AF251415

The South Atlantic in the Late Quaternary

Springer
Berlin
Heidelberg
New York
Hong Kong
London
Milan
Paris
Tokyo

G. Wefer • S. Mulitza • V. Ratmeyer (Eds.)

The South Atlantic in the Late Quaternary

Reconstruction of Material Budgets and Current Systems

With 315 Figures and 57 Tables

Springer

Editors:

PROFESSOR DR. GEROLD WEFER
DR. STEFAN MULITZA
DR. VOLKER RATMEYER

DFG Research Center Ocean Margins
University of Bremen
Klagenfurter Straße
28359 Bremen
Germany

ISBN 3-540-21028-8 Springer-Verlag Berlin Heidelberg New York

Library of Congress Cataloging-in-Publication Data

Bibliographic information published by Die Deutsche Bibliothek
Die Deutsche Bibliothek lists this publication in the Deutsche Nationalbibliografie;
detailed bibliographic data is available in the Internet at <http://dnb.ddb.de>.

Springer-Verlag Berlin Heidelberg New York
Springer-Verlag is a part of Springer Science+Business Media

springeronline.com

Typesetting: Camera-ready by C. Hayn, University of Bremen
Cover design: E. Kirchner, Heidelberg

Printed on acid-free paper SPIN 10983683 32/2132/AO 5 4 3 2 1 0

Preface

The South Atlantic plays a critical role in the coupling of oceanic processes between the Antarctic and the lower latitudes. The Antarctic Ocean, along with the adjacent southern seas, is of substantial importance for global climate and for the distribution of water masses because it provides large regions of the world ocean with intermediate and bottom waters. In contrast to the North Atlantic, the Southern Ocean acts more as an "information distributor", as opposed to an amplifier. Just as the North Atlantic is influenced by the South Atlantic through the contribution of warm surface water, the incoming supply of NADW – in the area of the Southern Ocean as Circumantarctic Deep Water – influences the oceanography of the Antarctic.

The competing influences from the northern and southern oceans on the current and mass budget systems can be best studied in the South Atlantic. Not only do changes in the current systems in the eastern Atlantic high-production regions affect the energy budget, they also influence the nutrient inventories, and therefore impact the entire productivity of the ocean. In addition, the broad region of the polar front is a critical area with respect to productivity-related circulation since it is the source of Antarctic Intermediate Water. Although the Antarctic Intermediate Water today lies deeper than the water that rises in the upwelling regions, it is the long-term source of nutrients that are ultimately responsible for the supply of organic matter to the sea floor and to sediments. Because particle flux in the source region, the polar front, greatly influences the mineral nutrient content of the intermediate water, the efficiency of low-latitude upwelling with respect to the removal of CO_2 from the atmosphere is determined in the Antarctic Intermediate Water. The South Atlantic therefore plays a significant role as a link between the Antarctic Ocean and the North Atlantic, both in terms of the heat budget of the North Atlantic and the circulation of the entire ocean.

The heat and mass exchange between the Antarctic Ocean and the South Atlantic during the Late Quaternary have been investigated over the past ten years, including their impact on world climate. This has required the study of present-day early diagenetic processes in the water column and sediments, as well as sediment properties that have a close relationship to environmental parameters ("proxies"), which can be used to decipher past conditions (temperature, salinity, productivity, etc.).

The interdisciplinary research project "The South Atlantic in the Late Quaternary - Reconstruction of material budgets and current systems" was a long-term scientific program, funded from 1989 to 2001 by the Deutsche Forschungsgemeinschaft as Sonderforschungsbereich 261 (SFB 261) at Bremen University. This program benefited from the sample material gained on several expeditions with the research vessels *Meteor* and *Polarstern*. This book presents the summarized results of the various topics of study in 30 articles arranged in seven sections.

The papers have benefited from detailed reviews by D. Archer, M. Arhan, S. Banerjee, R. Barman, W.H. Berger, J. Bijima, M. Dekkers, G. Delaygue, A. de Vernal, G. R. Dickens, T. Dokken, J.A. Flores, R. Francois, J. Giraudeau, F. Grousset, S. Hess, S. Hovan, R. A. Jahnke, F. Jorissen, S. Kanfoush, M. Kienast, D. Kirchman, W. Krijgsman, M. Kucera, C. Langereis, E. Laws, P. Le Grand, K. Matsumoto, H. Meggers, E. Michel, J. Milliman, A. Oschlies, M. Pagani, U. Passow, F. Peeters, D. Piper, M. Prins, C. J. Pudsey, O. Ragueneau, D. Rey, A. Rosell-Mele, C. Rühlemann, R. Schlitzer, M. Séranne, I. Snowball, G. Snyder, J. Thomson, T. Trull, M. Urbat, M. Voss, I. Wainer, M. Weinelt, H. Westphal, A. Winguth and R. Zahn.

Our research was supported by several international organizations, including the Joint Global Flux Study (JGOFS), Past Global Changes (PAGES) and Ocean Drilling Program (ODP) communities. Financial support was provided by the Deutsche Forschungsgemeinschaft and the University of Bremen as well as by the Senate of the Free and Hanseatic City of Bremen. Technical support was given by C. Hayn, W. Hale and B. Oelkers.

To each and all of those involved, our sincere thanks.

Gerold Wefer
Stefan Mulitza
Volker Ratmeyer   Bremen, September 2003

Contents

Early Diagenetic Processes and Preservation of Primary Signals

History of Upper Ocean Circulation

History of Bottom and Deep Water Circulation

Inverse Modeling of Particulate Organic Carbon Fluxes in the South Atlantic

R. Schlitzer[1*], R. Usbeck[1] and G. Fischer[2]

[1] Alfred Wegener Institute for Polar and Marine Research, Columbusstrasse,
D-27568 Bremerhaven, Germany
[2] Universität Bremen, Fachbereich Geowissenschaften, Postfach 33 04 40,
D-28334 Bremen, Germany
* corresponding author (e-mail): rschlitzer@awi-bremerhaven.de

Abstract: The biological production of particulate material near the ocean surface and its subsequent remineralization during sinking and after deposition on the seafloor strongly affect the distribution of oxygen, dissolved nutrients and carbon in the ocean. Dissolved nutrient distributions therefore reveal the underlying biogeochemical processes, and these data can be used to determine rates of production, remineralization and accumulation with the aid of inverse techniques. Here, an ocean circulation, biogeochemical model that exploits the existing large sets of hydrographic, oxygen, nutrient and carbon data is presented and results for the export production of particulate organic matter, vertical fluxes in the water column, and sedimentation rates are presented. In the model, the integrated export flux of particulate organic carbon (POC) in the South Atlantic amounts to about 1300 Tg C yr^{-1} (equivalent to 1.3 Gt C yr^{-1}), most of which occurring in the Benguela/Namibia upwelling region and in a zonal band following the course of the Antarctic Circumpolar Current (ACC). Remineralization of POC in the upper water column is intense, and only about 7% of the export reaches a depth of 2000 m. Comparison of modeled particle fluxes with sediment trap data suggests that shallow traps tend to underestimate the downward flux, whereas the deep traps seem to be affected by the lateral input of material and apparently overestimate the vertical flux. These findings are consistent with recent radionuclide studies. The rapid degradation of POC with depth produces geographical patterns of POC fluxes to the seafloor and POC accumulation in the sediment that are very different from the pattern of surface productivity, because of the modulation with varying bottom depth. Whereas there is significant surface production in deep-water, open-ocean regions, the benthic fluxes occur predominantly in coastal and shelf areas.

Introduction

Biogeochemical processes in the ocean act as a downward pump (Volk and Hoffert 1985) that produces low concentrations of nutrients and carbon in surface water and high values below. The low surface concentrations are caused by biological production in the sun-lit euphotic zone, which utilizes the dissolved nutrients in the formation of particulate material (primary production PP). One fraction of the particles sink out of the euphotic zone (export production EP), thereby effectively removing nutrients and carbon from the surface layer. In intermediate, deep and bottom layers nutrients are subsequently returned to the dissolved pool by particle decomposition (remineralization) while they sink in the water column, or after they settle on the seafloor. The decrease of surface water carbon concentrations due to the biological pump is a crucial factor that affects surface water $p\mathrm{CO_2}$ values and ultimately controls the air-sea $\mathrm{CO_2}$ exchange and atmospheric $\mathrm{CO_2}$ concentrations.

Because of the large potential impact on the atmospheric $\mathrm{CO_2}$ budget and global climate, deter-

From WEFER G, MULITZA S, RATMEYER V (eds), 2003, *The South Atlantic in the Late Quaternary: Reconstruction of Material Budgets and Current Systems.* Springer-Verlag Berlin Heidelberg New York Tokyo, pp 1-19

mining the strength of the biological pump, e.g. the magnitudes of export production and downward particle fluxes, has been a high-priority research goal for a long time. Three main experimental approaches have been developed: (1) the direct measurement of the vertical particle fluxes with moored or drifting sediment traps (Honjo et al. 1982; Wefer et al. 1982; Fischer et al. 1988; Deuser et al. 1995; Fischer et al. 1996; Fischer and Wefer 1996; Neuer et al. 1997; Fischer et al. 2000), (2) the estimation of primary or export production from surface-water chlorophyll concentrations obtained from satellites (Longhurst et al. 1995; Antoine et al. 1996; Behrenfeld and Falkowski 1997; Arrigo et al. 1998; Laws et al. 2000), and (3) the determination of oxygen utilization rates in tracer-dated thermocline, intermediate and deep water masses (Jenkins 1982; Jenkins and Goldman 1985; Jenkins 1987; Broecker et al. 1991).

Sediment trap measurements are point measurements and strongly influenced by the large temporal and spatial variability of production and particle flux events. Therefore, mean export production rates for larger regions or on global scale, as required for climate studies, are difficult to obtain from trap data. In addition, results from shallow traps (needed to estimate the export flux) have been questioned because of the suspected low trapping efficiency (see below). Satellite observations provide a global view, however, estimates of primary or export production from satellites have large uncertainties, because the conversion of chlorophyll concentrations into productivity rates and vertical particle fluxes is difficult and only relies on a very limited number of regional calibration studies. Furthermore, the optical sensors of the satellites only probe a thin surface layer of the ocean and deeper chlorophyll patches often found by towed instruments are missed. Recently, a systematic comparison of various satellite productivity algorithms has revealed significant random and systematic differences (Campbell et al. 2002).

Here, a very different approach is taken to estimate marine export production and downward particle fluxes through the water column and to the seafloor. The basic principle was proposed and applied by Riley (1951), who realized that the observed oxygen, nutrient, and carbon distributions in the ocean can be used to quantify the biogeochemical processes that shape the distributions. To give an example, Figure 1 shows phosphate and oxygen distributions along a meridional (AJAX; Nierenberg and Nowlin (1985)) and zonal (WOCE A7; WOCE Data Products Committee (2000)) section in the South Atlantic. In tropical and subtropical waters, biological consumption leads to a drawdown of surface phosphate to almost non-detectable levels. As shown in Figure 1a, this nutrient depleted surface layer is very thin (<100 m), except in the center of the subtropical gyre (ca. 30°S) where low-nutrient waters are found down to about 400 m due to Ekman pumping. In the sub-Antarctic and Antarctic zone, surface phosphate concentrations are much higher, but significant nutrient draw-down in a very thin surface layer seems to occur in this area as well, as indicated by the large vertical phosphate gradient found south of about 50°S (Figure 1a). Below the depleted surface layer, phosphate concentrations increase drastically with depth and reach maximum levels at about 500 m. These maximum phosphate concentrations a are most pronounced in the eastern tropical and subtropical Atlantic where biological productivity is high, and they are associated with strong oxygen minima (Fig. 1b). The anti-correlation between oxygen and phosphate provides additional evidence that the subsurface phosphate and oxygen signals are caused by biogeo-chemical processes, e.g. remineralization of sinking particulate material that brings nutrients back into solution and at the same time consumes dissolved oxygen.

It is important to note that the biogeochemically induced features of the nutrient, oxygen and carbon distributions have large amplitudes, and that these structures are very well resolved by the large amount of available data. Comparisons of repeated observations as long as 60 years apart show that these features are quite stable, and temporal changes are quite small even on decadal time-scales (Schlitzer 2000). This is in contrast to the large spatial and temporal variability seen in productivity maps from satellites and in sediment trap time series, and suggests that the surface nutrient depletion and the sub-surface nutrient maxima as well as the pronounced shallow oxygen minima are the integrated and averaged response of many

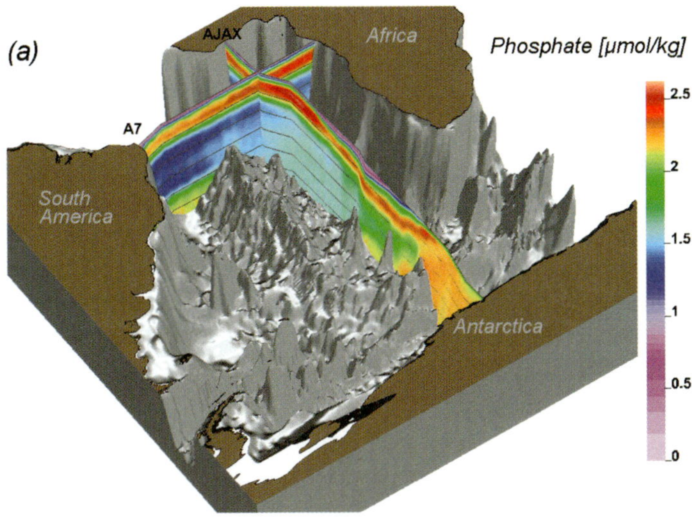

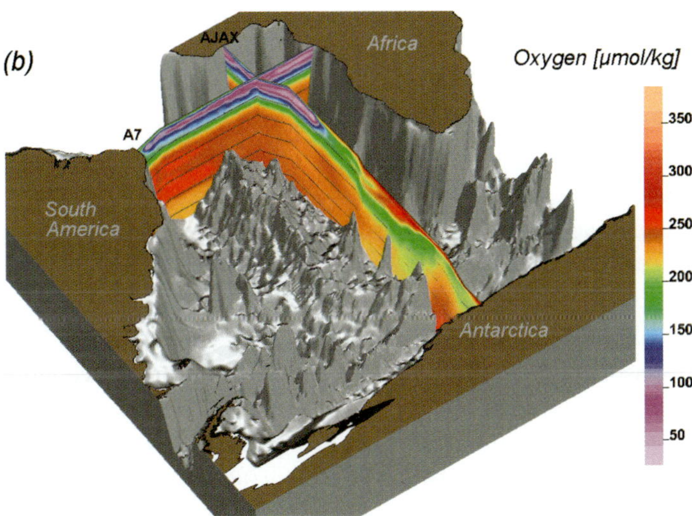

Fig. 1. Three-dimensional display of **a)** phosphate and **b)** oxygen distributions along two sections in the South Atlantic. Dark lines indicate depth levels in 1000 m intervals.

bloom and flux events from many years. Therefore, utilizing dissolved nutrients, carbon and oxygen fields promises to yield large-scale and long-term productivity and downward carbon flux estimates - quantities needed for global budget calculations and difficult to obtain from satellite or sediment trap observations alone.

In addition to the influence of biogeochemical processes on marine nutrient, carbon and oxygen fields, these distributions are also strongly affected by the circulation and water mass transport. This can be clearly seen in the western part of the A7 section in the Brazil Basin, where the vertical layering of waters with high phosphate concentrations at a depth of about 800 m, low phosphate levels between about 1500 and 4000 m, again high values below 4000 m, and near the bottom reflect the meridional overturning circulation consisting of northward moving Antarctic Intermediate Water (AAIW) and Antarctic Bottom Water (AABW), with southward flowing North Atlantic Deep Water (NADW) in between. A successful determination of biogeochemical flux rates from dissolved nutrient and oxygen fields therefore also requires to take physical transports into account. This is addressed by the approach described below.

In contrast to nutrient simulations using forward models, the estimation of physical or biogeochemi-cal rate constants from concentration data is commonly referred to as inverse calculations. Since Riley's (1951) work, the amount and quality of available data has increased tremendously, and advances in computer technology and numerical methods now allow detailed studies on a global scale dealing with many particle species and the large variety of measured properties simultaneously.

The model strategy described below employs the *adjoint method* to determine flows (physics) and production as well as remineralization parameters (biogeochemistry) that explain the global hydrography and the measured concentration fields of dissolved nutrients, carbon and oxygen best. The problem is formulated as a constrained optimization and the solution is found iteratively. It should be noted that in contrast to ecosystem models (Fasham et al. 1990) that attempt to describe the complicated interactions and

feedbacks of biological food-webs and community structures explicitly, in the present case the processes leading to the production, export and subsequent remineralization of particles are not resolved by the model, but parameterized as rate constants to be determined by the model. The unknown production or remin-eralization rates are only constrained by the requirement to reproduce the observed property distributions realistically. As described below, empirical flux-depth relations are applied, but biogeochemical a priori information about the underlying processes is not incorporated.

In this paper we present results from a model setup that includes particle production in the surface layer, remineralization in the water column and at the seafloor, and accumulation of material in the sediment. Input of material from rivers is included to account for losses in the sediment. Results from two model runs that account for dissolved organic matter and N_2-fixation and denitrification effects in the South Atlantic are also included. One of these experiments is further constrained by fitting it to observations of natural radiocarbon and chloro-fluorocarbons (CFC). Maps of organic carbon export fluxes, fluxes towards the bottom, and sediment accumulation rates are shown for the South Atlantic, and regional integral flux estimates are given. In addition, a detailed comparison of model particle fluxes with sediment trap measurements is included.

Inverse Model

As described above, determining particle fluxes and near-surface production rates from nutrient concentrations requires a biogeochemical model that also includes the ocean's 3-D circulation. Here, the global model of de las Heras and Schlitzer (1999) and Schlitzer (2000) is used that has already been fitted to global hydrographic, nutrient and oxygen fields. The general model strategy is based on Schlitzer (1993), and more details can be found there and in Schlitzer (1995, 1996, 2000, 2002). In the following, the model setup is briefly described and then the extensions with respect to biogeo-chemical cycles are explained.

Overall goal of the model calculations is to find a steady global ocean flow field (representing the climatological mean circulation) and mean export production of particulate organic carbon (POC) as well as POC fluxes throughout the water column and to the seafloor. The model run described here also includes the accumulation of organic material in the sediment. To be considered as optimal, the calculated velocities must have velocity shears that are close to geostrophic shear estimates and simulations of hydrographic and nutrient distributions must reproduce the respective observations accurately (see Schlitzer 1993 for more details). The *adjoint method* provides the means that drive the model to the desired state (see Schlitzer 2000 for more details).

The model is of global extent and uses a non-uniform grid with horizontal resolution ranging from 5x4 degrees longitude by latitude in open ocean areas to 1x1 degrees in regions with narrow currents (Drake Passage, Atlantic part of the Antarctic Circumpolar Current, Indonesian and Caribbean archipelagos), along coastal boundaries with strong currents (Florida Current, Gulf Stream, Brazil Current, Agulhas Current, Kuroshio), over steep topography (Greenland-Iceland-Scotland overflow region) and in areas with pronounced coastal up-welling or downwelling (Fig. 2a). In all cases the refinements are implemented in the direction of the strongest property gradients (usually perpendicular to fronts and currents) to better trace changes in ocean properties.

The model has 26 vertical layers, with thickness progressively increasing from 60 m at the surface to approximately 500 meters at 5000 m depth. Realistic topography is used, based on the U.S. Navy bathymetric data and averaged over grid-cells. Model depths over ridges and in narrow channels are adjusted manually to respective sill or channel depths. The model has three open boundaries, along which ocean properties and transports are prescribed in each model layer. They are located at the exit of the Mediterranean Sea, Red Sea and Persian Gulf. These three marginal seas are not modeled explicitly, but their impact on the global circulation is taken into account.

A schematic diagram of a vertical model column extending from the sea surface down to the respective bottom depth is shown in Figure 2b. Model property values are defined at the centers

of the grid-boxes whereas flows are defined on the interfaces (Arakawa C-grid). Formation of particulate material is occurring in the top two model layers with the top layer contributing 75% and the second layer contributing 25% of the total production. The bottom of the second model layer is considered the base of the euphotic zone (here: z_{EZ}=133 m). Particle fluxes below the euphotic zone are assumed to decrease with depth according to

$$j_P(z) = a \cdot (z / z_{EZ})^{-b} \qquad (1)$$

This functional relationship is commonly used for the depth dependence of the flux of particulate organic carbon (POC) (Suess 1980; Martin et al. 1987; Bishop 1989). In (1) a is the particle flux at the base of the euphotic zone, z_{EZ}, and represents the export production. The parameter b determines the shape of particle flux profile and thus controls the depth of remineralization. Large values for b correspond to steep particle flux decreases and thus large remineralization rates just below the euphotic zone, whereas values for b close to zero result in almost constant particle fluxes independent of depth, with little remineralization in the water column and most of the particle export reaching the ocean floor.

The particle flux to the bottom is given by $j_P(z_b) = a \cdot (z_b / z_{EZ})^{-b}$, where z_b is the bottom depth. In the model that includes sedimentation, a fraction of this flux reaching the bottom- is buried in the sediment $j_B = s \cdot j_P(z_b)$ (with $0 \le s \le 1$), while the remainder is remineralized in the bottom box. The globally integrated loss of material to the sediment is compensated in the model by input from rivers. Here, the thirteen largest rivers are considered. The sources are located in the surface layer of the model columns closest to the respective river mouths, and the strength of the material sources is set proportional to the water discharge and nutrient concentrations of the respective river (Lerman 1980; Berner and Berner 1996; Usbeck 1999). In the model runs without sedimentation all material reaching the seafloor is remineralized there.

Export production a, remineralization scale height b, and sedimentation fraction s may vary

(a)

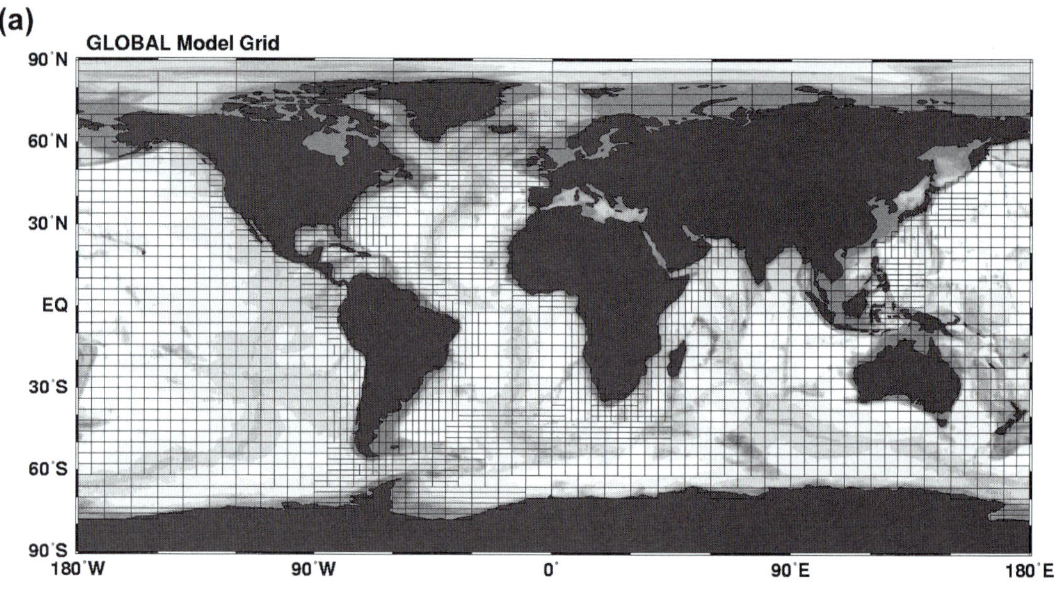

(b)

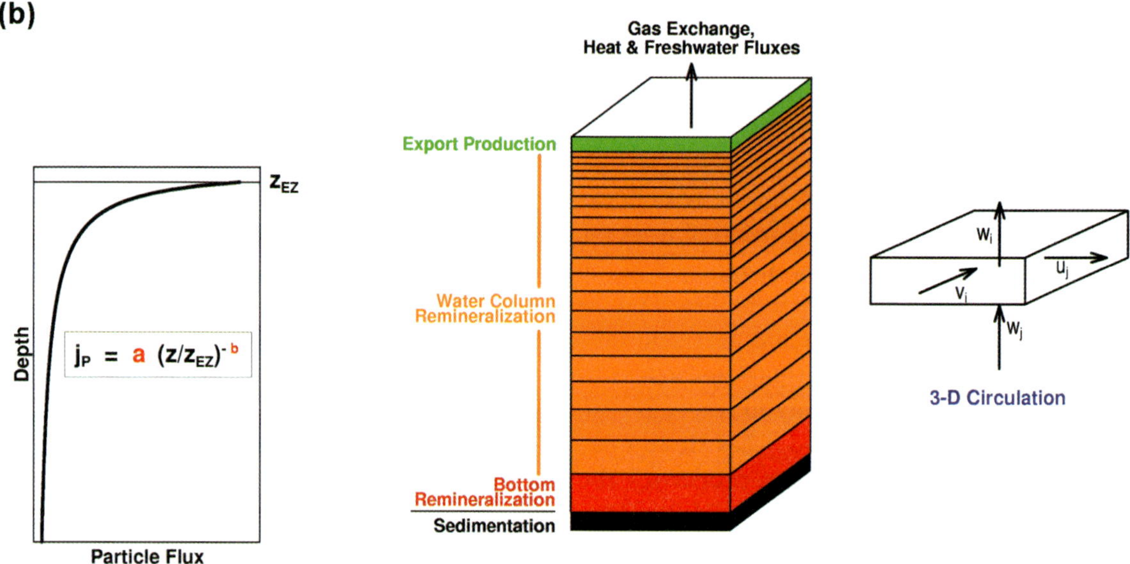

$$j_P = a \, (z/z_{EZ})^{-b}$$

Fig. 2. Horizontal **a)** and **b)** vertical model grid and definition of model parameters.

from grid column to grid column. Note that these parameters are not formally dependent on environmental factors such as nutrient availability, light intensity or density stratification, as implemented in other biogeochemical models (Maier-Reimer 1993; Yamanaka and Tajika 1996), but can be freely chosen by the model to achieve optimal agreement with measured oxygen, nutrient and carbon fields.

The computational strategy is explained in detail in Schlitzer (2000) and Schlitzer (2002), here only a brief summary will be given. The horizontal model flow field is initialized using geostrophic velocities calculated from original hydrographic data. Following suggestions from the literature (Reid 1986; Whitworth and Nowlin 1987; Rintoul and Wunsch 1991; Rintoul 1991; Reid 1997) the reference level for the geostrophic calculations varies geographically: $z_r = 1500$ m north of 25°S; $z_r = 3800$ m south of 45°S; linear transition between 25°S and 45°S.

The initial export fluxes of POM a are derived from the productivity map of Berger (1989), and the initial values of the remineralization parameters b are set to 1 (Suess 1980). Using this set of initial independent parameters, model vertical velocities as well as model fields for temperature, salinity, oxygen phosphate, nitrate, total inorganic carbon and alkalinity (dependent parameters) are calculated by applying steady-state conservation equations for mass, heat, salt, oxygen, nutrients, carbon, and alkalinity, respectively. This step of calculating the vertical velocities and model property fields is commonly referred to as a *simulation*.

Once calculated, the simulated property distributions can be compared with measurements. Usually, this model/data comparison is done subjectively, and attempts are made to correct potential causes for model/data misfits by modifying the velocity and biogeochemical parameters and then running new simulations. While this approach usually is successful for box-models with a limited number of parameters, for large 3D models like the present one, the manual adjustment of parameters is impractical and in most cases does not lead to desired improvements.

The so-called *adjoint method* (Hestenes 1975; Thacker and Long 1988; Schlitzer 2000) is an alternative to manual parameter-tuning, and allows treatment of optimization problems with a large number of adjustable parameters, like the present one. Here the evaluation of model/data misfits is performed automatically by using a suitably formulated cost function (see Schlitzer 2000 for a detailed discussion of cost function terms). A twin-model of the simulation step (the *adjoint model*) "learns" from the structure in the misfits and modifies the independent parameters (flows and biogeochemical parameters) in a systematic way that guarantees a better model/data agreement in the next simulation. By running the forward/adjoint steps repeatedly, the model is driven closer and closer to the observations. The calculations are terminated when the decrease of the cost function (improvement) per iteration is smaller than a pre-defined limit.

Because of the huge computational requirements of the optimization procedure described above, and because of a general lack of biogeochemical data that fully describe the seasonal cycle, the present model is formulated as a steady-state, annual-mean model. The optimized model flow field and particle flux patterns represent long-term means over the time periods covered by the data that are used by the model (ca. 1939 – 2002), and no attempt is made to resolve time-dependent and sporadic events, such as eddies or individual bloom events.

In the following, results from three different model runs are presented (Table 1). Run A includes accumulation of material in the sediment, but neither dissolved organic matter cycling nor N_2 fixation or denitrification. This experiment is fitted to global ocean temperature, salinity, oxygen, dissolved nutrient and carbon data and represents the reference experiment from which most conclusions are drawn here.

Experiments B and C disregard loss of material in the sediment but include cycling of semi-labile dissolved organic matter (Kirchman et al. 1991; Hansell and Carlson 1998), and N_2 fixation and denitrification as sources and sinks for nitrate. Nitrogen fixation and denitrification rates are not derived directly from measurements, but are specified using information on globally integrated fixation rates from the literature (Codispoti and Christensen 1985; Gruber and Sarmiento 1997). In the model, nitrogen fixation is evenly distributed

Experiment	Processes Included			Data Used		
	sedimentation	dissolved organic matter	N$_2$ fixation, denitrification	T, S	O$_2$, nutrients, carbon	^{14}C, CFC
A	yes	no	no	yes	yes	no
B	no	yes	yes	yes	yes	no
C	no	yes	yes	yes	yes	yes

Table 1. Model experiments used for the present study.

over those surface areas where dissolved nitrate is lower than 5 μmol kg⁻¹, which includes most of the low and mid-latitudes. Denitrification exactly matches total nitrogen fixation and is equally divided into water-column and benthic denitrification. The benthic denitrification is set to be proportional to the POC flux to the seafloor, and water-column denitrification is evenly spread over all volumes in which dissolved oxygen is below 20 μmol kg⁻¹. This places most of the denitrification in the Pacific and Indian Ocean at a depth between 100 and 800 m, which is in agreement with observational evidence (Codispoti and Christensen 1985; Codispoti et al. 1986; Howell et al. 1997).

Experiment C is fitted to natural radiocarbon data and chlorofluorocarbons (CFC) (WOCE Data Products Committee 2000) in addition to hydrography and dissolved oxygen, nutrients and carbon, as for the other runs. These experiments are performed to explore the range and uncertainties of model flux estimates. In most regions, reactive dissolved organic matter is confined approximately to the upper 200 m of the water column (Hansell and Carlson 1998; Anderson and le B. Williams 1999) and N$_2$ fixation and denitrification are more important in the Pacific as compared to the South Atlantic. Therefore, experiment A is considered a valid representation of the carbon cycle in the South Atlantic.

The automatic fitting procedure (*adjoint model*) described above has been applied successfully for all three experiments A to C, and the overall agreement between model simulated fields and observations is excellent and much better (misfit between modeled and measured values is about one order of magnitude smaller) than in other global biogeochemical models (Maier-Reimer 1993; Sarmiento et al. 2000). In the South Atlantic, at a depth of 500 m, where biogeochemical effects on dissolved nutrient and oxygen distributions are greatest (see Fig. 1), the mean model phosphate concentration is only 0.03 μmol kg⁻¹ higher than observed, and the root-mean-square (rms) deviation amounts to only ±0.16 μmol kg⁻¹. For oxygen in the same region, the deviations are -0.5±16 μmol kg⁻¹. Given typical measurement errors of 0.02 and 2 μmol kg⁻¹ for phosphate and oxygen, respectively, the model misfits are considered to be small, and experiments A to C all explain the observed distributions realistically.

Export of Particulate Organic Carbon

Figure 3 shows the model export fluxes of particulate organic carbon in the South Atlantic that is required for the successful reproduction of the measured property fields. This plot is for experiment A, but the geographical patterns as well as the magnitudes of the fluxes are similar for experiments B and C. According to the model, the downward carbon export in the South Atlantic predominantly occurs in two areas: (1) the Benguela-Namibia coastal upwelling region, and (2) a broad zonal band centered at about 50°S and roughly coinciding with the course of the Polar Front and the Antarctic Circumpolar Current (ACC). Here, the highest productivity is found in the western part, e.g. in northern Drake Passage, on the Patagonian shelf and eastward (downstream) over the North Scotia Ridge (ca. 45°W/53°S). Low export fluxes (< 1 mol C m⁻² yr⁻¹) are found in the center of the subtropical gyre at about 30°S, where surface

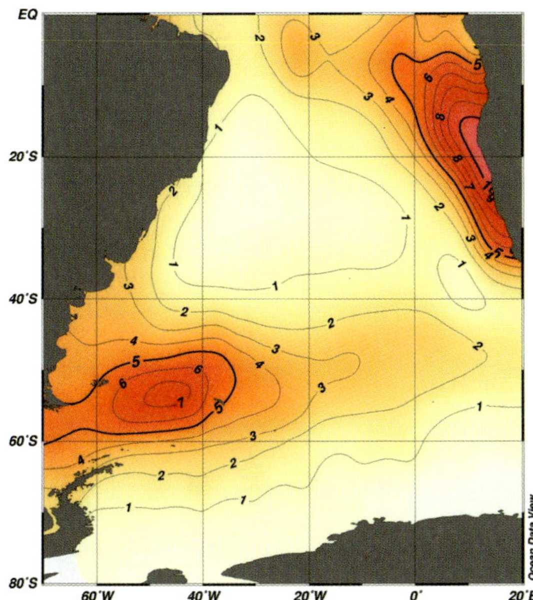

Fig. 3. Export flux of particulate organic carbon [mol C m^{-2} yr^{-1}] in 133 m depth for the model including accumulation of material in the sediment (experiment A).

waters are nutrient-depleted (Schlitzer 2002), and in the Weddell Sea, where permanent or seasonal ice coverage and presumably the unavailability of micronutrients such as iron (de Baar et al. 1995) limit biological productivity. Given the high phosphate and nitrate concentrations in ACC surface waters (PO$_4$ > 1 ¼ mol kg^{-1}; Schlitzer 2002), the export fluxes in the southwest Atlantic appear small compared to the Namibia coastal upwelling region. Again, iron limitation, strong vertical mixing and weak surface stratification are possible explanations for the relatively small productivity in these high nutrient/low chlorophyll (HNLC) regions.

A detailed comparison of inverse model export fluxes with estimates based on satellite productivity maps (Schlitzer 2002) has shown that the two methods agree reasonably well in most regions of the South Atlantic except south of 50°S, where satellite productivity algorithms indicate only weak biological productivity, but where the inverse model requires quite high downward carbon exports to reproduce the measured water column structure of oxygen, phosphate, nitrate and carbon. Schlitzer (2002) argues that the satellite productivity estimates south of 50°S might be too low because of a generally poor calibration in this region, and because of undetectable (by satellites) sub-surface chlorophyll patches frequently observed in the Southern Ocean. Comparisons of different satellite algorithms (Campbell et al. 2002) also have identified the Southern Ocean as the region where the results differ most and systematic offsets are largest.

Integrated export fluxes in experiment A amount to 9840 Tg C yr^{-1} (equivalent to 9.84 Gt C yr^{-1}) for the global ocean, 14% of which (equivalent to1340 Tg C yr^{-1}) occurring in the South Atlantic (see Table 2). Experiments B and C have slightly smaller globally integrated carbon exports (9440 and 9040 Tg C yr^{-1}, respectively), however, in both cases the POC export in the South Atlantic is almost identical with the result of run A (1320 and 1350 Tg C yr^{-1}, respectively). The partitioning of the South Atlantic carbon export into contributions from the Benguela/Namibia upwelling region and the high productivity region in the southwest Atlantic is quite different for the three experiments: while the Benguela/Namibia upwelling region is dominant in experiment A (558 versus 480 Tg C yr^{-1}), the southwest Atlantic is shown to be the most important area for downward carbon export in experiments B and C.

Flux of Particulate Organic Carbon to the Seafloor

Model particle fluxes at a depth of 2000 m (a level onto which sediment trap measurements are often extrapolated) are calculated using the model estimates for the parameters a and b and equation (1). As for the global ocean, the carbon fluxes at a depth of 2000 m in the South Atlantic and two subdomains turn out to be much smaller than the export fluxes (Table 2), confirming results from sediment trap measurements (Martin et al. 1987; Suess 1980) indicating that most of the exported POC is remineralized in the upper 2000 m of the water column. In experiment A, about 5% of the exported material arrive at 2000 m, whereas in experiments B and C this fraction is slightly larger, approximately amounting to 8%. Taking all three runs into consideration, the POC flux at a depth of 2000 m in the South Atlantic amounts to 96±26 Tg C yr^{-1}.

	A	**B**	**C**
Export Flux (133 m)			
Global Ocean	9840	9440	9040
South Atlantic	1340	1320	1350
Southwest Atlantic	480	673	718
Benguela/Namibia	558	507	331
Flux in 2000 m			
Global Ocean	643 (6.5)	911 (9.6)	647 (7.1)
South Atlantic	66 (5)	109 (8.3)	114 (8.4)
Southwest Atlantic	20 (4.2)	53 (7.9)	64 (8.9)
Benguela/Namibia	28 (5)	45 (8.9)	28 (8.5)
Flux to Seafloor			
Global Ocean	602 (6.1)	714 (7.6)	478 (5.3)
South Atlantic	62 (4.6)	95 (7.2)	74 (5.5)
Southwest Atlantic	30 (6.3)	58 (8.6)	39 (5.4)
Benguela/Namibia	20 (3.6)	31 (6.1)	19 (5.7)

Table 2. Integrated downward fluxes of particulate organic carbon (Tg C yr^{-1}) for the global ocean, the South Atlantic (as defined by the map domain of Fig. 3), the southwest Atlantic and the Benguela/Namibia high-productivity regions (delineated by 2 mol C m^{-2} yr^{-1} contour). Results are given for the three model experiments A, B, and C and for different depths. Values in parenthesis are percentages of respective export fluxes.

Values of POC fluxes to the seafloor have been calculated, again, using parameters a and b, and applying equation (1) with the bottom depth z_b of each model column. The geographical pattern of the resulting bottom fluxes is shown in Fig. 4, and the integral values for the domains defined above are given in Table 2. The POC fluxes to the seafloor in the South Atlantic (Fig. 4) are clearly dominated by relatively high values (between 0.5 and 3 mol C m^{-2} yr^{-1}) along the southwest African and South American coasts. In the open ocean South Atlantic, fluxes are much smaller and generally below 0.1 mol C m^{-2} yr^{-1}. In large regions, values are even smaller than 0.05 mol C m^{-2} yr^{-1}. The overall pattern of bottom POC fluxes differs markedly from the map of export fluxes from the euphotic zone shown in Fig. 3. While in shallow coastal waters and over shelves or the upper continental slopes large fractions of the exported material reaches the bottom, most of the exported POC is remineralized by the time the material reaches the floor in the deep waters of the open ocean. As a consequence, the pronounced high productivity band centered in the southwest Atlantic and stretching eastward is only weakly reflected in the bottom fluxes. The strong decoupling of surface fluxes and bottom signals observed here for POC should equally apply to other materials that are degraded significantly during sinking. Integrated POC fluxes to the seafloor in the South Atlantic amount to 77±17 Tg C yr^{-1} (Table 2). It is important to note that more than 80% of this flux occurs near the coast and on the shelves.

Accumulation of Particulate Organic Carbon in the Sediment

Model experiment A includes accumulation of particulate organic material in the sediments. It determines the burial fraction s (for definition see above) from signals in the distribution of dissolved oxygen, nutrients and carbon in deep and bottom waters. No a priori geological accumulation rate estimates are used for the present calculations. Comparison of model/data misfits of A with its parent experiment P (not shown; identical to A except that P does not include sedimentation but all material reaching the seafloor is remineralized

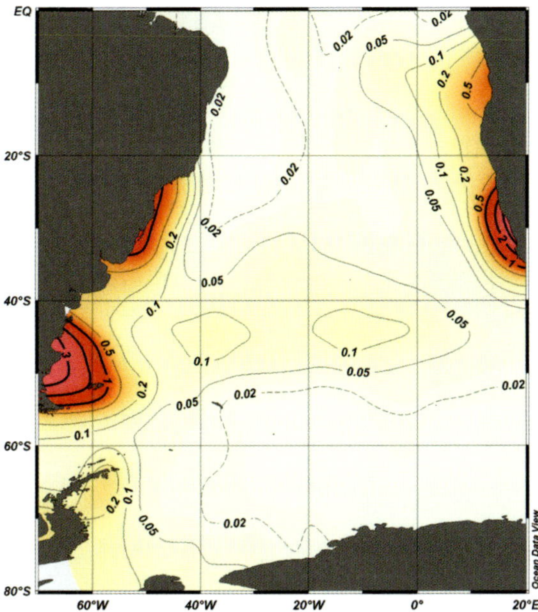

Fig. 4. Flux of particulate organic carbon [mol C m^{-2} yr^{-1}] to the seafloor for the model including accumulation of material in the sediment.

in the bottom box) shows that experiment A explains the data slightly better than P. However, this improvement is rather small, suggesting that removal of material in the sediments has a limited effect on the bottom water characteristics, probably because of the small magnitude of the burial fluxes.

Fig. 5 shows the geographical pattern of model POC accumulation in South Atlantic sediments. Comparison with Fig. 4 shows that the accumulation in the sediment is closely linked with the POC fluxes to the seafloor, with about 10% of the bottom-reaching flux actually being accumulated. Integrated over the entire South Atlantic, the POC accumulation amounts to 11 Tg C yr^{-1}, which represents only about 0.8% of the POC exported.

Comparison with Sediment Trap Data

Sediment trap deployments allow for the direct measurement of downward particle fluxes in the water column, and at the same time provide samples of the material for composition and structure analysis. Within the SFB216 project, a large number of moored sediment traps have been deployed for

long periods of time, in order to measure the seasonal cycle of downward particle fluxes and to catch sinking particles. These measurements have been used to estimate export fluxes of particulate organic carbon and nutrients from the euphotic zone into the deep ocean. Additionally, the decrease of particle fluxes with depth has been used to derive particle remineralization rates (Fischer et al. 1988; Fischer et al. 1996; Fischer and Wefer 1996; Neuer et al. 1997; Fischer et al. 2000; Fischer et al. 2002).

However, the reliability of vertical particle fluxes from sediment trap data has been questioned: Horizontal flow velocities in the upper ocean can be relatively high, so that tilting of the trap may occur and/or turbulences may prevent particles to settle into the trap ("under-sampling"; (Baker et al. 1988; Gust et al. 1994; Honjo and Manganini 1996). It has been shown for the deeper and near bottom traps that part of the material collected in the traps was not produced in the overlying water column, but obviously represented reworked suspended material that was transported laterally to the trap's position (Siegel and Deuser 1997; Fischer et al. 2000; Conte et al. 2001). This "over-sampling" appeared to be strongest in the vicinity of sloping bathymetry and close to the ocean bottom.

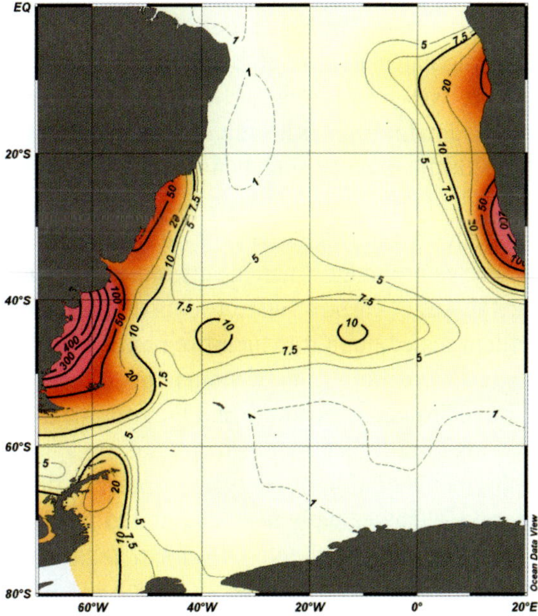

Fig. 5. Accumulation flux of particulate organic carbon [mmol C m^{-2} yr^{-1}] in the sediment.

Sampling efficiencies of sediment traps have been estimated using budgets of radionuclides (Bacon et al. 1985; Buesseler 1991; Buesseler et al. 1994) or aluminum fluxes (Walsh et al. 1988). At present, the calculation of ^{230}Th budgets seems to be the most reliable calibration method. ^{230}Th is produced in the water column by the decay of ^{234}U and is highly particle reactive. Sinking particles scavenge ^{230}Th, and the flux at a given depth must (on time scales of months to years) balance the production rate in the overlying water column. Therefore the ^{230}Th flux into a sediment trap should equal the amount of ^{230}Th produced above the trap. This quantity increases with depth and is well known, because the concentration of the parent nuclide ^{234}U in seawater is constant and well determined. If the measured ^{230}Th flux in a sediment trap is smaller than the theoretical value, part of the downward particle flux is obviously missed by the trap and it is therefore "under-trapping". If on the other hand, the measured ^{230}Th flux is larger than expected, the additional ^{230}Th can only originate from lateral input, and the trap obviously overestimates the vertical flux ("over-trapping"). Recently, Scholten et al. (2001) found that ^{230}Th fluxes indicate low trapping efficiencies in shallow traps (at the average 40%). In deep traps, they found ambiguous results, some traps showing larger as well as smaller particle fluxes than have been calculated from ^{230}Th budgets.

Here we compare direct flux measurements from sediment traps with the downward particle fluxes of the inverse model. As described above, the model POC fluxes are determined on the basis of long-term steady state oxygen, nutrient and carbon budgets, and thus provide estimates of downward fluxes of carbon and nutrients that are independent of sediment trap measurements. Figure 6 shows the positions of the sediment traps from the SFB216 project in the South and Equatorial Atlantic considered here. Most traps are long-term deployments of at least one year or more. Annual mean POC fluxes have been calculated from the original data (see Usbeck 1999 for details of the averaging procedure), values of which are listed in Table 3.

Figures 7 and 8 show model POC fluxes versus the annual average POC fluxes obtained from the

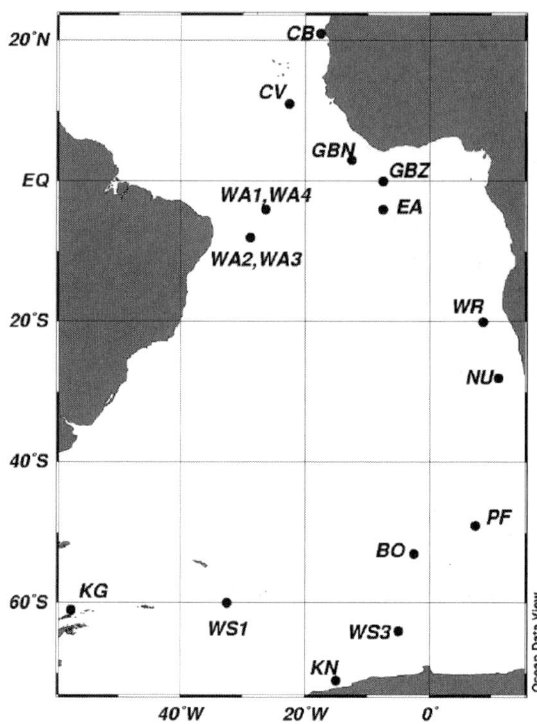

Fig. 6. Positions of sediment traps used for the present study.

sediment traps. Altogether, it is worth emphasizing that the model fluxes are of the same order of magnitude as the sediment trap measurements. However, systematic deviations of model fluxes from the trap measurements do exist. In the shallow traps (<1000m depth; Fig. 7), model POC fluxes are systematically higher than in measurements performed at all positions in the South- and Equatorial Atlantic. Model POC fluxes lie well above the 1:1-relation in Fig. 7. The ratios of trap fluxes relative to model fluxes range from 0.03 (WS1) to 0.85 (PF), with an average ratio of 0.46. Thus, on average, the model fluxes at the sediment trap positions are about a factor of 2 higher than the trap measurements. POC fluxes higher than 0.6 mol m^{-2} yr^{-1} are found in the model calculations, but no trap in the South Atlantic showed fluxes of this size.

For the deep traps (>1000m depth; Fig. 8), model POC fluxes are lower than measurements. With WR and CV being the only exceptions, all other model fluxes lie below the 1:1 relation. The ratios of trap and modeled POC fluxes range from 0.93 (WR) to 18.1 (GBZ), with an average of 5.1,

Trap	Lat (°N)	Lon (°E)	Depth (m)	Duration (days)	POC Flux
KN1	-71	-12	250	53	0.1583
WS3	-65	-3	360	368	0.1942
KG1	-62	-58	1600	360	0.3583
WS1	-62	-35	900	418	0.0025
BO1	-54	-3	453	460	0.225
PF3	-50	6	700	366	0.2808
NU2	-29	13	768	361	0.5762
WR2l	-20	9	1654	360	0.3142
WA3u	-8	-28	671	350	0.0792
WA3l	-8	-28	5031	350	0.0246
WA4l	-4	-26	4555	375	0.0813
EA8m	-6	-9	1833	296	0.1967
GBZ5l	-2	-10	3382	360	0.1917
GBN3l	2	-11	3965	361	0.181
CV1u,CV2u	11.5	-21	1000	679	0.2239
CV1l,CV2l	11.5	-21	4500	679	0.106
CB2,CB3	21	-20	3525	718	0.5042

Table 3. Long-term mean downward fluxes of POC (mol C m^{-2} yr^{-1}) from sediment trap measurements. Trap deployment was for one or two years in most cases.

i.e. measured POC fluxes are about a factor 5 larger than modeled fluxes. The systematic deviation of model fluxes and deep sediment trap data seems to be much larger than for the shallow traps, but the absolute deviations are in fact smaller. POC fluxes in the very deep ocean are relatively small, and additions of laterally transported, re-suspended material in the sediment trap can affect the ratios considerably, even if these additions are small in terms of absolute magnitude. Small systematic errors in the model deep ocean particle fluxes possibly due to deviations from the Martin-type functional relationship (1) in the real ocean, cannot be ruled out completely and provide an alternative explanation for the model/observation mismatch below the depth of 1000 m.

To investigate whether the systematic deviations between model and the sediment traps described above (traps lower than model in upper 1000 m of the water column; traps higher than model below 1000 m) are robust, an extension of the model was developed, which allows to drive the model particle fluxes towards measured fluxes. Various experiments were performed to test whether it is possible to obtain model particle fluxes closer to the sediment trap data, i.e. to reduce model/trap misfits in Figs. 7 and 8. The sediment trap data were included as soft constraints by adding a new term to the model cost function that penalizes deviations of model fluxes from trap data. Different model runs with different weight factors for the new cost function term were conducted. None, of these experiments was considered successful because systematic model/data misfits persisted in all cases. For the run with a very large weight on the model/trap penalty term, the model POC fluxes at the trap locations agreed rather well with the trap data, however, the model adjustment was local (singularity at the trap locations only) and the high model fluxes in the upper water column persisted at all other grid-points. Another experiment was conducted to enforce the trap data in the model over larger horizontal distances by applying smoothness constraints. This run showed that the smaller trap fluxes in the upper water column were inconsistent with the measured distributions of oxygen, nutrients and carbon in the ocean, because enforcing these fluxes over great areas led to un-

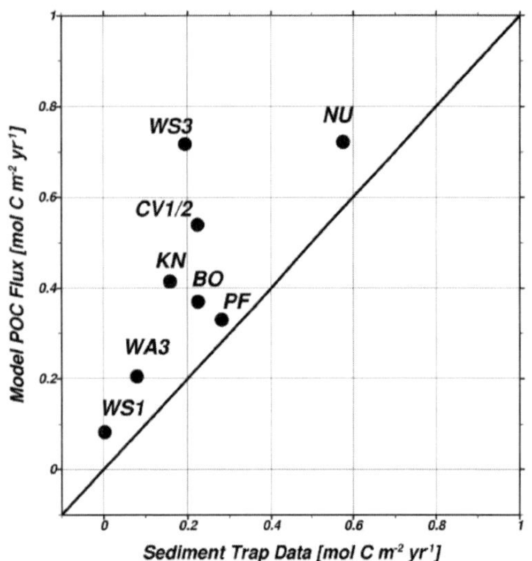

Fig. 7. POC model fluxes versus sediment trap data for traps shallower than 1000m. The solid line is the 1:1 relation.

realistic simulations for oxygen, nutrients and carbon (Usbeck et al. 2003).

Summary and Discussion

The present model calculations have shown that the strong and systematic traces of marine biogeochemical processes found in the distributions of dissolved oxygen, nutrients and carbon in the ocean can be used by a coupled physical/biogeochemical model to determine rate-constants as well as biogeochemical carbon and nutrient fluxes. The *adjoint method* has proven to be an efficient tool to drive the model towards the observations. The level of agreement between model simulation and observations achieved by the *adjoint method* is unmatched by any other biogeochemical model. Given that more data of even better quality will be available in the near future, the success of this study should motivate more inverse-type evaluations of marine biogeochemical cycles in the future.

For the global ocean, the three model experiments discussed here yielded integrated export fluxes of POC lying between 9040 and 9840 Tg C yr⁻¹. This is in relatively good agreement with export fluxes found in other global biogeochemical

models (8000 to 12,000 Tg C yr⁻¹; (Najjar et al. 1992; Anderson and Sarmiento 1995; Six and Maier-Reimer 1996; Yamanaka and Tajika 1997; Popova et al. 2000)) and with global estimates from satellite observations (12,000 Tg C yr⁻¹; (Laws et al. 2000)). Lower global export fluxes between 2000 and 4000 Tg C yr⁻¹ as obtained by simple box models (Broecker and Peng 1982; Sarmiento and Toggweiler 1984; Siegenthaler and Wenk 1984) are found to be incompatible with observed nutrient and oxygen fields (Schlitzer 2000).

The contribution of the South Atlantic (70°W to 20°E; 80°S to Equator; see map domain in Fig. 3) to the POC export amounts to 1350 Tg C yr⁻¹ (ca. 14% of global integral) and occurs mainly in the Benguela/Namibia coastal upwelling region and in a westward intensified zonal band that roughly follows the course of the ACC and Polar Front (ca. 52°S). Comparison with satellite derived productivity maps (Schlitzer 2002) has shown rather good agreement north of 50°S, however, there is a significant difference south of this line, with the model export fluxes being about twice as large as the satellite derived values. Because of unrealistic oxygen, nutrients and carbon simulations, the lower satellite estimates could not be reproduced by the

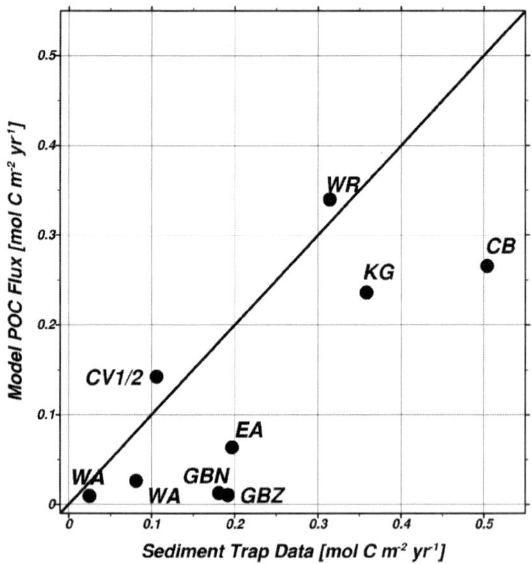

Fig. 8. POC model fluxes versus sediment trap data for traps deeper than 1000m. The solid line is the 1:1 relation.

model. Schlitzer (2002) argues that satellite based productivity estimates in the Southern Ocean might be too low, because satellites miss the productivity associated with frequently observed sub-surface chlorophyll patches and because of poor calibration with only few in- situ data in this region. This seems to be confirmed by the large variability and significant systematic offsets between different satellite productivity algorithms, found especially in the Southern Ocean (Campbell et al. 2002).

In the model, POC fluxes rapidly decrease with depth, and at a depth of 2000 m amount to 96 ± 26 Tg C yr^{-1} only. The integrated flux to the seafloor is of comparable magnitude, 77 ± 17 Tg C yr^{-1}, and concentrated in coastal and shelf regions. This pattern is very similar to the map of benthic oxygen consumption of Jahnke (1996), which also has highest rates along the southwest African coast and on the Patagonian shelf. Jahnke's map also shows the weak benthic imprint of the high open ocean surface productivity between 35°S and 50°S. Not only are the patterns similar, the numerical values also agree remarkably well (> 0.5 mol C m^{-2} yr^{-1} respectively mol O_2 m^{-2} yr^{-1} along the Namibian coast and <0.3 mol C m^{-2} yr^{-1} in the open ocean of the South Atlantic). It is noteworthy that, in the model, more than 90% (for experiment A; 100% for B and C) of the carbon flux to the seafloor is respired, leading to the consumption of oxygen as measured by Jahnke. A good agreement of model benthic fluxes is also obtained when comparing them with the release of benthic phosphate and nitrate as described by Hensen et al. (1998) and Zabel et al. (1998) (geographical pattern and numerical values; conversion using Redfield values C/P=106, C/N=6.6).

The globally integrated burial of POC resulting from experiment A (11 Tg C yr^{-1}) is about a factor of 1.5 higher than the estimate of Broecker and Peng (1993) for present-day carbon cycling. Sundquist (1985) summarized estimates from various sources varying by a factor of more than 100. The total accumulation of organic carbon derived from the *adjoint model* is well in the range of these estimates and comparably close to Broecker and Peng's (1993) most recently published value. The global mean relative (to export flux) accumulation rate, or preservation efficiency, of organic carbon

is within the range proposed by Berger (1989) (0.88% vs. 0.3 - 3%).

The comparison of model downward POC fluxes with direct measurements using sediment traps has revealed large, systematic discrepancies that could not be reconciled. At all shallow trap positions presented here, the mean model particle fluxes are higher as compared to the sediment trap data. This is consistent with independent results from ^{230}Th calibrations (Buesseler 1991; Scholten et al. 2001) and suggests that sediment traps tend to underestimate vertical particle fluxes at shallow water depths. For the deep traps, radionuclide calibrations indicate over-trapping as well as under-trapping (Scholten et al. 2001), whereas the model results shown in Fig. 8 suggest that overestimation of deep POC fluxes by traps might be a common problem. Lateral input of re-suspended material in the vicinity of sloping bathymetry or near the bottom is a process that could explain an over-sampling by traps. This has also been suggested by Fischer et al. (2000), Abelmann and Gersonde (1991), and Siegel and Deuser (1997).

Comparison of POC fluxes to the seafloor and POC burial in the sediment (Figs 4 and 5) with the POC export from the euphotic zone (Fig. 3) reveals significantly different patterns and shows that major surface productivity features (especially in open-ocean areas over deep water) are not, or only weakly, represented in bottom fluxes. This is attributable to the relatively rapid remineralization of POC with depth, that allows significant portions of the exported material to reach the seafloor in coastal and shelf regions, but leaves little material in deep-sea sediments. Reconstructing surface productivity from sediment accumulation rates, which involves the mapping of (small) bottom signals to much larger surface fluxes, therefore requires a good knowledge of the remineralization in the water column and sediment. The projection of sediment trap data to a common depth level, which is commonly performed to produce maps of downward particle fluxes, also relies on a known depth dependence of the POC flux. It is common practice to use formulations of POC flux versus depth from the literature (Bishop 1989; Martin et al. 1987; Suess 1980) for this purpose. However, these approaches employ different function types and parameters and yield projected fluxes that are significantly different, indicating that remineralization and depth-dependence of POC in the ocean is actually poorly known. Results of Usbeck et al. (2002) suggest that remineralization scale heights might vary regionally and that none of the published formulations are universally applicable.

Acknowledgements

We acknowledge inspiring scientific discussions with Richard Jahnke, Mike Behrenfeld, Ken Buesseler, Michiel Rutgers van der Loeff, Peter Müller and Susanne Neuer. A detailed and constructive review by Andreas Oschlies helped to improve the manuscript. This work was funded by the Deutsche Forschungsgemeinschaft (Sonderforschungsbereich 261, publication no. 376). Data are available under www.pangaea.de/Projects/SFB261.

References

Abelmann A, Gersonde R (1991) Biosiliceous particle flux in the Southern Ocean. Mar Chem 35: 503-536

Anderson LA, Sarmiento J (1995) Global ocean phosphate and oxygen simulations. Glob Biogeochem Cycl 9: 621-636

Anderson TR, le B Williams PJ (1999) A one-dimensional model of dissolved organic carbon cycling in the water column incorporating combined biological-photochemical decomposition. Glob Biogeochem Cycl 13: 337-349

Antoine D, Andre J-M, Morel A (1996) Oceanic primary production 2. Estimation at global scale from satellite (coastal zone color scanner) chlorophyll. Glob Biogeochem Cycl 10: 57-69

Arrigo KR, Worthen D, Schnell A, Lizotte MP (1998) Primary production in Southern Ocean waters. J Geophys Res 103: 15587-15600

Bacon MP, Huh C-A, Fleer AP, Deuser WG (1985) Seasonality in the flux of natural radionuclides and plutonium in the deep Sargasso Sea. Deep-Sea Res 32: 273-286

Baker ET, Milburn HB, Tennant DA (1988) Field assessment of sediment trap efficiency under varying flow conditions. J Mar Res 46: 573-592

Behrenfeld MJ, Falkowski PG (1997) Photosynthetic rates derived from satellite-based chlorophyll concentration. Limnol Oceanogr 42: 1-20

Berger WH (1989) Appendix. Global maps of ocean productivity. In: Berger WH, Smetacek VS, Wefer G (eds) Productivity of the Ocean: Present and Past. J Wiley & Sons, Chichester, pp 429-455

Berner E, Berner RA (1996) Global Environment: Water, Air, and Geochemical Cycles. Prentice Hall, New Jersey

Bishop JKB (1989) Regional extremes in particular matter composition and flux: Effects on the chemistry of the ocean interior. In: Berger WH, Smetacek VS, Wefer G (eds) Productivity of the Oceans: Present and Past. J Wiley & Sons, Chichester, pp 117-137

Broecker WS, Peng T-H (1982) Tracers in the Sea. Lamont-Doherty Geological Observatory, Columbia University, Palisades, NY

Broecker WS, Blanton S, Smethie W, Östlund G (1991) Radiocarbon decay and oxygen utilization in the deep Atlantic Ocean. Glob Biogeochem Cycl 5: 87-117

Broecker WS, Peng T-H (1993) Greenhouse puzzles. Lamont- Doherty Earth Observatory of Columbia University, Palisades, NY

Buesseler KO (1991) Do upper-ocean sediment traps provide an accurate record of particle flux? Nature 353: 420-423

Buesseler KO, Michaels AF, Siegel DA, Knap H (1994) A three dimensional time-dependent approach to calibrating sediment trap fluxes. Glob Biogeochem Cycl 8: 179-193

Campbell J, Antoine D, Armstrong R, Arrigo K, Balch W, Barber R, Behrenfeld M, Bidigare R, Bishop J, Carr M-E (2002) Comparison of algorithms for estimating ocean primary production from surface chlorophyll, temperature and irradiance. Glob Biogeochem Cycl 16: DOI 10.1029/2001GB001444

Codispoti LA, Christensen JP (1985) Nitrification, denitrification and nitrous oxide cycling in the eastern tropical south Pacific Ocean. Mar Chem 16: 277-300

Codispoti LA, Friederich GE, Packard TT, Glover HE, Kelly PJ, Spinrad RW, Barber RT, Elkins JW, Ward BB, Lipschultz F, Lostaunau N (1986) High nitrite levels off nothern Peru: A signal of instability in the marine denitrification rate. Science 233: 1200-1202

Conte MH, Ralph N, Ross EH (2001) Seasonal and interannual variability in deep ocean particle fluxes at the Oceanic Flux Program (OFP)/Bermuda Atlantic Time Series (BATS) site in the western Sargasso Sea near Bermuda. Deep-Sea Res Part II: Topical Studies in Oceanography 48: 1471-1505

de Baar HJW, de Jong JTM, Bakker DCE, Löscher BM, Veth C, Bathmann U, Smetacek V (1995) Importance of iron for plankton blooms and carbon dioxide drawdown in the Southern Ocean. Nature 373: 412-414

de las Heras M, Schlitzer R (1999) On the importance of intermediate water flows for the global ocean overturning. J Geophys Res 104: 15515-15536

Deuser WG, Jickells TD, King P, Commeau JA (1995) Decadal and annual changes in biogenic opal and carbonate fluxes to the deep Sargasso Sea. Deep-Sea Res 42: 1923-1932

Fasham MJR, Ducklow HW, McKelvie SM (1990) A nitrogen-based model of plankton dynamics in the oceanic mixed layer. J Mar Res 48: 591-639

Fischer G, Wefer G (1996) Long-term observation of particle fluxes in the eastern Atlantic: Seasonality, changes of flux with depth and comparison with the sediment record. In: Siedler G, Wefer G, Berger WH, Webb D (eds) The South Atlantic: Present and Past Circulation. Springer, Berlin, pp 325-344

Fischer G, Fütterer D, Gersonde R, Honjo S, Ostermann D, Wefer G (1988) Seasonal variability of particle flux in the Weddell Sea and its relation to ice cover. Nature 335: 426-428

Fischer G, Donner B, Ratmeyer V, Davenport R, Wefer G (1996) Distinct year-to-year particle flux variations off Cape Blanc during 1988-1991: Relation to δ^{18}O-deduced sea-surface temperatures and trade winds. J Mar Res 54: 73-98

Fischer G, Ratmeyer V, Wefer G (2000) Organic carbon fluxes in the Atlantic and the Southern Ocean: Relationship to primary production compiled from satellite radiometer data. Deep-Sea Res II 47: 1961-1997

Fischer G, Gersonde R, Wefer G (2002) Organic carbon, biogenic silica and diatom fluxes in the marginal winter sea-ice zone and in the Polar Front Region: interannual variations and differences in composition. Deep-Sea Res Part II: Topical Studies in Oceanography 49: 1721-1745

Gruber N, Sarmiento JL (1997) Global patterns of marine nitrogen fixation and denitrification. Glob Biogeo-chem Cycl 11: 235-266

Gust G, Michaels AF, Johnson RG, Deuser WG, Bowles W (1994) Mooring line motions and sediment trap hydromechanics: *In situ* intercomparison of three common deployment designs. Deep-Sea Res 41: 831-857

Hansell DA, Carlson CA (1998) Net community production of dissolved organic carbon. Glob Biogeochem Cycl 12: 443-453

Hensen C, Landenberger H, Zabel M, Schulz HD (1998) Quantification of diffusive benthic fluxes of nitrate, phosphate and silicate in the southern Atlantic Ocean. Glob Biogeochem Cycl 12: 193-210

Hestenes MR (1975) Optimization Theory. J Wiley & Sons, Inc, New York

Honjo S, Manganini SJ, Cole JJ (1982) Sedimentation of biogenic matter in the deep ocean. Deep-Sea Res 29: 609-625

Honjo S, Manganini SJ (1996) Annual biogenic particle fluxes to the interior of the North Atlantic Ocean studied at 34°N 21°W and 48°N 21°W. In: Milliman JD (ed) Topical Studies in Oceanography JGOFS: The North Atlantic Bloom Experiment. Pergamon Press, Oxford, pp 587-607

Howell EA, Doney SC, Fine RA, Olson DB (1997) Geochemical estimates of denitrification in the Arabian Sea and the Bay of Bengal during WOCE. Geoph Res Lett 24: 2549-2552

Jahnke RA (1996) The global ocean flux of particulate organic carbon: areal distribution and magnitude. Glob Biogeochem Cycl 10: 71-88

Jenkins WJ (1982) Oxygen utilization rates in north Atlantic subtropical gyre and primary production in oligotrophic systems. Nature 300: 246-248

Jenkins WJ, Goldman JC (1985) Seasonal oxygen cycling and primary production in the Sargasso Sea. J Mar Res 43: 465-491

Jenkins WJ (1987) 3H and 3He in the Beta Triangle: Observations of gyre ventilation and oxygen utilization rates. J Phys Oceanogr 17: 763-783

Kirchman DL, Suzuki Y, Garside C, Ducklow HW (1991) High turnover rates of dissolved organic carbon during a spring phytoplankton bloom. Nature 352: 612-614

Laws EA, Falkowski PG, Smith WO, Ducklow H, McCarthy JJ (2000) Temperature effects on export production in the open ocean. Glob Biogeochem Cycl 14: 1231-1246

Lerman A (1980) Controls on river water composition and mass balance of river systems. In: Martin JM (ed) River Inputs to Ocean Systems. J Wiley & Sons, New York, pp 1 - 12

Longhurst A, Sathyendranath S, Platt T, Caverhill C (1995) An estimate of global primary production in the ocean from satellite radiometer data. J Plankt Res 17: 1245-1271

Maier-Reimer E (1993) Geochemical cycles in an ocean general circulation model. Preindustrial tracer distributions. Glob Biogeochem Cycl 7: 645-677

Martin JH, Knauer GA, Karl DM, Broenkow WW (1987) VERTEX: Carbon cycling in the northeast Pacific. Deep-Sea Res 34: 267-285

Najjar RG, Sarmiento JL, Toggweiler JR (1992) Downward transport and fate of organic matter in the ocean: simulations with a general circulation model. Glob Biogeochem Cycl 6: 45-76

Neuer S, Ratmeyer V, Davenport R, Fischer G, Wefer G (1997) Deep water particle flux in the Canary Island region: seasonal trends in relation to long-term satellite derived pigment data and lateral sources. Deep-Sea Res I: Oceanographic Research Papers 44: 1451-1466

Nierenberg WA, Nowlin WD, Jr. (1985) Physical, chemical and *in situ* CTD data from the AJAX Expedition aboard RV KNORR. Scripps Inst. Oceanography, La Jolla, CA, SIO Reference 85-24

Popova EE, Ryabchenko VA, Fasham MJR (2000) Biological pump and vertical mixing in the Southern Ocean: Their impact on atmospheric CO_2. Glob Biogeochem Cycl 14: 477-498

Reid JL (1986) On the total geostrophic circulation of the South Pacific Ocean: Flow patterns, tracers, and transports. Prog Oceanogr 16: 1-61

Reid JL (1997) On the total geostrophic circulation of the Pacific Ocean: Flow patterns, tracers, and transports. Prog Oceanogr 39: 263-352

Riley GA (1951) Oxygen, phosphate, and nitrate in the Atlantic Ocean. Bull Bingham Oceanogr Coll 13: 1-124

Rintoul S (1991) South Atlantic interbasin exchange. J Geophys Res 96: 2675-2692

Rintoul S, Wunsch C (1991) Mass, heat, oxygen and nutrient fluxes and budgets in the North Atlantic Ocean. Deep-Sea Res 38 (suppl): 355-377

Sarmiento JL, Toggweiler JR (1984) A new model for the role of the oceans in determining atmospheric pCO_2. Nature 308: 621-624

Sarmiento JL, Monfray P, Maier-Reimer E, Aumont O, Murnane R, Orr J (2000) Sea-air CO2 fluxes and carbon transport: a comparison of three ocean general circulation models. Glob Biogeochem Cycl 14: 1267-1281

Schlitzer R (1993) Determining the mean, large-scale circulation of the Atlantic with the adjoint method. J Phys Oceanogr 23: 1935-1952

Schlitzer R (1995) An adjoint model for the determination of the mean oceanic circulation, air-sea fluxes and mixing coefficients. Alfred-Wegener-Institut, Bremerhaven, Ber Polarforsch 156

Schlitzer R (1996) Mass and heat transports in the South Atlantic derived from historical hydrographic data. In: Siedler G, Wefer G, Berger WH, Webb D (eds) The South Atlantic: Present and Past Circulation. Springer, Berlin, pp 305-323

Schlitzer R (2000) Applying the adjoint method for glo-bal biogeochemical modeling. In: Kasibhatla P, Heimann M, Hartley D, Mahowald N, Prinn R, Rayner P (eds) Inverse Methods in Global Biogeochemical Cycles. AGU Geophys. Monograph Series, Vol. 114, pp 107-124

Schlitzer R (2002) Carbon export fluxes in the Southern Ocean: Results from inverse modeling and comparison with satellite based estimates. Deep-Sea Res II 49: 1623-1644

Scholten JC, Fietzke J, Vogler S, Rutgers van der Loeff MM, Mangini A, Koeve W, Waniek J, Stoffers P, Antia A, Kuss J (2001) Trapping efficiencies of sediment traps from the deep Eastern North Atlantic: The 230Th calibration. Deep-Sea Res II: Topical Studies in Oceanography 48: 2383-2408

Siegel DA, Deuser WG (1997) Trajectories of sinking particles in the Sargasso Sea: Modeling of statistical funnels above deep-ocean sediment traps. Deep-Sea Res I 44: 1519-1541

Siegenthaler U, Wenk T (1984) Rapid atmospheric CO_2 variations and ocean circulation. Nature 308: 624-626

Six KD, Maier-Reimer E (1996) Effects of plankton dynamics on seasonal carbon fluxes in an ocean general circulation model. Glob Biogeochem Cycl 10: 559-583

Suess E (1980) Particulate organic carbon flux in the oceans-surface productivity and oxygen utilization. Nature 288: 260-263

Sundquist ET (1985) Geological perspectives on carbon dioxide and the carbon cycle. In: Sundquist E, Broecker W (eds) The Carbon Cycle and Atmospheric CO_2: Natural Variations Archean to Present. AGU Geophysical Monograph, Washington DC, vol 32, pp 5-59

Thacker WC, Long RB (1988) Fitting dynamics to data. J Geophys Res 93: 1227-1240

Usbeck R (1999) Modeling of marine biogeochemical cycles with an emphasis on vertical particle fluxes. Alfred Wegener Institute, Bremerhaven, Rep Polar Res 332, 105 p

Usbeck R, Loeff MRvd, Hoppema M, Schlitzer R (2002) Shallow remineralization in the Weddell Gyre. Geo-chem Geophys Geosyst 3: 10.1029/2001GC000182

Usbeck R, Schlitzer R, Fischer G, Wefer G (2003) Particle fluxes in the ocean: Comparison of sediment trap data with results from inverse modeling. J Mar Sys 39: 167-183

Volk T, Hoffert MI (1985) Ocean carbon pumps: analysis of relative strengths and efficiencies in ocean-driven atmospheric CO_2 changes. In: Sundquist ET,

Broecker WS (eds) The Carbon Cycle and Atmospheric CO_2: Natural Variations Archean to Present. AGU Geophysical Monograph 32, Washington DC, pp 99-110

Walsh I, Dymond J, Collier R (1988) Rates of recycling of biogenic components of settling particles in the ocean derived from sediment trap experiments. Deep-Sea Res 35: 43-58

Wefer G, Suess E, Balzer W, Liebezeit G, Müller PJ, Ungerer CA, Zenk W (1982) Fluxes of biogenic components from sediment trap deployment in circumpolar waters of the Drake Passage. Nature 299: 145-147

Whitworth T, III, Nowlin WD Jr (1987) Water masses and currents of the southern ocean at the Greenwich meridian. J Geophys Res 92: 6462-6476

WOCE Data Products Committee (2000) WOCE global data, version 2.0. WOCE Intern. Project Office, Southampton, UK, 171/00

Yamanaka Y, Tajika E (1996) The role of the vertical fluxes of particulate organic matter and calcite in the oceanic carbon cycle: Studies using an ocean biogeo-chemical general circulation model. Glob Biogeo-chem Cycl 10: 361-382

Yamanaka Y, Tajika E (1997) Role of dissolved organic matter in the marine biogeochemical cycle: Studies using an ocean biogeochemical general circulation model. Glob Biogeochem Cycl 11: 599-612

Zabel M, Dahmke A, Schulz HD (1998) Regional distributions of diffusive phosphate and silicate fluxes through the sediment-water interface: The eastern South Atlantic. Deep-Sea Res I 45: 277-300

Transfer of Particles into the Deep Atlantic and the Global Ocean: Control of Nutrient Supply and Ballast Production

G. Fischer[1*], G. Wefer[1], O. Romero[1], N. Dittert[2], V. Ratmeyer[1] and B. Donner[1]

[1] *Universität Bremen, Fachbereich Geowissenschaften, Klagenfurter Strasse, 28359 Bremen, Germany*
[2] *Institute Universitaire Européen de la Mer, Technopôle Brest-Iroise, Place Nicolas Copernic, F-29280 Plouzané, France*
* *corresponding author: gerhard.fischer@rcom-bremen.de*

Abstract: Particle fluxes from 20 trap sites in the Atlantic/Southern Ocean have been compiled to study the regional variations in comparison with important environmental variables. In turn, these results have been compared to other study sites from the world ocean, mainly regarding the relationship between bulk fluxes/various flux ratios to nutrient supply. It is shown that the supply of dissolved silicic acid to the surface waters (the 'silicate pump', Dugdale et al. 1995) plays a central role in opal fluxes, $BSi:C_{org}$ ratios, BSi:carbonate ratios, and thus carbon rain ratios. The mean annual $BSi:C_{org}$ ratio (mol/mol) normalized to 1000 m was 0.05 in the Atlantic, 0.4 in the Indian, 0.5 in the Pacific, and 0.1-3 in the Southern Ocean and follows the general path of the conveyor belt (Ragueneau et al. 2000). A shift in the primary producer community from coccolithophorids to diatoms, reflected by an exponential increase of the annual BSi:carbonate flux ratios, occurs above a molar $Si:N_{(250m)}$ nutrient threshold of about 1.7. The surface sediment opal:carbonate ratios (%) versus the $Si:N_{(250m)}$ nutrient values produce a threshold of 2-2.5, however, this value may be biased by opal dissolution during early diagenesis. We also tested the most recent findings about particle ballast which presume that carbonate is most important for the rapid downward transport of organic particles to bathypelagic depths. Our compilation of global flux data confirms such a general relationship. However, at certain sites and in particular years/seasons, other minerals may serve as ballast for organic carbon. Off NW Africa, for instance, lithogenic components were the major particle carriers. There, relationships between carbonate/lithogenic/total ballast fluxes versus daily organic carbon fluxes may even vary from year to year. Off Cape Blanc, the carbonate-C_{org}-relationship is highly significant during a strong coccolithophorid bloom in 1991, probably resulting in an efficient downward transfer of organic carbon. Interannual variation of fluxes was highest in high production systems combined with high seasonality of fluxes. We obtained ca. 20% variability in oligotrophic regions and up to 100% in the Southern Ocean where seasonality is most pronounced.

Introduction

A central goal of the Joint Global Ocean Flux Study (JGOFS) was to understand the physicochemical and biological conditions that control the regional variations of particle fluxes and the strength of the organic carbon pump which may drawdown oceanic and atmospheric CO_2. This drawdown is also dependent on the production of biogenic carbonate (calcium carbonate=calcite and aragonite) and, consequently, on the carbon rain ratio ($CRR= C_{org}/C_{carbonate}$ = POC/PIC). The CRR may be applied to describe the efficiency of the biological pump in the world oceans (e.g. Tsunogai and Noriki 1991). Basin-wide fluxes in the Atlantic Ocean and an 'effective carbon flux' (a measure of atmospheric CO_2 sequestration) below the euphotic zone and the depth of the mixed layer have been estimated by

From WEFER G, MULITZA S, RATMEYER V (eds), 2003, *The South Atlantic in the Late Quaternary: Reconstruction of Material Budgets and Current Systems.* Springer-Verlag Berlin Heidelberg New York Tokyo, pp 21-46

Antia et al. (2001): for the Atlantic Ocean between 65°N and 65°S, a value of 2.5 GT C yr^{-1} was calculated. This study also demonstrated that organic carbon flux differences show little variation at depths greater than 3000 m, despite very different primary production and export ratios at shallower depths. This was confirmed by modeling studies performed by Jackson and Burd (2002) and is consistent with bottom-near organic carbon fluxes estimated by oxygen uptake rates (Jahnke 1996). Antia et al. (2001) also pointed out the non-linear relationship between the export fraction and primary production which was also found by Betzer et al. (1984), Lampitt and Antia (1997) and Fischer et al. (2000).

Biogenic silica (BSi) or opal production and export, mostly provided by diatoms, is also crucial for the organic carbon export fluxes to the deep-sea and the CRR of settling particles (Dugdale et al. 1995; Pollock 1997) as was also demonstrated by the OPALEO group (Ragueneau et al. 2000). BSi production and flux show a strong dependency on the supply of dissolved silicic acid ('silicate pump', Dugdale et al. 1995). In addition, Ragueneau et al. (2000) found a relationship between the Si:N ratios in the source waters (250m) and the BSi:carbonate flux ratios, and emphasized that the flux ratio increases dramatically at a nutrient Si:N value of about 2. Fluxes of some components, such as organic carbon, BSi and carbonate, have also been used as productivity proxies (see Wefer et al. 1999, and references therein). Under certain circumstances modern relationships between organic carbon and calcium carbonate can be used to reconstruct the past primary production at sites in the oligotrophic Atlantic (Rühlemann et al. 1999; Brummer and van Eijden 1992). BSi:C$_{org}$ ratios of sinking particles, although highly variable in space and time (Ragueneau et al. 2000), are now much better understood (e.g. Ragueneau et al. 2002) and may be incorporated in biogeochemical models in the near future. Recent studies show the role of iron in the control of nutrient uptake ratios and BSi:C$_{org}$ and BSi:N ratios of diatoms (Hutchins and Bruland 1998; Takeda 1998).

Armstrong et al. (2002) emphasized that ballast minerals largely determine deep-water fluxes rather than primary production changes in the surface layer. According to these modeling approaches, two fractions of organic carbon exist, one associated with and probably protected by minerals, and another appearing to be more labile and degrading within the upper few 100 m of the water column. Following the study by Francois et al. (2002), bathy-pelagic fluxes are largely determined by the carbonate fluxes and their seasonality, resulting in different sinking velocities of the particles and changing the biodegradability of organic carbon. According to Francois et al. (2002), other ballast minerals such as lithogenic particles and biogenic opal, although much denser than organic particles, do not significantly affect the 'transfer efficiency'. Berelson (2002) found that the lithogenic particles do not increase sinking velocities, and thus do not reduce organic carbon degradation in the water column.

Long-term data on particle fluxes have become increasingly important (e.g. Deuser et al. 1995; Karl et al. 1997; Haake et al. 1996; Thunell 1998; Wong et al. 1999; Steinberg et al. 2001) but they are not available from all ocean areas. In particular, this holds true for the Southern Ocean; data from five years of sampling downward particulate matter from the Polar Front and a site located close to the mean winter sea-ice edge revealed a high inter-annual variability (Fischer et al. 2002). Long-term data may provide important information on potential effects of climate change on particle formation, and the composition and export of important primary producers. Antia et al. (2001) compiled the opal: carbonate flux ratios from the North Atlantic Ocean, including the values provided by Deuser et al. (1995). This data set indicates that a general decrease in the ratio of primary producers (diatoms vs coccolithophorids) may have occurred since about 1980. Applying the findings of Ragueneau et al. (2000), such a change might be explained by a decrease in the Si:N nutrient ratio of the source waters.

In this review paper we present results obtained during the last decade in the Atlantic and the Southern Ocean with respect to fluxes into the oceans' interior. We studied the latitudinal changes of fluxes and important flux ratios in the Atlantic Ocean. We also discuss the relationships between satellite-derived primary production and modeled export to a depth of 133 m (Schlitzer 2002, pers comm) and the organic carbon fluxes in 1000 m water depth. We incorporated recent findings about the flux of mineral ballast which obviously determines the transfer of organic carbon to the deep-sea to a major extent (Armstrong et al. 2002). We further rein-vestigated the relationship between nutrient supply and BSi:carbonate flux ratios (Ragueneau et al. 2000), to determine the Si:N nutrient threshold value where the production switches from carbonate-secreting organisms to diatoms. For comparison, we compiled data from surface sediments from the world ocean. Finally, long-term changes of particle fluxes will be briefly discussed.

Papers dealing with specific aspects of certain sites in the Atlantic Ocean were published by Ratmeyer et al. (1999a, b: dust fluxes at CI, CB, CV); Romero et al. (1999a, b, 2000, 2002a, b: diatoms at CB, WA, diatoms, foraminifera and coccolithophorids at NU); Treppke et al. (1996a, b: diatoms, silicoflagellates and alkenones at WR, diatoms at sites GBN, GBS); Lange et al. (1994, diatoms at GBN, GBS); Boltovskoy et al. (1993a, b, 1996: radiolarians at GBN, CB); Holmes et al. (2002: nitrogen isotopes at WR); Müller and Fischer (2001: alkenones at CB); Fischer et al. (1997, 1998: stable organic carbon isotopes); Fischer et al. (1999: fluxes of pteropods and foraminifera and their stable isotopes) and Walter et al. (2001: radionuclide fluxes in the Southern Ocean). Other papers in this volume related to this study deal with stable nitrate isotope ratios of sinking matter and surface sediments sampled in the Atlantic/Southern Ocean (Holmes et al. this volume) and the biogeochemical modeling of fluxes (Schlitzer et al. this volume). Alkenones as proxies for SST were tested using particle flux data (Müller and Fischer this volume).

Study Area

Our study sites were distributed over the Atlantic Ocean from about 30°N to 70°S (Fig. 1) and encompassed eutrophic, mesotrophic and oligotrophic zones with a wide range of mixed layer depths and nutrient supply (Fig. 2). According to Antoine et al. (1996), primary productivity ranged from about 40 in the Southern Ocean to about 328 gC m^{-2} yr^{-1} in the Namibian upwelling area. (Table 1). Several sites (CI, CB, WR, NU) were located in eastern boundary currents (Canary Current, Benguela Current). Another focus of the studies was set on the eastern equatorial Atlantic around 10°W (sites EA, GB, Table 1) and, for comparison, in the western equatorial Atlantic between about 23° and 28°W (sites WA, Table 1, Fig. 1). Both transects (EA and WA) differ in their oceanographic settings with a shallower thermocline and higher primary production in the eastern part of the tropical Atlantic.

Upon termination of our investigations within the special research project SFB 261, we deployed arrays in the Northern Brazil Basin, located at the northern edge of the subtropical gyre (site WAB, Table 1), to study the fluxes at an oligotrophic site which is presumably not influenced by Saharan dust particles. A more detailed description of the oceanographic background at the various sites has been published by Fischer et al. (2000, 2002); Romero et al. (2000, 2002); Holmes et al. (2002) and Neuer et al. (1997). For further information on the supply of Saharan dust to the three North Atlantic sites CI, CB and CV, we recommend the paper of Ratmeyer et al. (1999a). A comprehensive description of the different biogeochemical provinces in the Atlantic Ocean is given by Longhurst et al. (1995). The hydrography, ecology and sedimentation in the Antarctic Circumpolar Current is briefly described by Smetacek et al. (1997).

Material and Methods

We used moored, large-aperture time-series sediment traps mostly of the Kiel type with 20 cups and 0.5m^2 openings (Kremling et al. 1996). At several sites in the Southern Ocean, we also applied HONJO-type traps (12 and 22 cups) possessing 1.17 and 0.5 m^2 openings (Honjo and Doherty 1988). All types had comparable aspect ratios, and we assume no significant bias for particle collection. Sampling cups were poisoned with HgCl$_2$ and pure NaCl was used to increase the density in the cups (40‰). We did not correct fluxes for the dissolution of different components in the sampling cups after particle collection which means that our given fluxes underestimate the true values. After splitting, particles were wet-sieved in the home laboratory through a 1-mm mesh. All fluxes presented here refer to this size fraction. We did not

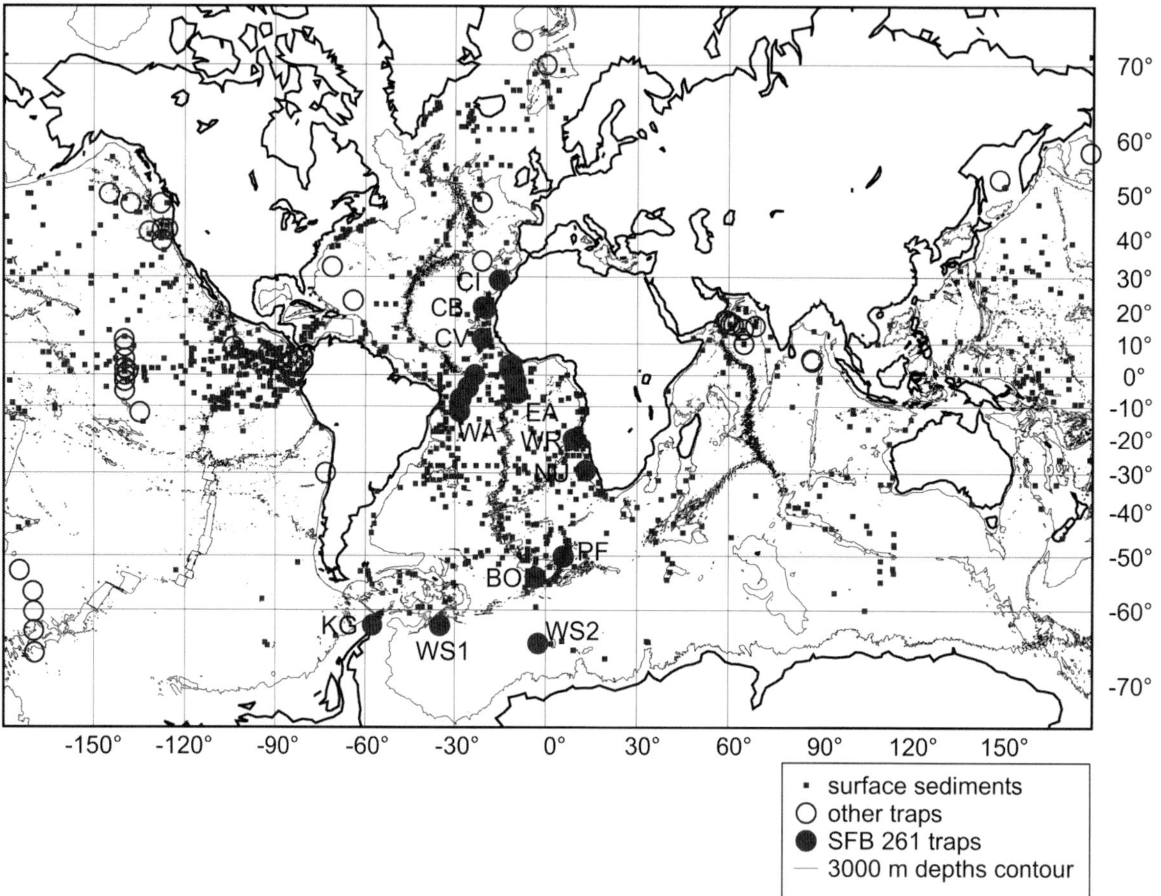

Fig. 1. Location of the sediment trap sites (SFB 261 traps in the Atlantic Ocean, other sites, modified data set from Ragueneau et al. 2000; Honjo et al. 2000), and the surface sediments (0-5 cm core depth).

pick swimmers by hand from the fraction < 1mm due to the generally deep deployment depths (Table 1). In almost all cases, the fraction of particles > 1mm is negligible (Fischer et al. 1996). Analysis of the fraction <1mm, using 1/4 and 1/5 wet splits, was performed according to Fischer and Wefer (1991). Organic carbon and calcium carbonate were measured with a CHN-Analyser (HERAEUS), biogenic opal was determined according to Müller and Schneider (1993). We did not include any water content (which is variable) in the opal calculations. Lithogenic fluxes were calculated according to: lith = total mass – carbonate opal – 2 *C_{org}. Ballast fluxes were calculated according to: ballast = total flux – 1.87*C_{org}.

We made a new selection of the data due to most recent findings based on the radionuclide studies of Yu et al. (2001) and Scholten et al. (2001) which indicate that the shallower flux data are mostly biased and seem to notably underestimate the true fluxes. Many data were from depth levels of around 1000 m (not shallower than 700 m), but the largest part originated from bathypelagic depths (Table 1). Thus, we were well below the zone important for the remineralization of organic carbon (Armstrong et al. 2002). Due to the different sampling depths, we made a normalization of organic carbon and total nitrogen fluxes to 1000 m, using the Martin function (Martin et al. 1987), however, with a variable b-value for organic carbon as suggested by Francois et al. (2002). On average, we applied a value of -0.86 (tropical-subtropical regions). We used b=-1.1 (Schlitzer 2002) for the Southern Ocean, a value which we also applied to

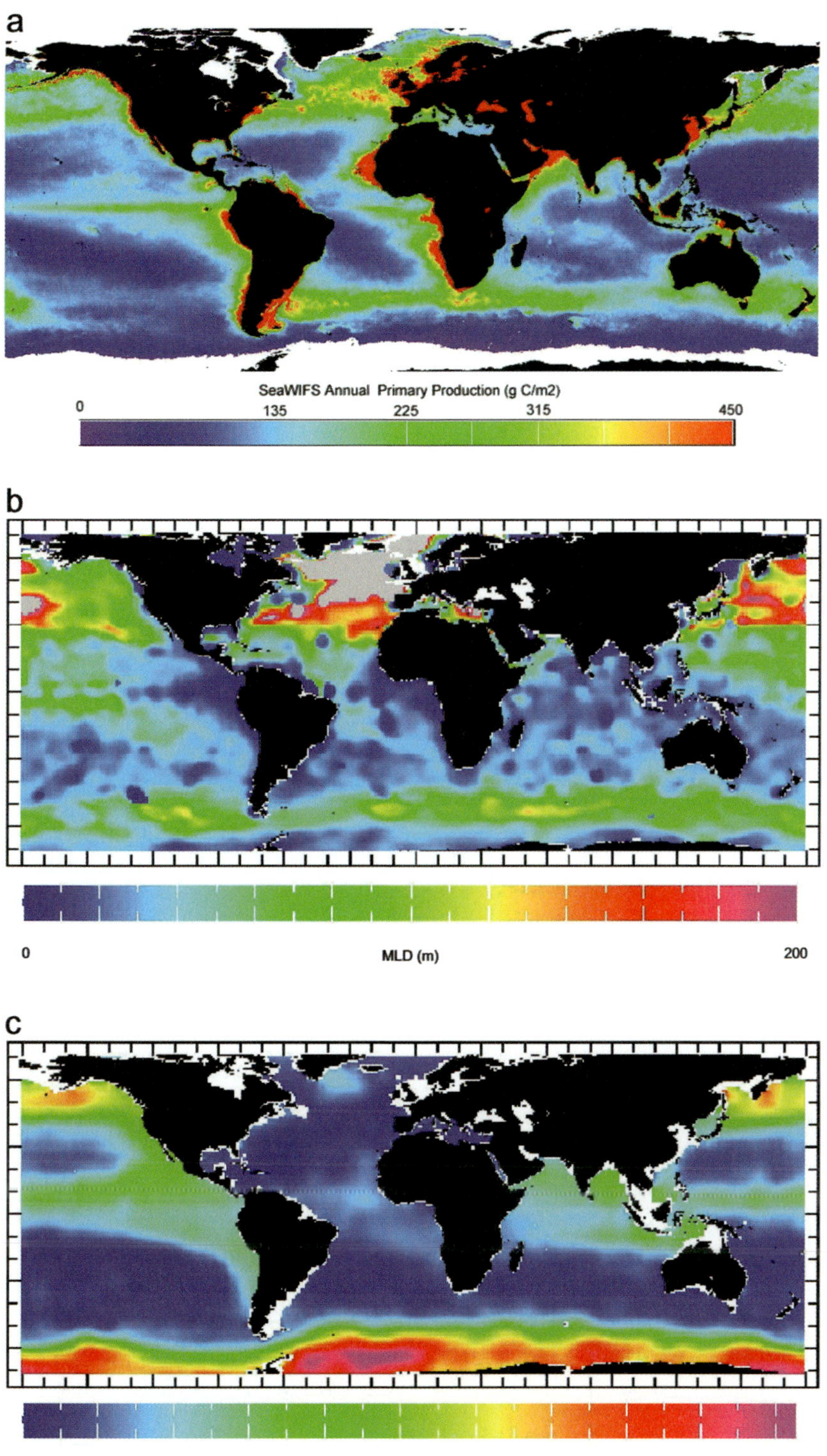

Fig. 2. a) SeaWIFS-derived primary production from 1988 (September) to 1999 (August) according to Behrenfeld and Falkowski (1997: http://marine.rutgers.edu/opp/swf/ Production/results/all2_swf.html). **b)** Mixed layer depths (=MLD, scale only to 200 m) in the world ocean extracted from Levitus et al. (1994, http://ingrid.ldeo.columbia.edu/ SOURCES/.LEVITUS94/.MONTHLY/) **c)** Dissolved silicic acid from 250 m water depths from the world ocean extracted from Levitus et al. (1994, http://ingrid.ldeo.columbia.edu/SOURCES/.LEVITUS94/.ANNUAL/).

Table 1.

Region	Site	Trap name	Trap depth (m)	Lat	Long	Interval from	to	total duration (days)	BULK FLUXES (g m-2) total	Corg	Corg (1000m)	Ntot	Ntot (1000m)	opal	CaCO3	Litho	MOLAR FLUX RATIOS C:N	BSi:Corg	BSi:N	Corg:Ccarb	BSi:CaCO3	NUTRIENTS (250m) H4SiO4 μmol	NO3- μmol	MOLAR FLUX RATIOS at 1000m C:N	BSi:Corg	BSi:N	Corg:Ccarb	Primary production g C m-2 yr-1	Export to 133 m g C m-2 yr-1	Ballast flux g m-2
NW Africa	CI	CII	1000	29.1	-15.3	25.11.91	25.09.92	305	4.92	0.37	0.37	0.05	0.05	0.12	1.43	2.63	7.96	0.06	0.51	2.17	0.14	5.3	8.1	7.96	0.06	0.51	2.17	113	41	4.2
	CB	CB1	2195	20.7	-19.7	22.03.88	08.03.89	351	64.28	2.62	5.14	0.31	0.68	5.23	27.58	26.22	9.73	0.40	3.89	0.79	0.32	11.1	19.6	8.79	0.20	1.79	1.55	285	85	59.4
		CB2	3502	21.2	-20.7	15.03.89	24.03.90	374	56.88	1.73	5.07	0.19	0.65	2.43	29.41	21.58	10.79	0.28	3.03	0.49	0.14	11.1	19.6	9.17	0.10	0.88	1.44	285	85	53.6
		CB3	3557	21.1	-20.7	29.04.90	08.04.91	344	48.05	2.05	6.10	0.25	0.87	1.94	23.32	18.68	9.66	0.19	1.83	0.73	0.14	11.1	19.6	8.19	0.06	0.52	2.18	285	85	44.2
		CB7	3568	21.3	-20.7	10.01.96	27.12.96	352	31.53	1.56	4.64	0.20	0.69	1.46	16.12	13.94	9.23	0.19	1.73	0.80	0.15	11.1	19.6	7.82	0.06	0.49	2.40	285	85	28.6
	CV	CV2	4435	11.5	-21.1	07.03.93	05.04.94	363	11.24	0.45	1.62	0.06	0.26	0.86	2.35	8.03	8.79	0.38	3.34	1.60	0.61	14.7	24.6	7.24	0.11	0.77	5.76	170	48	10.4
																														0.0
E'Equatorial Atlantic	EA 3°N	EA1	984	3.2	-11.3	13.04.91	29.11.91	230	22.49	2.33	2.30	0.32	0.31	2.13	9.13	6.56	8.59	0.18	1.58	2.12	0.39	12.4	24.4	8.63	0.19	1.60	2.10	218	32	18.1
		EA6	901	3.1	-11.9	03.12.91	06.10.92	308	13.58	1.22	1.13	0.16	0.14	1.21	4.56	5.28	8.90	0.20	1.76	2.23	0.44	12.4	24.4	9.17	0.21	1.96	2.07	218	32	11.3
	EA 2°N	GB2	880	1.6	-11.9	15.03.88	18.02.89	339	8.74	0.83	0.76	0.10	0.09	1.01	3.62	2.44	10.03	0.24	2.42	1.92	0.46	11.8	23.7	10.41	0.26	2.75	1.75	196	32	7.2
		GBn3	853	1.8	-11.1	01.03.89	16.03.90	380	29.23	3.16	2.83	0.39	0.34	3.55	14.98	4.39	9.38	0.22	2.11	1.76	0.39	11.8	23.7	9.82	0.25	2.47	1.57	196	32	23.3
		GBn6	859	1.8	-11.1	04.04.90	07.04.91	369	30.94	2.98	2.68	0.40	0.35	3.38	16.58	5.02	8.59	0.23	1.95	1.50	0.34	11.8	23.7	8.98	0.25	2.27	1.34	196	32	25.4
		EA2	953	1.8	-11.3	13.04.91	29.11.91	230	24.24	2.49	2.41	0.36	0.34	2.25	11.60	5.41	8.07	0.18	1.46	1.79	0.32	11.8	23.7	8.18	0.19	1.53	1.73	196	32	19.6
	EA 0°	EA3	1097	0.7	-10.8	13.04.91	29.11.91	230	11.40	0.97	1.03	0.14	0.16	0.60	7.17	1.77	7.88	0.12	0.97	1.12	0.14	11.5	23.7	7.67	0.12	0.89	1.20	219	18	9.6
		EA7	949	0.7	-10.8	05.12.91	06.10.92	307	32.07	2.84	2.73	0.38	0.36	3.06	16.76	6.57	8.73	0.22	1.88	1.41	0.30	11.5	23.7	8.86	0.22	1.98	1.36	219	18	26.8
		EA10	1280	0.0	-10.8	02.04.93	25.03.94	357	14.35	1.81	2.15	0.22	0.27	1.10	9.36	0.28	9.80	0.12	1.20	1.61	0.20	11.5	23.7	9.13	0.10	0.94	1.91	219	18	11.0
	EA 2°S	GBs4	696	-2.2	-9.9	01.03.89	16.03.90	380	9.80	1.13	0.88	0.13	0.09	0.76	5.51	1.27	10.06	0.13	1.34	1.71	0.23	11.5	25.4	11.17	0.17	1.92	1.33	180	38	7.7
		EA4	1068	-2.2	-10.1	13.04.91	29.11.91	230	11.00	0.96	1.01	0.13	0.13	0.65	7.23	0.94	8.99	0.13	1.21	1.11	0.15	11.5	25.4	8.82	0.13	1.13	1.16	180	38	9.2
	EA 4°S	EA5	947	-4.3	-10.3	13.04.91	29.11.91	230	11.92	0.79	0.76	0.12	0.11	0.78	7.81	1.48	8.04	0.20	1.59	0.85	0.17	11.9	27.7	8.17	0.21	1.68	0.81	176	57	10.4
	EA 6°S	EA8	1833	-5.8	-9.4	15.12.91	06.10.92	296	23.05	1.91	2.93	0.30	0.54	1.46	15.50	2.26	7.47	0.15	1.14	1.03	0.16	12.2	28.4	6.27	0.10	0.63	1.57	154	24	19.5
W'Equatorial Atlantic	WA 0°	WA5	3170	0.0	-23.5	30.03.93	05.04.94	371	26.21	1.11	2.99	0.16	0.51	1.60	18.69	3.71	7.90	0.29	2.27	0.50	0.14	11.8	24.4	6.80	0.11	0.73	1.33	151	16	24.1
		WA8u	718	0.0	-23.5	25.08.94	17.08.95	357	23.71	2.49	1.88	0.38	0.27	1.72	14.19	2.94	7.64	0.14	1.05	1.46	0.20	11.8	24.4	7.97	0.18	1.46	1.10	151	16	19.0
		WA11u	834	0.0	-23.4	27.03.96	26.02.97	336	24.26	3.47	2.97	0.45	0.37	0.96	13.82	3.87	9.03	0.06	0.50	2.09	0.12	11.8	24.4	9.25	0.06	0.60	1.79	151	16	17.8
		WA11l	3139	0.0	-23.4	27.03.96	26.02.97	336	21.56	1.18	3.14	0.15	0.45	1.83	12.58	4.79	9.41	0.31	2.92	0.78	0.24	11.8	24.4	8.11	0.12	0.94	2.08	151	16	19.4
		WA12	3173	0.0	-23.4	21.02.97	20.02.98	364	25.03	1.31	3.53	0.16	0.50	2.25	16.92	3.23	9.55	0.34	3.28	0.64	0.22	11.8	24.4	8.22	0.13	1.05	1.74	151	16	22.6
		WA15	3180	0.0	-23.4	28.05.98	14.05.99	352	33.97	1.89	5.10	0.23	0.72	3.31	21.61	5.26	9.63	0.35	3.37	0.73	0.26	11.8	24.4	8.28	0.13	1.08	1.97	151	16	30.4
	WA 4°S	WA4u	808	-4.0	-25.6	28.03.93	05.04.94	373	20.75	2.20	1.83	0.35	0.28	1.46	13.32	1.72	7.45	0.13	0.99	1.38	0.18	9.9	26.5	7.66	0.16	1.22	1.15	124	26	16.6
		WA4l	4555	-4.0	-25.6	28.03.93	05.04.94	373	18.98	0.94	3.44	0.12	0.52	1.64	13.40	2.06	9.43	0.35	3.30	0.58	0.20	9.9	26.5	7.75	0.10	0.74	2.14	124	26	17.2
		WA7u	854	-4.0	-25.7	20.08.94	17.08.95	362	21.96	1.69	1.47	0.25	0.21	1.52	15.35	1.77	7.87	0.18	1.42	0.92	0.17	9.9	26.5	8.04	0.21	1.66	0.80	124	26	18.8
		WA7l	4630	-4.0	-25.7	20.08.94	17.08.95	362	18.02	0.81	3.03	0.10	0.47	1.41	12.38	2.60	9.22	0.35	3.18	0.55	0.19	9.9	26.5	7.55	0.09	0.70	2.04	124	26	16.5
		WA10	4585	-3.9	-25.7	27.03.96	09.02.97	319	7.59	0.37	1.36	0.04	0.20	0.49	5.29	1.07	9.78	0.26	2.57	0.58	0.15	9.9	26.5	8.02	0.07	0.57	2.15	124	26	6.9
	WA 8°S	WA3u	671	-7.5	-28.0	26.03.93	05.04.94	375	10.00	0.95	0.67	0.14	0.09	0.41	6.22	1.20	8.08	0.09	0.69	1.27	0.11	9.8	24.9	8.51	0.12	1.03	0.90	99	15	8.2
		WA3l	5031	-7.5	-28.0	26.03.93	05.04.94	375	6.71	0.38	1.53	0.05	0.24	0.35	4.42	1.06	9.23	0.18	1.70	0.72	0.13	9.8	24.9	7.48	0.05	0.34	2.88	99	15	6.0
		WA6	4410	-7.5	-28.1	18.08.94	17.08.95	364	9.30	0.47	1.69	0.06	0.25	0.34	6.85	1.17	9.55	0.14	1.38	0.58	0.08	9.8	24.9	7.88	0.04	0.32	2.06	99	15	8.4
		WA9	4456	-7.5	-28.1	23.03.96	26.02.97	340	9.82	0.51	1.85	0.06	0.27	0.49	7.22	1.08	9.83	0.19	1.87	0.59	0.11	9.8	24.9	8.10	0.05	0.43	2.14	99	15	8.9
		WA13	4736	-7.5	-28.2	19.03.97	14.03.98	360	4.83	0.23	0.88	0.03	0.12	0.12	3.59	0.64	10.78	0.10	1.08	0.54	0.05	9.8	24.9	8.81	0.03	0.23	2.03	99	15	4.4
		WA14	4705	-7.5	-28.2	24.05.98	14.05.99	356	8.38	0.50	1.88	0.06	0.26	0.57	5.45	1.37	10.25	0.23	2.33	0.76	0.17	9.8	24.9	8.38	0.06	0.50	2.88	99	15	7.4
	WA 11°S	WAB1	4515	-11.5	-28.5	27.02.97	20.02.98	358	4.02	0.20	0.71	0.02	0.09	0.19	2.88	0.56	10.83	0.19	2.09	0.56	0.11	7.1	18.6	8.91	0.05	0.47	2.06	103	11	3.7
		WAB2	710	-11.6	-28.5	22.05.98	14.05.99	356	7.53	0.73	0.54	0.11	0.08	0.28	4.78	1.01	7.90	0.08	0.61	1.26	0.10	7.1	18.6	8.26	0.10	0.86	0.94	103	11	6.2

Table 1 continued

Region	Site	Trap name	Trap depth (m)	Lat	Long	Interval from	to	total duration (days)	BULK FLUXES (g m-2) total	Corg	Corg (1000m)	Ntot	Ntot (1000m)	opal	CaCO3	Litho	MOLAR FLUX RATIOS C:N	BSi:Corg	BSi:N	Corg:Ccarb	BSi:CaCO3	NUTRIENTS at 250m H4SiO4 (µmol)	NO3- (µmol)	MOLAR FLUX RATIOS at 1000m C:N	BSi:Corg	BSi:N	Corg:Ccarb	Primary production (g C m-2 yr-1)	Export to 133 m (g C m-2 yr-1)	Ballast flux (g m-2)
SW Africa	WR	WR1	1640	-20.1	9.2	04.03.88	16.03.89	376	60.84	6.26	9.57	0.67	1.09	8.48	33.86	5.99	10.92	0.27	2.96	1.54	0.42	11.9	28.2	10.24	0.18	1.81	2.35	328	68	49.1
		WR2	1648	-20.1	9.2	18.03.89	13.03.90	360	36.93	3.77	5.79	0.55	0.91	3.02	23.88	2.49	7.96	0.16	1.27	1.32	0.21	11.9	28.2	7.46	0.10	0.78	2.02	328	68	29.9
		WR3	1647	-20.1	9.2	25.03.90	09.04.91	380	24.63	2.86	4.38	0.41	0.67	2.87	12.86	3.79	8.17	0.20	1.64	1.85	0.37	11.9	28.2	7.65	0.13	1.00	2.84	328	68	19.3
		WR4	1717	-20.1	9.0	21.04.91	17.12.91	240	28.68	2.38	3.78	0.40	0.68	3.91	15.00	5.27	6.91	0.33	2.28	1.32	0.43	11.9	28.2	6.44	0.21	1.33	2.10	328	68	24.2
	NU	NU2	2516	-29.2	13.1	20.01.91	03.02.93	380	31.99	1.61	3.56	0.22	0.54	1.99	22.19	4.58	8.59	0.25	2.12	0.61	0.15	8.3	16.4	7.62	0.11	0.85	1.34	205	37	29.0
Southern Ocean (Atlantic S.)	PF	PF3	3196	-50.1	5.8	01.01.90	23.12.90	356	29.40	4.61	16.55	0.29	0.92	8.98	7.54	3.69	18.36	0.39	7.15	5.09	1.98	37.9	32.7	20.91	0.11	2.27	18.28	65	14	20.8
		PF7	3056	-50.1	5.8	19.04.94	29.12.94	255	3.19	0.40	1.37	0.07	0.20	1.55	0.43	0.42	6.97	0.78	5.40	7.75	6.01	37.9	32.7	7.89	0.23	1.79	26.48	65	14	2.4
		PF8	3110	-50.2	5.9	01.01.95	31.12.95	365	6.44	0.51	1.78	0.07	0.21	2.17	2.54	0.71	8.50	0.85	7.23	1.67	1.42	37.9	32.7	9.65	0.24	2.36	5.83	65	14	5.5
	BO	BO1	2194	-54.3	-3.3	01.01.91	31.12.91	365	28.82	0.51	1.21	0.09	0.19	17.49	1.76	8.55	6.84	6.86	46.91	2.41	16.56	66.8	32.7	7.47	2.89	21.58	5.73	40	15	27.9
		BO1/2	2188	-54.3	-3.3	01.01.92	31.12.92	307	2.60	2.90	6.86	0.10	0.21	2.90	1.45	0.31	35.61	0.20	7.12	16.66	3.33	66.8	32.7	38.88	0.08	3.29	39.42	40	15	0.2
		BO5	2251	-54.3	-3.3	01.01.95	31.12.95	365	7.38	0.13	0.32	0.03	0.06	5.03	0.16	1.93	6.07	7.74	46.95	6.77	52.40	66.8	32.7	6.64	3.17	21.06	16.52	40	15	7.1
	KG	KG11	1588	-62.3	-57.5	01.12.83	25.11.84	360	107.30	2.32	3.86	0.34	0.54	38.62	5.20	53.33	7.99	3.33	26.58	3.72	12.38	68.6	33.0	8.42	2.00	16.83	6.19	94	23	103.0
	WS1	WS1u	863	-62.4	-34.8	25.01.85	23.01.86	363	0.26	0.01	0.01	-		-	-	-						95.8	32.1					84	14	0.2
	WS2	WS2l	4556	-64.9	-2.5	20.01.87	20.11.87	304	7.92	0.17	0.92	-		3.97	1.04	2.38	-	4.56	-	1.40	6.38	96.9	34.4		0.86		7.41	68		7.6

Table 1. Positions, particle fluxes for the sampling interval (mostly around one year), various ratios in molar units, and major nutrients (annual means) at the Atlantic/Southern Ocean trapping sites. Annual primary production was derived from Antoine et al. (1996), mixed layer depths (=MLD) and nutrients from 250 m water depths from Levitus et al. (1994). Annual organic carbon export to 133 m water depths (EP$_{(133m)}$) resulted from inverse modeling studies (Schlitzer 2002, pers comm). For normalization of organic carbon fluxes to 1000 m depths, the 'Martin function' was applied (Martin et al. 1987), applying variable b-values (see 'Methods').

the North Pacific and the Nordic Seas, and b= -0.7 for the equatorial Pacific and the upwelling regions in the Guinea Basin and the Arabian Sea.

We used the primary production values of Antoine et al. (1996) which are generally in good agreement with other models (Oschlies and Garcon 1999; Schlitzer 2002). However, as emphasized by Schlitzer (2002), the values from the Southern Ocean seem to be drastically underestimated by a factor of 2-5 when the satellite approach is used. Values derived from inverse modeling studies made by Schlitzer (2002, pers comm) were included to calculate export production, i.e. the flux through 133 m. Annual average nutrients and mixed layer depths were taken from Levitus et al. (1994; http://ingrid.ldeo.columbia.edu/SOURCES/ .LEVITUS94/). Particle flux data from the world ocean used in this study (modified data set after Ragueneau et al. 2000; Honjo et al. 2000) are shown in Table 1. Surface sediment opal:carbonate ratios (0-5 cm core depth) and the corresponding nutrient values used here are found at http://www.pangaea.de/Projects/ORFOIS/Results. In order to avoid extreme values for the corresponding ratios, data from below 4500 m (average CCD in the world ocean) and above 200 m water depths were not included/considered, neither data with very low C_{org} (<0.01%), opal (<1%) and calcium carbonate (<1%) contents.

Results and Discussion

Latitudinal Changes of Fluxes and Nutrient Supply in the Atlantic Ocean

The latitudinal changes of annual total mass, carbonate, organic carbon and BSi fluxes are shown in Fig. 3. Very low to very high mass fluxes were obtained in the Southern Ocean (0.26-107.3 g m^{-2} yr^{-1}, Table 1). On average, both Atlantic coastal upwelling systems off NW and SW Africa, located in the eastern boundary currents, provided higher mass fluxes as compared to the equatorial Atlantic. An almost similar pattern can be found for organic carbon fluxes normalized to 1000 m water depth. Carbonate fluxes in the Southern Ocean were generally lower as compared to the coastal or equatorial Atlantic. The correlation between the organic carbon and carbonate fluxes in the tropical/subtropical Atlantic is highly significant (R^2=0.80, N=43) and the mean carbon rain ratio ($C_{org\ (1000m)}$:$C_{carbonate}$) for the tropical/subtropical Atlantic is at 1.8 (insert in Fig. 3b). This mean value appears to be rather high considering the relatively high carbonate fluxes. Deep water particulates in low fertility areas generally have lower values (e.g. 0.6-1.1 in the North Atlantic; Honjo and Manganini 1993). Our mean value of 1.8 at 1000 m depth resembles data from high productivity areas (e.g. Panama Basin: 2-2.5) and the average value derived from a compilation of Atlantic flux data ($CRR_{(1000m)} \approx 2$; Antia et al. 2001). In the Southern Ocean, the relationship between both components is weaker (R^2=0.67, N=8) and the mean carbon rain ratio is at 14.5 (insert in Fig. 3b), indicating a strong efficiency of the biological pump to draw-down CO_2 in this region, at least on shorter time-scales. This value is higher compared to the rain ratios ranging from 6-10 in the Drake Passage (Wefer et al. 1982) and those derived from the US-AESOPS sites MS-1-3 which ranged from 1.2 to 1.5 at a water depth of 1000 m (Honjo et al. 2000).

The most distinct latitudinal pattern is found for the BSi fluxes, dropping from the Southern Ocean to the North Atlantic site CI at around 30°N (Fig. 3d). As discussed by Ragueneau et al. (2000), this flux pattern is due to the silicic acid supply, i.e. the Si:N nutrient ratios from the source waters, confirming the model of the 'silicate pump' (Dugdale et al. 1995). Ragueneau et al. (2000) obtained very different BSi-C_{org} –relationships for the global ocean, following the path of the conveyor belt. The molar BSi:C_{org} ratios were around 2 in the Southern Ocean, but only 0.14 in the tropical-subtropical Atlantic Ocean (not normalized for C_{org}). In our compilation, we obtained the same value (BSi:$C_{org(1000m)}$=0.14, R^2=0.68, N=43; insert in Fig. 3d). In the Southern Ocean, the BSi:$C_{org\ (1000m)}$ ratios ranged from about 0.1 to about 3. From a recent study in the Atlantic Sector of the Southern Ocean, Fischer et al. (2002) also obtained large seasonal and interannual BSi:C_{org} variations. Five-year mean molar BSi-C_{org} ratios were at 1.3 at the Antarctic Polar Front (site PF), but 4.0 further south at the northernmost ice-edge in winter (site BO). This latitudinal distribution is consistent with

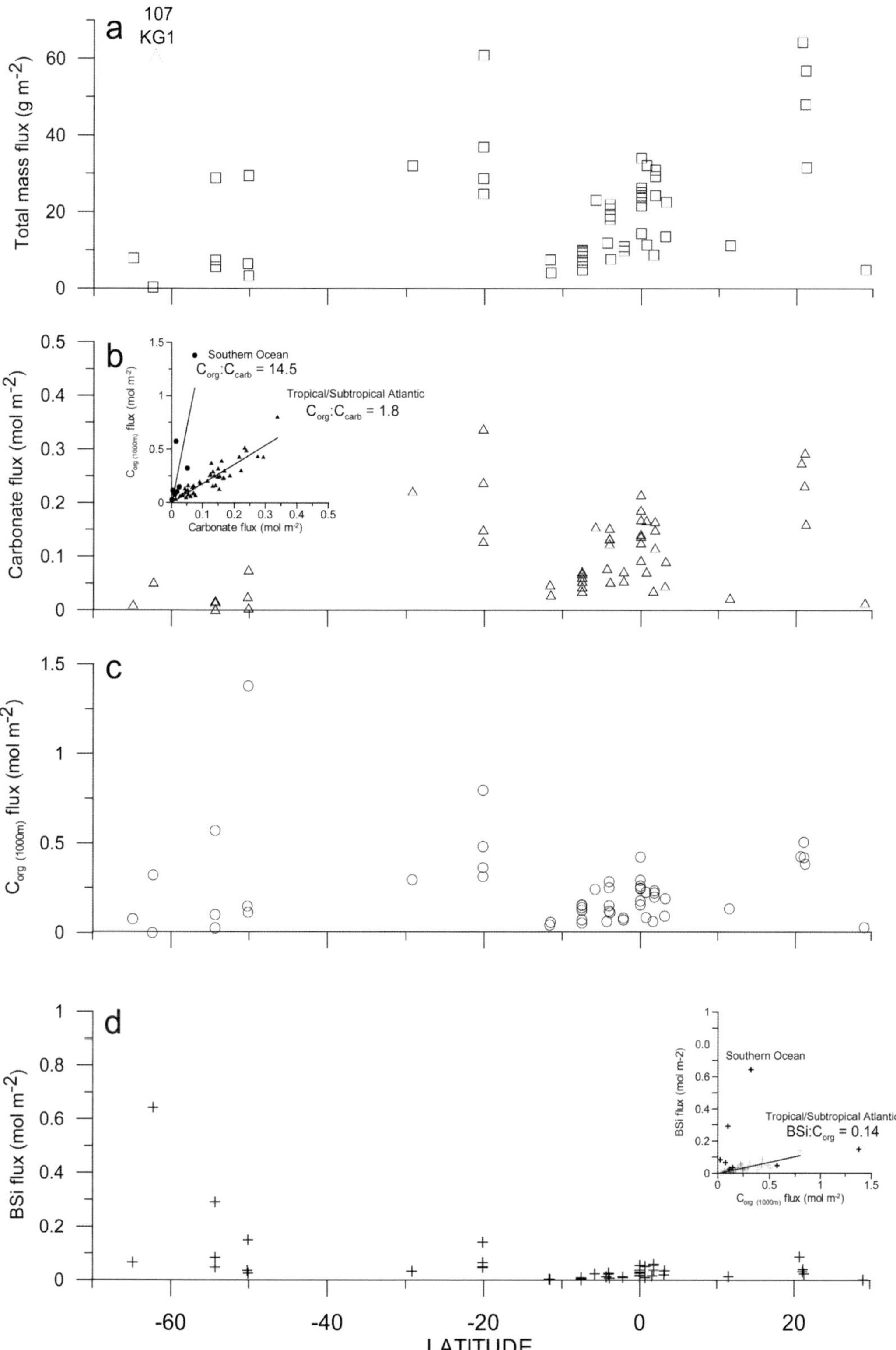

Fig. 3. Latitudinal variation of total mass **a)**, carbonate **b)**, C$_{org(1000m)}$ (**c**), normalized to 1000 m) and BSi fluxes **d)**. Note the latitudinal decrease of BSi fluxes from south to north. The mean annual molar BSi:C$_{org\,(1000\,m)}$ ratio is 0.14 in the tropical /subtropical Atlantic, and around two in the Southern Ocean (insert in d). Note the high correlation between C$_{org}$ and carbonate in the tropical/subtropical ocean (mean annual C$_{org(1000m)}$/C$_{carbonate}$ = 1.8; insert in b).

the general decrease of silicic acid availability in the surface waters from south to north, and with the concentrations of dissolved iron, additionally moderating the BSi:C$_{org}$ and BSi:N ratios (Hutchins and Bruland 1998; Takeda 1998).

To emphasize the importance of silicic acid supplied from the subsurface to the productive zone for the ratios of BSi:C$_{org}$ in sinking matter, the BSi:carbonate and the carbon rain, we plotted the latitudinal distribution of annual silicic acid/nitrate in the surface zone and at water depths of 250 m (Levitus et al. 1994). In addition, the nutrient differences between both depths (Δsilicic acid and Δnitrate), as well as the BSi:N ratios (N$_{total}$ was normalized to 1000 m; Martin et al. 1987) and BSi:carbonate molar flux ratios are shown (Fig. 4). The pattern indicates that the BSi:N$_{(1000m)}$ and BSi:carbonate ratios of sinking particles and the BSi fluxes clearly match with the latitudinal distribution of silicic acid availability in the surface (0 m) and subsurface waters (250 m) as well as Δsilicic acid $_{(250-0\,m)}$. Our study suggests a significant increase of BSi:N$_{(1000m)}$ of sinking particles above a silicic acid concentration of 40-60 µM. Plotting the Si:N nutrient ratio in 250 m versus the BSi:carbonate flux ratio also reveals no clear relationship over the entire range of nutrient ratios. Using a global data set, Ragueneau et al. (2000) discussed the existence of a Si:N nutrient threshold value at around 2, where the production of biogenic opal, mostly provided by diatoms, is exponentially increased as compared to carbonate production. This may occur, for instance, in the North Pacific and the Southern Ocean. Such a threshold value is supported by Egge and Aksens (1992) for diatom dominance over carbonate primary producers. Below, we compiled annual flux data from the world ocean to obtain a clearer picture where the threshold value may be situated in the present ocean and compared these results to values derived from surface sediments.

The two coastal upwelling systems, off NW and SW Africa, differ in their biological, chemical and atmospheric settings. Off SW Africa, a stronger seasonality and intensity in the SE- trade winds induces a stronger upwelling of cold subsurface waters, leading to higher primary productivity (Antoine et al. 1996). This is documented in the larger seasonal sea surface temperature range (see Treppke et al. 1996a). Significantly different macronutrient availability cannot be found in the source waters of upwelling (e.g. at 250 m water depth; SACW=South Atlantic Central Water; Levitus et al. 1994), but the ultimate source of upwelled waters, the Antarctic Intermediate Water (AAIW), is richer in silicic acid (35.9 µM) and nitrate (37.4 µM), and has a higher Si:N ratio (1.0) off SW Africa as compared to NW Africa (Si:N=0.8). The Atlantic Ocean off NW Africa is further characterized by a much higher dust and iron input (Donaghay et al. 1991) as compared to the SW Atlantic. Accordingly, daily BSi fluxes were generally higher at the WR (Fig. 6a), as compared to CB (Fig. 5a), resulting in higher BSi:C$_{org\,(1000m)}$ ratios (Table 1), this being in accordance with higher silicic acid concentrations delivered by the AAIW to the Namibian upwelling zone (Levitus et al. 1994). This underscores the importance of the 'silicate pump model' (Dugdale et al. 1995) for the BSi:C$_{org}$ ratios and the opal fluxes (Ragueneau et al. 2000). However, at both sites, the plot of organic carbon versus BSi displayed large scattering of values (Figs. 5a, 6d), when the data sets from 1988 to 1991 were used. Interestingly, there was a tendency to lower BSi:C$_{org(1000m)}$ ratios at site CB from 1988 to 1991 (Fig. 5d) and, less clearly, probably a reversed change at site WR (Fig. 6c). These variations were mostly due to changes in the BSi fluxes (Figs. 5a, 6a), and not the C$_{org}$ fluxes (Figs. 5c, 6b). The change in the BSi fluxes at site CB is accompanied by a decrease in the total number of diatoms (Fig. 5b; Romero et al. 2002a). Applying the silicate pump model and the findings of Ragueneau et al. (2000), the decrease in BSi:C$_{org\,(1000m)}$ at site CB would translate into a decrease in the silicic acid input, or the Si:N nutrient ratio of the source waters from 1988 to 1991. Such a change might be associated with a stronger dominance of the North Atlantic Central Water (NACW, less silicic acid) over the SACW (higher silicic acid) during this time period. We have indications that a decrease in upwelling intensity from 1988 to 1991 was not the reason for the decrease of BSi and diatom fluxes. Instead, alkenone studies suggest a decrease in

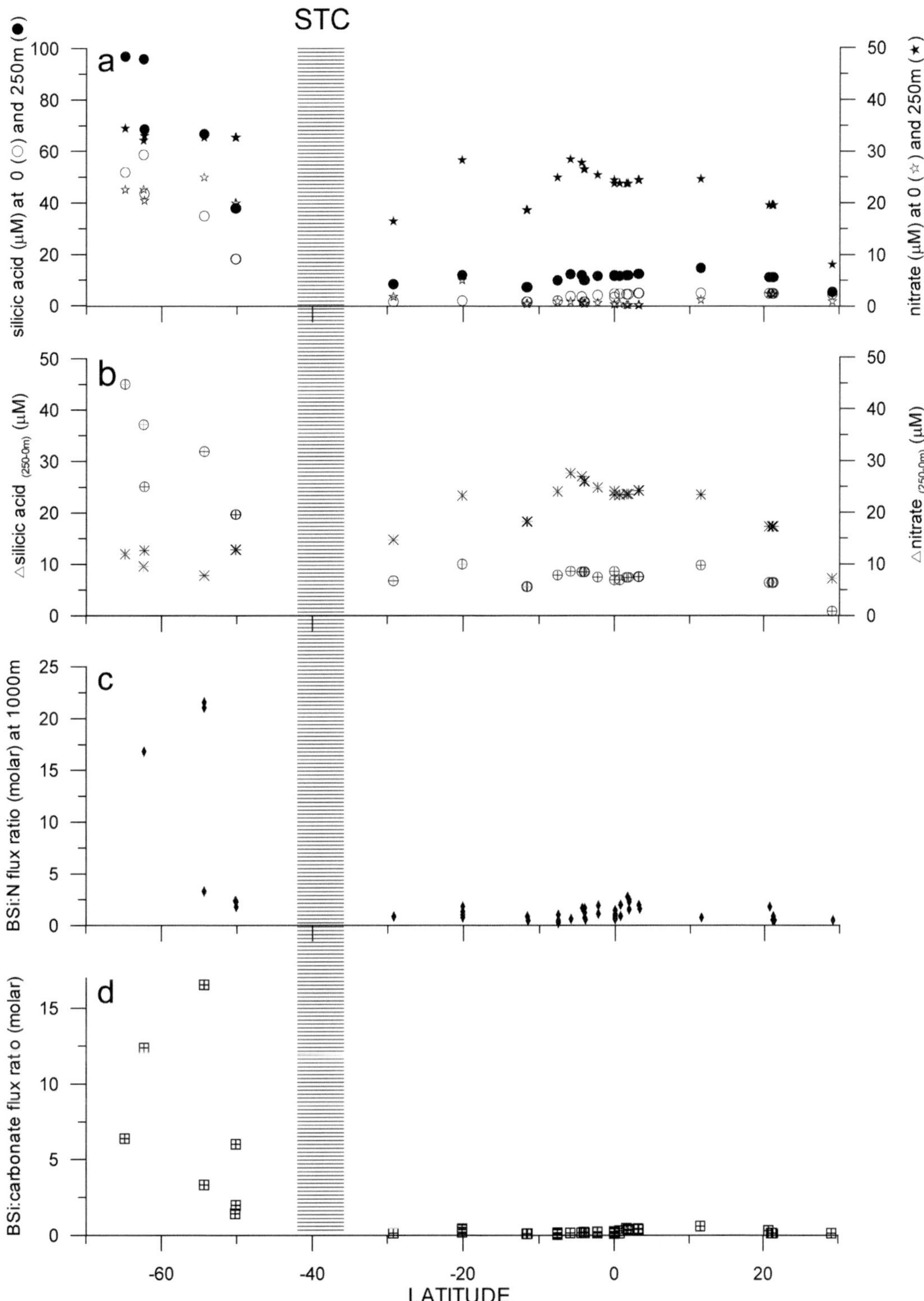

Fig. 4. Latitudinal distribution of nutrients at the surface (0 m) and in subsurface waters (250 m, **a)**) and difference in nutrients between both depths at the trapping sites **b)** (Levitus et al. 1994). The mean annual BSi:N$_{(1000m)}$ molar ratio of sinking particles **c)** and the molar BSi:carbonate flux ratios **d)** are also plotted versus latitude and show a distinct decrease from south to north in accordance with Δsilicic acid$_{(250-0m)}$ (STC=Subtropical Convergence). The extremely high BSi:carbonate flux ratio of 52.4 (Table 1) at BO5 was excluded in **d)**.

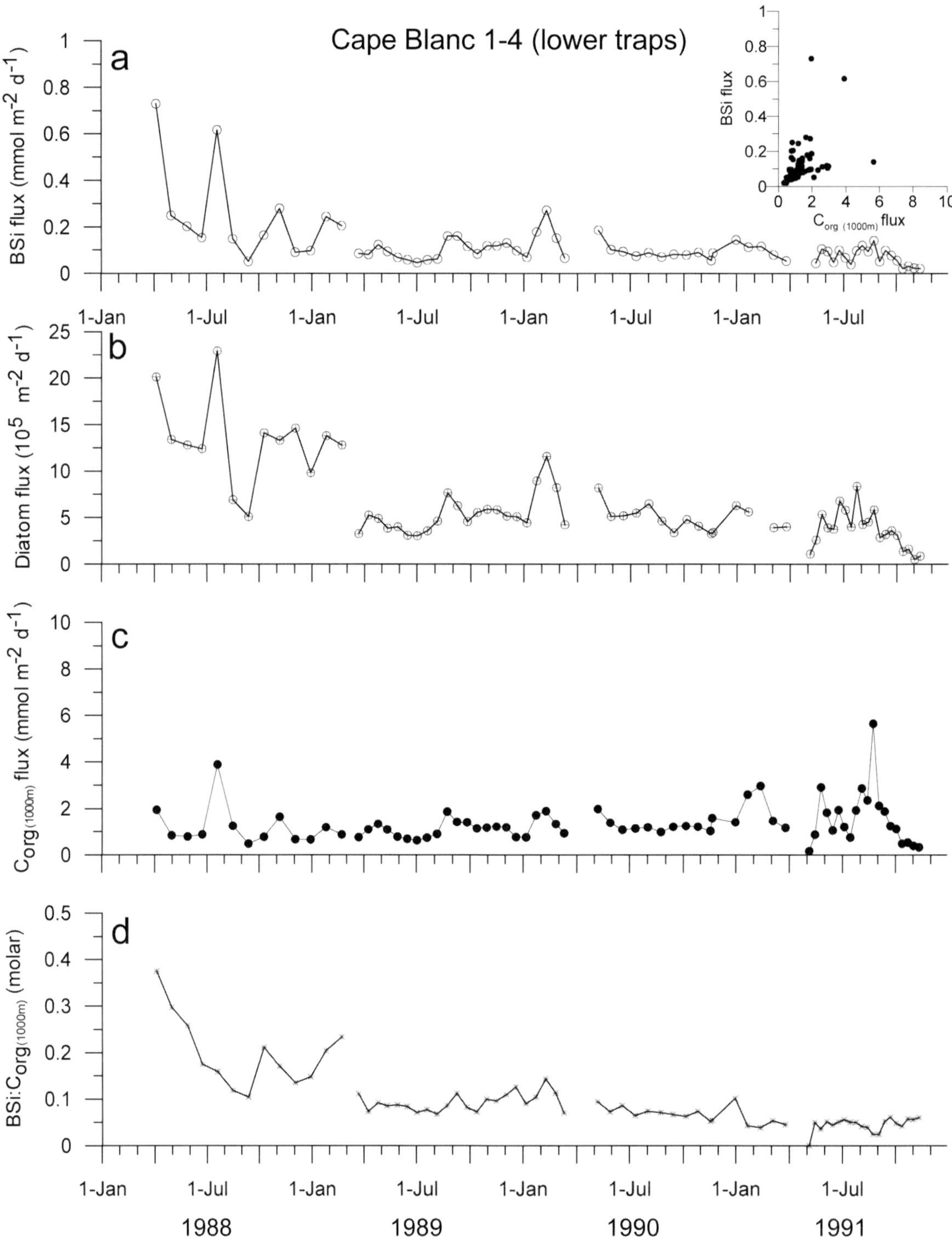

Fig. 5. Seasonal and interannual changes of the BSi **a)**, diatom **b)** and C$_{org\,(1000m)}$-fluxes (**c**), normalized to 1000 m) at the mesotrophic coastal upwelling site Cape Blanc (CB1-4, lower traps) from 1988 to 1991. Note that the molar BSi:C$_{org(1000\,m)}$ ratios decrease during this time period.

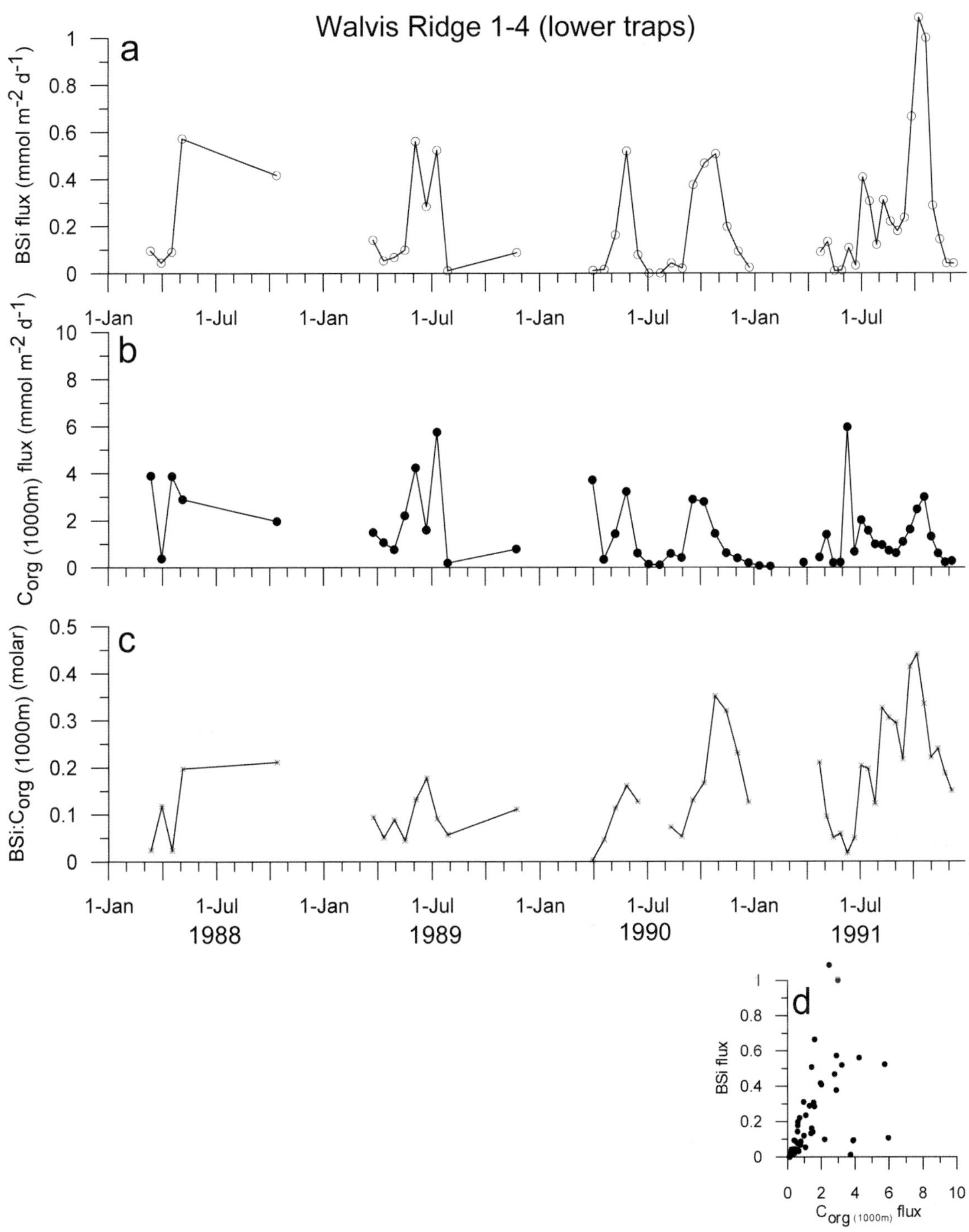

Fig. 6. Seasonal and interannual changes of the BSi **a)** and $C_{org\,(1000m)}$-fluxes **b)** (normalized to 1000 m) at the mesotrophic coastal upwelling site Walvis Ridge (WR1-4, lower traps) from 1988 to 1991. The molar $BSi:C_{org\,(1000\,m)}$ ratios **c)** appear to increase during this time period **d)**.

SST (Müller and Fischer 2001), presumably due to an increase in upwelling intensity from 1988 to 1991.

Silicic Acid Input and Carbonate versus Opal Production Systems in the Global Ocean

We have compiled a larger flux data set from sediment trap experiments (modified data set after Ragueneau et al. 2000; Honjo et al. 2000), in order to reinvestigate the relationship between annual C_{org} and BSi fluxes in the modern global ocean. As emphasized by Ragueneau et al. (2000), the molar $BSi:C_{org}$ flux ratios reflect the path of the conveyor belt and therefore strongly depend on the silicic acid supply from deeper waters. Low subsurface silicic acid concentrations (250 m) can be found in the Atlantic, intermediate values in the Indian Ocean, and high values in the North Pacific and the Southern Ocean (Codispoti 1983). Consequently, the modern Atlantic is a 'carbonate' ocean, and the North Pacific and the Southern Ocean are 'silicate oceans'. Using depth-normalized annual organic carbon fluxes, we obtained ratios of 0.05 in the Atlantic, 0.4 in the Indian, 0.5 in the Pacific and 0.1-3 in the Southern Ocean (Fig. 7a). All ratios are higher as compared to the results from Ragueneau et al. (2000), but the general picture is similar. These authors also examined at which Si:N nutrient ratio the production system shifted from carbonate (coccolithophorids) to biogenic opal (diatom) production. They found a linear increase of the BSi:carbonate flux ratios up to a nutrient Si:N threshold value of 2, above which the flux ratio increases exponentially. Our data indicate a rapid increase at a fairly lower Si:N value of about 1.7, but the relationship might be different for the various oceans (Fig. 7b). A general linear increase up to a threshold value of 2 cannot be inferred from our compilation. The correlation coefficients for the exponential relationships shown in Fig. 7 are only about 0.4 (R^2) for both lines.

In Fig. 8, we plotted the C_{org} versus the opal contents (a), and the opal:carbonate ratios (%) relative to the Si:N nutrient ratios at 250 M water depth (b). As for the water column fluxes (Fig. 7a), there is a large scatter of C_{org} and opal values measured in the sediment. However, we can still distinguish the three large ocean areas by means of the opal/C_{org} ratios. The data derived from the Atlantic show the largest scatter, although the ratios may be very low (='carbonate ocean'). In contrast, very high ratios are obtained for the Southern Ocean (='silicate ocean') similar to those for the water column (Figs. 7a, 8a). The values from the Pacific Ocean (='silicate ocean') are mostly high, but the Indian Ocean is clearly intermediate (='carbonate-silicate') with respect to the opal/C_{org} ratios.

The opal:carbonate ratios from the surface sediments (Fig. 8b) also show some similarities to the water column data discussed above (Fig. 7b), and offer a shift from carbonate to diatom producers at a Si:N nutrient threshold value of 2 to 2.5, slightly higher than the water column threshold value (Si:N≈1.7). However, the correlation coefficient for this exponential equation is low (R^2=0.27). The difference in the threshold value might be due to the diagenetic alteration of the opal:carbonate ratio, most probably due to opal dissolution at the sediment surface or at the water/sediment interface.

Intradecadal Variability and Nutrient Supply in the North Atlantic

So far, not many long-term records of particle fluxes in the world oceans are available to assess the interannual variations of probably periodic or episodic nature. Deuser et al. (1995) reported interannual variations of particle fluxes in the oligotrophic Sargasso Sea at 3200 m water depths; he attributed the observed opal:carbonate flux changes between 1978 and 1991 to subtle climate changes in the North Atlantic. From a few sites with particle flux records of three years or more, we calculated the 1σ-standard deviations of total fluxes and the relative (percentage) standard deviation (variability coefficient), in order to find out where the greatest interannual variation might be expected. Our data indicate that interannual changes are at least partly related to primary production and the seasonality of the production system ('seasonality index' of Berger and Wefer 1990). We obtained

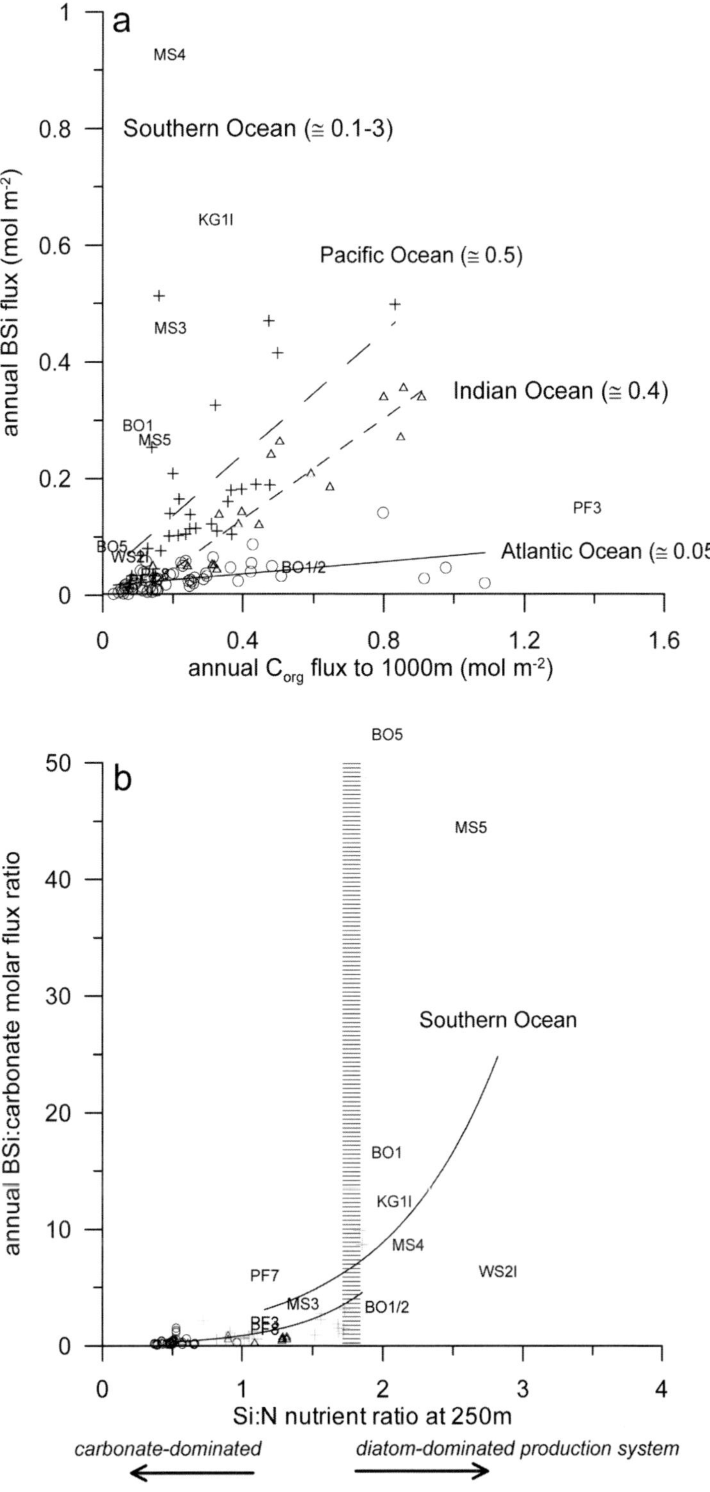

Fig. 7. Compilation of global annual organic carbon fluxes normalized to a water depth of 1000 m versus BSi fluxes **a)** (modified data set from Ragueneau et al. 2000; Honjo et al. 2000). Note the low molar ratios in the Atlantic (0.05) and the high values in the Pacific/Southern Oceans (0.5, 0.1-3). The Indian Ocean (0.4) is intermediate. **b)** Molar Si:N nutrient ratios of the source waters (250 m, Levitus et al. 1994) versus the annual molar BSi:carbonate flux ratios. Note the switch from carbonate to opal producers above a Si:N nutrient threshold value of 1.7 indicated as a vertical bar. Data can be found in Table 1.

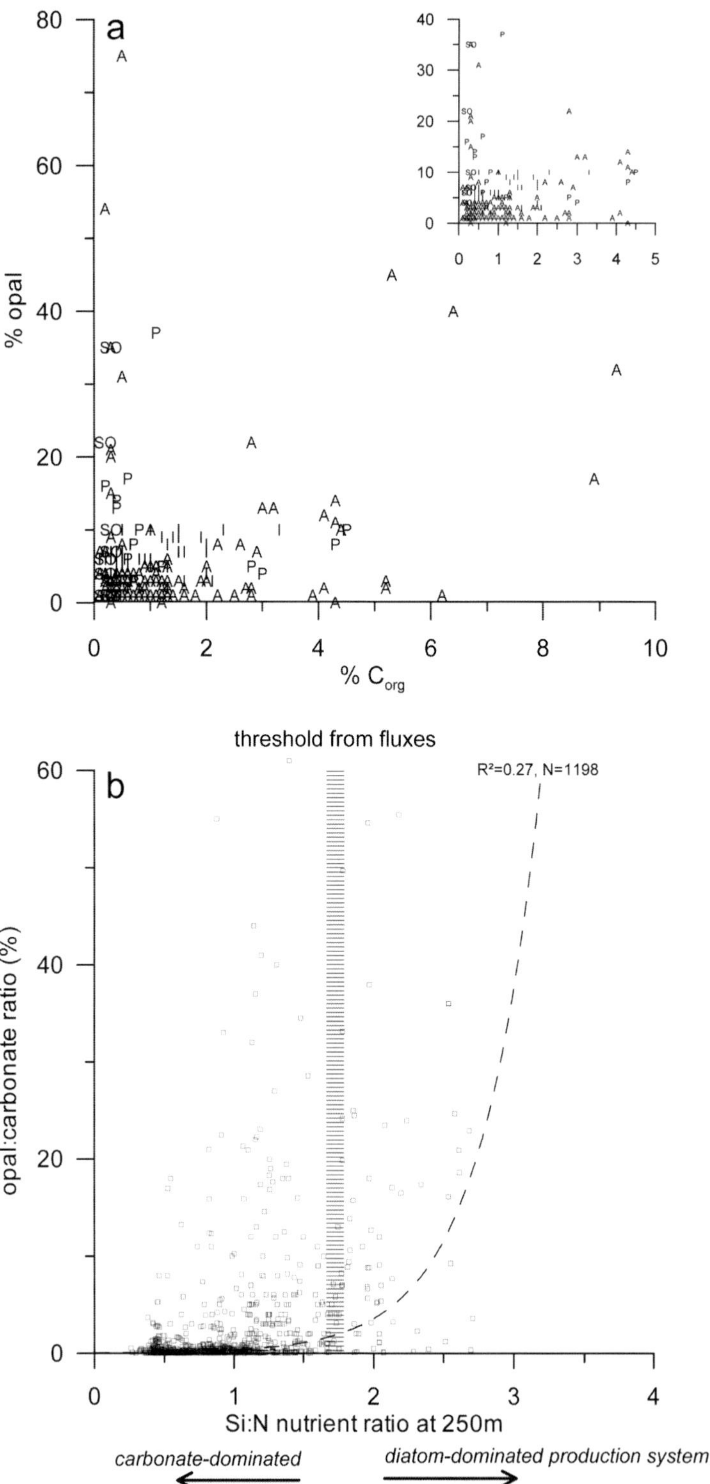

Fig. 8. Compilation of global %C_{org} versus %opal (**a**): SO=Southern Ocean, A=Atlantic, I=Indian Ocean, P=Pacific Ocean; insert: detail of %C_{org} vs %opal) and the opal:carbonate ratios (%) from surface sediments (core depth 0-5 cm, water depths: >200m to <4500 m) versus the molar Si:N$_{(250m)}$ nutrient ratios (**b**), Levitus et al. 1994). For criteria on data selection see 'Methods'. An exponential increase of the opal:carbonate ratios (%) occurs at a nutrient Si:N$_{(250m)}$ value of 2-2.5, which is slightly higher as compared to the threshold value derived from water column fluxes (Si:N$_{(250m)}$ » 1.7, see vertical bar in **b**). Data can be found at http://www.pangaea.de/Projects/ORFOIS/Results.

the lowest variations (17-35%) in the meso-trophic-oligotrophic western Atlantic (WA0°,4°S, 8°S) where primary production (according to Antoine et al., 1996) and seasonality of fluxes were the lowest (Fischer et al. 2000; Romero et al. 2000). Intermediate values (37-43%) can be calculated for the eastern equatorial upwelling zone (EA2°N, EA0°, EA2°S) and for the filamentous zones of coastal upwelling at the Walvis Ridge (WR), where primary production seasonality were higher ('sinusoidal to highly peaked production systems'; Berger and Wefer 1990; in Fischer and Wefer 1996). The highest variability was observed in the Southern Ocean at the Polar Front (90%) and in the zone influenced by winter sea-ice (site BO: 76%). At both latter sites, Fischer et al. (2002) obtained relative standard deviations rising up to 94% (site PF) and even 120% (site BO) using an extended data set. Both locations are characterized by a high degree of seasonality ('strongly seasonal to pulsed production systems', Berger and Wefer 1990) as described in Fischer et al. (2002). However, primary production values according to Antoine et al. (1996) are rather low at both sites (40-65 gC m^{-2} yr^{-1}), although probably underestimated by a factor of two or more (Schlitzer 2002). Even assuming such potential underestimation, primary production values in the Southern Ocean would not be as high as in the coastal or equatorial upwelling systems.

Two four-year records of particle fluxes from sites CB and WR have already been discussed and presented above (Figs. 5, 6). The decreasing opal and diatom fluxes measured at site CB match with the long-term flux pattern discussed by Deuser et al. (1995) for the North Atlantic. Antia et al. (2001) have compiled a larger data set from the North Atlantic plotting the opal:carbonate flux ratios since 1980. All data, including our results from site CB show a tendency of decreasing opal:carbonate ratios in the period from 1980 to 1997. As reported by Deuser et al. (1995), these changes were primarily due to a drop in opal fluxes, whereas the carbonate fluxes remained rather constant. The opal flux data from Cape Blanc confirm their findings and the diatom fluxes show a decreasing tendency from 1988 through 1991 (Fig. 5b; Romero et al. 2002a). At site EA2°N in the eastern equato-rial Atlantic which is under the influence of the NE trade wind belt (and the North Atlantic Oscillation, NAO), we also recorded a decrease of the annual opal:carbonate flux ratios from 0.28 to 0.19 between 1988 and 1991 (Table 1), which fits into the general pattern presented by Antia et al. (2001). These authors speculate that their observation may be common in the North Atlantic, probably reflecting changes in larger-scale physical forcing, such as the North Atlantic Oscillation.

Relationships between Primary Production, Export and Deep-Water Fluxes

The latitudinal distribution of organic carbon fluxes shows no clear pattern (Fig. 3c). Largest variations were found in the Southern Ocean, with values from almost 0 to 1 mol m^{-2}yr^{-1}. High values combined with larger variations were observed off SW Africa, whereas low to intermediate values were obtained in the tropical Atlantic and intermediate values. Less variation was found off NW Africa. The relationship between primary production (satellite-derived values from Antoine et al. 1996) and carbon export to 133 m water depths from biogeochemical modeling (Schlitzer 2002, pers comm) is shown in Fig. 9a. In the given range of primary production, a linear increase of organic carbon export to 133 m is obvious, resulting in an average f-ratio of 0.23 which seems to be reasonable (Eppley and Peterson 1979). This value, in turn, is consistent with the range of <0.1-0.38 given by Antia et al. (2001) over the range of primary production from 50-400 gC m^{-2} yr^{-1}. There also seems to be an increase of organic carbon fluxes normalized to 1000 m with increasing carbon export and primary production, despite a large scatter of the values (Fig. 9b, c). However, a hyperbolic relationship between primary production and depth-normalized carbon fluxes (to 2000 m), which was favored by Lampitt and Antia (1997), cannot be recognized from our data set. Instead, the plot reveals a similar broad increasing trend up to 300 g C m^{-2} yr^{-1} as shown by Francois et al. (2002). As a result from our compilation, about 2% of the organic carbon fixed by primary production in the Atlantic Ocean is exported to a

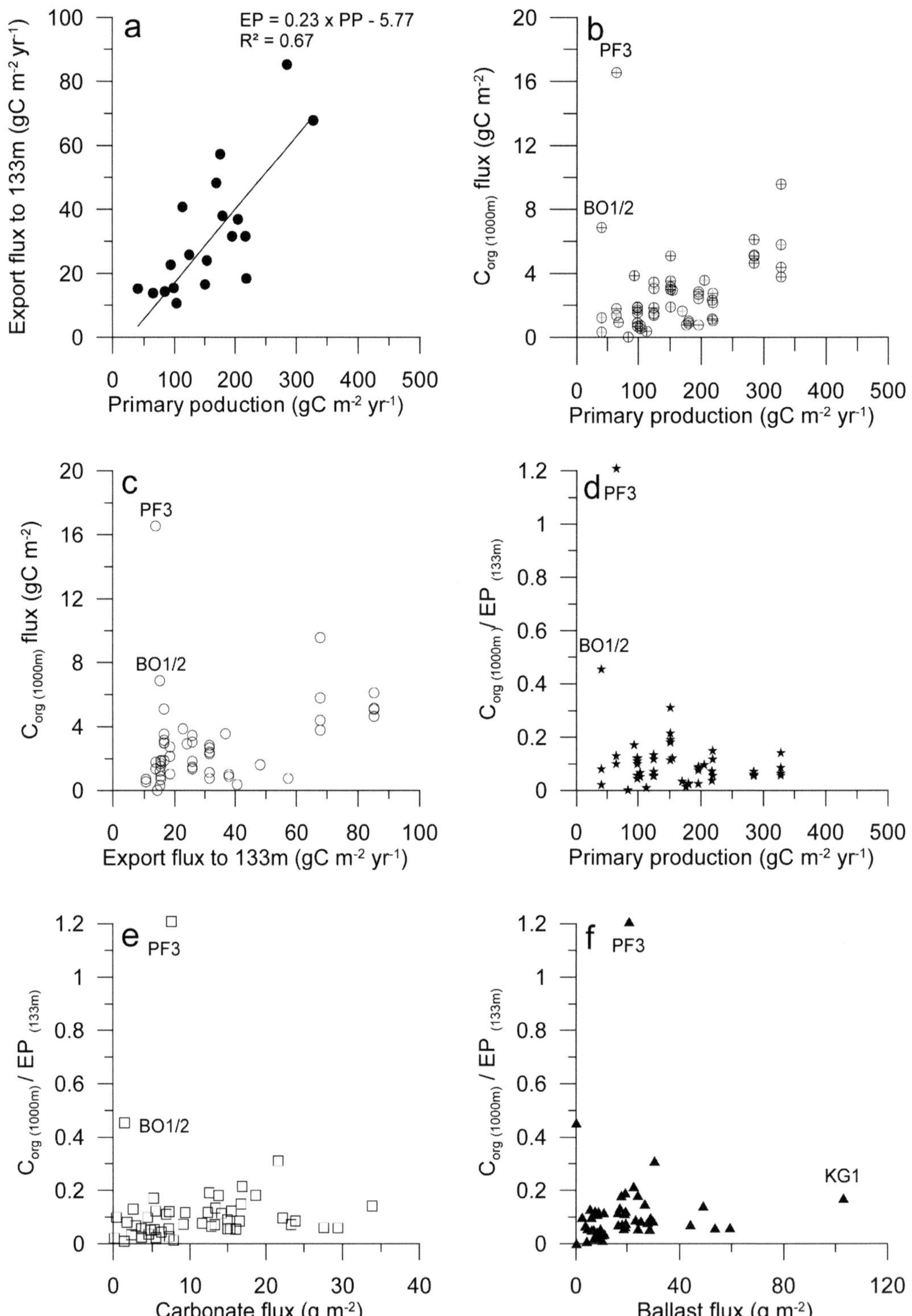

Fig. 9. Satellite annual primary production (Antoine et al., 1996) versus model-derived annual export production to 133 m water depth (**a**), Schlitzer 2002, pers comm) and annual organic carbon fluxes at 1000 m **b)** in the Atlantic/Southern Ocean (Table 1). A mean f-ratio of 0.23 was obtained from a. Organic carbon fluxes increase with export to 133 m **c)**; the 'transfer efficiency' (Francois et al. 2002; ($C_{org(1000m)}/EP_{(133m)}$) also increases slightly with primary production **d)**. A small linear increase between annual carbonate/ballast fluxes and $C_{org(1000m)}/EP_{(133m)}$ can be recognized (**e**), **f**)). Sites PF3 and BO1/2 do not comply with the general relationships.

water depth of 1000 m. This value appears to be quite high compared, for instance, to sites in the Equatorial Pacific, which have export ratios between 0.34 and 0.5% on average (Honjo et al. 1995). Lampitt and Antia (1997) obtained a global average of 1% with respect to the 2000 m depth level, being in good agreement with our results (2% at 1000 m). A simple relationship between normalized organic carbon fluxes to export fluxes cannot be concluded from Fig. 9c. As discussed by Francois et al. (2002), the slopes may vary largely between ocean areas and productivity systems, and a simple unique correlation should not be expected. As a consequence, organic carbon fluxes and sediment accumulation rates may not serve as a proxy for 'new production' when a simple unique equation is applied.

From a plot of annual primary production versus the ratio of $C_{org\,(1000m)}/EP_{(133m)}$ (='transfer efficiency' cf. Francois et al., 2002), we might deduct a slight increase (Fig. 9d, excluding BO1/2, PF3), but the whole picture remains unclear. Our findings support earlier models by Dugdale and Goering (1967) and Eppley and Peterson (1979) who showed that the export fraction and thus, the corresponding f-ratios, increase with primary production. In Fig. 9e we plotted the annual carbonate fluxes versus the ratio of $C_{org\,(1000m)}/EP_{(133m)}$. There seems to be indeed some coupling between the carbonate fluxes and this ratio. Plotting the total annual ballast fluxes versus the 'transfer efficiency' produces a slight increase (see also chapter below) at least in the lower range of ballast fluxes (Fig. 9f). But the relationships between both carbonate and ballast and the 'transfer efficiency' are not very convincing.

Organic Carbon Transfer to Greater Depths and Different Types of Ballast Minerals

Relationships between the most important flux components such as carbonate, biogenic opal and lithogenic components to annual organic carbon fluxes from the global ocean reveal that carbonate strongly determines organic carbon fluxes even though the scatter is rather large and the correlation coefficient is not very high (R^2=0.6, N=115, Fig. 10a). The mean global rain ratio to the deep ocean is ca. 1, close to the value found by Antia et al. (2001) applying to the deep Atlantic ($\approx$1 at 2000 m). As discussed above, the relationship between opal/BSi and organic carbon fluxes varies with the region, and no clear picture emerges when lithogenic versus organic carbon fluxes are plotted. A plot of total ballast versus organic carbon seems to indicate a linear increase in the lower range of ballast and organic carbon fluxes (insert in Fig. 10d) and the resulting annual mean C_{org}/ballast ratio is around 0.06 (R^2=0.48, N=61). In the higher range, the scattering of data points is very large. These data confirm conclusions obtained from a mechanistic model (Armstrong et al. 2002) showing that POC fluxes are tightly linked to the fluxes of ballast minerals in the deep ocean and that the POC:ballast ratios range from 0.044 to 0.065 .

How are fundamental environmental differences at the Cape Blanc (CB) and Walvis Ridge (WR) sites, both located at the edge of large coastal filaments (Fischer et al. 2000), documented in the bulk flux parameters and their ratios? A high dust input into the NW Africa upwelling area is documented by elevated lithogenic fluxes. Studies from Ratmeyer et al. (1999a, b) discussed the dust fluxes off NW Africa (Cape Blanc, Canary and Cape Verde Islands) in more detail. Plotting daily lithogenic fluxes versus organic carbon fluxes provides a good correlation at site CB (mainly in 1991), but not at site WR (Fig. 11c, f) where lithogenic fluxes were much lower. According to Antoine et al. (1996), primary production at the two study sites is almost indistinguishable. Possible effects of particle loading (Ittekkot 1993) by dust minerals off NW Africa should enhance the organic carbon fluxes. However, we found no evidence for such processes at our sites; instead, we recorded slightly higher mean organic carbon fluxes at site WR (Table 1), most likely due to a higher seasonality of production and export rates off Namibia (Berger and Wefer 1990; Fischer et al. 1996), and a higher input of silicic acid (Levitus et al. 1994) favoring diatom production.

In particular, low correlations of organic carbon with carbonate fluxes could be found at site WR (Fig. 11d), indicating that carbonate is not the major ballast material there. At site CB, carbonate versus organic carbon fluxes revealed a significant

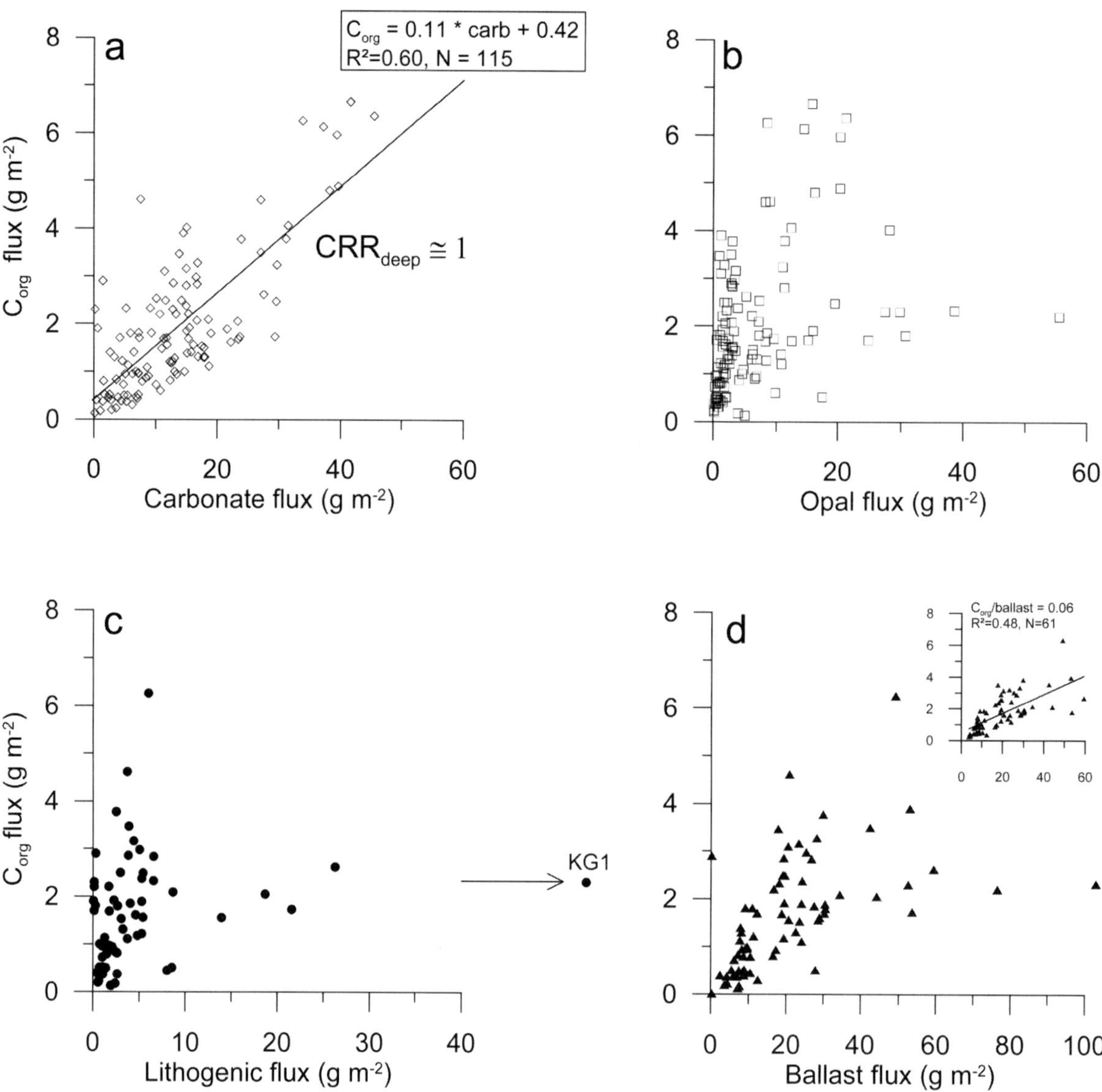

Fig. 10. Relationships between different bulk flux components (**a**) carbonate, **b**) opal, **c**) lithogenic, **d**) ballast (=total flux - $1.87 \cdot C_{org}$ flux) and organic carbon fluxes in the world ocean (modified data set after Ragueneau et al. 2000; Honjo et al. 2000). The mean global carbon rain ratio (=CRR=$C_{org}/C_{carbonate}$) into the deep ocean was ca. 1 (a), the mean global C_{org}:ballast flux ratio was 0.06 in the lower range of fluxes (insert in d).

correlation coefficient in 1991 (N=24, R^2=0.57; Fig. 11a), but not during 1988-1990. According to alkenone studies performed by Müller and Fischer (2001), the year 1991 was characterized by a high contribution of coccolithophorides. These rather heavy biogenic carbonate particles may serve as ballast for the lighter organic particles packed together in fecal pellets. We also plotted the daily ballast fluxes (which also include the lithogenic component) versus the organic carbon fluxes at site CB for the two different scenarios in 1988-1990 and 1991, respectively (Fig. 12). Now the relationships are significant for both time intervals, and the correlation for 1991 in the lower range of fluxes is even better. C_{org}/ballast ratios increased from 0.05 in 1988-1990 (a value close to the global value of 0.06, Fig. 10d) to 0.12 in 1991, that year being characterized by a strong coccolithophorid bloom.

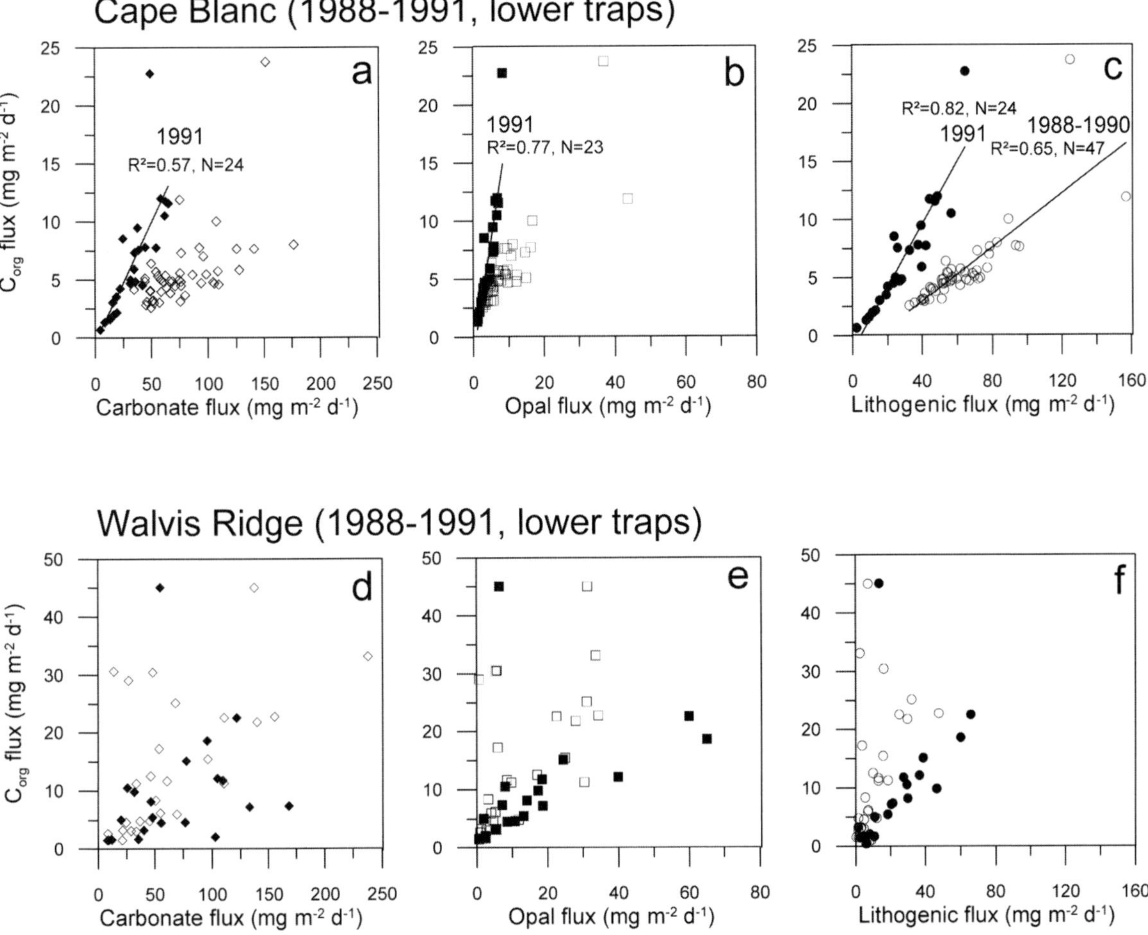

Fig. 11. Relationships between different flux bulk components (**a**), **d**) carbonate, **b**), **e**) opal, **c**), **f**) lithogenic material) versus the daily organic carbon fluxes at two coastal upwelling sites (CB, WR). Note the higher correlation coefficients at site CB, in particular, for lithogenic/carbonate versus organic carbon fluxes in 1991 (c). The relationship between carbonate and organic carbon is poorly established at both sites except at site CB in 1991 (a) during a strong coccolithophorid bloom (see Müller and Fischer 2001). Filled symbols indicate the values for 1991, open symbols for 1988-1990. Regression lines for site WR are not shown because of low linear correlation coefficients.

These differences in the C_{org}:ballast ratios (='transport ratios', cf. Armstrong et al., 2002) may be due to a difference in the relative proportions of ballast minerals as hypothesized by Armstrong et al. (2002).

Particle fluxes of all components in the western part of the tropical Atlantic clearly decreased from north to south, following the general decrease in primary production from the mesotrophic equatorial site (WA0°) to the oligotrophic site in the northernmost gyre of the Brazil Basin (WA11°S; Table 1; see Fischer et al. 2000). This tendency is documented at both trap levels. In general, fluxes were not higher in the eastern part of the tropical Atlantic as one might expect from the overall higher production (Fig. 2a), a shoaling of the thermocline (Houghton 1989), and a shallower mixed layer in boreal summer (Levitus et al. 1994). Organic carbon fluxes normalized to 1000 m derived from the shallower EA-traps in the Guinea Basin were almost similar to comparable latitudes in the western Atlantic. Surprisingly, the export fraction EF$_{1000}$ (=$C_{org(1000m)}$/PP) was generally higher in the western Atlantic (0.61-1.25%, Fischer et al. 2000), despite

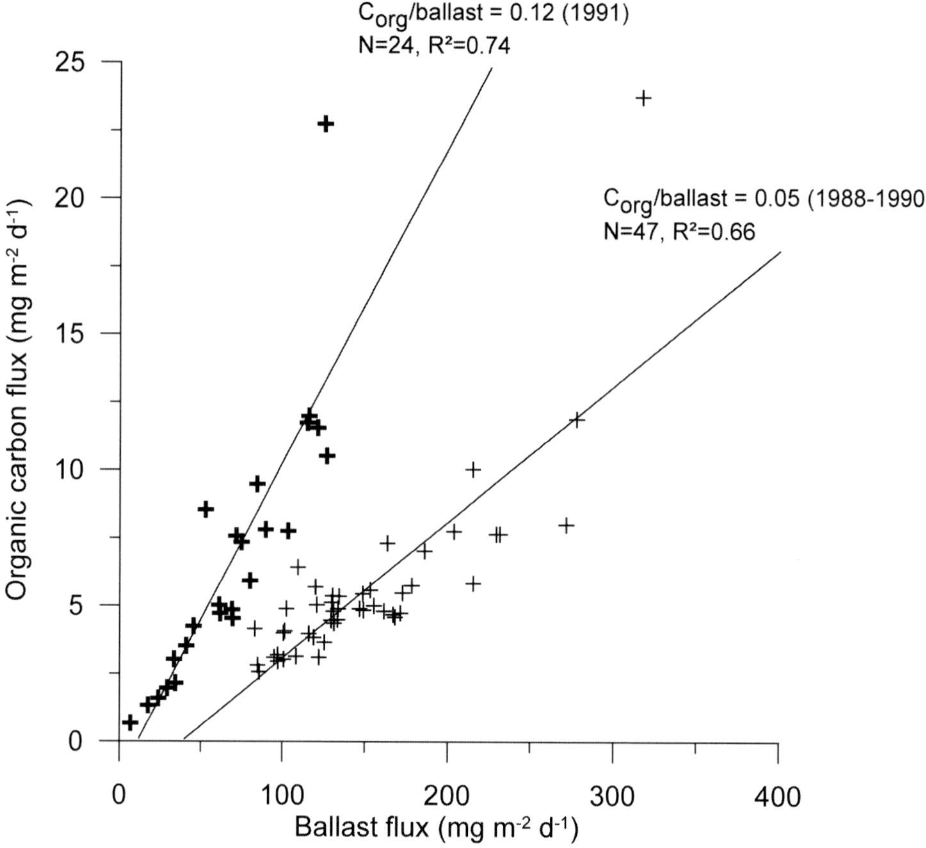

Fig. 12. Variable correlations between daily total ballast fluxes (=total flux - 1.87*C_{org} flux) and organic carbon fluxes at site CB off NW Africa. Note the significant correlation in 1991 when coccolithophorids accumulated in high quantities (Müller and Fischer 2001), leading to a high C_{org}/ballast ratio of 0.12. However, the correlation coefficient was nearly identical for 1988 to 1990 when lithogenic fluxes were highest (Fig. 11).

the lower primary production. An average value of only 0.4% was obtained for the equatorial Pacific (Honjo et al. 1995). Deep chlorophyll maxima (DCM) in the western Atlantic (Longhurst et al. 1995), not detected by satellite imagery, may be one explanation for the relatively high values. However, considering the effect of ballast minerals on the downward transport of organic carbon (Armstrong et al. 2002; Francois et al. 2002), we might also speculate that the 'transfer efficiency' in the western Atlantic is enhanced by higher carbonate production, which was most presumably provided by coccolithophorids. Our interpretation is supported by daily organic carbon versus carbonate fluxes from several study sites across the equatorial Atlantic: a much closer relationship is seen in

the western (sites WA1-2: R²=0.70, N=40) than in the eastern part (sites EA1-4: R²=0.26, N=64) (Fischer et al. 1995).

Summary and Conclusions

We have presented fluxes and various important flux ratios from the Atlantic Ocean including the Southern Ocean and compared these results to data from the global ocean. For reasons of comparison, we have also presented data from surface sediments to study the relationships between organic carbon versus biogenic opal as well as between opal:carbonate versus nutrient supply. Major findings from this comprehensive study are the following:

• A strong dependency of the BSi/opal fluxes upon the silicic acid supply from subsurface waters is observed. As suggested by Ragueneau et al. (2000), the highly variable annual $BSi:C_{org}$ ratios (0.05-3, mol/mol) mirror the general flow path of the conveyor belt.

• The relationship between the annual molar BSi:carbonate flux ratios and the nutrient $Si:N_{(250m)}$ ratios remains difficult to assess. There seems to be an exponential increase exceeding a $Si:N_{(250m)}$ nutrient threshold value of ca. 1.7, but this change, reflecting the switch from calcium carbonate to diatom production, may vary depending on the region. Comparing the opal:carbonate ratios (%) from surface sediments with the $Si:N_{(250m)}$ nutrient ratio provides a threshold value between 2 and 2.5, although the scattering of data is rather large.

• Interannual variability was lowest in oligotrophic areas, intermediate in coastal and equatorial upwelling regions, and highest in the Southern Ocean, where the seasonality of the fluxes is most pronounced. Long-term records of opal:carbonate flux ratios correspond well to the general pattern obtained for the North Atlantic Ocean (a decrease since 1980; see Antia et al. 2001; Deuser et al. 1995.

• We were able to demonstrate that there is an overall correlation of carbonate fluxes with the organic carbon fluxes for the Atlantic/Southern Ocean and the world ocean. These results suggest a ballast effect of carbonate on the 'transfer efficiency' of organic carbon to greater depths. The ballast fluxes and the organic carbon fluxes also show a positive linear relationship to a global mean C_{org}:ballast ratio of 0.06 (='ballast transport ratio; cf. Armstrong et al. 2002).

• According to Francois et al. (2002), lithogenic particles seem to be less important for the downward transfer of organic carbon on a global scale, as compared to carbonate particles. However, off NW Africa where the input of dust is exceptionally high, lithogenic particles may also constitute important particle carriers. We obtained variable positive linear relationships between the daily total ballast/lithogenic fluxes and organic carbon fluxes with C_{org}:ballast ratios of 0.06 and 0.12. During a strong coccolithophorid bloom in 1991, as follows from previous alkenone studies (Müller and Fischer 2001), the correlation between carbonate and C_{org} was also highly significant, stressing the importance of carbonate particles for the efficient transfer of organic carbon through the water column.

Acknowledgements

We thank the crews of the research vessels RV *Meteor*, RV *Polarstern*, RV *Polar Duke* (USA) for their work and assistance during deployments and recoveries of the mooring arrays. We are indebted to M. Scholz, V. Diekamp, M. Klann and C. Slickers for laboratory work. We also thank G. Ruhland, K. Dehning, U. Rosiak and D. Hebbeln for the logistical support during the cruises and the participation in the cruises. We are also grateful to the Alfred-Wegener-Institute for Polar and Marine Research in Bremerhaven for their support in deploying and redeploying the mooring arrays in the Southern Ocean with RV *Polarstern*. Surface sediment data were made available by PANGAEA/ WDC MARE. The comments of Oliver Ragueneau were helpful revising the manuscript. This research was funded by the Deutsche Forschungsgemeinschaft (Sonderforschungsbereich 261 at Bremen University). The study presented was partly supported by the European Commission, Grant EVK2-CT-2001-00100 ORFOIS. Data are available under www.pangaea.de/Projects/SFB261.

References

Antia AN et al. (2001) Basin-wide particulate carbon flux in the Atlantic Ocean: Regional export patterns and potential for atmospheric CO_2 sequestration. Glob Biogeochem Cycl 15: 845-862

Antoine D, Jean-Michel A, Morel A (1996) Ocean primary production 2. Estimation at global scale from satellite (Coastal Zone Colour Scanner) chlorophyll. Glob Biogeochem Cycl 10: 57-69

Armstrong RA, Lee C, Hedges JI, Honjo S, Wakeham SG (2002) A new, mechanistic model of organic carbon fluxes in the ocean based on the quantitative association of POC with ballast minerals. Deep-Sea Res II 49: 219-236

Behrenfeld MJ, Falkowski PG (1997) Photosynthetic rates derived from satellite-based chlorophyll concentration. Limnol Oceanogr 42: 1-20

Berelson WM (2002) Particle settling rates increase with depth in the ocean. Deep-Sea Res II 49: 237-251

Berger WH, Wefer G (1990) Export production:

seasonality and intermittency, and paleoceanographic implications. Palaeogeogr Palaeoclimatol Palaeoecol 89: 245-254

Betzer PR et al. (1984) Primary productivity and particle fluxes on a transect of the equator at 153 W in the Pacific Ocean. Deep-Sea Res 31: 1-12

Boltovskoy D, Alder VA, Abelmann A (1993a) Annual flux of radiolaria and other shelled plankters in the eastern equatorial Atlantic at 853 m: Seasonal variations and polycystine species-specific responses. Deep-Sea Res 40: 1863-1895

Boltovskoy D, Alder VA, Abelmann A (1993b) Radiolarian sedimentary imprint in Atlantic equatorial sediments: Comparison with the yearly flux at 853 m. Mar Micropal 23: 1-12

Boltovskoy D, Uliana E, Wefer G (1996) Seasonal variation in the flux of microplankton and radiolarian assemblage compositions in the northeastern tropical Atlantic at 2,195 m. Limnol Oceanogr 41: 615-635

Brummer GJA, van Eijden AJM (1992) 'Blue Ocean' paleoproductivitiy estimates from carbonate mass accumulation rates. Mar Micropal 19: 99-117

Codispoti LA (1983) On nutrient variability and sediments in upwelling areas. In: Suess E, Thiede J (eds) Coastal Upwelling, Part A. Plenum Press, pp 125-145

Deuser WG, Jickells TD, King P, Commeau JA (1995) Decadal and annual changes in biogenic opal and carbonate fluxes to the deep Sargasso Sea. Deep-Sea Res 42: 1923-1932

Donaghay PL et al. (1991) The role of episodic atmospheric nutrient inputs in chemical and biological dynamics of oceanic ecosystems. Oceanography 4: 62-70

Dugdale RC, Goering JJ (1967) Uptake of new and regenerated forms of nitrogen in primary productivity. Limnol Oceanogr 12: 196-296

Dugdale RC, Wilkerson FP, Minas HJ (1995) The role of a silicate pump in driving new production. Deep-Sea Res 42: 697-719

Egge JK, Aksens DL (1992) Silicate as regulating nutrient phytoplankton competition. Mar Ecol Progr Ser 83: 281-289

Eppley R, Peterson BJ (1979) Particulate organic matter flux and planktonic new production in the deep ocean. Nature 282: 677-680

Fischer G, Wefer G (1991) Sampling, preparation and analysis of marine particulate matter. In: Hurd DC, Spencer DW (eds) Marine Particles: Analysis and Characterization. American Geophysical Union, 63, Washington DC, pp 391-397

Fischer G, Wefer G (1995) Downward particulate matter fluxes in the eastern and the western Atlantic: a comparison. In: Tsunogai S, Iseki K, Koike I, Oba T (eds) Global Fluxes of Carbon and its related Substances in the Coastal-Ocean-Atmosphere System. Proceedings of the 1994-IGBP symposium, Sapporo, Japan, pp 317-331

Fischer G, Wefer G (1996) Long-term observations of particle fluxes in the Eastern Atlantic: Seasonality, changes of flux with depth and comparison with the sediment record. In: Wefer G, Berger WH, Siedler G, Webb DJ (eds) The South Atlantic: Present and Past Circulation. Springer, Berlin, pp 325-344

Fischer G, Donner B, Ratmeyer V, Davenport R, Wefer G (1996) Distinct year-to-year flux variations off Cape Blanc during 1988-1991: Relationship to δ^{18}O-deduced sea-surface temperatures and trade winds. J Mar Res 54: 73-98

Fischer G, Schneider RR, Müller PJ, Wefer G (1997) Anthropogenic CO_2 in Southern Ocean surface waters: Evidence from stable organic carbon isotopes. Terra Nova 9: 153-157

Fischer G, Müller PJ, Wefer G (1998) Latitudinal $\delta^{13}C_{org}$ variations in sinking matter and sediments from the South Atlantic: Effects of anthropogenic CO_2 and implications for paleo-PCO_2 reconstructions. J Mar Sys 17: 471-495

Fischer G, Kalberer M, Donner B, Wefer G (1999) Stable isotopes of pteropod shells as recorders of subsurface water conditions: Comparison with the record of *G.. ruber* and measurements. In: Fischer G, Wefer G (eds) Use of Proxies in Paleoceanography: Examples from the South Atlantic. Springer, Berlin, pp 191-206

Fischer G, Ratmeyer V, Wefer G (2000) Organic carbon fluxes in the Atlantic and the Southern Ocean: Relationship to Primary Production compiled from satellite radiometer data. Deep-Sea Res 47: 1961-1997

Fischer G, Gersonde R, Wefer G (2002) Organic carbon, biogenic silica and diatom fluxes in the marginal winter sea-ice zone and in the Polar Frontal Region: Interannual variations and differences in composition. Deep-Sea Res II 49: 1721-1745

Francois R, Honjo S, Krishfield R, Manganini S (2002) Factors controlling the flux of organic cabon in the bathypelagic ocean. Glob Biogeochem Cycl 16:1087, doi:10.1029/20016B001722, 2002

Haake B, Rixen T, Reemtsma T, Ramaswamy V, Ittekkot V (1996) Processes determining seasonality and interannual variability of settling particle fluxes in the deep Arabian Sea. In: Ittekkot, V, Schäfer P, Honjo S, Depetris PJ (eds) Particle Flux in the Ocean. J Wiley & Sons, Chichester, pp 251-268

Holmes E et al. (2002) Seasonal variability of δ^{15}N in sinking particles in the Benguela upwelling region. Deep-Sea Res I 49: 377-394

Honjo S, Doherty KW (1988) Large scale aperture time-series sediment traps; design, objectives, construction and application. Deep-Sea Res 35: 133-149

Honjo S, Manganini SJ (1993) Annual biogenic particle fluxes to the interior of the north Atlantic Ocean; studies at 34°N 21°W and 48°N 21°W. Deep-Sea Res I 40: 587-607

Honjo S, Dymond J, Collier R, Manganini SJ (1995) Export production of particles to the interior of the equatorial Pacific Ocean during the 1992 EqPac experiment. Deep-Sea Res II 42: 831-870

Honjo S, Francois R, Manganini S, Dymond J, Collier R (2000) Particle fluxes to the interior of the Southern Ocean in the Western Pacific sector along 170°W. Deep-Sea Res II 47: 3521-3548

Houghton RW (1989) Influence of local and remote wind forcing in the Gulf of Guinea. J Geophys Res 94: 4816-4828

Hutchins DA, Bruland KW (1998) Iron-limited diatom growth and Si:N uptake ratios in a coastal upwelling regime. Nature 393: 561-564

Ittekkot V (1993) The abiotically driven biological pump in the ocean and short-term fluctuations in atmospheric CO_2 contents. Glob Planet Change 8: 17-25

Jackson GA, Burd AB (2002) A model for the distribution of particle flux in the mid-water column controlled by subsurface biotic interactions. Deep-Sea Res II 49: 193-217

Jahnke R (1996) The global ocean flux of particulate carbon: Areal distribution and magnitude. Glob Biogeochem Cycl 10: 71-88

Karl DM et al. (1997) The role of nitrogen fixation in biogeochemical cycling in the subtropical North Pacific Ocean. Nature 388: 533-538

Kremling K, Lentz U, Zeitzschel B, Schulz-Bull DE, Duinker JC (1996) New type of time-series sediment trap for the reliable collection of inorganic and organic trace chemical substances. Rev Scient Instr 67: 4360-4363.

Lampitt RS, Antia AN (1997) Particle flux in deep-seas: Regional characteristics and temporal variability. Deep-Sea Res 44: 1377-1403

Lange CB, Treppke UF, Fischer G (1994) Seasonal diatom fluxes in the Guinea Basin and their relationship to trade winds, hydrography and upwelling events. Deep-Sea Res 41: 859-878

Levitus S., Burgett R, Boyer T (1994) World Ocean Atlas 1994 Volume 3: Nutrients. NOAA Atlas NESDIS 3, US Department of Commerce, Washington, DC

Longhurst A, Sathyendranath S, Platt T, Caverhill C (1995) An estimate of global primary production in the ocean from satellite radiometer data. J Plankt Res 17: 1245-1271

Martin J H, Knauer GA, Karl DM, Broenkow WW (1987) VERTEX: Carbon cycling in the northeast Pacific. Deep-Sea Res 34: 267-285

Müller PJ, Schneider R (1993) An automated leaching method for the determination of opal in sediments and particulate matter. Deep-Sea Res I 40: 425-444

Müller PJ, Fischer G (2001) A 4-year sediment trap record of alkenones from the filamentous upwelling region off Cape Blanc, NW Africa and a comparison with distributions in underlying sediments. Deep-Sea Res I 48: 1877-1903

Neuer S, Ratmeyer V, Davenport R, Fischer G, Wefer G (1997) Deep water particle flux in the Canary Island region: Seasonal trends in relation to long-term satellite derived pigment data and lateral sources. Deep-Sea Res I 44: 1451-1466

Oschlies A, Garçon V (1998) Eddy-induced enhancement of primary production in a model of the North Atlantic Ocean. Nature 394: 266-269

Pollock DE (1997) The role of diatoms, dissolved silicate and Antarctic glaciation in glacial/interglacial climatic change: A hypothesis. Glob Planet Change 14: 113-125

Ragueneau O et al. (2000) A review of the Si cycle in the modern ocean: Recent progress and missing gaps in the application of biogenic opal as a paleoproductivity proxy. Glob Planet Change 26, 317-365

Ragueneau, O, Dittert N, Pondaven P, Tréguer P, Corrin L (2002) Si/C decoupling in the world ocean: Is the Southern Ocean different? Deep-Sea Res II 49: 3127-3154

Ratmeyer V, Fischer G, Wefer G (1999a) Lithogenic particle fluxes and grain size distributions in the deep ocean off northwest Africa: Implications for seasonal changes of aeolian dust input and downward transport. Deep-Sea Res II 46: 1289-1337

Ratmeyer V, Fischer G, Wefer G (1999b) Seasonal impact of mineral dust on deep-ocean particle flux in the eastern subtropical Atlantic Ocean. Mar Geol 159: 241-252

Romero O, Lange CB, Fischer G, Treppke U, Wefer G (1999a) Variability in export production documented by downward fluxes and species composition of marine planktonic diatoms. Observations from the tropical and equatorial Atlantic. In: Fischer G, Wefer G (eds) Use of Proxies in Paleoceanography: Examples from the South Atlantic. Springer, Berlin, pp 365-392

Romero OE, Lange C B, Swap R, Wefer G (1999b) Eolian-transported freshwater diatoms and phytoliths across the equatorial Atlantic record: Temporal changes in Saharan dust transport patterns. J Geophys Res 104: 3211-3222

Romero OE, Fischer G, Lange C B, Wefer G (2000) Siliceous phytoplankton of the western equatorial Atlantic: Sediment traps and surface sediments. Deep-Sea Res II 47: 1939-1959

Romero OE, Lange CB, Wefer G (2002a) Interannual variability (1988-1991) of siliceous phytoplankton fluxes off northwest Africa. J Plankt Res 24: 1035-1046

Romero OE et al. (2002b) Seasonal productivity dynamics in the pelagic central Benguela System inferred from the flux of carbonate and silicate organisms. J Mar Sys 37: 259-278

Rühlemann C, Müller PJ, Schneider RR (1999) Organic carbon and carbonate as paleoproductivity proxies: Examples from high and low latitude productivity areas of the tropical Atlantic. In: Fischer G, Wefer G (eds) Proxies in Paleoceanography: Examples from the South Atlantic. Springer, Berlin, pp 315-344

Schlitzer R (2002) Carbon export fluxes in the Southern Ocean: Results from inverse modeling and comparison with satellite-based estimates. Deep-Sea Res II 49: 1623-1644

Scholten JC et al. (2001) Trapping efficiencies of sediment traps from the deep Eastern North Atlantic: the ^{230}Th calibration. Deep-Sea Res II 48: 2383-2408

Smetacek V, deBaar HJW, Bathmann UV, Lochte K, Rutgers van der Loeff MM (1997) Ecology and biogeochemistry of the Antarctic Circumpolar Current during austral spring: A summary of Southern Ocean JGOFS cruise ANT X/6 of RV Polarstern. Deep-Sea Res 44: 1-21

Steinberg DK et al. (2001) Overview of the US JGOFS Bermuda Atlantic Time-series Study (BATS): A decade-scale look at ocean biology and biogeochemistry. Deep-Sea Res II 48: 1405-1447

Takeda S (1998) Influence of iron availability on nutrient consumption ratio of diatoms in oceanic waters. Nature 393: 774-777

Thunell RC (1998) Seasonal and interannual variability in particle fluxes in the Gulf of California. Deep-Sea Res I 45: 2059-2083

Treppke UF et al. (1996a) Diatom and silicoflagellate fluxes at the Walvis Ridge: An environment influenced by coastal upwelling in the Benguela system. J Mar Res 54: 991-1016

Treppke UF, Lange CB, Wefer G (1996b) Vertical fluxes of diatoms and silicoflagellates in the eastern equatorial Atlantic, and their contribution to the sedimentary record. Mar Micropal 28: 73-96

Tsunogai S, Noriki S (1991) Particulate fluxes of carbonate and organic carbon in the ocean: Is the marine biological activity working as a sink of atmospheric carbon? Tellus, Ser A 43: 256-266

Walter HJ, Geibert W, Rutgers van der Loeff MM, Fischer G, Bathmann U (2001) Shallow vs. deep-water scavenging of ^{231}Pa and ^{230}Th in radionuclide enriched waters of the Atlantic sector of the Southern Ocean. Deep-Sea Res: 48: 471-493

Wefer G et al. (1982) Fluxes of biogenic components from sediment trap deployments in circumpolar waters of the Drake Passage. Nature 299: 145-147

Wefer G, Berger WH, Bijma J, Fischer G (1999) Clues to ocean history: A brief overview of proxies. In: Fischer G, Wefer G (eds) Use of Proxies in Paleoceanography: Examples from the South Atlantic. Springer, Berlin, pp 1-68

Wong CS et al. (1999) Seasonal and interannual variability in particle fluxes of carbon, nitrogen and silicon from time series of sediment traps at Ocean Station P, 1982-1993: Relationship to changes in subarctic primary productivity. Deep-Sea Res II 46: 2735-2760

Yu EF, Francois R., Honjo S, Fleer AP, Manganini SJ, Rutgers van der Loeff MM, Ittekkot V (2001) Trapping efficiency of bottom-tethered sediment traps estimated from the intercepted fluxes of ^{230}Th and ^{231}Pa. Deep-Sea Res I 48: 865-889

Radionuclides as Tracers for Particle Flux and Transport of Water Masses in the Atlantic Sector of the Southern Ocean

M. Rutgers van der Loeff[1,2], J. Friedrich[2*], W. Geibert[2], C. Hanfland[2], H. Höltzen, I. Vöge[2] and H.J. Walter

[1]RIKZ, P.O.Box 20907, 2500 Ex Den Haag, The Netherlands
[2]Alfred-Wegener-Institut für Polar- und Meeresforschung, Columbusstraße,
27515 Bremerhaven, Germany
* corresponding author (e-mail): jfriedrich@awi-bremerhaven.de

Abstract: The natural uranium decay series provide a suite of tracers to study transport processes in the ocean. We have used nuclides of the particle-reactive elements Th, Pa, Pb and Po for studies of particle flux in the Southern Ocean, whereas isotopes of the elements Ra and Ac served as tracers for the transport of water masses. Here we summarize the specific aspects of the behaviour of these nuclides in the Southern Ocean and give some examples of their application. We review the important influence of exchange between ocean basins by advection and upwelling on the long-lived nuclides. We show how the distribution of ^{234}Th in surface waters across the ACC represents the export production, whereas in the benthic nepheloid layer this tracer is used to illustrate how the resuspension regime in the ACC is linked to the position of the oceanographic fronts.

Introduction

U-series nuclides have been used extensively to study the rate of transport processes in the ocean (Broecker and Peng 1982; Cochran 1992; Rutgers van der Loeff 2001). The worldwide distribution of ^{226}Ra has been studied in detail in the GEOSECS program in search for a tracer for the conveyor belt circulation. Since then, the study of particle-reactive nuclides has allowed the modelling of scavenging processes and the use of these nuclides as tracers for particle flux. The understanding of the scavenging of ^{230}Th and ^{231}Pa as a reversible exchange process (Bacon and Anderson 1982; Nozaki et al. 1981) was the basis for the interpretation of the distribution of long-lived particle-reactive isotopes in sediment and water column. Particle fluxes on short time scales have been studied with shorter-lived particle-reactive nuclides. The scavenging in surface waters was studied with the tracer pair ^{210}Po/^{210}Pb (Bacon 1976; Bacon et al. 1976), whereas ^{234}Th gained

importance as a tracer on the time scale of weeks both in surface waters (Coale and Bruland 1985) and near the sea floor (Bacon and Rutgers van der Loeff 1989). In the Southern Ocean, the GEOSECS Program provided a strong basis for the description of radionuclides in relationship with the water mass circulation. The distribution of the long-lived nuclides of Uranium (Ku et al. 1977; Chen et al. 1986) and of ^{226}Ra (Chung 1981), the isotopes that are at the top of the decay chains, was now well established. In recent years, natural radionuclides have been used in the Southern Ocean to quantify present-day particle fluxes and to establish the role of the Southern Ocean in the glacial-interglacial changes in productivity and water mass circulation. Here we discuss recent advances in the interpretation of the distribution of ^{231}Pa and ^{230}Th in the Atlantic sector of the Southern Ocean and we show how the short-lived tracers have improved our understanding

From WEFER G, MULITZA S, RATMEYER V (eds), 2003, *The South Atlantic in the Late Quaternary: Reconstruction of Material Budgets and Current Systems.* Springer-Verlag Berlin Heidelberg New York Tokyo, pp 47-63

of particle fluxes on the time scale of weeks: export production and resuspension. We summarize data obtained with ^{234}Th with additional data from other tracers like ^{210}Po. Finally, we give examples of the advances that have been made in the application of the well-soluble tracers ^{228}Ra and ^{227}Ac to study water mass circulation.

^{226}Ra, ^{210}Pb and ^{210}Po

The distribution of ^{226}Ra in the ocean is controlled by biological cycling, input from bottom sediments and radioactive decay (Broecker et al. 1976; Cochran 1992). In the Southern Ocean, the upwelling of old and ^{226}Ra-enriched deep waters causes a high and relatively uniform distribution of ^{226}Ra of 0.21 dpm/L in the entire water column south of the Polar Front (Ku and Lin 1976; Chung and Applequist 1980; Chung 1981). North of the Polar Front, ^{226}Ra decreases to a constant value of 0.07dpm/L in Atlantic surface waters, a decrease that is correlated with silicate (Ku and Lin 1976; Broecker et al. 1976). ^{210}Pb is depleted relative to its parent ^{226}Ra in the entire water column. The remoteness of continental sources of ^{222}Rn to the atmosphere explains why surface waters in the Southern Ocean lack the ^{210}Pb excess that is present in all other oceans. In the Weddell Sea, the lowest depletion (highest ^{210}Pb concentrations) is found at the depth of the Warm Deep Water (WDW), whereas deep waters are strongly depleted in ^{210}Pb, down to a ^{210}Pb/^{226}Ra ratio of 0.3 in the Antarctic Bottom Water (AABW), corresponding to a mean residence time of only 14 years. This was explained by Chung (1981) as indication for scavenging by resuspended particulate matter. A detailed description of the distribution of ^{210}Pb and the ^{210}Pb/^{226}Ra ratio on the Greenwich meridian (AJAX section) across the Antarctic Circum-polar Current (ACC) was described by Farley and Turekian (1990).

^{210}Po, the short-lived (138 day half-life) and highly reactive daughter of ^{210}Pb, was studied by Friedrich (1997) as tracer of particle flux in the surface waters of the ACC. The depletion relative to its parent ^{210}Pb was used to quantify the export of ^{210}Po (cf. Bacon 1976), similar to the ^{234}Th-based procedure to calculate export production (see below). The seasonal changes in the scavenging of both parent (^{210}Pb) and daughter (^{210}Po) nuclide required a new non-steady state solution of the scavenging equation (Friedrich 1997). This dependence on two analyses causes the depletion and correspondingly the export values to be associated with appreciably larger errors than in case of the ^{234}Th-based export, which makes use of the conservative distribution of the parent ^{238}U (see below). The interesting additional information contained in the ^{210}Po data is related to the scavenging behaviour, which is different from that of ^{234}Th. Whereas Th adsorbs to any surface, Po tends to be accumulated by plankton in its cytoplasm, and even to accumulate in the food chain. This differential scavenging was exploited to make independent estimates of the export rates of Corg and of BSi (see below).

^{230}Th and ^{231}Pa

Based on studies in the Pacific and Atlantic Ocean, ^{230}Th and ^{231}Pa have been used extensively as tracers for particle flux and (paleo)productivity. Their use is based on the known production of these isotopes by decay of a well-soluble uranium parent nuclide. As a result of high reactivity and correspondingly short scavenging residence time of Th in the ocean (about 30 years), ^{230}Th is thought to be deposited within the same ocean area where it is produced, which makes it an ideal tracer to correct for horizontal sediment redistribution in the deep-sea (Bacon and Rosholt 1982; Suman and Bacon 1989; Francois et al. 1993; Frank et al. 1999). The somewhat longer scavenging residence time of Pa allows ^{231}Pa to be transported over larger distances before it is deposited in areas of enhanced particle flux. Consequently, the ^{231}Pa/^{230}Th ratio is interpreted to mirror (paleo)productivity (Anderson et al. 1983 a,b; Kumar et al. 1993, 1995; Walter et al. 1999). The application of these tracers has been reviewed recently by Francois et al. (2003)

The distribution of these isotopes in the ocean is not only controlled by the circulation and diffusion within a basin. Yu et al. (1996) showed that a large part of ^{231}Pa produced in the Atlantic Ocean is exported to the Southern Ocean before it is scavenged and removed to the seafloor. But even the distribution of the more reactive ^{230}Th is affected by ocean circulation, as was observed in the Weddell Sea (Rutgers van der Loeff and Berger 1993), in the North Atlantic (Vogler et al. 1998) and in the South Pacific (Chase 2001; Chase et al. 2003). The redistribution of ^{230}Th has to be taken into account when this isotope is used to correct sedimentation fluxes for effects of focussing and winnowing. To a first approximation, this correction can be based on the global scavenging/circulation model used by Henderson et al. (1999). In the Southern Ocean, and especially in the Weddell Sea, the correction predicted by the Henderson model is inadequate. The low particle fluxes in the deep Weddell Sea and the advection of this basin by upwelling Th-enriched deep waters causes ^{230}Th to be enriched in the Weddell Sea (Rutgers van der Loeff and Berger 1993) and exported to the ACC (Walter et al. 2000). Sedimentation rates that are corrected for focussing/winnowing using the conventional constant flux method would overestimate fluxes in the Weddell Sea and underestimate fluxes in the adjoining parts of the ACC by a factor of up to 2 (Fig. 1).

^{230}Th is widely used to correct fluxes in sediment traps for over- or undercollection. As a first approximation, the intercepted flux was compared with the production in the water column, but later the ocean boundary effect on the scavenging of ^{230}Th was taken into account. This could be quantified by the simultaneous measurement of ^{231}Pa (Bacon et al. 1985) and can in absence of ^{231}Pa data be estimated with the model of Henderson et al. (1999). Two recent compilations show a general undertrapping of shallow sediment traps (Scholten et al. 2001; Yu et al. 2001), studies that were corroborated by calculations based on inverse modelling (Usbeck 1999; Usbeck et al. 2003). In studies dealing with fluxes in the Weddell Sea or affected by Weddell Sea outflow, the specific Weddell-Sea effect mentioned above has to be taken into account (Fig. 1, Walter et al. 2000).

The use of ^{231}Pa/^{230}Th ratio as a proxy for (paleo)productivity (Kumar et al. 1993) is based on a constant fractionation of these nuclides during adsorption on suspended particles (e.g. Anderson et al. 1983 a,b). We have observed that the preferential adsorption of Th over Pa, by a factor of about 10 in most open oceans, collapses at high latitudes to a factor of only 1-2 (Rutgers van der Loeff and Berger 1993; Walter et al. 1997). This change in fractionation factor has been related to the chemical composition of the particles, especially the ratio of opal to clay (Anderson et al. 1992; Walter et al. 1997; Luo and Ku 1999). While measurements of Th and Pa isotope ratios cannot tell us whether the change is due to a reduced scavenging of Th or an enhanced scavenging of Pa, recent field observations (Chase 2001; Chase et al. 2002) and experiments (Geibert and Usbeck 2003) suggest it is primarily the former: a reduced scavenging of Th, which may be related to the scarcity of clay mineral particles in the remote Southern Ocean. The collapse of the fractionation factor strongly reduces the value of the ^{231}Pa/^{230}Th ratio as a proxy for (paleo)productivity in the Southern Ocean southward of approximately the southern Polar Front, and can at best be used in conjunction with other tracers (Kumar et al. 1993, 1995; Frank et al. 2000; Francois et al. 2003; Walter et al. 1999).

Finally, budget estimates of ^{231}Pa and ^{230}Th in an entire ocean basin have been used to estimate exchange rates between ocean basins, an approach that can also be used for time horizons (within the lifetime of these nuclides) like the Last Glacial Maximum. Yu et al. (1996) showed from a compilation of ^{231}Pa and ^{230}Th data from Atlantic sediments that export of Pa to the Southern Ocean, as it occurs now through North Atlantic Deep Water (NADW), was not significantly reduced during the Last Glacial Maximum (LGM). Recent modelling studies have shown, however, that ^{231}Pa/^{230}Th ratios in sediments from the Southern Ocean or from the south Atlantic are not very sensitive to past changes in thermohaline circulation (Asmus et al. 1999; Marchal et al. 2000). In the North and equatorial Atlantic the ratio is sensitive enough to provide useful constraints on paleocirculation (Marchal et al. 2000).

^{234}Th: Tracer for Export Production

This 24-day half-life daughter of ^{238}U is very suitable to quantify particle fluxes in the upper ocean and near the seafloor. From the depletion of ^{234}Th

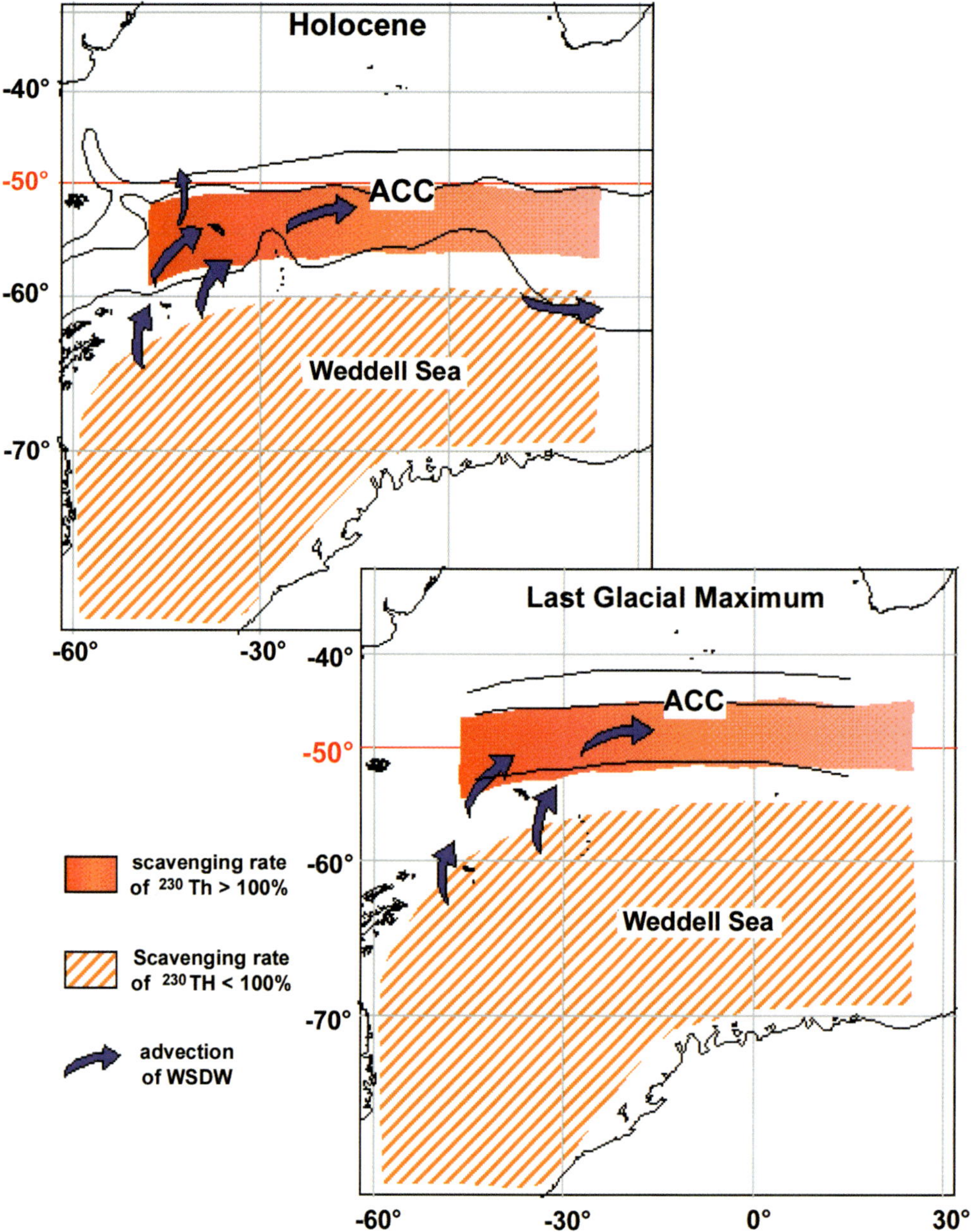

Fig. 1. [230]Th export from the Weddell Sea with implications for its rain rate to the sediment. (redrawn from Walter et al. 2000). WSDW: Weddell Sea Deep Water.

with respect to [238]U in the upper ocean the particulate flux or [234]Th out of the surface mixed layer or to depths below the euphotic zone can be calculated (Coale and Bruland 1985). The calculation is very simple when steady state can be assumed (i.e. if the cumulative depletion in the photic zone is constant with time, cf. Figs. 2 and 5), but can be adapted for non-steady state situation (Buesseler et al. 1992).

The improved methodology for [234]Th measurements (Rutgers van der Loeff and Moore 1999; Buesseler et al. 2001; Benitez-Nelson et al. 2001; Usbeck et al. 2002) has allowed us to routinely monitor the surface water distribution of [234]Th from

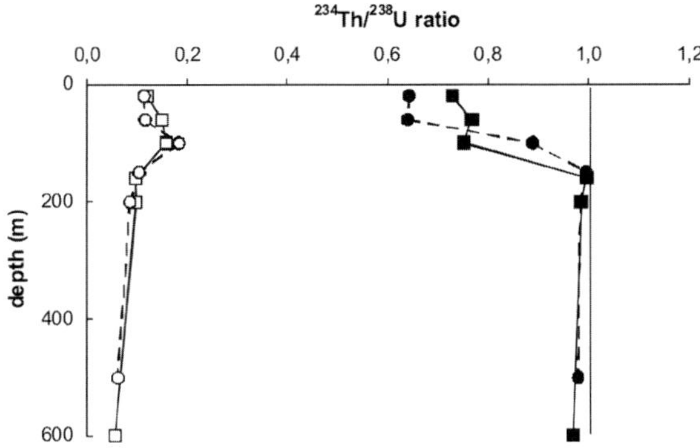

Fig. 2. Example of the distribution of ^{234}Th (expressed as ratio to its parent ^{238}U) in surface water near the Polar Front (49°50'S, 20°E) showing the depletion of total ^{234}Th (closed symbols) in the upper layer that must be balanced by an export on sinking particles to depth. The enhanced particulate activities (open symbols) in the surface water result from adsorption to plankton; the change in total activities between subsequent samplings (circles: 26 March; squares: 27 April) reflects a change in the mixed-layer depth (*Polarstern* Expedition ANT XVI/3; Rutgers van der Loeff and Westernströer 2000).

the ship's seawater supply on meridional transects across the ACC in increasing spatial resolution. The combination of these meridional transects with discrete profiles measured on stations has yielded a description of the seasonal development of export production. Apart from winter transects where no depletion of ^{234}Th was observed, transects later in the season revealed areas of high recent export rates, most often related to fronts.

Various approaches allow this flux to be converted to a flux of other components of the sinking material, e.g. POC or biogenic silica (see below). Export fluxes have been determined with this technique in JGOFS-related studies on the development of blooms in the Atlantic sector of the Southern Ocean in austral spring (ANT X/6) summer (ANT XIII/2) and autumn (ANT XVI/3), and during an iron enrichment experiment (ANT XVIII/2).

Seasonal Development of Export Fluxes:

Winter: By the end of the winter period, total ^{234}Th is found to be in equilibrium with ^{238}U in the sur-

face waters (Antarctic Peninsula: beginning of November; ANT VI/2, Rutgers van der Loeff and Berger 1991; Polar Frontal Zone in SE Atlantic: 11-31 October, ANT X/6, Rutgers van der Loeff et al. 1997; 26-31 October, ANT XVIII/2, Figs. 3,4). As the memory of the ^{234}Th distribution is several half-lives, this implies that around the Polar Front and southward, export production is negligible in the last several winter months up to the end of October.

Spring: In November a spring bloom develops near the Polar Front. In 1992 (ANT X/6) the bloom was probably induced by abundant Fe (De Baar et al. 1995; Löscher et al. 1997), which may have been related to a high iceberg density (van Franeker 1994; Dauelsberg et al. unpublished data). The tracer ^{234}Th showed the two phases of the bloom: first a build-up of biomass, which led to the transfer of Th to the particulate phase, and only in a second phase a removal of Th on sinking particles (Fig. 4; Rutgers van der Loeff et al. 1997).

Summer: In summer (ANT XIII/2, Dec 1995/Jan 1996), we observed a highly stable situation near the Polar Front (Figs. 5,6) with a constant small export flux. As this flux is sustained over an appreciable time, the resulting total flux is comparable to the event of a strong spring bloom (Rutgers van der Loeff et al. 2002).

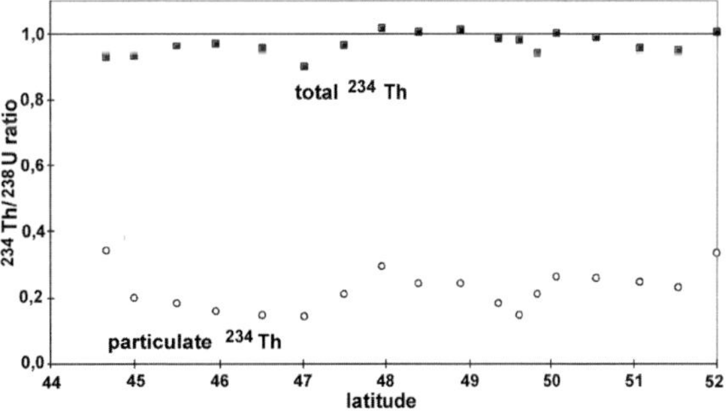

Fig. 3. Particulate (open symbols) and total (closed symbols) ^{234}Th /^{238}U ratio in surface water on the southward transect from Cape Town to the Polar Front, 26-31 Oct, 2000 (*Polarstern* Expedition ANT XVIII/2) showing that no export had occurred yet.

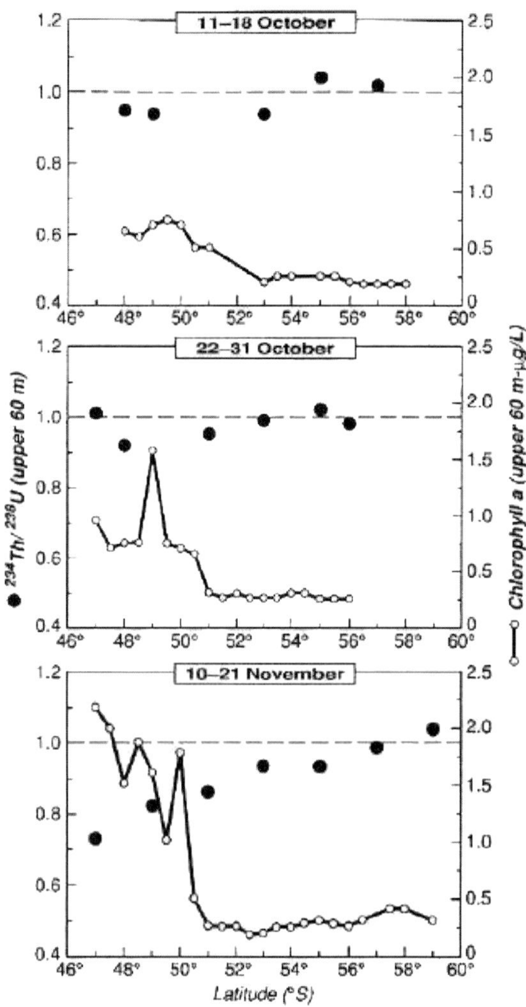

Fig. 4. Development of chlorophyll-a (line) and of ^{234}Th/ ^{238}U (dots) in surface waters in three successive transects across the ACC showing in the first phase the development of the bloom, followed in the second phase by export in the region near the Polar Front at about 50°S (Expedition ANT X/6, redrawn from Rutgers van der Loeff et al. 1997 and Bathmann et al. 1997 after Buesseler 1998).

Autumn: In autumn (ANT XVI/3, 1999) there was a continuing export with highest export rates associated with the oceanographic fronts (Fig. 7). The export near 52°S was related to a bloom that was subducted northward.

Conversion of ^{234}Th Export into the Export of Corg and BSi

Whereas the calculation of the export of ^{234}Th is straightforward, the conversion of the ^{234}Th flux to fluxes of other components, especially of POC and biogenic silica, is still a matter of debate. For this conversion we need to know the composition (i.e. the POC/^{234}Th and BSi/^{234}Th ratios) of the material settling out, a parameter that turns out to be difficult to determine. A first approach is to measure the composition of suspended material, which has a well-defined POC/^{234}Th ratio (Fig. 8) and assume that it settles out in the same composition.

The POC/^{234}Th ratio has been found to depend on particle size, with larger sizes usually having higher POC/^{234}Th ratios as one might expect from the surface to volume ratio (Buesseler et al. 1992, 1995; Cochran et al. 2000), but the opposite has been observed as well and the nature of this relationship is not yet well understood. An alternative approach is the analysis of material collected with sediment traps (e.g. NABE, Buesseler et al. 1992). Although perhaps the best approach available, even sediment traps are known to have biases in

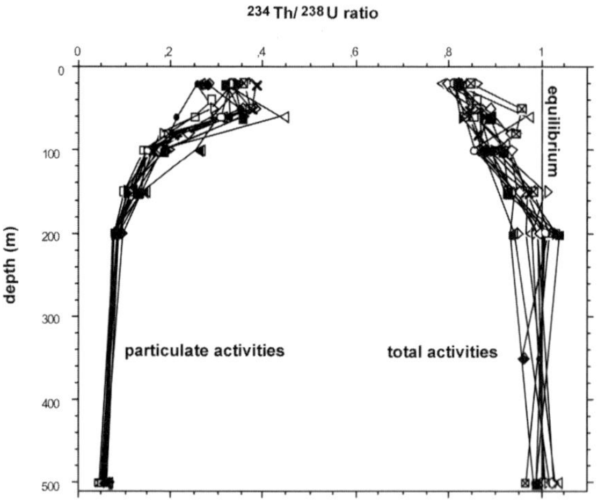

Fig. 5. Summer. Depth profiles of particulate (left) and total (right) ^{234}Th activities distributed over the period 25 Dec 1995 to 20 Jan. 1996 (ANT XIII/2) in a 150x130km wide box near the Polar Front, presented as ratio to the ^{238}U activity, showing the stable situation in summer.

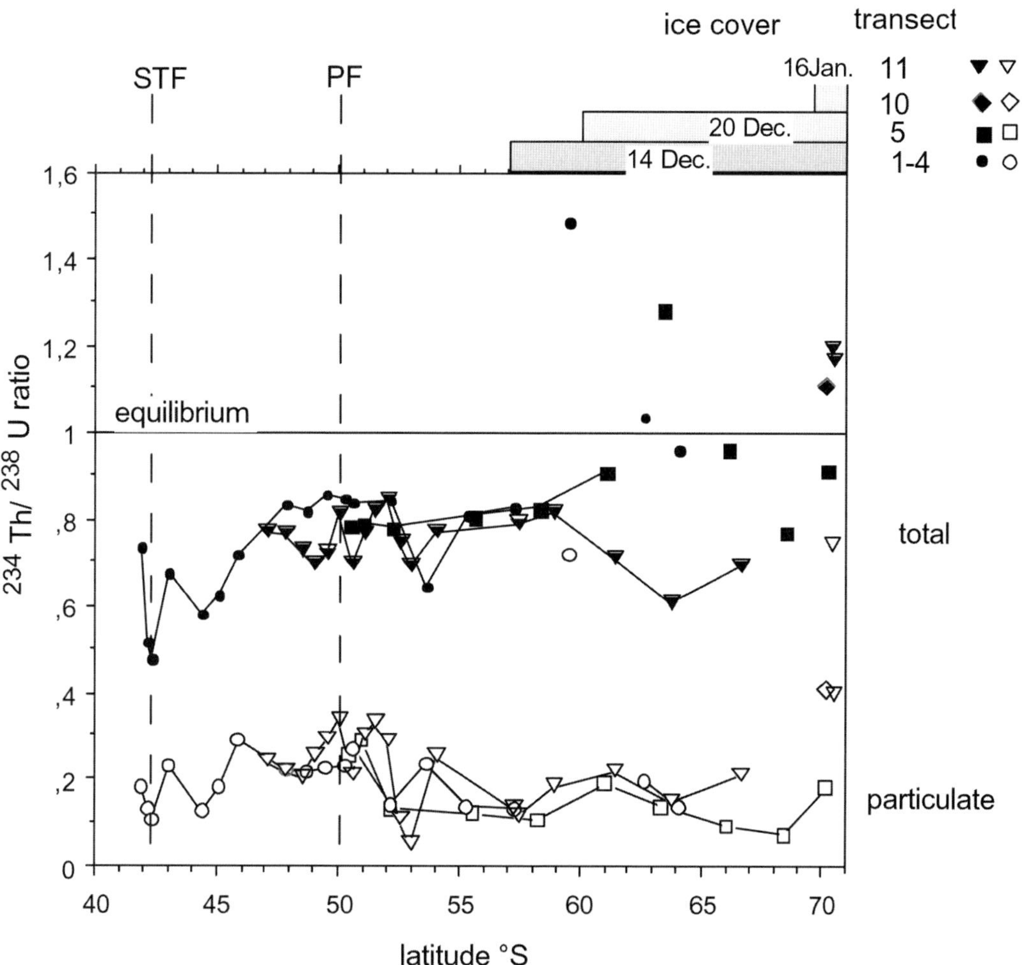

Fig. 6. Summer. Particulate (open symbols) and total (closed symbols) ^{234}Th in surface water, expressed as a ratio to ^{238}U, over various N-S transects during ANT XIII/2, Dec 1995/Jan1996, showing export near the Subtropical Front (STF) and moderate export near the Polar Front (PF) and further south. The erratic excess activities in ice-covered regions are ascribed to the continuous accumulation of ^{234}Th on ice algae that live attached to the percolated structures of the underside of the ice, and that are released as the ice is crushed by the ship's bow (Rutgers van der Loeff et al. 2002)

collection efficiency that must be size-dependent, which also implies a bias in POC/^{234}Th ratio. Moreover, sediment trap data can be obtained only for a relatively small number of stations. Andersson et al. (2000) showed a large discrepancy between these two approaches in the Baltic Sea. Ideally, the particles should be separated on sinking rate before analysis. A separation on particle size is not sufficient as the relationship between particle size and sinking rate is not straightforward: large particle aggregates may settle rapidly, but living plankton cells may not settle at all. A new approach (Gustafsson et al. 2000) uses experimental separation based on sinking rate, which should in principle allow determining the composition and corresponding POC/^{234}Th ratios dependent on sinking rate.

In a different approach we used an additional tracer, ^{210}Po, to separate two particle classes. It is well known that Po is strongly adsorbed to organic phases, whereas Th (and Pb) adsorbs strongly to any surface. Analysis of our data from the ANT X/

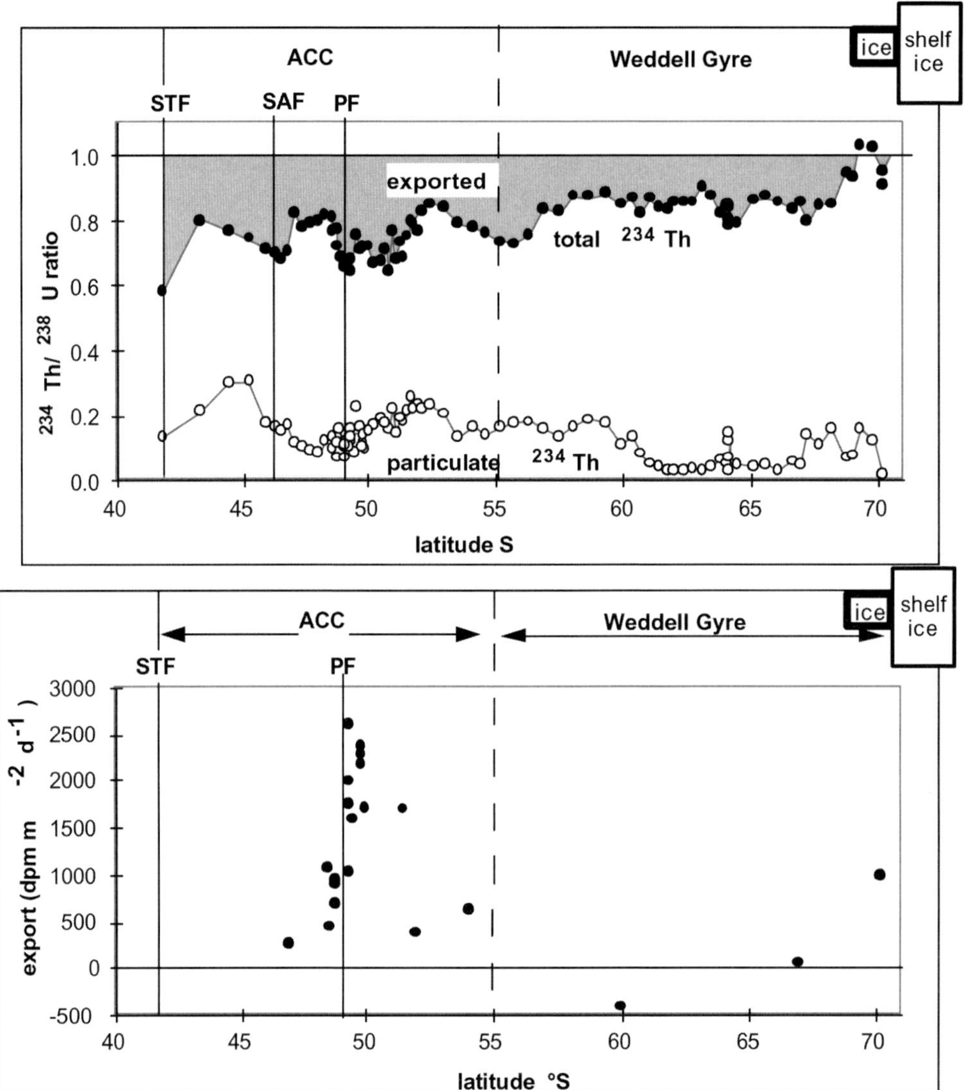

Fig. 7. Autumn. Above: particulate (open symbols) and total (closed symbols) ^{234}Th in surface water, expressed as a ratio to ^{238}U, during ANT XVI/3, March-April, 1999, showing export (hatched area) near the Subtropical Front (STF), Subantarctic Front (SAF) and south of the Polar Front (PF) and even in the Weddell Gyre (approx. 61-67°S) where particulate levels were extremely low. Below: Export rates of ^{234}Th calculated from ^{234}Th depletion at discrete stations along the same transect (assuming steady state, cf. Fig. 2). Redrawn from Usbeck et al. 2002.

6 expedition confirmed that the particulate Po was more strongly correlated with POC, whereas the adsorption of Th was more strongly related to BSi. In principle, this fractionation can be used to derive the export flux of POC and BSi separately from measurements of ^{234}Th and ^{210}Po depletion in the surface water (Friedrich and Rutgers van der Loeff 2002)

Surface Water Transects: The Prospects of Monitoring

Although the accurate calculation of export production of organic carbon requires both the determination of the time-dependent depth profile of the ^{234}Th depletion and the POC/^{234}Th ratio in the exported flux, much information can already be ob-

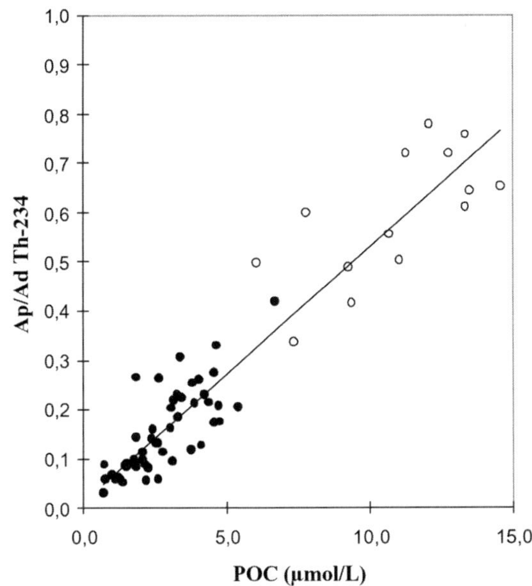

Fig. 8. Particulate ^{234}Th (A$_p$, normalised to the dissolved activity A$_d$) as a function of POC in suspended material collected near the Polar Front at 20-60m depth (open circles) and below 80m (closed circles), an example of the kind of data used to determine the POC/^{234}Th ratio on particles suspended at the depth where the export is to be determined. Summer expedition, ANT XIII/2 (Rutgers van der Loeff et al. 2002)

The Fate of Exported Particles:
Shallow Mineralization in the Weddell Sea

In an area of the Weddell Sea with extremely low suspended particle load (61-67°S, Fig. 7) we found for the first time a situation with export from the surface mixed layer that was balanced by a release at only 150-350m depth (Fig. 10) implying a very shallow mineralization. This situation is extraordinary, as it is associated with a net uptake of CO_2, which is in part exported to greater depths through the formation of AABW, whereas no record of this export is retained in the underlying sediments (Usbeck et al. 2002; Hoppema et al. 1997).

This shallow mineralization is also reflected in the distribution of dissolved nutrients (Whitworth and Nowlin 1987), as shown in detail by inverse modelling of available hydrographic and nutrient data sets from the literature (Usbeck 1999; Usbeck et al. 2002; Schlitzer 2000, 2002). It is also found in the distribution of oxygen and CO_2 (Hoppema et al. 1997), and in the distribution of another particle-reactive nuclide: ^{210}Pb (Farley and Turekian 1990). The mineralization depth recorded with ^{234}Th is somewhat shallower than expected from the distribution of these other tracers. This is probably related to the short half-life of ^{234}Th, which implies that the ^{234}Th release at 150-350m is asso-

tained from a high-resolution transect of total ^{234}Th in surface water alone as shown by the examples given above (Figs. 6,7). Another example is given by a transect from the SE Pacific (Fig. 9), where a strong export could be demonstrated near the ice edge, probably related to iron inputs (de Baar et al. 1999).

We conclude that the under-way analysis of total ^{234}Th in surface waters, collected at high resolution from the ship's seawater supply, is a powerful tool to describe the geographical distribution of export production. Such a monitoring, supported by occasional measurements of activity-depth profiles and/or of the mixed layer depth, and linked with a satellite-based (SeaWiFS) monitoring of the distribution of chlorophyll (calibrated with shipboard observations, e.g. Bathmann et al. 1997), would provide a tool to estimate export production on a basin-wide scale.

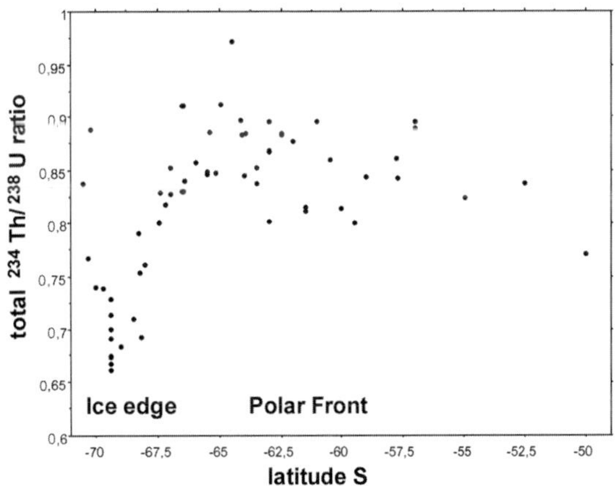

Fig. 9. April 1995: Total ^{234}Th/^{238}U ratios on a transect across the ACC in the SE Pacific (90°W, *Polarstern* Expedition ANT XII/4, redrawn from de Baar et al. 1999)

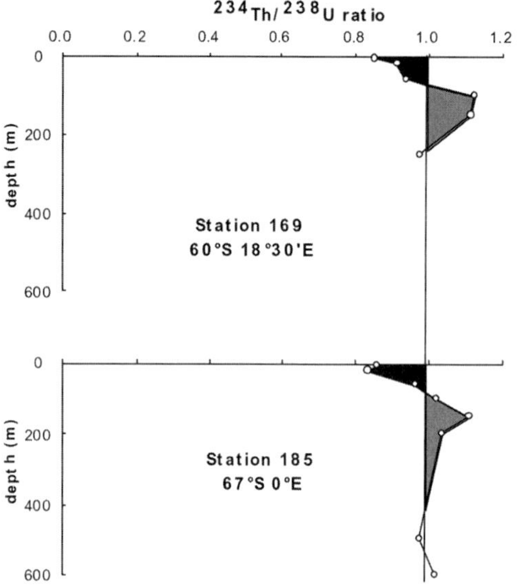

Fig. 10. Depth profiles of total ^{234}Th/^{238}U ratios, showing remineralization maxima and negligible net export production below 350m in the central Weddell Sea (ANT XVI/3, redrawn from Usbeck et al. 2002)

ciated with the decomposition of the most labile fractions, whereas the distribution of the other tracers is determined by the cumulative decomposition of a wider range of fractions.

^{234}Th in the Nepheloid Layer

Near the seafloor a situation exists that is very similar to that in the surface ocean. In the benthic nepheloid layer (BNL), ^{234}Th is adsorbed on resuspended particles. The resuspension-sedimentation cycle causes a depletion of total (dissolved + particulate) ^{234}Th in a layer that is on the order of 100m thick. This depletion can be used to quantify the particle exchange between sediment and bottom water (Bacon and Rutgers van der Loeff 1989; Rutgers van der Loeff and Boudreau 1997). We have measured the distribution of ^{234}Th in bottom waters on three transects across the Antarctic Circumpolar Current (ACC) in the southeast Atlantic (approx. 0° and 40°E, ANT XI/4) and in the southeast Pacific (90°W, ANT XII/4) (Fig. 11). The distribution of particulate ^{234}Th can be considered as a measure of particle load. Whereas in surface waters particulate ^{234}Th is well correlated with al-

gal biomass and chlorophyll, in deep waters it depicts the intensity of the nepheloid layer. Relative to the low particle loads in the far south (central Weddell Sea), we observe enhanced particle loads associated with the major fronts in the ACC, probably as a result of high bottom water currents and of rugged topography.

In the entire area we observed only minimal depletion of ^{234}Th with respect to ^{238}U in the bottom water. This implies that the residence time of particles in the nepheloid layer is generally longer than a few weeks, the time scale that can be measured with this tracer. One station in the central Enderby basin with high suspended load in the bottom water formed a conspicuous exception with a depletion of 13% (Fig. 12). Apparently, a strong resuspension and partial sedimentation of the opal rich sediments had occurred here.

An essential parameter of the budget of ^{234}Th in the BNL is the activity of ^{234}Th on the resuspended material. The interaction between resuspension and bioturbation can be described in a model that can be calibrated with ^{234}Th measurements. With this model (Rutgers van der Loeff and Boudreau 1997), turnover rates of other exchange processes at the sediment-water interface can be gauged with the ^{234}Th tracer.

Water Mass Tracers: ^{228}Ra, ^{227}Ac

Radium and Actinium are relatively mobile elements with isotopes that are produced in the uranium decay series from highly insoluble parents. Consequently, they are very well suited to trace water masses after their contact with sediment surfaces. In the remote oceans like the Southern Ocean it is of particular importance to have a tool to trace the advection of water masses with a potential input of terrigenous micronutrients. In the framework of investigations on the sources of iron in the South Atlantic we have studied the distribution of the isotopes ^{228}Ra and ^{227}Ac.

^{228}Ra: Tracer for Shelf Waters

^{228}Ra is produced by decay of ^{232}Th, which is ubiquitous in all sediments. ^{228}Ra is therefore released by all sediments, both in the deep-sea and near-

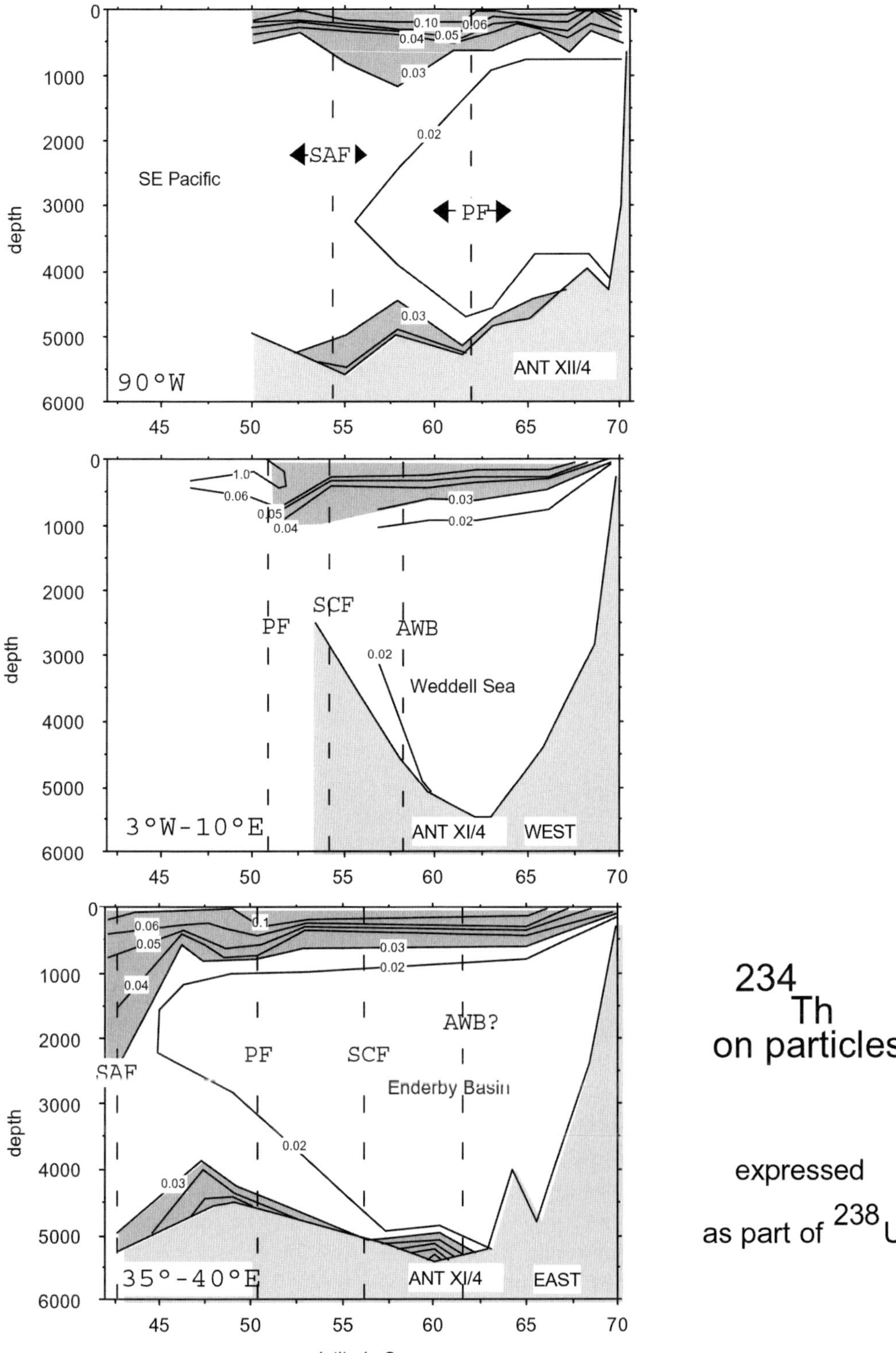

Fig. 11. Transects of particulate [234]Th on N-S sections across the ACC in the SE Pacific, the SE Atlantic and the SW Indian Ocean. South of the ACC, suspended loads in the bottom water are generally low.

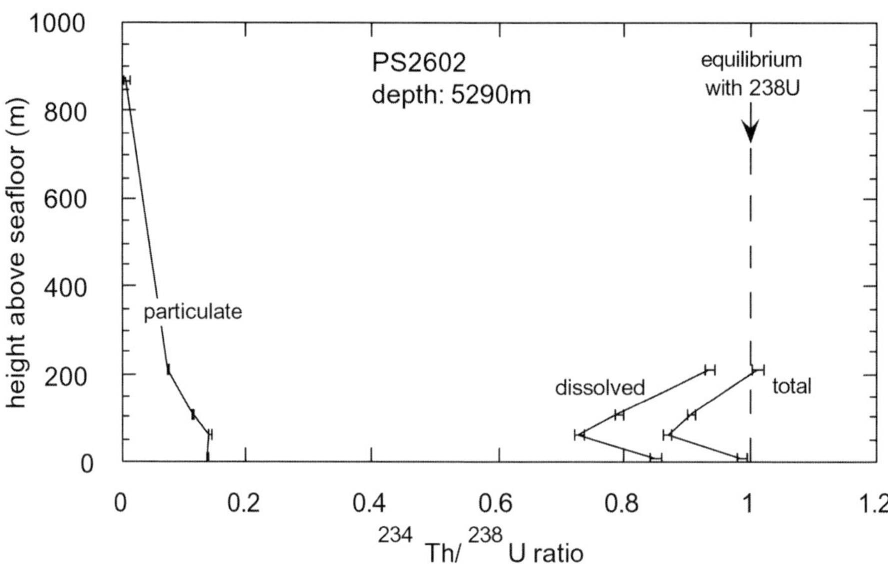

Fig. 12. An example of an exceptionally well-developed nepheloid layer in the center of the Enderby Basin. The 13% depletion of total ^{234}Th (2.8 dpm cm^{-2}) implies an intensive exchange with surface sediments with an average residence time of about 12 days.

shore. The extended residence time of coastal water masses in shallow shelf seas allows the tracer to build up here far higher activities than are reached in near-bottom waters in the deep-sea. ^{228}Ra is therefore especially well suited as a tracer of shelf waters. After the fundamental work of the GEOSECS expedition had established global distribution patterns of many radionuclide tracers including ^{228}Ra (Li et al. 1980), the first transect across the ACC towards the Antarctic continent (Rutgers van der Loeff 1994) confirmed the expected enrichment of ^{228}Ra on the Antarctic shelf. In the framework of the search for pathways of iron to the surface waters of the High Nutrient Low Chlorophyll (HNLC) Southern Ocean we have recently completed a detailed survey of ^{228}Ra in the Atlantic sector of the Southern Ocean (Hanfland 2002). Gamma spectroscopy of radium fractions collected from surface waters on MnO$_2$ adsorbers revealed the strong ^{228}Ra sources both near South Africa (in the Agulhas Current, carrying signals obtained on the east-African coast), near the Antarctic continent (both near Neumayer station and

near the Antarctic Peninsula) and near the south American continent. In the Argentinean Basin, enhanced ^{228}Ra activities were also found far offshore, indicating the far-reaching continental influence in the south Atlantic Gyre, in support of the studies of Li et al. (1980).

In the ACC proper, Hanfland (2002) mapped the distribution of ^{228}Ra in the surface water with the far more sensitive ^{228}Th ingrowth technique (after Li et al. 1980). The idea was, that if shelf-influenced water masses intrude into the frontal jets of the ACC, this ought to be visible in enhanced ^{228}Ra levels. Radioactive decay of this 5.8-y half-life tracer cannot be significant during the transit across the Atlantic sector of the ACC, which can be estimated to be on the order of several months. But on several high-resolution sections across the ACC at 0° to 20°E Hanfland observed mostly very low ^{228}Ra activities, discounting the role of this transport route for terrigenous material. As it can be expected that uptake and scavenging cause iron to be removed more rapidly than radium, Hanfland concluded that the shelf of South America, of the

Peninsula or of the south Sandwich Islands is not an important source for iron to alleviate the iron stress of plankton communities in the SE Atlantic (De Baar et al. 1995).

^{227}Ac: Tracer for Deep Upwelling

The application of this tracer is very similar to the previous tracer but with two major differences:
• the half-life of ^{227}Ac (21.8 y) is much longer (^{228}Ra 5.8 y). The tracer is therefore suitable for transport processes on a timescale of decades rather than years.
• the activity of the parent nuclide of ^{227}Ac (^{231}Pa) in sediments is dependent on water depth. As a consequence, deep-sea sediments are a far more important source for ^{227}Ac than shallow sediments.

Geibert (2001; Geibert et al. 2002) compared the release rate of ^{228}Ra and ^{227}Ac from marine sediments into the overlying water. This release rate depends not only on the activity of the mother nuclide in surface sediments, but also on the fraction of the daughter reaching the pore water, the adsorption equilibrium in the sediment pore water, and the bioturbation rate. Based on a model of Cochran and Krishnaswami (1980) for ^{228}Ra and an application of this model to ^{227}Ac by Nozaki et al. (1990) he shows that the ^{227}Ac/^{228}Ra release rate ratio is indeed highly distinctive, changing from 0.026 for shelf sediments to 1.4 for the deep-sea. Consequently, ^{227}Ac can be considered as a specific tracer for contact with deep-sea sediments.

The first ^{227}Ac profiles from the Southern Ocean (Geibert 2001) show indeed this bottom source of ^{227}Ac, and can in principle be used to quantify mixing rates in the deep-sea. Similar studies have been performed based on the distribution of ^{228}Ra, but can now be extended to the longer time scale of ^{227}Ac. It is remarkable that these profiles show significant excess ^{227}Ac activities (i.e. in excess over the amount supported by ^{231}Pa in the water column) in surface waters south of the Polar Front. This implies that ^{227}Ac can be used here as a tracer for upwelling of water masses from the deep-sea. Upwelling rates in the Weddell Sea as based on these ^{227}Ac data (approx. 55 ma^{-1}) are in line with earlier estimates based on heat budgets. The tracer appears to have a large potential for

studies of upwelling and diapycnal mixing rates in the ocean. The tracer may thus be used to monitor temporal changes in upwelling rates. If in the past two decades deep water production in the Weddell Sea had been smaller than usual, as has been hypothesised by Broecker et al. (1998), this would require a corresponding change in upwelling rate. If the circulation returned to normal, this might be detected as an increase in the ^{227}Ac levels in surface waters.

Concluding Remarks

Natural radionuclides help us to study the transport of particles and water masses in the southern Atlantic Ocean.

Water masses that flow over continental shelf regions obtain a strong ^{228}Ra signal. Mostly very low ^{228}Ra activities in the frontal jets of the ACC show that such water masses cannot be responsible for the local fertilization in the central ACC with terrigenous trace substances like iron.

Upwelling of deep water south of the Polar Front can be followed with ^{227}Ac, a promising new tracer for water masses that have been in contact with deep-sea sediments.

Scavenging of particle-reactive isotopes helps us to calibrate particle fluxes. Export production, as measured with ^{234}Th, is absent in winter, shows large pulses in spring and continues with moderate rates during summer and autumn, with highest rates associated with the fronts of the ACC. South of the ACC, in the Weddell Gyre, we find a significant export production from the euphotic zone, but a highly efficient shallow remineralization in the depth zone of 150-350m, leaving only extremely low particle fluxes below 350m.

The low particle flux south of the ACC causes inefficient scavenging of ^{230}Th, observed in low ^{230}Th inventories and an accumulation of ^{230}Th in the water column. ^{230}Th is thus exported with AABW to the ACC where the surplus is removed by scavenging.

In the ACC, strong currents down to the seafloor are responsible for resuspension of particles in the bottom waters, as observed by increases in turbidity and in particulate ^{234}Th. This causes a large-scale redistribution of sediments, with accumulation rates that can exceed the local rain rates by an order of magnitude. ^{230}Th is very suitable to correct for these focusing and winnowing effects, provided that the boundary scavenging of ^{230}Th in the water column is taken into account.

Boundary scavenging is much more pronounced for ^{231}Pa than for ^{230}Th, making the Southern Ocean a sink for ^{231}Pa produced further north in the Atlantic. The demonstration that this process continued during the last glacial (Yu et al. 1996) tells us that NADW transport continued during the LGM but not at what intensity. The lack of fractionation between Th and Pa in the opal-dominated Southern Ocean south of the Polar Front causes the ^{231}Pa$_{xs}$/^{230}Th$_{xs}$ ratio here to have very limited value as proxy for (paleo)productivity.

Acknowledgments

We are grateful for the assistance given by captains and crew of RV *Polarstern* on the many expeditions on which this paper is based. We wish to thank Dieter Fütterer for his full hearted support of the geochemistry group. M. Roy-Barman and R. Francois provided helpful reviews. This work was supported by the Deutsche Forschungsgemeinschaft (Sonderforschungsbereich 261). Data are available under www.pangaea.de/Projects/SFB261.

References

Anderson HL, Francois R, Moran SB (1992) Experimental evidence for differential adsorption of Th and Pa on different solid phases in seawater. EOS 73 (43S, fall meeting), 270 p

Anderson RF, Bacon MP, Brewer PG (1983a) Removal of ^{230}Th and ^{231}Pa at ocean margins. Earth Planet Sci Lett 66: 73-90

Anderson RF, Bacon MP, Brewer PG (1983b) Removal of ^{230}Th and ^{231}Pa from the open ocean. Earth Planet Sci Lett 62: 7-23

Andersson PS, Gustafsson O, Roos P, Broman D, Toneby A (2000) Particle mediated surface water export: comparison of estimates from ^{238}U-^{234}Th disequilibria and sediment traps in a continental shelf region. EOS AGU/ASLO Ocean Sciences Meeting, San Antonio, abstract only

Asmus T, Frank M, Koschmieder C, Frank N, Gersonde R, Kuhn G, Mangini A (1999) Variations of biogenic particle flux in the southern Atlantic section of the Subantarctic Front during the late Quaternary: Evidence from sedimentary ^{231}Pa$_{ex}$ and ^{230}Th$_{ex}$. Mar Geol 159: 63-78

Bacon MP (1976) Applications of Pb-210/Ra-226 and Po-210/Pb-210 disequilibria in the study of marine geo-chemical processes. PhD thesis, Woods Hole Oceanographic Institution

Bacon MP, Anderson RF (1982) Distribution of thorium isotopes between dissolved and particulate forms in the deep-sea. J Geophys Res 87: 2045-2056

Bacon MP, Rosholt JN (1982) Accumulation rates of ^{230}Th and ^{231}Pa and some transition metals on the Bermuda Rise. Geochim Cosmochim Acta 46: 651-666

Bacon MP, Rutgers van der Loeff MM (1989) Removal of Thorium-234 by scavenging in the bottom nepheloid layer of the ocean. Earth Planet Sci Lett 92: 157-164

Bacon MP, Spencer DW, Brewer PG (1976) ^{210}Pb/^{226}Ra and ^{210}Po/^{210}Pb disequilibria in seawater and suspended particulate matter. Earth Planet Sci Lett 32: 277-296

Bacon MP, Huh C-A, Fleer AP, Deuser WG (1985) Seasonality in the flux of natural radionuclides and plutonium in the deep Sargasso Sea. Deep-Sea Res 32: 273-286

Bathmann UV, Scharek R, Klaas C, Dubischar CD, Smetacek V (1997) Spring development of phytoplankton biomass and composition in major water masses of the Atlantic sector of the Southern Ocean. Deep-Sea Res II 44: 51-67

Benitez-Nelson C, Buesseler KO, Rutgers van der Loeff MM, Andrews J, Ball L, Crossin G, Charette MA (2001) Testing a new small-volume technique for determining thorium-234 in seawater. J Radioanalytical Nuclear Chem 248: 795-799

Broecker WS, Peng T-H (1982) Tracers in the Sea. Lamont- Doherty Geol Obs, Columbia University 690 p

Broecker WS, Goddard J, Sarmiento JL (1976) The distribution of ^{226}Ra in the Atlantic Ocean. Earth Planet Sci Lett 32: 220-235

Broecker WS, Peacock SL, Walker S, Weiss R, Fahrbach E, Schroeder M, Mikolajewicz U, Heinze C, Key R, Peng T-H, Rubin S (1998) How much deep water is formed in the Southern Ocean. J Geophys Res 103(C8): 15833-15843

Buesseler K O (1998) The decoupling of production and particulate export in the surface ocean. Glob Biogeochem Cycl 12: 297-310

Buesseler KO, Bacon MP, Cochran JK, Livingston HD

(1992) Carbon and nitrogen export during the JGOFS North Atlantic Bloom Experiment estimated from ^{234}Th:^{238}U disequilibria. Deep-Sea Res 39: 1115-1137

Buesseler KO, Andrews JA, Hartman MC, Belastock R, Chai F (1995) Regional estimates of the export flux of particulate organic carbon derived from thorium-234 during the JGOFS EqPac program. Deep-Sea Res II 42: 777-804

Buesseler KO, Benitez-Nelson C, Rutgers van der Loeff MM, Andrews J, Ball L, Crossin G, Charette MA (2001) A comparison of methods with a new small-volume technique for thorium-234 in seawater. Mar Chem 74: 15-28

Chase Z (2001) Trace elements as regulators (Fe) and recorders (U, Pa,Th, Be) of biological productivity in the ocean. PhD, Columbia University

Chase Z, Anderson RF, Fleisher MQ, Kubik P (2002) The influence of particle composition on scavenging of Th, Pa and Be in the ocean. Earth Planet Sci Lett 204: 215-229

Chase Z, Anderson RF, Fleisher MQ, Kubik P (2003) Scavenging of ^{230}Th, ^{231}Pa and ^{10}Be in the Southern Ocean (SW Pacific sector): The importance of particle flux and advection. Deep-Sea Res II 50: 739-768

Chen JH, Edwards LR, Wasserburg GJ (1986) ^{238}U, ^{234}U and ^{232}Th in seawater. Earth Planet Sci Lett 80: 241-251

Chung Y (1981) ^{210}Pb and ^{226}Ra distributions in the Circumpolar waters. Earth Planet Sci Lett 55: 205-216

Chung Y, Applequist MD (1980) ^{226}Ra and ^{210}Pb in the Weddell Sea. Earth Planet Sci Lett 49: 401-410

Coale KH, Bruland KW (1985) ^{234}Th:^{238}U disequilibria within the California current. Limnol Oceanogr 30: 22- 33

Cochran JK (1992) The oceanic chemistry of the Uranium and Thorium-series nuclides. In: Ivanovich M and Harmon RS (eds) Uranium-Series Disequilibrium: Applications to Earth, Marine, and Environmental Sciences. 2nd edition, Clarendon Press, pp 334-395

Cochran JK, Krishnaswami S (1980) Radium, Thorium, Uranium and ^{210}Pb in deep-sea sediments and sediment pore waters from the North Equatorial Pacific. Am J Sci 280: 849-889

Cochran JK, Buesseler KO, Bacon MP, Wang HW, Hirschberg DJ, Ball L, Andrews J, Crossin G, Fleer A (2000) Short-lived thorium isotopes (^{234}Th, ^{228}Th) as indicators of POC export and particle cycling in the Ross Sea, Southern Ocean. Deep-Sea Res II 47: 3451-3490

De Baar HJW, De Jong JTM, Bakker DCE, Löscher BM, Veth C, Bathmann U, Smetacek V (1995) Importance of iron for plankton blooms and carbon dioxide drawdown in the Southern Ocean. Nature 373: 412-415

De Baar HJW, de Jong JTM, Nolting RF, Timmermans KR, van Leeuwe MA, Bathmann U, Rutgers van der Loeff M, Sildam J (1999) Low dissolved Fe and the absence of diatom blooms in remote Pacific waters of the Southern Ocean. Mar Chem 36: 1-34

Farley KA, Turekian KK (1990) Lead-210 in the circumpolar South Atlantic. Deep-Sea Res 37(12): 1849-1860

Francois R, Bacon MP, Altabet MA, Labeyrie LD (1993) Glacial/Interglacial changes in sediment rain rate in the SW Indian sector of subantarctic waters as recorded by ^{230}Th, ^{231}Pa, U, and δ^{15}N. Paleoceanography 8: 611-629

Francois R, Frank M, Rutgers van der Loeff MM, Bacon MP (2003) 230Th-normalization: An essential tool for interpreting sedimentary fluxes during the late Quaternary. Paleoceanography, submitted

Frank M, Gersonde R, Mangini A (1999) Sediment redistribution, ^{230}Th$_{ex}$ - normalization and implications for the reconstruction of particle flux and export paleoproductivity. In: Fischer G, Wefer G (eds) Use of Proxies in Paleoceanography. Springer, Berlin, pp 409-426

Frank M, Gersonde R, Rutgers van der Loeff MM, Bohrmann G, Nürnberg CC, Kubik PW, Suter M, Mangini A (2000) Similar glacial and interglacial export bioproductivity in the Atlantic sector of the Southern Ocean: Multiproxy evidence and implications for glacial atmospheric CO_2. Paleoceanography 15: 642-658

Friedrich J (1997) Polonium-210 und Blei-210 im Südpolarmeer: Natürliche Tracer für biologische und hydrographische Prozesse im Oberflächenwasser des Antarktischen Zirkumpolarstroms und des Weddellmeeres. Ber Polarforsch 235: 1-155

Friedrich J, Rutgers van der Loeff MM (2002) A two-tracer (^{210}Po-^{234}Th) approach to distinguish organic carbon and biogenic silica export flux in the Antarctic Circumpolar Current. Deep-Sea Res I 49: 101-120

Geibert W (2001) Actinium-227 as tracer for advection and mixing in the Deep-Sea. Reports on Polar and Marine Research 385, 112 p

Geibert W, Usbeck R (2003) The adsorption of Thorium and Protactinium onto different particle types: Experimental findings. Geochim Cosmochim Acta, in press

Geibert W, Rutgers van der Loeff MM, Hanfland C, Dauelsberg H-J (2002) Actinium-227 as a Deep-Sea Tracer: Sources, Distribution and Applications. Earth Planet Sci Lett 198: 147-165

Gustafsson Ö, Düker A, Larsson J, Andersson P, Ingri

J (2000) Functional separation of colloids and gravitoids in surface waters based on differential settling velocity: Coupled cross-flow filtration - split flow-thin cell fractionation (CFF-SPLITT). Limnol Oceanogr 45: 1731-1742

Hanfland C (2002) Radium-226 and Radium-228 in the Atlantic sector of the Southern Ocean. Reports on Polar and Marine Research. PhD Thesis Univ Bremen, Germany 431, 135 p

Henderson GM, Heinze C, Anderson RF, Winguth AME (1999) Global distribution of the ^{230}Th flux to ocean sediments constrained by GCM modelling. Deep-Sea Res I 46: 1861-1893

Hoppema M, Fahrbach E, Schröder M (1997) On the total carbon dioxide and oxygen signature of the Circum-polar Deep Water in the Weddell Gyre. Oceanol Acta 20: 783-798

Ku TL, Lin MC (1976) ^{226}Ra distribution in the Antarctic Ocean. Earth Planet Sci Lett 32: 236-248

Ku TL, Knauss KG, Mathieu GG (1977) Uranium in the open ocean: Concentration and isotopic composition. Deep-Sea Res 24: 1005-1017

Kumar N, Gwiazda R, Anderson RF, Froelich PN (1993) ^{231}Pa/^{230}Th ratios in sediments as a proxy for past changes in Southern Ocean productivity. Nature 362: 45-48

Kumar N, Anderson RF, Mortlock RA, Froelich PN, Kubik P, Dittrich-Hannen B, Suter M (1995) Increased biological productivity and export production in the glacial Southern Ocean. Nature 378: 675-680

Li Y-H, Feely HW, Toggweiler JR (1980) ^{228}Ra and ^{228}Th concentrations in GEOSECS Atlantic surface waters. Deep-Sea Res 27A: 545-555

Löscher BM, de Jong JTM, de Baar HJW, Veth C, Dehairs F (1997) The distribution of Fe in the Antarctic Circumpolar Current. Deep-Sea Res II 44(1/2): 143-187

Luo S, Ku T-L (1999) Oceanic ^{231}Pa/^{230}Th ratio influence by particle composition and remineralization. Earth Planet Sci Lett 167: 183-195

Marchal O, Francois R, Stocker TF, Joos F (2000) Ocean thermohaline circulation and sedimentary ^{231}Pa/^{230}Th ratio. Paleoceanography 15: 625-641

Nozaki Y, Horibe Y, Tsubota H (1981) The water column distributions of thorium isotopes in the western North Pacific. Earth Planet Sci Lett 54: 203-216

Nozaki Y, Yamada M, Nikaido H (1990) The marine geochemistry of Actinium-227: Evidence for its migration through sediment pore water. Geophys Res Lett 17: 1933-1936

Rutgers van der Loeff MM (1994) ^{228}Ra and ^{228}Th in the Weddell Sea. In: Johannessen OM, Muench RD, Overland JE (eds) The Polar Oceans and their Role in Shaping the Global Environment: The Nansen Centennial Volume. Geophysical Monograph 85, American Geophysical Union, pp 177-186

Rutgers van der Loeff MM (2001) Uranium-Thorium decay series in the water column. In: Steele J, Thorpe S, Turekian K (eds) Encyclopedia of Ocean Sciences. Vol. MS 168, Academic Press

Rutgers van der Loeff MM, Berger GW (1991) Scavenging and particle flux: Seasonal and regional variations in the South Ocean (Atlantic sector) Mar Chem 35: 553-567

Rutgers van der Loeff MM, Berger GW (1993) Scavenging of ^{230}Th and ^{231}Pa near the Antarctic Polar Front in the South Atlantic. Deep-Sea Res I 40: 339-357

Rutgers van der Loeff MM, Boudreau BP (1997) The effect of resuspension on chemical exchanges at the sediment water interface - A modelling and natural radiotracer approach. J Mar Syst 11: 305-342

Rutgers van der Loeff MM, Moore WS (1999) Determination of natural radioactive tracers. In: Grasshoff K, Kremling K, Ehrhardt M (eds) Methods of Seawater Analysis. Wiley-VCH, Weinheim, pp 365-397

Rutgers van der Loeff MM, Westernströer U (2000) Export production measured through the ^{234}Th/^{238}U disequilibrium in surface waters. In: Bathmann U, Smetacek V, Reinke M (eds) The Expeditions ANTARKTIS XVI/3-4 of the RV *Polarstern* in 1999. Rep Polar Res 364, pp 114-118

Rutgers van der Loeff MM, Friedrich J, Bathmann UV (1997) Carbon export during the spring bloom at the southern Polar Front, determined with the natural tracer ^{234}Th. Deep-Sea Res II 44: 457-478

Rutgers van der Loeff MM, Buesseler KO, Bathmann U, Hense I, Andrews J (2002) Comparison of carbon and opal export rates between summer and spring bloom periods in the region of the Antarctic Polar Front, SE Atlantic. Deep-Sea Res II 49: 3849-3869

Schlitzer R (2000) Applying the Adjoint Method for Global Biogeochemical Modeling. In: Kasibhatla P, Heimann M, Hartley D, Mahowald N, Prinn R, Rayner P (eds) Inverse Methods in Biogeochemical Cycles. AGU, pp 107-124

Schlitzer R (2002) Carbon export fluxes in the Southern Ocean: Results from inverse modeling and comparison with satellite based estimates. Deep-Sea Res II 49: 1623-1644

Scholten JC, Fietzke J, Vogler S, Rutgers van der Loeff MM, Mangini A, Koeve W, Stoffers P, Antia A, Neuer S, Waniek J (2001) Trapping efficiencies of sediment traps from the deep waster north Atlantic: The [230]Th calibration. Deep-Sea Res II JGOFS North Atlantic Synthesis 48: 2383-2408

Suman DO, Bacon MP (1989) Variations in Holocene sedimentation in the North American Basin determined from [230]Th measurements. Deep-Sea Res 36: 869-878

Usbeck R (1999) Modeling of marine biogeochemical cycles with an emphasis on vertical particle fluxes. Ber Polarforsch 332, 105 p

Usbeck R, Rutgers van der Loeff MM, Hoppema M, Schlitzer R (2002) Shallow mineralization in the Weddell Gyre. Geochem Geophys Geosyst 3, 1: 10.1029/2001GC000182

Usbeck R, Schlitzer R, Fischer G, Wefer G (2003) Particle fluxes in the ocean: Comparison of sediment trap data with results from inverse modeling. J Mar Syst 39: 167-183

van Franeker J (1994) Sea-ice cover and icebergs. In: Bathmann UV, Smetacek V, de Baar HJW (eds) The Expedition Antarktis X/6-8 of the research vessel *Polarstern* in 1992/1993. Ber Polarforsch 135, AWI, Bremerhaven, pp 17-22

Vogler S, Scholten J, Rutgers van der Loeff M, Mangini A (1998) [230]Th in the eastern North Atlantic: The importance of water mass ventilation in the balance of [230]Th. Earth Planet Sci Lett 156: 61-74

Walter HJ, Rutgers van der Loeff MM, Höltzen H (1997) Enhanced scavenging of [231]Pa relative to [230]Th in the South Atlantic south of the Polar Front: Implications for the use of the [231]Pa/[230]Th ratio as a paleo-productivity proxy. Earth Planet Sci Lett 149: 85-100

Walter HJ, Rutgers van der Loeff MM, Francois R (1999) Reliability of the [231]Pa/[230]Th activity ratio as a tracer for bioproductivity of the ocean. In: Fischer G, Wefer G (eds) Use of Proxies in Paleoceanography - Examples from the South Atlantic. Springer, Berlin, pp 393-408

Walter HJ, Rutgers van der Loeff MM, Höltzen H, Bathmann U (2000) Reduced scavenging of [230]Th in the Weddell Sea: Implications for paleoceanographic reconstructions in the South Atlantic. Deep-Sea Res I 47: 1369-1387

Whitworth T, Nowlin WD (1987) Water masses and currents of the Southern Ocean at the Greenwich meridian. J Geophys Res 92(C6): 6462-6476

Yu E-F, Francois R, Bacon MP (1996) Similar rates of modern and last-glacial ocean thermohaline circulation inferred from radiochemical data. Nature 379: 689-694

Yu E-F, Francois R, Bacon MP, Honjo S, Fleer AP, Manganini SJ, Rutgers van der Loeff MM, Ittekot V (2001) Trapping efficiency of bottom-tethered sediment traps estimated from the intercepted fluxes of [230]Th and [231]Pa. Deep-Sea Res I 48: 865-889

Heterotrophic Particle-Associated Bacteria from the South Atlantic: A Community of Marine Microorganisms with a High Organic Carbon Degradation Potential

I. Berkenheger*, A. Heuchert, S. de Silva and U. Fischer

Universität Bremen, Fachbereich Biologie/Chemie, Zentrum für Umweltforschung und Umwelttechnologie (UFT), Abt. Marine Mikrobiologie, Leobener Straße, 28359 Bremen, Germany
** corresponding author (e-mail): imke@biotec.uni-bremen.de*

Abstract: Samples of the Equatorial Atlantic (EA) and the South Atlantic/Antarctica (SA) were taken during two cruises with RV *Meteor* in 1996 and 1997 and one cruise with RV *Polarstern* in 1998 in order to study the bacterial communities attached to organic particles. Ten heterotrophic bacterial strains, isolated from particles of the EA, and 11 strains, isolated from particles of the SA, were chosen for further investigations. All isolates are Gram-negative rods, which differ strongly in their ability to metabolize high and low molecular weight organic compounds (polysaccharides, di- and monosaccharides, organic acids). The phylogenetic composition of the bacterial communities on sinking particles was investigated by random amplified polymorphic DNA (RAPD), amplified ribosomal DNA restriction analysis (ARDRA), fluorescence *in situ* hybridization (FISH), and 16S rDNA-sequencing. In both investigation areas, members of the α- and γ-subclass of Proteobacteria and also of the *Cytophaga/Flavobacteria*-cluster were detected. The genus *Sulfitobacter* was present in both investigation areas, whereas other genera such as *Marinobacter* and *Psychrobacter* could each be found on only one sampling site.

Introduction

Macroscopic aggregates (marine snow, particles > 0.5 mm in size) are the dominant fraction involved in the transport of biogenic carbon from the surface water to the deep-sea bottom (Alldredge and Silver 1988).

Marine snow shows highly variable physical characteristics, appearance, and composition. The aggregates originate from two general pathways of formation (Alldredge and Silver 1988): (a) particles can be produced *de novo* by plants and mucus feeding webs of zooplankton, and (b) particles result from the biologically-enhanced physical aggregation of smaller particles (e.g. micro-aggregates, fecal pellets). In the senescent phase of the phytoplankton bloom, the cell surface becomes sticky due to the release of polysaccharides, and then single phytoplankton cells aggregate (Iriberri and Herndl 1995).

Meanwhile, several new classes of organic particles of smaller size (from 20 nm to hundreds of micrometers) have been discovered, hence changing the ideas of the physical and chemical nature of the particle field with which pelagic bacteria interact (Long and Azam 1996; Azam 1998). These smaller particles were 2 to 3 orders of magnitude more abundant than large particles and their components are polysaccharides, transparent exopolymer particles (TEP), or proteins (Alldredge et al. 1993; Long and Azam 1996).

TEP are highly significant components of marine aggregates, form the matrix of marine snow, and serve as substrates and microhabitats which provide attached bacteria with physical refuges from predators unable to feed on surfaces. Investigations of marine snow samples collected off coastal California showed that 24-68% of the bac-

From WEFER G, MULITZA S, RATMEYER V (eds), 2003, *The South Atlantic in the Late Quaternary: Reconstruction of Material Budgets and Current Systems.* Springer-Verlag Berlin Heidelberg New York Tokyo, pp 65-79

teria in seawater samples were attached to TEP (Alldredge et al. 1993).

All aggregates have higher concentrations of nutrients and show elevated microbial activities in comparison to the surrounding water (Shanks and Trent 1979; Caron et al. 1982; Azam et al. 1993). Their degradation is characterized by microbial succession in which auto- and heterotrophic bacteria and bacterivorous protozoa are involved (Biddanda and Pomeroy 1988; Iriberri and Herndl 1995). The particle-associated microbial communities undergo complex successional changes which significantly alter the chemical and biological properties of the particles. Rates of particle production and breakdown are difficult to predict flux and to understand biological community structure and transformations of matter and energy in the water column (Alldredge and Silver 1988). Aggregates provide enriched microenvironments of organic matter in the oligotrophic ocean and are in so far hotspots of microbial respiration which cause a fast and efficient respiratory turnover of particulate organic carbon in the sea (Caron et al. 1982; Azam 1998; Ploug et al. 1999). Marine snow is a microhabitat highly enriched in phytoplankton, bacteria, flagellates and detritus compared to the surrounding water (Ploug et al. 1999). In both habitats, different bacterial populations developed concerning their metabolic, morphological and biochemical properties. More rod-like and in general larger bacteria exist on the aggregates than in the surrounding water (Alldredge et al. 1986), and phylogenetic differences exist, as well. Associated heterotrophic bacteria could be classified as *Cytophaga/Flavo-bacteria*, *Planctomyces*, or γ-Proteobacteria, whereas most of the free-living bacteria could be classified as α-Proteobacteria and some as γ-Proteobacteria (DeLong et al. 1993; Rath et al. 1998; Schweitzer et al. 2001). Acinas et al. (1999) showed a very low diversity of attached bacteria-assemblages from the Mediterranean Sea, which consist only of α- and γ-Proteobacteria.

Fukami et al. (1981) found high abundances of species of the *Cytophaga/Flavobac-teria*-group on particles. Members of this branch are able to degrade a large spectrum of polymers (proteins, polysaccharides, chitin, nucleic acids) and some

species are well-known for their ability of surface-associated gliding motility and the production of hydrolytic exoenzymes. In the decomposition of marine phytoplankton, the change of the population structure correlates with the ability of degrading high-molecular organic components. Laboratory investigations on the microbial degradation of chitin over a long period have shown a decrease of attached bacteria, whereas the number of free-living bacteria increased (Kirchner 1995). While particle-associated bacteria are able to solubilize more dissolved organic matter (DOM) than they can actually consume, free-living bacteria can benefit from the DOM pool cleaved by the attached bacterial community (Karner and Herndl 1992).

It is known that, in spite of better cultivation conditions, usually only a small part of bacteria (often less than 1%) of a natural marine sample can be cultivated with conventional cultivation approaches. Molecular methods, such as sequencing of 16S rDNA or fluorescence *in situ* hybridization (FISH), may help to solve this problem. Because of better fluorescent dyes (e.g. CY3, a very sensitive carbocyanine-dye), FISH enables identification of routinely more than 50% of the cells even in oligotrophic aquatic samples in which the visualization of small cells with low numbers of ribosomes had been problematic so far (Amann et al. 1997). But there is still a need of classical methods for a better identification and comparison of microorganisms (Rüger 1993).

It was the aim of this study to describe and characterize the bacterial community attached to sinking particles in the Equatorial Atlantic and in the South Atlantic (Antarctica). Since there exists nearly no information about the examined habitats concerning bacterial abundances, it was necessary to determine the most densely populated layer first. Then, bacterial strains could be isolated from particles, representing typical microorganisms of these habitats. In order to study the microbial decomposition potential of sinking aggregates, which depends upon multiple interacting factors, the isolated strains were investigated with classical, modern microbiological, and molecular-biological methods concerning their physiological and phylogenetic properties.

Sampling Site

Seawater samples were taken in the Equatorial Atlantic (EA) and in the South Atlantic/Antarctica (SA) during two cruises with RV *Meteor* in 1996 and 1997 (stations 3906-3, 3907-3, 3908-10, 3925-4, 4301-10, 4302-2, 4303-3, 4305-5 and 4318-8) and one cruise with RV *Polarstern* in 1998 (stations 49/020/1 and 49/088/2). The sampling sites (Fig. 1) differ significantly in their grade of eutrophication. Stations 4301-10 (north of Gran Canaria), 4302-2 (Cape Blanc) and 4303-3 (south of Cape Verde Islands) represent habitats in an eutrophic upwelling zone in the eastern EA, stations 3906-3, 3907-3, 3908-10, and 4318-8 stand for oligotrophic areas in the western EA (continental slope off Northern Brazil). The sampling sites in the SA were chosen to investigate the influence of the water temperature on the composition of the attached bacterial community and physiological properties of single isolates.

Methods

Water sampling and cell count. Water samples were taken using 10 l Niskin-bottle water samplers (Hydrobios, Kiel, Germany) and portions of 20 ml were fixed with formaldehyde (2% v/v). Two ml of such preserved samples were stained with DAPI (4',6-diamidino-2-phenylindole) as described by Grossart (1995). Total cell counts were determined by epifluorescence microscopy (Zeiss Axiolab) equipped with a filterset for DAPI (filterset 02, Zeiss, Germany). For each sample, at least 100 fields (14,4 mm^2) were counted.

Isolation of bacteria. Water samples from 50 m waterdepth of stations 4301-10 and 4318-8 (EA) were filtered using Isopore membranes (2.5 cm in diameter, 10 µm pore-size, Millipore, Eschborn, Germany), which were then transferred into sterile seawater, gently shaken to suspend the aggregates, and stored at 4 °C on board of the RV and in the lab until cultivation. Dilution series up to 10^5 were made with these aggregate suspensions in order to isolate dominant species. Agar plates with ASN$_{III}$ medium (Rippka et al. 1979) were used as isolation medium (3.3-3.7% salinity) with the following modification: citric acid and EDTA were omitted, but 0,1% w/v yeast extract, trace elements SL10 (Trüper and Pfennig 1992), a 7 vitamin-solution (Vogt, 1997), 0.05% w/v actidione, and 2% agar were added. Plates were inoculated with material from the highest dilution rate and incubated at 19 °C (station 4301-10) and 27 °C (station 4318-8) up to two weeks.

Water samples from Antarctica were also taken using 10 l Niskin-bottle water samplers. Water samples from 20 m waterdepth were filtrated either through polycarbonate membranes with 10 µm pore size (station 49/020/1) or with 5 µm pore size (station 49/088/2). The filters were then transferred into sterile seawater and stored at -30 °C on board of the RV and at -18 °C in the lab until cultivation. 50 and 100 µl of each thawed sample were streaked onto ASN$_{III}$-agar plates and incubated at 27 °C and nearly 0 °C up to four weeks.

Colonies were randomly picked and purification of isolates was performed by streaking a single-cell colony onto new agar plates. This procedure was repeated several times in order to get pure isolates. Plates were incubated in the dark at the same temperatures as described above.

Utilization of substrates. Since aggregates consist of different polymers, enzymatic activities of cellulase, gelatinase, amylase, lecithinase and DNase were tested by streaking the isolates on artificial seawater agar plates (ASN$_{III}$ with 0.1% w/v yeast extract, supplemented with either 1% w/v cellulose, 0.4% w/v gelatin, 2% w/v starch, 5% w/v egg yolk emulsion or 0.2% w/v DNA and 0.005% w/v methylgreen, respectively). Cellulase activity of the EA isolates was tested by streaking them onto agar plates with ASN$_{III}$ and cellulose (1% w/v) as sole carbon source. The plates were incubated at 27 °C for 4-6 days. Cellulose hydrolysis resulted in clear zones (SA) or forming of colonies (EA) on the agar plates, while lecithinase was detectable by its appearance of cloudy precipitations (modified after Gerhardt et al. 1994). Starch hydrolysis was documented using Lugol's iodine solution forming colorless areas around the growth, and gelatin degradation by using picric acid to visualize unhydrolyzed polymers (modified after Süßmuth et al. 1999). DNA hydrolysis was analyzed treating the plates with 2 M HCl to reveal undigested DNA (modified after Smith et al. 1969).

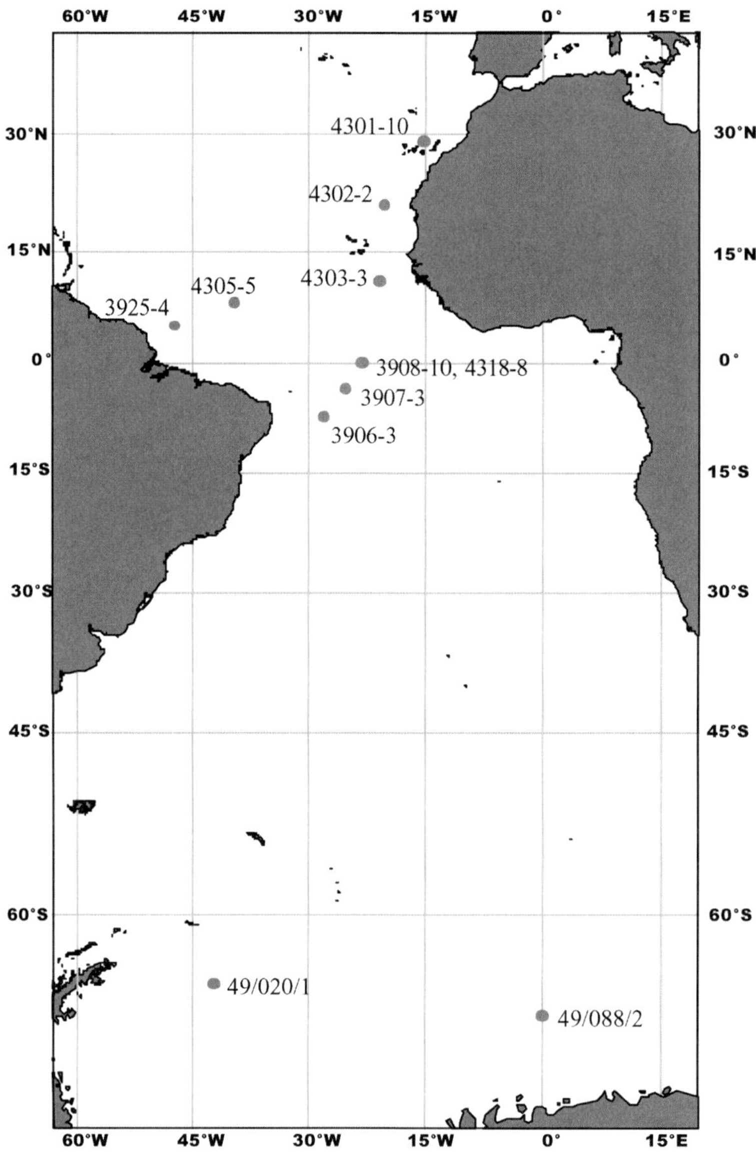

Fig. 1. Sampling sites of two cruises with RV *Meteor* in spring 1996 (stations 3906-3, 3907-3, 3908-10 and 3925-4) and spring 1997 (stations 4301-10, 4302-2, 4303-3, 4305-5 and 4318-8) and one cruise with RV *Polarstern* in spring 1998 (stations 49/020/1 and 49/088/2).

Utilization of low molecular weight organic compounds such as sugars, organic – or amino acids as sole carbon sources for growth was examined by following the increase of the optical density at 600 nm. ASN_{III}-medium without yeast and Fe-NH_4-citrate served as basic growth medium.

Activity of exoenzymes (amylase, DNase, and protease) in dependence of temperature was documented by measuring the clear zones on agar plates (methods as described above, protease was estimated as reported by Frazier and Rupp 1928 and modified after Brand 1939) after incubating at 5, 27, and 30 °C. The influence of salinity was investigated with NaCl concentrations ranging from 0.5 to 10%. The pH-optimum was detected by incubating the organisms between pH 5 and pH 10.

BIOLOG GN microtiter plates (Biolog, Hayward, USA) were inoculated with the following mixtures of strains (in equal amounts): i) all 10 EA isolates, ii) strains 4301-10/1, 4301-10/2, 4301-10/3, 4301-10/4, 4301-10/5, 4301-10/6 and iii) strains 4318-8/1, 4318-8/2, 4318-8/4, 4318-8/7. The plates were incubated at 27 °C for 72 h.

RAPD-PCR. DNA was isolated with Qiagen Genomic-tips 100/G as described in the Qiagen Genomic DNA Handbook (1997). The genomic DNA was amplified by using *Taq* Polymerase (1.25 U, REDTaq™, Sigma-Aldrich Co.). Bacterial primers CRA22 (5'-CCGCAGCCAA-3') and CRA23 (5'-GCGATCCCCA-3') were used for amplification (Neilan, 1995).

ARDRA. Genomic DNA (isolation as described above) was amplified using REDTaq™ and the bacterial primers 10-30F (5'-AGAGTTTGATC MTGGCTCGA-3') and 1542R (5'-AGAAAGGAG GTGATCCARCC-3'). The amplificational mixture consisted of 25-30 ng template DNA, 0.4 µM of each primer, 200 µM dNTP-Mix, 4 µl bovine serum albumin (2 mg/ml), 2 mM $MgCl_2$, 5 µl reaction buffer (100 mM Tris-HCl, 500 mM KCl, 11 mM $MgCl_2$, and 0.1% gelatin), 2 U REDTaq Polymerase and sterile water to a final volume of 50 µl. The PCR products (5 µl) were digested for 1 h at 37 °C with 3 U of the following restriction enzymes each, recognizing four bp sites: Hin6I, AluI, HpaII, Bsh1236I, BsuRI, and MboI, 1 µl reaction buffer (MBI Fermentas), and sterile water to a final volume of 10 µl. Amplification was performed with a Biometra Personal Cycler™. Thermal cycling conditions consisted of a hot start (95 °C for 5 min) and a cooling down to 85 °C, followed by 30 cycles of denaturing (93 °C, 30 s), annealing (52 °C, 30 s), and extension steps (72 °C, 60 s). Afterwards, a final polymerization for 5 min at 72 °C was performed. The patterns were analyzed in a 2% (w/v) agarose gel and stained with ethidium bromide.

FISH. Formaldehyde-fixed (7% final concentration) cells from a single colony of each isolate were transferred onto Teflon-coated microscope slides and immobilized by drying at 46 °C. After dehydration and fixation with 50, 80, and 96% (v/v) ethanol, slide-fixed cells were hybridized with oligonucleotide probes EUB338 (5'-GCTGCCT

CCCGTAGGAGT-3'), ARCH915 (5'-GTGCTCC CCCGCCAATTCCT-3'), CF319a (5'-TGGTCCG TGTCTCAGTAC-3'), ALF968 (5'-GGTAAGGT TCTGCGCGTT-3'), BET42a (5'-GCCTTCCCA CTTCGTTT-3'), and GAM42a (5'-GCCTTCCCA CATCGTTT-3') (modified after Eilers et al. 2000). All probes were CY3-labeled and synthesized by Interactiva (Ulm, Germany). The cells were counter-stained with DAPI (1 µg/ml) and incubated for 5 min in the dark. After washing (70% ethanol, aqua bidest.) and air drying, the slides were mounted with a 5:1 (v/v) mixture of Citifluor AF1 (Citifluor, London, UK) and Vectashield (Linaris, Wertheim, Germany). The evaluation was performed by an epifluorescence microscope (Zeiss Axiolab) with filtersets for DAPI (see above) and CY3 (Chroma HQ 41007).

16S rDNA sequencing. Isolation of DNA, bacterial primers and amplification were the same as described above. PCR amplification was confirmed via gel electrophoresis with a 1% agarose gel stained with ethidium bromide. PCR products were purified with QIAquick Gel Extraction Kits (Qiagen) and desalted by using 0.025 µm nitrocellulose filters (Millipore). Sequencing was conducted with the bacterial primers 10-30F (5'-AGAGTTTG ATCMTGGCTCGA-3') and 518F (5'-CCAGCAG CCGCGGTAAT-3') by MWG Biotech, Germany. For each sequence, the closest relative was determined by comparison to the NCBI database (http://www.ncbi.nlm.nih.gov).

Results and Discussion

Abundances of Free Living Bacteria

In order to define the layer with the highest microbial density, the abundances of free-living bacteria in the upper 250 m of the watercolumn were estimated by staining with DAPI. The results are summarized in Table 1. The highest abundances were found in a waterdepth of 20 m (0.95 to 2.89 x 10^5 cells ml[-1]), with the exception of station 4301-10. At this station, the most densely populated layer was found at 100 m depth (0.59 x 10^5 cells ml[-1]).

Only little is known about the abundances of free-living bacteria in the Equatorial Atlantic de-

Station No.	Sampling site (Latitude/Longitude)	Date	Depth (m)	Temperature[a] (°C)	Total cell numbers (10^5 ml^{-1})
3906-3	07°24.8′S/28°08.0′W	21.03.96	20	28.4	0.95 ± 0.3
			50	27.5	0.95 ± 0.4
			100	23.4	0.64 ± 0.2
			250	9.9	0.42 ± 0.2
3907-3	03°55.8′S/25°41.0′W	23.03.96	20	28.5	1.99 ± 0.5
			50	25.5	1.02 ± 0.4
			100	13.3	0.25 ± 0.2
			250	10.6	0.33 ± 0.2
3908-10	00°00.4′S/23°26.4′W	26.03.96	20	27.8	2.16 ± 0.9
			100	16.0	0.29 ± 0.1
			250	13.1	0.24 ± 0.3
3925-4	05°08.3′N/47°30.7′W	07.04.96	20	27.5	1.49 ± 0.5
			50	27.5	0.88 ± 0.5
			100	27.3	0.98 ± 0.4
			250	9.9	0.21 ± 0.2
4301-10	29°10.0′N/15°30.1′W	26.01.97	25	19.2	0.20 ± 0.1
			50	19.2	0.49 ± 0.3
			100	18.9	0.59 ± 0.3
			200	15.8	0.14 ± 0.1
4302-2	21°18.5′N/20°39.6′W	28.01.97	20	21.3	2.89 ± 0.7
			50	21.2	1.70 ± 0.6
			100	20.2	0.67 ± 0.3
			200	16.7	0.20 ± 0.1
4303-3	11°28.9′N/21°01.0′W	01.02.97	20	24.6	1.42 ± 0.5
			50	18.8	0.23 ± 0.2
			100	14.3	0.20 ± 0.2
			200	12.5	0.73 ± 0.3
4305-5	08°29.7′N/39°57.3′W	08.02.97	20	25.5	2.81 ± 0.6
			50	18.9	0.18 ± 0.2
			100	13.1	0.12 ± 0.1
			200	10.0	0.57 ± 0.1

[a] measured by CTD (Conductivity-Temperature-Depth) -profiler (Ratmeyer, 2002)

Table 1. Total cell numbers (DAPI cell counts) of free-living bacteria from seawater-samples of stations 3906-3, 3907-3, 3908-10 and 3925-4, taken in March/April 1996 (cruise M34/4) and stations 4301-10, 4302-2, 4303-3 and 4305-5, taken in January/February 1997 (cruise M38/1).

tected by direct counting. Eilers et al. (2000) found $1.2 – 10.5$ x 10^6 cells ml^{-1} in the North Sea near Helgoland, while Glöckner et al. (1999) detected $0.7 – 1.6$ x 10^6 cells ml^{-1} in the South Atlantic (Antarctica). Bacterial abundances in the Pacific range from 0.6 x 10^6 cells ml^{-1} (Oregon Coast, Suzuki et al. 1993) to 3 x 10^6 cells ml^{-1} (Playa del Rey, Glöckner at al. 1999). All these data show 10fold higher cell numbers which could attribute to higher nutrient contents.

Morphological, Physiological and Taxonomical Properties of Particle-Associated Bacteria

Twenty-one strains of heterotrophic bacteria were isolated from organic particles with sizes >10 µm (see Fig. 2). Ten strains originated from stations 4301-10 and 4318-8 (EA) and eleven strains were isolated from stations 49/020/1 and 49/088/2 (SA). These isolates were taken for a comparative characterization of particle-associated bacteria.

Morphology: All of the 21 isolates are Gram-negative, rodshaped (0.5 x 1-20 µm) or more or less coccoid (0.7 x 0.8-2 µm). Twelve strains showed motility on swarming agar. All isolates form colonies of similar shape (circular with convex or flat surface) on artificial seawater agar plates (ASN$_{III}$ plus 0.1% yeast extract). The pigmentation of the colonies varied from colorless to white, grayish, yellow or brown.

Utilization of substrates: The obtained isolates were examined with regard to their enzymatic equipment and ability to use different substances as sole carbon source (Tab. 2). All strains differ strongly in their capacity to metabolize high-molecular organic compounds such as starch and cellulose. As shown in Table 2, only one of the SA strains (strain 88/2-1) and three of the EA ones (strains 4301-10/1, 4301-10/4 and 4301-10/5) can use starch and gelatin, while three of the SA isolates and four of the EA isolates metabolize only either starch or gelatin. Seven SA and three EA isolates are not able to use neither starch nor gelatin. Cellulose is metabolized by seven of the ten EA isolates (strains 4301-10/1, 4301-10/2, 4301-10/5, 4301-10/6, 4318-8/2, 4318-8/4 and 4318-8/7). Likewise, the isolates vary in their ability to metabolize different lower molecular weight compounds (organic acids, mono- and disaccharides or amino acids), which were offered in different concentrations (1 mM or 5 mM). The majority of the isolates shows better growth with the higher concentration. This is in agreement to Unanue et al. (1999), who found two different uptake systems for carbohydrates and proteins. The authors suggested that free-living bacteria should be adapted to low concentrations, while attached bacteria could operate efficiently at very high substrate concentrations. Strains 4301-10/2, 4301-10/5, 4301-10/4, 4301-10/7, 20/1-2, 20/1-3, and 20/1-4 can use nearly all organic compounds tested, while strain 88/2-4 uses only the lower molecular weight ones. Some isolates are very restricted in their enzymatic equipment and are only able to use

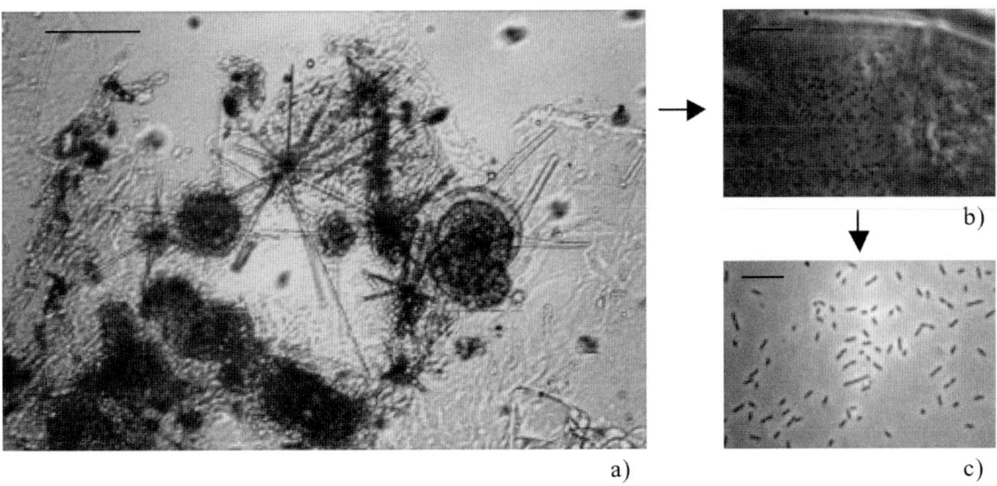

Fig. 2. Photomicrographs of aggregates and attached bacteria from the Equatorial Atlantic demonstrating isolation of a single bacterial strain from a mixed culture. **a)** organic particle, consisting of zoo- and phytoplankton debris, bar = 100 µm; **b)** particle colonised with a bacterial community consisting of different bacterial cell types, bar = 10 µm; **c)** pure culture of strain 4301-10/1, isolated from particles, bar = 10 µm.

	Equatorial Atlantic strains										South Atlantic (Antarctica) strains										
Properties	4301-10/1	4301-10/2	4301-10/3	4301-10/4	4301-10/5	4301-10/6	4318-8/1	4318-8/2	4318-8/4	4318-8/7	88/2-1	88/2-2	88/2-3	88/2-4	88/2-5	88/2-6	88/2-7	20/1-1	20/1-2	20/1-3	20/1-4
Enzymes — Catalase	+	+	+	+	+	+	+	+	+	+	-	+	+	+	+	(+)	+	+	+	+	+
Oxidase	-	-	+	+	-	+	+	+	-	-	-	+	+	+	+	-	+	-	+	+	+
Cellulase	+	+	-	-	+	+	-	(+)	(+)	+	n.t.	n.t.	-	-	n.t.	n.t.	-	n.t.	-	n.t.	n.t.
Gelatinase	+	-	-	+	+	-	+	-	-	-	+	-	-	-	-	-	-	-	+	n.t.	n.t.
Amylase	+	+	-	+	+	-	-	-	+	+	+	-	-	-	-	-	-	-	-	+	+
Lecithinase	+	+	+	-	+	+	+	+	+	+	n.t.	n.t.	+	-	n.t.	n.t.	+	n.t.	+	n.t.	n.t.
DNase	-	+	-	-	+	-	-	+	+	+	n.t.	n.t.	-	-	n.t.	n.t.	-	n.t.	+	+	+
Carbon sources — Acetate	-	+	+	-	+	+	+	(+)	+	+	-	+	-	+	(+)	n.t.	-	(+)	(+)	+	+
L-Alanine	-	+	+	-	+	+	+	-	+	+	-	-	-	+	-	n.t.	+	-	(+)	+	+
Cellobiose	+	+	-	-	+	+	-	+	+	+	-	-	-	+	+	+	-	+	+	n.t.	n.t.
Citrate	-	+	+	-	+	+	-	-	+	+	-	-	-	+	(+)	n.t.	-	(+)	+	+	+
Fructose	+	+	+	-	+	+	-	-	+	+	-	-	-	+	-	+	-	-	+	+	+
Glucose	+	+	+	-	+	+	-	-	+	+	-	-	-	+	+	+	-	+	+	+	+
Glycogen	+	+	+	-	+	+	-	-	+	+	n.t.	n.t.	n.t.	n.t.	n.t.	n.t.	n.t.	n.t.	n.t.	n.t.	n.t.
Maltose	-	+	-	-	+	-	-	-	+	+	-	-	-	-	-	-	-	-	-	+	+
Mannose	+	+	-	-	+	-	-	-	+	+	-	-	-	+	(+)	(+)	-	(+)	(+)	-	-
Propionat	n.t.	n.t.	n.t.	n.t.	n.t.	n.t.	n.t.	n.t.	n.t.	n.t.	-	-	-	+	+	+	-	+	+	+	+
Sucrose	+	-	-	-	-	-	-	-	-	-	-	-	-	+	+	+	-	+	-	+	+

- negative reaction, (+) weak reaction, + positive reaction, n.t. not tested

Table 2. Physiological investigations of 21 bacterial strains isolated from particles of *Meteor*-cruise M38/1 (EA 1997) and *Polarstern*-cruise ANTXV/4 (SA 1998).

one or two of the offered carbon sources: strains 4301-10/4 (gelatin and starch), 88/2-2 (acetate), 88/2-3 (lecithin), and 88/2-7 (lecithin and alanine). Only one EA isolate (4301-10/1) utilizes sucrose as sole carbon source, while the majority of SA isolates is able to degrade this disaccharide.

Production of exoenzymes (amylase, DNase, and protease) in dependence of time, temperature, salinity, and pH was investigated with SA isolates 20/1-3 and 20/1-4. Hydrolysis of starch depending on the incubation time is shown in Figure 3. Isolate 20/1-4 exhibits amylase activity immediately, while isolate 20/1-3 needs a lag-time of two days until starch can be degraded. Amylase seems to be constitutively available in isolate 20/1-4, while it is synthesized *de novo* in isolate 20/1-3. Both isolates exhibit highest amylase activity at 30 °C, while that of isolate 20/1-3 is strongly inhibited at 5 °C. DNase activity of both isolates is highest at 27 °C, but is inhibited at 5 °C. The psychrophilic bacterium *Pseudoalteromonas haloplanktis*, a member of the same genus as isolate 20/1-3, shows highest amylase activities also at relative high temperatures between 30 and 40 °C (Feller et al. 1994). The production of extracellular protease, amylase, and DNase depends on the sodium chloride concentration in the medium. For both isolates, the decomposition of the substrates by these enzymes occurs between 1 and 3.5% NaCl (salinity of seawater). No amylase and DNase activity were found in the presence of 10% NaCl. High sodium chloride concentrations inhibit the production of enzymes, because they induce an instability (Copeland 1996). While isolate 20/1-3 prefers weak alkaline pH values (7-8) for optimal amylase activity, the other isolate needs weak acidic conditions (pH 6-7).

Additionally to the investigation of the metabolic properties of each individual isolate, experiments with defined mixed cultures were carried out using the BIOLOG-method, in order to examine bacterial community behavior. BIOLOG-plates (micro-

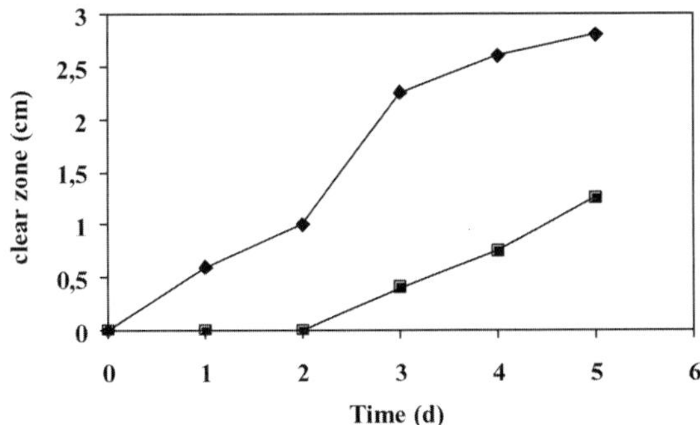

Fig. 3. Time dependent hydrolysis of starch by amylase excreted by the isolates 20/1-3 (■) and 20/1-4 (◆). Amylase activity is expressed by the clear zones (cm) formed (data taken from de Silva 2001).

titer plates with 95 different substrates, indicating a positive reaction by a change of color) were inoculated with three different mixed cultures: all six isolates from station 4301-10 (A), all four isolates from station 4318-8 (B), and all 10 isolates from both stations (C). It could be demonstrated that the culture-mixtures (A) and (B) had more positive reactions than single isolates and that the combined culture (C) had more positive reactions than the other two culture mixtures (A) and (B). These results suggest the existence of a synergetic behavior of aggregate-attached bacteria. Single species or groups of bacteria profit from each other and are able to degrade more substrates in a community than alone.

Salinity range: The isolates were grown in media containing 0-20% NaCl (except isolates 20/1-3 and 20/1-4, which were cultivated in the range of 0-16% NaCl). EA strains 4301-10/2, 4301-10/5, 4318-8/4 and 4318-8/7 and SA strain 88/2-4 grow in the presence of 0-20% NaCl and, therefore, show a halotolerant growth behavior (Brock et al. 2000). At low NaCl concentrations, growth of EA strains 4301-10/4 and 4318-8/1 (growth at 3.4-5% NaCl) and SA strains 88/2-1, 20/1-1 (growth at 2-4% NaCl) and 88/2-5 (growth at 2-8% NaCl) is inhibited. This is typical for marine organisms, which have a demand of sodium ions for their growth activities (Brock et al. 2000; see also chapter: Utilization of substrates).

Temperature range: Temperature dependency for growth was investigated by incubating the cultures at temperatures ranging from 5-40 °C for SA strains and 5-42 °C for EA strains. All strains grow well at temperatures between 5 and 30 °C. Some isolates (4301-10/2, 4301-10/5, 4318-8/4, 4318-8/7, 88/2-2 and 88/2-4) are even able to grow at relatively high temperatures (40-42 °C). Therefore, these isolates could be classified as psychrotroph (optimal growth temperature >20 °C, but growth can occur also at about 4 °C; Bowman 1997). Within the Antarctic strains, no psychrophilic strains were found, which are characterized by inhibition of growth above 20 °C. Bowman et al. (1997) as well as Delille (1992) have demonstrated that most of the Antarctic bacterial strains, isolated from seawater, must be considered as psychrotrophs and not as psychrophiles.

SA isolate 88/2-7 showed an unusual growth behavior when a 24 h old liquid culture, pregrown at 27 °C, was incubated for four days at 5 °C on a rotary shaker (Fig. 4). Under this low temperature condition, the cells of the culture developed large, spherical to elliptic shaped flocs (up to 1 mm in diameter), and the medium became clear. After transferring the culture back to 27 °C, the flocs disaggregated and the medium became turbid again. The observed aggregation could be a kind of adaptation mechanism to low temperatures, because such a strong reaction did not occur at 27 °C. Isolate 88/2-7 seems to produce exopolymeric substances (EPS), which were visible as small slime layers after staining with nigrosin. The EPS could be necessary for the attachment of the cells to other cells or particles.

Taxonomy: To divide the isolated bacterial strains into groups of very related ones and those which are nearly identical, different molecular-biological methods such as random amplified polymorphic DNA (RAPD), amplified ribosomal DNA restriction analysis (ARDRA), or fluorescence *in situ* hybridization (FISH) were used for a taxonomical characterization. RAPD produces DNA-fragments of varying nuclear size which can be used as fingerprints on agarose gels to distinguish between related individuals. Using this method, four of our EA isolates 4301-10/2, 4301-10/5, 4318-8/4, and 4318-8/7 can be summarized in one group, while the two isolates 4301-10/3 and 4301-10/6 form another distinct group (see Fig. 5). All other iso-

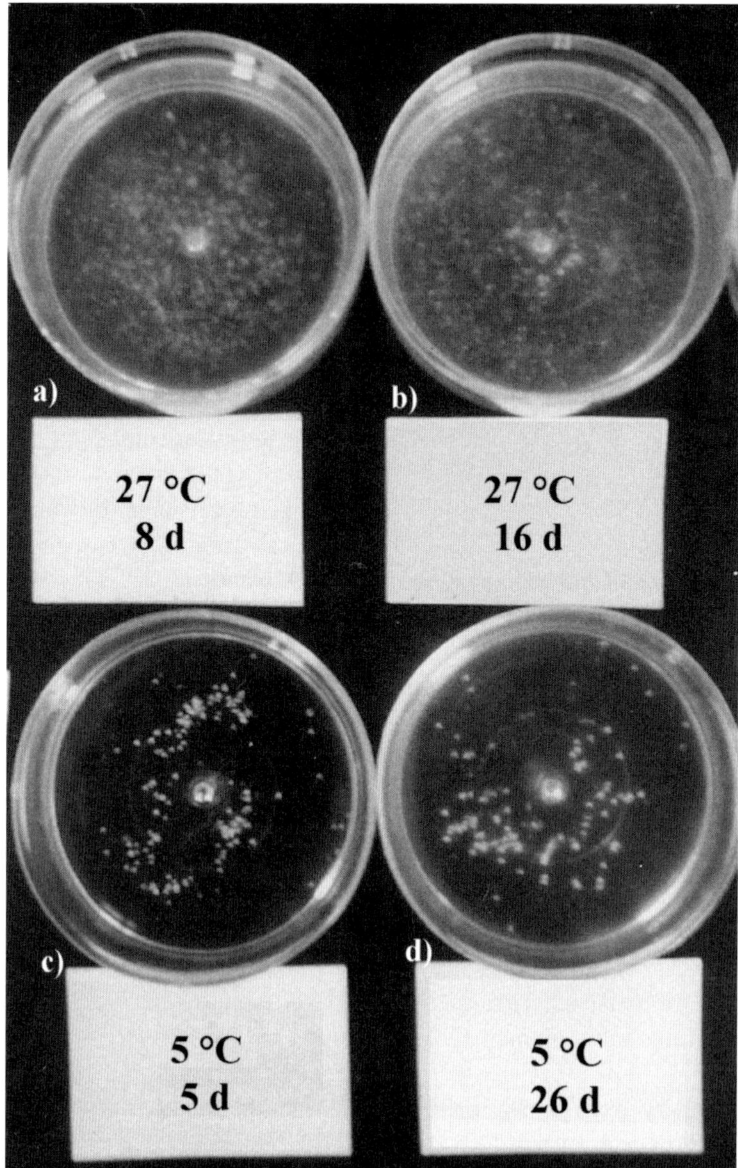

Fig. 4. Comparison of flocculation of isolate 88/2-7 at different temperatures and incubation times. **a)** and **b)** appearance of the culture after growth at 27 °C for 8 and 16 d, respectively. **c)** appearance of the culture after growth at 5 °C for 5 d (the culture was pregrown at 27 °C for 24 h). **d)** appearance of the culture after growth at 5 °C for 26 d (the culture was inoculated with a colony taken from a slant agar tube). Modified after Heuchert 1999.

lates are very different from each other and can therefore not be combined in one group. The SA-isolates seem to have a higher diversity in their RAPD PCR pattern than the EA-isolates. Only isolates 20/1-1 and 88/2-5 match in their DNA-fingerprints (data not shown).

Different restriction enzymes (e.g. Hin6I, AluI, HpaII, Bsh1236I, BsuRI and MboI) are applied in ARDRA, which is an efficient method to examine the diversity of a microbial community and which was used to reduce the number of isolated strains that were analyzed further. The isolates 88/2-5 and 88/2-6 show no difference in their PCR-patterns independent of the restriction enzymes used (data not shown). All other SA isolates show mismatch by the use of at least one restriction enzyme. The

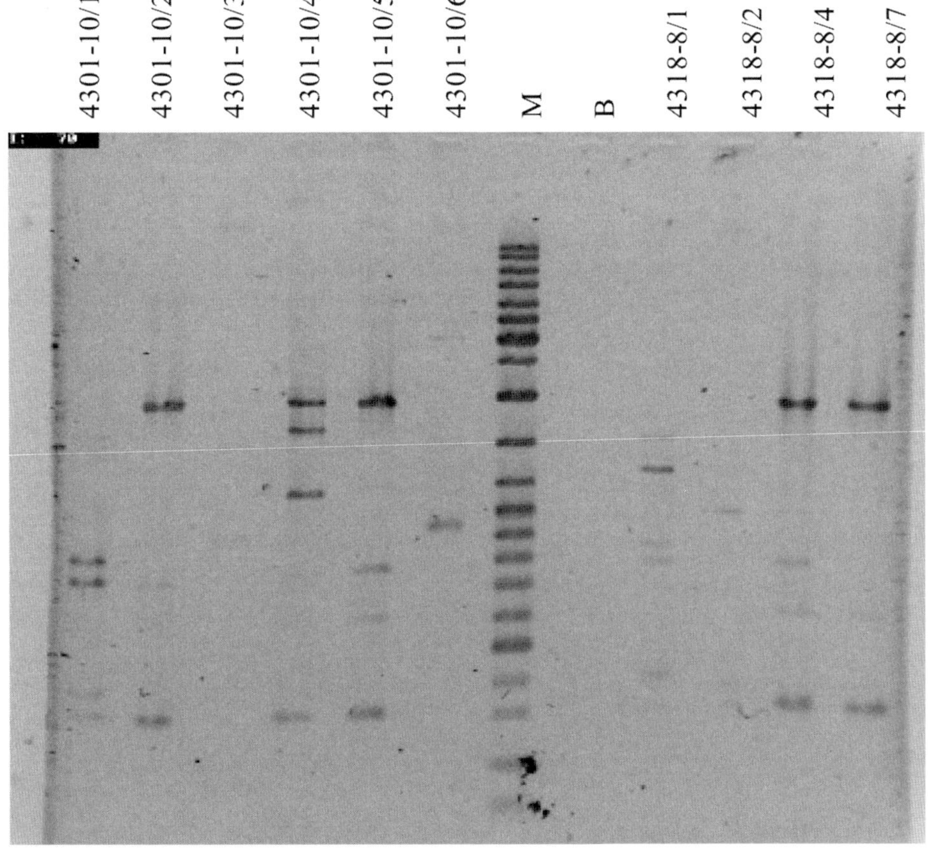

Fig. 5. RAPD-PCR of ten bacterial isolates from two sampling sites of the Equatorial Atlantic (B: control; M: DNA Ladder Mix).

discrepancies in the RAPD- and ARDRA-results do not contradict to each other, because the distinguishable ARDRA patterns may be derived from the same organism as reported by Dang and Lovell (2000).

FISH with different rRNA-targeted oligonucleotide probes for Bacteria, Archaea, *Cytophaga/Flavobacteria*, α-Proteobacteria, β-Proteobacteria, and γ-Proteobacteria was used for a systematic classification of our EA and SA isolates (see Tab. 3). As expected, all strains hybridized with the Bacteria probe and none with the probe for Archaea. Four strains of the EA and SA samples each could be detected with the probe for the γ-subclass of Proteobacteria (GAM42a). The probe, specific for most members of the *Cytophaga/Flavobacteria* group (CF319a), hybridized with two EA and one SA strains. This corresponds to the investigations of DeLong et al. (1993) with aggregates collected in the Santa Barbara Channel. The most abundant phylogenetic types detected in macroaggregate-associated bacterial populations fell within the groups of *Cytophaga/Flavobacteria*, *Planctomyces* and γ-Proteobacteria. *Cytophaga* was highly abundant in aggregate-associated bacterial communities in the Southern Californian Bight (Ploug et al. 1999). The probe for the α-subclass of Proteobacteria hybridized with four EA and three SA strains, while the probe for the β-subclass of the Proteobacteria did not hybridize with any of the isolates tested. Although β- and γ-Proteobacteria coexist in coastal environments, β-Proteobacteria are not found in the oligotrophic ocean but are abundant in freshwater habitats, where they seem to displace γ-Proteobacteria (Cottrell and Kirchman 2000). Representatives of

Investigation area and Isolate	Oligonucleotide probe					
	EUB338	ARCH915	CF319a	ALF968	BET42a	GAM42a
Equatorial Atlantic						
4301-10/1	+	-	+	-	-	-
4301-10/2	+	-	-	-	-	+
4301-10/3	+	-	-	+	-	-
4301-10/4	+	-	+	-	-	-
4301-10/5	+	-	-	-	-	+
4301-10/6	+	-	-	+	-	-
4318-8/1	+	-	-	+	-	-
4318-8/2	+	-	-	+	-	-
4318-8/4	+	-	-	-	-	+
4318-8/7	+	-	-	-	-	+
South Atlantic (Antarctica)						
88/2-1	+	-	+	-	-	-
88/2-3	+	-	-	-	-	+
88/2-4	+	-	-	+	-	-
88/2-5	+	-	-	+	-	-
88/2-6	+	-	-	+	-	-
88/2-7	+	-	-	-	-	+
20/1-3	+	-	-	-	-	+
20/1-4	+	-	-	-	-	+

EUB338, Bacteria; ARCH915, Archaea; CF319a, *Cytophaga-Flavobacteria*; ALF968, α-Proteobacteria; BET42a, β-Proteobacteria; GAM42a, γ-Proteobacteria
+, positive reaction; -, no reaction

Table 3. Fluorescence *in situ* hybridization with samples collected from the Equatorial and South Atlantic using different oligonucleotide probes.

the β-Proteobacteria group were found only within lake or river snow-associated bacteria and free-living bacteria in coastal oceans and freshwater habitats, whereas members of the α-Proteobacteria occur particle-associated as well as free-living in both, seawater and freshwater environments (Schweitzer et al. 2001; Crump et al. 1999; Rath et al. 1998; Acinas et al. 1999). Bowma et al. (1997) found bacterial isolates from Antarctic sea ice, which belong to the α- and γ-subdivisions of Proteobacteria and the *Flexibacter-Bacteroides-Cytophaga* phylum.

Based on the results obtained by RAPD, 16S rDNA-sequencing was carried out with one representative of each group of the EA samples and with all isolates of the SA samples. Sequence results correspond with the investigations obtained by FISH (Tab. 4). Only members of the α- and γ-subclass of Proteobacteria and of the *Cytophaga/Flavobacteria* group were found. The genus *Sulfitobacter* was present in both investigation areas, whereas other genera such as *Cytophaga*, *Roseobacter* and *Marinobacter* (EA) and *Polaribacter*, *Paracoccus*, *Pseudoalteromonas* and *Psychrobacter* (SA) were found only in one sampling site each. Members of the genus *Roseobacter* (α-Proteobacteria) are known as ubiquitously occurring organisms and rapid colonizers of surfaces in coastal environments (Dang and Lovell 2000), playing an important role in primary succession of surfaces. Members of the γ-subclass of Proteobacteria appeared to be involved in the very early stages of colonization as well. DeLong et al. (1993) reported that *Cytophaga/Flavobacteria*-species show highest abundances on particles. This finding differs from our observation, because only one representative of this group was detected in the EA (isolate 4301-10/1) and SA (isolate 88/2-1), each.

Bacterial groups/species	Isolates		16S rDNA-sequence similarity (%)
	Equatorial Atlantic	South Atlantic (Antarctica)	
Cytophaga/Flavobacteria			
Cytophaga sp. KT11ds2	4301-10/1		99
Polaribacter irgensii		88/2-1	97
α-Proteobacteria			
α-Proteobacterium QSSC9-5		88/2-5, 88/2-6, 20/1-1	97-98
Paracoccus sp. MBIC3966		88/2-4	98.1*
Roseobacter sp.DSS-8	4318-8/1		97
Sulfitobacter mediterraneus		88/2-2	97
Sulfitobacter pontiacus	4301-10/3; 4301-10/6		99
Marine bacterium SCRIPPS_101	4318-8/2		99
γ-Proteobacteria			
Marinobacter sp.	4301-10/2; 4301-10/5; 4318-8/4; 4318-8/7		99
Pseudoalteromonas sp. ER72M2		20/1-2, 20/1-3	99
Psychrophilic sea water bacterium PS16		20/1-4	99
Psychrobacter glacincola		88/2-7, 88/2-3	99.6*, 97

* Isolate 88/2-4 and 88/2-7 are completely sequenced (Heuchert personal communication)

Table 4. Percent similarity of the isolates originating from organic particles of the Equatorial and South Atlantic, respectively, with strains in BLAST based on their 16S rDNA-sequences.

Both of these species are able to use high-molecular weight compounds such as gelatin and starch, as mentioned above (Tab. 2). It is known that representatives of this group produce one ore more exoenzymes and are able to degrade a wide spectrum of polymeric compounds such as proteins, polysaccharides, chitin, and nucleic acids (DeLong et al. 1993).

Conclusions

All our bacterial strains, isolated from particles, can be classified as α- (42.85%) or γ-Proteobacteria (42.85%) or as a member of the *Cytophaga/Flavobacteria* group (14.3%). The isolates are all Gram-negative, psychrotrophic and differ strongly in their capacity to utilize the major polymeric components of sinking particles in the watercolumn (cellulose, starch, chitin) as well as the hydrolyzed products (di- and monosaccharides, organic acids). From this consumption behavior we presume that there must exist a complex synergy between the different species which is very interesting and demands further investigation. This assumption is confirmed by a study with seawater samples of the Delaware Bay estuary (Cottrell and Kirchman 2000). The authors detected that no phylogenetic group dominated the consumption of all DOM, suggesting that the participation of a diverse assemblage of bacteria is essential for the complete degradation of complex DOM in the oceans. Grossart (1995) demonstrated for Lake Constance that Lake Snow plays an important role in the supply of substrates for bacteria of the aphotic zone because of the significant DOM release even beyond the thermokline.

Acknowledgements

Helpful comments were provided by two anonymous referees. This study was funded by the Deutsche Forschungsgemeinschaft (Sonderforschungsbereich 261, Teilprojekt A1). Data are available under www.pangaea.de(Projects/SFB261.

References

Acinas SG, Antón J, Rodríguez-Valera F (1999) Diversity of free-living and attached bacteria on offshore western mediterranean waters as depicted by analysis of genes encoding 16S rRNA. Appl Environ Microbiol 65: 514-522

Alldredge AL, Cole JJ, Caron DA (1986) Production of heterotrophic bacteria inhabiting macroscopic organic aggregates (marine snow) from surface waters. Limnol Oceanogr 31: 68-78

Alldredge AL, Silver MW (1988) Characteristics, dynamics and significance of marine snow. Prog Oceanogr 20: 41-82

Alldredge AL, Passow U, Logan BE (1993) The abundance and significance of a class of large, transparent organic particles in the ocean. Deep-Sea Res 40: 1131-1140

Amann R, Glöckner FO, Neef A (1997) Modern methods in subsurface microbiology: *In situ* identification of microorganisms with nucleic acid probes. FEMS Microbiol Rev 20: 191-200

Azam F (1998) Microbial control of oceanic carbon flux: The plot thickens. Science 280: 694-696

Azam F, Martínez J, Smith D (1993) Bacteria-organic matter coupling on marine aggregates. In: Guerrero R and Pedrós-Alió C (eds) Trends in Microbial Ecology. Spanish Society for Microbiology, Barcelona, Spain, pp 410-414

Biddanda BA, Pomeroy LR (1988) Microbial aggregation and degradation of phytoplankton-derived detritus in seawater. I. Microbial succession. Mar Ecol Prog Ser 42: 79-88

Bowman JP, McCammon SA, Brown MV, Nichols DS, McMeekin TA (1997) Diversity and association of psychrophilic bacteria in Antarctic sea ice. Appl Environ Microbiol 63: 3068-3078

Brand H (1939) Die bakteriologische Untersuchung von Butter und ihre Auswirkungen. Molkereizeitung 9: 262-264

Brock TD, Madigan MT, Martinko JM, Parker J (2000) Biology of Microorganisms (9th ed) Prentice Hall International Limited, London

Caron DA, Davis PG, Madin LP, Sieburth JMcN (1982) Heterotrophic bacteria and bacterivorous protozoa in oceanic macroaggregates. Science 218: 795-797

Copeland RE (1996) Enzymes. VCH Publishers, Inc New York

Cottrell MT, Kirchman DL (2000) Natural assemblages of marine Proteobacteria and members of the *Cytophaga-Flavobacteria* cluster consuming low- and high-molecular-weight dissolved organic matter. Appl Environ Microbiol 66: 1692-1697

Crump BC, Armbrust EV, Baross JA (1999) Phylogenetic analysis of particle-attached and free-living bacterial communities in the Columbia River, its estuary, and the adjacent coastal ocean. Appl Environ Microbiol 65: 3192-3204

Dang H, Lovell CR (2000) Bacterial primary colonization and early succession on surfaces in marine waters as determined by amplified rRNA gene restriction analysis and sequence analysis of 16S rRNA genes. Appl Environ Microbiol 66: 467-475

de Silva S (2001) Physiologische und molekularbiologische Untersuchungen an zwei "Partikel-assoziierten" Bakterien aus dem Südpolarmeer mit extrazellulärer Enzymproduktion. Diploma-Thesis, Universität Bremen, Germany

Delille D (1992) Marine bacterioplankton at the Weddell Sea ice edge: Distribution of psychrophilic and psychrotrophic populations. Polar Biol 12: 205-210

DeLong EF, Franks DG, Alldredge AL (1993) Phylogenetic diversity of aggregate-attached vs. free-living marine bacterial assemblages. Limnol Oceanogr 38: 924-934

Eilers H, Pernthaler J, Glöckner FO, Amann R (2000) Culturability and *in situ* abundance of pelagic bacteria from the North Sea. Appl Environ Microbiol 66: 3044-3051

Feller G, Payan F, Theys F, Qian M (1994) Stability and structural analysis of alpha amylase from the antarctic psychrophile *Alteromonas haloplanktis*. Eur J Biochem 222: 441-447

Frazier WC, Rupp P (1928) Studies on the proteolytic bacteria of milk. 1, Medium for the direct isolation of caseolytic milk bacteria. J Bacteriol 16: 57-63

Fukami K, Simidu U, Taga N (1981) Fluctuation of the communities of heterotrophic bacteria during the decomposition process of phytoplankton. J Exp Mar Biol Ecol 55: 171-184

Gerhardt P, Murray RGE, Wood WA, Krieg NR (1994) Methods for general and molecular bacteriology. American Society for Microbiology, Washington DC

Glöckner FO, Fuchs BM, Amann R (1999) Bacterioplank-ton Compositions of Lakes and Oceans: A First Comparison Based on Fluorescence *in situ* Hybridization. Appl Environ Microbiol 65:3721-3726

Grossart HP (1995) Auftreten, Bildung und mikrobielle Prozesse auf makroskopischen organischen

Aggregaten (Lake Snow) und ihre Bedeutung für den Stoffumsatz im Bodensee. PhD-Thesis, Universität Konstanz, Germany

Heuchert A (1999) Isolierung und vergleichende Charakterisierung von zwei Partikel-assoziierten heterotrophen Bakterien aus dem Südatlantik (Antarktis). Diploma-Thesis, Universität Bremen, Germany

Iriberri J, Herndl GJ (1995) Formation and microbial utilization of amorphous aggregates in the sea: Ecological significance. Microbiologia Sem 11: 309-322

Karner M, Herndl GJ (1992) Extracellular enzymatic activity and secondary production in free-living and marine-snow-associated bacteria. Mar Biol 113: 341-347

Kirchner M (1995) Microbial colonization of copepod body surfaces and chitin degradation in the sea. Helgoländer Meeresuntersuchungen 49: 201-212

Long RA, Azam F (1996) Abundant protein-containing particles in the sea. Aquat Microb Ecol 10: 213-221

Neilan BA (1995) Identification and phylogenetic analysis of toxigenic cyanobacteria by multiplex randomly amplified polymorphic DNA PCR. Appl Environ Microbiol 61: 2286-2291

Ploug H, Grossart HP, Azam F, Jørgensen BB (1999) Photosynthesis, respiration, and carbon turnover in sinking marine snow from surface waters of Southern California Bight: Implications for the carbon cycle in the ocean. Mar Ecol Prog Ser 179: 1-11

Rath J, Wu KY, Herndl GJ, DeLong EF (1998) High phylogenetic diversity in a marine-snow-associated bacterial assemblage. Aquat Microb Ecol 14: 261-269

Rippka R, Deruelles J, Waterbury JB, Herdman M, Stanier RY (1979) Generic assignment, strain histories and properties of pure cultures of cyanobacteria. J Gen Microbiol 111: 1-61

Rüger HJ (1993) Isolierung und Identifizierung benthischer Bakterien. In: Meyer-Reil LA, Köster M (eds) Mikrobiologie des Meeresbodens. Gustav Fischer Verlag, Stuttgart, pp 121-143

Schweitzer B, Huber I, Amann R, Ludwig W, Simon M (2001) α- and β-Proteobacteria control the consumption and release of amino acids on lake snow aggregates. Appl Environ Microbiol 67: 632-645

Shanks AL, Trent JD (1979) Marine snow: Microscale nutrient patches. Limnol Oceanogr 24: 850-854

Smith PB, Hancock GA, Rhoden DL (1969) Improved methods for detecting deoxyribonuclease production by bacteria. Appl Microbiol 18: 991-993

Süßmuth R, Eberspächer J, Haag R, Springer W (1999) Mikrobiologisch-Biochemisches Praktikum. 2nd ed. Georg Thieme Verlag, Stuttgart

Suzuki MT, Sherr EB, Sherr BF (1993) DAPI direct counting underestimates bacterial abundances and average cell size compared to AO direct counting. Limnol Oceanogr 38:1566-1570

Trüper HG, Pfennig N (1992) The family Chlorobiaceae. In: Balows A, Trüper HG, Dworkin M, Harder W, Schleifer KH (eds) The Prokaryotes, 2nd Edition. Springer, New York, pp 3583-3592

Unanue M, Ayo B, Agis M, Slezak D, Herndl GJ, Iriberri J (1999) Ectoenzymatic activity and uptake of monomers in marine bacterioplankton described by a biphasic kinetic model. Microb Ecol 37: 36-48

Vogt C (1997) Untersuchungen zum Metabolismus von DMS und verwandter Verbindungen durch phototrophe Bakterien. PhD-Thesis, Universität Bremen, Germany

Contribution of Calcareous Plankton Groups to the Carbonate Budget of South Atlantic Surface Sediments

K.-H. Baumann[*], B. Böckel, B. Donner, S. Gerhardt, R. Henrich, A. Vink, A. Volbers, H. Willems and K.A.F. Zonneveld

Universität Bremen, Fachbereich Geowissenschaften, Postfach 330440, D-28334 Bremen, Germany
** corresponding author (e-mail) baumann@uni-bremen.de*

Abstract: A total of more than 400 surface sediment samples from the equatorial, central and subpolar South Atlantic Ocean were investigated for their carbonate content as well as for the carbonate contribution from various calcareous plankton groups. The modern pattern of marine carbonate production is exemplified by comparing two sediment traps located in different domains of the South Atlantic. In addition, this paper presents new carbonate calculations for the content of coccoliths, calcareous dinocysts, planktic foraminifera, and pteropods in surface sediments. In general, carbonate input of the different organism groups is highly variable although dominated by both planktic foraminifera and coccolithophorids. Whereas coccolith carbonate dominates the oligotrophic gyres of the South Atlantic, carbonate derived from planktic foraminifera is much more important in more fertile, mesotrophic areas, such as the equatorial divergence zone. In contrast, calcareous dinocysts only supply a minor proportion of calcium carbonate to the sediments. The aragonite content, mainly derived from pteropod shells, is of regional importance at the continental margin of the western South Atlantic. Here, aragonite contents of up to 50 wt-% of the total sediments were measured. Carbonate dissolution has a major effect below the lysocline depth, but also in highly productive areas (supralysoclinal dissolution). Foraminiferal carbonate is much more affected by dissolution than either coccolith or calcareous dinocyst carbonate. Preservation of pteropod shells is restricted to relatively shallow parts of the ocean distant from continental margins, as aragonite is much more susceptible to dissolution than calcite. As a result, the maximum aragonite content is observed at an intermediate depth, i.e. between 2000 to 3000m.

Introduction

The carbonate system is an important part of the global carbon cycle that controls the atmospheric CO_2 content. Since the ocean contains approximately 60 times more CO_2 than the atmosphere (Berger 1985; Broecker and Peng 1989), it plays a very important role in global CO_2 budgeting, in particular during the Quaternary climatic oscillations. It has been shown that atmospheric CO_2 concentrations were lower during the last glacial period (Barnola et al. 1987; Jouzel et al. 1993). Biological productivity, as driven by photosynthesis, is one of the primary mechanisms responsible of

partitioning carbon within the ocean (Berger et al. 1987). The efficiency of this biological pump is affected by the amount of carbon buried in the sediment as particulate organic carbon or in the form of carbonate. Generally, organic carbon is efficiently demineralised in the water column and on the sea-floor, so that only about 0.3 wt-% of the euphotic zone production is preserved in the sediment (Berger et al. 1989; Brummer and van Eijden 1992). Nevertheless, this preserved organic carbon has been widely used as an indicator of primary productivity, especially in high-productivity areas (e.g.

From WEFER G, MULITZA S, RATMEYER V (eds), 2003, *The South Atlantic in the Late Quaternary: Reconstruction of Material Budgets and Current Systems.* Springer-Verlag Berlin Heidelberg New York Tokyo, pp 81-99

Müller and Suess 1979; Sarnthein et al. 1992; Schneider et al. 1994). In contrast, biogenic carbonate is much better preserved in the sediments than the organic carbon accompanying it, however, the accumulation of carbonate is only rarely used as a reliable palaeo-productivity proxy (Brummer and van Eijden 1992; Rühlemann et al. 1996, 1999; van Krefeld et al. 1996).

The Atlantic Ocean is regarded as the largest present-day deep-sea carbonate sink as it serves as a huge carbonate depocenter, with an average deep calcite lysocline (Milliman 1993). The solubility of $CaCO_3$ in seawater increases with pressure, so that the ocean is typically supersaturated at shallow and intermediate depths, and undersaturated in the deepest parts. Only a fraction of $CaCO_3$ production is buried, and this proportion depends on the area of the seafloor that is shallower than the depth of calcite saturation. However, calculations and models that attribute the calcite lysocline to the critical undersaturation depth (hydrographic or chemical lysocline), and not to the depth at which significant calcium carbonate dissolution is observed (sedimentary calcite lysocline), strongly overestimate the preservation potential of calcareous deep-sea sediments. From hydrographic parameters alone, one would expect significant calcium- carbonate dissolution to begin below 5000 m in the deep Guinea and Angola Basin, and below 4400 m in the Cape Basin. However, carbonate preservation studies (Volbers and Henrich 2002; Henrich et al. this volume) clearly show that it starts already at much shallower depths.

Present-day production of $CaCO_3$ in the world's ocean is calculated to be at about 5 billion tons per year. Carbonate production ranges from about 2.5 g m^{-2} d^{-1} in the central ocean gyres to up to about 30 g m^{-2} d^{-1} in highly productive areas (Milliman 1993). It is assumed that about 60% accumulates in sediments, whereas the remaining 40% is dissolved (Milliman 1993). Calculation of the global oceans' carbonate budget includes carbonate production, accumulation, and dissolution, but published estimations of the carbonate budget vary widely as it is difficult to account for all input and output mechanisms of the system. On a longer time scale, for example, during the late Quaternary, calculation of the carbonate budget should even take the glacial and interglacial end-members into account. Carbonate production on continental slopes and especially in the pelagic open ocean is almost exclusively planktic. Carbonate producers such as coccolithophorids and foraminifera are primarily important for the long-term storage of carbon in open ocean sediments (e.g. Berger 1976, 1978). In addition, previous studies suggest that aragonitic pteropod production average at about 10 to 12% of the total $CaCO_3$ production (Berner and Honjo 1981; Fabry 1990; Fabry and Deuser 1991, 1992). Aragonite, a metastable polymorph of $CaCO_3$, dissolves at shallower water depths than calcite, as it is much more soluble in seawater (Morse et al. 1980). Approximately 90% of the aragonite flux is already remineralised in the upper 2200m of the water column (Betzer et al. 1984). However, aragonite can still contribute up to 10% of the total mass flux.

The distribution of the total calcium- carbonate content in sediments of the deep-sea is already well known (e.g. Archer 1996). However, detailed analyses of the carbonate fraction and partitioning of its various contributors are still missing. Total fluxes of matter are determined by chemical or physical measurements, but to test the biotic contribution to the complex marine carbonate system, it is required to breakdown fluxes, at least by organism groups. So far, this has been done only for individual groups, such as coccolithophorids (e.g. Broerse et al. 2000; Sprengel et al. 2000; Young and Ziveri 2000; Ziveri and Thunnell 2000), calcareous dinocysts (Broerse et al. 2000), and pteropods (Betzer et al. 1984; Kalberer et al. 1993).

In this paper we focus on the calcium carbonate content, the predominant biogenic material of pelagic oceans. Firstly, the modern pattern of marine carbonate production is exemplified by comparing two sediment traps located in different domains of the South Atlantic. In addition, this paper presents new carbonate calculations for coccoliths, calcareous dinocysts, planktic foraminifera, and pteropod contents in surface sediments. Principal distribution patterns of the carbonate portion of the major organism groups are described and discussed.

Hydrography and Production Characteristics of the Study Area

The surface current system in the South Atlantic is characterised by a northward transport of warm (>24°C) surface waters across the equator, ultimately feeding the Gulf Stream. This process is crucial for the global thermohaline circulation and especially for the heat transfer to the North Atlantic (Macdonald and Wunsch 1996). Since the circulation system has often been described in great detail (e.g. Peterson and Stramma 1991; articles in Wefer et al. 1996), only a brief summary of the main features is given in the following.

The upper level circulation in the South Atlantic is dominated by the subtropical anticyclonic gyre, and is closely coupled to lower atmospheric wind stress. Surface water (upper 50-100m) of the northwestward flowing South Equatorial Current (SEC) represents the northern part of the subtropical gyre (Fig. 1). The SEC itself consists of two branches, a mainstream flowing south of 10°S, and a smaller, faster flowing branch forced by the trade winds between 2° and 4°S (Peterson and Stramma 1991). In the equatorial area, these two branches are separated by the South Equatorial Counter Current (SECC) which pushes surface water eastward. At about 10°S off Brazil, the SEC splits into two branches, building the southward flowing Brazil Current (BC) and the northward flowing North Brazil Current (NBC) (Peterson and Stramma 1991). The latter contributes to the eastwards flowing North Equator Counter Current (NECC). Its interaction with the northern branch of the SEC

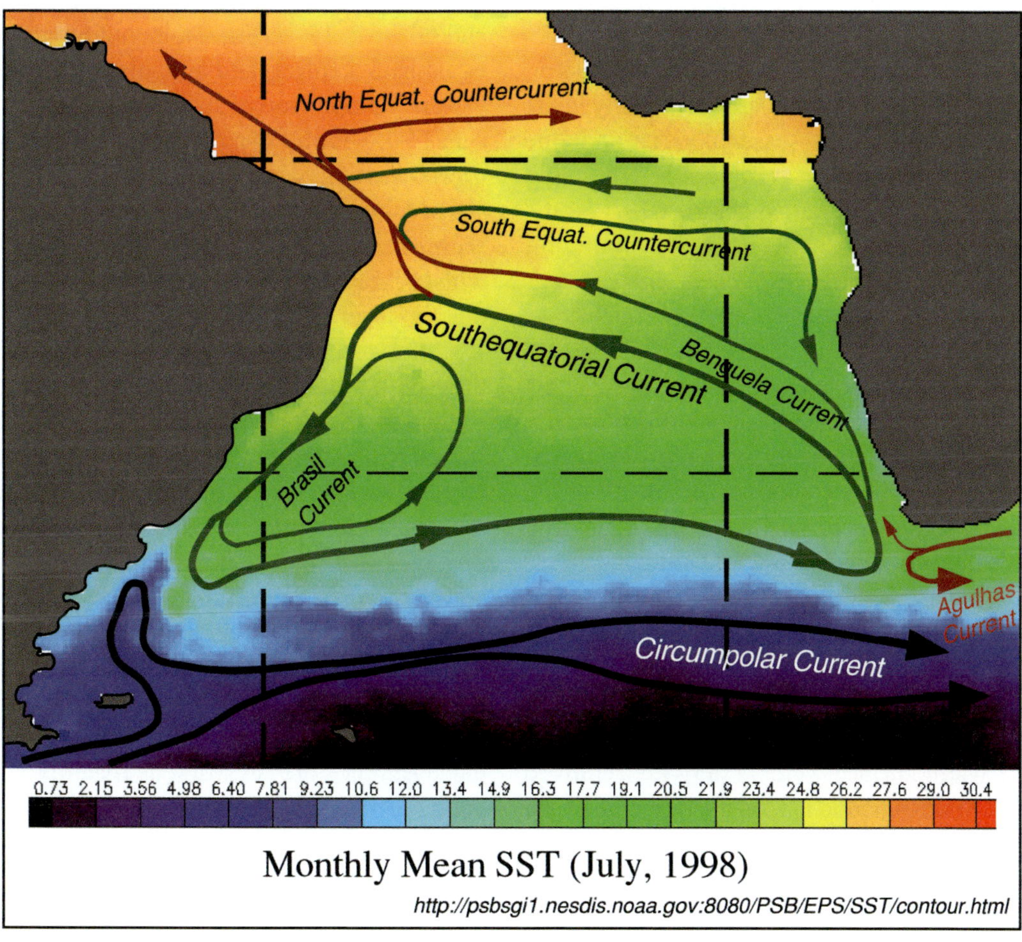

Fig. 1. Schematic representation of the large-scale upper-level circulation pattern in the South Atlantic (from various sources, see text) plotted on a map of annual mean sea-surface temperatures.

leads to a strong convergence of water masses in a mixing area at about 3° to 5°N. This results in downwelling of surface waters, which supports the eastward flowing Equatorial Undercurrent (EUC). This current is present at a depth of 50 to 125m along the entire equator and feeds surface currents off the African coast (Peterson and Stramma 1991). The contact zone of EUC and SEC forms the equatorial divergence where upwelling of colder water from around the thermocline depth occurs (Fig. 1). The source of the water that is upwelled is essential, as it determines the nutrient concentrations in the upper photic zone, where most of the phytoplankton productivity takes place. If, for example, water becomes upwelled from the EUC, which is an nutrient depleted "recycled" water mass, no significant increase in nutrients and associated phytoplankton production will be observed in the surface waters (Monger et al. 1997).

Thus, the mean sea surface temperatures (SST) are cooler at and a somewhat south of the equator, and warmest north of the equator, especially in the western Atlantic. Strong seasonal variation in the forcing winds, however, produces a fluctuating equatorial system. In boreal summer SST in the eastern equatorial Atlantic is at its minimum, whereas in boreal winter, SST is at its annual maximum and part of the equatorial surface water flows back as counter-currents. Thus, the SEC and associated features also have different seasonal aspects, such as the thermocline depth that is usually deeper in the west and relatively shallow in the east (Fig. 1). During boreal summer, the depth of the thermocline is lowered in the western equatorial Atlantic due to an increased westward transport of surface waters, which again is a consequence of the increased trade winds. Today, the thermo-cline depth is slightly shallow towards the east, allowing thermocline water to mix with warmer surface waters.

The warm surface water is actively removed from the South Atlantic to feed the heat sink of the North Atlantic and hence the global conveyor-belt, and is replaced by southern and subsurface waters. These cool waters supplanting the loss are generally rich in nutrients, wherefore the South Atlantic is relatively productive in upwelling areas. Of course, the present pattern of productivity as seen in pigment distribution (Fig. 2) represent an integration of a large number of dynamic properties including seasonal wind, geostrophic currents, and nutrient content of subsurface waters. Generally, nutrients are depleted in surface waters except in upwelling regions. Here, the primary production rate can reach >100g C m^{-2} a^{-1}, and is between 50 and 100g C m^{-2} a^{-1} in the equatorial divergence zone. In contrast, primary production is generally <50g C m^{-2} a^{-1} in the oligotrophic gyres (Berger 1989). Organic carbon in the surface sediments of the South Atlantic very well reflects this general production pattern of all organisms (Fig. 2).

Material and Methods

Surface Sediment Material

The study area stretches from about 20°N to 60°S and 60°W to 15°E (Fig. 3), including most of the equatorial, central, and subpolar South Atlantic Ocean. Comparison of sediment surfaces with recent hydrography is a certain method to test the usefulness of a paleoceanographic proxy. We thus studied the uppermost centimetre of more than 400 sediment surfaces (Fig. 3). Samples were collected with multicorers, box corers, or Van Veen grabbers during several ship expeditions. These coretop samples were retrieved during various cruises to the South Atlantic Ocean from water depths of <100m to >5500m. All sediment surfaces are assumed to be of Holocene age. However, only few of these surface sediments have been dated, nor is the exact sedimentation rate known at most sites. The ages of surface sediments may vary from decades to several hundreds, or even up to several thousands of years, depending on the local sedimentation rate. Nevertheless, considering that climatic and oceanic settings throughout the Holocene are comparable to present-day conditions, these samples are regarded as recently accumulated sediments.

In total, the data set includes 381 samples analysed for their bulk carbonate content and 316 samples analysed for their total organic carbon content. In addition, 204 samples were analysed for coccolith carbonate, 167 samples for calcareous dinocyst carbonate, 97 samples for planktic foraminiferal

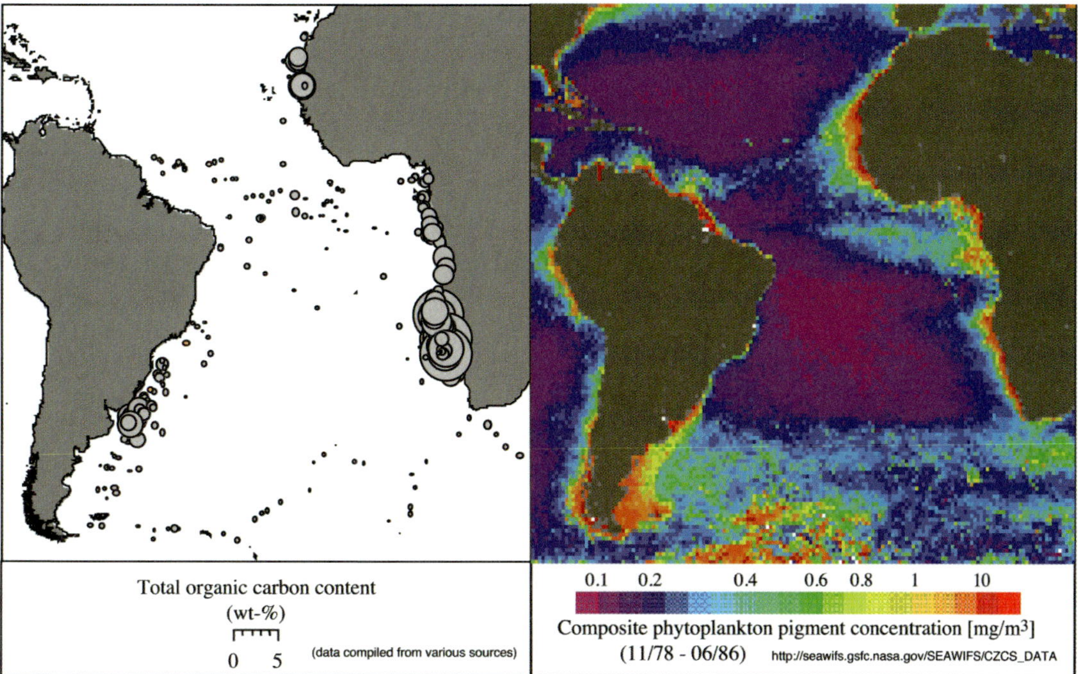

Fig. 2. Map showing the distribution of TOC in surface sediments (left), as well as the pigment distribution in surface waters inferred from colour scanning data aboard CZCS satellite (right).

carbonate, and 58 samples for their aragonite content. The sediments are from different regions with various oceanographic and environmental conditions, such as shelf regions displaying large amounts of benthic organisms and lithogenous material, continental slope regions with turbidites and con-tourites, deep-sea basins distinguished by strong carbonate dissolution beneath the compensation depth, and from various ridges and rises. The investigated sediments have accumulated in coastal and equatorial upwelling regions, as well as in oligotrophic regions and contain varying amounts of organic material

Determination of Bulk Carbonate, Aragonite, and Carbonate of the Various Planktonic Groups

Prior to the bulk geochemical analyses, sediment samples were freeze-dried and homogenised. *Bulk carbonate content.* Total carbon (TC) concentrations and total organic carbon (TOC) contents were measured using an LECO-CS 300 elemental analyser. This device measured the total carbon content. In a first step, 100 mg of the homogenised sediment were analysed (TC). In the subsequent analysis, HCl-treated samples (to remove calcium carbonate and determine TOC content) were measured in same way. Calculations were made according to the equation

$CaCO_3$ (wt-%) = 8.33 * (TC wt-% - TOC wt-%).

Analytical precision of the carbon measurements of carbonate standards and in replicates was regularly better than 2 %.

Planktic Foraminiferal Carbonate Content. Sediment samples for coarse fraction analyses were washed on a sieve with 63 mm mesh size. Two independent techniques were used to estimate the carbonate content of planktic foraminifera. (1) Carbonate of the >63 µm fraction was measured geochemically as described for the total carbonate content. (2) The carbonate content was estimated by the number of specimens counted in the weight-balanced sand fraction. Samples were split until at least 300 particles remained. Planktic foraminiferal species as well as various other particles (such as

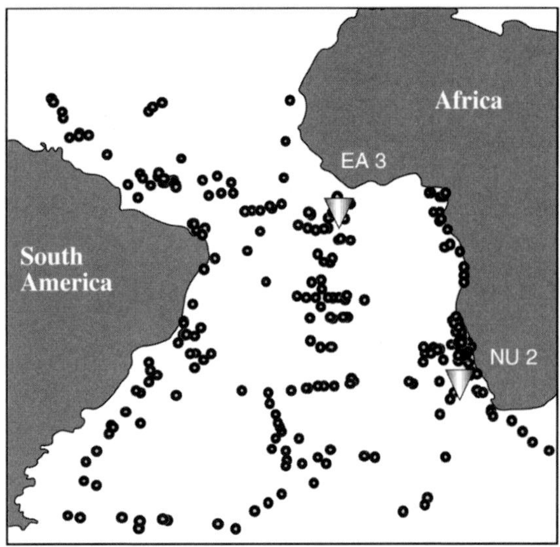

Fig. 3. Locations of the studied surface sediments (dots) and sediment traps (filled triangles).

benthic foraminifera, radiolarians, and quartz) were counted under the light microscope and the percentage of the foraminifera was determined. Absolute abundances expressed as specimens per weight percentage of dry sediment (wt-%), were taken as the carbonate contents.

The contribution of fragmentary planktic foraminifera tests to the carbonate content within the coarse silt fraction (10-63µm) was not taken into account. Therefore the estimations of the planktic foraminifera carbonate should be considered as minimum values. However, different means with both techniques indicate very similar results (Fig. 4) and therefore these data are assumed to represent a good approximation of the content of planktic foraminifera carbonate. The rather good comparison of the different means is affected by the occurrence of calcareous microfossils in some of the samples (Fig.4).

Coccolith Carbonate Content. Coccolith numbers calculated from surface sediments (see Vink et al. this volume) were converted into coccolith-carbonate contents on the basis of mean estimates of coccolith masses (Fig. 5). This is an approach that is already used routinely on sediment trap samples (e.g. Broerse et al. 2000; Giraudeau et al. 2000; Sprengel et al. 2000). Since there is an extremely wide range of coccolith sizes between different

coccolith species and consequently a great variation in mass estimates (e.g. Young and Ziveri 2000), specific carbonate masses for the most frequent species were used in this study.

The substantial errors in such calculations may be relatively high, since a single mean carbonate value does not account for the size variations of most common species (e.g. Knappertsbusch et al. 1997; Baumann and Sprengel 2000). To optimise carbonate calculations size ranges of the numerically most common species and the volumetrically most abundant species were examined (Baumann and Böckel unpubl. data). However, even with great caution, the calculations may produce errors of up to 50%. Nevertheless, the data given here seem very reasonable in comparison to the total carbonate data and is therefore expected to be a good approximation of the coccolith-carbonate content.

Calcareous Dinocyst Carbonate Content. The carbonate contents of calcareous dinocysts were calculated by the same method as

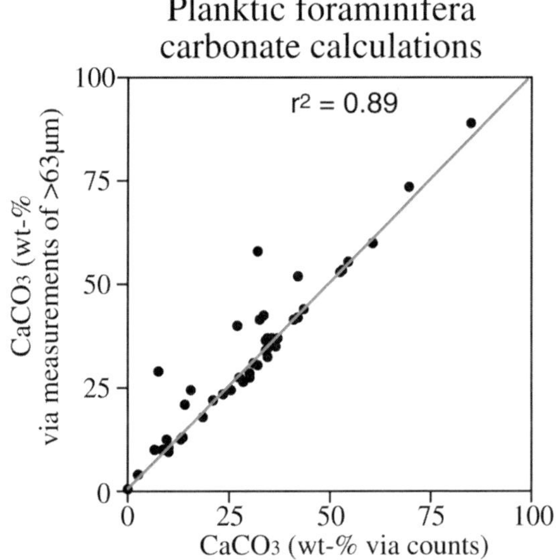

Fig. 4. Comparison of the measurements derived by the two different foraminifera carbonate estimations used in this study, indicating a fairly good correspondence of the data. Data points outside the 1:1 line are either from the continental margins or from the equatorial upwelling and include benthic foraminifera and pteropods.

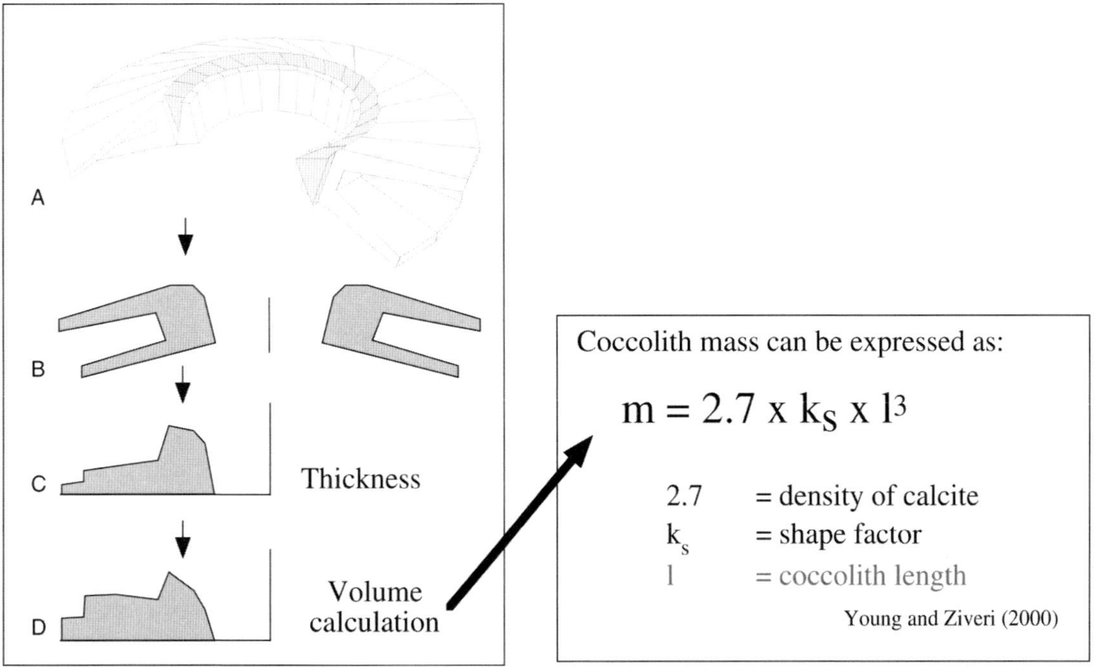

Fig. 5. Determination of species-specific coccolith carbonate contents by using the method introduced by Young and Ziveri (2000). Recommended shape values (ks) for each species together with own unpublished size measurements were used.

the cocco-lith carbonate content. The inner and outer diameters of the spheres of various species were measured in the SEM and their porosity estimated (Zonneveld unpubl. data). Numbers of calcareous dinocysts calculated in surface sediments (see Vink et al. this volume) were converted into carbonate on the basis of averaged individual species-specific mass estimates (Young and Ziveri 2000).

Aragonite content, calcite and aragonite x-ray diffraction analysis (XRD). After freeze drying, about 0.5 g bulk sediment of each sample was carefully ground to obtain a grain size <63 μm. All samples were measured with a Philips PW1830 goniometer (Crystallography, Bremen University) equipped with a fix divergence slit, using Cu Ka radiation (40 kV, 30 mA). The XRD measurements were carried out between 20-50° 2θ with a step size of 0.02° and a scan time of 2 seconds per step. Quantitative estimations were made by determining aragonite peak heights and calcite peak areas for each sample and subsequent comparison with

the respective calibration curves (after Milliman 1974).

Preparation of Sediment Trap Samples

Particle fluxes were determined using large aperture time-series sediment traps. Here, data of two sediment traps located in the eastern equatorial Atlantic (EA 3 at 0°N, 10.8°W) and off Namibia (NU 2 at 28.9°S, 14.6°E) were used. Trap depths were around 1000m (EA 3) and 2500m (NU 2). For sample preparation and detailed particle flux data see Fischer and Wefer (1996), Fischer et al. (2000), and Romero et al. (2002).

All planktic foraminifera tests and pteropod shells larger than 63 μm were identified microscopically and picked out by hand from a wet-split sample. The removed material was then washed briefly with fresh water, dried at 60°C and weighed to determine the organism fluxes and their carbonate proportions. Coccolith flux rates from sediment trap samples, which will be presented elsewhere, were converted into coccolith-carbonate fluxes following the same approach as for the surface sediments (see above).

Results and Discussion

Methodological Approach

In this paper, calculating organism-specific carbonate data for gathering information about past climates and ocean productivity from marine micro-paleontological records is unprecedented. Because of dissimilar analytical procedures for the different groups, we combined various analytical and calculation methods to determine the proportion of each group to the total carbonate content. There is neither any single geochemical approach allowing for this to be done, nor is it feasible to pick the small (<2-20 μm) coccoliths individually as it is routinely done with other microfossils. Despite the fact that the carbonate volumes of the specific plankton groups were calculated in rather different ways, the results invariably gave reasonable values for both the single organism groups and the sum of their total carbonate values.

Hence, one important methodological result of this work is that a carbonate budget of the various carbonate contributors is feasible and produces reasonable results. This approach is useful and, indeed, the good quality of the carbonate estimations may be surprising. As far as the foraminifera are concerned, this depends on the size fraction that is measured geochemically and the available species-specific size data applying to coccoliths and calcareous dinocysts. Therefore, it is important to select the most appropriate size classes and shape factors that cover best the average variability in the respective groups. Nevertheless, we have applied a relatively conservative approach for each group, which rather implies an underestimation of carbonate because only complete specimens were considered (coccoliths, calcareous dinocysts). Juvenile tests <63μm were neglected (foraminifera).

Bulk Carbonate and Composition of the Carbonate Fraction of Trapped Samples

The variability and magnitude of the overall carbonate flux pattern in the eastern equatorial Atlan-tic (EA 3) and off Namibia (NU 2) were rather comparable to each other. The composition of the material was generally dominated by carbonate, which constituted >65% in both traps (Wefer and Fischer 1993; Fischer and Wefer 1996; Romero et al. 2002). Total carbonate fluxes ranged from 2.0 to 100.5 mg m^{-2} d^{-1} in EA 3 and from 8.5 to 139.3 mg m^{-2} d^{-1} in NU 2 (Fig. 6). However, there were slight differences in the timing of peak carbonate fluxes. They were measured at EA 3 in June and July/August (collected for only 7.5 months) and at NU 2 in February and July/August.

The most striking difference in the carbonate flux pattern between the investigated traps is that the relative contribution of coccolith and foraminifera carbonate varied significantly (Fig. 6). The eastern equatorial Atlantic carbonate flux is dominated by planktic foraminifera (16-98%), whereas coccolithophorids and pteropods only contributed less than 30% of the total carbonate. Calculated absolute coccolith-carbonate fluxes were between 0.1 and 9.3 mg m^{-2} d^{-1}, and, on average, contributed 9.1% to the measured $CaCO_3$ fluxes. The input of aragonite-producing pteropods is approximately the same. Foraminifera constituted between 0.5 and 48.6 mg $m^{-2}$$d^{-1}$, with an average contribution of 46%. In contrast, much higher coccolith carbonate fluxes with maximum values of > 60 mg m^{-2} d^{-1} were measured in the Namibia upwelling zone (Fig. 6). The mean relative amount was high, varying between 12.6% and 66.2%, and mostly showing correlation with the total amounts of carbonate. The foraminiferal carbonate flux measured was 7.7 mg m^{-2} d^{-1}; the contribution of the planktic foraminiferal $CaCO_3$ fluxes to the measured total carbonate fluxes varied between 3.1 and 22.3%, with an average contribution of 10.8%. In addition, pteropods accounted for up to 23% of the total $CaCO_3$ fluxes, with a mean input of 2.4 mg m^{-2} d^{-1}.

As expected, the calculated carbonate fluxes of the calcareous organisms were lower than the measured total $CaCO_3$ fluxes, except for two values in the eastern equatorial Atlantic (EA 3). Here, collective $CaCO_3$ fluxes were slightly higher than the measured total carbonate fluxes, which probably is a calculation artefact. In addition, estimates of carbonate fluxes are prone to significant errors as a

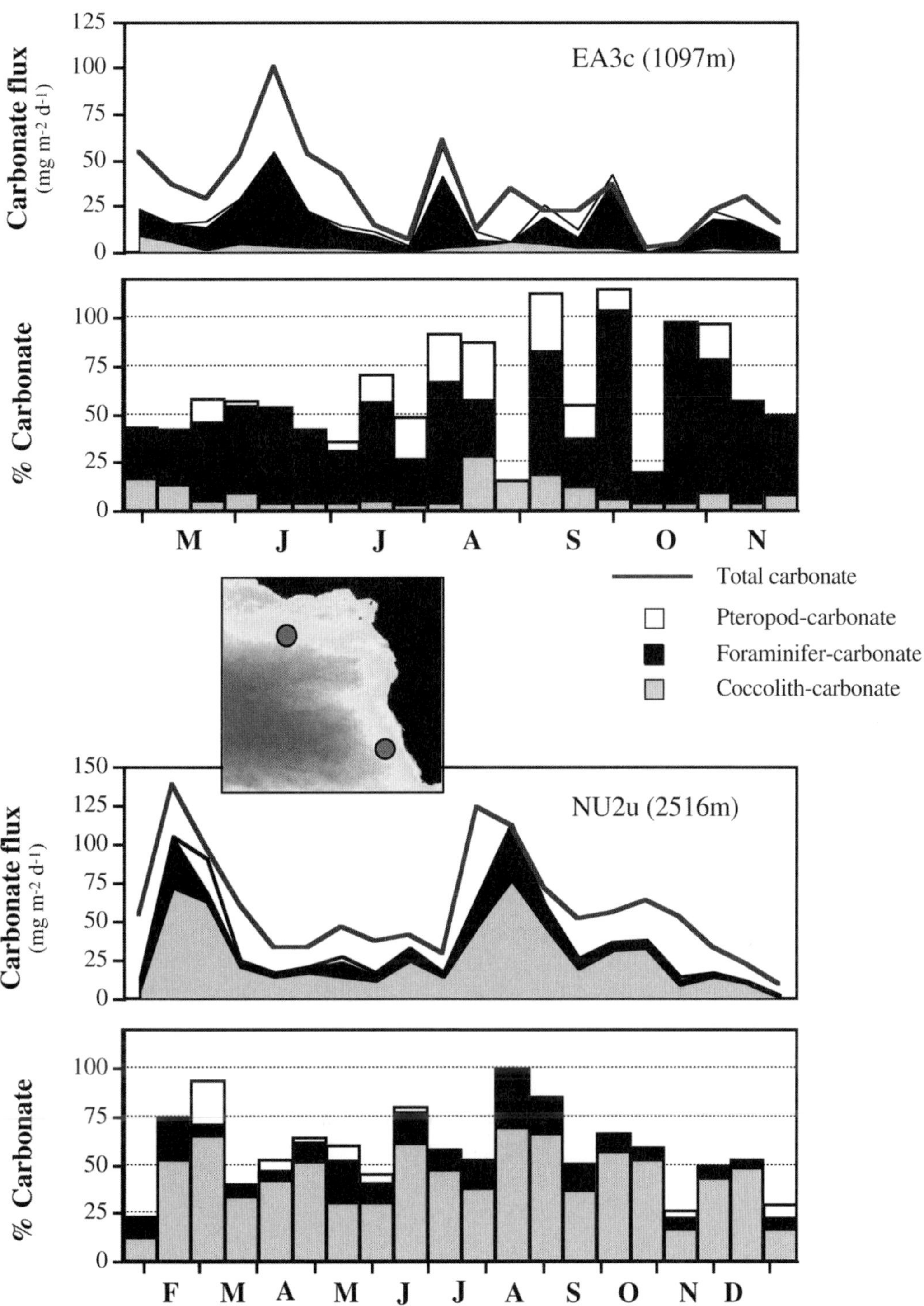

Fig. 6. Determination of organism-specific carbonate contents in the eastern equatorial Atlantic (EA 3) and off Namibia (NU 2). The most striking difference in the carbonate flux pattern between the investigated traps is that the relative contribution of coccolith and foraminifera carbonate varied significantly between the sites. Coccoliths overwhelmingly dominate the carbonate fraction close to the Namibia upwelling, whereas highest foraminiferal carbonate content occurs in the relatively fertile, mesotrophic waters of the equatorial divergence.

consequence of using the >150 μm fraction, breaking down tests, and problems related to accurate sample splitting (especially when the flux is extremely low). Generally, this resulted in a slight underestimation of the carbonate contents of the various carbonate contributors. Nevertheless, despite these uncertainties in absolute values, carbonate fluxes and the relative contribution of the various groups provided a valuable new perspective for the study of carbonate accumulation.

Distribution of Total Carbonate Contents and Carbonate Portion of the Organism Groups

To interpret the distribution of carbonate-bearing organisms (see Vink et al. this volume) and estimate lithogenic and organic particle dilution within the surface sediment, we need detailed information on the sediment carbonate content. The overall carbonate contents ranged from <2 wt-% at sites close to the southeastern continental margin off South America, but also from the deep basins and the western African continental margin, to >95 wt-% at sites of the Mid-ocean Ridge (Fig. 7). Highest carbonate contents in the surface sedi-ments were encountered more or less at the mid-Atlantic Ridge, from the equator to about 40°S, and the Walvis Ridge off southwestern Africa. In these oligotrophic areas carbonate values often exceeded 90wt-%. In contrast, nearshore sediments on continental shelves and slopes had considerably lower calcium carbonate contents, but also had much higher organic carbon contents (up to 8 wt-%; see Fig. 2). In addition, the calcium carbonate contents were quite low (<20 wt-%) in the deep Cape Basin, Argentine Basin, and Brazil Basin, close to and below carbonate the lysocline depth. The depth of the <20 wt-% carbonate contour did not change, most probably due to the fact that chemical erosion may reduce the actual amplitude of the fluctuations (see Henrich et al. this volume).

Nevertheless, calcium carbonate was fairly well preserved above the lysocline in the oligotrophic open ocean and most probably reflects the primary production of these areas. However, differences in the $CaCO_3$ contents between different areas may, at least, be partially due to different assemblage compositions of plankton. In fact, the carbonate input of the different organism groups was highly variable, although dominated by planktic foraminifera and coccolithophorids (Fig. 8). The weight-balanced coccolith carbonate contents rose to up to >80 wt-% in the mid-Atlantic Ridge sediments of the central South Atlantic. Obviously, coccoliths dominate the carbonate fraction in this area (60-70%), whereas they only play a minor role at the continental margin off South America, off eastern Africa, and in the equatorial divergence zone. The latter is characterised by carbonate derived from planktic foraminifera, which generally comprise between 30 and 55 wt-% of the total sediment. With a few exceptions (Walvis Ridge), foraminifera carbonate is moderately important in the other areas of the South Atlantic, but never exceeded 30 wt-% (Figs. 8).

In contrast, calcareous dinocysts only play a minor role as a supplier of calcium carbonate to the sediments. Carbonate contents of up to a maximum of 4 wt-% in the mid-Atlantic Ridge sediments were calculated, whereas less than 1 wt-% were recorded in the deep basins. The aragonite content, mainly derived from pteropod shells, is of local importance in the western South Atlantic especially off central South America. Here, the measured contents of aragonite amounted to up to 50 wt-% of the total sediments. In addition, some of the samples from the mid-Atlantic Ridge were characterised by medium-high contents of >20 wt-% (Fig. 8). It must be mentioned that the aragonite content was not measured in any sample recovered below a water depth of 4000m, This was due to fact that pteropods are highly susceptible to fragmentation than calcite particles at such water depths (see also Henrich et al. this volume).

Depth Related Carbonate Input

Coccolithophorids are by far the main contributors to the carbonate in the oligotrophic gyres of the South Atlantic. In fact, they predominate in the central South Atlantic, with highest coccolith carbonate contents encountered at a water depth of 3000 to 4700m (Fig. 9). Therefore, extremely high carbonate input derived from coccoliths is observed down to about the lysocline depth. Thus, at

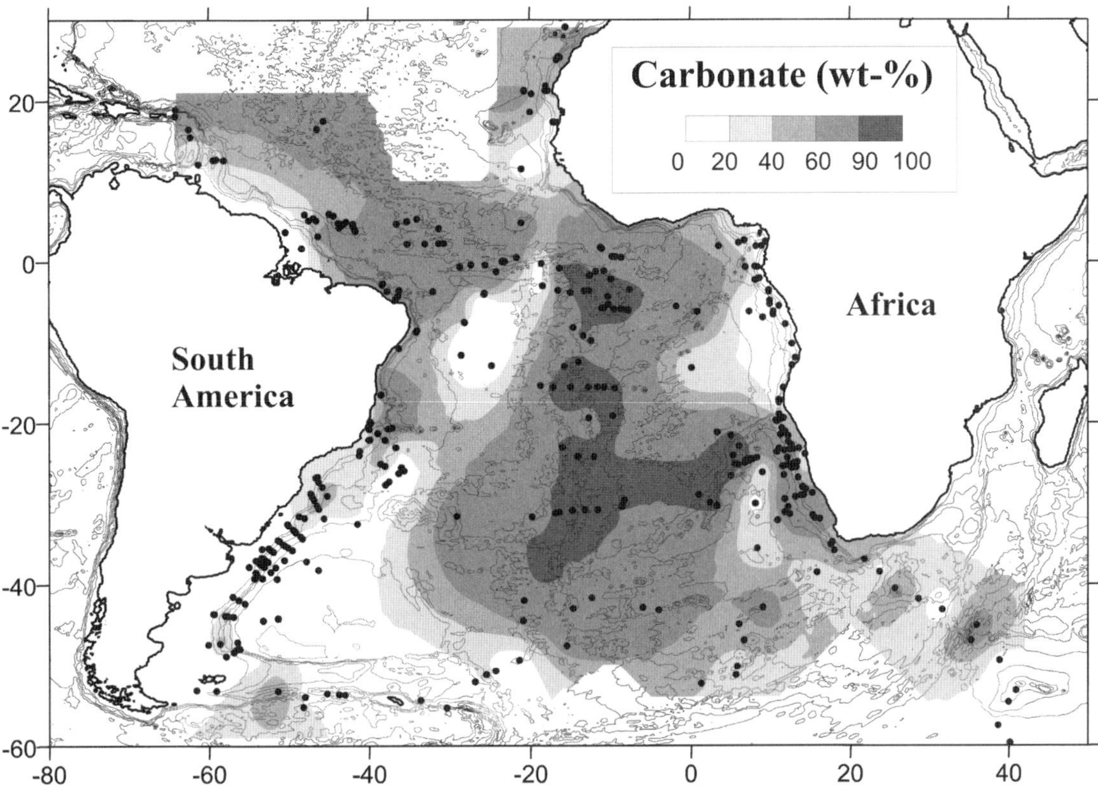

Fig. 7. Total carbonate content in surface sediments of the South Atlantic. Highest carbonate contents generally follow the mid-Atlantic Ridge, from the equator to about 40°S, and the Walvis Ridge off southwestern Africa. In contrast, nearshore sediments on continental shelves and slopes as well as sediments from the basins have considerably lower calcium carbonate contents.

least a number of coccolith species seem to be very resistant to dissolution. In contrast, the amount of foraminiferal carbonate decreases significantly below about 3500m. Highest foraminiferal carbonate comprises more than 50% of the total carbonate content in the relatively fertile, mesotrophic waters of the equatorial divergence. However, in terms of weight-balanced carbonate input, foraminifera only seldomly exceed the 50 wt-% and the decrease begins already far above the calcite lysocline. This could also be due to the relatively long-term exposure at the sediment/water interface leading to increased carbonate dissolution of the surface sediments (Volbers and Henrich 2002).

Contrary to planktic foraminifera, and despite the low absolute carbonate input into the total sedi-ment, the contribution of calcareous dinocysts to the carbonate fraction increased with depth. The highest relative abundances were measured close to the lysocline depth (Fig. 9). This was indicative of a relatively high preservation potential of these calcitic primary producers, which has also been shown for other regions (e.g. Zonneveld et al. 2001). In comparison, aragonitic pteropods are relatively abundant in the western South Atlantic, whereas only very few specimens have been found in the eastern Atlantic Ocean, particularly along the African continental margin (Gerhardt and Henrich 2001). The aragonite content displayed a s-shaped trend, with highest amounts occurring in intermediate water depth of 2000 to 3000m (Fig. 9). It has been shown that good preservation of pteropods correspond well to the increased arago-

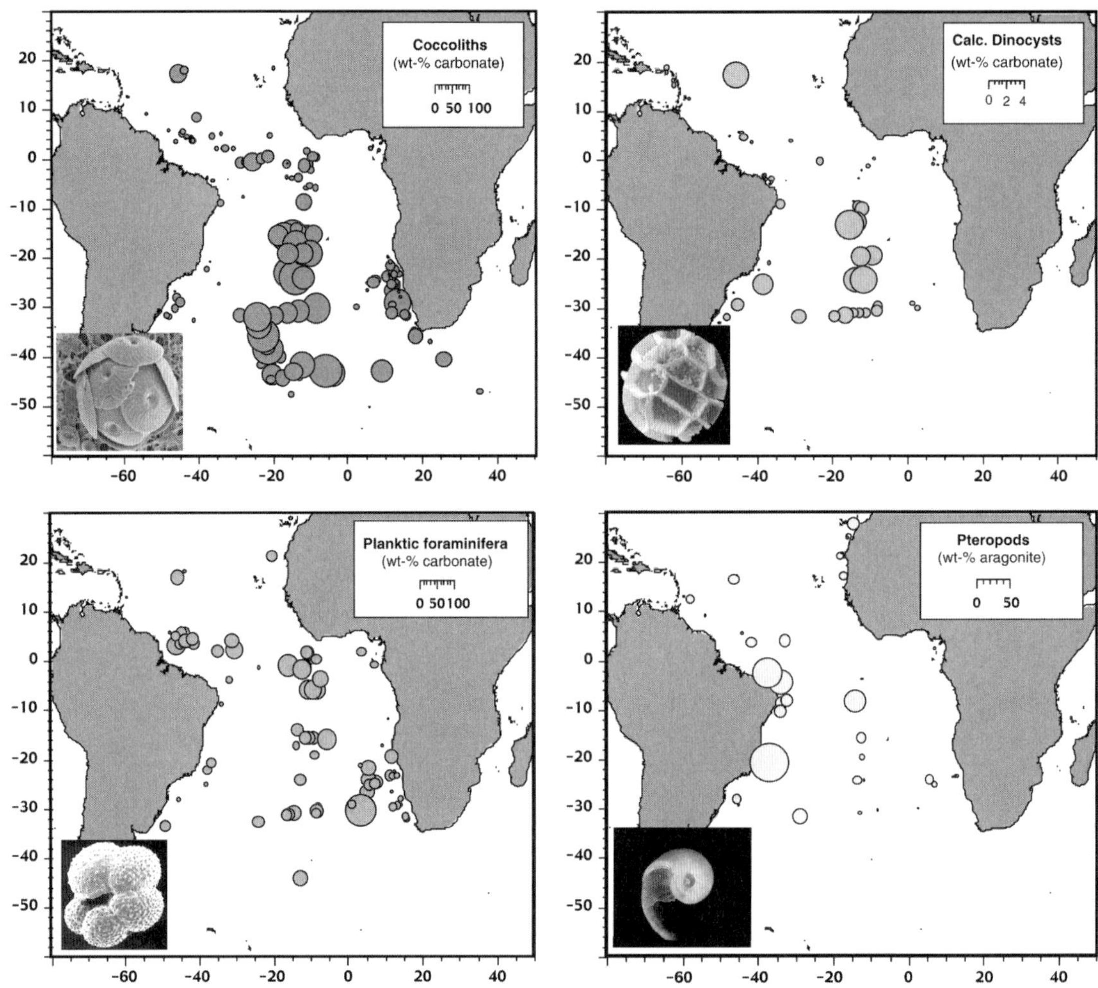

Fig. 8. Total carbonate contents (all weight-balanced) of the different organism groups in the studied surface sediment samples of the South Atlantic. In fact, carbonate input of the different organism groups is highly variable although dominated by planktic foraminifera and coccolithophorids, respectively. In contrast, aragonitic pteropods are of geographically restricted importance, whereas calcareous dinocysts contribute only minor to the carbonate.

nite content and vice versa (Gerhardt and Henrich 2001).

Factors Influencing Surface Sediment Carbonate Distribution

Surprisingly, the carbonate content in surface sediments hardly correlates with the upper level circulation of the South Atlantic as is for the case in other regions. For example, differences in sedimentary carbonate content have often been used to distinguish the different surface water masses in the Norwegian-Greenland Sea (e.g. Kellogg 1976; Baumann et al. 1993; Hebbeln et al. 1998; Henrich 1998). High carbonate contents are characteristic for the water masses of the warm Norwegian Current, whereas low contents and relatively high dissolution of $CaCO_3$ characterise the polar surface water masses (Huber et al. 2000). Of course, the actual carbonate content of deep-sea sediments is controlled by a complex interplay of production in overlying surface waters, dilution by non-carbonate phases and, especially, dissolution in the water column, at the sea floor, and in the sediment pore

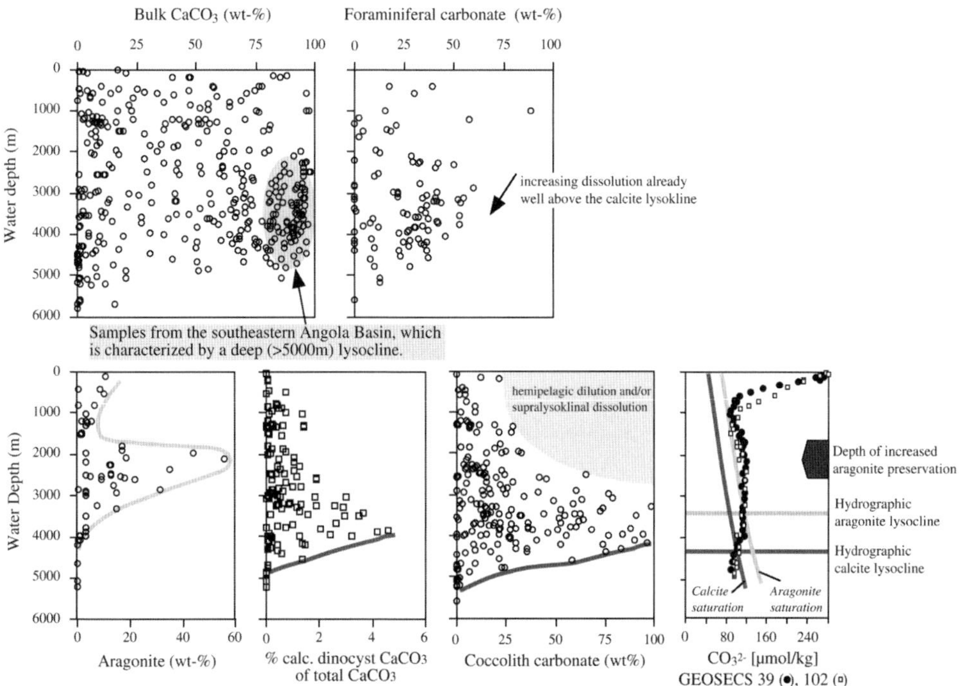

Fig. 9. Plot of the carbonate contents of the different organism groups versus water depths. High carbonate input derived from coccoliths is observed down to about the lysocline depth whereas foraminiferal carbonate already starts to decrease already far above the calcite lysocline. In addition, carbonate ion concentrations from GEOSECS stations 39 in the western equatorial Atlantic and from GEOSECS station 102 in the eastern South Atlantic (data from Bainbridge 1981), as well as saturation curves of CO_3^{2-} concentration for calcite and aragonite are indicated.

waters. As a result, it is often difficult to uniquely separate changes due to the carbonate saturation of deep water from other controlling factors.

The production of calcium carbonate by pelagic organisms in upper water levels is difficult to estimate. Due mainly to different time-scales involved (days vs. several hundreds of years), only a limited number of studies are available, which compare series of plankton data directly with spatial patterns of assemblages found in surface sediments (McIntyre and Bé 1967; Samtleben and Schröder 1992; Baumann et al. 2000). However, sediment traps are seen to be a valuable tool both to estimate carbonate production and to link the dynamics of pelagic plankton production in the photic zone, as represented by plankton samples, with the assemblages found in the underlying surface sediments (e.g. Samtleben et al. 1995). From studies such as these, it is known at least for coccolithophorids that

coccolith-carbonate fluxes and accumulation rates are in the same order of magnitude and show a remarkable correlation between the traps and sediments (Sprengel et al. 2002). In addition, determinations of aragonitic pteropod fluxes in the Pacific provide evidence of a patchy distribution of these organisms (Betzer et al. 1984). According to these sparse data, carbonate production varies significantly from oligotrophic to eutrophic surface waters (Milliman 1993), but is highly variable already in small areas. The presented carbonate estimates of the two sediment traps are also in good agreement with the findings in surface sediments. In the present study, coccolithophorids mainly thrived in oligotrophic waters, whereas planktic foraminifera become progressively more abundant in mesotrophic areas, especially when the nutrient content increases (Giraudeau et al. 2000). The concentration of planktic foraminifera is at least ten times higher in the fertile coastal and equatorial regions than in the gyres (Bé and Tolderlund 1971; Meggers et al. 2002). Nevertheless, absolute numbers of coccolithophorids also increased considerably in areas where ocean and upwelled water mix, as compared to the oligotrophic ocean (e.g. Klejine et al. 1989; Giraudeau and Bailey 1995). This is indeed in good accordance with the distribution of coccolith and foraminiferal carbonate especially in the surface sediments off southwestern Africa. Despite a relatively low contribution to the total $CaCO_3$ mass flux (mean of 10%), the foramini-feral carbonate content in the surface sediments covaries closely with the carbonate input by coccoliths (Fig. 10).

However, the mechanisms controlling the burial of carbonates in this area (as well as in other regions) are relatively complex and, of course, the carbonate content is influenced by dissolution (e.g. Dittert et al. 1999; Henrich et al. this volume). This is indicated by a relatively high proportion of unidentified carbonate in these sediments (reported as "Rest" in Fig. 10). This unidentifiable rest most probably comes from broken coccoliths, as well as fragmented foraminifers and pteropods, which were not taken into account in our calculations of the carbonate content. Recent dissolution studies suggest a strong influence of supralysoclinal dissolution with respect to aragonite and carbonate in the high-productivity areas of the eastern South

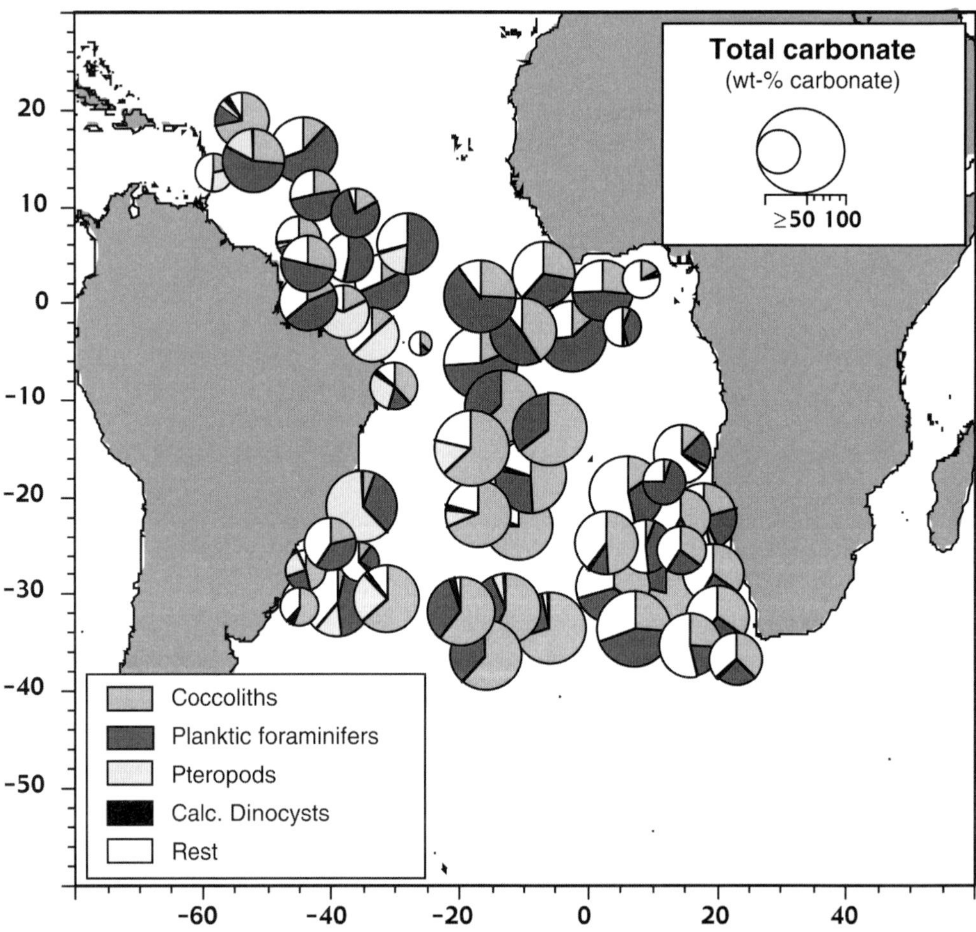

Fig. 10. Summary of the different organism group carbonate contents in surface sediments of the South Atlantic. A high proportion of unidentified carbonate, most probably originating from broken coccoliths, as well as fragmented foraminifers and pteropods, were not taken into account in our calculations and thus figured as "Rest".

Atlantic (Gerhardt and Henrich 2001; Volbers and Henrich 2002). This is due to the fact that degradation of organic carbon in sediments promotes the dissolution of calcium carbonate. Yet, a number of studies have demonstrated that calcite dissolution driven by metabolic CO_2 produced within the sediments forms a significant part of the diagenesis of sedimentary calcite even above the lysocline ("supralysoclinal dissolution"; e.g. Archer et al. 1989; Hales et al. 1994; Hales and Emerson 1997). Recently, Milliman et al. (1999) have suggested that considerable dissolution (perhaps as much as 60-80%) occurs even in the upper 500-1000 m of the ocean. In addition, preservation of pteropod shells is restricted to shallow parts of the seafloor (generally above 3500m water depth), i.e. shelf regions, continental slopes, ridges, and rises, as the aragonitic shells of pteropods are much more susceptible to solution than the calcitic remains of foraminifera and coccolithophorids. Thus, it appears reasonable to relate the absence of pteropod shells as well as higher fragmented planktic fora-minifera tests and coccoliths off southwestern Africa (see Dittert et al. 1999) to the local formation of more carbonate-dissolving bottom and pore waters. More detailed information on the preservation of carbonate is given in a number of papers, mostly dealing with the implications of such studies for oceanic and atmospheric carbon cycling (e.g. Howard and Prell 1994; Rühlemann et al. 1999; Hodell et al. 2001; Henrich et al. this volume).

Another factor influencing the weight-balanced carbonate contents is given by the dilution by non-carbonate phases and, thus, highly varying accumulation rates. Particularly, continental slope sediments are often dominated by various amounts of terrigenous sediments, discharged by nearby rivers, as well as relatively high contents of both organic carbon and biogenic opal (e.g. Schneider et al. 1997; Arz et al. 1999; Abrantes 2000). In addition, winds are well known to carry tremendous amounts of lithogenic dust from dry African source areas to the deep-sea (e.g. Tiedemann et al. 1989; Ruddiman 1997). Therefore, relative amounts of carbonate are highest on the mid-ocean ridge, where dilution (as well as dissolution) is at its minimum. This problem can only be ruled out on the basis of carbonate accumulation rates as a product of carbonate content, dry bulk density, and sedimentation rate. Unfortunately, only few of the samples analysed have been dated to obtain an exact age of the sediments surface, so that the sedimentation rate is not known. Fragmentary information on sedimentation rates in some of the areas, however, can be obtained from sediment cores nearby that indicate a rather high variation of sedimentation rates. They vary from 5-10 cm ka^{-1} on the Amazon fan to less than 1 cm ka^{-1} on the mid-Atlantic Ridge (e.g. Rühlemann et al. 1996). However, this calculation has so far not been done for the carbonate contents of the samples investigated and therefore will be subject of a future project.

Summary and Conclusions

A total of more than 400 surface sediment samples from the equatorial, central and subpolar South Atlantic Ocean were investigated for their carbonate content as well as the carbonate contribution of the calcareous plankton groups. These new carbonate calculations of coccolith, calcareous dinocyst, planktic foraminifera, and pteropod contents in surface sediments yielded the following results:

• A carbonate budget of the various carbonate contributors is possible and gives reasonable results. Indeed, the rather good quality of carbonate estimations is astonishing. It depends on the size fraction geochemically measured (foraminifera) and the group/species-specific size data available (coc-colithophorids, calcareous dinocysts). Nevertheless, the applied methods rather imply an underestimation of carbonate, because only unbroken specimens were considered (coccolithophorids, calcareous dinocysts) and juvenile tests <63μm neglected (foraminifera).

• Coccolithophorids and planktic foraminifers are the major components of pelagic carbonate. While coccolith carbonate dominates the oligotrophic gyres of the South Atlantic, carbonate derived from planktic foraminifera increased considerably in more fertile, mesotrophic areas, such as the equatorial divergence zone. In contrast, aragonitic ptero-pods are of geographically limited importance, whereas calcareous dinocysts (and probably benthic foraminifers) contribute only minor amounts to carbonate.

• Carbonate dissolution has a major effect below lysocline depth, but also in highly productive areas (supralysoclinal dissolution). Foraminiferal carbonate is much more affected by dissolution than coc-colith and calcareous dinocyst carbonate. The maximum aragonite content is observed in intermediate depth due to the fact that the aragonitic shells of pteropods are more susceptible to solution than calcitic tests.

Acknowledgements

We would like to thank all members of our Sedimentology and Palaeontolgy working groups for their general assistance and discussion. We appreciate the constructive comments of two anonymous reviewers. The useful suggestions of H. Meggers (Bremen) on an earlier version of the manuscript are also gratefully acknowledged. We are indebted to R. Henning, H. Heilmann, A. Brune, and A. Freesemann for laboratory work and we would also like to thank the captains, crews, and numerous colleagues for their help during several ship expeditions on board of the *RV Meteor*, *Polarstern*, and *Sonne*. This research was funded by the Deutsche Forschungsgemeinschaft (Sonderforschungsbereich 261 at Bremen University). Data are available under www.pangaea.de/Projects/SFB261.

References

Abrantes F (2000) 200 000 yr diatom records from Atlantic upwelling sites reveal maximum productivity during LGM and a shift in phytoplankton community structure at 185 000 yr. Earth Planet Sci Lett 176: 7-16

Archer DE (1996) An atlas of the distribution of calcium carbonate in sediments of the deep-sea. Glob Biogeochem Cycl 10: 159-174

Archer D, Emerson S, Reimers C (1989) Dissolution of calcite in deep-sea sediments: pH and O_2 microelec-trode results. Geochim Cosmochim Acta 53: 2831-2845

Arz HW, Pätzold J, Wefer G (1999) Climatic changes during the last deglaciation recorded in sediment cores from the northeastern Brazilian continental margin. Geo-Mar Lett 19:209-218

Bainbridge AE (1981) GEOSECS Atlantic Expedition, Hydrographic Data, 1972-1973. National Science Foundation, US Government Printing Office, Washington DC, pp 1-121

Barnola JM, Raynaud D, Korotkevich YS, Lorius C (1987) Vostok ice core provides 160 000-year record of atmospheric CO_2. Nature 329: 408-414

Baumann K-H, Lackschewitz KS, Erlenkeuser H, Henrich R, Jünger B (1993) Late Quaternary calcium carbonate sedimentation and terrigenous input along the east Greenland continental margin. Mar Geol 114: 13-36

Baumann K-H, Sprengel C (2000) Morphological variations of various coccolith species in a sediment trap north of the Canary Islands. J Nannoplankt Res 22: 185-193

Baumann K-H, Andruleit H, Samtleben C (2000) Cocco-lithophores in the Nordic Seas: Comparison of living communities with surface sediment assemblages. Deep-Sea Res II 47: 1743-1772

Bé AWH, Tolderlund DS (1971) Distribution and ecology of living planktonic foraminifera in the surface sediments of the Atlantic and Indian Oceans. In: Funnell B, Riedel WR (eds) Micropaleontology of the Oceans. Cambridge University Press, London, pp 105-149

Berger WH (1976) Biogenous deep-sea sediments: Production, preservation and interpretation. In: Riley JP, Chester R (eds) Chemical Oceanography. Academic Press, London, New York, San Francisco pp 265-389

Berger WH (1978) Deep-sea carbonate: Pteropod distribution and the aragonite compensation depth. Deep-Sea Res I 25: 447-452

Berger WH (1985) CO_2 increase and climate prediction: Clues from deep-sea carbonates. Episodes 8:163-168

Berger WH (1989) Global maps of ocean productivity. In: Berger WH, Smetacek VS, Wefer G (eds) Productivity of the Oceans: Present and Past. J Wiley & Sons, Chichester, pp 455-486

Berger WH, Fischer K, Lai C, Wu G (1987) Ocean productivity and organic carbon flux. Part I. Overview and maps of primary production and export production. University of California, San Diego, SIO Reference 87-30, pp 1-67

Berger WH, Smetacek VS, Wefer G (1989) Ocean productivity and paleoproductivity: An overview. In: Berger WH, Smetacek VS, Wefer G. (eds) Productivity of the Ocean: Present and past. J Wiley & Sons, Chichester, pp 1-34

Berner RA, Honjo S (1981) Pelagic sedimentation of aragonite: Its geochemical significance. Science 211: 940-942

Betzer PR, Byrbe RH, Acker JG, Lewis CS, Jolley RR,

Feely RA (1984) The oceanic carbonate system: A reassessment of biogenic controls. Science 226: 1074-1077

Brummer GJA, van Eijden AJM (1992) "Blue ocean" paleoproductivity estimates from pelagic carbonate mass accumulation rates. Mar Micropaleontol 19: 99-117

Broecker WS, Peng T-H (1989) The cause of the glacial to interglacial atmosheric CO_2 change: A polar alkalinity hypothesis. Glob Biogeochem Cycl 3: 215-239

Broerse ATC, Ziveri P, van Hinte JE, Honjo S (2000) Coccolithophore export production, species composition, and coccolith-$CaCO_3$ fluxes in the NE Atlantic (34°N 21°W and 48°N 21°W). Deep-Sea Res II 47: 1877-1905

Dittert N, Baumann K-H, Bickert T, Henrich R, Huber R, Kinkel H, Meggers H (1999) Carbonate dissolution in the deep-sea: Methods, quantification, and paleoceanographic application. In: Fischer G, Wefer G (eds) Use of Proxies in Paleoceanography: Examples from the South Atlantic. Springer, Berlin, pp 255-284

Fabry VJ (1990) Shell growth rates of pteropod and heteropod molluscs and aragonite production in the open ocean: Implications for the marine caronate system. J Mar Res 48: 209-222

Fabry VJ, Deuser WG (1991) Aragonite and magnesian calcite fluxes to the deep Sargasso Sea. Deep-Sea Res 38: 713-728

Fabry VJ, Deuser WG (1992) Seasonal changes in the isotopic compositions and sinking fluxes of euthecosomatous pteropod shells in the Sargasso Sea. Paleoceanography 7 (2): 195-213

Fischer G, Wefer G (1996) Long-term observations of particle fluxes in the eastern Atlantic: Seasonality, changes of flux with depth and comparison with the sediment record. In: Wefer G, Berger WH, Siedler G, Webb DJ (eds) The South Atlantic: Present and Past Circulation. Springer, Berlin, pp 325-344

Fischer G, Ratmeyer V, Wefer G (2000) Organic carbon fluxes in the Atlantic and the Southern Ocean: Relationship to primary production compiled from satellite radiometer data. Deep-Sea Res II 47: 1961-1997

Gerhardt S, Henrich R (2001) Shell preservation of *Limacina inflata* (Pteropoda) in surface sediments from the Central and South Atlantic Ocean: A new proxy to determine the aragonite saturation state of water masses. Deep-Sea Res I 48: 2051-2071

Giraudeau J, Bailey GW (1995) Spatial dynamics of coccolithophore communities during an upwelling event in the southern Benguela system. Cont Shelf Res 15: 1825-1852

Giraudeau J, Bailey GW, Pujol C (2000) A high-resolution time-series analses of particle fluxes in the northern Benguela coastal upwelling system: Carbonate record of changes in production and particle transfer processes. Deep-Sea Res II 47: 1999-2028

Hales B, Emerson S (1997) Calcite dissolution in sediments of the Ceara Rise: *In situ* measurements of porewater O_2, pH, and $C O_{2(aq)}$. Geochim Cosmochim Acta 61: 501-514

Hales B, Emerson S, Archer D (1994) Respiration and dissolution in the sediments of the western North Atlantic: Estimates from models of *in situ* microelec-trode measurements of porewater oxygen and pH. Deep-Sea Res I 41: 695-719

Hebbeln D, Henrich R, Baumann K-H (1998) Paleoceanography of the last interglacial/glacial cycle in the Polar North Atlantic. Quat Sci Rev 17: 125-153

Henrich R (1998) Dynamics of Atlantic water advection to the Norwegian-Greenland Sea - A time-slice record of carbonate distribution in the last 300 ky. Mar Geol 145: 95-131

Hodell DA, Charles CD, Sierro FJ (2001) Late Pleistocene evolution of the earth's carbonate system. Earth Planet Sci Lett 192: 109-124

Howard WR, Prell WL (1994) Late Quaternary $CaCO_3$ production and preservation in the Southern Ocean: Implications for oceanic and atmosheric carbon cycling. Paleoceanography 9: 453-482

Huber R, Meggers H, Baumann K-H, Henrich R (2000) Recent and Pleistocene carbonate dissolution in sediments of the Norwegian-Greenland Sea. Mar Geol 165: 123-136

Jouzel J, Barkov NI, Barnola JM, Bender M, Chappelaz J, Genthon C, Kotlyakov VM, Lipenkov V Lorius C, Petit JR, Raynaud D, Raisbeck G, Ritz C, Sowers T, Stievenard M, Yiou F, Yiou P (1993) Extending the Vostok ice-core record palaeoclimate to the penultimate glacial record. Nature 364: 407-412

Kalberer M, Fischer G, Pätzold J, Donner B, Segl M, Wefer G (1993) Seasonal sedimentation and stable isotope records of pteropods off Cap Blanc. Mar Geol 113: 305-320

Kellogg T.B (1976) Paleoclimatology and paleoceanography of the Norwegian and Greenland Seas: The last 450,000 years. Mar Micropaleontol 2: 235-249

Kleijne A, Kroon D, Zevenboom W (1989) Phytoplankton and foraminiferal frequencies in northern Indian Ocean and Red Sea surface waters. Netherlands J Sea Res 24: 531-539

Krefeld van SA, Knappertsbusch M, Ottens J, Ganssen GM, van Hinte JE (1996) Biogenic carbonate and ice-rafted debris (Heinrich layers) accumulation in deep-sea sediments from a northeast Atlantic piston core. Mar Geol 131: 21-46

Macdonald AM, Wunsch C (1996) An estimate of global ocean circulation and heat fluxes. Nature 382: 436-439

McIntyre A, Bé AWH (1967) Modern coccolithophoraceae of the Atlantic Ocean - I. Placoliths and Cyrtoliths. Deep-Sea Res I 14: 561-597

Meggers H, Freudenthal T, Nave S, Tragona J, Abranzes F, Helmke P (2002) Assessment of geochemical and micropaleontological sedimentary parameters as proxies of surface water properties in the Canary Islands region. Deep-Sea Res II 49:3631-3654

Milliman JD (1974) Marine Carbonates. Springer, New York, pp 1-363

Milliman JD (1993) Production and accumulation of calcium carbonate in the ocean: Budget of a non steady state. Glob Biogeochem Cycl 7: 927-957

Milliman JD, Troy PJ, Balch WM, Adams, AK, Li Y-H, Mackenzie FT (1999) Biologically mediated dissolution of calcium carbonate above the chemical lysocline? Deep-Sea Res I 46: 1653-1669

Monger B, McClain C, Murtugudde R (1997) Seasonal phytoplankton dynamics in the eastern tropical Atlantic. J Geophys Res 102: 12,389-12,411

Morse JW, Mucci A, Millero FJ (1980) The solubility of calcite and aragonite in seawater at various salinities, temperatures and 1 atmosphere total pressure. Geochim Cosmochim Acta 44: 85-94

Müller PJ, Suess E (1979) Productivity, sedimentation rate, and sedimentary organic matter in the oceans I. Organic carbon preservation. Deep-Sea Res I 26: 1347-1362

Peterson RG, Stramma L (1991) Upper-level circulation in the South Atlantic Ocean. Progr Oceanogr 26: 1-73

Romero O, Böckel B, Donner B, Lavik G, Fischer G, Wefer G (2002) Seasonal productivity dynamics in the pelagic central Benguela System inferred from the flux of carbonate and silicate organisms. J Mar Systems 37: 259-278

Ruddiman WF (1997) Tropical terrigenous fluxes since 25,000 yrs BP. Mar Geol 136: 189-207

Rühlemann C, Frank M, Hale W, Mangini A, Mulitza S, Müller PJ, Wefer G (1996) Late Quaternary productivity changes in the western equatorial Atlantic Evidence from 230^{Th}-normalized carbonate and organic accumulation rates. Mar Geol 135: 127-152

Rühlemann C, Müller PJ, Schneider RR (1999) Organic carbon and carbonate as paleoproductivity proxies: Examples from high and low productivity areas of the tropical Atlantic. In: Fischer G., Wefer G. (eds) Use of Proxies in Paleoceanography: Examples from the South Atlantic. Springer, Berlin, pp 315-344

Samtleben C, Schröder A (1992) Living coccolithophore communities in the Norwegian-Greenland Sea and their records in sediments. Mar Micropaleontol 19: 333-354

Samtleben C, Schäfer P, Andruleit H, Baumann A, Baumann K-H, Kohly A, Matthiessen J, Schröder-Ritzrau A (1995) Plankton in the Norwegian-Greenland Sea: From living communities to sediment assemblages - an actualistic approach. Geol Rundsch 84: 108-136

Sarnthein M, Pflaumann U, Ross R, Tiedemann R, Winn K (1992) Transfer functions to reconstruct ocean paleoproductivity: A comparison. In: Summerhayes CP, Prell WL, Emeis K-C (eds) Upwelling systems. Evolution since the early Miocene. Geol Soc Spec Publ 64: 411-427

Schneider RR, Müller PJ, Wefer G (1994) Late Quaternary paleoproductivity changes off the Congo deduced from stable carbon isotopes of planktonic foraminifera. Palaeogeogr Palaeoclimatol Palaeoecol 110: 255-274

Schneider RR, Price B, Müller PJ, Kroon D, Alexander I (1997) Monsoon related variations in Zaire (Congo) sediment load and influence of fluvial silicate supply on marine productivity in the east equatoial Atlantic during the last 200,000 years. Paleoceano-graphy 12: 463-481

Sprengel C, Baumann K-H, Neuer S (2000) Seasonal and interannual variation of coccolithophore fluxes and species composition in sediment traps north of Gran Canaria (29°N 15°W). Mar Micropaleontol 39: 157-178

Sprengel C, Baumann K-H, Hendericks J, Henrich R, Neuer S (2002) Modern coccolithophore and carbonate sedimentation along a productivity gradient in the Canary Islands region: Seasonal export production and surface accumulation rates. Deep-Sea Res II 49: 3577-3598

Tiedemann R, Sarnthein M, Stein R (1989) Climatic changes in the western Sahara: Aeolo-marine sediment record of the last 8 million years (sites 657-661). Proc ODP Sci Res 108: 241-277

Volbers ANA, Henrich R (2002) Present water mass calcium carbonate corrosiveness in the eastern South Atlantic inferred from ultrastructural breakdown of *Globigerina bulloides* in surface sediments. Mar Geol 186:471-486

Wefer G, Fischer G (1993) Seasonal patterns of vertical

particle flux in equatorial and coastal upwelling areas of the eastern Atlantic. Deep-Sea Res 40: 1613-1645

Wefer G, Berger WH, Siedler G, Webb DJ (1996) The South Atlantic: Present and Past Circulation. Springer, Berlin, 644 p

Young JR, Ziveri P (2000) Calculation of coccolith volume and its use in calbration of carbonate flux estimates. Deep-Sea Res II 47: 1679-1700

Ziveri P, Thunnell RC (2000) Coccolithophore export production in Guayamas Basin, Gulf of California: response to climate forcing. Deep-Sea Res II 47: 2073-2100

Zonneveld KAF, Versteegh GJM, de Lange GJ (2001) Palaeoproductivity and post-depositional aerobic organic matter decay reflected by dinoflagellate cyst assemblages of the eastern Mediterranean S1 sapropel. Mar Geol 172: 181-195

Coccolithophorid and Dinoflagellate Synecology in the South and Equatorial Atlantic: Improving the Paleoecological Significance of Phytoplanktonic Microfossils

A. Vink[*], K.-H. Baumann, B. Böckel, O. Esper, H. Kinkel, A. Volbers, H. Willems and K.A.F. Zonneveld

Universität Bremen, Fachbereich Geowissenschaften, Postfach 330 440, D-28334 Bremen, Germany
* corresponding author (e-mail): vink@micropal.uni-bremen.de

Abstract: Individual planktonic microfossil species, or assemblage groups of different species, are often used to, qualitatively and/or quantitatively, reconstruct past (sub)surface-water conditions of the world's oceans and seas. Until now, little information has been available on the surface sediment distribution patterns and paleoenvironmental reconstruction potential of coccolith, calcareous dinoflagellate cyst and organic-walled dinoflagellate cyst assemblages of the South and equatorial Atlantic, especially at the species level. This paper (i) summarizes the distributions of these three phytoplanktonic microfossil groups in numerous Atlantic surface sediments from 20°N–50°S and 30°E–65°W and determines their relationship with the physicochemical and trophic conditions of the overlying (sub)surface-waters, and (ii) determines the synecology of the three phytoplankton groups by carrying out statistical analyses (i.e. detrended and canonical correspondence analyses) on all groups simultaneously. Ecological relationships are additionally strengthened by statistically comparing the distribution patterns of the phytoplankton groups with those of planktonic foraminifera (Pflaumann et al. 1996; Niebler et al. 1998), as the ecological preferences of the latter are much better known. Many of the analyzed phytoplanktonic microfossil species or groups of species in the surface sediments do show restricted distributions which primarily reflect the environmental conditions of the upper water masses above them (e.g. sea-surface temperature, productivity, stratification). The acquired 'reference' data sets are large and diverse enough to allow future development of transfer functions for the reconstruction of past surface-water conditions, and show that there is still an enormous paleoenvironmental reconstruction potential concealed in many fossil coccolith and dinoflagellate cyst assemblages.

Introduction

Detailed quantitative and qualitative analyses of planktonic microfossil assemblages (e.g. foraminifera, diatoms, coccoliths, dinoflagellate cysts) have long been successfully applied in paleoenvironmental and -oceanographic reconstructions of the upper water column of the world's oceans. Such studies often focus on the analysis of one microfossil group only, in which the spatial and temporal fluctuations in abundance of particular assemblage groups, or even specific indicator species, provide important information on the changing environment and its possible causes. For example, seasonal but also long-term variations in thermocline and nutricline development and depth are shown by changes in the abundance of the coccolithophorid species *Florisphaera profunda* (Molfino and McIntyre 1990; Beaufort et al. 1997; Kinkel et al. 2000) or the calcareous dinoflagellate cyst (dinocyst)

From WEFER G, MULITZA S, RATMEYER V (eds), 2003, *The South Atlantic in the Late Quaternary: Reconstruction of Material Budgets and Current Systems.* Springer-Verlag Berlin Heidelberg New York Tokyo, pp 101-120

species *Calciodinellum albatrosianum* (Vink et al. 2002). Specific diatom associations can deliver important clues concerning position and intensity of both coastal and equatorial upwelling (Nelson et al. 1995), whereas the contrasting accumulation rates of organic-walled and calcareous dinocysts provide information on the state of eu-/oligotrophy and stratification in the upper water column (e.g. Höll et al. 1998; Vink et al. 2002). In recent decades, reliable algorithms for the calculation of paleotem-perature and paleoproductivity, derived from the species composition of planktonic foraminiferal associations in surface sediments, have been intensively developed and successfully applied in many oceanic areas (i.e. transfer function techniques [TFT: Imbrie and Kipp 1971; Mix et al. 1999]; modern analogue techniques [MAT: Prell 1985]). Similar transfer functions have recently been developed, although only regionally, for coccolith and organic-walled dinocyst associations, and appear just as promising (e.g. Giraudeau and Pujos 1990; de Vernal et al. 1993).

The general applicability and significance of species communities or individual species in sediments as proxy indicators for particular environmental conditions can, however, only be accurately determined when detailed knowledge of their exact autecology and synecology has been obtained. As numerous factors may influence the species associations between the times of production and deposition (e.g. lateral transport in the water column, [selective] dissolution/oxidation, redistribution at the sediment-water interface, bioturbation), the original production signal of most species is likely overprinted by other information once sedimentation occurs, leading to plankton–sediment mismatching. Any studies involving the traditional approach of comparing surface sediment distributions with the environmental conditions of the overlying (sub) surface-water masses to obtain paleoenvironmental information should thus involve as large a data set as possible, so that inconsistencies caused by any of the above-mentioned post-production processes are not weighted so heavily. Such larger data sets (e.g. for coccoliths and dinocysts) can consequently be used as the basis for developing transfer functions to quantitatively predict past (sub)surface-water conditions, although additional knowledge of the distributions of their living counterparts in the upper water column is still required to exactly pinpoint their preferred depth habitat. Furthermore, most paleoenvironmental reconstructions tend to focus on the changing temporal patterns in the assemblage of one microfossil group only, thus often leading to the publication of different and thus ambiguous paleoenvironmental scenarios when microfossil groups with totally different life cycles and habitats are investigated in the same core. Only few studies mainly from the North Atlantic have made first attempts at integrating the information from different plankton groups from surface-waters, sediment traps and surface sediments (e.g. Samtleben et al. 1995; Matthiessen et al. 2001; Schröder-Ritzrau et al. 2001). Instead of publishing contradictory results obtained from different microfossil groups, we should rather focus on using the information from one group for which trustworthy ecological data are already available to "calibrate" the other. Such calibration studies should occur through the spatial correlation of species distribution patterns, for which no or little environmental information is currently available, with those of other organisms for which the ecological preferences are much better known.

Here, the distribution patterns of individual species or assemblage groups of three phytoplanktonic microfossil groups (coccoliths, calcareous dinocysts and organic-walled dinocysts) are described throughout the South and equatorial Atlantic Ocean from 20°N–50°S and 30°E–65°W and their distributions are compared with the main physicochemical and trophic conditions of the overlying (sub) surface-waters, in order to create calibration/reference data sets which can potentially be used as input for the development of paleoenvironmental transfer functions at a later stage. Parts of the data sets used in this study have already been published elsewhere, i.e. coccoliths in Baumann et al. (1999), Shokati et al. (1999) and Kinkel et al. (2000); calcareous dinocysts in Vink et al. (2000a) and Zonne-veld et al. (2000); and organic-walled dinocysts in Vink et al. (2000b), Zonneveld et al. (2001a) and Esper and Zonneveld (2002). Furthermore, we have made a first attempt at determining the synecology of these three phytoplankton groups by carrying out detrended correspondence analyses (DCA) and canonical correspondence analyses (CCA) on reduced sample sets for which count data of all microfossil groups were available. Ecological relationships have been additionally strengthened by statistically comparing the distribution patterns of the phytoplankton groups with those of planktonic foraminifera (based on published data sets of Pflaumann et al. 1996 and Niebler and Gersonde 1998), as more detailed ecological information is generally available for this latter group. Despite offering a novel approach to the use of micropaleontological proxies, such methods have only rarely been carried out on different fossil groups so far (e.g. by Vénec-Peyré et al. 1997). They do, however, allow a comprehensive ecological calibration of species from different microfossil groups to each other and the modern (sub)surface-water conditions, and should contribute to a better understanding of the planktonic ecological community as a whole.

Current Systems and Hydrogeography of the South and Equatorial Atlantic

Upper-level circulation in the South and equatorial Atlantic Ocean is dominated by a system of gyres and by the equatorial and circumpolar current systems, as summarized in Fig. 1. The circulation system has often been described in great detail elsewhere (e.g. Peterson and Stramma 1991; Wefer et al. 1996). Therefore, only a summary of the main currents and their corresponding surface-water conditions is provided here.

Eastern boundary currents of the South Atlantic are dominated by the presence of the northward flowing Benguela Coastal Current (BCC), which forms the eastern part of the anticyclonic South Atlantic Subtropical Gyre. The BCC has a relatively low water temperature and high nutrient content, which is maintained by the effect of year-round trade wind action and pronounced coastal upwelling along the western coast of southern Africa (Namibia). The BCC is deflected westward at ca. 20°S and feeds into the broad and relatively uniform South Equatorial Current (SEC). The SEC flows westwards across the South Atlantic, steadily increasing its temperature and salinity until it bifurcates into the North Brazil Current (NBC) and

the southward flowing Brazil Current (BC) at the eastern promontory of South America. The BC thus transports relatively warm and saline surface-water back southwards along the eastern coast of South America, until it meets up with the cold and relatively fresh, northward-flowing Falkland Current (FC) at ca. 40°S and is deflected eastwards into the South Atlantic Current (SAC). The Subtropical Front distinguishes the relatively warm SAC from the cold, nutrient-rich and relatively fresh Antarctic Circumpolar Current (ACC) to the south.

In contrast, the equatorial Atlantic Ocean is a region characterized by generally warm surface-waters and a well developed thermocline. Currents are largely driven by the trade winds. The SEC and the NBC are at their strongest in the austral winter, when the SE trade winds are at their strongest. The strong surge of the trade winds and the rapid northward movement of the Intertropical Convergence Zone (ITCZ) leads to an exceptionally strong positive curl of the wind stress off northern Brazil, and the formation of the North Equatorial Countercurrent (NECC) during this season. Increased precipitation and fluvial runoff reduce salinities in the Amazon and Orinoco Fan areas during this season, and the NECC is fed with river-derived nutrients. Concurrently, the increased speed of the SEC leads to enhanced equatorial divergence and hence enhanced upwelling and productivity in the eastern equatorial Atlantic. During the austral summer, on the other hand, the ITCZ shifts southward and the SE trade winds are typically less strong, leading to a weakening, or even vanishing, of the NECC. Most of the NBC waters are transported through the Guyana Current (GUC) and the Caribbean directly into the North Atlantic. In addition, the weaker SEC leads to a reduced equatorial divergence. To the north of the equator, the NE trade winds and, consequently, the North Equatorial Current (NEC) are stronger, the latter transporting nutrient-poor, saline, cooler waters derived from the Canary Current into the tropics.

The NECC continues eastwards into the Guinea Current (GC). The combined effect of fluctuations in the intensity of the trade winds, and thus also in the intensity of the NECC and GC, and the strength of the West African SW-NE monsoon system, produces maximum seasonal variability in current sys-

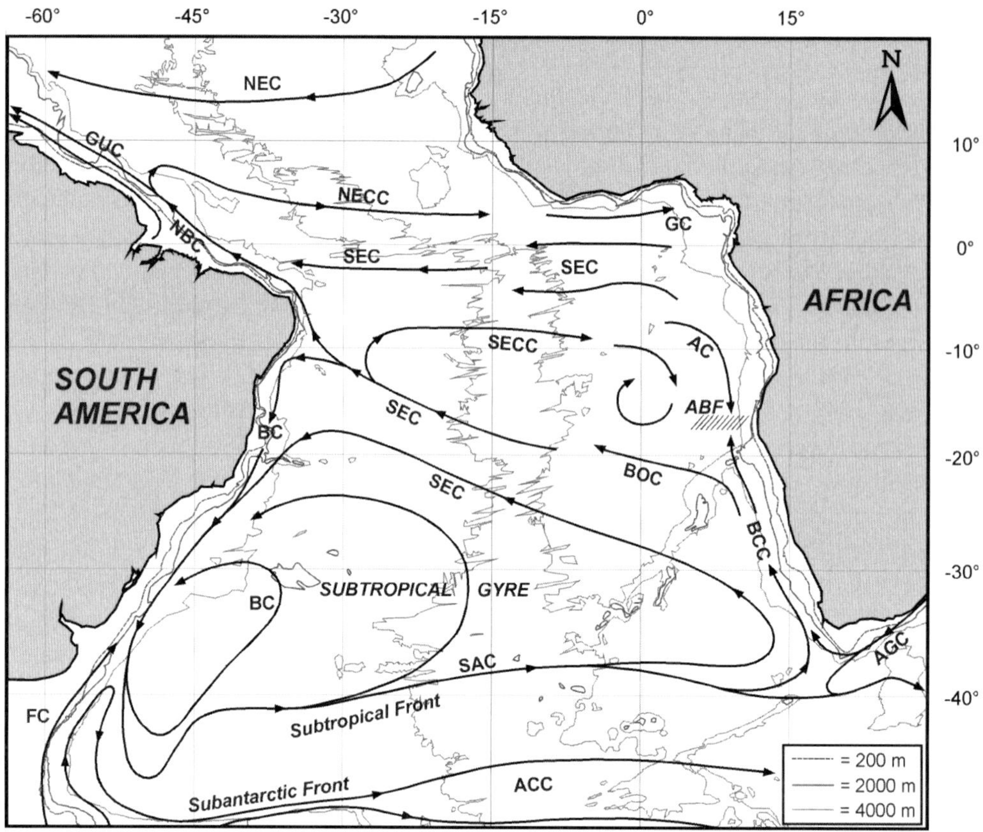

Fig. 1. Simplified sketch of the bathymetry and sea-surface currents of the South and equatorial Atlantic Ocean (adapted from Peterson and Stramma 1991; Schneider et al. 1996; Niebler and Gersonde 1998). NEC = North Equatorial Current; NECC = North Equatorial Countercurrent; SEC = South Equatorial Current; SECC = South Equatorial Countercurrent; BC = Brazil Current; NBC = North Brazil Current; GUC = Guyana Current; GC = Guinea Current; AC = Angola Current; BCC = Benguela Coastal Current; BOC = Benguela Oceanic Current; FC = Falkland Current; SAC = South Atlantic Current; ACC = Antarctic Circumpolar Current; AGC = Agulhas Current. *ABF* = Angola–Benguela Front.

tems and surface-water conditions in the Gulf of Guinea. In addition, fluvial runoff in the Gulf, notably through the Niger River, is responsible for annually low surface-water salinity and nutrient content in this area. In the eastern sector of the South Atlantic, below the Gulf, circulation is characterized by a cyclonic gyre circulation which includes the Benguela Oceanic Current (BOC), the South Equatorial Countercurrent (SECC) and the Angola Current (AC). Seasonal shoaling of nutrient-rich, shallow, subsurface-waters in the Angola gyre leads to an enhanced oceanic productivity. Somewhat south of the gyre, the convergence of warm equatorial AC waters with northward flowing, colder subtropical waters of the BCC at ca. 15°S produces the Angola-Benguela Front (ABF). The confrontation also brings nutrient-rich, shallow, subsurface-water into the photic zone.

Material and Methods

Material

The surface sediments of the South and equatorial Atlantic used for this study cover a broad area from 20°N–50°S and 30°E–65°W, and were obtained from the upper centimeter of multiple multicores and boxcores and, more rarely, by means of Van Veen grabbers, during expeditions of the RV *Meteor*, *Victor Hensen* and *Polarstern*. The data sets include 146 samples analyzed for coccoliths (Fig. 2a), 167 samples for calcareous dinocysts (Fig. 3a) and 96 samples for organic-walled dinocysts (Fig. 4a). These samples represent oceanographic environments varying from neritic (30 m water depth) to fully oceanic (5500 m), from cold (3°C) to warm (28°C) and from eutrophic to oligotrophic. Unfor-

tunately, only few of these surface sediments have been dated, neither do we know the exact sedimentation rates at most sites. However, many gravity- and multicores at, or in the vicinity of, the sample positions do exhibit relatively continuous late Quaternary sedimentation rates, in which the top sediments are at least allocated to the Holocene (based on the oxygen isotope stratigraphy, the occurrence of the interglacial foraminifer *Globorotalia menardii*, or AMS ^{14}C dates). Here, we have to make the basic assumption that all samples are of

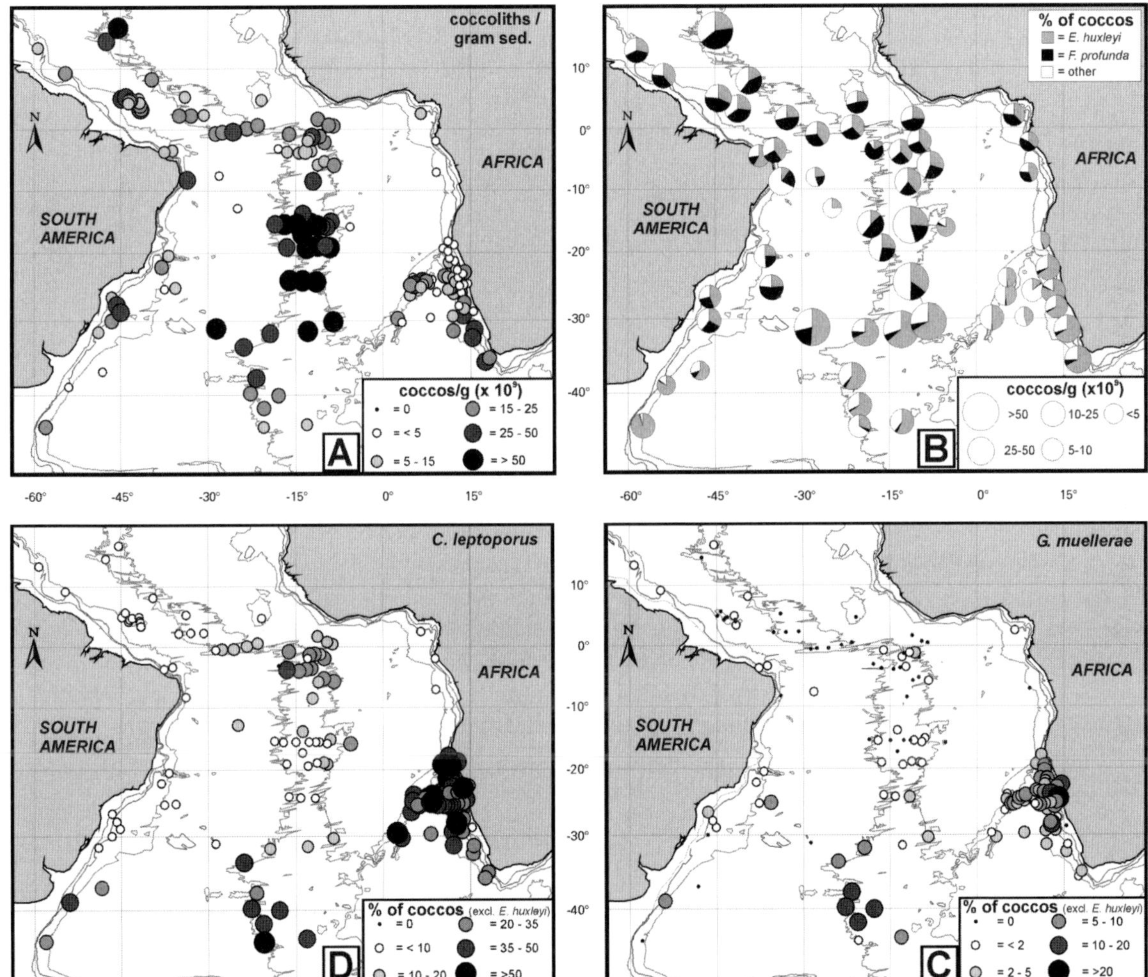

Fig. 2. Surface sediment distributions of selected coccolith species. **a)** Total number of coccoliths per gram of dry sediment; **b)** Pie charts showing the relative abundances (in % of total coccolith sum) of *Emiliania huxleyi* and *Florisphaera profunda* in selected, representative samples; **c)** Relative abundances of *Gephyrocapsa muellerae* (in % of total coccolith sum excluding the dominant species *E. huxleyi*); **d)** Relative abundances of *Calcidiscus leptoporus* (also in % of total coccolith sum excluding *E. huxleyi*).

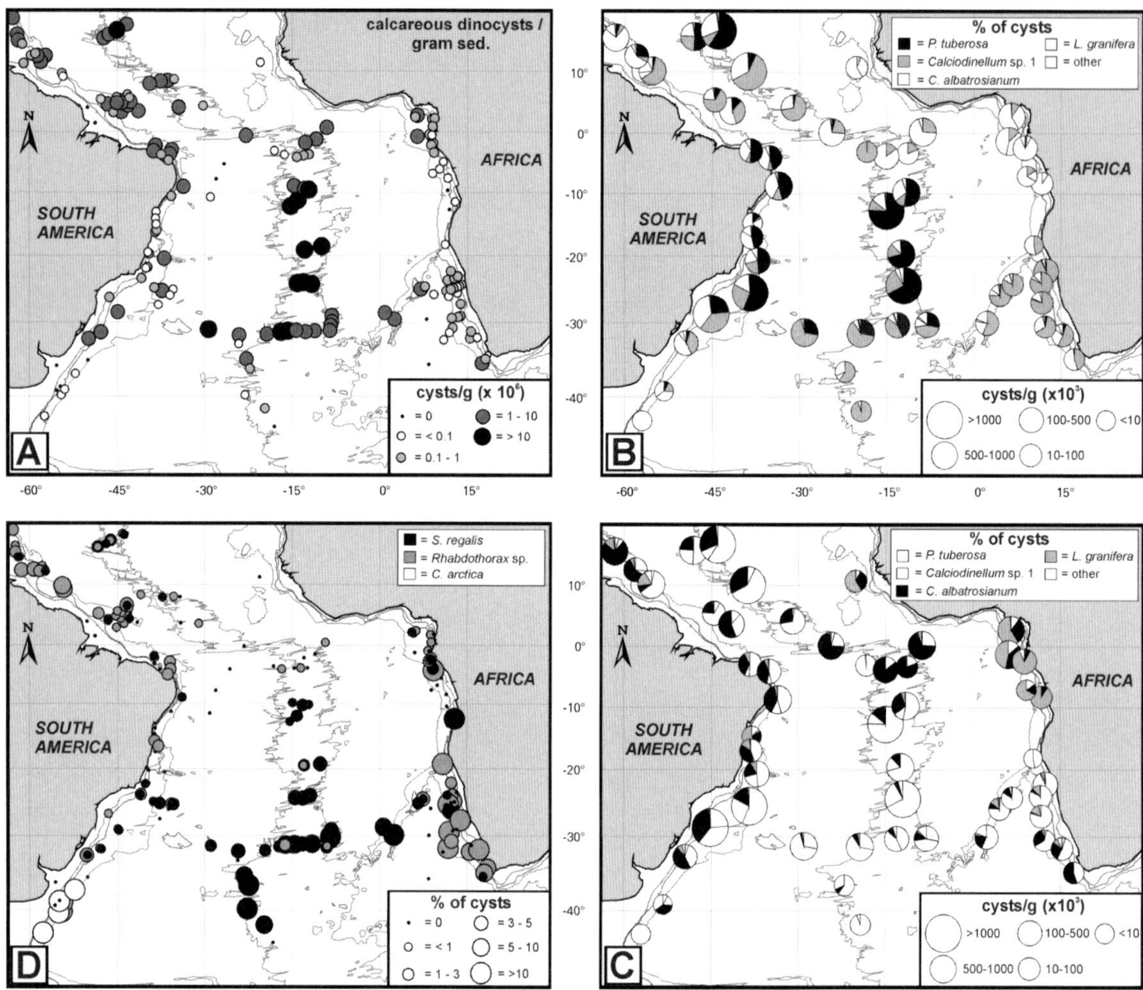

Fig. 3. Surface sediment distributions of calcareous dinocyst species. **a)** Total number of specimens (including the dominant species *Thoracosphaera heimii*) per gram of dry sediment; **b)** Pie charts showing the relative abundances (in % of total cyst sum excluding *T. heimii*) of *Pernambugia tuberosa* and *Calciodinellum* sp. 1 in selected, representative samples; **c)** Relative abundances of *Calciodinellum albatrosianum* and *Leonella granifera* (in % of total cyst sum excluding *T. heimii*); **d)** Relative abundances of the less common cyst species *Scrippsiella regalis*, *Rhabdothorax* sp. and *Caracomia arctica* (also in % of total cyst sum excluding *T. heimii*).

Holocene age (if not modern), and be aware of the limitations of the available data. Due to the different sedimentation rates and possible bioturbation processes, samples may contain several hundreds to several thousands of years of sediment. However, considering that climatic and oceanographic conditions during the late Holocene have remained relatively similar to those predominating today, we assume that the samples represent present-day plankton associations with an acceptable degree of accuracy.

Data Sources and Laboratory Preparation

The coccolith distributions in 38 surface sediment samples from the eastern South Atlantic used in this study have been published before by Baumann et al. (1999) and Shokati et al. (1999), and those in 27 western equatorial Atlantic samples were described by Kinkel et al. (2000). The data of the remaining 81 samples shown in Fig. 2 have not been published before. All samples were identically prepared according to the combined dilution/

filtering technique described in detail by Andruleit (1996) and Baumann et al. (1999). For species identification, the taxonomy of Jordan and Kleijne (1994) was applied.

Calcareous dinocyst distributions in 36 surface sediment samples derived mainly from the eastern South Atlantic were previously published by Zonne-veld et al. (2000), and those in 37 samples from the western equatorial Atlantic were described by Vink et al. (2000a). The data of the remaining 94 samples shown in Fig. 3 are published for the

first time. Two different preparation methods have been used. Zonneveld et al. (2000) used a very simple preparation method in which the diluted sediment was directly brought onto a microscopic slide and counted. This method was replaced by a more detailed preparation in which the sediment was sieved into two separate fractions (5–20 μm and 20–75 μm) in order to concentrate the generally abundant and small-sized coccoid species *Thoracosphaera heimii* into one fraction and all the larger cysts into the other (Vink et al. 2000a). Although

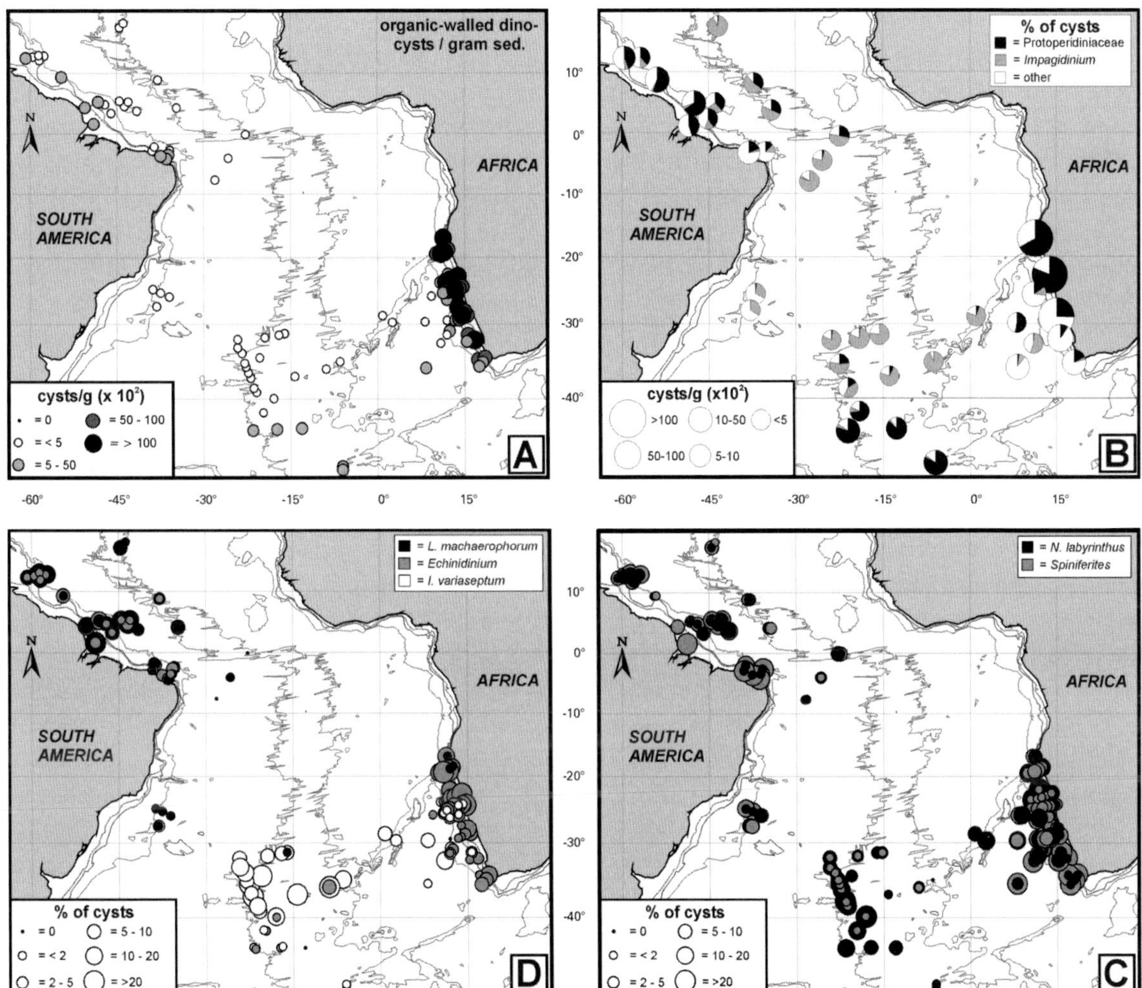

Fig. 4. Surface sediment distributions of organic-walled dinocyst species or groups of species. **a)** Total number of cysts per gram of dry sediment; **b)** Pie charts showing the relative abundances (in % of total cyst sum) of all protoperidiniaceaen species and all species belonging to the genus *Impagidinium* in selected, representative samples; **c)** Relative abundances of the species *Nematosphaeropsis labyrinthus* and all species belonging to the genus *Spiniferites*; **d)** Relative abundances of *Lingulodinium machaerophorum*, *Impagidinium variaseptum* and all species belonging to the genus *Echinidinium*.

absolute cyst concentrations are consistently higher using the sieving method, samples subjected to both methods were comparable in their relative cyst abundances and displayed similar trends in distribution (Esper et al. 2000; A. Vink, unpubl. data). For this reason, we used the sieving method for samples where-ever possible, but have added remaining samples from Zonneveld et al. (2000) when doubles were not counted. Taxonomy conforms to that listed in Vink et al. (2002).

With the exception of 7 surface sediment samples, all samples used for organic-walled dinocyst analysis in this study (Fig. 4) have been published elsewhere. The distributions in 41 samples derived from the eastern South Atlantic were described by Zonneveld et al. (2001a), those in 27 samples from the western equatorial Atlantic by Vink et al. (2000b), and those in 21 samples from the Subtropical Gyre and Subantarctic Front by Esper and Zonneveld (2002). Sample preparation and taxonomy were carried out as described by Vink et al. (2000b) and Esper and Zonneveld (2002).

Phytoplankton counts of all samples used in this study are available from the PANGAEA data base; http://www.pangaea.de/home/avink.

Statistical Methods

Geographic distribution maps were produced of the most common and/or ecologically significant species, or groups of species, of each of the three phytoplankton groups analyzed in the surface sediments recovered from the investigated area (Figs. 2–4). Relative instead of absolute abundances were used, because of the greatly fluctuating sedimentation rates in the studied regions, varying from <1 cm/ka over the mid-Atlantic Ridge to >20 cm/ka in the Benguela upwelling region and the Congo Fan. Problems associated with relative abundances and the so-called "closed-sum effect", i.e. many of the dominating species have widespread geographic ranges and tend to suppress the patterns shown by less abundant species, are assumed to be smaller than the artifacts which would arise in the absolute data because of variable sedimentary dilution. Counts of the cosmopolitan and eurythermal species *Thoracosphaera heimii* and *Emiliania huxleyi* have been removed from the percentage abundance calculations of calcareous dinocysts and coccoliths, respectively, in order to resolve difficulties arising from the underestimation of abundances of less common species.

In order to determine which species, or groups of species, of the different plankton groups show similar spatial distributions, and how these may be related to specific environmental gradients in the upper water column, detrended correspondence analyses (DCA) and canonical correspondence analyses (CCA) have been applied on (i) a data set of 33 samples for which relative abundances of species of all three phytoplankton groups are available, and (ii) a data set of 28 samples in which three calcareous plankton groups (i.e. planktonic foraminifera, coccoliths, calcareous dinocysts) have been examined. A DCA involving all four plankton groups was impossible as there are only a few common samples which have been analyzed for both planktonic foraminifera and organic-walled dinocysts. Samples located at depths >4700 m were excluded from the analyses as the calcareous associations have most likely been altered by dissolution. All analyses were carried out using the program CANOCO (**CANO**nical **C**ommunity **O**rdination: version 4.0 for Windows, authors C.J.F. Ter Braak and P. Šmilauer 1998, Wageningen, The Netherlands). DCA is a form of indirect gradient analysis in which species are arranged along axes in such a way that the first axis (or *x*-axis) represents an unknown, theoretical environmental gradient (or a combination of gradients) which causes the largest variation in species composition, and the second axis (or *y*-axis) represents the second-most important direction of variation. We ignore higher numbered DCA axes that explain only a small proportion of variance in the species data. Species which are plotted close to each other in such an analysis are likely to occur in the same set of samples and thus have similar distribution patterns, whereas species plotted far apart from each other tend towards different distribution patterns (Figs. 5 and 6). The perpendicular projection of a species point to an axis gives the position of the abundance optimum of that species on the axis. The length of the ordination axis is expressed in standard deviations. DCA and CCA are based on the assumption that species show a unimodal response in relation

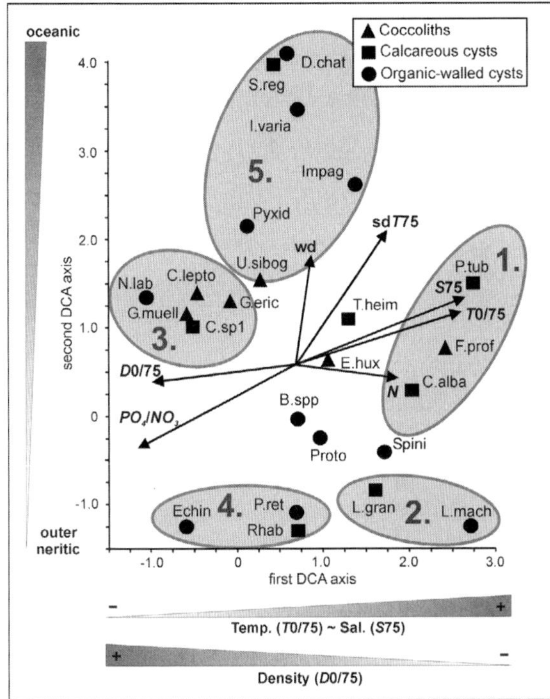

Fig. 5. DCA plot showing the variation within the species distribution of the three phytoplankton groups combined. Only the directions of significant environmental variables (variance $\geq$ 6%) have been added. See Table 1 for the explanation of variables. Five synecological phytoplankton groups are marked (their characteristics are summarized in Table 2 and are described in the discussion). Coccoliths: E. hux = *Emiliania huxleyi*; F. prof = *Florisphaera profunda*; G. muell = *Gephyrocapsa muellerae*; G. eric = *Gephyrocapsa ericsonii*; C. lepto = *Calcidiscus leptoporus*; U. sibog = *Umbilicosphaera sibogae*. Calcareous dinocysts: T. heim = *Thoracosphaera heimii*; L. gran = *Leonella granifera*; C. sp1 = *Calciodinellum* sp. 1; C. alba = *Calciodinellum albatrosianum*; P. tub = *Pernambugia tuberosa*; Rhab = *Rhabdothorax* sp.; S. reg = *Scrippsiella regalis*. Organic-walled dinocysts: D. chat = *Dalella chathamense*; Impag = all species of *Impagidinium*; I. varia = *Impagidinium variaseptum*; Pyxid = *Pyxidinopsis reticulata*; N. lab = *Nematosphaeropsis labyrinthus*; B.spp = *Brigantedinium* spp.; Echin = all species of *Echinidinium*; Proto = all protoperidiniaceans excluding species of *Brigantedinium* and *Echinidinium*; Spini = all species of *Spiniferites*; P. ret = cysts of *Protoceratium reticulatum*; L. mach = *Lingulodinium machaerophorum*.

to changing environmental gradients, which has been tested and confirmed for both data sets. To test which environmental variables may possibly play an important role in determining the nature of the species distribution, the species arrangements can be related to known environmental variables of the upper water column (e.g. temperature, salinity, nutrient concentration) through the linear regression of these variables onto the DCA-axes (e.g. Fig. 5). The direction and the length of the line representing a variable are indicative of its importance in determining the species composition: a long line running almost parallel to one of the two DCA axes is most likely to significantly influence the species distribution. Species that are plotted close to the center of a variable are either insensitive to that variable, or have their abundance optimum near the center of the gradient.

The species arrangements can be directly related to known environmental gradients in the overlying surface-waters by the use of CCA (i.e. direct gradient analysis). In this case the ordination axes are linear aggregates of those environmental variables which best explain the variation in the species data. Variables explaining <5% of the variation in the data are not significant at the 99% confidence interval (*P*-values >0.01). The environmental data which were digitally available for our statistical analyses were sample water depth (wd), mean annual temperatures (T), salinities (S), densities (D), phosphate (PO_4) and nitrate (NO_3) concentrations, and the standard deviations of monthly temperatures (sd$T$: hereafter referred to as seasonality), at 0 and 75 m water depth. Seasonal data were generally not used due to the great amount of covariation which arose between seasons. Values of the above-mentioned parameters were obtained for the last 92 years for one degree latitude and longitude square blocks from the World Ocean Atlas 1994 Data Set, National Oceanographic Data Center, Washington D.C. (see http://ingrid.Idgo.columbia.edu/SOURCES/.LEVITUS94/). Furthermore, Brunt-Väisälä frequencies (N: in rad/s) were calculated from D as a measure of the stratification of the upper water column between 0 and 75 m water depth:

$$N = \sqrt{(9.8 * \delta D)/(1026 * \delta z)}$$

where δD represents the density difference over the distance δz (in this case 75 m). Higher values of N represent more stratified water conditions between 0 and 75 m water depth.

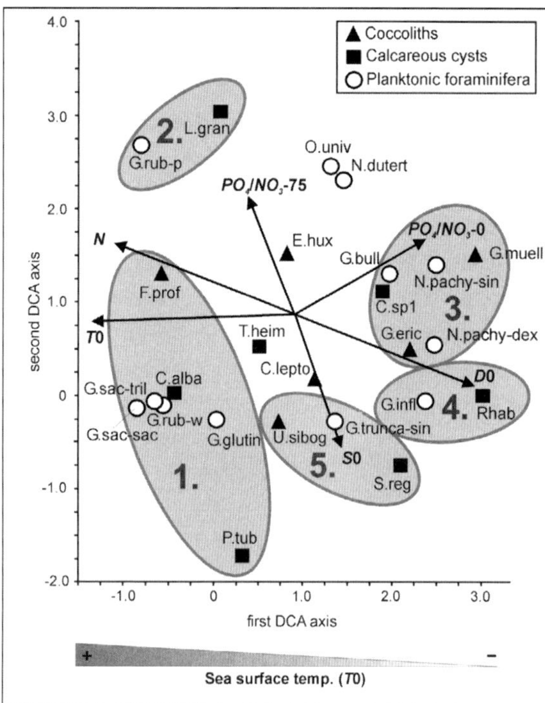

Fig. 6. DCA plot showing the variation within the species distribution of the three calcareous plankton groups combined. Only the directions of significant environmental variables (variance $\geq 6\%$) have been added, of which SST is the most significant. See Table 1 for the explanation of variables. The same five synecological groups as in Fig. 5 are marked. Abbreviations of coccolith and calcareous dinocyst species as in Fig. 5. Planktonic foraminifera: G.rub-p = *Globigerinoides ruber* pink; G.rub-w = *Globigerinoides ruber* white; O.univ = *Orbulina universa*; N.dutert = *Neogloboquadrina dutertrei*; N.pachy-sin = *Neogloboquadrina pachyderma* sinistral; N.pachy-dex = *Neogloboquadrina pachyderma* dextral; G.bull = *Globigerina bulloides*; G.infl = *Globorotalia inflata*; G.trunca-sin = *Globorotalia truncatulinoides* sinistral; G.glutin = *Globigerinita glutinata*; G.sac-sac = *Globigerinoides sacculifer sacculifer*; G.sac-tril = *Globigerinoides sacculifer trilobus*.

Modern Phytoplanktonic Microfossil Assemblages of the South and Equatorial Atlantic and their Paleoenvironmental Applications

Surface Sediment Distributions of Phytoplanktonic Species: Main Trends

Most species of each of the three phytoplanktonic microfossil groups analyzed show distinct distribution patterns which are clearly related to the environment, regardless whether their occurrence is widespread throughout the South and equatorial Atlantic or restricted to more regional or even local areas within the complex hydrographic system (Figs. 2–4). The species' composition in surface sediments does not, of course, depend on the physicochemical status of the photic layer only, but also on a number of abiotic factors such as lateral transport by ocean currents, resuspension, dissolution processes, aerobic decay, or bioturbation. Additionally, differences in distribution between the micro-fossil groups are caused by the variable sample coverage of the different groups, a factor which must be considered when making inter-group comparisons. Despite all of these limitations which can clearly complicate interpretations, we believe that the size of the collected data sets should at least filter out the erroneous samples and reflect the main, large-scale environmental trends of the species under consideration.

As coccoliths are the most important calcitic primary producers in the world's oceans, it is not surprising that their absolute abundances (Fig. 2a) greatly exceed those of calcareous dinocysts (Fig. 3a) in most areas, a fact which also manifests itself in the quantitatively important vs. negligible contributions of these two phytoplankton groups to the sedimentary carbonate, respectively (see Bau-mann et al. this volume). We believe that these markedly contrasting sedimentary concentrations are not due to the calcareous dinocysts' greater sensitivity to dissolution, as we would assume similar, perhaps even stronger, dissolution effects in coccoliths due to their comparable calcite chemistry albeit much smaller size. Instead, the primary production of coccolithophorids appears to be several dimensions higher in most oceanic areas as compared to calcareous dinocysts. Figure 2a also shows that coccoliths tend to occur in high concentrations in surface sediments encompassing a broad geographic range, mostly because of the capability of selected coccolithophorid species to thrive in cold and eutrophic areas, in contrast to most calcareous dinocyst species (Fig. 3). In the statistical analyses (Figs. 5 and 6), the six most prominent coccolithophorid species of the South and equatorial Atlantic are plotted far apart on the first axis, but close together along the second axis, suggesting that not all environmental parameters used in the analyses equally influence the distributions of the species. *Emiliania huxleyi* is the dominant species in most of the surface samples, often forming more that 50% of the association (Fig. 2b). Not considering *Emiliania huxleyi*, *Florisphaera profunda* is the most common species in warm, oligo-trophic-to-mesotrophic equatorial regions, whereas *Calcidiscus leptoporus* and *Gephyrocapsa muellerae* dominate in cool, eutrophic, southern South Atlantic regions (Fig. 2).

Generally speaking, calcareous dinocysts show highest concentrations in the sediments underlying the relatively warm, oligotrophic-to-mesotrophic, oceanic surface-waters north of ca. 40°S (Fig. 3a). The distributions of the individual species do, however, show striking differences which are reflected in Fig. 3 as well as by the fact that they are plotted far apart from each other in the statistical analyses (Figs. 5 and 6). Disregarding *Thoracosphaera heimii*, whose different reproductive strategy and higher annual reproduction rates do not allow direct quantitative comparison with dinocyst production (e.g. Tangen et al. 1982), *Calciodinellum albatrosianum* and *Calciodinellum* sp. 1 (previously called? *Sphaerodinella tuberosa* var. 2: e.g. Vink et al. 2000a; Zonneveld et al. 2000) are by far the most common and widely distributed species in the South and equatorial Atlantic (Fig. 3). The distribution of *Calciodinellum albatrosianum* shows a clear relationship with warm, oligotrophic-to-mesotrophic surface-waters, whereas the distribution of *Calciodinellum* sp. 1 appears to be genuinely cosmopolitan. *Pernambugia tuberosa* may form up to 75% of the association in nutrient-poor regions.

In contrast to the calcareous dinocysts, organic-walled dinocysts are produced in relatively low concentrations throughout the South and equatorial Atlantic and appear in highest concentrations in more shelfward, nutrient-rich areas such as the Benguela upwelling area (Fig. 4a). As such, the recent distribution patterns of these two dinocyst groups show little overlap, illustrating that most of the dinoflagellate species producing organic-walled cysts occupy different surface-water habitats as compared to those producing calcareous cysts. Considering that over 40 species of organic-walled dinocysts are found in the South and equatorial Atlantic, species of the same genera and/or with similar ecological affinities have been grouped together in the statistical analyses and in Fig. 4 for reasons of better clarity. The geographic distributions of these groups of species, but also of individual species, vary significantly, as is shown in Fig. 4 as well as by the fact that they are plotted far apart from each other in the statistical analysis (Fig. 5). The main group of heterotrophic species belonging to the family of Protoperidiniaceae generally dominates the assemblages in high productivity regions, whereas the phototrophic species belonging to the genus *Impagidinium* dominate, despite their very low absolute concentrations, in the oligotrophic, oceanic regions (Fig. 4b).

Statistical Grouping of Planktonic Microfossil Species: Inter-Group Ecological Calibration and Relationships with Surface-water Conditions

DCA and CCA carried out on the data set of all three phytoplankton groups (Fig. 5) shows that $T0/75$, $D0/75$ and $S75$ are the most significant environmental variables which may determine species variation along the first axis (accounting for 21.3% of the total variation in the data set), whereas the second axis (accounting for 14.1% of variation) is interpreted to reflect an outer neritic–oceanic trend in the data set (wd). Explanations of the abbreviated environmental variables are given in Table 1. PO_4 and NO_3 at 0 and 75 m water depth show so much covariation that they have been taken as a single parameter, which turns out to be significant in determining the species'

Variable	Abbreviation	Phytoplankton Groups (Fig. 5)	Calcareous Plankton Groups (Fig. 6)
SST (°C)	$T0$	15.96 – 27.58	16.63 – 27.62
temperature at 75 m (°C)	$T75$	13.48 – 26.23	14.14 – 21.12
seasonality in monthly SST	$sdT0$	0.44 – 2.80	1.28 – 2.28
seasonality in monthly temp. at 75 m	$sdT75$	0.13 – 0.89	0.7 – 1.27
SSS (p.s.u.)	$S0$	34.29 – 36.64	35.25 – 36.59
salinity at 75 m (p.s.u.)	$S75$	35.12 – 36.64	35.17 – 36.49
sea surface density (g/cm^3 % 1000)	$D0$	21.97 – 25.66	19.92 – 25.58
density at 75 m (g/cm^3 % 1000)	$D75$	24.34 – 26.57	25.82 – 26.51
stratification between 0 and 75 m (rad/s)	N	0.0109 – 0.0185	0.0096 – 0.0286
sea surface phosphate conc. (µmolar)	$PO_4\text{-}0$	0.109 – 0.648	0.090 – 0.593
phosphate conc. at 75 m (µmolar)	$PO_4\text{-}75$	0.147 – 1.222	0.143 – 1.183
sea surface nitrate conc. (µmolar)	$NO_3\text{-}0$	0.166 – 5.441	0.166 – 4.975
nitrate conc. at 75 m (µmolar)	$NO_3\text{-}75$	0.392 – 15.333	0.392 – 18.360
sample water depth (m)	**wd**	773 – 4675	1004 – 5675

Table 1. Minimum and maximum mean annual values of environmental variables used in the respective statistical analyses.

distributions in the DCA, as do N and $sdT75$. All other parameters appear to be insignificant (i.e. fall below the 5% significance level in the CCA). Species plotted close to the centroid of the environmental parameters are most likely unaffected by these parameters (e.g. *Emi-liania huxleyi* and *Brigantedinium* spp. in Fig. 5), whereas species which are plotted far away from the centroid and in the direction of change of a significant environmental variable are most likely positively related to that variable. Similarly, species of different microfossil groups plotted close to each other show the same geographic, and probably environmental, preferences.

DCA and CCA carried out with the data set of the three calcareous plankton groups (Fig. 6) reflect a slightly different pattern of species' variation. The first axis accounts for 33.1% of the total variation in the data set and likely represents a SST ($T0$) gradient, whereas the second axis, accounting for 13.6% of the variation, may represent a combination of $PO_4/NO_3\text{-}75$ gradients and SSS ($S0$). The other environmental variables which relate significantly to the species' distributions are $D0$, $PO_4/NO_3\text{-}0$ and N.

The environmental preferences and inter-group relationships of many of the microfossil species can be approximately estimated by combining information about the species' ecology gathered from (i) the literature (where available), (ii) the surface sediment distribution patterns (Figs. 2–4), and (iii) the DCA and CCA analyses (Figs. 5 and 6). We make the basic assumption that groups of species with similar geographic distributions probably react to the same environmental gradients in the upper water column, and may thus exhibit comparable ecological preferences. Clearly, such assumptions should be handled with caution as they tend to oversimplify complex ecosystem structures: factors such as variable depth habitats within the upper water column, or the significance of different parts of the life cycles and/or symbionts for survival and production, are largely ignored. Nevertheless, more general environmental conditions promoting a high abundance of species can be determined. Five synecological groups of species with similar relationships to environmental conditions became apparent (Figs. 5 and 6; Table 2):

Group 1. – Warm, stratified, nutrient-poor environments: In the phytoplankton DCA (Fig. 5), the two calcareous dinocyst species *Calciodi-nellum albatrosianum* (Fig. 3c) and *Pernambugia tuberosa* (Fig. 3b), and the coccolithophorid species *Florisphaera profunda* (Fig. 2b) are plotted close

Group Nr.	Species / Species Group	wd	T0/75	sdT75	S0/75	D0/75	PO_4/NO_3	N
1.	*Calciodinellum albatrosianum*		+		+	−	−	+
	Florisphaera profunda		+	+	+	−	−	+
	Pernambugia tuberosa		+	+	+	−	−	+
	Globigerinoides sacculifer sacculifer		+				−	+
	Globigerinoides sacculifer trilobus		+				−	
	Globigerinoides ruber white		+				−	
	Globigerinita glutinata						−	
2.	*Globigerinoides ruber* pink		+		−	−		+
	Leonella granifera	−			−	−		+
	Lingulodinium machaerophorum	−	+			−		+
3.	*Gephyrocapsa muellerae*		−		−	+	+	−
	Nematosphaeropsis labyrinthus		−		−	+	+	−
	Calciodinellum sp. 1		−		−	+	+	
	Gephyrocapsa ericsonii		−			+		−
	Neogloboquadrina pachyderma dex.		−			+		−
	Calcidiscus leptoporus					+		−
	Neogloboquadrina pachyderma sin.		−				+	
	Globigerina bulloides		−				+	
4.	*Echinidinium* spp.	−	−	−	−		+	+
	Rhabdothorax sp.	−		−	+	+	+	−
	Globorotalia inflata		−		+	+		−
	Protoceratium reticulatum	−		−			+	
5.	*Scrippsiella regalis*	+		+	+	+	−	−
	Dalella chathamense	+		+	+		−	−
	Impagidinium spp.	+		+	+		−	
	Impagidinium variaseptum	+		+			−	
	Globorotalia truncatulinoides sin.				+	+	−	−
	Umbilicosphaera sibogae				+		−	
	Pyxidinopsis reticulata	+						
- -	*Thoracosphaera heimii*							
	Emiliania huxleyi							
	Brigantedinium spp.							
	Protoperidiniaceae[a]							
	Spiniferites spp.							+

Table 2. Main synecological groups of species as summarized from surface sediment distribution patterns and the statistical analyses. For explanation of environmental variables and their minimum and maximum mean annual values, see Table 1. + = positive correlation (variance ≥ 6%); − = negative correlation (variance ≥ 6%); no sign = insignificant correlation. [a] excludes species of *Brigantedinium* and *Echinidinium*.

together and relate to relatively high temperatures and salinities, relatively stratified subsurface-waters and nutrient-poor conditions (as summarized in Table 2). Absolute values of these environmental variables are given in Table 1. *Calciodinellum albatrosianum* is additionally plotted very close to the foraminiferal species *Globigerinoides sacculifer* and *Globigerinoides ruber* (white) in the plankton DCA (Fig. 6), both of which are well-known for thriving in warm (26–28°C), relatively oligotrophic, stratified surface-waters (e.g. Thunell and Reynolds 1984; Ravelo and Fairbanks 1990; Hale and Pflaumann 1999). Furthermore, *Globigerinoides ruber* (white) often occurs in high abundances in the Equatorial Undercurrent, where salinities are >36‰ (Kemle-von Mücke and Oberhänsli 1999). Their calcification depths are assumed to lie between 0 and 50 m (Kemle-von Mücke and Oberhänsli 1999), suggesting that their environmental preferences could well overlap with those of *Calciodinellum albatrosianum*. The foraminiferal species *Globigerinita glutinata*, although often occurring at greater depths within the upper water column, is also known for its distribution in relatively warm and saline waters (Oberhänsli et al. 1992; Kemle-von Mücke and Oberhänsli 1999) and is plotted close to the three species mentioned above. It is interesting that *Florisphaera profunda* is plotted close to *Calciodinellum albatrosianum*, although it apparently prefers a somewhat greater stratification and more nutrients at 75 m water depth (Fig. 6), which is in conformity with the data from the literature which describes this species as a typical thermocline-dwelling species in nutrient-poor areas (Molfino and McIntyre 1990; Beaufort et al. 1997). These distributions confirm the suggestion made by Vink et al. (2002) that *Calciodinellum albatrosianum*, like *Florisphaera profunda*, requires a well developed thermocline for optimal production. *Pernambugia tuberosa* is the only phytoplankton species which shows a very strong negative relation with phosphate and nitrate parameters at both 0 and 75 m water depth (i.e. throughout the whole mixed layer) in both DCAs. We conclude from these statistical results and the clear distribution pattern in nutrient-poor regions (Fig. 3b) that abundances of *Pernambugia tuberosa* are primarily controlled by the nutrient content. The clear breaks in its distribution pattern, from common in the oligotrophic South Atlantic Gyre to rare in the nutrient-rich NECC and back to common in the oligotrophic NEC, show how well this species reflects hydrographic fronts in the South and equatorial Atlantic.

Group 2. – (Outer-)neritic, warm, low-density, stratified environments: The calcareous dinocyst species *Leonella granifera* (Fig. 3c) and the organic-walled species *Lingulodinium machaerophorum* (Fig. 4d) relate to relatively warm, low-density and stratified surface-water conditions in the more shelfward regions of the South and equatorial Atlantic (Fig. 5). In addition, *Leonella granifera* relates to extremely low surface-water salinity and is plotted in the upper left hand corner of Fig. 6, together with the foraminiferal species *Globigerinoides ruber* (pink). The latter occurs in highest abundances in oceanic regions where warm, nutrient-poor and especially stratified conditions prevail in the upper 50 m of the water column. In contrast to *Globigerinoides ruber* (pink), however, *Leonella granifera* shows a clear preference for the more shelfward regions, where riverine influence may play an important role (e.g. see the higher abundances of *Leonella granifera* in the vicinities of the Niger, Amazon and Orinoco Rivers, as well as along the coast of eastern Brazil, where river input is substantial: Fig. 3c). In the DCA, both species also correlate with high nutrient concentrations at 75 m water depth, but do not relate to surface-water nutrient concentrations. As we still do not know where exactly *Leonella granifera* lives in the upper water column, it is difficult to say whether the species reacts to the low SSS and extreme stratification in the uppermost layer of the water column, or whether it prefers the deeper parts of the euphotic zone where nutrients are available. An additional possibility could be that this species requires particular, river-derived nutrients for optimal production. *Lingulodinium machaerophorum* has often been described as a species which generally prefers regions with relatively high SST, slightly reduced SSS (i.e. greater water-column stability) and where nutrients are available (Lewis and Hallett 1997). The distributions in this study support this description.

Group 3. – Cool, dense, well-mixed, nutrient-rich environments: The three coccolithophorid species *Gephyrocapsa muellerae* (Fig. 2c), *Gephyrocapsa ericsonii* and *Calcidiscus leptoporus* (Fig. 2d), the calcareous dinocyst species *Calciodinellum* sp. 1 (Fig. 3b) and the organic-walled species *Nematosphaeropsis labyrinthus* (Fig. 4c) are not so much influenced by the outer neritic–oceanic gradient but correlate with relatively low temperatures and salinities, high densities, and well-mixed, nutrient-rich conditions, such as those prevailing in the upwelling areas (Fig. 5). In Fig. 6, *Gephyrocapsa muellerae* is plotted close to *Neogloboquadrina pachyderma* (sin.), which is a foraminiferal species known to occur in high concentrations in cold surface-water environments (e.g. Niebler and Gersonde 1998) as well as in upwelling areas such as offshore Namibia, most likely due to the higher nutrient concentrations there (Ufkes et al. 2000). *Gephyrocapsa muellerae* thus appears to be a temperate species which requires nutrient-rich upper surface-waters for optimal production. The only real cold water (calcareous) dinocyst species is *Caracomia arctica*, found in only four samples underlying the cold, nutrient-rich and relatively fresh Falkland Current (Fig. 3d; not included in the DCA sample set). Its absence in the Benguela upwelling region indicates that this species may have a more polar distribution than either *Gephyrocapsa muellerae* or *Neogloboquadrina pachyderma* (sin.). *Calciodinellum* sp. 1 is plotted close to the foraminiferal species *Globigerina bulloides*, which has been reported to occur in maximum abundances in relatively cool regions (15–20°C) that are affected by coastal or oceanic upwelling (e.g. Niebler and Gersonde 1998), or in front systems when nutrient concentration is moderately elevated (e.g. Oberhänsli et al. 1992). However, judging from the surface sediment distributions, *Calciodinellum* sp. 1 shows a more widespread distribution (Fig. 3b), thus tolerating less fertile conditions and higher surface temperatures than *Globigerina bulloides*. As such, *Calciodinellum* sp. 1 appears to be the most cosmopolitan of all the calcareous cyst species. The extremely high relative abundances of *Calciodinellum* sp. 1 in the samples of the Benguela upwelling region (often forming >75% of the calcareous dinocyst association) suggests that this species is more resistant to dissolution related to organic matter degradation and metabolic CO_2 production than the others. However, Wendler et al. (2002) revealed that this species is probably the most sensitive of all to dissolution associated with remineralisation reactions in the Arabian Sea. We therefore believe that at least the larger part of the Atlantic signal reflects high cyst production in nutrient-rich environments. *Gephyrocapsa ericsonii* is plotted close to the two foraminiferal species

Neogloboquadrina pachyderma (dextral) and *Globorotalia inflata* in Fig. 6. Both foraminiferal species are known to have their highest abundances in cool environments where seasonal upwelling may occur, but neither are assumed to be typical upwelling species (e.g. Niebler and Gersonde 1998). In addition, Boltovskoy et al. (1996) argue that the distribution of *Globorotalia inflata* is not directly temperature-related, but that this species rather reflects transitional environments. *Gephyrocapsa ericsonii* has often been described as a species which is characteristic for eutrophic environments (e.g. Winter and Siesser 1994). Its distribution shows that temperature cannot be the most important controlling factor, as its highest abundances occur in the cooler regions of the South Atlantic, but it definitely still thrives in tropical areas (Kinkel et al. 2000). We suggest that this species requires meso- to eutrophic (sub)surface-water conditions for optimal production. *Nematosphaeropsis labyrinthus* is an opportunistic species which can tolerate a wide range of temperatures and salinities and has been observed in Arctic as well as tropical domains (e.g. Wall et al. 1977). However, most studies, including this one, show that this species tends towards higher abundances in nutrient-rich environments, where it could be very useful as a paleoproductivity proxy, considering its great resistance against bacterial degradation (Zonneveld et al. 2001b). Although the distribution pattern of *Calcidiscus leptoporus* shows an obvious affinity for nutrient-rich environments (Fig. 2d), this species shows no significant relationship with the PO_4/NO_3 parameter in the DCA (Fig. 5). This discrepancy probably occurs because of the similar directions of change of the nutrient and temperature parameters, whereby the statistically insignificant effect of temperature on distribution has been weighted more than the positive effect of nutrients.

Group 4. – (Outer-)neritic, cool, dense, well-mixed, nutrient-rich environments: This group contains the calcareous dinocyst species *Rhabdothorax* sp. (Fig. 3d) and the two organic-walled dinocyst species *Echinidinium* spp. (Fig. 4d) and *Protoceratium reticulatum*. It shows much overlap with Group 3. described above, but differs mainly due to its strong preference for shelfward regions (Fig. 5). Like *Gephyrocapsa ericsonii*, *Rhabdothorax* sp. is plotted close to the two foraminiferal species *Neogloboquadrina pachyderma* (dextral) and *Globorotalia inflata* in Fig. 6 and it thus shows a similar preference for meso- to eutrophic (sub)surface-water conditions. However, *Gephyrocapsa ericsonii* and *Rhabdotho-rax* sp. are plotted far apart from each other on the second axis in Fig. 5, as *Gephyrocapsa ericsonii* does not show a relation with the outer neritic–oceanic gradient, whereas *Rhabdothorax* shows a clear preference for outer neritic environments. The *Echinidinium* group is plotted well to the left of Fig. 5 and may be the best indicator for coastal/neritic, high nutrient concentrations (most likely associated with coastal upwelling), although care must be taken in interpreting their distributions quantitatively as these species, together with the other protoperi-diniacean cyst species, are generally quite sensitive to oxygen-dependent bacterial decay (Zonneveld et al. 2001b). *Protoceratium reticulatum* is generally considered to be a cosmopolitan species which tolerates large fluctuations in temperature and salinity, but increased cyst abundances have often been reported from nutrient-rich upwelling areas and during phases of surface-water eutrophication (e.g. Bolch and Hallegraeff 1990). The results of this study also show a tendency of this species towards higher cyst abundances in nutrient-rich areas.

Group 5. – Oceanic, saline, well-mixed, nutrient-poor environments: The calcareous dino-cyst species *Scrippsiella regalis* (Fig. 3d), the four organic-walled dinocyst species *Dalella chathamense, Impagidinium variaseptum* (Fig. 4d), *Impagidinium* spp. (Fig. 4b), *Pyxidinopsis reticulata* and the coccolithophorid species *Umbilicosphaera sibogae* correlate to a greater or lesser extent (see Table 2) with greater water depths, relatively high seasonality and salinity, and unstratified, nutrient-poor conditions (Fig. 5). *Scrippsiella regalis* shows a mainly oceanic distribution in cool and oligotrophic environments and is plotted close to the foraminiferal species *Globorotalia truncatulinoides* (sin.) on the opposite side of *Leonella granifera* and *Globigerinoides ruber* (pink) in the DCA (Fig. 6). *Globorotalia truncatulinoides* (sin.) is a deep-dwelling, subtropical species which is found in its highest abundances in regions where cool and well-mixed upper water masses predominate (e.g. Niebler and Gersonde 1998). We assume similar environmental affinities for *Scrippsiella regalis*. The distribution of *Scrippsiella regalis* is similar to that of *Dalella chathamense* and *Impagi-dinium variaseptum*, which is not so much influenced by T, S and D but shows a tendency towards higher abundances in oligotrophic, oceanic regions where seasonality may be important (Fig. 5). However, distributions of *Impagidinium variaseptum* are restricted to the southern hemisphere, whereas both *Dalella chathamense* and *Scrippsiella regalis* are present, though perhaps in low absolute abundances, in northern hemisphere sediments such as those underlying the NEC. Most other species of the genus *Impagidinium* (Fig. 4b) have comparable distributions but often occur in warmer environments as well. It is difficult to pinpoint the main ecological characteristics of the coccolitho-phorid species *Umbilicosphaera sibogae*. It shows a positive relationship with $S0$ and relates slightly negatively to the phosphate and nitrate parameters at 0 and 75 m water depth in the calcareous plankton DCA (Fig. 6), but does not significantly relate to these parameters or indeed any others in the phytoplankton DCA (Fig. 5). These relationships suggest that it may not always be correct to place this species within the group characteristic for eutrophic environments (Baumann et al. 1999). However, living specimens show highest abundances in water samples from below the thermocline in the subtropical gyre (Baumann et al. 1999), illustrating that they may just be reacting to higher nutrient levels in the deep euphotic zone. An additional problem could be that it is impossible to distinguish between two different (sub)species of *Umbilicosphaera sibogae* in sediments, making it necessary to deal with them as a single group, although they probably have different ecological preferences. In this study, the distributions of *Umbilicosphaera sibogae* and the organic-walled cyst species *Pyxidinopsis reticulata* appear quite similar (Fig. 5), although *Pyxidinopsis reticulata* tends towards more nutrient-poor, oceanic environments with a significant temperature seasonality. However, *Pyxidinopsis reticulata* still has an affinity for slightly higher nutrient levels than, for example, most of the *Impagidinium* species.

Several species show no obvious affinities to the given environmental variables (Table 2). This group consists of the cosmopolitan species *Thoraco-sphaera heimii* and *Emiliania huxleyi* (Fig. 2b), but also includes the organic-walled dinocyst species belonging to *Brigantedinium* spp., the family of the Protoperidiniaceae (excluding species of *Echinidinium* and *Brigantedinium*: Fig. 4b), and *Spiniferites* spp. (Fig. 4c). It is interesting that *Brigantedinium* spp. and the group of the Proto-peridiniaceae do not relate with the nutrient parameters or indeed with any of the other environmental variables in the DCA, despite their long tradition of application in paleoproductivity reconstructions. We believe that these results are not due to the fact that such species have a cosmopolitan distribution (Fig. 4b shows that this is probably not the case), but rather that their sensitivity to oxidation/organic-matter degradation (Zonneveld et al. 2001b) may have altered their distribution patterns and overprinted the original relationships with environmental parameters. This stands in contrast to the distributions of *Thoracosphaera heimii* and *Emiliania huxleyi*, which really do have cosmopolitan distributions. Both species are assumed to have a very broad range of ecological tolerance (e.g. Karwath 2000; Schröder-Ritzrau et al. 2001).

Conclusions and Future Perspectives

Many of the phytoplanktonic microfossil species, or groups of species, found in South and equatorial Atlantic surface sediments show restricted distributions, or specific regions of highest abundance, which primarily reflect the environmental conditions of the upper water masses directly above them. Of the three phytoplankton groups analyzed in this study (i.e. coccoliths, calcareous dinocysts, organic-walled dinocysts), the coccolith species clearly show the most widespread distributions throughout the South and equatorial Atlantic, generally being present in the entire sub-polar to tropical sample range. The six species documented in this study do not much discriminate between outer neritic and oceanic areas, but temperature and nutrient content

appear to be the two most important factors controlling their abundances on a regional scale. Several calcareous dinocyst species similarly show widespread distributions, but they generally have their highest abundances in relatively warm, oligotrophic and mesotrophic regions and are related to a greater spectrum of environmental factors than is the case for most of the coccolithophorid species, i.e. they are probably less tolerant to seasonally or inter-annually changing environmental conditions at one particular site. In turn, the organic-walled dinocyst group is the most diverse, containing opportunistic, cosmopolitan species (e.g. *Nematosphaeropsis labyrinthus*) as well as species showing a restricted habitat (e.g. species of *Echinidinium*). As such, we could only very roughly distinguish between coccolithophorids as predominant r-strategists (i.e. pioneering/opportunistic organisms which usually respond to nutrient enrichment with enhanced growth rates or productivity), calcareous cyst-producing dinoflagellates as predominant K-strategists (i.e. adapted to relatively low nutrient levels in a stable environment), and organic-walled cyst-producing dinoflagellates as partly r-strategists, partly K-strategists and partly specialists (i.e. adapted to particular combinations of environmental conditions only).

High abundances of individual microfossil species or particular species assemblage groups can be used as indicators for specific environmental conditions in the uppermost water column. For example, typical species for relatively high SSTs in the South and equatorial Atlantic are *Calciodinellum albatrosianum* and *Lingulodinium machaerophorum*, whereas low SSTs are indicated by *Gephyrocapsa muellerae* and *Cara-comia arctica*. Similarly, nutrient-rich upper water conditions can be reconstructed from relatively high abundances of *Calcidiscus leptoporus*, *Gephyrocapsa ericsonii*, *Rhabdothorax* sp., *Nematosphaeropsis labyrinthus* and species of *Echinidinium*, whereas nutrient-poor conditions are reflected by high abundances of *Pernambugia tuberosa* and species of *Impagidinium*. Based on their similar distribution patterns, it was possible to distinguish between five different synecological groups of coccolithophorids and dinoflagellates in the South and equatorial Atlantic, whose exact relationships with (sub)surface-water conditions are summarized in Table 2.

Although organic-walled dinocyst associations are increasingly being used in the development of transfer functions in the North Atlantic (e.g. de Vernal et al. 1993), the present data set in the South and equatorial Atlantic is still too small and patchy to allow a similar development there. More surface samples must be analyzed before such an attempt is made. However, the coccolith and calcareous dinocyst data sets are large and cover environmentally/ecologically diverse areas, and we have shown that many of the species react sensitively to changing environmental conditions. This suggests that these two microfossil groups may be just as adequate as planktonic foraminifera for paleotemperature and -productivity reconstructions based on transfer functions in this region. In fact, one could argue that they may be even more suitable for surface-water reconstructions, as the distributions of their living counterparts are restricted to the euphotic zone and vertical migration through the water column is much less likely (which often leads to submergence, displacement and survival of foraminifera in the deeper parts of moving water masses over many hundreds of kilometers away from the production area, e.g. Boltovskoy et al. 1996). However, it is important that we gather more information on the regional depth distribution patterns and reproduction/calcification depths of living species in the upper water column in order to pinpoint their exact depth habitats prior to TFT development (especially for the calcareous dinocysts, for which so little information is presently available). In addition, more detailed information from sediment trap studies should improve our understanding of the transformation of living phyto-plankton communities into fossil assemblages in deep-sea sediments. Despite these restrictions, the results of surface sediment analysis clearly accentuate the paleoenvironmental reconstruction potential hidden in many fossil coccolith and dinocyst assemblages, and these should be further exploited in future paleoenvironmental and -oceanographic research.

Acknowledgements

All members of the working groups of Sedimentology and Historical Geology/Palaeontology of Bremen University are thanked for their general assistance and openness to discussion. We thank A. Bammann, A. Brune, A. Freesemann, E. Freytag, F. Heidersdorf, S. Hüneke and C. Wienberg for their technical assistance. T. Jennerjahn is thanked for kindly providing the samples of the eastern Brazilian continental margin. We would also like to thank the crews and numerous colleagues of the RV *Meteor*, *Polarstern* and *Victor Hensen* for their help during ship expeditions. Special thanks are due to J. Giraudeau and an anonymous reviewer for their constructive comments on a previous version of this manuscript. This research was funded by the Deutsche Forschungsgemeinschaft through the special research project Sonderforschungsbereich 261 "Der Südatlantik im Spätquartär: Rekonstruktion von Stoffhaushalt und Stromsystemen". This is SFB contribution 345. Data are available under www.pangaea.de/Projects/SFB261.

References

Andruleit H (1996) A filtration technique for quantitative studies of coccoliths. Micropaleontology 42: 403-406

Baumann K-H, Cepek M, Kinkel H (1999) Coccolithophores as Indicators of Ocean Water Masses, Surface-Water Temperature, and Paleoproductivity – Examples from the South Atlantic. In: Fischer G, Wefer G (eds) Use of Proxies in Paleoceanography – Examples from the South Atlantic. Springer, Berlin, pp 117-144

Beaufort L, Lancelot Y, Camberlin P, Cayre O, Vincent E, Bassinot F, Labeyrie L (1997) Insolation cycles as a major control of Equatorial Indian Ocean primary production. Science 278: 1451-1454

Bolch CJ, Hallegraeff GM (1990) Dinoflagellate cysts in recent sediments from Tasmania, Australia. Bot Mar 33: 173-192

Boltovskoy E, Boltovskoy D, Correa N, Brandini F (1996) Planktic foraminifera from the southwestern Atlantic (30°–60°S): Species-specific patterns in the upper 50 m. Mar Micropaleontol 28: 53-72 de Vernal A, Rochon A, Hillaire-Marcel C, Turon J-L, Guiot J (1993) Quantitative reconstruction of sea-surface conditions, seasonal extent of sea-ice cover and meltwater discharges in high latitude marine environments from dinoflagellate cyst assemblages. In: Peltier WR (ed) Ice in the Climate System. NATO ASI Ser I (12), pp 611-621

Esper O, Zonneveld KAF (2002) Distribution of organic-walled dinoflagellate cysts in surface sediments of the Southern Ocean (eastern Atlantic sector) between the Subtropical Front and the Weddell Gyre. Mar Micropaleontol 46: 177-208

Esper O, Zonneveld KAF, Höll C, Karwath B, Kuhlmann H, Schneider RR, Vink A, Weise-Ilho I, Willems H (2000) Reconstruction of palaeoceanographic conditions in the South Atlantic Ocean at the last two Terminations based on calcareous dinoflagellate cysts. Int J Earth Sci 88: 680-693

Giraudeau J, Pujos A (1990) Fonction de transfert basee sur les nannofosilles calcaires du Pleistocene des Caraibes. Oceanol Acta 13: 453-469

Hale W, Pflaumann U (1999) Sea-surface temperature estimations using a modern analog technique with foraminiferal assemblages from Western Atlantic Quaternary sediments. In: Fischer G, Wefer G (eds) Use of Proxies in Paleoceanography – Examples from the South Atlantic. Springer, Berlin, pp 69-90

Höll C, Zonneveld KAF, Willems H (1998) On the ecology of calcareous dinoflagellates: The Quaternary Eastern Equatorial Atlantic. Mar Micropaleontol 33: 1-25

Imbrie J, Kipp NG (1971) A new micropaleontological method for quantitative paleoclimatology: Application to a late Pleistocene Caribbean core. In: Turekian KK (ed) The Late Cenozoic Glacial Ages. Yale University Press, New Haven, pp 71-181

Jordan RW, Kleijne A (1994) A classification system for living coccolithophores. In: Winter A, Siesser WG (eds) Coccolithophores. Cambridge University Press, pp 83-106

Karwath B (2000) Ecological studies on living and fossil calcareous dinoflagellates of the equatorial and tropical Atlantic Ocean. Ber Fachber Geowiss, Univ Bremen 152, 175 p

Kemle-von Mücke S, Oberhänsli H (1999) The Distribution of Living Planktic Foraminifera in Relation to Southeast Atlantic Oceanography. In: Fischer G, Wefer G (eds) Use of Proxies in Paleoceanography – Examples from the South Atlantic. Springer, Berlin, pp 91-115

Kinkel H, Baumann K-H, Èepek M (2000) Coccolithophores in the equatorial Atlantic Ocean: Response to seasonal and Late Quaternary surface water variability. Mar Micropaleontol 39: 87-112

Lewis J, Hallett R (1997) *Lingulodinium polyedrum* (*Gonyaulax polyedra*) A blooming dinoflagellate. Oceanogr and Mar Biol: Annual Review 35: 97-161

Matthiessen J, Baumann K-H, Schröder-Ritzrau A, Hass C, Andruleit H, Baumann A, Jensen S, Kohly A, Pflaumann U, Samtleben C, Schäfer P, Thiede J (2001) Distribution of calcareous, siliceous and organic-walled planktic microfossils in surface sediments of the Nordic Seas and their relation to surface-water masses. In: Schäfer P, Ritzrau W, Schlüter M, Thiede J (eds) The Northern North Atlantic: A Changing Environment. Springer, Berlin, pp 105-127

Mix AC, Morey AE, Pisias NG, Hostetler SW (1999) Foraminiferal faunal estimates of paleotemperature: Circumventing the no-analog problem yields cool ice age tropics. Paleoceanography 14: 350-359

Molfino B, McIntyre A (1990) Precessional forcing of nutricline dynamics in the equatorial Atlantic. Science 29: 766-769

Nelson DM, Tréguer P, Brzezinski MA, Leynaert A, Quéguiner B (1995) Production and dissolution of biogenic silica in the ocean: Revised global estimates, comparison with regional data and relationship to biogenic sedimentation. Glob Biogeochem Cycl 9: 359-372

Nieber H-S, Gersonde R (1998) A planktic foraminiferal transfer function for the southern South Atlantic Ocean. Mar Micropaleontol 34: 213-234

Oberhänsli H, Bénier C, Meinecke G, Schmidt H, Schneider R, Wefer G (1992) Planktonic foraminifers as tracers of ocean currents in the eastern South Atlantic. Paleoceanography 7: 607-632

Peterson RG, Stramma L (1991) Upper-level circulation in the South Atlantic Ocean. Prog Oceanogr 26: 1-73

Pflaumann U, Duprat J, Pujol C, Labeyrie LD (1996) SIMMAX: A modern analog technique to deduce Atlantic sea surface temperatures from planktonic foraminifera in deep-sea sediments. Paleoceanography 11: 15-35

Prell WL (1985) The stability of low latitude sea surface temperatures: An evaluation of the CLIMAP reconstruction with emphasis on positive SST anomalies. US DOE Rep TR 025, 60 p

Ravelo AC, Fairbanks RG (1990) Reconstructing tropical Atlantic hydrography using planktonic foraminifera and an ocean model. Paleoceanography 5: 409-431

Samtleben C, Schäfer P, Andruleit H, Baumann A, Baumann K-H, Kohly A, Matthiessen J, Schröder-Ritzrau A (1995) Plankton in the Norwegian-Greenland Sea: From living communities to sediment assemblages – an actualistic approach. Geol Rundsch 84: 108-136

Schneider RR, Müller PJ, Ruhland G, Meinecke G, Schmidt H, Wefer G (1996) Late Quaternary surface temperatures and productivity in the east-equatorial South Atlantic: Response to changes in trade/monsoon wind forcing and surface water advection. In: Wefer G, Berger WH, Siedler G, Webb DJ (eds) The South Atlantic: Present and Past Circulation. Springer, Berlin, pp 527-551

Schröder-Ritzrau A, Andruleit H, Jensen S, Samtleben C, Schäfer P, Matthiessen J, Hass HC, Kohly A, Thiede J (2001) Distribution, export and alteration of fossilizable plankton in the Nordic Seas. In: Schäfer P, Ritzrau W, Schlüter M, Thiede J (eds) The Northern North Atlantic: A Changing Environment. Springer, Berlin, pp 81-104

Shokati I, Baumann K-H, Èepek M, Henrich R (1999) Zur Sedimentation von Coccolithophoriden und ihrer Akkumulation in spätpleistozän-holozänen Sedi-menten des östlichen Südatlantiks. Zbl Geol Paläont Teil I: 859-876

Tangen K, Brand LE, Blackwelder PL, Guillard RRL (1982) *Thoracosphaera heimii* (Lohmann) Kamptner is a dinophyte: observations on its morphology and life cycle. Mar Micropaleontol 7: 193-212

Thunell RC, Reynolds LA (1984) Sedimentation of planktonic foraminifera: Seasonal changes in species flux in the Panama Basin. Micropaleontol 30: 241-260

Ufkes E, Jansen JHF, Schneider RR (2000) Anomalous occurrences of *Neogloboquadrina pachyderma* (left) in a 420-ky upwelling record from Walvis Ridge (SE Atlantic). Mar Micropaleontol 40: 23-42

Vénec-Peyré M-T, Caulet JP, Grazzini CV (1997) Glacial/interglacial changes in the equatorial part of the Somali Basin (NW Indian Ocean) during the last 355 kyr. Paleoceanography 12: 640-648

Vink A, Zonneveld KAF, Willems H (2000a) Distributions of calcareous dinoflagellate cysts in surface sedi-ments of the western equatorial Atlantic Ocean, and their potential use in palaeoceanography. Mar Mi-cropaleontol 38: 149-180

Vink A, Zonneveld KAF, Willems H (2000b) Organic-walled dinoflagellate cysts in western equatorial Atlantic surface sediments: distributions and their relation to environment. Rev Paleobot Palynol 112: 247-286

Vink A, Brune A, Höll C, Zonneveld KAF, Willems H (2002) On the response of calcareous dinoflagellates to oligotrophy and stratification of the upper water column in the equatorial Atlantic Ocean. Palaeogeogr Palaeoclimatol Palaeoecol 178: 53-66

Wall D, Dale B, Lohmann GP, Smith WK (1977) The environmental and climatic distribution of dinoflagellate cysts in modern marine sediments from regions in the North and South Atlantic Oceans and adjacent seas. Mar Micropaleontol 2: 121-200

Wefer G, Berger WH, Siedler G, Webb DJ (1996) The South Atlantic: Present and Past Circulation. Springer, Berlin, 729 p

Wendler I, Zonneveld KAF, Willems H (2002) Oxygen availability effects on early diagenetic calcite dissolution in the Arabian Sea as inferred from calcareous dinoflagellate cysts. Glob Planet Change 34: 219-239

Winter A, Siesser W (1994) Coccolithophores. Cambridge University Press, Cambridge, 242 p

Zonneveld KAF, Brune A, Willems H (2000) Spatial distribution of calcareous dinoflagellate cysts in surface sediments of the Atlantic Ocean between 13°N and 36°S. Rev Palaeobot Palynol 111: 197-223

Zonneveld KAF, Hoek R, Brinkhuis H, Willems H (2001a) Geographical distributions of organic-walled dino-flagellate cysts in surficial sediments of the Benguela upwelling region and their relationship to upper ocean conditions. Prog Oceanogr 48: 25-72

Zonneveld KAF, Versteegh GJM, de Lange GJ (2001b) Palaeoproductivity and post-depositional aerobic organic matter decay reflected by dinoflagellate cyst assemblages of the Eastern Mediterranean S1 sapropel. Mar Geol 172: 181-195

The South Atlantic Oxygen Isotope Record of Planktic Foraminifera

S. Mulitza[*], B. Donner, G. Fischer, A. Paul, J. Pätzold,
C. Rühlemann and M. Segl

Universität Bremen, Fachbereich Geowissenschaften, Klagenfurter Strasse,
D-28359 Bremen, Germany
* corresponding author (e-mail): smulitza@uni-bremen.de

Abstract: This paper reviews the recording of oxygen isotope ratios in planktic foraminifera and summarizes recent results of the application of oxygen isotopes in paleoceanographic studies of the South Atlantic. The most important factors controlling the $\delta^{18}O$ of planktic foraminifera are temperature, the $\delta^{18}O$ and the pH of ambient seawater. Seasonal and vertical calcification weight the mean $\delta^{18}O$ of a foraminiferal population towards the hydrographic conditions in the preferred ecological niche. After deposition, the $\delta^{18}O$ signal is affected by bioturbation and dissolution. Despite many influence factors, the composition of oxygen isotopes in fossil tests of planktic foraminifera provides important constraints on variations of the surface water hydrography of the South Atlantic and the Southern Ocean throughout the past 20,000 years. During the last glacial maximum, the Polar Front remained close to its modern position or shifted only slightly towards the north. In the tropics, oxygen isotopes indicate only a moderate glacial cooling of 2-3°C. During deglaciation, oxygen isotope ratios in the eastern boundary currents of the subtropical South Atlantic decreased asynchronously relative to those in the eastern North Atlantic, with the highest interhemispheric contrasts during the Younger Dryas and the Heinrich Event 1. This pattern is consistent with a redistribution of heat within the Atlantic Ocean in response to a weakening of the thermohaline circulation. The slowdown of deglacial overturning was associated with a southward displacement of the thermal equator and the Intertropical Convergence.

Introduction

The ratio of the stable oxygen-isotopes ^{18}O and ^{16}O is one of the most important tools in paleoclimatology. The importance of this proxy results, to a large extent, from the fact that oxygen isotopes can be used in interpretations of marine (e.g. Guilderson et al. 1994; Wolff et al. 1998), terrestrial (e.g. Schwalb et al. 1999), and cryospheric records (e.g. Johnsen et al. 1972; Dansgaard et al. 1993). Furthermore, oxygen isotopes circulate in the main climatic subsystems, namely in the ice, in the atmosphere, and in the ocean, and are fractionated whenever a phase transition between or within the reservoirs occurs. For these reasons, oxygen isotopes contribute significantly to the understanding of past climates.

In marine sediments, the oxygen isotopic composition of planktic foraminifera is primarily used for stratigraphy (e.g. Shackleton and Opdyke 1973; Imbrie et al. 1984). Additionally, the $\delta^{18}O$ of planktic foraminifera provides information on the temperature (e.g. Emiliani 1955) and the isotopic composition of seawater in which calcification took place (e.g. Duplessy et al. 1991).

Currently, a data base comprising more than 100.000 oxygen isotope measurements on planktic foraminifera is available (see PANGAEA, www.pangaea.de). The use of this huge and rapidly growing paleoceanographic archive, however, is limited by a number of factors that are critical to the application of oxygen isotopes as a paleoceano-

From WEFER G, MULITZA S, RATMEYER V (eds), 2003, *The South Atlantic in the Late Quaternary: Reconstruction of Material Budgets and Current Systems.* Springer-Verlag Berlin Heidelberg New York Tokyo, pp 121-142

graphic tool. This paper reviews the results of the SFB 261 project with respect to the recording of $\delta^{18}O$ by planktic foraminifera, and the late Quaternary oxygen isotope record of these organisms in the tropical and South Atlantic.

Oxygen Isotopes

Isotopes are variants of an element containing different numbers of neutrons. There are three stable isotopes of oxygen: ^{16}O, ^{17}O and ^{18}O. ^{16}O is the most abundant (99.76%) whereas ^{17}O and ^{18}O only comprise 0.04% and 0.2% of the total oxygen. Most studies concentrate on the ratio of the most abundant isotopes ^{18}O and ^{16}O. The isotopic composition of a sample being measured is expressed as $\delta^{18}O$:

$$\delta^{18}O(\%o) = \left[\frac{(^{18}O/^{16}O)_{\text{sample}} - (^{18}O/^{16}O)_{\text{standard}}}{(^{18}O/^{16}O)_{\text{standard}}} \right] \cdot 1000$$

which represents the difference in the $^{18}O/^{16}O$ ratios between the sample and the standard, expressed as parts per thousand (per mil). Generally, carbonate samples are measured relative to the PDB-standard (Cretaceous belemnite formation at Peedee in South Carolina, USA), whereas water samples refer to SMOW (Standard Mean Ocean Water). This nomenclature has recently been changed to V-PDB and V-SMOW (Coplen 1996). To convert the V-SMOW scale into the V-PDB scale, a small correction of -0.27 ‰ is necessary (Hut 1987) on account of the different preparation techniques of carbonate and water samples. The basic principles of mass spectrometry and the preparation of carbonate and water samples are described by Hoefs (1987) and Rohling and Cooke (2001).

Oxygen Isotopes in the Hydrological Cycle

Whenever a water parcel undergoes a phase transition (e.g. evaporation or condensation), a temperature-dependent kinetic fractionation of oxygen isotopes occurs. Hence, water evaporating from the sea surface is depleted in heavy isotopes relative to ocean water, while rain precipitating from a cloud is enriched relative to the cloud's moisture.

The tropical oceans are the major source of atmospheric water vapor. Poleward transport of this water results in a gradual rainout and thus in a depletion of oxygen-18 in the remaining moisture. Hence, the isotopic composition of precipitation varies strongly with latitude, altitude and continentality.

Compared to the atmosphere, the variation in the isotopic composition of seawater (δ_w) is relatively small. The distributions of both salinity and oxygen isotopes in seawater are mainly controlled by precipitation and evaporation. For this reason, salinity and δ_w of surface waters are linearly related, displaying high δ_w values (high salinities) in low latitudes, and low δ_w values (low salinities) in high latitudes (Fig. 1). However, the slope of the δ_w-salinity relationship varies between 0.1 for tropical surface waters and 1 for high-latitude surface waters (e.g. Paul et al. 1999). This reflects the fact that precipitation is much more depleted of oxygen-18 in high latitudes than in low latitudes. Hence, salinity and δ_w label the surface waters in different ways. Mixing of water masses that originate from the sea surface creates the characteristic vertical sections known from GEOSECS (Birchfield 1987; Östlund et al. 1987). The isotopic composition of sea water is also affected by the formation of sea ice, but the fractionation effect is so small that the freezing process leads to an increase in salinity, with essentially no observable influence on δ_w. This gives a zero slope in a δ_w-salinity diagram (Craig and Gordon 1965). As a result, the surface waters of the Southern Ocean show a wide range of salinities, although δ_w is nearly constant at about -0.3 ‰ (Jacobs 1985). In present-day oceanography, δ_w allows to distinguish between different freshwater sources. Thus, it is possible to investigate the influence of local P - E, river-runoff and glacial meltwater on a water mass of interest (e.g. Weiss et al. 1979; Jacobs et al. 1985).

Recording of $\delta^{18}O$ by Planktic Foraminifera

Palaeotemperature Equations and Disequilibrium Effects

Another type of isotope fractionation is the so-called equilibrium (or thermodynamic) fractionation.

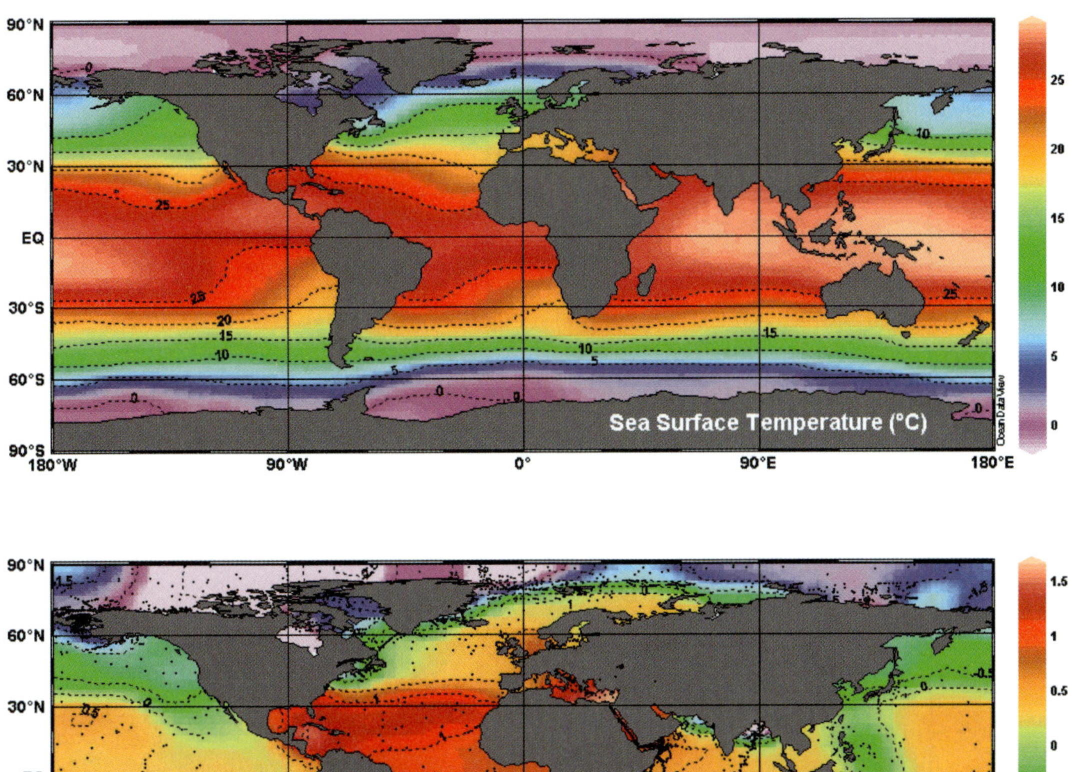

Fig. 1. Distribution patterns of temperature (upper panel, data from Levitus and Boyer 1994) and $\delta^{18}O$ of surface waters (lower panel). $\delta^{18}O$ data are from the 0-50 m level from the data base of Schmidt et al. 1999, and have been objectively interpolated to a 5° grid with the method described in Levitus and Boyer (1994).

This is essentially the isotope effect involved in temperature-dependent equilibrium reactions, for example, in the system CO_2-H_2O-$CaCO_3$. The temperature-dependence of the equilibrium fractionation has been determined by laboratory experiments (e.g. Kim and O'Neil 1997). Urey (1947) first proposed that the composition of oxygen isotopes in carbonate fossils could be used to reconstruct temperatures. During the past decades many empirically derived paleotemperature equations have been published, which often produce results that differ by several degrees, when applied to the same data set (see Bemis et al. 1999 for review). For this reason, a thorough understanding of the governing processes and an accurate calibration of the different fossil groups is still necessary.

Oxygen isotope measurements on living planktic foraminifera caught in surface waters provide the opportunity to test paleotemperature equations if both temperature and $\delta^{18}O$ of the surrounding seawater is known. Fig. 2 shows a compilation of $\delta^{18}O$ measurements on the species *Globigerinoides ruber*, *Globigerinoides sacculifer*, *Globigerina bulloides* and *Neogloboquadrina pachyderma*

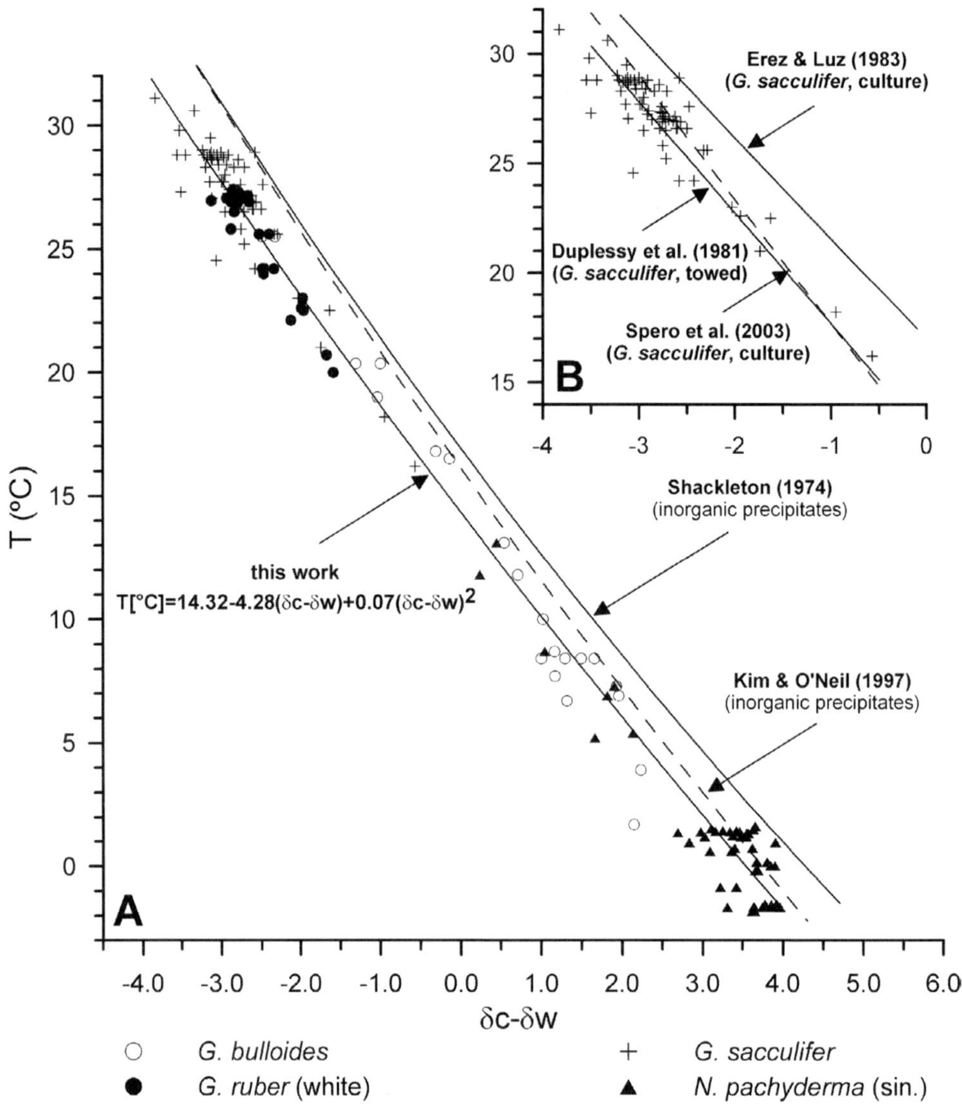

Fig. 2. a) Variation of the difference between the oxygen isotope composition of foraminiferal calcite (δc) and ambient seawater (δw) versus temperature for living specimens of *G. ruber*, *G. sacculifer*, *G. bulloides* and *N. pachyderma* and comparison with the paleotemperature equations of Shackleton (1974) and Kim and O'Neil (1997). The plot also includes published data from Duplessy et al. (1981), Kahn and Williams (1981), Ganssen (1983), Bauch et al. (1997) and Mulitza et al. (2003). **b)** Comparison of the oxygen isotope composition of living *G. sacculifer* with the paleotemperature equations of Duplessy et al. (1981) Erez and Luz (1983) and Spero et al. (2003).

done at the University of Bremen and data from published sources (see references in Mulitza et al. 2003). These four species were chosen because their geographical distribution covers the entire present-day temperature range, they all belong to the shallowest dwelling species, and because they are the most frequently species used in paleoceanographic investigations. For all measurements, $\delta^{18}O$

of ambient seawater was available, either measured directly or derived from salinity or atlas values (Mulitza et al. 2003).

It is significant that all species displayed lower $\delta^{18}O$ values than indicated by the established paleotemperature equations of Shackleton (1974), Kim and O'Neil (1997) and Erez and Luz (1983). The latter equation is based on measurements of *G.*

sacculifer cultured at temperatures between 14 and 30 °C. The difference between culture grown *G. sacculifer* and *G. sacculifer* grown in a natural environment strongly suggests that the offset is due to an environmentally induced effect other than temperature or $\delta^{18}O$ of ambient seawater.

Two important factors influencing the stable oxygen isotopic composition of planktic foraminifera besides temperature and $\delta^{18}O$ of seawater have been identified: (1) the carbonate ion- or pH-effect (Spero et al. 1997; Bijma et al. 1999; Zeebe 1999), and (2) the photosynthetic activity of sym-biontic algae (Spero and Lea 1993). Both increased pH and increased photosynthetic activity result in a decrease of $\delta^{18}O$ values in foraminiferal shells. A photosynthetic effect, however, seems unlikely to be the reason for the negative deviation because the nonsymbiontic species *G. bulloides*, and *N. pachyderma* fall on the same scale as the symbiontic species *G. ruber* and *G. sacculifer*.

It seems reasonable that the observed offset is due to different carbon chemistries in the culture experiments and the present-day open ocean. As already mentioned by Bemis et al. (1998), the carbonate system was not controlled in the experiments of Spero and Lea (1993) and Erez and Luz (1983). Hence, it is possible that atmospheric carbon dioxide migrated into the culture vessel during the experiments and decreased pH. The published calibrations for the species discussed here, in which the pH was controlled to be close to the oceanic pH (Bemis et al. 1998; Spero et al. 1997) are very close to the plankton tow data (Fig. 2b), and strongly suggests the pH effect as the primary mechanism explaining the differences between the equations.

The pH effect can also explain the deviations from the synthetic carbonate equations. Zeebe (1999) calculated that the critical pH, i.e. the pH at which carbonate begins to precipitate, had been at 7.8 in the experiments of Kim and O'Neil (1997). Open ocean pH values in the tropics are close to 8.2. Assuming a decrease of 1.1‰ per unit pH increase (Zeebe 1999), this difference of 0.4 units alone could account for 0.4-0.5‰ of the observed 0.6‰ difference between Kim and O'Neil (1997) and our equation at the warm end. This would also be consistent with the observation that the differences between the two equations decrease towards

cold temperatures, since pH values in high latitudes and deep waters are slightly lower than in tropical surface waters (Brewer et al. 1995).

It seems that variations in ambient pH can explain some of the differences between laboratory cultures and field observations, and even differences between different benthic species (Bemis et al. 1998). However, we must also take into account that the pH in the microenvironment of the shell can deviate considerably from the oceanic pH (Jørgensen et al. 1985; Wolf-Gladrow et al. 1999). Hence, the pH or carbon-ion effect might also be partly regarded as a vital effect.

Effects of Seasonality and Vertical Migration

Seasonality. A number of studies have demonstrated that a significant seasonal variability of both $\delta^{18}O$ and flux occurs in many species of planktic foraminifera (e.g. Williams et al. 1981; Deuser and Ross 1989; Thunell et al. 1999). To give an example, Fig. 3 shows the fluxes of several planktic species at sediment trap station CB (off Mauretania, Cape Blanc) over the period from 1988 to 1991, together with the oxygen isotope composition of *G. ruber*. It is evident that seasonal variations in the oxygen isotope ratio of *G. ruber* are a direct function of surface- water temperature as predicted by the regression equation of the field data in Fig. 2a. This proves that the surface- water temperature signal is transported into the sediment without any significant time lag.

The different species peak at different times of the year, with an apparent succession of abundance peaks throughout the year. Theoretically, the oxygen isotope composition of a foraminiferal population in the sediment is the flux-weighted mean of all oxygen isotope values. In Fig. 3g we calculated the theoretical mean flux-weighted sea- surface temperature recorded in each of the individual species. Apparently, *G. ruber* and *G. sacculifer* record a summer signal at the position of CB, while *O. universa* records annual mean conditions, and *G. bulloides* and *G. inflata* are shifted towards the winter. This evidence suggests that the $\delta^{18}O$ of different species might be a useful proxy of the seasonal temperature contrast. However, although some attempts have been made to reconstruct

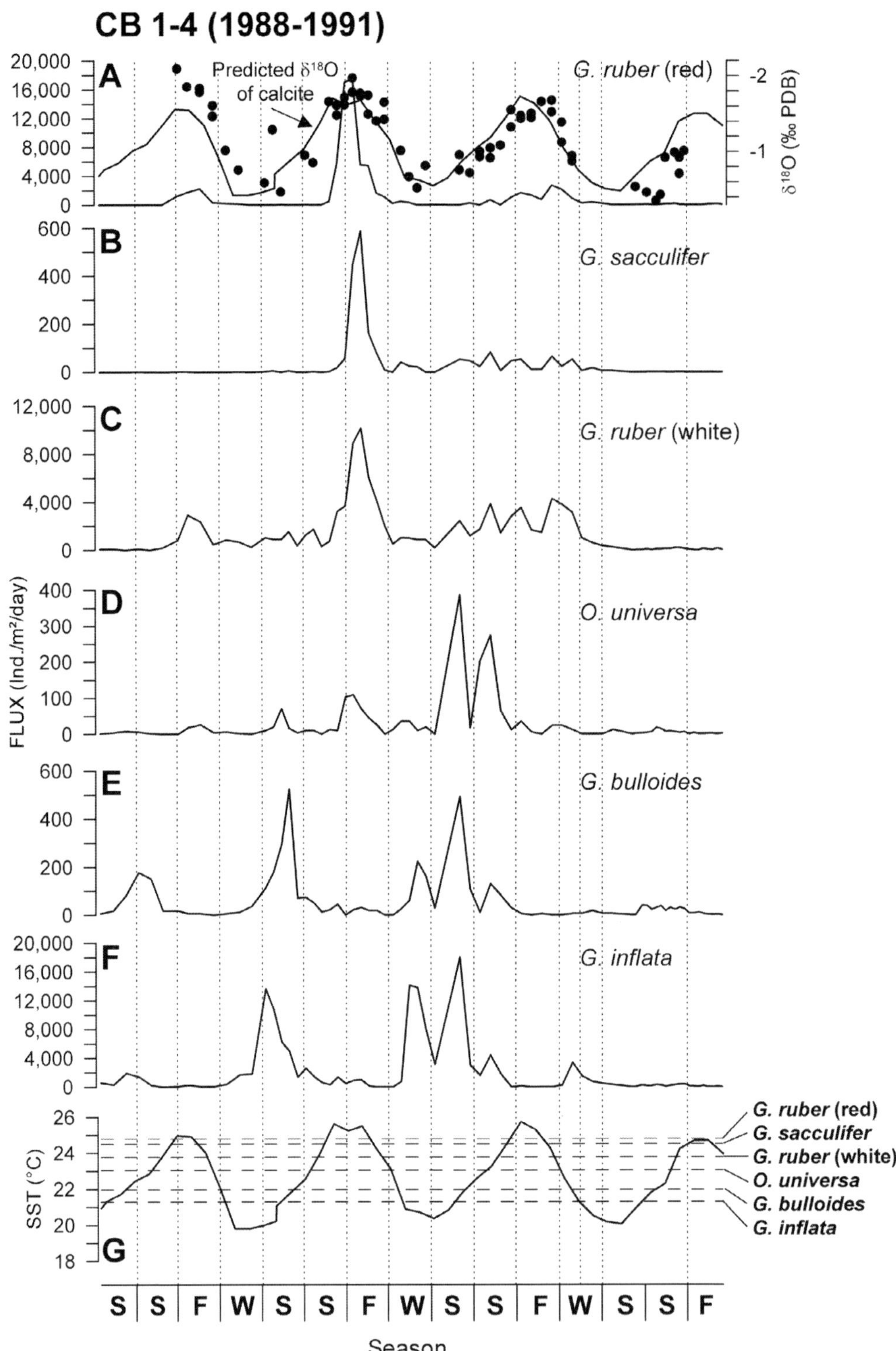

Fig. 3. (a - f) Seasonal succession of foraminiferal fluxes at sediment trap station CB off Cape Blanc (Mauretania) and **a)** seasonal variations of the oxygen isotope composition of *G. ruber* (black dots, Fischer et al. 1999). Predicted δ18O of foraminiferal calcite **a)** has been calculated from the regression equation in Fig. 2 and seasonal variations in measured sea surface temperature **g)**. Dashed lines **g)** represent the flux weighted (recorded) temperatures of the different species.

seasonality form oxygen isotope values measured in planktic foraminifera (e.g. Deuser 1978; Mulitza et al. 1998), it seems to be difficult to quantify seasonality because most winter species are not surface-dwelling foraminifera and the seasonal flux pattern is unlikely to be constant through time.

Vertical Migration. Investigating plankton tows has revealed that planktic foraminifera live vertically dispersed in the upper water column (e.g. Fairbanks et al. 1980; Kemle -von Mücke and Oberhänsli 1999) and that the depth habitat can also have significant influence on the oxygen isotope composition of planktic foraminifera. Many planktic foraminiferal species deposit their shells in the chlorophyll maximum zone (Fairbanks et al. 1982; Kohfeld et al. 1997), a major food source exploited by planktic foraminifera. However, the isotope data of primarily shallow-dwelling species indicate that foraminifera calcify in depth zones that are significantly narrower than the overall vertical distribution of these species implies (Fairbanks et al. 1980). For example, Fig. 4 shows the vertical distributions of *G. sacculifer* and *G. ruber* (white) at the plankton tow station GeoB 1402, located in the eastern equatorial Atlantic, together with the oxygen isotope compositions of the species from the different towing intervals. Although both species can be found in significant amounts down to a water depth of several hundred meters, the isotope values indicate that calcification occurs within the photic zone, where both species have their highest abundance. However, it is also evident that the $\delta^{18}O$ values of *G. sacculifer* increase within the seasonal thermocline, whereas the $\delta^{18}O$ of *G. ruber* is relatively constant throughout the entire water column. This implies a much smaller and shallower depth range of calcification in *G. ruber* (white) as compared to *G. sacculifer*.

Since planktic foraminifera live in different depth habitats, their $\delta^{18}O$ differences are a function of water column stratification and mixed-layer depth. This observation has been used to reconstruct the stratification of past oceans from $\delta^{18}O$ differences of several species (e.g. Williams and Healy-Williams 1980; Mulitza et al. 1997; Rühlemann et al. 2001), which is a parameter independent of global ice volume variations. This approach, however, is complicated by the fact that foramini-

fera are not restricted to a certain depth level, but can change their depth habitat. For example, Sauter and Thunell (1991) have shown that both *N. dutertrei* and *N. pachyderma* follow isothermal ranges, migrating to shallower waters during upwelling and subsequently descending after upwelling. Due to both, depth stratification and seasonal occurrence, planktic foraminifera have the potential to be used for the reconstruction of the vertical and seasonal variability of the hydrography within the photic zone. However, it must be taken into account that the signal carrier itself is sensitive to environmental changes, which can lead to a bias in the oxygen isotope record.

$\delta^{18}O$ from Core Top Foraminifera

Foraminifera from core- top sediments contain the integrated signal of all processes discussed in the previous paragraphs and are therefore important upon assessing the potential of foraminifera to record surface water conditions. Fig. 5 shows a comparison of the expected $\delta^{18}O$ of foraminiferal calcite with the observed $\delta^{18}O$ values from a global core- top compilation (see Schmidt and Mulitza 2002 and references therein). Although oxygen isotope values from core top foraminifera are generally much higher than indicated for surface waters, most of the lateral hy-drographic gradients of present-day surface waters are remarkably well preserved. For example, note the effects of the eastern and western boundary currents, or the input of freshwater into the north-eastern Indian Ocean.

Fig. 6 shows species-specific histograms of the deviations from annual mean surface-water $\delta^{18}O$. It is evident that most of the foraminiferal shells from the sediment show oxygen isotope values that are approximately 0.5-1 ‰ higher than indicated for annual mean surface waters. As outlined above, two primary mechanisms may account for this observation: (1) a distortion of the foraminiferal flux towards seasons or periods that were colder than the annual mean, or (2) subsurface calcification occurring at lower temperatures. *G. bulloides* is most abundant in the upwelling season (Hebbeln et al. 2000). Hence the bias towards higher values might reflect a seasonal distortion to colder up-welling conditions. The highest flux of the other three

GeoB 1402-2

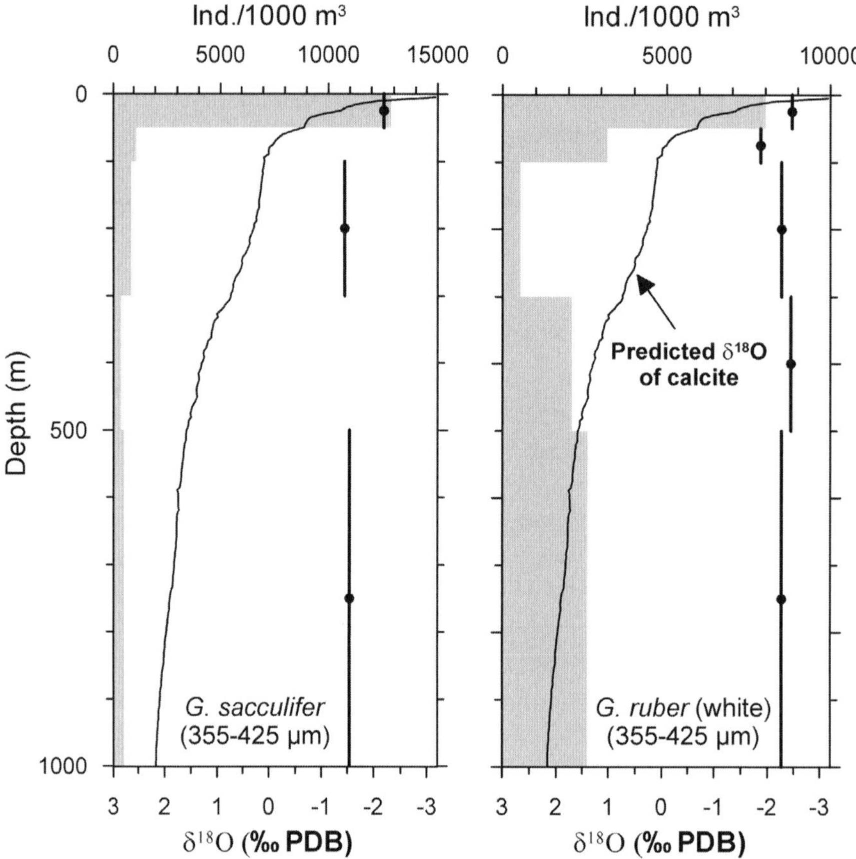

Fig. 4. Vertical distribution (shaded bars) and oxygen isotope composition of *G. sacculifer* and *G. ruber* (white) at the plankton tow station GeoB 1402-2 in the eastern equatorial Atlantic. Data are from Kemle-von Mücke and Oberhänsli (1999).

species is primarily during the summer (Deuser and Ross 1989; Donner and Wefer 1994; cf. Fig. 3). Therefore, a distortion of the foramini-feral population in the sediment towards colder seasons seems unlikely to be the reason for the positive deviations in *G. sacculifer*, *G. ruber* and *N. pachyderma*. It is well known that *G. sac-culifer* and *N. pachyderma* (sin.) add calcite in deeper and cooler water masses (Duplessy et al. 1981; Lohmann 1995; Kohfeld et al. 1997). Interestingly, the deviations of both *G. ruber* and *G. sacculifer* have a pronounced bimodal distribution, with one maximum close to surface-water conditions, and one at about +0.6 and +1.0 (Fig. 6). This is consistent with observations that the shells of *G. ruber* and *G.*

sacculifer are essentially a mixture of two end-member calcites, the initial shell and the calcite crust (Duplessy et al. 1981; Lohmann 1995). Thus the bimodal distribution might represent high abundances of both crusted and uncrusted shells.

Most of the deviations from surface-water $\delta^{18}O$ in the core-top measurements examined here are in the same range as the differences between the classical equations and the equation derived in this work. For example, the equations of Shackleton (1974), or Erez and Luz (1983), would indicate calcification close to equilibrium at the sea surface for most of the core-top samples, or negative off-sets due to vital effects. For paleoceanographic applications, it has to be considered that the $\delta^{18}O$

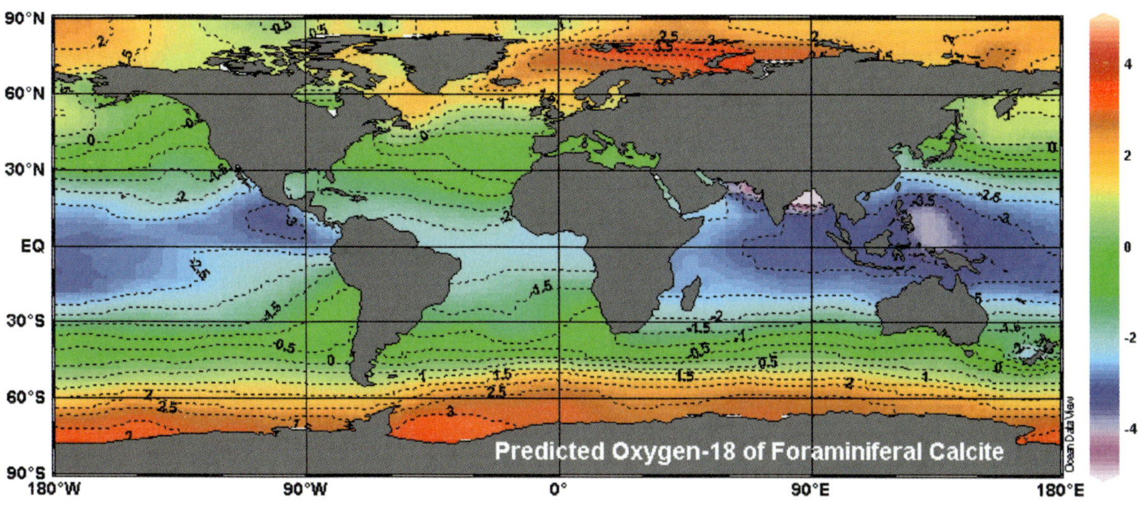

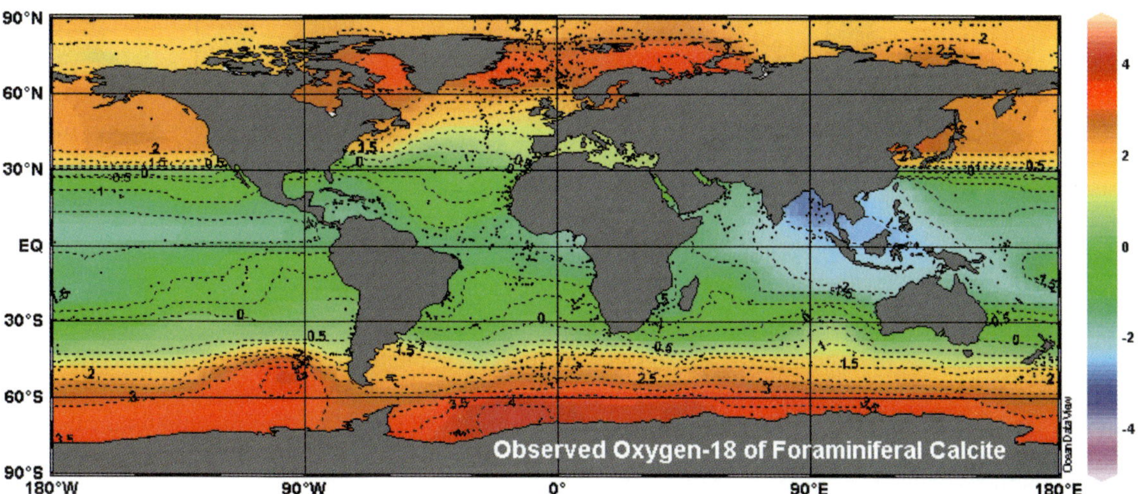

Fig. 5. Upper panel: Distribution of predicted $\delta^{18}O$ of calcite calculated from the $\delta^{18}O$ and temperature of surface waters as displayed in Fig. 1 with the equation derived from living planktic foraminifera (see Fig. 2a). Lower panel: distribution of coretop foraminiferal $\delta^{18}O$ of the species *G. ruber*, *G. sacculifer*, *G. bulloides* and *N. pachyderma* (data sources in Schmidt and Mulitza 2002, http://www.pangaea.de/home/smulitza/).

composition of most foraminiferal populations, even the shallowest dwelling foraminifera, is not a pure surface-water signal, but a mixture of shells or shell parts calcified at different water depths.

Postdepositional Effects

Two processes may alter the isotopic signal of the mean foraminiferal population in the sediment af-

ter deposition: bioturbation and calcite dissolution. The main effect of bioturbation on oxygen isotope records is a truncation of the glacial/Interglacial amplitude. Fig. 7a shows the oxygen isotope difference between the core top and the LGM versus the sedimentation rate of sediment cores recovered from the tropical and South Atlantic ocean by the SFB 261 project. Obviously, the glacial/interglacial amplitude decreases considerably below a

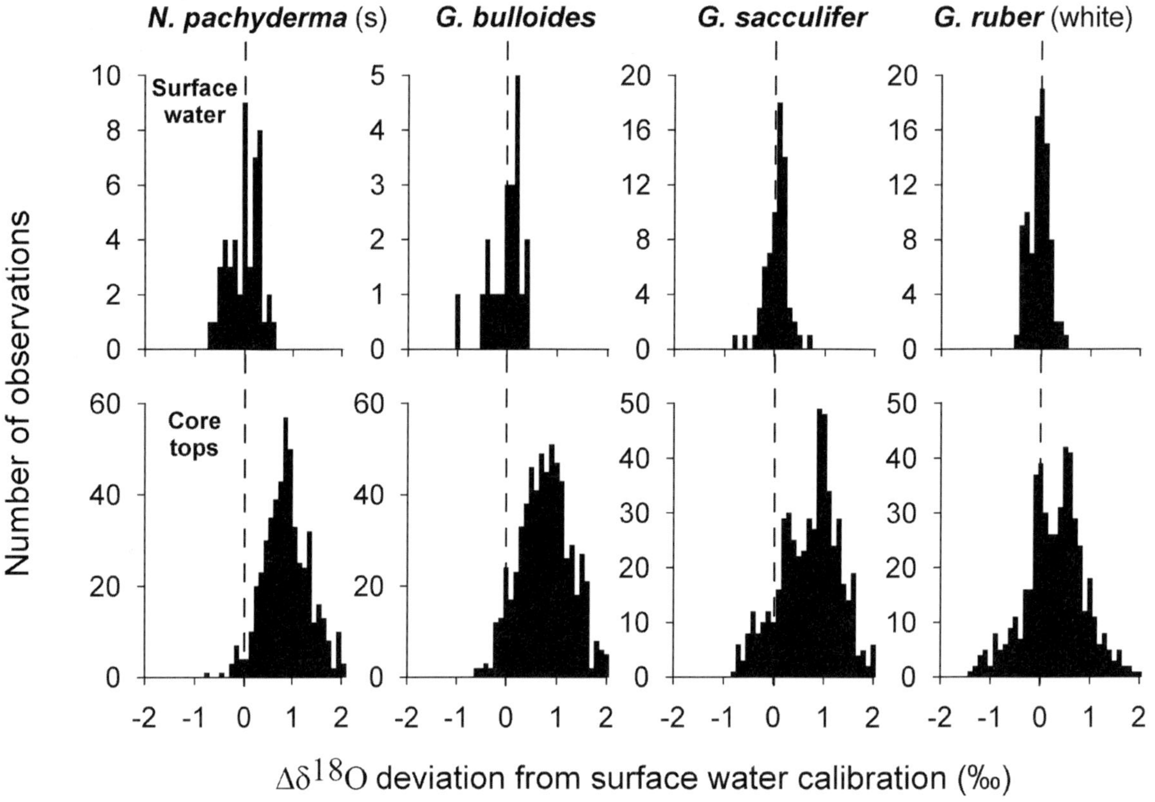

Fig. 6. Frequency distributions of the deviations between annual mean equilibrium calcite for surface waters and foraminiferal carbonate from core tops. Core top data are from a compilation (Schmidt and Mulitza 2002, http://www.pangaea.de/home/smulitza/). Positive deviations mean that $\delta^{18}O$ is higher in core top samples than indicated by the relationship derived from living planktic foraminifera (Fig. 2a). $\delta^{18}O$ of foraminiferal calcite has been calculated by solving the quadratic equation derived in this study for calcite, and by using annual mean temperature and $\delta^{18}O$ of seawater derived from atlas values.

threshold of about 2 cm/ka. This pattern is consistent with a continuous mixing of shells (Broecker 1986; Mix 1987). Below a sedimentation rate of 2 cm/ka, shells from both glacial and interglacial sediments mix within a sample and the $\delta^{18}O$ amplitude consequently decreases. Another effect of bioturbation is the shift of the maximum and minimum values in an oxygen isotope record, when considerable changes in the abundance of the measured species occur. As more shells are transported from maximum to minimum abundances than vice versa, peaks in the oxygen isotope record can be considerably shifted along the core depth, if they occur at minimum abundance.

Dissolution can also alter the isotopic composition of foraminiferal shells after deposition. The effect of dissolution clearly depends on the homogeneity of the foraminiferal shells. Many foraminifera accumulate different kinds of calcite during their life cycle. For example *G. sacculifer* usually adds a thick calcite crust in deeper and colder waters (Duplessy et al. 1981; Lohmann 1995). Thus the $\delta^{18}O$ of the shell of *G. sacculifer* is a mixture of two end-member isotope values: that of the initial shell, and that of the calcite crust. If the initial shell is preferentially dissolved, the $\delta^{18}O$ of the remainder will be shifted towards higher values. For this reason, $\delta^{18}O$ values of *G. sacculifer* from the Ontong Java Plateau increase with increasing water depth and increasing calcite dissolution (Wu and Berger 1989, Fig. 7b).

Late Quaternary $\delta^{18}O$ Records

The Ice Volume Effect

As shown in the previous paragraphs, the composition of oxygen isotopes in planktic foraminifera mainly depends on the temperature and the $\delta^{18}O$ of the surrounding seawater. The latter was subject to large variations throughout the Quaternary due to variations in ice volume. Continental ice generally has a very low $\delta^{18}O$. For this reason the waxing and waning of the ice sheets had a huge effect on the isotopic composition of the ocean. Estimates of the mean "ice effect" range between 1.3 ‰ (Fairbanks 1989) and 1 ‰ (Labeyrie et al. 1987; Schrag and DePaolo 1993) for the last deglaciation. Hence, typically two- thirds of the isotope signal recorded in deep- sea cores is attributable to variations in continental ice volume. This

global change in seawater $\delta^{18}O$ has to be considered, when absolute temperatures are reconstructed from oxygen isotopes of planktic fora-minifera. However, it must also be noted that the oxygen isotope composition of seawater at a single site is determined by local effects, e.g. changes in circulation or moisture balance and global ice volume. Even the glacial interglacial $\delta^{18}O$ change in deep waters is not uniformly distributed. For example, Schrag et al. (2002) and Adkins et al. (2002) reported changes between 0.7 and 1.1 ‰ for different deep water sites in the Atlantic.

Although the temporal evolution of global sea-level changes is well constrained through coral terraces and concomitant U/Th datings (e.g. Chappell and Shackleton 1986; Fairbanks 1989), some uncertainty exists about the timing of global $\delta^{18}O$ changes. Since the $\delta^{18}O$ composition of the ice sheet is not homogenous, a change in ice volume must not lead to a proportional change in $\delta^{18}O$ of sea water. Mix and Ruddimann (1985) estimated that the effect of changing isotopic composition of glacier ice may induce a lag of oceanic $\delta^{18}O$ of

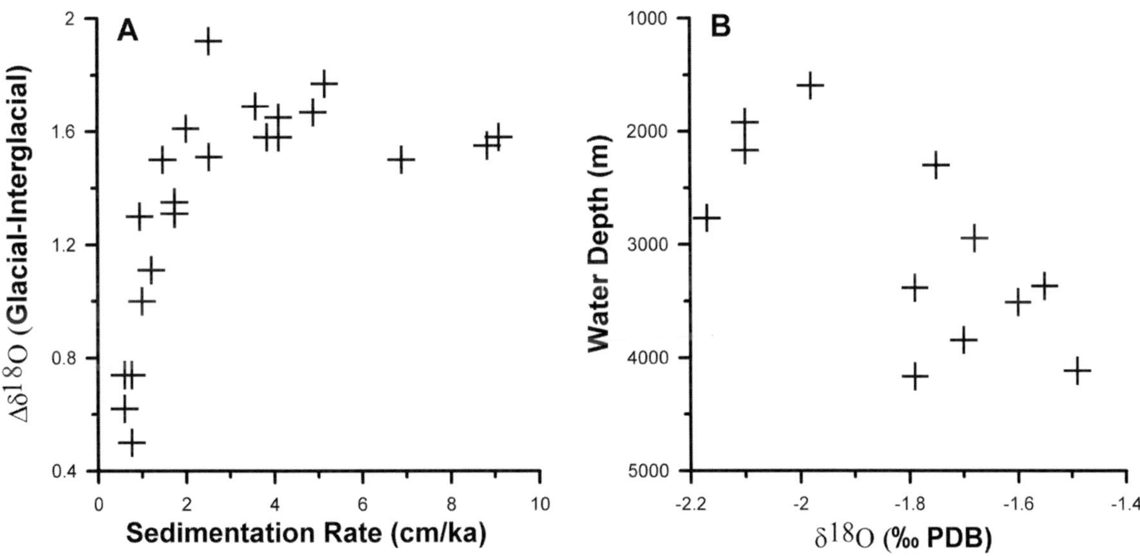

Fig. 7. a) Glacial interglacial $\delta^{18}O$ difference of various species from cores in the tropical and South Atlantic. $\delta^{18}O$ amplitudes decrease with decreasing sedimentation rate suggesting that sediment mixing affects the oxygen isotope record below a threshold of about 2 cm/ka. **b)** Oxygen isotope composition of *G. sacculifer* from coretops of the Ontong Java Plateau (Wu and Berger 1989). $\delta^{18}O$ values increase with water depth and suggest an effect of calcite dissolution on the oxygen isotope composition of foraminiferal carbonate.

about 1 to 3 kyrs behind the real ice volume change. In the light of these uncertainties, the best strategy to use oxygen isotopes in paleoceanography seems to consist in looking at regional gradients (e.g. Arz et al. 1999) instead of absolute changes.

The Last Glacial Maximum in the Tropical and South Atlantic

Atlantic Sector of the Southern Ocean. The Southern Ocean is the ventilation area for much of the world ocean, and many bottom, deep and intermediate water masses are derived from Ant-arctic surface waters. For this reason, the history of the surface hydrography is critical for the understanding of past circulation changes. Oxygen isotopes in planktic and benthic foraminifera have been used to constrain the Southern Ocean hydrography during the LGM (Charles and Fairbanks 1990; Matsumoto et al. 2001). Fig. 8 shows a compilation of Holocene and LGM $\delta^{18}O$ data for the planktic foraminifer *N. pachyderma* and benthic *Cibicidoides* on a latitudinal transect (Matsumoto et al. 2001). The Holocene *N. pachyderma* $\delta^{18}O$ agrees with the predicted $\delta^{18}O$ at a water depth of 200 m. The northward decrease of $\delta^{18}O$ reflects the temperature gradient between the cold polar and warm subtropical subsurface waters, and occurs almost entirely to the north of the present-day Polar Front. Poleward of 50°S, *N. pachyderma* shows nearly constant values. $\delta^{18}O$ of Holocene *Cibicidoides* is nearly constant across the latitudinal transect, because it mostly reflects the approximately constant temperature and salinity of Circumpolar Deepwater (CDW). South of the present-day Polar Front, benthic and planktic $\delta^{18}O$ values are almost similar, because they both calcify in the CDW, which is upwelled south of the Polar Front. Hence the position of the Polar Front is marked by the region where benthic and planktic $\delta^{18}O$ values diverge.

As indicated in Fig. 8b, the structure of the latitudinal distribution of glacial *N. pachyderma* and *Cibicidoides* is very similar to the Holocene. This implies that the position of the Polar Front was not very different from today (Matsumoto et al 2001). Indeed, results from other proxies also suggest only slight northward shifts of the Polar Front during

the LGM. For example, Brathauer and Abelmann (1999) reconstructed a northward shift of the Polar Front of about 3°. Considering the rather sparse data for the LGM close to the Polar Front, a slight northward shift does not seem to contradict glacial $\delta^{18}O$ distribution. However, more oxygen isotope data from the frontal area of the Southern Ocean are needed to further specify the paleoposition of the Polar Front during the LGM.

Tropical Atlantic. The magnitude of glacial cooling in the tropics is an important parameter to assess the sensitivity of the climate system to radiative perturbations, because the tropics are remote from any cooling effect of ice (Crowley 2000). CLIMAP (1981) reconstructed very small

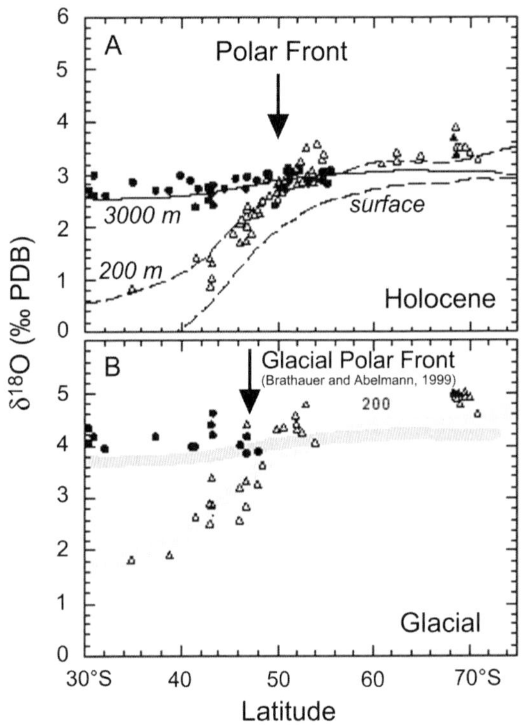

Fig. 8. South Atlantic Holocene **a)** and LGM **b)** latitudinal $\delta^{18}O$ distributions of *N. pachyderma* (open triangles), *Cibicidoides* species (circles), and *E. exigua* (solid triangles). Lines represent equilibrium $\delta^{18}O$ for calcite at the surface, at 200 m and at 3000 m water depth. The approximate position of the Polar Front is indicated by an arrow. Position of glacial Polar Front is from Brathauer and Abelmann (1999). Modified from Matsumoto et al. (2001).

temperature changes in the order of 1-2°C in the western tropical Atlantic warm pool. In contrast, subsequent estimates from corals off Barbados point to temperature changes of 5-6°C (Guilderson et al. 1994). Foraminiferal oxygen isotopes, however, do not support such a strong cooling of tropical surface waters. Fig. 9 represents a compilation of glacial-interglacial $\delta^{18}O$ values (Wolff et al. 1998). They were obtained by averaging samples within 10 specific regions, where several cores cluster and both surface sediments and LGM samples were available. Generally, glacial-interglacial differences in $\delta^{18}O$ values ranged between 1.8 and 2.0‰ throughout the study area, with highest values around 2.1‰ in the north-western tropical Atlantic. Lowest values of about 1.6‰ were obtained off the Brazilian Coast at about 8°S. Wolff et al.

(1998) compared the oxygen isotope differences with the CLIMAP estimates by assuming that a change in temperature of 1°C is equivalent to a oxygen isotope change of 0.22‰ (cf. Fig. 2). In the eastern and southern parts of the study area, oxygen isotope signatures are in good agreement with CLIMAP. In the northwestern tropical Atlantic warm pool, however, the differences between CLIMAP and our estimates increases, reaching a maximum of 0.66 ‰ at the Ceara Rise. Considering local changes in the $\delta^{18}O$ of seawater (see Wolff et al. 1998 for details), the observed oxygen isotope differences would require an additional change of sea- surface temperatures of 1-2°C compared to CLIMAP, which would amount to a total glacial cooling of 2-3°C in the north-western tropical Atlantic. A temperature change of 2-3°C

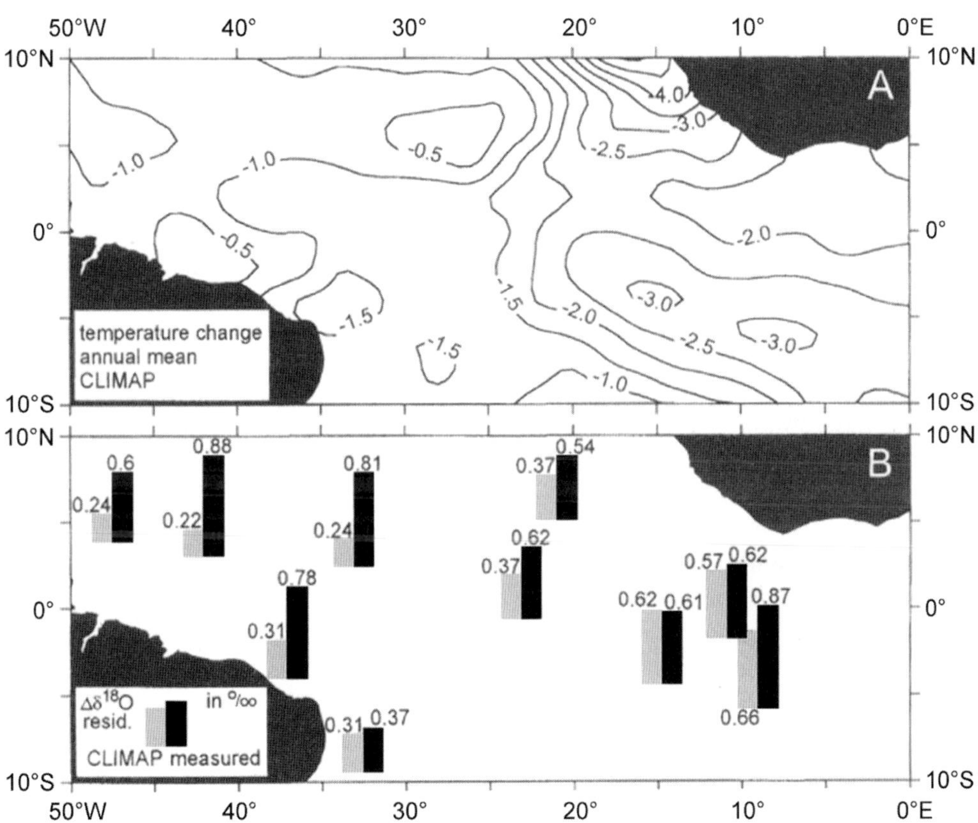

Fig. 9. a) Glacial to interglacial differences in annual mean temperature according to CLIMAP (1981) (we used mean of February and August). **b)** $\delta^{18}O$ residuals after subtracting an ice effect 1.2 ‰ (right bar) as opposed to the theoretical $\delta^{18}O$ residuals (left bar) computed from CLIMAP temperature differences in a. Redrawn from Wolff et al. (1998).

as depicted in the study conducted by Wolff et al. (1998), represents an intermediate estimate between the results of CLIMAP (~1°C) and the results derived from corals (5-6°C). Recent studies using coupled ocean-atmosphere models (Ganopolski et al. 1998) and alkenones (Rühlemann et al. 1999), also suggest a cooling of about 3°C in the tropical Atlantic. Altogether, these results are indicative of an intermediate sensitivity of the tropical climate to radiative forcing.

The Last Deglaciation in the Tropical and South Atlantic

Interhemispheric Comparison of Oxygen Isotope Records. The deglaciation offers the opportunity to study the response of the climate system to perturbations like changes in sea level, atmospheric greenhouse gases, or insolation. In the high northern latitudes, the last deglaciation was interrupted by several brief returns to near-glacial climate (Dansgaard et al. 1993; Bond et al. 1993; Dokken and Jansen 1999). The most prominent of these events was the Younger Dryas cold period. Temperatures over Greenland, as seen in the GRIP ice-core $\delta^{18}O$ record (Fig. 10a), are essentially controlled by heat released from the surface of the North Atlantic Ocean. This heat is delivered from the warm water reservoir in the western tropical Atlantic by the northward transport of warm upper-layer waters, which are imported to compensate for the southward flow of North Atlantic Deep Water (NADW) formed in the Norwegian-Greenland and Labrador seas. As a result of this cross-equatorial exchange of water, the South Atlantic exports heat to the North Atlantic (Crowley 1992). Modeling experiments predict an abrupt Younger Dryas cooling in response to a freshwater-induced reduction in northward heat transport and a concomitant warming over much of the South Atlantic, with a maximum warming in the eastern boundary currents (e.g. Rind et al. 2001). An asynchronous coupling of the two hemispheres caused by changes of thermohaline circulation has been inferred from temperature records of Arctic and Antarctic ice cores (Blunier et al. 1998, Fig. 10a and d), and from alkenone and oxygen-isotope records which display warm surface temperatures

in the southern Atlantic during periods of reduced NADW production (Charles et al. 1996; Rühlemann et al. 1999; Mulitza and Rühlemann 2000). A seesaw pattern of sea surface temperatures is also indicated by foraminiferal oxygen isotopes. Fig. 10b is a comparison of two high- resolution oxygen isotope records of *G. bulloides*, one from the Benguela upwelling region at 20°S (GeoB 1023-5, Schneider et al. 1992), the other from the Iberian Continental Margin (SU81-18, Bard et al. 1989; Bard et al. 2000) at 38°N. Both records have been extensively dated with AMS ^{14}C of foraminiferal calcite. It is evident that the $\delta^{18}O$ difference between both sites is highest during deglaciation, namely during the Younger Dryas and the Heinrich Event 1. Since both time slices are associated with strong reductions in deep water formation (e.g. Boyle and Keigwin 1987), and cooling in the northern North Atlantic, it is likely that the seesaw pattern is due to a reduced heat transport into the northern hemisphere, and the concomitant heat storage and warming in the South Atlantic. It must be noted, however, that there is recent evidence for an early response of sea surface temperatures in other oceans (Lea et al. 2000; Herbert et al. 2001). Since the Atlantic is the only Ocean with a net northward heat transport, the seesaw mechanism cannot explain the early warming in other oceans. Future work might show that the temperature signal displayed in the Antarctic Byrd ice core reflects the global temperature pattern, and that the North Atlantic lagged behind because of the huge northern continental ice sheets supplying large amounts of melt water into the polar North Atlantic, which led to a disruption in the formation of North Atlantic Deep Water and thermohaline circulation.

Influence of the Seesaw on Atmospheric Circulation and Precipitation. The Holocene climatic evolution of tropical Africa is characterized by drastic changes in precipitation patterns. Early Holocene lake levels rose to a maximum across much of Africa between about 12,000 and 5,000 ^{14}C yr B.P. (Butzer et al. 1972; Street and Grove 1979) indicating strong monsoon precipitation. Generally, the early Holocene humid phase is thought to be the result of higher-than-present summer insolation in the Northern Hemisphere

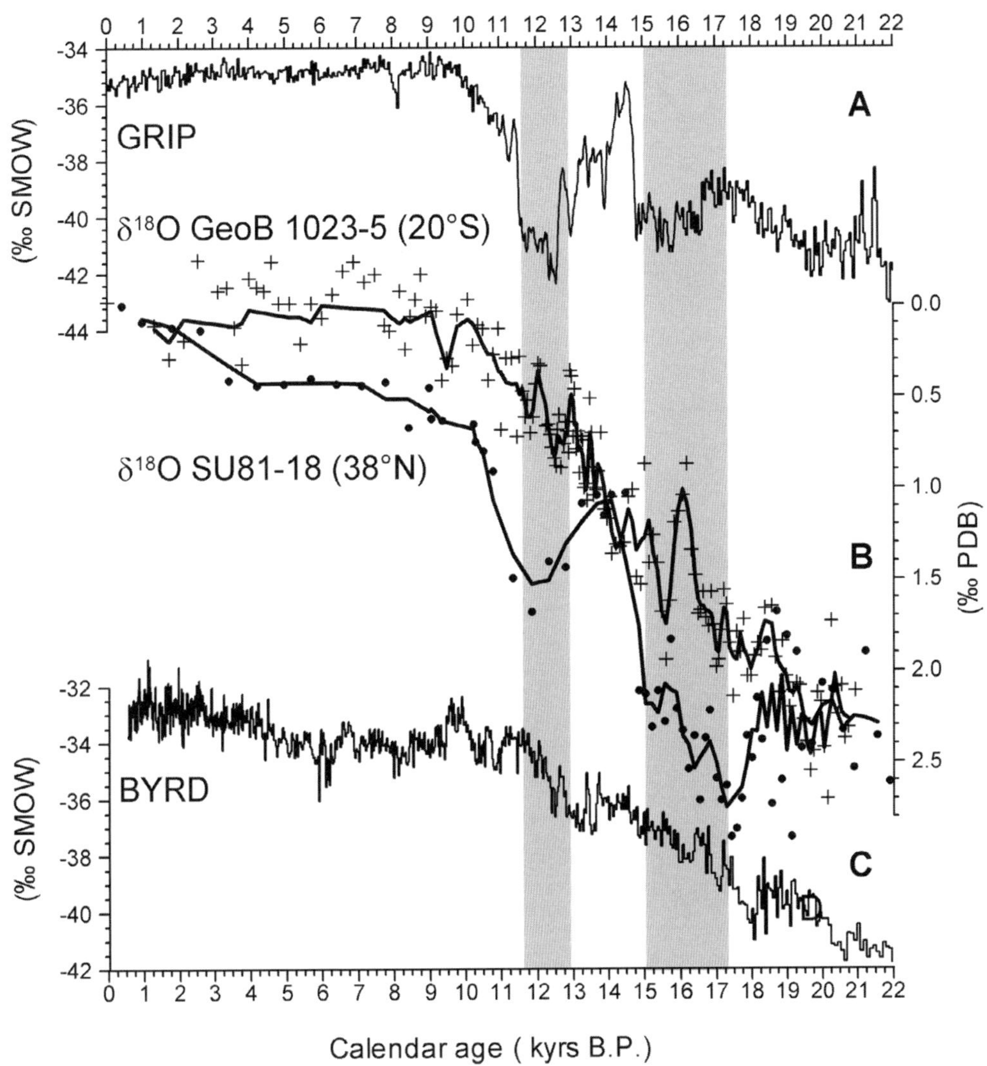

Fig. 10. Interhemispheric comparison of oxygen isotope records of *G. bulloides* of for cores Su81-18 (Bard et al 1989; Bard et al. 2000) and GeoB 1023-5 (Schneider et al. 1992) **b)** over the last 22 kyrs and oxygen isotope records of BYRD **a)** and GRIP **c)** (Johnson et al. 1972; Dansgaard et al. 1993).

(Kutzbach 1981; Kutzbach and Otto-Bliesner 1982; Kutzbach and Liu 1997). Several rapid regressions of African lake levels during the last deglaciation, however, cannot simply be explained by insolation changes. Street-Perrott and Perrott (1990) suggested that the deglacial seesaw pattern in sea surface temperatures (Berger and Vincent 1986; Mix et al. 1986; Fairbanks 1989; Berger 1990) was responsible for rapid climatic shifts in the tropics during the last deglaciation. Again, oxygen isotope records combined with other proxies provide clues for the response of the tropical circulation to rapid

climate changes. Fig. 11 shows the oxygen isotope records of *G. ruber* and *U. peregrina* from core GeoB 1007-4, which was recovered from the upper part of the Congo deep-sea fan at a water depth of 1502 m. To distinguish the regional deviations from the global $\delta^{18}O$ signal, we plotted the $\delta^{18}O$ curve on the same scale with the global $\delta^{18}O$ record calculated from sea- level variations (Fairbanks 1989), by matching both records in the late Holocene (0–8,000 yr B.P.). The $\delta^{18}O$ record of *G. ruber* shows three negative excursions from the global signal during the last deglaciation at about 12,600

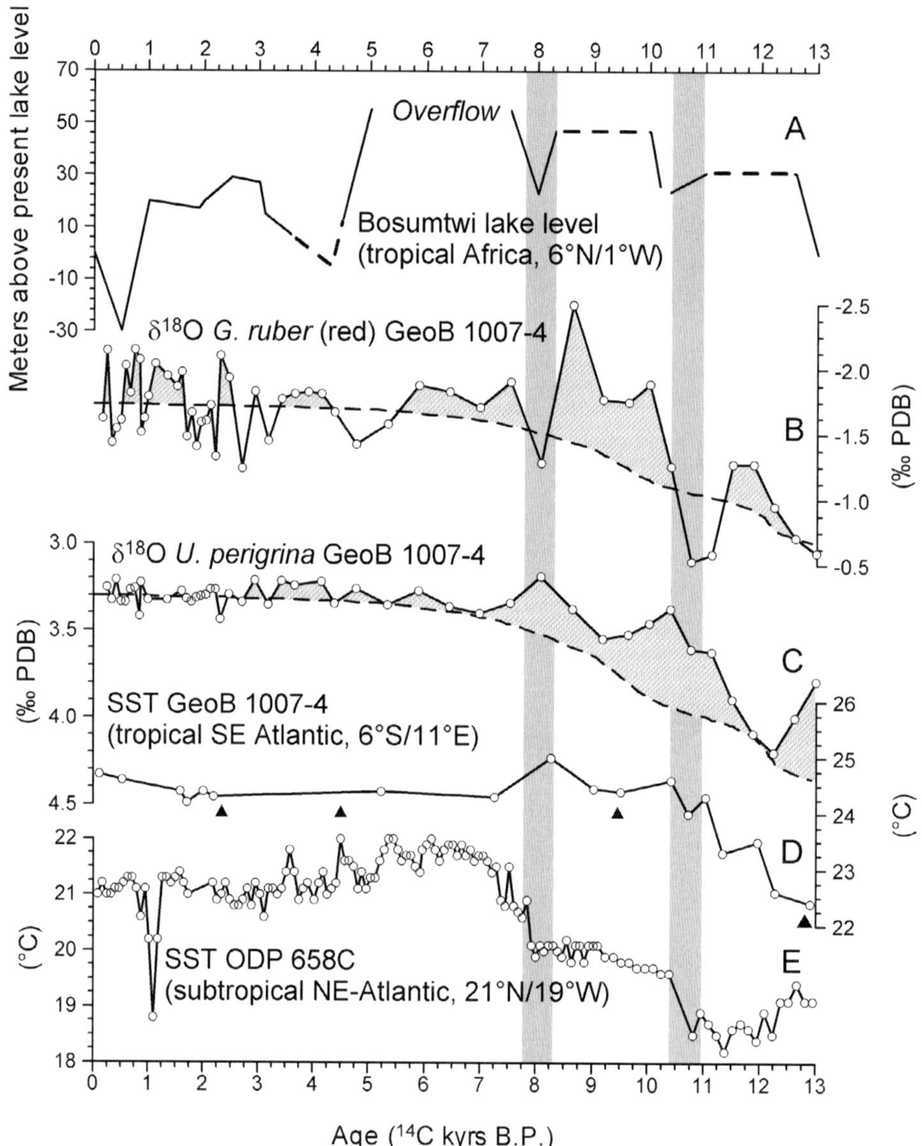

Fig. 11. a) Fluctuations in lake level, Lake Bosumtwi (Ghana) (Talbot et al. 1984; Street-Perrott and Harrison 1984). **b)** Oxygen-isotope record of the shallow-dwelling planktic foraminifer *G. ruber* (red) in sediment core GeoB 1007-4. The shaded area denotes negative deviations from the global δ¹⁸O signal (Fairbanks 1989). **c)** Oxygen-isotope record of bottom-dwelling benthic foraminifer *U. peregrina*. Shaded areas as for *G. ruber* in Fig. 2c. **d)** Alkenone-derived sea-surface temperature of core GeoB 1007-4. Triangles indicate age control points which are based on ¹⁴C dating of total organic carbon. **e)** Alkenone-derived sea-surface temperatures of ODP core 658C (Zhao et al. 1995). Stippled areas represent times of increased interhemispheric temperature gradients coinciding with lake-level lowstands in tropical Africa. Redrawn from Mulitza and Rühlemann (2000)

–11,300, 10,400–8200, and 7,800–5,500 yr B.P. At 8,600 yr B.P., the δ¹⁸O values of *G. ruber* are up to 1‰ more negative than to be expected from the global signal. As foraminiferal δ¹⁸O decreases by about 0.22‰ with each degree of warming, a de-cline of 1‰ would require an SST increase of about 4°C relative to the mean for the late Holocene. However, as indicated by the alkenone record measured along the same core (Fig. 2d), early Holocene temperatures were only slightly higher, or even colder than during the late Holocene. Hence, it appears that the negative $\delta^{18}O$ deviations are due to regionally lower $\delta^{18}O$ of surface waters, instead of temperature. Pastouret et al. (1978) and Giresse et al. (1982) suggested that early Holocene excursions in the $\delta^{18}O$ of *G. ruber*, observed in sediment cores from the Niger and Congo deep-sea fans, were due to the increased discharge of isotopically light river water. Support for this hypothesis comes from the water- level record of Lake Bosumtwi (Street-Perrott and Harrison 1984; Talbot et al. 1984) (Fig. 11a), which is representative of western Africa south of the Sahara (Street-Perrott and Perrott 1990). Early Holocene lake high stands, indicating greater precipitation, are coeval with low $\delta^{18}O$ residuals in GeoB 1007-4 owing to higher freshwater discharge (Fig. 11a, b). This is in excellent agreement with the results of circulation models that predict stronger monsoonal precipitation due to higher thermal land–sea contrasts in response to increased summer insolation in the Northern Hemisphere (Kutzbach 1981; Kutzbach and Otto-Bliesner 1982). The early Holocene "wet period", however, was interrupted by two shorter dry periods at about 11,000 –10,000 and 8,000–7,000 yr B.P (Fig. 10a). The 11,000 –10,000 yr drought corresponds to the well-known Younger Dryas period (e.g. Wohlfarth 1996) marked by positive residuals in the *G. ruber* record of GeoB 1007-4 and low water levels of Lake Bosumtwi. As indicated by the alkenone record of ODP site 658C (Fig. 11e, Zhao et al. 1995), temperatures in the eastern North Atlantic remained relatively low throughout deglaciation until 8,000 yr B.P. The strongest interhemispheric temperature contrasts occur during the Younger Dryas (11,000 –10,000 yr B.P.) and at about 8,000 yr B.P. (Fig. 11d, e). As already mentioned above, this indicates that a smaller fraction of the heat received by the South Atlantic during deglaciation was exported into the North Atlantic.

The weather systems that produce sub-Saharan rainfall receive most of their moisture from the tropical Atlantic. The extent of the moist southwesterly airflow from the tropical Atlantic into North Africa is controlled by the location of the near-equatorial low-pressure trough (Lamb 1978), the Intertropical Convergence. A comparison with the temperature data of site 658C suggests that the

thermal equator and the associated Intertropical Convergence was displaced southward relative to its present position during deglaciation, reaching extremes at 11,000 –10,000 and at 8,000 yr B.P. As observed for the modern ocean (Lamb 1978), this configuration may have reduced the moisture-bearing southwest monsoonal flow into northwestern Africa. The opposite climatic response occurred during the mid-Holocene (7,000– 5,000 yr B.P.), when the thermal equator was displaced far north of its modern position. At that time, several north African lakes (e.g. Lake Bosumtwi, Fig. 2a) attained their highest levels, reflecting intense monsoonal rains. Modeling studies with prescribed modern SSTs suggest that the monsoon reached its greatest intensity in the early Holocene, about 9,000 yr B.P. (Kutzbach 1981; Kutzbach and Street-Perrott 1985), because summer insolation in the Northern Hemisphere was at a maximum. The lagged response of Northern Hemisphere SSTs during the last deglaciation might explain why lake levels and precipitation were still high in the mid-Holocene when summer insolation was already declining. Although the importance of insolation changes for the African climate is unquestionable, changes in the tropical SST distribution related to the global conveyor circulation might explain some of the rapid changes in precipitation recorded in African lake levels. More comprehensive time-slice SST reconstructions for the early and middle Holocene are necessary to assess this hypothesis with atmospheric models.

Conclusions and Future Work

The ratio of oxygen isotopes in foraminiferal calcite is an outstanding proxy for paleoceanographic applications because it has a huge lateral and temporal coverage through much of Earth's history and can in many cases even be measured in extreme environments like the polar oceans or in isolated basins like the Red Sea. Hence, it is worthwhile to undertake more efforts to understand and tackle some of the problems associated with this proxy. First, we must be aware that planktic foraminifera have an environmental sensitivity that may distort the oxygen isotope composition of a foraminiferal population in the sediment towards the hydrographic conditions within the ecological niche of the particular species. Hence, it is necessary to gain more information about the nature and the magnitude of the environmental sensitivity to develop models which enable us to predict seasonal flux, abundance and $\delta^{18}O$ composition of a foraminiferal population in the sediment. Furthermore, the factors that determine the incorporation of oxygen isotopes in living foraminiferal shells must be better understood. Although laboratory cultures have contributed much to the understanding of "vital effects", there is still the need for both laboratory and field studies to obtain information about the influence of the microenvironment and to improve calibrations. Also, the distribution of oxygen isotopes in the present-day hydrological cycle must be better documented by more measurements of sea water and precipitation. There is still an order of magnitude more measurements available for fossil foramini-fera than for modern sea water. Finally, oxygen isotopes should be incorporated into coupled climate model experiments to be able to predict the $\delta^{18}O$ of calcite directly rather than deconvolving the $\delta^{18}O$ record with independent temperature estimates. Once these problems have been solved, the oxygen isotope composition of foraminifera will provide a powerful tool to test coupled climate models under boundary conditions different from the present climate state.

Acknowledgements

The preparation of this paper was supported by the Bundesministerium für Bildung und Forschung (DEKLIM, Grant 01 LD 0019) and the Deutsche Forschungsgemeinschaft (SFB 261 and RCOM). We thank Jelle Bijma and Trond Dokken for careful reviews and Helge Arz and Claire Waelbroeck for discussion and comments. Thanks to P. Grootes and staff of the Leibniz-Labor in Kiel for providing AMS- ^{14}C datings. Data are available under www.pangaea.de/Projects/SFB261.

References

Adkins JF, McIntyre K, Schrag DP (2002) The salinity, temperature, and $\delta^{18}O$ of the glacial deep ocean. Science 298: 1769-1773

Arz H, Pätzold J, Wefer G (1999) The deglacial history of the western tropical Atlantic as inferred from high resolution stable isotope records off northeastern Brazil. Earth Planet Sci Lett 167: 105-117

Bard E, Fairbanks R, Arnold M, Maurice P, Duprat J, Moyes J, Duplessy JC (1989) Sea-level estimates during the Last Deglaciation based on $\delta^{18}O$ and Accelerator Mass Spectrometry ^{14}C ages measured in *Globigerina bulloides*. Quat Res 31: 381-391

Bard E, Rostek F, Turon JL, Gendreau S (2000) Hydrological impact of Heinrich Events in the subtropical Northeast Atlantic. Science 289: 1321-1324

Bauch D, Carstens J, Wefer G (1997). Oxygen isotope composition of living *Neogloboquadrina pachyderma* (sin.) in the Arctic Ocean. Earth Planet Sci Lett 146: 47-58

Bemis BE, Spero HJ, Bijma J, Lea DW (1998) Reevaluation of the oxygen isotopic composition of planktonic foraminifera: Experimental results and revised paleotemperature equations. Paleoceanography 13: 150-160

Berger WH, Vincent E (1986) Sporadic shutdown of North Atlantic deep water production during the Glacial-Holocene transition? Nature 324: 53-55

Berger WH (1990) The Younger Dryas cold spell-a quest for causes. Palaeogeogr Palaeoclimatol Palaeoecol 49: 189-206

Bijma J, Spero HJ, Lea DW (1999) Reassessing foraminiferal stable isotope geochemistry: Impact of the oceanic carbonate system (experimental results). In: Fischer G and Wefer G (eds) Use of Proxies in Paleoceanography. Springer, Berlin, pp 489-512

Birchfield GE (1987) Changes in deep-ocean water $\delta^{18}O$ and temperature from last glacial maximum to the present. Paleoceanography 2: 431-442

Blunier T, Chappellaz J, Schwander J, Dällenbach A, Stauffer B, Stocker TF, Raynaud D, Jouzel J, Clausen HB, Johnson SJ (1998) Asynchrony of Antarctic and Greenland climate change during the last glacial period. Nature 391: 739-743

Bond G, Broecker W, Johnsen S, McManus J, Labeyrie L, Jouzel J, Bonani G (1993) Correlations between climate records from North Atlantic sediments and Greenland ice. Nature 365: 143-147

Boyle EA, Keigwin L (1987) North Atlantic thermohaline circulation during the past 20,000 years linked to high-latitude surface temperatures. Nature 330: 35-40

Butzer KW, Isaac GL, Richardson JA, Washbourn-Kamau C (1972) Radiocarbon dating of East African lake levels. Science 175: 1069

Brathauer U, Abelmann A (1999) Late Quaternary variations in sea surface temperatures and their relationship to orbital forcing recorded in the Southern Ocean (Atlantic Sector) Paleoceanography 14: 135-148

Brewer PG, Glover DM, Goyet C, Shafer DK (1995) The pH of the North Atlantic Ocean: Improvements to the global model for sound absorption in seawater. J Geophys Res 100: 8761-8776.

Broecker WS (1986) Oxygen isotope constraints on surface temperatures. Quat Res 26: 121-134

Chappel J, Shackleton NJ (1986) Oxygen isotopes and sea level. Nature 324: 134-140

Charles CD, Fairbanks RG (1990) Glacial to interglacial changes in the isotopic gradients of the Southern Ocean surface water. In: Bleil U, Thiede J (eds) Geological History of the Polar Oceans: Artic versus Antartic. Kluwer, Dordrecht, pp 519-538

Charles CD, Lynch-Stieglitz J, Ninnemann US, Fairbanks RG (1996) Climate connections between the hemispheres revealed by deep-sea sediment core/ice core correlations. Earth Planet Sci Lett 142: 19-27

CLIMAP (1981) Seasonal reconstructions of the Earth's surface at the last glacial maximum. GSA Map and Chart Ser MC-36

Coplen TB (1996) Editoral: More uncertainty than necessary. Paleoceanography 11: 369-370

Craig H, Gordon LI (1965) Deuterium and oxygen-18 variations in the ocean and the marine atmosphere. In: Tongiorgi E (ed) Stable Isotope in Oceanographic Studies and Paleotemperatures. Spoleto, Consiglio Naz Delle Ricerche, Laboratorio di Geol Nuc, Pisa, pp 9-130

Crowley TJ (1992) North Atlantic Deep Water cools the southern hemisphere. Paleoceanography 7: 489-497

Crowley TJ (2000) CLIMAP SSTs re-revisited. Clim Dyn 16: 241-255

Dansgaard W, Johnsen SJ, Clausen HB, Dahl-Jensen D, Gundestrup NS, Hammer, CU, Hvidberg CS, Steffensen JP, Sveinjörnsdottir AE, Jouzel J, Bond G (1993) Evidence for general instability of past climate from 250-kyr ice-core record. Nature 364: 218-220

Deuser WG (1978) Stable-isotope paleoclimatology: A possible measure of past seasonal contrast from foraminiferal tests. DSIR Bull 220: 55-60

Deuser WG, Ross EH (1989) Seasonally abundant planktonic foraminifera of the Saragossa Sea: Succession, deep-water fluxes, isotopic compositions, and paleoceanographic implications. J Foram Res 19: 268-293

Dokken TM, Jansen E (1999) Rapid changes in the mechanism of ocean convection during the last glacial period. Nature 401: 458-461

Donner B, Wefer G (1994) Flux and stable isotope composition of *Neogloboquadrina pachyderma* and other planktonic foraminifers in the Southern Ocen (Atlantic Sector) Deep-Sea Res 41: 1733-1743

Duplessy JC, Blanc PL, Bé AWH (1981) Oxygen-18 enrichment of planktonic foraminifera due to gametogenic calcification below the euphotic zone. Science 213: 1247-1250

Duplessy JC, Labeyrie L, Juillet-Leclerc A, Maitre F, Duprat J, Sarnthein M (1991) Surface salinity reconstruction of the North Atlantic Ocean during the last glacial maximum. Oceanol Acta 14: 311-324

Emiliani C (1955) Pleistocene temperatures. J Geol 63: 538-578

Erez J, Luz B (1983) Experimental paleotemperature equation for planktonic foraminifera. Geochim Cosmochim Acta 47: 1025-1031

Fairbanks RG, Wiebe PH, Bé AWH (1980) Vertical distribution and isotopic composition of living planktonic foraminifera in the western North Atlantic. Science 207: 61-63

Fairbanks RG, Sverdlove M, Free R, Wiebe PH, Bé AWH (1982) Vertical distribution and isotopic fractionation of living planktonic foraminifera from the Panama Basin. Nature 298: 841-844

Fairbanks RG (1989) A 17,000-year long glacio-eustatic sea level record: Influence of glacial melting rates on the Younger Dryas event and deep-ocean circulation. Nature 342: 637-642

Fischer G, Kalberer M, Donner B, Wefer G (1999) Stable isotopes of pteropod shells as recorders of sub-surface water conditions: Comparison to the record of G. ruber and to measured values. In: Fischer G and Wefer G (eds) Use of Proxies in Paleoceanography. Springer, Berlin, pp 191-206

Ganopolski A, Rahmstorf S, Petoukhov V, Claussen M (1998) Simulation of modern and glacial climates with a coupled global model of intermediate complexity. Nature 391: 351-356

Ganssen G (1983) Dokumentation von küstennahem Auftrieb anhand stabiler Isotope in rezenten Foraminiferen vor Nordwest-Afrika. "*Meteor*"-Forschungsergebnisse 37C: 1-46

Giresse P, Bongo-Passi G, Delibrias G, Duplessy JC (1982) La lithostratigraphie des sédiments hémipelagiques du delta profond du fleuve Congo et ses indications sur les paéoclimats de la fin du Quat Bull Soc Géol. France 7: 803-815

Guilderson TP, Fairbanks RG, Rubenstone JL (1994)

Tropical temperature variations since 20,000 years ago: Modulating interhemispheric climate change. Science 263: 663-665

Hebbeln D, Marchant M, Wefer G (2000) Seasonal variations of the particle flux in the Peru-Chile current at 30°S under normal' and El Nino conditions. Deep-Sea Res 47: 2101-2128

Herbert TD, Schuffert JD, Andreasen D, Heusser L, Lyle M, Mix A, Ravelo AC, Stott LD, Herguera JC (2001) Collapse of the California Current during glacial maxima linked to climate change on land. Science 293: 71-76

Hoefs J (1987) Stable isotope geochemistry. Springer, Berlin, 241 p

Hut G (1987) Consultants group meeting on stable isotope reference samples for geochemical and hydrological investigations, Int At Energy Agency, Vienna, 42 p

Imbrie J, Hays JD, Martinson DG, McIntyre A, Morley JJ, Pisias NG, Prell WL, Shackleton NJ (1984) The orbital theory of Pleistocene climate: Support from a revised chronology of the marine $\delta^{18}O$ record. In: Berger A, Imbrie J, Hays J, Kukla G, Saltzman B (eds) Milankovitch and Climate, Part I. D Reidel Publishing Co, pp 269-305

Jacobs SS, Fairbanks RG, Horibe Y (1985) Origin and evolution of water masses near the antarctic continental margin: Evidence from $H_2^{18}O/H_2^{16}O$ ratios in seawater. Ant Res Ser 43: 59-85

Johnsen SJ, Dansgaard W, Clausen HB, Langway CC (1972) Oxygen isotope profiles through the Antarctic and Greenland ice sheets. Nature 235: 429-434

Jørgensen BB, Erez J, Revsbech NP, Cohen Y (1985) Symbiotic photosynthesis in a planktonic foraminiferan, *Globigerinoides sacculifer* (Brady), studied with microelectrodes. Limnol Oceanogr 30: 1253-1267

Kahn M, Williams DF (1981) Oxygen and carbon isotopic composition of living planktonic foraminifera from the northeast Pacific Ocean. Palaeogeogr Palaeo-climatol Palaeoecol 33: 47-69

Kemle-von Mücke S, Oberhänsli H (1999) The distribution of living planktic foraminifera in relation to southeast Atlantic oceanography. In: Fischer G, Wefer G (eds) Use of Proxies in Paleoceanography: Examples from the South Atlantic. Springer, Berlin, pp 91-115

Kim ST, O'Neil JR (1997) Equilibrium and nonequilibrium oxygen isotope effects in synthetic carbonates. Geochim Cosmochim Acta 61: 3461-3475

Kohfeld KE, Fairbanks RG, Smith SL, Walsh ID (1997) *Neogloboquadrina pachyderma* (sinistral coiling) as paleoceanographic tracers in polar oceans: Evidence from Northeast Water Polynya plankton tows, sediment traps, and surface sediments. Paleoceanography 11: 679-699

Kutzbach JE (1981) Monsoon climate of the Early Holo-cene. Science 214: 59-61

Kutzbach JE, Otto-Bliesner BL (1982) The sensitivity of the African-Asian monsoonal climate to orbital parameter changes for 9000 years BP in a low-resolution general circulation model. J Atmos Sci 39: 1177-1188

Kutzbach JE, Street-Perrott FA (1985) Milankovitch forcing of fluctuations in the level of tropical lakes from 18 to 0 kyr BP. Nature 317: 130-134

Kutzbach JE, Liu Z (1997) Response of the African Monsoon to Orbital Forcing and Ocean Feedbacks in the Middle Holocene. Science 278: 440-442

Labeyrie LD, Duplessy JC, Blanc PL (1987) Variations in mode of formation and temperature of oceanic deep waters over the past 125,000 years. Nature 327: 477-482

Lamb PJ (1978) Case studies of tropical Atlantic surface circulation patterns during recent sub-Saharan weather anomalies: 1967 and 1968. Mon Weather Rev 106: 482-491

Lea DW, Pak DK, Spero HJ (2000) Climate Impact of Late Quaternary Equatorial Pacific Sea Surface Temperature Variations. Science 289: 1719-1724

Levitus S, Boyer T (1994) World Ocean Atlas, Vol. 4: Temperature. US Gov Printing Office, Washington DC

Lohmann GP (1995) A model for variation in the chemistry of planktonic foraminifera due to secondary calcification and selective dissolution. Paleoceanography 10: 445-457

Matsumoto K, Lynch-Stieglitz J, Anderson RF (2001) Similar glacial and Holocene Southern Ocean hydro-graphy. Paleoceanography 16: 1-10

Mix AC, Ruddiman WF (1985) Structure and timing of the last deglaciation: Oxygen isotope evidence. Quat Sci Rev 4: 59-108

Mix AC, Ruddiman WF, McIntyre A (1986) Late Quaternary paleoceanography of the tropical Atlantic, 1: Spatial variability of annual mean sea-surface temperatures, 0-20,000 years B.P. Paleoceanography 1: 43-66

Mix AC (1987) The oxygen-isotope record of glaciation. In: Ruddiman WF, Wright HEJ (eds) North America and Adjacent Oceans during the Last Deglaciation. The Geology of North America, vol. K-3, Geol Soc Am, pp 111-135

Mulitza S, Dürkoop A, Hale W, Wefer G, Niebler HS (1997) Planktonic foraminifera as recorders of past surface-water stratification. Geology 25: 335-338

Mulitza S, Wolff T, Pätzold J, Hale W, Wefer G (1998) Temperature sensitivity of planktonic foraminifera and its influence on the oxygen isotope record. Mar Micropaleontol 33: 223-240

Mulitza S, Rühlemann C (2000) African monsoonal precipitation modulated by interhemispheric temperature gradients. Quat Res 53: 270-274

Mulitza S, Boltovskoy D, Donner B, Meggers H, Paul A, Wefer G (2003) Temperature: $\delta^{18}O$ relationships of planktic foraminifera collected from surface waters. Palaeogeogr Palaeoclim Palaeoecol 202: 143-152

O'Neil JR, Clayton RN, Mayeda TK (1969) Oxygen isotope fractionation in divalent metal carbonates. J Chem Phys 51: 5547-5558

Östlund HG, Craig C, Broecker WS, Spencer D (1987) GEOSECS Atlantic, Pacific, and Indian Ocean Expedition, Shorebased data and graphics. GEOSECS Atlas Ser vol 7. US Government Printing Office

Pastouret L, Chamley H, Delibrias G, Duplessy JC, Thiede J (1978) Late Quaternary climatic changes in western tropical Africa deduced from deep-sea sedimentation off the Niger delta. Oceanol Acta 1: 217-232

Paul A, Mulitza S, Pätzold J, Wolff T (1999) Simulation of oxygen isotopes in a global ocean model. In: Fischer G, Wefer G (eds) Use of Proxies in Paleoceanography. Springer, Berlin, pp 655-686

Rind D, Russell GL, Schmidt GA, Sheth S, Collins D, DeMenocal P, Teller J (2001) Effects of glacial meltwater in the GISS Coupled atmosphere-Ocean Model: Part II: A bi-polar seesaw in Atlantic Deep Water production. J Geophys Res 106: 27355-27366

Rohling EJ, Cooke S (2001) Stable oxygen and carbon isotope ratios in foraminiferal carbonate. In: Gupta S (ed) Modern Foraminifera. Kluwer Academic, Dordrecht, The Netherlands, pp 239-258

Rühlemann C, Mulitza S, Müller PM, Wefer G, Zahn R (1999) Warming of the tropical Atlantic Ocean and slowdown of thermohaline circulation during the last deglaciation. Nature 402: 511-514

Rühlemann C, Diekmann B, Mulitza S, Frank M (2001) Late Quaternary changes of western equatorial Atlantic surface circulation and Amazon lowland climate recorded in Ceará Rise deep-sea sediments. Paleoceanography 10: 293-305

Sautter LR, Thunell RC (1991) Seasonal variability in the $\delta^{18}O$ and $\delta^{13}C$ of planktonic foraminifera from an upwelling environment; Sediment trap results from the San Pedro Basin, Southern California. Paleoceanography 6: 307-34

Schmidt GA, Bigg GR, Rohling EJ (1999) Global Seawater Oxygen-18 Database. http://www.giss.nasa.gov/data/o18data/

Schmidt GA, Mulitza S (2002) Global calibration of ecological models for planktic foraminifera from coretop carbonate oxygen-18. Marine Micropaleontol 44: 125-140

Schneider R, Dahmke A, Kölling A, Müller PJ, Schulz HD, Wefer G (1992) Strong deglacial minimum in the $\delta^{13}C$ record from planktionic foraminifera in the Benguela upwelling region: Palaeoceanographic signal or early diagenetic imprint? In: Summerhayes CP, Prell WL, Emeis KC (eds) Upwelling Systems: Evolution Since the Early Miocene. Geological Society Special Publication, pp 285-297

Schrag DP, DePaolo DJ (1993) Determination of $\delta^{18}O$ of seawater in the deep ocean during the last glacial maximum. Paleoceanography 8: 1-6

Schrag DP, Adkins JF, McIntyre K, Alexander JL, Hodell DA, Charles CD, McManus JF (2002) The oxygen isotopic composition of seawater during the Last Glacial Maximum. Quat Sci Rev 21: 331-342.

Schwalb A, Burns SJ, Kelts K (1999) Holocene environments from stable isotope stratigraphy of ostracodes and authigenic carbonates in chilean altiplano lakes. Palaeogeogr Palaeoclim Palaeoecol 148: 153-168

Shackleton NJ, Opdyke ND (1973) Oxygen isotope and paleomagnetic stratigraphy of equatorial Pacific core V 28-238: Oxygen isotope temperatures and ice volumes on a 10^5 year scale. Quat Res 3: 39-55

Shackleton N (1974) Attainment of isotopic equilibrium between ocean water and the benthonic foraminifera genus Uvigerina: Isotopic changes in the ocean during the last glacial. Colloq Int CNRS 219: 203-209

Spero HJ, Lea DW (1993) Intraspecific stable isotope variability in the planktic foraminifera *Globigerinoides sacculifer*: Results from laboratory experiments. Mar Micropaleontol 22: 221-234

Spero HJ, Bjima J, Lea DW, Bemis BB (1997) Effect of seawater carbonate concentration on foraminiferal carbon and oxygen isotopes. Nature 390: 497-500

Spero HJ, Mielke KM, Kalve EM, Lea DW, Pak DK (2003) Multispecies approach to reconstructing eastern equatorial Pacific thermocline hydrography during the past 360 kyr. Paleoceanography 18: doi:10.1029/2002PA000814.

Street FA, Grove AT (1979) Global maps of lake-level

fluctuations since 30,000 years BP. Quat Res 12: 83-118

Street-Perrot FA, Harrison SP (1984) Temporal variations in lake levels since 30,000 yr BP - an index of the global hydrological cycle. Geophys Monogr 29: 118-129

Street-Perrott FA, Perrott A (1990) Abrupt climate fluctuations in the tropics: The influence of Atlantic Ocean circulation. Nature 343: 607-616

Talbot MR, Livingstone PG, Palmer PG, Maley J, Melack JM, Delibrias G, Gulliksen S (1984) Preliminary results from sediment cores from Lake Bosumtwi, Ghana. Paleoecol Afr 16: 173-192

Thunell R, Tappa E, Pride C, Kincaid E (1999) Sea-surface temperature anomalies associated with the 1997-1998 El Nino recorded in the oxygen isotope composition of planktonic foraminifera. Geology 27: 843-846

Urey HC (1947) The thermodynamic properties of isotopic substances. Jour Chem Soc 562-581

Weiss RF, Östlund HG Craig H (1979) Geochemical studies of the Wedell Sea. Deep-Sea Res 26: 1093-1120

Williams DF, Healy-Williams N (1980) Oxygen isotopic-hydrographic relationships among recent planktonic foraminifera from the Indian ocean. Nature 283: 848-852

Williams DF, Be AWH, Fairbanks RG (1981) Seasonal stable isotopic variations in living planktonic foraminifera from Bermuda plankton tows. Palaeogeogr Palaeoclim Palaeoecol 33: 71-102

Wohlfarth B (1996) The chronology of the Last Termination: a review of radiocarbon-dated, high-resolution terrestrial stratigraphies. Quat Sci Rev 15: 267-284

Wolf-Gladrow DA, Bijma J, Zeebe RE (1999) Model simulation of the carbonate chemistry in the microenvi-ronment of symbiont bearing foraminifera. Mar Chem 64: 181-198

Wolff T, Mulitza S, Arz H, Pätzold J, Wefer G (1998) Oxygen isotopes versus CLIMAP (18 ka) temperatures: a comparison from the tropical Atlantic. Geology 26: 675-678

Wu G, Berger WH (1989) Planktonic foraminifera: Differential dissolution and the Quaternary stable isotope record in the West Equatorial Pacific. Paleoceano-graphy 4: 181-198

Zeebe RE (1999) An explanation of the effect of seawater carbonate concentration on foraminiferal oxygen isotopes. Geochim Cosmochim Acta 63: 2001-2007

Zhao M, Beveridge NAS, Shackleton NJ, Sarnthein M, Eglinton G (1995) Molecular stratigraphy of cores off northwest Africa: Sea surface temperature history over the last 80 ka. Paleoceanography 10: 661-675

Nitrogen Isotopes in Sinking Particles and Surface Sediments in the Central and Southern Atlantic

M.E. Holmes, G. Lavik, G. Fischer[*] and G. Wefer

Universität Bremen, Fachbereich Geowissenschaften, Klagenfurter Strasse,
28359 Bremen, Germany
** corresponding author (e-mail): gerhard.fischer@rcom-bremen.de*

Abstract: This manuscript provides an overview of sedimentary nitrogen isotope records in the tropical and southern Atlantic Ocean. Sedimentary $\delta^{15}N$ in most of this region reflects the extent of surface water nitrate depletion. Nitrogen isotopes in sediments from the coastal upwelling regions of the Angola Basin and Benguela region off of Africa ranged from 5 to 12‰ and were negatively correlated with averaged near surface (0 - 50 m) historical nitrate concentrations. Coincidence of low $\delta^{15}N$ in sinking particles (2 - 5‰) with low sea surface temperatures and high fluxes confirm the importance of relative nitrate utilization for the isotopic composition of organic matter in the Benguela during modern times. Off the Brazilian coast, isotope ratios were 5 - 7‰ and showed weak correspondence to surface nitrate concentrations, which are low. The expected relationship between nitrate and $\delta^{15}N$ may be obscured here and in other oligotrophic regions because the $\delta^{15}N$ of the small pool of nitrate can be readily altered by the advection or diffusion of even low levels of nitrate with a different isotopic composition. In the tropical and central South Atlantic, sediment isotopic values were between 6 and 11‰. The lack of any apparent relationship between $\delta^{15}N$ and surface nitrate may be partly due to a paucity of nitrate concentration data, but $\delta^{15}N$ near the equator also may be influenced by nitrogen fixation. In these oligotrophic waters, the input of iron via aeolian transport of African dust may support the fixation of N_2 into organic matter, thereby lowering $\delta^{15}N$. Higher input of dust to the surface waters north of the equator is hypothesized to be the cause of the southward $\delta^{15}N$ increase in sinking particles revealed by a north-south transect of moored sediment traps. In the southern South Atlantic, relative nitrate utilization is evidently the main control on sedimentary nitrogen isotopes between around 35°S and 50°S, where average near surface nitrate concentrations were strongly correlated with $\delta^{15}N$. South of the Polar Front, at around 50°S, this relationship is not observable in our data and there is an apparent switch from nitrate-based primary production to production based on ammonium. Sinking particles at the Polar Front are enriched in ^{15}N in austral winter and show $\delta^{15}N$ minima when fluxes are high, but because of the consistently low relative nitrate utilization in the Southern Ocean, this pattern is likely caused by changes in the plankton community or to increased degradation during times of low flux. $\delta^{15}N$ values throughout the tropical and southern Atlantic are correlated with sediment organic carbon content and also generally mirror primary productivity patterns, with low $\delta^{15}N$ associated with areas of high productivity and vice versa. Sediment trap data indicate that sediments are enriched in ^{15}N relative to sinking particles by up to 4‰. The offset (-2.0 to 4.3‰) does not vary greatly between the Polar Front, the productive Benguela upwelling area and the oligotrophic tropical Atlantic, in spite of the vastly different environmental conditions prevailing in these three regions.

Introduction

The spectre of global warming has focused attention on the role of the oceans in the carbon and nutrient cycles. Because the past holds the key to the present, sediment and ice cores have been examined and much has been learned about past environmental variations, especially glacial-interglacial changes.

From WEFER G, MULITZA S, RATMEYER V (eds), 2003, The South Atlantic in the Late Quaternary: Reconstruction of Material Budgets and Current Systems. Springer-Verlag Berlin Heidelberg New York Tokyo, pp 143-165

Ice cores show that atmospheric CO_2 was lower during the last glacial period (Berner et al. 1980). The reason for this is still unclear, but there are a multitude of theories regarding the mechanisms by which the world's oceans may regulate CO_2 levels in the atmosphere. Broecker and Peng (1989) suggested that increased alkalinity in surface waters could have been sufficient to lower glacial atmospheric CO_2 without an increase in nutrient concentrations. Alternatively, higher export production at high latitudes due to increased biological removal of nutrients and CO_2 from surface waters may have led to lower glacial CO_2 in the atmosphere (Knox and McElroy 1984). More recently, Francois et al. (1997) proposed that surface-water stratification in the Southern Ocean greatly reduced the loss of CO_2 from the ocean to the atmosphere, leading to the lowered atmospheric PCO_2.

Photosynthesis plays an important role in the global climate because of its connection to CO_2 and nutrients. Due to the capacity of nitrogen to limit primary production in some parts of the ocean, it is essential that we expand our knowledge of the biogeochemistry of this nutrient. The usefulness of nitrogen isotopes in reconstructing changes in the marine nitrogen cycle has been repeatedly demonstrated. For example, Holmes et al. (1997) and Martinez et al. (2000) examined past changes in relative nitrate utilization and productivity off the west coast of Africa (to 300 kyr) based on $\delta^{15}N$ in sediments. $\delta^{15}N$ is defined as $\{[(^{15}N/^{14}N)_{sample}/(^{15}N/^{14}N)_{atmosphere}] - 1\}*1000$ and $\delta^{15}N$ of atmospheric nitrogen is 0‰. Glacial-interglacial changes in nutrient utilization in the equatorial Pacific were examined by Farrell et al. (1995) and Ganeshram et al. (2002) inferred from bulk sediment $\delta^{15}N$ that nitrogen fixation was lower during glacial times. Bertrand et al. (2000) employed $\delta^{15}N$ to demonstrate the effect of sea level changes during deglaciation on nutrient cycling in coastal upwelling areas. Sedimentary $\delta^{15}N$ was also used by Altabet et al. (2002) and Suthhof et al. (2001) to study millenial-scale changes in denitrification in the Arabian Sea and Ganeshram et al. (1995) determined that the low $\delta^{15}N$ values they measured in glacial sediments reflect decreased denitrification. Kienast (2000), on the other hand, inferred from unchanged $\delta^{15}N$ in sediment organic matter from the South China Sea that the nitrogen isotopic composition of western Pacific subsurface nitrate did not change on a glacial-interglacial scale.

The basis for the application of $\delta^{15}N$ in organic matter is the fractionation that occurs during various processes involving nitrogen in the ocean. Estimation of nutrient utilization makes use of one of these processes - the assimilation of dissolved inorganic nitrogen (DIN) by phytoplankton. Plankton may take up nitrate (NO_3^-), nitrite (NO_2^+) or ammonium (NH_4^+) during photosynthesis, but in productive regions such as the eastern subtropical Atlantic and much of the Southern Ocean, nitrate is the main inorganic nutrient source (Chapman and Shannon 1985; de Baar et al. 1997). Fractionation during NO_3^- uptake occurs because of the more rapid reaction rate associated with the lighter isotope (^{14}N). Phytoplankton thus preferentially utilizes ^{14}N, producing particulate nitrogen (PN) that is enriched in ^{14}N relative to the nitrate (Wada and Hattori 1978). Estimates for the fractionation factor associated with nitrate uptake are around 5‰ (Wu et al. 1997; Sigman et al. 1999a). In a simplification of the real world, assuming no nitrate is added or removed except by phytoplankton uptake, the $\delta^{15}N$ of the remaining nitrate increases according to Rayleigh fractionation kinetics as isotopically light NO_3^- is removed (Cifuentes et al. 1988; Waser et al. 1998). Plankton that take up this NO_3^-, which is isotopically heavy relative to NO_3^- from the deep-sea (deep-sea nitrate is around 5‰; Sigman et al. 1997) is correspondingly also heavy. For example, Montoya and McCarthy (1995) and Wu et al. (1997) found an inverse relationship between the $\delta^{15}N$ of PN and nitrate concentrations. Regenerated nitrogen, or NH_4^+, may be used as a substrate under nitrate-limited conditions, as well as in parts of the Southern Ocean where iron may be limiting (Kioke et al. 1986; Glibert et al. 1982). NH_4^+ assimilation is associated with isotopic fractionation ranging from as low as 0‰ under N-depleted conditions (Waser et al. 1999) to 9‰ in eutrophic systems (Cifuentes et al. 1989; Montoya et al. 1991).

Denitrification and nitrogen fixation are two additional processes that may affect the isotopic composition of the nitrate available in surface water for assimilation. In suboxic marine waters, the fractionation factor associated with NO_3^- reduction may be around 27‰ (Brandes et al. 1998), leaving the remaining NO_3^- depleted in ^{14}N relative to the mean deep-sea $\delta^{15}N$ (Cline and Kaplan 1975). The elevated $\delta^{15}N$ signal is reflected in organic matter in regions where nitrate reduction occurs, such as the Arabian Sea and the eastern tropical North Pacific (Altabet et al. 1995; Ganeshram et al. 1995). Denitrification does not have an important influence at the present time on the isotopic signature of nitrate in most parts of the southern Atlantic, although it may have some influence where anoxia occurs, e.g. on the upper shelf off Namibia. Nitrogen fixation, in contrast, occurs mainly in oligotrophic regions of the ocean. Nitrogen fixing organisms can convert nitrogen gas into a biologically usable form of nitrogen. There is little or no isotopic fractionation associated with the fixation of N_2, so the fixed nitrogen has a $\delta^{15}N$ value near that of atmospheric air (0‰) (Minagawa and Wada 1986; Carpenter et al. 1997). These two processes do not affect only the isotopic composition of nitrate, but they are also the main controls on the amount of biologically available nitrogen in the global ocean.

Several recent investigations have furthered our knowledge of the patterns of isotopic alteration in organic matter during sinking through the water column and burial on the seafloor, although the exact processes behind these $\delta^{15}N$ shifts are not yet clearly understood. Early studies suggested that isotopic fractionation during remineralization left the residual organic matter enriched in ^{15}N (Saino and Hattori 1980; Altabet and McCarthy 1985). On the other hand, Altabet et al. (1991) and Nakatsuka et al. (1997) found that $\delta^{15}N$ in sinking particles decreased with depth, ascribing this trend to degradation of compounds rich in ^{15}N. The conflicting reports indicate that competing processes occur during early diagenesis. Nakatsuka et al. (1997) found that another characteristic of $\delta^{15}N$ diagenetic change was an increase at the sediment-water interface. Incubation experiments have demonstrated that $\delta^{15}N$ in particulate organic matter (POM) increases due to fractionation during bacterial degradation under oxic conditions (Libes and Deuser 1988; Holmes et al. 1999; Lehmann et al. 2002). However, preferential removal of isotopically heavy compounds such as amino acids can lead to decreases in $\delta^{15}N$ in organic material (Sweeney and

Kaplan 1980; De Lange et al. 1994). Anaerobic incubations of particulate matter from a lake showed that $\delta^{15}N$ decreased due to the loss of ^{15}N-enriched nitrogen and/or to an increase in the contribution of ^{15}N-depleted bacterial biomass (Lehmann et al. 2002). Libes and Deuser (1988) and Holmes et al. (1999) also reported $\delta^{15}N$ decreases in PN incubated under suboxic or anoxic conditions. The removal of ^{15}N-enriched compounds during early diagenesis of bulk sedimentary $\delta^{15}N$, as well as the presence of DIN in sediments, counteracts the increase in residual organic matter $\delta^{15}N$ caused by fractionation during bacterial remineralization because DIN is isotopically light (Freudenthal et al. 2001a). These authors found that the counteracting effect of DIN can be more important in suboxic sediments due to ammonium fixation and in organic-poor sediments. In spite of the apparently complicated processes altering the $\delta^{15}N$ of settling PN, most studies have found diagenetic increases of 1-5‰ between the surface-produced PN signal and the underlying sediments, with values at the lower end of this range found in regions of moderate to high fluxes (Francois et al. 1992; Altabet and Francois 1994; Farrell et al. 1995; Altabet et al. 1999; Holmes et al. 2002).

We present $\delta^{15}N$ values of sinking particles and surface sediments in the southern part of the Atlantic Ocean. The data cover a large area, from south of the Antarctic Circumpolar Current to north of the equator and from the coast of Africa to South America. The broad coverage of the data allows comparisons of ecosystems under different environmental constraints – from upwelling-influenced ecosystems in the Benguela region to the high nutrient, low chlorophyll conditions in the Southern Ocean. We will present evidence that relative nitrate utilization dictates isotopic composition in much of this ocean, with some interesting exceptions. Our ability to understand $\delta^{15}N$ variations in sediments and PN is hampered by the absence of water column NH_4^+, $\delta^{15}N$-NH_4^+ and $\delta^{15}N$-NO_3^- profiles. Attempts to decipher nitrogen isotope records in sediments and especially in PN in the water column would certainly profit from comparison with such water column data. Despite these limitations on our interpretation of $\delta^{15}N$ variations, the data presented here contribute to our understanding of the marine nitrogen cycle in the Atlantic Ocean.

Materials and Methods

Surface sediments were sampled during several expeditions to the Atlantic Ocean using either box corers or multiple corers (Table 1). The upper 0.5 or 1.0 centimeter of sediment was removed immediately upon core recovery and stored at 4°C in either plastic syringes or in petri dishes sealed with electrical tape. The sediment was later freeze-dried and homogenized in a shore-based laboratory. Most $\delta^{15}N$ measurements were performed on bulk sediment using a Finnigan Delta Plus mass spectrometer coupled with a Carlo/Erba NC2500 elemental analyzer. For samples from the Angola Basin and Benguela region, off the west coast of Africa, $\delta^{15}N$ determinations were made on sediment that had been decalcified with 1M HCl using a Finnigan MAT 252 mass spectrometer equipped with a trapping box. Experiments carried out in our laboratory and at the Institut fuer Ostseeforschung in Warnemuende on a working standard sediment showed no difference in isotopic values between sediments that were decalcified with HCl and those that were untreated (Holmes et al. 1999). Standard error was $< \pm 0.4‰$ for samples from the Angola Basin and Benguela region and $< \pm 0.2‰$ for all other samples.

Ship	Cruise	Date
RV *Meteor*	M6/6	February/March 1988
RV *Meteor*	M9/4	February/March 1989
RV *Meteor*	M12/1	March/April 1990
RV *Meteor*	M16/1	March/April 1991
RV *Meteor*	M16/2	April/May 1991
RV *Meteor*	M20/2	December 1991/February 1992
RV *Meteor*	M22/1	September/October 1992
RV *Meteor*	M23/1	February 1993
RV *Meteor*	M23/2	February/March 1993
RV *Meteor*	M23/3	March/April 1993
RV *Sonne*	SO84	January/February 1993
RV *Victor Hensen*	JOPSII-6	March 1995
RV *Polarstern*	ANT VIII/3	November 1989
RV *Polarstern*	ANT XI/2	December 1993 – January 1994
RV *Polarstern*	ANT XI/4	April - May 1994

Table 1. Cruises during which samples presented in this work were retrieved.

Sinking particles were collected using sediment traps moored in the eastern tropical Atlantic, over the Walvis Ridge and at the Polar Front in the eastern South Atlantic (Table 2). The Kiel-type traps had $0.5m^2$ collecting areas and samples were preserved with $HgCl_2$. Pure NaCl was added to the cups to increase salinity and "swimmers" were removed prior to geochemical analyses. For WR2-upper samples, $\delta^{15}N$ was measured on GF/F-filtered material. $\delta^{15}N$ determination on all other samples was carried out on homogenized, freeze-dried material. Isotopic measurements of WR2-upper samples were made using the same MAT 252 as described above for sediment samples from the Angola Basin and Walvis Bay (standard error $< \pm 0.4‰$). The other sediment trap samples were measured using the Delta Plus mass spectrometer and Carlo Erba NC2500 elemental analyzer (standard error $< \pm 0.2‰$).

Results and Discussion

Coastal Areas: Angola Basin, Benguela Upwelling Region, Brazilian Shelf

In general, $\delta^{15}N$ in surface sediments of the Angola Basin and Benguela upwelling region reflect the extent of nitrate utilization in surface waters, as demonstrated by Holmes et al. (1996, 1998). Sedimentary $\delta^{15}N$ ranged from 4.6 to 11.8‰ (Fig. 1), with lower values inshore, increasing with distance from the coast. The coastal upwelling off southwest Africa has been well documented (Calvert and Price 1971; Chapman and Shannon 1985; Lutjeharms and Meeuwis 1987). Nitrate concentrations are high near shore and decrease with distance from the coast, as phytoplankton takes up the nutrients. The range of nitrate concentrations in the surface waters overlying our samples was 0.5 to 8.3μM (Conkright et al. 1998). The strong influence of relative nitrate depletion can be seen when sedimentary $\delta^{15}N$ is plotted against the natural log of the nitrate concentrations (Fig. 2a and b). Although it is not nitrate concentration per se that controls $\delta^{15}N$, high concentrations do lead to lower relative utilization, and vice versa. This is clear from studies that show inverse relationships between surface $[NO_3^-]$ and $\delta^{15}N$-NO_3^-

Sample	Location	Deployment	Trap/Water depth (m)	$\delta^{15}N$ sinking particles (‰)	$\delta^{15}N$ surface sediment (‰)	$\Delta\delta^{15}N$
WR2 upper	20°03'S/9°09'E	3/89-3/90	599/2196	5.5	7.1	1.6
WR2/3/4 lower	20°03'S/9°09'E	3/89-12/91	1648/2196	6.9	7.1	0.2
EA1	3°10'N/11°15'W	4/91-11/91	984/4524	3.4	6.1	2.7
EA2	1°47'N/11°15'W	4/91-11/91	953/4399	4.3	7.6	3.3
GBN3	1°48'N/11°08'W	3/89-3/90	853/4481	4.2	7.6	3.4
GBN6	1°48'N/11°08'W	4/90-4/91	859/4522	4.6	7.6	3.0
EA3	0°05'S/10°46'W	4/91-11/91	1097/4141	5.2	8.2	3.0
EA4	2°11'S/10°06W	4/91-11/91	1068/3906	5.7	8.7	2.9
EA5	4°20'S/10°16'W	4/91-11/91	948/3490	5.4	9.4	4.3
EA8a	5°47'S/9°26'W	12/91-9/92	598/3450	6.1	9.6	3.5
PF3 upper	50°08'S/5°50'E	11/89-12/90	614/3750	2.8	5.2	2.4
PF3 lower	50°08'S/5°50'E	11/89-12/90	3196/3750	7.2	5.2	-2.0

Table 2. $\delta^{15}N$ in sinking particles and surface sediments in the southern Atlantic Ocean. Data for the Walvis Ridge (WR2), eastern tropical Atlantic (EA and GBN) and South Atlantic (PF) are from Holmes et al. (2002), Lavik (2001) and Fischer et al. (2003), respectively.

(Altabet 2001) and surface $[NO_3^-]$ and $\delta^{15}N$ in POM and (Altabet and McCarthy 1985; Wu et al. 1997). With the exception of the area of middle Angola (Fig. 2a), the nitrogen isotope ratios in surface sediments are negatively correlated with ln $[NO_3^-]$ but the relationships differ in slope and intercept between different ocean areas.

A sediment trap study of sinking particles over the Walvis Ridge demonstrated the link between upwelled nutrients and $\delta^{15}N$ (Holmes et al. 2002) on shorter timescales. Low nitrogen isotope ratios (2.9‰ in November/December) in sinking particles collected at 599m water depth coincided with low sea surface temperatures (SST) and high nitrogen fluxes (Fig. 3). Low SST indicates the presence of cold, nutrient-rich water upwelled at the coast. It would be helpful to have $[NO_3^-]$ and $\delta^{15}N$- NO_3^- measurements from each trap collection interval, but in the absence of these data, we surmise that the high nutrient concentrations in this water supported increased productivity, as seen by the high fluxes, and led to low $\delta^{15}N$ values due to the relatively unutilized, recently upwelled nitrate pool with $\delta^{15}N$ values close to that of new nitrate (~5‰). $\delta^{15}N$ in the 3 samples from October to December preceding the $\delta^{15}N$ minimum in December were

elevated relative to what would be expected in view of the high fluxes and the decrease in SST. The cause of the higher than expected isotopic values is unclear, but possible explanations include the uptake of ^{15}N-enriched nitrate during photosynthesis, a reduction in fractionation by plankton during nitrate incorporation, or a temporary change in trophic structure. With no $\delta^{15}N$- NO_3^- values, it is impossible to determine the cause of this apparent discrepancy between isotopic value, temperature and flux in October – December. Results from a sediment trap mooring in the equatorial Pacific at 2°N and 5°N revealed a similar pattern of $\delta^{15}N$ change in sinking particles over a one year period during which a transition from El Niño (low $[NO_3^-]$, elevated $\delta^{15}N$) to non-El Niño (higher $[NO_3^-]$, lower $\delta^{15}N$) conditions occurred (Altabet 2001). Isotope values there were comparable to those in sinking particles at the Walvis Ridge and ranged from ~5 to 10‰. Neuer et al. (2002) also found a relationship between total flux and sinking particle $\delta^{15}N$ in an upwelling-influenced zone off northwest Africa (trap depth 700 m). The range in measured $\delta^{15}N$ was 2‰ to nearly 6‰, with higher values of >8‰ found at sediment trap moorings farther offshore in oligotrophic locations. These

                                   Holmes et al.

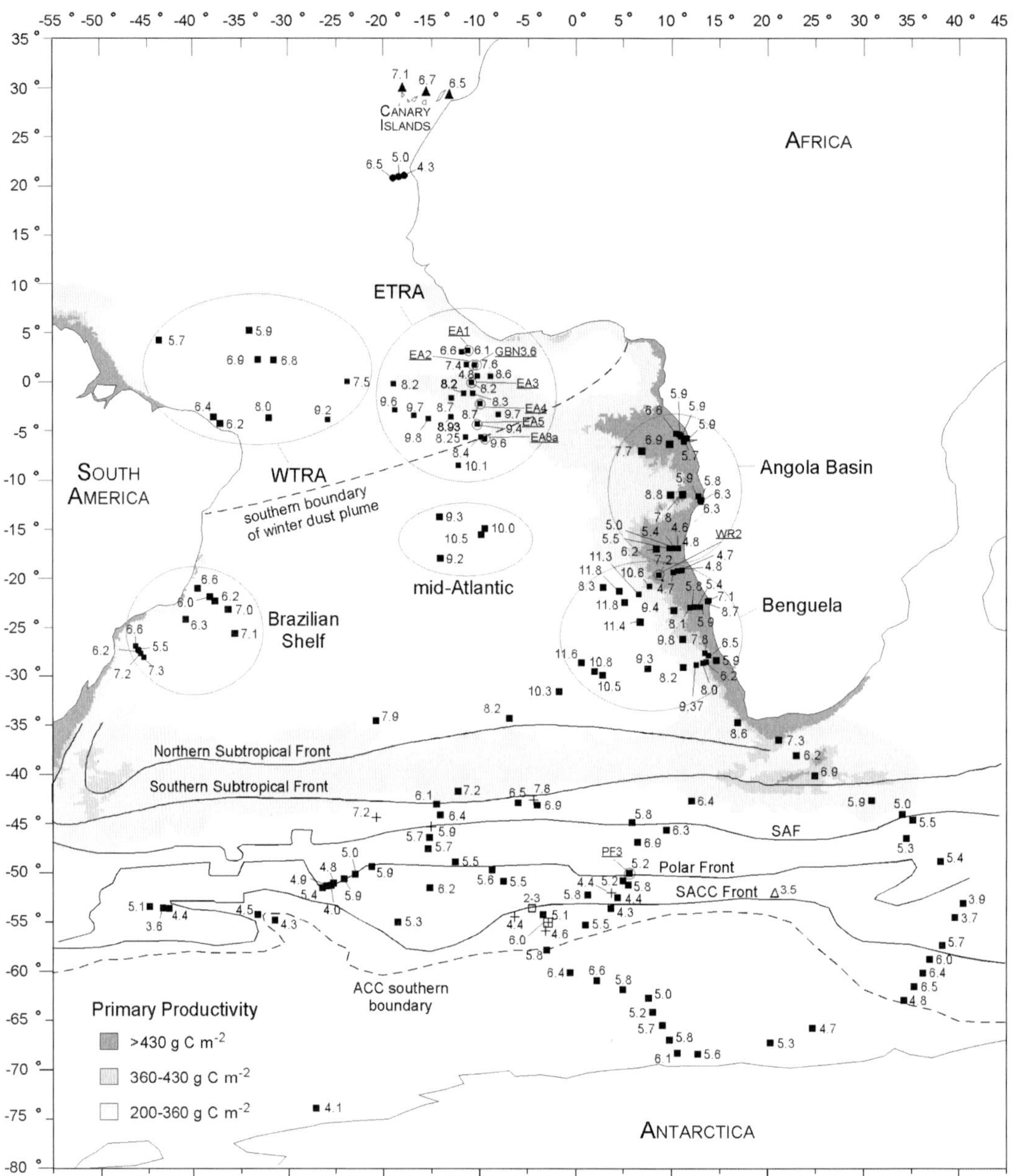

Fig. 1. Locations of sediment traps (open circles) and surface sediment samples from the southern Atlantic Ocean (filled squares). Where samples were located too close together for visual separation, the $\delta^{15}N$ values shown are averages. Primary productivity derived from satellite-based chlorophyll is represented by shaded areas (Behrenfeld and Falkowski 1997) and was obtained from the Ocean Primary Production Team at Rutgers University (www.marine.rutgers.edu). No productivity levels are shown for the region north of 10°N. Holocene sediment $\delta^{15}N$ data from the literature are shown in the southern South Atlantic as crosses (Francois et al. 1997), open triangle (diatom-bound organic matter $\delta^{15}N$, Crosta and Shemesh 2002), cross inside a box (Sigman et al. 1999b) and open square (Rau and Froelich 1993) and off northern Africa as filled triangles (Freudenthal et al. 2001b) and filled circles (Martinez et al. 2000). Angola Basin, Benguela and southern South Atlantic $\delta^{15}N$ values are from Holmes et al. (1996, 1998) and Fischer et al. (2003). $\delta^{15}N$ values from the tropical Atlantic and two samples from the Namibian shelf are from Lavik (2001). The positions of the Northern and Southern Subtropical Fronts (NSTF and SSTF) are from Belkin (1993) and Belkin and Gordon (1996). The positions of the Subantarctic (SAF), Polar (PF), southern Antarctic Circumpolar Current (SACC) Fronts and the southern boundary of the ACC are from Orsi et al. (1995).

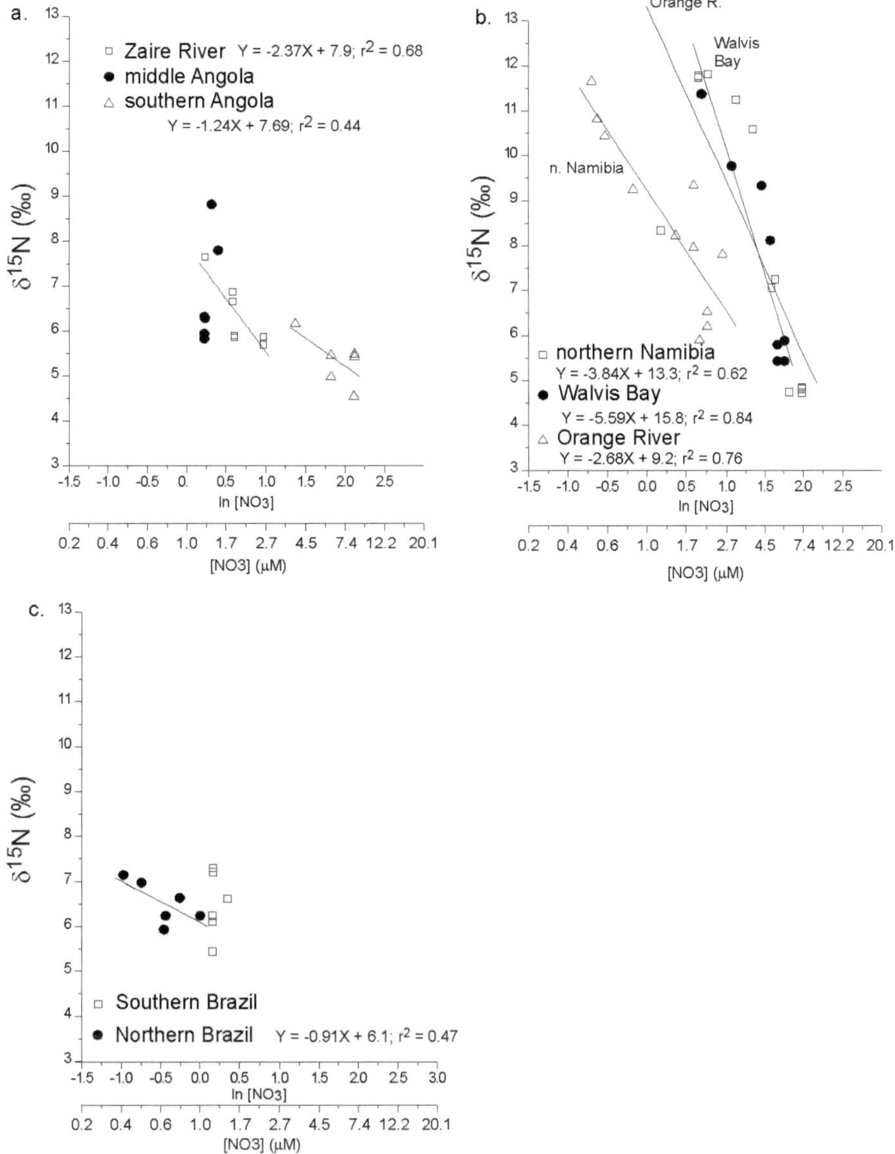

Fig. 2. Linear regression of sedimentary $\delta^{15}N$ versus ln [NO_3^-] (from Conkright et al. 1998) for the Angola Basin **a)**, Benguela upwelling region **b)** and Brazilian shelf **c)**.

authors concluded that the observed seasonal isotopic variability was due to upwelling-related enhanced plankton biomass. In contrast, Altabet et al. (1999) found that variations in sinking particle $\delta^{15}N$ (6 to around 11‰) in traps from various locations in the Eastern North Pacific were not tightly coupled with maxima in PN flux. Their traps were at depths ranging from 450 to 665 m and were located in areas seasonally influenced by coastal upwelling and where sub-euphotic zone $\delta^{15}N$-NO_3^- is elevated relative to the oceanic average.

The sediments over the Brazilian shelf exhibit a lower isotopic range (5.0 – 7.3‰) compared to those from near Africa. The northern transect shows a similar negative correlation between $\delta^{15}N$ and ln [NO_3^-] (Fig. 2c), but there is no apparent

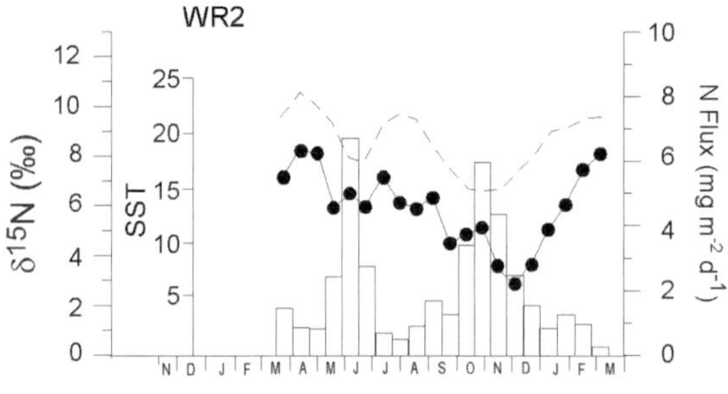

Fig. 3. δ^{15}N in sinking particles (circles), nitrogen fluxes (bars) and alkenone-derived sea surface temperatures (SST - dashed line) at the Walvis Ridge in the Benguela upwelling region. SST values are from Treppke et al. (1996), δ^{15}N values are from Holmes et al. (2002) and flux data are from Wefer and Fischer (1993).

relationship between δ^{15}N and ln [NO$_3$-] in the southern transect. In surface waters overlying this southern transect, as well as in the transect from middle Angola, average nitrate concentrations were within a narrow range (1.2 – 1.4 µM for southern Brazil and 1.3 – 1.4µM for middle Angola). The lack of correlation between these two parameters in middle Angola and the southern Brazil transect could be caused by a paucity of surface nitrate measurements because the fewer data points that are available, the less likely the data are to accurately represent annual and interannual nitrate concentrations. Since surface sediments represent hundreds to thousands of years of accumulated material, long-term nitrate concentrations are necessary to make meaningful comparisons with sedimentary δ^{15}N.

When ln [NO$_3^-$] is plotted against δ^{15}N, the slope of the regression line should ideally represent the fractionation factor (α) for nitrate uptake (Altabet and Francois 1994). The correlations between ln [NO$_3^-$] and sedimentary δ^{15}N in Fig. 2 are reasonably good, but the slopes of the lines are lower than expected from field estimates of α. These field estimates are between 4 and 9‰ (compilations by Montoya 1994 and Waser et al. 1998). In the Angola Basin and Benguela region, we obtained slopes as low as 1.2 and off Brazil, the slope of the regression line was only 0.9. Because most of the ocean surface is an open system with respect to

DIN, local vertical or horizontal input of nitrate to the surface tends to disrupt the linear relationship expected between δ^{15}N and ln [NO$_3^-$] (Altabet and Francois 1994). This is especially important where surface nitrate concentrations are low since dissolved inorganic nitrogen from lateral advection/ diffusion or precipitation becomes relatively more important and may alter the δ^{15}N. As pointed out by Freudenthal et al. (2001b), the sedimentary record in regions with strong productivity gradients, such as this, may be influenced by a mixture of autochthonous particles and particles originating from the nearby high productivity areas. This would serve to lower the δ^{15}N signal in the oligotrophic region, decreasing the slopes of the lines in Fig. 3. Denitrification in shallow shelf waters may elevate δ^{15}N-NO$_3^-$ off Namibia, further dampening the slopes of the regressions by raising δ^{15}N in organic matter. Lavik (2001) measured δ^{15}N of 7.1 and 8.7‰ in two surface sediment samples from around 60 km from the coast in 98 and 125 m water depth in a continuation of our "Walvis Bay" transect. These values are significantly higher (by 1.5 to >3‰) than sediments further offshore. Oxygen levels in the water column above one of these sites are <0.2mL L^{-1}, leading Lavik (2001) to conclude that denitrification was the cause of the high δ^{15}N.

In the Arabian Sea, one of the most important sites of denitrification, nutrients are completely utilized on an annual basis and changes in relative nitrate depletion do not alter the δ^{15}N signal of denitrification in sediments (Suthhof et al. 2001). This is in contrast to the limited effect found in the Atlantic of nitrate reduction in the water column on shallow shelf sediments. Studies of δ^{15}N in other upwelling regions have yielded results comparable to ours in the South Atlantic. For example, Martinez et al. (2000) found that sedimentary δ^{15}N on the northwest African slope increased from 4.3‰ near the shore to 6.5‰ further out to sea and ascribed this to increasing relative utilization of nitrate as it is advected away from the region of upwelling. In a study near the Canary Islands by Freudenthal et al. (2001b), sedimentary δ^{15}N was 6.5‰ at an upwelling-influenced site and 7.1‰ further offshore in an oligotrophic region. δ^{15}N values reported in these two papers are similar in value to the isotope ratios we measured. While Altabet (2001) also found that δ^{15}N in POM in the equatorial Pacific increased with decreasing near-surface [NO$_3^-$] and distance from the equatorial upwelling area, he reported a greater range of δ^{15}N values than we found. δ^{15}N in POM near the equator was around 2-4‰ and increased to ~16‰ at around 12°. The relationship between δ^{15}N and nitrate utilization holds true in the Arctic Ocean as well, where Schubert and Calvert (2001) found that sedimentary δ^{15}N in organic matter ranged from 4.6 to 13.1‰ and was inversely correlated with near surface [NO$_3^-$], which was between 3 and 6μmol/kg.

Eastern and Western Tropical Atlantic/Mid-Atlantic

In the Eastern Tropical Atlantic (ETRA), δ^{15}N ranged from 4.8 to 10.1‰, in the Western Tropical Atlantic (WTRA) from 5.7 to 9.2‰ and in the mid-Atlantic from 9.2 to 10.5‰ (Lavik 2001). In the ETRA and WTRA, sedimentary δ^{15}N generally increases from north to south (Figs. 1 and 4a). Due to the increasing distance from the African continent, the north-south trend in the ETRA might be influenced by terrestrial input from coastal runoff or by aeolian transport of dust from the Sahara and Sahel deserts on the African continent. C/N ratios can provide evidence as to whether significant amounts of land-derived organic material is present in the sediments because C/N ratios generally are lower in marine-derived organic matter (4 - 10) and higher in terrestrial matter (>20) due to the nitrogen-depleted nature of plant material (Westerhausen et al. 1993; Talbot and Johannessen 1992). Indeed, C/N ratios decrease from north to south in the ETRA (Fig. 4a) from 9.5 in the northernmost sample to between 6 and 7 in the southern part of the transect. This trend could be due to decreasing levels of terrigenous input away from the coast, but even the slightly elevated C/N ratios in the north are not high enough to be considered non-marine. More evidence against the presence of significant amounts of terrestrial matter in these sediments comes from sediment trap studies nearby in which less than 10% of the diatoms were non-marine (Lange et al. 1994). Even if some terrestrially derived nitrogen is present in these

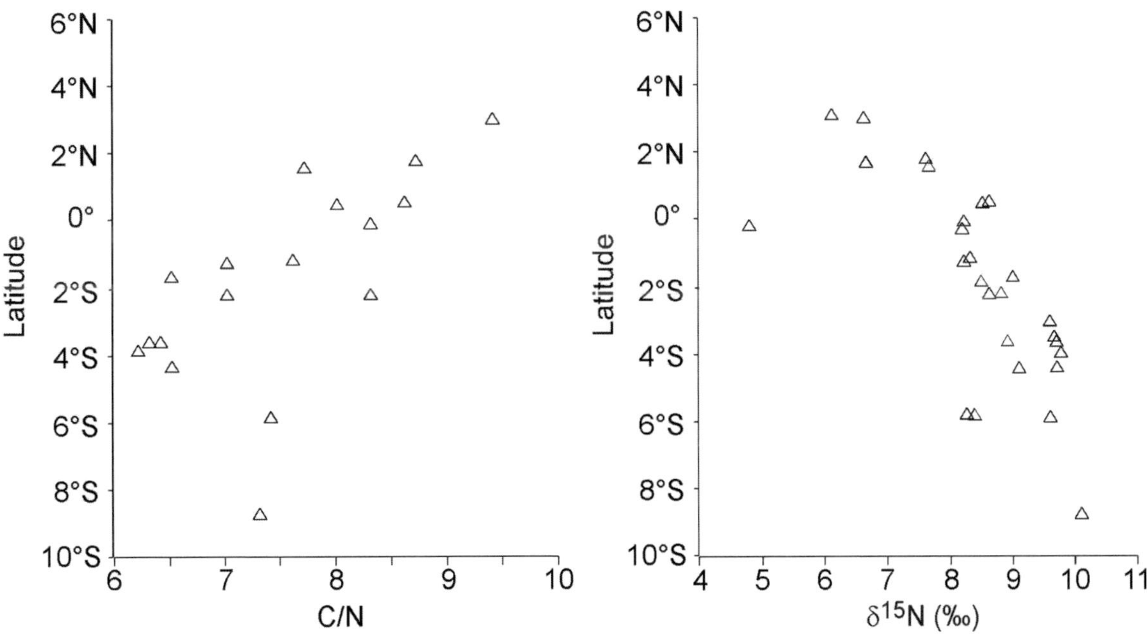

Fig. 4. Latitude versus C/N ratios **a)** and δ^{15}N **b)** for sediment samples from the eastern tropical Atlantic. Note the general increase of C/N to the north and the corresponding decrease of δ^{15}N.

sediments, the effect on the isotopic composition is probably small, considering that the $\delta^{15}N$ of terrestrial organic matter (5.0 – 9.0‰ for modern lake sediments in western Africa; Talbot and Johannessen 1992) is not particularly light and thus does not likely alter the sedimentary $\delta^{15}N$ significantly.

The large increase in sedimentary $\delta^{15}N$ from 5 - 6.5‰ in the north to ~10‰ in the south in the ETRA (Fig. 4b) can also not be satisfactorily explained by nitrate concentrations or relative nitrate utilization. In contrast to the coastal areas discussed above, the nitrogen isotopic composition of sediments from the eastern and western tropical Atlantic as well as from the mid-Atlantic is not correlated with [NO_3^-] (Fig. 5). Trade wind-induced upwelling at the equator results in high phytoplankton biomass and production during boreal summer between ~3°N and 7°S (Longhurst 1993). Sediment traps show that total flux is highest north of the equator and decreases to the south (Fischer and Wefer 1996; Romero et al. 1999), indicating that there is a productivity gradient, albeit weaker than in the coastal upwelling areas. In the tropical and middle regions of the Atlantic, annually averaged nitrate concentrations are low (mostly < 1 µM), but changes in the similarly low [NO_3^-] in the northern Brazil transect were sufficent to explain some of the sedimentary $\delta^{15}N$ variations observed there. One explanation for the apparent absence of a link between $\delta^{15}N$ and [NO_3^-] could be that water column [NO_3^-] measurements may be scarce in this region and the average concentration data may not represent the actual annual concentrations, as discussed above for middle Angola and Brazil. However, in order to explore other scenarios that might explain the observed isotopic variability, we assume for now that these nitrate concentrations are indeed representative of annual averages.

N_2 gas can be converted into biologically usable substrates by certain organisms, such as cyanobacteria, providing a DIN source. N_2-fixing organisms can flourish under low nitrate conditions such as those found in the tropical Atlantic (Carpenter 1983; Gruber and Sarmiento 1997; Capone et al. 1997), producing isotopically light organic matter. Montoya et al. (2002) reported blooms of *Trichodesmium* (an important N_2-fixing cyanobacterium)

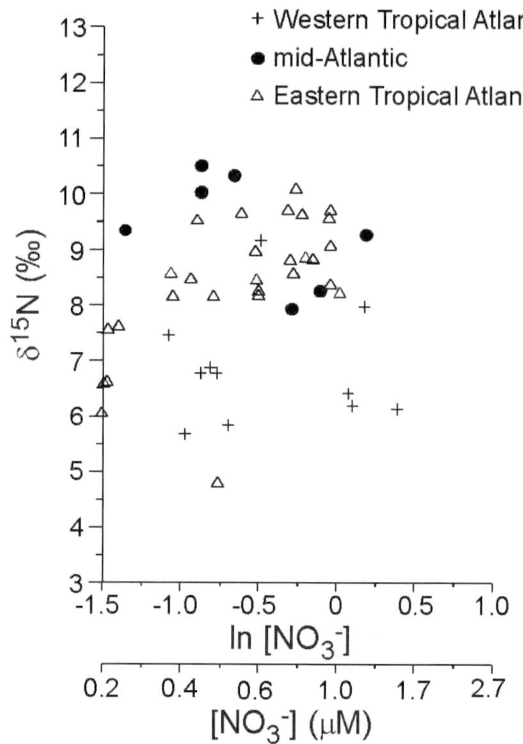

Fig. 5. Scatter plot of sedimentary $\delta^{15}N$ versus ln [NO_3^-] (from Conkright et al. 1998) for the eastern tropical, western tropical and middle Atlantic.

in the tropical North Atlantic and measured isotopic values of –1 to –2‰ in these cyanobacteria. $\delta^{15}N$ in *Trichodesmium* from the subtropical North Pacific exhibited similarly low values of 0.41‰ (Karl et al. 1997). These authors determined that the annual mean $\delta^{15}N$ of sinking material at the same location was 3.1‰. Montoya et al. (2002) determined that the low $\delta^{15}N$ values they measured in suspended particles and zooplankton where there were significant diazotroph populations were caused by the input of new nitrogen by N_2 fixation. They also determined that this process was more important in the western part of the Atlantic than the east. According to Gruber and Sarmiento (1997) N_2 fixation is highest north of ~10°N and decreases southward. This pattern of decreasing N_2 fixation from the north to the south fits the sedimentary $\delta^{15}N$ trend of lower values in the north (due to the incorporation of isotopically light N derived from N_2 fixation) and higher $\delta^{15}N$ in the south. Haug et al. (1998) determined that light sedimentary $\delta^{15}N$ (~2 - 3‰) in the Cariaco Basin resulted from increased N_2 fixation during interglacial periods (driven by a decrease in the N/P ratio of the nutrient supply to surface waters) and heavier values (~5‰) suggest that N_2 fixation was unimportant during glacials. Similarly, Struck et al. (2001) surmised that low $\delta^{15}N$ values (-1 - 1‰) in a sapropel (128 – 123 kyr) from the Mediterranean Sea were caused by extensive N_2 fixation supplying much of the nitrogen to near surface waters. The isotopic ratios these authors measured in surface sediments (2.6 to 5.6‰) indicate that marine nitrate is the main DIN source for primary productivity at the present time. Our lowest $\delta^{15}N$ values from the tropical Atlantic are only 5 to 6‰, significantly higher than those measured by these other authors, indicating that N_2 fixation supplies only part of the DIN in the tropical Atlantic.

Iron can limit N_2 fixation and Lavik (2001) proposed that low $\delta^{15}N$ values in sinking particles north of the equator in the ETRA were due to N-fixation fueled by increased iron supplied by wind-blown dust. Dust plumes from the Sahara and Sahel deserts on the northern African continent have been shown to extend far across the Atlantic (Moulin et al. 1997). Several model studies show that atmospheric iron input to the Atlantic Ocean is high between the equator and ~25°N and decreases toward the south and west (compilation by Moore et al. 2002). According to Moulin et al. (1997), the Saharan dust plumes are centered around 5°N in the winter and move north to around 20°N during the summer. If dust provides iron for N-fixation, the effect should be strongest in the northern part of the tropical Atlantic. Surface sediment $\delta^{15}N$ is, in fact, lower in the north and increases to the south as discussed above (Fig. 4b). Sinking particles may also reflect the stronger influence of aeolian iron deposition farther north, since flux-weighted average $\delta^{15}N$ values are lower north of the equator (3.4 – 4.3‰) than to the south (5.2 – 6.1‰) (Table 2). Similarly, Romero et al. (1999) found higher amounts of lithogenic particles and freshwater diatoms in sediment traps EA1 and EA2 (north of the equator) than EA3, EA4 or EA5 (at or south of the equator) and attributed this to Saharan dust transport patterns.

Sediment traps allow us to examine temporal variations in $\delta^{15}N$ in sinking particles, which might help elucidate the causes of the observed patterns (Table 3). In the northern part of the tropical Atlantic sinking particles should have lower $\delta^{15}N$ values in the winter compared to the summer if dust input is an important influence on N_2-fixation in this region and this temporal difference should decrease to the south. Although we have good coverage of the summer sinking particle flux, few data were collected between December and April. We attempted to compare summer with winter $\delta^{15}N$ (Table 3) by averaging values from April through September (summer) and October through March (winter). At 3°N (EA1) and at 2°N in 1991 (EA2) winter $\delta^{15}N$ was higher than in the summer, but no samples were collected from December until April. In 1989 and 1990 at 2°N, when summer and winter sinking particles are more equally represented, winter $\delta^{15}N$ values were lower (by ~0.7‰). At the equator (EA3), winter $\delta^{15}N$ was also lower than in summer (by 1.2‰), but again, winter samples are underrepresented. At 2°S (EA4) no sinking particles were collected in the winter and at 4°S (EA5), the one available wintertime sample was 6.7‰, compared to the average summertime value of 5.2‰. In the southernmost sediment trap, where both seasons were well sampled, winter $\delta^{15}N$ was also higher than in summer, but the aeolian deposition of dust should be lower in the south. On the other hand, evidence refuting the idea that input of iron from dust plumes supports increased N_2-fixation and that this is reflected in $\delta^{15}N$ comes from a study by Sañudo-Wilhelmy et al. (2001) who found that *Trichodesmium* sampled in April from 0 to 6°N; 50 to 28°W in the Atlantic Ocean were not iron-limited, but rather phosphorous-limited. Other field studies have also suggested that iron limitation does not play a strong role in the tropical Atlantic (Martin et al. 1993; Sunda 1997). If iron is not limiting throughout our study area in the tropical Atlantic, perhaps it is not iron, but DIN from aeolian sources that may be contributing to the low bulk $\delta^{15}N$ values in the northern part of our study. Lavik (2001) found that nitrogen in dust (3 - 8‰; Lavik 2001) collected just off the coast was iso-topically similar to that of marine organic matter in this region. Agusti et al. (2001) reported that atmospheric N sources are

Site	Location trap/water depth	Sampling Year	Season (boreal)	N Flux (mg m^{-2}day^{-1})	δ^{15}N (‰)
Northern Guinea Basin					
EA1	3°10'N/11°10'W	1991	Summer	1.6	2.8 (n=15)
	984 m / 4524 m		Winter	0.8	3.6 (n=5)
EA2	1°47'N/11°15'W	1991	Summer	1.9	4.1 (n=15)
	953 m / 4399 m		Winter	0.6	5.1 (n=5)
GBN3	1°48'N/11°08'W	1989-	Summer	1.2	4.4 (n=10)
	853 m / 4481 m	1990	Winter	1.0	3.7 (n=6)
GBN3	1°47'N/11°08'W	1990-	Summer	0.8	4.5 (n=4)
GBN6	859 m / 4522 m	1991	Winter	0.7	3.9 (n=4)
Equatorial Guinea Basin					
EA3	0°05'S/10°46'W	1991	Summer	0.9	5.5 (n=13)
	1097 m / 4141 m		Winter	0.2	4.3 (n=3)
Southern Guinea Basin					
EA4	2°11'S/10°06'W	1991	Summer	0.9	5.4 (n=13)
	1068 m / 3906 m		Winter	n.a.	n.a.
EA5	4°11'S/10°16'W	1991	Summer	0.9	5.2 (n=10)
	948 m / 3490 m		Winter	1.4	6.7 (n=1)
EA8a	5°47'S/9°26'W	1991-	Summer	1.3	6.2 (n=13)
	598 m / 3450 m	1992	Winter	1.0	6.9 (n=7)

Table 3. N-Flux and δ^{15}N in sinking particles in the tropical Atlantic (Lavik 2001).

almost an order of magnitude too small to make up the N deficit they calculated for the area between 5 and 13°N and ~20°W, so although this may partly explain our δ^{15}N data, it does not seem likely, based on the amount estimated to enter the ocean and its isotopic composition, to be the main factor in determining δ^{15}N in PN.

Alternatively, the light δ^{15}N north of the equator may be partly caused by the contribution of dissolved inorganic nitrogen from rain associated with the Intertropical Convergence Zone. Due to a paucity of data, the magnitude of this effect cannot be quantified, but the wide range of δ^{15}N values measured in rain water (-15 to +15‰; Kendall 1998 and references therein) precludes ignoring rainwater as a potential source of light nitrogen. Another possibility is that ammonium-based primary production could explain the lack of δ^{15}N correlation with [NO$_3^-$]. Le Bouteiller (1986) reported that at the equator at 4°W in the Atlantic Ocean, NH$_4^+$ was

the principal source of nitrogen in the euphotic zone. On the other hand, Waser et al. (2000) measured δ^{15}N values in suspended particulate matter (SPM) of 4-5‰ in the fall in the oligotrophic subtropical gyre of the North Atlantic. These authors determined that changes in nitrate supply are the primary control on δ^{15}N in SPM in the surface ocean and that production based on NH$_4^+$ had little effect on δ^{15}N. There are many factors influencing the nitrogen isotopic composition of organic matter in the tropical Atlantic and we cannot make a convincing case for ruling any of them out as the cause of the pattern we observe in δ^{15}N. Direct measurements of iron, ammonium and nitrate concentrations in combination with δ^{15}N-NO$_3^-$, δ^{15}N-NH$_4^+$ or δ^{15}N-N$_2$ would be invaluable in sorting out the δ^{15}N record in the tropical and mid-Atlantic. For example, Brandes et al. (1998) used isotopic variations of DIN in the Arabian Sea to trace mixing and nitrogen cycles and could determine that a significant amount of primary productivity in that region, known for its denitrification, was fueled by N_2 fixation. In spite of the absence of these data, we suggest that based on the widespread N_2 fixation occurring in this region, this probably has some influence on $\delta^{15}N$, whether or not input of iron or DIN by dust deposition have any affect.

South Atlantic

The South Atlantic (defined here as the region south of 35°S) is considered to play an important role in the atmospheric CO_2 cycle and global climate change. The South Atlantic is a high nutrient, low chlorophyll region where surface nitrate is incompletely utilized by phytoplankton. The region between the Subantarctic Front (SAF) in the north and the Polar Front (PF) (also known as the Antarctic Convergence) to the south is the Polar Frontal Zone (PFZ). The PF is a jet within the Antarctic Circumpolar Current (ACC) and exhibits strong gradients in temperature, density and other oceanographic properties (Moore et al. 1997). As discussed by Fischer et al. (2003) the SAF and PF delineate changes in sedimentary $\delta^{15}N$ (Fig. 6). North of the SAF, $\delta^{15}N$ ranged from 5.0 to 7.3‰, with little variation among the 3 transects. In the PFZ, $\delta^{15}N$ dropped to 5.3 to 5.9 in the eastern and western transects and to 6.3 – 6.9 in the middle transect. The PF marks an even more pronounced decrease in sedimentary $\delta^{15}N$, with average values south of the PF around 1‰ lighter than in the PFZ. Holocene sedimentary $\delta^{15}N$ in the Southern Ocean reported previously by other authors is similar to ours. For example, Francois et al. (1993) found values of 4-5‰ at the top of a core from the southwest Indian sector of the Southern Ocean near the SAF. Bulk sedimentary $\delta^{15}N$ measured by Francois et al. (1997) ranged from 4.4 to 7.8‰ and is in agreement with our data (Fig. 1). Diatom-bound organic nitrogen (which may by less likely affected by diagenetic alteration than bulk sedimentary $\delta^{15}N$) near the sediment surface north of the SACC Front was 3.5‰ as reported by Crosta and Shemesh (2002). Sigman et al. (1999b) found a value of 6‰ for bulk surface sediment $\delta^{15}N$ in the Atlantic sector south of the PF. Rau and Froelich (1993) obtained a value of 2-3‰ for surface sedi-

ment near the SACC Front, which is almost 2‰ lower than the isotopic ratios we measured near that area.

Nitrate utilization appears to play a major role in determining the nitrogen isotopic composition of surface sediments north of the PF. In the eastern and western South Atlantic, ln [NO_3^-] and $\delta^{15}N$ are significantly negatively correlated north of the PF (Fig. 7). The absence of a strong correlation in the middle South Atlantic may be due to the low number of samples. Unfortunately, we do not have direct measurements of [NO_3^-] or $\delta^{15}N$-[NO_3^-] and must make use of the historically averaged data of Conkright et al. (1998). Francois et al. (1997) determined that increasing surface sediment $\delta^{15}N$ from south of the PF (around 4‰ in the Indian sector) to north of this boundary (to ~9‰ at around 40°S) was due to the higher $\delta^{15}N$-[NO_3^-] supplied to subantarctic surface water by the northward drift of partially depleted nitrate. These authors also estimated that 30% of the nitrate in the euphotic zone is utilized by phytoplankton south of the PF at the present time while during the last glacial, there was 70% utilization.

Sinking particles in the PF region exhibit a pattern similar to that seen at the Walvis Ridge, with low $\delta^{15}N$ when N fluxes were high and high $\delta^{15}N$ during times of low flux (Fig. 8). $\delta^{15}N$ was around 1‰ in March and in December when N flux was >3 mg m^{-2} d^{-1}. The high austral winter $\delta^{15}N$ values occurred simultaneously with low fluxes, but are not likely due to increased relative nitrate utilization because of the high levels of this nutrient in the Southern Ocean. Fischer et al. (2003) proposed that either a change in the main plankton species or increased decomposition due to longer residence time in the water column may have led to the elevated isotope ratios. Changes in relative nitrate utilization were also determined by Francois et al. (1993) to have led to elevated $\delta^{15}N$ in sediments of the Last Glacial Maximum south of the SAF in the Indian sector of the Southern Ocean.

In contrast, south of the PF, there is no relationship between $\delta^{15}N$ and ln [NO_3^-]. The correspondence between nitrate and $\delta^{15}N$ north of the PF may be made possible due to the relatively high iron concentrations found in the PF and to the north of it (Kumar et al. 1995; de Baar et al. 1995; Loescher

                                    Holmes et al.

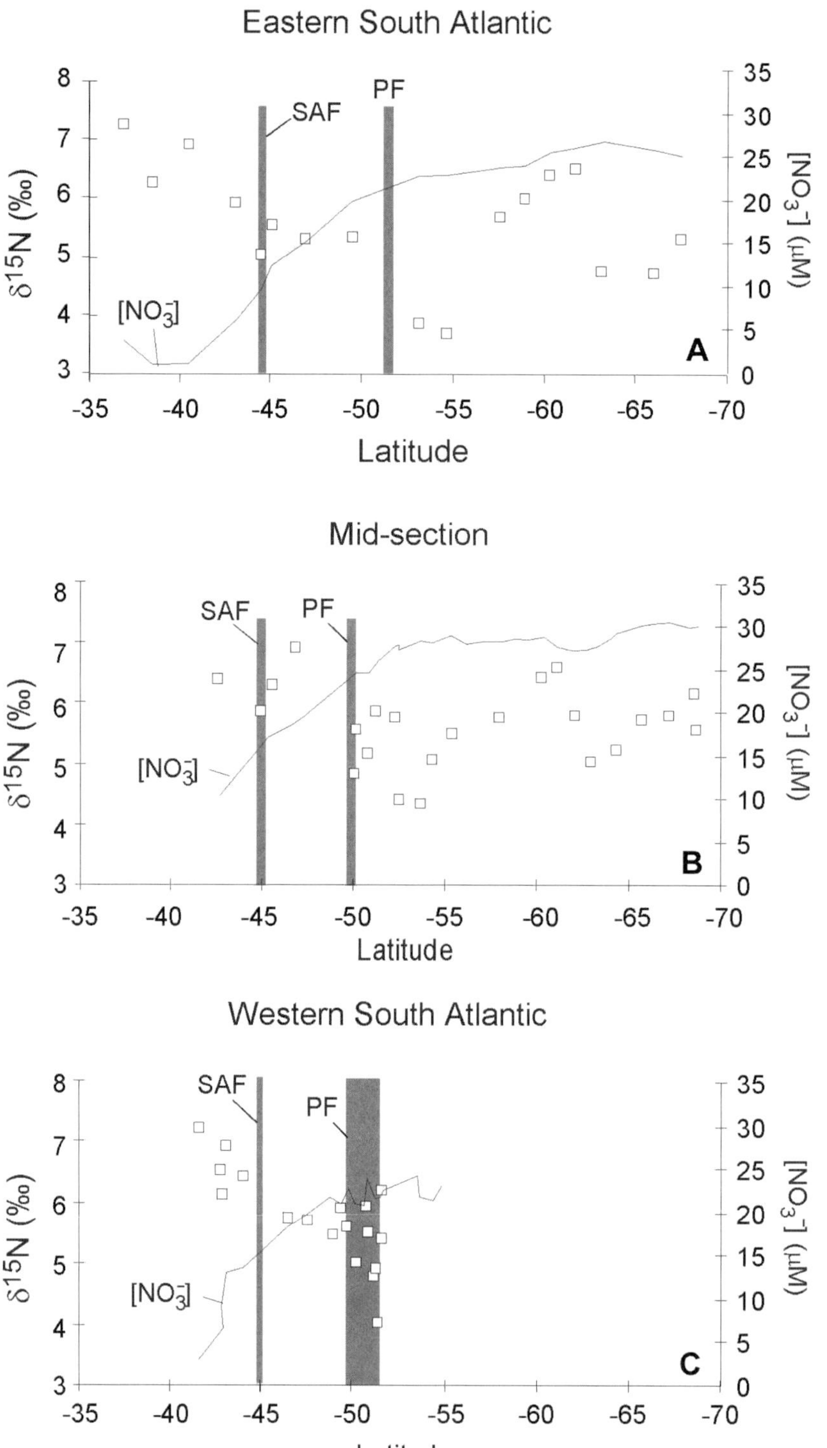

Fig. 6. Latitudinal variations in sedimentary $\delta^{15}N$ in the eastern **a)**, mid-section **b)** and western **c)** regions of the South Atlantic. Fronts are denoted by gray bars. The solid line shows average austral spring $[NO_3]$ from 0-50m depth (from Conkright et al. 1998). Redrawn from Fischer et al. (2003).

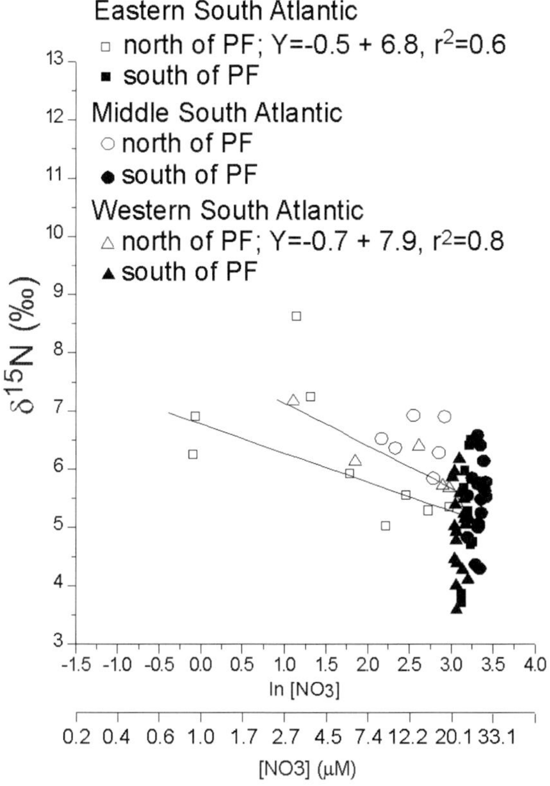

Fig. 7. Sedimentary $\delta^{15}N$ versus ln [NO_3^-] (from Conkright et al. 1998) in the southern South Atlantic. Samples from north of the Polar Front (PF) are represented by open symbols and those from south of the PF by filled symbols. No regression line is shown for the samples in the area of the middle South Atlantic.

et al. 1997). The PF could supply the iron from continental sources or it may be deposited via aeolian transport. Iron is needed not only for N-fixation as discussed earlier, but also for nitrate uptake by phytoplankton and its presence positively effects productivity in much of the iron-limited Antarctic (Martin et al. 1990; Hutchins and Bruland 1998; Takeda 1998). South of the PF, dissolved iron levels are low and because ammonium assimilation does not require extra iron, primary productivity there is probably based on ammonium (de Baar et al. 1997). Although we have no ammonium concentration data to compare with our $\delta^{15}N$ values, an ammonium-based production system would explain the lack of correlation between nitrate and $\delta^{15}N$. The findings of Rau et al. (1991), that $\delta^{15}N$

in Weddell Sea POM was correlated to [NH_4^+] support this hypothesis, as do reports of ammonium utilization in the Scotia and Weddell Seas (Glibert et al. 1982; Kioke et al. 1986; Smith and Nelson 1990). Another possible explanation for the variation in nitrogen isotopic compositions south of the PF is the contribution of POM from melting sea ice to the sediments, which Rau et al. (1991) found to have a high $\delta^{15}N$ signature. On the other hand, the lack of correlation between $\delta^{15}N$ and historically averaged [NO_3^-] may be due to sparse [NO_3^-] measurements from the southern South Atlantic. The findings of Crosta and Shemesh (2002) may be evidence of this, since diatom-bound organic nitrogen $\delta^{15}N$ south of the PF was found to have varied due to changes in relative nitrate utilization in both the Atlantic and Indian sectors of the Southern Ocean during the past 240 kyr.

Productivity and $\delta^{15}N$

The relationship between $\delta^{15}N$ and nitrate is shown for all the areas discussed above (Fig. 9). The large scatter of the data demonstrates the variations in nitrogen biogeochemistry among different regions. The link between nitrogen isotopes and productivity is emphasized by the close correspondence between $\delta^{15}N$ and total organic carbon (TOC).

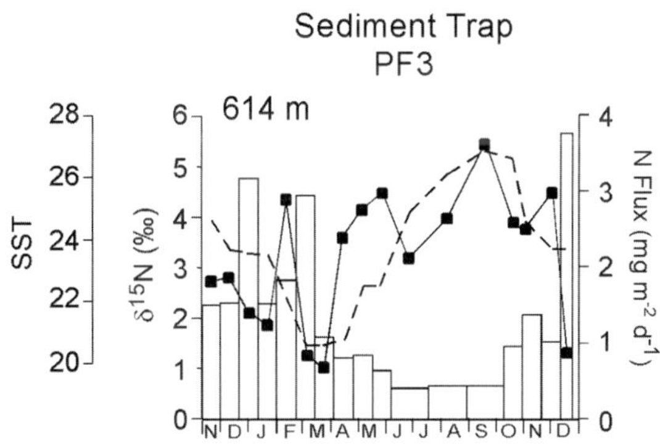

Fig. 8. $\delta^{15}N$ in sinking particles (squares), nitrogen fluxes (bars) and [NO_3^-] (dashed line) at the Polar Front. $\delta^{15}N$ values are from Fischer et al. (2003) and [NO_3^-] is from Conkright et al. (1998).

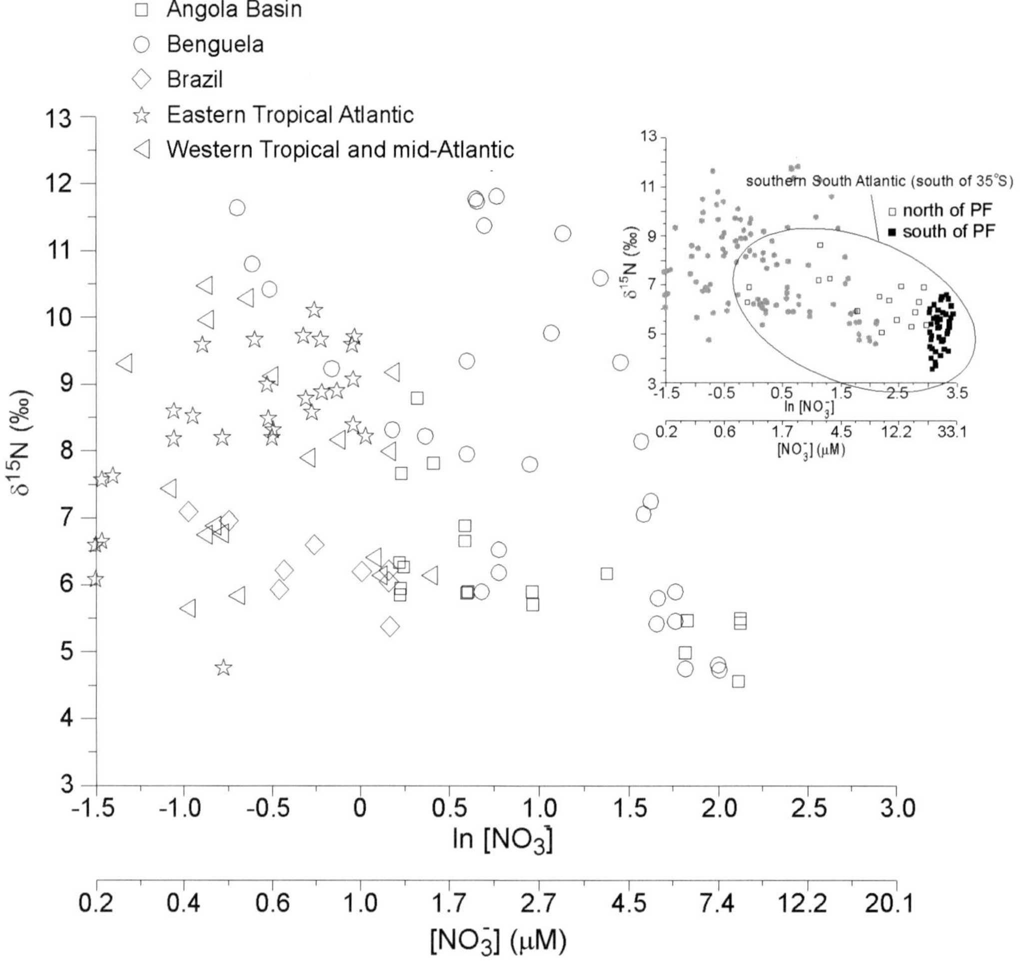

Fig. 9. $\delta^{15}N$ versus ln $[NO_3^-]$ (from Conkright et al. 1998) for all sediment surface samples from the Atlantic Ocean north of 35°S. Inset is the same plot with southern South Atlantic samples (south of 35°S) included. In inset, filled squares represent samples from north of the PF and unfilled squares those from south of the PF.

TOC content is negatively correlated with $\delta^{15}N$ in sediments in parts of the Atlantic Ocean north of 35°S (Fig. 10a-f). Strong correlations are observed in the Benguela, Orange River and the western tropical Atlantic. Surprisingly, the relationship is weak in the Angola Basin and the eastern tropical Atlantic. $\delta^{15}N$ in sediments south of 35°S also shows no relationship to TOC (Fig. 10g). Here, despite low $\delta^{15}N$ values, especially south of the PF, TOC is uniformly low (less than 1%). This is not surprising in a high nutrient, low chlorophyll region, because uptake of nitrate from a relatively unde-pleted pool will lead to light $\delta^{15}N$ although overall productivity may be low to moderate (Antoine et al. 1996; Behrenfeld and Falkowski 1997). We cannot explain the weak correlation between the two parameters in the Angola Basin and eastern tropical Atlantic. The general link between $\delta^{15}N$ and annual primary productivity is seen in figure 1, where the shaded areas on the map indicate various levels of primary production derived from satellite-based chlorophyll concentrations (Behrenfeld and Falkowski 1997).

Decomposition Processes and $\delta^{15}N$

The overall $\delta^{15}N$ offset between sinking particles and sediments is 2.8‰ in the southern Atlantic

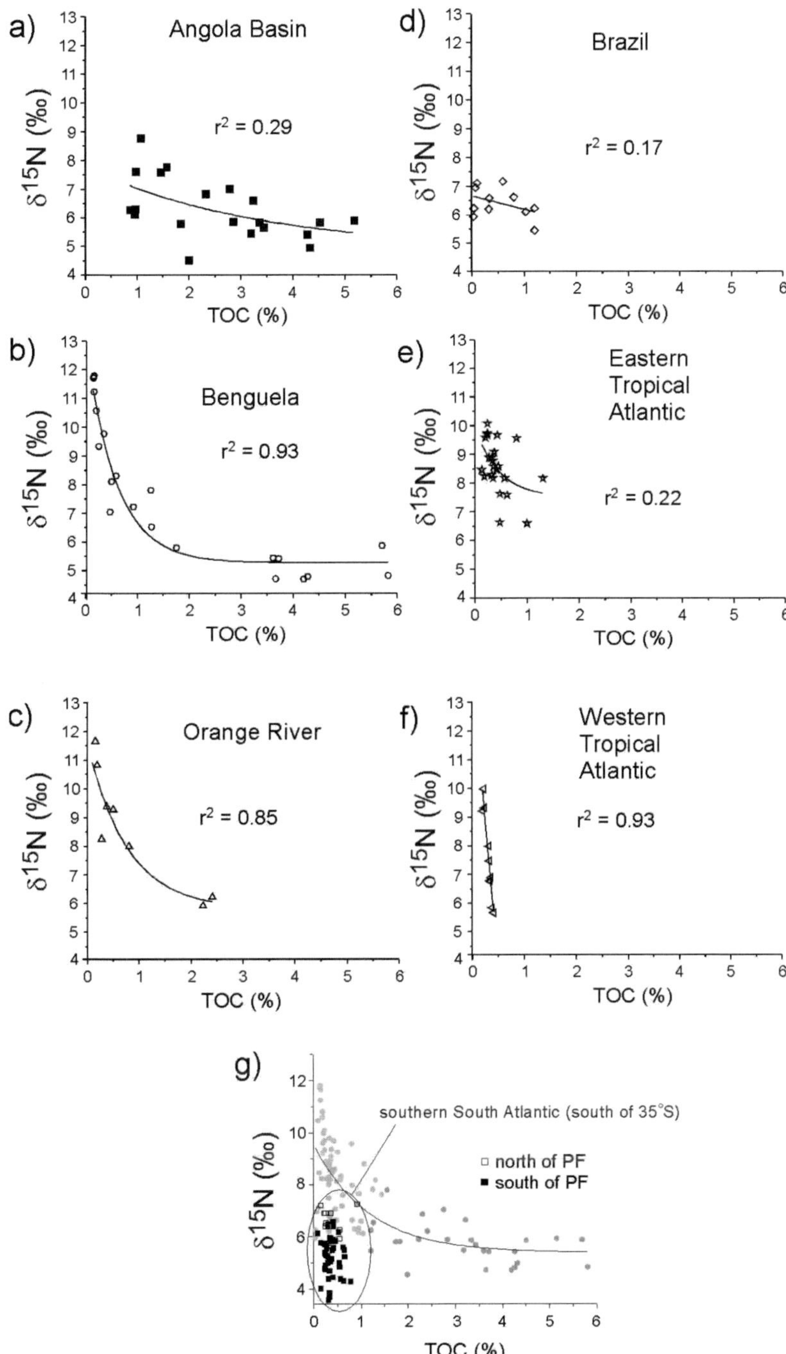

Fig. 10. $\delta^{15}N$ versus total organic carbon (TOC) for the different regions of the Atlantic Ocean north of 35°S **a)-f)**. Lines represent logarithmic decay fits. **g)** is $\delta^{15}N$ versus total organic carbon (TOC) for all sediment surface samples from the Atlantic Ocean with southern South Atlantic samples included. Filled squares represent samples from north of the PF and unfilled squares those from south of the PF. TOC values for samples from the Angola Basin and Benguela are from Müller et al. (1994) and Holmes et al. (1998).

sediment traps (Table 2). The deep traps at the Walvis Ridge and the Polar Front likely contain resuspended sediment and are therefore not included in this calculation. There are regional variations in this offset; in the eastern tropical Atlantic the difference is somewhat higher (3.3‰) than at the Polar Front (2.4‰) or the Walvis Ridge (1.6‰). The small difference at the Walvis Ridge between sinking particles and sediments may be caused by the high flux rates in this upwelling region. In such systems, organic matter is better preserved in comparison with oxic deep-sea sediments and the isotope ratios may be less altered by bacterial degradation (Altabet et al. 1999). Additionally, particles are exposed to remineralization during sinking to the sea floor for a shorter time than in the eastern tropical Atlantic or at the Polar Front because the water depth is much greater in both of the latter regions. In general, δ^{15}N at all three trap locations indicates that remineralization increases the isotopic signature of sedimented organic material.

This observation is in line with several other studies, which have come to the conclusion that the net effect of decomposition on the sedimentary isotopic signal is enrichment in ^{15}N. For example, Francois et al. (1992) and Altabet and Francois (1994) reported a 5 - 7‰ isotopic enrichment in surface sediments compared to the surface signal due to decomposition. More recently, Francois et al. (1997) and Lavik (2001) found diagenetic shifts of 3 - 4‰ between sinking particles and surface sediments and a study by Sigman et al. (1999b) comparing diatom-bound nitrogen with bulk Southern Ocean sediments revealed an increase of ~3‰ due to degradation. Freudenthal et al. (2001a) found that early diagenesis in sediments of the eastern subtropical Atlantic led to variability in δ^{15}N of <1‰. Since sedimentary isotopic variations due to denitrification, nitrate fixation and utilization and paleoceanographic changes among others, are often larger than those caused by decomposition, this proxy is proving to be a valuable tool in the study of nitrogen biogeochemistry.

Summary

δ^{15}N records are complex, recording mainly relative nitrate utilization in some regions, nitrogen fixation in others and non-nitrate (ammonium) based primary production elsewhere. It is clear that in some types of marine systems, especially upwelling regions and other areas where nitrate is the main limiting factor for primary productivity, sedimentary δ^{15}N is a good proxy for surface nitrate depletion. It is also obvious that caution must be taken in the interpretation of the sedimentary nitrogen isotope record because δ^{15}N does not reflect relative nitrate utilization in all ocean regions. This is especially true for the South Atlantic south of the Polar Front, where iron may be largely limiting and where NH_4^+ may be the main inorganic nitrogen source for photosynthesis. We found no significant correlation between surface $[NO_3^-]$ and δ^{15}N in the eastern and western tropical Atlantic (between 5°N and around 10°S) and the mid-Atlantic and surmise that N_2-fixation is a source of isotopically light POM. Denitrification, although insignificant on a broad scale in the southern Atlantic Ocean at the present time, does influence the δ^{15}N record in parts of the world's oceans and may have done in the past in the southern Atlantic. The extensive coverage of our isotope data from the Atlantic Ocean highlight the validity of nitrogen isotopes as a proxy, especially for relative nitrate utilization. Our work also points out the aspects of this tool in which interpretation is not straightforward, for example, 1) in regenerated nitrogen-based production systems, 2) when N_2-fixation occurs or 3) when particles are transported some distance from their origin.

Acknowledgments

We gratefully acknowledge Monika Segl and Tim Freudenthal for their help in the isotope laboratory. Thanks are also due to Marco Klann and Volker Diekamp who assisted with sample processing and measurement. We also wish to thank Marianne Huettel, whose support during the writing of this manuscript is much appreciated. The comments of Maren Voss and an anonymous reviewer were helpful in revising the manuscript. Access to nutrient data was facilitated by the LDEO/IRI Data Library (http://ingrid.ldgo.columbia.edu). This research was funded by the Deutsche Forschungsgemeinschaft (Sonderforschungsbereich 261 at Bremen University, Contribution No. 368) and by the Bundesminister für Bildung, Wissenschaft, Forschung and Technologie (BMBF). Data are available under www.pangaea.de/Projects/SFB261.

References

Agusti S, Duarte CM, Vaqué D, Hein M, Gasol JM, Vidal M (2001) Food-web structure and elemental (C, N and P) fluxes in the eastern tropical North Atlantic. Deep-Sea Res 48: 2295-2321

Altabet (2001) Nitrogen isotopic evidence for micronutrient control of fractional NO_3^- utilization in the equatorial Pacific. Limnol Oceanogr 46: 368-380

Altabet MA, McCarthy J (1985) Temporal and spatial variation in the natural abundance of ^{15}N in PON from a warm-core ring. Deep-Sea Res 32: 755-772

Altabet M, Francois R (1994) Sedimentary nitrogen isotopic ratio as a recorder for surface ocean nitrate utilization. Glob Biogeochem Cycl 8: 103-116

Altabet M, Deuser WG, Honjo S, Stienen C (1991) Seasonal and depth-related changes in the source of sinking particles in the North Atlantic. Nature 354: 136-139

Altabet MA, Francois R, Murray DW, Prell WL (1995) Climate-related variations in denitrification in the Arabian Sea from sediment $^{15}N/^{14}N$ ratios. Nature 373: 506-509

Altabet MA, Pilskaln C, Thunell R, Pride C, Sigman D, Chavez F, Francois R (1999) The nitrogen isotope biogeochemistry of sinking particles from the margin of the Eastern North Pacific. Deep-Sea Res I 46: 655-679

Altabet MA, Higginson MJ, Murray DW (2002) The effect of millenial-scale changes in Arabian Sea denitrification on atmospheric CO_2. Nature 415: 159-162

Antoine D, André J-M, Morel A (1996) Oceanic primary production-2. Estimation at global scale from satellite (coastal zone colour scanner) chlorophyll. Glob Biogeochem Cycl 10: 57-69

de Baar HJW, de Jong JTM, Bakker DCE, Loescher BM, Veth C, Bathmann U, Smetacek V (1995) Importance of iron for plankton blooms and carbon dioxide drawdown in the Southern Ocean. Nature 373: 412-415

de Baar HJW, van Leeuwe MA, Scharek R, Goeyens L, Bakker KMJ, Fritsche P (1997) Nutrient anomalies in *Fragilariopsis kerguelensis* blooms, iron deficiency and the nitrate/phosphate ratio (AC

Redfield) of the Antarctic Ocean. Deep-Sea Res 44: 229-260

Behrenfeld MJ, Falkowski PG (1997) Photosynthetic rates derived from satellite-based chlorophyll concentration. Limnol Oceanogr 24: 1-20

Belkin IM (1993) Frontal structure of the South Atlantic (in Russian). In: Voronina NM (ed) Pelagischeskie Ekosistemy Yuzhnogo Okeana (Pelagic Ecosystems of the Southern Ocean). Nauka, Moscow, pp 40-53

Belkin IM, Gordon AL (1996) Southern Ocean fronts from the Greenwich meridian to Tasmania. J Geophys Res 101: 3675-3696

Berner W, Oeschger H, Stauffer B (1980) Information on the CO_2 cycle from ice core studies. Radiocarbon 22: 227-235

Bertrand P, Pedersen TF, Martinez P, Calvert S, Shimmield G (2000) Sea level impact on nutrient cycling in coastal upwelling areas during deglaciation: Evidence from nitrogen isotopes. Glob Biogeochem Cycl 1: 341-355

Brandes JA, Devol AH, Yoshinari T, Jayakumar DA, Naqvi SWA (1998) Isotopic composition of nitrate in the central Arabian Sea and eastern tropical North Pacific: A tracer for mixing and nitrogen cycles. Limnol Oceanogr 43: 1680-1689

Broecker W, Peng TH (1989) The cause of the glacial to interglacial atmospheric CO_2 change: A polar alkalinity hypothesis. Glob Biogeochem Cycl 3: 215-239

Calvert SE, Price NB (1971) Upwelling and nutrient regeneration in the Benguela Current, October, 1968. Deep-Sea Res 18:505-523

Capone DG, Zehr JP, Pearl HW, Bergmann B, Carpenter EJ (1997) *Trichodesmium*, a globally significant cyanobacterium. Science 276: 1221-1229

Carpenter EJ (1983) Nitrogen fixation by marine *Oscilla-toria* (*Trichodesmium*) in the world's oceans. In: Carpenter EJ (ed) Nitrogen in the Marine Environment. Academic Press, London, New York, pp 65-103

Carpenter EJ, Harvey HR, Fry B, Capone DG (1997) Bio-geochemical tracers of the marine cyanobacterium, *Trichodesmium*. Deep-Sea Res 44: 27-38

Chapman P, Shannon L (1985) The Benguela ecosystem Part II. Chemistry and related processes. Oceanogr Mar Biol Ann Rev 23: 183-251

Cifuentes LA, Sharp JH, Fogel ML (1988) Stable carbon and nitrogen isotope biogeochemistry in the Delaware estuary. Limnol Oceanogr 33: 1102-1115

Cifuentes LA, Fobel ML, Pennock JR, Sharp JH (1989) Biogeochemical factors that influence the stable nitrogen isotope ratio of dissolved ammonium in the Delaware Estuary. Geochim Cosmochim Acta 53: 2713-2721

Cline JD, Kaplan IR (1975) Isotopic fractionation of dissolved nitrate during denitrification in the eastern tropical North Pacific Ocean. Mar Chem 3: 271-299

Conkright M, Levitus S, O'Brien T, Boyer TP, Antonov J, Stephens C (1998) World Ocean Atlas 1998 CD-ROM Data Set Documentation. Tech Rep 15: NODC Internal Report, Silver Spring, MD, 16 p

Crosta X, Shemesh A (2002) Reconciling down core anticorrelation of diatom carbon and nitrogen isotopic ratios from the Southern Ocean. Paleoceanography 17: 1010, doi: 10.1029/2000PA000565

De Lange GJ, van Os B, Pruysers PA, Middelburg JJ, Castradori D, van Santvoort P, Müller PJ, Eggenkamp H, Prahl FG (1994) Possible early diagenetic alteration of palaeo proxies. In Zahn R, Pedersen TF, Kaminski MA, Labeyrie L (eds) Carbon Cycling in the Glacial Ocean: Constraints on the Ocean's Role in Global Change. Springer, NATO ASI Series, Vol. 1 (17), pp 225-258.

Farrell JW, Pedersen TF, Calvert SE, Nielsen B (1995) Glacial-interglacial changes in nutrient utilization in the equatorial Pacific Ocean. Nature 377: 514-517

Fischer G, Wefer G (1996) Long-term observation of particle fluxes in the Eastern Atlantic: Seasonality, changes of flux with depth and comparison with the sediment record. In: Wefer G, Berger WH, Siedler G, Webb DJ (eds) The South Atlantic: Present and Past Circulation. Springer, Berlin, pp 325-344

Fischer G, Holmes ME, Wefer G (2003) $\delta^{15}N$ in plankton, sinking particles and surface sediments in the southern Atlantic Ocean. Deep-Sea Res, submitted

Francois R, Altabet MA, Burckle LH (1992) Glacial to interglacial changes in surface nitrate utilization in the Indian sector of the Southern Ocean as recorded by sediment $\delta^{15}N$. Paleoceanography 7: 589-606

Francois R, Bacon MP, Altabet MA, Labeyrie LD (1993) Glacial/interglacial changes in sediment rain rate in the SW Indian sector of Subantarctic waters as recorded by ^{230}Th, ^{231}Pa, U, and $\delta^{15}N$. Paleoceanography 8: 611-629

Francois R, Altabet MA, Yu E-F, Sigman DM, Bacon MP, Frank M, Bohrmann G, Bareille G, Labeyrie LD (1997) Contribution of Southern Ocean surface-water stratification to low atmospheric CO_2 concentrations during the last glacial period. Nature 389: 929-935

Freudenthal T, Wagner T, Wenzhöfer, Zabel M, Wefer, G (2001a) Early diagenesis of organic matter from sediments of the eastern subtropical Atlantic: Evidence from stable nitrogen and carbon isotopes. Geochim Cosmochim Acta 65: 1795-1808

Freudenthal T, Neuer S, Meggers H, Davenport R, Wefer G (2001b) Influence of lateral particle advection and organic matter degradation on sediment accumulation and stable nitrogen isotope ratios along a productivity gradient in the Canary Islands region. Mar Geol 177: 93-109

Ganeshram RS, Pedersen TF, Calvert SE, Murray JW (1995) Large changes in oceanic nutrient inventories from glacial to interglacial periods. Nature 376: 755-758

Ganeshram RS, Pedersen TF, Calvert SE, Francois R (2002) Reduced nitrogen fixation in the glacial ocean inferred from changes in marine nitrogen and phosphorus inventories. Nature 415: 156-159

Glibert MG, Biggs DC, McCarthy JJ (1982) Utilization of ammonium and nitrate during austral summer in the Scotia Sea. Deep-Sea Res 29: 837-850

Gruber N, Sarmiento JL (1997) Global patterns of marine nitrogen fixation and denitrification. Glob Biogeo-chem Cycl 11: 235-266

Haug GH, Pedersen TF, Sigman DM, Calvert SE, Nielsen B, Peterson LC (1998) Glacial/interglacial variations in production and nitrogen fixation in the Cariaco Basin during the last 580 kyr. Paleoceanography 13: 427-432

Holmes ME, Mueller PJ, Schneider RR, Segl M, Paetzold J, Wefer G (1996) Stable nitrogen isotopes in Angola Basin surface sediments. Mar Geol 134: 1-12

Holmes ME, Schneider RR, Mueller PJ, Segl M, Wefer G (1997) Reconstruction of past nutrient utilization in the eastern Angola Basin based on sedimentary $^{15}N/^{14}N$ ratios. Paleoceanography 12: 604-614

Holmes ME, Müller PJ, Schneider RR, Segl M, Wefer G (1998) Spatial variations in euphotic zone nitrate utilization based on $\delta^{15}N$ in surface sediments. Geo-Mar Lett 18: 58-65

Holmes ME, Eichner C, Struck U, Wefer G (1999) Reconstruction of surface ocean nitrate utilization using stable nitrogen isotopes in sinking particles and sediments. In: Fischer G, Wefer G (eds) Use of Proxies in Paleoceanography: Examples from the South Atlantic. Springer, Berlin, pp 447-468

Holmes E, Lavik G, Fischer G, Segl M, Ruhland G, Wefer G (2002) Seasonal variability of $\delta^{15}N$ in sinking particles in the Benguela upwelling region. Deep-Sea Res 49: 377-394

Hutchins DA, Bruland KW (1998) Iron-limited diatom growth and Si:N uptake ratios in a coastal upwelling regime. Nature 393: 561-564

Karl D, Letelier R, Tupas L, Dore J, Christian J, Hebel D (1997) The role of nitrogen fixation in the biogeochemical cycling in the subtropical North Pacific Ocean. Nature 388: 533-538

Kendall C (1998) Tracing nitrogen sources and cycling in catchments. In: Kendall C, McDonnell JJ (eds) Isotope Tracers in Catchment Hydrology. Elsevier Sciences BV Amsterdam, pp 519-576

Kienast M (2000) Unchanged nitrogen isotopic composition of organic matter in the South China Sea during the last climatic cycle: Global implications. Paleoceanography 15: 244-253

Kioke I, Holm-Hanson O, Biggs DC (1986) Inorganic nitrogen metabolism by Antarctic phytoplankton with special reference to ammonium cycling. Mar Ecol Prog Ser 30: 105-116

Knox F, McElroy MB (1984) Changes in atmospheric CO_2: Influence of the marine biota at high latitude. J Geophys Res 89: 4629-4637

Kumar N, Anderson RF, Mortlock RA, Froelich PN, Kubik, PW, Dittrich-Hannen B Suter M (1995) Increased biological productivity and export production in the glacial Southern Ocean. Nature 378: 675-680

Lange CB, Treppke, UF, Fischer G (1994) Seasonal diatom fluxes in the Guinea Basin and their relationship to trade winds, hydrography and upwelling events. Deep-Sea Res 41: 859-878

Lavik G (2001) Nitrogen isotopes of sinking matter and sediments in the South Atlantic. PhD Thesis, Ber Fachber Geowiss Univ Bremen 169, 155 p

Le Bouteiller A (1985) Environmental control of nitrate and ammonium uptake by phytoplankton in the Equatorial Atlantic Ocean. Mar Ecol Prog Ser 30: 167-179

Lehmann MF, Bernasconi SM, Barbieri A, McKenzi JA (2002) Preservation of organic matter and alteration of its carbon and nitrogen isotopic composition during simulated and *in situ* early sedimentary diagenesis. Geochim Cosmochim Acta 66: 3573-3584

Libes SM, Deuser WG (1988) The isotopic geochemistry of particulate nitrogen in the Peru Upwelling Area and the Gulf of Maine. Deep-Sea Res 35: 517-533

Loescher BM, de Baar HJW, de Jong JTM, Veth C, Dehairs F (1997) The distribution of Fe in the Antarctic Circumpolar Current. Deep-Sea Res 44: 143-187

Longhurst A (1993) Seasonal cooling and blooming in tropical oceans. Deep-Sea Res I 40: 2145-2165

Lutjeharms JRE, Meeuwis JM (1987) The extent and variability of southeast Atlantic upwelling. South African J Mar Sci 5: 51-62

Martin JH, Fitzwater SE, Gordon RM (1990) Iron deficiency limits phytoplankton growth in the Antarctic waters. Glob Biogeochem Cycl 4: 5-12

Martin JH, Fitzwater SE, Gordon RM, Hunter CN, Tanner SJ (1993) Iron, primary production and carbon-nitrogen flux studies during the JGOFS North Atlantic bloom experiment. Deep-Sea Res 40: 115-134

Martinez P, Bertrand P, Calvert SE, Pedersen TF, Shimmield GB, Lallier-Vergès E, Fontugne MR (2000) Spatial variations in nutrient utilization, poroduction and diagenesis in the sediments of a coastal up-welling regime (NW Africa): Implications for the paleoceanographic record. J Mar Res 58: 809-835

Minagawa M, Wada E (1986) Nitrogen isotope ratios of red tide organisms in the East China Sea: A characterization of biological nitrogen fixation. Mar Chem 19: 245-249

Moore JK, Abbott MR, Richman JG (1997) Variability in the location of the Antarctic Polar Front (90° - 20°W) from satellite sea surface temperature data. J Geophys Res 102: 27,825-27,833

Moore JK, Doney SC, Glover DM, Fung IY (2002) Iron cycling and nutrient-limitation patterns in surface waters of the World Ocean. Deep-Sea Res 49: 463-507

Montoya JP, Horrigan SG, McCarthy JJ (1991) Rapid, storm-induced changes in the natural abundance of ^{15}N in a planktonic ecosystem. Chesapeake Bay, USA. Geochim Cosmochim Acta 55: 3627-3638

Montoya JP (1994) Nitrogen isotope fractionation in the modern ocean: implications for the sedimentary record. In: Zahn R, Pedersen TF, Kaminski MA, Labeyrie L (eds) Carbon Cycling in the Glacial Ocean: Constraints on the Ocean's Role in Global Change. Springer, Berlin, pp 259-279

Montoya JP, McCarthy JJ (1995) Isotopic fractionation during nitrate uptake by phytoplankton grown in continuous cultures. J Plankton Res 17: 436-464

Montoya JP, Carpentar EJ, Capone DG (2002) Nitrogen fixation and nitrogen isotope abundances in zooplankton of the oligotrophic North Atlantic. Limnol Oceanogr 47: 1617-1628

Moulin C, Lambert CE, Dulac F, Dayan U (1997) Control of atmospheric export of dust from North Africa by the North Atlantic Oscillation. Nature 387: 691-694

Müller PJ, Schneider R, Ruhland G (1994) Late Quaternary pCO2 variations in the Angola Current: evidence from organic carbon $\delta^{13}C$ and alkenone temperatures. In: Zahn R, Kaminski MA, Labeyrie L, Pedersen TF (eds) Carbon Cycling in the Glacial Ocean: Constraints on the Ocean's Role in Global

Change. NATO ASI Series. Springer, Berlin, Vol. I 17, pp 343-366

Nakatsuka T, Handa N, Harada N, Sugimoto T, Imaizumi S (1997) Origin and decomposition of sinking par-ticulate organic matter in the deep water column inferred from the vertical distributions of its $\delta^{15}N$, $\delta13C$ and $\delta^{14}C$. Deep-Sea Res 44: 1957-1979

Neuer S, Freudenthal T, Davenport R, Llinás O, Rueda M-J (2002) Seasonality of surface water properties and particle flux along a productivity gradient of NW Africa. Deep-Sea Res 49: 3561-3576

Orsi AH, Whitworth TW, Nowlin WDJ (1995) On the meridional extent and fronts of the Antarctic Circum-polar Current. Deep-Sea Res 42: 641-673

Rau GH, Sullivan CW, Gordon LI (1991) $\delta^{15}N$ and $\delta^{13}C$ variations in Weddell Sea particulate organic matter. Mar Chem 35: 355-369

Rau GH, Froelich P (1993) A 450 kyr record of bulk sediment $\delta^{13}C_{org}$ and $\delta^{15}N$ at 54°S; significant glacial-interglacial changes in Southern Ocean CO_2 supply/demand? EOS, Transactions, AGU 74; 43, Suppl, 345 p

Romero OE, Lange CB, Fischer G, Treppke UF, Wefer G (1999) Variability in export production documented by downward fluxes and species composition of marine planktic diatoms: Observations from the tropical and equatorial Atlantic. In: Fischer G, Wefer G (eds) Use of Proxies in Paleoceanography: Examples from the South Atlantic. Springer, Berlin, pp 365-392

Saino T, Hattori A (1980) ^{15}N natural abundance in oceanic suspended particulate matter. Nature 283: 752-754

Sañudo-Wilhelmy SA, Kustka AB, Gobler CJ, Hutchins DA, Yang M, Lwiza K, Burns J, Capone DG, Raven JA, Carpenter EJ (2001) Phosphorus limitation of nitrogen fixation by *Trichodesmium* in the central Atlantic Ocean. Nature 411: 66-69

Schubert CJ, Calvert SE (2001) Nitrogen and carbon isotopic composition of marine and terrestrial organic matter in Arctic Ocean sediments: Implications for nutrient utilization and organic matter composition. Deep-Sea Res 48: 789-810

Sigman DM, Altabet MA, Michener R, McCorkle DC, Fry B, Holmes RM (1997) Natural abundance-level measurement of the nitrogen isotopic composition of oceanic nitrate: An adaptation of the ammonia diffusion method. Mar Chem 57: 227-242

Sigman DM, Altabet MA, McCorkle DC, Francois R, Fischer G (1999a) The $\delta^{15}N$ of nitrate in the Southern Ocean: Consumption of nitrate in surface waters. Glob Biogeochem Cycl 13: 1149-1166

Sigman DM, Altabet MA, Francois R, McCorkle DC, Gaillard J-F (1999b) The isotopic composition of diatom-bound nitrogen in Southern Ocean sediments. Paleoceanography 14: 118-134

Smith WO, Nelson DM (1990) Phytoplankton growth and new production in the Weddell Sea marginal ice zone during austral spring and autumn. Limnol Oceanogr 35: 809-821

Struck U, Emeis K-C, Voss M, Krom MD, Rau GH (2001) Biological productivity during sapropel S5 formation in the eastern Mediterranean Sea: Evidence from stable isotopes of nitrogen and carbon. Geochim Cosmochim Acta 65: 3241-3258

Sunda WG (1997) Control of dissolved iron concentrations in the world ocean: A comment. Mar Chem 57: 169-172

Suthhof A, Ittekkot V, Gaye-Haake B (2001) Millennial-scale oscillation of denitrification intensity in the Arabian Sea during the late Quaternary and its potential influence on atmospheric N_2O and global climate. Glob Biogeochem Cycl 15: 637-649

Sweeney RE, Kaplan IR (1980) Natural abundances of ^{15}N as a source indicator for near-shore marine sedimentary and dissolved nitrogen. Mar Chem 9: 81-94

Takeda S (1998) Influence of iron availability on nutrient consumption ratio of diatoms in oceanic waters. Nature 393: 774-777

Talbot MR, Johannessen T (1992) A high resolution palaeoclimatic record for the last 27,500 years in tropical West Africa from the carbon and nitrogen isotopic composition of lacustrine organic matter. Earth Planet Sci Lett 110: 23-37

Treppke UF, Lange CB, Donner B, Fischer G, Ruhland G, Wefer G (1996) Diatom and silicoflagellate fluxes at the Walvis Ridge: An environment influenced by coastal upwelling in the Benguela system. J Mar Res 54: 991-1016

Wada E, Hattori A (1978) Nitrogen isotope effects in the assimilation of inorganic nitrogenous compounds by marine diatoms. Geomicrobiol J 1: 85-101

Waser NAD, Harrison PJ, Nielsen B, Calvert SE, Turpin DH (1998) Nitrogen isotope fractionation during the uptake and assimilation of nitrate, nitrite, ammonium and urea by a marine diatom. Limnol Oceanogr 43: 215-224

Waser NAD, Yu Z, Yin K, Nielsen B, Harrison PJ, Turpin DH, Calvert SE (1999) N isotopic fractionation during a simulated diatom spring bloom: Importance of N-starvation in controlling fractionation. Mar Ecol Prog Ser 179: 291-296

Waser NAD, Harrison WG, Head EJH, Nielsen B, Lutz VA, Calvert SE (2000) Geographic variations in the nitrogen isotope composition of surface particulate nitrogen and new production across the North Atlantic Ocean. Deep-Sea Res I 47: 1207-1226

Wefer G, Fischer G (1993) Seasonal patterns of vertical particle flux in equatorial and coastal upwelling areas of the eastern Atlantic. Deep-Sea Res 40: 1613-1645

Westerhausen L, Poynter J, Eglinton G, Erlenkeuser H, Sarnthein M (1993) Marine and terrigenous origin of organic matter in modern sediments of the equatorial East Atlantic; the $\delta^{13}C$ and molecular record. Deep-Sea Res 40: 1087-1112

Wu JP, Calvert SE, Wong CS (1997) Nitrogen isotope variations in the subarctic northeast Pacific - relationships to nitrate utilization and trophic structure. Deep-Sea Res 44: 287-314

C$_{37}$-Alkenones as Paleotemperature Tool:
Fundamentals Based on Sediment Traps and Surface Sediments from the South Atlantic Ocean

P. J. Müller[*] and G. Fischer

Universität Bremen, Fachbereich Geowissenschaften, Postfach 33 04 40,
D-28334 Bremen, Germany
[*] corresponding author (e-mail): pmueller@uni-bremen.de

Abstract: The alkenone paleotemperature method has gained wide acceptance, but questions remain concerning the water depth and seasonality of alkenone production or the temperature calibration of the $U^{K'}_{37}$ unsaturation index. In this paper, we summarize alkenone results from the South Atlantic Ocean which were obtained within the scope of the collaborative research project SFB 261 at Bremen University. We present sediment trap time-series from the eastern Equatorial Atlantic, the Northern and Southern Benguela, the Polar Frontal Zone and the Antarctic Zone, and compare the $U^{K'}_{37}$ records to concurrent temperature variations in the surface waters (Reynolds and Smith 1994). To convert $U^{K'}_{37}$ into temperature, we used the *Emiliania huxleyi* calibration of Prahl et al. (1988). In addition, we recapitulate surface sediment results and provide an update of the global core-top calibration. Our sediment trap results confirm earlier conclusions deduced from surface sediments that $U^{K'}_{37}$ principally reflects mixed-layer temperatures in the eastern South Atlantic. A shallow alkenone source is indicated, for example, by coinciding SST and $U^{K'}_{37}$ records, comparable temperature amplitudes and identical flux-weighted SST and $U^{K'}_{37}$ values within $\pm 1°C$. The sediment traps further reveal that seasonal variations in alkenone production have little effect on the overall $U^{K'}_{37}$ signal exported out of the euphotic zone. Canonical spring-autumn blooms as observed in the Northern Benguela and episodic flux events prevailing in filamentous upwelling regions produce average $U^{K'}_{37}$ signals not significantly different from the annual mean SST. Additional interannual variations weaken seasonal effects. In the Polar Frontal Region, where the dominant alkenone flux occurred in late winter and spring, the flux-weighted $U^{K'}_{37}$ signal was lower by about $1°C$ compared to the mean SST in the collection period. Only at site BO1 south of the Polar Front, did the $U^{K'}_{37}$ time series fail to reproduce the annual SST cycle. Relatively low alkenone temperatures (-0.4° to 0°C) obtained for the productive summer season at this site may be attributed to the calibration, although other factors cannot be ruled out. Altogether, our sediment trap and sediment results suggest that $U^{K'}_{37}$ reflects the mean annual temperature of the mixed layer in most regions of the South Atlantic. An exception is the western Argentine Basin, where the sedimentary $U^{K'}_{37}$ ratios appear to be biased by offshore and northward redistribution processes. An update of the global core-top calibration (n=518) using annual mean SST data of World Ocean Atlas 1994 yields exactly the same relationship as before ($U^{K'}_{37}$ = 0.033 SST + 0.044; Müller et al. 1998). A slightly different equation is obtained using temperature data of World Ocean Atlas 1998 ($U^{K'}_{37}$ = 0.032 SST + 0.073) but both relationships yield similar temperature estimates (within 1°C) as the Prahl et al. (1988) calibration. Core-top as well as sediment trap results do not indicate a systematic deviation from linearity at the warm and cold ends of the calibrations. The linear relationships may therefore be used to determine paleotemperatures in the range from 0 to 29°C, with an uncertainty of about $\pm 1°C$. They also produce reasonable temperature estimates for periods that predate the first occurrence of *E. huxleyi*, suggesting that the *Gephyrocapsa* species contributing alkenones to Quaternary and Pliocene sediments responded similarly to temperature changes.

From WEFER G, MULITZA S, RATMEYER V (eds), 2003, *The South Atlantic in the Late Quaternary: Reconstruction of Material Budgets and Current Systems.* Springer-Verlag Berlin Heidelberg New York Tokyo, pp 167-193

Introduction

Sea-surface temperature (SST) is one of the major variables in paleoceanographic studies. Established micropaleontological methods for estimating SST use sedimentary abundances of planktonic microfossils (foraminifera, radiolarians, and diatoms) and their isotopic ($\delta^{18}O$) or elemental (Mg/Ca) compositions (e.g. Abelmann et al. 1999; Mix et al. 1999; Nürnberg et al. 2000). In the mid-eighties, Brassell et al. (1986) published an intriguing article suggesting that past SST could also be derived from the analysis of specific organic compounds preserved in marine sediments. These compounds are long-chain (C_{37}-C_{39}) unsaturated ketones (alkenones) which are synthesized by a limited number of planktonic algae and whose extent of unsaturation (number of double bonds) depends on the temperature of the water in which the algae grow. Since then, an increasing number of biogeochemical and paleoceanographic studies dealt with the occurrence, fate and applications of these unique organic compounds. In this introduction, we only give prominence to a few milestones in the evolution of the alkenone paleotemperature method. For comprehensive reviews, the reader is referred to Brassell (1993) and the position papers and reports of a recent workshop on alkenone-based paleoceanographic indicators (Grimalt et al. 2000; Mix et al. 2000; Prahl et al. 2000a; Sachs et al. 2000; Volkman 2000; Bard 2001; Bijma et al. 2001; Eglinton et al. 2001; Herbert 2001; Schneider 2001).

Long-chain alkenones were first recognized in Pleistocene and Miocene sediments from the Walvis Ridge (Boon et al. 1978). De Leeuw et al. (1980) confirmed their structure as C_{37}-C_{39} methyl and ethyl ketones, and Volkman at al. (1980) identified them in extracts of *Emiliania huxleyi*, an ubiquitous marine coccolithophorid alga. Subsequently, these compounds have been identified in some other haptophyte microalgae, namely *Gephyrocapsa oceanica* and a few related coastal forms (Conte et al. 1995; Volkman et al. 1995; Volkman 2000). *E. huxleyi* and *G. oceanica* are the most common sources of alkenones in the modern ocean while other yet unknown species of

the family Gephyro-capsae were the likely sources in the Cenozoic ocean (Marlowe et al. 1990).

In order to permit a direct comparison between water temperature and alkenone unsaturation, Brassell et al. (1986) introduced the unsaturation index U^{K}_{37} which is calculated from the relative abundances of C_{37} methyl alkenones containing 2-4 double bonds. However, the $C_{37:4}$ alkenone is not generally present, especially being absent in samples from warmer waters, and seems to be controlled by factors other than temperature (e.g. Bard et al. 2000; Rosell-Melé et al. 2002; Sikes and Sicre 2002). Therefore, a simplified version of the unsaturation index is usually employed:

$$U^{K'}_{37} = [C_{37:2}]/[C_{37:2} + C_{37:3}] \qquad (1)$$

Prahl and co-workers provided the first generally accepted temperature calibrations of the $U^{K'}_{37}$ index, based on laboratory cultures of *E. huxleyi* and verified by analyses on water column particulates:

$$U^{K'}_{37} = 0.033T + 0.043 \qquad (2)$$
$$\text{(Prahl and Wakeham 1987)}$$

$$U^{K'}_{37} = 0.034T + 0.039 \qquad (3)$$
$$\text{(Prahl et al. 1988)}$$

The validity of these culture equations for paleoceanographic use has been confirmed by sediment-based calibrations (reviews by Müller et al. 1998; Prahl et al. 2000a; Herbert 2001) and by comparison with other paleotemperature proxies (Sikes and Keigwin 1994; Müller et al. 1997; Marlow et al. 2000; Bard 2001).

Despite the wide acceptance of the alkenone paleotemperature method, the exact meaning of temperature values derived by alkenone analysis is still a matter of concern (e.g. Volkman 2000; Bijma et al. 2001). In oceanic settings with significant production below the mixed layer, the overall temperature recorded by alkenones will be lower than the SST (Prahl et al. 1993, 2001; Ternois et al. 1997; Ohkouchi et al. 1999). Systematic seasonal differences in alkenone production may also bias the $U^{K'}_{37}$ signal (Conte et al. 1992; Sikes et al. 1997; Ternois et al. 1998; Rosell-Melé et al. 2000; Bijma et al.

2001). Another question still unresolved is whether U$^{K'}_{37}$ varies linearly with temperature over the entire range of oceanic temperatures. There is some evidence for systematic deviations from linearity especially at water temperatures above 24°C (Sonzogni et al. 1997a, b; Goni et al. 2001; Pelejero and Calvo 2003) and below about 5°C (Sikes and Volkman 1993; Ternois et al. 1998).

In the present study, we summarize alkenone results from the South Atlantic Ocean and its Antarctic sector which were obtained within the scope of the collaborative research project SFB 261 at Bremen University. In the first part, we present largely unpublished sediment trap results which give further insight into the seasons and water depths of alkenone production in the eastern Atlantic (see also Müller and Fischer 2001). In the second part, we summarize calibration studies based on surface sediments (Müller et al. 1998; Benthien and Müller 2000) and provide an update of the global core-top calibration. Finally, we briefly address studies which indicate that calibrations established in the modern ocean also apply to older sediments (Müller et al. 1997; Marlow et al. 2000). We do not present paleotemperature records from the South Atlantic in this paper which have been evaluated elsewhere (Müller et al. 1994, 1997; Schneider et al. 1995, 1996, 1999; Kirst et al. 1999; Rühlemann et al. 1999; Schneider 2001; Kim et al. 2002a). The stable carbon isotopic composition of alkenones as a potential proxy for carbon dioxide in surface waters is addressed in another study of this volume (Schulte et al.).

Material and Methods

The sediment trap samples of this study represent subsets of a series of long-term sediment trap deployments in the South Atlantic and its Antarctic sector (Fischer and Wefer 1996; Fischer et al. 2000, 2002). The samples were collected with cone-shaped multi-sample sediment traps (collection area 0.5 m^2) attached to moored arrays. Prior to the deployment, NaCl was added to the sampling cups to reach a final salinity of about 40 ‰. The cups were poisoned before and after deployment with mercury chloride. Samples were wet-sieved through a 1 mm screen to remove larger shells and swimmers, split into aliquots, freeze-dried and homogenized. The deployment data are summarized in Table 1. Full details of the sampling and splitting procedures are given elsewhere (Fischer et al. 2000, 2002).

The procedures employed in our laboratory to determine long-chain alkenones in sediments and sediment trap samples have been described elsewhere (Müller et al. 1998; Benthien and Müller 2000; Müller and Fischer 2001). To convert the U$^{K'}_{37}$ time series into water temperatures, we used equation 3 as suggested by Prahl et al. (2000a). These values are referred to as alkenone or U$^{K'}_{37}$ temperatures in the following.

Actual SST measurements for the collection periods of our traps were obtained from the NOAA-CIRES Climate Diagnostics Center, Boulder, Colorado, from their web site at http://www.cdc.noaa.gov/. These data represent filtered

Trap	Position	Water depth (m)	Trap depth (m)	Collection period considered for alkenone analysis	Sampling interval (days)
EA8a	05°47.1'S 09°25.5'W	3450	598	15.12.91-06.10.92	15.4
WR2	20°02.8'S 09°09.3'E	2196	599	18.03.89-13.03.90	18
WR3	20°02.4'S 09°09.7'E	2208	1648	25.03.90-09.04.91	19
WR4	20°07.5'S 08°57.6'E	2263	1717	21.04.91-17.12.91	12
NU2	29°12.0'S 13°07.0'E	3055	2516	20.01.92-03.02.93	19
PF3	50°07.6'S 05°50.0'E	3785	614	10.11.89-23.12.90	21
BO1	54°20.3'S 03°22.6'W	2734	450	28.12.90-08.12.91	23

Table 1. Station data of sediment traps from the South Atlantic used for alkenone analysis.

and bias-corrected weekly mean sea-surface temperatures based on *in situ* and satellite measurements (Reynolds and Smith 1994). For core-top calibrations, we used the annual mean SST values of the World Ocean Atlas 1994 (WOA94, Levitus and Boyer 1994) and 1998 (WOA98, Conkright et al. 1998), which are also available at the website cited above.

Sediment Trap Results

The station data for the sediment traps of this study are listed in Table 1. For space limitations, we did not tabulate the analytical results in this paper but archived them in the Pangaea data base (http://www.pangaea.de/PangaVista; search for Müller PJ Fischer G 2003). Bulk parameters (organic carbon, carbonate, opal) have been reported elsewhere

(Fischer et al. 2000, 2002). Below, we will present the major features of each time series and compare them to the variations in SST during the collection periods. To facilitate this comparison, we reduced the collection dates of the trap data for the average settling time of the particles as indicated in the captions of the relevant figures. The time lag between the sea surface and the trap was estimated by cross-correlating the weekly SST and $U^{K'}_{37}$ records (see also Müller and Fischer 2001). To give an example, this approach is illustrated in Fig. 7 for trap PF3 from the Polar Frontal Zone.

Eastern Equatorial Atlantic (EA8a)

Trap EA8a was located in the eastern tropical Atlantic about 6° south of the equator (Fig. 1). The trap was deployed in a water depth of 598 m and

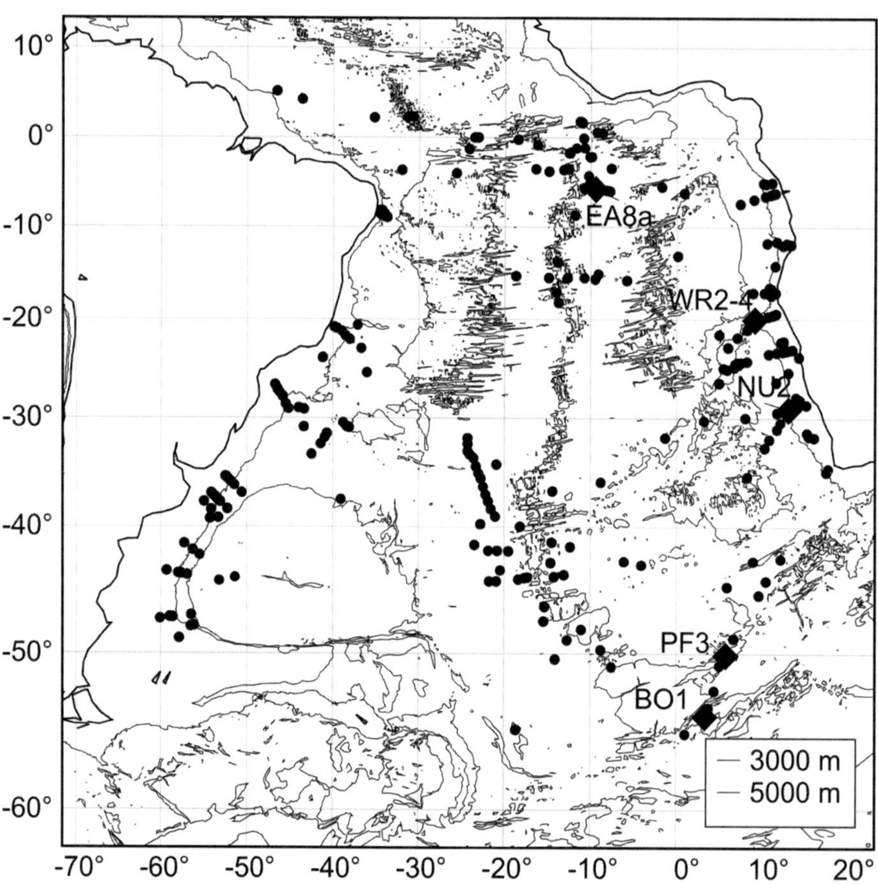

Fig. 1. Locations of sediment traps (labeled diamonds) and surface sediments (dots) analyzed for alkenones in the South Atlantic and the Antarctic sector.

collected particles over a period of 296 days from December 1991 to October 1992 (Table 1). Figure 2 compares the alkenone results to the annual SST and primary productivity cycles at this site. To compensate for the settling time of the particles, we reduced the collection dates by 45 days in this figure. This correction is based on an optimum correlation between SST and $U^{K'}_{37}$ for a lag of three collection periods (r = 0.934, Fig. 2a) suggesting a relatively low sinking rate of 13 m d^{-1} for particles containing alkenones. Besides, we sorted the samples from January to December in order to simulate a normal annual cycle. The last five samples in Fig. 2 thus actually represent the beginning of the time series in November to December 1991.

The flux of C$_{37}$ alkenones into trap EA8a (Fig. 2d) varied from 0.5 to 13.5 µg m^{-2} d^{-1}, with the exception of one higher value in early November (42 µg m^{-2} d^{-1}). However, the latter value is based on a shorter collection interval (3 days) than the other samples (15 days) and may document a short-lived flux event. In general, elevated alkenone fluxes (>5 µg m^{-2} d^{-1}) occurred in January, March and November. The fluxes co-varied fairly well with the flux of TOC (Fig. 2c), but did not reflect the seasonal primary productivity patterns (Fig. 2b). Primary productivity is generally enhanced during austral winter (June to October) in this region, when the thermocline shoals in response to intensified SE trade winds in the western Atlantic (Longhurst 1993; Antoine et al. 1996). However, as this prominent productivity peak in winter was not completely sampled, our time series may have missed the season of highest alkenone production. This is indicated by high alkenone fluxes in November just at the end of the period of high primary productivity, and by high alkenone concentrations relative to TOC in the particles of this month (4800 and 2200 µg gC^{-1}, Fig. 2e). A comparable enrichment of C$_{37}$ alkenones (up to 4100 µg gC^{-1}) in periods of exceptionally high alkenone fluxes was also observed in the filamentous upwelling regime off Cape Blanc, NW Africa (Müller and Fischer 2001).

The $U^{K'}_{37}$ ratios of the EA8a time series varied between 0.893 and 0.980, closely mirroring the annual SST cycle (Fig. 2a). In terms of temperature, however, the range of the $U^{K'}_{37}$ values (25.1-27.7°C)

was clearly smaller than the range of the weekly SST measurements (22.7-28.2°C). As shown in Fig. 3a, both temperature records were in good agreement (±0.5°C) in summer and autumn (Feb-May), but disagreed by about 2°C in the colder seasons. This discrepancy cannot be explained by a change in the slope of the $U^{K'}_{37}$-temperature relationship as observed by Sonzogni et al. (1997b) for water temperatures >24°C in the Indian Ocean ($U^{K'}_{37}$ = 0.23T + 0.316). This relationship would indicate slightly higher temperatures than the calibration of Prahl et al. (1988) but a comparable low seasonal temperature range (Fig. 3a).

Another possible explanation for the diverging SST and $U^{K'}_{37}$ records at site EA8 is lateral particle advection from the western equatorial Atlantic. Siegel and Deuser (1997) demonstrated that particles collected by sediment traps can have surface departure points many hundreds of kilometers apart, depending on the ocean flow field, particle sinking velocities and the depth of the trap. Assuming, for example, that the particles intercepted by trap EA8a originate from 20°W, the two temperature records would largely synchronize (Fig. 3b). This hypothesis is supported by the general upper-level circulation in this region. Trap EA8 was positioned at 5.8°S just south of the eastward flowing South Equatorial Undercurrent (SEUC). This strong countercurrent is typically located between 3°S and 5°S and transports waters from the region off Brazil all across the Atlantic into the Angola Basin. It has a core depth of 150-200 m and maximum velocities of 30-60 cm s^{-1} (Peterson and Stramma 1991; and references therein). Assuming an average sinking rate of 13 m d^{-1}, as estimated above, and an eastward current velocity of 10-20 cm s^{-1}, we calculate a possible displacement of particles by 400-800 km in the upper 600 m of the water column. Hence, small and slowly sinking particles produced in the central equatorial South Atlantic could have easily been transported to the position of our trap.

Northern Benguela (WR2-4)

The eastern edge of the Walvis Ridge was chosen as a favorable site to study seasonal and interannual variations in alkenone production in the northern

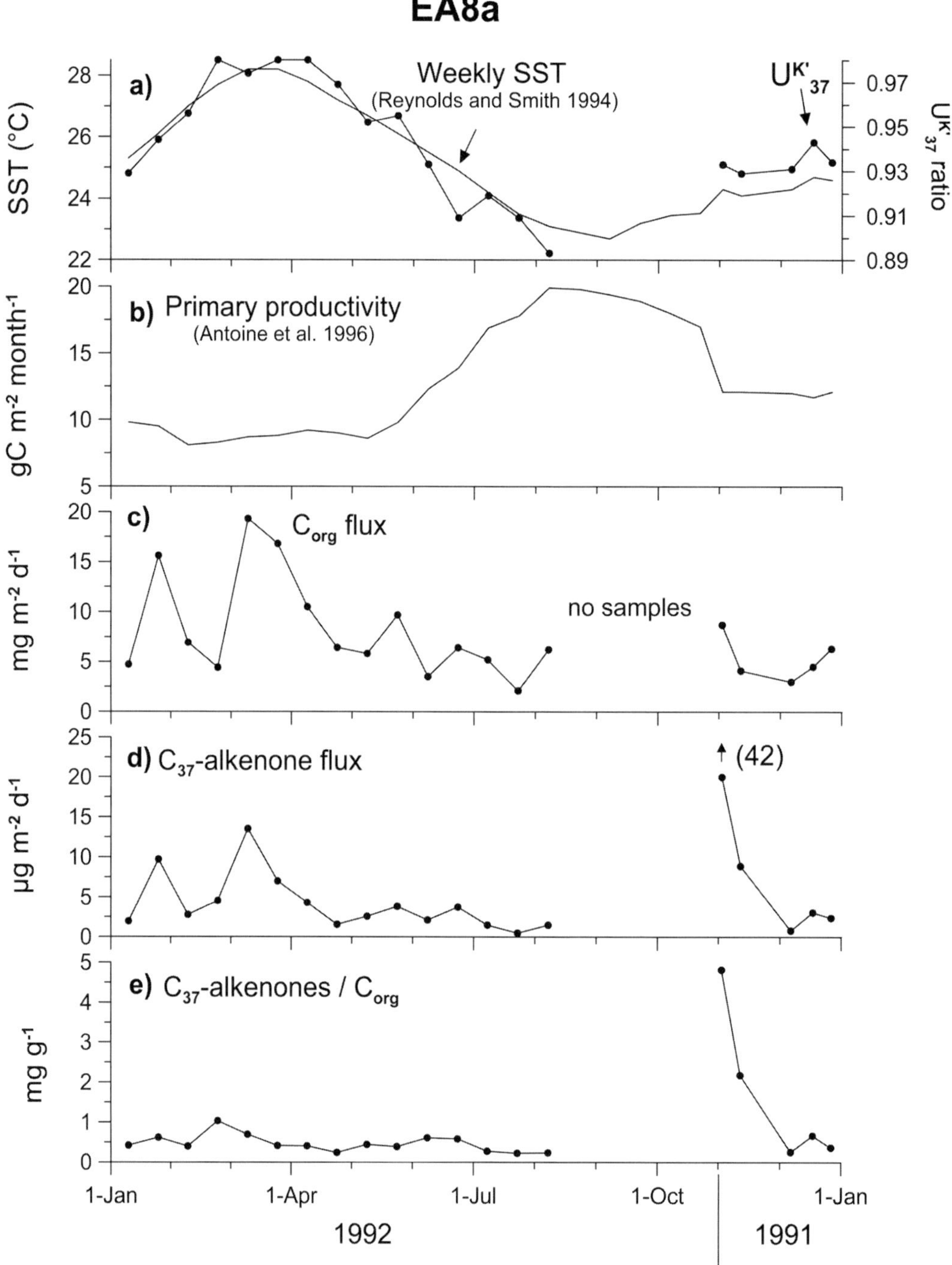

Fig. 2. Time series data for sediment trap EA8a in the eastern equatorial Atlantic. Data points represent the centers of the collection intervals. They have been reduced by 45 days to compensate for the settling time of the particles and were rearranged to simulate an annual cycle from January to December (see text). **a)** $U^{K'}_{37}$ ratios of particles intercepted at a water depth of 598 m compared to the temperature changes at the sea surface (Reynolds and Smith 1994); **b)** monthly mean primary production rates based on satellite (coastal zone color scanner) chlorophyll data (Antoine et al. 1996); **c)** flux of total organic carbon; **d)** flux of C_{37} alkenones; **e)** concentration of C_{37} alkenones relative to total organic carbon. Enhanced alkenone fluxes and high carbon-normalized alkenone concentrations just at the end of the period of high primary production in austral spring indicate that we may have missed the season of highest alkenone production at this site.

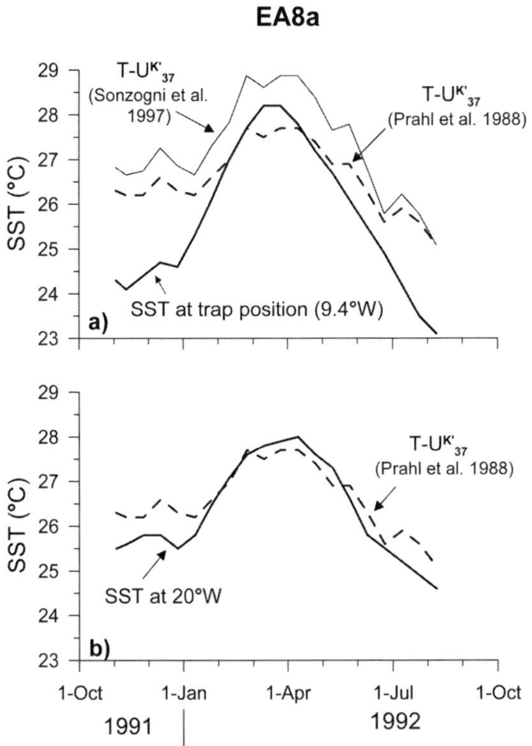

Fig. 3. Alkenone temperature records (T-U$^{K'}_{37}$) for trap EA8a, using two different calibrations, compared to the annual SST cycle (Reynolds and Smith 1994) at **a)** the trap position and **b)** about 10° west of the trap. The calibrations used are the culture equation of Prahl et al. (1988) (eq. 3 in the text) and the warm-end relationship of Sonzogni et al. (1997b) (U$^{K'}_{37}$ = 0.23T + 0.316). The collection dates for the trap data were reduced by 45 days, but not rearranged as in Fig. 2. The better agreement of U$^{K'}_{37}$ temperatures with SST values from 20°W implies that the alkenone record was biased by particle advection from the (warmer) central Atlantic.

Benguela upwelling regime. Our traps were located about 200 nautical miles offshore Namibia within the filamentous mixing area providing a three-year alkenone record from March 1989 to December 1991 (Table 1, Fig. 1). The first trap (WR2) was deployed at a water depth of 599 m within the water mass of the Antarctic Intermediate Water, and the two other traps (WR3-4) at 1648 and 1717 m within the North Atlantic Deep Water about 550 m above the sea floor (Fischer and Wefer 1996). A CTD profile from this station and the U$^{K'}_{37}$ record of trap WR2 have been published by Treppke et al. (1996).

The most prominent feature of the alkenone time series from the Walvis Ridge is a pronounced seasonality in the fluxes with maximum values in austral autumn (Apr-Jun) and spring (Sep-Nov), and low rates for most of the summer and winter seasons (Fig. 4a). The flux of C$_{37}$ alkenones varied by more than two orders of magnitude over the three-year cycle (0.04-13.6 µg m^{-2} d^{-1}) and co-varied with that of TOC (r = 0.818, Fig. 4b) and carbonate (r = 0.683, Fig. 4c). A good correlation also exists between the flux of alkenones and the fluxes of total coccoliths (r = 0.758) and *E. huxleyi* (r = 0.737), based on countings reported by Cepek (1996) for the WR2 time series. The dashed curve in Fig. 4a shows the flux record for *E. huxleyi*, the principal alkenone producer in this region. According to Cepek (1996), *E. huxleyi* constituted 52-75 % of the total coccoliths in the samples of trap WR2, whereas *G. oceanica*, another alkenone-producing species, was only of secondary importance (0.5-11%). The average annual fluxes of C$_{37}$ alkenones into the shallow and deep traps were 3.5 and 1.6 µg m^{-2} d^{-1}, respectively. Hence about 50% of the alkenones appeared to be lost by decomposition or disaggregation processes during settling, or were transported laterally further offshore. A comparable reduction of alkenones in the water column was observed off Cape Blanc, NW Africa, based on contemporaneous shallow and deep deployments (Müller and Fischer 2001).

The U$^{K'}_{37}$ values of the WR2-4 time series ranged from 0.535 to 0.842 and showed a clear seasonality which is in general accordance with the weekly SST record (Fig. 5a). Regressions between SST and the U$^{K'}_{37}$ measurements of the two deeper traps indicate a time lag of about 5 weeks and a sinking rate of about 48 m d^{-1}. The collection dates in Figs. 4 and 5 were corrected accordingly. In terms of temperature, the U$^{K'}_{37}$ range was 14.6-23.6°C, somewhat greater than the SST amplitude for the collection period (16.3-23.0°C, Fig. 5a). Overall, the two temperature records showed identical mean (± σ) values (traps: 19.1 ± 2.3 °C; weekly SST: 18.9 ± 2.0 °C). The highest U$^{K'}_{37}$ temperatures occurred from late summer to early autumn (Mar-Apr). They were associated with low alkenone fluxes (Fig. 5b) and a relatively shallow surface mixed layer (20-30 m, Fig. 5c). The low-

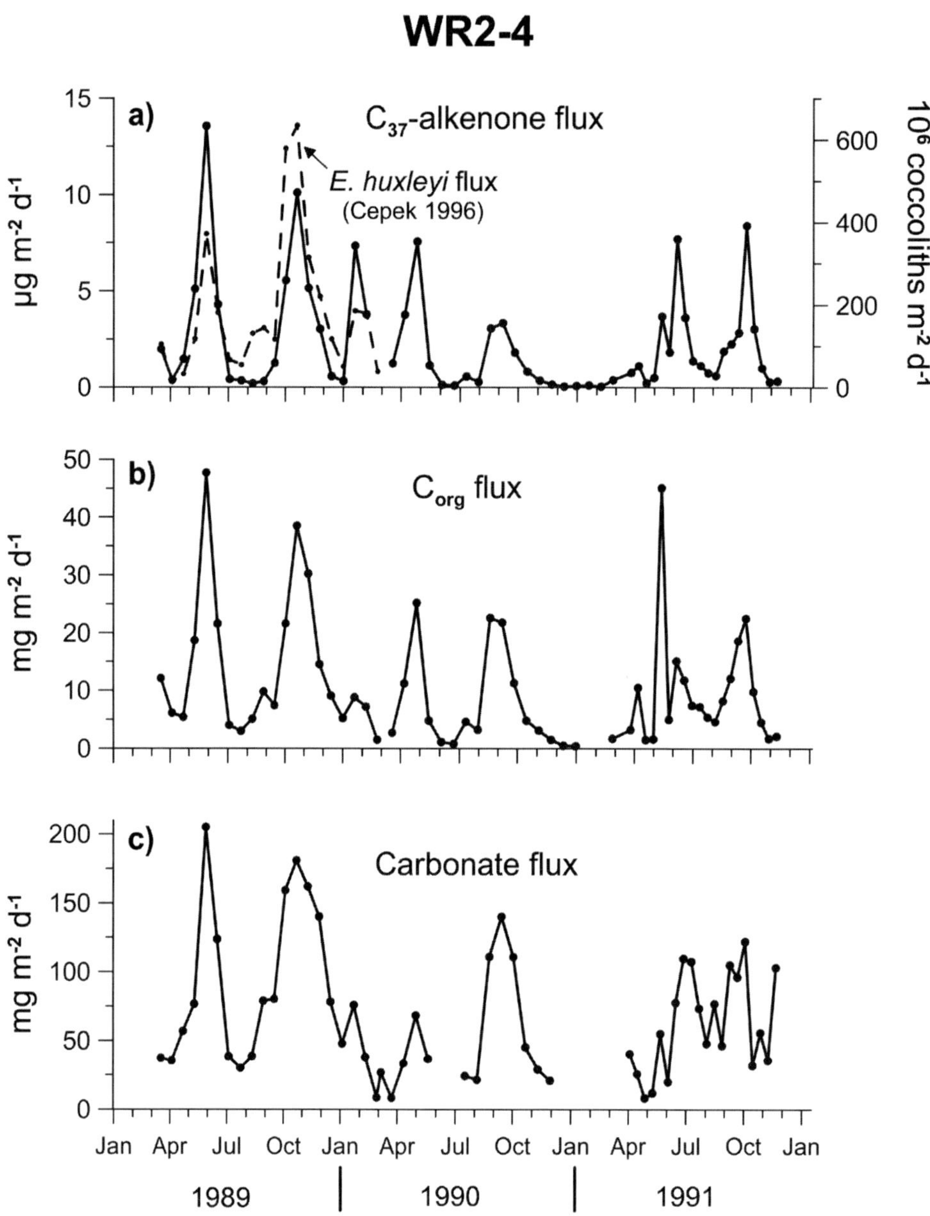

Fig. 4. Three-year record for fluxes of C$_{37}$ alkenones, total organic carbon and carbonate at the Walvis Ridge. Also shown is the WR2 flux record for coccoliths of *E. huxleyi* (dashed line in a) as reported by Cepek (1996). The collection dates of the trap data have been reduced by 2 weeks (WR2) and 5 weeks (WR3-4) to compensate approximately for the lag due to particle settling. The time series reveals a pronounced seasonality in alkenone production with maximum values in austral autumn (Apr-Jun) and spring (Sep-Nov), and low rates for most of the summer and winter seasons (see also Fig. 5).

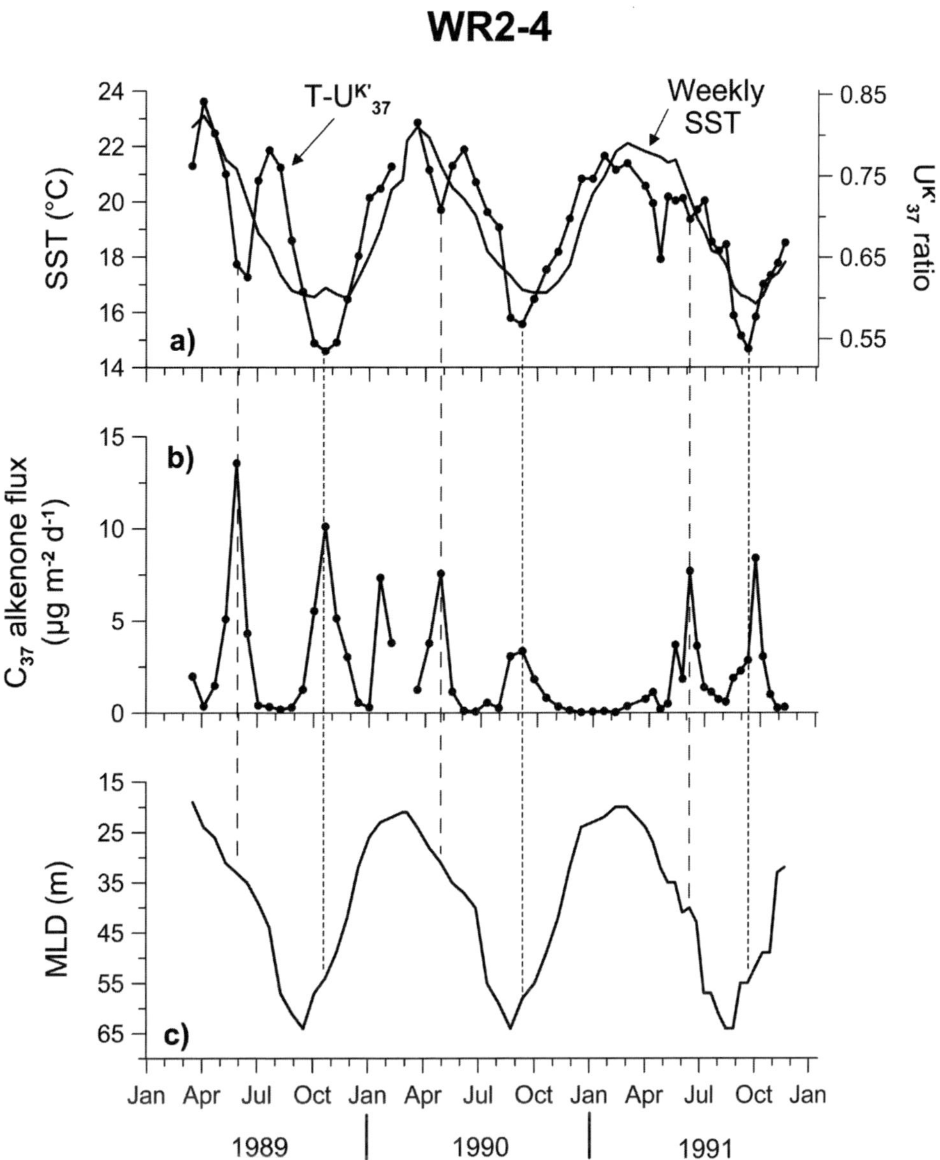

Fig. 5. Three-year record of alkenone temperatures (T-U$^{K'}_{37}$) at the Walvis Ridge compared to the seasonal variations in **a)** weekly SST (Reynolds and Smith 1994), **b)** alkenone fluxes and **c)** mixed layer depths (MLD). The latter was estimated from monthly mean temperature distributions in the uppermost water column (Levitus and Boyer 1994) and defined as the depth in which the temperature changes by 0.5°C relative to the surface value (Longhurst 1995). The collection dates of the trap data have been reduced as indicated in the caption of Fig. 4. Peak alkenone fluxes were recorded in autumn when the stratification broke down (dashed vertical lines) and in spring, when the water column recovered from deep winter mixing (dotted vertical lines).

est temperatures were recorded during periods of high alkenone production in spring (Oct-Nov), when the mixed layer recovered from deep winter mixing (dotted vertical lines in Fig. 5). A second, less pronounced $U^{K'}_{37}$ minimum, associated with high alkenone fluxes as well, occurred in late autumn (May-June, dashed vertical lines in Fig. 5). This temperature minimum is most clearly documented in the shallow record of trap WR2 and also indicated in the two deeper records of the subsequent years, but not evident in the weekly SST measurements (Fig. 5a).

In summary, and with reference to the Coastal Domain Model 6 of Longhurst (1995), the seasonal cycle of alkenone production at the Walvis Ridge can be characterized as canonical spring-autumn blooms. Accordingly, the flux of alkenones is relatively low during summer stratification, when the productivity is mainly fuelled by regenerated nutrients. Progressive breakdown of the stratification and deepening of the mixed layer in autumn induces renewed alkenone productivity fuelled by upwelled nutrients that have been accumulated below the mixed layer in summer. The production slows down again under winter conditions, when the mixed layer extends below the photic depth down to about 60 m (Fig. 5c). Finally, a spring bloom in alkenone production develops from accumulated winter nutrients, when the water column begins to stratify. Interestingly, Prahl et al. (2000b) observed a similar relationship between mixed-layer depth and alkenone export in the central Arabian Sea. The major export events occurred when the mixed layer was changing while minimum fluxes appeared during inter-monsoon and mid-monsoon periods which are characterized by shallow or deep mixing, respectively.

Southern Benguela (NU2)

Mooring site NU2 was located in the southern part of the Benguela upwelling region (Fig. 1) within the filamentous mixing area west of the Namaqua upwelling cell (Lutjeharms and Stockton 1987; Shannon and Nelson 1996). The trap used for alkenone analysis was fixed at a water depth of 2516 m, approximately 540 m above the sea floor, capturing particles over a period of 380 days (Ta-

ble 1). The alkenone results are illustrated in Fig. 6. We corrected the trap dates in this figure for a settling time of 8 weeks based on an optimum correlation between SST and $U^{K'}_{37}$ (r = 0.879) for a lag of 3 sampling intervals. The corresponding sinking rate is 45 m d^{-1}.

Like in the northern Benguela, we observed a closer correlation of alkenone fluxes with fluxes of TOC (r = 0.798) rather than to carbonate (r = 0.686). However, the seasonal extremes in production were less pronounced than at the Walvis Ridge (cp. Fig 4) and the average annual flux of C_{37} alkenones was lower by 40% (1.0 µg m^{-2} d^{-1}, range: 0.15-3.0 µg m^{-2} d^{-1}). A major pulse in alkenone sedimentation was recorded in December/January, but peak fluxes also occurred in March/April, June/July, and September-November (Fig. 6b). Thus, the seasonal variations in alkenone production had no discernible effect on the average $U^{K'}_{37}$ temperature of the time series (Table 2). Both, the $U^{K'}_{37}$ and weekly SST records (Fig. 6a) showed identical mean ($\pm \sigma$) values ($U^{K'}_{37}$: 17.8 $\pm$ 1.9 °C; weekly SST: 18.0 $\pm$ 2.0 °C) and comparable seasonal ranges (7.1°C and 6.1°C, respectively).

According to a recent biological study of the NU2 time series (Romero et al. 2002), the coccolithophorid assemblage was dominated by *E. huxleyi* (41-79%). The second most important coccolithophorids were *Calcidiscus leptoporus* (4-11.5%) and *Florisphaera profunda* (1.5-13%) which probably do not produce alkenones. *G. oceanica*, another alkenone-producing species, was only of minor importance and not reported in the paper. The flux record for coccoliths of *E. huxleyi* is displayed in Fig. 6c. It is characterized by a prominent flux maximum at the autumn to winter transition, and a secondary one in late spring and early summer, both being in accordance with elevated fluxes of alkenones, organic carbon and carbonate. However, the maximum fluxes for alkenones and *E. huxleyi* did not occur in the same season. While alkenones peaked in December, *E. huxleyi* was most abundant in June/July. According to Romero et al. (2002), the summer flux maximum of the NU2 time series is characterized by relatively high fluxes of neritic (coastal) diatoms and the right-coiled planktonic foraminifera *Neoglobo-quadrina pachyderma* considered to be

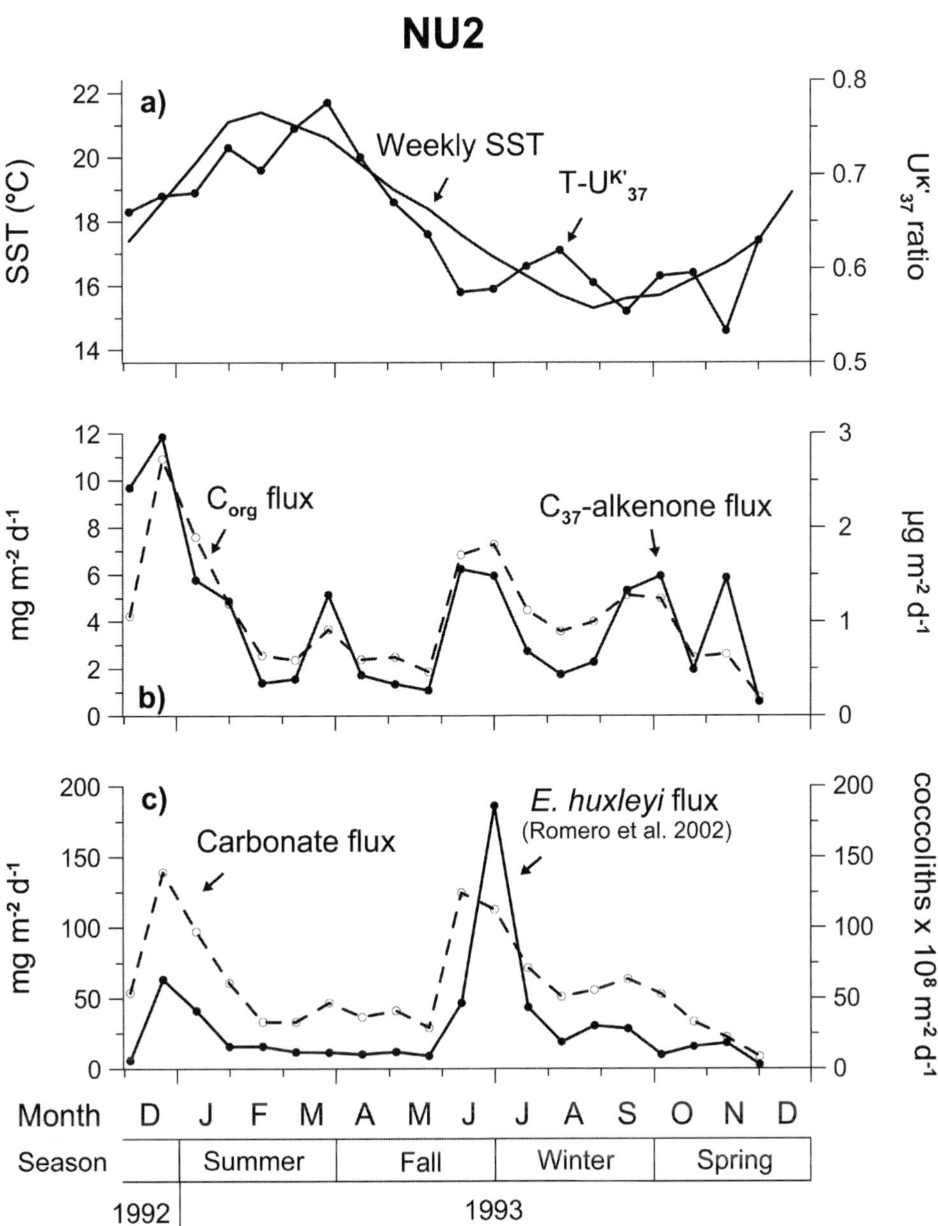

Fig. 6. Time series data for sediment trap NU2 from the southern part of the Benguela upwelling region. The collection dates of the trap data have been reduced by 8 weeks to compensate for the lag due to particle settling. **a)** U$^{K'}_{37}$ temperatures for particles intercepted at a water depth of 2516 m compared to the temperature changes at the sea surface (Reynolds and Smith 1994); **b)** comparison between the fluxes of total organic carbon and C$_{37}$ alkenones; **c)** comparison between the flux records for total carbonate and coccoliths of *E. huxleyi* (Romero et al. 2002).

an up-welling indicator. Hence, the peak flux of alkenones recorded in December appears to be related to an increased coastal upwelling and off-shore spreading of nutrient-rich filaments. On a yearly basis, however, the relative abundance of neritic diatoms was only about 20%, suggesting that the time series was dominated by *in situ* production with subordinate coastal contributions.

Polar Frontal Zone (PF3)

Trap PF3 was located in the eastern Atlantic at 50.1°S (Fig. 1) within the Antarctic Polar Frontal Zone as defined by Read et al. (2002). The trap was deployed in a water depth of 614 m and put in operation from November 1989 to December 1990 (Table 1). To estimate the average settling time of the particles, we first compared the $U^{K'}_{37}$ record to the weekly SST variations at this site. A cross-correlation between the two records revealed no significant relationship (r = 0.210) based on the original collection dates of the trap (Fig. 7a) and the best correlation (r = 0.909) for a lag of 10 weeks (Fig. 7b). This suggests a relatively low sinking rate of 9 m d^{-1} which is considerably lower than the rates obtained in the Benguela region (40-50 m d^{-1}, see above) and off Cape Blanc, NW Africa (60-280 m d^{-1}, Müller and Fischer 2001). Much higher sinking rates for alkenone-containing particles (145-290 m d^{-1}) were also reported from the western Pacific off Japan (Sawada et al. 1998).

The low sinking rate obtained in the region of the Polar Front may be explained by lateral advection of particles with the eastward-flowing frontal jet. Assuming, for instance, a speed of 15-30 cm s^{-1} (Strass et al. 2002), a particle advected with the jet could be displaced by a maximum distance of 900-1800 km within 10 weeks. However, due to the small zonal SST variation in this region, the $U^{K'}_{37}$ record of trap PF3 was apparently not biased. The alkenone temperatures ranged from 0.6° to 5.4°C (Fig. 8a), fairly in accordance with the weekly SST variations (1.2-4.9°C). In the warm season from January to May, both records showed identical mean (±σ) values ($U^{K'}_{37}$: 4.4 ± 0.7 °C; weekly SST: 4.4 ± 0.6 °C). In the more productive cold season, the alkenone temperatures were slightly lower by 0.6-1.5°C. The flux-weighted

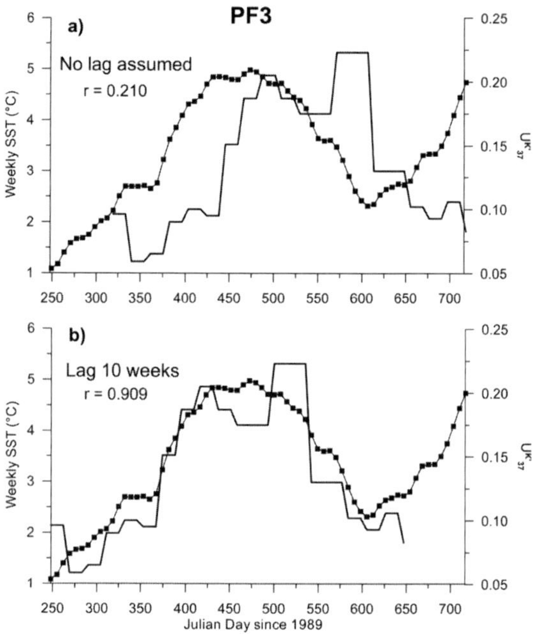

Fig. 7. $U^{K'}_{37}$ unsaturation record (solid line) for trap PF3 from the Polar Frontal Zone in the eastern Atlantic compared to the seasonal variations in weekly SST (squares, Reynolds and Smith 1994). A cross-correlation between the two records suggests a time lag of 10 weeks for the trap data, and a rather low sinking rate of 9 m d^{-1}, indicating advection of particles with the eastward-flowing frontal jet.

temperature signal of the PF3 time series was thus lower by 0.9°C, as compared to the predicted SST value (Table 2).

As depicted in Fig. 8b, the PF3 time series reveals a clear seasonality in alkenone production. The highest fluxes of C_{37} alkenones (up to 2.8-3.8 µg m^{-2} d^{-1}) occurred between winter and spring 1990 (Aug-Oct). In the preceding year, the alkenone flux in spring was significantly lower (0.8 µg m^{-2} d^{-1}) but still enhanced, compared to the warmer season (Jan-Jun; 0.2 µg m^{-2} d^{-1}). Overall, about 87% of the annual alkenone flux was recorded in winter and spring at this site (Table 2). Total organic carbon (Fig. 8c) and carbonate (Fig. 8d) showed enhanced fluxes in this season as well, but the correlation to alkenones was less close (r = 0.324 and 0.585, respectively) than at the subtropical sites (Figs. 4 and 6). Curiously, perhaps caused by particle aggregation, we observed the closest

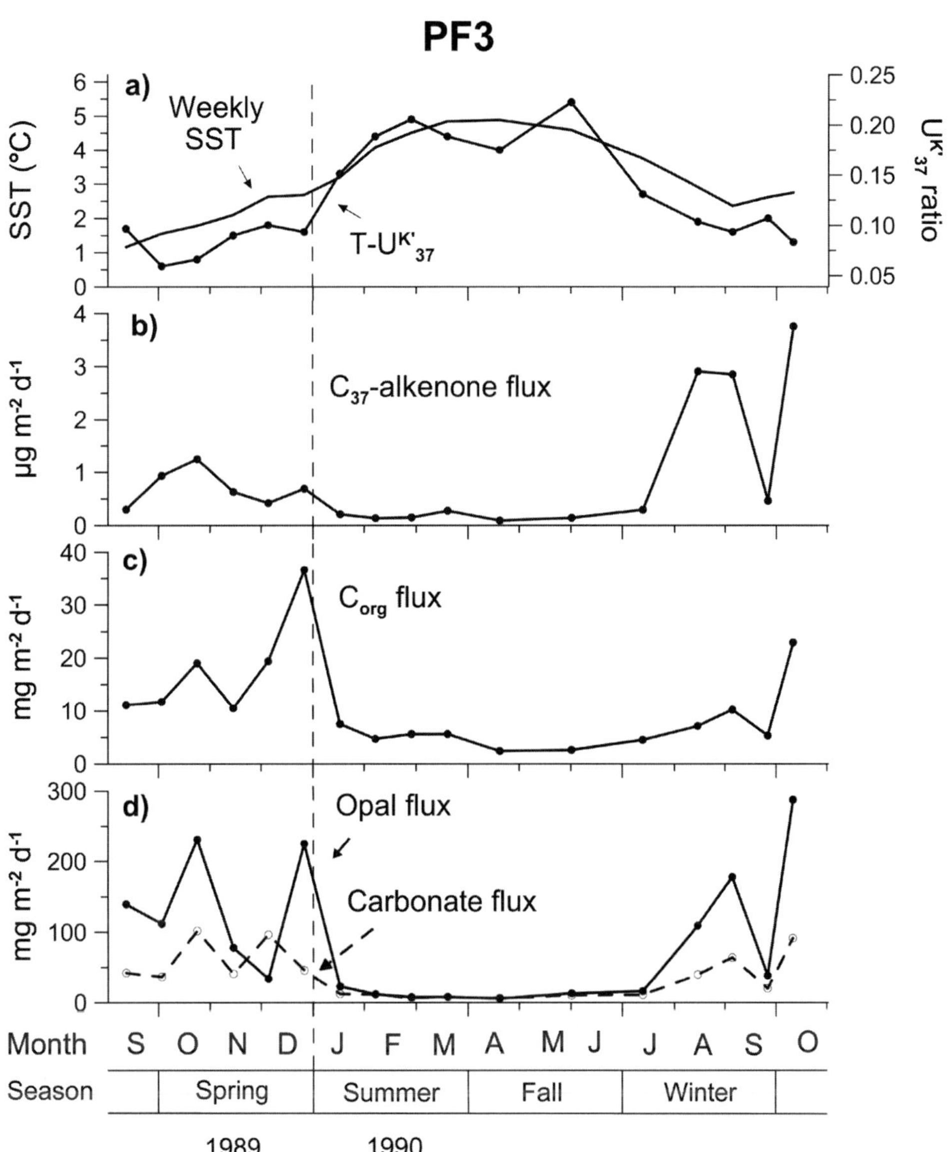

Fig. 8. Time series data for sediment trap PF3 from the Polar Frontal Zone in the eastern Atlantic. The collection dates for the trap data have been corrected for a lag of 10 weeks (see Fig. 7). The dominant alkenone export occurred in winter and spring (Aug-Dec) in accordance with increased fluxes of organic carbon, carbonate and opal. The U$^{K'}_{37}$ temperatures were in agreement with the weekly SST in the warm season but lower by on average 0.8°C in the cold season.

correlation (r = 0.720) between alkenone fluxes and fluxes of opal, the dominant biogenic component (Fig. 8d).

Antarctic Zone (BO1)

Sediment trap BO1 was located in the Antarctic Zone west of Bouvet Island, about 4° south of the Polar Front. The trap was deployed within the upper Circumpolar Deep Water in a water depth of 450 m and provided an alkenone record of 345 days for the year 1991 (Table 1, Fig. 1). As illustrated in Fig. 9, alkenones as well as bulk biogenic components (opal, carbonate, organic carbon) showed a pronounced seasonality in production, with elevated fluxes in austral summer and autumn. Much lower fluxes were recorded in winter and spring, when the sea ice approached the mooring location (Gersonde and Zielinski 2000; Fischer et al. 2002). Biogenic silica was the predominating component, accounting for about 80% of the annual mass flux (Fig. 9a). A distinct difference in the timing of peak fluxes points to a succession in primary producers. Diatoms providing the bulk of opal (Gersonde and Zielinski 2000) bloomed in February and alkenone-producing algae apparently one to two months later. This is indicated by coinciding peak fluxes of carbonate and alkenones in April (Fig. 9b,c) and by enhanced carbon-normalized C_{37} alkenone concentrations in the particles of this month (up to 1250 $\mu g\ gC^{-1}$). As a composite, the organic carbon record (Fig. 9d) reflects the contributions of both siliceous and calcareous producers.

As to be expected for a position south of the Polar Front, the $U^{K'}_{37}$ ratios of the BO1 time series (0.027 to 0.121, Fig. 9e) were clearly lower than at site PF3 (cp. Fig. 8a). In terms of temperature, the $U^{K'}_{37}$ values ranged from -0.4° to 2.4°C. However, unlike all other sediment trap records of this study, the BO1 record did not reproduce the annual SST cycle (Fig. 9e). The alkenone temperatures were systematically lower during the warm season (Jan-Apr) and higher during the cold season (Jun-Nov), in each case by an average of 1.7°C. A cross-correlation between the two records did not yield an unambiguous result. We therefore considered no lag for the data of this trap in Figures 9 and 12.

The high $U^{K'}_{37}$ values obtained for the cold season at site BO1 may be explained by the advection of particles from the nearby Polar Frontal Zone. This hypothesis is corroborated by increased proportions of the diatom *Thalassionema nitzschioides* in the particles collected in late winter and spring (Fischer et al. 2002). Considering the very low alkenone fluxes recorded during the cold season at this site (0.05-0.17 $\mu g\ m^{-2}\ d^{-1}$, Fig. 9c), even small allochthonous contributions of alkenones could have significantly biased the $U^{K'}_{37}$ values. However, the overall $U^{K'}_{37}$ signal of the BO1 time series (0.3°C) was determined by the relatively low $U^{K'}_{37}$ values recorded during summer and early autumn, when 98% of the annual alkenone flux occurred (Table 2).

The phytoplanktonic source of alkenones in the present-day Antarctic Ocean is not definitely known. However, it is likely that the cosmopolitan coccolithophorid *E. huxleyi* is the main producer. According to a study from the southern Indian Ocean, this taxon dominated the nanoflora in sediment trap material from the Polar Frontal Region and contributed most of the total phytoplanktonic cells in the upper 100 m of the water column (Ternois et al. 1998). *Phaeocystis pouchetii*, a common bloom-forming haptophyte in the Southern Ocean, probably does not synthesize alkenones (Volkman 2000).

Water Depth and Seasonality of Alkenone Production

In order to further assess the influence of water depth and seasonality of alkenone production on the overall $U^{K'}_{37}$ signal exported from the euphotic zone, we partitioned the annual alkenone fluxes between the four major seasons and weighted the $U^{K'}_{37}$ ratios according to the measured alkenone fluxes (Table 2). Also included in this table are the results of a 4-year time series from the upwelling region off Cape Blanc, NW Africa (CB1-4; Müller and Fischer 2001). The $U^{K'}_{37}$ ratios were converted into temperature values using the calibration of Prahl et al. (1988), and compared to the flux-weighted and mean weekly SST values for each (lag-corrected) collection period. The weighted

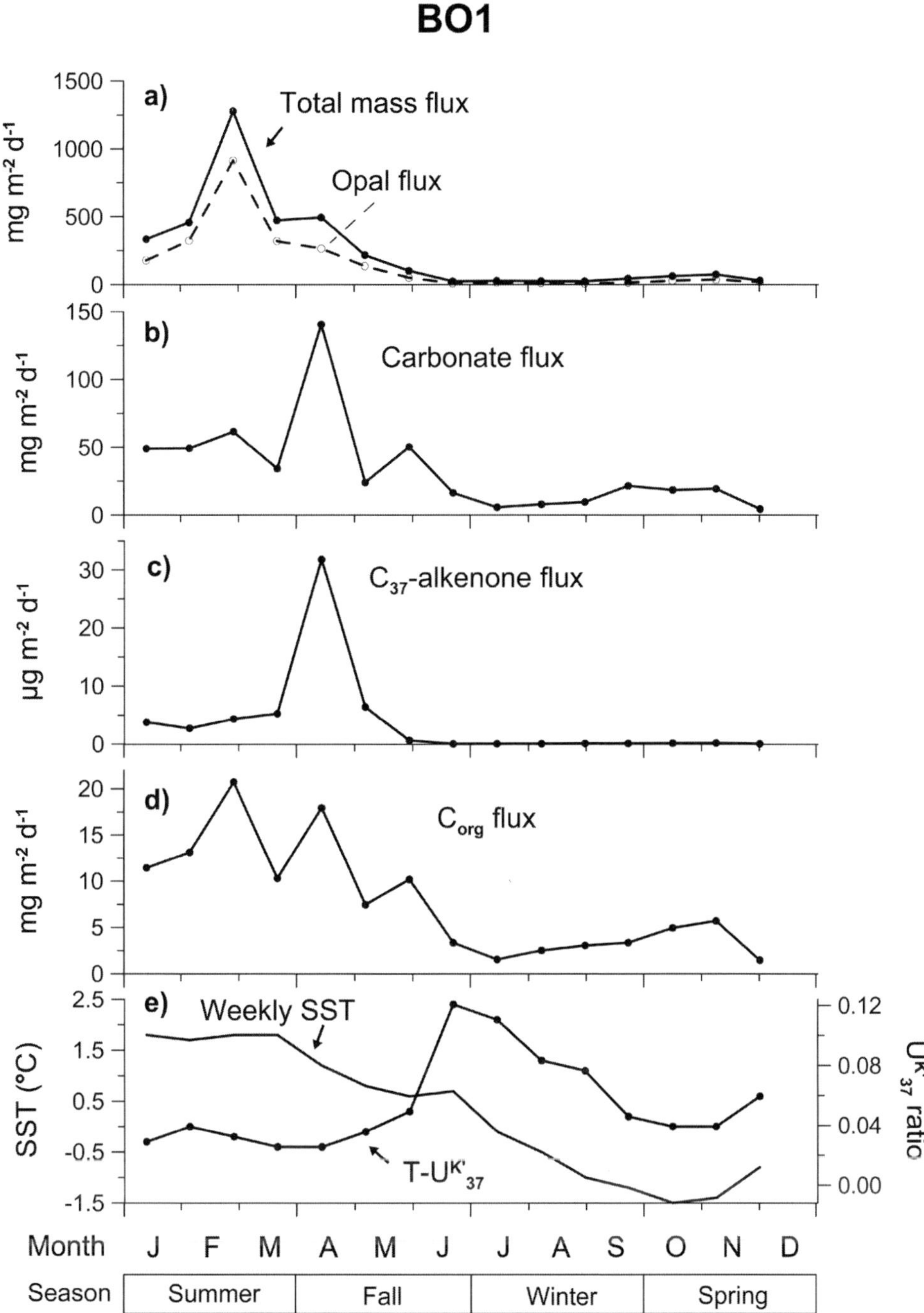

Fig. 9. Time series data for sediment trap BO1 from the Antarctic Zone in the eastern Atlantic about 4° south of the Polar Front. A distinct difference in the timing of peak fluxes points to a succession in siliceous and calcareous primary producers. About 98% of the annual alkenone export occurred in the warm season from January to April (Table 2). Unlike all other sediment traps of this study, the U$^{K'}_{37}$ record of this trap did not reproduce the weekly SST cycle. The U$^{K'}_{37}$ values for winter and spring were probably biased by advection of "warmer" material from the Polar Frontal Zone, whereas the relatively low U$^{K'}_{37}$ temperatures obtained for the warm season may be attributed to the calibration (see text and Fig. 12).

Trap Sediment	Annual flux of C_{37} alkenones	Seasonal partition (%)				Weighted $U^{K'}_{37}$ ratio	Weighted $U^{K'}_{37}$ temperature	Weighted and mean SST [1]	Difference[2]
	$\mu g\ m^{-2}\ y^{-1}$	Sum	Aut	Win	Spr		°C	°C	°C
CB1-4	1170	30	21	32	17	0.790	22.1	21.9	0.2
6 cores						0.796	22.3	21.7	0.6
EA8a	1535	39	19	7[3]	35	0.947	26.7	25.7	1.0
GeoB1613-10						0.932	26.3	25.5	0.8
WR2-4	849	14	42	21	23	0.651	18.0	18.6	-0.6
GeoB1028-2						0.696	19.3	18.9	0.4
NU2	372	22	19	22	37	0.640	17.7	17.9	-0.2
GeoB1721-4						0.674	18.7	17.6	1.1
PF3	240	8	5	48	39	0.100	1.8	2.7	-0.9
PS1759-1						0.149	3.2	2.9	0.3
BO1	1244	29	69	1	1	0.030	-0.3	1.3	-1.6
no core						---	---	---	---

$U^{K'}_{37}$ temperatures calculated from $T = (U^{K'}_{37} - 0.039)/0.034$ (Prahl et al. 1988)

[1] Flux-weighted weekly SST (Reynolds and Smith 1994) for the lag-corrected collection periods of the traps and annual mean SST (0-10 m) (lower lines) according to WOA98 (Conkright et al. 1998)

[2] Difference between $U^{K'}_{37}$ temperatures and SST measurements

[3] probably underestimated (see Fig. 2)

Table 2. Seasonal alkenone fluxes and flux-weighted $U^{K'}_{37}$ and SST values for sediment trap time-series from the South Atlantic and off Cape Blanc (CB1-4), NW Africa (Müller and Fischer 2001). Also shown are the $U^{K'}_{37}$ values for surface sediments (lower lines) from the trap locations (Müller et al. 1998).

SST values are predictions which should agree with the weighted $U^{K'}_{37}$ temperatures, if the alkenones had originated from the mixed layer and the calibration was suitable. To demonstrate the general agreement between sinking particles and underlying sediments, we additionally show the $U^{K'}_{37}$ values for surface sediments from the trap positions (Müller et al. 1998) in comparison to the mean annual SST according to World Ocean Atlas 1998 (Conkright et al. 1998).

From Table 2 it is evident that the $U^{K'}_{37}$ temperature signal of our traps predominantly derived from the surface mixed layer. At the three subtropical locations (CB, WR, NU), flux-weighted SST and $U^{K'}_{37}$ values agreed within ±0.6°C. The 1°C deviation observed at site EA8a in the eastern equatorial Atlantic may be related to the advection of alkenones from warmer waters west of the trap (see above). In any case, the $U^{K'}_{37}$ record of this trap can be explained reasonably only with a shallow water source of the alkenones (Fig. 3). The seasonal amplitudes of $U^{K'}_{37}$ temperatures in the Benguela region (Figs. 5a and 6a) and off Cape Blanc (Müller and Fischer 2001) also argue against significant alkenone contributions from deeper waters. To look at further details of this aspect, Fig. 10 compares the monthly averaged U$^{K'}_{37}$ temperatures for traps WR3-4 and NU2 to the seasonal temperature variations in the upper 75 m of the water column (Conkright et al. 1998). In this figure it is obvious that the U$^{K'}_{37}$ amplitudes (5-6°C) at both sites can only be explained with a relatively shallow alkenone source. At a depth of 50 m, the seasonal temperature variation is only about 2°C. A shallow alkenone source is also indicated by plankton studies which showed the highest cell numbers of *E. huxleyi* generally above the thermocline (Giraudeau and Bailey 1995; Baumann et al. 1999).

Table 2 reveals further that seasonal differences in alkenone production had little effect on the overall U$^{K'}_{37}$ signals at tropical-subtropical locations. The

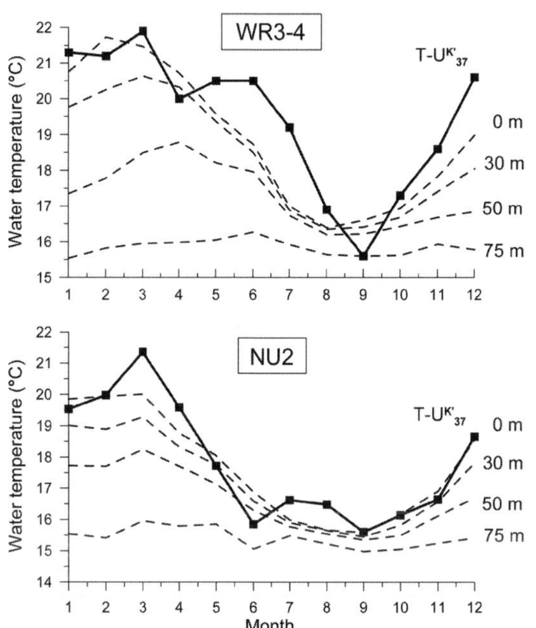

Fig. 10. Monthly averaged U$^{K'}_{37}$ temperature records for traps WR3-4 and NU2 from the northern and southern Benguela region (filled squares) as compared to the average seasonal temperature variations in the upper 75 m of the water column (dashed lines, Conkright et al. 1998). The collection dates of the trap have been corrected by 5 and 8 weeks, respectively. The absolute U$^{K'}_{37}$ temperatures and their seasonal amplitudes (5-6°C) can best be explained by alkenones originating from a mixed-layer source.

production of alkenones is perennial in these regions, and the differences between cold and warm seasons are apparently small. Canonical autumn-spring blooms as observed in the Northern Benguela at the Walvis Ridge (Figs. 4 and 5b) and episodically varying alkenone fluxes in filamentous mixing areas of upwelling regions produce average U$^{K'}_{37}$ signals not significantly different from the annual mean SST. Examples for the latter situation are trap NU2 from the Southern Benguela (Fig. 6b) and the 4-year time series analyzed off Cape Blanc, NW Africa (Müller and Fischer 2001). Interannual variations and long-term averaging during sediment formation will further suppress seasonal effects. As a result sediments in general show U$^{K'}_{37}$ temperatures close to the annual mean SST (Table 2).

For the Southern Ocean, our time series revealed a distinct seasonality in alkenone production, but different timings in the Polar Frontal Region (PF3) and the Antarctic Zone south of the front (BO1). At site PF3, about 87% of the annual alkenone flux occurred in late winter and spring (Table 2), coinciding with elevated fluxes of opal, carbonate and TOC (Fig. 8), and in accordance with the general production cycle in this region. Phytoplankton blooms develop in spring, when the water column recovers from deep winter mixing (Longhurst 1995; Smetacek et al. 1997; Abbott et al. 2000). Chlorophyll concentrations are usually highest in the mixed layer which rarely shoals above the photic depth (Smetacek and Passow 1990; Longhurst 1995; Strass et al. 2002). High phytoplankton stocks observed deeper than 40-70 m in the region of the Polar Front can probably be attributed to downwelling rather than *in situ* growth (Bathmann et al. 1997; Smetacek et al. 1997; Strass et al. 2002).

In marked contrast to the Polar Frontal Region, the export of alkenones was almost completely confined to the warm season at site BO1 in the Antarctic Zone, where about 98% of the annual flux occurred from January to May (Fig. 9c, Table 2). However, the U$^{K'}_{37}$ temperatures obtained for the warm season were clearly lower than the weekly SST (Fig. 9e). The overall temperature signal of the time series thus deviated from the predicted value by -1.6°C (Table 2). Since deep production is an unlikely cause in this region, we must consider that the calibration published by Prahl et al. (1988) and comparable relationships (Table 3) slightly underestimate the temperature of waters colder than about 3°C (Figs. 8a, 9e and 12). However, our sediment trap results do not support the existence of a non-linear relationship at the lower end of the $U^{K'}_{37}$-temperature calibration as has been implied by a study on suspended matter (Sikes and Volkman 1993, see the dashed curve in Fig. 12).

The very high alkenone fluxes observed in summer and early autumn at site BO1 are noteworthy and somewhat unexpected. As can be seen from Table 2, the annual alkenone flux at this location was about 5 times higher than at the Polar Front, and comparable to the fluxes recorded in the up-welling regions off Cape Blanc and at the Walvis Ridge. The peak flux measured in April 1991 (31.8 μg m^{-2} d^{-1}, Fig. 9c) was even an order of magnitude larger than the highest flux recorded at the Polar Front (Fig. 8b). However, these high alke-none fluxes are probably exceptional and not representative for longer time periods. This is indicated by considerably lower mass fluxes observed in the following years at the BO position (Fischer at al. 2002). The annual organic carbon flux, for instance, was lower by a factor of 5 in the years 1992 and 1995, and by more than one order of magnitude in the years 1993 and 1994. Large interannual variations were also observed at site PF in the Polar Frontal Region on the basis of a 5-year record.

Core-Top Calibrations

The primary aim of our core-top studies in the South Atlantic was to establish a sediment-based $U^{K'}_{37}$-temperature calibration and to define the most suitable relationship for paleotemperature reconstructions. In the first instance, we analyzed surface sediments from the eastern South Atlantic and correlated the $U^{K'}_{37}$ values to seasonal, annual, and production-weighted annual mean temperatures of the overlying surface waters (Müller et al. 1998). The best correlations were obtained using temperatures from 0-10 m water depth, suggesting that $U^{K'}_{37}$ predominantly reflects mixed-layer temperatures in this part of the ocean. Moreover, regional variations in the seasonality of primary production had only a minor effect on the sedimentary $U^{K'}_{37}$

temperature signal (<0.5°C). Sonzogni et al. (1997b) made similar observations in the tropical-subtropical Indian Ocean. The most suitable relationship for the eastern South Atlantic thus used the mean annual temperature in the upper 10 m of the water column:

$$U^{K'}_{37} = 0.033 \text{ SST} + 0.069 \quad (n = 149, r = 0.990) \tag{4}$$

The standard deviation of the residuals for this relationship is ±0.032 $U^{K'}_{37}$ units or ±1.0°C.

Comparison of these results with published core-top calibrations from other oceanic regions, e.g. the North Atlantic (Rosell-Melé et al. 1995), Indian Ocean (Sonzogni et al. 1997), North Pacific (Doose et al. 1997), South China Sea (Pelejero and Grimalt 1997) and Southern Ocean (Sikes et al. 1997) revealed a high degree of accordance. Müller et al. (1998), therefore, established a global core-top calibration based on 370 sites between 60°S and 60°N and the mean annual SST derived from the World Ocean Atlas 1994:

$$U^{K'}_{37} = 0.033 \text{ SST} + 0.044 \quad (n = 370, r = 0.978) \tag{5}$$

Subsequently, Benthien and Müller (2000) evaluated the applicability of this calibration in the western South Atlantic. Surface sediments from the Brazil Current region and the Tropical Gyre north of 32°S were found to have $U^{K'}_{37}$ values, which were in agreement with annual mean atlas SSTs and equation 5. Most of the unsaturation values from the western Argentine Basin, however, were lower than expected. Particularly, sediments from the region of the Brazil-Malvinas Confluence (35-39°S) and the Argentine continental slope between 41° and 48°S showed lower alkenone temperatures (by 3-6°C) compared to annual mean atlas SSTs. Only the $U^{K'}_{37}$ values of sediments from the upper continental slope at 47-48°S and from the deep Argentine Basin below a water depth of about 4500 m were found to match the surface water data.

As pointed out by Benthien and Müller (2000), the oceanographic characteristics of the Malvinas Current region (shallow photic depth, high nutrient concentrations in the mixed layer, perennial cocco-lithophore production with spring-summer maximum, minor subsurface production) suggest that the primary (unbiased) U$^{K'}_{37}$ signal in this region should reflect the annual mean temperature of the mixed layer. The low alkenone temperatures in the western Argentine Basin can therefore not be explained by production below the mixed layer or during the winter season. Differences in nutrient availability and growth rate (e.g. Epstein et al. 1998; Popp et al. 1998) are also unlikely causes because these cannot explain the simultaneous occurrence of biased and

unaffected U$^{K'}_{37}$ ratios in closely neigh-bored sediments underlying the same production system. Instead, it appears that lateral displace-ments of suspended particles and sediments, caused by northward-directed surface and bottom currents, benthic storms, and downslope processes were responsible for the deviating U$^{K'}_{37}$ temperatures in this region.

For this reason, we did not consider data from the Argentine Basin in updating the global core-top calibration (Fig. 11). We also rejected the low core-top U$^{K'}_{37}$ values from mid-latitudes (19°-35°N) and

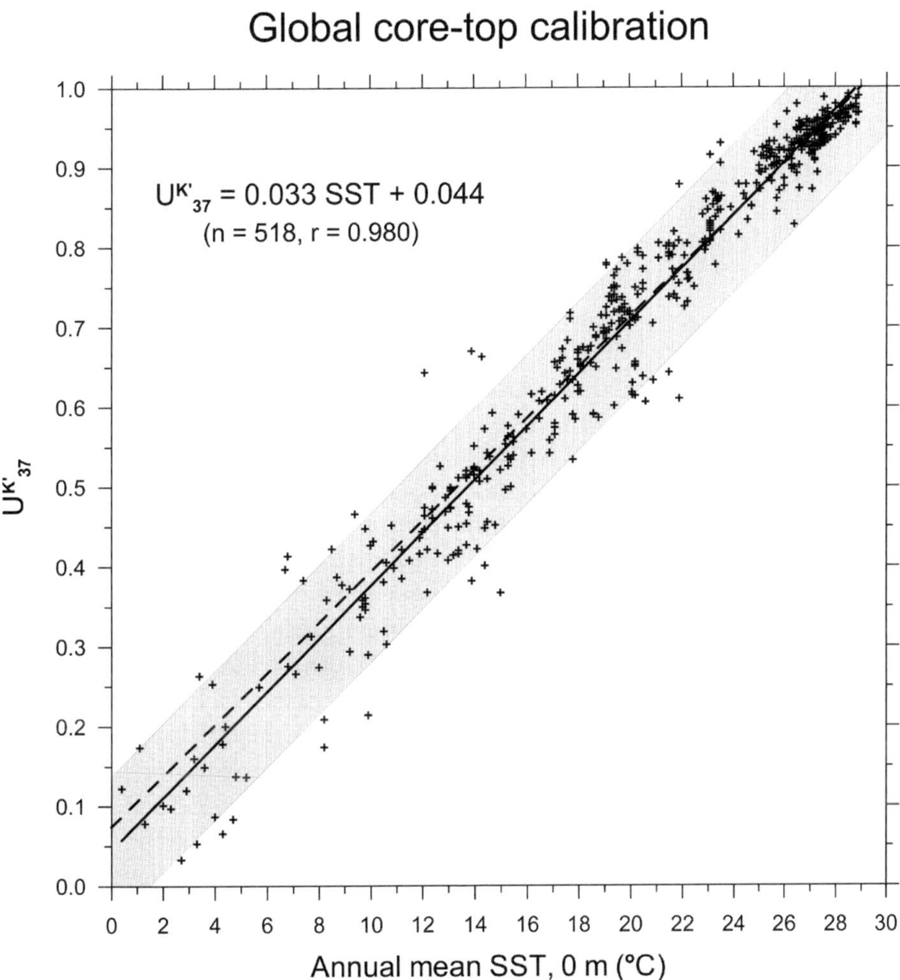

Fig. 11. Update of the global core-top calibration of Müller et al. (1998) based on 518 locations between 60°N and 60°S and SST data of World Ocean Atlas 1994 (Levitus and Boyer 1994). See text for new references. The regression equation is exactly the same as before, and only slightly different if SST data of World Ocean Atlas 1998 are applied (dashed line, eq. 7 in the text). Temperature estimates based on these core-top relationships agree within 1°C with the *E. huxleyi* culture calibrations of Prahl and Wakeham (1987) and Prahl et al. (1988). See Table 3 for a comparison.

the equatorial region (5-8°S) in the Central Pacific claimed to be biased by alkenone production below the mixed layer (Ohkouchi et al. 1999). Yet another possibility is that these red clay deposits, which typically show very low sedimentation rates in the order of a few mm/1000 years or even less (e.g. Chester and Aston 1976), were contaminated by glacial material admixed from underneath. Finally, we did not consider reported $U^{K'}_{37}$ index values of unity to avoid detection limit problems (Grimalt et al. 2001; Pelejero and Calvo 2003). The relationship shown in Fig. 11 is thus based on 518 locations between 60°S and 60°N, including new core-top values from the Atlantic (Benthien and Müller 2000; Müller and Fischer 2001; Müller, unpubl. data), the Pacific (Herbert et al. 1998; Ohkouchi et al. 1999; Kim et al. 2002b), and the Arabian Sea (Budziak 2001). Interestingly, the enlarged data set yields exactly the same regression of $U^{K'}_{37}$ to annual mean SST (WOA94) as equation 5, with a slightly improved correlation coefficient:

$$U^{K'}_{37} = 0.033 \,(\pm\, 0.0006)\, SST + 0.044 \,(\pm\, 0.012)$$
$$(n = 518,\ r = 0.980) \tag{6}$$

Recalculating this sediment calibration with annual mean SST values of World Ocean Atlas 1998 yields a slightly different relationship with similar regression statistics:

$$U^{K'}_{37} = 0.032 \,(\pm\, 0.0005)\, SST + 0.073 \,(\pm\, 0.011)$$
$$(n = 518,\ r = 0.982) \tag{7}$$

The errors given in brackets for equations 6 and 7 represent the 95% confidence intervals for the slope and intercept values. Accordingly, both relationships show comparable temperature sensitivities but somewhat different intercept constants. The standard deviation of the residuals, expressed as temperature, is ±1.3°C for both regressions.

The main reason for the slightly different regressions is that WOA98 provides lower SST values in some regions, especially in mid- and high latitudes. Consequently, equation 7 yields slightly lower temperature estimates than equation 6. However, as shown in Fig. 11 and Table 3, the differences are negligible (<0.3°C) for warm waters and only

	Emiliania huxleyi cultures		Global core-tops	
			WOA94	WOA98
	Eq. 2	Eq. 3	Eq. 6	Eq. 7
$U^{K'}_{37}$	°C	°C	°C	°C
1.0	(29.0)	(28.3)	29.0	29.0
0.9	26.0	25.3	25.9	25.8
0.8	22.9	22.4	22.9	22.7
0.7	19.9	19.4	19.9	19.6
0.6	16.9	16.5	16.8	16.5
0.5	13.8	13.6	13.8	13.3
0.4	10.8	10.6	10.8	10.2
0.3	7.8	7.7	7.8	7.1
0.2	(4.8)	(4.7)	4.7	4.0
0.1	(1.7)	(1.8)	1.7	0.8
0.0	(-1.3)	(-1.1)	-1.3	-2.3

Values outside calibration range given in brackets

Table 3. Comparison of $U^{K'}_{37}$ temperatures based on *E. huxleyi* cultures and global core-top calibrations (see text for equations and references)

significant (0.6-1°C) for waters colder than about 10°C. Overall, the core-top calibrations yield similar temperature estimates (within 1°C) as the culture calibrations of Prahl and Wakeham (1997) and Prahl et al. (1988), again attesting the general applicability of these relationships. Formally, these culture calibrations are constrained to growth temperatures between 8° and 25°C and it has been suggested that the linearity breaks down at lower temperatures (Sikes and Volkman 1993; Ternois et al. 1998). However, our core-top results from the eastern South Atlantic (e.g. Fig. 4 in Müller et al. 1998) and the global relationship depicted in Fig. 11 rather support a linear relationship. A linear relationship is also indicated by the sediment trap results from the Antarctic sector in the Atlantic, as illustrated in Fig. 12. The $U^{K'}_{37}$ values of the particles intercepted by traps PF3 and BO1 scatter within the 4σ-range of the global core-top calibration and deviate only slightly from the relationship of Prahl et al. (1988) (dashed extrapolated line) at water temperatures <3°C. For reasons of comparison, the dashed curve in Fig. 12 shows the non-linear relationship of Sikes and Volkman (1993) which is based on particulate matter samples from the Southern Ocean south of Tasmania ($U^{K'}_{37} < 0.35$) and the field samples reported by Prahl and Wakeham (1987).

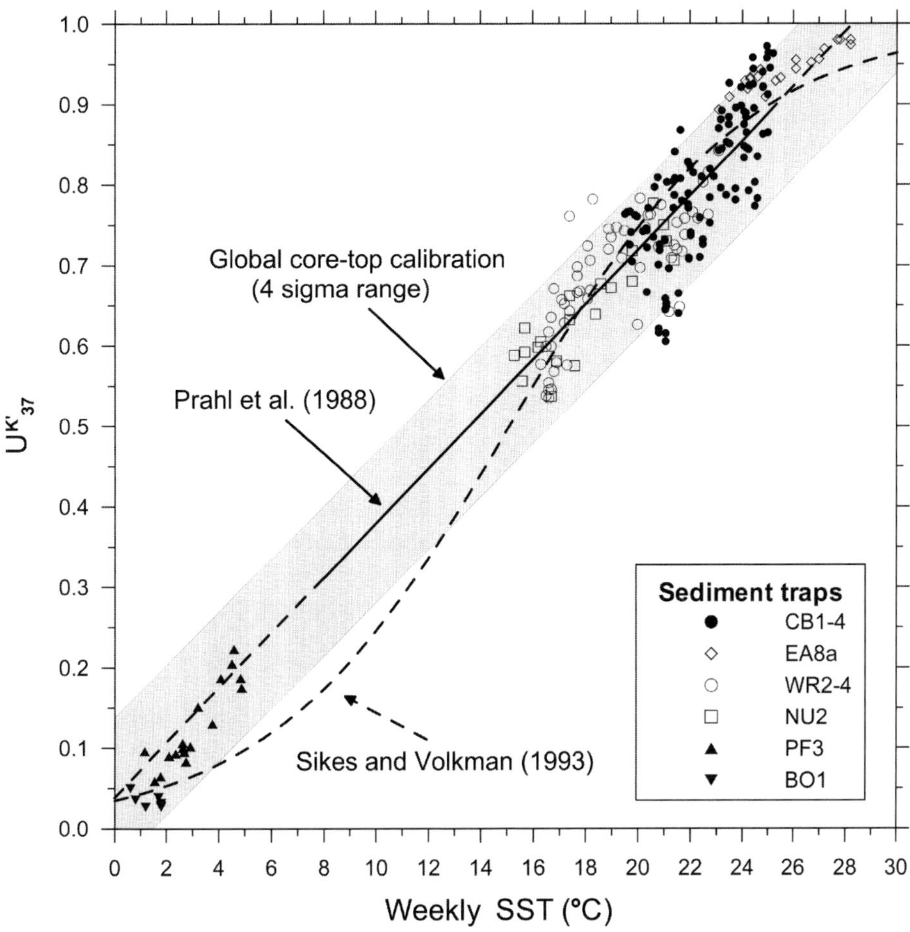

Fig. 12. Relationship between U$^{K'}_{37}$ and weekly SST (Reynolds and Smith 1994) for the sediment traps of this study, omitting the abnormal winter-spring values of the BO1 record (Fig. 9e). The trap data were corrected for the average time-lags due to particle settling as indicated above. The U$^{K'}_{37}$ values scatter within the 4σ range of the core-top calibration (shaded area, cp. Fig. 11), but tend to be slightly low for SST values <3°C as compared to the calibration of Prahl et al. (1988) (dashed ends indicate extensions outside the calibration range). The non-linear particulate matter relationship of Sikes and Volkman (1993) is shown for comparison (dashed curve).

Paleotemperature Assessment

The use of calibrations based on cultures of *E. huxleyi*, sea-water particulates or modern sediments for paleotemperature determinations assumes that this taxon or related haptophyte algae similarly responding to temperature changes were the principal alkenone producers in the past. *E. huxleyi* first appeared about 268 kyr ago during oxygen-isotope stage 8 and became the dominant coccolithophorid between 85 and 65 ka depend-

ing on the latitude (Hay 1977; Thierstein et al. 1977). Alkenones, on the other hand, have been found in much older sediments, e.g. throughout the Quaternary and Pliocene (Marlow et al. 2000), in the Eocene (Marlowe et al. 1984), and even in the Cretaceous (Farrimond et al. 1986). It is important, therefore, to verify that calibrations established in the modern ocean also apply to ancient sediments.

Müller et al. (1997) addressed this problem in a late Quaternary sediment core from the Walvis Ridge. To trace possible effects of past species changes on the $U^{K'}_{37}$ temperature signal, they determined alkenones and coccolithophorid species abundances and compared the results to SST estimates extracted from the $\delta^{18}O$ record of a planktonic foraminifer (*Globigerinoides ruber*). Alkenones and isotopes were determined over the entire 400-kyr record of the core, whereas the coccolithophorid study was confined to the last 200 kyr when the most pronounced changes in alkenone content occurred. Species of the genus *Gephyrocapsa* were found to be the predominating coccolithophorids throughout oxygen-isotope stages 6 and 5. *E. huxleyi* systematically increased in relative abundance since the stage 5/4 transition (about 70 kyr B.P.), became dominant over *Gephyrocapsa* spp. in stage 3, and reached the highest abundances in the Holocene. The carbon-normalized alkenone contents were inversely related to the abundances of *E. huxleyi*, and directly related to that of *Gephyrocapsa* spp., suggesting that species of this genus were the principal alkenone contributors to the pre-Holocene sediments.

Even though, $U^{K'}_{37}$ temperatures based on the *E. huxleyi* calibration of Prahl et al. (1988) matched well with the isotope temperatures, whereas a culture relationship established for a strain of *G. oceanica* (Volkman et al. 1995) produced unrealistically high values. Interestingly, the calibration based on *E. huxleyi* produced reasonable paleotemperatures even in periods that predate the first appearance of this species. Recent results from the Central North Atlantic point in the same direction (Villanueva et al. 2002). These authors determined three major reversals between *Gephyrocapsa* and *Emiliania* species over the last 290,000 years and found no evidence for any significant changes in the $U^{K'}_{37}$-temperature relationship. Hence there is growing evidence that late Quaternary *Gephyrocapsa* species synthesizing alkenones responded similarly to temperature changes as *E. huxleyi* in the present ocean. Furthermore, temperature reconstructions based on longer ODP records from the Equatorial Pacific (Emeis et al. 1995), the North and South Atlantic (Herbert and Schuffert 1998; Marlow et al. 2000), and the Mediterranean Sea (Emeis et al. 1998) indicate that the $U^{K'}_{37}$ paleothermometer can also be applied to pre-Quaternary periods, at least over the last 6 million years.

As a final point, it is difficult to evaluate the uncertainty of $U^{K'}_{37}$ paleotemperature estimates. Recent intercalibration results indicate an overall intralaboratory variance (repeatability) of <1.6°C, and maximum $U^{K'}_{37}$ temperature differences of up to 2.1°C between any two laboratories (Rosell-Melé et al. 2001). Analytical uncertainties may thus also explain most of the scattering of values in the global core-top relationship (Fig. 11) which is based on about 15 laboratories and various analytical techniques. Processes such as resuspension, lateral advection and mixing of sediments, and the general assumption of annual mean SST for the calibration, may have additionally contributed to scattering. Thus, the relatively high standard error of estimate obtained for this relationship (±1.3°C) should not be considered as the general error of the $U^{K'}_{37}$ paleotemperature method.

The analytical precision of a particular laboratory is often much better, in the order of ±0.01 $U^{K'}_{37}$ units or ±0.3°C (e.g. Sonzogni et al. 1997a; Herbert et al. 1998; Kim et al. 2002a). Downcore sediment samples from defined climatic periods with relatively low natural temperature variations also display less scattering than the global core-top data set. For example, the standard deviations of $U^{K'}_{37}$ temperatures determined in our laboratory for the Last Glacial Maximum (LGM) ranged between 0.2° and 0.4°C, based on sediment cores with 3-23 samples per LGM time slice (e.g. Schneider et al. 1995, 1996; Müller et al. 1997; Kirst et al. 1999; Kim et al. 2002a). Moreover, sample to sample differences for LGM core sections with a relatively high resolution (n ≥ 4) were <0.5°C in most cases. Thus, considering also the differences between the calibrations (Table 3), an error of 1°C for the alkenone paleotemperature method might be a realistic estimation.

Acknowledgements

We thank the masters, crews and scientists aboard RV *Meteor* and RV *Polarstern* for their help during trap deployments and recovery and sediment sampling. We are also grateful to Hella Buschhoff,

Dietmar Grotheer and Ralph Kreutz for careful analyses in the home laboratory and to A. Rosell-Melé and T. Trull for constructive reviews of the paper. This research was funded by the Deutsche Forschungsgemeinschaft (Sonderforschungsbereich 261 at Bremen University, Contribution No. 373) and the European Community (Contract ENV4-CT97-0564). Data are available under www.pangaea.de/Projects/SFB261.

References

Abbott MR, Richmann JG, Letelier RM, Bartlett JS (2000) The spring bloom in the Antarctic Polar Frontal Zone as observed from a mesoscale array of bio-optical sensors. Deep-Sea Res 47: 3285-3314

Abelmann A, Brathauer U, Gersonde R, Sieger R, Zielinski U (1999) Radiolarian-based transfer function for the estimation of sea surface temperatures in the Southern Ocean (Atlantic sector). Paleoceanography 14: 410-421

Antoine D, André J-M, Morel A (1996) Oceanic primary production. 2. Estimation at global scale from satellite (coastal zone color scanner) chlorophyll. Glob Biogeochem Cycl 10: 57-69

Bard E (2001) Comparison of alkenone estimates with other paleotemperature proxies. Geochem Geophys Geosyst 2: doi: 10.1029/2000GC000050

Bard E, Rostek F, Turon J-L, Gendreau S (2000) Hydrological impact of Heinrich events in the subtropical northeast Atlantic. Science 289: 1321-1324

Baumann K-H, Cepek M, Kinkel H (1999) Coccolithophores as indicators of ocean water masses, surface-water temperature, and paleoproductivity - Examples from the South Atlantic. In: Fischer G, Wefer G (eds) Use of Proxies in Paleoceanography. Examples from the South Atlantic. Springer, Berlin, pp 117-144

Benthien A, Müller PJ (2000) Anomalously low alkenone temperatures caused by lateral particle and sediment transport in the Malvinas Current region, western Argentine Basin. Deep-Sea Res I 47: 2369-2393

Bijma J, Altabet M, Conte MH, Kinkel H, Versteegh GJM, Volkman JK, Wakeham SG, Weaver PP (2001) Primary signal: Ecological and environmental factors - Report from Working Group 2. Geochem Geophys Geosyst 2: doi: 10.1029/2000GC000051

Boon JJ, van de Meer FW, Schuyl OJW, de Leeuw JW, Schenck PA, Burlingame AL (1978) Organic geochemical analyses of core samples from site 362, Walvis Ridge, DSDP Leg 40. Initial Reports of the Deep-Sea Drilling Project 40: Washington (US Government Printing Office), pp 627-637

Brassell SC (1993) Applications of biomarkers for delineating marine paleoclimatic fluctuations during the Pleistocene. In: Engel MH, Macko SA (eds) Organic Geochemistry: Principles and Applications. Plenum Press, New York, London, pp 699-738

Brassel SC, Eglinton G, Marlowe IT, Pflaumann U, Sarnthein M (1986) Molecular stratigraphy: A new tool for cimatic assessment. Nature 320: 129-133

Budziak D (2001) Late Quaternary monsoonal climate and related variations in paleoproductivity and alkenone-derived sea-surface temperatures in the western Arabian Sea. Ber Fachber Geowiss Univ Bremen 170, 114 p

Cepek M (1996) Zeitliche und räumliche Variationen von Coccolithophoriden-Gemeinschaften im subtro-pischen Ost-Atlantik: Untersuchungen an Plankton, Sinkstoffen und Sedimenten. Ber Fachber Geowiss Univ Bremen 86, 160 p

Chester R, Aston SR (1976) The geochemistry of deep-sea sediments. In: Riley JP, Chester R (eds) Chemical Oceanography, vol. 6. Academic Press, London, pp 281-390

Conkright M, Levitus S, O'Brien T, Boyer T, Antonov J, Stephens C (1998) World Ocean Atlas. Available online and on CD-ROM Data Set Documentation. Techn Rep 15, NODC Internal Report, Silver Spring, MD: 16 p

Conte MH, Eglinton G, Madureira LAS (1992) Long-chain alkenones and alkyl akenoates as palaeotemperature indicators: their production, flux and early sedimentary diagenesis in the Eastern North Atlantic. In: Eckart CB, Larter SR (eds) Advances in Organic Geochemistry 1991. Org Geochem 19: 287-298

Conte MH, Thompson A, Eglinton G, Green JC (1995) Lipid biomarker diversity in the coccolithophorid *Emiliania huxleyi* (Prymnesiophyceae) and the related species *Gephyrocapsa oceanica*. J Phycol 31: 272-282

De Leeuw JW, van der Meer FW, Rijpstra WIC, Schenck PA (1980) On the occurrence and structural identification of long chain unsaturated ketones and hydrocarbons in sediments. In: Douglas AG, Maxwell JR (eds) Advances in Organic Geochemistry 1979. Pergamon Press, Oxford, pp 211-217

Doose H, Prahl FG, Lyle MW (1997) Biomarker temperature estimates for modern and last glacial surface waters of the California Current system between 33° and 42°N. Paleoceanography 12: 615-622

Eglinton TI, Conte MH, Eglinton G, Hayes JM (2001) Proceedings of a workshop on alkenone-based paleoceanographic indicators. Geochem Geophys Geosyst 2: doi: 10.1029/2000GC000122

Emeis K-C, Doose H, Mix AC, Schulz-Bull D (1995) Alkenone sea-surface temperatures and carbon burial at Site 846 (Eastern Equatorial Pacific Ocean): The last 1.3 MY. In: Pisias NG, Mayer LA, Janecek TR, Palmer-Julson A, van Andel TH (eds) Proc ODP Sci Results 138. College Station, TX, pp 605-613

Emeis K-C, Schulz H-M, Struck U, Sakamoto T, Doose H, Erlenkeuser H, Howell M, Kroon D, Paterne M (1998) Stable isotope and alkenone temperature records of sapropels from Sites 964 and 967: Constraining the physical environment of sapropel formation in the eastern Mediterranean Sea. In: Robertson AHF, Emeis K-C, Richter C, Camerlenghi A (eds) Proc ODP Sci Results 160. College Station, TX, pp 309-331

Epstein BL, D'Hondt S, Quinn JG, Zhang J (1998) An effect of dissolved nutrient concentrations on alkenone-based temperature estimates. Paleoceanography 13: 122-126

Epstein BL, D'Hondt S, Hargraves PE (2001) The possible metabolic role of C_{37} alkenones in *Emiliania huxleyi*. Org Geochem 32: 867-875

Farrimond P, Eglinton G, Brassell SC (1986) Alkenones in Cretaceous black shales, Blake-Bahama Basin, western North Atlantic. In: Leythaeuser D, Rullkötter J (eds) Advances in Organic Geochemistry 1985. Org Geochem 10, pp 897-903

Fischer G, Wefer G (1996) Long-term observations of particle fluxes in the Eastern Atlantic: Seasonality, changes of flux with depth and comparison with the sediment record. In: Wefer G, Berger WH, Siedler G, Webb DJ (eds) The South Atlantic: Present and Past Circulation. Springer, Berlin, pp 325-344

Fischer G, Ratmeyer V, Wefer G (2000) Organic carbon fluxes in the Atlantic and the Southern Ocean: Relationship to primary production compiled from satellite radiometer data. Deep-Sea Res 47: 1961-1997

Fischer G, Gersonde R, Wefer G (2002) Organic carbon, biogenic silica and diatom fluxes in the marginal winter sea-ice zone and in the Polar Front Region: Inter-annual variations and differences in composition. Deep-Sea Res 49: 1721-1745

Gersonde R, Zielinski U (2000) The reconstruction of late Quaternary Antarctic sea-ice distribution - the use of diatoms as a proxy for sea-ice. Palaeogeogr Palaeo-climatol Palaeoecol 162: 263-286

Giraudeau J, Bailey GW (1995) Spatial dynamics of coccolithophore communities during an upwelling event in the Southern Benguela system. Cont Shelf Res 15: 1825-1852

Goni MA, Hartz DM, Thunell RC, Tappa E (2001) Oceanographic considerations for the application of the alkenone-based paleotemperature $U^{K'}_{37}$ index in the Gulf of California. Geochim Cosmochim Acta 65: 545-557

Grimalt JO, Rullkötter J, Sicre M-A, Summons R, Farrington J, Harvey HR, Goni M, Sawada K (2000) Modifications of the C_{37} alkenone and alkenoate composition in the water column and sediment: Possible implications for sea surface temperature estimates in paleoceanography. Geochem Geophys Geosyst 1: doi: 10.1029/2000GC000053

Grimalt JO, Calvo E, Pelejero C (2001) Sea surface paleo-temperature errors in $U^{K'}_{37}$ estimation due to alken-one measurements near the limit of detection. Paleoceanography 16: 226-232

Hay WW (1977) Calcareous nannofossils. In: Ramsay ATS (ed) Oceanic Micropaleontology. Academic Press, London, pp 1055-1200

Herbert TD (2001) Review of alkenone calibrations (culture, water column, and sediments). Geochem Geophys Geosyst 2: doi: 10.1029/2000GC000055

Herbert TD, Schuffert JD (1998) Alkenone unsaturation estimates of Late Miocene through Late Pliocene sea-surface temperatures at site 958. Proc ODP Sci. Results 159T: 17-21

Herbert TD, Schuffert JD, Thomas D, Lange C, Weinheimer A, Peleo-Alampay A, Herguera J-C (1998) Depth and seasonality of alkenone production along the California margin inferred from a core top transect. Paleoceanography 13: 263-271

Kim J-H, Schneider RR, Müller PJ, Wefer G (2002a) Interhemispheric comparison of deglacial sea-surface temperature patterns in Atlantic eastern boundary currents. Earth Planet Sci Lett 194: 383-393

Kim J-H, Schneider RR, Hebbeln D, Müller PJ, Wefer G (2002b) Last deglacial sea-surface temperature evolution in the Southeast Pacific compared to climate changes on the South American continent. Quat Sci Rev 21: 2085-2097

Kirst GJ, Schneider RR, Müller PJ, von Storch I, Wefer G (1999) Late Quaternary temperature variability in the Benguela Current system derived from alkenones. Quat Res 52: 92-103

Levitus S, Boyer TP (1994) World Ocean Atlas. Vol 4: Temperature. NOAA Atlas NESDIS 4, pp 1-117

Longhurst A (1993) Sesonal cooling and blooming in tropical oceans. Deep-Sea Res 40: 2145-2165

Longhurst A (1995) Seasonal cycles of pelagic production and consumption. Prog Oceanogr 36: 77-167

Lutjeharms JRE, Stockton PL (1987) Kinematics of the upwelling front off Southern Africa. In: Payne AIL,

Gulland JA, Brink KH (eds) The Benguela and Comparable Ecosystems. S Afr J mar Sci 5: 35-49

Marlow JR, Lange CB, Wefer G, Rosell-Melé A (2000) Upwelling intensification as part of the Pliocene-Pleistocene climate transition. Science 290: 2288-2291

Marlowe IT, Brassell SC, Eglinton G, Green JC (1984) Long chain unsaturated ketones and esters in living algae and marine sediments. Org Geochem 6: 135-141

Marlowe IT, Brassell SC, Eglinton G, Green JC (1990) Long-chain alkenones and alkyl alkenoates and the fossil coccolith record of marine sediments. Chem Geol 88: 349-375

Mix AC, Morey AE, Pisias NG (1999) Foraminiferal faunal estimates of paleotemperature: Circumventing the non-analog problem yields cool ice age tropics. Paleoceanography 14: 350-359

Mix AC, Bard E, Eglinton G, Keigwin LD, Ravelo AC, Rosenthal Y (2000) Alkenones and multiproxy strategies in paleoceanographic studies. Geochem Geophys Geosyst 1: doi: 10.1029/2000GC000056

Müller PJ, Fischer G (2001) A 4-year sediment trap record of alkenones from the filamentous upwelling region off Cape Blanc, NW Africa and a comparison with distributions in underlying sediments. Deep-Sea Res I 48: 1877-1903

Müller PJ, Schneider R, Ruhland G (1994) Late Quaternary PCO$_2$ variations in the Angola Current: Evidence from organic carbon δ^{13}C and alkenone temperatures. In: Zahn R, Pedersen TF, Kaminski MA, Labeyrie L (eds) Carbon Cycling in the Glacial Ocean: Constraints on the Ocean's Role in Global Change. NATO ASI Series I, Vol 17, Springer, Berlin, pp 343-366

Müller PJ, Cepek M, Ruhland G, Schneider RR (1997) Alkenone and coccolithophorid species changes in late Quaternary sediments from the Walvis Ridge: Implications for the alkenone paleotemperature method. Palaeogeogr Palaeoclimatol Palaeoecol 135: 71-96

Müller PJ, Kirst G, Ruhland G, Von Storch I, Rosell-Melé A (1998) Calibration of the alkenone paleotemperature index U$^{K'}_{37}$ based on core-tops from the eastern South Atlantic and the global ocean (60°N-60°S). Geochim Cosmochim Acta 62: 1757-1772

Nürnberg D, Müller A, Schneider RR (2000) Paleo-sea surface temperature calculations in the equatorial east Atlantic from Mg/Ca ratios in planktic foraminifera: A comparison to sea surface temperature estimates from U$^{K'}_{37}$, oxygen isotopes, and foraminiferal transfer function. Paleoceanography 15: 124-134

Ohkouchi N, Kawamura K, Kawahata H, Okada H (1999) Depth ranges of alkenone production in the central Pacific Ocean. Glob Biogeochem Cycl 13: 695-704

Pelejero C, Grimalt JO (1997) The correlation between the U$^{K}_{37}$ index and sea surface temperatures in the warm boundary: The South China Sea. Geochim Cosmochim Acta 61: 4789-4797

Pelejero C, Calvo E (2003) The upper end of the U$^{K'}_{37}$ temperature calibration revisited. Geochem Geophys Geosyst 4: 1014, doi: 10.1029/2002GC000431

Peterson RG, Stramma L (1991) Upper-level circulation in the South Atlantic Ocean. Progr Oceanogr 26: 1-73

Popp BN, Kenig F, Wakeham SG, Laws EA, Bidigare RR (1998) Does growth rate affect ketone unsaturation and intracellular carbon isotopic variability in *Emiliania huxleyi*? Paleoceanography 13: 35-41

Prahl FG, Wakeham SG (1987) Calibration of unsaturation patterns in long-chain ketone compositions for palaeotemperature assessment. Nature 330: 367-369

Prahl FG, Muehlhausen LA, Zahnle DL (1988) Further evaluation of long-chain alkenones as indicators of paleoceanographic conditions. Geochim Cosmochim Acta 52: 2303-2310

Prahl FG, Collier RB, Dymond J, Lyle M, Sparrow MA (1993) A biomarker perspective on prymnesiophyte productivity in the northeast Pacific Ocean. Deep-Sea Res 40: 2061-2076

Prahl F, Herbert T, Brassell SC, Ohkouchi N, Pagani M, Repeta D, Rosell-Mele A, Sikes E (2000a) Status of alkenone paleothermometer calibration: Report from working group 3. Geochem Geophys Geosyst 1: doi: 10.1029/2000GC000058

Prahl FG, Dymond J, Sparrow MA (2000b) Annual biomarker record for export production in the central Arabian Sea. Deep-Sea Res 47: 1581-1604

Prahl FG, Pilskaln CH, Sparrow MA (2001) Seasonal record for alkenones in sedimentary particles from the Gulf of Maine. Deep-Sea Res 48: 515-528

Read JF, Pollard RT, Bathmann UV (2002) Physical and biological patchiness of an upper ocean transect from South Africa to the ice edge near the Greenwich Meridian. Deep-Sea Res II 49: 3713-3733

Reynolds RW, Smith TM (1994) Improved global sea surface temperature analyses using optimum interpolation. J Clim 7: 929-948

Romero O, Boeckel B, Donner B, Lavik G, Fischer G, Wefer G (2002) Seasonal productivity dynamics in the pelagic central Benguela System inferred from the flux of carbonate and silicate organisms. J Mar Syst 37: 259-278

Rosell-Melé A (1998) Interhemispheric appraisal of the value of alkenone indices as temperature and salinity proxies in high-latitude locations. Paleoceanography 13: 694-703

Rosell-Mele A, Eglinton G, Pflaumann U, Sarnthein M (1995) Atlantic core-top calibration of the $U^{K'}_{37}$ index as a sea-surface palaeotemperature indicator. Geochim Cosmochim Acta 59: 3099-3107

Rosell-Melé A, Comes P, Müller PJ, Ziveri P (2000) Alkenone fluxes and anomalous $U^{K'}_{37}$ values during 1989-1990 in the Northeast Atlantic (48°N 21°W). Mar Chem 71: 251-264

Rosell-Melé A, Bard E, Emeis K-C, Grimalt JO, Müller P, Schneider R, Bouloubassi I, Epstein B, Fahl K, Fluegge A, Freeman K, Goñi M, Güntner U, Hartz D, Hellebust S, Herbert T, Ikehara M, Ishiwatari R, Kawamura K, Kenig F, de Leeuw J, Lehman S, Mejanelle L, Ohkouchi N, Pancost RD, Pelejero C, Prahl F, Quinn J, Rontani J-F, Rostek F, Rullkötter J, Sachs J, Blanz T, Sawada K, Schulz-Bull D, Sikes E, Sonzogni C, Ternois Y, Versteegh G, Volkman JK, Wakeham S (2001) Precision of the current methods to measure the alkenone proxy $U^{K'}_{37}$ and absolute alkenone abundance in sediments: Results of an interlaboratory comparison study. Geochim Geophys Geosyst 2: doi: 10.1029/2000GC000141

Rühlemann C, Mulitza S, Müller PJ, Wefer G, Zahn R (1999) Warming of the tropical Atlantic Ocean and slowdown of thermohaline circulation during the last deglaciation. Nature 402: 511-514

Sachs JP, Schneider RR, Eglinton TI, Freeman KH, Ganssen G, McManus JF, Oppo DW (2000) Alkenones as paleoceanographic proxies. Geochem Geophys Geosyst 1: doi: 10.1029/2000GC000059

Sawada K, Handa N, Nakatsuka T (1998) Production and transport of long-chain alkenones and alkyl alkenoates in a sea water column in the northwestern Pacific off central Japan. Mar Chem 59: 219-234

Schneider R (2001) Alkenone temperature and carbon isotope records: Temporal resolution, offsets, and regionality. Geochem Geophys Geosyst 2: doi: 10.1029/2000GC000060

Schneider RR, Müller PJ, Ruhland G (1995) Late Quaternary surface circulation in the east equatorial South Atlantic: Evidence from alkenone sea surface temperatures. Paleoceanography 10: 197-219

Schneider RR, Müller PJ, Ruhland G, Meinecke G, Schmidt H, Wefer G (1996) Late Quaternary surface temperatures and productivity in the east-equatorial South Atlantic: Response to changes in trade/monsoon wind forcing and surface water advection.

In: Wefer G, Berger WH, Siedler G, Webb DJ (eds) The South Atlantic: Present and Past Circulation. Springer, Berlin, pp 527-551

Schneider RR, Müller PJ, Acheson R (1999) Atlantic alkenone sea-surface temperature records: Low versus mid latitudes and differences between hemispheres. In: Abrantes F, Mix A (eds) Reconstructing Ocean History: A Window into the Future. Plenum Publishers, New York, pp 33-55

Shannon LV, Nelson G (1996) The Benguela: Large Scale Features and Processes and System Variability. In: Wefer G, Berger WH, Siedler G, Webb DJ (eds) The South Atlantic: Present and Past Circulation. Springer, Berlin, pp 163-210

Siegel DA, Deuser WG (1997) Trajectories of sinking particles in the Sargasso Sea: modeling of statistical funnels above deep-ocean sediment traps. Deep-Sea Res 44: 1519-1541

Sikes EL, Volkman JK (1993) Calibration of alkenone unsaturation ratios ($U^{K'}_{37}$) for paleotemperature estimation in cold polar waters. Geochim Cosmochim Acta 57: 1883-1889

Sikes EL, Keigwin LD (1994) Equatorial Atlantic sea surface temperature for the last 30 kyr: A comparison of $U^{K'}_{37}$, $\delta^{18}O$ and foraminiferal assemblage temperature estimates. Paleoceanography 9: 31-45

Sikes EL, Sicre M-A (2002) Relationship of the tetra-unsaturated C_{37} alkenone to salinity and temperature: Implications for paleoproxy applications. Geochem Geophys Geosyst 3: 1063, doi: 10.1029/2002GC000345

Sikes EL, Volkman JK, Robertson LG, Pichon J-J (1997) Alkenones and alkenes in surface waters and sedi-ments of the Southern Ocean: Implications for paleo-temperature estimation in polar regions. Geochim Cosmochim Acta 61: 1495-1505

Smetacek V, Passow U (1990) Spring bloom initiation and Sverdrup's critical depth model. Limnol Oceanogr 23: 1256-1263

Smetacek V, De Baar HJW, Bathmann UV, Lochte K, Rutgers van der Loeff MM (1997) Ecology and bio-geochemistry of the Antarctic Circumpolar Current during austral spring: a summary of Southern Ocean JGOFS cruise ANT X/6 of RV *Polarstern*. Deep-Sea Res 44: 1-21

Sonzogni C, Bard E, Rostek F, Lafont R, Rosell-Mele A, Eglinton G (1997a) Core-top calibration of the alkenone index vs sea surface temperature in the Indian Ocean. Deep-Sea Res 44: 1445-1460

Sonzogni C, Bard E, Rostek F, Dollfus D, Rosell-Mele A, Eglinton G (1997b) Temperature and salinity effects on alkenone ratios measured in surface sediments from the Indian Ocean. Quat Res 47: 344-355

Strass VH, Garabato ACN, Pollard RT, Fischer HI, Hense I, Allen JT, Read JF, Leach H, Smetacek V (2002) Mesoscale frontal dynamics: shaping the environment of primary production in the Antarctic Circum-polar Current. Deep-Sea Res 49: 3735-3769

Ternois Y, Sicre M-A, Boireau A, Conte MH, Eglinton G (1997) Evaluation of long-chain alkenones as paleo-temperature indicators in the Mediterranean Sea. Deep-Sea Res 44: 271-286

Ternois Y, Sicre M-A, Boireau A, Beaufort L, Miquel J-C, Jeandel C (1998) Hydrocarbons, sterols and alkenones in sinking particles in the Indian Ocean sector of the Southern Ocean. Org Geochem 28: 489-501

Thierstein HR, Greitzenauer KR, Molfino B (1977) Global synchroneity of late Quaternary coccolith datum levels: Validation by oxygen isotopes. Geology 5: 400-404

Treppke UF, Lange CB, Donner B, Fischer G, Ruhland G, Wefer G (1996) Diatom and silicoflagellate fluxes at the Walvis Ridge: An environment influenced by coastal upwelling in the Benguela system. J Mar Res 54: 991-1016

Villanueva J, Flores JA, Grimalt JO (2002) A detailed comparison of the U^{k}_{37} and coccolith records over the past 290 kyears: Implications to the alkenone paleotemperature method. Org Geochem 33: 897-905

Volkman JK (2000) Ecological and environmental factors affecting alkenone distributions in seawater and sediments. Geochem Geophys Geosyst 1: doi: 10.1029/ 2000GC000061

Volkman JK, Barrett SM, Blackburn SI, Sikes EL (1995) Alkenones in Gephyrocapsa oceanica: Implications for studies of paleoclimate. Geochim Cosmochim Acta 59: 513-520

Volkman JK, Eglinton G, Corner EDS, Sargent JR (1980) Novel unsaturated straight-chain C_{37}-C_{39} methyl and ethyl ketones in marine sediments and a cocco-lithophore *Emiliania huxleyi*. In: Douglas AG, Max-well JR (eds) Advances in Organic Geochemistry 1979. Pergamon Press, Oxford, pp 219-227

Stable Carbon Isotopic Composition of the $C_{37:2}$ Alkenone: A Proxy for CO_2(aq) Concentration in Oceanic Surface Waters?

S. Schulte[1,2,*], A. Benthien[1,3], N. Andersen[1,4], P.J. Müller[1], C. Rühlemann[1] and R.R. Schneider[1]

[1]Universität Bremen, Fachbereich Geowissenschaften, Postfach 330 440, D-28334 Bremen, Germany
[2]Institut für Chemie und Biologie des Meeres (ICBM), Universität Oldenburg, Postfach 2503, D-26111 Oldenburg, Germany
[3]Alfred Wegener Institut für Polar- und Meeresforschung, Postfach 120161, D-27515 Bremerhaven, Germany
[4]ETH Zürich, Geologisches Institut, CH-8092 Zürich, Switzerland
* corresponding author (e-mail): sschulte@uni-bremen.de

Abstract: We tested the applicability of the carbon isotopic composition of $C_{37:2}$ alkenones ($\delta^{13}C_{37:2}$) as a proxy for dissolved carbon dioxide CO_2(aq) in oceanic surface waters. For this purpose we determined $\delta^{13}C_{37:2}$ in suspended particulate organic matter (POM) and surface sediments from the South Atlantic. In opposite of what would be expected from a diffusive CO_2 uptake model for marine algae we observed a positive correlation between $1/[CO_2(aq)]$ and the isotopic fractionation (ε_p) calculated from $\delta^{13}C_{37:2}$. This clearly demonstrates that CO_2(aq) is not the primary factor controlling ε_p at the sites studied. On the other hand we found a negative correlation between ε_p and the phosphate concentration in the surface waters (0-10 m) supporting the assumption of Bidigare et al. (1997) that ε_p is primarily related to nutrient-limited algal growth rather than to $[CO_2(aq)]$. Reconstructing past CO_2(aq) levels from $\delta^{13}C_{37:2}$ thus requires additional proxy information in order to correct for the influence of haptophyte growth on the isotopic fractionation. In the eastern Angola Basin, we previously used $\delta^{15}N$ of bulk organic matter as proxy for nutrient-limited growth rates. As an alternative the Sr/Ca ratio of coccoliths has been recently suggested as growth-rate proxy which should be tested in future studies.

Introduction

From air trapped in the Vostok ice core (Antarctica) it is well known that the concentration of atmospheric carbon dioxide (CO_2) varied during the past 420,000 years in concert with late Quaternary climatic cycles (Petit et al. 1999). CO_2 is one of the so-called greenhouse gases controlling the global heat budged and there is growing evidence that the increase in the atmospheric CO_2 level observed since the beginning of the 19th century is contributing to world wide warming (e.g. Crowley 2000). However, predictions of future changes of Earth's climate are difficult since the relationship between atmospheric CO_2 and long-term climatic cycles is very complex and poorly understood (Stouffer et al. 1994).

Atmospheric CO_2 levels depend on the balance of CO_2 between the world's oceans and terrestrial ecosystems. To a first approximation, equatorial regions of the modern ocean are supersaturated in CO_2 with respect to the atmosphere while sub-polar regions approach air-sea equilibrium and polar regions are undersaturated (Feely et al. 2001). To

From WEFER G, MULITZA S, RATMEYER V (eds), 2003, *The South Atlantic in the Late Quaternary: Reconstruction of Material Budgets and Current Systems.* Springer-Verlag Berlin Heidelberg New York Tokyo, pp 195-211

define better the mechanisms by which oceanic and atmospheric levels of CO_2 have changed over geological time scales, paleoceanic sources and sinks of carbon dioxide must be delineated. Therefore paleo-indicators (proxies) for past CO_2 concentrations in the surface oceans are required.

Empirical relationships between the carbon isotopic composition of suspended organic matter ($\delta^{13}C_{org}$) and the concentration of dissolved carbon dioxide ($[CO_2(aq)]$) in oceanic surface waters imply that $\delta^{13}C_{org}$ of marine phytoplankton varies as a function of ambient $[CO_2(aq)]$ (Degens et al. 1968; Rau et al. 1989, 1992; Francois et al. 1993; Rau 1994; Fischer et al. 1997, 1998; Bentaleb et al. 1998). Based on these observations it has been suggested that the isotopic composition of sedimentary organic carbon can be used as proxy for $[CO_2(aq)]$ in ancient surface waters. Respective paleoceanographic reconstructions of CO_2 used the isotopic composition of bulk sedimentary organic carbon (Arthur et al. 1985; Fontugne and Calvert 1992; Müller et al. 1994; Bentaleb et al. 1996) as well as of individual biomarker compounds such as geoporphyrins (Popp et al. 1989) and long-chain alkenones (e.g. Jasper et al. 1994; Andersen et al. 1999; Pagani et al. 1999).

However, recent laboratory and field experiments as well as theoretical considerations indicate that the isotopic fractionation (ε_p) of carbon during photosynthesis, and consequently also the sedimentary $\delta^{13}C$ of organic matter, could also be influenced by physiological processes and environmental factors such as active carbon uptake (e.g. Laws et al. 1997), direct bicarbonate utilisation (Burns and Beardall 1987; Keller and Morel 1999), cell geometry and membrane permeability (e.g. Popp et al. 1998; Burkhardt et al. 1999). Moreover, growth rate and the growth rate limiting resources (e.g. temperature, nutrient supply, irradiance) were recently identified as additional factors (e.g. Bidigare et al. 1997, 1999; Riebesell et al. 2000a; Laws et al. 2001; Gervais and Riebesell 2001; Benthien et al. 2002).

Popp et al. (1998) demonstrated that some of the above mentioned factors can be evaded by using the isotopic composition of C_{37} alkenones. This biomarker is exclusive to haptophyte algae such as the coccolithophorid *Emiliania huxleyi* and species of the genus *Gephyrocapsa* (Volkman et al.

1989; Volkman et al. 1995; Conte et al. 1995), which both have a limited range in cell size and geometry. Assuming that alkenone-producing algae assimilate CO_2 mainly by passive diffusion, the relation between ε_p and $[CO_2(aq)]$ can be expressed as follows (Jasper et al. 1994):

$$\varepsilon_p = \varepsilon_f - \frac{b}{[CO_2(aq)]} \qquad (1)$$

where ε_f is the enzymatic fractionation during carbon fixation and b is an arbitrary, empirically derived parameter accounting for all physiological factors influencing the carbon isotope discrimination (e.g. Rau et al. 1996).

Some laboratory and field studies have focused on the isotopic fractionation in alkenone producing algae and the $\delta^{13}C$ signal of C_{37} alkenones (e.g. Bidigare et al. 1997; 1999; Popp et al. 1999; Riebesell et al. 2000b; Gonzales et al. 2001; Benthien et al. 2002). The results of these studies indicate that the fractionation of the $C_{37:2}$ alkenone depends not solely on ambient $[CO_2(aq)]$ but on a variety of factors (for a recent review see Laws et al. 2001). Bidigare et al. (1997) invoked that most variations in ε_p result from variations in growth rate related to nutrient availability rather than to $[CO_2(aq)]$. In their laboratory experiments they used nitrate limited chemostat cultures with continuous light and temperature conditions. This finding is consistent with the results of Riebesell et al. (2000b) who demonstrated by using N-repleted, light controlled batch cultures, that the effect of $[CO_2(aq)]$ on isotope fractionation in *E. huxleyi* is small compared to potential changes of ε_p due to growth rate variations and factors that affect the growth rate (μ). However, considerable discrepancies were found in absolute values of ε_p as well as in the slope of ε_p vs. $\mu/[CO_2(aq)]$. Probably these discrepancies are an effect of differences in the phytoplankton culture conditions (for a detailed discussion about the influence of culture experiment design see Laws et al. 2001).

In the present paper, we summarize isotopic results obtained from alkenones in the South Atlantic within the scope of the Collaborative Research Project 261 (Andersen et al. 1999; Benthien et al. 2002, and unpubl. data). In order to establish relationships between ε_p and environmental conditions characteristic for the present surface ocean, we analysed suspended particulate organic matter (POM) and surface sediment samples for the isotopic composition of $C_{37:2}$ alkenones and compared the results to CO_2 and phosphate concentrations in the overlying surface waters (0-10 m). In accordance with literature data, our results indicate only a weak effect of $[(CO_2(aq)]$ on ε_p and $\delta^{13}C_{alkenones}$ but a strong influence of nutrient-related growth rate. Finally, we compare late Quaternary paleo-PCO_2 reconstruc-tions from the eastern Angola Basin based on the isotopic composition of total organic carbon (Müller et al. 1994) and C_{37} alkenones (Andersen et al. 1999) and discuss them in the light of the new extended calibration results.

Material and Methods

Samples

Sediment samples were collected during several cruises of RV *Meteor*, RV *Sonne*, and R/V *Victor Hensen* (Andersen et al. 1999; Benthien et al. 2002; Fig. 1). Surface sediments were obtained using a giant box corer and a multiple corer (Tab. 1), except for site GeoB1016-3 where the surface sediment sample was taken from the top of the gravity corer. For the surface sediments a Holocene age was confirmed by the occurrence of the planktonic foraminifera *Globorotalia menardii* and/or by $\delta^{18}O$ and ^{14}C measurements on fora-minifera (for details see Benthien et al. 2002 and references therein). The stratigraphy of gravity core GeoB1016-3 is based on foraminiferal $\delta^{18}O$ (Schneider et al. 1995). After core recovery, subsamples from gravity and giant box cores were taken using plastic syringes (10 ml) and stored at 4°C until analysis. The gravity core was re-sampled 6 years after recovery for isotopic analyses on alkenones (Andersen et al. 1999). Sediments collected by the multiple corer were sectioned into 1 cm slices and frozen at –18°C until analysis.

Particulate organic matter was obtained by filtering 40 l of surface sea water through pre-combusted Whatman GF/F (nominal pore size 0.7μm) glass-fibre filters. Water for filtering was obtained aboard RV *Meteor* (cruise M46/3, Tab. 1)

using a membrane pump situated at about 5 m water depth at the front of the vessel. The filters were stored at –18°C until analysis.

Alkenone Analysis

Detailed descriptions of analytical procedures can be found elsewhere (Schneider et al. 1995; Müller et al. 1998; Benthien and Müller 2000). Briefly, alkenones were extracted from freeze-dried and homogenised sediment with a UP 200H ultrasonic disrupter probe using three successively less polar mixtures of methanol and dichloromethane. For this purpose filters with POM were cut into small pieces after freeze drying. Thereafter the extracts where purified by passing them over a silica gel cartridge and then saponified to remove possibly interfering esters. Alkenone unsaturation ratios were determined by gas chromatography.

The carbon isotopic analysis of $C_{37:2}$ alkenones was performed using a HP 5890 II gas chromatograph (GC) coupled via a combustion interface to a Finnigan MAT 252 mass spectrometer (for analytical details see Andersen et al. 1999 and Benthien et al. 2002). The isotopic composition of the $C_{37:2}$-alkenone was determined relative to PDB by comparison with co-injected *n*-alkanes (C_{34}, C_{36}, C_{37}, C_{38}) and a standard gas (CO_2) of known isotopic composition. Generally each sample was measured two to four times revealing an analytical precision better than 0.3‰.

Surface Water CO_2 and Nutrients

In contrast to other studies (e.g. Bidigare et al. 1997; Popp et al. 1999) we did not determine the concentrations of CO_2 and nutrients simultaneously with the sampling of POM. Instead we used integrated values of nutrients and CO_2 available from data collections or ocean atlases. Since core-top sedi-ments in general represent at least a few hundred years a comparison with annually integrated values of surface water parameters is probably the best approach.

Unfortunately, for CO_2 no annually integrated values are available. We therefore estimated the partial pressure of CO_2 in the surface waters (PCO_2) from numerous expeditions carried out in the South Atlantic Ocean during different seasons

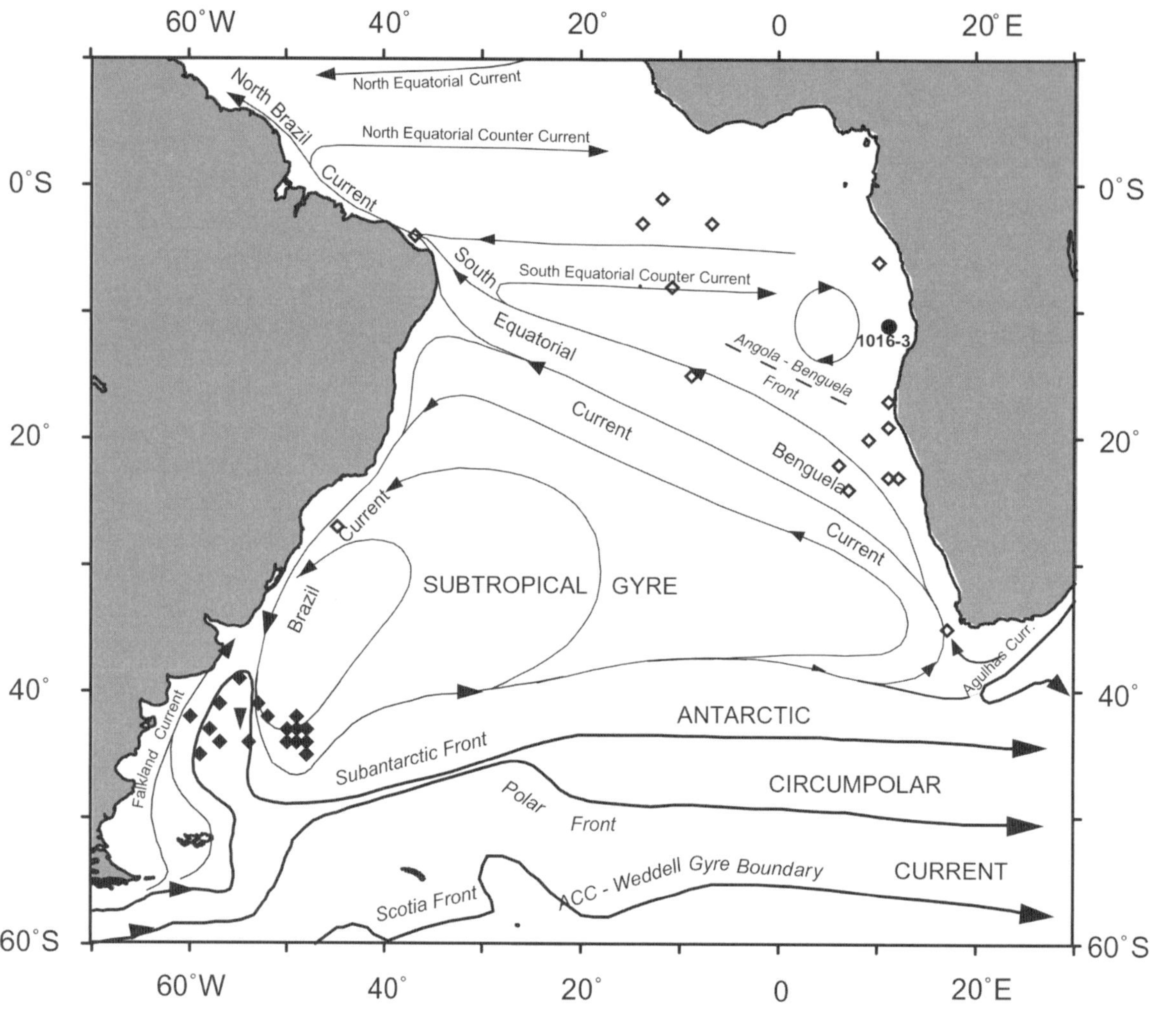

Fig. 1. Map of the South Atlantic Ocean showing sample locations and main surface currents: open diamonds represent locations of core top sediment, the black dot that of the gravity core GeoB1016-3 and filled diamonds those of particulate organic matter filtered from the water column.

and years (Weiss et al. 1992; Groupe CITHER 1, 1994; Johnson et al. 1995, 1998; extracted from the CDIAC (Carbon Dioxide Information Analysis Centre, web-page: http://cdiac.esd.ornl.gov) and converted the values into concentrations of dissolved carbon dioxide ([CO_2(aq)]) applying Henry's Law (Fig. 2; Benthien et al. 2002). Pre-industrial CO_2(aq) concentrations were calculated in the same manner after subtracting the industrial increase of 70 ppmv (Takahashi et al. 1992). Such a correction is probably not appropriate for areas where strong upwelling of cold CO_2-rich sub-surface waters results in a distinct sea-air imbalance

(e.g. Tans et al. 1990). Consequently the modern anthropogenic CO_2 increase has no significant influence on the surface water PCO_2 in such regions (Lee et al. 1997). Therefore, we did not correct the CO_2(aq) concentrations for the anthropogenic influence at core sites of the Angola and Benguela upwelling cells (Fig. 1). This problem was discussed in detail by Benthien et al. (2002). Finally, average CO_2(aq) concentrations were computed for each sample location (Fig. 2, Benthien et al. 2002) using the Kriging method (Davis 1986).

The annual mean concentration of phosphate in the surface water (0-10 m) at each location was

Core	Cruise	Lat.	Long.	SST [°C]	$\delta^{13}C$ [‰]	ε_p [‰]	[CO$_2$(aq)] [µmol/L]	b-value [‰µmol/L]	[PO$_4$]$^{3-}$ [µmol/L]
					Sediment samples				
1016-3*	M6/6	11°46S	11°40E	24.7	-23.5	13.3	11.5	135	0.32
1008-6*	M6/6	6°34S	10°19E	25.5	-23.1	12.9	11.8	142	0.28
1028-4#	M6/6	20°06S	9°11E	20.6	-22.9	12.1	12.2	157	0.52
1032-2#	M6/6	22°54S	6°02E	19.8	-24.3	13.5	10.1	117	0.31
1041-1#	M6/6	3°28S	7°36W	26.1	-23.6	13.5	9.2	106	0.18
1105-3#	M9/4	1°39S	12°25W	25.7	-23.4	13.2	8.9	105	0.18
1117-3#	M9/4	3°48S	14°53W	26.2	-23.6	13.5	9.5	109	0.17
1214-2#	M12/1	24°41S	7°14E	20.6	-23.4	12.7	10	124	0.32
1413-1#	M16/1	15°40S	9°27W	24.8	-23.6	13.4	9	105	0.23
1501-1*	M16/2	3°40S	32°00W	26.2	-25.4	15.3	9.5	92	0.16
1503-2*	M16/2	2°18N	30°38W	26.7	-25	15	8.4	84	0.13
1505-3*	M16/2	2°16N	33°00W	26.3	-25.1	15.1	8.4	82	0.16
1508-1*	M16/2	5°20N	34°01W	27.1	-25.1	15.2	8.7	86	0.19
1515-2*	M16/2	4°14N	43°40W	26.6	-25	15	8	79	0.24
1703-5*	M20/2	17°27S	11°01E	19.6	-23.8	13	13.2	158	0.66
1706-1*	M20/2	19°33S	11°10E	18.1	-23.5	12.5	12.9	161	0.71
1710-2#	M20/2	23°25S	11°41E	18.9	-23	12.1	12.2	157	0.59
1711-5*	M20/2	23°18S	12°22E	18.3	-23.9	12.9	12.2	147	0.63
1712-2#	M20/2	23°15S	12°48E	18.2	-22.5	11.5	13.2	178	0.64
1713-6#	M20/2	23°13S	13°00E	17.9	-22.6	11.5	13.2	178	0.66
1719-5*	M20/2	28°55S	14°10E	17.5	-24.8	13.7	8.5	96	0.36
2102-1*	M23/2	23°59S	41°12W	24.7	-25.6	15.4	8.2	79	0.13
2109-3#	M23/2	27°54S	45°52W	23.9	-24	13.6	8.4	96	0.13
2125-2*	M23/2	20°49S	39°51W	25.2	-25.8	15.7	8.3	77	0.11
2215-8*	M23/3	0°00N	23°29W	26.7	-24.5	14.5	8.3	87	0.11
2204-1#	M23/3	8°31S	34°01W	26.5	-23.6	13.5	8.1	93	0.16
3603-1#	M34/1	35°07S	17°32E	19.9	-23.6	12.8	8.9	109	0.43
1903-1*	SO84	8°40S	11°50W	24.9	-25.7	15.6	8.7	82	0.19
3117-3#	JOPS II	4°17S	37°05W	27.5	-24.2	14.3	8.8	94	0.2
					Particulate organic matter (POM)				
	M46/3	39°41S	55°82W	20.5	-29.7	15.8	13.9	128	0.75
	M46/3	41°85S	53°20W	19.9	-22.7	11	13.7	192	0.6
	M46/3	42°08S	52°70W	19.6	-25	11	13.7	194	0.53
	M46/3	42°84S	49°65W	18.5	-24.8	13.2	13.5	159	0.49
	M46/3	44°05S	48°68W	16.5	-24.9	13.2	13.3	157	0.55
	M46/3	45°58S	48°26W	15.7	-24.6	12.3	14.1	179	0.65
	M46/3	44°01S	49°94W	19.7	-25.9	13.8	13.6	152	0.52
	M46/3	44°38S	50°17W	19.8	-28.9	16.9	13.6	110	0.52
	M46/3	43°68S	49°68W	16.1	-25.1	13.3	13.6	159	0.52
	M46/3	43°80S	50°36W	18.8	-27.2	15.7	13.6	127	0.5
	M46/3	43°93S	48°95W	17.6	-25	12.8	13.3	161	0.59
	M46/3	44°23S	50°93W	18.7	-27.6	15.9	13.5	123	0.53
	M46/3	44°90S	54°45W	14.5	-22.7	10.5	13.6	197	0.66
	M46/3	45°20S	59°09W	12.3	-22.8	8.5	15.2	250	1.02
	M46/3	44°32S	57°50W	15.8	-22.7	10.1	14.4	215	0.9
	M46/3	44°34S	57°87W	15.9	-24.6	12.3	14.4	183	0.9
	M46/3	43°74S	58°99W	13.5	-21.8	9.6	15.2	234	0.95
	M46/3	42°30S	60°27W	19.2	-26.4	14	13.6	150	0.89
	M46/3	41°03S	57°53W	19.5	-23.6	9.4	14.8	232	0.85

Table 1. Isotopic composition and fractionation data (both in ‰) of the $C_{37:2}$ alkenone determined in South Atlantic particulate organic matter (POM) and core top sediments. Data marked with * are from Andersen et al. (1999) and data marked with # are from Benthien et al. (2002). Data for POM are from this study. Sea-surface temperature (in °C) for the core tops are derived from alkenone analyses [U_{37}^K, Müller et al. 1998; Benthien and Müller 2000] applying the calibration of Prahl et al. 1988. Sea-surface temperature of POM samples are measured at sample station. Also given are the calculated concentrations of surface water carbon dioxide (Benthien et al. 2002) as well as the surface water concentration of phosphate (Conkright et al. 1994 extracted from the web site: http://ferret.wrc.noaa.gow/fbin/climate_server).

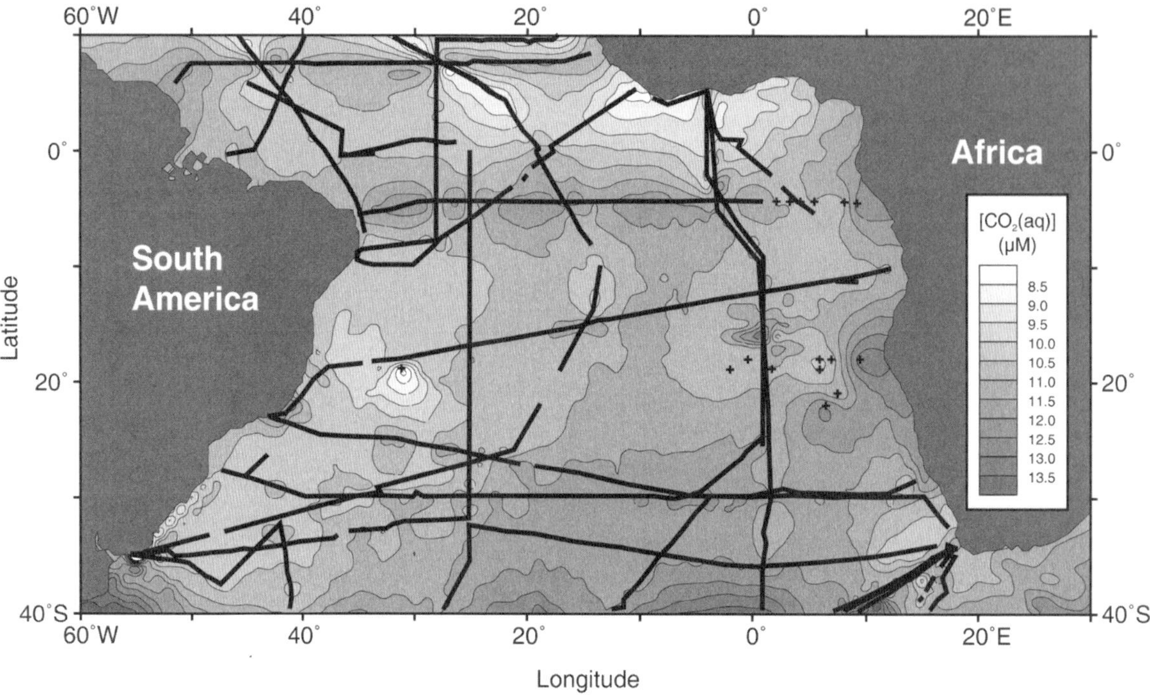

Fig. 2. Modern surface-water concentration of CO_2(aq) in the equatorial and South Atlantic Ocean. Concentrations were calculated from surface water partial pressure of carbon dioxide measured continuously during several expeditions (Weiss et al. 1992, Groupe CITHER 1 1994; Johnson et al. 1995, 1998). The tracks of this expeditions are indicated as solid thick lines. Single locations of measurements are shown as crosses. [CO_2(aq)] isolines were created with the Kridging method (redrawn from Benthien et al. 2002).

obtained from the World Ocean Atlas 1994 (http:/ /ferret.wrc.noaa.gov/fbin/climate_server; Conkright et al. 1994).

Calculation of ε_p

The isotopic fractionation ε_p of alkenones associated with photosynthetic fixation of carbon was calculated using the following equation (Freeman and Hayes 1992), where δ_d is the isotopic composition of dissolved carbon in CO_2(aq) and δ_p that of photosynthate carbon:

$$\varepsilon_p = (\frac{\delta_d + 1000}{\delta_p + 1000} - 1)\, 1000 \qquad (2)$$

δ_p was derived from the isotopic composition of $C_{37:2}$ alkenones and corrected for the compound-specific fractionation $\varepsilon_{alkenone}$ between the $C_{37:2}$ alkenones and the biomass of alkenone-producing organisms. We adopted 4.2‰ as the value for $\varepsilon_{alkenone}$ (Popp et al. 1998):

$$\delta_p = \delta_{C37:2} + \varepsilon_{alkenone}\, (1+\frac{\delta C37:2}{1000}) \qquad (3)$$

δ_d was calculated from $\delta_{\Sigma CO2}$ following the equation of Rau et al. (1996) based on Mook et al. (1974):

$$\delta_d = \delta_{\Sigma CO2} + 23.644 - \frac{9701.5}{T} \qquad (4)$$

where T is the temperature in Kelvin. Temperatures for sediment samples were obtained from the alkenone unsaturation index (U_{37}^K) using the calibration of Prahl et al. (1988). For POM samples directly measured temperatures were available. For both sample types we assumed a constant pre-industrial $\delta_{\Sigma CO2}$-value of 2.5‰ (Kroopnick 1985). In core GeoB1016-3 $\delta_{\Sigma CO2}$ was calculated from the $\delta^{13}C$ record of the surface-dwelling planktonic foraminifer *Globigerinoides ruber* (pink) (Andersen et al. 1999). The tests of *G. ruber* are depleted in ^{13}C relative to the isotopic composition of total dissolved CO_2. To consider this we assumed an offset of 0.5‰ (Fairbanks et al. 1982; Curry and Crowley 1987). Based on propagating error calculations Andersen et al. (1999) determined an absolute error of +/-0.42‰ for the calculation of ε_p.

Results and Discussion

Modern Situation (Attempt of Calibration)

Since laboratory data cannot perfectly mimic natural environments, which undergo changes in physical (e.g. temperature, mixing), chemical (e.g. major and trace nutrients) and biological (e.g. grazing pressures, species competition) processes, extrapolation to the field must be done with caution (Laws et al. 2001). In addition, laboratory and water column studies cannot completely imitate the sum of environmental conditions and physiological processes which may influence the sedimentary isotopic signal of alkenones. Therefore it is necessary to test hypotheses like the potential of an influence of CO_2(aq) concentrations on the fractionation of $C_{37:2}$ alkenones also by sediment-based studies.

Such a calibration requires knowledge of the water depth were the $\delta^{13}C$ signal of alkenones ($\delta^{13}C_{37:2}$) was produced. Hapthophytes use sunlight for photosynthesis and thus live in the photic zone. Dependent on suspended particles in the water column, the photic zone depth varies from about 20 m in eutrophic to about 120 m in oligotrophic regions (Morel and Berthon 1989; Longhurst 1993). Our core-top studies in the South Atlantic showed that for most hydrographic regions the best correlation between the sedimentary $U_{37}^{K'}$ signal and sea-surface temperature (SST) is obtained when modern atlas values from 0-10 m water depth are used (Müller et al. 1998; Benthien and Müller 2000). It was further demonstrated that seasonal changes in primary production had only a negligible effect on the sedimentary $U_{37}^{K'}$ signal in this region (Müller et al. 1998). Recent sediment trap studies from the tropical-subtropical eastern Atlantic also suggest that $U_{37}^{K'}$ principally records the annual average of mixed-layer temperature in this region (e.g. Müller and Fischer 2001; PJ Müller unpubl. data). We therefore used surface water

properties (upper 10 m) to "calibrate" the alkenone $\delta^{13}C$ signal in POM and surface sediments.

In the low to mid latitude South Atlantic $\delta^{13}C_{37:2}$ values for the POM range from −29.7‰ to − 21.8‰ and for surface sediments from -25.8‰ to −22.5‰. The corresponding ε_p-values (calculated from equation 1) range from 8.5‰ to 15.8‰ and 11.5‰ to 15.7‰, respectively (Tab. 1). Figure 3a shows the relationship between ε_p of the $C_{37:2}$-alkenone and the reciprocal of [CO_2(aq)]. If the concentration of [CO_2(aq)] was the major factor controlling the carbon isotopic fractionation of the $C_{37:2}$ alkenone, a negative correlation between ε_p and the $1/[CO_{2(aq)}]$ would be expected (e.g. Rau et al. 1992, Eq. 1). Instead, a positive, but not identical relationship is observed for both, the POM and core top sediments (Fig. 3a; R = 0.72 and R = 0.77, respectively). This result is in accordance with the relationship between ε_p of alkenones and $1/[CO_{2(aq)}]$ found by Popp et al. (1999) in POM of the Southern Ocean and indicates that isotopic fractionation in alkenones is influenced by factors other than [CO_2(aq)].

In spite of these regionally observed correlations between ε_p and $1/[CO_2$(aq)] (this study; Popp et al. 1999), no general relationship between both parameters is apparent if additional literature data are considered (Fig. 3b, Laws et al. 2001). This also implies that factors other than [CO_2(aq)] influence the isotopic fractionation.

In the case of taxon-specific biomarkers such as alkenones the effects of variation in cell size and membrane permeability are accounted to be relatively small (e.g. Bidigare et al. 1997; Popp et al. 1998), especially for core-top sediments which typically represent at least a few hundred years. A constant value for the maximum fractionation ε_f attributed to enzymatic carbon fixation can also be assumed, since it is likely to be species-dependent (Bidigare et al. 1997) and thus not a problem for taxon-specific biomarkers. We adopted a value of 25‰ which was reported from culture experiments of two different strains of *E. huxleyi* (Bidigare et al. 1997). In addition, variations of growth rate and the growth rate influencing factors (e.g. Riebesell et al. 2000a; Gervais and Riebesell 2001) as well as different carbon acquisition mechanisms and inorganic carbon sources (i.e. CO_2 and HCO_3^-)

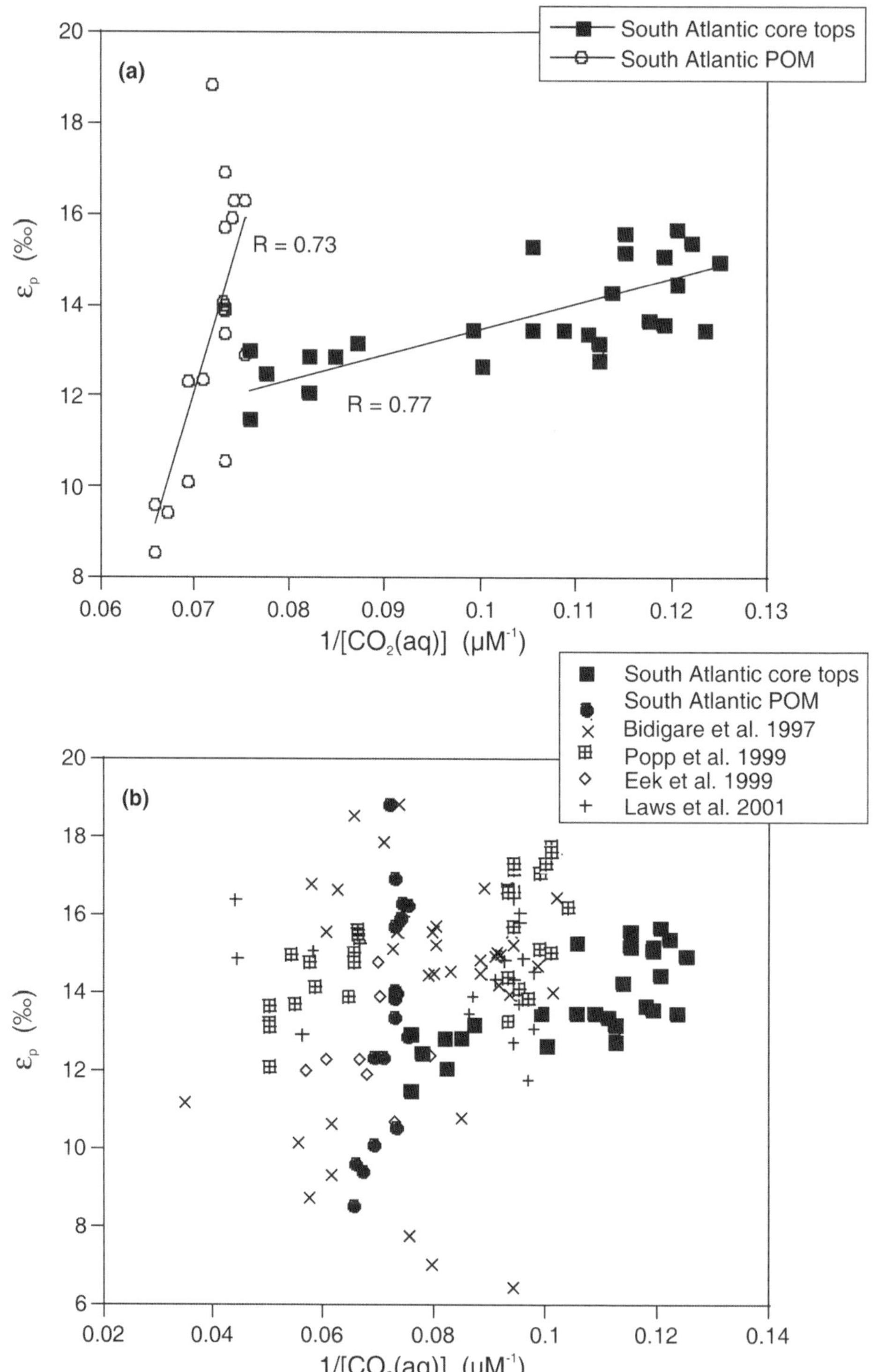

Fig. 3. a) Carbon isotopic fractionation (ε_p) of the $C_{37:2}$ alkenone determined in particulate organic matter (POM; open circles) and core top sediments (filled squares) from the South Atlantic in relation to $1/[CO_{2(aq)}]$. **b)** Relationship between ε_p and $1/[CO_2(aq)]$ in POM (this study and literature data (Laws et al. 2001)) and core top sediments (this study). Linear regression of all data (n = 145) yields no significant correlation (R = 0.12).

may further influence the isotopic fractionation of alkenones (Rau et al. 1996; Keller and Morel 1999; Gonzales et al. 2001; Benthien et al. 2002).

The ε_p values of $C_{37:2}$ alkenones in POM and surface sediments of the South Atlantic correlate well with ambient surface water concentrations of phosphate which is in accordance with results of Popp et al. (1999) and Andersen et al. (1999). Both POM and surface sediments exhibit a negative relationship between ε_p and surface water phosphate, but the correlations have different intercepts and slopes (Fig 4a). A compilation of all currently published ε_p values of $C_{37:2}$ alkenones from POM together with the data of the present study reveals only a weak negative correlation (R = 0.46) between ε_p and the surface water phosphate concentration (Fig. 4b). Bidigare et al. (1997) pointed out that for environments with non-zero concentrations of phosphate the growth rate of alkenone-producing algae is linearly related to $[PO_4^{3-}]$. Based on a predictive model, Rau et al. (1996) proposed, that ε_p decreases with increasing growth rates. Apparently, the variations of ε_p observed in Figure 4b are related to variations in growth rate. When the growth rate increases, the diffusive CO_2 flux through the cell membrane decreases relative to carbon fixation. At first, due to the discrimination against ^{13}C, more ^{12}C is consumed. This subsequently leads to an enrichment of the internal carbon pool in ^{13}C, which is then consumed by carbon fixation. As a result of this mechanism, $\delta^{13}C_{37:2}$ values increase while those for ε_p decrease (Keller and Morel 1999).

The influence of growth rate on the isotopic fractionation in alkenones is further supported by a significant correlation between the calculated *b*-values and phosphate concentrations (Fig 5a, R = 0.88). We added our results to the compilation of Laws et al. (2001) and got a nearly identical relationship (Fig. 5a). It was reasoned by Bidigare et al. (1997) that this correlation may be caused by growth-rate limiting concentrations of micronutrients (e.g. Fe, Zn, Co) rather than by phosphate concentrations since *E. huxleyi* has a low phosphorus requirement. This interpretation is supported by the fact that the phosphate concentrations in the studied areas are much higher than the half–saturation constant for growth determined for *E.*

huxleyi (Riegman et al. 2000). In addition, if growth rate would be limited by phosphate it should follow Michaelis-Menten saturation kinetics rather than a linear relationship (Fig. 5a). Consequently, it seems more reasonable that micronutrients such as iron, zinc and cobalt are the growth-rate limiting factors and that the phosphate concentration is closely related to that of micronutrients (see Bidigare et al. 1997 for a detailed discussion).

A positive relationship between the *b*-value and phosphate concentration could also be simply the result from a correlation between phosphate and dissolved CO_2 as pointed out by Bidigare et al. (1997). Indeed, a high correlation is observed between these two parameters (R = 0.89; Fig. 5b). Popp et al. (1999) examined the effect of $[PO_4^{3-}]$ on ε_p at relatively constant concentrations of dissolved CO_2 and found a significant correlation. By contrast, no correlation between $[CO_2(aq)]$ and ε_p was observed. They thus concluded that the correlation between the *b*-value and $[PO_4^{3-}]$ is not caused by the relationship between $[CO_2(aq)]$ and $[PO_4^{3-}]$ in surface waters. This is further supported by the positive correlation observed here between ε_p and $1/[CO_2(aq)]$ in POM and core top sediment from the South Atlantic (Fig 3a) as well as in POM from the subantarctic Southern Ocean (Popp et al. 1999). As mentioned above such a correlation is opposite of what is expected from a $[CO_2(aq)]$ controlled, diffusive uptake model. This strongly suggests, that the isotopic fractionation in haptophyte algae is more sensitive to variations of growth rate and/or growth rate limiting factors (e.g. Riebesell et al. 2000a; Rost et al. 2002) and micronutrient availability (Bidigare et al. 1997) than to varying concentration of $[CO_2(aq)]$.

Paleo-PCO_2 Estimates in the Angola Current

In this section we briefly summarize the results of earlier paleo-PCO_2 reconstructions in the South Atlantic (Müller et al. 1994; Andersen et al. 1999) and evaluate them in the light of the above results. These studies are based on a late Quaternary sediment core (GeoB1016-3) from the eastern Angola Basin covering the last 200,000 years (Fig. 6). In the first study, we used the isotopic composition of bulk sedimentary organic carbon ($\delta^{13}C_{org}$), pub-

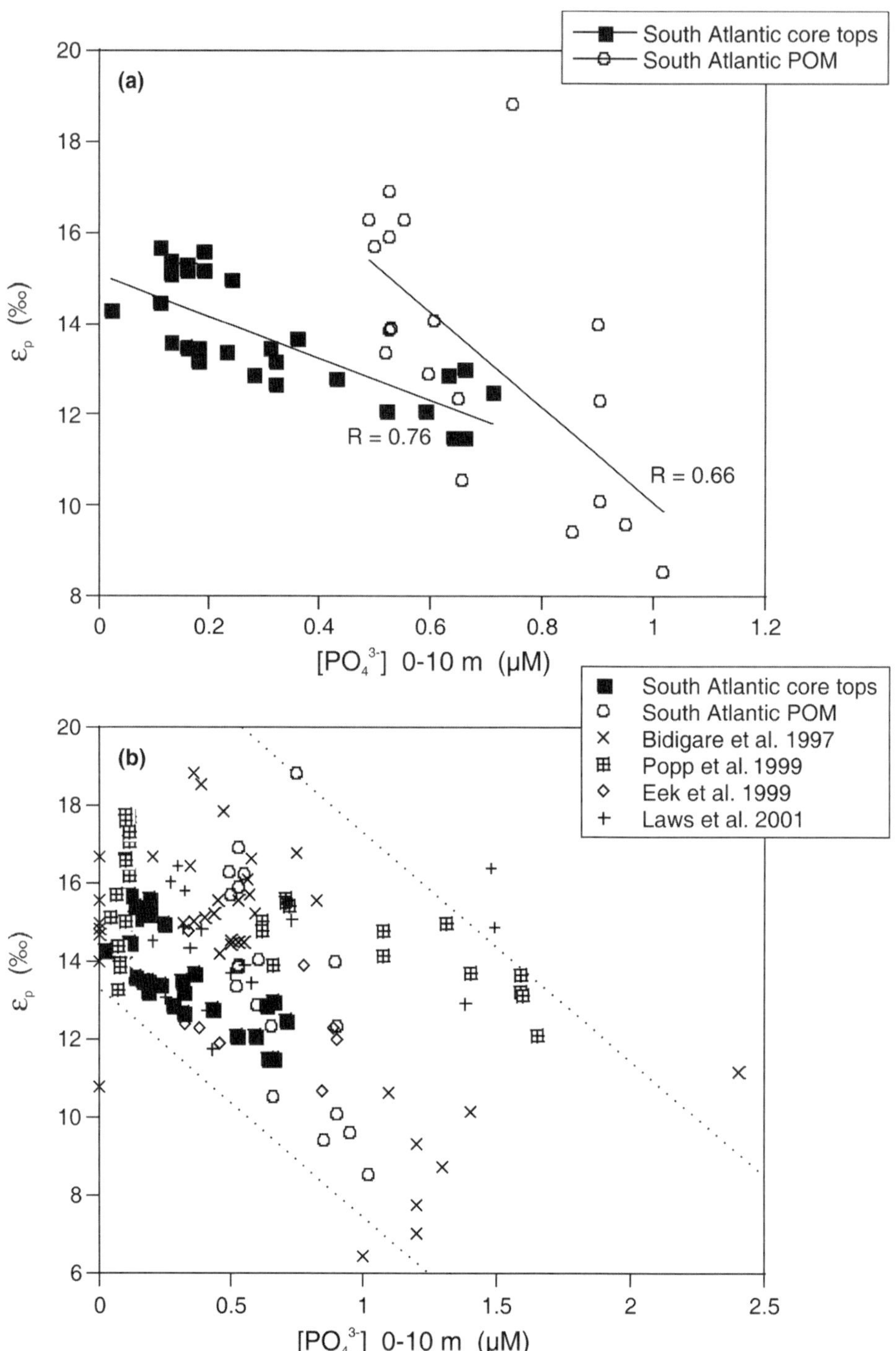

Fig. 4. a) Relationship between the isotopic fractionation (ε_p) of the $C_{37:2}$ alkenone measured in POM (open circles) and core top sediments (filled squares), and the annual mean surface water concentration of phosphate (0-10 m) in the South Atlantic. The lines represent the linear correlation between the two parameters. **b)** Relationship between ε_p and phosphate. Compilation of literature data for POM and data obtained in the present study (POM and sediments). Stippled lines are for better visualization of the weak correlation observed between ε_p and phosphate ($R = 0.46$; all data $n = 145$).

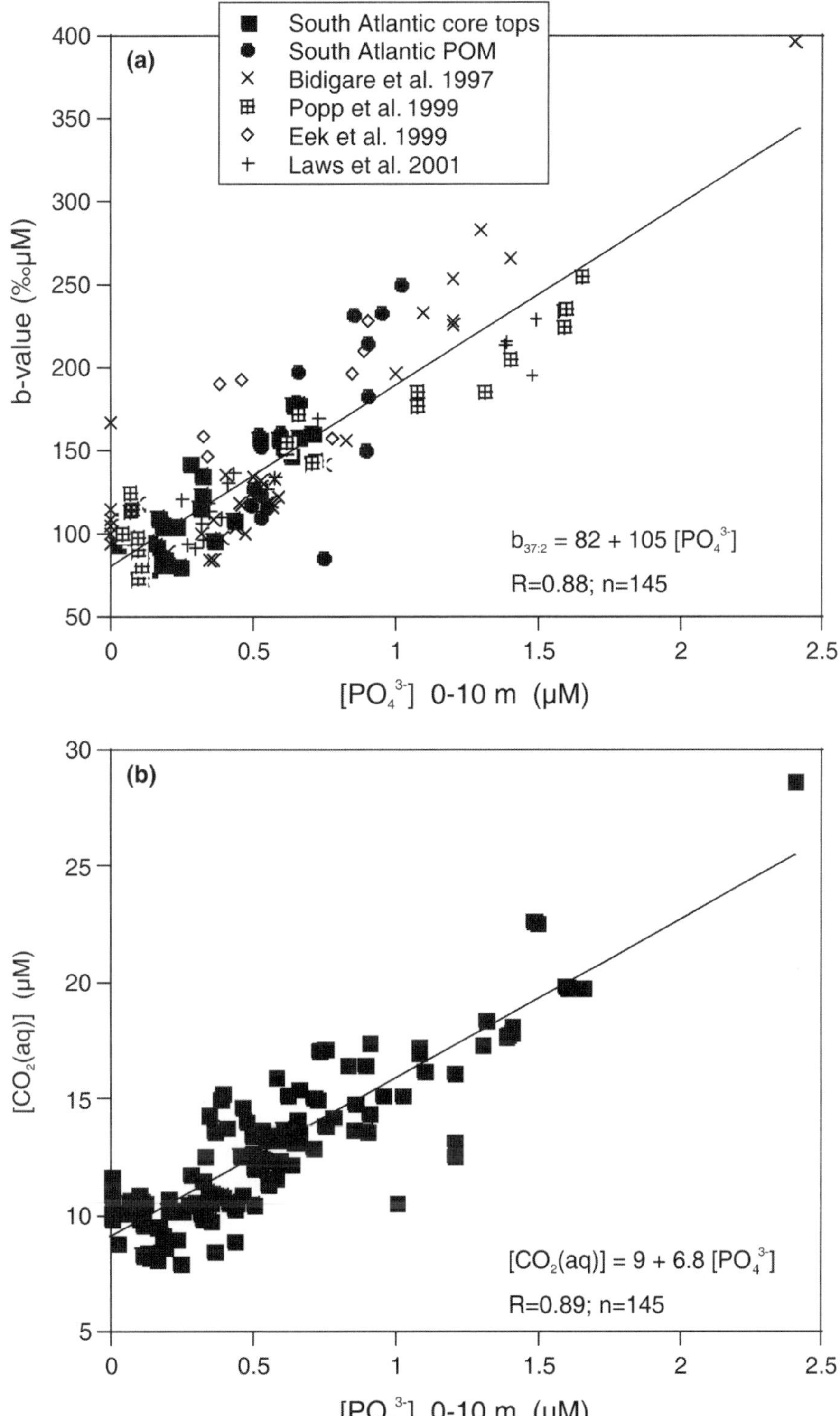

Fig. 5. a) Relationship between b-value and phosphate concentration of data of the present study plus literature data (Laws et al. 2001). The relationship found is nearly identical with that reported by Laws et al. (2001): $b_{(37:2)}$ = 79 + 120[PO_4^{3-}]. **b)** Relationship between dissolved CO_2 and surface water phosphate concentration for the whole data set (present study plus data published in Laws et al. 2001).

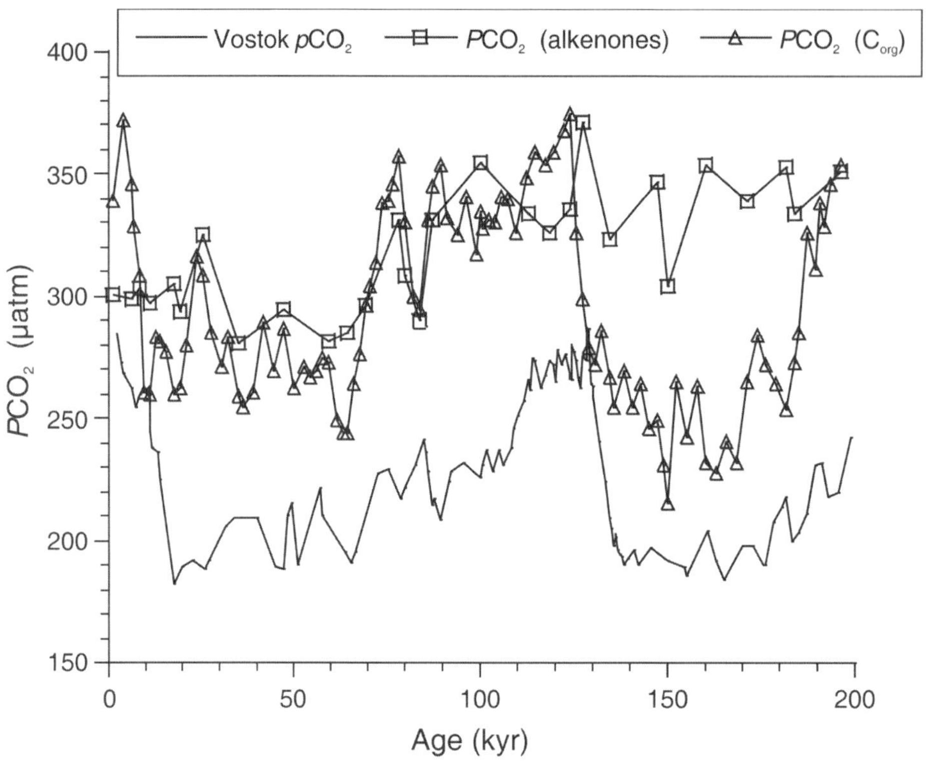

Fig. 6. Reconstructions of paleo-PCO_2 in core GeoB1016-3 versus age using carbon isotopic fractionation of $C_{37:2}$ alkenone (triangles, from Andersen et al. 1999) and bulk organic matter (squares, from Müller et al. 1994). The line represents atmospheric CO_2 concentration recorded in Vostok ice core (Jouzel et al. 1993). The age model of the marine record is based on foraminiferal $\delta^{18}O$ that was tuned to the ice core chronology (for details see Müller et al. 1994).

lished empirical relationships between $\delta^{13}C_{org}$ and $[CO_2(aq)]$, and alkenone-derived ($U_{37}^{K'}$) temperatures to estimate past PCO_2 levels (Müller et al. 1994). The results indicate that the Angola Current has generally acted as a source region for atmospheric CO_2 throughout the last two glacial-interglacial cycles, consistent with the modern situation. However, the difference between our surface-water estimate and atmospheric PCO_2 (Vostok ice core, e.g. Jouzel et al. 1993) appeared to be smaller in glacial periods suggesting a reduced CO_2 output into the atmosphere in periods of lowered sea-surface temperatures and enhanced biological productivity in this region (Schneider et al. 1996).

In a subsequent study, Andersen et al. (1999) used a different approach based on the isotopic composition of the $C_{37:2}$ alkenone in conjunction with that of bulk sedimentary nitrogen ($\delta^{15}N$). The latter was considered as indirect proxy for nutri-

ent-limited growth rates and used to calculate the growth-rate dependent parameter b in equation 1. This application was based on the observation that $\delta^{15}N$ in surface sediments from eutrophic regions is positively correlated with the phosphate concentration in overlying surface waters. For more details on the methods used in the above two studies the reader is referred to the original publications.

Figure 6 compares the paleo-PCO_2 records from the Angola Basin to the atmospheric record of the Vostok ice core. In general, the two oceanic reconstructions agree quite well and show generally higher PCO_2 values compared to the Vostok record supporting the initial contention of a permanent source for atmospheric CO_2 over the past 200,000 years. However, significantly different PCO_2 levels were obtained for the penultimate glacial, the marine isotope stage 6 (128-186 kyr in Fig. 6). While the PCO_2 record derived from bulk

$\delta^{13}C_{org}$ shows a glacial to interglacial pattern parallel to the atmospheric CO_2 variations indicated by the ice core, the biomarker approach yields high values for glacial stage 6 comparable to the level observed for interglacials. We have no definite explanation for this deviation and must consider several effects.

Reconstructions based on $\delta^{13}C$ of bulk organic matter may suffer from species-specific differences in isotope fractionation among marine photosynthetic organisms (e.g. Burkhardt et al. 1999), shifts in isotope composition of buried organic matter during degradation, and contamination by terrestrial organic matter (Jasper and Hayes 1990). Based on a strong positive relationship between C/N and $\delta^{13}C_{org}$ values in core GeoB1016-3, Müller et al. (1994) argued that differing proportions of marine and terrigenous organic matter can be ruled out as a cause for the $\delta^{13}C_{org}$ variability in this core. However, strictly speaking this is only valid for organic matter derived from C3 plants (e.g. Meyers 1997). A higher proportion of isotopically heavier C4 plant material in climatic periods with increased savannah vegetation (e.g. Partridge et al. 1999) could have led o generally higher bulk $\delta^{13}C_{org}$ values resulting in the lower PCO_2 estimates obtained for glacial stage 6. A critical review addressing problems and advantages associated with the use of bulk sedimentary $\delta^{13}C_{org}$ as tracer of dissolved carbon dioxide in the surface ocean was recently presented by Kienast et al. (2001).

The phytoplanktonic organic matter preserved in marine sediments represents a mixture derived from various groups, e.g. diatoms, dinoflagellates and coccolithophorids, which may grow at different rates and may show different isotopic signatures. It is well known that rapidly growing diatoms have ^{13}C-rich isotopic compositions. Fry and Wainright (1991), for example, reported that diatoms in spring blooms were relatively rich in ^{13}C, while other phytoplankton had ^{13}C-depleted values. A significant contribution of diatoms or other rapidly growing phytoplankters to the bulk sedimentary organic matter deposited during glacial stage 6 could therefore also explain the different PCO_2 levels obtained by the bulk C_{org} and alkenone methods. Such an interpretation is also compatible with recent sediment trap results from the Southern Ocean which show a coupling of $\delta^{13}C_{org}$ with the flux of opal and that of $\delta^{13}C_{alkenones}$ with the carbonate flux (S. Schulte, unpublished results).

In principle, the isotopic composition of alkenones is a less ambiguous proxy parameter for surface-water CO_2 than bulk $\delta^{13}C_{org}$ because alkenones are synthesized by a limited number of haptophytes only. However, their isotopic composition is also influenced by more than one factor as demonstrated above by the POM and core-top results. In the modern South Atlantic, the isotopic fractionation of the $C_{37:2}$ alkenone appears to be mainly influenced by growth rate rather than CO_2. Variations in growth rate must therefore also be taken into account in paleoceanographic studies. There is indeed evidence for significant variations in paleoproductivity and, by inference, growth rate at site GeoB1016-3 (Schneider et al. 1996; Andersen et al. 1999). As described above, the influence of the growth rate on the isotopic fractionation of alkenones is included in the b-value (see equation 1) which may be derived from bulk sedimentary $\delta^{15}N$ values, at least in eutrophic regions (Andersen et al. 1999). However, $\delta^{15}N$ is also not unambiguous because a variety of environmental processes, independent from nutrient-growth rate relationships, can influence the nitrogen cycle and the isotopic signal in the sediments (e.g. mineralization, nitrification, denitrification, diagenesis; Altabet et al. 1999).

More reliable reconstructions of past dissolved CO_2 concentration and partial pressure in the surface ocean could be obtained if an independent proxy for variations in algal growth rate would be available. The Cd/Ca and $\delta^{13}C$ ratios of planktonic foraminifera have been suggested as nutrient proxies and useful indirect parameters to constrain the effects of growth rate on the isotopic composition of alkenones (e.g. Popp et al. 1999). However, the $\delta^{13}C$ of planktonic foraminiferal calcite is also controlled by a variety of other factors which complicates its interpretation as a proxy for dissolved phosphate (e.g. Spero et al. 1997; Bijma et al. 1999). Although benthic foraminiferal Cd/Ca ratios have proven useful as nutrient proxy in deep waters, the planktonic Cd/Ca ratio is no genuine alternative for surface waters because the incorporation of Cd into foraminiferal calcite is temperature sensitive (Rickaby and Elderfield 1999).

Recently it has been suggested that the Sr/Ca ratio in coccoliths is mainly controlled by growth and calcification rates, although temperature also exerts some influence on this ratio (Stoll and Schrag 2000; Rickaby et al. 2002; Stoll et al. 2002). Sr/Ca measurements on coccoliths thus provide a promising alternative to constrain the effect of growth rate on the isotopic fractionation during photosynthesis. In combination with the $\delta^{13}C$ of alkenones, this would permit more reliable estimations of $[CO_2(aq)]$ and PCO_2 in ancient oceanic surface waters.

Summary and Conclusions

The alkenone ε_p record in POM and surface sediments of the South Atlantic Ocean was compared with overlying surface-water concentrations of phosphate and dissolved carbon dioxide. The results demonstrate that regardless of the ε_p-controlling mechanism, the observed changes in the carbon isotopic composition of alkenones are linked to variations in phosphate concentrations rather than to $[CO_2(aq)]$. Although an unambiguous evidence for a specific mechanism cannot be presented at the moment, the relationship between ε_p and $[PO_4^3]$ found in our South Atlantic field studies suggest, in agreement with the theory of Bidigare et al. (1997), that a strong relationship exists between isotopic fractionation in haptophytes and nutrient-limited growth rate. Furthermore, the fact that the sediment data fit excellently with the water column data clearly indicates that the isotopic record of sedimentary alkenones reflects surface water conditions.

Estimates of ancient dissolved CO_2 concentrations based on compound specific $\delta^{13}C$-values must consider the effect of changing growth rates of certain algae species which determine the CO_2 concentration gradient between ambient water and the phytoplanktonic cell. The bulk sediment $\delta^{15}N$ signal may be useful in some restricted areas of the ocean (e.g. the eastern Atlantic Ocean) as a paleo proxy for the *b*-value. The Sr/Ca-ratio in coccolithophorids might provide an independent record of past changes in coccolithophorid growth rates, which in combination with data on the carbon isotopic fractionation in coccolithophorid organic matter may permit more reliable calculations of past dissolved CO_2 in the surface ocean.

Acknowledgements

We thank the officers, crews and scientists aboard RV *Meteor*, RV *Sonne* and RV *Victor Hensen* for their help during coring and sampling operations. The technical assistance during laboratory work of Hella Buschhoff, Wolfgang Bevern, Dietmar Grotheer, Birgit Meyer-Schack and Monika Segl is appreciated. The paper benefited from helpful reviews of Edward Laws and Mark Pagani. This research was funded by the Deutsche Forschungsgemeinschaft (Sonderforschungsbereich 261 at Bremen University, contribution No. 378). Data are available under http://www.pangaea.de/Projects/SFB261.

References

Andersen N, Müller PJ, Kirst G, Schneider RR (1999) Alkenone $\delta^{13}C$ as a proxy for past PCO_2 in surface waters: Results from the Late Quaternary Angola current. In: Fischer G, Wefer G (eds) Use of Proxies in Paleoceanography: Examples from the South Atlantic. Springer, Berlin, pp 469-488

Altabet MA, Pilskahn C, Thunell R, Pride C, Sigman D, Chavez F, Francois R (1999) The nitrogen isotope biogeochemistry of sinking particles from the margin of the Eastern North Pacific. Deep-Sea Res 46: 655-679

Arthur MA, Dean WE, Claypool GE (1985) Anomalous C-13 enrichment in modern marine organic carbon. Nature 315: 216-218

Bentaleb I, Fontugne M, Descolas-Gros C, Girardin C, Mariotti A, Pierre C, Brunet C, Poisson A (1996) Organic carbon isotopic composition of phytoplankton and sea-surface pCO_2 reconstructions in the Southern Indian Ocean during the last 50,000 yr. Organic Geochem 24: 399-410

Bentaleb I, Fontugne M, Descolas-Gros C, Girardin C, Mariotti A, Pierre C, Brunet C, Poisson A (1998) Carbon isotopic fractionation by plankton in the Southern Indian Ocean: Relationship between $\delta^{13}C$ of particulate organic carbon and dissolved carbon dioxide. J Mar Sys 17: 39-58

Benthien A, Müller PJ (2000) Anomalously low alkenone temperatures caused by lateral particle and sediment transport in the Malvinas Current region, western

Argentine Basin. Deep-Sea Res 47: 2369-2393

Benthien A, Andersen N, Schulte S, Müller PJ, Schneider RR, Wefer G (2002) Carbon isotopic composition of the $C_{37:2}$ alkenone in core-top sediments of the South Atlantic Ocean: Effects of CO_2 and nutrient concentrations. Glob Biogeochem Cycl 16: 1012: doi 10.1029/2001GB001433

Bidigare RR, Flügge A, Freeman KH, Hanson KL, Hayes JM, Hollander D, Jasper JP, King LL, Laws EA, Milder J, Millero FJ, Pancost R, Popp BN, Steinberg PA, Wakeham SG (1997) Consistent fractionation of ^{13}C in nature and in the laboratory: Growth-rate effects in some haptophyte algae. Glob Biogeochem Cycl 11: 279-292

Bidigare RR, Flügge A, Freeman KH, Hanson KL, Hayes JM, Hollander D, Jasper JP, King LL, Laws EA, Milder J, Millero FJ, Pancost R, Popp BN, Steinberg PA, Wakeham SG (1999) Correction to "Consistent fractionation of ^{13}C in nature and in the laboratory: Growth-rate effects in some haptophyte algae". Glob Biogeochem Cycl 13: 251-252

Bijma J, Spero HJ, Lea DW (1999) Reassessing foraminiferal stable isotope geochemistry: Impact of the oceanic carbonate system (experimental results). In: Fischer G, Wefer G (eds) Use of Proxies in Paleocea-nography: Examples from the South Atlantic. Springer, Berlin, pp 489-512

Burkhardt S, Riebesell U, Zondervan I (1999) Effects of growth rate, CO_2 concentration, and cell size on the stable carbon isotope fractionation in marine phytoplankton. Geochim Cosmochim Acta 63: 3729-3741

Burns BD, Beardall J (1987) Utilization of inorganic carbon by marine microalgae. J Experiment Mar Biol Ecol 107: 75-86

Conkright ME, Levitus S, Boyer TP (1994) World Ocean Atlas 1994 Vol. 1: Nutrients, NOOA Atlas NESDIS 1. US Government Printing Office

Conte MH, Thompson A, Eglinton G, Green JC (1995) Lipid biomarker diversity in the coccolithophorid *Emiliania huxleyi* (Prymnesiophycae) and the related species *Gephyrocapsa oceanica*. J Phycol 31: 272-282

Crowley TJ (2000) Causes of climate change over the past 1000 years. Science 289: 270-277

Curry WB, Crowley TJ (1987) The $\delta^{13}C$ of equatorial Atlantic surface waters: Implications for ice age pCO_2 levels. Paleoceanography 2: 489-517

Davies JC (1986) Statistics and Data Analysis in Geology. J Wiley & Sons, New York, 646 p

Degens ET, Behrendt M, Gotthardt B, Reppmann E (1968) Metabolic fractionation of carbon isotopes in marine plankton – II. Data on samples collected off the coasts of Peru and Ecuador. Deep-Sea Res 15: 11-20

Eek MK, Whiticar MJ, Bishop JKB, Wong CS (1999) Influence of nutrients on carbon isotope fractionation by natural populations of *Prymnesiophyte algae* in NE Pacific. Deep-Sea Res II 46: 2863-2876

Fairbanks RG, Sverdlove M, Free R, Wiebe PH, Bé AWE (1982) Vertical distribution and isotopic fractionation of living planktonic foraminifera from the Panama basin. Nature 298: 841-844

Feely RA, Sabine CL, Takahashi T, Wannikhof R (2001) Uptake and storage of carbon dioxide in the ocean: The global CO_2 survey. Oceanography 14: 18-32

Fischer G, Schneider R, Müller PJ, Wefer G (1997) Anthropogenic CO_2 in Southern Ocean surface waters: Evidence from stable organic carbon isotopes. Terra Nova 9: 153-157

Fischer G, Müller PJ, Wefer G (1998) Latitudinal $\delta^{13}C_{org}$ variations in sinking matter and sediments from the South Atlantic: Effects of anthropogenic CO_2 and implications for paleo-PCO_2 reconstructions. J Mar Sys 17: 471-495

Fontugne MR, Calvert SE (1992) Late Pleistocene variability of the carbon isotopic composition of organic matter in the eastern Mediterranean: Monitor of changes in carbon sources and atmospheric CO_2 concentrations. Paleoceanography 7: 1-20

Francois R, Altabet MA, Goericke R, McCorkle DC, Brunet C, Poisson A (1993) Changes in the $\delta^{13}C$ of surface water particulate organic matter across the subtropical convergence in the SW Indian Ocean. Glob Biogeochem Cycl 7: 627-644

Freeman K, Hayes JM (1992) Fractionation of carbon isotopes by phytoplankton and estimates of ancient CO_2 levels. Glob Biogeochem Cycl 6: 185-198

Fry B, Wainright SC (1991) Diatom sources of ^{13}C-rich carbon in marine food webs. Mar Ecol Prog Ser 76: 149-157

Gervais F, Riebesell U (2001) Effect of phosphorus limitation on elemental composition and stable carbon isotope fractionation in a marine diatom growing under different CO_2 concentrations. Limnol Oceanogr 46: 497-504

Gonzáles EL, Riebesell U, Hayes JM, Laws EA (2001) Effects of biosynthesis and physiology on relative abundances and isotopic compositions of alkenones. Geochem Geophys Geosyst 2: doi 2000GC 000052

Groupe CITHER 1 (1994) Recueil des données, campagne CITHER 1, NO L'Atalante (2 janvier-19 mars 1993). Volume 1 : Mesures «en route» - Cou-ranto-métrie ADCO et Pegasus, Documents Scientifiques, OP 14, 161 pp, Centre ORSTOM, Cayenne

Jasper JP, Hayes JM (1990) A carbon isotope record of CO_2 levels during the late Quaternary. Nature 347: 462-464

Jasper JP, Hayes JM, Mix AC, Prahl FG (1994) Photosynthetic fractionation of ^{13}C and concentrations of dissolved CO_2 in the central equatorial Pacific during the last 255,000 years. Paleoceanography 9: 781-798

Johnson KM, Wallace DWR, Wilke RJ, Goyet C (1995) Carbon dioxide, hydrographic, and chemical data obtained during RV *Meteor* cruise 15/3 in the South Atlantic Ocean (WOCE Section A9, Feburary-March 1991), ORNL/CDIAC-82, NDP-051, Carbon Dioxide Information Analysis Centre, Oak Ridge National Laboratory, Oak Ridge, Tennessee

Johnson KM, Schneider B, Mintrop L, Wallace DWR, Kozyr A (1998) Carbon dioxide, hydrographic, and chemical data obtained during RV *Meteor* cruise 22/5 in the South Atlantic Ocean (WOCE Section A10, December 1992, January 1993), ORNL/CDIAC-113, NDP-066, Carbon Dioxide Information Analysis Centre, Oak Ridge National Laboratory, Oak Ridge, Tennessee

Jouzel J, Barkov NI, Barnola JM, Bender M, Chappelaz J, Genthon C, Kotlyakov VM, Lipenkov V, Lorius C, Petit JR, Raynaud D, Raisbeck G, Ritz C, Sowers T, Stievenard M, Yiou F, Yiou P (1993) Extending the Vostok ice-core record of paleoclimate to the penultimate glacial record. Nature 364: 407-412

Keller K, Morel FMM (1999) A model of carbon isotopic fractionation and active carbon uptake in phyto-plankton. Mar Ecol Prog Ser 182: 295-298

Kienast M, Calvert SE, Pelejero C, Grimalt JO (2001) A critical review of marine sedimentary $\delta^{13}C_{org}$-pCO_2 estimates: New paleorecords from the South China Sea and a revisit of lower-latitude $\delta^{13}C_{org}$-pCO_2 records. Glob Biogeochem Cycl 15: 113-127

Kroopnick PM (1985) The distribution of ^{13}C of CO_2 in the world oceans. Deep-Sea Res 32: 57-84

Laws EA, Thompson PA, Popp BN, Bidigare RR (1997) Effect of growth rate and CO_2 concentration on carbon isotopic fractionation by the marine diatom Phaeodactylum tricornutum. Limnol Oceanogr 42: 1552-1560

Laws EA, Popp BN, Bidigare RR, Riebesell U, Burkhardt S, Wakeham SG (2001) Controls on the molecular distribution and carbon isotopic composition of alkenones in certain haptophyte algae. Geochem Geophys Geosyst 2: doi 2000GC000057

Lee K, Millero FJ, Wanninkof R (1997) The carbon dioxide system in the Atlantic Ocean. J Geophys Res 102 (C7): 15693-15707

Longhurst A (1993) Seasonal cooling and blooming in tropical oceans. Deep-Sea Res 40: 2145-2165

Meyers PA (1997) Organic geochemical proxies of paleo-oceanographic, paleolimnologic, and paleoclimatic processes. Organic Geochem 27: 213-250

Mook WG, Bommerson JC, Staverman WH (1974) Carbon isotope fractionation between dissolved bicarbonate and gaseous dioxide. Earth Planet Sci Lett 2: 169-176

Morel A, Berthon JF (1989) Surface pigments, algal biomass profiles, and potential production of the euphotic layer: Relationships reinvestigated in view of remote-sensing applications. Limnol Oceanogr 34: 1545-1562

Müller PJ, Schneider RR, Ruhland G (1994) Late Quaternary PCO_2 variations in the Angola current: evidence from organic carbon $\delta^{13}C$ and alkenone temperatures. In: Zahn R, Pedersen TF, Kaminski MA, Labeyrie L (eds) Carbon Cycling in the Glacial Ocean: Constrains on the Ocean's Role in Global Change. NATO ASI Series I 17, Springer, Berlin, pp 343-366

Müller PJ, Kirst G, Ruhland G, von Stroch I, Rosell-Melé A (1998) Calibration of the alkenone paleotemperature index based on core tops from the eastern South Atlantic and the global ocean (60°N-60°S). Geochim Cosmochim Acta 62: 1757-1772

Müller PJ, Fischer G (2001) A 4-year sediment trap record of alkenones from the filamentous upwelling region off Cape Blanc, NW Africa and a comparison with distributions in underlying sediments. Deep-Sea Res 48: 1877-1903

Pagani M, Arthur MA, Freeman KH (1999) Miocene evolution of atmospheric carbon dioxide. Paleoceanography 14: 273-292

Partridge TC, Scott L, Hamilton JE (1999) Synthetic reconstructions of southern African environments during the Last Glacial Maximum (21-18 kyr) and the Holocene altithermal (8-6 kyr). Quat Int 57/58: 207-214

Petit JR, Jouzel J, Raynaud D, Barkov NI, Barnola J-M, Basile I, Bender M, Chappellaz J, Davis M, Delaygue G, Delmotte M, Kotlyakov VM, Legrand M, Lipenkov VY, Lorius C, Pépin L, Ritz C, Saltzmann, Stievenard M (1999) Climate and atmospheric history of the past 420,000 years from the Vostok ice core, Antarctica. Nature 399: 429-436

Popp BN, Takigiku R, Hayes JM, Louda JW, Baker EW (1989) The post-Paleozoic chronology and mechanism of ^{13}C depletion in primary marine organic matter. J Ameri Sci 289: 436-454

Popp BN, Laws EA, Bidigare RR, Dore JE, Hanson KL, Wakeham SG (1998) Effect of phytoplankton cell geometry on carbon isotopic fractionation. Geochim Cosmochim Acta 62: 69-77

Popp BN, Hanson KL, Dore JE, Bidigare RR, Laws EA, Wakeham SG (1999) Controls on the carbon isotopic composition of phytoplankton. In: Abrantes F, Mix AC (eds) Reconstructing Ocean History: A Window into the Future. Kluwer Academics/ Plenum Publishers, New York, pp 381-398

Prahl FG, Muehlhausen LA, Zahnle DL (1988) Further evaluation of long-chain alkenones as indicators of paleoceanographic conditions. Geochim Cosmochim Acta 52: 2303-2310

Rau GH, Takahashi T, DesMarais DJ (1989) Latitudinal variations in plankton $\delta^{13}C$: implications for CO_2 and productivity in past oceans. Nature 341: 516-518

Rau GH, Takahashi T, DesMarais DJ, Repeta DJ, Martin JH (1992) The relationship between the $\delta^{13}C$ of organic matter and c_e in ocean surface water: Data from a JGOFS site in the northeast Atlantic Ocean and a model. Geochim Cosmochim Acta 56: 1413-1419

Rau GH (1994) Variations in sedimentary organic carbon $\delta^{13}C$ as a proxy for past changes in ocean and atmospheric CO_2 concentrations. In: Zahn R, Pedersen TF, Kaminski MA, Labeyrie L (eds) Carbon Cycling in the Glacial Ocean: Constrains on the Ocean's Role in Global Change. NATO ASI Series I 17, Springer, Berlin, pp 307-321

Rau GH, Riebesell U, Wolf-Gladrow D (1996) A model of photosynthetic ^{13}C fractionation by marine phytoplankton based on diffusive molecular CO_2 uptake. Mar Ecol Prog Ser 133: 275-285

Rickaby REM, Elderfield H (1999) Planktonic Cd/Ca: Paleonutrients or Paleotemperature? Paleoceanography 14: 293-303

Rickaby REM, Schrag DP, Zondervan I, Riebesell U (2002) Growth rate dependence of Sr incorporation during calcification of *Emiliania huyleyi*. Glob Biogeochem Cycl 16: doi 10.1029/2001GB001408

Riebesell U, Burkhardt S, Dauelsberg A, Kroon B (2000a) Carbon isotope fractionation by a marine diatom: Dependence on the growth rate limiting resource. Mar Ecol Prog Ser 193: 295-303

Riebesell U, Revill AT, Holdsworth DG, Volkman JK (2000b) The effects of varying CO_2 concentration on lipid composition and carbon isotope fractionation in *Emiliania huxleyi*. Geochim Cosmochim Acta 64: 4179-4192

Riegman R, Stolte W, Noordeloos AAM, Slezak D (2000) Nutrient uptake and alkaline phosphatase (EC 3:1:3:1) activity of *Emiliania huxleyi* (Prymnesiophyceae) during growth under N and P limitation in continuous cultures. J Phycol 36: 87-96

Rost B, Zondervan I, Riebesell U (2003) Light-dependent carbon isotope fractionation in the coccolithophorid *Emiliania huxleyi*. Limnol Oceanogr, in press

Schneider RR, Müller PJ, Ruhland G (1995) Late Quaternary surface circulation in the east equatorial south Atlantic: Evidence from alkenone sea-surface temperatures. Paleoceanography 10: 197-219

Schneider RR, Müller PJ, Meinecke G, Schmidt H, Wefer G (1996) Late Quaternary surface temperatures and productivity in the east-equatorial South Atlantic: Response to changes in trade/ monsoon wind forcing and surface water advection. In: Wefer G, Berger WH, Siedler G, Webb DJ (eds) The South Atlantic: Present and Past Circulation. Springer, Berlin, pp 527-551

Spero H, Bijma J, Lea DW, Bemis BE (1997) Effect of seawater carbonate concentration on foraminiferal carbon and oxygen isotopes. Nature 390: 497-500

Stoll HM, Schrag DP (2000) Coccolith Sr/Ca as a new indicator of coccolithophorid calcification and growth rate. Geochem Geophys Geosyst 1: doi 1999 GC000015

Stoll HM, Rosenthal Y, Falkowski P (2002) Climate proxies from Sr/Ca of coccolith calcite: Calibrations from continuous culture of *Emiliania huxleyi*. Geochim Cosmochim Acta 66: 927-936

Stouffer RJ, Manabe S, Vinnikov KY (1994) Model assessment of the role of natural variability in recent global warming. Nature 367: 634-636

Takahashi T, Tans PP, Fung I (1992) Balancing the budget – carbon dioxide sources and sinks, and effects of the industry. Oceanus 35: 18-28

Tans PP, Fung IY, Takahashi T (1990) Observational constrains on the global atmospheric CO_2 budget. Science 247: 1431-1438

Volkman JK, Eglinton G, Corner EDS, Sargent JR (1989) Novel unsaturated straight-chain methyl and cthyl kctones in marine sediments and a coccolthophore *Emiliania huxleyi*. In: Douglas AG and Maxwell JR (eds) Advances in Organic Geochemistry. Pergamon, New York, pp 219-227

Volkman JK, Barrett SM, Blackburn SI, Sikes EL (1995) Alkenones in Gephyrocapsa oceanica: Implications for studies of paleoclimate. Geochim Cosmochim Acta 59: 513-520

Weiss RF, van Woy FA, Salameh PK (1992) Surface water and atmospheric carbon dioxide and nitrous oxide observations by shipboard automated gas chromatography: Results from expeditions between 1977 and 1999. Scripps Institution of Oceanography Reference, SIO 92-11, 124pp, University of California, San Diego

Late Quaternary Terrigenous Sedimentation in the Western Equatorial Atlantic South American versus African Provenance Discriminated by Magnetic Mineral Analysis

U. Bleil* and T. von Dobeneck

Universität Bremen, Fachbereich Geowissenschaften, Postfach 33 04 40,
D-28334 Bremen, Germany
* corresponding author (e-mail): bleil@geomarin.uni-bremen.de

Abstract: Magnetic mineral accumulation at the Ceará Rise has been studied with the aim to discriminate and reconstruct fluvial South American and eolian African terrigenous fluxes to the late Quaternary western Equatorial Atlantic. Seven sediment series recovered along two bathymetric transects were investigated with standard environmental magnetic techniques. Climatically controlled fluctuations in continental detrital discharge and marine biogenic carbonate fluxes strongly modulate the susceptibility records. Their coherent precessional and higher-frequent signal components could be used to establish a high-resolution age framework for these sediments. According to a partial susceptibility analysis, on average 79 % of the susceptibility signal originates from magnetite of different grain size, 13 % from hematite and 8 % from paramagnetic matrix compounds. In terms of absolute concentrations this implies that hematite is almost twenty times more abundant than magnetite, because of its orders of magnitude lower intrinsic susceptibility. The longitudinal gradients of their respective accumulation rates document a delivery from two major sources characterized by largely different magnetite to hematite ratios (about 1:12 versus 1:50). A mixing model of this scenario provided detailed insight into the past variability of the separate magnetic mineral fluxes and their most probable provenance. Overall about 56 % of hematite and 84 % of magnetite were transported in the Amazon fluvial load. Their accumulation is closely related to sea level changes, reaching highest (lowest) rates, when most South American shelf areas fell dry (were flooded) before and after Termination I and II. Hematite and magnetite of African provenance, 44 and 16 %, respectively, follow a distinctly different accumulation pattern with prominent maxima during cold intervals of glacial periods. By statistically linking these trace minerals to total lithogenic fluxes, we find that during the last 200 kyr, on average 79 % of total terrigenous material in the Ceará Rise area originates from South American sources in the Amazon River catchment, while African dust sources contributed 21 %.

Introduction

The delivery of terrigenous materials to the world oceans still remains inadequately constrained in various respect, in spite of substantial research efforts in recent decades. Quantifying the composition and fluxes of lithogenic continental weathering products in the marine realm is a mandatory prerequisite to identifying their sources, transport media and pathways as well as the evolution of climate, vegetation and tectonic regimes on land.

An attractive target region for sediment source tracking is the Ceará Rise, a submarine elevation in the western Equatorial Atlantic located some 900 km northwest off the Amazon River mouth. This supralysoclinal plateau situated in the trajec-

From WEFER G, MULITZA S, RATMEYER V (eds), 2003, *The South Atlantic in the Late Quaternary: Reconstruction of Material Budgets and Current Systems.* Springer-Verlag Berlin Heidelberg New York Tokyo, pp 213-236

tories of the distal Amazon discharge and Sahara dust plumes is covered by clay-rich, calcareous sediments. The discrete terrigenous contributions of western and eastern provenance have not yet been precisely quantified, in particular with respect to their variability over the late Quaternary glacial/interglacial cycles.

A general outline of the terrigenous sedimentation on the Ceará Rise since late Paleocene has been published by Dobson et al. (1997, 2001). On the basis of chemical characteristics they could identify three different South American sources that successively dominated through Tertiary time. The presently active pattern prevailed during the last about 10 Myr, when a ten-fold increase of terrigenous accumulation rates took place which is attributed to the Andean uplift and a related intensified Amazon River flow. Due to their limited temporal resolution, the data sets do not give any conclusive information about late Quaternary variations in terrigenous inflow. Moreover, the authors consider eolian dust deposits as negligible if present at all in the Ceará Rise region.

Analyzing late Quaternary sediment sequences, Zabel et al. (1999) more recently came to a similar conclusion. Using Al:Ti ratios to differentiate between terrigenous material supplied by the Amazon River discharge and the Sahara dust plume, a very predominant South American influence was traced from the Ceará Rise to the western flanks of the mid-Atlantic Ridge at about 5°N, conforming with the regional distribution of clay mineral compositions in the surface sediments (Petschick et al. 1996). However, obvious problems remain particularly regarding a reliable South American Al:Ti source signature and possible modifications of the Al:Ti ratios during transport. A notably different interpretation of the data, namely a non-negligible eolian influx of African terrigenous material to the western equatorial Atlantic, cannot be excluded, therefore. Rühlemann et al. (1996) also reported that during the late Quaternary eolian deposits have been of very minor importance at the Ceará Rise. On the other hand, a significant present-day long-range westward transport of terrigenous material from the African deserts is well known to reach the Ameri-cas in the mid-tropospheric Saharan Dust Layer (Prospero et al. 1981; Swap et al. 1992).

In this study environmental magnetism methods have been applied for a more detailed understanding of the terrigenous sedimentation in the western equatorial Atlantic. The regional setting of the Ceará Rise is particularly appropriate for this approach as the sediment magnetic mineral inventories are very predominantly of continental origin (Petermann and Bleil 1993; Petermann 1994).

Investigations of magnetic signals in marine sedimentary deposits have been notably advanced in recent years, focusing on a variety of different applications. As other physical, chemical and mineralogical sediment properties, magnetic attributes typically reflect a succession of interdependent, mostly climate-driven processes. To decipher and understand this network, specifically its timing and rates of change, an accurate age control is indispensable. The now popular use of the magnetic susceptibility (among other bulk physical parameters) as a cyclostratigraphic proxy has allowed to greatly refine and extend astronomically tuned time scales even beyond the Neogene (for a comprehensive review see Langereis and Dekkers 1999 and references therein). This technique was also very successful here, using available oxygen iso-tope data as initial dating scheme (von Dobeneck and Schmieder 1999). Relative paleointensity, which is attracting broad interest for regional to global correlations and high-resolution chronostratigraphies (e.g. Channell et al. 2000; Laj et al. 2000), yielded disappointing results, though. The Ceará Rise sedimentary records entirely failed to match widely accepted target patterns, most likely due to specific lock-in problems (Bleil and von Dobeneck 1999).

Prime objectives in analyzing the Ceará Rise sediment series by environmental magnetic methods was to (1) isolate the contributions of terrigenous magnetic mineral components to the susceptibility signal from modulations by diluting biogenic carbonate, to (2) determine regional characteristics of the magnetic mineral assemblages, in the first place variations of concentration, composition and grain size spectra in time and space that could yield diagnostic hints to (3) infer their provenance.

To quantify the highly variable accumulation of the tracer minerals magnetite and hematite the 'partial susceptibilities' concept (von Dobeneck

1998; Frederichs et al. 1999) was applied to isothermal remanent magnetization, anhysteretic remanent magnetization and frequency dependent susceptibility measurements supplemented by avail-able data sets of density, carbonate and iron con-tents. On these grounds, a two-source model could be evaluated outlining magnetic mineral and total terrigenous material fluxes from South America and Africa to the western equatorial Atlantic during the late Quaternary.

Materials and Methods

Ceará Rise Sediments

The Ceará Rise volcanic edifice, now located about 900 km to the northeast of the Amazon Delta (Fig. 1), formed some 80 Ma ago at the mid-Atlantic Ridge. Several NW to SE aligned platforms reach up to approximately 3200 m water depth. To the southwest they are bordered by steep, to the northeast by relatively gentle slopes. The Ceará Rise is draped with a thick sequence (> 1000 m) of lithogenic and biogenic sediments (Shackleton et al. 1997). Their uppermost late Quaternary layers were probed with a gravity corer on two bathymetric transects across the northeastern flanks during R/V *Meteor* Cruise M 16/2 (Schulz et al. 1991). Water depths at the coring sites varied between about 3100 and 4200 m, core lengths from about 660 to 820 cm (Table 1).

The sediments consist of alternating predominantly calcareous and terrigenous layers. Biogenic carbonate content, in variable amounts composed of coccolithophorid and foraminiferal tests, ranges from about 20 to 80 % documenting changes in paleoproductivity with carbonate maxima occurring during warm substages of glacials and

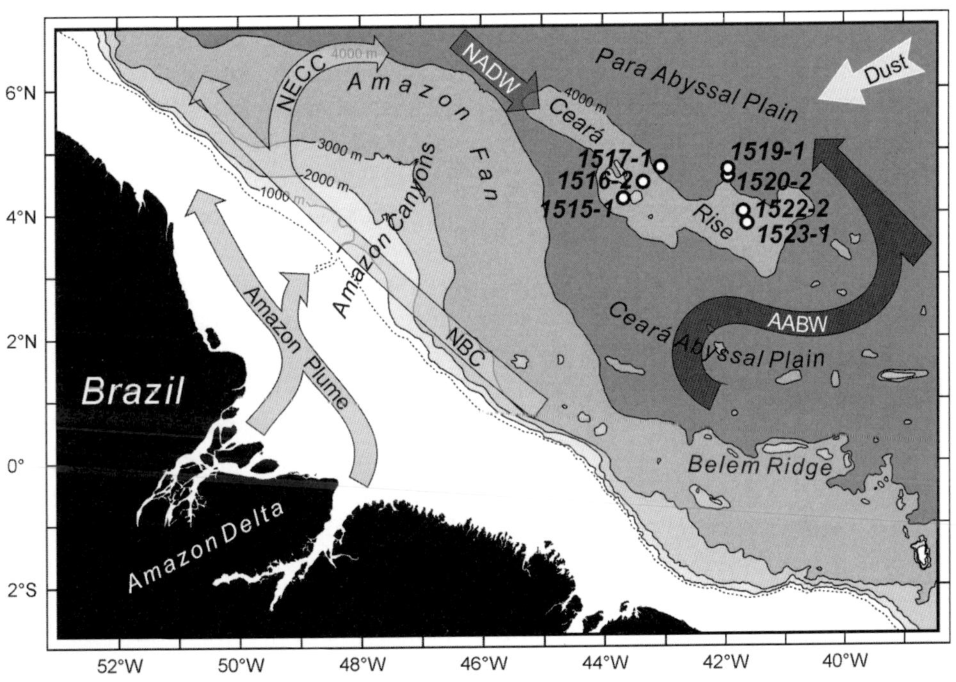

Fig. 1. Locations of gravity core transects on the northeastern flanks of the Ceará Rise and present-day regional system of surface (light grey) and deep (dark grey) currents in the study area. The retroflection of the North Brazil Current (NBC) into the North Equatorial Counter Current (NECC) persists from June to January (Muller-Karger 1988). North Atlantic Deep Water (NADW) flows directly over the Ceará Rise, while Antarctic Bottom Water (AABW) encompasses the structural high. Trajectories of the African dust plume reach the Ceará Rise region during the boreal winter season (e.g. Prospero et al. 1981).

Coring Transect	Core	Latitude [N]	Longitude [W]	Water Depth [m]	Core Length [cm]
NW	GeoB 1515-1	04°14.3'	043°40.0'	3119	658
	GeoB 1516-2	04°29.9'	043°20.2'	3582	692
	GeoB 1517-1	04°44.2'	042°02.8'	4001	686
SE	GeoB 1523-1	03°49.9'	041°37.3'	3292	728
	GeoB 1522-2	04°01.5'	041°41.0'	3481	780
	GeoB 1520-2	04°35.4'	041°56.0'	3915	817
	GeoB 1519-1	04°42.3'	041°56.0'	4196	813

Table 1. Ceará Rise gravity cores analyzed in this study.

interglacials (Rühlemann et al. 1996). Geochemical data hint at oxic to slightly suboxic conditions throughout the sediment columns and suggest that a similar environ-ment prevailed over the entire time interval of their deposition (Schulz pers. comm.). This view is sup-ported by magnetic hysteresis measurements at 2 cm intervals at the shallowest core location of the eastern transect, GeoB 1523, that revealed on average slightly coarser magnetic minerals below around 3.5 m, possibly reflecting some minor diage-netic corrosion. However, there is no apparent evidence for a chemically or biologically mediated precipitation of magnetic minerals at depth (Bleil and von Dobeneck 1999). A limited influx of reactive organic matter obviously sustained only a moderate diagenesis in the late Quaternary Ceará Rise sediments which did not substantially affect their magnetic mineral assemblage.

Analytical Procedures

Magnetic susceptibility logs were measured with a *Bartington M.S.2.F* spot reading probe operating the *M.S.2* meter in high precision mode. An experi-mentally determined calibration factor of 16.1 was applied to scale volume susceptibilities in 10^{-6} SI. Frequent base line readings were taken and linearly interpolated to account for generally small instru-mental drifts. For measurements at a spacing of 1 cm the sensor was directly lowered onto the split surface of the archive core halves. The sediments were covered with a thin transparent foil to prevent contamination.

For detailed magnetic measurements working halves of the cores were sampled with 6.2 cm^3 plastic boxes at on average 5 cm intervals. After a complete paleomagnetic analysis (Bleil and von Dobeneck 1999) various mineral magnetic parame-ters have been determined of which the following are the basis of this study:
- An anhysteretic remanent magnetization σ_{ar} acquired in alternating fields decaying from 0.1 T peak amplitude and a 40 µT bias direct field was measured on an automated *2G Enterprises 755R* DC SQUID magnetometer with built-in AC and DC coil systems.
- Isothermal remanent magnetizations σ_{ir} of 0.1, 0.3 and 2.5 T were measured on the same instrument using its built-in pulse magnetizer up to 0.3 T and a separate *2G Enterprises* pulse magnetizer for higher fields.
- Frequency dependent susceptibility χ_{fd} was determined with a *Bartington M.S.2* meter operating the *M.S.2B* single sample sensor in the high precision mode at 470 and 4700 Hz alternating fields.

Dry bulk densities were determined by weighting wet and freeze dried sediment sampled at 5 cm intervals. For core GeoB 1522-1 no density data are available and only volume susceptibility has been included in the chronostratigraphic analysis of the core collection. Inorganic carbon content of cores GeoB 1515-1, 1520-2 and 1523-1 was meas-ured at 5 cm spacing by the standard coulometric technique. For the same cores total iron concentra-tions were determined at 10 cm

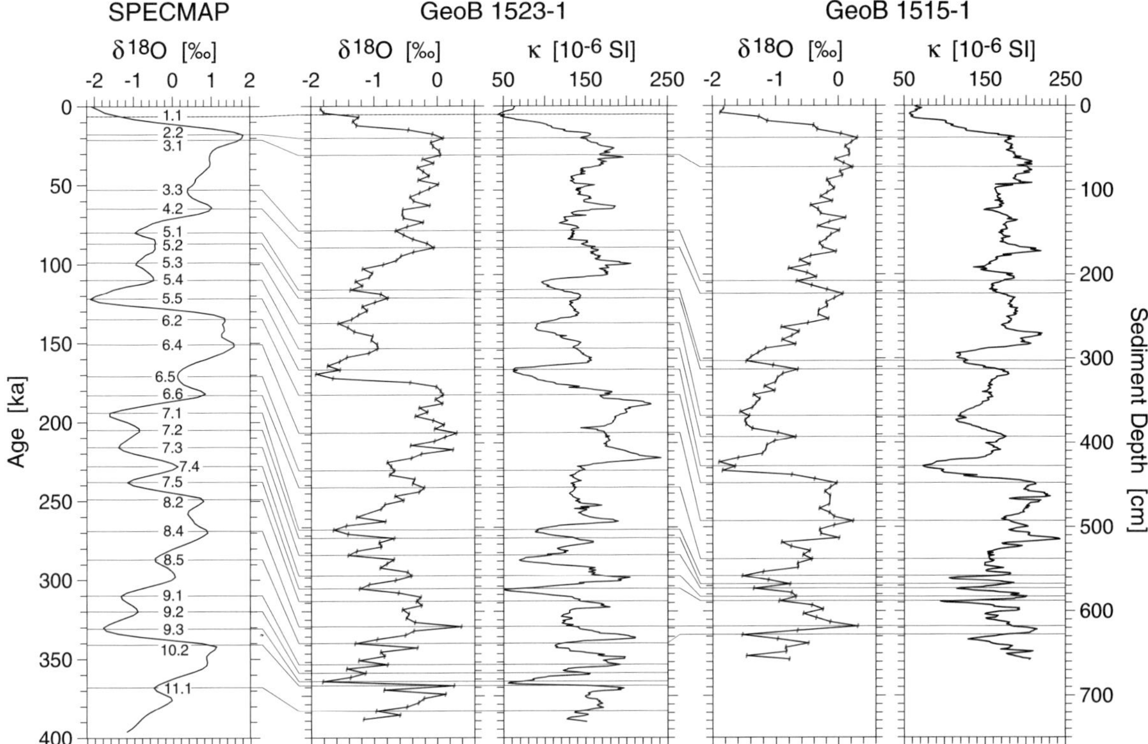

Fig. 2. Initial age model of the Ceará Rise sediment series. *Globigerinoides sacculifer* $\delta^{18}O$ records of cores GeoB 1523-1 (Mulitza 1994) and GeoB 1515-1 (Rühlemann et al. 1996) were graphically correlated to the SPECMAP oxygen isotope stack of Imbrie et al. (1984) providing 27 (19) age tie points for the respective susceptibility logs. Note incoherent leads and lags of κ to $\delta^{18}O$ in several cold substages.

intervals. The procedure described by Kotz et al. (1972) was used for pressure digestion of the samples and inducti-vely coupled plasma atomic emission spectrometry (ICP-AES) applied to quantify the iron contents. All the data sets of this study are stored in the *PANGAEA* database (www.pangaea.de/projects/SFB261).

Chronostratigraphy

Oxygen Isotopes

For the two shallowest sites of both coring transects (GeoB 1515 and 1523) stable oxygen isotope data of planktic foraminiferal species *Globigerinoides sacculifer* are available at 5 cm intervals (Mulitza 1994; Rühlemann et al. 1996). Following Prell et al. (1986), these authors developed an age scheme by graphic correlation

of characteristic isotopic events in the $\delta^{18}O$ records (19 and 27 tie points, respectively) to the SPECMAP stack of Imbrie et al. (1984) which is used as a reference frame in this study (Fig. 2).

Magnetic Susceptibility

The striking similarities of $\delta^{18}O$ and susceptibility records and their obvious agreement with the SPECMAP standard (Fig. 2) confirm the climatic dependence of both parameters during the late Quaternary glacial/interglacial cycles. In view of the depositional regime at the Ceará Rise it is interesting to note that the synchrony of oxygen isotope and susceptibility signals is distorted by variable leads and lags or sections of obvious pattern dissimilarities. These effects are too large to be simply explained by the higher spatial resolution of the magnetic (1 cm) versus the isotopic (5 cm) data sets. Both $\delta^{18}O$ spectra exhibit well defined and similar, albeit not identical orbital characteristics. The susceptibility spectrum has considerably more precessional power as well as some additional peaks. A detailed analysis (von Dobeneck and Schmieder 1999) revealed that the susceptibility record of core GeoB 1523-1 lags precessional forcing by 1.77 kyr and leads $\delta^{18}O$ by 3.90 kyr. On average, the delay between precession index (PI) and $\delta^{18}O$ adds up to a plausible mean of 5.67 kyr.

By graphic correlation of prominent features in the susceptibility records (48 tie points) the GeoB 1523-1 $\delta^{18}O$ age model was transferred to all other cores (Fig. 3a). Based on this chronostratigraphic scheme, precessional tuning is a straightforward, but typically repetitive procedure (Fig. 3b). Each step starts with filtering the time series in an extended precession band of 15 - 28 kyr to include compressed or stretched signal sections, proceeds with identification and correlation of clearly developed maxima and minima (33 tie points) to the shifted precession signal (lag 1.77 kyr) and ends with adopting the improved age model. Resulting age modifications were altogether small averaging 1.3 - 2.2 ka for the first and 0.4 - 0.9 ka for the second tuning step. While the initial correlation pattern is based on matching individual susceptibility peaks, the orbital tuning procedure refers to the maxima and minima of a filtered signal to which entire record sections contribute. Consequently, the tuned age model relies more on the continuous evolution than on singularities. On the other hand, narrow peaks and steep slopes have a strong impact on filter-extracted signals and therefore also their dating. The origin of these signal components is unknown *a priori*. In the present case convincing links to known high-frequency climate variations such as Heinrich events (Grousset et al. 1993; Robinson et al. 1995) and in particular Bond (Bond et al. 1993) or Daansgard-Oeschger cycles (Dans-gaard et al. 1993) could be established (von Doben-eck and Schmieder 1999).

Extracted and emphasized by highpass filtering in the sub-Milankovitch band (< 15 kyr), the high-frequency susceptibility patterns of the Ceará Rise sediment collection could be graphically correlated at an average spacing of about 3 kyr between 90 tie points (Fig. 3c). Each tie point was set to a common mean age assuming that relative delay times within the study area are negligible. The graphic correlation of high-frequency signal components results in mean age shifts of 0.8 - 1.5 ka. At this refinement stage, the age model is synchronous with the external precession signal to an estimated absolute precision of about 2 ka and internally coherent to relative age discrepancies of less than 1 ka between sites (von Dobeneck and Schmieder 1999). While our initial age model relies on the precision of the SPECMAP oxygen isotope stack, the final version exclusively depends on precessional tuning of the susceptibility signals and subsequent correlation of their high-frequency features extracted from the sub-Milankovitch band. Note that with the above accuracy estimates accu-mulation rates based on this age model are still of limited resolution in detail.

Susceptibility Records

The susceptibility signal at the Ceará Rise reflects highly variable deposition of both biogenic and lithogenic sediment components (Figs. 2, 4). The main factors controlling glacial/interglacial contrasts are shifts in carbonate production (Rühle-mann et al. 1996) and concurrent changes in terrigenous inflow involving several pathways (Fig. 1). Of major importance is the sea level fall in glacial periods causing erosion of the South American shelves (Richter et al. 1997), channeling of the Amazon discharge directly into the Amazon Canyons (Milliman et al. 1975; Damuth 1977) and intensification of mass flow events on the slopes of the Amazon Fan (Maslin and Mikkelsen 1997). These different processes enhance the suspended particle loads at different water depths. In recent times they reach the Ceará Rise either by surface transport via retroflection of the North Brazil Current into the North Equatorial Countercurrent (Johns et al. 1990) or by nepheloid transport via the southeastward directed NADW flow (Kumar and Embley 1977; Francois and Bacon 1991). Increased eolian input from the African continent (Sarnthein et al. 1981; Matthewson et al. 1995) and a larger sediment yield of the Amazon during glacial periods presumably further contribute to the susceptibility variations.

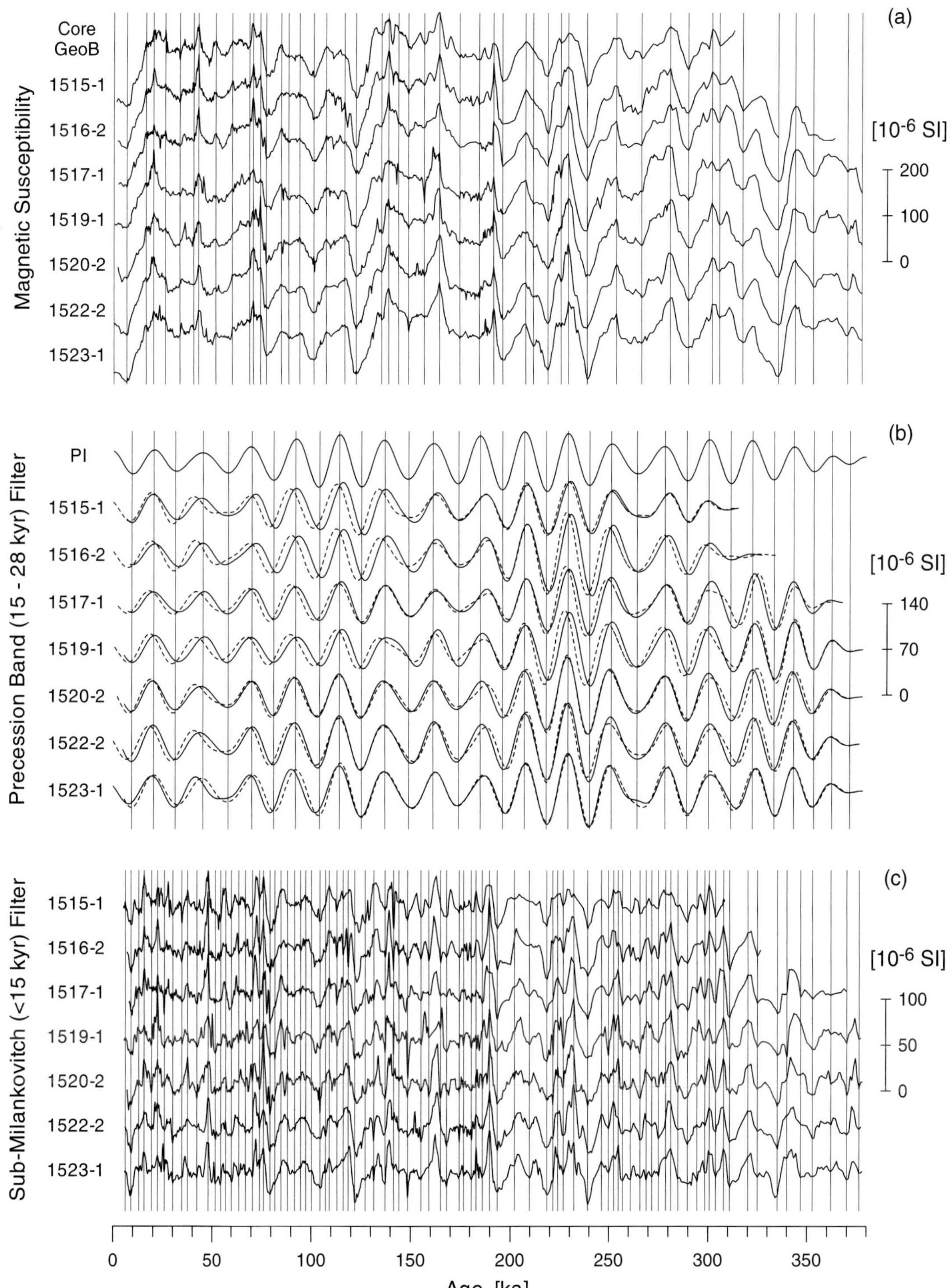

Fig. 3. Three step age model refinement of the Ceará Rise sediment series. **a)** Graphic correlation of susceptibility records to the $\delta^{18}O$ age model of core GeoB 1523-1 (Mulitza 1994, 48 tie points). **b)** Initial (dashed) and subsequent (solid) stage of synchronizing bandpass filtered susceptibility to the astronomical precession index (PI, lag 1.77 kyr, 33 tie points). **c)** Graphic correlation of sub-Milankovitch band (< 15 kyr) susceptibility characteristics to their common mean age (90 tie points).

The Marine Magnetic Signal

The magnetic mineral inventory in late Quaternary Ceará Rise sediments is very predominantly associated with the terrigenous sediment fraction, i.e. it does not involve significant amounts of authigenic bacterial magnetite (Petermann and Bleil 1993; Petermann 1994), nor has it been notably affected by diagenetic processes (Bleil and von Dobeneck 1999). The initial step to discern climatically control-led susceptibility changes resulting from fluctuating biogenic carbonate deposition versus variations in terrigenous influx is to eliminate potentially inter-fering effects of porosity. The mass-specific sus-ceptibility χ simply refers to dry bulk density ρ_{db} as pore water diamagnetism ($< - 0.9 \cdot 10^{-8}$ m^3/kg, Dunlop and Özdemir 1997) can be neglected. The κ (Fig. 4a) and χ (Fig. 4c) records correlate closely (Pearson's r = 0.95) and have essentially identical morphologies, because overall downcore increase and also individual pronounced short-term shifts in dry bulk density are limited to about 15 % (Fig. 4b). Carbonate contents, available for three cores, GeoB 1515-1, 1520-1 and 1523-1, range from less than 20 to over 75 % of dry mass (Fig. 4d). The carbonate-free mass-specific susceptibility χ_{cf} calculated on the basis of these data (Fig. 4e)

$$\chi = \chi_{Carb} \cdot \frac{\%CaCO_3}{100} + \chi_{cf} \cdot \left(1 - \frac{\%CaCO_3}{100}\right) \quad (1a)$$

$$\chi_{cf} = \left(\chi - \chi_{Carb} \cdot \frac{\%CaCO_3}{100}\right) \Big/ \left(1 - \frac{\%CaCO_3}{100}\right) \quad (1b)$$

primarily represents the susceptibility signal of the ferrimagnetic and antiferromagnetic mineral components, but also includes paramagnetic portions of the terrigenous sediment matrix. Contributions of carbonate diamagnetism ($\chi_{Carb} \approx - 0.5 \cdot 10^{-8}$ m^3/kg, Dunlop and Özdemir 1997) are of very minor importance. Hence, the actual '*marine magnetic sus-ceptibility signal*' is the factor $(1 - \%CaCO_3/100)$ in the second term of equation (1a) which is equi-valent to the ratio of terrigenous to total sediment accumulation. Its correlation to χ (r = 0.87) is distinctly different from that to χ_{cf} (r = 0.20),

apparently due to the presence of two independent controls. First and foremost, both the carbonate and the non-carbonate content vary by approximately a factor of 4 as compared to a factor of less than 2 for χ_{cf}. Moreover, the two time series have different frequency contents and are clearly not in phase for most of the last 380 kyr. Influx of materials from the continents apparently did not substantially influence the marine productivity in the Ceará Rise region. The χ and κ records thus essentially document the balance of a fluctuating marine carbonate flux, mirroring climatically mediated oceanic productivity and calcite preservation (Martin and Sayles 1996), with an also highly variable terrigenous flux. This flux balance strongly modulates the equally variable, but less indicative terrigenous susceptibility signal. Mainly for this reason, the volume susceptibility logs could be successfully used to develop a detailed age model for the sediment series starting from their correlation to the SPECMAP oxygen isotope stack as discussed above. At the Ceará Rise, close to the South American continent and the Amazon River mouth, an intense (Milliman and Meade 1983) and presumably highly fluctuating source of magnetic components, this is a rather unexpected result. It exemplifies, however, why in most mid- to low-latitude open-ocean settings magnetic susceptibility can be very effectively applied as a cyclostratigraphic parameter (e.g. Bloemendal and deMenocal 1989; Shackleton and Crowhurst 1997; Schmieder et al. 2000). As generally stated in such studies, susceptibility is indeed a measure of magnetic mineral concentration in the sediments, yet in such environments it primarily reflects past marine surface or deep water milieus and can hence be applied as an adequate proxy for carbonate productivity or preservation. On the other hand, the susceptibility records will provide little or no valuable information about provenance and characteristics of the magnetic mineral assemblage without additional, more elaborate magnetic analyses.

The Continental Magnetic Signal

For a more comprehensive assessment of the susceptibility data, the 'partial susceptibility analysis' (von Dobeneck 1998; Frederichs et al. 1999) has been applied. Equations (2) and (3) outline the basic

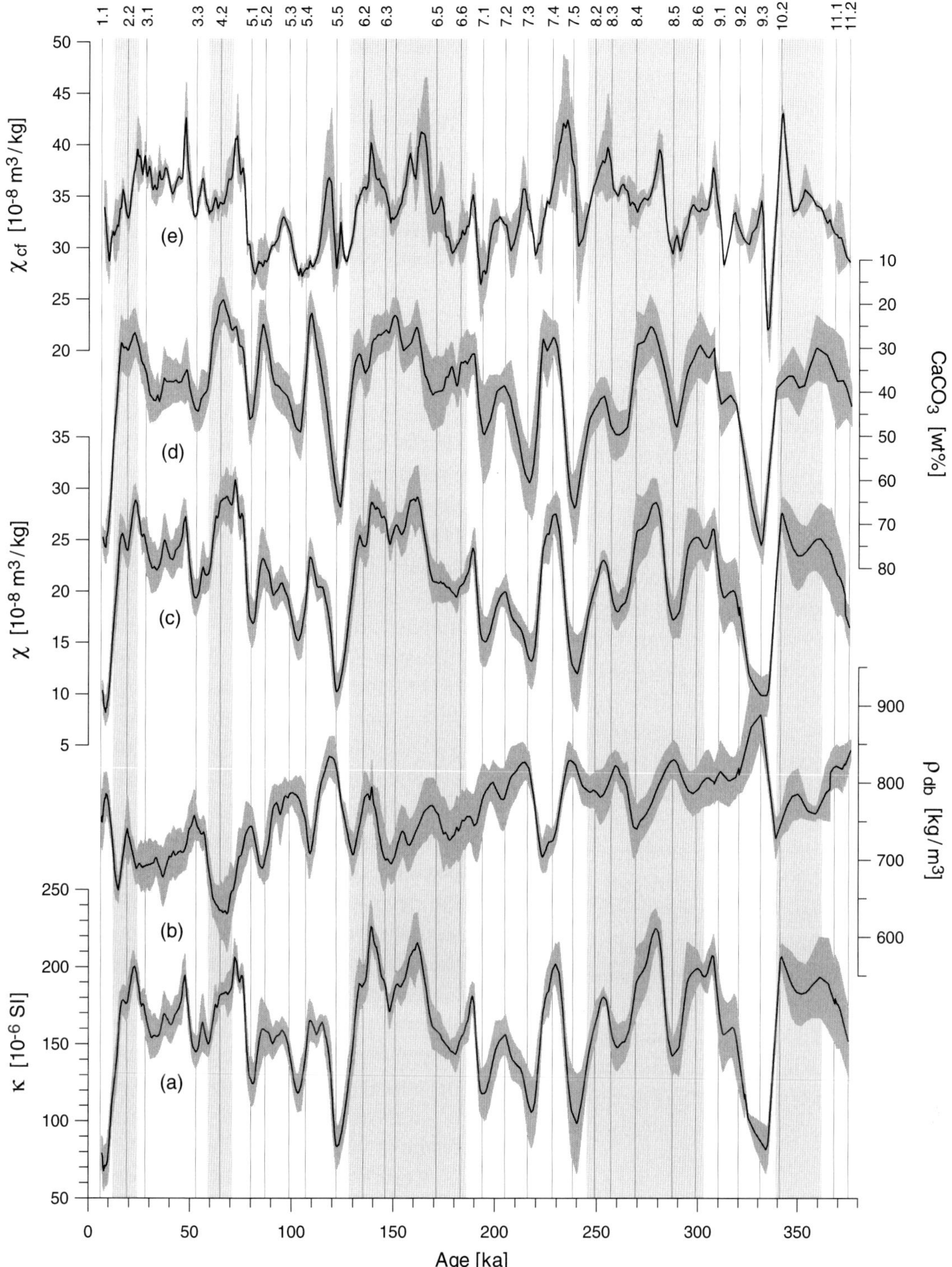

Fig. 4. Stacks of **a)** magnetic volume susceptibility κ, **b)** dry bulk density ρ_{db} and **c)** mass-specific susceptibility χ (6 cores each), **d)** carbonate content (note reversed axis) and **e)** carbonate-free mass-specific susceptibility χ_{cf} (3 cores each). Using individual chronostratigraphic models of the cores (Fig. 3), the data have been interpolated to a 0.2 ka interval age scheme for averaging. Grey shading denotes ± one standard deviation from the mean. Vertical light grey bars outline glacial periods, vertical lines oxygen isotope events as labeled on top (Imbrie et al. 1984).

concept. The susceptibility signal is assumed to be composed of contributions from a magnetically 'soft' ferrimagnetic magnetite components, refer-red to as χ_{Mag}, and a magnetically 'hard' antiferromagnetic hematite component, referred to as χ_{Hem}, supplemented by paramagnetic χ_{para} and diamagnetic χ_{dia} components of the sediment matrix

$$\chi = \chi_{Mag} + \chi_{Hem} + \chi_{para} + \chi_{dia} \qquad (2)$$

The magnetite component χ_{Mag} is taken to be proportional to a linear combination of low field (0.1 T) isothermal remanent magnetization σ_{ir} (Fig. 5c), anhysteretic remanent magnetization σ_{ar} (acquired in 0.1 T peak amplitude alternating field and 40 µT bias direct field, Fig. 5b) and frequency dependent susceptibility χ_{fd} (Fig. 5a), representing the coarser grained so-called multi-domain (MD), finer grained single-domain (SD) and ultra-fine grained superparamagnetic (SP) magnetite fractions, respectively (Frederichs et al. 1999). Analogously, the hematite fraction χ_{Hem} is taken to be proportional to the high-field (2.5 - 0.3 T) remanence which may also comprise contributions of goethite (Balsam and Deaton 1991; Harris and Mix 1999) or other high coercivity components (Fig. 5d).

While χ_{dia} of the carbonate portion can be neglected, measurements directly corresponding to χ_{para} are not available for the entire core collection. For iron-bearing minerals, the by far most important paramagnetic compounds, susceptibility is proportional to the Fe content (Bleil and Petersen 1982). Iron concentrations from geochemical analyses of three cores, GeoB 1515-1, 1520-1 and 1523-1, range between about 0.9 to 4.4 wt% (Schulz pers. comm). The three core average Fe wt% (Fig. 5e) was used for the remaining three cores to calculate the partial susceptibilities applying equation (3).

$$\chi_{est} = a_1 \cdot \sigma_{ir} + a_2 \cdot \sigma_{ar} + a_3 \cdot \chi_{fd}$$
$$+ a_4 \cdot \sigma_{hir} + a_5 \cdot wt\% \ Fe \qquad (3)$$

The first three terms on the right hand of equation (3) represent the different magnetite grain size fractions, the fourth term represents hematite and the fifth term the paramagnetic terrigenous matrix. To quantify their respective portions of total sus-

ceptibility, the coefficients a_i have been determined by multiple linear regression analysis yielding

$$\chi_{est} = 7.9 \cdot 10^{-5} \sigma_{ir} / M_0 + 12.1 \cdot 10^{-5} \sigma_{ar} / M_0$$
$$+ 1.8 \cdot \chi_{fd} + 7.3 \cdot \sigma_{hir} / M_0 + 0.6 \cdot 10^{-8} \ wt\% \ Fe \qquad (4)$$

where susceptibilities are scaled to 10^{-8} m^3/kg, remanent magnetizations to 10^{-3} Am^2/kg divided by a unit magnetization of $M_0 = 1$ A/m.

These partial susceptibility estimates delineate the regional late Quaternary Ceará Rise sediment magnetic characteristics. Individual coefficients conform reasonably well with those obtained in the central equatorial Atlantic (von Dobeneck 1998; Frederichs et al. 1999) implying similar overall depositional regimes. The coarser grained magnetite fraction strongly dominates the susceptibility signal (Fig. 6e). Compared to its 65.0 % contribution, all others, hematite (13.1 %), superparamagnetic magnetite (8.8 %), paramagnetic matrix minerals (7.7 %) and fine grained magnetite (5.4 %) are of minor importance. Also remarkable is the small amount and limited variability of $\chi_{Mag\ SD}$. No apparent close relation of the fine grained magnetite fraction exists to any of the other magnetic components. A biogenic origin, unrelated to the influx of terrigenous material could be possible, therefore. Also of sparse information is the χ_{para} record, mainly because sampling resolution of Fe data was only about half that of the magnetic measurements. Nevertheless, a general relationship of the paramagnetic matrix fraction to the magnetite and hematite mineral inventory clearly exists. The relatively small partial susceptibility coefficient of $0.6 \cdot 10^{-8}$ indicates that a measurable portion of the total iron does not reside in paramagnetic minerals.

Correlation between all other partial susceptibility parameters, r = 0.72 to r = 0.75, are all statistically significant. However, a closer inspection of the records reveals some interesting peculiarities, e.g. almost all prominent minima of the χ_{Hem} and $\chi_{Mag\ MD}$ records coincide exactly, whereas a number of prominent maxima show conspicuous lags and leads (see also σ_{ir} and σ_{hir} data sets in Figure 5). These features imply that the magnetomineral assemblage originates from at least two

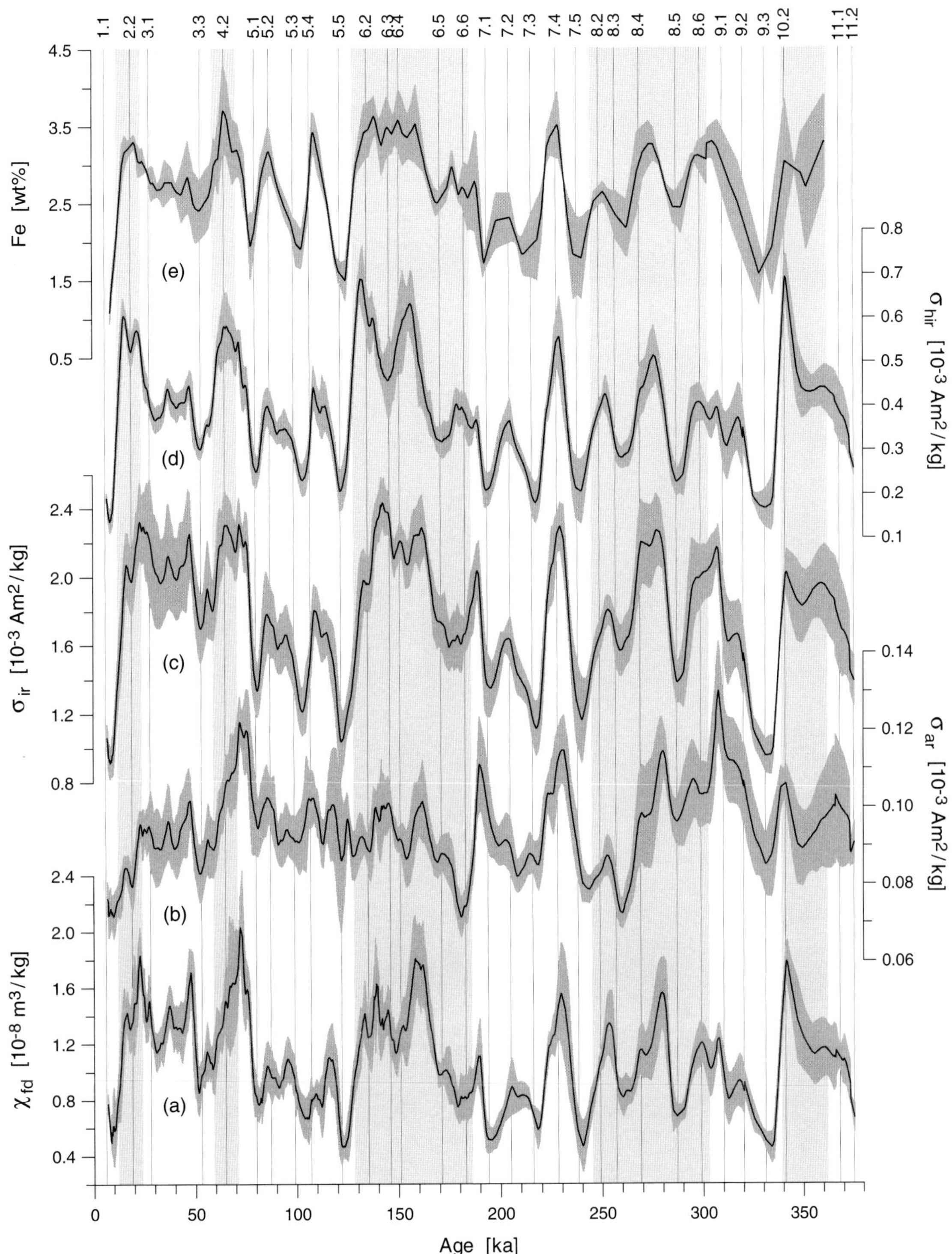

Fig. 5. Six-core stacks of **a)** frequency dependent susceptibility χ_{fd}, **b)** anhysteretic remanent magnetization σ_{ar} acquired in 0.1 T peak amplitude alternating field and 40 µT bias direct field, **c)** 0.1 T isothermal remanent magnetization σ_{ir} depicting the ultra-fine grained (SP), fine grained (SD) and coarser grained (MD) fractions of the magnetically 'soft' magnetite component, respectively. **d)** High-field (2.5 - 0.3 T) isothermal remanent magnetization σ_{hir} represents the magnetically 'hard' hematite component. **e)** Three-core stack of dry mass iron concentration. See Figure 4 for further explanations.

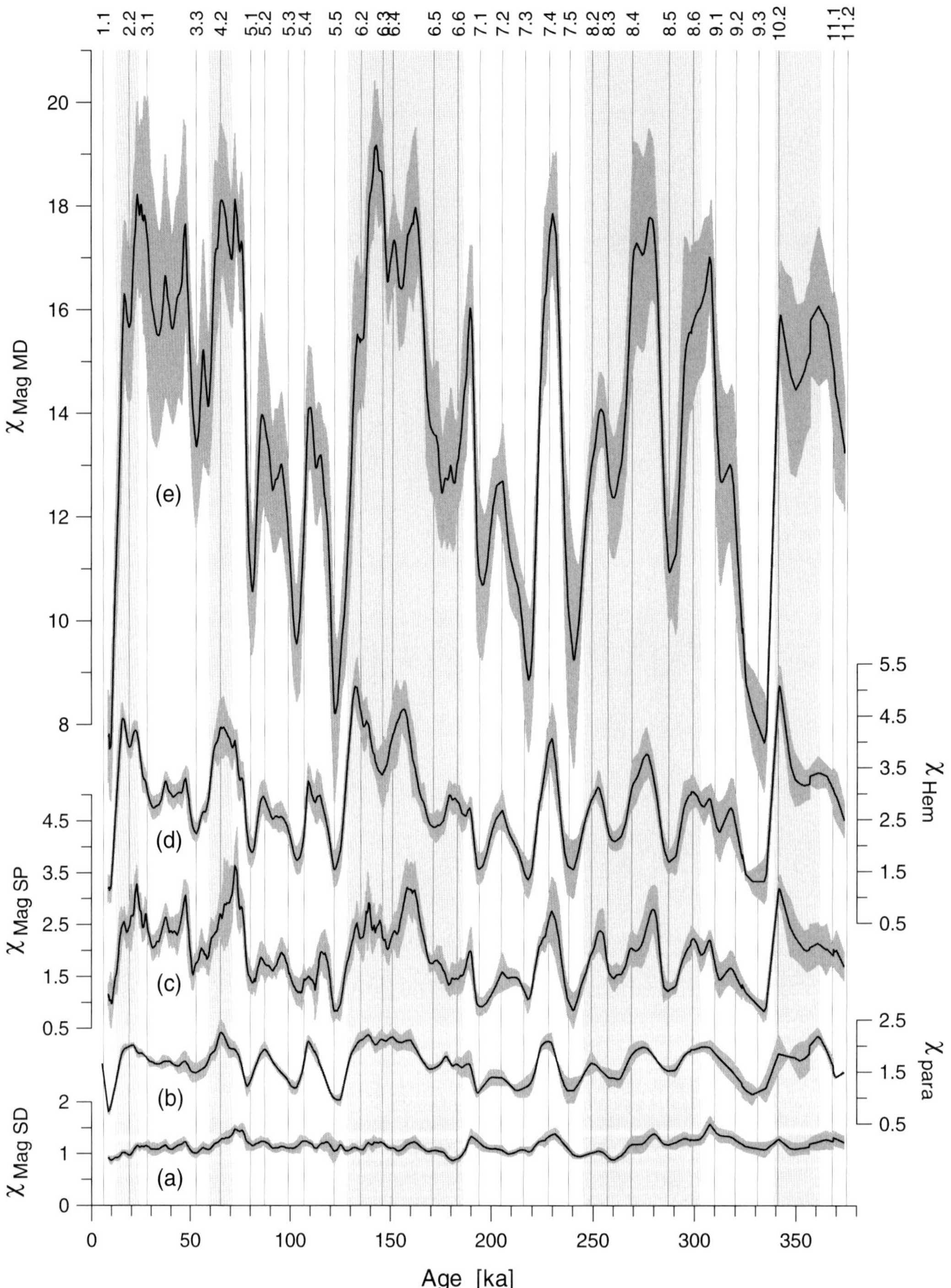

Fig. 6. Results of the partial susceptibility analysis as defined by equation (4) using the mass-specific susceptibility record (Fig. 4c) and magnetic data sets shown in Figure 5. Identically scaled (in 10^{-8} m³/kg) estimates of **a)** the fine grained magnetite component $\chi_{\text{Mag SD}}$, **b)** the sediment matrix paramagnetism χ_{para}, **c)** the ultra-fine grained magnetite component $\chi_{\text{Mag SP}}$, **d)** the hematite component χ_{Hem} and **e)** the coarser grained magnetite component $\chi_{\text{Mag MD}}$ are arranged in order of their relative contributions to total susceptibility. See Figure 4 for further explanations.

different terrigenous sources which clearly contrast in magnetic signatures. Their supply varies asynchronously over time. Obviously, one source has a distinctly higher hematite to magnetite ratio than the other. At this stage it cannot be decided whether a single pathway is involved, e.g. the Amazon discharge mostly represents an Amazon Basin signal, while during specific climatic conditions a strong Andean signal is superimposed (Harris and Mix 1999) or different sources contribute through alternative transport routes and media. The least biased measure to attain more insight into these questions are the magnetite and hematite accumulation rates as they will allow to quantify variations in component fluxes.

Terrigenous Fluxes

Magnetic Minerals

An obvious choice to determine magnetite and hematite concentrations (weight percentages of dry mass) are the partial susceptibility estimates. Unlike magnetite remanence properties, susceptibility is fairly constant over a very wide grain size range ($\approx 6 \cdot 10^{-4}$ m^3/kg) and only increases by about a factor of 2 for the ultra-fine grained SP fraction (Heider et al. 1996). With these data an average magnetite content of 0.028 ± 0.006 wt% was calculated as the sum of $\chi_{\mathrm{Mag\ SP}}$ (Fig. 6c), $\chi_{\mathrm{Mag\ SD}}$ (Fig. 6a) and $\chi_{\mathrm{Mag\ MD}}$ (Fig. 6e). For late Quaternary Ceará Rise sediments recovered at Ocean Drilling Program Leg 154 Site 929 somewhat lower magnetite concentrations of 0.009 ± 0.007 wt% (Richter et al. 1997) have been determined from measurements of saturation magnetization at 1.5 T. Nevertheless, as these data refer to wet bulk density ρ_{wb}, ($\rho_{\mathrm{wb}} \approx 2 \times \rho_{\mathrm{db}}$) and a much smaller number of samples, the agreement with the present results appears quite acceptable. The same is true for the contribution of magnetite to susceptibility estimated from low-field and high-field susceptibilities of the ODP sediments to 65 ± 13 % (Richter et al. 1997). No attempt has been made to also quantify their hematite (or high coercivity mineral) contents, though.

For our sediment series the $S_{-0.3\,\mathrm{T}}$ parameter (Bloemendal et al. 1992), a widely used, however coarse, measure to assess concentrations of high versus low coercivity minerals, ranges from 0.757 to 0.907 (Fig. 7a) approximately corresponding to magnetite to hematite ratios between 1:50 and 1:10. Applying a recently published hematite susceptibility of $\approx 120 \cdot 10^{-8}$ m^3/kg (Kletetschka and Wasilewski 2002) to evaluate the hematite wt% from χ_{Hem} (Fig. 6d) results in magnetite to hematite ratios of 1:180 to 1:50. To adequately compensate for this large difference, a susceptibility of $600 \cdot 10^{-8}$ m^3/kg is appropriate to calculate the hematite wt% from χ_{Hem} yielding an average hematite content of 0.509 ± 0.151 wt%. Such a high intrinsic hematite susceptibility has been reported (Collinson 1983), yet it seems more likely that a mineral component with higher magnetic stability than magnetite and considerably higher susceptibility than hematite is responsible. Pyrrhotite complies very well with these requirements (Dunlop and Özdemir 1997; Thompson and Oldfield 1986) and, most importantly, Solheid et al. (1997) found convincing evidence that it is present in the late Quaternary Ceará Rise sediments. Accordingly, on average about 12 % of the sediments total iron is contained in hematite, the remainder, apart from comparatively tiny magnetite and pyrrhotite fractions, in paramagnetic minerals.

In Figure 7 time series of hematite and magnetite concentrations are compared to total accumulation rates. While the magnetic mineral contents of both coring transects show similar, apparently climatically controlled variations around essentially stable means, distinctly different trends exist for the accumulation rates. As summarized in Table 2, total accumulation rate (and hence also the hematite and magnetite accumulation rates) on average have remained fairly constant over the last 200 kyr for the SE transect. In contrast, the average accumulation rate of the NW transect more or less continuously increased during this period. Normalized to these average trends, the amplitudes of climate-driven shifts in accumulation rates are quite similar on both transects for this interval, but noticeable decline in sediments older than about 200 ka on the NW transect and older than about 300 ka on the SE transect (Fig. 7d). These older deposits have all been recovered in the deepest about 1 m of the cores which appear to be considerably shortened,

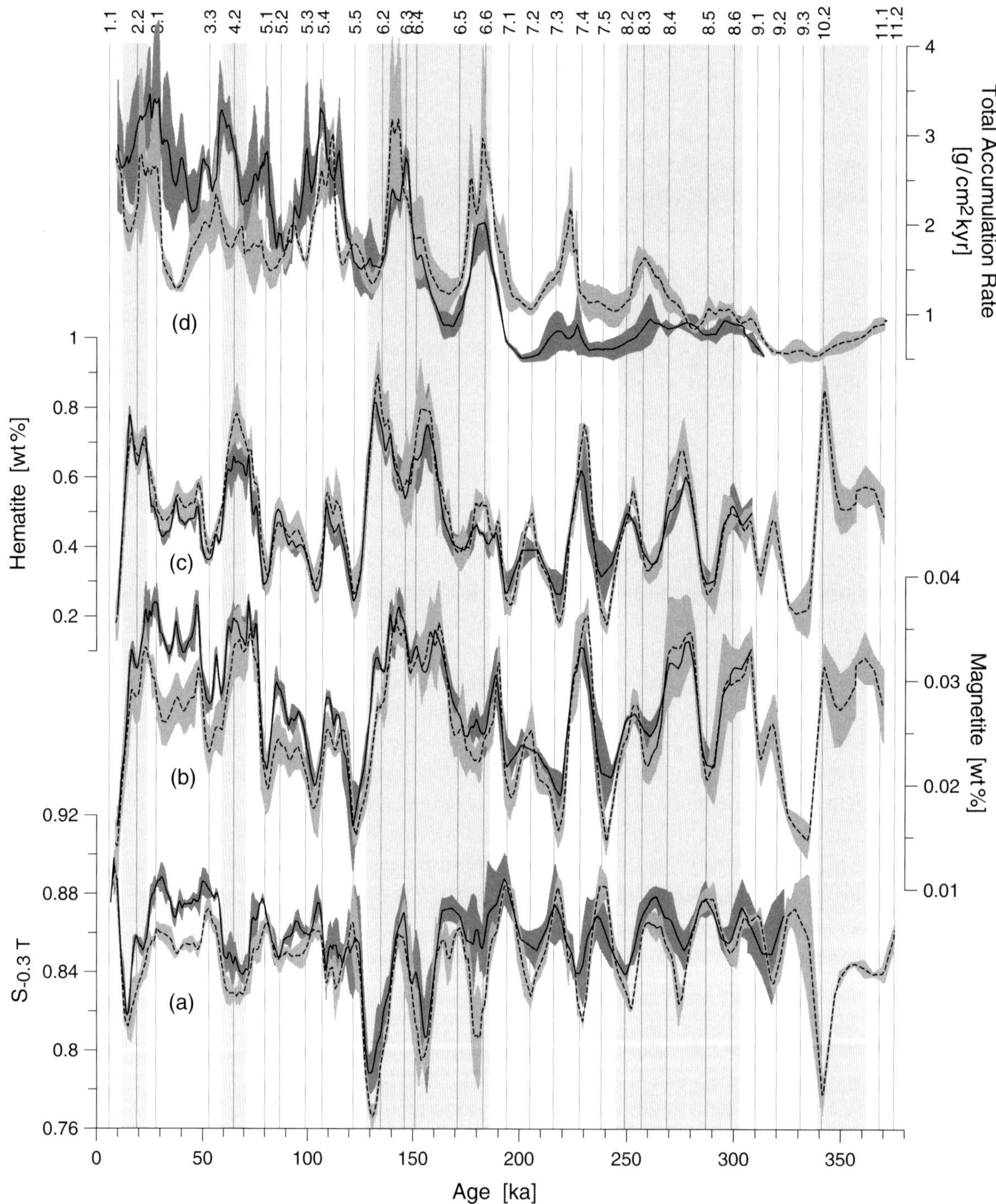

Fig. 7. a) $S_{-0.3\,T}$ parameter outlining the magnetite to hematite ratio, **b)** magnetite and **c)** hematite concentrations, **d)** total accumulation rates of sediment series from the northwestern (solid lines, dark shading) and southeastern (dashed lines, light shading) Ceará Rise coring transects. Note that small scale accumulation rate shifts may result from uncertainties in the age model (about ± 2 kyr). See Figure 4 for further explanations.

Accumulation Rate	Total [g/cm^2·kyr]		Hematite [g/cm^2·kyr]		Magnetite [g/cm^2·kyr]	
	NW Transect	SE Transect	NW Transect	SE Transect	NW Transect	SE Transect
0 – 200 ka	2.12 ± 0.28	1.88 ± 0.25	104 ± 14	96 ± 15	6.2 ± 0.9	5.0 ± 0.7
0 – 100 ka	2.59 ± 0.39	1.87 ± 0.24	129 ± 19	97 ± 15	8.0 ± 1.2	5.0 ± 0.7
100 – 200 ka	1.69 ± 0.19	1.89 ± 0.25	81 ± 10	95 ± 16	4.6 ± 0.6	4.9 ± 0.7

Table 2. Average total, magnetite and hematite accumulation rates (± one standard deviation) on the Ceará Rise northwestern and southeastern coring transects. Standard deviations refer to the mean between site variability, not to the climatically controlled fluctuations of the entire time series.

a well known effect of gravity coring. As in the deeper parts of ice sheets, the plastic deformation does not substantially modify the physical or other properties, but strongly reduces the apparent accumulation rates. Ocean Drilling Project Leg 154 hydraulic piston cores from the same area did not show a comparable decrease in sedimentation rates (e.g. Curry and Cullen 1997), confirming that the present observations indeed result from a coring artifact. The following discussion is therefore restricted to the last 200 kyr.

Mean total accumulation rate during the last 100 kyr on the NW transect was 28 % higher than on the SE transect, but 11 % lower from 200 to 100 ka (Table 2). Because of this remarkably different depositional history on the two transects, these time intervals are distinguished in the following, the boundary set at 100 ka being somewhat arbitrary, however (Fig. 7). The respective data for hematite (+ 25 % and – 17 %) and magnetite (+ 38 % and – 7 %) accumulation rates are plausibly explained if magnetic minerals from the predominant source, most likely the Amazon fluvial discharge, reached the eastern slopes of the Ceará Rise via an overall prevailing oceanographic transport system from the northwest during the last 100 kyr, similar to the present situation (Fig. 1). To also account for the notably different gradients in hematite and magnetite deposition between the two transects, additional hematite must have been supplied (at least) to the SE transect from a second, yet to be identified source, which contained much less magnetite than the Amazon hinterland. As suggested by the almost unchanged amounts and ratios of magnetite and hematite accumulation on the SE transect, both

source regions and their magnetic mineral characteristics, i.e. their magnetite to hematite ratios, were essentially the same during the foregoing 200 - 100 ka interval. The marked shifts on the NW transect, 37 % less hematite and 43 % less magnetite, should then be the result of a considerably modified system of oceanic currents that led Amazon material primarily from the southwest (?) to the Ceará Rise. The observed gradients in hematite and magnetite deposition between the two transects could on principle also result from additional magnetite of a second (northern or western) source predominately reaching the NW transect. As there is no independent oceanographic or atmospheric transport system available, this option has been discarded.

More insight into the short-term variability of magnetite and hematite deposition on both coring transects is obtained from time series of their accumulation rates (Fig. 8) which have been normalized to respective means of unity to allow for a direct comparison. Overall, the covariance is quite obvious. Maxima typically occur around cold periods of glacial stages (2.2, 4.2, 6.2, 6.4, 6.6) and also interglacial period 5.4. As indicated by the ratios of normalized accumulation rates, an enhanced influx is received from the hematite enriched source during these intervals. Minima correspond to warm periods of interglacial stages (1.1, 3.1, 3.3, 5.1, 5.3, 5.5 7.1) and also glacial periods 6.3 and 6.5. At these times, an increased amount of material from the relatively more magnetite containing source is transported to the Ceará Rise region. To quantitati-vely separate the shifting influences of the two magnetic mineral sources, the following simple model has been evaluated.

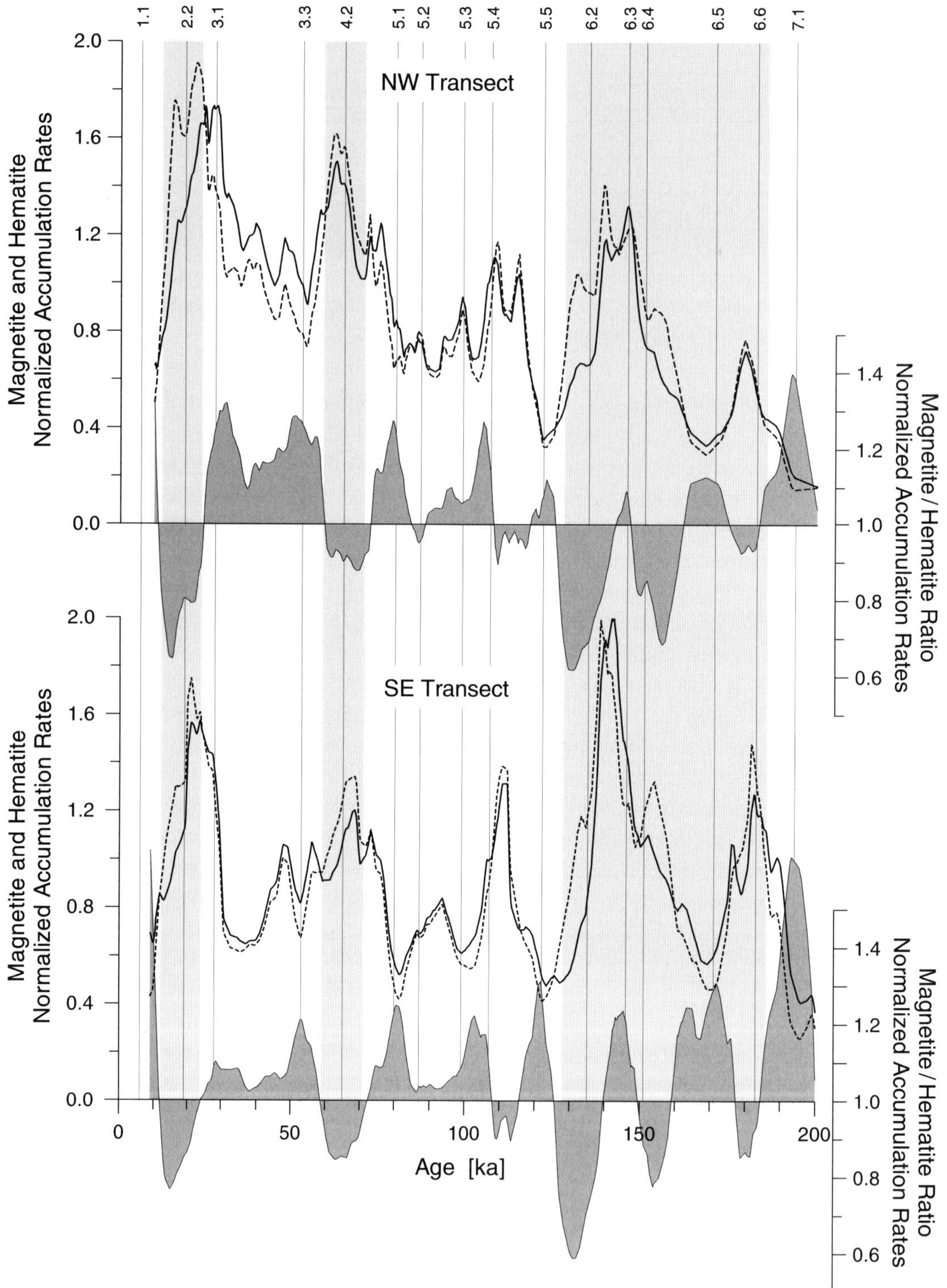

Fig. 8. Normalized magnetite (solid lines) and hematite (dashed lines) accumulation rates in sediment series of the NW (upper) and SE (lower) Ceará Rise coring transects. The ratios of normalized accumulation rates (= normalized content ratios, shaded) reveal distinct variations, apparently dominated by obliquity and precession, implying that the terrigenous magnetic mineral assemblages originate from two main sources with clearly different magnetic signatures. See Figure 4 for further explanations.

The magnetic characteristics of sources A and B are described by their magnetite to hematite ratios $C_A = Mag_A / Hem_A \gg C_B = Mag_B / Hem_B$ assumed to be fairly constant over the last 200 kyr. At any given point in time the hematite Hem (t) and magnetite Mag (t) accumulation rates (AR) on the NW and SE coring transects will then be composed of

$$AR\ Hem_{NW}(t)$$
$$= AR\ Hem_{A,NW}(t) + AR\ Hem_{B,NW}(t) \qquad (5a)$$

$$AR\ Hem_{SE}(t)$$
$$= AR\ Hem_{A,SE}(t) + AR\ Hem_{B,SE}(t) \qquad (5b)$$

$$AR\ Mag_{NW}(t)$$
$$= AR\ Mag_{A,NW}(t) + AR\ Mag_{B,NW}(t)$$
$$= C_A \cdot AR\ Hem_{A,NW}(t) + C_B \cdot AR\ Hem_{B,NW}(t)$$

$$AR\ Mag_{SE}(t)$$
$$= AR\ Mag_{A,SE}(t) + AR\ Mag_{B,SE}(t)$$
$$= C_A \cdot AR\ Hem_{A,SE}(t) + C_B \cdot AR\ Hem_{B,SE}(t)$$
$$(6a,\ b)$$

Hence, hematite and magnetite contributions of the individual sources amount to

$$AR\ Hem_{A,NW,SE}(t)$$
$$= \frac{AR\ Mag_{NW,SE}(t) - C_B \cdot AR\ Hem_{NW,SE}(t)}{C_A - C_B}$$

$$AR\ Hem_{B,NW,SE}(t)$$
$$= \frac{C_A \cdot AR\ Hem_{NW,SE}(t) - AR\ Mag_{NW,SE}(t)}{C_A - C_B}$$
$$(7a,b)$$

$$AR\ Mag_{A,NW,SE}(t) = C_A \cdot AR\ Hem_{A,NW,SE}(t)$$

$$AR\ Mag_{B,NW,SE}(t) = C_B \cdot AR\ Hem_{B,NW,SE}(t)$$
$$(8a,\ b)$$

The critical, *a priori* unknown parameters are the magnetite to hematite ratios C_A and C_B in the source regions. C_A were directly estimated from the measured sediment magnetite to hematite ratios (weight percent or accumulation rates, Figure 9). They reach almost identical maxima at around oxygen isotope event 7.1 on both transects. The delivery of magnetic minerals from source B must have been minimal at this stage. Assuming that it essentially vanished, yields a magnetite to hematite ratio of $C_A = 0.083$ for the Amazon source region. Note that this ratio of about 1:12 is an absolute lower limit as any smaller C_A would occasionally result in negative hematite accumulation rates on both coring transects.

The lowest magnetite to hematite ratio of the 876 individual samples was 0.027. To some extend it should overestimate the magnetite content of source B as the Amazon source supply presumably never entirely ceased anywhere in the study area. A C_B of 0.020 has therefore been used in the model calculations.

With such estimates of magnetic source characteristics, late Quaternary records of magnetite and hematite fluxes from the Amazon and source B have been inferred from equations (7) and (8). During the last 200 kyr on average 84 % of magnetite and 56 % of hematite originated from the Amazon. On the whole, these proportions, 82 % versus 53 % from 200 to 100 ka and 86 % versus 59 % during the last 100 kyr, were relatively stable. Mean hematite and magnetite accumulations rates from individual sources on the two coring transects are listed in Table 3. Assuming that source B did not contribute any magnetite ($C_B = 0$) increases the Amazon hematite fraction to on average 67 %. Both sources delivered about equal amounts of hematite for $C_B = 0.027$ and for $C_B > 0.027$ most hematite came from source B. Note that W-E trends observed in magnetic mineral fluxes of different provenance to the two coring transects (Table 3) are independent of the magnetite to hematite ratio inferred for source B.

Deposition of Amazon source components on average decreased from W to E during the last 100 kyr and was roughly the same on both transects from 200 to 100 ka. Over the entire last 200 kyr more hematite and magnetite from source B was deposited on the SE transect. This persistent gradient indicates that source B should be to the east of the Ceará Rise. Most likely, the respective

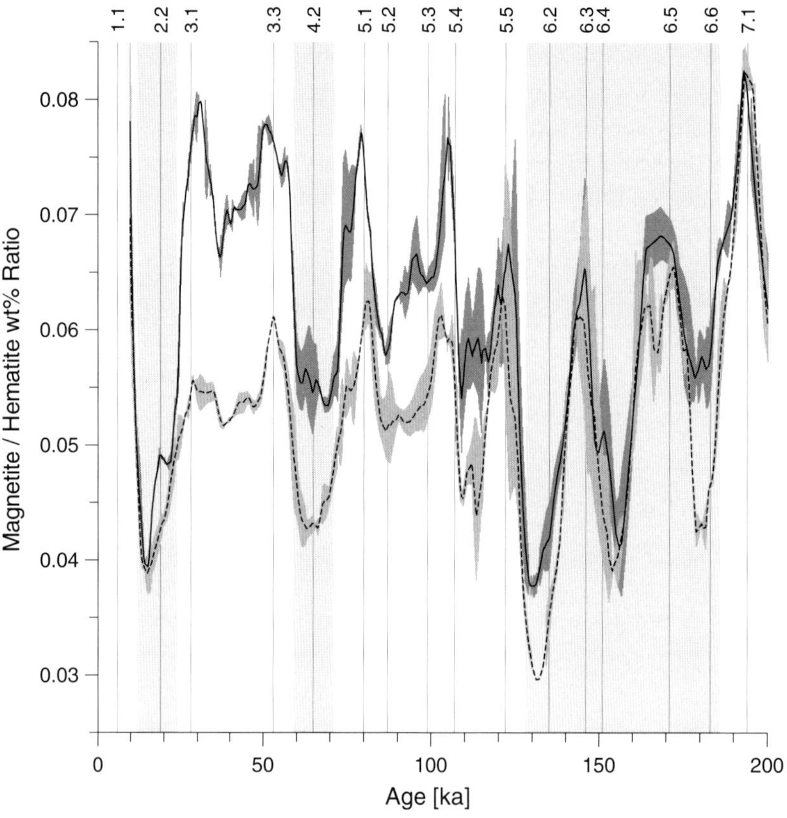

Fig. 9. Magnetite to Hematite weight percent ratios in sediments of the northwestern (solid line) and southeastern (dashed line) Ceará Rise coring transects. See Figure 4 for further explanations.

Accumulation Rate [g/cm^2·kyr]	NW Transect				SE Transect			
	Amazon Source		Sahara Source		Amazon Source		Sahara Source	
	Hematite	Magnetite	Hematite	Magnetite	Hematite	Magnetite	Hematite	Magnetite
0 – 200 ka	66 ± 10	5.5 ± 0.8	38 ± 8	0.8 ± 0.2	48 ± 7	4.0 ± 0.6	51 ± 11	1.0 ± 0.2
0 – 100 ka	87 ± 13	7.2 ± 1.1	42 ± 8	0.8 ± 0.2	49 ± 7	4.0 ± 0.6	52 ± 9	1.0 ± 0.2
100 – 200 ka	47 ± 7	3.9 ± 0.6	34 ± 8	0.7 ± 0.2	48 ± 7	3.9 ± 0.6	50 ± 12	1.0 ± 0.2

Table 3. Late Quaternary hematite and magnetite accumulation rates (± one standard deviation) from South American and African sources on the Ceará Rise northwestern and southeastern coring transects. Standard deviations refer to mean between site variability, not to the climatically controlled fluctuations of the entire time series.

magnetic mineral fractions originate from the African continent and have been transported in the Saharan dust plume.

More conclusive information about this central question can be obtained from Figure 10 showing time series of hematite and magnetite fluxes from the Amazon and Sahara/Sahel sources to the Ceará Rise coring transects. Among others, a most interesting feature is the conspicuous covariance of the Amazon source accumulation rates with the δ^{18}O sea level proxy of Bickert and Mackensen (this volume). This evidently reflects climatically controlled cyclic transgressions and regressions on the South American shelf. Maxima (minima) of hematite and magnetite accumulation do not directly coincide with relative sea level low (high) stands, but seem to be associated with relative sea level falls (rises) reaching highest (lowest) rates, when large areas of the shelves fell dry (were flooded). Ultimately, the material must have been transported by the Amazon River and its main source regions are in the Andes (Meade 1994). The Andean magnetic signal may have been variably modified in the Amazon Basin depending on climate conditions and residence time. Obviously, there are also other complications to interpret the Amazon source record in more detail without further information. In particular, the apparent differences between the two coring transects hint at fluctuating systems of ocean currents carrying the Amazon materials to the Ceará Rise region.

The influx from the Sahara and Sahel source is obviously controlled by totally different mechanisms. Noteworthy are the pronounced accumulation rate peaks in cold climate intervals of glacial stages (2.2, 4.2, 6.2, 6.4 and 6.6) and also interglacial period 5.4. From interglacials to glacials hematite and magnetite accumulation rates typically rise by a factor of more than 10. Various studies in the eastern equatorial and tropical Atlantic (e.g. Sarnthein et al. 1981; Pokras and Mix 1985; Matthewson et al. 1995; Ruddiman 1997) have documented much more vigorous glacial wind systems and at the same time a shift of their trajectories to south (similar to today's boreal winter situation). Combined with enhanced continental aridity this caused a considerably higher eolian transport of African terrigenous material to the oceans. In late Quaternary sediment series from the Senegal continental slope and Sierra Leone Rise region Bloemendal et al. (1988) found lowest magnetite to hematite ratios of about 1:50 in glacial periods. Although this $S_{-0.3\,T}$ parameter estimate is subject some uncertainties, it perfectly complies with the African source signature of ($C_B = 0.02$) applied in our model calculations. The roughly symmetrical shape of the African glacial signals implies a rapid increase and decrease of dust fluxes. This does not provide conclusive evidence, however, about the controlling component of the climate system, African aridity as opposed to wind intensity and routes. Our data confirm observations of Matthewson et al. (1995) in the Cape Verde area that conditions in interglacial substage 5.4 (not so much in substage 5.2) were similar to those in full glacials. On the other hand, a sawtooth pattern they report, i.e. a sudden rise of dust fluxes and gradual transitions into the next cycle, is clearly not recorded in the magnetic mineral accumulations at the Ceará Rise.

Terrigenous Fraction

The magnetic mineral accumulation rates may be used as proxies to assess the fluxes of bulk terrigenous materials from South America and Africa to the western equatorial Atlantic. Terrigenous accumulation rates, simply taken as the difference of total and carbonate accumulation rates, are assumed to be proportional to a linear combinationof AR Hematite (or Magnetite) of Amazon and Sahara/Sahel provenance. Similar to the partial susceptibility concept, the respective coefficients were determined by multiple linear regression analysis. The results are shown in Figure 11. On average, 79 % of terrigenous fluxes originate from South America, 21 % from Africa. Depositional trends between the two coring transects as well as changes from the 200 to 100 ka interval to the last 100 kyr essentially correspond to those seen in the magnetic mineral fluxes. The eolian transport system of African dust clearly varied at a precessional climate rhythm (23 kyr) as repeatedly observed in the eastern equatorial Atlantic (e.g. Tiedemann et al. 1989; Matthewson et al. 1995). In this respect, influx of the Amazon River fluvial discharge shows a much more complex pattern of superimposed sea level changes and variable ocean current systems that may completely mask the immediate impact of climate shifts on the South American continent. A suspicious, so far unexplained feature in both the eolian and fluvial fluxes are twin peaks separated by several thousand years at numerous relative accumulation rate maxima.

Conclusions

1. Susceptibility records of late Quaternary Ceará Rise sediments are largely controlled by variations in carbonate content modulating the primary mag-

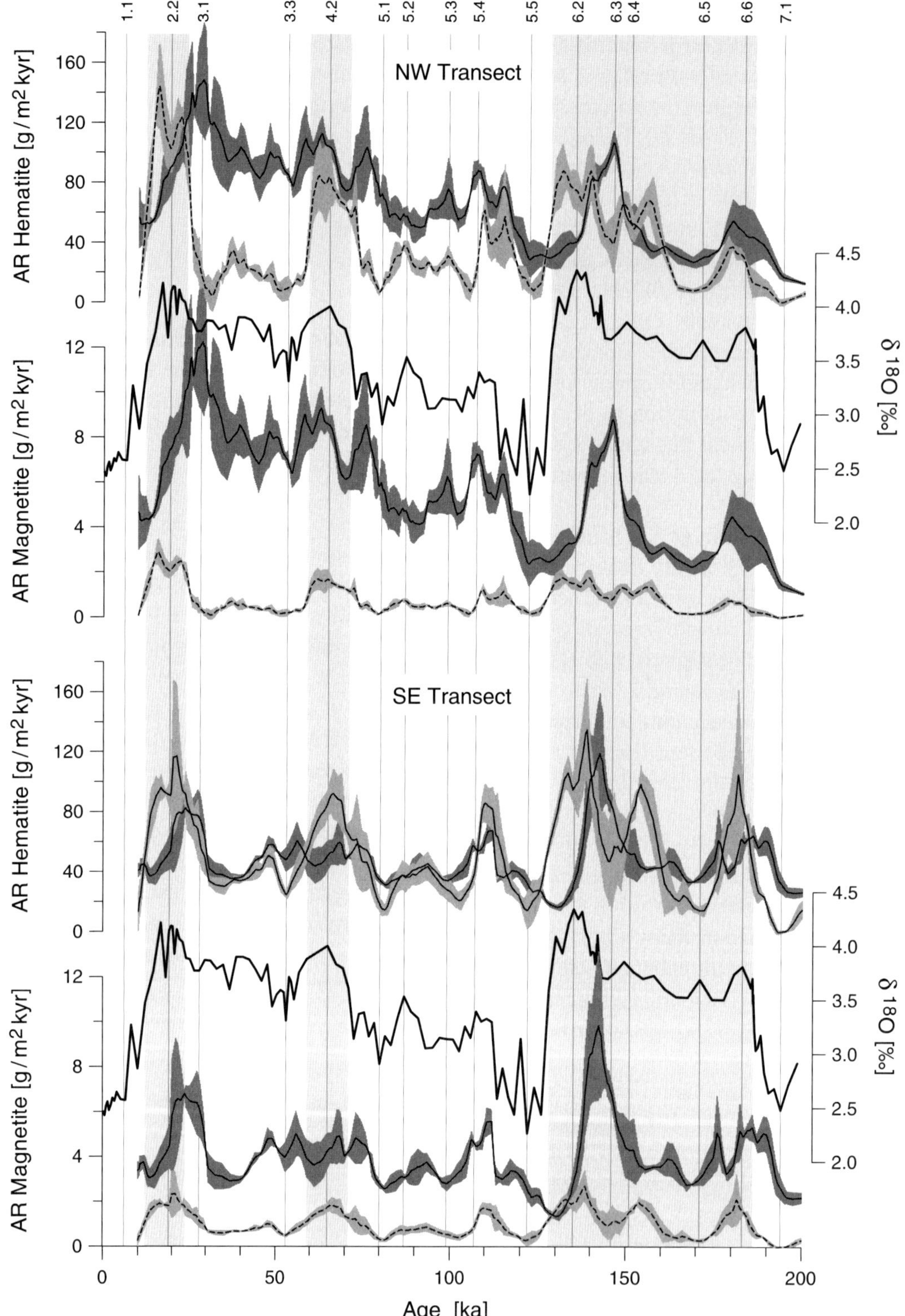

Fig. 10. Magnetite and hematite accumulation rates on the northwestern (upper) and southeastern (lower) Ceará Rise coring transects differentiating components of South American (Amazon, solid lines) and African (Sahara/Sahel, dashed lines) origin. Minima (maxima) in the $\delta^{18}O$ record (thick solid lines) of Bickert and Mackensen (this volume) indicate relative sea level high (low) stands. See Figure 4 for further explanations.

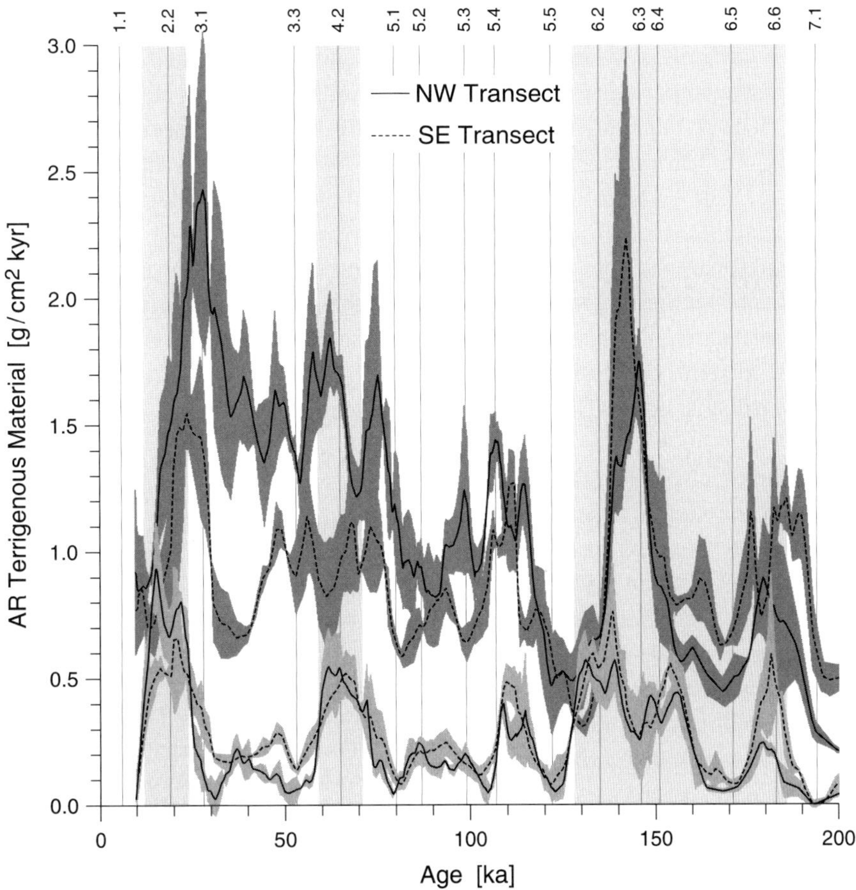

Fig. 11. Accumulation rates of bulk terrigenous material on the Ceará Rise coring transects differentiating components of South American (Amazon, dark shading) and African (Sahara, light shading) origin. See Figure 4 for further explanations.

netic signal which very predominantly resides in a magnetic mineral assemblage of lithogenic, conti-nental origin.

2. To delineate regional sediment magnetic attributes the 'partial susceptibility concept' has been successfully applied. On average 65 % of susceptibility originate from a relatively coarse grained magnetite fraction. Hematite (13.1 %), ultrafine grained magnetite (8.8 %), paramagnetic matrix minerals (7.7 %) and fine grained magnetite (5.4 %) contribute comparatively minor portions. Because of the extreme differences in intrinsic susceptibilities, corresponding mean hematite concentrations (0.509 wt%) are substantially higher than magnetite contents (0.028 wt%).

3. Time series of magnetite and hematite accumulation rates reveal that their supply varied asynchronously over time implying that the magneto-

mineral inventories results from two main terrigenous sources, one being distinguished by a notably higher magnetite to hematite ratio (1:12) than the other (1:50). This scenario was evaluated in detail with a two-source model.

4. Taking into account persistent EW depositional gradients as well as literature data of magnetic sediment characteristics in the western and eastern equatorial Atlantic, the Saharan dust plume could be identified as the second, hematite-rich source besides the obvious magnetite-rich Amazon River fluvial load. During the last 200 kyr on average 84 % of magnetite and 56 % of hematite were of South American provenance versus 16 % of magnetite and 44 % of hematite from the North African desert areas.

5. The inflow from the two sources varied at distinctly different climatically mediated modes. Mag-

netic minerals accumulations from African dust fluxes to the western equatorial Atlantic during peak cold substages of glacial periods are typically more than ten times higher than those of warm inter-glacial periods. Conditions around cold interglacial substage 5.4 should have been comparable to full glacials. Both increase and decrease of African dust fluxes were obviously very rapid, an observa-tion which provides no unequivocal clue, however, to identify the initial trigger of the climate system, African aridity or intensity and southward shift of the Saharan Air Layer trajectories.

Deposition of magnetic minerals from the Amazon discharge at the Ceará Rise is largely constrained by intermittent transgressions and regressions on the South American shelf. Maxima (minima) of hematite and magnetite accumulation correlate to relative sea level rises (falls) reaching highest (lowest) rates, when most shelf areas fell dry (were flooded). Another main control is the regional ocean current system transporting Amazon material to the Ceará Rise. There are clear indications that it was substantially modified around 100 kyr ago.

6. During the last 200 kyr, on average 79 % of total terrigenous material fluxes to the Ceará Rise area originate from South American sources in the Amazon River catchment. African dust sources contributed 21 %. The eolian transport very pre-dominantly varied at precessional climate rhythm (23 kyr). The much more pronounced contrast between South American (1:12) and African (1:50) magnetite to hematite source characteristics compared to the respective Al:Ti signatures of 1:20 versus 1:12 may be partly responsible that, different from the present results, Zabel et al. (1999) could not document significant African dust deposits in the western equatorial Atlantic on the basis of these element ratios. Presumably more important to explain this inconsistency are modifications of the initial Al:Ti ratio, mainly a loss of Al during oceanic transport, where the composition of magnetic mineral assemblages will remain largely unchanged.

Acknowledgements

We thank captain and crew of RV *Meteor* for their efficient support during cruise M 16/2. Liane Brück coordinated parts of the laboratory work. Various discussions with Karl Fabian and constructive reviews by Wolfgang Berger and Wout Krijgsman greatly helped to improve the manuscript. The study was funded by the Deutsche Forschungs-gemeinschaft (Sonderforschungsbereich 261, Contribution No. 384). TvD presently enjoys a visiting research fellowship at Utrecht University supported by the Netherlands Research Center for Integrated Solid Earth Science (ISES).

References

Balsam WL, Deaton BC (1991) Sediment dispersal in the Atlantic Ocean: evaluation by visible light spectra. Rev Aqua Sci 4: 411-447

Bleil U, Petersen N (1982) Magnetic properties of natural minerals. In: Angenheister G (ed) Numerical Data and Functional Relationships in Science and Tech-nology. Springer, Berlin, Landolt-Börnstein V/1b: pp 308-365

Bleil U, von Dobeneck T (1999) Geomagnetic events and relative paleointensity records – clues to high-reso-lution paleomagnetic chronostratigraphies of late Quaternary marine sediments? In: Fischer G, Wefer G (eds) Use of Proxies in Paleoceanography, Exam-ples from the South Atlantic. Springer, Berlin, pp 635-654

Bloemendal J, deMenocal P (1989) Evidence for a change in the periodicity of tropical climate cycles at 2.4 Myr from whole-core magnetic susceptibility measure-ments. Nature 342: 897-900

Bloemendal J, King JW, Hall FR, Doh S-J (1992) Rock magnetism of late Neogene and Pleistocene deep-sea sediments: relationship to sediment source, diagenetic processes, and sediment lithology. J Geophys Res 97: 4361-4375

Bond G, Broecker W, Johnsen S, McManus J, Labeyrie L, Jouzel J, Bonati G (1993) Correlations between climate records from North Atlantic sediments and Greenland ice. Nature 365: 143-145

Channell JET, Stoner JS, Hodell DA, Charles CD (2000) Geomagnetic paleointensity for the last 100 kyr from the sub-Antarctic South Atalantic: A tool for inter-hemispheric correlation. Earth Planet Sci Lett 175: 145-160

Collinson DW (1983) Methods in Rock Magnetism and Palaeomagnetism, Techniques and Instrumentation. Chapman Hall, London

Curry WB, Cullen JL (1997) Carbonate production and dissolution in the western equatorial Atlantic during the last 1 my. In: Shackleton NJ, Curry WB, Richter

C, Bralower TJ (eds) Proceedings of the Ocean Drilling Program. Scientific Results 154: 189-199

Damuth JE (1977) Late Quaternary sedimentation in the western equatorial Atlantic. Geol Soc Am Bull 88: 695-710

Dansgaard W, Johnsen SJ, Clausen HB, Dahl-Jensen D, Gundestrup NS, Hammer CU, Hvidberg CS Steffensen JP, Sveinbjornsdottir AE, Bond G (1993) Evidence for general instability of past climate from a 250-kyr ice-core record. Nature 364: 218-220

Dobson DM, Dickens GR, Rea DK (1997) Terrigenous sedimentation at the Ceará Rise. In: Shackleton NJ, Curry WB, Richter C, Bralower TJ (eds) Proceedings of the Ocean Drilling Program. Scientific Results 154: 465-473

Dobson DM, Dickens GR, Rea DK (2001) Terrigenous sedimentation on Ceará Rise: A Cenozoic record of South American orogeny and erosion. Palaeogeogr Palaeoclim Palaeoecol 165: 215-229

Dunlop DJ, Özdemir Ö (1997) Rock Magnetism, Fundamentals and Frontiers. Cambridge University Press, Cambridge

Frederichs T, Bleil U, Däumler K, von Dobeneck T, Schmidt A (1999) The magnetic view on the marine paleoenvironment: parameters, techniques, and potentials of rock magnetic studies as a key to paleoclimatic and paleoceanographic changes. In: Fischer G, Wefer G (eds) Use of Proxies in Paleoceanography, Examples from the South Atlantic. Springer, Berlin, pp 575-599

Francois R, Bacon MP (1991) Variations in terrigenous input into the deep equatorial Atlantic during the past 24,000 years. Science 251: 1473-1475

Grousset FF, Labeyrie L, Sinko JA, Cremer M, Bond G, Duprat J, Cortijo E, Huon S (1993) Patterns of ice-rafted detritus in the glacial North Atlantic (40 − 55 °N). Paleoceanography 8: 175-192

Harris SE, Mix AC (1999) Pleistocene precipitation balance in the Amazon Basin recorded in deep-sea sediments. Quat Res 51: 14-26

Heider F, Zitzelsberger A, Fabian K (1996) Magnetic susceptibility and remanent coercive force in grown magnetite crystals from 0.1 μm to 6 mm. Phys Earth Planet Int 93: 239-256

Imbrie J, Hays JD, Martinson DG, McIntyre A, Mix AC, Morley JJ, Pisias NG, Prell WL, Shackleton NJ(1984) The orbital theory of Pleistocene climats: support from a revised chronology of the marine $\delta^{18}O$ record. In: Berger AL et al. (eds) Milankovitch and Climate Part I. Reidel, Dordrecht, pp 269-305

Johns WE, Schott FA, Zantopp RJ, Evans RH (1990) The North Brazil Current retroflection: Seasonal structure and eddy variability. J Geophys Res 95: 22103-22120

Kletetschka G, Wasilewski PJ(2002) Grain size limit for SD hematite. Phys Earth Planet Int 129: 173-179

Kotz L, Kaiser G, Tschöpel P, Tölg G (1972) Aufschluß biologischer Matricen für die Bestimmung sehr nied-riger Spurenelementgehalte bei begrenzter Einwaage mit Salpertersäure unter Druck in einem Teflongefäß. Z anal Chem 260: 207-209

Kumar N, Embley RW (1977) Evolution and origin of Ceará Rise: An aseismic rise in the western equatorial Atlantic Geol Soc Am Bull 88: 683-694

Laj C, Kissel C, Mazaud A, Channell JET, Beer J (2000) North Atlantic paleointensity stack since 75 ka (NAPIS) and the duration of the Laschamp event. Phil Trans R Soc London A358: 1009-1025

Langereis CG, Dekkers MJ (1999) Magnetic cyclostratigraphy: High-resolution dating in and beyond the Quaternary and analysis of periodic changes in diagenesis and sedimentary magnetism. In: Maher BA, Thompson R (eds) Quaternary Climates, Environments and Magnetism. Cambridge University Press, Cambridge, pp 352-382

Martin WR, Sayles FL (1996) CaCO3 dissolution in sediments of the Ceará Rise, western equatorial Atlantic. Geochim Cosmochim Acta 60: 243-263

Maslin M, Mikkelsen N (1997) Amazon Fan mass-transport deposits and underlying interglacial deposits: age estimates and fan dynamics. In: Flood RD, Piper DJW, Claus A, Peterson LC (eds) Proceedings of the Ocean Drilling Program. Sci Res 155: 353-365

Matthewson AP, Shimmield GB, Kroon D, Fallick AE (1995) A 300 kyr high-resolution aridity record of the North African continent. Paleoceanography 10: 677-692

Meade RH (1994) Suspended sediments of the modern Amazon and Orinoco rivers. Quat Intern 21: 29-39

Milliman JD, Meade RH (1983) World-wide delivery of river sediment to the oceans. J Geol 91: 1-21

Milliman JD, Summerhayes CP, Barretto HT (1975) Quaternary sedimentation on the Amazon continental margin: A model. Geol Soc Amer Bull 88: 610-614

Muller-Karger FE, McClain CR, Richardson PL (1988) The dispersal of the Amazon's water. Nature 333: 56-59

Mulitza S (1994) Spätquartäre Variationen der oberflächennahen Hydrographie im westlichen äquatorialen Atlantik. Ber Fachber Geowiss, Univ Bremen 57, 97 p

Petermann H (1994) Magnetotaktische Bakterien und ihre Magnetosome in Oberflächensedimenten des Südatlantiks. Ber Fachber Geowiss, Univ Bremen 56, 134 p

Petermann H, Bleil U (1993) Detection of live magneto-tactic bacteria in South Atlantic deep-sea sediments. Earth Planet Sci Lett 117: 223-228

Petschick R, Kuhn G., Gingele F (1996) Clay mineral distribution in surface sediments of the South Atlantic: sources, transport, and relation to oceanography. Mar Geology 130: 203-229

Pokras EM, Mix AC (1985) Eolian evidence for spatial variability of Late Quaternary climates in tropical Africa. Quat Res 24: 137-139

Prell WL, Imbrie J, Martinson DG, Morley JJ, Pisias NG, Shackleton NJ, Streeter HF (1986) Graphic correlation of oxygen isotope stratigraphy: Application to the late Quaternary. Paleoceanography 1: 137-162

Prospero JM, Glaccum RA, Nees RT (1981) Atmospheric transport of soil dust from Africa to South America. Nature 289: 570-572

Richter C, Valet J-P, Solheid PA (1997) Rock magnetic properties of sediments from Ceará Rise (Site 929): implications for the origin of the magnetic suscep-tibility signal. In: Shackleton NJ, Curry WB, Richter C, Bralower TJ (eds) Proceedings of the Ocean Drilling Program. Scientific Results 154: 169-179

Robinson SG, Maslin MA, McCave IN (1995) Magnetic susceptibility variations in Upper Pleistocene deep-sea sediments of the NE Atlantic: Implications for ice rafting and paleocirculation at the last glacial maximum. Paleoceanography 10: 221-250

Ruddiman WF (1997) Tropical Atlantic terrigenous fluxes since 25,000 yrs BP. Mar Geol 136: 189-207

Rühlemann C, Frank M, Hale W, Mangini A, Mulitza S, Müller PJ, Wefer G (1996) Late Quaternary produc-tivity changes in the western equatorial Atlantic: evidence from ^{230}Th-normalized carbonate and organic carbon accumulation. Mar Geol 135: 127-152

Sarnthein M, Tetzlaff G, Koopmann B, Wolter K, Pflau-mann U (1981) Glacial and interglacial wind regimes over the eastern subtropical Atlantic and north-west Africa. Nature 293: 193-196

Schulz HD, Cruise Participants (1991) Bericht und erste Ergebnisse über die *Meteor*-Fahrt M 16/2. Ber Fachber Geowiss, Univ Bremen, 149 p

Schmieder F, von Dobeneck T, Bleil U (2000) The mid-Pleistocene climate transition as documented in the deep South Atlantic Ocean: Initiation, interim state and terminal event. Earth Planet Sci Lett 179: 539-549

Shackleton NJ, Curry WB, Richter C, Bralower TJ (1997) Proceedings of the Ocean Drilling Program. Scientific Results 154: 552

Shackleton NJ, Crowhurst S (1997) Sediment fluxes based on an orbitally tuned time scale 5 Ma to 14 Ma, Site 926. In: Shackleton NJ, Curry WB, Richter C, Bralower TJ (eds) Proceedings of the Ocean Drilling Program. Scientific Results 154: 69-82

Solheid PA, Banerjee SK, Richter C, Valet J-P (1997) High-resolution rock-magnetic study of Ceará Rise sedi-ments at Site 529. In: Shackleton NJ, Curry WB, Richter C, Bralower TJ (eds) Proceedings of the Ocean Drilling Program. Scientific Results 154: 181-186

Swap R, Garstang M, Greco S, Talbot R, Kallberg P (1992) Saharan dust in the Amazon Basin. Tellus 44B: 133-149

Tiedemann R, Sarnthein M, Stein R (1989) Climate chan-ges in the western Sahara: Aeolo-marine sedi-ment record of the last 8 million years (Sites 657-661). In: Ruddiman W, Sarnthein M, Baldauf J et al. (eds) Pro-ceedings of the Ocean Drilling Program. Scientific Results 108: 241-277

Thompson R, Oldfield F (1986). Environmental Mag-netism. Allen Unwin, London

von Dobeneck T (1998) The concept of 'partial suscepti-bilities'. Geol Carpath 49: 228-229

von Dobeneck T, Schmieder F (1999) Using rock mag-netic proxy records for orbital tuning and extended time series analyses into the super- and sub-Milan-kovitch bands. In: Fischer G, Wefer G (eds) Use of Proxies in Paleoceanography, Examples from the South Atlantic. Springer, Berlin, pp 601-633

Zabel M, Bickert T, Dittert L, Haese RR (1999) Signifi-cance of the sedimentary Al:Ti ratio as an indicator for variations in the circulation patterns of the equa-torial North Atlantic. Paleoceanography 14: 789-799

Integrated Rock Magnetic and Geochemical Quantification of Redoxomorphic Iron Mineral Diagenesis in Late Quaternary Sediments from the Equatorial Atlantic

J.A. Funk[1*], T. von Dobeneck[1,2] and A. Reitz[3]

[1]*Universität Bremen, Fachbereich Geowissenschaften, Postfach 33 04 40, D-28334 Bremen, Germany*
[2]*Paleomagnetic Laboratory 'Fort Hoofddijk', Faculty of Earth Sciences Utrecht University, Budapestlaan 17, 3584 CD Utrecht, The Netherlands*
[3]*Geochemistry Department, Faculty of Earth Sciences Utrecht University, PO Box 80 021, 3508 TA Utrecht, The Netherlands*
* *corresponding author (e-mail): funk@geomarin.uni-bremen.de*

Abstract: Rock magnetic and geochemical data logged by fast, non-destructive X-ray fluorescence and susceptibility half core scanning techniques have been combined to create high-resolution records of redoxomorphic iron mineral diagenesis in suboxic marine sediments. The great potential of this approach and advantage to standard single sample methods is demonstrated on two Late Quaternary sequences from the central Equatorial Atlantic (GeoB 2908-7 and 4317-2). Reductive dissolution of ferric minerals, most prominently magnetite (Fe_3O_4) and hematite (Fe_2O_3), induced by organic carbon degradation is shown to represent a gradual, mineral- and grain-size selective process. Proportionality of Fe, Ti and magnetite concentrations in the unaltered sections lead us to define proxy parameters for magnetite depletion (Fe/κ_{nd}) below and precipitation (κ_{nd}/Ti) above the modern and numerous fossil redox boundaries, while iron relocation was detected on basis of the Fe/Ti ratio. By calibrating all three ratios internally, we reconstruct and quantify primary deposition and secondary change of both, magnetite and total Fe profiles. Fine-scaled C_{org} variations (0.1 to 0.6 %) and susceptibility losses (up to $200 \cdot 10^{-6}$ SI) show high signal resemblance and appear to be equivalent signatures of cyclic productivity pulses in the study area. Some minor suboxic events are still expressed in the rock magnetic proxy signal, but are not accompanied by residual C_{org} enrichments.

Introduction

Numerous paleoenvironmental studies on suboxic to anoxic sediments have combined rock magnetic and chemical data (e.g. Vigliotti et al. 1999) to identify mineral alterations due to *redoxomorphic diagenesis* (Dapples 1962), a "collective noun for processes of early diagenesis involving both reductive and oxidative stages" (Robinson et al. 2000). Redox reactions including dissolution, depletion, relocation and precipitation of iron occur in and around sapropels (Passier et al. 1998), turbiditic sequences (Robinson et al. 2000) and organically enriched sediments deposited under conditions of upwelling (Tarduno 1994), reduced bottom water circulation (Calvert and Pedersen 1993) or eutrophication (Snowball 1993). The cited studies were concerned with reconstructions of paleo-productivity and sediment accumulation, with alteration of primary minerals and diagenetic overprinting of paleo- and rock magnetic information (Dekkers et al. 1994), and with the establishment of cyclostratigraphic age models based on orbital rhythms (Langereis and Dekkers 1999; van Santvoort et al. 1997).

The geochemical settings for redoxomorphic diagenesis are alike in all mentioned situations: Bacterially mediated oxidation of embedded organic

From WEFER G, MULITZA S, RATMEYER V (eds), 2003, *The South Atlantic in the Late Quaternary: Reconstruction of Material Budgets and Current Systems.* Springer-Verlag Berlin Heidelberg New York Tokyo, pp 237-260

238   Funk et al.

matter follows a declining energy yield sequence of terminal electron acceptors from interstitial oxygen and nitrate to Mn (IV) oxides, Fe (III) oxides, and sulfate (Froelich et al. 1979). At the stage of iron reduction this degradation leads to gradual dissolution of magnetic primary ferric iron minerals such as magnetite, maghemite and hematite. This process is illustrated (Fig. 1) by exemplary rock magnetic and geochemical data of an active iron redox boundary located at about 45 cm depth (Fig. 1a) in an Equatorial Atlantic sediment core. Characteristic features of organically enriched layers (Fig. 1b) are local minima in magnetic mineral concentration (Fig. 1c) and a relative decrease of the finer vs. coarser magnetite fractions (Fig. 1d) and of ferrimagnetic vs. antiferromagnetic oxides, which is due to grain size and mineral selectivity of reductive mineral dissolution. The resulting ferrous iron anoxically precipitates *in situ* as paramagnetic phase or diffuses upwards to the active iron redox boundary to form authigenic, generally paramagnetic Fe^{3+} oxihydroxides as well as biogenic magnetite (Smirnov and Tarduno 2000). Relocation of iron

(Fig. 1e) is shown by the element ratio of (mobile) iron and (stable) aluminum (Thomson et al. 1999).

The most easily measurable and universal, hence widely used magnetic parameter susceptibility κ (Fig. 1f) cumulates all iron mineral concentrations with pronounced emphasis on ferrimagnetic species. Two sharp signal minima of susceptibility deviating from the general trend of other climate proxy signals, e.g. $\delta^{18}O$, are indicative, but not specific of diagenesis effects. This is unfortunate as core scanning techniques generate high resolution susceptibility logs at rates of about 100 data points per hour.

At half that speed runs a new logging device capable of measuring iron concentrations, the X-ray fluorescence (XRF) core scanner (Jansen et al. 1998). This fully automatic instrument developed and built at the Netherlands Institute for Sea Research (NIOZ, Texel) is also available at the Department of Geosciences, University of Bremen. For most of the exemplary sediment section Fe counts (Fig. 1g) mimic respective susceptibility signals as both parameters largely delineate terri-

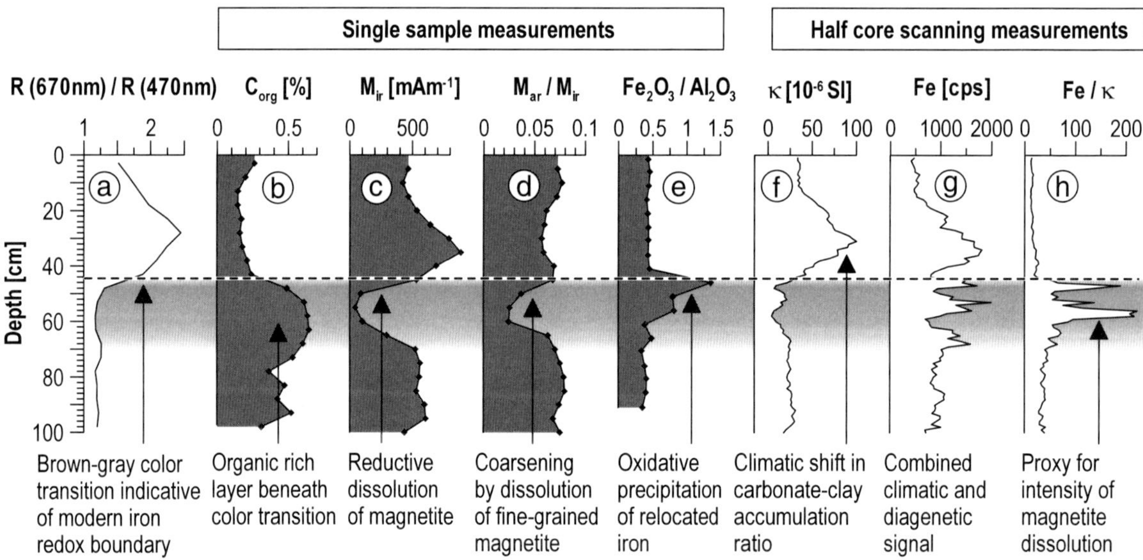

Fig. 1. Characteristic signatures of redoxomorphic iron mineral diagenesis at the active iron redox boundary (dotted line) exemplified by equatorial Atlantic gravity core GeoB 2908-7. Gray curve fillings represent single sample, white fillings half core scanning measurements. The horizontal band symbolizes the magnetite dissolution layer. Curves represent **a)** ratio of red and blue reflectance, **b)** organic carbon content, **c)** isothermal remanent magnetization, **d)** magnetogranulometric ratio, **e)** iron/aluminum ratio, **f)** magnetic susceptibility, **g)** iron content, **h)** iron/susceptibility ratio.

genous content. However, as XRF data are largely unbiased by mineralogy, they do not follow susceptibility variations due to alterations converting strongly magnetic into weakly magnetic iron species. The Fe/κ ratio of both logs (Fig. 1h) therefore highlights exactly those intervals, where such shifts occur. The double spikes correspond primarily to the two susceptibility minima of Fig. 1f, but partly also to a Fe precipitation layer at the Fe^{2+}/Fe^{3+} redox zone (Fig. 1g). Both are not resolved in the single sample measurements of Figs. 1b-d, which are followed by the broader underlying peak in Fig. 1h. We are not aware of any preceding examples where magnetic parameters were explicitly normalized by Fe and Ti content to define quantitative proxies for magnetic mineral diagenesis. However a good step in this direction was taken by Rosenbaum et al. (1996) by utilizing crossplots of concentratio-nal magnetic parameters against iron and titanium content to detect diagenetic magnetite and hematite loss in limnic sediments .

By combining both scanning techniques in the above depicted way, diagenetically affected sediment sections can be detected at a hitherto unprecedented speed and precision - provided that the primary composition of the terrigenous sediment fraction remains fairly constant. It is the aim of this contribution to investigate implications and prospects of this new technique and to discuss its results in the context of established element and rock magnetic analytics.

Materials

Quaternary sediment sequences of two gravity cores recovered from marginal positions of the Equatorial Atlantic Divergence Zone (Fig. 2) provide suitable conditions to demonstrate the full potential of the new methods described here. Both show organic carbon enrichments in glacial periods and related reductive iron mineral diagenesis of various intensities and frequencies.

Gravity core GeoB 2908-7 was recovered during RV *Meteor* cruise M 29/3 from 3809 m water depth in the western Equatorial Atlantic (00°06.4'N, 03°19.6'W) near the Central Equatorial Fracture Zone, gravity core GeoB 4317-2 during cruise M 38/1 (Fischer et al. 1998) from 3507 m water depth on the eastern slope of the Mid-Atlantic Ridge (04°21.3'N, 30°36.2'W). These sediments span the last 360,000 years and 500,000 years, respectively. The foraminiferal/nannofossil oozes

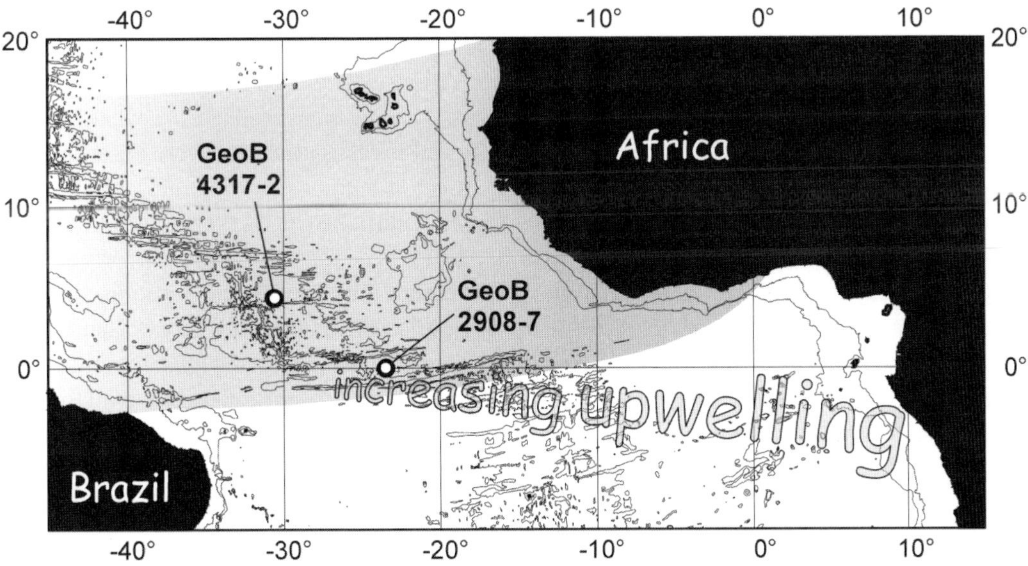

Fig. 2. Study area with sites of gravity cores GeoB 4317-2 and 2908-7. Saharan dust fall area is marked in gray. Eastwards increasing upwelling at the Equatorial Divergence enhances productivity and organic carbon accumulation.

feature continuous glacial/interglacial cycles manifested in oxygen isotopes, carbonate content and organic carbon content. The sampling sites are located within the trajectory of the boreal winter NE trade wind, represented by the Passat and Harmattan wind systems. They receive huge amounts of terrigenous material from the Saharan dust plume (Ruddiman et al. 1989), which is the dominant source of detrital iron and magnetic minerals in the Equatorial Atlantic (Bloemendal et al. 1992). Rainout within the Intertropical Convergence Zone is the major particle removal process from the atmosphere. The surface currents in this region are equally controlled by trade winds, generating the Equatorial Divergence upwelling system, an enhanced productivity region.

Methods

Single Sample Rock Magnetic Measurements

The work halves of both cores were sampled onboard at 5 cm intervals for paleo- and rock magnetic measurements (oriented 6.2 cm^3 cube samples) and later resampled at the same positions for powder XRF element analyses.

After paleomagnetic analysis, the cube samples were investigated by standard remanence-based measurements carried out with the automated *2G* SQUID rock magnetometer of the Marine Geophysics Division at the Department of Geosciences, University of Bremen. Anhysteretic remanent magnetization (M_{ar}), a grain-size selective parameter primarily quantifying submicron magnetite (single/pseudo-single domain size) was imparted in an alternating field decaying from 250 mT with a superimposed constant field of 0.04 mT. Isothermal remanent magnetization (M_{ir}), acquired in a pulsed field of 250 mT, equally quantifies magnetite, but with considerably less influence of particle size. The M_{ar}/M_{ir} ratio (Maher 1988) is therefore a well-established magnetite grain-size index. Saturation isothermal remanent magnetization (M_{sir}) acquired in a pulsed field of 2.5 T was subsequently overprinted by a back field of -0.3 T to evaluate the relation of low coercive (magnetite) to high coercive (hematite, goethite) mineral concentrations, the

so-called S-ratio (Bloemendal et al. 1992) defined as

$$S_{-0.3T} = 0.5 \cdot \left(1 + M_{-0.3T} / M_{sir}\right)$$

Kruiver and Passier recently demonstrated (2001) the presence of considerable internal variability and overlap within the coercivity distributions of both magnetite and hematite, which may mo-dulate this parameter independent of actual relative mineral concentrations. The above given standard interpretation of $S_{-0.3 T}$ should therefore be treated with caution when dealing with drastically changing grain sizes or mineral phases of various (lithogenic, authigenic, biogenic) origin.

The viscous loss of the initial M_{sir} in 24 hours was termed M_{vr}. 'Magnetic viscosity' is due to metastable single domain magnetite particles near the superparamagnetic (SP) threshold grain-size of approximately 20 nm (Butler and Banerjee 1975). Similarly fine SP hematite particles may act in the same sense (Banerjee 1971). The ratio M_{vr}/M_{sir} therefore quantifies the relative contribution of ferri-and antiferromagnetic particles with grain sizes near the SP/SSD threshold. Its general tendency compares to the more customary, but less sensitive frequency-dependent susceptibility $\kappa_{fd\%}$, which is commonly used to quantify the ultra-fine magnetite fraction in environmental materials (Dearing et al. 1996). Because of instrumental limitations of the used *Bartington* Susceptometer, the $\kappa_{fd\%}$, data measured on the weakly magnetic material investigated here were regarded as being to noisy to support detailed interpretation and are therfore not shown. Nevertheless, all measured $\kappa_{fd\%}$, data sets are, within their error limits, in full agreement with the M_{vr}/M_{sir} ratio used here.

Single sample susceptibility (κ) data of all samples are also not shown, as they coincide very well with scanning susceptibility data available at much higher resolution. They were, however, used to determine the carbonate free dry mass susceptibility χ_{cfdm}. This parameter is derived from single sample volume susceptibility measurements by mathematically eliminating the effects of porosity, pore water and carbonate dilution. It refers exclusively to fluctuations of magnetic mineral linked to diagenesis or terrigenous sedimentation.

For hysteresis measurements miniature samples < 50 mg have been prepared using a technique described by von Dobeneck (1996). Measurements were carried out with a *PMC* M2900 alternating gradient force magnetometer. By processing with the HYSTEAR program (von Dobeneck 1996), we derived basic hysteresis parameters such as the specific saturation magnetization σ_s and remanent magnetization σ_{rs}, the magnetogranulometric M_{rs}/M_s ratio by Day et al. (1977), the coercive field B_c, and the median field B_{rh} of the remanent (symmetric) hysteresis component. The equally hysteresis-based susceptibility parameter χ_{nf} quantifies contributions of paramagnetic and diamagnetic sediment matrix constituents, while χ_{tot}, the slope of the induced (antisymmetric) hysteresis component at zero field, cumulates all induced magnetizations. The mass specific ferrimagnetic susceptibility χ_{fer} is their difference:

$$\chi_{fer} = \chi_{tot} - \chi_{nf}$$

Single Sample Element Analyses

The organic carbon content was determined with a *LECO* CS-300 infrared analyzer after removal of carbonate by 6 % HCl. Analytical accuracy was checked using a standard every 10 to15 sample.

Element analyses of the first meter of core GeoB 2908-7 were performed at 1 cm resolution by Inductively Coupled Plasma Atomic Emission Spectroscopy (ICP-AES). After total digestion of 50 mg freeze-dried sediment in a mixture of 3 ml HNO_3 (65% s.p.), 2 ml HF (48% s.p.) and 2 ml HCl (30% s.p.) at 200°C and evaporating the solution, using a microwave device, the residual was homogenized with a solution of 0.5 ml HNO_3 (65% s.p.) and 4.5 ml MilliQ-water, also in the microwave. For each rank, one blank and a reference standard (MAG-1, USGS; Gladney and Roelandts 1988) have been treated like the samples. The resulting solutions were analyzed with a *Perkin-Elmer* optima 3300 RL ICP-AES system. The relative standard deviation is less than 3%.

Pressed powder tablets of selected sediment layers from both studied cores were also analyzed for absolute concentrations of major and minor elements using a wavelength dispersive *Philips* PW 1400 X-ray fluorescence spectrometer. Prior to analysis, the samples were disintegrated by ultrasonic treatment, dialyzed and washed to remove pore water, ball-milled and pressed.

These two very precise element analytics are far too laborious to be generally applied for high resolution studies of full sediment cores or core collections. They were only applied in specific sections and replaced by the much faster XRF scanning for the remaining core lengths. The comparison of absolute single sample and relatice scanning methods provides a basis for calibration and error estimate for the latter.

In this study mainly the terrigenous elements Al, Fe and Ti are of interest. To detect relocations of redox sensitive elements, dilution effects resulting from changes in biogenous accumulation must be compensated, which is commonly done by normalizing their concentrations to Al or Ti. According to Thomson et al. (1999), this procedure implies the following inherent assumptions:

Al and Ti are conservative elements which do not suffer appreciable diagenesis (Thomson et al. 1998). Carbonate and opal have insignificant contents of redox sensitive elements. Each terrigenous element under study is assumed to be deposited at a constant ratio relative to Al and Ti (same source material). Aluminum resides naturally in alumosilicates associated with the silt and clay fraction of the sediment. Thus, it is considered as a surrogate of fine-grained material (Mudroch and Azcue 1995). During continental weathering Al, Ti, Fe, and Mn are relatively immobile and are therefore considered as refractory elements (Canfield 1997).

Scanning Susceptibility Measurements

The magnetic volume susceptibility κ was determined with a custom-made, automated split-core susceptibility scanner employing a *Bartington* MS2 susceptometer with a high resolution MS2F (Ø15 mm) spot sensor (Fig. 3.a). Its lateral sensitivity is approximated by a Gaussian distribution with a half width of 12 mm. Vertically, a layer of 10 mm contributes over 90 % to the signal (Nowaczyk and Antonow 1997; Lecoanet et al. 1999). Measurements were taken at 1 cm spacing on archive halves in the sensitive mode. Instrument drift was controlled after each step by a zero measurements in air. The raw data were specified in SI units using an empirically derived calibration factor of 18.1. With a digital precision of 0.1 scale units, the numerical resolution is therefore in the order of $\pm 1 \cdot 10^{-6}$ SI.

Strictly speaking, magnetic susceptibility is not directly proportional to ferri- or paramagnetic mineral content, because many matrix minerals (e.g. calcite, opal) and pore water have a weakly negative diamagnetic background susceptibility shifting κ to lower, sometimes negative values. This effect can create numerical artefacts in ratios using κ as denominator. An effective counter-strategy used here is to simply subtract a flat value κ_{dia} for diamagnetism from κ yielding a 'non-diamagnetic' susceptibility κ_{nd}:

$$\varkappa_{nd} = \varkappa - \overline{\varkappa}_{dia} \approx \varkappa + 15 \cdot 10^{-6}$$

This value is a minimum estimate based on typical values for pure calcite, opal and water (Thompson and Oldfield 1986). A more specific value for κ_{dia} could well be determined on the basis of known porosities and diamagnetic mineral contents. In that case, reliable susceptibility estimates for all matrix components (calcite, opal, silicates) would be required, but are typically not at hand for non-stoichiometric mineral phases. Circumventing these problems, the proposed flat value for κ_{dia} can simply be justified by the fairly uniform background diamagnetism inherent to every atom, which is even present in the case of a para- or ferrimagnetic component.

Scanning XRF Measurements

Relative element contents of potassium (K), calcium (Ca), titanium (Ti), manganese (Mn), iron (Fe), copper (Cu) and strontium (Sr) were determined at a depth resolution of 1 cm using a *NIOZ* XRF core scanner (Fig. 3.b). This computer controlled logging system has been developed for fast, non-destructive major element analysis on split sediment cores. The output data represent relative element concentrations, which are given in counts per second [cps]. The core halves are fixed by a

pneumatic sample holder and passed at preset positions along the source and detection unit by a stepping motor. The sensor averages over an area of 1 cm². The response depth ranges from tenths to hundreds of a μm for the above mentioned element range. The instrument and its applications are described by Jansen et al. (1998) and Röhl and Abrams (2000). Absolute detection limits depend on measurement time, surface and lithology and are therefore not easily expressed. The limitations of this instrument are clearly demonstrated by comparative diagrams of single sample and scanning data in the following chapter.

Results

Combined Geochemical and Rock Magnetic Stratigraphy of Core GeoB 4317-2

The organic carbon content of core GeoB 4317-2 varies rhythmically between 0.08 and 0.5 wt.% with a mean of 0.20 wt.% (Fig. 4a). The changes observed are attributed to a glacial-interglacial cyclicity of productivity (Lyle 1988) with enhanced C_{org} deposition during cold periods, particularly marine oxygen isotope stages (MIS) 6, 10 and 12. According to the definition by Kidd et al. (1978), discrete layers thicker than 1 cm containing 0.5 to 2.0 wt.% of organic carbon are denoted as *sapropelic*, with over 2.0 wt.% as *sapropels*. With presently almost oligotrophic conditions and less than *sapropelic* C_{org} contents in glacial sections, the sediments were nevertheless zonally affected by suboxic diagenesis. The following data clearly demonstrate that sequences with a relatively high organic carbon content correlate with zones of marked magnetite dissolution (gray shaded horizons in Fig. 4). The magnetogranulometric ratio M_{ar}/M_{ir} (grain-size proxy of ferrimagnetic minerals, Fig. 4b) is sensitive both to variations in the primary terrigenous input and to iron oxide dissolution. Fine grained magnetite is more susceptible to dissolution than coarser grained magnetite because of its higher surface-to-volume ratio, resulting in a shift to coarser grain-size distributions (Karlin and Levi 1983). Most parts of the M_{ar}/M_{ir} signal of core GeoB 4317-2 show a cyclic change of magnetite

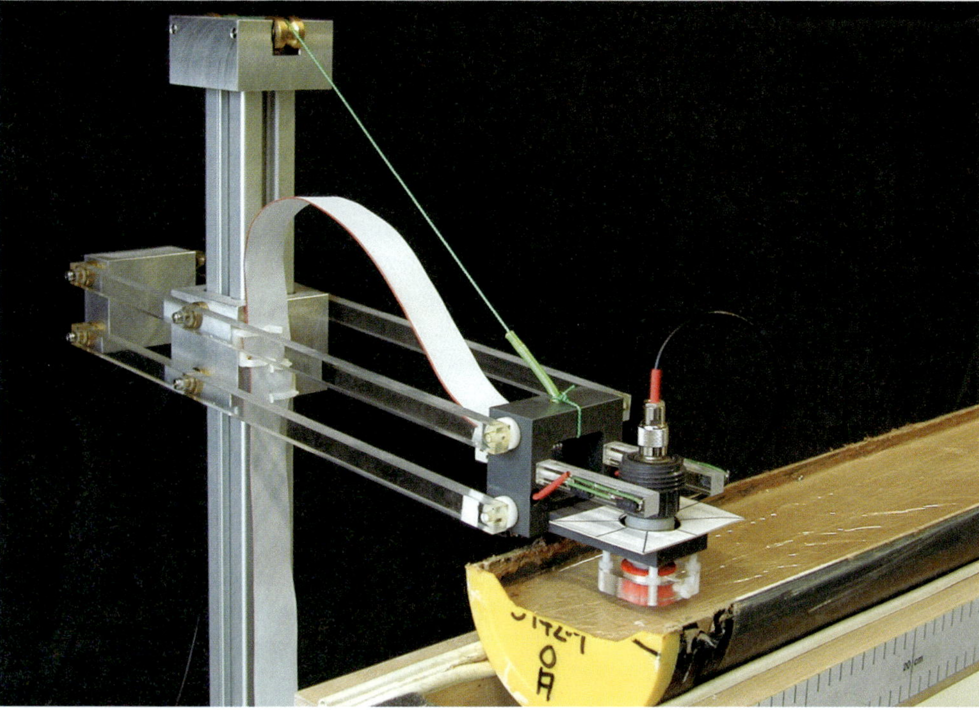

Fig. 3. a) Automated magnetic susceptibility half core logger with a self-leveling Bartington MS2F spot sensor mounted on optically controlled lever.

Fig. 3. b) Automated X-ray fluorescence half core scanner for non-destructive element analysis with X-ray source (left), prism attached to core surface (center, core not shown) and X-ray detector (right).

grain-size related to climatic variations of wind intensity and/or to productivity related effects of a mild reductive diagenesis. Magnetite dissolution layers (DL) I and more clearly II and III deviate from this pattern featuring an abrupt coarsening at or slightly below the C_{org} maxima and a peak of relative fining due to reprecipitation directly above. The lower pervasiveness of magnetite reduction in DL I as compared to DL II and III may result from the shorter duration of suboxic conditions, ie. from the respective burial history. Mild forms of reduction affect primarily ultra-fine (SP) magnetic ferric minerals. They are therefore less apparent in remanence-based parameter records such as M_{ar}/M_{ir}, but still prominent in parameters derived from induced magnetizations such as χ_{nf}/χ_{tot} (Fig. 4h)

The $S_{-0.3T}$ index (Fig. 4c) mirrors the ratio of high- to low-coercive components, in general hematite and magnetite (Bloemendal et al. 1992). This parameter shows distinct minima, relative hematite enrichments, in the dissolution layers, suggesting that reductive diagenesis has a lesser effect on hematite than on magnetite. This observation will be further substantiated and discussed at the end of this chapter.

Pronounced $CaCO_3$ variations ranging from 35 to 84 wt.% reflect mainly cycles of carbonate dissolution due to glacial emplacement of corrosive Circumpolar Deep Water (de Menocal et al. 1993; Bickert and Wefer 1996). In absence of significant opal contents the non-carbonate sediment fraction (Fig. 4e) is largely representative of the terrigenous content (Schmieder et al. 2000). In most parts it matches the magnetic susceptibility record κ in great detail. However, both signal clearly diverge in the three major dissolution layers I, II and III, where susceptibility (Fig. 4d) is far too low for glacial conditions (Bloemendal et al.1988). As reductive diagenesis involves transformation of magnetic into non-magnetic iron phases, the correlation of rock magnetic logs to climatic cycles is strongly compromised in these intervals. In interglacial layers, carbonate maxima resulting from much better preservation conditions correspond reasonably well to susceptibility lows.

The non-ferrimagnetic susceptibility (χ_{nf}, Fig. 4f) is the slope of the linear outer branch of a saturation hysteresis loop (von Dobeneck 1996). This parameter quantifies contributions of paramagnetic phases plus a small, nearly constant offset due to diamagnetic matrix minerals. Relative Fe counts measured with the XRF scanner (Fig. 4g) and χ_{nf} closely mimic each other throughout the sediment sequence implying that most iron is located in paramagnetic mineral phases like iron-bearing silicates. The much higher level of conformity of non-$CaCO_3$ with total Fe records compared to magnetic susceptibility underlines this statement.

The ratio of non-ferrimagnetic to total susceptibility (χ_{nf}/χ_{tot}, Fig. 4h) delineates unequivocally where magnetite dissolution takes place. The prevailing stable baseline value (~ 0.15) represents an average 15 % contribution of paramagnetism to total susceptibility, a characteristic source signature. This parameter value doubles or even triples at the peaks of reductive dissolution implying that about two thirds of the primary magnetite have vanished. The ratio Fe/κ_{nd} (Fig. 4i), based exclusively on fast logging measurements, follows the χ_{nf}/χ_{tot} trends in most every aspect, but provides much higher spatial resolution and detail at a fraction of measurement time. Both parameters image magnetite depletion layers as distinctive, internally subdivided sawtooth patterns with upwards increasing dissolution levels. For the three major dissolution layers, the diagenetic features stretch vertically over about 100 cm. The slightly elevated Fe/κ_{nd} signals at numerous other mildly reductive layers are much narrower and also more symmetric features.

For reasons elaborated in the methods section, redox sensitive elements as Fe are often normalized to the stable terrigenous element Al (Karlin 1990a, b). As this light element is not within the range of the XRF scanner (K to Sr), we use here the immobile terrigenous element Ti as normalizer assuming that the sedimentary iron and titanium minerals are identical or have a common origin. From a fairly constant background level the Fe/Ti ratio (Fig. 4j) shows distinct multiple enrichment peaks immediately above the lower two dissolution layers, while the effect above dissolution layer I is smaller. A sharp subsurface peak at 11 cm depth, a broader peak at 540 cm and many less developed peaks pinpoint additional faint iron precipitation horizons associated with underlying layers of partial magnetite dissolution.

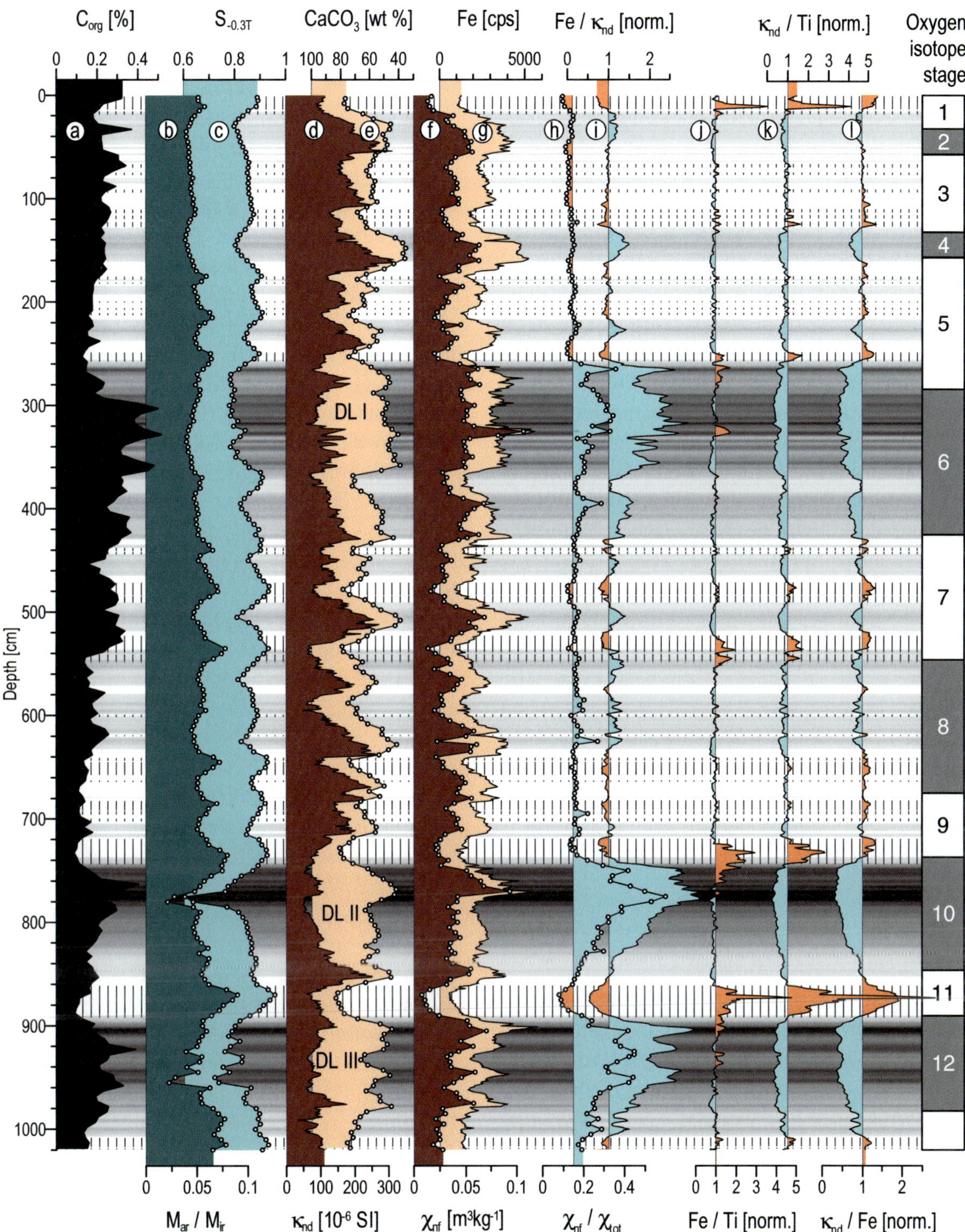

Fig. 4. Geochemical and rock magnetic profiles of core GeoB 4317-2. Labels DL I to III indicate major magnetite dissolution layers. Background shading indicates degree of magnetite dissolution based on Fe/κ_nd. Magnetic mineral precipitation zones based on κ_nd/Ti are shown as hatched areas.

An alternative and related, but magnetic mineral selective Fe precipitation proxy is κ_{nd}/Ti (Fig. 4k), also derived by combining susceptibility and XRF logs. Other than the Fe/Ti signal, κ_{nd}/Ti has no stable baseline as it also registers magnetite depletion (inversely to Fe/κ_{nd}). The iron residing in nonmagnetic minerals is either not bioavailable, immobile or ferrous, while the ferric iron in all magnetic minerals is generally bioavailable (Lovley et al. 1987). Subtle distinctions of the lower slopes of the Fe/Ti and κ_{nd}/Ti peaks, particularly at the upper boundary of magnetite dissolution layer II, probably result from cancellation effects of superimposing magnetite depletion and subsequent precipitation.

Magnetic mineral precipitation cannot be seen in susceptibilities, only the M_{ar}/M_{ir} and $S_{-0.3T}$ signals indicate enrichment by a very fine, probably bio-genic (Karlin et al. 1987; Robinson et al. 2000; Tarduno et al. 1998) magnetite phase. Obviously, this secondary magnetite has been precipitated after reductive dissolution of primary magnetite during partial burn-down of the initially broader C_{org} enrichment layer.

The κ_{nd}/Fe ratio (Fig. 4.l) indicates magnetic mineral depletion as well as enrichment effects and is therefore a bivalent proxy. However, as magnetic and total iron enrichment often go along, this parameter may not identify precipitation correctly. Defining a baseline value for standardization is also less evident.

Combined Geochemical and Rock Magnetic Stratigraphy of Core GeoB 2908-7

Core GeoB 2908-7, recovered about 450 km southeast of core GeoB 4317-2, has on average 35 % higher C_{org} contents with a mean of 0.27 wt.% and variations between 0.1 and 0.66 wt.% (Fig. 5a). The C_{org} signal carries a typical equatorial Atlantic climate signature, modulated by orbitally forced changes in trade wind zonality (Verardo and Mc Intyre 1994). Productivity is more pronounced at this site due to stronger upwelling and higher productivity towards the Equatorial Divergence Zone. More frequent productivity pulses cause higher sedimentation rates and greatly increase the number of organically enriched layers and of

resulting magnetite dissolution and precipitation horizons. While in core GeoB 4317-2 dissolution zones are mainly found at major terminations, core GeoB 2908-7 carries consecutive diagenetic features also at stadials (Fig. 5).

The M_{ar}/M_{ir} ratio of core GeoB 2908-7 (Fig. 5b) also shows much stronger overprint of the primary signal. Ten layers of intense dissolution appear as deep, notchlike grain-size shifts (abrupt magnetite coarsening) in the record. In combination with climatically unexpected susceptibility minima (Fig. 5d), these peaks indicate massive loss of fine-grained magnetite, whereas high-coercive iron mineral phases ($S_{-0.3T}$, Fig. 5c) are less affected. A synopsis of C_{org}, dissolution and precipitation proxies suggests that also many minor M_{ar}/M_{ir} and $S_{-0.3T}$ shifts may be related to partial magnetite dissolution in organically enriched horizons. It is not totally clear from these data, however, to what extent these signatures still carry traits of grain-size and mineral variations of primary detrital iron oxides.

The wide range of susceptibilities κ_{nd} from 0.9 to $100 \cdot 10^{-6}$ SI documents two superimposing influences: The eolian iron input from the African continent is systematically lower at this more southern site (Ruddiman et al. 1989) and frequently overprinted by reductive dissolution. The large scale variations in $CaCO_3$ content from 51 to 90 wt.% mainly reflect intensified carbonate dissolution due to glacial advances of Circumpolar Deep Water (Bickert and Wefer 1996). This cyclical dilution of the terrigenous (non-$CaCO_3$) phase (Fig. 5e), is inversely correlated to magnetic susceptibility.

Over most of the sediment column this initial accumulation pattern is also conserved in the very similar non-ferrimagnetic susceptibility (χ_{nf}, Fig. 5f) and bulk iron (Fe Fig. 5g) records. Distinct narrow discrepancies from 40 to 55 cm, at 440 cm and at 590 cm, all can be attributed to precipitation of non-magnetic iron phases. This interpretation is clearly supported by enhanced Fe/κ_{nd} ratios (Fig. 5i), indicating that this parameter does not only identify dissolution of ferrimagnetics (decreasing denominator κ_{nd}), but also precipitation of paramagnetics (increasing numerator Fe).

Fine-scaled variability is observed in the ratios of ferrimagnetic, non-ferrimagnetic and total Fe

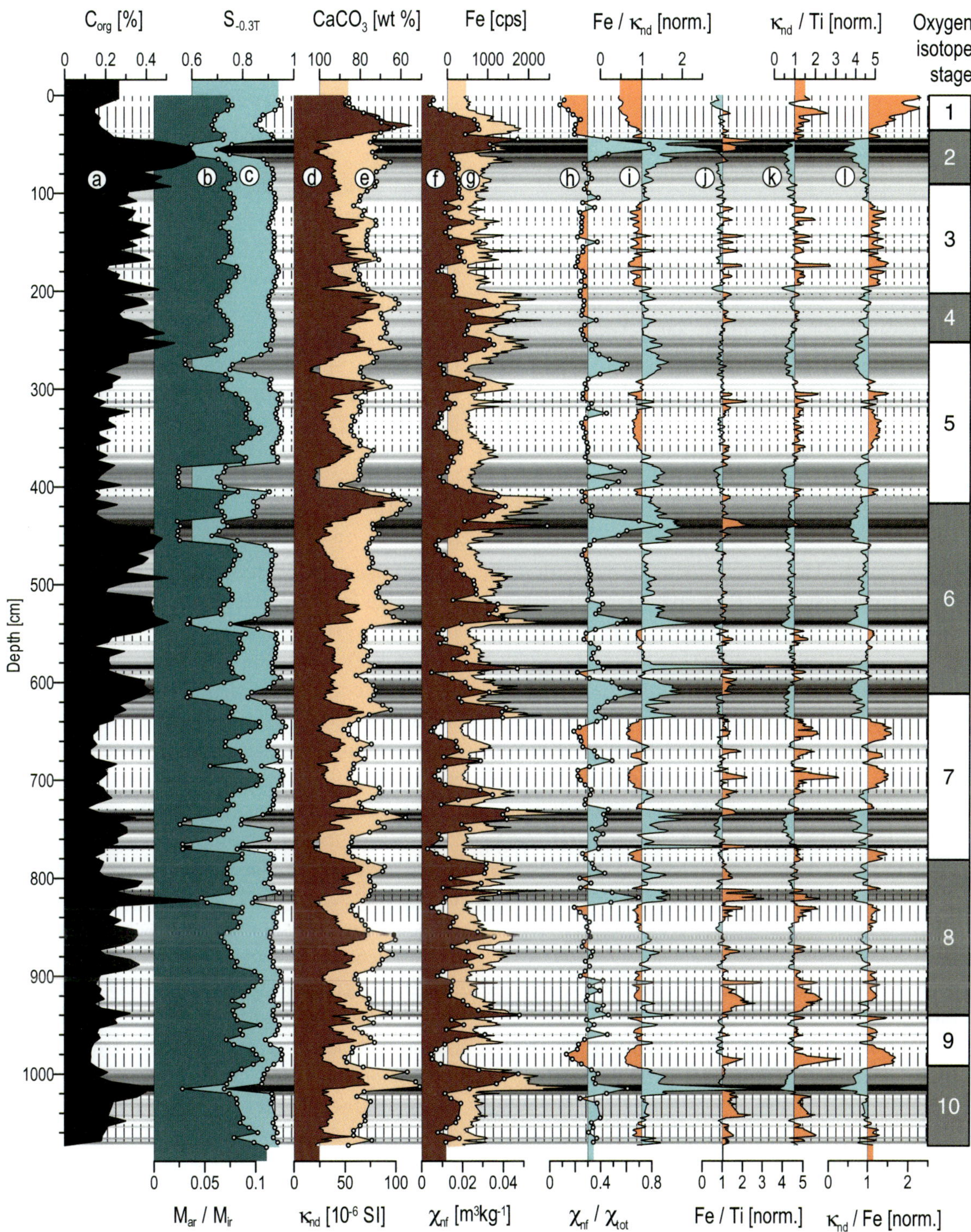

Fig. 5. Geochemical and rock magnetic profiles of core GeoB 2908-7. Background shading indicates magnetite dissolution stage based on Fe/κ_{nd}. Magnetic mineral precipitation zones based on κ_{nd}/Ti are shown as hatched areas.

(Fig. 5h,i,l) and respective relations to Ti (Fig. 5j,k). In spite of lower signal levels, particularly regarding Ti counts, and therefore poorer signal-to-noise ratios, complex suites of alternating fine-scale depletion and enrichment layers are recognized. Some precipitations situated within dissolution zones are purely paramagnetic (e.g. at 445 and 820 cm) and do not appear in the κ_{nd}/Ti ratio.

Detailed Analysis of the Active Fe Redox Boundary in Core GeoB 2908-7

For a better definition of redoxomorphic processes and proxy parameters, additional and more detailed rock magnetic and geochemical analyses of the subsurface sediments encompassing the present-day iron redox boundary are presented in Figure 6. The same section already shown in the introductory Figure 1 and as top of the core GeoB 2908-7 logs in Figure 5 is now presented with an extended parameter spectrum to illustrate and discuss alternative methods. The 28 selected rock magnetic and geochemical data sets are displayed in a three-part compilation of concentration (Fig. 6.I), dissolution (Fig. 6.II) and precipitation (Fig. 6.III) signatures. A comparison of data sets from low (5 cm spacing, circles) and high resolution (1 cm spacing, dots) single sample measurements with core logging data (no symbols) shows assets and drawbacks of the traditional versus scanning methods. High-resolution magnetic hysteresis (dark gray) and ICP-AES element measurements (hatched) covering the top 100 cm of core GeoB 2908-7 in 1 cm steps allow to verify and investigate some fine-scaled structures (horizontal lines) visible in the combined XRF/susceptibility scanning data, to calibrate XRF logging data and to determine their noise floor.

The organic carbon enrichment at 44-78 cm (Fig. 6.I a) triggers reductive dissolution of magnetite (Fig. 6.I b) and, to a minor extent, hematite (Fig. 6.I c) affecting visibly the total susceptibility signal (Fig. 6.I d), but not its paramagnetic component (Fig. 6.I e). The latter log however resembles the total Fe content (Fig. 6.I f,g) and shows three conspicuous peaks within the reductive layer. These are neither found in the Ti (Fig. 6.I h,i) nor in the magnetic iron mineral (Fig. 6.I b,c) records and must therefore be regarded as non magnetic

Fe precipitation. In contrast, the broad iron enrichment peak from 18 to 42 cm is shared by Ti and κ logs and represents a primary relative maximum of terrigenous accumulation. Comparing Fe and Ti data sets (Fig. 6.I f-i), determined both by scanning and single sample methods, it is apparent that in case of Fe (Fig. 6.I f,g) even the fine-scaled features largely coincide, while for the rarer element Ti (Fig. 6.I h,i), the noise floor of the scanning XRF measurement obviously disturbs the signal.

The next series of plots (Fig. 6II) groups relational, i.e. concentration independent active iron redox boundary over a width of about 20 cm (coarse residue fraction). The coercive field (Fig. 6.II c) shows a superposition of two effects: a low in the reduction zone (loss of magnetically hard SD/PSD particles), but also a noticeable increase in coercivity between 35 to 45 cm. This gradient delineates the irreversible loss of ultra-fine grained (< 30 nm), magnetically very soft SP magnetite. As SP particles have considerably higher specific susceptibilities than all other magnetite grain-size fractions (Heider et al. 1996), their depletion is equally reflected in a sharp decline of χ_{cfdm} (Fig. 6.II f). Losses are at maximum in C_{org}-enriched layers, where reactive carbon and magnetite nanoparticles coexist in intimate contact enabling or facilitating redox reactions between both solid phases.

As earlier mentioned, the $S_{-0.3\,T}$ parameter (Fig. 6.II d) mirrors the reductive layer as a relative enrichment of hematite compared to magnetite. It is not sure, though, whether this actually implies a higher dissolution resistance of hematite. From a geochemical viewpoint, the hematite reduction is energetically favorable to magnetite reduction and has a much higher reaction rate constant (Canfield et al. 1992; Haese 2000). This argument regards only isolated grains but not inclusions in silicate minerals that are protected against reductive dissolution (Walden and White 1997). But a purely rock magnetic argument may be more decisive: coarse and fine hematite have similar remanence carrying capacities (both are SD particles), while coarse magnetite (MD) particles hold much less remanence than fine (SD/PSD). Consequently, the S-Ratio relates a grain-size independent hematite content to a measure of magnetite

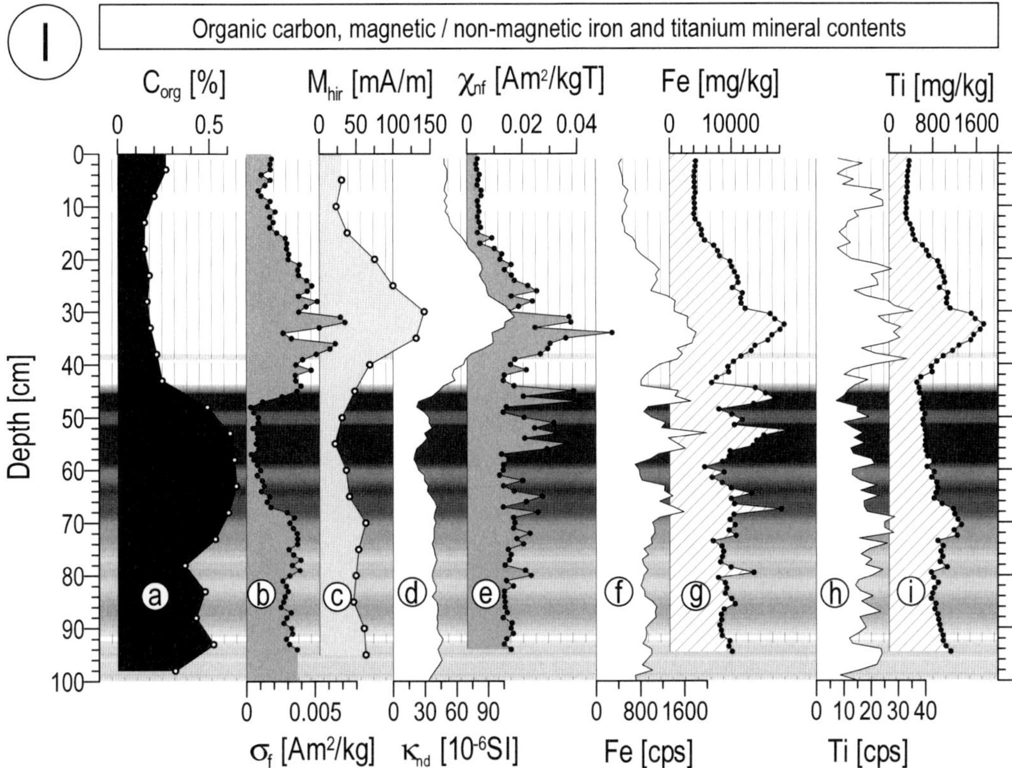

Fig. 6. I. Magnetomineral and element concentration profiles (core GeoB 2908-7, 0-100 cm) encompassing the active Fe redox boundary. The shown parameters quantify **a)** organic carbon, **b)** magnetite, **c)** hematite, **d)** ferri- and paramagnetism, **e)** paramagnetic iron, **f, g)** total iron and **h, i)** titanium. The latter element data were acquired by XRF scanning (f, h) and single sample ICP-AES (g, i) measurements, respectively. Background shading indicates the degree of magnetite dissolution based on Fe/κ_{nd}. Magnetic mineral precipitation zones according to κ_{nd}/Ti are shown as hatched areas. Plot symbols and fillings refer to sampling density and method (see legend Fig. 6.III).

content, which is heavily biased towards small (sub-micron) particles. Any diagenetic process reducing the sub-micron fractions of both minerals proportionally would therefore effectively shift the $S_{-0.3\,T}$ parameter to smaller values. This grain-size effect introduced by the SD remanence and coercivity of residual coarse hematite equally explains the other-wise surprising increase of the remanent coercive field B_{rh} (Fig. 6.II e) over the magnetite depletion layer.

As shown in Figure 5 with less spatial resolution, the two magnetite dissolution proxies χ_{nf}/χ_{tot} (Fig. 6.II g) and Fe/κ_{nd} (Fig. 6.II h) clearly delimit the diagenetically affected zone by multiple overlapping peaks protruding from a stable plateau level. The Fe/χ_{fer} ratio (Fig. 6.II i), combining high resolution XRF and hysteresis measurements, delineates this even more impressively, because the de-

nominator measures strictly ferrimagnetic iron and can therefore approach near zero values boosting the ratio to high numbers. In case of the χ_{nf}/χ_{tot} and Fe/κ_{nd} ratios, the denominators have positive lower limits in paramagnetic susceptibility. It could, however, be argued that peaks in all three ratios may not result from decreasing denominators, but rather from an increase of the numerator χ_{nf} or Fe, i.e. non-magnetic iron precipitation. Since the primary accumulation of Fe and Ti is highly proportional (Fig. 6III a,d) and Ti is minimally affected by diagenetic mobilization (Brown et al. 2000), the latter element should be an unambiguous normalizer for susceptibility. Two alternatives to Fe/κ_{nf} and Fe/χ_{fer} are therefore the Ti-based ratios Ti/κ_{nd} (Fig. 6II j) and Ti/χ_{fer} (Fig. 6II k). These AGFM and ICP-AES derived data convincingly document the high degree of coincidence between Fe- and

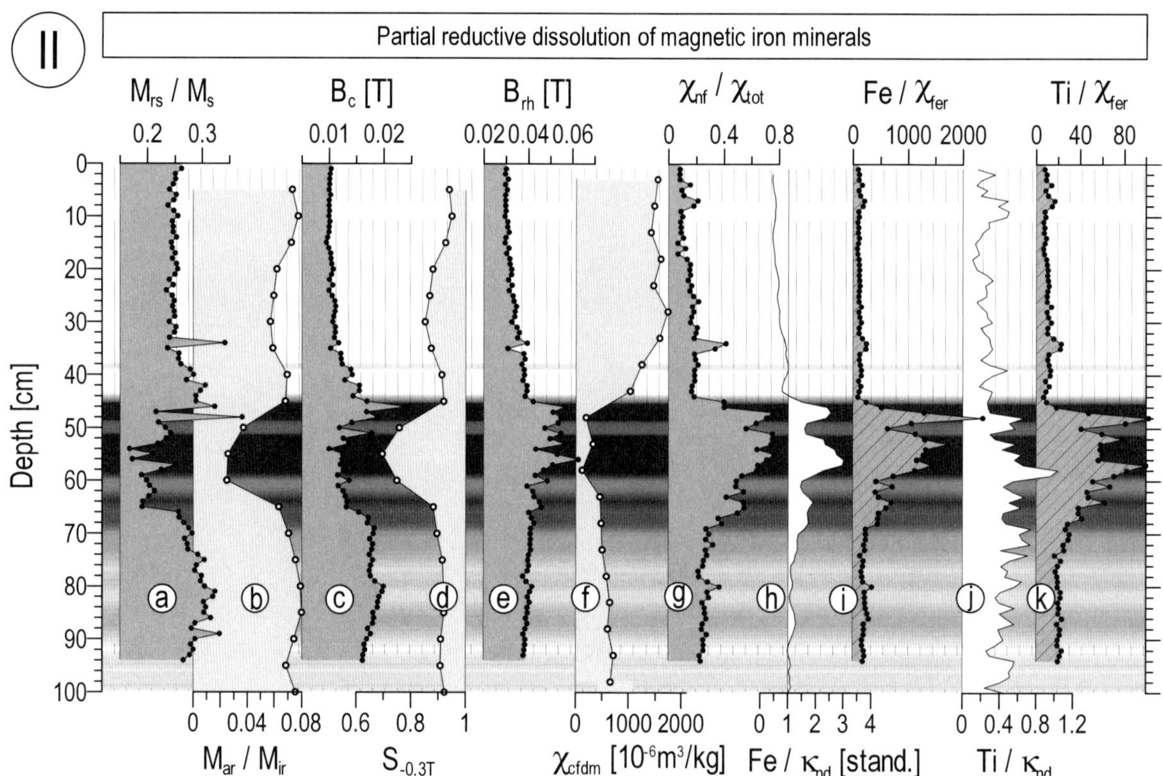

Fig. 6. II. Integrated rock magnetic and geochemical profiles (core GeoB 2908-7, 0-100 cm) indicating reductive iron diagenesis below the active Fe redox boundary. The shown parameters measure **a)** magnetite grain-size (domain state) after Day et al. (1977), **b)** relative contribution of fine grained magnetite, **c)** coercive field, **d)** S-ratio (values below 1 denote increasing influence of high-coercive magnetic iron oxides, primarily hematite), **e)** hysteresis derived remanence coercive field, **f)** carbonate-free dry mass susceptibility, **g)** paramagnetic contribution to susceptibility. Newly proposed magnetite dissolution proxies relating **h, i)** Fe content or **j, k)** Ti content to nondiamagnetic or ferrimagnetic susceptibility. The ratios are based on XRF and susceptibility scanning measurements (h, j), and on single sample ICP-AES and hysteresis data (i, k); normalization is explained in chapter 5.2. Background shading indicates the magnetite dissolution stage based on Fe/κ_{nd}. Magnetic mineral precipitation zones according to κ_{nd}/Ti are shown as hatched areas. Plot symbols and fillings refer to sampling density and method (see legend Fig. 6.III).

Ti-related magnetic mineral dissolution records and dispel doubts on major Fe precipitation effects in the present case. The scanner-based equivalents (Fig. 6II h,j) generally agree, but the poor signal to noise level of the Ti measurement seriously reduces the interpretability of the Ti/κ_{nf} ratio leaving Fe/κ_{nf} without alternative.

The third series of plots (Fig. 6 III) deals with Fe relocation and precipitation effects and starts out by comparing the inert terrigenous normalizers Ti and Al. As in other marine records (van Santvoort et al. 1996) the Al-normalized Ti profile (Fig. 6 III a) remains almost constant and distinctly different from the Fe/Ti (Fig. 6 III b,c) and Fe/Al

(Fig. 6 III d) ratios. It is therefore clear that the sharply defined maxima in logs III b-d at and below the Fe redox front can only result from diagenetic iron enrichment (König et al. 1997) which more than doubles the primary detrital Fe content. Interestingly, complementary minima (sources) visible in the Fe$_{excess}$ record (Fig. 6 III e), defined as

$$Fe_{excess} = Fe_{tot} - \left(Al_{tot} \cdot [Fe/Al]_{average}\right)$$

where Fe$_{tot}$ (Al_{tot}) refers to the measured element content per sample, are much shallower and broader than the maxima (sinks) implying that the mobi-

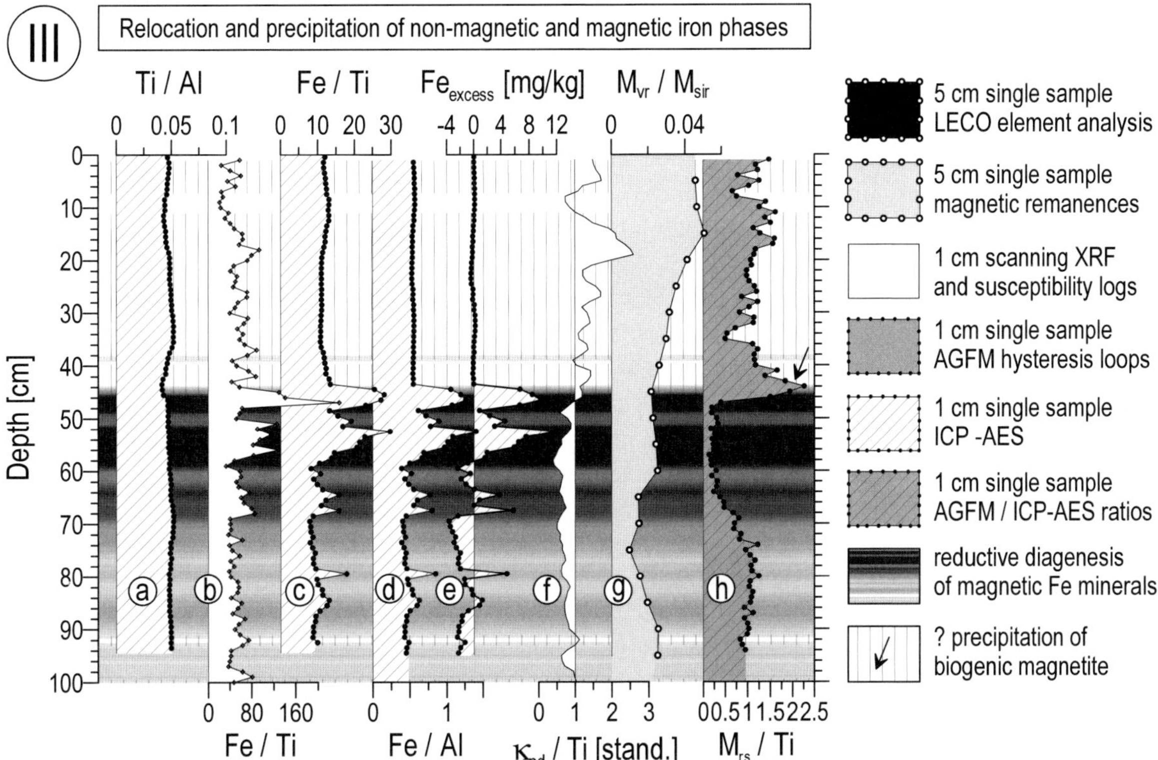

Fig. 6. III. Integrated rock magnetic and geochemical profiles (core GeoB 2908-7, 0-100 cm) indicating iron relocation and precipitation in and above the active Fe reduction zone. The shown parameters delineate **a)** equivalence of Ti and Al as normalizers of terrigenous input, iron enrichment detected by element ratios of **b)** Fe/Ti (XRF scanner), **c)** Fe/Ti (ICP-AES) and **d)** Fe/Al (ICP-AES), **e)** Fe excess based on Fe/Al ratio of $0-40$ cm. The newly proposed proxies quantify authigenic enrichment by **f)** ferrimagnetic Fe minerals (scanner data), **g)** ultrafine (SP) magnetite particles (single sample remanence data), **h)** fine (SD) magnetic Fe mineral particles (hysteresis and ICP-AES data), most likely bacterial magnetosomes. Background shading indicates the degree of magnetite dissolution based on Fe/κ_{nd}. Magnetic mineral precipitation zones according to κ_{nd}/Ti are shown as hatched areas. Plot symbols and fillings refer to sampling density and method (see legend).

lized and precipitated iron was drawn from an iron depletion zone largely exceeding the directly underlying magnetic mineral dissolution zone. This can also be seen by extra-polating the very stable primary Fe/Al ratio (Fig. 6 III d) of the oxygenated top layer downwards.

Three individual precipitation layers, each approximately 5 cm wide, can be discerned. They are located below, within and above the organically enriched and magnetically depleted zone and separated by magnetic mineral dissolution maxima. From geochemical considerations, two of these enrichments have to be regarded as relict peaks from former redox boundaries during prolonged non-steady state conditions. The theoretical ques-

tion (Burdige 1993), whether the upper (upward shift of the Fe redox boundary due to Holocene sedimentation) or the middle (downward shift of the Fe redox boundary to due to enhanced Holocene bottom water oxygenation) peak represents the active Fe precipitation zone is clearly answered by the sediment's color (Fig. 1a): The brown-green color transition (Lyle 1983) is located precisely at 45 cm depth.

In comparison to Fe/Al (Fig. 6 III d) and Fe/Ti (Fig. 6 III b,c) the κ_{nd}/Ti ratio (Fig. 6 III f) remains at a fairly stable low level over the precipitation zones implying that the iron relocated here primarily resides in nonmagnetic phases. The Ti-based scanner data are again too noisy to allow an interpretation of minor features. However, it is obvious that there is a marked upward increase in κ_{nd}/Ti from the Fe redox boundary (46 cm) culminating at a depth of 20 cm and sharply decaying above. Although no pore water data are available for this core, this interval is very likely to represent the nitrate zone, where magnetite biomineralization by anaerobic, nitrate respirating dissimilatory iron-reducing microorganisms (Lovley et al. 1987) and magnetotactic bacteria (Bazylinski et al. 1988; Blakemore 1975; Petermann and Bleil 1993) widely occurs. The shape of the κ_{nd}/Ti profile is paralleled by the M_{vr}/M_{sir} ratio (Fig. 6 III g) sensing ultra-fine, magnetically viscous magnetite particles (SP/SD transition; grain-sizes around 20 nm); (Worm and Jackson 1999). This magnetite fraction carries a two- to threefold higher susceptibility than all coarser fractions (Heider et al. 1996) and is therefore over-represented in susceptibility-based parameters. Its presence throughout the nitrate zone indicates a bacterially mediated reduction of amorphous ferric iron yielding superparamagnetic magnetite as metabolic byproduct and electron acceptor. As noted above, the minute size of these SP particles makes them a first victim of even mild reductive dissolution (Frederichs et al. 1999) and explains for their depletion from deeper sediment layers.

The M_{rs}/Ti ratio (Fig. 6 III h) focuses mainly on relative enrichments of slightly larger, magnetically stable magnetite particles (stable SD; grain-sizes 30-100 nm) and excludes SP particles. Its profile clearly marks the magnetite dissolution zone (46-70 cm), topped by a narrow, but clearly defined peak (40-46 cm, black arrow) just above the Fe redox boundary. This is probably an enrichment of fossil magnetosomes as described by Tarduno and Wilkison (1996). The producers, magnetotactic bacteria, are nitrate reducers and assimilate and oxidize ferrous iron available near the Fe redox boundary. Although potentially significant for paleo- and rock magnetic data (Tarduno et al. 1998), biogenic magnetite precipitation appears to play a negligible role in terms of absolute iron concentrations here as it is completely absent in the Fe/Al record (Fig. 6 III d).

Discussion and Conclusions

Efficiency, Data Quality and Calibration of Scanning Techniques

These studies demonstrate that combining rock magnetic and geochemical methods is very effective both in detecting redoxomorphic diagenesis of magnetic and nonmagnetic iron minerals and in distinguishing between variations in primary input and secondary overprint. At the present, traditional single sample methods still offer a greater variety, accuracy and selectivity of procedures and proxy parameters. The great advantage of the newly proposed proxies based on scanning techniques is therefore the efficiency of data acquisition. With the now available non-destructive XRF core scanner combined with well established high-resolution spot measurements of magnetic susceptibility, the data basis can be established in a small fraction of the time required for conventional single sample hysteresis and element analyses.

Sample preparation and data acquisition rates for the two alternative methods employed here are compiled in Table 1. Scanning methods are about 20 to 50 times faster than single sample techniques. Diagenetic layers are often fine-scaled and typically require centimeter spacing records for ample resolution. Only the scanning techniques permit to study long and multiple sediment cores at such high-resolutions in affordable time.

The price to pay for the enormous gain in speed is a loss in sensitivity. The necessary technical compromises accepted by transforming XRF spectrometry into a surface scanning technique clearly impair the quantification of minor elements such as Ti and Mn (Fig.6 I f-i). The theoretical, but not necessarily influential error sources of XRF analyses are manifold: grain-size, surface and matrix effects, instability of source and detector, sample roughness, inhomogeneity and porosity, count statistics, not to forget the imponderables of an automated algorithm transforming X-ray emission spectra into relative element concentrations. These effects make it in some instances questionable to transform XRF counts into absolute concentrations. As a detailed error assessment is complicated and beyond the scope of this paper, we simply ar-

Analytic method	Sample preparation procedures	Preparation time	Measurement time
XRF element analysis, powder measurement	Sampling, drying, grinding, weighting, pressing	60 min per sample	50 min per sample
ICP – AES element analysis	Sampling, freeze drying, homogenizing, digestion	35 min per sample	5 min per sample
Magnetic remanence measurements	Sampling	5 min per sample	10-60 min per sample
Magnetic hysteresis measurement	Sampling, drying, weighting, impregnating	20 min per sample	10-20 min per sample
XRF scanning element analysis	Cleaning of split core surface	5 min per 1 m core section	2 min per data point
Susceptibility scanning measurement	Cleaning of split core surface	5 min per 1 m core section	1 min per data point

Table 1. Comparison of single sample and scanning analytical methods.

gue on basis of repeatability and consistence with calibrated, precise single sample measurements.

In Figure 7 the XRF scanner data of Fe and Ti [wet bulk sample, counts per second] of core GeoB 4317-2 are correlated with calibrated conventional XRF measurements [wt.%] of single samples obtained from identical core depths. The linear regression lines and equations, the 95% confidence ranges and Pearson's correlation coefficients altogether indicate that for Fe contents of 1-5 wt.%, XRF counts can be reliably calibrated (Fig. 7a). In case of the lower Ti contents of 0.1-0.5 wt.%, the scatter and zero-offset of scanning XRF analyses is considerably larger, yet there is sufficient correlation to interpret at least first order variations quantitatively (Fig. 7b).

For the scanning susceptibility measurements the potential error sources are fewer and less problematic. Unevenness and inhomogeneity of the core surface may cause minor signal deviations as do fluctuating sample and sensor temperatures. Typically, measurements can be repeated within some 5–10% and systematically reproduced by single sample data. The lateral shape (but not the vertical extension!) of the sediment volumes contributing respectively to XRF and susceptibility scanner measurements are quite similar in area. Therefore, the error resulting from combining these parameters in a single ratio remains small.

Normalization of Proposed Diagenesis Proxies

The ratios Fe/κ_{nd}, κ_{nd}/Fe, κ_{nd}/Ti and Fe/Ti do not quantify the progression of iron mineral diagenesis by their absolute value since they also depend on local magnetic mineral source characteristics. Site specific references for the terrigenous input can usually be inferred from the stable signal platforms observed in sediment sections with low C_{org} contents (see Figs. 4 and 5). These baseline levels are used here to normalize the continuous records. This approach implies that the initial proportion of accumulated magnetic and nonmagnetic Fe and Ti minerals did not extensively vary through time. In the equatorial Atlantic as in other open ocean settings with a single predominant terrigenous sediment source this assumption is probably acceptable. It may not be fulfilled, however, if contributions from geologically differing sources strongly alternate within the record.

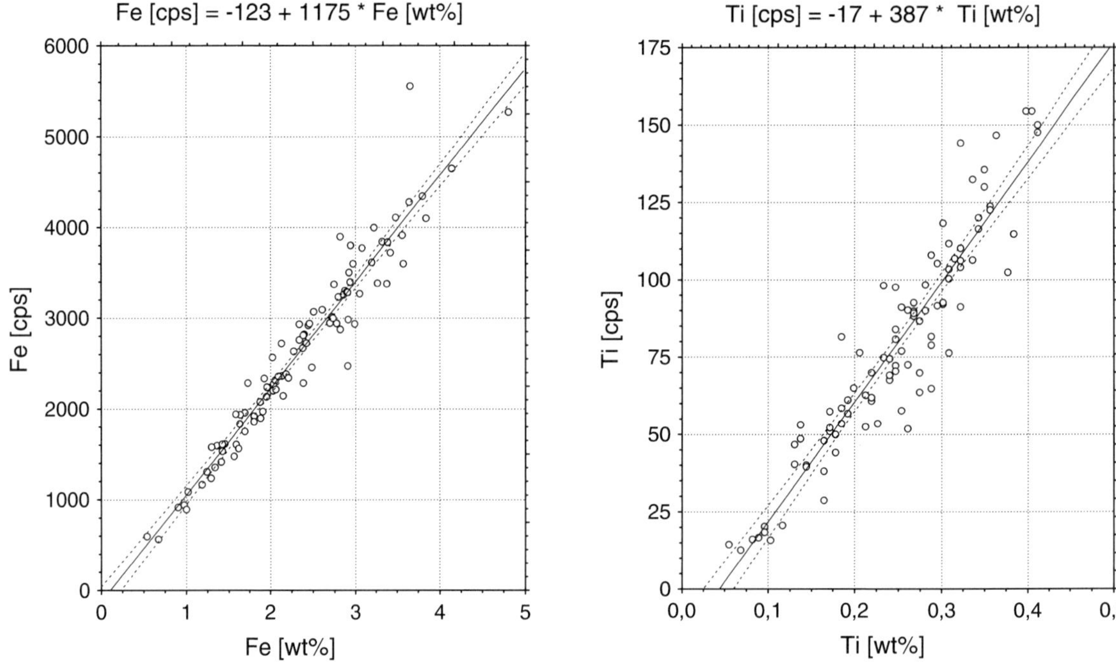

Fig. 7. Calibration of relative **a)** Fe and **b)** Ti XRF scanner data [cps] of core GeoB 4317-2 by conventional single sample XRF analyses [wt % oxide]. Shown here are Pearson's correlation coefficients r, linear regression equations, regression lines and their 95 % confidence ranges.

By normalizing these ratios, positive deviations from unity quantify the grade of magnetic mineral dissolution and authigenesis as well as iron depletion and enrichment, respectively. In the following the proxy parameters shall be symbolized as

$$\left[Fe/\varkappa_{nd}\right]_n = \frac{Fe/\varkappa_{nd}}{\left(Fe/\varkappa_{nd}\right)_{base}};$$

$$\left[\varkappa_{nd}/Fe\right]_n = \frac{\varkappa_{nd}/Fe}{\left(\varkappa_{nd}/Fe\right)_{base}};$$

$$\left[\varkappa_{nd}/Ti\right]_n = \frac{\varkappa_{nd}/Ti}{\left(\varkappa_{nd}/Ti\right)_{base}};$$

$$\left[Fe/Ti\right]_n = \frac{Fe/Ti}{\left(Fe/Ti\right)_{base}};$$

It is of no relevance, whether the XRF element data are inserted as raw or calibrated values. The problem of discontinuities in the Fe/κ record due to zero or negative values of κ is resolved by using the strictly positive non-diamagnetic susceptibility κ_{nd}. Sections with very low Ti contents can result in unrealistic, noise related positive peaks in Ti-based ratios. This problem should be avoidable in the future by normalizing with Al. This element will be measurable with the next generation of XRF core scanners.

Quantification of Depletion and Enrichment Processes

Prior to further considerations, it shall be recalled that the terms 'depletion' and 'enrichment' have fundamentally different meanings, when being applied to magnetic minerals or iron content: to deplete or enhance a layer magnetically, it suffices to dissolve, alter or precipitate magnetic minerals *in situ*; iron relocation is not required. Ferrimagnetism itself has no balanced budget, as it is easily exchanged into or from the (paramagnetic) total Fe reservoir. However, depletion and enrichment of paramagnetic iron in marine sediments is always balanced and involves dissolution and relocation by diffusive transport in the pore water soluble ferrous form. The ferrimagnetic mineral content, (titano-) magnetite or maghemite with a susceptibility equivalent of 10-100 ppm, is typically several orders of magnitude smaller than the nonmagnetic iron mineral content. Consequently, the magnetic fraction is very subordinate in the total iron budget, but largely dominates the rock magnetic properties.

The integration of concentration dependent rock magnetic and geochemical parameters permits to quantitatively reconstruct diagenetic changes in magnetic and nonmagnetic iron budgets as will be demonstrated for core GeoB 4317-2.

With suitable threshold values for the ratios $[Fe/\kappa_{nd}]_n$ and $[\kappa_{nd}/Ti]_n$, the sediment column can be subdivided into magnetically pristine ($[Fe/\kappa_{nd}]_n \approx 1$, $[\kappa_{nd}/Ti]_n \approx 1$), enriched ($[\kappa_{nd}/Ti]_n > 1$), weakly ($2 > [Fe/\kappa_{nd}]_n > 1$) and strongly ($[Fe/\kappa_{nd}]_n > 2$) depleted sections. In Figure 4, 5 and 9 this classification has already been implicitly introduced: background patterns are shaded (magnetically depleted) and hatched (magnetically enriched) to correlate and clarify the specific signatures of diagenesis in all parameter records. In Figure 8 the same categories were used as grouping variables in cross-plots of concentrational rock magnetic and geo-chemical parameters of core GeoB 4317-2. The samples assumed as pristine (black dots) show an almost perfect mutual proportionality of susceptibility, Fe and Ti contents. In all three cases the respective regression lines (dashed; Fig. 8a-c) approximately pass through the origin.

The regression equations have been used to define the primary (predicted) susceptibility and Fe content of all modified samples on basis of their Ti contents. The measured (Fig. 9c,e, thick lines) and the Ti-derived (Fig. 9c,e, thin lines) parameters were subtracted to quantify the diagenesis related excess (orange) and deficiency (cyan) of κ_{nf} and Fe in scaled, absolute values (Fig. 9b,d, loss/gain). Together, Figures 8 and 9 describe and link the rock magnetic and geochemical perspective and put it into a joint stratigraphic framework. The range and

correlation of magnetic and nonmagnetic Fe loss and gain is visualized in crossplot Figure 8d.

From the data point distribution of the magnetically depleted and enhanced samples (Fig. 8) some fundamental characteristics and trends of redoxomorphic iron diagenesis in the central Equatorial Atlantic can be derived:

• Loss of susceptibility, i.e. dissolution of magnetite, by reductive diagenesis is limited to C_{org} and Fe rich layers deposited during glacials and their terminations and amounts up to $200 \cdot 10^{-6}$ SI (75% of the initial κ). Enhancement of susceptibility, probably by bacterial biomineralization, takes place around C_{org} and Fe poor carbonate maxima overlying the magnetically depleted zones. It reaches up to $50 \cdot 10^{-6}$ SI and may almost double the Ti- derived pristine value. Susceptibility losses greatly exceed the gains.

• The Ti-derived pristine susceptibility signal (Fig. 9c, light gray shaded curve) shows only a weak positive correlation with C_{org} (Fig. 9a) while the budget curve of diagenetic loss and gain (Fig. 9b) is almost an inverse mirror image of C_{org}. Assuming a proportionality of mineralized C_{org} and magnetite dissolution, this would imply that some magnetic depletion peaks during cold oxygen (sub)stages 2, 4, 5.4, 7.4 and 8 are no longer present in the C_{org} record. Very likely, these 'minor' C_{org} maxima were eliminated by a downward progressing oxidation front. As the susceptibility loss is permanent, it may - in analogy to barium - represent a more authentic archive of past C_{org} accumulation than C_{org} content itself, at least under the favorable conditions encountered in the equatorial Atlantic.

• With total Fe contents ranging between 1 to 5 wt % (Fig. 9e, thick black line) the average depletions accounting for less than 0.5 wt.% can be considered as moderate. Susceptibility is greatly reduced in all iron-depleted zones. The iron enrichments (Fig. 9 e, orange) may enhance pristine contents by up to 3 wt %. Total iron precipitation correlates positively with susceptibility and hence magnetic iron precipitation (Fig. 8d triangles). Yet, the highest Fe enrichments are found around paleo-oxidation fronts, where susceptibility is at loss (Fig. 8 d, diamonds and Fig.9 b,d).

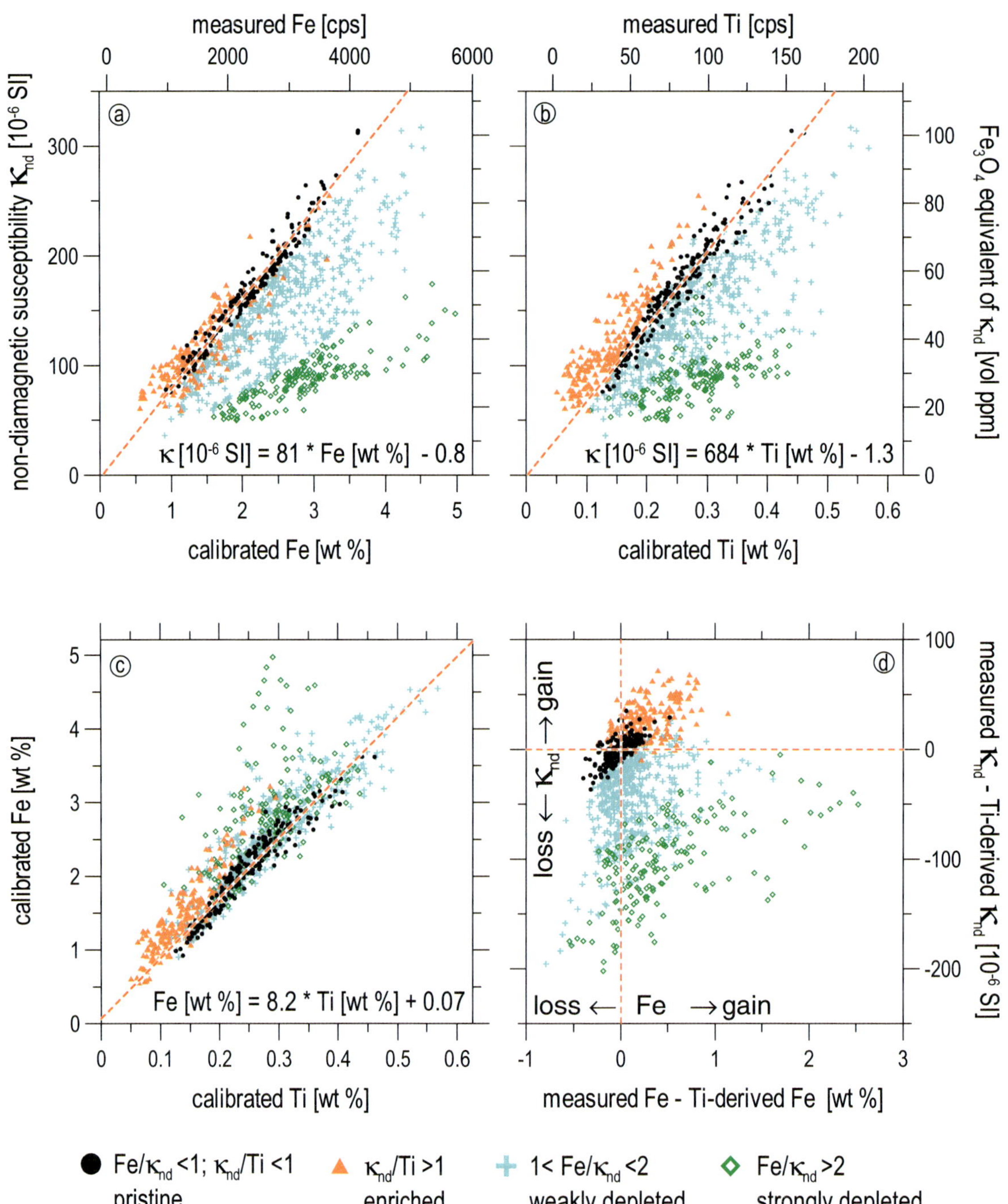

Fig. 8. Crossplots of non-diamagnetic susceptibility, Fe and Ti contents of core GeoB 4317-2 quantifying diagenetic loss and gain of magnetic and total Fe. The newly introduced proxies κ_{nd}/Ti and Fe/κ_{nd} are used as criteria to categorize magnetically pristine (black dots), enriched (orange triangles) and weakly (cyan crosses) to strongly (green diamonds) depleted sections (see legend). Diagenetically unaffected sediments show high mutual proportionality of susceptibility, Fe and Ti contents. Their linear regression lines (dashed red lines) pass very near the origin **a-c**). The regression equations can be used to derive pre-diagenetic κ_{nd} and Fe values for altered sediments on basis of the unaffected Ti contents. In **d**), the differences between measured and Ti-derived pristine susceptibilities as well as Fe contents are cross-plotted. The data scatter over the four quadrants visualizes range and linkage of magnetic and non-magnetic Fe loss and gain in the course of redoxomorphic diagenesis.

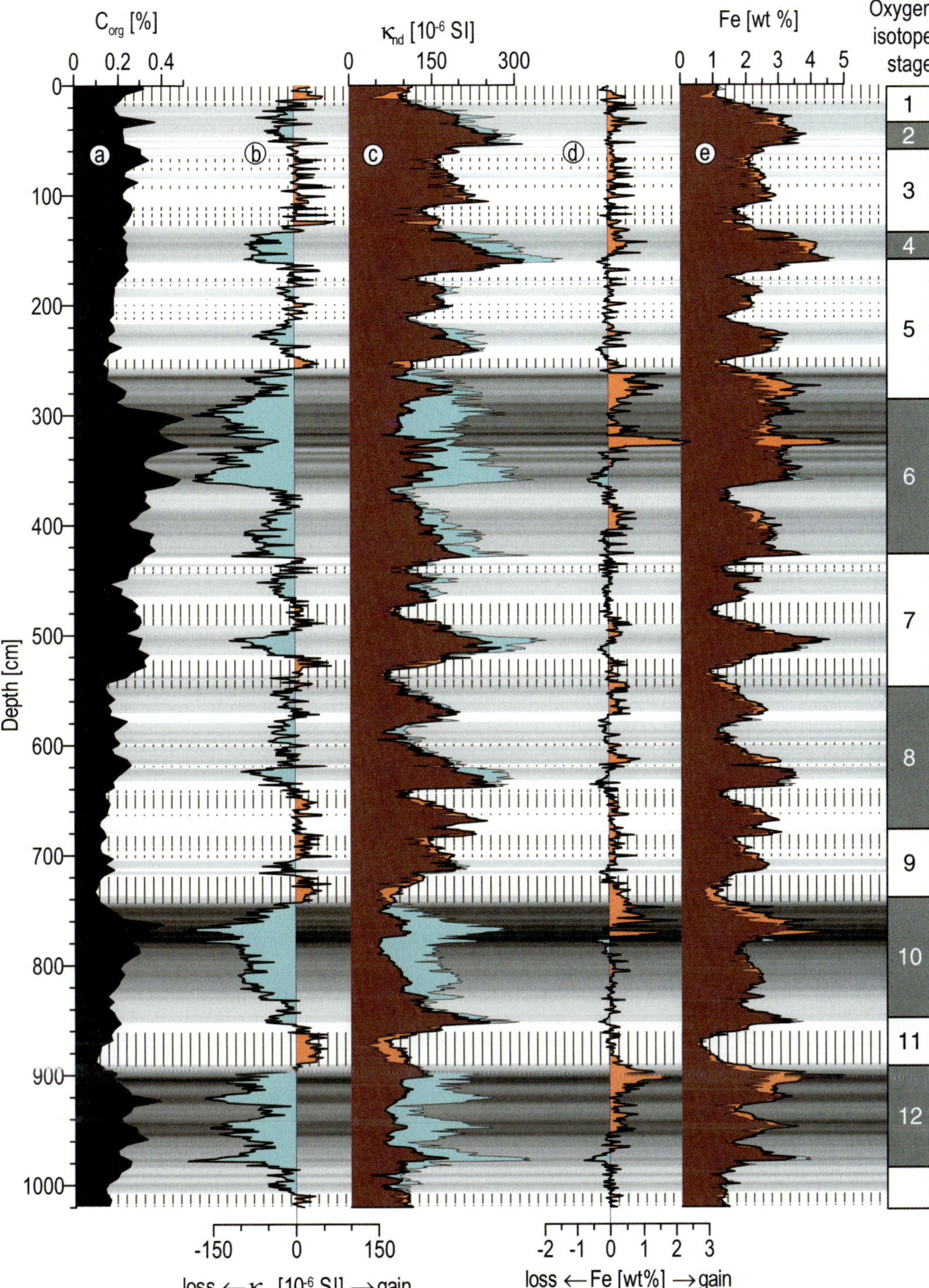

Fig. 9. Stratigraphic compilation of C_{org} content **a)** with measured (thick lines) and Ti-derived pristine (thin lines) susceptibility **c)** and Fe **e)** records of core GeoB 4317-2. The calculated loss (cyan) and gain (orange) of magnetic **b)** and non-magnetic **d)** iron is scaled accordingly to curves (c) and (e). Note the reverse patterns of (a) and (b) and the much higher relative losses of magnetic minerals compared to total iron. Background shading indicates the degree of magnetite dissolution based on Fe/κ_{nd}. Magnetic mineral precipitation zones according to κ_{nd}/Ti are shown as hatched areas.

Acknowledgements

We thank officers and crews of RV *Meteor* for their efficient support during cruises M 29/3 and M 38/1. Arguments and constructive criticism raised by U. Bleil and the two peer reviewers, M. Dekkers and S. Banerjee, have greatly im-proved this manuscript and are gratefully acknow-ledged. T. Wagner provided C_{org} data of core GeoB 4317-2. U. Röhl made the XRF scanner measurements possible and was very helpful. This study was funded by the Deutsche Forschungs-gemeinschaft (Sonderforschungsbereich 261 at Bremen University, contribution No. 385). J.A. Funk was supported by the Deutsche Forschungs-gemeinschaft in the framework of Graduierten-kolleg 221. T. von Dobeneck is a visiting research fellow at Utrecht University supported by the Neth-erlands Research Center for Integrated Solid Earth Science (ISES). Data are availalble under www. pangaea.de/Projects/SFB261.

References

Banerjee SK (1971) Decay of marine magnetic anomalies by ferrous ion diffusion. Nature: Physical Science 229: 181–183

Bazylinski DA, Frankel RB, Jannasch HW (1988) Anaerobic magnetite production by a marine, magnetotactic bacterium. Nature 334: 518–519

Bickert T, Wefer G (1996) Late Quaternary deep water circulation in the South Atlantic: Reconstruction from carbonate dissolution and benthic stable isotopes. In: Wefer G, Berger WH, Siedler G, Webb DJ (eds) The South Atlantic: Present and Past Circulation. Springer, Berlin, pp 599-620

Blakemore RP (1975) Magnetotactic bacteria. Science 190: 377-379

Bloemendal J, Lamb B, King J (1988) Paleoenvironmental implications of rock - magnetic properties of Late Quaternary sediment cores from the eastern Equa-torial Atantic. Paleoceanography 3: 61-87

Bloemendal J, King JW, Hall FR, Doh S-J (1992) Rock magnetism of late Neogene and Pleistocene deep-sea sediments: Relationship to sediment source, diagenetic processes, and sediment lithology. J Geophys Res 97: 4361-4375

Brown ET, Le Callonnec L, German CR (2000) Geochemical cycling of redox-sensitive metals in sediments from Lake Malawi: A diagnostic paleotracer for epi-sodic changes in mixing depth. Geochim Cosmochim Acta 64: 3515-3523

Burdige DJ (1993) The biochemistry of manganese and iron reduction in marine sediments. Earth Sci Rev 35: 249-284

Butler RF, Banerjee SK (1975) Theoretical single-domain grain size range in magnetite and titano-magnetite. J Geophys Res 80: 4049–4058

Calvert SE, Pedersen TF (1993) Geochemistry of recent oxic and anoxic marine sediments: Implications for the geological record. Mar Geol 113: 67-88

Canfield DE, Raiswell R, Bottrell S (1992) The reactivity of sedimentary iron minerals toward sulfide. Am J Sci 292: 659-683

Canfield DE (1997) The geochemistry of river particles from the continental USA: Major elements. Geochim Cosmochim Acta 61: 3349-3365

Dapples EC (1962) Stages of diagenesis in the development of sandstones. Bull Geol Soc Am 73: 913-934

Day R, Fuller M, Schmidt VA (1977) Hysteresis properties of titanomagnetites: Grain-size and compositional dependence. Phys Earth Planet Int 13: 260-267

Dearing JA, Dann RJL, Hay K, Lees JA, Loveland PJ, Maher BA, O'Grady K (1996) Frequency dependent susceptibility measurements of environmental materials. Geophys J Int 124: 228-240

Dekkers MJ, Langereis CG, Vriend SP, van Santvoort PJM, de Lange GJ (1994) Fuzzy c-means cluster analysis of early diagenetic effects on natural remanent magnetisation acquisition in a 1.1 Myr piston core from the Central Mediterranean. Phys Earth Planet Int 85: 155-171

deMenocal PB, Ruddiman WF, Pokras EM (1993) Influences of high- and low- latitude processes on African terrestrial climate: Pleistocene eolian records from equatorial Atlantic ocean drilling program Site 663. Paleoceanography 8: 209-242

Fischer G and cruise participants (1998) Report and preliminary results of *Meteor* cruise M38/1. Ber Fachber Geowiss, Univ Bremen 94, 178 p

Frederichs T, Bleil U, Däumler K, von Dobeneck T, Schmidt A (1999) The magnetic view on the marine paleoenvironment: Parameters, techniques, and potentials of rock magnetic studies as a key to paleoclimatic and paleoceanographic changes. In: Fischer G and Wefer G (eds) Use of Proxies in Paleoceanography: Examples from the South Atlantic. Springer, Berlin, pp 575-599

Froelich PN, Klinkhammer GP, Bender ML, Luedtke NA, Heath GR, Cullen D, Dauphin P, Hammond D, Hartman B, Maynard V (1979) Early oxidation of organic matter in pelagic sediments of the eastern equatorial Atlantic: Suboxic diagenesis. Geochim Cosmochim Acta 43: 1075-1090

Gladney ES, Roelandts I (1988) 1987 compilation of elemental concentration data for USGS BHVO-1, MAG-1, QLO-1, RGM-1, SCO-1, SDC-1, SGR-1 and STM-1. Geostandards Newsletters 12: 253-362

Haese RR (2000) The reactivity of iron. In: Schulz HD and Zabel M (eds) Marine Geochemistry. Springer, Berlin, pp 233-262

Heider F, Zitzelsberger A, Fabian F (1996) Magnetic susceptibility and remanent coercive force in grown magnetite crystals from 0.1 μm to 6 mm. Phys Earth Planet Int 93: 239-256

Jansen JHF, Van der Gaast SJ, Koster AJ (1998) CORTEX, a shipboard XRF-scanner for element analyses in split sediment cores. Mar Geol 151: 143-153

Karlin R (1990a) Magnetic mineral diagenesis in suboxic sediments at Bettis Site W-N, NE Pacific Ocean. J Geophys Res 95: 4421-4436

Karlin R (1990b) Magnetite diagenesis in marine sediments from the Oregon continental margin. J Geophys Res 95: 4405-4419

Karlin R, Levi S (1983) Diagenesis of magnetic minerals in recent haemipelagic sediments. Nature 303: 327-330

Karlin R, Lyle M, Heath GR (1987) Authigenic magnetic formation in suboxic marine sediments. Nature 326: 490-493

Kidd RB, Cita MB, Ryan WBF (1978) Stratigraphy of eastern Mediterranean sapropel sequences recovered during DSDP Leg 42A and their paleoenvironmental significance. In: Hsü KJ and Montadert L et al. (eds) Initial Reports of the Deep-Sea Drilling Project 42A. US Government Printing Office, Washington, pp 421-443

Kruiver PP, Passier HF (2001) Coercivity analysis of magnetic phases in sapropel S1 related to variations in redox conditions, including an investigation of the S ratio. Geochem Geophys Geosyst 14 December: 1-21

König I, Drodt M, Suess E, Trautwein AX (1997) Iron reduction through the tan - green color transition in deep-sea sediments. Geochim Cosmochim Acta 61: 1679-1683

Langereis CG, Dekkers MJ (1999) Magnetic cyclostratigraphy: High resolution dating in and beyond the Quaternary and analysis of periodic changes in diagenesis and sedimentary magnetism. In: Maher B and Thompson R (eds) Quaternary Climates, Environments and Magnetism. Cambridge University Press, pp 352-382

Lecoanet H, Lévêque F, Segura S (1999) Magnetic susceptibility in environmental applications: Comparison of field probes. Phys Earth Planet Int 115: 191-204

Lovley DR, Stolz JF, Nord GL Jr, Phillips EJP (1987) Anaerobic production of magnetite by a dissimilatory iron-reducing microorganism. Nature 330: 252–254

Lyle M (1983) The brown-green color transition in marine sediments: A marker of the Fe(III)-Fe(II) redox boundary. Limnol Oceanogr 28: 1026-1033

Lyle M (1988) Climatically forced organic carbon burial in equatorial Atlantic and Pacific Oceans. Nature 335: 529-532

Maher BA (1988) Magnetic properties of some synthetic submicron magnetites. Geophys J 94: 83–96

Mudroch A, Azcue JM (1995) Manual of aquatic sediment sampling. Lewis Publishers, Boca Raton, 219 p

Nowaczyk NR, Antonow M (1997) High-resolution magnetostratigraphy of four sediment cores from the Greenland Sea I. Identification of the Mono Lake excursion, Laschamp and Biwa I / Jamaica geomagnetic polarity events. Geophys J Int 131: 310-324

Passier HF, Dekkers MJ, de Lange GJ (1998) Sediment chemistry and magnetic properties in an anomalously reducing core from the eastern Mediterranean Sea. Chem Geol 152: 287-306

Petermann H, Bleil U (1993) Detection of live magnetotactic bacteria in South Atlantic deep-sea sediments. Earth Planet Sci Lett 117: 223-228

Robinson SG, Sahota JTS, Oldfield F (2000) Early diagenesis in North Atlantic abyssal plain sediments characterized by rock magnetic and geochemical indices. Mar Geol 163: 77-107

Röhl U, Abrams LJ (2000) High-resolution downhole and nondestructive core measurements from Sites 999 and 1001 in the Caribbean Sea: Application to the Late Paleocene thermal maximum. In: Leckie RM, Sigurdsson H, Acton GD, Draper G (eds) Proc ODP Sci Results 165. College Station, TX, pp 191-203

Rosenbaum JG, Reynolds RL, Adam DP, Drexler J, Sarna-Wojcicki AM, Whitney GC (1996) Record of middle Pleistocene climate change from Buck Lake, Cascade Range, southern Oregon - evidence from sediment magnetism, trace-element geochemistry, and pollen. Bull Geol Soc Am 108: 1328-1341

Ruddiman WF, Janecek TR (1989) Pliocene-Pleistocene biogenic and terrigenous fluxes at equatorial Atlantic Sites 662, 663, and 664. In: Ruddiman W and Sarnthein M (eds) Proc ODP Sci Results 108. College Station, TX, pp 211-240

Schmieder F, von Dobeneck T, Bleil U (2000) The Mid-

Pleistocene climate transition as documented in the deep South Atlantic Ocean: Initiation, interim state and terminal event. Earth Planet Sci Lett 179: 539-549

Smirnov AV, Tarduno JA (2000) Low-temperature magnetic properties of pelagic sediments (Ocean Drilling Program Site 805C): Tracers of maghemitization and magnetic mineral reduction. J Geophys Res 105: 16,457-16,471

Snowball IF (1993) Geochemical control of magnetite dissolution in subarctic lake sediments and the impli-cations for environmental magnetism. J Quat Sci 8: 339-346

Tarduno JA (1994) Temporal trends of magnetic disso-lution in the pelagic realm: Gauging paleoproduc-tivity? Earth Planet Sci Lett 123: 39-48

Tarduno JA, Tian W, Wilkison S (1998) Biogeochemical remanent magnetization in pelagic sediments of the western equatorial Pacific Ocean. Geophys Res Lett 25: 3987-3990

Tarduno JA, Wilkison SL (1996) Non-steady state magnetic mineral reduction, chemical lock-in, and delayed remanence acquisition in pelagic sediments. Earth Planet Sci Lett 144: 315-326

Thompson R, Oldfield F (1986) Environmental Mag-netism. Allen & Unwin, London, UK, 227 p

Thomson J, Jarvis I, Green DRH, Green D (1998) Oxi-dation fronts in Madeira abyssal plain turbidites: Per-sistence of early diagenetic trace-element enrich-ments during burial, Site 950. In: Weaver PPE, Schmincke H-U, Firth JV, Duffield W (eds) Proc ODP Sci Results 157. College Station, TX, pp 559-571

Thomson J, Mercone D, de Lange GJ, van Santvoort PJM (1999) Review of recent advances in the interpre-tation of eastern Mediterranean sapropel S1 from geochemical evidence. Mar Geol 153: 77-89

Verardo DJ, McIntyre A (1994) Production and des-truction: Control of biogenous sedimentation in the tropical Atlantic 0-300,000 years BP. Paleocean-ography 9: 63-86

Vigliotti L, Capotondi L, Torii M (1999) Magnetic pro-perties of sediments deposited in suboxic - anoxic environments: Relationships with biological and geochemical proxies. In: Tarling DH and Turner P (eds) Paleomagnetism and Diagenesis in Sediments. Geol Soc Spec Publ London, 151, pp 71-83

von Dobeneck T (1996) A sytematic analysis of natural magnetic mineral assemblages based on modelling hysteresis loops with coercivity-related hyberbolic basis functions. Geophys J Int 124: 675-694

Walden J, White K (1997) Investigation of the con-trols on dune colour in the Namib Sand Sea using mineral magnetic analyses. Earth Planet Sci Lett 152: 187-201

Worm H-U, Jackson M (1999) The superparamag-netism of Yucca Mountain Tuff. J Geophys Res B: Solid Earth 104: 25,415-25,425

Magnetic Signals in Plio-Pleistocene Sediments of the South Atlantic: Chronostratigraphic Usability and Paleoceanographic Implications

F. Schmieder

Universität Bremen, Fachbereich Geowissenschaften, Postfach 33 04 40,
D-28334 Bremen, Germany
e-mail: schmiede@uni-bremen.de

Abstract: The origin of the magnetic signals used to build age models for marine sediments recovered in the framework of the long-term Quaternary South Atlantic research project SFB 261 is twofold. Conventional magnetostratigraphy makes use of well-dated polarity reversals of the Earth's magnetic field recorded in the natural remanent magnetization (NRM) of the sediments. In addition, magnetic cyclostratigraphy has been successfully established as a very efficient dating tool for marine sediment sequences during recent years. In the oligotrophic South Atlantic, confirmation of orbital forcing of magnetic susceptibility records made it possible to establish high-resolution age models, by tuning the respective components to astronomical variations. A set of twelve individually tuned and well-correlated Pleistocene magnetic susceptibility records were stacked within the stratigraphic network SUSAS and can now be used as a correlation reference for other cores recovered in this region. The suitability of this target curve for age control is tested against paleomagnetic ages. Pattern correlation is possible for nine of ten selected cores recovered in the oligotrophic South Atlantic between 15°S and 35°S, but seems to be only partly successful for sediments from the Congo Basin. In the Pleistocene sequences, the magnetic age models provide further evidence for the simultaneous deposition of previously reported unusual diatom ooze layers between 23°S and 33°S at approximately 540 – 530 ka, at the end of the Mid-Pleistocene climate transition (MPT). The age models also indicate enhanced carbonate dissolution during the MPT interim state (920 – 640 ka). The concept of tuning magnetic susceptibility records to orbital variations is extended to the late Pliocene and reveals characteristics obviously related to the rearrangement of ocean circulation as Northern Hemisphere glaciation intensified. Enhanced carbonate preservation since approximately 3.0 Ma and the establishment of obliquity-driven dissolution cycles since about 2.5 Ma document increasing influx of North Atlantic Deep Water (NADW) into the subtropical South Atlantic. In a deep core from the Rio Grande Rise area, an abrupt change from red deep-sea clay to carbonaceous sediments is recorded at 2.73 Ma, exactly the time proposed for the major intensification of Northern Hemisphere glaciation.

Introduction

A crucial requirement for understanding the dynamics of geological and paleoenvironmental processes documented in sediments is age control. Knowledge of the timing of the reversals of the Earth's magnetic field and the fact that these reversals are frequently recorded in the natural remanent magnetization (NRM) of marine sediments enables dating via a comparison of normally and reverse magnetized intervals to a geomagnetic polarity timescale (GPTS). Magnetostratigraphy provides quite reliable age control, although its value is limited by the number of reversals documented in the studied material. For the Pleistocene, the frequently used GPTS of Cande and Kent (1992, 1995; Fig. 1) contains three such reversals. Additionally, the authors included two short geomagnetic events (therein called "cryptochrons") in their template, the Emperor (504 to 493 ka) and the Cobb Mountain event (1212 to 1201 ka).

With regard to high-resolution dating, astronomical tuning has proven to be one of the most powerful tools during the last two decades. It has

From WEFER G, MULITZA S, RATMEYER V (eds), 2003, *The South Atlantic in the Late Quaternary: Reconstruction of Material Budgets and Current Systems.* Springer-Verlag Berlin Heidelberg, pp 261-277

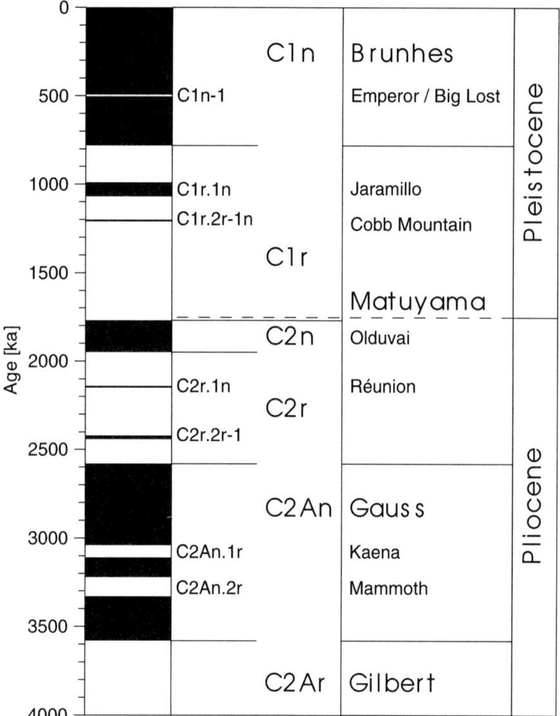

Fig. 1. Geomagnetic polarity timescale (GPTS) for the last 4000 ka (left) after Cande and Kent (1992, 1995) and the commonly used indications of reversals (right).

been successfully used for late Pliocene and Pleistocene sections, but it is by no means restricted to these epochs (e.g. Schwarzacher 1993a, b; Shackleton et al. 1995; Lourens et al. 1996; D'Argenio et al. 1998). Oxygen isotopes are by far the most frequently used and best studied proxy regarding cyclostratigraphy. However, because of much less laboratory effort and proven suitability, other climatically induced signals have also become popular and are now accepted as useful candidates (e.g. Langereis and Dekkers 1999). In this context, physical properties are of special interest because they can be measured relatively rapidly and with high stratigraphic resolution. Their usefulness has been proven by numerous chronostratigraphies (e.g. Shackleton et al. 1995; Shackleton and Crowhurst 1997; Bickert et al. 1997; Heslop et al. 2000).

In many marine environments magnetic susceptibility can be used to refine the age models for sedimentary depositions due to certain, regionally varying, orbital linkages. As a function of concentration and composition of the magnetic mineral

fraction, the magnetic susceptibility of marine sediments is in the first instance a measure of the content of (titano-) magnetite (Thompson and Oldfield 1986). Most magnetite reaches the sediment as part of the terrigenous input (an important exception in some regions is bacterial magnetite, e.g. Petersen et al. 1986), and in the majority of cases the magnetic susceptibility record reflects the variable ratio of biogenic to lithogenic components (Robinson 1990). Variations in the magnetic susceptibility records of marine carbonate sediments can be influenced by carbonate dissolution, varying terrigenous input, dilution of a constant terrigenous input by varying carbonate production, or a combination of these often climatically controlled processes. Consequently, magnetic susceptibility can serve as a proxy for the carbonate record which mirrors paleoclimatic history. Therefore, magnetic susceptibility records can be tuned to orbital variations and can additionally aid in paleoceanographic reconstructions.

Magnetic susceptibility records from twelve GeoB cores recovered in the framework of SFB 261 in the South Atlantic were used by von Dobeneck and Schmieder (1999) to construct the Subtropical South Atlantic Susceptibility (SUSAS) stack by tuning characteristic Pleistocene Milankovitch-style patterns to orbital obliquity and precession. The SUSAS cores are located between the Rio Grande Rise in the west and the Walvis Ridge in the east and form an ocean-wide transect between 22°S and 34°S. In the South Atlantic the ice age rhythm causes cyclic variations of the sediment $CaCO_3$ content with increased carbonate dissolution during glacial periods (Volat et al. 1980; Tappa and Thunell 1984; Balsam and McCoy 1987) because glacial reduction of North Atlantic Deep Water (NADW) gives way for a vertical expansion of more corrosive Lower Circumpolar Deep Water (LCDW) (e.g. Bickert and Wefer 1996). The resulting dissolution cycles are mirrored in magnetic susceptibility with higher values during glacial periods.

Oxygen isotope records exist for four SUSAS cores, but their stratigraphic interpretability is limited by very low sediment accumulation rates, generally as low as 0.5 - 1.5 cm/kyr. Nevertheless, astro-nomical tuning of the magnetic susceptibility

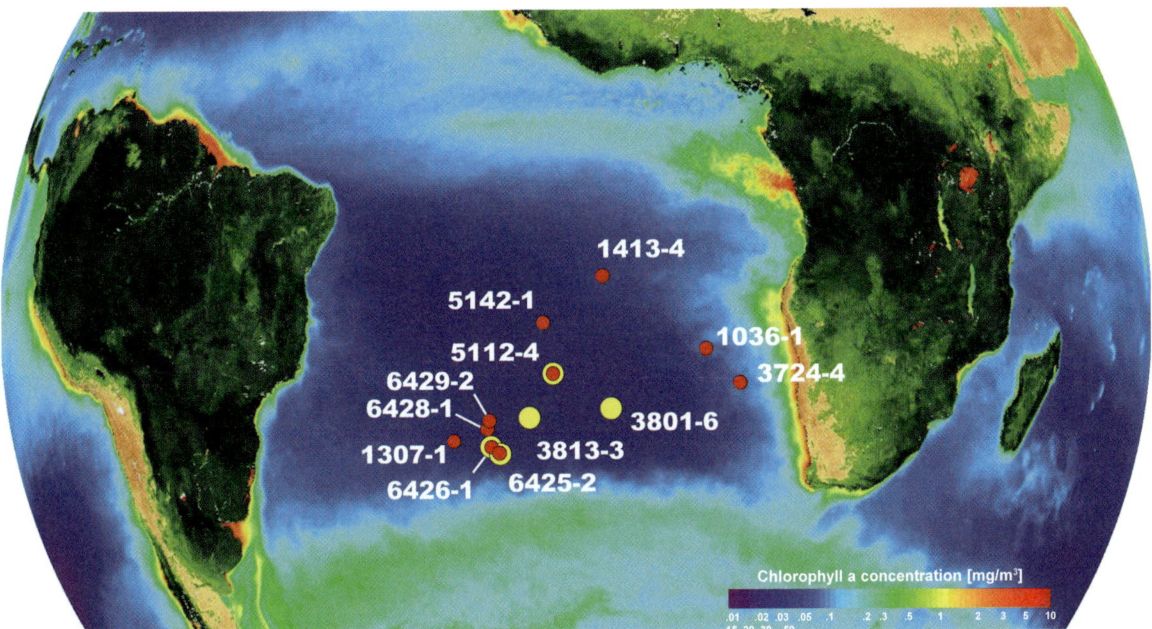

Fig. 2. Locations of GeoB cores investigated (red dots). SeaWIFS image (SeaWiFS Project, NASA Goddard Space Flight Center) depicts primary productivity as visualization of chlorophyll a content in the surface water, values increase from dark blue to red (data collected over the time period September 1997 through August 1998). All cores are located in the oligotrophic South Atlantic. Yellow dots mark locations where an unusual diatom ooze layer of *Ethmodiscus rex* has been found in the sediments. GeoB 3801-6 and 3813-3 are SUSAS cores containing this layer.

records could be established based on magnetostratigraphic tie points. The applied phase lag was determined by cross spectral analysis of $\delta^{18}O$ and magnetic susceptibility. It yielded for the 100 and 41 kyr cycles a coherence of 0.99 and 0.96 and reasonable phase angles (von Dobeneck and Schmieder 1999). In the obliquity band, magnetic susceptibility lags the orbital forcing by 4.5 kyr.

The individually tuned and well-correlated SUSAS records form a master-curve, and can be used as a target record for other cores in this region. The remarkably good similarity of magnetic susceptibility profiles in the whole oligotrophic South Atlantic has been used to construct preliminary age models for sediments recovered between 20°S and 38°S during RV *Meteor* cruises M 41/3 (Frederichs at al. 1999) and M 46/4 (Donner and Schmieder 2001) by comparing typical patterns to the SUSAS stack. The post-cruise spot sensor magnetic susceptibility data measured in the laboratory for these and other cores allow for a more detailed signal definition and hence improved age modeling. By subsequently tracing the magnetic signal back to its

paleoceanographic origin it can be interpreted in terms of modifications of deep water chemistry.

Material and Methods

From several hundreds of cores recovered between 1988 and 2000 in the South Atlantic within the framework of Sonderforschungsbereich (SFB) 261 at the University of Bremen, ten cores were chosen to test the suitability of the SUSAS stack as a target curve to construct high-resolution age models. A main criterion for the selection was the recording of at least one paleomagnetic reversal to provide an independent age control. As gravity coring on board RV *Meteor* is limited to 15 - 20 m, only cores from regions with relatively low sediment accumulation rates of less than 2.5 cm/kyr reach the youngest well-defined polarity reversal, the Brunhes/ Matuyama boundary at 780 ka (Fig. 1). Therefore, regions with enhanced productivity are excluded from this study. The chosen cores were all recovered from the oligotrophic South Atlantic (Fig. 2) in water depths between 3789 and 5073 m during several RV *Meteor* expeditions. The carbonaceous sediments are mainly composed of nannofossil ooze. Table 1 summarizes the position, water depth, core length, and the sampling interval for paleomagnetic analyses of the cores, and lists the respective cruise reports. The sediments were recovered using a gravity corer with an internal diameter of 12 cm. Subsampling for paleomagnetic measurements was carried out on board, generally at a 10 or 5 cm sampling interval (Table 1), using cubic plastic boxes with a volume of 6.2 cm^3.

The natural remanent magnetization (NRM) was measured using two different three axis SQUID magnetometers. A stepwise alternating field (AF) demagnetization was performed for all samples to isolate the characteristic remanent magnetization (ChRM). The different demagnetiza-tion schemes used are listed in Table 1. For the samples of GeoB 1036-1, 1307-1, and 1413-4 (Thießen 1993) a *Cryogenic Consultants* GM 400 and a single axis demagnetizer *2G Enterprises* 2G 600 were used. In 1998, the laboratory of Mar-ine Geophysics in the Department of Geosciences at Bremen University was equipped with a new *2G Enterprises* magneto-meter (Model 755R) with an integrated demag-netizing unit. This fully automated magnetometer was used for all other samples. Directions of the ChRM were calculated by using the principal component analysis (Kirschvink 1980) and at least three successive demagnetization steps.

Magnetic susceptibility was measured post-cruise on archive halves using a *Bartington Instruments* spot sensor (F) at a 1 cm spacing. Before measuring, the archive halves, which are stored at 4°C, were warmed up to room temperature to avoid temperature effects. Each measurement was corrected with a separate background reading.

Results and Discussion

Magnetostratigraphy

In most cases the ChRM could be determined unambiguously. Stable ChRM directions were isolated from the linear sections of the demagnetiza-tion curves. Figure 3 summarizes ChRM inclina-tions and declinations of all cores. Depths of all identified reversals and the corresponding ages referring to Cande and Kent (1992, 1995) are listed in Table 2. Superimposed trends on some declina-tion values, e.g. between 10 and 13.5 m in GeoB 6426-1, point to rotation during the core recovery. As they do not complicate the identifica-tion of geomagnetic polarity, these intervals are plotted unchanged.

By containing the Gauss/Gilbert boundary (3580 ka), core GeoB 1307-1 contains the oldest sediments of this study. In GeoB 6429-2, the Matuyama/Gauss boundary (2581 ka) is recorded, all the other cores end within the Matuyama Chron.

METEOR Expedition	GeoB core	Position		Water Depth [m]	Core Length [m]	Sampling interval [cm]	Demag. scheme	MTS published in	Cruise report
		Latitude	Longitude						
M 6/6	1036-1	21°10.4'S	03°19.7'E	5073	10.76	10 (5)	A	Thießen '93	Wefer et al. (1988)
M 15/2	1307-1	33°36.1'S	27°39.9'W	4017	6.77	10	B	this paper	Pätzold et al. (1993)
M 16/1	1413-4	15°40.8'S	9°27.3'W	3789	10.10	10	A	Thießen '93	Wefer et al. (1991)
M 34/2	3724-4	26°08.2'S	08°55.6'E	4759	11.41	10	C	this paper	Schulz et al. (1996)
M 41/3	5112-4	23°49.5'S	16°15.5'W	3842	5.59	10	C	this paper	Pätzold et al. (1999)
	5142-1	19°05.4'S	17°08.7'W	3946	5.64	5	C	this paper	
M 46/4	6425-2	33°49.5'S	23°35.2'W	4352	10.73	5	C	this paper	Wefer et al. (2001)
	6426-1	33°30.0'S	24°01.5'W	4385	13.41	5	C	this paper	
	6428-1	32°30.6'S	24°14.9'W	4015	7.26	5	C	this paper	
	6429-2	31°57.0'S	24°14.9'W	4335	7.44	5	C	this paper	

Table 1. Water depth, position, core length, and sampling interval of the GeoB gravity cores studied in this paper. MTS: Magnetostratigraphy; Demag.: Demagnetization. Demagnetization was performed at A: 10, 20, 30, 40, 50, (70, 100) mT, B: 5, 10, 20, 30, 40, 50, 70, 100 mT, C: 5, 10, 15, 20, (25), 30, (35), 40, (45), 50, 65, 80, (95, 100, 110, 120) mT.

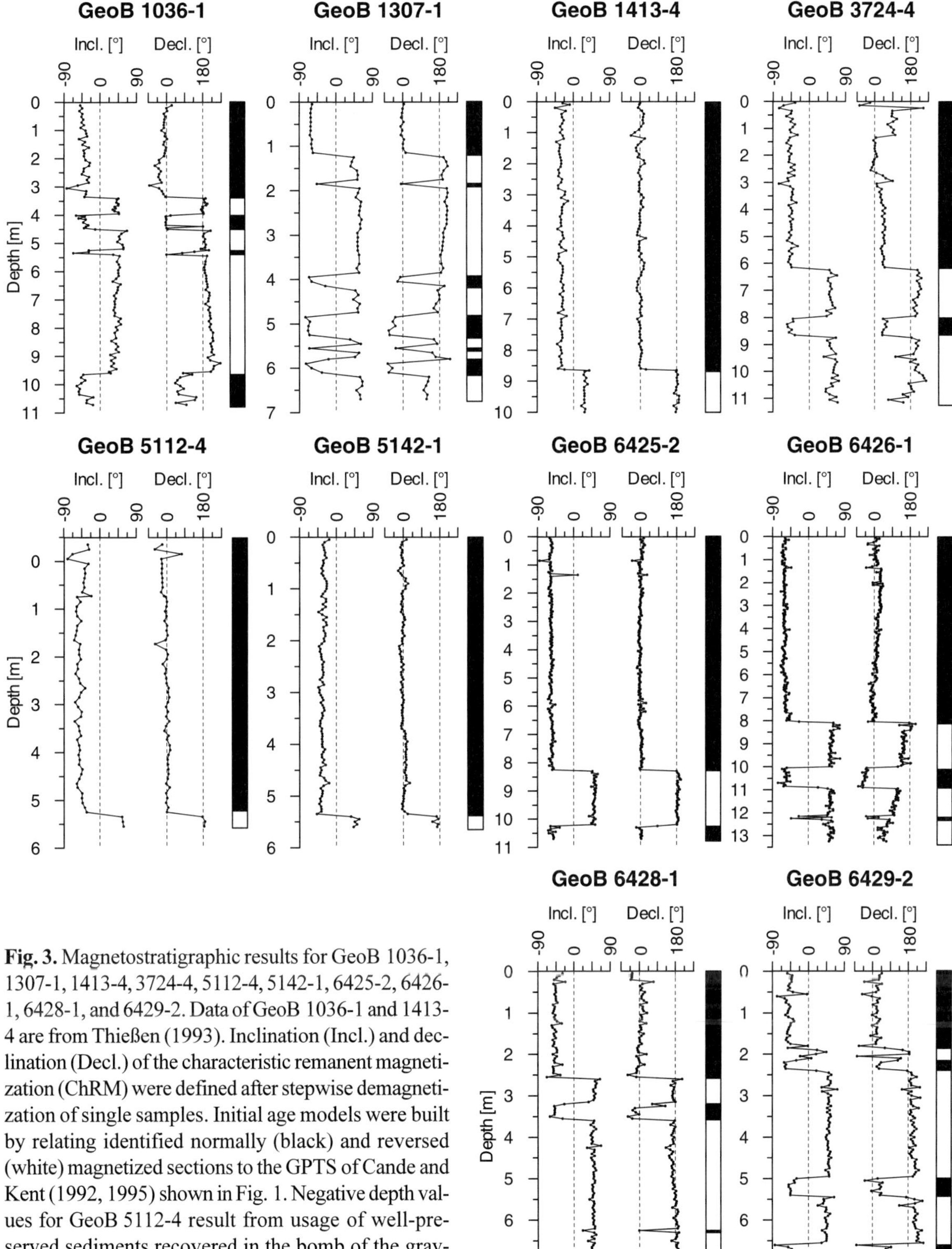

Fig. 3. Magnetostratigraphic results for GeoB 1036-1, 1307-1, 1413-4, 3724-4, 5112-4, 5142-1, 6425-2, 6426-1, 6428-1, and 6429-2. Data of GeoB 1036-1 and 1413-4 are from Thießen (1993). Inclination (Incl.) and declination (Decl.) of the characteristic remanent magnetization (ChRM) were defined after stepwise demagnetization of single samples. Initial age models were built by relating identified normally (black) and reversed (white) magnetized sections to the GPTS of Cande and Kent (1992, 1995) shown in Fig. 1. Negative depth values for GeoB 5112-4 result from usage of well-preserved sediments recovered in the bomb of the gravity corer.

No short polarity events could be identified unambiguously within the Brunhes Chron. Only one of all analyzed samples in the Brunhes Chron recorded a positive inclination, the one from 1.35 m depth in GeoB 6425-2 (+10.4°; α_{95} = 0.4°). The subsequent refinement of the age model (see below) results in an age of 135 ka for this sample. The nearest potential event, the Blake event, has been dated by Langereis et al. (1997) at 110 - 120 ka. In these oligotrophic environments, a combination of very low sediment accumulation rates and bioturbation is likely to hinder the recording of very short geomagnetic reversals or excursions.

However, under certain conditions a duration of about 10 kyr seems to be sufficient for a recording, since within the Matuyama Chron the short Cobb Mountain event is clearly documented in two of the six cores which reach the necessary age. In GeoB 1036-1, three samples with inverse inclination and/or declination are found between 5.23 m and 5.38 m, and in GeoB 6426-1 between 12.13 m and 12.28 m (Fig. 3). The well-defined inverse declination at 6.25 m (α_{95} = 3.2°) in GeoB 6428-1 does not correspond to the Cobb Mountain event. Relating the normal interval at the bottom of this core to the Olduvai event results in an age of 1602 ka for the depth of this sample. Subsequent refinement of the age model (see below) produces

an age of 1582 ka in marine isotope stage (MIS) 54. It can be assumed that this feature corresponds to a short event first found by Clement and Kent (1987) in North Atlantic sediments from ODP Site 609. By retuning the ODP 609 data to the record of ODP Site 677 (Shackleton et al. 1990) C. Langereis and D. Heslop dated this "Stage 54 Event" to 1580 ka (pers. comm.).

In Figure 4 age/depth relations based on identified geomagnetic reversals (open symbols) are shown for all cores. When more than one tie point is available these results indicate more or less constant sediment accumulation rates during the Pleistocene. Slightly lower rates are calculated between the top of the Jaramillo event at 990 ka and the Brunhes/Matuyama boundary at 780 ka in GeoB 1036-1, 6428-1, and 6429-2 (Fig. 4 b). This result is in agreement with the finding that the time interval between 920 and 640 ka is characterized by markedly lowered carbonate accumulation in the subtropical South Atlantic as a result of enhanced influence of corrosive bottom waters of southern origin (Schmieder et al. 2000).

The paleomagnetic analyses indicate much lower sediment accumulation rates in the lowest ~ 2 m of GeoB 1307-1 and 6429-2. At a first glance, this change could be assumed to result from sediment compaction during the core recovery. But

CK '95	GeoB 1036	GeoB 1307-1	GeoB 1413-4	GeoB 3724-4	GeoB 5112-4	GeoB 5142-1	GeoB 6425-2	GeoB 6426-1	GeoB 6428-1	GeoB 6429-2
Age [ka]	Depth [m]									
780	3.38	1.2	8.645	6.2	5.3	5.375	8.275	8.075	2.575	1.875
990	3.98	1.8	-	8.2	-	-	10.225	10.025	3.175	2.125
1070	4.525	1.9	-	8.7	-	-	-	10.875	3.575	2.375
1201	5.225	-	-	-	-	-	-	12.125	-	-
1211	5.375	-	-	-	-	-	-	12.275	-	-
1770	9.625	3.9	-	-	-	-	-	-	7.175	4.975
1950	-	4.2	-	-	-	-	-	-	-	5.425
2581	-	4.8	-	-	-	-	-	-	-	6.575
3050	-	5.3	-	-	-	-	-	-	-	-
3110	-	5.5	-	-	-	-	-	-	-	-
3220	-	5.6	-	-	-	-	-	-	-	-
3330	-	5.775	-	-	-	-	-	-	-	-
3580	-	6.15	-	-	-	-	-	-	-	-

Table 2. Depths of identified geomagnetic reversals and ages corresponding to the geomagnetic polarity timescale of Cande and Kent (1995).

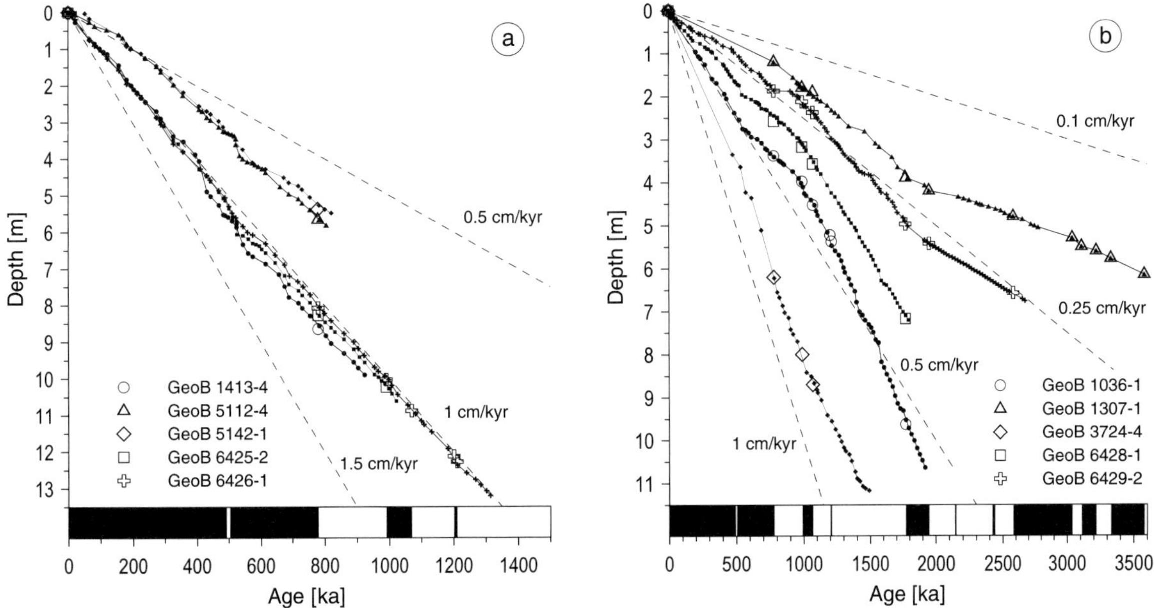

Fig. 4. Age/depth relations for GeoB cores resulting from the magnetostratigraphy shown in Fig. 3 (open symbols) and refined age models after correlation to SUSAS stack and obliquity tuning shown in Figs. 5, 6, and 7 (small symbols). Dotted lines depict different sediment accumulation rates. Note different age and depth scales on both plots.

remarkably, the change occurs in both cores exactly at the same time, between top and bottom of the Olduvai event (1950 - 1770 ka). Additionally, a potential compaction of the lower part of the core is not observed for any of the other cores which do not reach the top of the Olduvai, except for the lowermost ~ 40 cm of GeoB 3724-4, where the refined age model described below (small symbols in Fig. 6) could be interpreted in terms of compaction (or strongly lowered sediment accumulation rates). Furthermore, seven late Pliocene magnetostratigraphic tie points in GeoB 1307-1 denote no further sediment accumulation rate reduction prior to the Olduvai which would be expected in case of compaction as a result of an increasing effect. Instead, nearly constant sediment accumulation rates of about 0.15 cm/kyr are observed during the whole late Pliocene sequence, except for the 70 kyr long Kaena subchron wherein they even increase to 0.29 cm/kyr.

For these reasons, we rule out sediment compaction and conclude that indeed sediment accumulation rates increased significantly in the western subtropical South Atlantic during the Olduvai, thus sometime between 1950 and 1770 ka. Average values calculated for the Pleistocene are

in both cores approximately 40% higher than in the Pliocene sections. In GeoB 1307-1 they shift from 0.15 to 0.21 cm/kyr and in 6429-2 from 0.18 to 0.26 cm/kyr. The subsequent refinement of the age models (small symbols in Fig. 4) results in the identification of hiatuses in the Pleistocene sections of these two cores. Taking these gaps into account would even slightly enhance the average sediment accumulation rates calculated for the Pleistocene.

Magnetic Susceptibility: Tuning and Interpretation

Within the SUSAS Stack (Pleistocene)

Based on the magnetostratigraphic tie points a straightforward top to bottom correlation even of small-scale characteristics in magnetic susceptibility to the SUSAS stack was feasible for the sediments of GeoB 1413-4 (Fig. 5 a), 5142-1 (b), 5112-4 (c), 6425-2 (d), and 6426-1 (e). The correlation indicates that MIS 1 to 3 are missing at the top of GeoB 5142-1 (Fig. 5 b). MIS 1 to 4 are missing at the top of GeoB 5112-4 but in accordance with Frederichs et al. (1999) the well-preserved sedi-

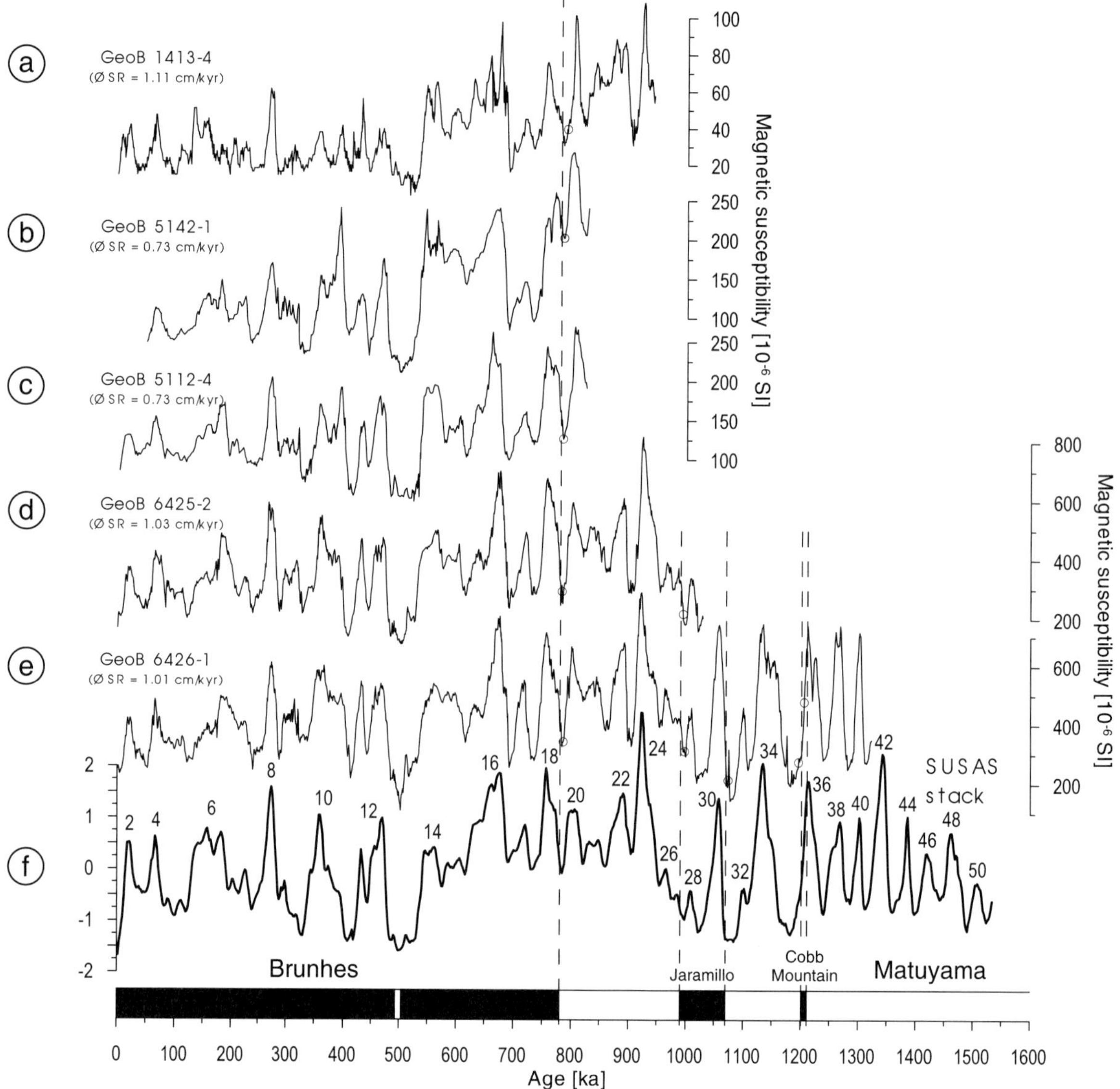

Fig. 5. Combined age models for GeoB **a)** 1413-4, **b)** 5142-1, **c)** 5112-4, **d)** 6425-2, and **e)** 6426-1 based on paleomagnetic tie points and correlation to **f)** SUSAS stack. Average sediment accumulation rates (Ø SR) range between 0.7 and 1.1 cm/kyr. Identified geomagnetic reversals are marked by dots. As a consequence of the lock-in process they are generally located deeper in the cores than the corresponding reference ages of Cande and Kent (1992, 1995) which are marked by vertical dotted lines. Labels at the SUSAS stack indicate even numbered cold oxygen isotope stages.

ments recovered in the bomb of the gravity corer have been added to complete the record as far as possible (Fig. 5 b). This approach is justified by a very good correlation of the resulting magnetic susceptibility pattern to the SUSAS stack and neighboring cores. Mean magnetic susceptibility values of the cores, calculated only for the Brunhes chron to enable a comparison, increase from north to south (Table 3). A lower contribution of biogenic

components in the deeper southern cores, e.g caused by enhanced influence of corrosive southern source deep waters, is unlikely, since sediment accumulation rates are similar or even higher in this area (see also Fig. 5). Consequently, it can be assumed that the trend to higher magnetic susceptibilities in the south indicates enhanced terrigenous input.

Apart from a good resemblance of characteristic Milankovitch style patterns with the main orbital periods of 41 kyr in the older part of the stack and approximately 100 kyr during the late Quaternary, high-frequency features are as well recorded as the longer-term changes. A good example is the 'MPT interim state' (Schmieder et al. 2000), between 920 and 640 ka (Fig. 5). The evolution of climate variability within the Mid-Pleistocene climate transition (MPT) causes much higher magnetic susceptibilities during this time period. These higher values are presumed to result from strongly reduced carbonate accumulation caused by enhanced influence of southern source deep waters. The discovery of considerably raised magnetic susceptibilities during the MPT interim state in the more northern cores GeoB 1413-4, 5112-4, and 5142-1 (Fig. 5) is further clear evidence for an ocean-wide change of sedimentation conditions in the middle Pleistocene.

Within the generally carbonaceous sequences, an unusual diatom layer was found in cores GeoB 5112-4, 6425-2, and 6426-1 (Pätzold et al. 1999, Wefer et al. 2001). All three layers are predominantly composed of the giant diatom *Ethmodiscus rex* (O. Romero pers. comm.) and the preliminary age estimations made on board by comparing the magnetic susceptibility records to the SUSAS stack (Frederichs et al. 1999; Donner and Schmieder 2001) suggest that they have been deposited at the same time as comparable laminated diatom ooze layers in the SUSAS cores GeoB 3801-6 and 3813-3 (Fig. 2). Together with unusual lithological features in other cores, the latter have been interpreted by Schmieder et al. (2000) as a documentation of the terminal event of the MPT related to the rearrangement of ocean circulation after a period of reduced NADW production and the onset of late Quaternary 100 kyr

climatic cycles. This event has been dated at 540 - 530 ka and the diatom layers were assumed to have been deposited during a time interval of 10 kyr. A prominent Ba/Al peak is associated with the diatom layer in GeoB 3813-3 indicating that the layers recorded a period of increased organic carbon flux to the sea floor (Gingele and Schmieder 2001).

In the case of GeoB 5112-4 and 6425-2, the top and bottom of the diatom ooze layer can be clearly detected by strongly enhanced P-wave velocities (Frederichs et al. 1999, Dillon et al. 2001) measured using an automated full waveform logging system (Breitzke and Spieß 1993) and are visible by eye. Therefore, the relevant section could easily be removed from the record before it was correlated to the stack. Unlike the dating procedure used within the SUSAS chronology for GeoB 3801-6 and 3813-3, no time interval has been attributed to this layer. Assuming a very rapid deposition of the diatoms (< 1 kyr) the correlation of the resulting magnetic susceptibility pattern to the SUSAS stack results in an age for the event of 524 ka in GeoB 5112-4 and of 528 ka in GeoB 6425-2. In core GeoB 6426-1 the diatom zone is not as clearly detectable as in the other two cores and could not be removed. The age model relates the sequence containing fragments of *Ethmodiscus rex* to the time interval 548 - 530 ka. The detection of this diatom layer in sediments from more northern and southern positions in the South Atlantic (Fig. 2) and the evidence for a simultaneous deposition strengthen the hypothesis that a major paleoceanographic event took place at the end of the MPT interim state.

In contrast to the cores shown in Fig. 5, GeoB 3724-4 from the Congo Basin displays stronger deviations from the SUSAS pattern (Fig. 6). The shift to higher magnetic susceptibilities during the

Core	Latitude	Mean susceptibility [10^{-6} SI] (780 – 0 ka)	Standard deviation [%]	Sediment accumulation rate [cm ka^{-1}] (780 – 0 ka)
GeoB 1413-4	15°40.8'S	33.1	48.3	1.11
GeoB 5142-1	19°05.4'S	124.9	43.2	0.73
GeoB 5112-4	23°49.5'S	128.7	34.9	0.73
GeoB 6425-2	33°49.5'S	347.7	34.5	1.06

Table 3. Mean magnetic susceptibilities, standard deviation and avearge sediment accumulation rates during the Brunhes chron for four investigated GeoB cores which form a transect from north to south.

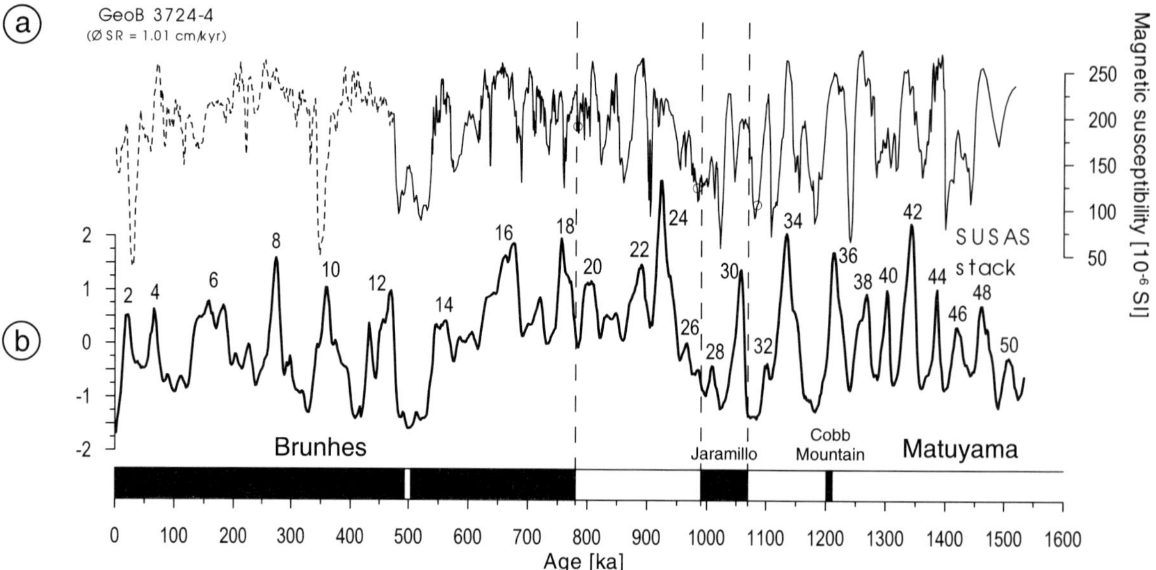

Fig. 6. Combined age model for **a)** GeoB 3724-4 based on paleomagnetic tie points (dots) and correlation to **b)** SUSAS stack. This correlation is not as good as those shown in Fig. 5 and not at all feasible for the youngest part of the record plotted with a dotted line. Reference ages of Cande and Kent (1992, 1995) are marked by vertical dotted lines. Labels at the SUSAS stack indicate even cold oxygen isotope stages.

MPT interim state is still visible, but a correlation of small scale patterns, especially during the early Pleistocene, is only feasible in some intervals. No correlation has been performed after MIS 13, where two minima with extremely low magnetic susceptibility values probably indicate significant disturbances.

Beyond the SUSAS Stack (Pliocene)

According to the magnetic reversal stratigraphy (Figs. 3 and 4), sediment cores GeoB 1036-1, 1307-1, 6428-1, and 6429-2 reach farther back in time than the 1534 ka old SUSAS stack. This results mainly from very low average sediment accumula-tion rates, ranging from 0.20 to 0.54 cm/kyr. The reduced temporal resolution makes it more difficult to relate magnetic susceptibility variations un-ambiguously to specific SUSAS features or orbital cycles, as demonstrated for the cores shown in Figure 5. To get a better impression of pattern variability, a simultaneous correlation strategy was applied. This parallel core processing makes use of all available magnetostratigraphic tie points, including those of the oldest, low-resolution cores as well as the better preserved magnetic suscep-

tibility patterns in the cores with higher sediment accumulation rates. Subsequent to paleomagnetic dating the logs were correlated to the SUSAS stack as far as possible (Fig. 7). Generally, this was more difficult in the youngest part of the records. In GeoB 1036-1, no unambiguous correlation was possible between MIS 2 and 7. The correlation results in a hiatus in GeoB 1307-1 between MIS 9 and 21 and in GeoB 6429-2 between MIS 19 and 23. These hiatuses result in higher average sedi-ment accumulation rates for the Pleistocene sect-ions than those calculated above, based on the magnetostratigraphic tie points alone. Although GeoB 6428-1 and 6429-2 are located near the region where several cores indicate a synchronous deposition of diatom ooze at approximately 530 ka this layer could not be detected in these two cores. Probably, sediment accumulation rates were too low at these locations to enable preservation of the diatoms.

In the older Early Pleistocene and late Pliocene sections of these cores, variations with a typical 41 kyr periodicity are better preserved because of higher sediment accumulation rates. Explicit evidence for obliquity forcing of magnetic sus-ceptibility comes from an additional tie point beyond the SUSAS stack provided by detection of the top of the Olduvai subchron at 1770 ka in GeoB 6428-1 and 1036-1. Six nearly equidistant magnetic susceptibility cycles recorded in both cores between this reversal and the last applied SUSAS tie point at 1508 ka result in a dominant period of approximately 44 kyr. Spectral analysis of the magnetostratigraphically dated magnetic susceptibility record of GeoB 6428-1 also verifies strong obliquity forcing (Fig. 8). These cycles were used to refine the paleomagnetic age models by tuning to the respective orbital variations (Fig. 7 c). In accordance to the procedure utilized during the construction of the SUSAS stack, obliquity variations calculated by Berger and Loutre (1991) shifted by -4.5 kyr were used as target curve. During the tuning procedure, low obliquity values, indicating colder conditions, were related to mag-netic susceptibility maxima. This strategy was substantiated during the construction of the SUSAS stack and is also supported by several magnetostratigraphic tie points in the Pleistocene and Pliocene (Fig. 7 a and b). Afterwards, the tuned sequences of GeoB 1036-1 and 6428-1 helped to adjust the record of GeoB 6429-2 (Fig. 7 b). In the older, late Pliocene section of this core, tuning of the filtered record to the shifted obliquity target curve was performed as far as possible. Below the last tie point at 2673.5 ka, the record was extrapolated with the last calculated sediment accumulation rate. Below the SUSAS stack, the obliquity tuning results only in small deviations from the magnetostratigraphic age model (Fig. 4 b). The average differ-ence of allocated ages for all measured data points of GeoB 6429-2 is 12.2 ± 12.4 kyr. Extremely low sediment accumulation rates in core GeoB 1307-1 (Fig. 7 a) restricted identification and correlation of particular magnetic susceptibility patterns. The SUSAS stack has been used as target record as far as possible, in the older part obvious similarities to GeoB 6429-2 were used to adjust the record.

In the SUSAS cores, magnetic susceptibility is a reliable proxy for carbonate content (Schmieder et al. 2000). The remarkable similarity of these magnetic records to Pleistocene climate variations results from glacial/ interglacial carbonate dis-solution cycles induced by the interplay of NADW and more corrosive bottom waters of southern origin. The SUSAS pattern is dominated by typical 100 kyr cyclicity in the late Quaternary and distinct 41 kyr cycles in the older part back to 1500 ka (von Dobeneck and Schmieder 1999). When applying this interpretation as a proxy for carbonate preser-vation, magnetic susceptibility in the older portions of the presented cores appears to be closely linked to the late Pliocene reorganization of ocean circula-tion.

This reorganization is coupled to the intensi-fication of Northern Hemisphere glaciation which took place between 3100 and 2500 ka with a major step at about 2750 - 2730 ka (Maslin et al. 1996; Maslin et al. 1998; Haug et al. 1999). According to Turnau and Ledbetter (1989), NADW was much reduced or even absent in the Rio Grande Rise region before the onset of Northern Hemisphere glaciation at about 3100 ka (ages adjusted to the new timescales of Shackleton et al. 1990 and Hilgen 1991). A first incursion of NADW to this region near 30°S occurred at approximately 3000 ka. The interval from 3000 to 2700 ka was dominated by climatic cooling and increasing influx of deep waters originating in the North Atlantic. During this interval, NADW expanded and deepened in the South Atlantic (Turnau and Ledbetter 1989).

Very low carbonate contents in the oldest sedi-ments of cores GeoB 1307-1 and 6429-2 (Pätzold et al. 1993, Wefer et al. 2001) are indicated by high magnetic susceptibilities (Fig. 7 b). During these times the absence of NADW in the subtropical South Atlantic distinctly hampered carbonate pre-servation at greater water depths. Extremely high magnetic susceptibility values around $800 \cdot 10^{-6}$ SI are found in the time range between 4000 and 3100 ka in GeoB 1307-1, two peak values of 2800 and $3500 \cdot 10^{-6}$ SI are dated to 3720 and 3400 ka respectively. In striking contrast to the otherwise carbonaceous sequences, the lowermost 0.5 m of GeoB 6429-2 consists of dark colored red deep-sea clay (Eberwein et al. 2001; see Fig. 7b). The combined magneto- and cyclostratigraphic age model relates the depth of this remarkable change in sediment composition to an age of 2730 ka, exactly the time proposed by Haug et al. (1999) for the abrupt major intensification of Northern Hemi-sphere glaciation. The southward expansion and deepening of NADW (Turnau and Ledbetter 1989)

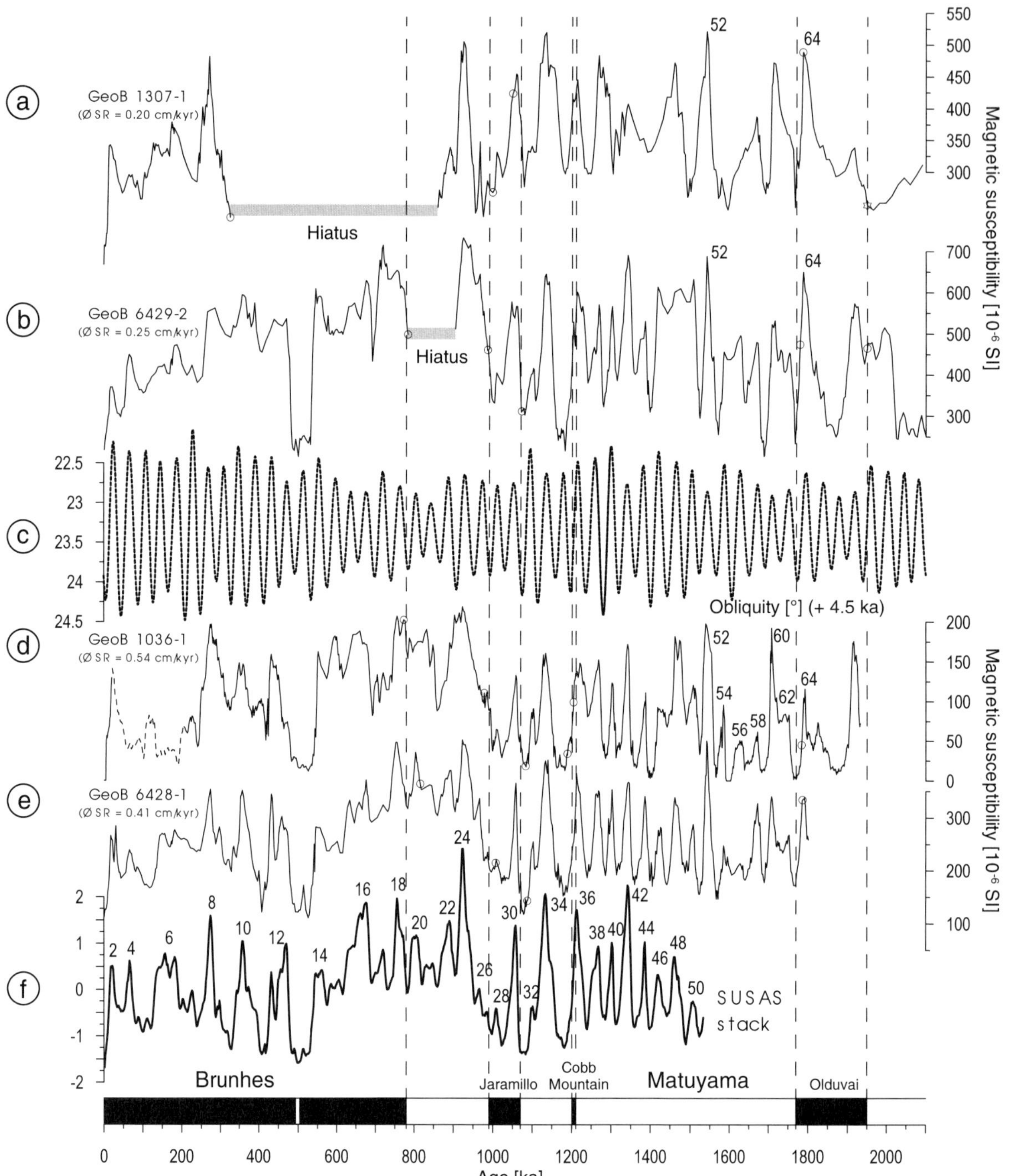

Fig. 7. Combined age models for **a)** GeoB 1307-1, **b)** 6429-2, **d)** 1036-1, and **e)** 6428-1 based on paleomagnetic tie points and correlation to **f)** SUSAS stack and tuning of extracted components to **c)** lagged obliquity variations calculated by Berger and Loutre (1991). Susceptibility data of GeoB 1036-1 are from Thießen (1993). The cores are characterized by very low average sediment accumulation rates (Ø SR). Identified geomagnetic reversals are marked by stars for positions kept in the final age model and by dots for those adjusted during the correlation and tuning procedure. As a consequence of the lock-in process the latter are generally located deeper in the cores than the corresponding reference ages of Cande and Kent (1992, 1995) which are marked by vertical dotted lines. Labels at the SUSAS stack indicate even numbered cold oxygen isotope stages. NHG: Northern Hemisphere glaciation.

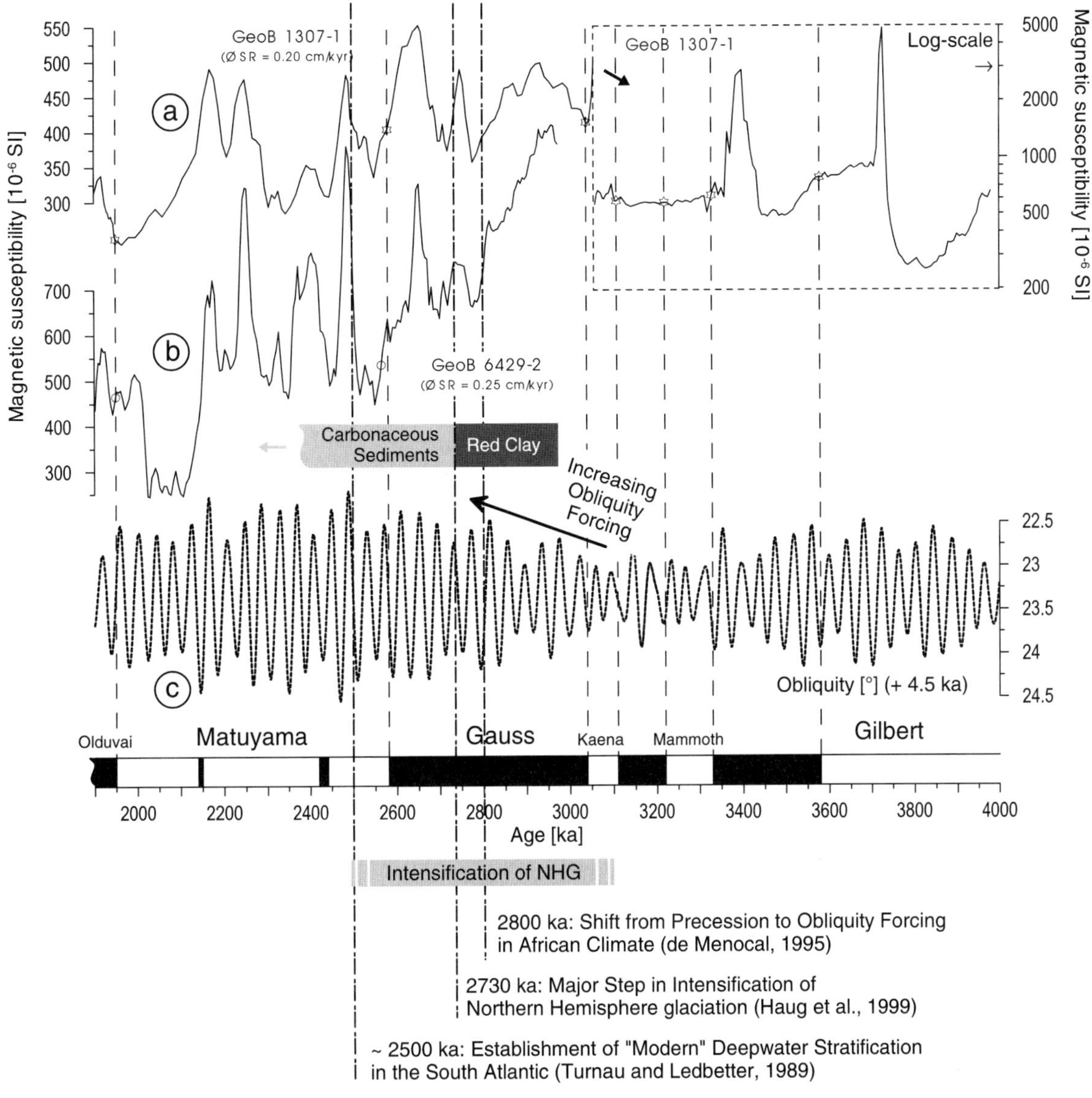

Fig. 7. continued (note 200 kyr overlap with left side).

apparently reached the location of GeoB 6429-2 in a water depth of 4335 m not before 2730 ka. Afterwards, it resulted in enhanced carbonate preservation at this location. The red clay is not found in GeoB 1307-1, probably because in this slightly shallower water depth of 4017 m at the eastern flank of the Rio Grande Rise, the carbonaceous shells were somewhat better preserved. Nevertheless, the lowermost part of this core is also characterized by a very low carbonate content (Donner 1993).

After the onset of the intensification of Northern Hemisphere glaciation at approximately 3100 ka, increasing carbonate content in both cores is indicated by decreasing mean magnetic susceptibility values. This trend continues until approximately 2100 ka and is superimposed on developing obliquity-driven cycles. The 41 kyr cyclicity is established in GeoB 6429-2 between 2800 and 2500 ka on a nearly constant background level of 617 ($\pm$ 128) $\cdot$ 10⁻⁶ SI (arithmetic mean $\pm$ standard deviation). At that time, a deepwater stratification similar to that of today was established in the Rio Grande Rise area as a result of southward expanding NADW (Turnau and Ledbetter 1989). The Northern Hemisphere glaciation

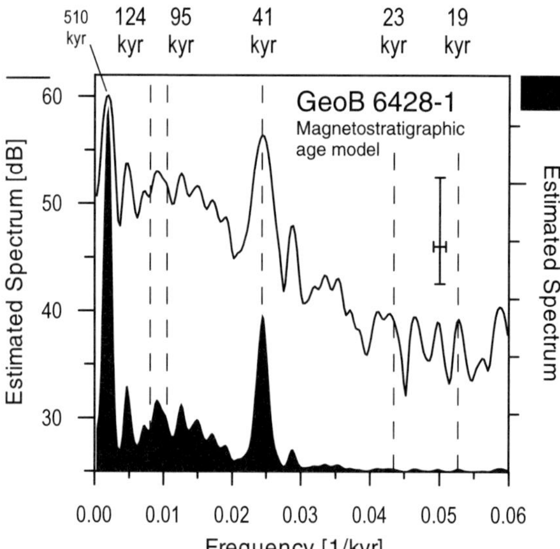

Fig. 8. Spectral analysis of magnetostratigraphically dated magnetic susceptibility of GeoB 6428-1 on a linear (solid black) and logarithmic (line) scale. Main orbital periods are marked by vertical dotted lines. Note the strong obliquity forcing (41 kyr) of the (untuned) record. The spectral peak near 500 kyr results from strongly enhanced susceptibilities during the MPT interim state. The cross is related to the logarithmic scale and depicts 6 dB bandwidth and 90% confidence interval. Spectral analysis were done using the program SPECTRUM (Schulz and Stattegger 1997).

intensified between 3100 and 2500 ka (e.g. Shackleton et al. 1984; Tiedemann et al. 1994) with a major step at 2750 - 2730 ka (Maslin et al. 1996; Maslin et al. 1998; Haug et al. 1999). African climate variability proxies document enhanced obliquity forcing since approximately 2500 ka (e.g. Bloemendal and de Menocal 1989; de Menocal 1995). After a long-term minimum between 4500 and 3100 ka, obliquity forcing of solar insolation intensified considerably between 3100 and 2500 ka (Fig. 7 b). This increase in obliquity amplitudes is asserted by several authors as an important additional factor (Maslin et al. 1995, 1998; Haug and Tiedemann 1998) within the scope of explanations given for the cause of the initiation of Northern Hemisphere glaciation (for a summary see Maslin et al. 1998), like the uplift of the Tibetan-Himalayan plateau (Raymo et al. 1988) or the gradual closure of the Isthmus of Panama (Haug and Tiedemann

1998). The latter led to a changed ocean circulation which supplied more moisture to, and hence more precipitation in the Northern Hemisphere. More obtuse angles of the Earth rotational axis occurred periodically and prevented the increased snow cover of the winters from melting during summers. Within this time period the magnetic susceptibility records of GeoB 1307-1 and 6429-2 indicate significant changes in deep water chemistry in the subtropical South Atlantic. Well-preserved obliquity cycles since 2500 ka can be assumed to result from periodically enhanced influence of NADW in this region.

The record of GeoB 6429-2 implies a further major change at about 2150 ka when the background level of magnetic susceptibility drops to the much lower early Pleistocene level of 429 ($\pm$ 112) $\cdot$ 10^{-6} SI (between 2150 and 1000 ka). Like a reduced influence of NADW during the MPT interim state caused much higher magnetic susceptibilities (Schmieder et al. 2000), the late Pliocene decline is interpreted as a further abrupt intensification of NADW influx to the subtropical South Atlantic. Most likely this major increase of NADW influx and the consequential better carbonate preservation accounts for the above discussed 40% increase of sediment accumulation rates near the Pliocene/Pleistocene boundary.

Conclusions

• The chronostratigraphic validity of the SUSAS stack (von Dobeneck and Schmieder 1999) as a target record has been demonstrated for ten previously magnetostratigraphically dated GeoB cores recovered in the oligotrophic South Atlantic between 15.5 and 33.5°S.

• Obliquity forcing of magnetic susceptibility is documented since 2500 ka in the late Pliocene and has been used to refine magnetostratigraphic age models in sections older than the SUSAS stack.

• Further evidence for a major paleoceanographic event at the end of the Mid-Pleistocene Transition at 540 – 530 ka (Schmieder et al. 2000) comes from the concordant dating of unusual diatom layers consisting of the giant diatom *Ethmodiscus rex* in three more cores (GeoB 5112-4, 6425-2, 6426-1).

• Enhanced magnetic susceptibilities and hence lowered carbonate accumulation rates during the MPT interim state are also documented at more northern locations and strengthen the hypothesis of an oceanwide enhanced influence of corrosive bottom waters of southern origin during that interval.

• Paleomagnetic age models indicate that sediment accumulation rates in the Rio Grande Rise area during the early Pleistocene are higher by approximately 40% than during the late Pliocene.

• In addition to the paleoceanographic history recorded in the Pleistocene SUSAS sequences the late Pliocene and early Pleistocene rearrangement of ocean circulation is mirrored in older magnetic susceptibility records. In the Rio Grande Rise area, obliquity forcing of sedimentation evolved simultaneously with the intensification of Northern Hemisphere glaciation between 3100 and 2100 ka. In the deep core GeoB 6429-2, the major intensification of the glaciation is documented in an abrupt change from dark colored red deep-sea clay to carbonaceous sediments at 2730 ka. Inflow of NADW to the Rio Grande Rise region increased steadily during the late Pliocene with major steps at approximately 2700 – 2600 and at 2150 ka.

Acknowledgements

Liane Brück provided technical assistance during NRM measurements of GeoB 1307-1, 5112-4, and 5142-1. Thomas Frederichs, Christian Hilgenfeldt and Karl Fabian automated the cryogenic magnetometer and the magnetic susceptibility measurement system in Bremen and developed various software to analyze the data. Thanks to Ian Snowball and Cor Langereis for thoughtful reviews and helpful comments on the manuscript and to David Heslop for improving the English. Captains and crews of RV *Meteor* are acknowledged for their efficient support during the cruises M 6/6, M 15/2, M 16/1, M 34/2, M 41/3, and M 46/4. This study was funded by the Deutsche Forschungsgemeinschaft (Sonderforschungsbereich 261, contribution no. 383). Data are available under www.pangaea.de/Projects/SFB261.

References

Balsam WL, McCoy Jr FW (1987) Atlantic sediments: Glacial/interglacial comparisons. Paleoceanography 2: 531-542

Berger A, Loutre MF (1991) Insolation values for the climate of the last 10 million years. Quat Sci Rev 10: 297-317

Bickert T, Wefer G (1996) Late Quaternary deep water circulation in the South Atlantic: Reconstruction from carbonate dissolution and benthic stable isotopes. In: Wefer G, Berger WH, Siedler G, Webb DJ (eds) The South Atlantic: Present and Past Circulation. Springer, Berlin, pp 599-620

Bickert T, Curry WB, Wefer G (1997) Late Pleistocene to Holocene (2.6-0 Ma) western equatorial Atlantic deep-water circulation: Inferences from benthic stable isotopes. In: Proc ODP Sci Results 154, pp 239-254

Bloemendal J, deMenocal P (1989) Evidence for a change in the periodicity of tropical climate cycles at 2.4 Myr from whole-core magnetic susceptibility measurements. Nature 342: 897-900

Breitzke M, Spieß V (1993) An automated full waveform logging system for high-resolution P-wave profiles in marine sediments. Mar Geophys Res 15: 297-321

Cande SC, Kent DV (1992) A new geomagnetic polarity time scale for the Late Cretaceous and Cenozoic. J Geophys Res 97: 13917-13951

Cande SC, Kent DV (1995) Revised calibration of the geomagnetic polarity timescale for the Late Cretaceous and Cenozoic. J Geophys Res 100: 6093-6095

Clement BM, Kent DV (1987) Short polarity intervals within the Matuyama: Transition field records from hydraulic piston cored sediments from the North Atlantic. Earth Planet Sci Lett 81: 253-264

D'Argenio BD, Fischer AG, Richter GM, Longo G, Pelosi N, Molisso F, Duarte Morais ML (1998) Orbital cyclicity in the Eocene of Angola: Visual and image-time series analysis compared. Earth Planet Sci Lett 160: 147-161

deMenocal P (1995) Plio- Pleistocene African Climate. Science 270: 53-59

Dillon M, von Dobeneck T, Schmieder F (2001) Physical Properties Studies. In: Wefer G and cruise participants, Report and preliminary results of *Meteor*-Cruise M 46/4, Mar del Plata - Salvador. Ber Fachber Geowiss, Univ Bremen 173, pp 77-82

Donner B (1993) Stratigraphie. In: Pätzold J and cruise participants, Bericht und erste Ergebnisse über die *Meteor*-Fahrt M 15/2, Rio de Janeiro - Vitória, 18.1. - 7.2.1991. Ber Fachber Geowiss, Univ Bremen 17, pp 22-29

Donner B, Schmieder F (2001) Preliminary stratigraphic correlations: Magnetic susceptibility and reflectance data. In: Wefer G and cruise participants, Report and preliminary results of *Meteor*-Cruise M 46/4, Mar del Plata - Salvador. Ber Fachber Geowiss, Univ Bremen 173, pp 82-84

Eberwein A, Mollenhauer G, Romero O (2001) Core description and smear slide analysis. In: Wefer G and cruise participants, Report and preliminary results of *Meteor*-Cruise M 46/4, Mar del Plata - Salvador. Ber Fachber Geowiss, Univ Bremen 173, pp 19-21

Frederichs T, Fabian K, Funk J (1999) Physical Properties Studies. In: Pätzold J and cruise participants, Report and preliminary results of *Meteor*-Cruise M 41/3, Vitória - Salvador. Ber Fachber Geowiss, Univ Bremen 129, pp 122-129

Gingele FX, Schmieder F (2001) Anomalous South Atlantic lithologies confirm global scale of unusual mid-Pleistocene climate excursion. Earth Planet Sci Lett 186: 93-101

Haug GH, Tiedemann R (1998) Effect of the formation of the Isthmus of Panama on Atlantic Ocean thermohaline circulation. Nature 393: 673-676

Haug GH, Sigman DM, Tiedemann R, Pedersen TF, Sarnthein M (1999) Onset of permanent stratification in the subarctic Pacific Ocean. Nature 401: 779-782

Hilgen FJ (1991) Extension of the astronomically calibrated (polarity) time scale to the Miocene/Pliocene boundary. Earth Planet Sci Lett 107: 349-368

Heslop D, Langereis CG, Dekkers MJ (2000) A new astronomical timescale for the loess deposits of Northern China. Earth Planet Sci Lett 184: 125-139

Kirschvink JL (1980) The least-sqare line and plane and the analysis of paleomagnetic data. Geophys J Roy Astr Soc 62: 699-718

Langereis CG, Dekkers MJ (1999) Magnetic cyclostratigraphy: High-resolution dating in and beyond the Quaternary and analysis of periodic changes in diagenesis and sedimentary magnetism. In: Maher BA, Thompson R (eds) Quaternary Climates, Environments and Magnetism. Cambridge Univ Press, pp 138-158

Langereis CG, Dekkers MJ, de Lange GJ, Paterne M, van Santvoort PJM (1997) Magnetostratigraphy and astronomical calibration of the last 1.1 Myr from an eastern Mediterranean piston core and dating of short events in the Brunhes. Geophys J Int 129: 75-94

Lourens LJ, Antonarakou A, Hilgen FJ, Van Hoof AAM, Vergnaud-Grazzini C, Zachariasse WJ (1996) Evaluation of the Plio-Pleistocene astronomical time scale. Paleoceanography 11: 391-413

Maslin MA, Haug GH, Sarnthein M, Tiedemann R, Erlenkeuser H, Stax R (1995) Northwest pacific Site 882: The initiation of Northern Hemisphere glaciation. In: Proc ODP Sci Results 145: 315-329

Maslin MA, Haug GH, Sarnthein M, Tiedemann R (1996) The progressive intensification of Northern Hemisphere glaciation as seen from the North Pacific. Geol Rundsch 85: 452-465

Maslin MA, Li XS, Loutre MF, Berger A (1998) The contribution of orbital forcing to the progressive intensification of Northern Hemisphere glaciation. Quat Sci Rev 17: 411-426

Pätzold J and cruise participants (1993) Bericht und erste Ergebnisse über die *Meteor*-Fahrt M 15/2, Rio de Janeiro - Vitória, 18.1. - 7.2.1991, Ber Fachber Geowiss, Univ Bremen 17, 46 p

Pätzold J and cruise participants (1999) Report and preliminary results of *Meteor* cruise M 41/3, Vitória – san Salvador de Bahia, 18.4.1998 - 15.5.1998, Ber Fachber Geowiss, Univ Bremen 129, 46 p

Petersen N, von Dobeneck T, Vali H (1986) Fossil bacterial magnetic in deep-sea sediments from the South Atlantic Ocean. Nature 320: 611-615

Raymo ME, Ruddiman WF, Froelich PN (1988) Influence of late Cenozoic mountain building on ocean geochemical cycles. Geology 16: 649-653

Robinson SG (1990) Applications for whole-core magnetic susceptibility measurements of deep-sea sediments. In: Proc ODP Sci Results 115: 737-771

Schmieder F, von Dobeneck T, Bleil U (2000) The Mid-Pleistocene climate transition as documented in the deep South Atlantic Ocean: Initiation, interim state and terminal event. Earth Planet Sci Lett 179: 539-549

Schulz HD and cruise participants (1996) Report and preliminary results of *Meteor*-Cruise M 34/2, Walvis Bay - Walvis Bay. Ber Fachber Geowiss, Univ Bremen 78, 133 p

Schulz M, Stattegger K (1997) SPECTRUM: Spectral analysis of unevenly spaced paleoclimatic time series. Comp Geosci 23: 929-945

Schwarzacher W (1993a) Cyclostratigraphy and the Milankovitch theory. Elsevier, Amsterdam, 225 p

Schwarzacher W (1993b) Milankovitch cycles in the pre-Pleistocene stratigraphic record: A review. In: Hailwood EA, Kidd RB (eds) High resolution stratigraphy. Geol Soc Spec Publ 70: 187-194

Shackleton NJ, Backman J, Zimmerman H, Kent DV, Hall A, Roberts DG, Schnitker D, Baldauf JG, Desprairies A, Homrighausen R, Huddlestun P, Keene JB, Kaltenback AJ, Krumsiek KAO, Morton AC, Murray JW, Westberg-Smith J (1984) Oxygen isotope calibration of the onset of ice-rafting and history of glaciation in the North Atlantic region. Nature 307: 620-623

Shackleton NJ, Berger A, Peltier WR (1990) An alternative astronomical calibration of the lower Pleistocene timescale based on ODP Site 677. Trans Royal Soc Edinburgh, Earth Sci 81: 251-261

Shackleton NJ, Crowhurst S, Hagelberg T, Pisias NG, Schneider DA (1995) A new Late Neogene time scale: application to Leg 138 sites. In: Proc ODP Sci Results 138: 73-101

Shackleton NJ, Crowhurst S (1997) Sediment fluxes based on an orbital tuned time scale 5 Ma to 14 Ma, site 926. In: Proc ODP, Sci Results 154: 69-82

Tappa E, Thunell R (1984) Late Pleistocene glacial/interglacial changes in planktonic foraminiferal biofacies and carbonate dissolution patterns in the Vema Channel. Mar Geol 58: 101-122

Thießen W (1993) Magnetische Eigenschaften von Sedimenten des östlichen Südatlantiks und ihre paläozeanographische Relevanz. PhD Thesis, Ber Fachber Geowiss, Univ Bremen 41, 170 p

Thompson R, Oldfield F (1986) Environmental magnetism. Allen and Unwin, London, 227 p

Tiedemann R, Sarnthein M, Shackleton NJ (1994) Astronomic timescale for the Pliocene Atlantic $\delta^{18}O$ and dust flux records of Ocean Drilling Program site 659. Paleoceanography 9: 619-638

Turnau R, Ledbetter MT (1989) Deep circulation changes in the South Atlantic Ocean: response to initiation of Northern Hemisphere glaciation. Paleoceanography 4: 565-583

Volat J-L, Pastouret L, Vergnaud-Grazzini C (1980) Dissolution and carbonate fluctuations in Pleistocene deep-sea cores: A review. Mar Geol 34: 1-28

von Dobeneck T, Schmieder F (1999) Using rock magnetic proxy records for orbital tuning and extended time series analyses into the super- and sub-Milankovitch bands. In: Wefer G, Fischer G (eds) Use of Proxies in Paleoceanography: Examples from the South Atlantic. Springer, Berlin, pp 601-633

Wefer G and cruise participants (1988) Kurzbericht über die *Meteor*-Expedition Nr. 6, Hamburg - Hamburg, 28. Oktober 1987 - 19. Mai 1988. Ber Fachber Geowiss, Univ Bremen 4, 29 p

Wefer G and cruise participants (1991) Bericht und erste Ergebnisse der *Meteor*-Fahrt M 16/1, Pointe Moire - Recife, 27.3. – 25.4.1991. Ber Fachber Geowiss, Univ Bremen 18, 120 p

Wefer G and cruise participants (2001) Report and preliminary results of *Meteor*-Cruise M 46/4, Mar del Plata - Salvador. Ber Fachber Geowiss, Univ Bremen 173, 136 p

Congo Fan Neogene and Quaternary Sedimentation: Interplay of Riverine and Current Induced Deposition

G. Uenzelmann-Neben[*] and H. Miller

Alfred-Wegener-Institut für Polar- und Meeresforschung, Postfach 120161, 27515 Bremerhaven, Germany
** corresponding author (e-mail): uenzel@awi-bremerhaven.de*

Abstract: The incorporation of information regarding sedimentation rates and lithology from ODP Leg 175 Sites 1075, 1076 and 1077 into the analysis and interpretation of high-resolution seismic reflec-tion data led to the revision and refinement of a depositional model for the upper Congo Fan area presented earlier by Uenzelmann-Neben (1998). For four time slices since the Eocene (Late Oligocene - Miocene/Pliocene, Pliocene - 600 ky, 600 ky - ~160 ky, ~160 ky - Recent) the main sed-iment contributor to the upper fan was determined. Thus we can say that in the Late Paleogene input of sediments from the north dominated the area by either a south setting current or the Kouilou/ Niari River. This situation continued to the period Pliocene - 600 ky when southern sedi-ment sources (the Congo River and upwelling) became dominant with the material being deflected to the north by the Benguela Current. Upwelling as a sediment source on the upper fan became even more important after 600 ky while the main sediment load of the Congo River is guided to the middle and lower fan via the Congo Canyon.

Introduction

The Congo Fan is the largest postsalt depocentre along the West African margin south of the Niger Delta. Since the Congo River is the only water drainage from the Congo Basin, the Congo Fan appears to be an ideal location to study the varia-tions of the terrigenous input into the ocean over time. This might help to understand the climatic development in this area.

The West African margin is characterised by upwelling cells whose evolution is connected to the development of the Benguela Current. In the present-day eastern South Atlantic, the South Equatorial Countercurrent in the north feeds the warm southward-flowing waters of the Angola Current (AC) (Fig. 1). At ~15°-17°S, these waters converge with the cold northward-flowing Ben-guela Coastal Current and are deflected to the northwest. A sharp frontal zone, the Angola-Ben-guela Front (ABF), is established and extends to a distance of up to 1000 km off the coast (Meeuwis and Lutjeharms 1990). The ABF effec-tively forms a barrier for surface ocean transport, as does the Walvis Ridge for deep water.

The cyclonic gyre in the Angola Basin is overlain by the low-salinity plume of the Congo River at ~5° S. ODP Leg 175 Sites 1075, 1076 and 1077 are lo-cated below the Congo plume. Riverine input, coastal upwelling, and incursions of open waters influence this environment (Schefuß et al. 2001). Site 1076 is the shallowest location and should record the strongest interaction of river discharge and coastal upwelling. Sites 1075 and 1077 are both at larger water depths and are presumably more influenced by open-ocean conditions.

The reconstruction of the Late Neogene devel-opment of the Benguela Current and its associated upwelling regimes was the goal of ODP Leg 175. Three sites (1075, 1076 and 1077) were drilled on the Congo Fan which contain the record of sedi-ment supply by the Congo River, intercalated with the oceanic record (Wefer, Berger, Richter et al. 1998). Unfortunately, the drilled record goes back only to the Late Pliocene (2 Ma, Wefer, Berger, Richter et al. 1998), and no light was shed on Miocene/Pliocene times and the early phase of the development of the Benguela Current.

From WEFER G, MULITZA S, RATMEYER V (eds), 2003, *The South Atlantic in the Late Quaternary: Reconstruction of Material Budgets and Current Systems.* Springer-Verlag Berlin Heidelberg New York Tokyo, pp 279-293

Palynological records from Site 1075 tell us about a change in environment in equatorial Africa ~1.05 Ma, which was followed by a decrease in Congo River discharge (Dupont et al. 2001). Stable isotopes, magnetic characterization analysis, and organic geochemistry carried out at material of Sites 1076 and 1077 also report a gradual decrease in terrestrial input from 0.9 to 0.4 Ma for the Congo Fan (Durham et al. 2001). In contrast to those observations, the terrigenous organic matter of Site 1075 shows no indications for major changes around 1 Ma (Hoeltvoeth et al. 2001). Furthermore, the mapping of the Holocene sedimentary record showed that recent sediments supplied by the Congo River have been distributed to the northwest by the north setting Benguela Current (Jansen et al. 1974). Thus, the information of the three ODP Leg 175 sites on the Congo Fan may not truly document the amount of river discharge but instead may reflect the distribution of the material.

On the southern Gabon margin (between the Ogooué and Congo Fans), elongated drifts formed by landward bottom-current flow have been identified (Séranne and Nzé Abeigne 1999). This bottom-current flow has been interpreted to be the result of upwelling mostly from Antarctic Intermediate Waters due to an Ekman induced northwest surface current. This process is inferred to have started already in Oligocene times.

So, we see that several factors are involved in transporting and shaping the sediments on the southern Gabon-Congo margin. It is not yet known, in which chronology the processes developed nor whether river discharge has been more important than biogenic productivity as a result of upwelling. Using depot centres and seismic structures of the sediments we want to help deciphering the chronological development of Congo River input and Benguela Current productivity.

A model for the Neogene/Quaternary was presented by Uenzelmann-Neben (1998). Using the information provided by ODP Leg 175 Site 1075 - 1077 that model is revised in this paper and an improved model will be presented for the deposition since Late Oligocene times which will consider both material entered by rivers (riverine sedimentation, Kouilou/Niari River and Congo River) and due to upwelling (biogenic debris, current induced sedi-

mentation) and the deflection of the material while being transported by the Benguela Current (current induced sedimentation).

To give a short summary on the general geologic evolution: Mesozoic rifting in the South Atlantic produced a succession of five distinct tectonic regimes: Prerift, Synrift (I and II), Postrift and Regional Subsidence (Brice et al. 1982). Up to 1000 m of evaporites were deposited during the Aptian. This was followed by initial movement of the salt in Albian time (Leyden et al. 1972; Emery et al. 1975; Emery and Uchupi 1984; Uchupi 1992). The seaward edge of the salt diapirs is formed by the prominent Angola Escarpment (Emery and Uchupi 1984). Crossing the escarpment, the seafloor drops sharply by up to 1000 m (Fig. 1). During the middle Cretaceous (Aptian-Turonian), black shales and bituminous sandstones were deposited on top of the evaporites (Bolli et al. 1978; Brice et al. 1982; Emery and Uchupi 1984; Uchupi 1992). Erosion associated with the uplift of the African continent in Late Cretaceous/Tertiary times was most likely the reason for a considerable build-up of sediments in the basins off West Africa during the Late Paleogene/Neogene (Valle et al. 2001). Deposition of turbidites started in the Oligocene (Brice et al. 1982; Uchupi 1992). The Benguela Current, the major oceanic current in the eastern South Atlantic, commenced in the Late Miocene (Dean et al. 1984). A large amount of organic material was produced by associated upwelling and was deposited quickly, which resulted in slope instabilities and led to mass movement (slumping, sliding, debris-flows) in the southern Angola Basin (Emery and Uchupi 1984; Uchupi 1992).On the southern Gabon margin (north of the Congo Fan) upwelling is inferred to have started already in Oligocene/Early Miocene times leading to bottom-current flow towards the shelf (Séranne and Nzé Abeigne 1999). Elongate drifts were thus formed of material supplied by the Congo Fan, which was transported north.

Data Acquisition and Processing

About 1200 km of high-resolution reflection seismic data were gathered on the Congo Fan during a joint study made by the Alfred-Wegener-Institute

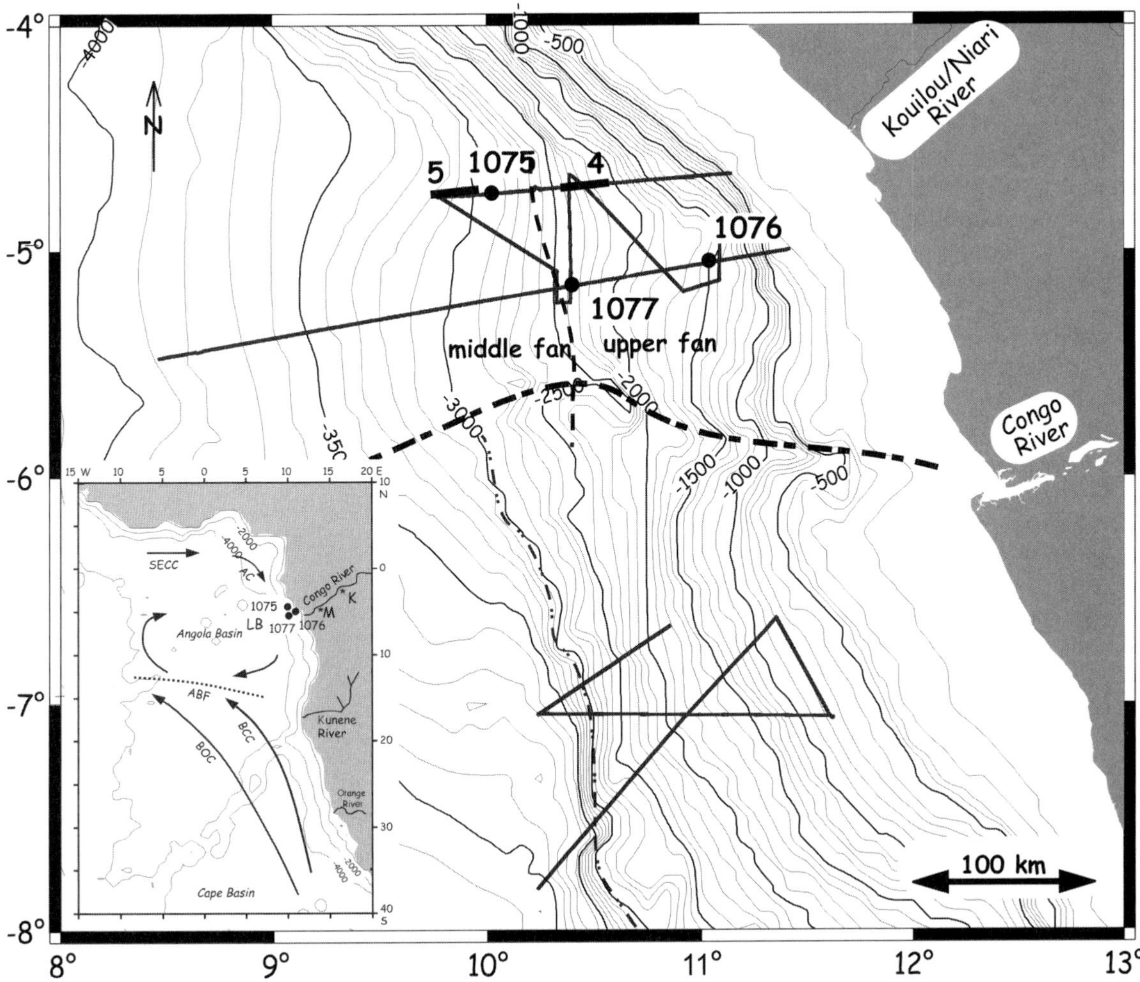

Fig. 1. Location of seismic profiles shot on the Northern Congo Fan. Bathymetry according to ETOPO-5 (National Geophysical Data Center, 1988). The italicized numbers refer to the parts of the lines displayed in Figs. 4 and 5. The dots show the locations of the recently drilled Sites 1075, 1076 and 1077 of ODP leg 175. The dot-dashed line marks the position of the Congo Canyon according to Shepard and Emery (1973). The double-dot-dashed line shows the position of the Angola Escarpment. The parts of the fan referred to in the text as middle and upper fan are separated by the dashed line. Modified from Uenzelmann-Neben (1998). The insert map shows the generalized surface hydrography (after Wefer, Berger, Richter et al. 1998). ABF= Angola-Benguela Front, AC= Angola Current, BCC= Benguela Coastal Current, BOC= Benguela Oceanic Current, K= Kinshasa, LB= Lower Congo Basin, M= Matadi, SECC= South Equatorial Counter Current.

and the University of Bremen (Fig. 1). A single GI-gun™ in the true GI mode (an airgun with two chambers: one to generate the seismic pulse and a second one triggered delayed to destroy the bubble pulse) was fired every 25 m as seismic source generating frequencies up to 300 Hz. This provided a vertical resolution of ~2.5 m. The data were re-ceived employing an 800 m streamer with 24 channels in a 600 m active section and recorded on a EG&G ES 2420 multichannel digital recording system. Positioning was by GPS (Global Positioning System).

In parallel with the seismic measurements, continuous records were acquired with the Parasound™ system, a narrow beam echo sounder of only 4° beam width with high vertical and lateral resolution. This system was run with a signal of 4 kHz and 250 µs pulse length. The shot interval varied depending on water depth, but generally was between 1 and 2 seconds. This system has a vertical resolution of up to 10 cm, while penetrating the upper 50-150 m of the sedimentary col-umn. Thus, high quality images of recent sedimentation and the surface structures were also collected (Bleil et al. 1994).

The processing of the multichannel seismic data comprised demultiplexing, definition of the geometry invoking the navigation data (CDP interval 25 m) and CDP sorting. Due to the very good seismic signal (short double pulse, bubble signal suppressed) and a negligible ocean bottom multiple, no deconvolution was applied. Despite the fact that the small streamer length gave only a relatively small move out, a detailed velocity analysis (every 50 CDP) was carried out. This was required because of the complex subbottom morphology. After correcting both spherical divergence and normal move out and stacking, a finite difference omega-x migration was applied to the data. This algorithm was chosen because it is accurate for dips up to 65 degrees, while simultane-ously being able to handle lateral velocity variations (Stolt 1978; Kjarstansson 1979; Yilmaz 2001). This method was especially useful in regions with steeply dipping reflectors due to salt diapirism. For display, no gain (i.e. AGC) or filtering except for an anti-aliasing-filter (360 Hz during acquisition) were applied to the data. The amplitudes shown on the profiles therefore represent values relative to the maximum amplitude of the whole section.

Stratigraphic Concept

An initial seismostratigraphic model was defined by Uenzelmann-Neben et al. (1997). Using preliminary results of ODP Leg 175 Sites 1075 - 1077 (Berger et al. 1998), Uenzelmann-Neben (1998) slightly modified the stratigraphic model. In general, this modified stratigraphic model is followed in this paper. New information from ODP Leg 175 Sites 1075 - 1077 as presented by Wefer et al. (2001)

regarding e.g. sedimentation rates shall be integrated.

The sites of ODP Leg 175 located within our target area (see Fig. 1) have all penetrated the seabed only to a subbottom depth of 202 - 207 m (Fig. 2, Wefer, Berger, Richter et al. 1998). This lies still within unit CF2 and hence nothing could be learned about the nature of the deeper sedimentary units or interfaces. The sites show relatively homogeneous material of Pliocene-Pleistocene age. A biostratigraphy, sedimentation rates and physical properties were derived for all sites and a sonic log was obtained at Site 1077 (Wefer, Berger, Richter et al. 1998).

Uenzelmann-Neben et al. (1997) defined four seismostratigraphic units according to their reflection characteristics ranging in age from Eocene to Quaternary (Fig. 2). They interpreted the uppermost unit CF1 (Late Pliocene, 80-140 m thick) to consist of deep marine hemipelagic clays and oozes (Figs. 2 and 3). Within unit CF1 a strong continuous reflection CF-A-int can be observed (Figs. 4 and 5). This reflector is very smooth and shows no indications of erosion, or of trans- or regression or non- depositional (toplap) character. Using sedimentation rates derived from biostratigraphy for ODP Leg 175 Sites 1075, 1076 and 1077 (19.6 cm/ky, 30.8 cm/ky, 20.5 cm/ky, resp., Wefer, Berger, Richter et al. 1998) the age of the reflector was found to be 160 ky, 112 ky and 160 ky, resp. Age-depth models based on the analysis of phytoplankton (Uliana et al. 2001) and δ^{18}O-data (Durham et al. 2001) are of higher resolution but of only limited usefulness since they have been established for only selected sites and depth ranges (phytoplankton: Site 1077 0-65 mbsf, Uliana et al. 2001; δ^{18}O-data: Site 1076 0-120 mbsf and Site 1077 0-160 mbsf, Durham et al. 2001). Unfortunately, those age-depth models are not consistent with biostrati-graphy or paleomagnetic results and also show strong age deviations between the sites (190 ky at Site 1076 vs 237 ky at Site 1077 for reflector CF-A-int). The strong deviation at Site 1076 is related to the unusual high sedimentation rates compared to both Sites 1075 and 1077 which may be a result of mass wasting on the continental slope and thus may not be representative for the whole area. Hence, we will use the age determined at Sites

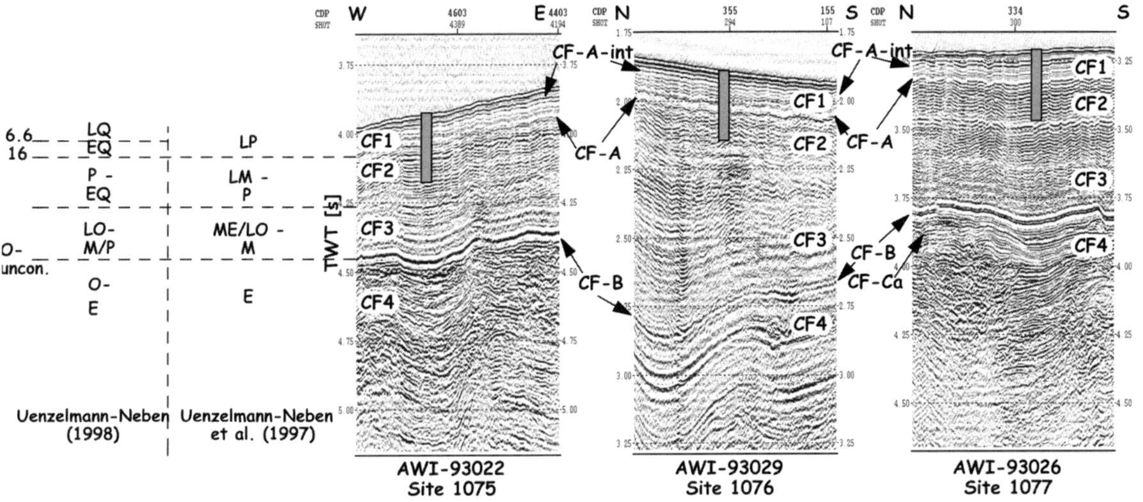

Fig. 2. Correlation of information from ODP Leg 175 Sites 1075 - 1077 (Wefer, Berger, Richter et al. 1998) with the seismic data. For location see Fig. 1. The hatched blocks show the location and penetration of the drill sites within the seismic lines. The seismostratigraphy according to Uenzelmann-Neben et al. (1997) and Uenzelmann-Neben (1998) has been included with CF1 - CF4 referring to the seismic units and CF-A-int - CF-Ca to the seismic reflectors. E= Eocene, EQ= Early Quaternary, LM= Late Miocene, LO= Late Oligocene, LP= Late Pliocene, LQ= Late Quaternary, M= Miocene, ME= Middle Eocene, O= Oligocene, P= Pliocene. The italic numbers and letters refer to the age of prominent reflectors. 6.6= marine isotope stage 6.6, 16= marine isotope stage 16, O-uncon.= Oligocene unconformity. Modified from Uenzelmann-Neben (1998).

1075 and 1077. Sites 1075 and 1077 which are located in larger waterdepths show a similar age which corresponds to marine isotope stage 6.6. Paleomagnetic data also suggest marine isotope stage 6 and not 5 (Wefer, Berger, Richter et al. 1998). The core itself does not show anything, which would help understand the origin of reflector CF-A-int. No unconformities were reported, physical properties (P-wave velocity and density) show no significant changes except for a drop in P-wave velocity above reflector CF-A-int at Site 1077, nor does the lithology change. Reflector CF-A-int was thus interpreted as a climatically- enhanced condensed zone related to a shift in the Angola-Benguela Front (Uenzelmann-Neben 1998). So, the upper part of unit CF1 is supposed to represent the Late Quaternary.

The correlation of the seismic data with both the biostratigraphic (Wefer, Berger, Richter et al. 1998) and the δ^{18}O-age-depth-model (Durham et al. 2001) shows reflector CF-A to be 600 ky and 500

ky old, resp. This corresponds to marine isotope stage 16 or 15. Paleomagnetic data from Site 1077 suggest MIS 16 (Wefer, Berger, Richter et al. 1998) for the depth of reflector CF-A. Except for a strong change in sedimentation rates from biostratigraphy at Site 1077 (from 19.2 cm/ky below to 6.9 cm/ky above) and less pronounced change at Site 1076 (5.6 cm/ky below, 7.2 cm/ky above) no direct indications for the origin of reflector CF-A have yet been found in the core material (Wefer, Berger, Richter et al. 1998). Durham et al. (2001) report a decrease in productivity and a minimum in terrigenous input for 600 ky, which could well explain the change in reflec-tion characteristics observed from unit CF1 to unit CF2.

The build-up of the Benguela Current (Late Miocene-Pliocene, Dean et al. 1984; Wefer, Berger, Richter et al. 1998) was considered to be represented by unit CF2 (80-140 mbsf, 130-170 m thick, Uenzelmann-Neben et al. 1997), which is characterised by a number of continuous, often

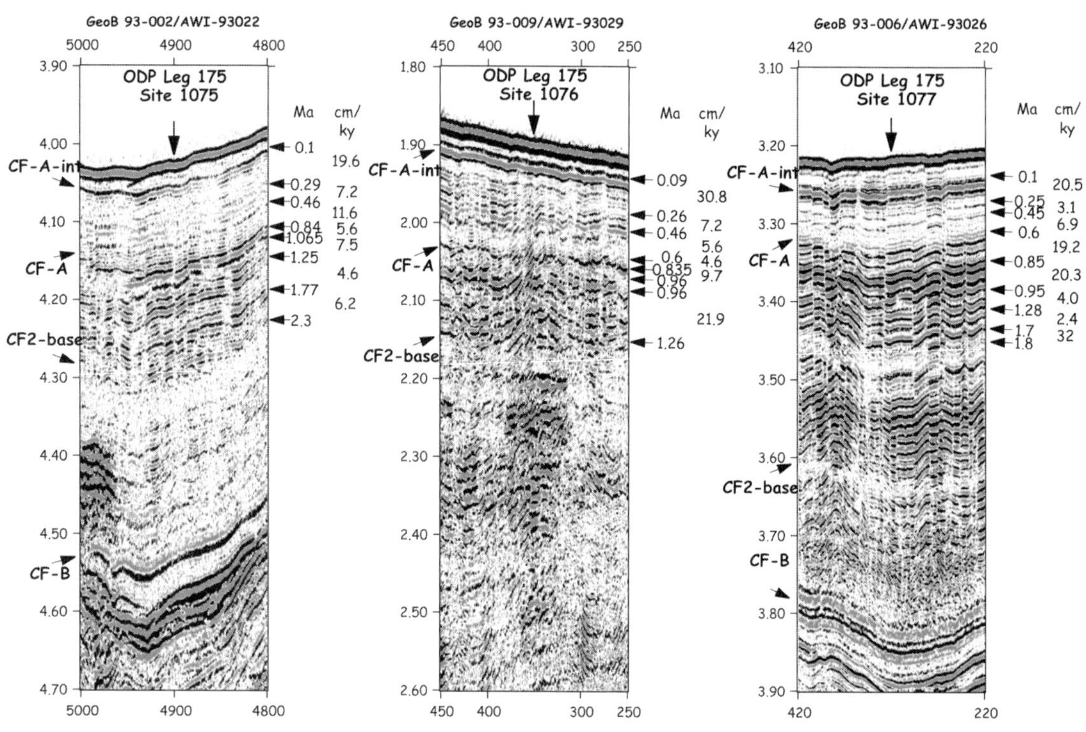

Fig. 3. Correlation of biostratigraphic sedimentation rates deduced for ODP Leg 175 Sites 1075, 1076 and 1077 (Wefer, Berger, Richter et al. 1998) with seismic data. The ages are given in million years before present (MA) and the sedimentation rates in cm per 1000 years (cm/ky).

undulating subparallel internal reflectors. A correlation of the seismic unit with lithological and biostratigraphic information reveals the unit to be of Pliocene-Early Pleistocene age (Fig. 2, Wefer, Berger, Richter et al. 1998).

Unit CF3 (Eocene/Late Oligocene-Miocene, 210-310 mbsf, 160-220 m thick, Uenzelmann-Neben et al. 1997) shows weak internal reflections. Since ODP Leg 175 Sites 1075, 1076 and 1077 only penetrated the sediments to a subbottom depth of 202-207 m (Wefer, Berger, Richter et al. 1998) which still lies within unit CF2 and recovered only Pleistocene-Pliocene material the depth of the Miocene/Pliocene boundary could not be determined. Extrapolated biostratigraphic sedimentation rates suggest that this boundary is located within unit CF3 thus modifying the top of the unit to be of Pliocene age. The base of unit CF3 is formed by a prominent reflector CF-B of inverse polarity (Fig. 4). Although reflector CF-B does not appear

as an erosional feature where imaged, it has been correlated via its reflection character and depth with reflector A of Emery et al. (1975, line 62; see Uenzelmann-Neben et al. 1997). Reflector A was interpreted as an Eocene/ Oligocene hiatus due to erosion (Emery and Uchupi 1984) resulting from the first major occurrence of an Antarctic icecap which removed large quantities of water from the oceans (Séranne 1999) and increased influx of Antarctic Bottomwater (AABW) in Late Oligocene times (Siesser 1978). Thus, unit CF3 is redated to cover the Late Oligocene to Miocene/ Pliocene.

The lowermost unit CF4 (Eocene-Oligocene, 370-530 mbsf, 130-215 m thick) is characterised by strong internal reflections and is interpreted to document an increasing terrigenous input (Uenzelmann-Neben et al. 1997). Below this unit at least two further units CF5 and CF6 can be recognised separated by strong continuous reflections. Since

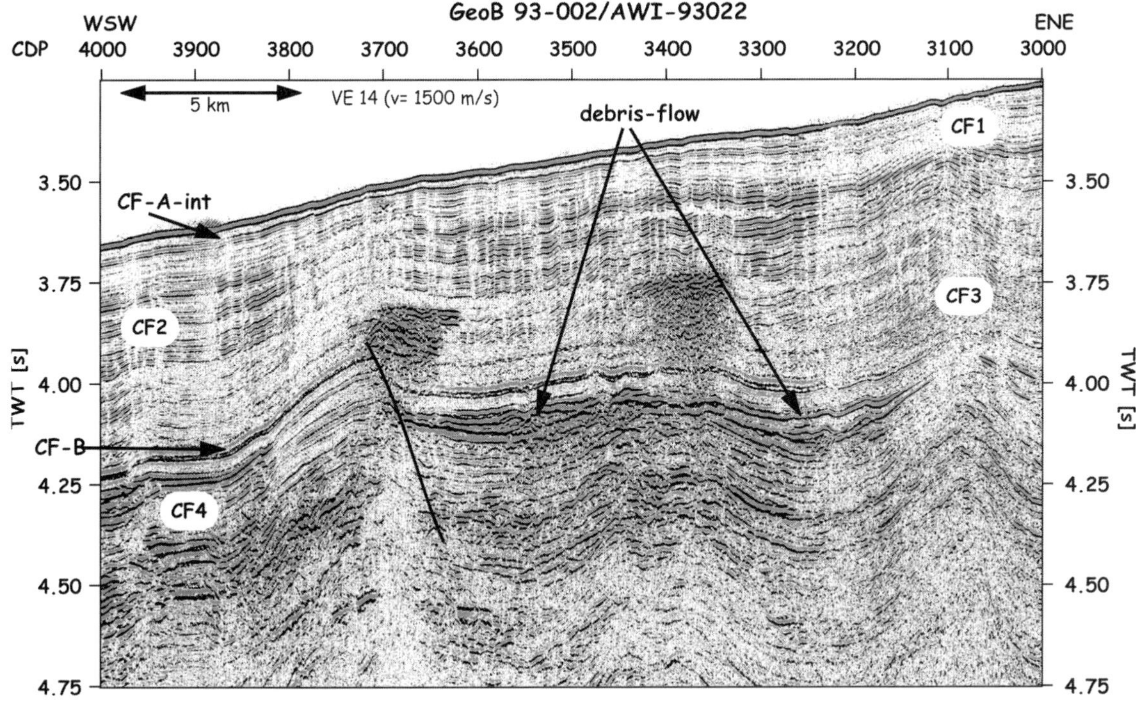

Fig. 4. Central part of line GeoB 93-002/AWI-93022. Note the debris-flows in unit CF4 and the reflectors of unit CF2 which are characterised by small-scale faulting due to differential conpaction.

we know very little about the nature of those units and, furthermore, they cannot be mapped within the whole target area due to disturbances as a result of salt tectonics the detailed analysis was limited to units CF1 to CF4.

Neogene/Quaternary Depositional Environments

The seismic data were analysed with respect to changes in reflection pattern, sedimentation structures such as slides and debris flows and the distribution of the seismic units. Changes in those parameters are the result of modifications in the depositional environment which in turn is due to variations in the currents systems and in the terrigenous supply of sediment.

Unit CF4 is the oldest unit investigated. Since the base reflector could not be mapped in the whole target area no isopach map was compiled. The unit comprises Eocene-Oligocene and possibly older

sediments (Uenzelmann-Neben 1998), which are imaged as strong parallel reflectors (Fig. 4). This indicates generally coarser material which is deposited proximal to its source. Only few channels were observed within this unit but especially in the northern part of our target area unit CF4 is characterised by a number of debris flows (Fig. 4, CDPs 3150-3700). Here, we are in front of the Kouilou/Niari River estuary. There are three possibilities for the transport mechanism of the sediments:

1. Input by the Congo River, entrainment in a north setting current and deposition on the northern fan. This is a difficult scenario because, since we found indications for coarse material in the reflection characteristics, this requires a strong current for transportation and no north setting current has yet been reported for the Eocene. The Benguela Current evolved only in Late Miocene times (Wefer, Berger, Richter et al. 1998). 2. Rapid deposition of material entered by the Kouilou/ Niari River

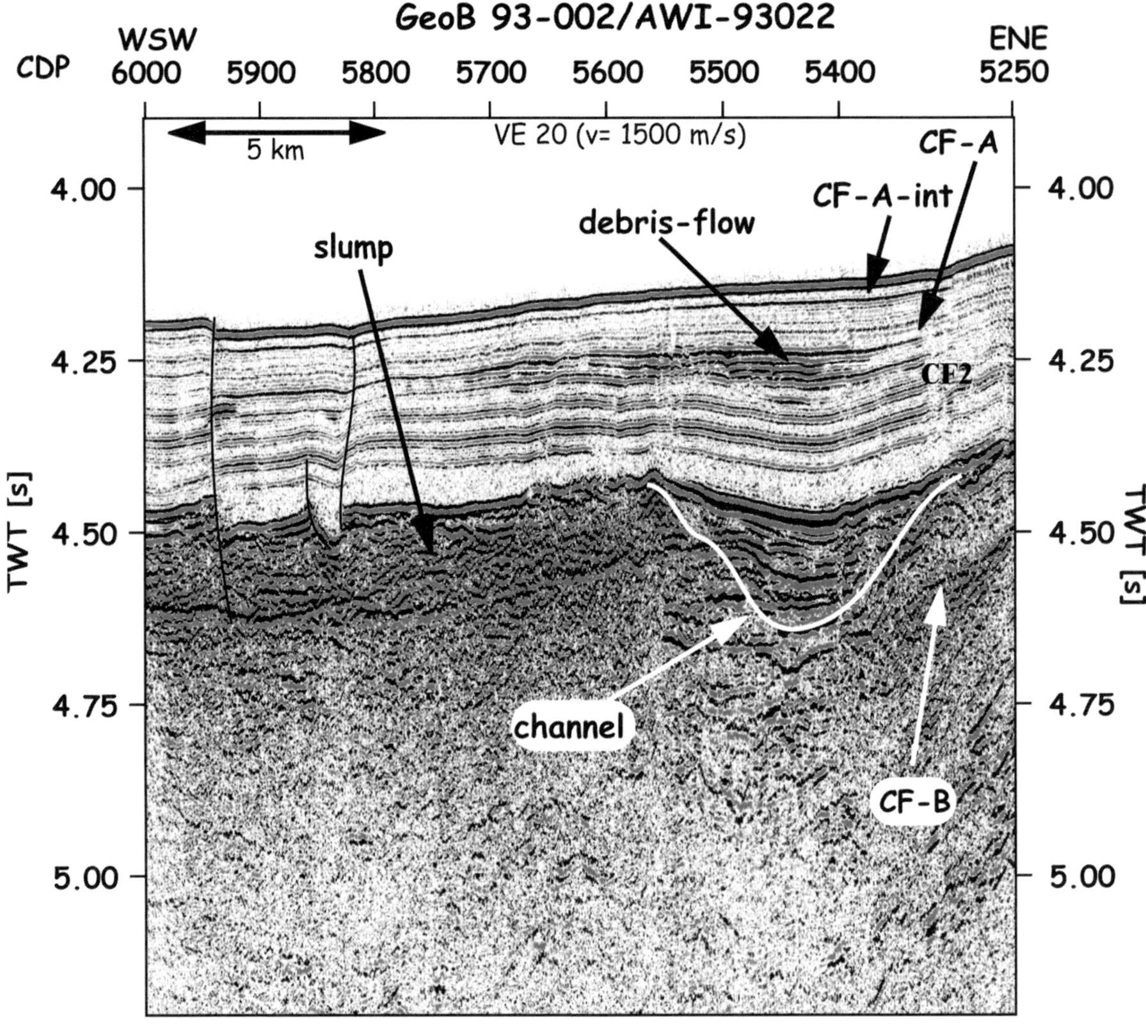

Fig. 5. Western part of line GeoB 93-002/AWI-93022. Note the slump in unit CF3 indicating mass transport due to rapid sedimentation on the slope and the channel. A debris-flow can be identified near the top of unit CF2.

led to instabilities at the shelf edge/continental slope which resulted in slope failure and debris flows. Droz et al. (1996) mapped three palaeochannels north of the Congo Fan, which they attributed to the effect of the Kouilou/Niari Turbidite System, which ceased in Post-Eocene times. Savoye et al. (2000) identified a system of palaeochannels both north and south of the current channel (more than 80 alone for the Quaternary), which can be re-grouped into three main structures that succeeded have each other in time: north, axial and south with the axial being the most recent one.

Savoye et al. (2000) have difficulties determining the relative chronology of the other two structures. 3. Input by a river in the north, e.g. the Ogooué River, entrainment in the south setting Angola Current and deposition on the northern fan. Again, the Angola Current needed to be fairly strong in order to transport the observed coarse material. Anderson et al. (2000) also give indications for contributions from a turbidite system north of the Congo Fan. This in combination with the observed locations of debris flows and channels in our data favours the second or third option.

A change in reflection characteristics occurs across reflector CF-B (Late Oligocene unconformity). While unit CF4 shows strong parallel reflectors, weaker reflections can be observed within unit CF3. In the lower 100-125 ms of the unit more continuous reflectors are found (Fig. 4, e.g. CDPs 3420-3620). Above this, unit CF3 often appears transparent indicating homogeneous material deposited under 'calm' constant conditions. This change in internal reflection pattern points towards a modification in the depositional environment, which could have been initiated by the onset of the northward flow of Antarctic Intermediate Water in Early Miocene (23 Ma) as was reported for the southern Gabon margin by Séranne (1999). This would give a sedimentation rate of 1.25 cm/ky for the lower part of unit CF3 which falls well into the values reported for the continental rise/ deep-sea (Michels pers. communication).

The number of channels observed in unit CF3 is larger than in unit CF4 (Uenzelmann-Neben et al. 1997). Furthermore, debris flows identified in the seismic data are located slightly farther in the west. Unit CF3 shows a big thick (max 750 ms TWT) depocentre near the mouth of the Kouilou/ Niari River whereas the sediments south of the Congo Canyon are relatively thin with only a thickening between salt domes (Fig. 6a). Again, there are two possible explanations for this distribution of sedi-ments: 1. Terrigenous sediment input by a river in the north, e.g. the Ogooué River or the Kouilou/Niari River as suggested by Droz et al. (1996), Uenzelmann-Neben et al. (1997) and Uenzelmann-Neben (1998). This material was then deposited close to the source. 2. Séranne and Nzé Abeigne (1999) suggested that upwelling already commenced in the Oligocene (a palaeo-Benguela Current?). If this was the case this would have led to a north setting surface current and thus a northward transport of material input by the Congo River. Upwelled watermasses always are rich in nutrients, which then results in high productivity. The low sedimentation rates we observe argue against a high biogenic component and favour either a northern sediment source or a low productivity current. The sediments were then deposited near the shelf edge, this led to instabilities and slope failure, and debris flows were transported via a

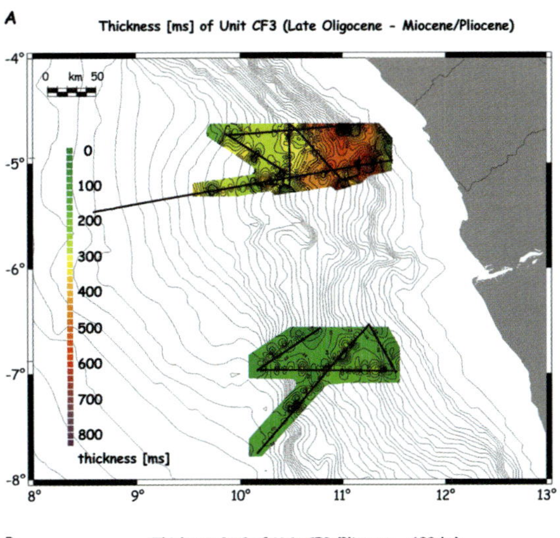

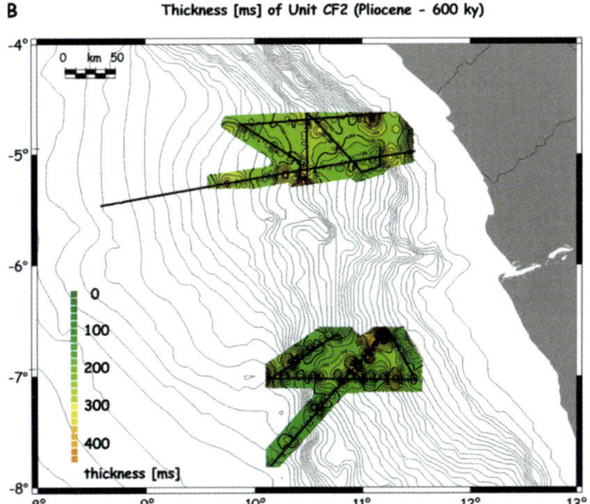

Fig. 6. a) Isopach map of unit CF3. Note the big depocentre in the northeast while south of the Congo Canyon little thickness variations occur. Modified from Uenzelmann-Neben (1998). **b)** Isopach map of unit CF2. Here, two smaller depocentres can be identified in the north which are not as distinct as the depocentre in unit CF3.

number of channels towards the middle and outer fan. Later on, the channels were filled with coarser material (Figs. 4 and 5).

The source of the sediment probably is the African continent where doming and uplift occurred in the Neogene (Hartmann et al. 1998; Frostick 1997), e.g. leading to a kinking of the Atlantic Hinge Zone in Oligocene times as the shelf and coastal plain rose by 2 to 3 km (Cramez and Jackson 2000).

No reflector marks the boundary of units CF3/CF2 but a change in reflection characteristics (Figs. 4 and 5). This implies that sedimentation continued and mainly the depositional environment changed. Unit CF2 is characterised by strong short continuous, often undulating, subparallel internal reflections and small-scale faulting. Biostratigraphic sedimentation rates are markedly higher within this unit with a strong increase in the upper part (Fig. 3, Wefer, Berger, Richter et al. 1998). The sediment drilled by ODP Leg 175, which cover the upper unit, were found to be mainly fine grained (Pufahl et al. 1998). The high sedimentation rates are the result of increased productivity due to the onset of the Benguela Current. The biogenic material was decomposed by bacteria leading to an increased gas content. Indications for this can be found both in the inversed polarity of reflector CF-A-int and the occurrence of bottom simulating reflectors (Uenzelmann-Neben et al. 1997, Uenzelmann-Neben 1998 and her Fig. 5).

Rapid sedimentation of fine grained material as the result of high productivity can easily create a lid which prevents the continuous release of fluids and gas. A result of this is differential compaction which is documented in small-scale faulting affecting mainly units CF2 and CF1.

Two depocentres of unit CF2 can be found north of the Congo Canyon (Fig. 6b). In the south, the unit shows thickness variations in tune with salt diapirs. No small-scale faulting can be observed, and the unit is generally thinner than in the north. This suggests that here in the south only little of the provided material was deposited. On the other hand, the depocentres in the north are located farther west than in unit CF3 (Late Oligocene - Miocene/Pliocene). Obviously, less sediment was supplied from the north. The Congo River started to play a more significant role in supplying material. It is the only exit for water from the Congo Basin. Doming associated with rifting in the East African Rift in Miocene times led to a strong pickup of the Congo River drain-age away from the East African Rift (Frostick and Reid 1989; Frostick 1997). Moreover, the western perimeter uplands (Crystal Mountains) of the Congo Basin provide a close source for fan material. From just below Stanley Pool (Kinshasa) the river drops 11 m per

14 km in its 350 km journey to Matadi near the coast (O'Brien pers. communication).

The material hence supplied by the Congo River together with the increased biogenic sediment supplied by the evolving Benguela Coastal Current was deflected to the north thus leaving only little material for the area south of the Congo Canyon.

Another change in reflection pattern occurs across reflector CF-A. While unit CF2 shows strong short reflections, unit CF1 is characterised by weak continuous reflections. This indicates a calm depositional environment with only few disturbances. Sedimentation rates are lower than for unit CF2 emphasising this (Fig. 3, Wefer, Berger, Richter et al. 1998). According to the biostratigraphic age model (Wefer Berger, Richter et al. 1998) this change in reflection pattern occurred about 0.84 Ma at Sites 1075, 1076 and 1077, according to δ^{18}O-data it occurred about 0.62 Ma at Site 1076 and about 0.77 Ma at Site 1077 (Durham et al. 2001). It is difficult to determine, which age is correct since both 0.84 and 0.62 Ma are linked to significant environmental changes. Dupont et al. (2001) observe a distinct environmental change in palynological records from Site 1075, which occurred in equatorial Africa 1.05 Ma. This led to a decrease in Congo River discharge and river-induced upwelling (Dupont et al. 2001). These environmental changes in tropical Africa precede the mid-Pleistocene climatic shift by 100 ky (Dupont et al. 2001). Following those changes, Durham et al. (2001) observe a decrease in terrigenous input (~ 0.92 Ma) and at 0.8 Ma an increase in productivity. Those environmental changes most certainly found their expression in a modification of reflection characteristics as observed from unit CF2 to unit CF1.

Durham et al. (2001) infer a minimum in terrigenous input and a maximum in productivity for 0.6 Ma. But we consider that already the onset of modifications in input/productivity will lead to different reflection characteristics and thus favour a move towards a more current controlled sedimentation as the origin of our observations.

The unit is divided into two subunits (lower and upper unit CF1) by a strong conformal reflector CF-A-int (Figs. 4 and 5, Uenzelmann-Neben 1998). The depth of this reflector correlates well with a renewed increase in sedimentation rates as reported by Wefer, Berger, Richter et al. (1998). Uliana et al. (2001) identified a decrease in Congo River discharge at the marine isotope stage 5/6 boundary while Durham et al. (2001) found indications for a reduced productivity at ~170 ky. Those observations correspond to the age of reflector CF-A-int. Furthermore, sedimentation within the Lower Congo Basin is dominated by rain-out of suspended clay derived from the Congo River to the southeast and by pelagic settling of biogenic debris (Pufahl et al. 1998). The coarse load of the Congo River is trapped within the river/canyon system and distributed to the outer fan. This is reflected by the size and the location of the depocentres of lower unit CF1 and upper unit CF1. The size of the depocentre shrank significantly and seems to document mainly that little sediment was transported to the northern fan, which may reflect the reduced terrigenous input and the decrease in productivity. Furthermore, the main load of the sediment is transported to the outer fan by the Benguela Current and in the Congo Canyon whereto our seismic lines do not extend.

The Model

The analysis of the seismic data led to the development of a scenario for the part both riverine sediment input and biogenic debris as a result of a modified current regime played in the Congo Fan area since Late Oligocene times Fig. 7 and Table 1). Of major importance in this context is not only the input of the material but also the distribution. This model refines the ideas of Uenzelmann-Neben (1998) which are based on the work of Droz et al. (1996) and Uenzelmann-Neben et al. (1997). Fig. 7 shows the different stages of the model from unit CF3 (Late Oligocene - Miocene/Pliocene) to upper unit CF1 (160 ky - Recent). The shaded area show the depocentres while the Dashed arrows represent the contributions from the Kouilou/Niari River and the Congo River. The black arrows mark material transported by the currents.

In Late Oligocene - Miocene/Pliocene the main sediment source appeared to lie in the north. This could have been either farther in the north with the material being transported into our area of interest by the south setting Angola Current or the Kouilou/Niari River (Fig. 7a). The isopach map of unit CF3 shows the main pile of sediments in the northeastern part of our target area (Fig. 6a). A large amount of material was deposited in front of the Kouilou/Niari River estuary as a result of an uplift of the coastal plains (Cramez and Jackson 2000) and redeposited via a number of debris-flows. We cannot exclude contributions from the north by a south setting current (Angola Current?). But this would probably result in a less restricted depocentre.

In Pliocene - 600 ky the depositional system was modified (Fig. 7b). Less material was contributed from the north (either by the Angola Current or the Kouilou/Niari River), and the Congo River took up its role as a major sediment source. Here, the western uplands of the Congo Basin supplied the sediment. Furthermore, the evolving Benguela Coastal Current led to an increase in biogenic debris and additionally deflected the material to the north. There, two smaller depocentres were built-up which is documented in the isopach of unit CF2 (Fig. 6b). Thus, this period saw a combination of riverine and current induced sedimentation on the upper Congo Fan.

The influence from the north decreased in the period 600 ky - ~160 ky (Fig. 7c). The depocentre now is located farther away from the estuary of the Kouilou/Niari River and lies more in the path of material deflected to the north by the Benguela Coastal Current. Only insignificant quantities were deposited south of the canyon. Sedimentation rates are lower than in the Period Pliocene - 600 ky although the productivity was still high due to upwelling (Wefer, Berger, Richter et al. 1998). And the size of the depocentre has shrunk indicating that now more and more of the sediment contributed by the Congo River is captured by the canyon and transported to the middle and lower fan. It appears that on the upper fan the deposition was mainly controlled by the current system.

Since about 160 ky the depocentre observable in our data further shrank. Less material was contributed from the north and most of the sediment contributed by the Congo River, which now was much reduced (Uliana et al. 2001), was captured by the Congo Canyon and thus distributed to the middle and lower fan. A small component as well

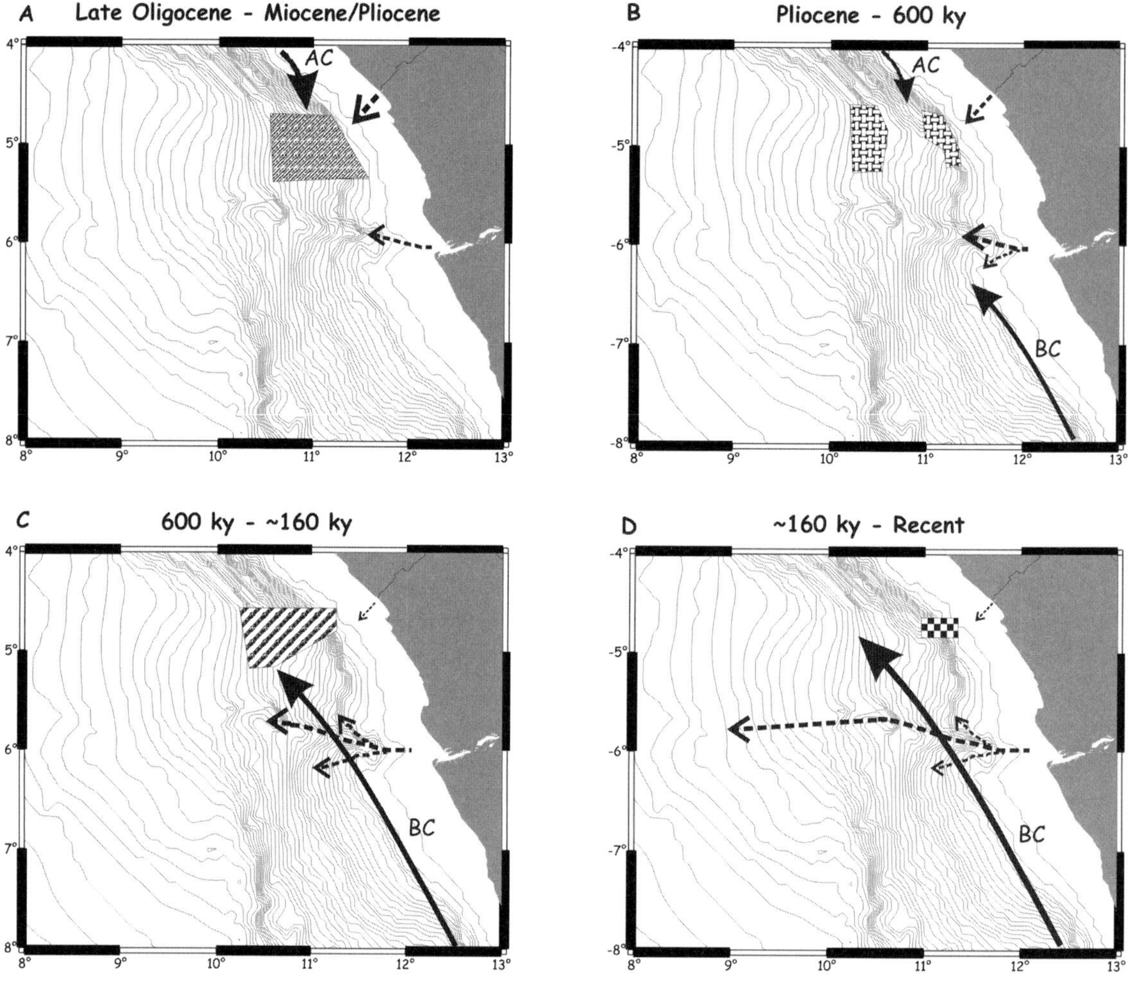

Fig. 7. Model for the evolution of the deposition in the Congo Fan since the Late Oligocene. The shaded areas show the depocentres for **a)** unit CF3 (Late Oligocene - Miocene/Pliocene), **b)** unit CF2 (Pliocene - 600 ky), **c)** lower unit CF1 (600 ky - ~160 ky), and **d)** upper unit CF1 (~160 ky - Recent). The dashed arrows represent the distributions by the Kouilou/Niari River and the Congo River, and the black arrows marks the material contributed and transported by the Angola (AC) and Benguela Currents (BC).

as the biogenic debris due to upwelling has again been deflected to the north by the Benguela Coastal Current and deposited on the northern flank of the fan (Fig. 7d). This would explain the observed distribution of upper unit CF1. Since the amount of sediment seized by the Benguela Coastal Current is less than during the period 600 ky - ~160 ky (lower unit CF1) the size of the depocentre is further reduced while its location has moved upslope. So, the most recent period shows both riverine and

current controlled deposition with the current (biogenic debris due to upwelling, sediment transport and deflection) being the most important on the upper fan.

Conclusions

In order to add to the understanding of the development of the depositional environments in the area of the northern Congo Fan in the Neogene seismic

seismo-stratigraphic unit	estimated age	main contributor to upper Congo Fan	
CF4	Eocene/ Oligocene	Kouilou/ Niari River or Ogooué River in the north	sedimentation on the slope, instabilities led to mass transport or material deflected from the north by the Angola Current
CF3	Late Oligocene - Miocene/ Pliocene	Kouilou/ Niari River or Ogooué River in the north	sedimentation on the slope, instabilities led to mass transport or material deflected from the north by the Angola Current
CF2	Pliocene - 600 ky	Congo River, upwelling	material input by Congo River and biogenic debris are deflected to the north by the Benguela Coastal Current
lower CF1	600 ky - ~160 ky	(Congo River), upwelling	less material input by Congo River and increased biogenic debris are deflected to the north by the Benguela Coastal Current
upper CF1	~160 ky - Recent	upwelling	material input by the Congo River is guided to the middle and lower fan, a small component and mainly biogenic debris are deflected to the north by the Benguela Coastal Current

Table 1. The Sedimentation model.

data were closely tied to geological information supplied by ODP Leg 175 Sites 1075, 1076 and 1077. This led to a model, which shows that in Late Oligocene - Miocene/Pliocene times sediment was transported into the target area mainly from the north. We are unable to strictly distinguish between a current or a river as the main sediment contributor. This situation changed to a sediment source in the south with the main distributor being the evolving Benguela Current. Thus, since the Pliocene the depositional environment of the upper Congo Fan has been characterised by a combined terrigenous (Congo River) and biogenic (Benguela Current) sediment source with the dominant carrier being a current (Benguela Current).

Acknowledgements

We are grateful for the support and assistance of chief scientist Prof. Bleil and the officers and crew of RV *Sonne* during the expedition. Many thanks go to M. Séranne and an anonymous reviewer for their constructive comments. This is Alfred-Wegener-Institut contribution No awi-n10336. The expedition was funded by the German Federal Ministry of Research and Technology (BMBF) under contract No 03R427. The authors are responsible for the contents of this paper.

References

Anderson JE, Cartwright J, Drysdall SJ, Vivian N (2000) Controls on turbidite sand deposition during gravity-driven extension of a passive margin: Examples from Miocene sediments in Block 4, Angola. Mar Pet Geol 17: 1165-1203

Berger W, Wefer G, Richter C et al. (1998) Leg 175 preliminary report. Ocean Drilling Program Preliminary Report 175 College Station TX (Ocean Drilling Program)

Bleil U, Breitzke M, Däumler K, Dittert L, Ewert J, Gohl K, Heesemann B, Heinitz WD, Helmke P, Keil H, Merker T, Pioch B, Potozki F, Rosiak U, Schneider R, Schwenk T, Spiess V, Uenzelmann-Neben G (1994) Report and preliminary results of *Sonne* cruise SO86 Buenos Aires - Capetown. Ber Fachber Geowiss, Univ Bremen 51, 116 p

Bolli HM, Ryan WBF et al. (1978) Initial Reports of the Deep-sea Drilling Project, 40, Washington (US Govt Printing Office).

Brice SE, Cochran MD, Pardo G, Edwards AD (1982) Tectonics and Sedimentation of the South Atlantic Rift Sequence: Cabinda, Angola. In: Watkins, Drake (eds) Studies in Continental Margin Geology. AAPG Mem 34: 5-18

Cramez C, Jackson MPA (2000) Superposed deformation straddling the continental-oceanic transition in deep-water Angola. Mar Pet Geol 17: 1095-1109

Dean WE, Hay WW, Sibuet J-C (1984) Geologic evolution, sedimentation, and paleoenvironment of the Angola Basin and adjacent Walvis Ridge: Synthesis of results of Deep-sea Drilling Project Leg 75. In: Hay WW, Sibuet J-C et al. (1984) Init. Rept. DSDP 75: 509-544 Washington (US Govt Printing Office)

Droz L, Rigaut F, Cochonat P, Tofani R (1996) Morphology and recent evolution of the Zaire Turbidite System (Gulf of Guinea). Geol Soc Am Bull 108: 253-269

Dupont LM, Donner B, Schneider R, Wefer G (2001) Mid-Pleistocene environmental change in tropical Africa began as early as 1.05 Ma. Geology 29: 195-198

Durham EL, Maslin MA, Platzman E, Rosell-Melé A, Marlow JR, Leng M, Lowry D, Burns SJ, ODP Leg 175 Shipboard Scientific Party (2001) Reconstructing the climatic history of the western coast of Africa over the past 1.5 My: A comparison of proxy records from the Congo Basin and the Walvis Ridge and the search for evidence of the Mid-Pleistoce Revolution. In: Wefer G, Berger WH, Richter C (eds) Proc ODP. Sci Results, 175: 1-46 [online]

Emery KO, Uchupi E (1984) The Geology of the Atlantic Ocean. Springer, New York

Emery KO, Uchupi E, Phillips J, Bowin C, Mascle J (1975) Continental margin off western Africa: Angola to Sierra Leone. AAPG Bull 59: 2209-2265

Frostick LE (1997) The East African Rift Basins. In: Selley R (ed) African Basins. Sedimentary Basins of the World. Elsevier, Amsterdam, pp 187-209

Frostick LE, Reid I (1989) Is Structure the Main Control of River Drainage and Sedimentation in Rifts? J Afr Earth Sci 8:165-182

Hartmann DA, Swanson WA, Smith PR, Goulding FJ, Kelly CA (1998) Structural development of the continental margin of Congo and northern Angola. AAPG abstract 370-371

Hoeltvoeth J, Wagner T, Horfield B, Schubert CJ, Wand U (2001) Late-Quaternay supply of terrigenous organic matter to the Congo deap-sea fan (Odp Site 1075): Implications for equatorial African paleoclimate. Geo Mar Lett 21: 23-33

Jansen JHF, Giresse P, Moguedet G (1974) Struvtural and sedimentary geology of the Congo and southern Gabon continental shelf; a seismic and acoustic reflection survey. Neth J Sea Res 17: 364-384

Kjarstansson E (1979) Modelling and migration by the monochromatic 45-degree equation. Stanford Exploration Project Report No 15 Stanford University

Leyden R, Bryan G, Ewing M (1972) Geophysical Reconnaissance on African Shelf: 2. Margin Sediments from Gulf of Guinea to Walvis Ridge. AAPG Bull 56: 682-693

Meeuwis JM, Lutjeharms JRE (1990) Surface thermal characteristics of the Angola-Benguela Front. S Afr J Mar Sci 9: 261-279

National Geophysical Data Center (1988) ETOPO-5 Bathymetry/Topography Data, Data Announce. 88-MGG-02 Natl Oceanic and Atmos Admin, US Dept Comm, Boulder Colorado

Pufahl PK, Maslin MA, Anderson L, Büchert V, Jansen F, Lin H, Perez M, Vidal L, Shipboard Scientific Party (1998) Lithostratigraphic summary for Leg 175: Angola-Benguela upwelling system. In: Wefer G, Berger W, Richter C (eds) Proceedings of the Ocean Drilling Program Initial Report 175: 533-542

Savoye B, Cochonat P, Apprioual R, Bain O, Baltzer A, Bellec V, Beuzart P, Bourillet JF, Cagna R, Cremer M, Crusson A, Dennielou B, Diebler D, Droz L, Ennes JC, Floch G, Guiomar M, Harmegnies F, Kerbrat R, Klein B, Kuhn H, Landuré JY, Lasnier C, Le Drezen E, Le Formal JP, Lopez M, Loubrieu B, Marsset T, Migeon S, Normand A, Nouzé H, Ondréas H, Pelleau P, Saget P, Séranne M, Sibuet JC, Tofani R, Voisset M (2000) Structur and recent evolution of the Zaire deep-sea fan: Preliminary results of the ZaïAngo 1 & 2 cruises (Angola-Congo margin). C R Acad Sci Paris, Earth and Planetary Sciences 331: 211-220

Schefuß E, Versteegh GJM, Jansen JHF, Sinninghe Damsté JS (2001) Marine and terrigenous lipids in Southeast Atlantic sediments (Leg 175) as paleoenvironmental indicators: Initial results. In: Wefer G, Berger WH, Richter C (eds) Proc ODP. Sci Results 175: 1-34 [online]

Séranne M (1999) Early Oligocene stratigraphic turnover on the west African continental margin: A signature of the Tertiary greenhouse-to-icehouse transition? Terra Nova 11: 135-140

Séranne M, Nzé Abeigne C (1999) Oligocene to Holocene sediment drifts and bottom currents on the clope of Gabon continental margin (west Africa) - consequences for sedimentation and southeast Atlantic upwelling. Sed Geol 128: 179-199

Shepard FP, Emery KO (1973) Congo Submarine Canyon and Fan Valley. AAPG Bull 57: 1679- 1691

Siesser WG (1978) Leg 40 results in relation to continental shelf and onshore geology. In: Bolli H, Ryan WB et al. (eds) Initial Reports DSDP 40: 965-979 Washington (US Govt Printing Office)

Stolt RH (1978) Migration by Fourier Transform. Geophysics 43: 23-48

Uchupi E (1992) Angola Basin: Geohistory and Construction of the Continental Rise. In: Poag and de Graciansky (eds) Geologic Evolution of Atlantic Continental Rifts. Van Nostrand Reinhold, New York, pp 77-99

Uenzelmann-Neben G (1998) Neogene sedimentation history of the Congo Fan. Mar Pet Geol 15: 635-650

Uenzelmann-Neben G, Spiess V, Bleil U (1997) A seismic reconnaissance survey of the northern Congo Fan. Mar Geol 140: 283-306

Uliana E, Lange CB, Donner B, Wefer G (2001) Siliceous phytoplankton productivity fluctuations in the Congo Basin over the past 460,000 years: Marine vs riverine influence, ODP Site 1077. In: Wefer G, Berger WH, Richter C (eds) Proc ODP. Sci Results 175: 1-32 [online]

Valle PJ, Gjelberg JG, Helland-Hansen W (2001) Tectono-stratigraphic development in the eastern Lower Congo Basin, offshore Angola, West Africa. Mar Pet Geol 18: 909-927

Wefer G, Berger W, Richter C et al. (1998) Leg 175 initial report. Ocean Drilling Program Intitial Report 175 College Station TX (Ocean Drilling Program)

Wefer G, Berger WH, Richter C (2001) Proc ODP. Sci Results 175 [online]

Yilmaz Ö (2001) Seismic data analysis. Soc Explor Geo-phys Invest Geophys 10 Tulsa

Terrigenous Signals in Sediments of the Low Latitude Atlantic –
Implications for Environmental Variations during the Late Quarternary:
Part I: Organic Carbon

T. Wagner[1,2]*, M. Zabel[2], L. Dupont[2], J. Holtvoeth[2] and C.J. Schubert[3]

[1]*Woods Hole Oceanographic Institution, Marine Chemistry and
Geochemistry Department, USA*
[2]*Universität Bremen, Fachbereich Geowissenschaften, Klagenfurter Strasse,
28359 Bremen, Germany*
[3]*AWAG, Limnological Research Center, Seestrasse 79, 6047 Kastanienbaum, Switzerland*
* corresponding author (e-mail): twagner@whoi.edu*

Abstract: The established view of a marine-dominated organic signature of modern and late
Quarternay deep ocean sediments is challenged by recently performed organic geochemical,
petrolocical, and palynological investigations. This study reviews multidisciplinary concepts that
were developed over the last decade in Bremen and have been successfully applied to modern and
late Quarternary sediments from the low latitude Atlantic. Relative proportions and compositional
variations of terrigenous OM are deduced from macerals (organic particles), freshwater diatoms,
phytoliths, pollen grains, lignin signatures, and carbon isotopic compositions of bulk organic matter as well as from higher plant-derived long-chain n-alkanes. For their variety of depositional
settings and their close location next to each other the dust-influenced central Equatorial Atlantic
and the West-African continental margin are examined. To assess environmental variations during
the late Quarternary, terrigenous organic records from the central Atlantic to the low latitude West-
African continental margin and the Congo deep-sea fan are discussed with regard to the paleoclimatic
evolution of central African dust source areas, continental run off and vegetational changes in the
Congo catchment area. Additionally, the influence of degradation processes and/or selective
preservation, both on short and long time scales, of non-reactive (mostly terrigenous) organic matter is investigated.

Introduction

One of the still poorly constrained elements in the
global carbon cycle concerns the importance and
fate of terrigenous organic matter (OM) in the deep
ocean. Flux estimates of terrestrial particulate OM
in front of major rivers suggest that about 65% of
the total particulate organic matter (POM) that enters
the ocean is refractory (Ittekkot 1988) probably
being more resistant to degradation. Knowledge of
quantitative proportions and composition of terrestrial OM in oceanic settings is essential since
climate-driven fluctations in total sedimentary organic carbon (TOC) have been widely used as a
proxy for paleoproductivity, especially beneath oceanic and continental upwelling regions (see

Sarnthein et al. 1992 for review). This fundamental
application appears to be supported by the general
statement that only a very small fraction of the OM
preserved in marine deep-sea sediments is land-derived (Hedges et al. 1997). On the other hand,
terrigenous OM in organic-carbon-poor marine environments remains difficult to quantify
mainly because of its broadly dispersed distribution,
its potential sub-fossil age and its low analytical
response (see Tyson 1995; Hedges et al. 1997;
Wagner and Dupont 1999 for discussion).

The established view of a marine-dominated
organic signature of modern and late Quaternay
deep ocean sediments is challenged by improve-

*From WEFER G, MULITZA S, RATMEYER V (eds), 2003, The South Atlantic in the Late Quaternary: Reconstruction of Material Budgets
and Current Systems.* Springer-Verlag Berlin Heidelberg New York Tokyo, pp 295-322

ments from collaborative geoscientific research at Bremen (e.g. Rühlemann et al. 1996; Schlünz et al. 1999; Wagner and Dupont 1999; Huang et al. 2000; Wagner 2000; Holtvoeth et al. 2001; for a complete list of contributions from Bremen with regard to the topic of this review see Wagner and Zabel, this volume) and from many other investigations reported in the literature (e.g Prahl and Muehlhausen 1989; Stein 1991; Schubert and Stein 1996; Eglinton et al. 1997; Bauer and Druffel 1998; Goñi et al. 1998, 2000; Knies and Stein 1998; Stein and Fahl 2000; Raymond and Bauer 2001).

Based on a broad variety of analytical techniques, all these studies recommend a re-evaluation of the assumption that the deposition of terrigenous organic carbon is restricted to deltas and continental margins, and rarely occurs beyond the shelf-slope boundary (e.g. Hedges and Keil 1995). The most important mechanisms invoked for the entrainment of terrigenous OM to the deep ocean are (I) eolian supply (Gagosian and Peltzer 1985; Gagosian et al. 1987; Huang et al. 2000; Wagner 2000), (II) sea level controlled displacement of main depositional areas in front of major rivers (Schlünz et al. 1999; Stein and Fahl 2000; Bauch et al. 2001), (III) redistribution and gravitational downslope transport from continental margins (Bauer and Druffel 1998) likely associated with long-range lateral advection (Walsh 1989; Gough et al. 1993), and (IV) sea ice transport at high latitudes (Wagner and Hölemann 1995; Schubert and Stein 1996).

In this study we review multidisciplinary concepts that were developed over the last decade in Bremen and were successfully applied to modern and late Quarternary sediments from the low latitude Atlantic. Due to the variety of depositional settings and their proximity to one another, this review will focus on the dust-influenced central Equatorial Atlantic and the West-African continental margin, where riverine supply has build up extensive deep-sea fans in front of the Congo and Niger. Related aspects concerning the export of lithogenic matter to the low latitude Atlantic are reviewed in a seperate contribution (Zabel and Wagner this volume).

The Low Latitude Atlantic: A key Area to Study Lateral and Spatial Processes Controlling Organic Carbon Deposition

One key area in environmental research is the low latitude Atlantic. Here the interaction of atmospheric and oceanic circulation is well documented and controls on organic carbon production and preservation are reasonably well understood. Numerous marine records from the central and eastern low latitude Atlantic as well as the corresponding Indian Ocean document that modern climatic and oceanographic boundary conditions have repeatedly changed in response to orbitally-forced Quarter-nary glacial-interglacial cycles (e.g. Sarnthein et al. 1994; DeMenocal 1995; Verardo and McIntyre 1994; Raymo et al. 1997; Schubert et al. 1998) to produce distinct cycles and spatial relationships in many proxy parameters. Funk et al. (this volume), for example, present spatial and temporal trends in various continental and marine proxy records covering large areas of the low latitude Atlantic that convincingly demonstrate the close interaction of African dust export and marine carbon sedimentation in association with late Quarternary African aridity-humidity cycles and the history of oceanic upwelling along the equator.

There is a distinct asymmetry between the western and central to eastern sectors of the low latitude Atlantic with regard to source areas, transport mechanisms and distances, and flux rates. Areas of special interest are the dust-influenced oceanic upwelling zone along the Equatorial Divergence and the fluvial-induced high productivity areas in front of the Congo, Niger, and Amazon rivers (Fig. 1). These three river systems supply about 10% of the annual global terrigenous organic carbon run-off into the Equatorial Atlantic and 28% of the total annual run-off for the Atlantic Ocean (Ludwig et al. 1998). Giant deep-sea fans seaward of these rivers testify to the local burial of huge amounts of terrigenous matter on the continental margins.

In the central low latitude Atlantic, in contrast, terrigenous OM is mainly introduced by eolian transport. The trade and Harmattan wind systems are well known to carry huge amounts of lithogenic dust from arid African source areas to the deep

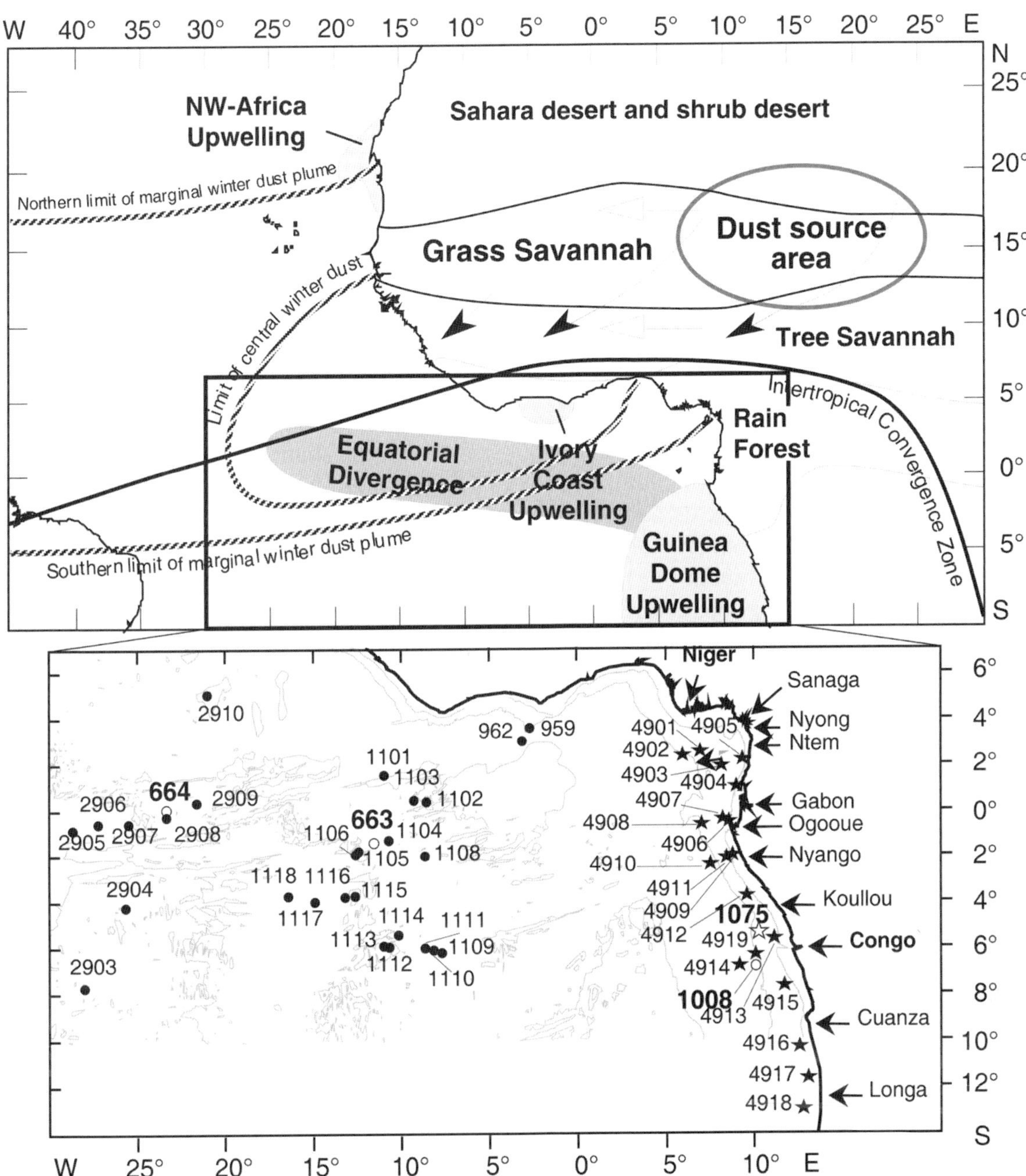

Fig. 1. Generalized map of the Equatorial Atlantic showing main coastal and oceanic upwelling areas, major African vegetation zones, and approximate boreal winter positions of source and corresponding depositional areas for atmospheric dust. Black arrows indicate surface trade winds, open arrows the mid-level African Easterly Jet (AEJ) and the Saharan Air Layer. Position of surface sediments and West-African rivers between 14°S and 6°S are shown in the inset map (filled circles mark locations along the open ocean E-W transect, stars of surface sediments from the near-continental N-S transect). Cores discussed in this review are highlighted in bold and indicated by open symbols.

Atlantic Ocean (e.g. Sarnthein et al. 1982; de Menocal 1995; Ratmeyer et al. 1999). Indicators of wind-borne OM in pelagic low latitude Atlantic sediments are pollen and spores (e.g. Dupont et al. 1998; Frédoux 1994), phytoliths (Pokras and Mix 1985; Ruddiman and Janecek 1989), fresh water diatoms (Pokras and Mix 1985; Pokras and Ruddiman 1989; deMenocal et al. 1993), charcoal (sensu Verardo and Ruddiman 1996), and fragments of higher plant tissue (terrigenous macerals; Stein et al. 1989; Wagner 1998, 1999, 2000). Geochemical studies have utilized the ratio of organic carbon to total nitrogen (Stein et al. 1989; Westerhausen et al. 1993; Verardo and McIntyre 1994; Wagner 1998), the hydrocarbon yield from kerogen as determined by Rock-Eval pyrolysis (Stein et al. 1989; Wagner and Dupont 1999; Wagner 2000), the isotopic signature of bulk organic carbon (Westerhausen et al. 1993; Müller et al. 1994; Rühlemann et al. 1996; Wagner and Dupont 1999; Wagner 2000), the amount and isotopic composition of elemental carbon (Verardo and Ruddiman 1996; Bird and Cali 1998) and other terrigenous biomarkers (e.g. concentration and isotopic signatures of long-chain n-alkanes, sterols, n-alkanols and lignin; Westerhausen et al. 1993; Huang et al. 2000; Eglinton et al. 2002; Schefuß 2002) to separate allochthonous from autochthonous organic carbon. Mixing of wind-blown with riverine terrestrial OM is documented oceanward of the main African rivers (Westerhausen et al. 1993). For an introduction to concepts, limitations and perspectives of each individual proxy parameter we refer to Wagner and Dupont (1999).

Modern Distribution and Sources of Sedimentary Terrigenous Organic Matter - Contrasts between the Central Pelagic and Eastern Near-Continental Equatorial Atlantic

The first comprehensive investigation on the modern distribution of terrigenous OM along the Equatorial African margin was presented by Westerhausen et al. (1993). Employing elemental, isotopic, and molecular records they quantified terrigenous organic carbon and further distinguished an eolian- and river-borne fraction. They concluded that terrigenous organic carbon comprise >60% of bulk TOC on the modern shelves off Liberia, Ivory Coast and Gabon whereas proportions drastically decrease below 20% along the upper continental slope.

Subsequent studies extended further north and west using samples from eolian dust, sediment traps and the ocean floor from as far as 35° north and 32° west (e.g. Hooghiemstra et al. 1986; Dupont and Agwu 1991; Wagner and Dupont 1999; Huang et al. 2000; Dupont and Wyputta 2002; Romero et al. 1999; Zhao et al. 2003). In these studies, relative proportions and compositional variations of terrigenous OM were deduced from macerals (organic particles), freshwater diatoms, phytoliths, pollen grains, lignin signatures, bulk isotopic data ($\delta^{13}C_{org}$), and carbon isotopes derived from long-chain n-alkanes. Application of such a multiparam-eter approach confirm differences in quantity and composition of terrigenous organic matter between near-continental and open ocean settings of the central and eastern Equatorial Atlantic and further north (Huang et al. 2000; Romero et al. 1999).

Organic Petrology and Stable Carbon Isotopic Composition of Bulk OM

Accordingly, terrigenous OM in surface and Quarternary deposits of the low latitude Atlantic is composed of the two main groups huminite/vitrinite and inertinite, both originating from vascular land plants, whereby inertinite is stronger oxidized and less reactive being indicative of plant burning and more arid climatic conditions ("fusinite" or "charcoal", Taylor et al. 1998; Fig. 2). The distribution of the inertinite-vitrinite ratio (I/V) in the low latitude Atlantic is displayed in Figure 3a. Apparently, I/V ratios separate the two contrasting depositional settings with low values along the African continental margin and a gradual advance with increasing distance to the continent. This observation is in favour with preferential long-range eolian transport of pre-oxidized terrigenous plant matter from African grassland-covered dust source areas (Wagner and Dupont 1999; Wagner 2000).

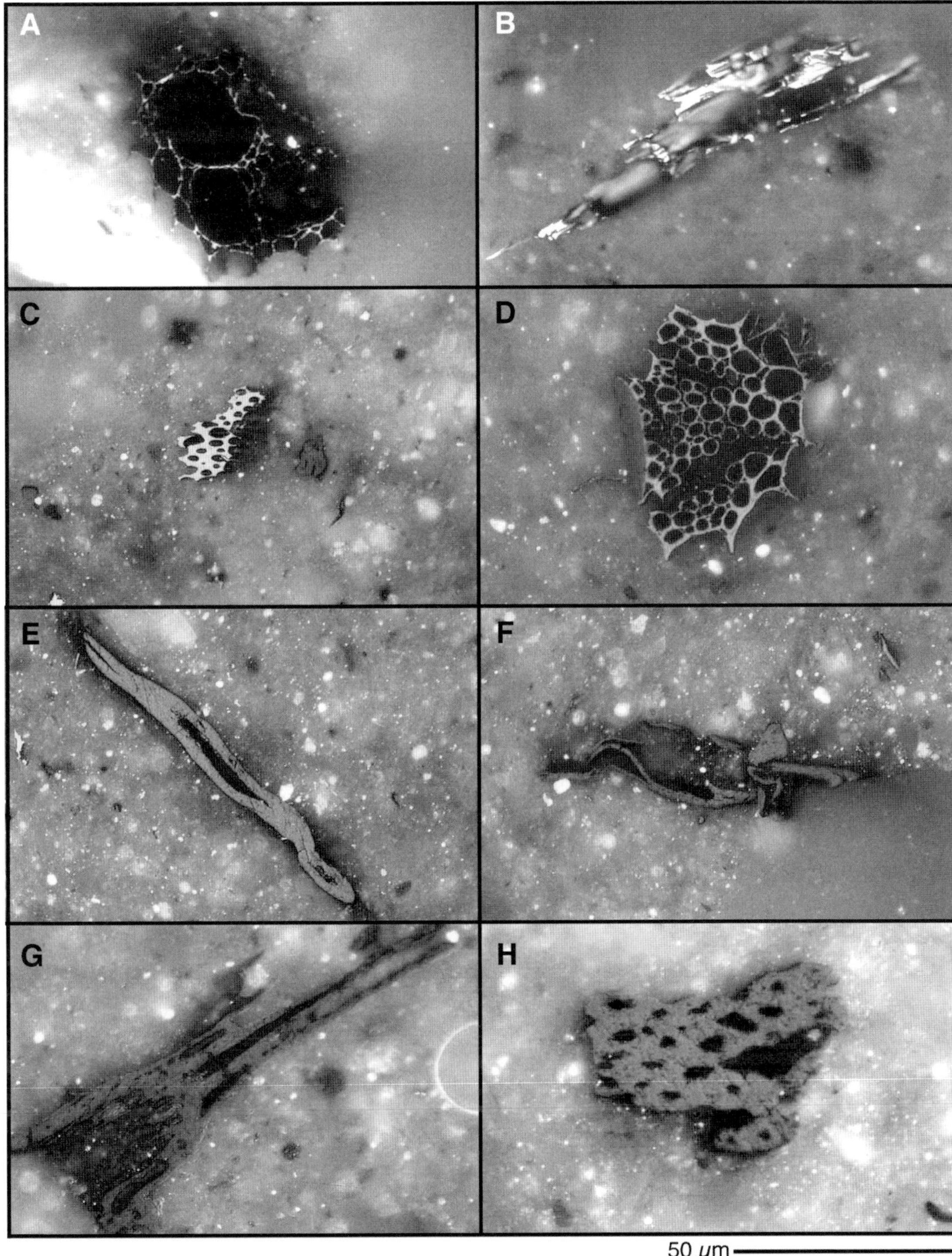

Fig. 2. Terrigenous organic matter (macerals) in modern sediments from the Equatorial West-African continental margin. **(a-d)** Various types of highly oxidized organic matter (inertinite, fusinite) revealing elongated, thick-walled, often-pitted structures of vascular bundles (tracheids). Severe oxidation may be caused by terrestrial plant burning resulting in "charcoal-type" OM. **(e-h)** Non-oxidized terrigenous organic particles (huminite/vitrinite) with partly filled cell lumina documenting cross-sections through parts of plant vascular systems (tracheids). 200-500 fold magnification on polished surfaces under white incident light (oil). **((a-b)** GeoB 4901; **(c-f)** Geob 4917; **g)** GeoB 4907; **h)** GeoB 4918).

Along the West-African continental margin between 4°N and 13°S (Cameroon to northern Angola), contribution of terrigenous organic carbon is dominated by river discharge including channeled downslope transport on the Congo deep-sea fan. The modern distribution of terrigenous organic carbon as depicted from $\delta^{13}C_{org}$ in comparison to estimates based on maceral analysis is shown in Figure 3b,c. Application of a simple two-component mixing model to bulk $\delta^{13}Corg$ signatures of marine deposits has been extensively applied to estimate marine and terrigenous fractions of organic carbon, although various other factors may influence the bulk isotopic signal (see Tyson 1995; Wagner and Dupont 1999 for reviews). In fact, near-continental settings in the Equatorial Atlantic need to be considered at least as three-component systems, taking into account that supply of sub-fossil organic matter or dead carbon from continental peat and swamp areas and erosion of older strata, all of them with barely unknown and probably wide-ranging isotopic compositions, may considerably modify the isotopic composition of the bulk organic matter in the marine sediments. For the main types of African vegetation, the contrasting isotopic signatures of C3 and C4 land plants are of special relevance. Terrigenous OM derived from vascular C3 plants (representing more than 95% of the present day terrestrial biomass) reveal depleted values between -25.5 to -29.3‰ with an average endmember signature of -27‰, whereas ^{13}C-enriched carbon from terrestrial C4 plant matter, including many grasses, range from -8 to -19‰ (average -12‰, Gearing 1988). Accordingly, temporal variations in admixture of C4 plant matter may influence the bulk $\delta^{13}C_{org}$ signature of marine sediments. Elevated proportions of C4 plant matter would result in higher ^{13}C values which, in turn, would lead to an underestimation of the terrigenous organic fraction if a simple two-endmember mixing model (marine versus C3 plants) is applied.

fraction based on petrographical assessment at the base of the Congo fan (3970 m water depth, Fig. 3c). A pronounced northward decrease in organic carbon concentrations along the continental slope (from 2.5% off northern Angola to 0.63% on the Niger fan) is mainly attributed to high productivity conditions within and south of the Congo river plume. To derive additional information on the type and source organic matter, lignin chemistry was investigated (Holtvoeth et al. 2001 in press.). Lignin, a principal component of terrestrial plant

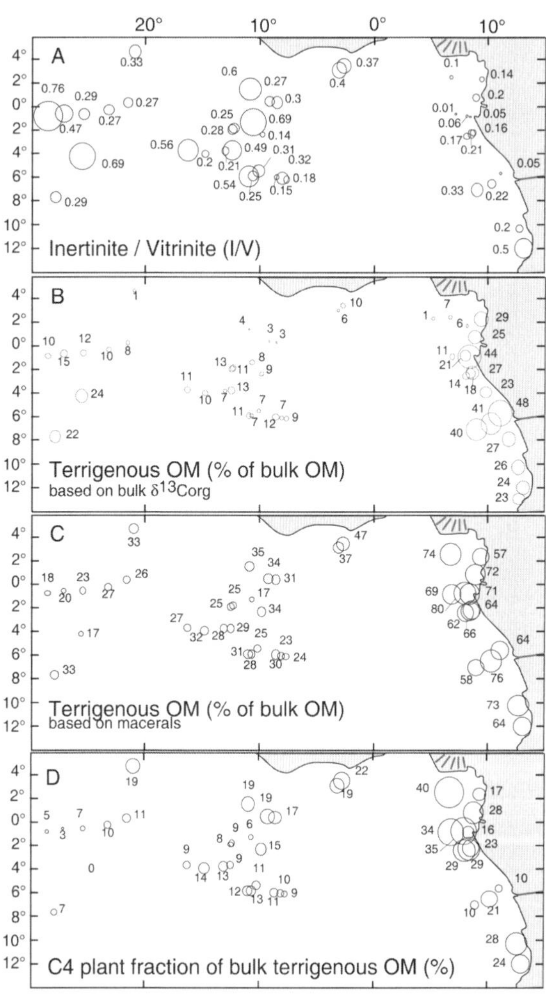

Fig. 3. Distribution of **a)** the ratio of oxidized to non-oxidized terrigenous organic matter (Inertinite/Vitrinite), terrigenous organic carbon estimates based on **b)** bulk $\delta^{13}C_{org}$ and **c)** maceral analysis and **d)** C4 plant fractions of bulk sedimentary organic matter in surface sediments of the central and eastern Equatorial Atlantic.

Lignin Phenols

The importance of deep-sea fans in front of major rivers with respect to the entrainment of allochtonous organic carbon and deep-sea carbon budgets is evident from an almost 60% terrigenous organic tissue, has been used to assess the contribution and composition of terrigenous OM in riverine suspended particles as well as modern and late Quarternary deposits from various lacustrine and marine environments (e.g. Hedges et al. 1986; Miltner and Emeis 2000; Goñi et al. 1998, 2000; Keil et al. 1998). Following the fundamental studies by Hedges and Parker (1976) and Hedges and Mann (1979a,b) quantification of eight specific phenols allows for characterization of plant tissues that contribute to the terrigenous organic compound of bulk organic carbon. The ratios of syringyl phenols to vanillyl phenols (S/V) and of cinnamyl phenols to vanillyl phenols (C/V) are commonly employed to distinguish gymnosperm or angiosperm, woody or non-woody vascular plant tissues.

In the modern low latitude Equatorial Atlantic results from lignin chemistry and organic petrology demonstrate a support non-oxidized (huminitic/vitrinitic) angiosperm origin of terrigenous organic matter along the West-African continental margin. Here the sum of the eight specific lignin phenols (Λ) broadly scatter from 0.07-0.95 mg/100gTOC depending on the distance from the various river mouths (Fig. 4a). Notably, the Ogooue river north of Cape Lopez at about 0.5°S results in the most pronounced lignin signature followed by the Congo river which apparently supplies considerable amounts of lignin and bulk terrigenous OM (76-58% according to maceral analysis) as far as the distal base of the deep-sea fan. Similar or higher lignin phenol concentrations are reported from various other continental shelf and slope environments, e.g from the Washingon coast (0.81-3.19 mg/100gTOC, Hedges and Mann 1979b; Keil et al. 1998), the NW-Atlantic Shelf (Louchouarn et al. 1999) or the Beauford Shelf in the Canadian Arctic (0.44-1.22 mg/100gTOC, Goñi et al. 2000; Table 1). The corresponding concentration of lignin phenols in the central Equatorial Atlantic is much lower (0.04-0.12mg/100mgTOC) but is similar to pelagic sediments from the North Atlantic (0.01-0.12 according to Gough et al. 1993, Table 1). The lignin phenols in African continental margin sedi-ments are dominated by a mixture of woody angio-sperm tissue (mainly barks) and non-woody angio-sperm tissue (e.g. leaves) as indicated by the ratios of Syringyl vs. Vannillyl (S/V) and Cinnamyl

vs. Vannillyl (C/V) (Fig. 5). Studies of the lignin chemistry of Amazon river suspension load exhibit comparable compositions of higher plant tissue (see filled circle in Figure 5, Hedges et al. 1986) and apparently characterizes terrigenous plant matter from both tropical Equatorial Atlantic areas. A southward increase in S/V ratios along the African continental slope is attributed to a gradually increased admixture of leaf plant material, which is likely to be related to a stronger release of eolian dust from southern Africa. Low C/V ratios along the African continental margin exclude any major supply of non-woody material, except at the southermost and northermost sectors of the transect, where C/V ratios approach slightly elevated values (0.34 on the Niger deep-sea fan and 0.35 off northern Angola). In these settings, riverine

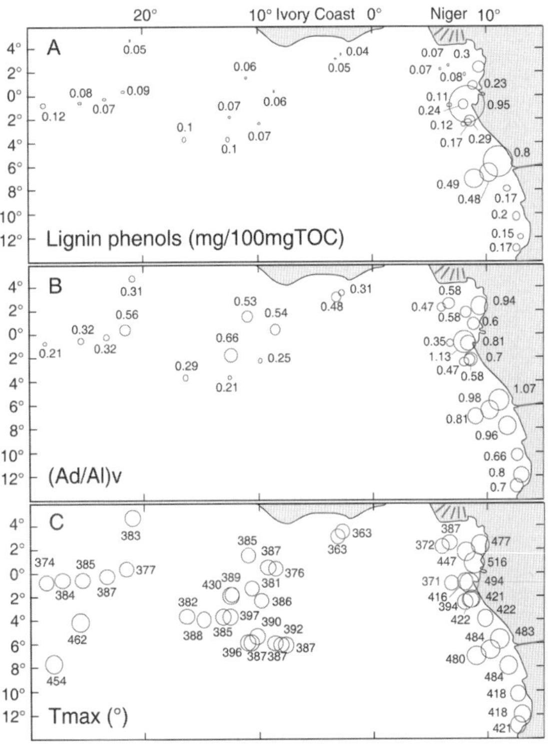

Fig. 4. Distribution of **a)** lignin phenol concerntration normalized to total organic carbon (in mg/100mgTOC), **b)** the vanillyl acid to aldehyde ratio (Ad/Al)v, and **c)** Tmax signatures (in °C) as obtained from Rock-Eval Pyrolysis in surface sediments of the central and eastern Equatorial Atlantic.

Setting	Area	Sample	Λ^* (mg/100mgOC)	S/V	C/V	$(Ad/Al)_V$	$(Ad/Al)_S$	Reference
Open ocean	NE-Atlantic	1	0.01-0.12	0.14-0.70	0.09-0.77	0.37-2.92	0.46-2.44	Gough et al. 1993
	Central Equat. Atlantic	1	0.04-0.12	0.32-1.98	0.27-0.84	0.21-0.66		this study
Marginal ocean basin	S-Baltic Sea	1	2.1-7.6	0.42-0.65	0.16-0.3	0.29-0.37	0.3-0.42	Miltner and Emeis 2000
	Mediterranean Sea	1	0.5-3.0	0.16-0.55	0.04-0.32	0.37-2.98	0.33-3.72	Gough et al. 1993
	Gulf of Mexico (US)	1	0.4-1.4	1.6 av.	0.5 av.			Goñi et al. 1998
Shelf/Slope	Beauford Shelf	1	0.44-1.22	0.54-1.17	0.15-0.24	0.55-0.80	0.46-1.33	Goñi et al. 2000
	Atlantic Shelf (Canada)	1	0.6	0.92	0.83	2.09-		Louchouarn et al. 1997, 1999
	Middle Atlantic Bay	1	0.05-0.11	0.37-0.49	0.24-1.1	0.52-1.45		Mitra et al. 2000
	Lake Washington (USA)	1	1.05 av.	0.44 av.	0.21 av.			Hedges and Mann 1979b
	Washington Coast (USA)	1	0.23-6.5	0.16-0.31	0.03-0.49			Hedges and Mann 1979b
	Washington Coast (USA)	1	3.19 (mid-shelf)/ 0.81 (slope)	0.29/0.32	0.06/0.12	0.35/0.64	0.24/0.38	Keil et al. 1998
	SW-African Margin (14°S-2°N)	1	0.07-0.95	0.36-1.15	0.06-0.35	0.35-1.13		this study
	Congo fan (ODP 1075)	3	0.07-0.37 (16=0.06-0.32)	0.47-1.38	0.15-0.39	0.47-1.74	0.26-1.94	Holtvoeth et al. in press
	Peru shelf	2	0.58	2.66-	0.49	1.0		Bergamashi et al. 1997
	Pakistan Margin Arabian Sea	3	0.03-0.20	0.66-2.49	0.01-0.66	0.09-0.66	0.01-0.68	Schubert unpubl. results
	Pakistan Margin (OMZ)	3	0.06-0.26	0.33-2.84	0.05-0.81	0.05-0.55	0.06-0.74	Schubert unpubl. results
Estuarine/Deltaic	Mackenzie Delta (Canada)	4	0.54-1.43	0.42-0.52	0.13-0.15	0.46-0.80	0.44-1.23	Goñi et al. 2000
	Saanich Inlet (Canada)		857-2707**					Hamilton and Hedges 1988
	Rhone Delta (France)	1	1.29-2.08	0.45-0.66	0.08-0.20	0.25-0.43	0.3-0.46	Gough et al. 1993
	Caeté Estuary (Brazil)	5	12.9***	1.25-	0.16	0.56	0.4	Dittmar and Lara 2001
Riverine	Arctic rivers	4	0.2-5.2	0.1-0.77	0.07-0.58	0.08-86		Lobbes et al. 2000
	US rivers	4	1.33 av.	0.82 sv.	0.12 sv.	0.5 av.		Onstad et al. 2000
	Amazon River (Brazil)	6	7.31 (coarse)/ 2.15 (fine)	0.78/0.85	0.07/0.1	0.23/0.44	0.19/0.25	Hedges et al. 1986
Lacustrian	Lake Baikal (Sibiria)	3	30-1143****	0.1-1.2	0-0.8	0-12		Orem et al. 1997
	Lake Biwa (Japan)	3	330-1570					Ishiwatari and Uzaki 1982

Table 1. Compilation og lignin data from various modern and Late Quaternary terrigenous and marine environments (sample material: 1=surface sediment; 2=near-surface sediment (10-15cm); 3=Quaternary sediment; 4=suspended POM; 5=mangrove sediment; 6=river suspension; *=S+C+V; **=(µg/L); ***=mmol/molOC; ****=V+S+C+P).

supplies from the Niger river and eolian transport from savannah-covered vegetation belts in southern Africa are probably responsible for the enhanced entrainment of non-woody plant matter.

Supply of Aged Terrestrial Organic Matter as Suggested by Degradation and Reactivity Proxies

Apart from the main equatorial African rivers, the Congo and Niger, there is a number of smaller rivers, e.g. the Sanaga in Cameroon or the Gabon and Oyooue in Gabun, that contribute to the bulk export of terrigenous matter to the African shelf (Fig. 1). Since most of these tropical African rivers drain extensive swamp and peat areas in the hinterland, it is reasonable to expect that they supply different types of terrigenous OM, i.e. fresh and reactive plant litter representing the modern vegetation cover of the catchment area and diageneti-

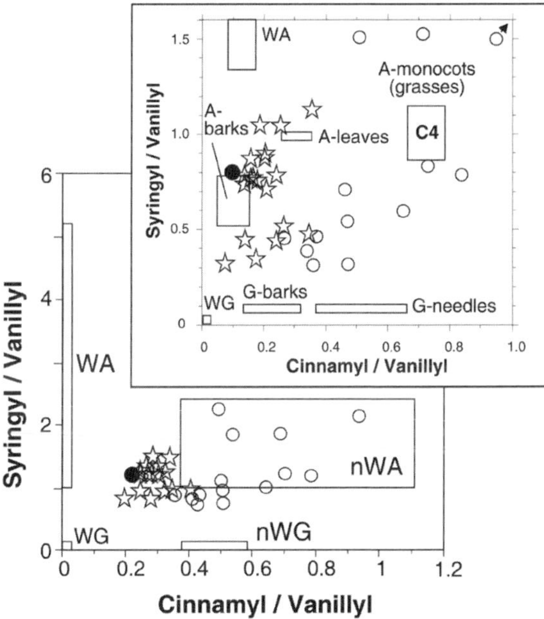

Fig. 5. Plot of cinnamyl to vannillyl (C/V) versus syringyl to vannillyl (S/V) phenol ratios of surface sediments from the central and eastern Equatorial Atlantic. Compositional ranges of major vascular plant tissues according to Goñi and Hedges (1992). Tissue abbreviations: WG gymnosperm wood; WA angiosperm woods; nWG non-woody gymnosperm; nWA non-woody angiosperm.

cally altered, older OM originating from eroded swamp and peat environments. The presence of a labile organic fraction associated with river discharge is indicated by the non-oxidized character of particulate OM (huminite/vitrinite; in Figure 3a expressed as low inertinite-vitrinite ratios) and steeply enhanced hydrogen indices (up to 560 mgHC/gTOC, Wagner personal communication) in surface sediments of the Congo fan and at shallow sites off Gabun and Cameroon. By comparison, the TOC-enriched sediments off northern Angola receive higher amounts of oxidized terrigenous OM (I/V ratio up to 0.5, Fig. 3a) and reveal considerably lower hydrogen indices (240-305 mgHC/gTOC, Wagner personal communication). The importance of supply and redistribution of allochthonous swamp and peat deposits along the coast of deltaic and estuarine systems has been documented for the Orinoco, Congo, Amazon, Mississippi and Mahakam Deltas (Stach et al. 1975; McCabe 1984; Wagner and Pfefferkorn 1997) and in southern Florida (e.g. Cohen 1970). The occurrence of such reworked terrigenous OM in marine sediments should be detectable e.g. by degradation and reactivity proxies. Enhanced contribution of non-reactive, sub-fossil OM is notable along the shallow water corridor by elevated T_{max} signatures >480°C as obtained from Rock-Eval pyrolysis (T_{max} depicts the temperature of maximum hydrocarbon generation from the kerogen fraction (S2) being a measure of reactivity and thermal maturity of sedimentary OM), which follow the African continental margin north of the Congo river and across the Congo fan to the deep-sea (Fig. 4c). This conclusion is further supported by overall elevated $(Ad/Al)_V$ ratios of lignin components which scatter around 1.0 north of the Congo river and approach highest levels of 1.13 on the slope north of the Ogooue river (Fig. 4b). Such high ratios suggest large input of degraded particle-bound OM to the modern slope and deep-sea fan deposits. Since the most efficient lignin-degrading organisms (e.g. soft-rot fungi, Nelson et al. 1995) are confined to terrestrial ecosystems (e.g soils, Erikson et al. 1990), it is most probable that lignin-decay took place prior to river-discharge and marine deposition. The range of $(Ad/Al)_V$ ratios along the modern African continental margin is comparable to or slightly higher than $(Ad/Al)_V$ ratios reported from other modern shelf-slope environments (Table 1), e.g. the Middle Atlantic Bay (Mitra et al. 2000), the Peru shelf (Bergamashi et al. 1997), or the Beauford Shelf (Goñi et al. 2000), and corroborates the recent implication of enhanced riverine export of aged terrestrial OM to the deep ocean (Eglinton et al. 1997; Ludwig 2001; Raymond and Bauer 2001).

In the central Equatorial Atlantic, by comparison, the consistently low T_{max} signatures and $(Ad/Al)_V$ ratios in surface sediments suggest that sub-fossil OM is not present at higher proportions, except for the northern Brazil Basin, where some advection of non-reactive OM from southern (probably South-American) source areas is assumed.

African Dust Source Areas and Support for C4 Plant Contribution

As noted above, transport of terrigenous OM to the central Equatorial Atlantic is mainly attributed to eolian supply from African source areas. Due to the very fine grain size of eolian dust in that remote pelagic area (modal grain size of terrigenous silt ranges from 6-9μm, Rath et al. 1999) it remains difficult to distinguish laterally advected from wind-born material. The composition of terrigenous OM in open ocean surface sediments is different from that on the African continental margin. The S/V and C/V ratios of lignin phenols and results from organic petrology confirm enhanced contribution of oxidized (inertinite), non-woody gymnosperm tissue (S/V 0.32-1.98, C/V 0.27-0.84; Fig. 4). Taking Lake Chad and the Sahel Zone as source areas of eolian dust (Ruddiman and Janecek 1989; Bonifay and Giresse 1992) terrigenous OM probably derives partly from C4 plant grass vegetation. Under modern climatic conditions, this area is mainly covered by savannah which is dominated by C4 grass plants (Giresse et al. 1994; Figure 1). Terrestrial C4 plants, including many grasses, exhibit optimum photosynthesis at temperatures about 10° higher than vascular C3 plants and are thus common in warm tropical and dry to arid climates. Due to their specific $\delta^{13}C_{org}$ signatures, the contribution of C4 plant matter to marine sediments needs to be

assessed as precise as possible usually requiring a combination of methods including palynology and carbon isotopes on land plant waxes.

Maceral analysis, bulk $\delta^{13}C_{org}$, (Figs. 3 b,c) and palynology were employed to assess and evaluate the potential influence from C4 plant matter in the low latitude Atlantic results. Accordingly, maceral-based estimates for proportions of terrigenous OM range from 62-80% along the African continental margin and 18-35% in the central Equatorial Atlantic. The continental-slope estimates are consistent with former approximations by Westerhausen et al. (1993). Corresponding proportions derived from bulk $\delta^{13}C_{org}$ are generally lower ranging from 6-48% along the African continental slope and 1-13% in the central Equatorial Atlantic. The largest discrepancies between maceral-based and isotope-based estimates are evident along the continental margin off Cameroon and Nigeria (up to 67%) but gradually decrease south and west to approach smallest values (5-9%) in the central Equatorial Atlantic at 29°W. Considering admixture of C4 plant matter to be responsible for these observed differences it is estimated that 3-40% (1-30%) of the terrigenous (bulk) organic fraction is composed of C4 plant matter (Fig. 3d) with strongest contribution on the Niger deep-sea fan and the adjacent continental margin off Cameroon and Gabon.

Recent biomarker and pollen studies on dust and sediment trap material and surface sediments from the eastern low latitude Atlantic support the presence of significant C4 plant material (Huang et al. 2000; Eglinton et al. 2002). Mapping the abundance of ^{13}C in leaf-wax components in surface sediments recovered from the seafloor off northwest Africa (0-35°N) reveals a clear pattern of $\delta^{13}C$ distribution, indicating systematic changes in the proportions of terrestrial C3 and C4 plant input (Fig. 6). They conclude that C4 plant derived terrigenous leave-waxes achieve highest proportions (>50%) off the NW-African coast at about 20°N. Notably, C4 plants still contribute about 40-50% of total leaf-waxes to the central Equatorial Atlantic. These proportions gradually decrease towards the tropical African continental margin to account for less than 20% off Niger and Cameroon. High C4 contributions, apparently carried by January trade winds, also extend far into the

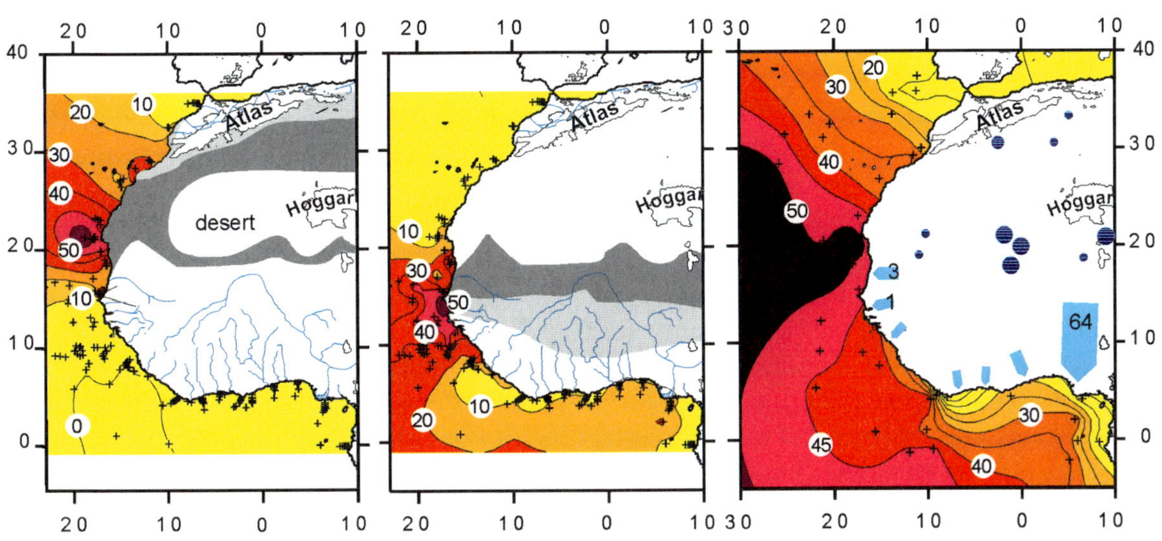

Fig. 6. (adapted after Huang et al. 2000): Source areas of C$_4$ plant material distributed in the eastern tropical Atlantic. Left. Isopol contours (percentage of total pollen) of ChenoAms pollen in marine surface sediment samples (Hooghiemstra et al. 1986) and the main source areas of Chenopodiaceae and Amaranthaceae (ChenoAms) pollen on the continent (Sahara desert and semi-desert areas). Middle. Isopol contours of Gramineae pollen in marine surface sediments (Dupont and Agwu 1991; Hooghiemstra et al. 1986) and the main source areas of grass pollen (Gramineae) on the continent (Sahel and savannah). Right. Distribution of C$_4$ plant wax (% of total C$_3$ and C$_4$) in marine surface sediment samples, calculated from values of $\delta^{13}C_{n\text{-}alkanes}$ using a two-component mixing equation with end member values of -34 and -19 ‰, respectively (Rieley et al. 1993; Collister et al. 1994). Circles indicate major source areas for dust deflation, such as Holocene lake deposits (N.B. not shown but to east of map is another major source area around Lake Chad, centred on ca. 14° N 11° E). Arrows mark major river inputs of particulate terrigenous organic carbon (10^4 tonnes·y^{-1}) (Ludwig et al. 1996). (+) Sampling sites.

equatorial Gulf of Guinea. Similar distributions are obtained if summed pollen counts for the Chenopodiaceae-Amaranthaceae and the Poaceae are used as an independent C4 proxy. Huang et al. (2000) conclude that the specificity of the latitudinal distribution of vegetation in North West Africa and the pathways of the trade and Sahara Air Layer wind systems are responsible for the isotopic patterns observed in the surface sedi-ments. Using a comparable analytical approach, Schefuß (2002) recently determined the relative amounts of C4 plant lipids of the bulk eolian lipid yield in dust samples that were collected along a transect from 33°N to 12°S along the African west coast. Employing endmember values of -36.0‰ and -21.5‰ for C3

and C4 plant leaf lipids, respectively, eolian C4 contribution was determined to be around 50% or higher off NW-Africa and northern Angola (8-10°S). Along the equatorial African coast (10°N to 7°S) these concentrations decrease to 30-35% due to the greater contribution of aerosols from C3 plant matter.

All these lines of evidence indicate that deposition of C4 plant matter in the central and eastern low latitude Atlantic may well be substantial. It may therefore be expected that the impact of this specific type of terrigenous OM on quantitative estimates of marine versus terrigenous OM is also of relevance for other low latitude, dust-influenced ocean basins.

Environmental Variations during the Late Quarternary

The Terrigenous Organic Record of the Central Equatorial Atlantic: Implications for the Paleoclimatic Evolution of African Dust Source Areas

Various paleoclimatic studies have shown that fluctuations in African aridity corresponded to changes in high latitude ice volume and North Atlantic sea surface temperatures (SST, Sarnthein et al. 1982; Street-Perrott and Perrott 1990; Tiedemann et al. 1994). In addition, DeMenocal et al. (1993) proposed that African terrestrial climate responded most sensitively to high latitude SSTs during peak glacial conditions when precessional forcing in the tropics was lowest. Both controls were considered to promote stronger African aridity and enhanced dust transport due to an intensification of the trade winds at low latitudes. As a consequence, African climate should be preconditioned for aridity during high latitude ice growth and full glacial periods.

Aridity/humidity cycles in terrestrial African climate were inferred from palynological (Pokras and Mix 1985; Hooghiemstra 1989; Dupont and Hooghiemstra 1989; Lezine 1991; Hooghiemstra et al. 1992; DeMenocal et al. 1993; Frédoux 1994; Giresse et al. 1994; Dupont et al. 2000) and geochemical (e.g. Verardo and Ruddimann 1996; Bird and Cali 1998; Zabel et al. 1999) data. Most of them note enhanced eolian supply and an equatorward displacement of African vegetation zones during full glacial conditions. The one million year terrigenous organic records from Site 663 in the central Equatorial Atlantic provides supporting evidence which supports fluctuations in African aridity (Fig. 7, from Wagner 2000). Accordingly, estimates for terrigenous OM range between 26-55% based on petrologic studies and 0-46% from the isotopic signal. Both records suggest that terrigenous OM supply not simply responded to glacial-interglacial cycles but was restricted to short time intervals during glacial periods (except for two exceptions during interglacial stages 15 and 7) when terrigenous fluxes were almost twice as high compared to glacial background and interglacial conditions.

At site 663, changes in the composition of the terrigenous organic fraction were also deduced from fluctuations in C4 supply, the I/V ratio, the percentage of the detrital fraction (<5 μm) of the bulk terrigenous OM (% detr/bulk Terr OM), and the flux of phytoliths (DeMenocal et al. 1993). Typically, maximum glacial accumulation of terrigenous OM coincided with peaks in C4 debris and the inertinite/vitrinite ratio, whereas the relative proportion of detrital terrigenous particles approached lowest percentages. Wagner (2000) suggested that such constellations likely represent short time periods when eolian supply of larger and more highly oxidized organic particles containing an elevated portion of C4 debris was enhanced. Glacial background and interglacial sections, in contrast, reveal almost entirely detrital terrigenous OM with very low dilution of C4 plant debris or oxidized OM.

Wind-blown inertinite may be regarded as an indicator of vegetation fires in African source areas, equivalent to charcoal (Verardo and Ruddiman 1996), phytoliths (Pokras and Mix 1985; DeMenocal et al. 1993), or black/dead carbon (Bird and Cali 1998). Pokras and Mix (1985) proposed that African hyperarid conditions were established due to enhanced glacial wind speeds which, in turn, caused and fueled extensive natural bush or savannah fires. Huge amounts of burned plant material were injected into the atmosphere and transported to the tropical Atlantic. The terrigenous organic record of Site 663 supply supporting evidence for changes in aridity of source areas over the past one million years. Increased glacial supply of C4 debris to the tropical Atlantic is also indicated by Gramineae and Cyperaceae pollen records from the Gulf of Guinea (Frédoux 1994; Dupont and Weinelt 1996; Dupont et al. 1998; Jahns et al. 1998), and has been proposed based on the carbon isotope signature of eolian black carbon for the Sierra Leone Rise (Bird and Cali 1998). The general covariance of C4 plant estimates with the inertinite/vitrinite record of Sites 663 favour a common aridity-driven control (Wagner 2000). It should be emphasized, however, that the proposed C4 plant supply only contributes a small fraction, never ex-

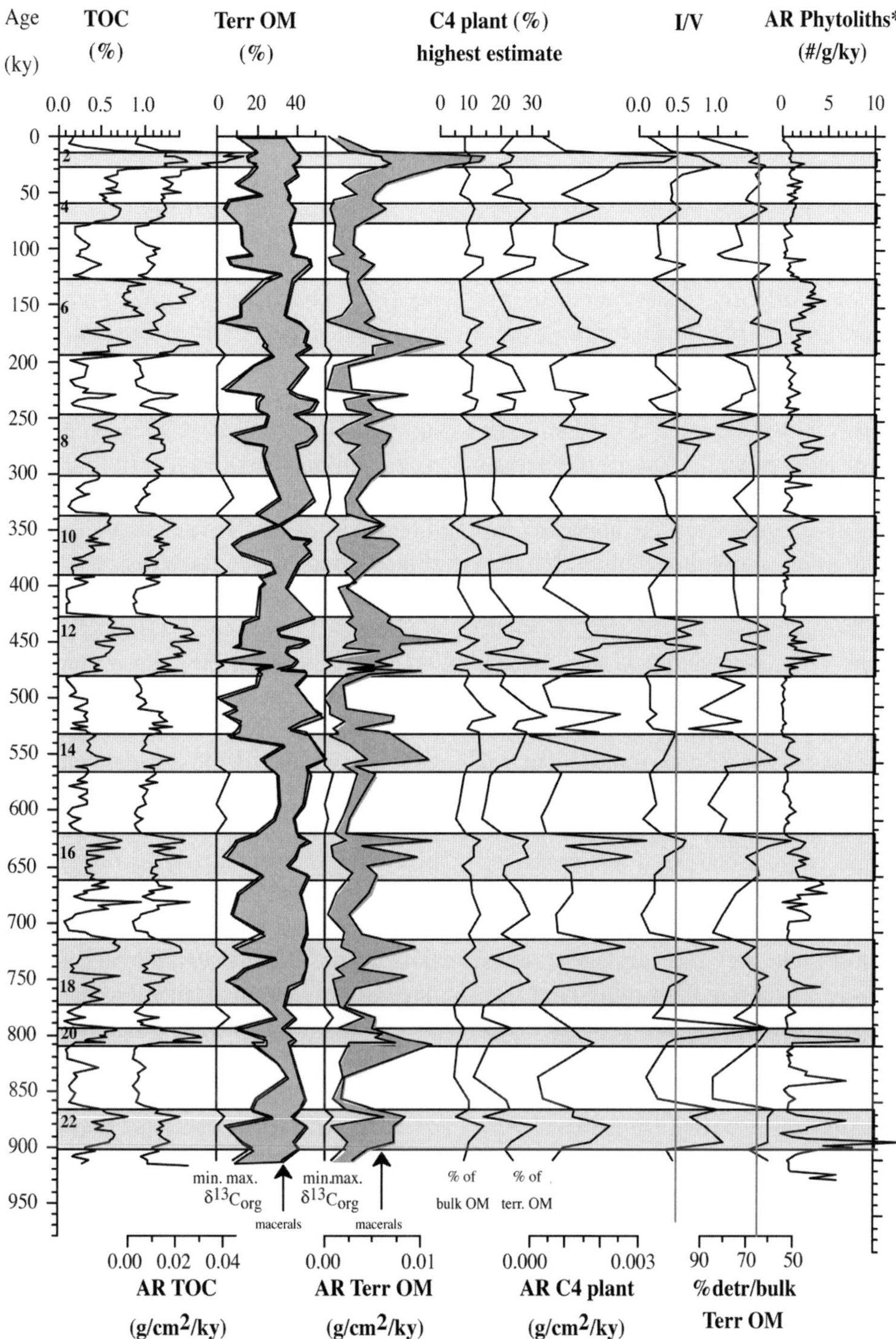

Fig. 7. Time series of ODP site 663 percentages and accumulation rates of organic carbon (TOC), terrigenous OM as determined by maceral analysis and isotopic measurements, and highest C4 plant estimates (shown as percentages of bulk and terrigenous OM. Elevated I/V ratios are attributable to periods of enhanced aridity in dust source areas and vegetation fires similar to the phytholith record (from deMenocal et al. 1993). Late Quaternary variations in the percentage of the detrital terrigenous organic fraction (< 5μm) relative to the total terrigenous organic fraction document changes in the transport energy of dust-carrying winds (from Wagner 2000).

ceeding one quarter of the bulk terrigenous organic fraction (14% with respect to the total organic matter).

Enhanced glacial dryness likely stimulated the expansion of the African grass-savannah covered vegetation belt and should be documented in marine palynological records. For example, in a recent study, Dupont et al. (2000) used marine and terrestrial palynological records from the eastern low latitude Atlantic to draw detailed vegetation maps of equatorial West Africa for eleven time slices over the last 150,000 years (Fig. 8). Accordingly, rain forest was widespread during interglacial stages 1 and 5, but strongly reduced during glacial stages 3 and 4 and especially during 2 and 6 when open, grass-rich vegetation prevailed. During the extreme glacial conditions of stage 6 savanna and open dry forest covered large areas along the northern coast of the Gulf of Guinea and the southern Saharan desert reached far to the south and the Namib desert far to the north. The area of rain forest was restricted to the southwest of the Guinean mountains and along the eastern coast of the Gulf of Guinea, while mangrove samps were strongly reduced and the area of *Podocarpus* forest fluctuated. In contrast, rain forest was widespread and mangroves were extensive along the coast during the last interglacial (Substage 5e). *Podocarpus* forest area strongly extended during Substages 5d and 5b. In Substages 5c and 5a, rain forest reclaimed areas it had lost in the previous substages (5d and 5b respectively). Mangrove swamps were less widespread in the later substages of Stage 5 than during Substage 5e. During Stage 4, the rain forest area was again strongly reduced, and recovered only slightly during the following Stage 3. Also the mangrove swamp area was reduced except along the Ivory Coast and along the northwestern coast of the Gulf of Guinea. *Podocarpus* forest only occurred in Angola and may be in Congo during Stage 4. Again forest was much reduced during Stage 2 and open vegetation types covered large parts of equatorial West Africa. Mangrove swamps must have been rare. At the beginning of the Holocene, mangrove swamps had recovered and reached their largest extension. Also the rain forests area increased in the early Holocene and Guinean and Congolian rain forests were probably not separated by a savanna corridor now exsisting in Togo and Benin.

Towards a Glacial-Interglacial Balance of OM Accumulation along the Central and Near-Continental Low Latitude Atlantic

It has been shown that surface water production beneath the Equatorial Divergence and dust supply from African source areas created late Quarternary changes in OM sedimentation. These relations allow the assessment of more general sedimentation patterns that link pelagic areas of the low latitude Atlantic to the African continent over the past 1 millon years. Wagner (2000) presented average glacial and interglacial flux rates and glacial-interglacial balances of main sediment components (Fig. 9a) and marine and terrigenous OM (Fig. 9b) for sites 663, 664 and 959 (for location see Fig. 1)

Late Quarternary glacial-interglacial patterns of calcium carbonate, biogenous opal, terrigenous matter, and TOC clearly distinguish the central Equatorial Divergence (site 663) from the other areas. Here, glacial-interglacial balances are consistently positive, reaching highest offsets for biogenic opal (+40%) and TOC (+80%) closely coupled to enhanced supply of terrigenous matter (+20%). Carbonate accumulation, in contrast, is not enhanced during glacials (+5%) probably reflecting the influence of moderate and mainly productivity-driven calcite dissolution (see Funk et al. this volume; and Henrich et al. this volume for further discussion on carbonate dissolution). Stronger carbonate dissolution due to enhanced contribution of corrosive deep waters from the southern ocean (Raymo et al. 1997) may be assumed from negative balances for carbonate accumulation (-10%) further west at site 664. The positive balance in TOC accumulation rates (+20%) probably document slightly enhanced glacial upwelling in the western Equatorial Divergence. Off the equatorial West-Africa (site 959), glacial-interglacial accumulation was continuously low and constant throughout the past 1 Ma. The minor decrease in glacial deposition and accumulation (-9% and -3%, respectively) probably relate to persistent winnowing on top of the exposed transform margin

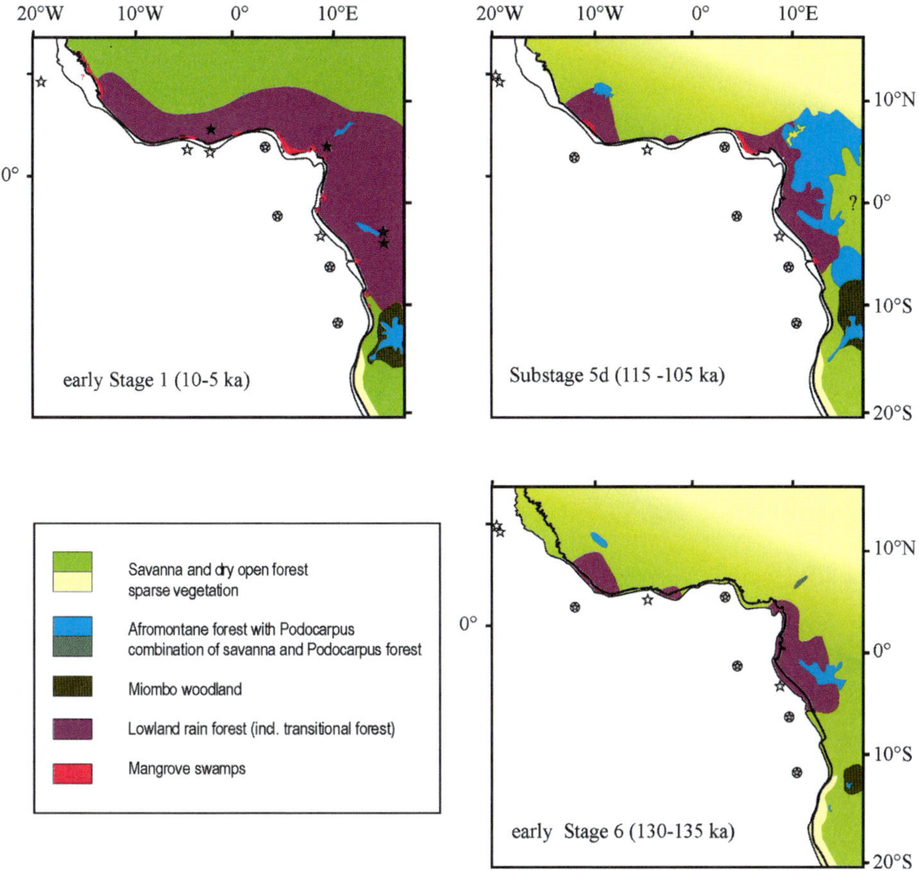

Fig. 8. Time slice reconstructions of African vegetation (adapted after Dupont et al. 2000).

(Wagner 1998; Giresse et al. 1998; Giresse and Wiewióra 2001).

Distinct changes in accumulation of marine and terrigenous OM are evident when glacial average flux rates are considered. Due to enhanced dust supply to the central Equatorial Atlantic terrigenous organic components at site 663 consistently reveal pronounced positive balances during glacials that amount to a 1.6-fold increase compared to interglacial periods. The proposed enhancement of African dust source aridity and/ or wind strenght during glacials is reinforced by almost 2.5-fold higher accumulation rates of inertinite and a doubling of the estimated C4 plant matter. A comparable change in glacial supply of terrigenous OM is not observed at sites 664 and 959. Probably stimulated by enhanced eolian deposition, glacial accumulation of marine OM

increased by about 20% to 60% below the Equatorial Divergence in comparison to interglacial conditions. Assuming that these pronounced changes in marine OM accumulation are mainly related to productivity rather than to perservation, they suggest considerably stronger oceanic upwelling during late Quarternary glacial periods.

The Terrigenous Organic Record of Marine Deep-Sea Fans: Land-Ocean Interaction and Influence of Selective Degradation – Aspects from the Congo Fan

In contrast to the broad dispersal of terrigenous OM by eolian transport, the export of organic carbon through rivers is more confined to continental margins (e.g. Hedges et al. 1997). Ludwig et al.

Main sediment components

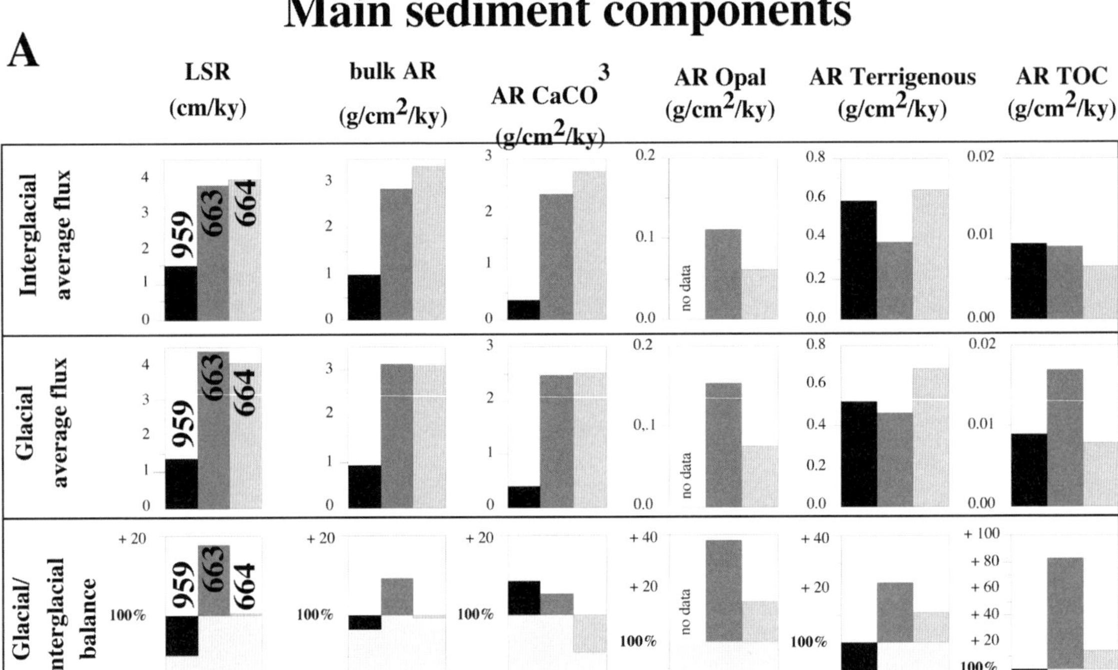

Marine and terrigenous organic matter

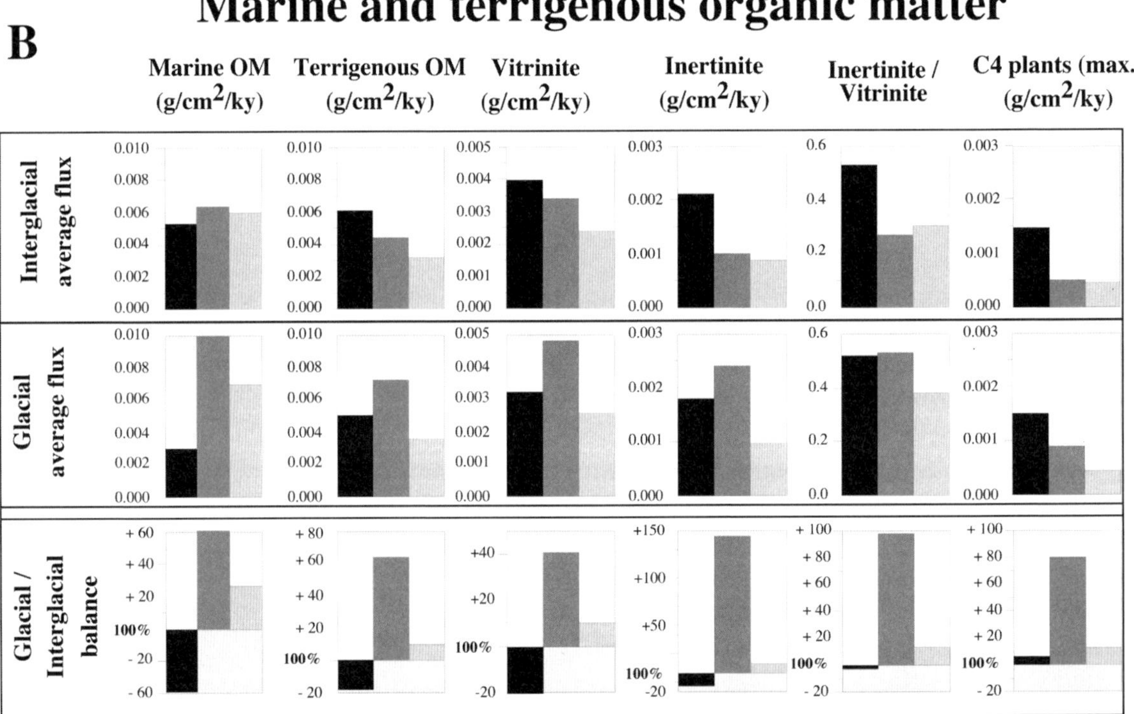

Fig. 9. Late Quaternary average interglacial and glacial flux rates (950 ka) and glacial-interglacial balances of **a)** main sediment components and **b)** marine and terrigenous OM from ODP sites 664, 663 and 959 (from Wagner 2000). Balances illustrate the relative increase or decrease of each individual component between glacial and interglacial periods at each position. Carbonate and biogenous opal data according to deMenocal et al. (1993; site 663) and Ruddiman and Janecek (1989, site 664). Data from site 959 as reported in Wagner (1998, 1999).

(1996) estimated the average global river carbon flux to the ocean to account for $392*10^{12}$g per year ($205*10^{12}$g/y dissolved organic carbon, DOC, and $187*10^{12}$g/y particulate organic carbon, POC).

On a global scale, organic carbon in modern continental shelf sediments may be composed of nearly the same amounts of marine and terrestrial organic carbon (Schlünz and Schneider 2000) although this view is challenged by other studies (e.g. see Hedges and Keil 1995; Hedges et al. 1997 for reviews). Results from the Amazon Fan have shown the dramatic effect of glacial sea level lowstands on the downslope transport of riverine organic carbon to distal areas of the deep-sea fan (Schneider et al. 1997b; Schlünz et al. 1999; Schlünz and Schneider 2000) and the adjacent western Equatorial Atlantic (Rühlemann et al. 1996). Estimates based on isotopic data from this area suggest that 15-40% of the total OM is of terrigenous origin. At present it is not well constrained how much of that riverine OM is laterally advected by ocean currents to settle over vast areas of the deep Atlantic basins (for shallow current advection see implications by Zabel et al. 1999). Goñi et al. (1998) emphasized the importance and sources of land-derived OM in surface sediments of the Gulf of Mexio using bulk and molecular level sedimentary compositions. They conclude that the terrigenous organic fraction may have been significantly underestimated in earlier studies (e.g. Hedges and Mann 1979b; Prahl et al. 1994) and that more than 50% of the lignin from land-derived organic carbon originated from C4 plants being eroded from grasslands of the Mississippi river drainage basin. These results imply that the terrigenous organic fraction may be underestimated in pelagic sediments where river-induced C4 plant matter contribute to the bulk organic carbon pool, especially during past glacial conditions (Bird et al. 1994).

Considering that vegetation zones covering the catchment areas of the two main tropical African rivers, i.e. the Congo and the Niger, are intimately linked to precessional-forced fluctuations in the West-African monsoon system it appears appropriate to expect considerable fluctuations in the supply and composition of river-borne OM to the easternmost Equatorial Atlantic. The importance of monsoon-driven variations in sediment load and fluvial nutrient supply on late Quarternary paleoproductivity within the Congo and Niger river plumes and the adjacent areas has been demonstrated by Schneider et al. (1997a,b) and Zabel et al. (2001). To further address this topic the following discussion will focus on three aspects, (i) general controls on organic sedimentation including a distinct pyrolytic feature (Holtvoeth et al. 2001), (ii) evidence from palynology and (iii) implications from lignin chemistry for selective degradation as preserved in sediments from the Congo fan (sites 1008 and 1075, Fig. 1).

Applying detailed organic geochemical investigations on the nature and reactivity of OM, Holtvoeth et al. (2001) recognized high amplitude fluctuations for various geochemical proxy records which are linked to the record of summer insolation at 15°N (Fig. 10). This can be illustrated with a simple method of data processing. For each sample the specific time interval to the next insolation maximum in the core profile is determined and converted to an equivalent within an idealized insolation cycle that covers 20 kyrs and is subdevided into 10 classes (corresponding to 2 kyrs intervals). Mean values of TOC content and accumulation rates of all samples within each individual class are calculated and compiled in a diagram (41 samples per class on average). The close link of TOC concentration and accumulation to the insolation cycle showing maximum values during periods of enhanced trade wind intensity, dry African climate and low continental run-off (insolation minima) is clearly recognizable (Fig. 11). The observed amplitudes for total organic carbon (TOC) within the profile (Fig. 10) range from 0.8 to 4.1% and from 0.2 to 4 for the ratio of low-mature to high mature OM (lm/hm). The latter parameter was recently established as new OM-reactivity proxy and is based on the relative share of hydrocarbon yields from low-mature (reactive) and high-mature (less-reactive) organic fractions during Rock-Eval pyrolysis. Based on the data, Holtvoeth et al. (2001) distinguish three general types of OM, (i) a low-mature marine source, (ii) a low-mature terrigenous source (lm; probably to some extend corresponding to the huminite/vitrinite fraction), and (iii) a high-mature terrigenous fraction (hm; related to

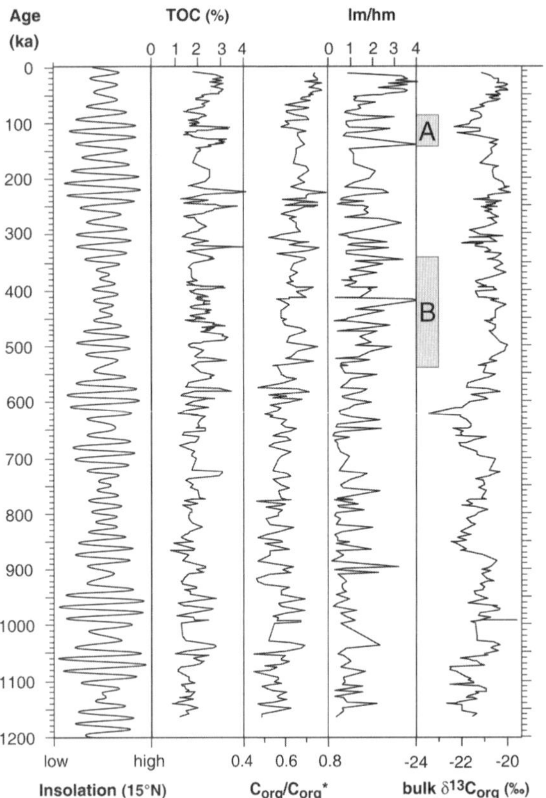

Fig. 10. Organic geochemical records of bulk organic matter from late Quaternary sediments of the Congo deep-sea fan (ODP Site 1075, from Holtvoeth et al. 2001). High amplitude flucuations of bulk geochemical record are linked to the summer insolation curve at 15°N. TOC maxima are dominated by marine organic carbon. Downcore trends point to a relative enrichment of a more stable (terrigenous) organic compound with a remarkable shift between 700 and 600 ka in TOC, lm/hm, and bulk $\delta^{13}C_{org}$ records. A and B indicate sections selected for more detailed lignin analysis (see Figures 15 and 16).

inertinite), although it remains a target of further studies to better assess chemical and physical properties of the low-mature terrigenous fraction. Combined with quantitative organic petrology, the lm/hm ratio may serve to improve the reconstruction of changes in the relative contributions of these different types of terrigenous OM. Figure 10 further shows the degradation rate profile of bulk OM (C_{org}/C_{org}*). Assuming that microbial sulfate reduction is the main degradation process for sedimentary organic matter in the Congo fan deposits, a minimum estimate of the original content of or-

ganic carbon (C_{org}*) was calculated according to Littke et al. (1997) following the equation C_{org}* = C_{org} + 2S * M_C/M_S (S = measured sulfur content, M_C = molecular mass of carbon, M_S = molecular mass of sulfur). The degradation rate of organic carbon is expressed as the ratio of the measured to the estimated original organic carbon contents (C_{org}/C_{org}*). Finally, bulk $\delta^{13}C_{org}$ signatures are presented which range from –20 to –23.5‰ throughout the past 1.2 million years at site 1075 (Fig. 10). Application of a simple binary mixing model considering marine and terrigenous end-member values of –18‰ and –27‰, respectively, result in terrigenous organic fraction in the order of 25-30%. It should be noted, however, that this approach ignores any influence from C4 plant matter and aged terrigenous OM, and thus most likely underestimates of the terrigenous organic fraction. To further address general fluctuations in supply from African vegetation belts including the admixture of C4 plant matter to the Congo Fan some key palynological results are discussed.

Palynological Results from the Congo Deep-Sea Fan

Pollen and spores found in the Congo Fan deposits have their sources in the Congolian rain forest area

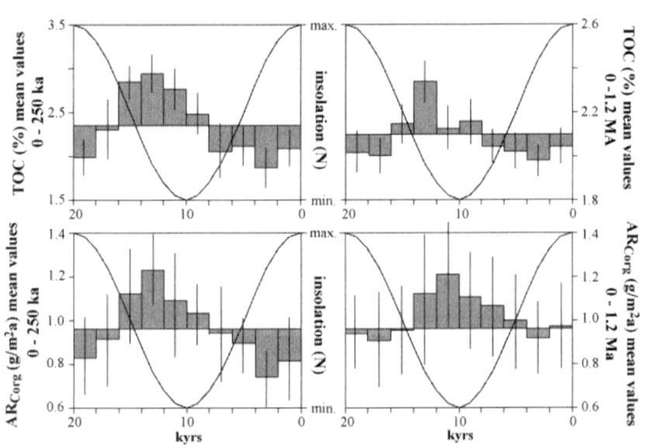

Fig. 11. Mean values of TOC contents and accumulation rates from 0 to 250 ka and from the whole record of ODP site 1075 within idealized cycle. For the last 250 kyrs insolation can be clearly identified as the driving force for organic sedimentation at the Congo deep-sea fan. Over the whole time span of 1.2 Ma the relation becomes blurred due to long term diagenesis and changes in the system control (e.g. mid Pleistocene transition).

and are mainly transported by the river. Additional aeolian pollen from the Angolan highlands are found (Dupont and Wyputta 2002). The record of ODP Site 1075 (5°S 10°E) located north of the Congo deep-sea canyon shows slightly more aeolian influence than the one of GeoB Site 1008 (7°S10°E) taken south of it (Van Iperen et al. 1987). Relative abundances of pollen of rain forest elements, *Podocarpus* (from the Afromontane forest), Poaceae (grasses and bamboos), and Cyperaceae (sedges, rushes, papyrus, and others), record changes in the lowland rain forest, woodland, Afromontane forest, swamp forest, and open swamps of the Congo Basin (Fig. 12).

The early Pleistocene pollen record of ODP Site 1075 shows a strong rise in the percentages of *Podocarpus* pollen concomitant with a reduction of woodland and lowland forest pollen, 1.05 Ma ago (Dupont et al. 2001). It indicates the extension of the Afromontane forest belt toward lower altitudes and suggests lower temperatures during (even-numbered) glacial stages. From the pollen record, it is inferred that *Podocarpus* forest was common during the glacial stages between 1.05 and 0.6 Ma, suggesting that the climate in equatorial Africa became cooler, but remained humid. Interglacials (odd-numbered isotope stages) between 0.9 and 0.6 Ma, in contrast, exhibit large variations in African vegetation comprising lowland rain forest, woodland, swamps, and some mountain forest. A trend to more open swamps is indicated within those mid-Pleistocene interglacial periods. The development of swamps in the Congo basin might be the result of reduced lake areas, which would suggest rather dry interglacial conditions.

The late Pleistocene pollen record of GeoB Site 1008 indicates the expansion of more open vegetation types, such as open swamps with grasses and Cyperaceae, during glacial Stages 6 and 2 (Jahns 1996). This pattern specifically holds true during periods in which the insolation at the Northern Hemisphere tropics increased. During periods with

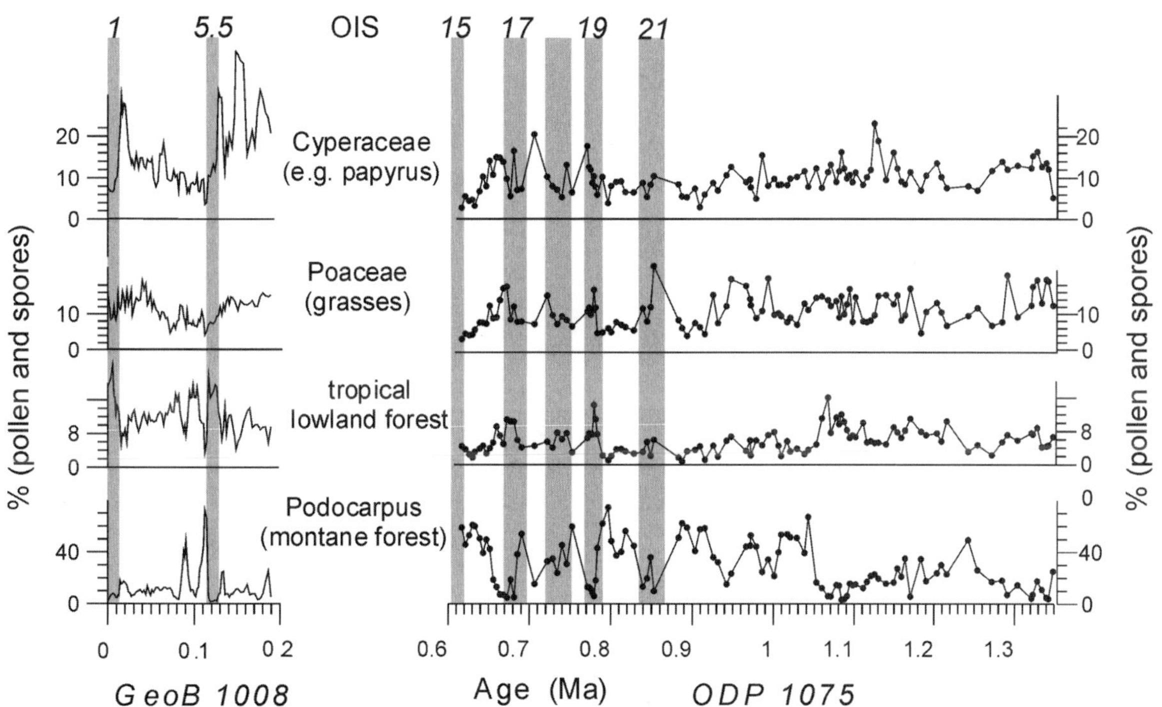

Fig. 12. Selected pollen taxa of Sites GeoB 1008 (left; Jahns 1996) and Ocean Drilling Program Site 1075 (right; Dupont et al. 2001) expressed in percent of the total of pollen and spores. Warm stages are shaded; odd-numbered interglacial isotope stages are indicated on top. OIS is Oxygen Isotope Stage.

open woodland (more grass pollen) and open swamps (more Cyperaceae pollen), the proportion of the C4 plants in the vegetation cover must have been larger than during periods with dense forests. However, in terms of biomass, the C3 trees probably remained the dominant source of terrigenous organic matter.

During the late Pleistocene climate cycles, the influence of the precession becomes more obvious. Several short-time fluctuations during Stages 6 and 5 underline the importance of precessional variation of the monsoon climate corroborating findings of Kutzbach (1987), Schneider et al. (1997a), and others. Spectral analysis of the GeoB 1008 record indicate that the response of the rain forest follows the pace of the Northern Hemisphere precession with a time lag of about 90°, equivalent to about 5 ka (Dupont et al. 1999). The timing of the relative abundance of selected pollen types within the precession dominated insolation cycle is illustrated in Figure 13.

Primary Supply Versus Selective Degradation of Organic Matter: Implications from Lignin Chemistry

It is generally accepted that terrigenous OM is less susceptible to microbial attack than marine OM. Therefore, variations in initial composition of bulk OM should result in a changing susceptibility to OM degradation. In that respect the records of degradation of bulk OM ($C_{org}/C_{org}*$), the lm/hm ratio, and bulk $\delta^{13}C_{org}$ at site 1075 are of interest (Fig. 10) since there are distinct, short term fluctuations in all parameters, which likely are attributable to changes in primary supply. Typically, insolation minima are followed with a delay of 2-7 kyrs by peak TOC (Fig. 11), minimum degradation rates, high lm/hm ratios, and generally heavier isotopic signatures. These trends suggest enhanced marine productivity and sedimentation (and/or preservation) of labile (mainly marine) OM in response to stronger wind-driven upwelling. This pattern is reversed during insolation maxima, when humid climate conditions over tropical Africa resulted in enhanced river discharge but lower marine produc-

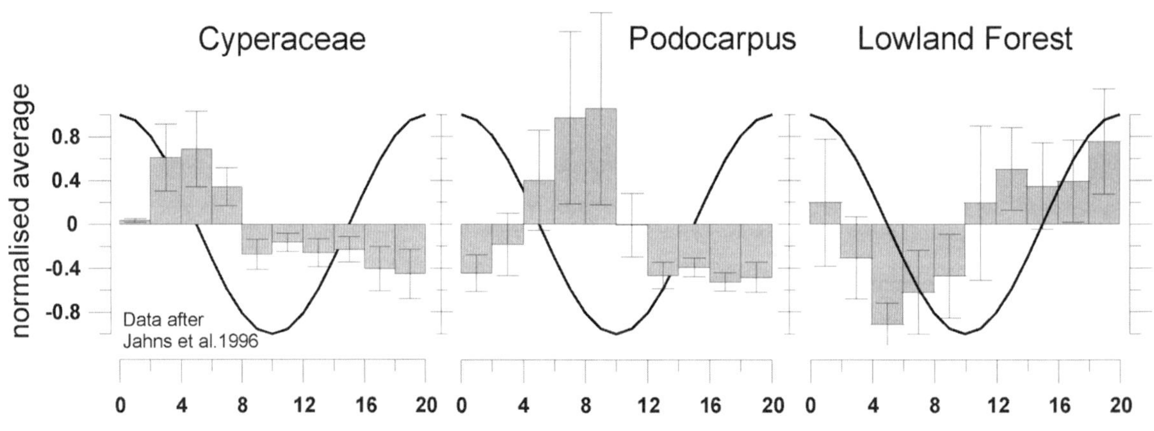

Fig. 13. Samples of GeoB 1008 (dated between 0 and 190 ka) are grouped according to the time lag after the last insolation maximum (insolation at 15°N) in classes of 2 ka. All cycles are normalised to 20 ka length. For each class the normalised average pollen percent of Cyperaceae (left), *Podocarpus* (middle), and sum of lowland forest elements (right) is plotted. The error bars are calculated by dividing the standard deviation of the class samples by the standard deviation of all samples. The line schematically depicts the insolation variation.

tivity due to reduced up-welling and lower organic carbon deposition. Such conditions apparently favoured the preservation of a less reactive (probably more terrigenous) OM in the sediments. Superimposed on these short-term fluctuations in primary composition of sedimenatry OM, a long-term down-core trend is evident (Fig. 10). Accordingly, C_{org}/C_{org}* and lm/hm ratios decrease with burial depth supporting the general conclusion that a more stable (and probably terrigenous) organic fraction is selectively enriched with increasing burial depth. In addition, an offset towards lower values is observed between 600 and 700 ka, most clearly documented in the TOC and lm/hm records. At the same time, bulk $\delta^{13}C_{org}$ signature shifts to slightly lighter values. Up to now it is not clear whether these shifts preserve direct response of the ecosystem in response to climatic change or secondary (sedimentary or diagenetic) adjustments of the depositional system.

To investigate variations in terrigenous OM composition and selective degradation two time slices (150-85 ka (A) and 540-340 ka (B), see indication in Figure 10) were studied in detail. The projection of the S/V and C/V ratios obtained from lignin chemistry according to Goñi and Hedges (1992) shows that terrigenous OM in late Quarternary deposits of the Congo fan mainly originates from angiosperm leaves and barks with variable admixture of angiosperm wood and grass tissues (Fig. 14). Apparently, there is no clear distinction in lignin composition that supports the existence of two simple terrestrial endmembers that reflect glacial or interglacial depositional conditions.

In addition to the general positive correlation of insolation minima to maxima in TOC and lm/hm ratios, the lignin parameters Λ and C/V reveal a negative correlation, as for example observed between 97-140 ka (time slice A, Figure 15). This pattern may either be explained by a better preservation of woody plant tissues (indicated by low C/V ratios) compared to non-woody tissues, by enhanced primary supply of woody material, or by processes related to hydrodynamic sorting (Goñi et al. 1997). Flucuations in primary supply is likely to have occurred during humid (interglacial) climates when the African tropical rain forest expanded over most of the Congo catchment area. Palynological

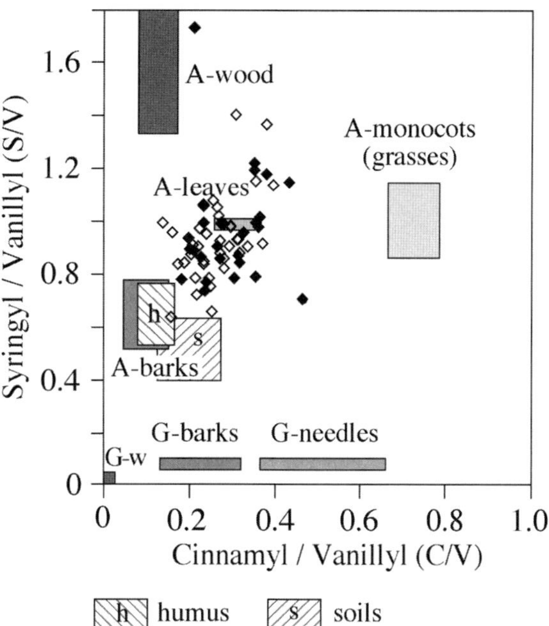

Fig. 14. S/V and C/V ratios obtaines from lignin chemistry of selected samples from site 1075 (from Holtvoeth et al. in press). Terrigenous OM in the sediments of the Congo fan mainly derives from barks and leaves with admixture of angiosperm wood and grass. Gw = gymnosperm wood, open/filled symbols: samples from insolation minima/maxima. Assignement of soil and humus OM according to Farella et al. (2001).

data compared to the C/V ratios from four samples between 111 and 139 ka indicate that higher amounts of woody terrigenous OM occur during (warm and humid) stages with extended lowland rainforest whereas supply of mainly non-woody terrigenous OM is concomitant with enhanced pollen records from montane forests. It may therefore be assumed that tropical grasslands expanded together with montane forests during the latter (cooler) climate stages. These observations clearly favour a varying primary composition causing the C/V record although diagenetic and transport effects can not be ruled out. To address this alternative mechanism, proxies indicative for degradation (C_{org}/C_{org}*), reactivity (lm/hm) and composition (bulk $\delta^{13}C_{org}$, Λ) of the OM are considered. Correlation of these parameters reveal two general groups of samples (Fig. 16), one reflecting the relative enrichment of high-mature terrigenous OM due

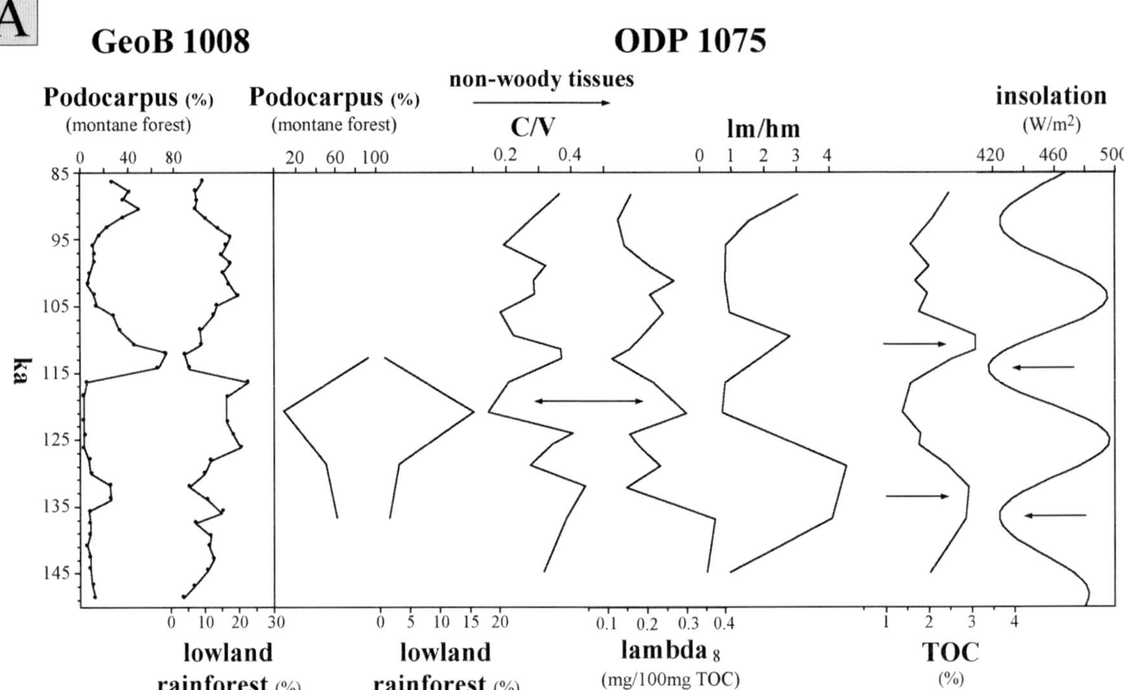

Fig. 15. Lignin data compared to bulk organic parameters (150 - 85 ka, from Holtvoeth et al. 2003), insolation record, and palynological data (GeoB 1008 from Jahns et al. 1996) for time interval A see Fig. 10. A significant negative correlation of C/V ratios and Λ values between 140 and 97 ka either indicates enhanced supply of woody plant tissues during humid stages with extended rainforest (supported by palynological results) and reduced primary production or a better preservation potential of woody plant tissues.

to selective diagenesis (group I, degradation line) and one representing mainly differences in primary composition of the OM (group II, combined mixing/degradation line).

Group I samples have low lm/hm signatures indicating generally lower proportions of reactive OM. Therefore, decreasing $C_{org}/C_{org}*$ and increasing Λ document the progressive enrichment of a high mature terrigenous organic component due to diagenetic processes. In comparison, group II samples reveal a broad range for all proxies reflecting a combined mixing/degradation line according to their variable primary nature. Since reactive OM may originate from marine and terrigenous sources their relative affiliation to the degree of degradation may help to distinguish the two groups. Enhanced proportions of reactive, terrigenous OM may thus be indicated by high Λ and lm/hm but low $C_{org}/C_{org}*$ signatures, assuming that low mature terrigenous OM is more resistant to degradation

compared to its marine counterpart. This model includes selective degradation of the most labile (marine) organic fraction, which likely was deposited in conjunction with fresh terrigenous OM. Enhanced supply of reactive terrigenous OM probably represents stronger river discharge during humic climatic conditions when tropical rain forests covered the catchment area. A schematic model summarizing the observed time relationships of insolation, organic carbon burial and the three organic components with indication of potential C4 plant and degradadtion influence is presented in Figure 17.

Concluding Remarks

Various organic geochemical and palynological studies performed on modern and late Quarternary sediments from the central to eastern Equatorial

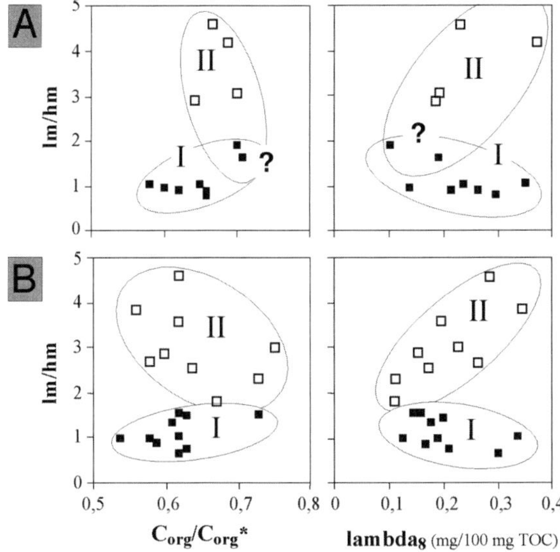

Fig. 16. Cross plots of proxies indicative for degradation (C_{org}/C_{org}*), reactivity (lm/hm) and composition (Λ) of bulk OM in late Quaternary sediments from the Congo Fan (site 1075) for time intervals A and B see Fig. 10. Assignment of group I reflects the relative enrichment of high-mature terrigenous OM due to selective diagenesis while group II mainly documents differences in primary composition of OM (from Holtvoeth et al. 2003).

Atlantic address the fundamental question of how terrigenous OM can be traced and quantified in TOC-poor deep-sea deposits. The application of multiparameter concepts combining macerals (organic particles), freshwater diatoms, phytoliths, pollen grains, lignin phenols, bulk isotopic data ($\delta^{13}C_{org}$), and carbon isotopes derived from long-chain n-alkanes to modern and late Quarternary near-continental and open ocean settings of the central and eastern Equatorial Atlantic provide evidence for spatial and temporal variations in quantity and composition of terrigenous organic matter. Most important, a growing body of data support the presence of significant and variable quantities of isotopically heavy C4 plant-derived OM from arid, grass-covered African source areas in deposits from the central Equatorial Atlantic and along the NW-African continental margin. The spatial and temporal distribution of C4 plant input is determined by climate-driven fluctuations in eolian dust export from the African continent. Terrigenous organic

matter in sediments form the Congo deep-sea fan, in contrast, mainly originates from the C3 plant dominated catchment area of the Congo River, although palynological data indicate a minor riverine supply of C4 plant to the deep-sea fan during past dry climate conditions. To validate these results and to improve the data base, additional research efforts utilizing molecular isotopic approaches are suggested.

Based on results from combined pyrolysis, lignin and isotope chemistry three general types of organic matter were distinguished in Congo fan sediments, i.e. (i) a highly reactive (low-mature) marine source, (ii) a reactive (low-mature) terrigenous source (probably associated to the huminite/vitrinite fraction), and (iii) a low-reactive (high-mature) terrigenous fraction (related to inertinite). Differences in selective degradation of the two reactive components apparently have a strong influence on organic geochemical attributes of individual samples and may lead to considerable fluctuation in the relative shares of the three organic components. Such diagenetic processes compete with primary fluctuations in supply of organic matter and may therefore complicate the interpretation of organic geochemical proxy records in a paleoenvironmental context. At present, the inter-

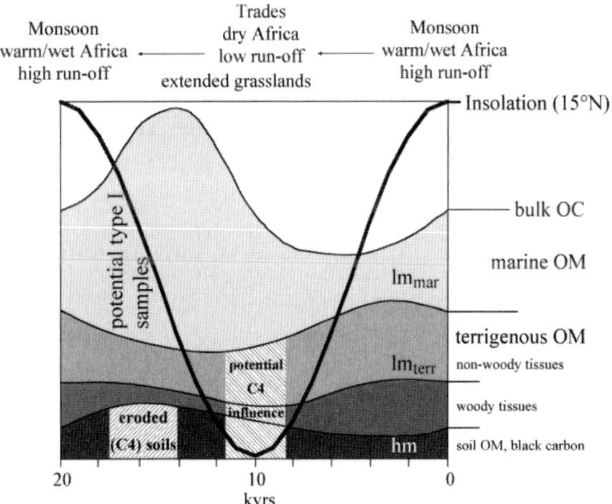

Fig. 17. Scematic model summarizing time relationships of insolation, organic carbon burial and the three organic components with indication of potential C4 plant and degradation influence as documented in late Quaternary sediments from the Congo fan (ODP Site 1075).

play of these complex processes are not well understood. It therefore remains a primary target for further studies to better assess chemical and physical properties of reactive organic matter, both of terrigenous and marine origin.

Acknowledgements

We thank R. Stein (AWI-Bremerhaven) and T. Eglinton (WHOI) for their critical and instructive comments on a former version of this manuscript. This research was funded by the Deutsche Forschungsgemeinschaft (Sonderforschungsbereich 261 at Bremen University, Contribution 381, and Ocean Drilling Program grants Wa 1036/3 and 5). Data are available under www.pangaea.de/ Projects/SFB261.

References

Bauch HA, Kassens H, Naidina OD, Kunz-Pirrung M, Thiede J (2001) Composition and flux of Holocene sediments on the Laptev Sea Shelf, Arctic Sibiria. Quat Res 55: 344-351

Bauer JE, Druffel ERM (1998) Ocean margins as a significant source of organic matter to the deep open ocean. Nature 392: 482-485

Bergamaschi BA, Tsamakis E, Keil RG, Eglinton TI, Montlucon DB, Hedges JI (1997) The effect of grain size and surface area on organic matter, lignin and carbonhydrate concentration, and molecular compostions in Peru Margin sediments. Geochim Cosmochim Acta 61: 1247-1260

Bird MI, Cali JA (1998) A million-year record of fire in sub-Saharan Africa. Nature 394: 767-769

Bird MI, Giresse P, Chivas AR (1994) Effect of forest and savanna vegetation on the carbon-isotope composition of sediments from the Sanaga River, Cameroon. Limnol Oceanogr 39: 1845-1854

Bonifay D, Giresse P (1992) Middle to Late Quarternary sediment flux and post- depositional processes between the continental slope off Gabon and the Mid-Guinean margin. Mar Geol 106: 107-129

Cohen AD (1970) An allochthonous paet deposit from south Florida. Geol Soc Am Bull 81:2477-2482

DeMenocal PB (1995) Plio-Pleistocene African climate. Science 270: 53-59

DeMenocal PB, Ruddiman WF, Pokras EM (1993) Influences of high- and low-latidude processes on African terrestrial climate: Pleistocene eolian records

from equatorial Atlantic Ocean Drilling Program site 663. Paleoceanography 8: 209-242

Dittmar T, Lara RJ (2001) Molecular evidence for lignin degradation in sulfate-reducing mangrove sediments (Amazonia, Brazil). Geochim Cosmochim Acta 65: 1417-1428

Dupont LM, Hooghiemstra H (1989) The Saharan-Sahelian boundary during the Brunhes chron. Acta Bot Neerland 38: 405-415

Dupont LM, Agwu COC (1991) Environmental control of pollen grain distribution patterns in the Gulf of Guinea and offshore NW- Africa. Geol Rundsch 80: 567-589

Dupont LM, Weinelt M (1996) Vegetation history of the savanna corridor between the Guinean and the Congolian rain forest during the last 150,000 years. Veget Hist Archaeobot 5: 273-292

Dupont LM, Wyputta U (2002) Comparing Atmospheric Trajectories and the Distribution of Pollen in Marine Sediments along the South-Western African Coast. Quat Sci Rev 22: 157-174

Dupont LM, Marret F, Winn K (1998) Land-sea correlation by means of terrestrial and marine palynomorphs from the equatorial East Atlantic:Phasing of SE trade winds and ocean productivity. Palaeogeogr Palaeoclimatol Palaeoecol 142: 51-84

Dupont LM, Schneider R, Schmüser A, Jahns S (1999) Marine terrestrial interaction of climate changes in west Equatorial Africa of the last 190,000 years. Palaeoecol Africa 26: 61-84

Dupont LM, Jahns S, Marret F, Ning S (2000) Vegetation change in equatorial West Africa: time-slices for the last 150 ka. Palaeogeogr Palaeoclimatol Palaeo-ecol 155: 95-122

Dupont LM, Donner B, Schneider R, Wefer G (2001) Mid-Pleistocene environmental change in tropical Africa began as early as 1.05 Ma. Geology 29: 195-198

Eglinton TI, Benitez-Nelson BC, Pearson A, McNichol AP, Bauer JE, Druffel ERM (1997) Variability in radiocarbon ages of individual organic compounds from marine sediments. Science 277: 796-799

Eglinton TI, Eglinton G, Dupont L, Sholkovitz ER, Montluçon D, Reddy CM (2002) Composition, Age and Provenance of Organic Matter in NW African Dust over the Atlantic Ocean. Geochem Geophys Geosys 3: doi 10.1029/2001GC000269

Eriksson KEL, Blanchette RA, Ander P (1990) Microbial and Enzymatic Degradation of Wood and Wood Components. Springer, Berlin

Farella N, Lucotte M, Louchouarn P, Roulet M (2001) Deforestation modifying terrestrial organic transport in the Rio Tapajós, Brazilian Amazon. Org Geochem 32: 1443-1458

Frédoux A (1994) Pollen analysis of a deep-sea core in the Gulf of Guinea: Vegetation and climatic changes during the last 225,000 years BP. Palaeogeogr Palaeoclimatol Palaeoecol 109: 317-330

Gagosian RB, Peltzer ET (1985) The importance of atmospheric input of terrestial organic material to deep-sea sediments. Organic Geochem 10: 661-669

Gagosian RB, Peltzer ET, Merrill JT (1987) Long-range transport of terrestrially derived lipids in aerosols from the South Pacific. Nature 325: 800-804

Gearing JN (1988) The use of stable isotope ratios for fracting the Nearshore - Offshore exchange of organic matter. In: Jansson BO (ed) Coastal-Offshore Ecosystem Interactions. Springer, Berlin, pp 69-101

Giresse P, Maley J, Brenac P (1994) Late Quarternary palaeoenvironments in the Lake Barombi Mbo (West Cameroon) deduced from pollen and carbon isotopes of organic matter. Palaeogeogr Plaeoclimatol Palaeoecol 107: 65-78

Giresse P, Gadel F, Serve L, Barusseau JP (1998) Indicators of climate- and sediment source-variations at Site 959: Implications for the reconstruction of paleo-environments in the Gulf of Guinea trough Pleisto-cene times. In: Mascle J, Lohman GP, Moullard M (eds) Ocean Drilling Program Scientific Results 159. Ocean Drilling Program, College Station, Texas, pp 585-604

Giresse P, Wiewióra A (2001) Stratigraphic condensed deposition and diagenetic evolution of green clay minerals in deep water sediments on the Ivory coast - Ghana Ridge. Mar Geol 179: 51-70

Goñi MA, Hedges JI (1992) Lignin dimers: Structures, distribution, and potential geochemical application. Geochim Cosmochim Acta 56: 4025-4043

Goñi MA, Ruttenberg KC, Eglinton TI (1997) Sources and contribution of terrigenous organic carbon to surface sediments in the Gulf of Mexico. Nature 389: 275-278

Goñi MA, Ruttenberg KC, Eglinton TI (1998) A reassessment of the sources and importance of land-derived organic matter in surface sediments of the Gulf of Mexico. Geochim Cosmochim Acta 62: 3055-3075

Goñi MA, Yunker MB, MacDonald RW, Eglinton TI (2000) Distribution and sources of organic biomarkers in arctic sediments from the Mackenzie River and Beaufort Shelf. Mar Chem 71: 23-51

Gough MA, Fauzi R, Mantoura C, Preston M (1993) Terrestrial plant biopolymers in marine sediments. Geochim Cosmochim Acta 57: 945-964

Hamilton SE, Hedges JI (1988) The comparative geochemistries of lignins and carbohydrates in an anoxic fjord. Geochim Cosmochim Acta 52: 129-142

Hedges JI, Parker PL (1976) Land-derived organic matter in the surface sediments from the Gulf of Mexico. Geochim Cosmochim Acta 40: 1019-1029

Hedges JI, Keil RG (1995) Sedimentary organic matter preservation: An assessment and speculative synthesis. Mar Chem 49: 81-115

Hedges JI, Keil RG, Benner R (1997) What happens to terrestrial organic matter in the ocean? Org Geochem 27: 195-212

Hedges JI, Mann DC (1979a) The characterization of plant tissues by their lignin oxidation products. Geochim Cosmochim Acta 43: 1803-1807

Hedges JI, Mann DC (1979b) The lignin geochemistry of marine sediments from the southern Washington coast. Geochim Cosmochim Acta 43: 1809-1818

Hedges JL, Clark WA, Quay PD, Richey JE, Devol AH, Santos U (1986) Composition and fluxes of particulate organic material in the Amazon River. Limnol Oceanogr 31: 717-738

Holtvoeth J, Wagner T, Schubert CJ, Horsfield B, Mann U (2001) Late Quarternary supply of terrigenous organic matter to the Congo Deep-Sea Fan (ODP Site 1075): Implications for equatorial African paleo-climate. Geo-Mar Lett 21: 23-33

Holtvoeth J, Wagner T, Schubert C (2003) Organic matter in river-influenced continental margin sediments: The land-ocean and climate linkage at the Late Quaternary Congo fan (ODP Site 1075). Geochem Geophys Geosyst, in press

Hooghiemstra H (1989) Variations of the NW African trade wind regime during the last 140,000 years: Changes in pollen flux evidenced by marine sediment records. In: Leinen M and Sarnthein M (eds) Paleo-climatology and Paleometeorology: Modern and past patterns of global atmospheric transport. NATO ASI Series C 282: 733-770

Hooghiemstra H, Agwu COC, Beug HJ (1986) Pollen and spore distribution in recent marine sediments: A record of NW-African seasonal wind patterns and vegetation belts. "*Meteor*"Forschungs-Ergebnisse C40, pp 87-135

Hooghiemstra H, Stalling H, Agwu COC, Dupont LM (1992) Vegetational and climatic changes at the northern fringe of the Sahara 250,000-5000 years BP: Evidence from 4 marine pollen records located between Portugal and the Canary Islands. Rev Palaeobot Palynol 74:1-53

Huang Y, Dupont L, Sarnthein M, Hayes JM, Eglinton G (2000) Mapping of C4 plant input from North-

West Africa into North-East Atlantic sediments. Geochim Cosmochim Acta 64: 3505-3513

Ishiwatari R, Uzaki M (1987) Diagenetic changes of lignin compounds in a more than 0.6 million-year-old lacustrine sediment (Lake Biwa, Japan). Geochim Cosmochim Acta 51: 321-328

Ittekkot V (1988) Global trends in the nature of organic matter in river suspensions. Nature 332: 436-438

Jahns S (1996) Vegetation history and climatic changes in West Equatorial Africa during the Late Pliocene and Holocene, based on a marine pollen diagram from the Congo fan. Veg Hist Archaeobot 5: 207-213

Jahns S, Hüls M, Sarnthein M (1998) Vegetation and climate history of west equatorial Africa based on a marine pollen record off Liberia (site GIK 16776) covering the last 400,000 years. Rev Palaeobot Palynol 102: 277-288

Keil RG, Tsamakis E, Giddings JC, Hedges JI (1998) Biochemical distributions (amino acids, neutral sugars, and lignin phenols) among size-classes of modern marine sediments from the Washington coast. Geochim Cosmochim Acta 62: 1347-1364

Knies J, Stein R (1998) New aspects of organic carbon deposition and its paleoceanographic implications along the northern Barents Sea margin during the last 30,000 years. Paleoceanography 13: 384-394

Kutzbach JE (1987) The changing pulse of the monsoon. In: Fein JS, Stephens P (eds) Monsoons. J Wiley & Sons, Chichester, pp 247-267

Lézine A-M (1991) West African paleoclimates during the last climatic cycle inferred from an Atlantic deep- sea pollen record. Quat Res 35: 456-463

Littke R, Lückge A, Welte DH (1997) Quantification of organic matter degradation by microbial sulfate reduction for Quaternary sediments from the Northern Arabian Sea. Geowissenschaften 84: 312-315

Lobbes JM, Fitznar HP, Kattner G (2000) Biogeochemical characteristics of dissolved and particulate organic matter in Russian rivers entering the Arctic Ocean. Geochim Cosmochim Acta 64: 2973-2983

Louchouarn P, Lucotte M, Canuel R, Gagné JP, Richard LF (1997) Sources and early diagenesis of lignin and bulk organic matter in the sediments of the Lower St. Lawrence Estuary and the Saguenay Fjord. Mar Chem 58: 3-26

Louchouarn P, Lucotte M, Farella N (1999) Historical and geographical variations of sources and transport of terrigenous organic matter within a large-scale coastal environment. Org Geochem 30: 675-699

Ludwig W (2001) The age of river carbon. Nature 409: 466

Ludwig W, Amiotte-Suchet P, Munhoven G, Probst JL (1998) Atmospheric CO_2 consumption by continental erosion: Present-day controls and implications for the Last Glacial Maximum. Glob Planet Change 16/17: 107-120

Ludwig W, Probst JL, Kempe S (1996) Predicting the oceanic input of organic carbon by continental erosion. Glob Biogeochem Cycl 10: 23-41

McCabe A (1984) Depositional environments of coal and coal-bearing strata. In: Rahmani RA, Flores RM (eds) Sedimentology of coal and coal-bearing sequences. Spec Pub Int Ass Sedimentol 7: 13-42

Miltner A, Emeis KC (2000) Origin and transport of terrestrial organic matter from the Oder lagoon to the Arkona Basin, Southern Baltic Sea. Org Geochem 31: 57-66

Mitra S, Bianchi TS, Guo L, Santschi PH (2000) Terrestrially derived dissolved organic matter in the Chesapeake Bay and the Middle Atlantic Bight. Geochim Cosmochim Acta 64: 3547-3557

Müller PJ, Schneider RR, Ruhland G (1994) Late Quaternary PCO_2 variations in the Angola current: evidence from organic carbon $\delta^{13}C$ and alkenone temperatures. In: Zahn R, Pedersen TF, Kaminski MA, Labeyrie L (eds) Carbon Cycling in the Glacial Ocean: Con-strains on the Ocean's Role in Global Change. NATO ASI Series I 17, Springer, Berlin, pp 343-366

Nelson BC, Goñi MA, Hedges JI, Blanchette RA (1995) Soft-rot fungal degradation of lignin in 2700 year old archeological woods. Holzforschung 49: 1-10

Onstad GD, Canfield DE, Quay PD, Hedges JI (2000) Sources of particulate organic matter in rivers from the continental USA: Lignin phenol and stable carbon isotope composition. Geochim Cosmochim Acta 64: 3539-3546

Orem WH, Colman SM, Lerch HE (1997) Lignin phenols in sediments of Lake Baikal, Siberia: Application to paleoenvironmental studies. Org Geochem 27: 153-172

Pokras EM, Mix AC (1985) Eolian evidence for spatial variability of Late Quarternary climates in tropical Africa. Quat Res 24: 137-149

Pokras EM, Ruddiman WF (1989) Evolution of south Saharan/Sahelian aridity based on freshwater diatoms (genus Melosira) and opal pytholiths: Sites 657-661. In: Ruddiman WF, Sarnthein M, Baldauf J, Backmann J, Bloemendal J, Curry W, Farrimond P, Faugeres JC, Janecek TR, Katsura Y, Manivit H, Mazzullo J, Mienert J, Pokras EM, Raymo ME, Schultheiss PJ, Steinn R, Tauxe L, Valet JP, Weaver P, Yasuda H (eds) Proceedings of the Ocean Drilling Program. Scientific Results 108: 143-148

Prahl FG, Ertel JR, Goñi MA, Sparrow MA, Eversmeyer B (1994) Terrestrial organic carbon contributions to sediments of the Washington margin. Geochim Cosmochim Acta 58: 3053-3048

Prahl FG, Muelhausen LA (1989) Lipid biomarkers as geochemical tools for paleoceanographic study. In: Berger WH, Smetacek VS, Wefer G (eds) Productivity of the Ocean: Present and Past. J Wiley & Sons, Chinchester, pp 271-289

Rath S, Wagner T, Henrich S (1999) Pleistozäne Schwankungen im Staubeintrag und in der Karbonaterhaltung im äquatorialen Atlantik (ODP Site 663) zwischen 480 und 390 Ka: Hinweise aus Korngrößenanalysen. Z Geol Paläontol Teil I (H 7-10): 1-19

Ratmeyer V, Fischer G, Wefer G (1999) Lithogenic particle fluxes and grain size distributionsin the deep ocean off northwest Africa: Implications for seasonal changes of aeolian dust input and downward transport. Deep-Sea Res 46: 1289-1337

Raymo ME, Oppo DW, Curry W (1997) The mid-Pleisto-cene climate transition: A deep-sea carbon isotopic perspectiv. Paleoceanography 12: 546-559

Raymond PA, Bauer JE (2001) Riverine export of aged terrestrial organic matter to the North Atlantic Ocean. Nature 409: 497-500

Romero O, Lange CB, Swap RJ, Wefer G (1999) Eolian-transported freshwater diatoms and phytoliths across the equatorial Atlantic record temporal changes in Saharan dust transport patters. J Geophys Res 104: 3211-3222

Ruddiman WF, Janecek TR (1989) Pliocene-Pleistocene biogenic and terrigenous fluxes at equatorial Atlantic Sites 662, 663 and 664. In: Ruddiman WF, Sarnthein M, Baldauf J, Backmann J, Bloemendal J, Curry W, Farrimond P, Faugeres JC, Janecek TR, Katsura Y, Manivit H, Mazzullo J, Mienert J, Pokras EM, Raymo ME, Schultheiss PJ, Steinn R, Tauxe L, Valet JP, Weaver P, Yasuda H (eds) Proceedings of the ODP, Scientific Results 108. Ocean Drilling Program, College Station, Texas, pp 221-240

Rühlemann C, Frank M, Hale, W, Mangini A, Mulitza S, Müller PJ, Wefer G (1996) Late Quarternary productivity changes in the western equatorial Atlantic from ^{230}Th-normalized carbonate and organic carbon accumulation rates. Mar Geol 135: 127-152

Sarnthein M, Pflaumann U, Ross R, Tiedemann R, Winn K (1992) Transfer functions to reconstruct ocean paleoproductivity: A comparison. In: Summerhayes CP, Prell WL, Emeis KC (eds) Upwelling Systems: Evolution Since the Early Miocene. Geol Soc Spec Pub 64: 411- 427

Sarnthein M, Thiede J, Pflaumann U, Erlenkeuser H, Fütterer D, Koopmann B, Lange H, Seibold E (1982) Atmospheric and oceanic circulation patterns off Northwest Africa during the past 25 million years. In: von Rad U, Hinz K, Sarnthein M, Seibold E (eds) Geology of the Northwest African Continental Margin. Springer, Berlin, pp 545-604

Sarnthein M, Winn K, Jung SJA, Duplessy JC, Labeyrie L, Erlenkeuser H, Ganssen G (1994) Changes in east Atlantic deepwater circulation over the last 30,000 years: Eight time slice reconstructions. Paleoceanography 9: 209-267

Schefuß E (2002) Paleo-environmental effects of the Mid-Pleistocene transition in the tropical Atlantic. Geologica Ultraiectina 223- Mededelingen van de Faculteit Aardwetenschappen Universiteit Utrecht, Utrecht, 183 p

Schlünz B, Schneider RR (2000) Transport of terrestrial organic carbon to the oceans by rivers: Re-estimating flux- and burial-rates. Int J Earth Sci 88: 599-606

Schlünz B, Schneider RR, Müller PJ, Showers WJ, Wefer G (1999) Terrestrial organic carbon accumulation on the Amazon Deep-Sea Fan during the last glacial sealevel lowstand. Chem Geol 159: 263-281

Schneider RR, Müller PJ, Kroon D, Price B, Alexander I (1997a) Monsoon related variations in Zaire (Congo) sediment load and influence of fluvial silicate supply on marine productivity in the east equatorial Atlantic during the last 200,000 years. Paleoceano-graphy 12: 463-481

Schneider RR, Müller PJ, Schlünz B, Segl M, Showers WJ, Wefer G (1997b) Late Quarternary Western Atlantic paleoceanography and terrigenous sedimentation on the Amazon Fan: A view from δ^{18}O and δ^{13}C of planktonic foraminifera and bulk organic matter δ^{13}C. In: FloodR, Piper D, Klaus A, Peterson LC (eds) Proceedings of the Ocean Drilling Program. Scientific Results 155: 319-333

Schubert CJ, Stein R (1996) Deposition of organic carbon in Arctic Ocean sediments: Terrigenous supply vs marine productivity. Org Geochem 24: 421-436

Schubert CJ, Villanueva J, Calvert SE, Cowie GL, von Rad U, Schulz H, Berner U, Erlenkeuser H (1998) Stable phytoplankton community structure in the Arabian Sea over the past 200,000 years. Nature 394: 563-566

Stach E, Mackowsky MT, Teichmüller M, Taylor GH, Chandra D, Teichmüller R (1982) Stach's Textbook of Coal Petrology. Bornträger, Berlin

Stein R (1991) Accumulation of Organic Carbon in Marine Sediments. Lecture Notes in Earth Sciences 34: 1-216

Stein R, Fahl K (2000) Holocene Accumulation of Organic Carbon atthe Laptev Sea Continental Margin (Arctic Ocean): Sources, pathways, and sinks. Geo-Mar Lett 20: 27-36

Stein R, ten Haven HL, Littke R, Rullkötter J, Welte DH (1989) Accumulation of marine and terrigenous organic carbon at upwelling Site 658 and nonupwelling Sites 657 and 659: Implications for the reconstruction of paleoenvironments in the eastern subtropical Atlantic through late Cenozoic times. In: Ruddiman WF, Sarnthein M, Baldauf J, Backmann J, Bloemendal J, Curry W, Farrimond P, Faugeres JC, Janecek TR, Katsura Y, Manivit H, Mazzullo J, Mienert J, Pokras EM, Raymo ME, Schultheiss PJ, Steinn R, Tauxe L, Valet JP, Weaver P, Yasuda H (eds) Proceedings of the Ocean Drilling Program. Scientific Results 108: 361-386

Street-Perrott FA, Perrott RA (1990) Abrupt climate fluctuations in the tropics: The influence of Atlantic Ocean circulation. Nature 343: 607-612

Taylor GH, Teichmüller M, Davis A, Dissel CFK, Littke R, Robert P (1998) Organic Petrology. Stuttgart, Schweitzerbart, 707 p

Tiedemann R, Sarnthein M, Shackleton NJ (1994) Astronomic timescale for the pliocene Atlantic $\delta^{18}O$ and dust flux records of Ocean Drilling Program site 959, Paleoceanography 9: 619-638

Tyson RV (1995) Sedimentary organic matter: Organic facies and palynofacies. Chapman and Hall, London

van Iperen JM, van Weering TEC, Jansen JHF, van Bennekom AJ (1987) Diatoms insurface sediments of the Zaire deep-sea fan (SE Atlantic Ocean) and their relation to overlying water masses. Neth J Sea Res 21: 203-217

Verardo DJ, McIntyre A (1994) Production and destruction: Control of biogenous sedimentation in the tropical Atlantic 0-300,000 years BP. Paleoceanography 9: 63-86

Verardo DJ, Ruddiman WF (1996) Late Pleistocene charcoal in tropical Atlantic deep-sea sediments: Climatic and geochemical significance. Geology 24: 855-857

Wagner T (1998) Pliocene-Pleistocene carbonate and organic carbon fluxes off Ivory Coast/Ghana (ODP Site 959): Paleoenvironmental implications for the eastern Equatorial Atlantic. In: Mascle J, Lohman GP, Moullard M (eds) Ocean Drilling Program Scientific Results 159. Ocean Drilling Program, College Station, Texas, pp 557-574

Wagner T (1999) Petrology of organic matter in modern and Late Quarternary deposits of the Equatorial Atlantic: Climatic and oceanographic links. Int J Coal Geol 39: 155-184

Wagner T (2000) Control of organic carbon accumulation in the late Quarternary Equatorial Atlantic (ODP Sites 664, 663): Productivity versus terrigenous supply. Paleoceanography 15: 181-199

Wagner T, Hölemann JA (1995) Deposition of organic matter in the Norwegian- Greenland Sea during the past 2.7 million years. Quat Res 44: 255-266

Wagner T, Dupont LM (1999) Terrestrial organic matter in marine sediments: Analytical approaches and eolian-marine records from the central Equatorial Atlantic. In: Fischer G, Wefer G (eds) The Use of Proxies in Paleoceanography: Examples from the South Atlantic. Springer, Berlin, pp 547-574

Wagner T, Pfefferkorn HW (1997) Tropical peat occurrences in the Orinoco Delta: Preliminary assessment and comparison to Carboniferous coal deposits. Proceedings of the XIII International Congress on the Carboniferous and Permian 2: 161-168

Walsh JJ (1989) How much shelf production reaches the Deep-Sea? In: Berger WH, Smetacek VS, Wefer G (eds) Productivity of the Ocean: Present and Past. J Wiley & Sons, Chinchester, pp 175-191

Westerhausen L, Poynter J, Eglinton G, Erlenkeuser H, Sarnthein M (1993) Marine and terrigenous origin of organic matter in modern sediments of the equatorial East Atlantic: The $\delta^{13}C$ and molecular record. Deep-Sea Res 40: 1087-1121

Zabel M, Bickert T, Dittert L, Haese RR (1999) The significance of sedimentary Al/Ti ratio as indicator for reconstructions of the terrestrial input to the Equatorial Atlantic. Paleoceanography 14: 789-799

Zhao M, Dupont LM, Eglinton G., Teece M (2003) n-Alkane and pollen reconstruction of terrestrial climate and vegetation for NW Africa over the last 160 kyr. Org Geochem 34: 131-143

Terrigenous Signals in Sediments of the Low-Latitude Atlantic - Indications to Environmental Variations during the Late Quaternary: Part II: Lithogenic Matter

M. Zabel[1*], T. Wagner[1] and P. deMenocal[2]

[1] Universität Bremen, Fachbereich Geowissenschaften, Postfach 330 440, D-28334 Bremen, Germany
[2] Lamont Doherty Earth Observatory, Columbia University, PO Box 1000, Palisades NY 10964
* corresponding author: mzabel@uni-bremen.de

Abstract: The inorganic terrigenous fraction of marine sediments offers a great number of different and well established proxy parameters to investigate the development of Earth's climate. This study presents a synthesis of multidisciplinary investigations which have been applied to late Quaternary sediments recovered from the low-latitude Atlantic during the Bremen Special Reseach Project 261. In the equatorial Atlantic terrigenous matter is supplied by eolian and fluvial pathways. In addition to the dust input from African deserts, the catchment areas of the three major rivers Amazon, Niger and Zaire (Congo) are the dominant sources. Small river systems are of local importance. Terrigenous records from near-continental and open pelagic depositional settings are discussed. The main questions we focused on are a) the control of climate change and b) the identification and timing of rapidly occurring events. Results from the low- latitude Atlantic support the suggestion that both high-latitude and low-latitude forcing influence tropical climate and marine sedimentation. Apparently, the frequency of climate variability in the tropics during the late Quaternary is controlled by the precessional insolation cycle, whereas amplitudes and timing of climate change are mainly determined by the high latitudes in the Northern Hemisphere. Within the phase relationships, however, regional differences arise. Furthermore, there is evidence for climate instability during glacials and interglacials which probably occurred on decadal to centennial time scales.

Introduction

In manifold ways the equatorial Atlantic represents a particularly suitable key position for investigations on climatic development, its fluctuations and land-ocean interrelations. This region represents a key corridor for the global water mass circulation, the dominant process of the interhemisphere and latitudinal heat transfer on Earth. On the other hand, the equatorial Atlantic is characterised by the contact of meteorological cycles between Northern and Southern Hemisphere at the Intertropical Convergence Zone (ITZC). In a very simplistic description, there are mainly the contrasting temperatures between land and ocean, or on global scale, between polar regions and low latitudes, combined with air pressure gradients which drive the atmospheric circulation, and therefore control climate conditions and marine sedimentation. As simple as this system may appear, its function is highly complex because interacting sub-processes strongly influence this system. During the last decades numerous studies

From WEFER G, MULITZA S, RATMEYER V (eds), 2003, *The South Atlantic in the Late Quaternary: Reconstruction of Material Budgets and Current Systems.* Springer-Verlag Berlin Heidelberg New York Tokyo, pp 323-345 on marine and lacustrine sediments have improved our knowledge about these processes, their cause, and their effect on climate change (e.g. Parkin and Shackleton 1973; Street and Grove 1976; Sarnthein 1978; Kolla et al 1979; Kutzbach 1981; Pokras and Mix 1985; Mix et al. 1986; McIntyre et al. 1989; deMenocal et al. 1993; Foley 1994; Hughen et al. 1996; Ganopolski et al. 1998; Zabel et al. 1999; Mulitza and Rühlemann 2000). One emphasis of this research was placed on the investigation of variations in the African monsoon system which primarily controls the transport of water vapor onto the West African continent. Despite the large scientific progress in this field, there are still open and intensively discussed questions which concern, for example a) the link between fluctuations in the African climate and evaporation in the tropical Atlantic, b) the interplay between high and low latitude forcing of African climate variability, and c) the cause of rapid climate changes, especially during late Quaternary.

Sediments from the equatorial Atlantic represent outstanding archive material. Although terrestrial records offer more direct, yet only temporally limited insights into regional climatic developments, continuously settled marine deposits permit reviews back to the Mesozoic (e.g. Wagner 2002). The region of the tropical Atlantic is one of the most important depot centers for the input of terrestrial source material. Particles are supplied to this area by eolian and fluvial transport from both adjacent continents (Fig. 1). Related to the affected sea-floor area, the eolian dust input from the African Sahara and Sahel regions may possess the greatest importance for the composition of the marine sediments. Estimates for the terrestrial dust deposition rate range between 100-400 Mt/yr (Prospero 1981). The largest quantity is, however, supplied by the Amazon River. Its discharge of suspended par-ticulate material is estimated to amount approximately to 1200 Mt/yr (Gaillardet et al. 1997). Depending on the ocean current pattern at present, the bulk of this material is carried parallel to the coast-line in northwestern direction to the Caribbean (Milliman et al. 1975). Nevertheless, there is evidence that Amazon suspensates may constitute the dominant portion of the terrigenous fraction as far as the Mid- Atlantic Ridge (Zabel et al. 1999). However, in contrast to this tremendous input, the wash- outs of the Niger River and the Zaire River are comparably small (each approx. 40 Mt/yr; Gaillardet et al. 1999). Apart from these four main sources, some smaller rivers of local importance additionally supply their load of suspended material.

With this synthesis paper we summarize some collaborative geoscientific studies on variations of the inorganic terrigenous fraction in marine sediments which were conducted between 1997 and 2001 within the previous German Special Research Project 261. On this occasion, we particularly want to put out their contribution for an improvement of the understanding of the climate history within the low-latitude Atlantic and the adjacent continents. First, short overviews are given on the most common methods to investigate the lithogenic fraction and its input to the ocean and on the state of knowledge from outside studies. Related aspects concerning the terrigenous organic matter in sediments from the low-latitude Atlantic are reviewed in a separate contribution (Wagner et al. this volume). In addition, further studies on the terrigenous fraction from other regions of the Atlantic Ocean are summarized by Diekmann et al. (this volume).

Methods to Investigate the Inorganic Terrigenous Fraction

There is a large number of well established approaches to investigate the inorganic terrigenous fraction in marine sediments. Although it is not the purpose of this paper to review them all in detail, we briefly turn our attention to some of these methods, or rather proxy parameters, because they may differ substantially in their force of expression. In extreme cases, two different proxies, applied to the same samples, can lead to apparent contradictions. Besides, individual approaches may occasionally contain risks of misinterpretation.

The first group of research methods uses concentrations or accumulation rates (AR) of specific particulate sediment constituents which are considered to be exclusively associated with the terrigenous fraction. The variety of parameters essentially concerns individual minerals or characteristic mineral suits. For example, feldspar, quartz, iron

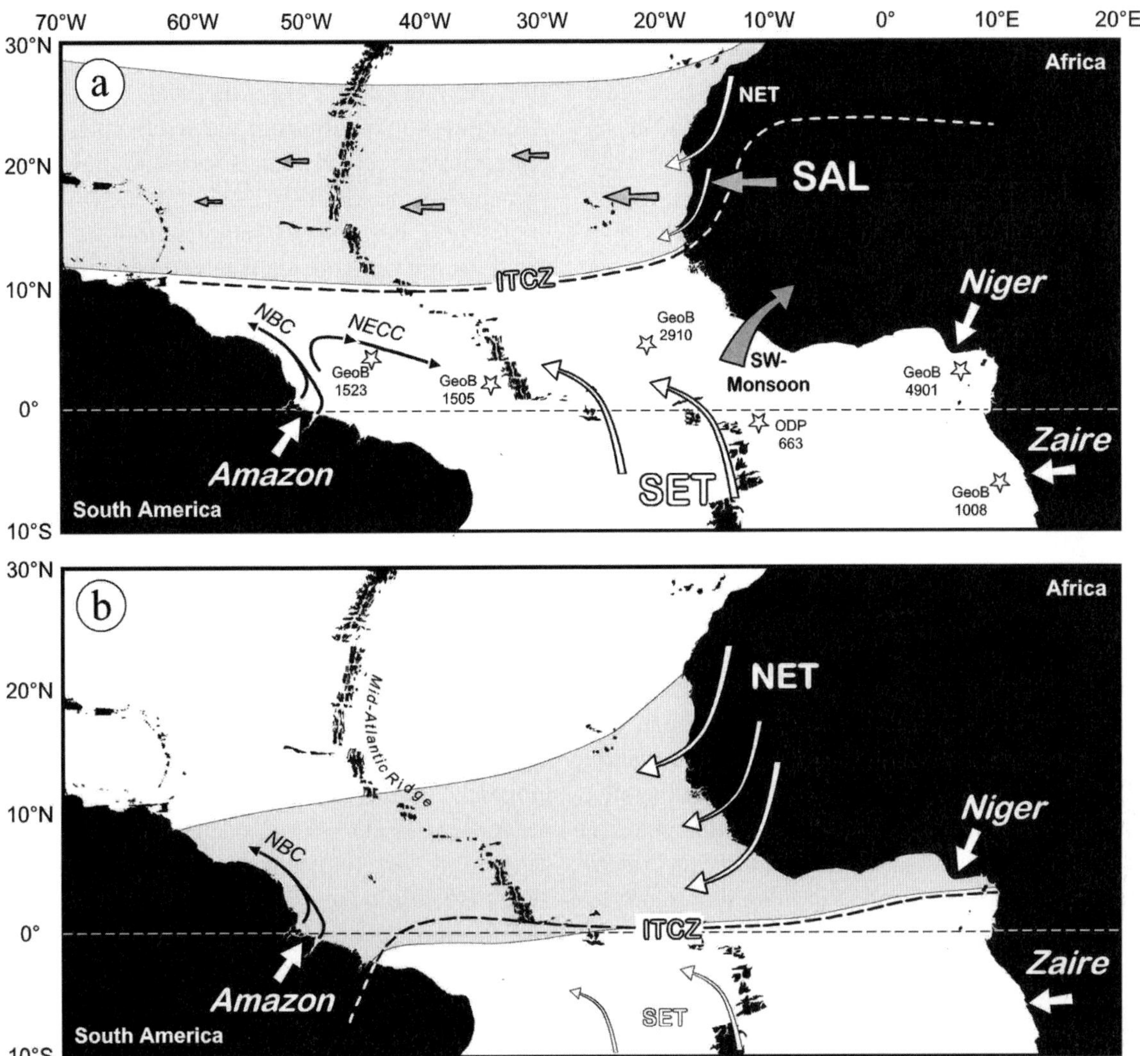

Fig. 1. The equatorial Atlantic and source areas of terrigenous input. Current low, mid-tropospheric and underlying wind regimes and main river systems are additionally shown. Shaded areas indicate the seasonal positions of the dust plums (modified after Sarnthein et al. 1981). Stars mark locations for which records are presented in Figs. 2 to 4. During the boreal summer season **a)** strong SE trade winds (SET) move the Intertropical Convergence Zone (ITCZ) to about 10°N. African dust is mainly transported by the Saharan Air Layer (SAL). Because the enhanced retroflection of the North Brazil Current (NBC) which accelerates the North Equatorial Countercurrent (NECC), suspended matter supplied by the Amazon can drift eastwards (Zabel et al. 1999). During the boreal winter season **b)** dust- loaded NE trade winds (NET) are dominant and the ITCZ is located close to the equator. River-suspended matter supplied by the Amazon is mainly transported in northwesterly direction.

oxide, certain clay minerals or clay mineral assemblages have been used to record the terrigenous input (e.g. Delany et al. 1967; Damuth and Fairbridge 1970; Parkin and Shackleton 1973; Windom 1975; Kolla et al. 1979; Lange 1982; Baslam et al. 1995; Schneider et al. 1997; Gingele et al. 1998; Harris and Mix 1999; Rühlemann et al. 2001; Diekmann et al. this volume). Quartz and feldspar contents may be the most reliable indicator of these proxies, because clay mineral assemblages or iron oxide contents are often much more difficult to interpret. On account of several factors, like the very slow soil formation rates, clay mineral assemblages may provide only integrated records of overall climatic impacts, from which it follows that their distribution patterns rather reflect the great, supra-regional changes than variations in specific local climate conditions (Thiry 2000). Also not to be neglected is the possible influence of a geochemical alteration of the primary signal due to authigenic mineral formation and/or dissolution. Hence, clay mineral distribution has been used more often to trace ocean currents (e.g. Biscaye 1965; Petschick et al. 1996; Diekmann et al. 1999).

Quite similar to the use of specific minerals, individual element concentrations have been interpreted as characterizing terrigenous constituents (e.g. Matthewson et al. 1995; Schneider et al. 1997; Zabel et al. 1999, 2001; Haug et al. 2001; Bozzano et al. 2002). One uncertainty of this assumption is that bulk chemical analyses of sediment do not naturally reveal which sedimentary phases add to the concentration of a particular element (e.g. van der Weijden 2002). Another risk stems from the possible distortion of the primary terrigenous signal by secondary processes during particle settling through the water column, like scavenging effects (e.g. Orians and Bruland 1985). We will return to this point in the next paragraph. However, interpretations of the element contents in marine sediments are manifold. While in some studies elements like Al, K, Rb, Ti, or Zr are considered to be indicators for terrigenous matter derived from a defined area and reflect the climate conditions therein (e.g. Matthewson et al. 1995; Schneider et al. 1997), others also use the elemental composition of bulk sediments to identify terrigenous quantities from different source areas, providing that the raw materials differ significantly from each other (e.g. Zabel et al. 1999, 2001). A description of the technical and analytical methods to determine element concentrations in marine sediments which were also applied for this study is given in Zabel et al. (1999).

A frequently used modification of this method focuses on element ratios. Besides the previously mentioned restriction regarding the allocation of single elements to specific mineral phases, element ratios may be suitable especially for reconstructions of the sediment's provenance and the paleoclimate

(e.g. Boyl 1983; Matthewson et al. 1995; Schneider et al. 1995; Jansen et al. 1998; Arz et al. 1998, 1999; Martinez et al. 1999; Zabel et al. 1999, 2001). According to conditions in the sedimentation regime, they can reflect climate variations more sensitively than the AR of terrigenous particles (Zabel et al. 1999). Nevertheless, studies from the equatorial Pacific have shown that caution is required when marine particles dominate the sedimentation flux and the terrigenous input is low. However, sedimentary Ti/Al ratios, for example, can extremely deviate from the natural composition of terrestrial source material. The underlying enrichment of Al has been interpreted in terms of scavenging on biogenic particles (e.g. Murray et al. 1993; Dymond et al. 1997). As a result, element/Al records can reflect variations in paleoproductivity instead of such related to the terrigenous signal.

When specific measurements are not available, the quantity of the non-biogenic, terrigenous fraction ($Terr_{tot}$) is often calculated as the residual from biogenic carbonate and opal analyses (e.g. deMenocal et al. 1993; Tiedemann et al. 1994; deMenocal 1995; Rühlemann et al. 1996; Ruddiman 1997). But due to the mutual dilution of the marine and the terrigenous fractions, this generalized method can at least result in an estimation of the latter, however, only producing reciprocal values of the marine components in summary. Additionally, effects due to variations in carbonate production or dissolution would have inevitable consequences for the interpretation of the terrestrial record (e.g. Bloemendal and deMenocal 1989). Information inherent in the terrigenous material itself as investigated by the specific methods described above, may not be observed. In contrast to individual concentrations and element ratios, AR records are much more sensitive with regard to the age model.

An unspecific but nevertheless expressive sedimentological approach is to look at the lithogenic grain-size distribution as representative for the energy which transported the terrigenous particles (e.g. Parkin and Shackleton 1973; Sarnthein et al. 1981; Matthewson et al. 1995; Grousset et al. 1998). The relation is simple: The coarser the particles, the higher the wind velocity or the discharge of river water.

Recently, geophysical sediment parameters were also used for the investigation and interpretation of the terrigenous fraction (Frederichs et al. 1999; Schmidt et al. 1999; Funk et al. this volume). First results obtained from the magnetic characterization of marine sediments are very promising and indicative of the high potential of this new approach. A future task of great importance lies in the calibration of the complex geophysical measurements using geochemical and mineralogical data.

State of Knowledge from Previous Studies

Fundamental work investigating the export of African dust into the equatorial Atlantic originates from the end of the sixties and beginning of the seventies. First indication of the extent to which African dust is carried across the Atlantic Ocean was made by Delany et al. (1967). Three years later, Prospero et al. (1970) identified considerable amounts of African dust particles within the Caribbean as well. Details of the meteorology associated with the transport mechanisms were elucidated by Carlson and Prospero (1972) and Prospero and Carlson (1972). Accordingly, long range transport occurs by the mid-tropospheric African easterly jet stream (Saharan Air Layer - SAL) during boreal summer season, whereas winter dust plumes are predominantly carried by the northeast trade wind system (Sarnthein et al. 1981; Fig. 1). However, these studies have not only shown the spatial and quantitative dimensions of the dust transport from West African deserts for the first time, but can also be seen as an important inspiration for the paleoceanographic and paleoclimatic studies of the last 30 years. Today, the input introduced by the NE trade wind during winter is described as the dominant pathway for the lithic fraction in sediments of the northeastern tropical Atlantic (Chester et al. 1972; Chiapello et al. 1997; Ratmeyer et al. 1999). Against earlier speculation (Kolla et al. 1979), there is a lot of evidence that the position of the trade-wind belt has not shifted substantially between glacials and interglacials (e.g. Sarnthein et al. 1981; Tiedemann et al. 1989; Baslam et al. 1995; Ruddiman 1997; Grousset et al. 1998).

Indication for Low-Latitude Forcing of the West African Monsoon System

The general availability of erodible soils depends, first of all, on the humidity of the respective climate. In Central and West Africa precipitation is controlled by the West African monsoon which drives moisture into the continent during boreal summer (Fig. 1). Apart from the highly seasonal pattern of this system, there is also plenty of evidence that its intensity has periodically varied responding to low-latitude insolation on the precessional frequency bands (19 and 23 kyr; e.g. Kutzbach 1981; Pokras and Mix 1985; Tiedemann et al. 1989). Consequently, orbitally induced variations of the low-latitude monsoon intensity are described as the primary driving force for the West African terrestrial climate (e.g. deMenocal et al. 1993; Matthewson et al. 1995). It inevitably follows that dry episodes must have been related to a general attenuation of the monsoon. Furthermore, it has been stated that monsoon intensity and precipitation over Central Africa is strongest, when the boreal summer perihelion coincides with the maximum summer insolation on the Northern Hemisphere (deMenocal et al. 1993 and references therein). However, this interferential effect indicates that, apart from precessional insolation, eccentricity, i.e. the variation in the elliptical course of the Earth around the sun, may also at least occasionally influence the tropical climate (cf. McIntyre and Molfino 1996).

The concept of enhanced aridity on the African continent during glacial periods was first developed for terrestrial samples by Fairbridge (1964). Meanwhile, this concept has been confirmed by a large number of mineralogical, geochemical, faunal, and palynological studies on marine and lacustrine paleoclimate records (e.g. Damuth and Fairbridge 1970; Parkin and Shackleton 1973; Williams 1975; Street and Grove 1976; Sarnthein 1978; Kolla et al. 1979; Sarnthein et al. 1981; Pokras and Mix 1985; Tiedemann et al. 1989; Street-Perrott and Perrott 1990; deMenocal et al. 1993; Tiedemann et al. 1994; Leroy and Dupont 1994; Matthewson et al. 1995; Ruddiman 1997; Gasse 2000 and references therein). The approximately simultaneous decrease in precessional forcing in the tropics is suggested as one reason for the dominance of dry conditions during glacials (e.g. deMenocal et al. 1993).

Indication for High-Latitude Forcing of African Aridity/Humidity Cycles

Apart from the frequently documented link between terrigenous input and variations in the monsoon system, there are various studies on marine and lacustrine records which indicate that fluctuations in the Saharan and Sahelian aridity, as well as the precipitation cycles in tropical South America, were essentially synchronous with cold events at high latitudes (e.g. Parkin and Shackleton 1973; Kolla et al. 1979; Sarnthein et al. 1981; Stein 1985; Gasse et al. 1989; Street-Perrott and Perrott 1990; Tiedemann et al. 1994). Based on fluctuations in benthic oxygen isotope records which are sensitive to ice volume variations in the Northern Hemisphere, this global climate signal is dominated by the 100-kyr orbitally eccentricity cycle since about 1Ma (deMenocal 1995) and is associated with an increased variance at the precession bands (e.g. Imbrie et al. 1984). According to Harris and Mix (1999) this nonlinear amplification of insolation forcing drives latitudinal shifts of the ITCZ which could explain the dominant 100-kyr component. However, comparison with the previous paragraph makes it clear that the control mechanisms of African aridity are still being controversially discussed.

Indications Derived from Sea Surface Temperatures

Valuable information to fathom this apparent contradiction result from reconstructions of variations in sea- surface temperature (SST). As mentioned above, the thermal land-sea contrast and accompanying pressure gradients are important factors for climatic development and variability. For example, results of circulation models predict greater monsoonal precipitation due to larger land-ocean pressure gradients acting in response to increased boreal summer insolation (e.g. Kutzbach 1981; Prell and Kutzbach 1987). SSTs themselves depend on the thermohaline circulation of the ocean which is clearly influenced by the production of North Atlantic Deep Water (e.g. Zhao et al. 1995; Manabe and Stouffer 1997; Mulitza and Rühlemann 2000). Hence, meltwater induced changes in the strength of the North Atlantic thermohaline circulation have been proposed quite early to explain fluctuations of the African monsoon (e.g. Mix et al. 1986; Street-Perrott and Perrott 1990; deMenocal et al. 1993). Accordingly, arid conditions prevail when North Atlantic SSTs are relatively cool and the South Atlantic, or in this case tropical, SSTs are relatively warm (e.g. Lough 1986; deMenocal and Rind 1993). Consequently, the African climate seems preconditioned for aridity during ice growth in high latitudes and full glacial periods. The buildup of heat in the equatorial region was clearly established by comparing alkenone-derived temperature records from ODP site 658 (NW Africa) and core GeoB 1007 recovered from the Congo fan (Zhao et al. 1995; Mulitza and Rühlemann 2000). Similar to other paleoclimate proxies, equatorial SST records show strong precessional variance with lower amplitudes in the west (McIntyre et al. 1989). However, variations in the strength of deep water ventilation and therefore in the cross-equatorial heat transport certainly may help to understand changes in monsoonal precipitation over Africa, but terrestrial surface fresh water indicatorscast doubt on the fact that changes in the thermohaline circulation can exclusively explain all the periods that evidence weakened monsoon. Especially during the Holocene, there still are considerable discrepancies in the number, timing and duration of the century-scale dry episodes (Guo et al. 2000; see below).

Rapid Climate Changes

A significant criterion for climate control issues is represented by the timing and amplitudes of climate changes. In particular, short-living and abrupt climatic changes are of special interest. At least the climatic instability in low latitudes during the Holocene and late Pleistocene were first proven by retreating lake levels in arid, semiarid and equatorial Africa (e.g. Street and Grove 1976; Talbot et al. 1984; Street-Perrott and Perrott 1990 and references therein). These high-amplitude events lasted over decades and occasionally over several centuries and are comparable with climatic changes which are documented in high-latitude marine sediments (e.g. Bond et al. 1993) and Greenland ice cores (e.g. GRIP 1993; Dansgaard et al. 1993). Equivalent variations in high-resolution marine records from the tropical Atlantic region are sparsely documented and were only observed for the most part in most recent times (Peterson et al. 1991; Hughen et al. 1996; deMenocal et al. 2000a, b; Maslin and Burns 2000; Haug et al. 2001; Marret et al. 2001). So, in the eastern subtropical Atlantic (ODP site 658) the Holocene cooling events were described as synchronous with changes in the SSTs of the subpolar North Atlantic which document a link between high- and low-latitude climate (deMenocal et al. 2000a). Similar correlations were also documented for the laminated deposits of the Cariaco Basin off Venezuela (ODP site 1002; Haug et al. 2001). In contrast to Maslin and Burns (2000), who have postulated an approximately constant increase of the effective moisture at the Amazon Basin throughout the Holocene, Cariaco Basin deposits revealed a period of enhanced precipitation between the global Younger Dryas event and about 5.4 kyr BP. This has been attributed to a northward shift of the ITCZ (Haug et al. 2001) which generally corresponds to palynological studies form this region (e.g. Ledru 1993), although the forest of the Amazon Basin does not seem to be replaced by an extensive savanna vegetation during glacial periods (Haberle and Maslin 1999). Additional support for Amazonian climate change is given by the large-scale variations in the Atlantic thermohaline circulation which could also be documented in western tropical Atlantic sediments for the last 10 kyr (Arz et al. 2001). However, its a fact that the instability of the Holocene climate is a global phenomenon and that slow orbital insolation forcing alone cannot explain the abruptness of the aridity/ humidity changes. In this context, deMenocal et al. (2000b) discussed a climate-threshold response as being responsible for the timing of climatic transitions.

In summary, the essence of all these studies is that the tropical Central African and South American climate is sensitive to both high- and low-latitude forcing. Therefore, frequently formulated questions remaining are a) how does the climate engine work and b) what is the cause for abrupt climate variations?

Climate Forcing in the Low-Latitude Atlantic as Inferred by Variations in the Lithogenic Fraction - Results from SFB 261

In this section we want to compare and discuss time series of terrigenous matter from six cores which cover the whole tropical Atlantic, resolving the last 250 kyr (Fig. 1). A first generalized view on the records indicates that the accumulation rates of the estimated total terrigenous fraction ($Terr_{tot}$ AR) and Ti flux rates (Ti AR) are expectedly almost parallel (Fig. 2). This definitively does not apply to time series of $Terr_{tot}$ AR and the Al/Ti ratio (Fig. 3). The records seem to be largely independent of each other. Figure 7 documents that this cannot be attributed to an age model effect. Records of the total terrigenous content ($Terr_{tot}$) and Al/Ti are nearly anticyclical, especially in cores that are influenced by a dust input from North Africa and suspensates from the Amazon River (GeoB2910, 1505, 1523). For sediments off the African River estuaries (GeoB1008 and 4901), this relationship reveals an inconsistency which could indicate changes in the interactions among underlying processes. However, the significant differences between the signals of terrigenous parameters consequently imply that their individual variations must have different causes. In the following, cycles and phase relationships in the lithogenic material of the individual cores will be discussed.

Variations in the African Dust Input (ODP site 663, GeoB2910)

Based on $Terr_{tot}$ from ODP site 663 (Figs. 2a and 4a), deMenocal et al. (1993) have deduced that AR records of Sahelian dust vary predominantly at orbital periodicities of 100-kyr and 41-kyr and that spectral phase estimates would implicate high-latitude forcing. Provided that sediments from this locations and core GeoB2910 are influenced by the same wind field or by rather continental climate conditions, parallel oscillations in $Terr_{tot}$AR and Ti

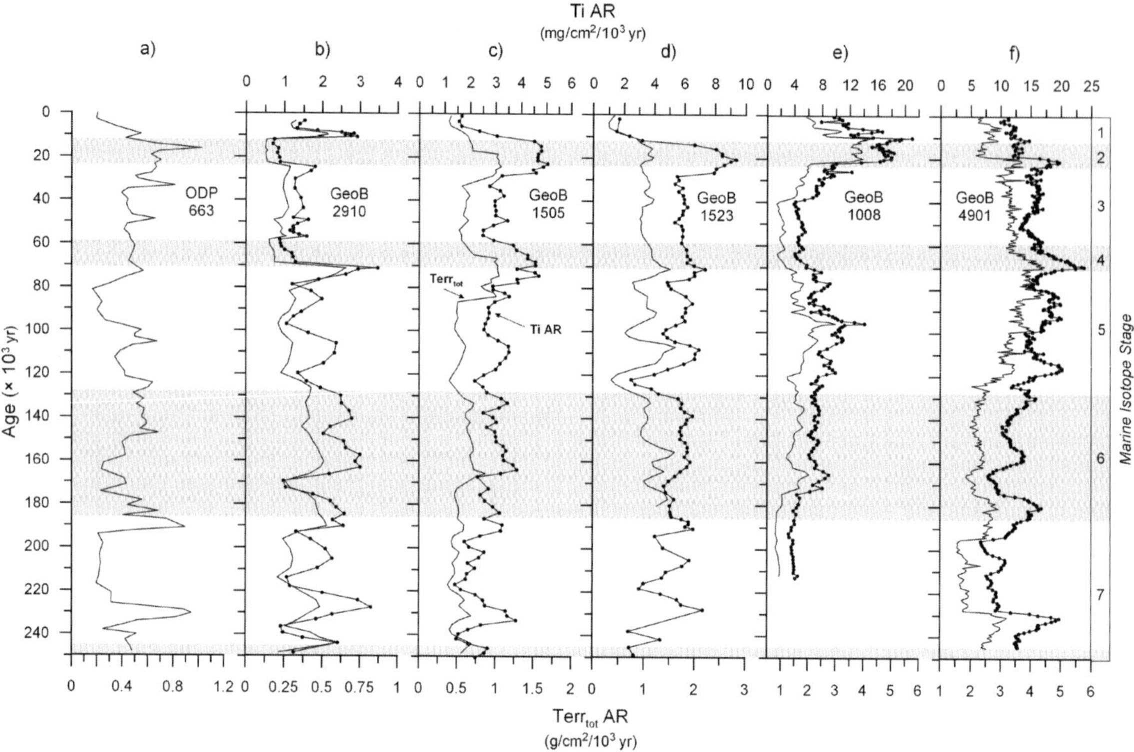

Fig. 2. Records of accumulation rates (AR) of the total terrigenous matter (Terr$_{tot}$) and Ti from six cores of the equatorial Atlantic. Terr$_{tot}$ was estimated by subtraction of biogenic carbonate, opal, and organic carbon contents from the total sediment. All age models are based on δ^{18}O-stratigraphy. **a)** ODP site 663 - western equatorial Atlantic, 1°12'S, 11°53'W, (deMenocal et al. 1993); **b)** GeoB2910 - Sierra Leone Rise (Zabel et al. 1999; cf. Fig. 4, Tab.1); **c)** GeoB1505 - central equatorial Atlantic (Zabel et al. 1999; cf. Fig. 4, Tab. 1); **d)** GeoB1523 - Ceara Rise (Zabel et al. 1999; cf. Fig. 4, Tab.1) Terr$_{tot}$AR based on ^{230}Th-normalization; **e)** GeoB1008 - Zaire deep-sea fan, 6°35'S, 10°19'E (cf. Schneider et al. 1997); **f)** GeoB4901 - Niger deep-sea fan (Zabel et al. 2001; cf. Fig. 4, Tab. 1)

AR of core GeoB2910 generally contradict this observation (Fig. 2b). Resembling the variations observed in the Al/Ti ratio (Fig. 3b), these records reveal a much stronger precessional 23-kyr component which is nearly in phase with the high-latitudinal insolation signal in the δ^{18}O record of the planktonic foraminifera *G. sacculifer* (Zabel et al. 1999). By contrast, the marine rain rate (total org. carbon/carbonate) in core GeoB2910 clearly fluctuates in tune with orbital eccentricity and subordinate obliquity cyclicity. Due to the potential and the reciprocal dilution effect of the terrigenous and biogenic marine fractions mentioned above, it seems obvious that the assumed domination of the eolian transport to the tropical western Atlantic at the 100-kyr periodicity is rather

an artifact. Nevertheless, the Al/Ti record in GeoB 2910 gives strong evidence to an eccentricity modulation of the prevailing precessional term. These overlying fre-quencies were also documented by the occurrence of the freshwater diatom *Melosia* in ODP site 663 (deMenocal et al. 1993) and in other cores re-covered from this region (Pokras and Mix 1985). In this context it is very interesting that *Melosira* variations, in contrast to the Al/Ti ratio, lag precession by about 9 kyr, which was interpreted on account of a lake desiccation in response to climate transitions rather than arid conditions in general (Pokras and Mix 1985). However, the results presented clearly indicate that while the frequency of African dust input and its composition during late Quaternary depends on

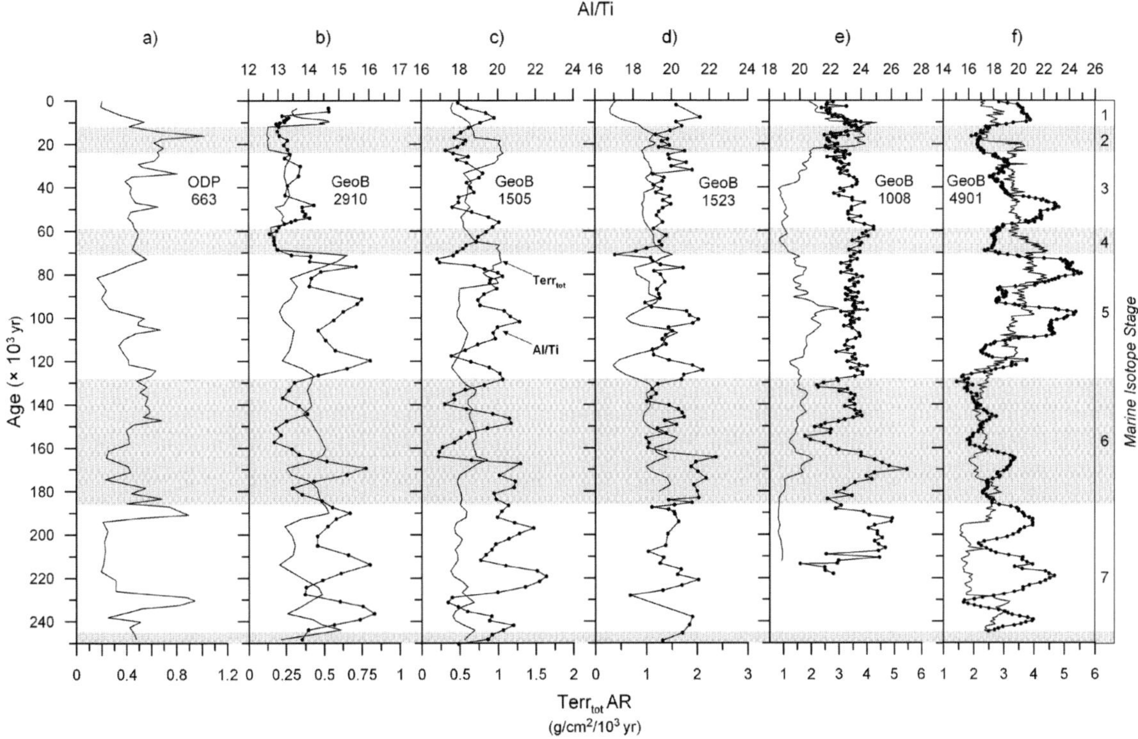

Fig. 3. Records of Terr$_{tot}$AR and the Al/Ti ratio (For information on the cores see caption of Fig. 2)

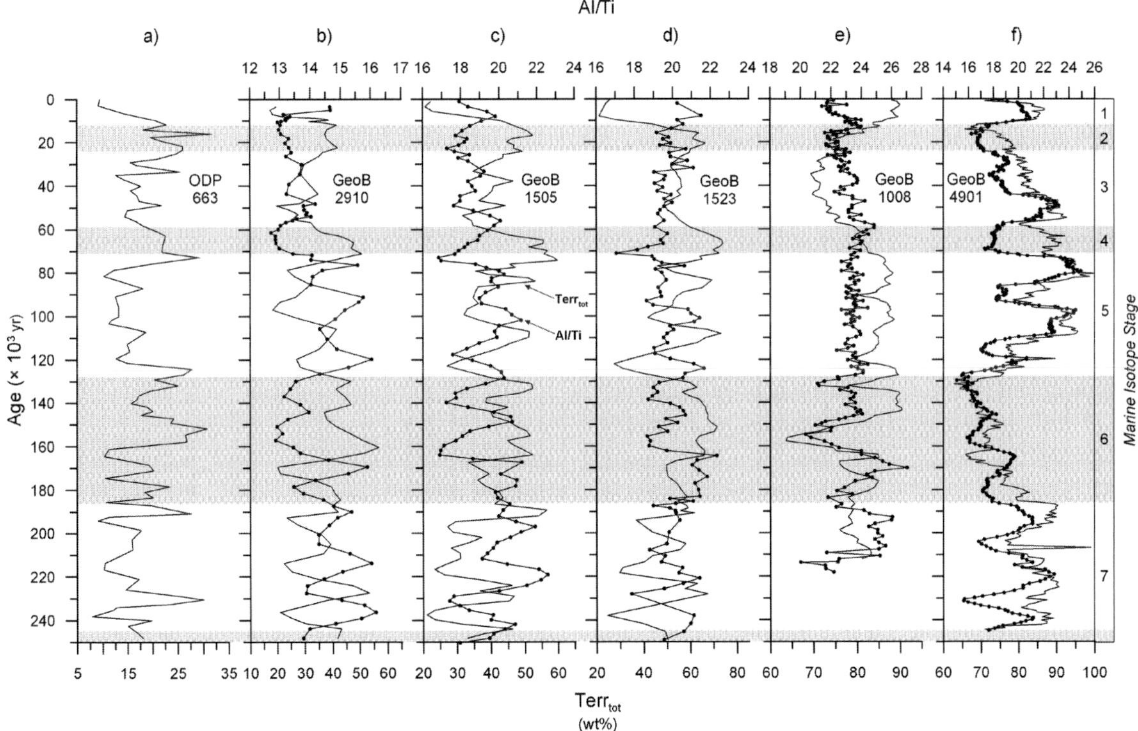

Fig. 4. Records of Terr$_{tot}$ and the Al/Ti ratio (For information on the cores see caption of Fig. 2)

Northern Hemisphere insolation, the amplitudes are governed by variations in the global ice volume.

Variations in the Amazon River Input (GeoB1523 and 1505)

The terrigenous fraction on the Ceará Rise (GeoB1523) consists almost exclusively of river-suspended matter (RSM) that is supplied by the Amazon River (Zabel et al. 1999; Rühlemann et al. 2001). Both, the total terrigenous input ($Terr_{tot}AR$) to the Ceará Rise and its composition as indicated by the Al/Ti ratio are dominated by precessional periodicity (Figs. 2d and 3d). In comparison, the clay mineralogy (illite/smectite ratio) in core GeoB1523 shows a inconsistent and partly divergent picture (Fig. 5), which is probably due to the artifacts on clay mineral assemblages previously discussed (cf. above). However, despite the last glacial amplitudes are most pronounced during interglacials as is particularly documented in the flux records. Therefore, Rühlemann et al. (2001) attributed changes in the supply of terrigenous matter to global sea level variations. Accordingly, the flat topography of the shelf off the Amazon estuary causes the main effect of shelf erosion and direct transport of RSM to the Ceará Rise via canyons to occur when sea-level oscillations only vary between present-day level and such being 40-50 m lower than today. This implies that the flux of Amazon suspensates to the tropical Atlantic Ocean indicates a major influence of high-latitudinal forcing. On the other hand, changes in the surface circulation patterns are described as being additionally responsible for variations in the terrigenous supply to the Ceará Rise and further to the east (Zabel et al. 1999; Rühlemann et al. 2001). Zabel et al. (1999) have therefore deduced an intensification of the North Equatorial Counter Current (NECC) (Fig. 1) as a result of strengthened SE trade winds which explain the relatively high Ti AR and corresponding low Al/Ti ratios in core GeoB1505 at the western flank of the Mid-Atlantic Ridge during glacials and cold interstadials (Figs. 2c and 3c). Significant leads of the terrigenous records against the precessional signal in oxygen isotope records as documented by phase shifts in cores GeoB1523 ($Terr_{tot}AR$: 1.4 kyr, Al/Ti: 1.9 kyr)

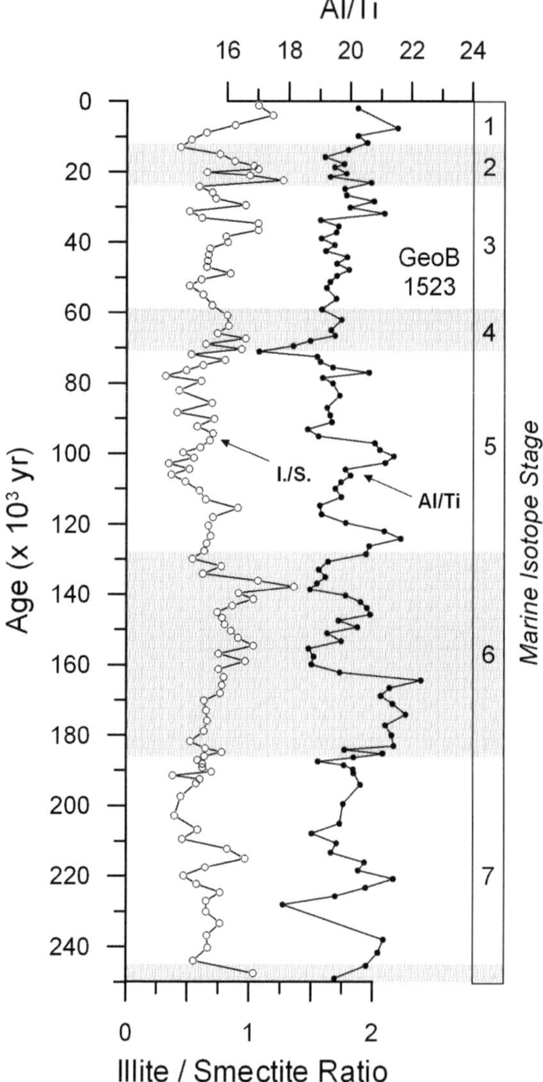

Fig. 5. Comparison between records of the illite/smectite ratio and the Al/Ti ratio in sediment core GeoB1523. Clay data were taken from Rühlemann et al. 2001.

and GeoB1505 (Al/Ti: 3.0 kyr) may additionally support previous studies which demonstrated that sequences of the Southern Hemisphere lead the northern circulation by 2-3 kyr (e.g. Hays et al. 1976; Imbrie et al. 1989; Zabel et al. 1999). However, interpreting the Al/Ti ratio as an indicator of the climate conditions in the source area (see below), terrigenous records would also support the assumption of Harris and Mix (1999), who have argued that Amazonian aridity is to be assigned to the chain of events leading to ice ages, rather than being a response to glacier oscillations.

Variations in the Zaire River Output (GeoB1008)

RSM output by the Zaire River was discussed by Schneider et al. (1997) and Gingele et al. (1998). Element ratios like K/Al and mineralogical investigations like the kaolinite/feldspar ratio, both to be interpreted as geochemical weathering indices, indicate to the expected larger input of kaolinite as compared with feldspar during warm and humid periods (cf. van der Gaast and Jansen 1984; Bonifay and Giresse 1992; Gingele 1996). Correspondingly, variations in the composition of the lithogenic fraction and its constituents and hence in the Central African climate are again dominated by orbital precession (Schneider et al. 1997). At least during the last 130 kyr, this signal in the 1/23 kyr^{-1} frequency band is less pronounced in the Al/Ti ratio (Fig. 3e) and completely absent in the Ti AR record as well as in the collective parameters Terr$_{tot}$ AR (Fig. 2e) and Terr$_{tot}$ (Fig. 4e). While potential problems with the total terrigenous fraction were already discussed before as a convincing indicator for climate conditions, the lesser significance of single element proxies could be connected with the composition of the soils in the source area. However, the strong precessional cyclicity in the terrestrial signals was described as having been more or less in phase with changes in Southern Hemisphere SST and boreal summer insolation in low latitudes, leading changes in Northern Hemisphere SST and continental ice volume (Schneider et al. 1997). Accordingly, as with the Amazon input, climate variability in the Zaire catchment area seems to be governed by the SE trade wind system which is known to lead the circulation in the Northern Hemisphere (e.g. Zabel et al. 1999). This scenario is consistent with results of clay mineralogy studies that were carried out on sediments from core GeoB1401 (6°56'S, 9°00'E; Gingele et al. 1998; Fig. 6). Smectite crystallinity and illite chemistry revealed that the freshwater discharge of the Zaire River is fostered by an intensified SW African monsoon system during times of maximal insolation in the Northern Hemisphere.

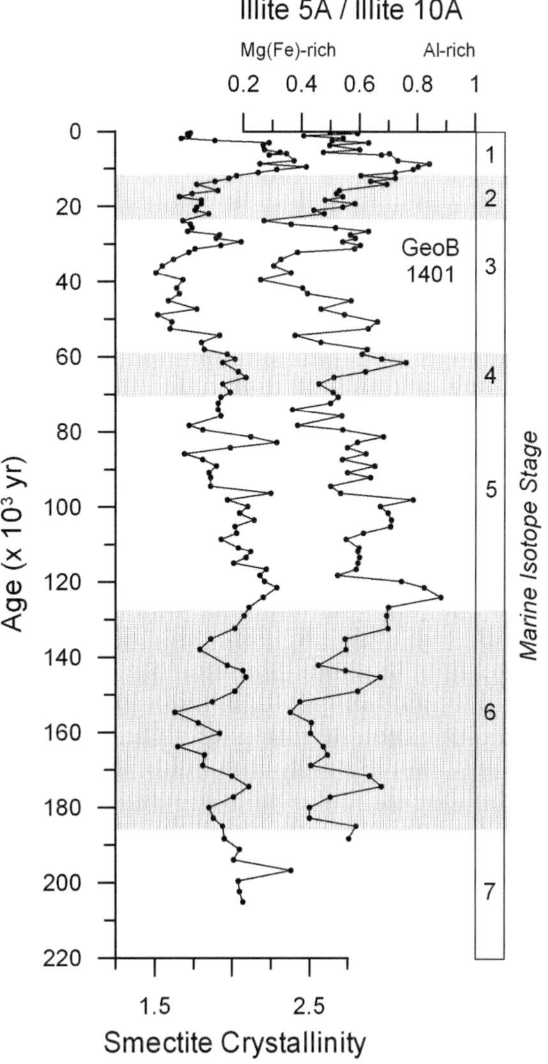

Fig. 6. Records of the smectite crystallinity and of the chemical character of illite (Fe, Mg- or Al-rich) in core GeoB1401 recovered from the Zaire River deep-sea fan at 6°56'S, 9°00'E. The clay mineralogy was analysed by Gingele et al. (1998). Whereas well crystalled smectite originates mainly from the continental shelf, poorly pedogenic crystalls are rather supplied to this region by RSM of the Zaire River. The chemistry of illites reflect the intensity of chemical weathering in the source area which is recently high (Al-rich) in the drainage area. Mg, Fe-rich illites can be traced to the eolian pathway from SW-African deserts. Therefore both parameters indicate to increased runoff during interglacials 1, 5 and 7 and record a high-latitude forcing of river runoff at 100 kyr periodicities reflecting glacial aridity.

Variations in the Niger River Input (GeoB4901)

Time series in the composition of the terrigenous fraction at the Niger fan (GeoB 4901) also showed considerable oscillations (Figs. 2f, 3f, 4f), but the connection to global climate conditions is much more pronounced than for Zaire sediments (Zabel et al. 2001). Consequently, the wrapping curve of the Al/Ti record, indicating relative de-creases in Al concentrations during cold inter-stadials of both penultimate interglacials and glacials, already reveal the typical saw- tooth pattern (Fig. 3f). These oscillations additionally denote that the heavy minerals (the Ti-carrier) are relatively reduced in sediments from warm periods. Based on the fact that the formation of kaolinite (Al) is favored by intensive chemical weathering in a humid climate, Al/Ti ratios corroborate the frequently documented response of African aridity to changes in glacial boundary conditionsdisplaying dry conditions during cold periods. However, based on spectral estimates for harmonic variances, Ti AR and Ti/Al time series are strongly controlled by precession-modulated insolation. In this clarity nothing similar can be determined for the bulk parameters $Terr_{tot}$ AR and $Terr_{tot}$ (Figs. 2f and 4f).

To study the link between orbital climate forcing and its documentation by terrestrial proxies Zabel et al. (2001) have applied cross-spectral analysis to the marine and terrigenous records from the Niger fan (GeoB 4901). Figure 7 shows the result of the relation between Ti AR, Ti/Al, maximum boreal summer insolation at 20°N as potentially reflecting the variability of the monsoon intensity in central equatorial Africa (e.g. Pokras and Mix 1985; Dupont and Leroy 1995), and the precessional signal in the oxygen isotope record of the benthic foraminifera Cibicidoides wuellerstorfi as representing high-latitude climate change. The latter was extracted from the raw data by digital filtering. Zabel et al. (2001) discovered a significant lag of fluctuations in the terrigenous fraction against variations in low-latitude solar radiation (Fig. 7c and e). Although delays are not constant over time, this observation implies that a direct response of terrigenous supply and composition to the monsoon cycle does not exist. Terrestrial signals fluctuate closer in tune with the 23-kyr orbital portion of global ice volume cyclicity. This concerns both chronology and magnitude of the amplitudes (Fig. 7d and f). Considering the well-known time lag of the global isotope response to insolation forcing (Imbrie et al 1984), sediments from the Niger fan confirm the frequently postulated considerable influence of high-latitude forcing on tropical African climate. Additional strong evidence for this concept is given by variations in the phase shifts (Fig. 7g). They show a clear modulation of the precessional period by the 100-kyr cycle which dominates high-latitude climate change and associated processes like the thermohaline circulation of the ocean. In general, this overlay of frequencies may be applicable to the tropical South American climate, where a dominant 100-kyr cyclicity in compositional variations of Amazon clay minerals has been documented (Rühlemann et al. 2001).

In summary, the results presented underline that the terrigenous input to the tropical Atlantic and its composition generally fluctuates during late Quaternary on the 23-kyr orbital cycle. A lot of studies have given evidence that the same applies to the African monsoon system, which mainly controls the eastern supply of terrigenous matter to the tropical Atlantic (cf. above). Consequently, the strong influence of the high-latitude insolation component on the terrigenous fraction of marine sediments at low latitudes reveals that precessional climate variations in tropical Africa and South America of late Quaternary were modulated by mechanisms operating at the subpolar North Atlantic, which are dominated by the 100-kyr eccentricity cycle. Time series going back further substantiate that the orbital constellation and associated interferences are subjected to nonlinear temporal variations (e.g. deMenocal 1995). This may indicate that the question, whether tropical climate were mainly influenced by high or low latitude processes, could only be answered for limited intervals.

The Period from the Last Glaciation to the Holocene

To consider the last glacial-interglacial change in more detail, we examined the upper sections of nine sediment cores recovered from the tropical

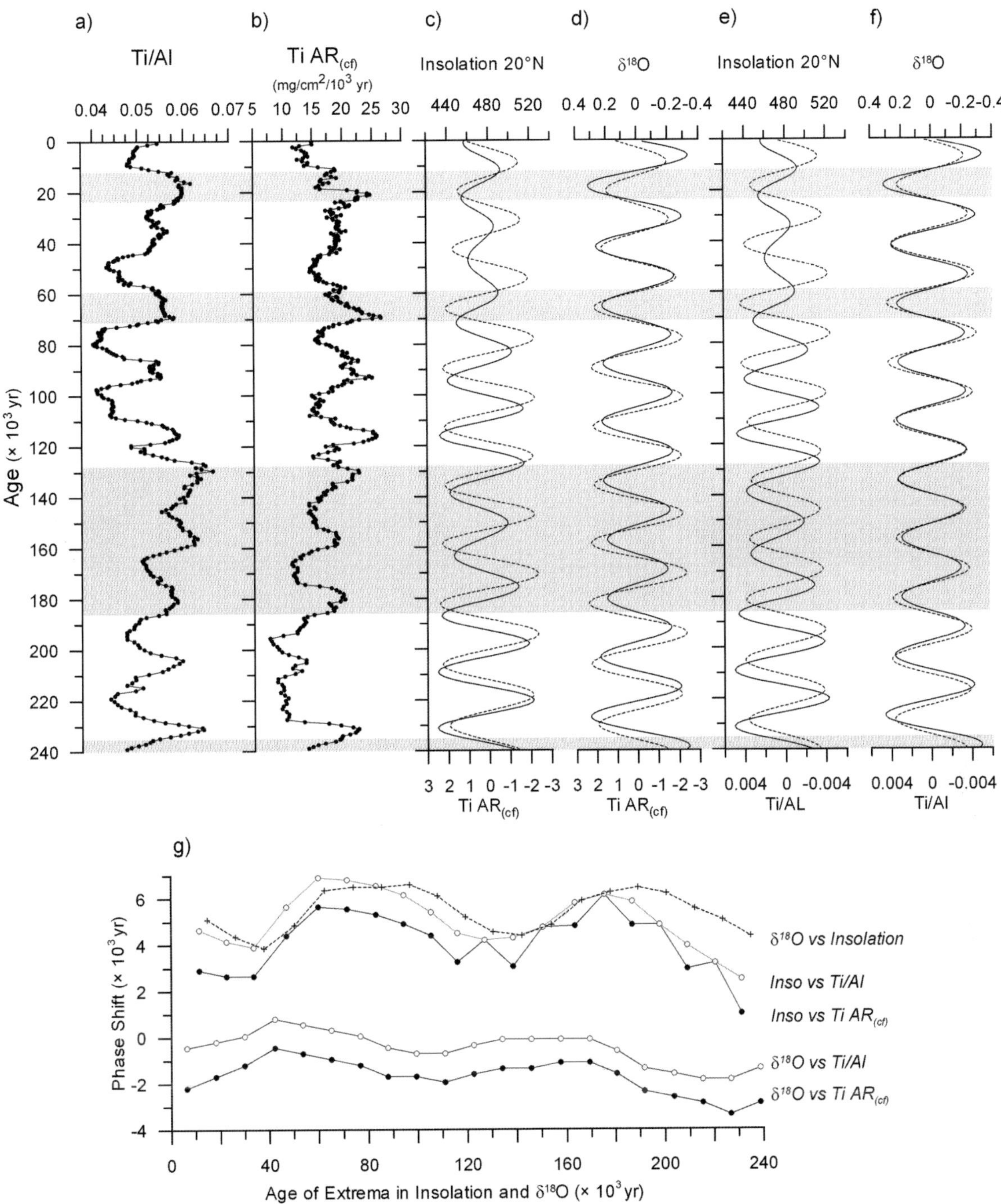

Fig. 7. Results from cross-spectral analysis on core GeoB4901 records. Comparison of the δ¹⁸O record of the benthic foraminifera *C. wuellerstorfi*, the insolation cycle for 20°N (W/m²), Ti/Al, and Ti AR at a 23-kyr periodicity. Dashed lines trace changes in the elemental variables. Phase and magnitude of amplitudes indicate that variations in the terrigenous input show a better match with changes in the Northern Hemisphere ice volume than with low-latitude solar radiation. Variations in the phase shifts reflect the overlay of the 100-kyr periodicity. (after Zabel et al. 2001)

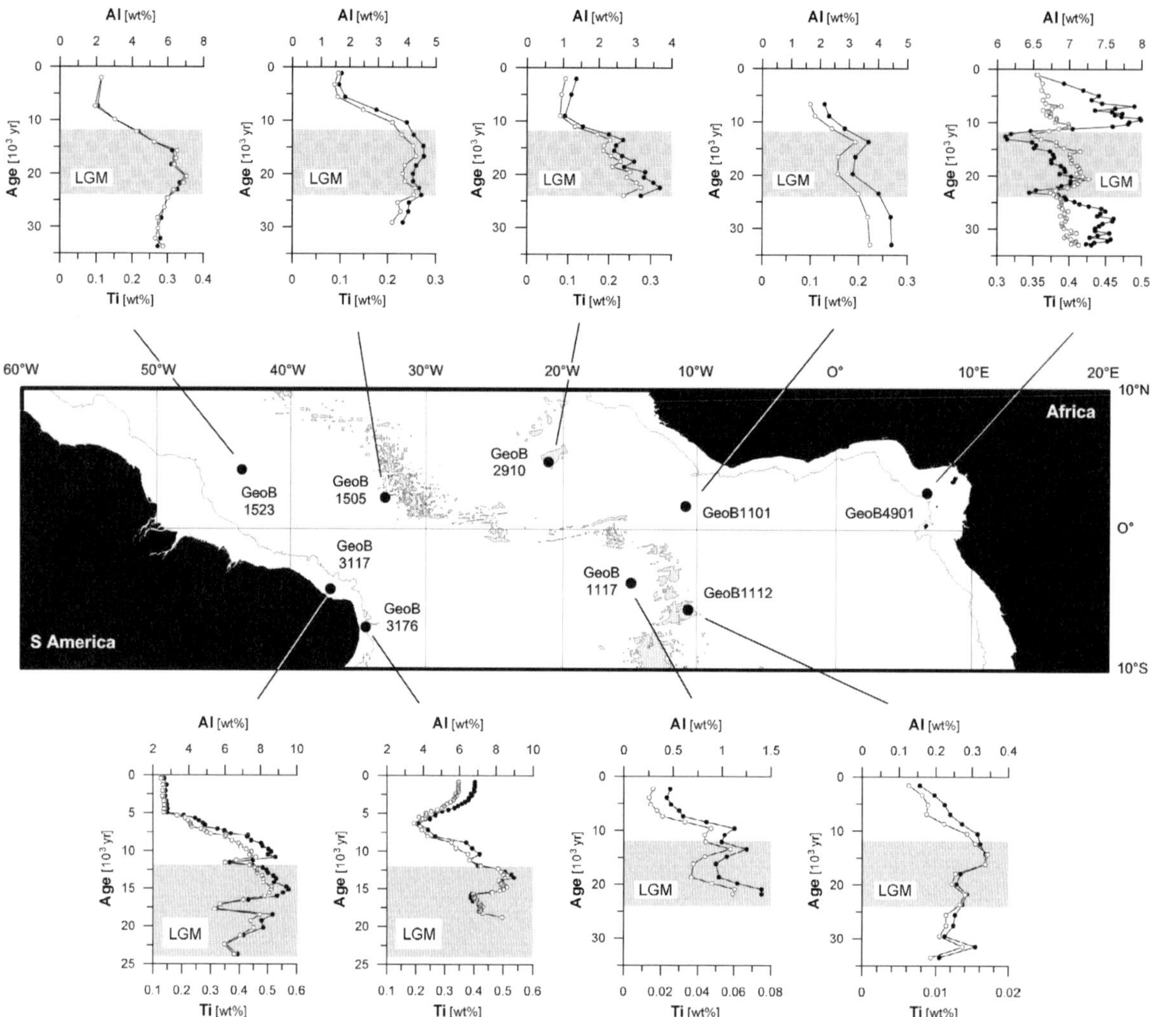

Fig. 8. Records of Al and Ti concentrations in 9 cores analyzed for this study. Al - closed cycles, Ti - open cycles. The terrigenous components in these sediments originated from different source areas adjacent to the equatorial Atlantic (Tab.1) The period of last glaciation is hatched (LGM – last glacial maximum). Age models are based on $\delta^{18}O$-stratigraphy from monospecific samples, transfer of accelerator mass spectrometry (AMS), radio carbon dating of monospecific samples from adjacent cores by correlation using records of several parameters, or new AMS data (Tab. 1).

Atlantic. Figure 8 shows the individual locations. All sediments are well dated by monospecific $\delta^{18}O$ records, a close correlation with adjacent ^{14}C-dated cores or by radio carbon measurements (Tab. 1). With the exception of the Niger sediments (GeoB 4901), all concentration profiles show significantly increased contents of the terrigenous source elements Al and Ti during last glaciation, as compared with the Holocene. Based on isotope stage averages, the enrichment factors range from 1.3 to 2.2 (abs. 1.2-3.6). Highest concentrations were found

at near-shore locations (GeoB 3117, 3176, 4901), where the terrigenous fraction is dominated by fluvial input (Arz et al. 1999; Zabel et al. 2001).

Calculation of the element AR agrees with this observation in nearly all cores (Fig. 9). Distinctive maximum flux rates can be noted in most records during Last Glacial Maximum (LGM) and close to deglaciation. This is consistent with the concept of predominantly arid glacials (cf. above). High AR can be associated with some interacting processes. Depending on the source area and transport

	Longitude	Latitude	Water depth [m]	Stratigraphy	Main Source Area
Core					
GeoB1101-5	01°40'N	10°59'W	4588	δ^{18}O - *C. wuellerstorfi* [1]	Sahara [6]
GeoB1112-4	05°47'S	10°45'W	3125	δ^{18}O - *C. wuellerstorfi* [1]	African Deserts
GeoB1117-2	03°49'S	14°54'W	3984	δ^{18}O - *C. wuellerstorfi* [1]	African Deserts
GeoB1505-2	02°16'N	33°01'W	3706	δ^{18}O - *C. wuellerstorfi* [2]	Amazon Basin [2]
GeoB1523-1	03°50'N	41°37'W	3292	δ^{18}O - *G. sacculifer* [3]	Amazon Basin [2,3]
GeoB2910-1	04°51'N	21°03'W	2703	δ^{18}O - *G. sacculifer* * [2]	Sahel Zone [2,6]
GeoB3117-1	04°11'S	37°08'W	930	AMS ^{14}C ** [4]	Local Brazilian Rivers [4]
GeoB3176-1	07°01'S	34°27'W	1385	AMS ^{14}C ** [4]	Local Brazilian Rivers [4]
GeoB4901-8	02°41'N	06°43'E	2184	AMS ^{14}C [5]	Niger Catchment Area [7]

[1] Bickert and Wefer 1996; [2] Zabel et al. 1999; [3] Rühlemann et al. 1996, 2001; [4] Arz et al. 1999; [5] Zabel unpubl. [6] Frederichs et al. 1999; [7] Zabel et al. 2001

* via CaCO$_3$ correlation; ** via δ^{18}O (*G. sacculifer* and *G. ruber*), XRF data, and color correlation

Table 1. Core locations, depths, methods for stratigraphy, and main source areas of the terrigenous fraction.

mechanisms they indicate (a) intense deflation of dust favored by strong winds and a reduced vegetation coverage due to low precipitation rates, (b) low retention capacity of soils under fluvial erosion which also depends mainly on the extent of root penetration, (c) an intensification of ocean currents when these serve as carriers of the terrigenous fraction. Absolute enhancement factors of flux rates during oxygen isotope stage 2 in comparison to the Holocene range from 1.5 to 13.5, where maximum differences are documented for the African dust input at station GeoB 2910 and for the local RSM input at station GeoB 3117. However, similar flux patterns were also reported for Terr$_{tot}$ in equatorial Atlantic sediments by Ruddiman (1997). But, his theory according to which stronger NE trade winds are more important than glacial hyperaridity for the dust influx to the tropical Atlantic can neither be corroborated nor disproved by our data set. However, special interesting features are the differences in the occurrence of the maximum ARvalues for dust. While the northernmost coring site GeoB 2910 is located under the center of the African dust plume, this is definitely not the case for the two cores south of the equator (GeoB 1112 and 1117; cf. Fig. 1). But, similar records and amplitudes in flux rates still indicate that GeoB 1101 and 1117

may be supplied by the same trajectory. In contrast to GeoB 2910, the other three records show significant variations in the dust input which took place during the last glaciation. Although distinct maxima at the transition to the Holocene and the lead of the prominent peak in core GeoB 1112 could be interpreted as caused by fluctuations of position of the ITCZ, the geochemical composition of the terrigenous fraction is rather supportive of variations in the zonal wind strength (cf. Sarnthein et al 1981 and references given above). So, maximum AR are parallel to minima in Al/Ti (Fig. 10) which indicate strongest atmospheric circulation. However, the interstage extrema substantiate that variations in the corresponding supply of terrestrial material to the low-latitude Atlantic, and more in general terms climate variations in the adjacent source areas, cannot sufficiently be explained by a simple glacial-interglacial contrast (cf. above). For example, on core GeoB 4901 Zabel et al. (2001) was able to demonstrate that the terrigenous input from the Niger River is mainly governed by the intensity of precipitation and the associated vegetation coverage. Nevertheless, interpretation of AR records with regard to the terrestrial input seems difficult. For example, AR maxima in the near-shore deposits GeoB 3117, 3176, and 4901 appear similar to the

Zabel et al.

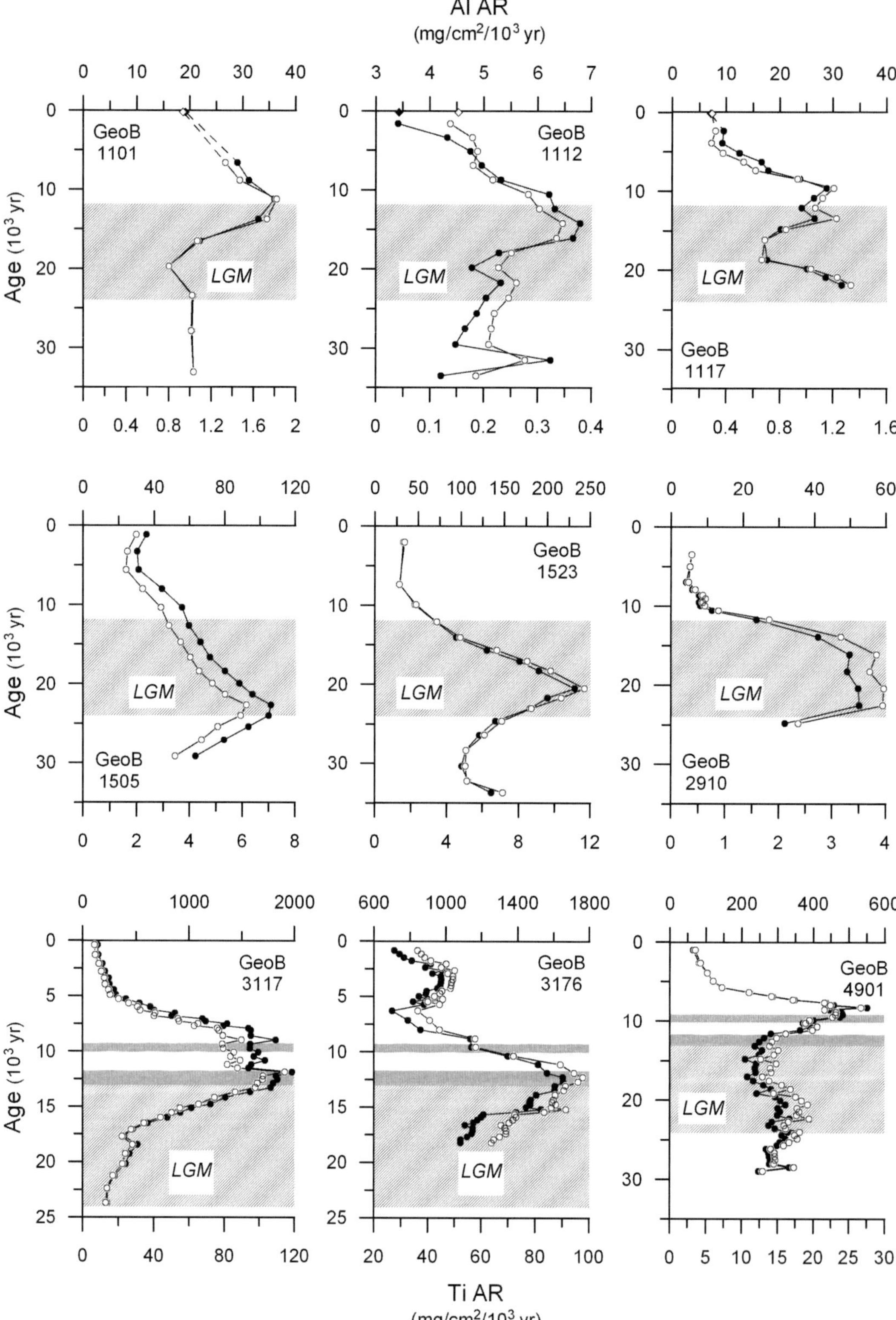

Fig. 9. Records of Al and Ti accumulation rates (AR). For signs and symbols see Fig. 8. Bars mark North Atlantic meltwater events mwpIA and mwpIB (Fairbanks 1989). Dry bulk densities for the estimation of AR were calculated via weight/volume by PJ Müller (unpubl. data).

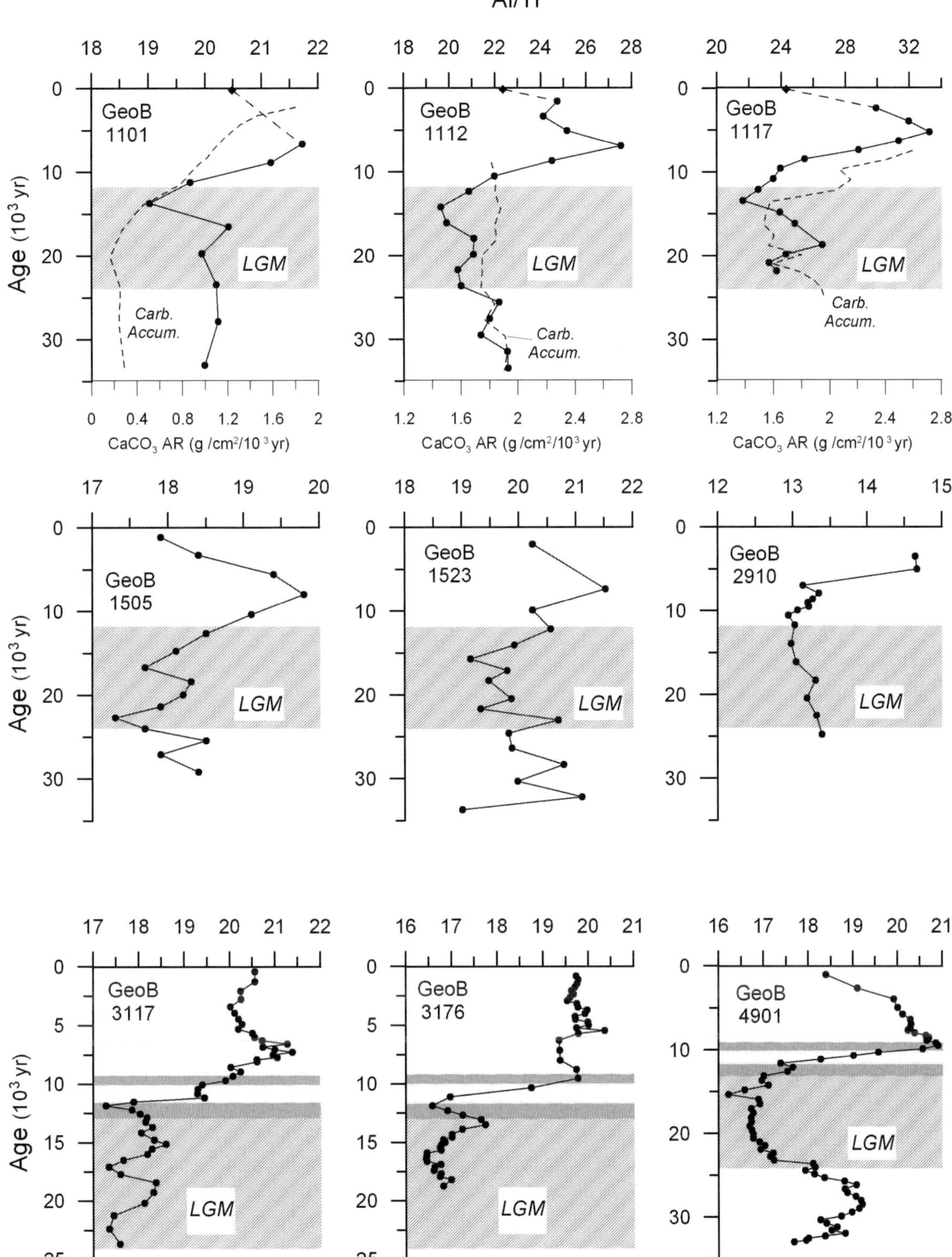

Fig. 10. Records of Al/Ti ratios. For signs and symbols see Figs. 9 and 10. Carbonate contents were measured by Bickert and Wefer (1996).

prominent North Atlantic melt-water discharge events recorded by Fairbanks (1989) on coral reefs off Barbados. This permits the assumption that temporally limited inputs due to shelf erosion during periods of rapid sea-level rise have affected these records significantly (Arz et al. 1999). Variations in the composition of the terrigenous fraction support the occurrence of this effect at least in both cores off NE Brazil (Fig. 10). While distinct decreases in the Al/Ti ratio during the first meltwater peak (mwp-IA) at around 12 kyr BP could be caused by redeposition of older shelf sediments to the continental slope, the second steep sea-level rise (mwp-IB) is documented by relatively constant ratios which would point to the focusing of material with similar geochemical signatures. An alternative interpretation of this period in Al/Ti records off Brazil is given by pollen analysis, indicating exceptionally strong rainfall between 15.5 and 11.8 kyr BP (Behling et al. 2000), which would correspond to the nearly coinciding local Al/Ti maxima (see below).

Despite these possible artifacts due to sea-level change, records of element ratios are less sensitive to age models and may thus elucidate differences in climate conditions easier than AR. Based on the well established assumption that the Al/Ti signature of dust is an indicator for the wind strength (e.g. Zabel et al. 1999), representative records of African dust input (GeoB 1101, 1112, 1117, 2910) indicate stronger atmospheric circulation during last glacial by significantly lower ratios. Al/Ti ratios at the Sierra Leone Rise (GeoB 2910) resemble the composition of African dust samples (Zabel et al. 1999 and references therein), whereas carbonate-rich off-shore sediments (up to 95 wt% $CaCO_3$) reveal relative enrichments of Al. Decoupling with carbonate AR records indicates, however, that the potential influence of Al-scavenging (Murray et al. 1993) may be of minor importance. The concept of predominant aridity during glacials is also supported by sediments in which RSM dominates (GeoB 1505, 1523, 3117, 3176, 4901). According to Zabel et al. (2001) the Al/Ti ratio of RSM can be interpreted as a chemical weathering index, representing the product of the interplay of precipitation rate, vegetation coverage, type and intensity of weathering, and erosion capacity. Due to preferentially chemical weathering during warm

and humid climates, when percolated water permits hydrolytical processes, somewhat more Al (kaolinite) than Ti (heavy minerals) is eroded, a fact which is expressed by high Al/Ti ratios during interglacial periods. In contrast to the eolian pathway, transport energy might be of subordinate importance for the composition of RSM, when sediment discharge and runoff are inversely correlated due to the retention effect of vegetation. However, following our argumentation it becomes clear why the terrigenous AR records and Al/Ti ratios presented here show roughly inverse patterns. The slight shifts between maxima and corresponding minima may be mainly the result of the particular response time in plant growth and decay to changes in rainfall intensity.

In summary, all records provide strong evidence that generally dry conditions prevailed during last glacial. Additionally, they reveal that significant changes in terrestrial flux rates often occurred on relatively short time scales. Depending on depth resolution and age model, some of these changes lasted only a few centuries. This agrees with the general opinion that fluctuation in orbital insolation cannot be the only driving force for the climate variability in the tropics.

Conclusions

In accordance with most outside studies, the presented results concerning the variations of terrigenous signatures in sediments from the tropical Atlantic reveal a strong precessional cyclicity in general, which is indicative of a close link between continental aridity fluctuations and evaporation in the ocean. But apart from this dependence on changes in orbitally produced solar radiation, paleo-climate records also clearly indicate that oscillations in high latitudes are necessary to explain the major features of late Quaternary climate change at centurial, millennial or longer time scales. Beyond that, the highly complex interplay between high- and low-latitude forcing is subject to regional differences. In particular, land-ocean interactions may be of crucial importance for this observation. These are mainly controlled by the thermocline circulation which could be one explanation for the predominately latitudinal differences. The additional sensitivity of the monsoon system to changes in the vegetation cover (e.g. Foley et al. 1994; Ganopolski et al. 1998; Claussen et al. 1999) implies that the timing of climate change is also influenced by the response of vegetation to significant changes in precipitation. However, the close connection between SSTs, temperature and pressure gradients, and the orbital insolation cycle may explain why the precessional frequency of terrigenous input variations and material compositions revealed a modulation by ice volume fluctuations in the Northern Hemisphere. At least during late Quaternary, the high-latitudinal 100-kyr periodicity determines the amplitudes of fluctuations in the terrigenous climate proxies, and therefore seems to have a dominant influence on the sum of all local parameters controlling climate change each probably working on different time scales.

Implications and Perspectives for Future Research

Owing to the effectiveness of high-resolution studies on marine sediments, future approaches should rely on a much closer association between terrestrial and marine investigations. These should not only include conventional comparisons between time series from lacustrine and marine deposits, but also interdisciplinary projects on such specific terrigenous source materials like soils and dust particles. At least today's interpretations of most terrigenous climate proxies which are based on causal connections, may be well-founded, but they are not really understood in detail. This is definitively the case in regard to regional differences and for processes of particle transport. Studies concentrating on the geochemical and mineralogical alteration of the particle load from their primary signature to the imbedding at the sea floor are extremely rare. This also includes investigations on the influences of lateral particle drift in the water column and redistribution processes near to the sea floor. In this context, an important but only sparsely discussed question concerns the preservation potential of distinct short-lasting peaks in marine sedi-ments related to benthic bioturbation.

Due to the highly complex connections among all the processes involved, further improvements of our understanding of the climate control from the results of coupled climatic models are to be expected. Last but not least, little is known about the anthropogenic disturbances taking an effect on natural climate variations.

Acknowledgements

At first we would like to thank all the domestic and foreign colleagues, who contributed to our investigations in the context of the DFG-Sonderforschungsbereich 261. Thank you all for your support and the very good co-operation during the last years. Additionally we are also indebted to Rüdiger Stein for constrictive comments that helped to improve the manuscript. This work was funded by the Deutsche Forschungsgemeinschaft (DFG). Data are available under www.pangaea.de/Projects/SFB261.

References

Arz HW, Pätzold J, Wefer G (1998) Correlated millennial-scale changes in surface hydrography and ter-rigenous sediment yield inferred from last-glacial marine deposits off northeastern Brazil. Quat Res 50: 157-166

Arz HW, Pätzold J, Wefer G (1999) Climate changes during the last deglaciation recorded in sediment cores from the northeastern Brazilian continental margin. Geo-Mar Lett 19: 209-218

Arz HW, Gerhardt S, Pätzold J, Röhl U (2001) Millennial-scale changes of surface- and deep-water flow in the western tropical Atlantic linked to Northern Hemisphere high-latitude climate during the Holocene. Geology 29: 239-242

Balsam WL, Otto-Bliesner BL, Deaton BC (1995) Modern and last glacial maximum eolian sedimentation patterns in the Atlantic Ocean interpreted from sediment iron oxid content. Paleoceanography 10: 493-507

Behling H, Arz HW, Pätzold J, Wefer G (2000) Late Quaternary vegetational and climate dynamics in northeastern Brazil, inferences from marine core GeoB 3104-1. Quat Sci Rev 19: 981-994

Bickert T, Wefer G (1996) Late Quaternary deep water circulation in the South Atlantic: Reconstruction from carbonate dissolution and benthic stable isotopes. In: Wefer G, Berger WH, Siedler G, Webb DJ (eds) The South Atlantic – Present and Past Circulation. Springer, Berlin, pp 599-620

Biscaye PE (1965) Mineralogy and sedimentation of recent deep-sea clay in the Atlantic ocean and adjacent seas and oceans. GSA Bull 76: 803-832

Bloemendal J, deMenocal PB (1989) Evidence for a change in the periodicity of tropical climate cycles at 2.4 Myr drom whole-core magnetic susceptibility measurements. Nature 342: 897-900

Bond G, Broecker W, Johnson S, McManus J, Labeyrie L, Jouzel J, Bonani G (1993) Correlations between climate records from North Atlantic sediments and Greenland ice. Nature 365: 143-147

Bonifay D, Giresse P (1992) Middle to late Quaternary sediment flux and post-depositional processes between the continental slope off Gabon and the Mid-Guinean margin. Mar Geol 106: 107-129

Boyl EA (1983) Chemical accumulation variations under the Peru Current during the last 130,000 years. J Geophys Res 88: 7667-7680

Bozzano G, Kuhlmann H, Alonso B (2002) Storminess control over African dust input to the Moroccan Atlantic margin (NW Africa) at the time of maxima boreal summer insolation: A record of the last 220 kyr. Palaeogeogr Palaeoclimat Palaeoecol 183: 155-168

Carlson TN, Prospero JM (1972) The large-scale movement of Saharan air outbreaks over the northern Equatorial Atlantic. J Appl Meterology 11:283-297

Chester R, Elderfield H, Griffin JJ, Johnson LR, Padgham RC (1972) Eolian dust along the eastern margins of the Atlantic Ocean. Mar Geol 13: 91-105

Chiapello I, Bergametti G, Chatenet B, Bousquet P, Dulac F, Santos Suarez E (1997) Origins of African dust transported over the north-eastern tropicl Atlantic. J Geophys Res 102: 13,701-13,709

Claussen M, Kubatzki C, Brovkin V, Ganopolski A, Hoelzmann P, Pacchur H-J (1999) Simulation of an abrupt change in Saharan vegetation in the mid-Holocene. Geophys Res Lett 26: 2037-2040

Damuth JE, Fairbridge RW (1970) Equatorial Atlantic deep-sea arkosic sands and ice-age aridity in tropical South America. GSA Bull 81: 189-206

Dansgaard W, Johnsen SJ, Claussen HB, Dahl-Jensen D, Gundestrup NS, Hammer CU, Hvidberg CS, Steffensen JP, Sveinbjörnsdottir AE, Jouzel J, Bond G (1993) Evidence for general instability of past climate from a 250-kyr ice-core record. Nature 364: 218-220

Delany AC, Parkin DW, Griffin JJ, Goldberg ED, Reimann BEF (1967) Airborne dust collected at Barbados. Geochim Cosmochim Acta 31: 885-909

deMenocal PB, Ruddiman WF, Pokras EM (1993) Influence of high- and low-latitude processes on African terrestrial climate: Pleistocene eolian records from equatorial Atlantic Ocean drilling program site 663. Paleocenography 8: 209-242

deMenocal PB, Rind D (1993) Sensitivity of Asian and African climate to variations in seasonal insolation, glacial ice cover, sea surface temperature, and Asian orography. J Geophys Res 98: 7265-7287

deMenocal PB (1995) Plio-Pleistocene African climate. Science 270: 53-59

deMenocal, P, Ortiz J, Guilderson T, Sarnthein M (2000a) Coherent High- and Low-Latitude climate variability during the Holocene warm period. Science 288: 2198-2202

deMenocal P, Ortiz J, Guilderson T, Adkins J, Sarnthein M, Baker L, Yarusinsky M (2000b) abrupt onset and termination of the African Humid Period: Rapid climate responses to gradual insolation forcing. Quat Sci Rev 19: 347-361

Diekmann B, Kuhn G, Mackensen A, Petschick R, Fütterer DK, Gersonde R, Rühlemann C, Niebler HS (1999) Kaolinite and chlorite as tracers of modern and late Quaternary deep water circulation in the South Atlantic and the adjoining Southern Ocean. In: Fischer G, Wefer G (eds) Use of Proxies in Paleoceanography – Examples from the South Atlantic. Springer, Berlin, pp 285-313

Dymond J, Collier R, McManus J, Honjo S, Manganini S (1997) Can the aluminium and titanium contents of the ocean sediments be used to determine the paleoproductivity of the oceans. Paleoceanography 12: 586-593

Fairbanks RG (1989) A 17,000-year glacio-eustatic sea level record: Influence of glacial melting rates on the Younger Dryas event and deep-ocean circulation. Nature 342: 637-642

Fairbridge RW (1964) African ice-age aridity. In: Nairn EM (ed) Problems in Paleoclimatology. J Wiley & Sons, New York, pp 356-363

Foley JA (1994) The sensitivity of the terrestrial biosphere to climatic change: A simulation of the middle Holocene. Glob Biogeochem Cycles 8: 505-525

Frederichs T, Bleil U, Däumler K, von Dobeneck T, Schmidt A (1999) The magnetic view on the marine paleoenvironment: Parameters, techniques, and potentials of rock magnetic studies as a key to paleoclimatic and paleoceanographic changes. In: Fischer G, Wefer G (eds) Use of Proxies in Paleoceanography: Examples from the South Atlantic. Springer, Berlin, pp 575-599

Gaillardet J, Dupré B, Allègre CJ (1997) Chemical and physical denudation in the Amazon river basin. Chem Geol 142: 141-173

Gaillardet J, Dupré B, Allègre CJ (1999) Geochemistry of large river suspended sediments: Silicate chemical weathering or recycling tracers? Geochim Cosmo-chim Acta 63: 4037-4051

Ganopolski A, Kubatzki C, Claussen M, Brovkin V, Petoukhov V (1998) The influence of vegetation-atmosphere-ocean interaction on the climate during the Mid-Holocene. Science 280: 1916-1919

Gasse F, Lédée V, Massault M, Fontes J-C (1989) Water-level fluctiations of lake Tanganyika in phase with oceanic changes during the last glaciation and deglaciation. Nature 342: 57-59

Gasse F (2000) Hydrological changes in the African tropics since the Last Glacial Maximum. Quat Sci Rev 19: 189-211

Gingele FX (1996) Holocene climatic optimum in southwest Africa – Evidence from the marine clay mineral record. Palaeogeogr Palaeoclimatol Palaeoecol 122: 77-87

Gingele FX, Müller PM, Schneider RR (1998) Orbital forcing of freshwater input in the Zaire Fan area – clay mineral evidence from the last 200 kyr. Palaeogeogr Palaeoclimat Palaeoecol 138: 17-26

Greenland Ice-core Project (GRIP) Members (1993) Climate instability durino the last interglacial period recorded in the GRIP ice core. Nature 364: 203-207

Grousset FE, Parra M, Bory A, Martinez P, Bertrand P, Shimmield G, Ellam RM (1998) Saharan wind regimes traced by the Sr-Nd isotopic composition of subtropical Atlantic sediments: Last Glacial Maximum vs today. Quat Sci Rev 17: 395-409

Guo Z, Petit-Maire N, Kröpelin S (2000) Holocene non-orbital climatic events in present-day arid areas of northern Africa and China. Glob Planet Change 26: 97-103

Haberle SG, Maslin MA (1999) Late Quaternary vegetation and climate change in the Amazon Bsin on a 50,000 year pollen record from the Amazon Fan, ODP site 932. Quat Res 51: 27-38

Harris SE, Mix AC (1999) Pleistocene precipitation balance in the Amazon Basin recorded in deep-sea sediments. Quat Res 51: 14-26

Haug GH, Hughen KA, Sigman DM, Peterson LC, Röhl U (2001) Southward migration of the Intertropical Convergence Zone through the Holocene. Science 293: 1304-1308

Hays JD, Imbrie J, Shackleton NJ (1976) Variations in the Earth's orbit: Pacemaker of the ice ages. Science 194: 1121-1132

Hughen KA, Overpeck JT, Peterson LC, Trumbore S (1996) Rapid climate changes in the tropical Atlantic region during the last deglaciation. Nature 380: 51-54

Imbrie J, Hays JD, Martinson DG, McIntyre A, Mix AC, Morley JJ, Pisias NG, Prell WL, Shackleton NJ (1984) The orbital theory of pleistocene climate: Support from a revised chronology of the marine $\delta^{18}O$ record. In: Berger LA, Imbrie J, Hays JD, Kukla J, Saltzman J (eds) Milankovich and Climate. Part I, Reidel, Dordrecht, pp 269-305

Imbrie J, McIntyre A, Mix A (1989) Oceanic response to orbital forcing in the late Quaternary: Observational and experimental strategies. In: Berger A (ed) Climate and Geoscience. Kluwer Acad, Norwell, Mass, pp 121-164

Jansen JHF, Van der Gaast SJ, Koster B, Vaars AJ (1998) CORTEX, a shipboard XRF-scanner for element analyses in split sediment cores. Mar Geol 151: 143-153.

Kolla V, Biscaye PE, Hanley AF (1979) Distribution of quartz in late Quaternary Atlantic sediments in relation to climate. Quat Res 11: 261-277

Kutzbach JE (1981) Monsoon climate of the early Holocene. Science 214: 59-61

Lange H (1982) Distribution of chlorite and kaolinite in eastern Atlantic sediments off North Africa. Sedimentology 29: 427-431

Ledru M-P (1993) Late Quaternary environmental and climate changes in central Brazil. Quat Res 39: 90-98

Leroy S, Dupont L (1994) Development of vegetation and continental aridity in northwestern Africa during Late Pliocene: The pollen record of ODP Site 658. Palaeogeogr Palaeoclimat Palaeoecol 109: 295-316

Lough J (1986) Tropical Atlantic sea surface temperatures and rainfall variations in sub-Saharan Africa. Monthly Weather Rev 114: 561-570

Manabe S, Stouffer RJ (1997) Coupled ocean-atmosphere model response to freshwater input: Comparison to Younger Dryas event. Paleoceanography 12: 321-336

Marret F, Scourse JD, Versteegh G, Jansen JHF, Schneider R (2001) Integrated marine and terrestrial evidence for abrupt Congo River palaeodischarge fluctuations during the last deglaciation. J Quat Sci 16: 761-766

Martinez P, Bertrand P, Shimmield GB, Cochrane K, Jorissen FJ, Foster J, Dignan M (1999) Upwelling intensity and ocean productivity changes off Cape Blanc (northwest Africa) during the last 70,000 years: geochemical and micropaleontological evidence. Mar Geol 158: 57-74

Maslin MA, Burns SJ (2000) Reconstruction of the Amazon Basin effective moisture availability over the past 14,000 years. Science 290: 2285-2287

Matthewson AP, Shimmield GB, Kroon D, Fallick A (1995) A 300kyr high-resolution aridity record of the North African continent. Paleoceanography 10: 677-692

McIntyre A, Ruddiman WF, Karlin K, Mix AC (1989) Surface water response of the equatorial Atlantic Ocean to orbital forcing. Paleoceanography 4: 19-55

McIntyre A, Molfino B (1996) Forcing of Atlantic equatorial and subpolar millenial cycles by precession. Science 274: 1867-1870

Milliman JD, Summerhayes CP, Barretto HT (1975) Quaternary sedimentation on the Amazon continental margin: A model. Geol Soc Am Bull 86: 610-614

Mix AC, Ruddiman WF, McIntyre A (1986) Late Quaternary paleoceanography of the tropical Atlantic, 1: Spatial variability of annual mean sea-surface temperatures, 0-20,000 years BP. Paleoceanography 1: 43-66

Mulitza S, Rühlemann C (2000) African monsoonal precipitation modulated by interhemispheric temperature gradients. Quat Res 53: 270-274

Murray RW, Leinen M, Isern AR (1993) Biogenic flux of Al to sediment in the central equatorial Pacific Ocean: Evidence for increased productivity during glacial periods. Paleoceanography 8: 651-670

Orians KJ, Bruland KW (1985) Dissolved aluminium in the central North Pacific. Nature 316: 427-429

Parkin DW, Shackleton NJ (1973) Trade wind and temperature correlation down a deep-sea core off the Sahara coast. Nature 245: 455-457

Peterson LC, Overpeck JT, Kipp NG, Imbrie J (1991) A high-resolution late Quaternary upwelling record from the anoxic Cariaco Basin, Venezuela. Paleoceanography 6: 99-119

Petschick R, Kuhn G, Gingele F (1996) Clay mineral distribution in surface sediments of the South Atlantic: Sources, transport, and relation to oceanography. Mar Geol 130: 203-229

Pokras EM, Mix AC (1985) Eolian evidence for spatial variability of late Quaternary climates in tropical Africa. Quat Res 24: 137-149

Prell WL, Kutzbach JE (1987) Monsson variability over the past 150,000 years. J Geophys Res 92: 8411-8425

Prospero JM (1981) Eolian transport to the world ocean. In: Emiliani C (ed) The Sea. Vol. VII. J Wiley & Sons, New York, pp 801-874

Prospero JM, Carlson TN (1972) Vertical and areal distribution of Sahara dust over the western equatorial North Atlantic Ocean. J Geophys Res 77: 5255-5265

Prospero JM, Bonatti E, Schubert C, Carlson TN (1970) Dust in the Caribbean atmosphere traced to an African dust storm. Earth Planet Sci Lett 9: 287-293

Ratmeyer V, Balzer W, Bergametti G, Chiapello I, Fischer G, Wyputta U (1999) Seasonal impact of mineral dust on deep-ocean particle flux in the eastern subtropical Atlantic Ocean. Mar Geol 159: 241-252

Ruddiman WF (1997) Tropical Atlantic terrigenous fluxes since 25,000 yrs BP. Mar Geol 136: 189-207

Rühlemann C, Frank M, Hale W, Mangini A, Mulitza S, Müller PJ, Wefer G (1996) Late Quaternary productivity changes in the western equatorial Atlantic. Evidence from ^{230}Th normalized carbonate and organic carbon accumilation. Mar Geol 159: 127-152

Rühlemann C, Diekmann B, Mulitza S, Frank M (2001) Late quaternary changes of western equatorial Atlantic surface circulation and Amazon lowland climate recorded in Ceará Rise deep-sea sediments. Paleoceanography 16: 293-305

Sarnthein M (1978) Sand deserts during glacial maximum and climatic optimum. Nature 272: 43-46

Sarnthein M, Tetzlaff G, Koopmann B, Wolter K, Pflaumann U (1981) Glacial and interglacial wind regimes over the eastern subtropical Atlantic and North-West Africa. Nature 293: 193-196

Schmidt AM, von Dobeneck T, Bleil U (1999) Magnetic characterization of Holocene sedimentation in the South Atlantic. Paleoceanography 14: 465-481

Schneider RR, Müller PJ, Ruhland G (1995) Late Quaternary surface circulation in the east equatorial Atlantic: Evidence from alkenone sea surface temperatures. Paleoceanography 10: 197-219.

Schneider RR, Price B, Müller PJ, Kroon D, Alexander I (1997) Monsoon related variations in Zaire (Congo) sediment load and influence of fluvial silicate supply on marine productivity in the east equatorial Atlantic during the last 200,000 years. Paleoceano-graphy 12: 463-481

Stein R (1985) Late Neogene changes of paleoclimate and paleoproductivity off Northwest Africa (DSDP Site 397). Palaeogeogr Palaeoclimatol Palaeoecol 49: 47-59

Street FA, Grove AT (1976) Environmental and climatic implications of late Quaternary lake-level fluctuations. Nature 261: 385-390

Street-Perrott FA, Perrott RA (1990) Abrupt climate

fluctuations in the tropics: The influence of Atlantic Ocean circulation. Nature 343: 607-612

Talbot MR, Livingstone DA, Palmer DG, Maley J, Melack JM, Delibrias G, Gulliksen J (1984) Preliminary results from sediments core from lake Bosumtwi, Ghana. Palaeoecol Afr 16: 173-192

Thiry M (2000) Palaeoclimatic interpretation of clay minerals in marine deposits: an outlook from the continental origin. Earth-Sci Rev 49: 201-221

Tiedemann R, Sarnthein M, Stein R (1989) Climatic changes in the western Sahara: Aeolo-marine sediment record of the last 8 million years (Sites 657-661). Proc Ocean Drill Prog Sci Res 108: 241-277.

Tiedemann R, Sarnthein M, Shackleton NJ (1994) Astronomic timescale for the Pliocene Atlantic $\delta^{18}O$ and dust flux records of Ocean Drilling Program site 659. Paleoceanography 9: 619-638

van der Gaast SJ, Jansen JHF (1984) Mineralogy, opal, and manganese of middle and late Quaternary sediments of the Zaire (Congo) deep-sea fan: Origin and climatic variation. Neth J Sea Res 17: 313-341

van der Weijden CH (2002) Pitfalls of normalization of marine geochemical data using a common divisor. Mar Geol 184: 167-187

Wagner T (2002) Late Cretaceous to early Quaternary organic sedimentation in eastern Equatorial Atlantic. Palaeogeogr Palaeoclimatol Palaeoecol 179: 113-147

Williams MAJ (1975) Late Pleistocene tropical aridity synchronous in both hemispheres? Nature 253: 617-618

Windom HL (1975) Eolian contribution to marine sediments. J Sediment Petrol 45:520-529

Zabel M, Bickert T, Dittert L, Haese RR (1999) Significance of the sedimentary Al:Ti ratio as an indicator for variations in the circulation patterns of the equatorial North Atlantic. Paleoceanography 14: 789-799

Zabel M, Schneider RR, Wagner T, Adegbie AT, de Vries U, Kolonic S (2001) Late Quaternary Climate Changes in Central Africa as Inferred from Terrigenous Input to the Niger Fan. Quat Res 56: 207-217

Zhao M, Beverridge NAS, Shackleton NJ, Sarnthein M, Eglington G (1995) Molecular stratigraphy of cores off northwest Africa: Sea surface temperature history over the last 80 ka. Paleoceanography 10: 661-675

Surface Sediment Bulk Geochemistry and Grain-Size Composition Related to the Oceanic Circulation along the South American Continental Margin in the Southwest Atlantic

M. Frenz[*], R. Höppner, J.-B.W. Stuut, T. Wagner and R. Henrich

*Universität Bremen, Fachbereich Geowissenschaften, Klagenfurter Strasse,
28359 Bremen, Germany*
** corresponding author (e-mail): mfrenz@uni-bremen.de*

Abstract: Surface sediments from the South American continental margin surrounding the Argentine Basin were studied with respect to bulk geochemistry ($CaCO_3$ and C_{org}) and grain-size composition (sand/silt/clay relation and terrigenous silt grain-size distribution). The grain-size distributions of the terrigenous silt fraction were unmixed into three end members (EMs), using an end-member modelling algorithm. Three unimodal EMs appear to satisfactorily explain the variations in the data set of the grain-size distributions of terrigenous silt. The EMs are related to sediment supply by rivers, downslope transport, winnowing, dispersal and re-deposition by currents. The bulk geochemical composition was used to trace the distribution of prominent water masses within the vertical profile. The sediments of the eastern South American continental margin are generally divided into a coarse-grained and carbonate-depleted southwestern part, and a finer-grained and carbonate-rich northeastern part. The transition of both environments is located at the position of the Brazil Malvinas Confluence (BMC). The sediments below the confluence mixing zone of the Malvinas and Brazil Currents and its extensions are characterised by high concentrations of organic carbon, low carbonate contents and high proportions of the intermediate grain-size end member. Tracing these properties, the BMC emerges as a distinct north-south striking feature centered at 52-54°W crossing the continental margin diagonally. Adjacent to this prominent feature in the southwest, the direct detrital sediment discharge of the Rio de la Plata is clearly recognised by a downslope tongue of sand and high proportions of the coarsest EM. A similar coarse grain-size composition extends further south along the continental slope. However, it displays better sorting due to intense winnowing by the vigorous Malvinas Current. Fine-grained sedimentary deposition zones are located at the southwestern deeper part of the Rio Grande Rise and the southern abyssal Brazil Basin, both within the AABW domain. Less conspicuous winnowing/accumulation patterns are indicated north of the La Plata within the NADW level according to the continental margin topography. We demonstrate that combined bulk geochemical and grain-size properties of surface sediments, unmixed with an end-member algorithm, provide a powerful tool to reconstruct the complex interplay of sedimentology and oceanography along a time slice.

Introduction

The southwestern part of the South Atlantic Ocean holds a key position in global ocean circulation, since southern surface, intermediate and deep waters are introduced into the thermohaline cycle (e.g. Gordon 1986). Due to the position at the western boundary of the South Atlantic Ocean, the circulation is focussed to (Deep) Western Boundary (Under-) Currents (Stommel 1958). The increased flow velocities are expected to have a major impact on sedimentary processes, i.e. conducting deposition, winnowing and lateral advection of sedimentary particles.

From WEFER G, MULITZA S, RATMEYER V (eds), 2003, *The South Atlantic in the Late Quaternary: Reconstruction of Material Budgets and Current Systems.* Springer-Verlag Berlin Heidelberg New York Tokyo, pp 347-373

Modern oceanography differentiates more than eight water masses along the western boundary of the South Atlantic (e.g. Arhan et al. 1999; Larqué et al. 1997; Mémery et al. 2000). Since sedimentological methods are not sensitive enough to distinguish whether the sedi-ments have been in contact to varieties of the same water mass, we apply a model essentially based on Piola and Matano (2001). This model manages the complex vertical stratification structure of the water masses in the western South Atlantic with four main water levels (Fig. 1). Upper and intermediate water levels are occupied by Subantarctic (SAW) and subtropical Thermocline Waters (TW; both < 500 m water depth and admixed with South Atlantic Central Water) and Antarctic Intermediate Water (AAIW, 500-1000 m water depth). As one unit, SAW (branch of the Antarctic Circumpolar Current) and AAIW enter the study area from the south, referred to as the Malvinas (Falkland) Current (MC), flowing NE. Its counterpart in the northeast is the south-westward flowing Brazil Current (BC). It contains both, TW as well as AAIW and marks the north-western closure of the subtropical gyre. Both current systems collide in front of the mouth of Rio de la Plata at 38°S, called the Brazil Malvinas Confluence (BMC; Gordon and Greengrove 1986). At the BMC both currents are deflected to the south and experience intense mixing that also affects deeper water levels. The BMC system is further complicated by the discharge from the Rio de la Plata estuary, which releases 470 km^3 a^{-1} of freshwater into the South Atlantic (Milliman and Meade 1983). In comparison, the two deep water levels reveal a much simpler structure. North of the BMC, water depths between 2000 and 4000 m are characterised by the southward flow of North Atlantic Deep Water (NADW). South of the BMC the NADW splits the northward flowing CDW into an upper and lower layer (UCDW and LCDW). Below 4000 m, Antarctic Bottom Water (AABW) flows northward.

Although the oceanographic configuration along the eastern continental margin of South America has been investigated in much detail, our knowledge on the variability of the sedimentary regimes in this area is rather poor and their driving forces not fully understood. Previous studies have elucidated local sediment dispersal mechanisms (e.g. Höppner and Henrich 1997; Benthien and Müller 2000; Hensen et al. 2000; Michaelovitch et al. 2002; Ledbetter 1984). They were based on specific parameters only. In addition, the discrimination of large-scale sediment bodies and their internal structure was investigated by echo sound properties (Mello et al. 1992; Flood and Shor 1988; Ledbetter 1986). Hence, a complementary multiproxy sedimentary study covering the entire region is still lacking.

This study is intended to fill this gap. We present bulk geochemical, bulk sedimentological and terrigenous silt grain-size distribution data from the eastern South American continental margin between 20 and 50°S latitude covering water depths from the outer shelf to the deep-sea. This detailed sedimentological data set is unmixed into subpopula-tions, using an end-member algorithm (Weltje 1997), which are subsequently interpreted in terms of sediment transport mechanisms and depositional processes. In particular, we focus the discussion on the complex interferences of biological and sedi-mentological processes with variations in physical and chemical properties of the different water masses.

Methods

The present study relies on grain-size investigations of 111 surface samples from the southwestern South Atlantic. In addition, 106 bulk geochemical data were used [CaCO$_3$ and C$_{org}$ data from this study, Mollenhauer et al. (2003) and Hease et al. (2000); see www.pangaea.de for accurate references]. The samples were recovered during 6 cruises of RV *Meteor* between 1993 and 2001. Where available, the uppermost cm (0-1 cm) of multi- and box-core samples was used for analysis. At 16 locations the core tops had already been sampled. For these core positions the second cm (1-2 cm) was used for granulometric analysis (www.pangaea.de).

The freeze-dried bulk samples were split into coarse (sand) and fine fractions by wet sieving using a 63 μm mesh. Subsequently, the fine fraction was split into silt and clay fractions according to their settling velocities after Stokes' Law, assuming a sample density of 2.65 g cm^{-3}. The bulk silt

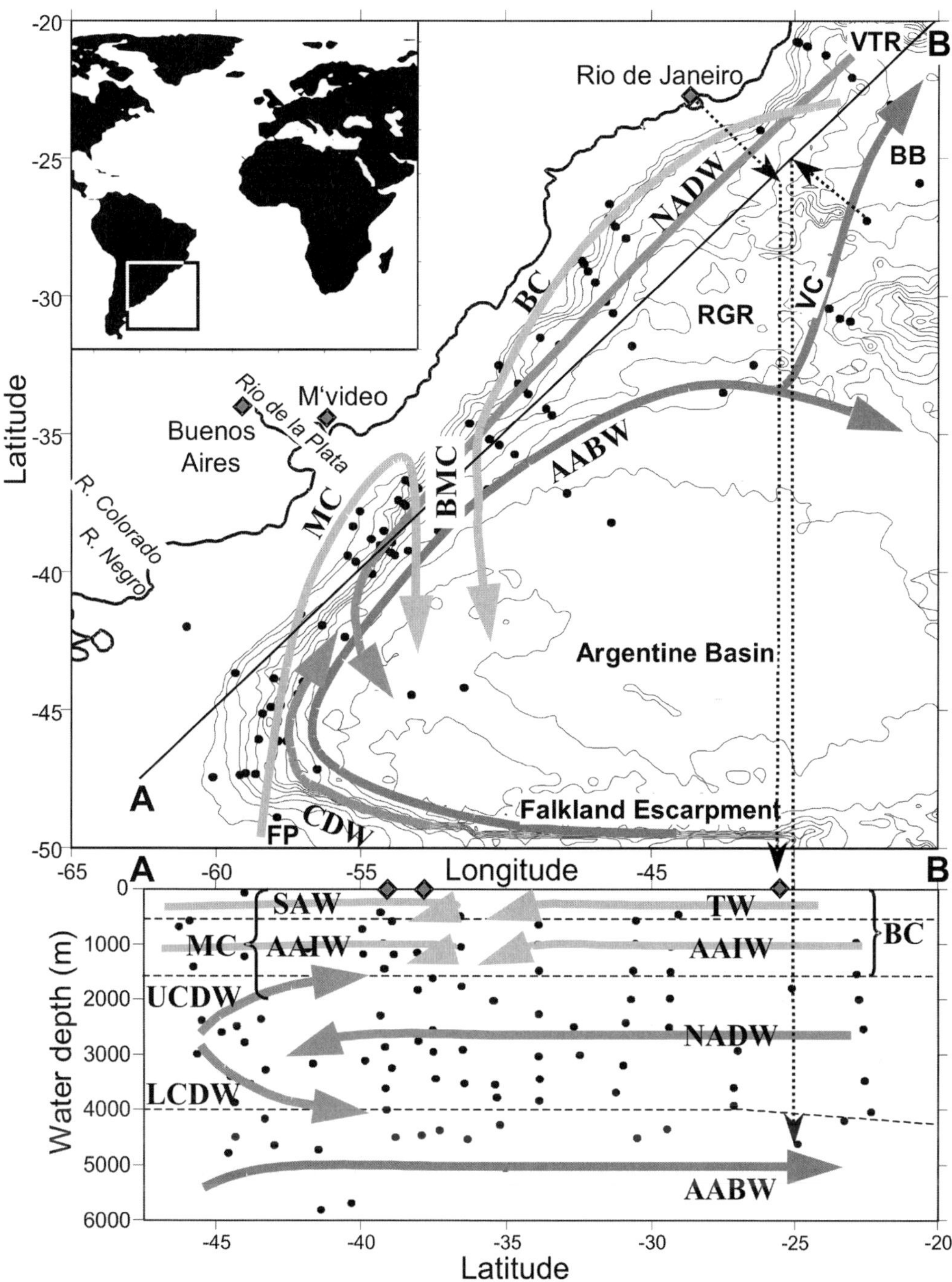

Fig. 1. Bathymetry according to the ETOPO 2'-grid (top) of the South American continental margin between 20 and 50°S. Main geographic features are indicated (M'video - Montevideo, VTR - Vitória Trindade Ridge, BB - Brazil Basin, RGR - Rio Grande Rise, VC - Vema Channel, FP - Falkland Plateau). Isobaths are 0 m and multiples of 500 m. Arrows show generalised main flow paths of water masses along the western boundary [MC - Malvinas Current, BC - Brazil Current, SAW - Subantarctic Water, TW - Thermocline Water, AAIW - Antarctic Intermediate Water, (U/L)CDW - (Upper/Lower) Circumpolar Deep Water, NADW - North Atlantic Deep Water, AABW - Antarctic Bottom Water]. BMC indicates the Brazil Malvinas Confluence and its extension. Line A-B shows the position of the projection plain of the part below. The dotted arrows indicate the projection path for two exemplary locations, Rio de Janeiro and a sample position from the abyssal Brazil Basin.

fraction was dried and weighed, decalcified using excess HCl, then washed, dried and weighed again. The grain-size distribution of the carbonate-free silt fraction was analysed using a Micromeritics Sedi Graph 5100. The measurements were limited to 95 samples, where more than 1 g carbonate-free sample material was left after decalcification. The SediGraph data were converted into fifty size classes from 9-4 Φ (bin width 0.1 Φ each) using the SediMac applet written by G. Kuhn (AWI, Bremerhaven). The statistical parameters [arithmetic mean, modal size (size class containing highest amount), sorting (standard deviation) and skewness] of the size-frequency distributions are based on momentum calculations according to Krumbein (1936). The terrigenous silt fraction comprises a size window that is highly sensitive for sediment sorting processes (e.g. McCave et al. 1995; Bianchi and McCave 2000; Bianchi et al. 2001; Gröger and Henrich 2003) that are assumed to have a great impact on the investigated sediments (Ledbetter 1984, 1986).

Endmember modelling of the carbonate-free silt distributions was carried out using the algorithm of Weltje (1997). This end-member algorithm has been successfully applied to unravel data sets of rodent palaeocommunity successions (Van Dam and Weltje 1999) and particle-size distributions of deep-sea sediments from the Arabian Sea (Prins and Weltje 1999; Prins et al. 2000), the North Atlantic Ocean (Prins et al. 2001, 2002), and the eastern South Atlantic (Stuut et al. 2002). The basic assumption underlying end-member studies is the fact that the data set consists of a number of sub-populations. For each set of end members, the degree to which the composition can be reconstructed is expressed as the coefficient of determination. The appropriate amount of end members is subsequently chosen at the threshold where the best fit is reached with the least amount of end members. The relative proportion of each end member for all samples can then be calculated.

Organic and total carbon were measured in homogenised sub-samples using a Leco CS-300 elemental analyser (precision of measurement $\pm$ 3 %). For the determination of organic carbon, calcium carbonate was removed by repeatedly adding aliquots of 0.25 N HCl. The carbonate con-

tent was calculated from the difference between total and organic carbon and expressed as calcite [CaCO$_3$ = (C$_{tot}$-C$_{org}$) * 8.33]. To evaluate the reproducibility of carbon data duplicates were measured routinely.

To enable a 3-D impression of the spatial distribution, each parameter is shown as kriging-interpolated isoplots over the ETOPO 2' bathymetric grid, and as a depth-related projection on a vertical plane oriented parallel to the continental slope. As is the nature of any interpolation, areas of higher sample coverage denote higher reliability compared to scarcely covered areas.

Following Benthien and Müller (2000), Holocene ages are assumed for our sample grid that is partly derived from identical stations and from positions in between of those.

Results

The bulk geochemical and grain-size composition of surface sediments along the South American continental margin varies from carbonate-rich and fine-grained in the northeast to carbonate-poor and coarse-grained in the southwest. The transition zone between these two modes coincides with the position of the BMC and is located at about 52° to 54°W. Overall, grain sizes generally decrease with increasing water depth. In contrast, the carbon-distribution patterns are preferentially related to specific water depths (carbonate) and geographic location (organic carbon).

Carbonate Distribution

One characteristic feature in the bulk carbonate distribution is a continuous level of elevated values all along the continental rise at water depths between 2000 and 4000 m (Fig. 2). Typical carbonate contents between 40 and 70 % are observed from 20 to 35°S and from 43 to 48°S, approaching maximum concentrations at about 2500 m water depth. Exceptionally high concentrations are observed at exposed sites, i.e. on top of the Vitória Trindade Ridge (VTR, CaCO$_3$ > 90 %) and on the eastern slope of the Vema Channel (VC, CaCO$_3$ > 80 %). In contrast to these high concentrations, the carbonate contents of the carbonate-rich level is considerably reduced (CaCO$_3$ < 20 %) in a broad

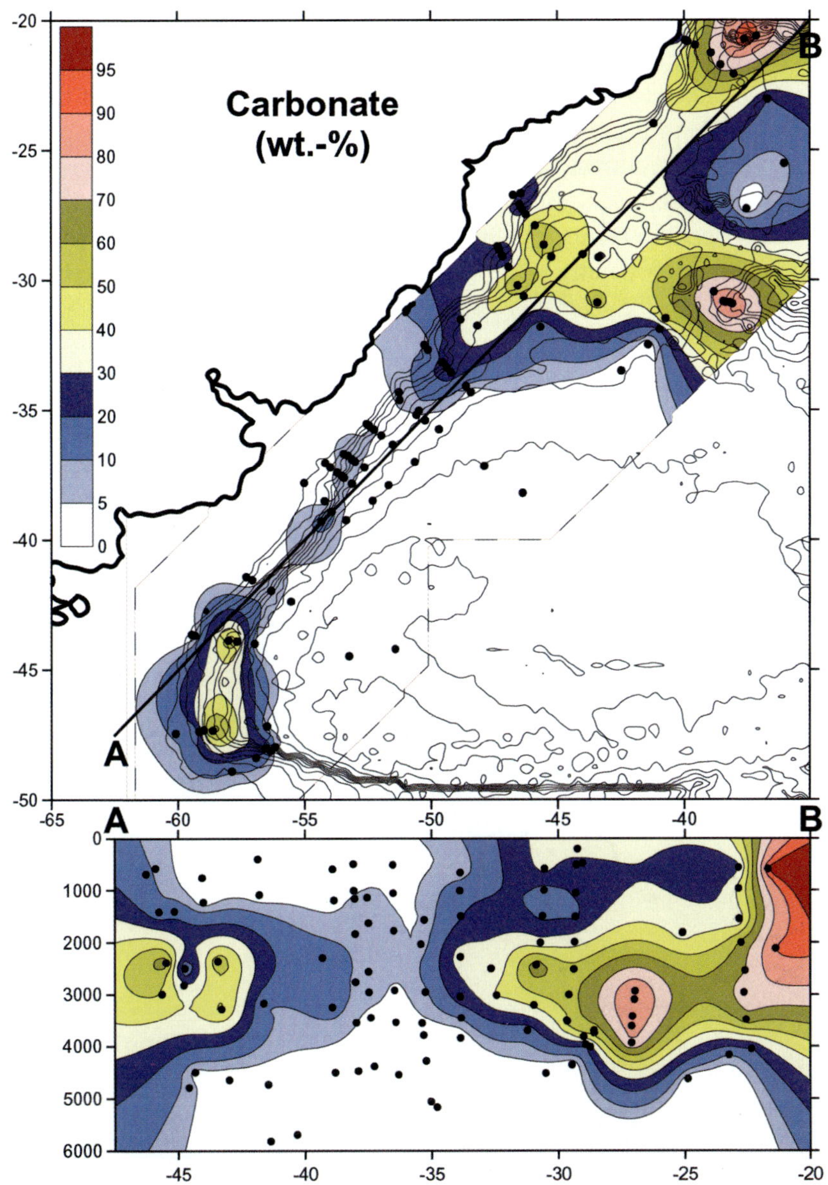

Fig. 2. Carbonate distribution of surface samples along the eastern South American continental margin (n = 106). NADW is indicated by elevated concentrations down to 48°S. The signal is weaker in a broad sector off the La Plata mouth. Sediments under the influence of AABW and Malvinas Current lack carbonate.

sector from about 35 to 42°S (= 50-57°W), which includes the La Plata mouth. Within this sector of depletion the carbonate-rich band is almost interrupted ($CaCO_3$ < 10 %) by a narrow corridor at about 52-54°W. At about 48°S (57°W) the continental margin turns east to build the Falkland Escarpment. Here, and at water depths corresponding to the carbonate-rich level, carbonate contents decrease to values below 20 %.

In the Argentine Basin samples from below 4000 m water depth are almost completely barren in calcium carbonate ($CaCO_3$ < 5 %), whereas in the southwestern Brazil Basin samples down to 4200 m still reveal carbonate contents of 10 to 20 % before they drop to almost zero below 4500 m water depth.

The continental slope (< 2000 m water depth) offshore Uruguay and Argentina is clearly distinguished from corresponding water depth levels in the northeast offshore Brazil on the basis of carbonate contents. In the southwest, carbonate generally remains below 10 % to decrease to < 5 % above 1200 m water depth. In contrast, sediments from the northeast contain about 20 % between 2000 and 1000 m water depth to increase to 30 % carbonate at about 600 m water depth.

Organic Carbon Distribution

The most striking feature of the modern organic carbon distribution along the south-eastern South American continental margin is a narrow N-S striking corridor off the Rio de la Plata mouth at about 52-54°W (Fig. 3). Here, the concentrations are up to three times higher compared to the adjacent ar-

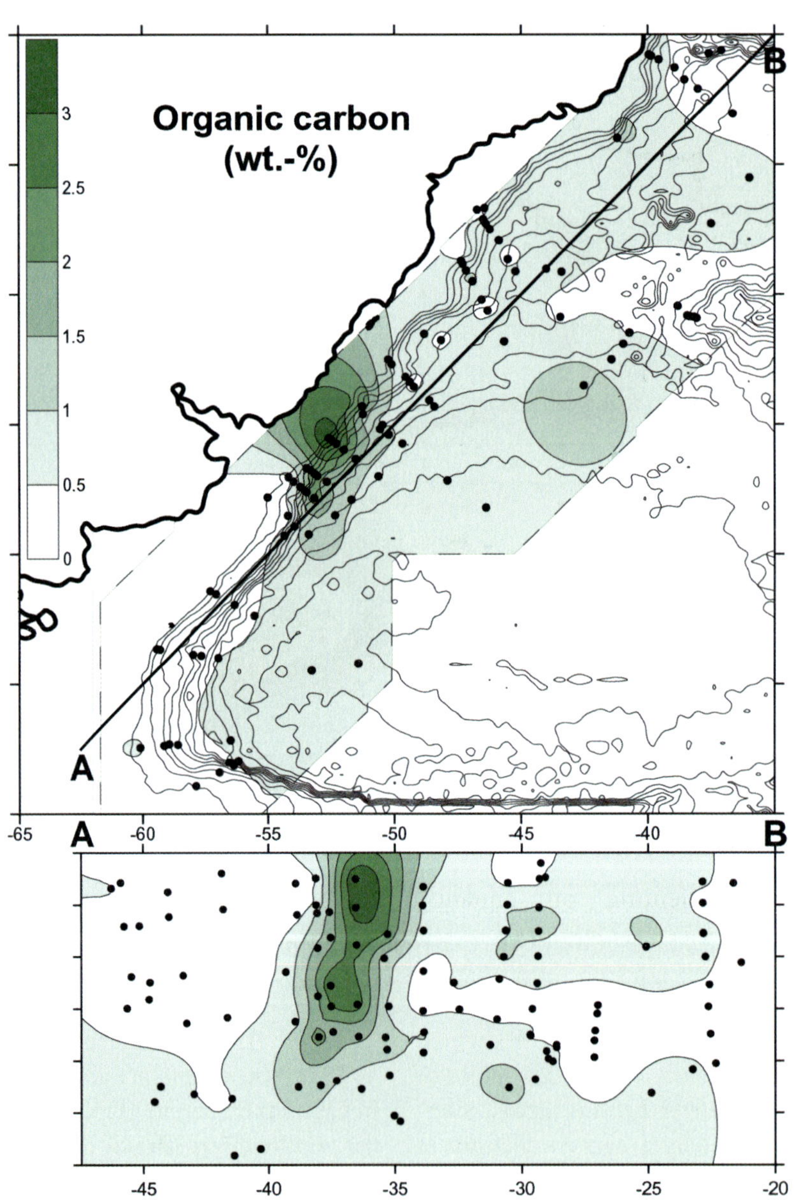

Fig. 3. Organic carbon concentrations of surface sediment samples along the eastern South American continental margin (n = 106). Note the high-C_{org} corridor (starting at the outer shelf heading south between 52 and 54°W) tracing the Brazil Malvinas Confluence and its extensions.

eas. In this corridor, organic carbon exceeds 2.5 % at water depths between 500 and 3500 m to gradually decrease to still considerable 1 % at 4500 m water depth. This corridor coincides with the almost carbonate-barren corridor at 52-54°W (Figs. 2, 3). The organic carbon contents remain low southwest of this corridor, in the order of 0.5 % and less, with values being highest below a water depth of 4000 m. The northeast of the study area indicates a stratification of the organic carbon contents which is related to the water depth. Above 700 m and between 2500-3500 m water depth the sediments hardly contain any organic carbon (C_{org} < 0.5 %). In between and below these two depleted levels the organic carbon contents are elevated in the order of 1.5-0.5 %.

Sediments enriched in organic carbon in front of the La Plata and below the BMC bear a mixed marine and terrigenous organic signature. Supply of riverine organic matter is traced by the presence of higher plant tissue, identified by lignin phenols and organic particles (macerals) of terrigenous origin (vitrinite, inertinite), whereas bodies of marine algae, dinoflagellate cysts and detrital fragments represent organic matter of marine origin (see Wagner et al. this volume and references therein for terrigenous organic matter in South Atlantic sediments). In front of the La Plata and below the BMC, the content of reactive and hydrogen-rich organic matter preserved in surface sediments gradually decreases with increasing distance to the continent. Accordingly, hydrogen indices obtained from Rock-Eval Pyrolysis decreased from 350 mg HC/g C_{org} at shallow sites to slightly above 100 mg HC/g C_{org} at about 4500 m water depth, documenting an enhanced remineralisation of reactive organic matter at greater water depths. First estimates of marine and terrigenous organic fractions based on maceral analysis suggest a 40-50 % contribution from continental sources in front of the La Plata, showing no clear trend over the continental slope.

Granulometry I: Sand/Silt/Clay Relationship of the Bulk Sediment

Measurements performed by Romero and Hensen (2002) revealed that the contents of biogenic opal in surface sediments of the region of interest are very low (< 1.3 %). Their sample set overlaps to a great extent with the one used in this study. Therefore biogenic opal can be neglected in the sand/silt/ clay relations, and the sediments can be regarded to consist of biogenic carbonate and terrigenous components only. Hence, the carbonate contents have to be taken into account when interpreting the relation of these bulk size fractions.

A clear separation of the northeastern and southwestern parts of the continental margin is evident on account of the sand/silt/clay distribution patterns, and projections of these basic granulo-metric data in a ternary plot (Figs. 4, 5). Accordingly, at approximately 52-54°W, there is a transition zone that separates relatively coarse-grained, sandy sediments in the southwest from relatively fine-grained, muddy deposits in the northeast. Furthermore, higher clay proportions at otherwise comparable sand contents are observed in surface sediments from the northeast in comparison to the southwest. In both areas the sand contents consistently decrease with increasing water depth.

In the southwest, the Argentine shelf and slope sediments contain more than 90 % of sand-sized material at water depths less than 1500 m. This signature persists towards the Falkland Escarpment in the south. The sand contents decrease with increasing water depth across the Argentine continental rise approaching 40 % at 4700 m, and even 20-30 % in the southwesternmost basin exceeding 3500 m water depth. Below a water depth of 3300 m , the fine fraction of all samples from the southwest is mainly composed of silt (> 40 % silt, < 20 % clay). The silt contents gradually increase across the continental rise towards the continent, most pronounced in front of the La Plata mouth and close to the Falkland Escarpment.

Comparable to the carbonate and organic carbon distribution patterns the sand, silt and clay contents of the sediments in the northeast show a 4-level structure related to water depth. Sediments from carbonate-rich layers, i.e. shallow (< 700 m water depth) and intermediate (2000-3500 m water depth) water levels are characterised by a very variable size composition (10-60 % sand, 20-50 % silt, 20-50 % clay). Between and below these levels the sand content is reduced to < 10 % at con-

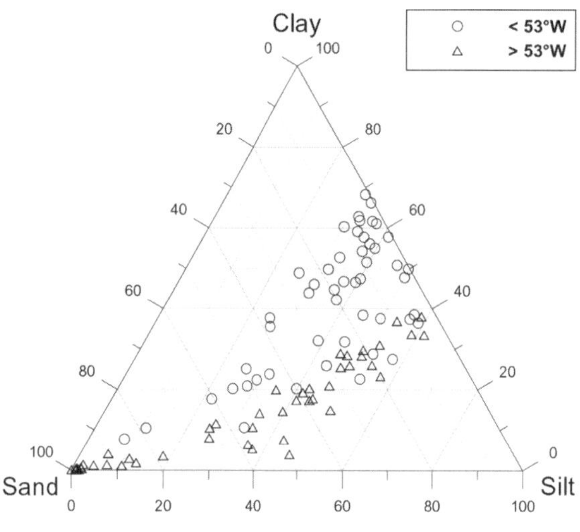

Fig. 4. Ternary bulk sand-silt-clay relations (in wt.-%) of surface sediments from the eastern South American continental margin. Fine-grained northeastern samples (< 53°W, circles) can be distinguished from coarse-grained southwestern samples (> 53°W, triangles).

stant silt (30-40 %) and high clay contents (> 50 %). Exceptionally coarse sediment compositions are observed at exposed sites (eastern slope of VC, VTR and beyond). Here, the sand contents exceed 60 %, whereas silt and clay each approach a relative value of about 20 %.

In the deep-sea, below 4000 m water depth, the marked coarse/fine size transition at about 53°W is shifted to the northeast. Although grain size generally decreases with increasing water depth, abyssal sediments in the southwest are relatively coarse-grained compared to those in the northeast. 20-30 % sand and 40-70 % silt are observed as far as 35°S (50°W), indicating that the coarse/fine transition is shifted by about 3° towards the north and the east, respectively, which corresponds to about 400 km.

Granulometry II: Terrigenous Silt Grain-Size Distributions

To test if the relative increase of opal contents due to the separation of the carbonate-free silt fraction (removal of sand, clay and carbonate) has a sig-

nificant effect on its size distribution, some additional analyses were carried out. Light microscopy of the sand fraction of samples with elevated opal contents (opal > 1 %, according to Romero and Hensen 2002) revealed some sponge spicules and radiolarians that have, due to their size, evaded the silt fraction. Spot tests of the carbonate-free silt fraction revealed maximum opal concentrations of 13 % in areas vulnerable for elevated opal contents like the southwestern most Argentine Basin and the low-$CaCO_3$/high-C_{org} corridor. Although for two cases this means a tenfold enrichment in the carbonate-free silt, compared to the bulk sample, the biogenic opal content is rather low and its impact on the grain-size distribution of the carbonate-free silt can be neglected. Therefore, the grain-size distributions of the carbonate-free silt fraction are considered to represent terrigenous silt (TS).

The content of TS of the bulk surface sediments (not shown) reveals maximum values of 70 % of the bulk sample. Generally, the TS proportions increase with increasing water depth. At the Argentine shelf and slope its contribution is insignificant because the sediments are predominantly sandy, i.e. they consist of more than 90 % terrigenous sand. On the other hand, the silt deposition is reduced and additionally diluted by high carbonate sand contents at exposed sites.

The general statistics for the 96 TS grain-size frequency distributions (minimum, maximum, average) are given in figure 6. It verifies the inhomogeneous and highly variable size composition of the sediments from the eastern South American continental margin. Maximum size variability is revealed close to the coarse and the fine end of the silt size range.

Although conclusions drawn from the statistical parameters of the TS distributions are restricted by artefacts that are introduced by the empirical size limits of the silt fraction (see Excursion), the general trends can be deduced from their spatial distribution pattern. The spatial variability of the commonly used statistical parameters is illustrated in figure 7. As expected from bulk granulometry (sand/silt/clay), the size parameters mean, mode and skewness confirm the observed differences between southwestern coarse and northeastern fine sediments. The transition of both modes of

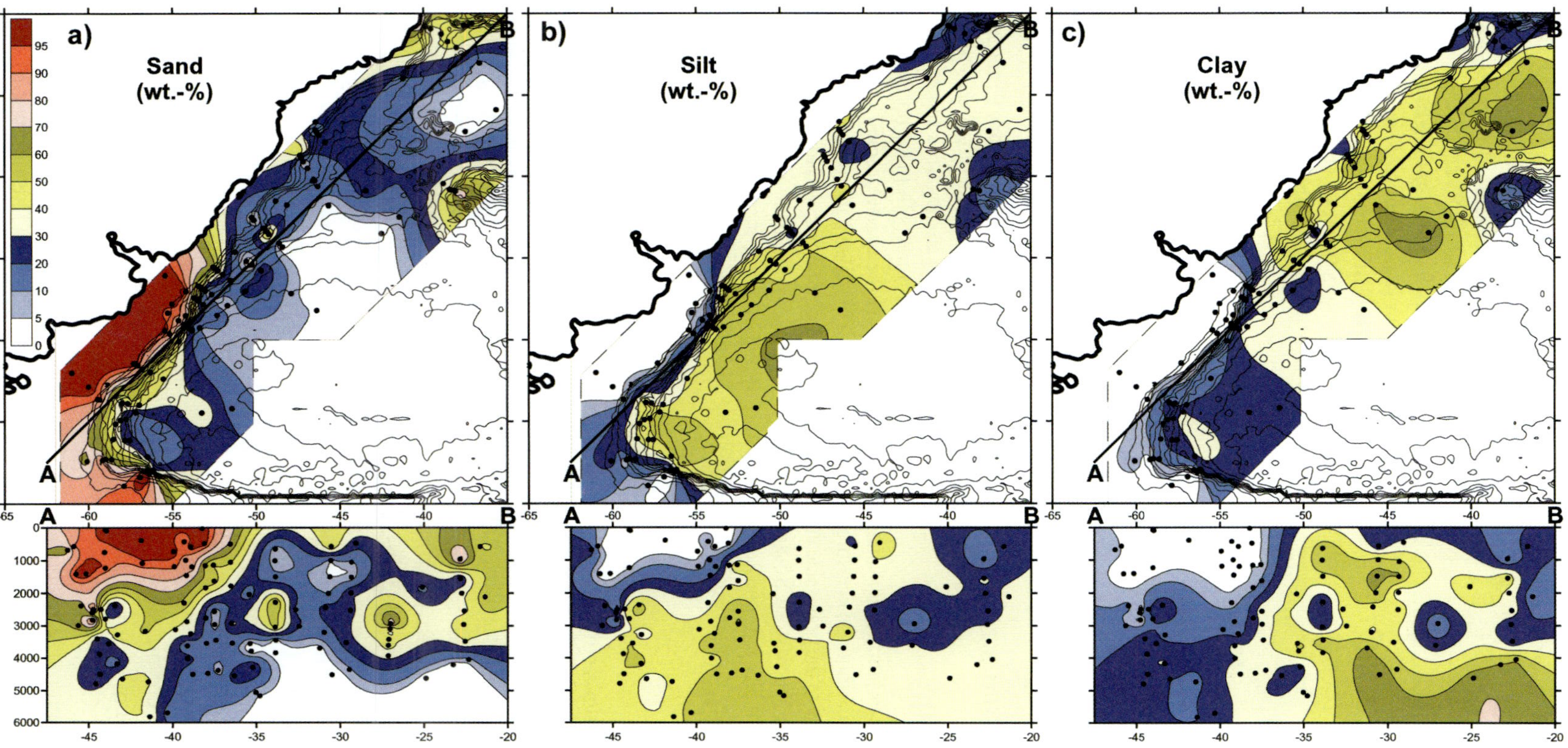

Fig. 5. Contribution of bulk size fractions to the surface sediments along the eastern South American continental margin. **a)** bulk sand (size fraction > 63 μm, n = 111) **b)** bulk silt (size fraction 63 2 μm, n = 98) **c)** bulk clay (size fraction < 2 μm, n = 98). Note the sharp break at about 53°W between southwestern coarse (silty and sandy) and northeastern fine sediment mode (mainly clayey). High terrigenous (compare Fig. 2) sand concentrations are observed at the Argentine outer shelf and slope.

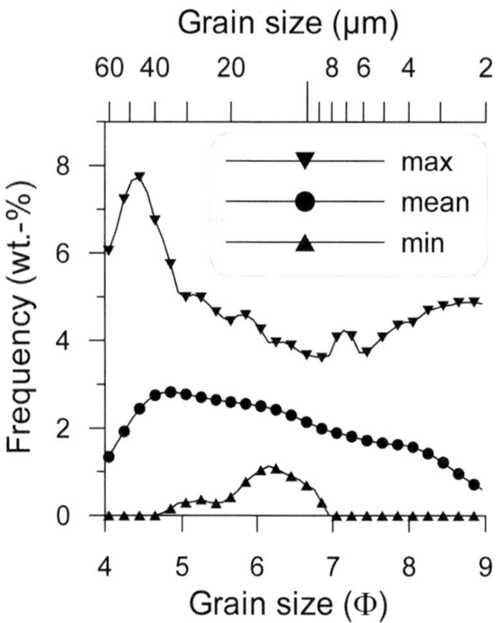

Fig. 6. Brief statistics (maximum, average and minimum) for the 95 TS grain-size distributions measured. The plot reveals the high variability of the grain-size composition of terrigenous silt along the eastern South American continental margin. Variability is highest above 32 µm (very coarse silt) and below 4 µm (very fine silt).

Fig. 7. right Spatial distribution of the statistical parameters calculated from the grain-size distributions of terrigenous silt (TS) according to Krumbein (1936), n = 95. **a)** TS mean grain size in µm (left of scale bar) and Φ-units (right of scale bar). **b)** TS modal grain size (size class of highest amount) in µm (left of scale bar) and Φ-units (right of scale bar). Very coarse TS grain size is indicated along the Argentine continental slope (contourite) and down the slope and rise off the la Plata mouth (turbidite). In contrast, accumulation of very fine sediments are found at deeper sites in the northern Argentine and the southern Brazil Basin. **c)** TS Sorting (standard deviation of the size frequency distribution). Good sorting (< 1) characterises the contourites off Argentina along the slope whereas the coarse la Plata downslope tongue shows poor sorting. Sediments in the area congruent to the low-CaCO$_3$/high-C$_{org}$/EM2 corridor (compare Figs. 2 and 9) and in the deeper northeastern parts are well sorted. **d)** TS skewness. The calculation reveals negative (coarse) skewness for fine grained sediments from the northeast and positive (fine) skewness for coarse grained sediments from the southwest. This false pattern (see Excursion) is the result of the artificial upper and lower size limits of the silt fraction. Most negative values occur at fine-grained accumulation sites, most positive off the mouth of the La Plata.

sediment composition again coincides with the low-CaCO$_3$/high-C$_{org}$ corridor at about 52-54°W.

In the southwest, the coarsest TS composition is observed at the continental slope at 1000-1500 m water depth and off the la Plata mouth down to water depths of > 3000 m (mean 32-20 µm, mode > 32 µm). With increasing water depths, the grain size of silt drops to minimum values in the southwesternmost rim of the basin below 4000 m water depth (mean and mode 16-11 µm). These sites are also characterised by symmetric distributions (skewness ~ 0) but poor sorting (> 1.3). For TS samples from the remaining area a positive skewness is calculated, most prominent off the La Plata. Sorting is best (< 0.9) at continental slope sites south of La Plata.

In the northeast, grain size decreases rapidly from shallow (< 1000 m) to intermediate and deep sites. TS mean values vary from 20-15 µm at the continental slope to 7-6 µm at the lower rise and in the southern Brazil Basin. Compared to the mean values, the corresponding modal values generally reveal a similar pattern but a greater variability and contrast, i.e. the water-depth related modal size decrease covers 45-3 µm. In the same way as size decreases, skewness decreases with water depth to negative values, showing minimum values (< -0.7) at the lower rise and southern Brazil Basin. Good sorting (< 1) is observed along a transect slightly north of the La Plata, at the eastern slope of the VC and in the southern Brazil Basin.

End-Member Modelling

The silt particle-size data set was unmixed into subpopulations using the end-member algorithm of Weltje (1997). The number of end members (dimensions of mixing space) is determined on the

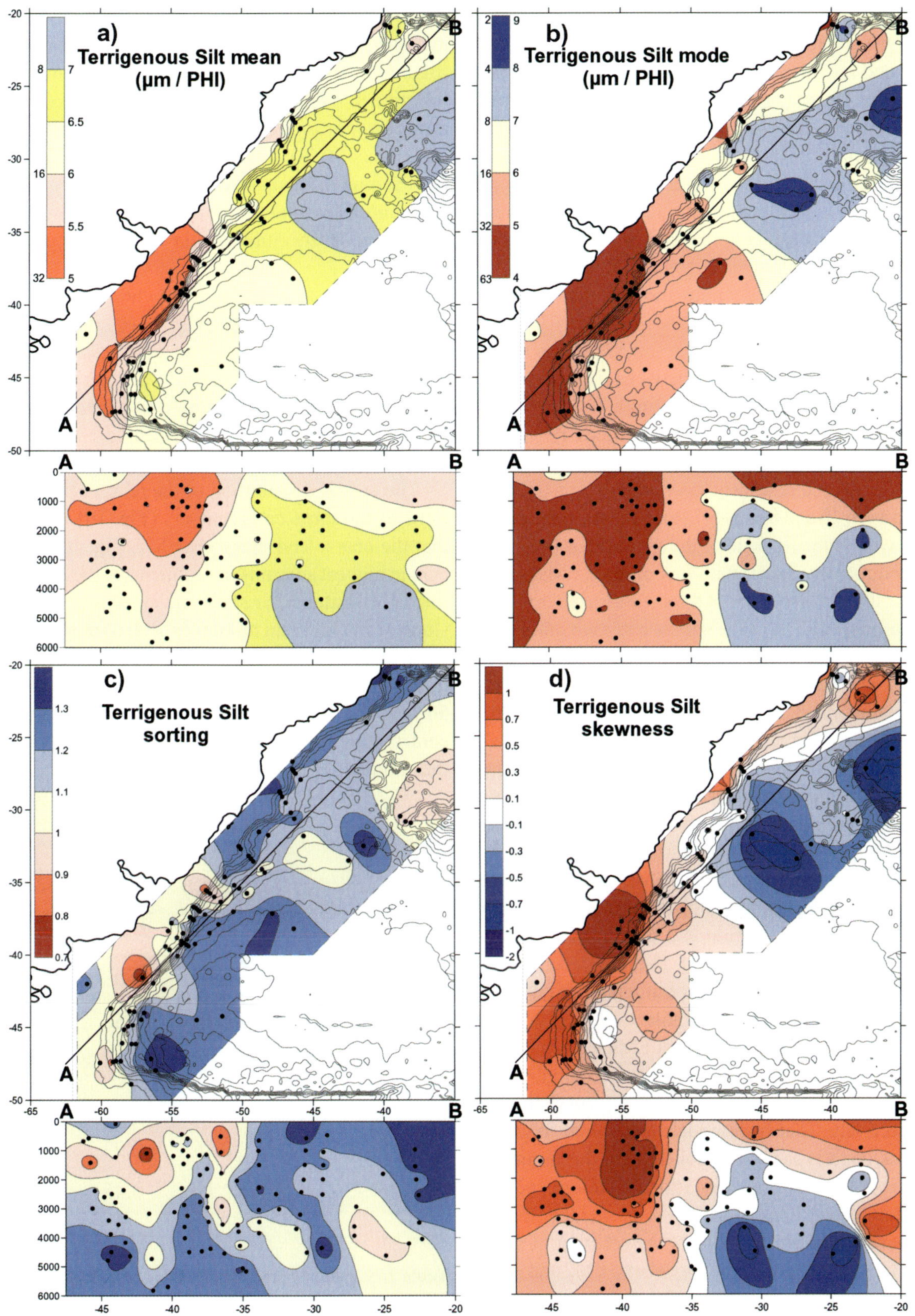
a)
Terrigenous Silt mean
(μm / PHI)
8   7
6.5
16   6
5.5
32   5
A
B

b)
Terrigenous Silt mode
(μm / PHI)
2   9
4   8
8   7
16   6
32   5
63   4
A
B

c)
Terrigenous Silt
sorting
1.3
1.2
1.1
1
0.9
0.8
0.7
A
B

d)
Terrigenous Silt
skewness
1
0.7
0.5
0.3
0.1
-0.1
-0.3
-0.5
-0.7
-1
-2
A
B

basis of goodness-of-fit statistics (Prins and Weltje 1999), i.e. the maximum fit with the least amount of end members. The end-member algorithm calculates the goodness of fit (coefficient of determination) per size class and per set of end members. For each set of end members, the mean coefficient of determination is used to determine the appropriate amount of end members. The mean coefficient of determination (Fig. 8a) increases with the increasing number of end members. Figure 8b shows the coefficients of determination per size class for models with 2 to 5 and 10 end members. The two-end-member model (mean $r^2 = 0.59$) shows low r^2 values in the size ranges between 32 and 10 μm (~5-6.5 Φ, $r^2 < 0.6$) and < 4 μm (> 8 Φ, $r^2 < 0.6$). The three-end-member model (mean $r^2 = 0.87$) shows low r^2 values in the size range < 2.7 μm (> 8.5 Φ) and relatively low r^2 values ($r^2 = 0.8$-0.9) between 16 and 8 μm (6-7 Φ). The latter size range as well as the fine end of the spectrum is better reproduced by models with four or more EMs (mean $r^2 \geq 0.93$). On average, a three-end-member model reproduces 87 % of the variance in each grain-size class compared to 93 % reproducibility of a four-end-member model. The mean coefficient of determination increases only slightly in models with more than four EMs (e.g. five-end-member model, $r^2 = 0.95$). In conclusion, the goodness-of-fit statistics suggest that a three- or a four-end-member model provides a reasonable choice with respect to contradictory requirements, i.e. reproducibility versus minimal number of EMs. The grain-size frequency distributions of the end members for both, the three and four end-member models are shown in figures 8c and 8d. The statistical parameters of the frequency distributions of the EMs in both models are given in Table 1.

The coarsest and the finest EMs in each model are almost identical, showing prominent modes at approximately 47 μm (4.4 Φ) and 4 μm (8 Φ), respectively. That both models are similar in goodness-of-fit is partly consequential to the similarity of their coarsest and finest EMs. Therefore, the two intermediate EMs of the four-end-member model are regarded as the result of the splitting of EM2 of the three-end-member model into EM2 and EM3 in the four-end-member model with minor contributions of the coarse and fine EMs. The marginal

improvement in goodness-of-fit from a three-end-member model to a four-end-member model is attributed to one of the EMs of the four-end-member model, namely EM3, that contributes only a minor proportion to the compositional variation (maximum 30 %). Therefore, it is concluded that the TS grain-size composition of the South American continental margin (Atlantic side) can be regarded as a mixture of three end members (Fig. 8c).

The distribution of the three end members in the ternary mixing space (Fig. 9) reveals hardly any concurrence of the coarse and the fine EMs. Samples from the southwest (> 53°W) are predominantly mixtures of medium and coarse EMs, whereas sediments from the northeast are mixtures of the medium and fine EMs.

The spatial distribution of the relative proportions of the three end members (Fig. 10) reveals that the coarse end member (EM1, Fig. 10a) occurs in highest amounts in sediments at the Argentine continental outer shelf and slope and downslope of the La Plata mouth, as well as at shallow sites in the northeast (40-90 %). The fine end member (EM3, Fig. 10c) contributes more than 50 % over a wide area of the northeast. Maximum contributions are observed at the lower rise and in the southern Brazil Basin (> 80 %). 75 of the 95

	three-end-member model		
	EM1	EM2	EM3
Mean (Φ)	4.92	6.01	7.44
Mean (μm)	33.14	15.50	5.76
Mode (Φ)	4.55	5.65	8.05
Mode (μm)	42.69	19.92	3.77
Sorting	0.76	0.93	0.95
Skewness	2.10	0.45	-0.75

Table 1. Statistical parameters of the size frequency distributions for the three end-member model calculated according to Krumbein (1936).

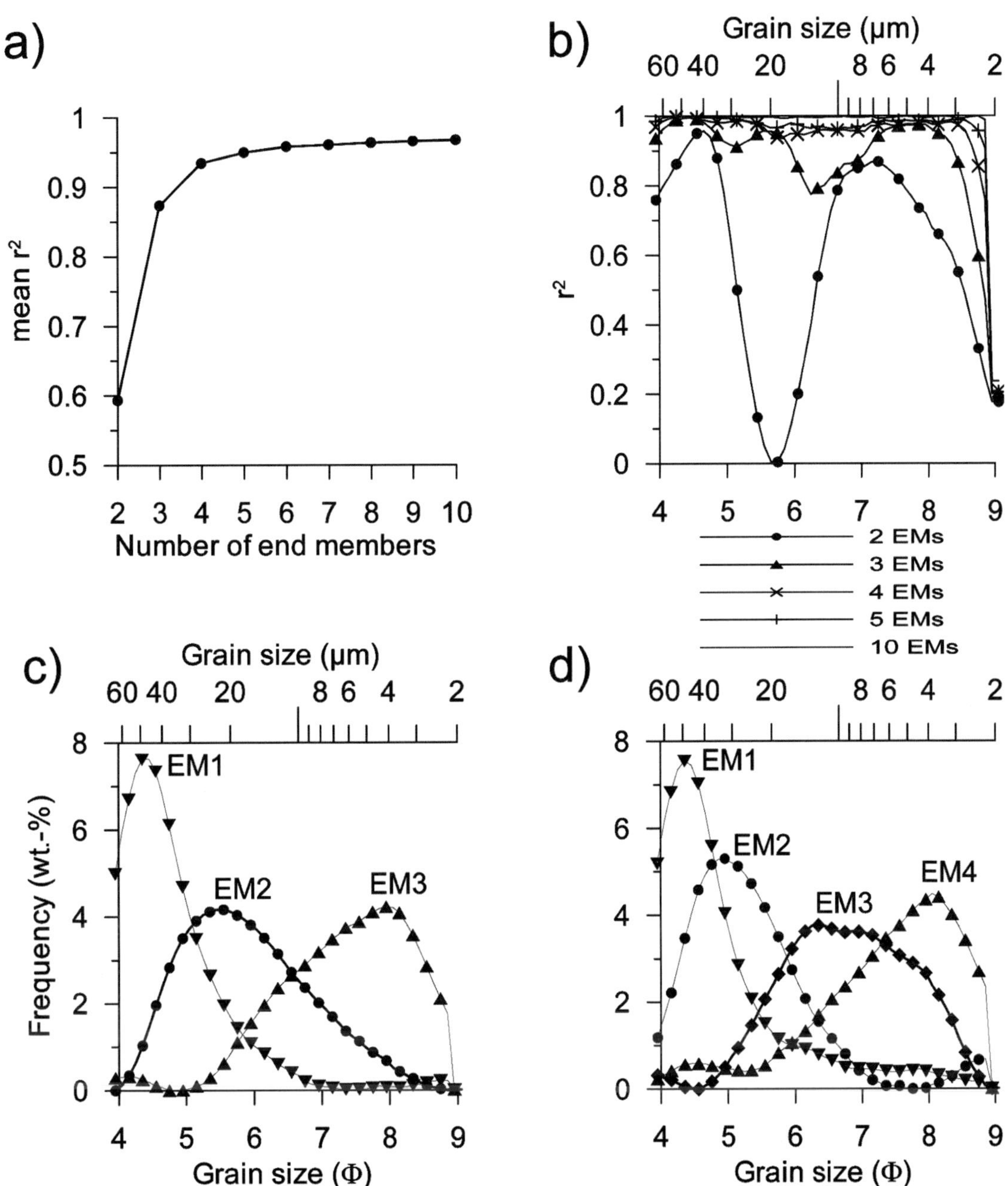

Fig. 8. Goodness-of-fit statistics to estimate the number of end members. **a)** Mean coefficient of determination for models with 1 to 10 end members. **b)** Coefficients of determination for each size class for models with 2-5 and 10 end members. **c)** and **d)** Modelled end members of terrigenous silt from the eastern South American continental margin (size frequency distributions of the end members) for the three-end-member model (c) and the four-end-member model (d), respectively.

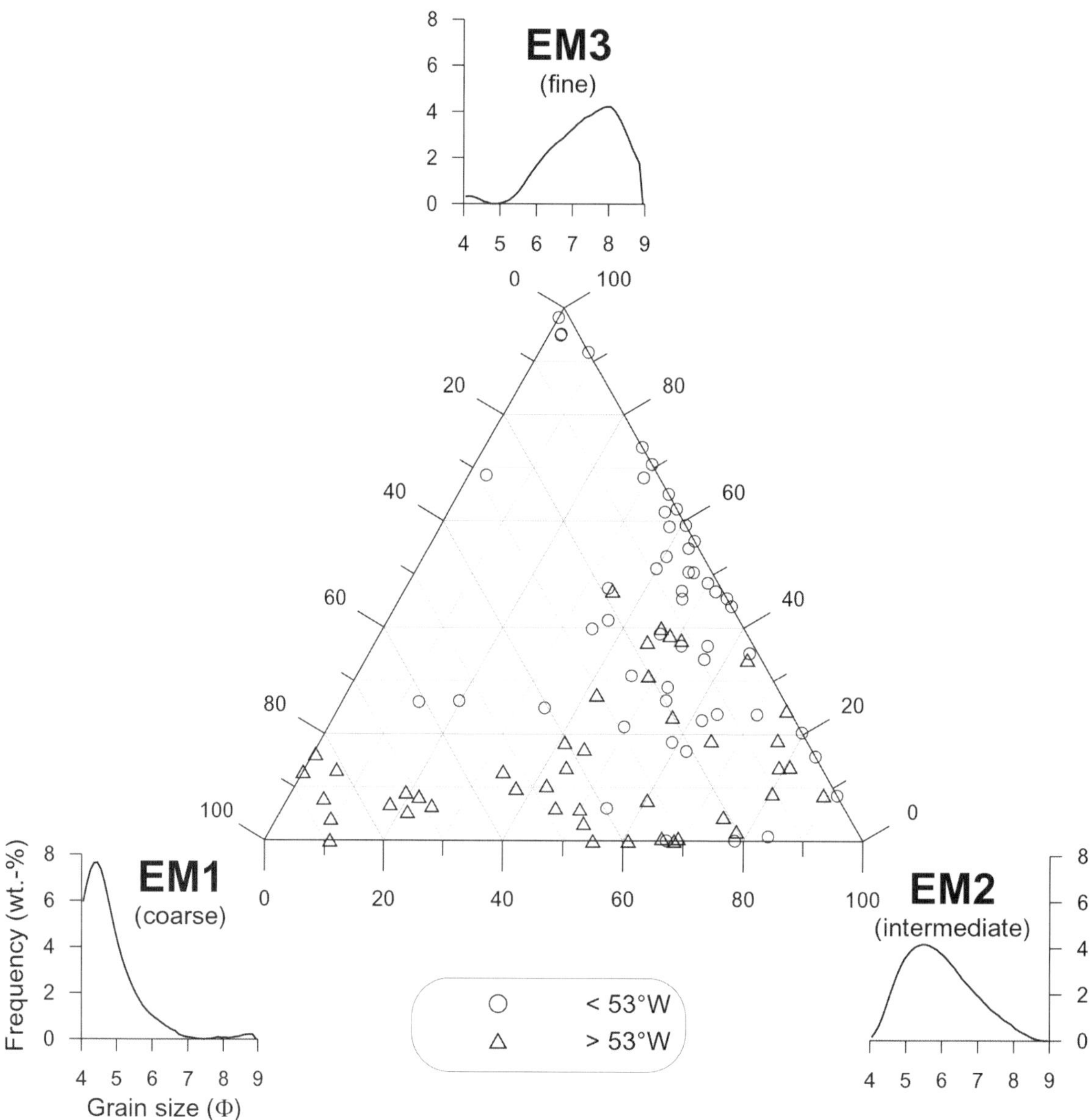

Fig. 9. Ternary mixing diagram for the terrigenous silt of surface sediments from the eastern South American continental margin and the size-frequency distributions of the three end members. Circles represent samples from the northeast (< 53°W) that show mainly binary mixing of the fine and the intermediate end members (EM3-EM2). Triangles denote samples from the southwest (> 53°W) that reveal predominantly binary mixing of the coarse and the fine end members (EM1-EM2).

samples consist by more than 30 % of the intermediate end member (EM2, Fig. 10b). High concentrations of EM2 (> 70 %) are found in a narrow corridor off the La Plata mouth that coincides with the low-$CaCO_3$/high-C_{org} corridor. Another area of higher contributions of EM2 (60-80 %) is located at the southwestern rim of the Argentine Basin.

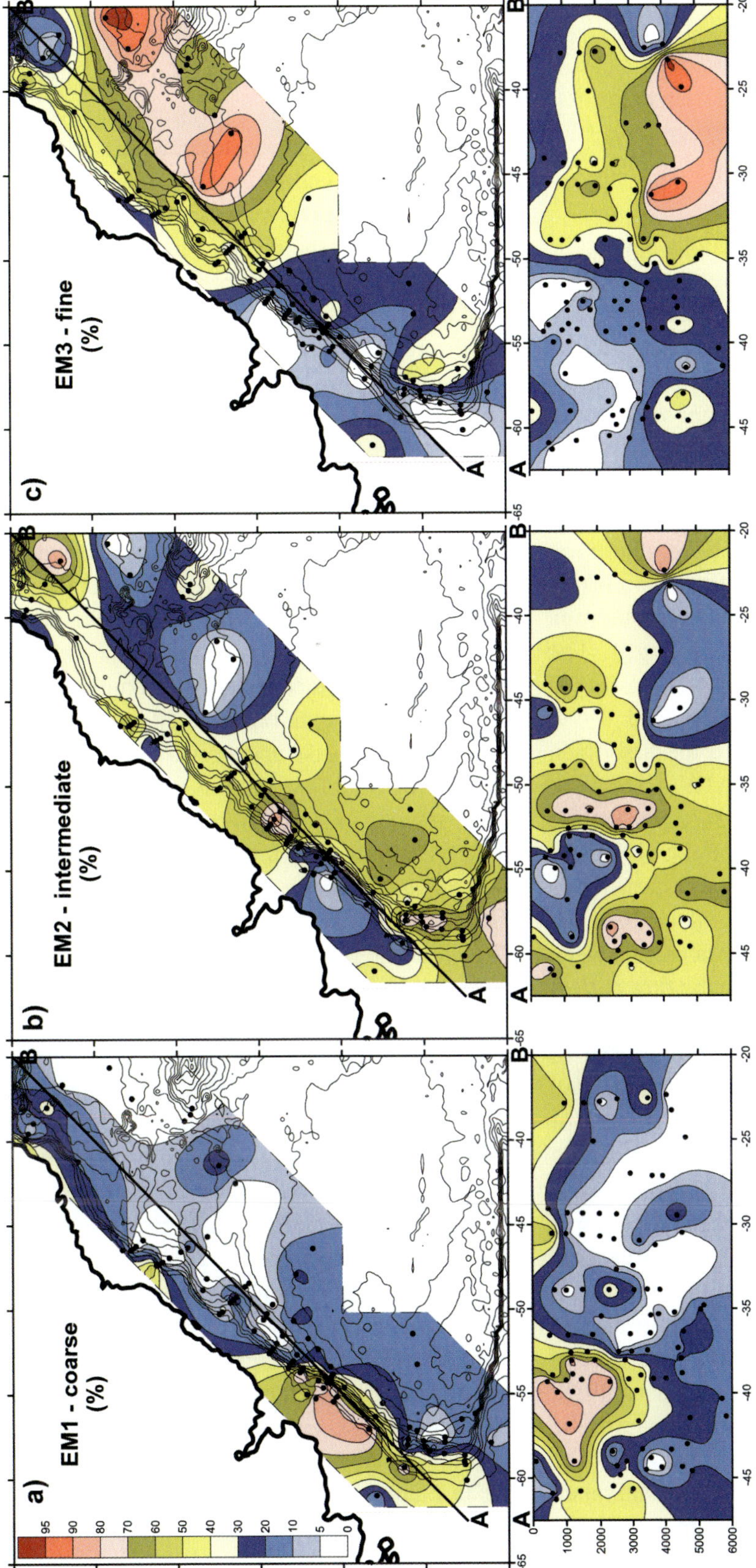

Fig. 10. Spatial distribution of the proportion of the three end members (EM). **a)** Highest amounts of the coarse EM1 are observed downslope the La Plata and along the Argentine continental slope corresponding to coarse TS grain size (Fig. 6a, b). High proportions of this end member indicate winnowing by strong currents and/or direct coarse river sediment discharge. **b)** Highest amounts of the intermediate EM2 are observed at the area congruent with the high C_{org}/low $CaCO_3$ corridor (Fig. 2) as the imprint of the BMC. The silt-sized terrigenous fraction picked up by the MC off the mouth of the La Plata is released from the high energetic mixing zone of thermocline and subtropical waters because the energy drops due to increasing water depths along its southward extension at about 52–54°W. High proportions of EM2 at the most southwestern rim of the Argentine Basin are attributed to the release of coarse suspended matter from the MC that was probably eroded on its way across the Falkland Plateau. **c)** Spatial distribution of the proportion of EM3 once more indicating the contrast between coarse southwestern and fine northeastern sediments. High amounts of this finest end member reveal accumulation processes where current velocities are strongly reduced. Reference areas are the southwestern deep part of the RGR where AABW is dammed up and the abyssal southern Brazil Basin where AABW broadens again behind the jet passage through the Vema Channel.

Discussion

Distribution of Carbonate and Organic Carbon as Proxies for Water Masses and Main Oceanographic Features

Apart from the dilution by terrigenous inputs, the preservation of marine sediment compounds is mainly governed by complex interactions between productivity and the physical and chemical properties of the overlying water mass. The NADW represents a relatively warm and saline water mass, compared to the overlying UCDW and AAIW, and the underlying LCDW and AABW. Most critical for the carbonate preservation is the carbonate-ion saturation level. In the Argentine Basin, the intermediate and deep water masses of southern origin (AAIW, CDW, AABW) display an undersaturation of carbonate ions ($CO_3^{2-} < 90$ µmol kg^{-1}). The NADW, in contrast, is oversaturated with respect to CO_3^{2-} ($CO_3^{2-} > 110$ µmol kg^{1}, Dittert et al. 1999; Broecker and Peng 1982). Therefore, the dissolution of carbonate is increased in AAIW, CDW and AABW water levels, whereas NADW fosters enhanced carbonate preservation. Following these basic considerations, the continuously elevated carbonate contents in surface sediments along the South American continental rise at water depths between 2000 and 4000 m clearly document the flow path of the low corrosive NADW – at least north of the BMC. This is supported by previous studies of Gerhardt and Henrich (2001) and Volbers and Henrich (2002), who used ultrastructures of pteropods and foraminifer tests to trace preservation/dissolution patterns in South Atlantic sediments.

According to the commonly accepted model of recent water mass distribution in the Argentine Basin (e.g. Peterson and Whitworth 1989; Maamaatuaiahutapu et al. 1992; Stramma and England 1999), the NADW travels as a deep western boundary current south as far as 39°S, where it turns away from the continent towards the east. At the continental rise south of the BMC, the NADW is displaced by the upper and lower branch of the CDW (Fig. 1). These modern hydrographic observations are in sharp contrast to the sedimentologic findings. South of the BMC (43-48°S), at a water depth between 2000 and 3500 m, high carbonate contents in the sediments (50-70 %) comparable to those north of the BMC (Fig. 2) are indicative of good carbonate preservation. A considerable decrease in carbonate contents is found only east of 57°W, where the Falkland Escarpment shows an increased steepness, in agreement with the presence of the CDW. The contradiction is based on the contrasting presence of the carbonate under-saturated CDW and good carbonate preservation. The conclusions on the basis of sedimentology are apparent: The NADW extended further south (down to 48°S) during the younger oceanographic history. The disagreement in results of the two scientific disciplines may be caused by the different time windows observed when analysing their specific samples. The set of hydrographic samples comprise the last few decades at the most, i.e. they are snap shots of the most recent oceanographic situation. In contrast to these short-term observations, surface sediments comprise a mean record over the time represented by the sampled interval, in this case up to several hundred years. Therefore possible variations (production and remote arrival) of northern deep-water flux to the south (Bianchi and McCave 1999) can cause the state of carbonate preservation observed south of the BMC, but is undetectable hydrographically. Consequently, the retroflexion of the CDW occurred already at the Falkland Escarpment near 57°W. From the sedi-mentological point of view, the whole horizontal band of elevated carbonate contents between 2000 and 3500 m water depth can be related to the presence of NADW along the eastern South American continental margin.

The variations of sediment carbonate concentrations within this main NADW flow path, on the other hand, are attributed to different origins. The particularly high carbonate values observed at the eastern slope of the VC and shallow sites of VTR relate to a reduced terrigenous supply at exposed and/or distal positions, likely supplemented by good carbonate preservation (Gerhardt and Henrich 2001). Alternatively, reduced carbonate contents off the la Plata mouth may result from two effects, namely terrigenous dilution in response to high river discharges and/or an enhanced supralysoclinal dissolution of carbonate due to high organic-carbon flux rates which are associated with the river-induced high productivity at the Brazil Malvinas Confluence zone. The latter is supported by previous studies that found high nutrient concentrations in surface waters (Benthien and Müller 2000) and sediments (Hensen et al. 2000) off the La Plata.

The presence of hydrogen-enriched organic matter of mixed marine-terrigenous origin at shallow sites in front of the river system and below the BMC, and its general trend to lower contributions in deeper marine settings supports the assumption of an enhanced carbonate dissolution in relation to the remineralisation of settling reactive organic matter. The present data on the origin of sedimentary organic matter from that area, however, are not sufficient to finally rule out a marine or mixed marine-terrigenous source of the reactive organic matter.

The lower boundary of the NADW is the interface to the AABW, which determines the position of the modern calcite lysocline (compare Henrich et al. this volume). According to the carbonate distribution presented here, this boundary is located at about 4000 m water depth in the Argentine Basin and about 200 m deeper in the Brazil Basin. These estimates are in good agreement with Thunell (1982) who reports identical water depths.

Decreasing carbonate contents between 2000 and 1500 m water depth all along the continental margin record the transition from the NADW to the UCDW and AAIW. In the southwest, four principal processes interact to cause the quasi absence of carbonate along the Argentine continental slope and outer shelf: 1) very low carbonate test production in the cold and turbid surface currents, 2) strong dilution by high terrigenous river discharge, and 3) increased carbonate corrosivity along the outer continental shelf and slope, that are occupied by the UCDW and AAIW.

In the northeast of the study area, the flow of the UCDW and the AAIW constitutes the western closure of the subtropical gyre, i.e. here the currents flow south-westwards (Boebel et al. 1997; Stramma and England 1999). In this part, the carbonate contents are only depressed at the intermediate water layer (~1500-500 m water depth). This points to the UCDW and the AAIW, although they produce a lower corrosive signal compared to the

southwest. To a certain extent, this is attributed to an advanced alteration of these water masses, after running a complete cycle through the South Atlantic. The main reason for the comparably higher carbonate contents, however, likely relates to Thermocline Water masses, which foster primary production supported by coastal and shelf-break upwelling (Piola et al. 2000). This induces higher carbonate test production that is not completely dissolved by the UCDW and the AAIW. As compared to the southwest, good preservation and minor terrigenous dilution in the surface water layer, enable higher carbonate contents (compare Gerhardt and Henrich 2001).

The corridor of distinctly elevated organic carbon contents (Fig. 3) marks a narrow zone of preferred deposition of organic matter, which coincides with a key position of upper level (surface and intermediate) circulation: the confluence of the Brazil and the Malvinas Currents in the vicinity of the mouth of Rio la Plata estuary (BMC). The collision of the contrasting MC and BC results in a southward deflection of both currents and a corridor of intense mixing that can affect even the deep water layer (Piola and Matano 2001; Olson et al. 1988; Vivier and Provost 1999). This intense mixing causes an enrichment in nutrients that is also stimulated by a high mean freshwater discharge of the La Plata [14,900 $m^3 s^{-1}$ (Milliman and Meade 1983) to 23,300 $m^3 s^{-1}$ (Piola et al. 2000)]. The latter carries 7.2 10^{12} g organic carbon annually to the South Atlantic (Depetris and Paolini 1991). These factors result in a high productivity area at the BMC and along its southward extension. The organic matter produced in the BMC drops along its pathway and traces it in the sediments. Therefore, the exact configuration of this specific oceanographic feature is nicely portrayed by the organic carbon contents of the surface sediments. Even though oceanographers like to define the BMC at 38°S (e.g. Maamaatuaiahutapu et al. 1998; Vivier and Provost 1999; Piola and Matano 2001), the high-C_{org} trace in the sediments has a more north-south alignment at about 52-54°W. This position is in good agreement with oceanographic observations of the MC and BC extensions following the BMC on the shelf (Piola et al. 2000) and over the deep-sea (Olson et al. 1988).

The organic carbon variations observed in the surface sediments northeast and southwest of the BMC corridor provide another tracer for water-mass properties. Below a water depth of 4000 m, the complete dissolution of carbonate leads to a relative enrichment of organic carbon. The preservation of organic carbon in the AABW is supported by low oxidation rates of organic matter due to the low oxygen contents of the AABW.

The core of NADW represented by high carbonate concentrations in the surface sediments at 2000-3500 m water depth in the northeast is characterised by very low concentrations of organic carbon. These are related to the high oxygen content of the NADW and thus its low preservation potential for organic matter. Elevated contents of organic carbon north of the BMC at 1000-2000 m water depth clearly reflect the presence of the UCDW with its very low oxygen concentrations. In the upper level in the northeast, a minor terrigenous dilution and a good carbonate preservation result in a slight decrease in organic carbon. In the southwest, the lack of considerable amounts of organic carbon in the intermediate and upper levels is the result of massive terrigenous dilution as well as of high oxygen concentrations in the formation region of the AAIW (Stramma and England 1999).

Bulk Size Composition and its Origin

The strong contrasts between southwestern coarse- and northeastern fine-grained deposits, obvious from all parameters related to particle size, are regarded as the product of the combined effects of geological, sedimentological and oceanographic conditions that act on provenance (availability/supply), transport, sorting and accumulation of source components.

Potential source areas for sediments deposited on the eastern South American margin range from the tropics in the north to the subpolar zone in the south. Hence, different weathering regimes act on a large variety of source rocks and soils. Chemical weathering is the most prominent process at low latitudes producing very fine-grained alteration products. In contrast, physical weathering (temperature and frost) becomes increasingly important

towards higher latitudes resulting in coarser grained and less chemically altered products.

Physical maps of South America (e.g. GEBCO 1984, chart 5-12) reveal that the central part of the continent is drained by the Rio Paraguay/Rio Paraná system, which is situated in the Paraná Basin that covers a drainage area of $2.83\ 10^6\ km^2$. This extensive river system is focussed at the broad river mouth of the Rio de la Plata, which discharges $470\ km^3\ a^{-1}$ of freshwater, supplying a sediment load of $92\ 10^6\ t\ a^{-1}$ (Milliman and Meade 1983). South of the La Plata mouth, other large rivers drain onto the extended Argentinean shelf (e.g. Rio Colorado, Rio Negro, Rio Chubut). In contrast, north of the La Plata mouth, the number of rivers draining into the South Atlantic is low and their size is small. According to this general difference in the presence (SW) and absence (NE) of rivers, an overall higher and more variable terrigenous input with a great variety of grain sizes is expected on the southern shelf.

The Argentine shelf is situated within and beyond the "roaring forties" which are known for their extraordinary high storm potential. During storms, the influence of waves can extend down to a water depth of 200 m (Chamley 1990). Hence, re-mobilisation processes have a major impact on modern sediments from the continental shelf and slope. Once sediment particles are taken up in suspension, the fine components are likely to be removed and advected downslope or laterally with the Malvinas Current, leaving a coarse residual type of sediment.

Even though we are not able to quantify the contribution of each of these processes, the expected northeast-southwest (i.e. fine-coarse) contrast is clearly recognised from the simple sand/silt/clay distribution patterns and their relative proportions (Figs. 4, 5).

Grain-Size Composition of the Terrigenous Silt Fraction

The eastern continental margin of South America corresponds to the western boundary of the South Atlantic Ocean. We therefore assume that western boundary currents have a major impact on sedimentary processes. Theoretical considerations on grain-size frequency distribution patterns lead to the assumption that three basic types of current-influenced sediments can be distinguished: 1) low energy depositional (accumulation) sediments, 2) erosional or residual sediments and 3) well-sorted sediments (Michels 2000). According to this model, depositional sediments are characterised by a positive (i.e. fine) skewness due to the lack of coarse grain size. They occur when current velocities are slowed down enough to allow even fine particles to settle. In contrast, the fine proportion of the residual type sediments is lacking due to the winnowing effect of strong currents. These deposits display a negative (i.e. coarse) skewness. Both accumulation and residual deposits are rather poorly sorted. The third distribution type, which intermediates between the first two, is well-sorted, i.e. it displays a narrow range of grain sizes. Here, both coarse and fine components are absent, due to frequent re-suspension/deposition cycling, which ideally results in a very good sorting and a symmetrical frequency distribution. Hence, for the interpretation of grain-size distributions its symmetry or "shape" matters more than mere "size". On the other hand, size is an indicative parameter to trace energetic regimes and the distance to the source. For example, waves and strong currents leave behind coarser sediments than lower energetic regimes (e.g. McCave et al. 1995; Ledbetter and Ellwood 1980), and the grain size of volcanic ash or atmospheric dust decreases downwind (e.g. Walker 1971; Prins and Weltje 1999).

Excursion: Some General Aspects on the Problem of Grain-Size Analysis limited to the Silt Fraction and Characterisation of the EMs

Before discussing the measured grain-size distributions in terms of sedimentary processes, a major problem of analysing the grain sizes of silt has to be addressed: the upper and lower size boundaries of the silt fraction. These artificial borders have a major impact on the character of the silt frequency distributions and hence on the resulting statistical parameters and their interpretation. As demonstrated in figure 5, the sediments along the South American continental margin show a great variety in general size composition ranging from almost pure sand to almost pure mud. For example, the muddy sediments consist of considerable amounts of clay (30-80 % of bulk sample), besides a major proportion of fine silt. Particularly the carbonate-poor sandy sediments on the Argentine shelf and slope (sand > 90 %) contain only a minor proportion of silt (< 10 %) and clay (< 2 %). These general size relations imply that the terrigenous silt fraction analysed may represent only a minor proportion of the whole size range of the sediment. Hence, conclusions only derived from the TS size frequency distributions and their statistical parameters, may cause misinterpretations of the actual processes governing sediment dispersal and deposition.

To discuss these potential problems in more detail the end members EM1 and EM3 are used as extreme examples for grain size distributions in the areas of their preferential occurrence. The coarse EM1 predominantly occurs on the Argentine continental shelf and slope and downslope of the La Plata mouth where the terrigenous sand content exceeds 90 %. In this case any grain size less than 8 μm (~7 Φ) can be neglected [$\Sigma(< 8\ \mu m) < 2\ \%$]. For EM1 the statistical calculations (Table 1) reveal a strong positive (fine) skewness, which is indicative for accumulation processes. On the other hand, very coarse mean and modal grain size and good sorting relate to winnowing by variable albeit strong currents. Considering that these attributes of EM1 are only valid for less than one-tenth of the carbonate-free sediment (Fig. 11a), the rather limited significance of parameters deduced from TS size frequency distributions is obvious. But from the high sand contents we know that the terrigenous bulk spectral mean and modal sizes are actually situated in the sand fraction, i.e. they become even coarser than calculated for TS. The strong positive skewness for TS is attributed to the sharp size cut at 63 μm (4 Φ) that produces a steep slope at the coarse end and therefore a relatively gentle slope towards the fine end of the spectrum. Since the size frequency distribution of the sand fraction is unknown, the question of "true" skewness is open, i.e. it can remain positive or even become negative in the bulk size range. Therefore, EM1 can represent a very coarse accumulation pattern as well as a residual sediment produced due to winnowing and reworking by strong currents. Important evidence allowing us to distinguish between theses converse processes represented by EM1 is provided by differences in TS sorting and skewness and by general environmental considerations (current velocities, sediment sources; see also next chapter). At comparable proportions of TS (Figs. 2, 4) and EM1 (Fig. 10a) the TS parameters downslope of the La Plata mouth reveal poorer sorting and even stronger positive skewness compared to alongside the slope further south. Therefore, the sediments downslope of the La Plata are finer grained, making a "true" positive (fine) skewness more reliable. Additionally, the spatial distribution of this pattern as a downslope tongue proximal to the La Plata mouth suggests a direct link to this sediment source. In contrast, the EM1 pattern further south is aligned straight along the Argentine continental slope. The TS from these locations is very well sorted and it can be assumed that the sharp size cut at 63 µm (4 Φ) will be converted into a more gentle slope leaving the slope at the fine end of the size spectrum as the steeper one, i.e. inverting skewness from positive (fake) to negative. That means that here, EM1 experiences a complete reversal in interpretation from a pretended accumulative to an actually erosional deposit. This type of sediment and its rather horizontal distribution is linked to the winnowing processes of relatively strong currents.

In contrast to EM1, for EM3 the size cut at 2 µm (9 Φ) results in a questionable negative (coarse) skewness. Samples with high proportions of EM3 mostly occur below 4000 m water depth (Fig. 10c) where carbonate is almost completely absent (Fig. 2) and the sand contents can be neglected (Fig. 5a). These sediments display silt/clay ratios which are at about 30/70 (Fig. 5b, c). In these cases it can be assumed that the 70 % clay thin out towards the fine end to reverse the fake negative (coarse) skewness to a positive (fine) one and therefore, reveal an accumulation type of sediment. Additionally, the generally fine grain size of these sediments suggests a low energetic deposition rather than current sorting or even winnowing.

In contrast to the two other complementary end-members, in areas of its frequent occurrence EM2 represents a high proportion of the complete

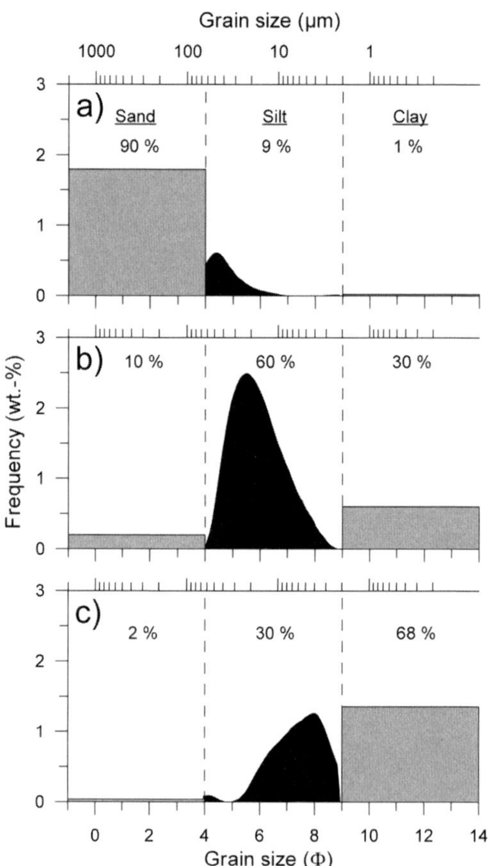

Fig. 11. Visualisation of the problem of grain-size measurements limited to the silt fraction by means of the three end-member examples (a-c). The grain-size distributions of terrigenous silt (black) are rescaled to the relative silt proportion. The sand and clay proportions are divided by 50 size classes and then inserted as one box because their internal grain-size distribution is unknown. Since the EM examples originate from positions with low carbonate contents (EM1 - Argentine continental slope, EM2 - low-$CaCO_3$/high-C_{org} corridor, EM3 - SW foot of RGR below a water depth of 4000 m.) the sand and clay contents are regarded as terrigenous. **a)** EM1 represents samples with very high amounts of sand: solely silt is not very representative for such samples, the size cut at the sand border results in a calculated falsely positive (fine) skewness indicating both, accumulation and winnowing. **b)** EM2 represents samples with high amounts of silt: interpretation of size distributions are correct, true negative skewness indicates accumulation. **c)** EM3 represents samples containing high amounts of clay: the falsely negative (coarse) skewness suggests that the samples originate from winnowing processes, however, they resulted from deposition at low current velocities.

size spectrum (Fig. 11b) exceeding 60 % of the carbonate-free sample. Therefore, its character can be derived directly from its frequency distribution in the silt-size spectrum. According to its positive (fine) skewed distribution we infer accumulation as the main sedimentation process. In contrast to EM3, the relative coarse grain size indicates a higher energetic regime within the seawater suspensions from which the particles settle.

In summary: all three end members indicate sediment accumulation attributed to different energetic regimes or proximity to source, except EM1, which can indicate residual sediments as well.

Spatial Distribution of End-Member Proportions Complemented by Parameter Distribution – the Relation to Sedimentary Processes

EM1 – Very Coarse, Accumulation and/or Residual deposit: In the northeast only few locations along the shelf break and slope off southern Brazil show considerable proportions of the coarse EM1 (> 50 %). The two other EMs at these positions provide about 25 % each. The relative high proportion of EM1 in this area is related to current sorting (winnowing) under the influence of the Brazil Current. The resulting relative coarse composition is supplemented by the relative proximity to the rare terrestrial sources due to a narrow shelf, which enables a distinct contribution of accumulative EM1 and EM2. However, compared to the southwest, the winnowing signal is rather weak indicating the BC to be less strong than its southern counterpart. Current measurements revealed mean speeds of 10-20 cm s^{-1} close to the slope in the intermediate water level (Boebel et al. 1999; Hogg et al. 1999). This velocity range is apparently sufficient to produce a winnowing pattern, since the clay contents (< 30 %) and therefore the cohesive behaviour of these sediments is reduced.

Even though variations are subtle, slightly elevated proportions of EM1 occur along a transect at about 34°S between 2000 and 3000 m water depth and at the lower part of the next northern transect at about 3000 m water depth (Fig. 10a). This pattern is likely to be caused by the flow of the NADW, which is focused at two topographic

ledges, whereas samples from other water depths of these transects and the same water depths of neighbouring transects lack this feature. There, the samples are situated at topographic inlets where current speeds drop again at lee sides. Elevated sand contents, and the resulting fake fine skewness (see above) at these sites support these assumptions.

In the southwest, there are two major axes of high EM1 concentrations (> 70 %), one along the continental slope and outer shelf, and one downslope off the La Plata mouth. The continental slope axis is clearly related to the strong Malvinas Current, leaving a residual type of sediment. Contourites have been detected in this area by means of parasound patterns (Segl et al. 1994; Bleil et al. 2001) confirming our conclusions. The downslope axis shows a weakening trend towards the south but an abrupt break to the east. Notably, this La Plata tongue is situated west of the low-CaCO$_3$/high-C$_{org}$/EM2 corridor (see below). It is assumed that the coarse La Plata tongue is superimposed by this adjacent EM2 corridor and therefore capped in the east to finally display an oblique downslope orientation. The equally high proportions of EM1 in the axes alongside the slope and downslope are divided into 1) residual sediments left by winnowing of vigorous MC (contourite) and 2) direct sandy la Plata discharge prograding downslope mainly as bed load, by differences in TS skewness and sorting (see above). Additionally, oceanographic current velocity data (Boebel et al. 1999; Peterson 1992; Maamaatuaiahutapu et al. 1998; Weatherly 1993; Vivier and Provost 1999) reveal high nearbottom current speeds of > 30 cm s^{-1} for the MC along the continental slope (800-1500 m water depth) which drop to < 5 cm s^{-1} directly off the la Plata mouth. The low current velocity off the La Plata is continued downslope towards the deep-sea allowing finer-grained sediment particles to settle. This increases an admixture of fine grain sizes and consequently results in the observed poorer sorting of the sediments downslope of the La Plata mouth.

EM3 – Very Fine, Accumulation Deposits: To emphasise the contrast to the other two end members we continue with the discussion of the fine EM3. Wide areas of the northeastern part are covered with sediments containing proportions of EM3 above 50 %. There are two areas where this EM3 signature occurs as a quasi-single feature (both other EMs < 5 %), these are situated at the deeper parts of the southern Rio Grande Rise (RGR) and in the southern Brazil Basin. The southern part of the RGR, below 4000 m water depth is situated where the AABW flows north-eastwards. Since the RGR represents a natural obstacle for the AABW, this results in a stowage effect that causes the current velocity to drop considerably, upon which even the finest sediment load is released.

After the jet passage of the Vema Channel (no data, but compare e.g. Ledbetter 1984) the AABW enters the southern Brazil Basin where there is plenty lateral space to spread. This again results in a rapid drop of current speed and release of fine load picked up further south. Low TS mean and modal grain sizes, as well as the (fake) negative skewness and poor sorting of about 1 for both EM3 reference areas support this assumption.

Another area of high proportions of EM3 occurs on the eastern slope of the VC at water depths between 3000 and 4000 m. Therefore, relatively low current velocities of the NADW are suggested to prevail at this site. This assumption is explained by the position at the eastern flank of the VC where minor current focussing occurs. Additionally, lower primary velocities are assumed for the NADW offshore compared to surface currents and the AABW in the deepest parts of the VC (Ledbetter 1984).

Other high EM3 contributions in the northeast are found at downslope transects that are situated within or across topographic inlets. They represent the leeward side of current focussing ledges, and are therefore places of accumulation as indicated by a negative (fake) skewness. The WOCE current metre velocities (Hogg et al. 1999) within one inlet at 2000 to 4000 m water depth (100 m above seafloor) revealed velocities of 6-7 cm s^{-1}, which is obviously low enough to allow fine grain sizes to settle, taking into account an even lower near-bottom speed and assuming a decreasing current profile towards the bottom.

In the southwest, there is only one area with a considerable concentration of the fine EM3 (30-50 %). This is in the deepest southwesternmost part of the Argentine Basin in the domain of the AABW (> 4000 m water depth). Here the AABW flows along the Falkland Escarpment from the east and collides with the lower continental rise before it turns northeastwards. Similar to the southern RGR, the rise represents an obstacle, causing obstruction, drop of flow velocity, and finally the settling of suspended matter. Complicating this view, intermediate amounts of EM2 (30-60 %) are also observed here, causing a symmetric to slight positive skewness and very poor sorting altogether.

EM2 – Intermediate, Coarse Accumulation Deposits: EM2 reveals medium proportions in a large part of the study area (Fig. 10b). One area of highest EM2 concentrations is congruent to the low-$CaCO_3$/high-C_{org} corridor. The coincidence of organic carbon and EM2 reveals that both, organic matter and terrigenous medium to coarse silt have settled together along this corridor. The organic matter originates from the La Plata river input and marine high production along the BMC mixing zone. First estimates based on optical investigation (maceral analysis) suggest a dominance of the marine organic fraction, although with considerable contribution (about 40-50 %) from non-woody angiosperm terrigenous plant tissue. The supply of both, oxidised and non-oxidised terrigenous organic matter (inertinite, vitrinite) presumably indicates erosion and subsequent riverine transport of differentially altered plant matter, either from peat deposits along the adjacent lowlands of the La Plata and/or from fossil, C_{org}-bearing strata in the hinterland (Wagner et al. this volume).

Our data suggest that the Malvinas Current on the shelf flowing north-eastwards picks up the suspended river discharge from the La Plata containing amounts of terrigenous detritic constituents and terrigenous organic material. Immediately after the passage across the river mouth the MC collides with the BC to form the BMC. The MC, including its suspension load, is deflected to the south, building the MC Return Flow that is intensely mixed with the thermocline water of the BC. Downstream, the velocity and energy regimes decrease due to the increasing water depth, which allows the particles of the suspension to settle starting with the coarse proportion (coarse silt). Scavenging of autochthonous marine organic matter (settling of algae blooms) might support this hemipelagic mechanism and include finer particle sizes.

The location of the low-$CaCO_3$/high-C_{org}/EM2 corridor between 52 and 54°W coincides almost exactly with the recent modal positions of the extension of the Brazil (53.5°W) and Malvinas Currents (54.5°W) of Olson et al. (1988). The subtropical and subantarctic fronts, respectively, are situated east and west of these extensions. The high energetic mixing zone of the BMC extension occupies the space in between. The slight shift of the position of the BMC (about 1° towards the east) is attributed to differences in spatial and temporal resolution between oceanographical and sedimentological methods. As discussed before, oceanographic studies display high spatial but low temporal resolution, i.e. they are snapshots of the modern situation. Surface sediments, in contrast, may have a lower spatial resolution but represent mean values of corresponding parameters over a longer period of time (decadal to centurial timescales). The sedimentologically traced BMC therefore, represents the mean position of a highly variable oceanographic feature that shifts to the northeast in winter and to the southwest in summer. South of 40°S the organic-carbon corridor is below the detection limit of the surrounding background of C_{org} = 0.5-1 %. This indicates that the highest proportion of organic matter introduced by the La Plata, or produced at the confluence, has been deposited.

The second area of high EM2 concentration extends north-south at an intermediate water depth (2000-4000 m) along the southwestern rim of the Argentine Basin. A combination of processes is suggested to be responsible, i.e. 1) drop of sediment load eroded from the Falkland Plateau within the MC, 2) drop of load picked up by NADW from downslope processes and/or 3) graded succession from erosion (EM1, slope) via coarse accumulation (EM2, upper rise) to fine accumulation (EM3, lower rise/abyss) according to decreasing current velocity.

Data of Peterson (1992) and Boebel et al. (1999) indicate high near-bottom speeds (> 30 cm s[-1]) of the MC along the continental slope off southern Argentina that is strong enough to cause erosion on and along the Falkland Plateau. When flowing north/northeast, the MC faces deeper waters and

therefore slows down to release the coarsest load first. Due to the similarity to the EM2 corridor at the BMC, we favour coarse silt deposition from the MC to be the important mechanism to produce this pattern that is supplemented by the downslope decreasing sediment grain size.

In the previous paragraphs we have discussed characteristic examples of end-member occurrence and accompanying parameters in terms of sedimentation processes. South of the VTR a downslope transect of samples shows no apparent affinity to sorting processes. Grain-size related parameters as well as the distribution of end-member proportions show an irregular pattern. According to studies of Mello et al. (1992) this transect is situated across a system of turbidite channels that feed the Columbia and Carioca Channels in the deep-sea. The observed irregularity of the distribution pattern of the sediment grain-size properties, therefore, is strongly attributed to mixing with turbidites.

Summary and Conclusions

This study presents a synthesis of the results of geochemical ($CaCO_3$ and C_{org}) and grain-size investigations (sand/silt/clay proportions and terrigenous silt grain-size distributions) on a set of surface sediment samples from the western South Atlantic that covers the South American continental margin from 20 to 50°S, and from the outer shelf to abyssal water depths. Bulk geochemical and grain-size distribution patterns reveal a multifarious composition of sediments along the eastern South American continental margin. These patterns are primarily attributable to differences in climatic conditions (weathering), geographic settings (presence of rivers) and marine production, and are secondary strongly modified by modern ocean circulation.

The bulk geochemical parameters mainly serve as tracers for water masses. Over the period of time represented by the surface sediment samples (100s of years?), elevated carbonate contents between 2000 and 4000 m water depth indicate NADW in this level down to 48°S before it leaves the continental margin. The almost complete absence of carbonate in sediments from below 4000 m water depth in the Argentine Basin and below 4200 m water depth in the southern Brazil Basin indicates the position of the NADW/AABW interface that simultaneously marks the modern calcite lysocline in these areas. Weaker differences in carbonate and organic carbon contents enable us to distinguish Subantarctic/Thermocline Waters, AAIW and UCDW.

In general, the grain-size investigations reveal sharp contrasts between northeastern fine-grained and southwestern coarse-grained sediments. The transition between both modes is situated at about 53°W. Sediment accumulation occurs in three different manifestations which are indicated with the three end members from the application of an end-member algorithm to the terrigenous silt grain-size distributions. The coarsest depositional pattern is observed in front of the mouth of the Rio de la Plata. This direct river discharge is traced as a straight downslope tongue of moderately sorted sandy terrigenous sediments down to 4000 m water depth (EM1).

Deposition of mainly silty sediments (EM2) occurs along a corridor striking north-south between 52 and 54°W, in combination with high accumulation rates of organic carbon and low accumulation rates of carbonate. We propose the following model for the hemipelagic sedimentation within this corridor: 1) incorporation of suspended Rio de la Plata discharges (mainly silt and clay sized terrigenous detritus and organic matter) by the MC in front of the La Plata mouth, 2) distribution of the suspended matter along the pathway of the MC that collides with the BC at the BMC and experiences intense mixing along the BMC extension to the south, 3) decrease in energy regime along the BMC extension due to increasing water depth and release of the suspended matter starting with the coarsest material (coarse silt). As a result the low-$CaCO_3$/high-C_{org}/EM2 corridor is the direct sedimentological imprint of the prominent oceanographic feature of the BMC and its southward extension.

Accumulation of fine-grained sediments (EM3) occurs at the southwestern Rio Grande Rise, at the southwesternmost rim of the Argentine Basin, and in the southern Brazil Basin, all in the AABW domain. The deposition of silty clay at these locations is attributed to a decrease in current velocity of the AABW in the course of stowage effects at bathymetric obstacles (RGR, lower Argentine continental rise) or due to re-spreading of this water mass following the jet passage of the Vema Channel (southern Brazil Basin).

The sediments along the Argentine continental outer shelf and slope reveal a similar coarse size composition compared to that of the sandy downslope tongue off the La Plata mouth (high sand contents, high proportions of EM1). In contrast to the La Plata tongue, however, the Argentine slope sediments show a less strong fine (fake) skewness and an extremely good sorting. From these differences a strong winnowing effect of the MC with its high near-bottom current velocities is inferred to be responsible for the contourite sediments at the Argentine outer shelf and slope. In this case, the end-member modelling and sedimentological considerations revealed one EM (here EM1) to indicate two different deposits: contourites (winnowing) and continuous bed load sediments (downslope accumulation).

South of VTR a sample transect crosses a system of turbidity channels, clearly documented by a completely unordered shallow-deep sedimentologi-cal sequence. As the conclusions from sedimento-logical data are in good agreement with oceanographic observations, except for this transect south of VTR, we assume that downslope mass wasting processes have a minor impact on sediment grain-size composition of surface sediments.

Methodically, this study tackled two basic aspects: We have demonstrated that the combination of bulk geochemical data and grain-size properties of surface sediments, unmixed with an end-member algorithm forms a powerful tool to reconstruct oceanographic conditions along a time slice. The use of discrete size fractions (i.e. separation of sand and/or clay) will affect the results concerning grain-size frequency distributions. It is recommended to have a look beyond these size limits in order to reveal the proper characteristics of a deposit and hence, to deduce appropriate interpretation.

Acknowledgements

We thank the members of the shipboard and scientific crews of the RV *Meteor*. G. J. Weltje kindly provided the end-member algorithm applet. M. Klann is acknowledged for carrying out the opal measurements. We like to thank M.A. Prins and an anonymous reviewer whose critical remarks improved the manuscript. This research was funded by the Deutsche Forschungsgemeinschaft (Sonderforschungsbereich 261 at the University of Bremen). This is SFB contribution no. 352. All data are available under www.pangaea.de/Projects/SFB261.

References

Arhan M, Heywood KJ, King BA (1999) The deep waters from the Southern Ocean at the entry to the Argentine Basin. Deep-Sea Res 46: 475-499

Benthien A, Müller PJ (2000) Anomalously low alkenone temperatures caused by lateral particle and sediment transport in the Malvinas Current region, western Argentine Basin. Deep-Sea Res 47: 2369-2393

Bianchi GG, McCave IN (1999) Holocene periodicity in North Atlantic climate and deep-ocean flow south of Iceland. Nature 397: 515-517

Bianchi GG, McCave IN (2000) Hydrography and sedimentation under the deep western boundary current on Björn and Gardar Drifts, Iceland Basin. Mar Geol 165: 137-169

Bianchi GG, Vautravers M, Shackleton NJ (2001) Deep flow variability under apparently stable North Atlantic Deep Water production during the last interglacial of the subtropical NW Atlantic. Paleoceanography 16: 306-316

Bleil U, cruise participants (2001) Report and preliminary results of *Meteor*-cruise M46/3. Ber Fachber Geowiss Univ Bremen 172, 161 p

Boebel O, Schmid C, Zenk W (1997) Flow and recirculation of Antarctic Intermediate Water across the Rio Grande Rise. J Geophys Res 102 (C9): 20967-20986

Boebel O, Schmid C, Zenk W (1999) Kinematic elements of Antarctic Intermediate Water in the western South Atlantic. Deep-Sea Res 46: 355-392

Broecker WS, Peng TH (1982) Tracers in the Sea. Lamont Doherty Geol Obs Publ. Columbia University, New York, 690 p

Chamley H (1990) Sedimentology. Springer, Berlin, 628 p

Depetris PJ, Paolini JE (1991) Beiogeochemical aspects of South American rivers: The Parana and the Orinoco. In: Degens ET, Kempe S, Richey JE (eds) Biogeochemistry of major world rivers. J Wiley & Sons, Chichester, pp 323-348

Dittert N, Baumann KH, Bickert T, Henrich R, Huber R, Kinkel H, Meggers H (1999) Carbonate dissolution in the deep-sea: Methods, quantification and paleoceanographic application. In: Fischer G, Wefer G (eds) Use of Proxies in Paleoceanography: Examples from the South Atlantic. Springer, Berlin, pp 255-284

Flood RD, Shor AN (1988) Mud waves in the Argentine Basin and their relationship to regional bottom circulation patterns. Deep-Sea Res 35: 943-971

GEBCO (1984) General bathymetric chart of the Oceans. Department of Fisheries and Oceans, Ottawa, Canada

Gerhardt S, Henrich R (2001) Shell preservation of *Limacina inflata* (pteropoda) in surface sediments from the Central and South Atlantic Ocean: A new proxy to determine the aragonite saturation state of water masses. Deep-Sea Res 48: 2051-2071

Gröger M, Henrich H (2003) Deep-water circulation during the Pleistocene (0.8-0.25 Ma): Inferences from near-bottom-current flow variability and deep-water chemistry in the western equatorial atlantic. Mar Geol, submitted

Gordon AL (1986) Interocean exchange to thermocline water. J Geophys Res 91(C4): 5037-5046

Gordon AL, Greengrove C (1986) Geostrophic circulation of the Brazil-Falkland Confluence. Deep-Sea Res 36: 359-384

Haese RR, Schramm J, Rutgers van der Loeff MM, Schulz HD (2000) A comparative study of iron and manganese diagenesis in continental slope and deep-sea basin sediments off Uruguay (SW Atlantic). Int J Earth Sci 88: 619-629

Hensen C, Zabel M, Schulz H (2000) A comparison of benthic nutrient fluxes from deep-sea sediments off Namibia and Argentina. Deep-Sea Res 47: 2029-2050

Höppner R, Henrich R (1997) Kornsortierungsprozesse am Argentinischen Kontinentalhang anhand von Siltkorn-Analysen. Zentralblatt Geol Pal I 7-9: 897-905

Hogg N, Siedler G, Zenk W (1999) Circulation and variability at the southern boundary of the Brazil Basin. J Phys Oceanogr 29: 145-157

Krumbein WC (1936) Application of logarithmic moments to size frequency distributions of sediments. J Sed Petrol 6: 35-47

Larqué L, Maamaatuaiahutapu K, Garçon V (1997) On the intermediate and deep water flows in the South Atlantic Ocean. J Geophys Res 102 (C6): 12425-12440

Ledbetter MT (1984) Bottom-current speed in the Vema Channel recorded by particle size of sediment fine-fraction. Mar Geol 58: 137-149

Ledbetter MT (1986) Bottom-current pathways in the Argentine Basin revealed by mean particle size. Nature 321: 423-425

Ledbetter MT, Ellwood BB (1980) Spatial and temporal changes in bottom water velocity and direction from analysis of particle size and alignment in deep-sea sediment. Mar Geol 38: 245-261

Maamaatuaiahutapu K, Garcon VC, Provost C, Boulahdid M, Osiroff AP (1992) Brazil-Malvinas Confluence: Water mass composition. J Geophys Res 97(C6): 9493-9505

Maamaatuaiahutapu K, Garçon VC, Provost C, Mercier H (1998) Transports of the Brazil and Malvinas Currents at their confluence. J Mar Res 56: 417-438

McCave IN, Manighetti B, Beveridge NAS (1995) Circulation in the glacial North Atlantic inferred from grain-size measurements. Nature 374: 149-152

Mello GA, Flood RD, Orsi TH, Lowrie A (1992) Southern Brazil Basin: Sedimentary processes and features and implications for continental-rise evolution. In: Poag CW, de Graciansky PC (eds) Geologic Evolution of Atlantic Continental Rises. Van Nostrand Reinhold, New York, pp 189-213

Mémery L, Arhan M, Alvarez-Salgado XA, Messias M-J, Mercier H, Castro CG, Rios AF (2000) The water masses along the western boundary of the south and equatorial Atlantic. Prog Oceanogr 47: 69-98

Michaelovitch de Mahiques M, Almeida da Silveira IC, de Mello e Sousa SH, Rodrigues M (2002) Post-LGM sedimentation on the outer shelf-upper slope of the northernmost part of the São Paulo Bight, southeastern Brazil. Mar Geol 181: 387-400

Michels KH (2000) Inferring maximum current velocities in the Norwegian-Greenland Sea from settling-velocity measurements of sediment surface samples: methods, application, and results. J Sed Res 70: 1036-1050

Milliman JD, Meade RH (1983) World-delivery of river sediment to the oceans. J Geol 91: 1 21

Mollenhauer G, Jennerjahn T, Müller PJ, Schneider RR, Wefer G (2003) Spatial distribution and accumulation of organic carbon in the South Atlantic Ocean: Its modern and glacial contribution to the carbon cycle. Glob Planet Change, submitted

Olson DB, Podestá GP, Evans RH, Brown OB (1988) Temporal variations in the separation of Brazil and Malvinas Currents. Deep-Sea Res 35: 1971-1990

Peterson RG (1992) The boundary currents in the western Argentine Basin. Deep-Sea Res 39: 623-644

Peterson RG, Whitworth III T (1989) The subantarctic and polar fronts in relation to deep water masses through the southwestern Atlantic. J Geophys Res 94: 10817-10838

Piola AR, Campos EJD, Möller OO Jr, Charo M, Martinez C (2000) Subtropical shelf front off eastern South America. J Geophys Res 105(C3): 6565-6578

Piola AR, Matano RP (2001) Brazil and Falklands (Malvinas) Currents. In: Steele JH, Thorpe SA, Turekian KK (eds) Encyclopedia of Ocean Sciences. Academic Press, San Diego, pp 340-349

Prins MA, Bouwer LM, Beets CJ, Troelstra SR, Weltje GJ, Kruk RW, Kuijpers A, Vroon PZ (2002) Ocean circulation and iceberg discharge in the glacial North Atlantic: Inferences from unmixing of sediment size distribution. Geology 30: 555-558

Prins MA, Postma G, Weltje GJ (2000) Controls on terrigenous sediment supply to the Arabian Sea during the late Quaternary: The Makran continental slope. Mar Geol 169: 351-371

Prins MA, Troelstra SR, Kruk RW, Borg van der K, Jong de AJ, Weltje GJ (2001) The Late Quaternary sediment record from Reykjanes Ridge, North Atlantic. Radiocarbon 43: 939-947

Prins M, Weltje G (1999): End-member modelling of siliciclastic grain-size distributions: The late Quaternary record of eolian and fluvial sediment supply to the Arabian Sea and its paleoclimatic significance. SEPM Spec Publ 62: 91-111

Romero O, Hensen C (2002) Oceanographic control of biogenic opal and diatoms in surface sediments of the Southwestern Atlantic. Mar Geol 186: 263-280

Segl M, cruise participants (1994) Report and preliminary results of Meteor-cruise M29/1. Ber Fachber Geowiss Univ Bremen 58, 94 p

Stramma L, England M (1999) On the water masses and mean circulation of the South Atlantic Ocean. J Geophys Res 104(C9): 20863-20883

Stommel H (1958) The abyssal circulation. Deep-Sea Res 5: 80-82

Stuut JB, Prins MA, Schneider RR, Weltje GJ, Jansen JHF, Postma G. (2002) A 300 kyr record of aridity and wind strength in southwestern Africa: Inferences from grain-size distributions of sediments on Walvis Ridge, SE Atlantic. Mar Geol 180: 221-233

Thunell RC (1982) Carbonate dissolution and abyssal hydrography in the Atlantic Ocean. Mar Geol 47: 165-180

Van Dam JA, Weltje GJ (1999) Reconstruction of the Late Miocene climate of Spain using rodent palaeocom-munity successions: an application of end-member modelling. Palaeogeogr Palaeoclimat Palaeoecol 151: 267-305

Vivier F, Provost C (1999) Direct velocity measurements in the Malvinas Current. J Geophys Res 104(C9): 21083-21103

Volbers A, Henrich R (2002) Present water mass calcium carbonate corrosiveness in the eastern South Atlantic inferred from ultrastructural breakdown of *Globigerina bulloides* in surface sediments. Mar Geol 186: 471-486

Walker GPL (1971) Grain-size characteristics of pyroclastic deposits. J Geol 79: 696-741

Weatherly GL (1993) On deep-current and hydrographic observations from a mudwave region and elsewhere in the Argentine Basin. Deep-Sea Res 40: 939-961

Weltje G (1997) End-member modelling of compositional data: Numerical-statistical algorithms for solving the explicit mixing problem. J Mat Geol 29: 503-549

Terrigenous Sediment Supply in the Polar to Temperate South Atlantic: Land-Ocean Links of Environmental Changes during the Late Quaternary

B. Diekmann[1,2*], D.K. Fütterer[1], H. Grobe[1], C.D. Hillenbrand[1], G. Kuhn[1], K. Michels[1], R. Petschick[2] and M. Pirrung[1]

[1]*Alfred-Wegener-Institut für Polar- und Meeresforschung, Columbusstraße, 27515 Bremerhaven, Germany*
[2]*Alfred-Wegener-Institut für Polar- und Meeresforschung, Forschungsstelle Potsdam,Telegrafenberg A43, 14473 Potsdam, Germany*
[3]*Geologisch-Palaeontologisches Institut, Universität Frankfurt,Senckenberganlage 32-34, 60054 Frankfurt/Main, Germany*
** corresponding author (e-mail): bdiekmann@awi-potsdam.de*

Abstract: Terrigenous sediment parameters in modern sea-bottom samples and sediment cores of the South Atlantic are used to infer variations in detrital sources and modes of terrigenous sediment supply in response to environmental changes through the late Quaternary climate cycles. Mass-accumulation rates of terrigenous sediment and fluxes of ice-rafted detritus are discussed in terms of temporal variations in detrital sediment input from land to sea. Grain-size parameters of terrigenous mud document the intensity of bottom-water circulation, whereas clay-mineral assemblages constrain the sources and marine transport routes of suspended fine-grained particulates, controlled by the modes of sediment input and patterns of ocean circulation. The results suggest low-frequency East Antarctic ice dynamics with dominant 100-kyr cycles and high rates of Antarctic Bottom Water formation and iceberg discharge during interglacial times. In contrast, the more subpolar ice masses of the Antarctic Peninsula also respond to short-term climate variability with maximum iceberg discharges during glacial terminations related to the rapid disintegration of advanced ice masses. In the northern Scotia Sea, increased sediment supply from southern South America points to extended ice masses in Patagonia during glacial times. In the southeastern South Atlantic, changes in regional ocean circulation are linked to global thermohaline ocean circulation and are in phase with northern-hemispheric processes of ice build-up and associated formation of North Atlantic Deep Water, which decreased during glacial times and permitted a wider extension of southern-source water masses in the study area.

Introduction

The South Atlantic plays an important role in the global climate system, as it represents the central junction box of ocean currents and water masses (Figs. 1, 2) that drive interhemispheric heat exchange (Keir 1988; Broecker and Denton 1989; Berger and Wefer 1996). The southern sector of the South Atlantic forms part of the circumpolar Southern Ocean, which provides the major conduit between the world oceans and maintains the thermal isolation of Antarctica. Marine sedimentary 'proxy' records in the South Atlantic document palaeoceanographic changes related to regional and

From WEFER G, MULITZA S, RATMEYER V (eds), 2003, *The South Atlantic in the Late Quaternary: Reconstruction of Material Budgets and Current Systems.* Springer-Verlag Berlin Heidelberg New York Tokyo, pp 375-399

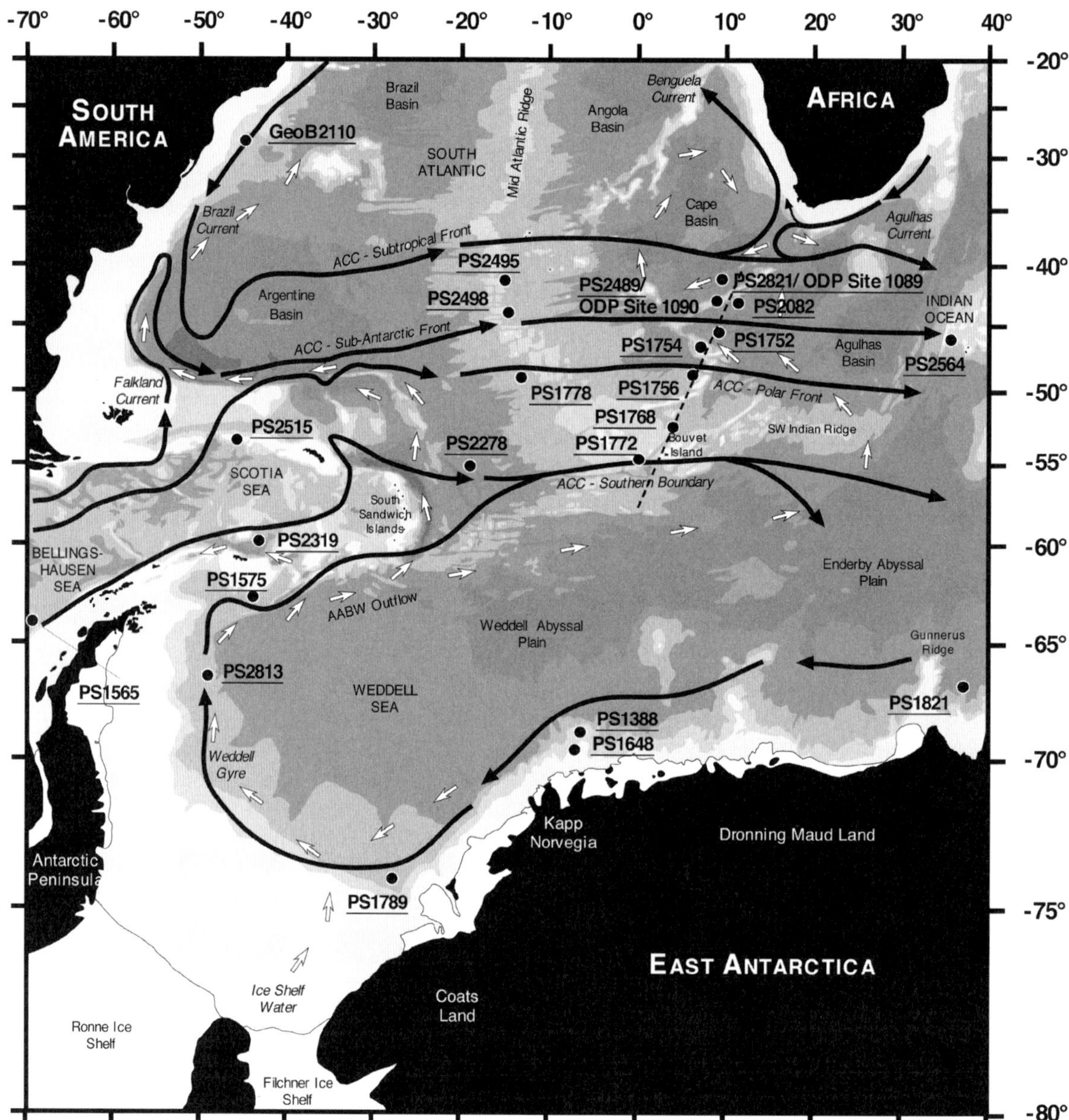

Fig. 1. Polar to temperate South Atlantic with locations of studied sediment cores. Black arrows show surface currents and the frontal system of the Antarctic Circumpolar Current (ACC) (Peterson and Stramma 1991; Orsi et al. 1995). The flow pattern of Antarctic Bottom Water (AABW), indicated by the white arrows, in many places appears independently from upper-level circulation (Georgi 1981; Tucholke and Embley 1984; Locarnini et al. 1993). Dashed line shows position of vertical water-mass profile depicted in Fig. 2. Sea floor below 2000 m is gradually shaded in steps of 1000 m.

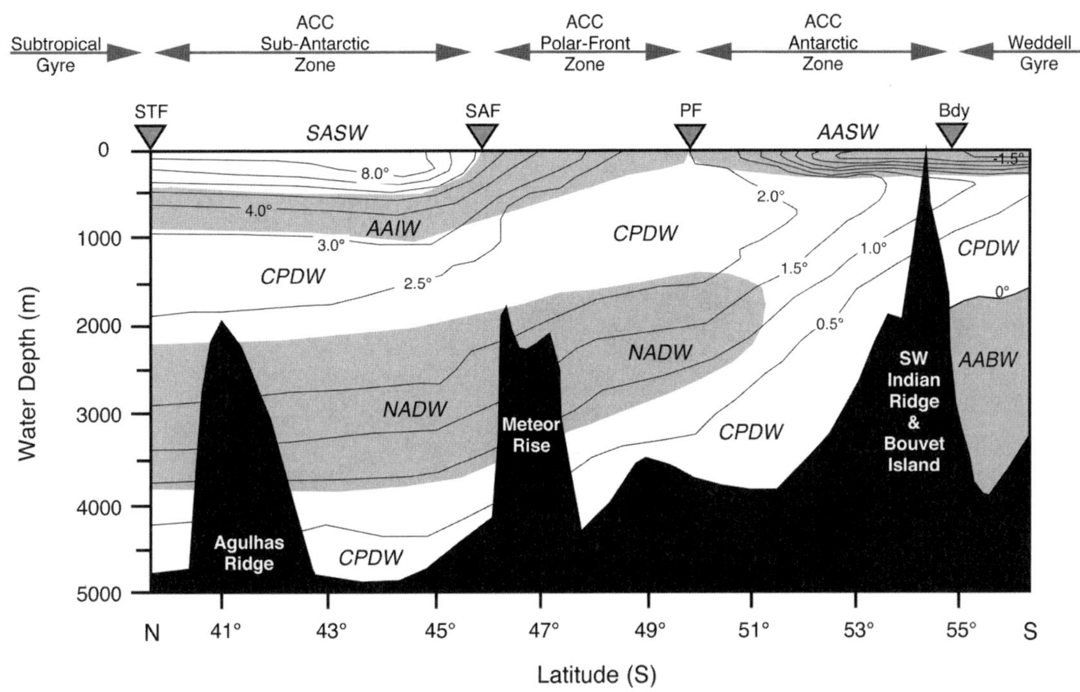

Fig. 2. Vertical distribution of water masses and potential temperatures on a north-south transect across the frontal system of the Antarctic Circumpolar Current (ACC) in the southeastern South Atlantic (see profile line in Fig. 1): (STF) Subtropical Front, (SAF) Subantarctic Front, (PF) Polar Front, (Bdy) boundary between ACC and Weddell Gyre, (SASW) Subantarctic Surface Water, (AASW) Antarctic Surface Water, (AAIW) Antarctic Intermediate Water, (NADW) North Atlantic Deep Water, (CPDW) Circumpolar Deep Water, (AABW) Antarctic Bottom Water. Modified from Gersonde et al. (1999).

global climate variability (Wefer et al. 1999). Moreover, they provide insights into Antarctic ice dynamics as a land-based factor of the southern-hemispheric climate system (Ehrmann 1994; Barrett 1996).

In this contribution, we will characterize terrigenous sediment supply to the temperate to polar South Atlantic and will highlight how terrigenous sediment parameters can be used to decipher the palaeoenvironment through the late Quaternary ice ages and interglacials. For a comprehensive overview of all aspects of terrigenous sedimentation, the reader is referred to the detailed compilations in respective textbooks (Chamley 1989; Lisitzin 1996; Anderson 1999). Since terrigenous sediments originate from terrestrial sources and are deposited in the ocean, they ideally record land-ocean links of climate processes. We will address fluxes of ter-

rigenous mud and ice-rafted detritus as well as the granulometric and mineralogical properties of the detrital particles. The terrigenous sediment parameters will be reviewed in terms of Antarctic and Patagonian ice dynamics and regional ocean circulation.

Material and Methods

We refer to sea-bottom samples and sediment cores recovered in the South Atlantic and adjoining seas with research vessels '*Polarstern*' (PS-samples) and '*Meteor*' (GeoB-samples) during the last decades (Fig. 1, Table 1). Moreover, we present data from Site 1089 and Site 1090 of Leg 177 of the Ocean Drilling Program, which were drilled in the southeastern South Atlantic (Fig. 1, Table 1). The Site 1089 record was used to extend the PS2821

Sediment Core	Gear	Latitude	Longitude	Water Depth	Age Model	Reference
GeoB2110-4	Gravity Corer	-28.650°	-45.521°	3003 m	Plankt. Foram. Biostrat./δ^{18}O	Gingele et al. (1999)
PS1388-3	Gravity Corer	-69.033°	-5.883°	2517 m	Plankt./Benth. Foram. δ^{18}O, Lithostrat.	Grobe and Mackensen (1992)
PS1565-2	Gravity Corer	-63.924°	-69.544°	3427 m	Lithostratigraphy	Hillenbrand (1994)
PS1575-1	Gravity Corer	-62.852°	-43.332°	3452 m	Litho-/Chemostratigraphy	Bonn et al. (1998)
PS1648-1	Gravity Corer	-69.740°	-6.524°	2531 m	Lithostratigraphy	Grobe and Mackensen (1992)
PS1752-1	Gravity Corer	-45.622°	9.597°	4519 m	Radiolarian Biostratigraphy	Brathauer and Abelmann (1999)
PS1754-1	Gravity Corer	-46.770°	7.611°	2471 m	Planktonic Foraminifera δ^{18}O	Niebler (1995), Frank et al. (1996)
PS1756-4	Gravity Corer	-48.899°	6.714°	3787 m	Organic Carbon δ^{13}C, Radiol. Biostrat.	Frank et al. (1996), Abelmann (unpubl.)
PS1768-8	Gravity Corer	-52.593°	4.475°	3270 m	Plankt. Foram. δ^{18}O/AMS ^{14}C	Niebler (1995), Frank et al. (1996)
PS1772-8	Gravity Corer	-55.458°	1.164°	4135 m	Diatom Biostratigraphy	Frank et al. (1996)
PS1778-5	Gravity Corer	-49.012°	-12.697°	3380 m	Radiolarian Biostratigraphy	Brathauer and Abelmann (1999)
PS1789-1	Gravity Corer	-74.241°	-27.300°	2411 m	Planktonic Foraminifera AMS ^{14}C	Weber et al. (1994)
PS1821-6	Gravity Corer	-67.064°	37.479°	3981 m	Lithostratigraphy	Bonn et al. (1998)
PS2082-1	Gravity Corer	-43.220°	11.738°	4610 m	Benthic Foraminifera δ^{18}O (80-250 ka)	Mackensen et al. (1994)
PS2278-3	Piston Corer	-55.970°	-22.195°	4415 m	Diatom Biostratigraphy	Gersonde (unpublished)
PS2319-1	Gravity Corer	-59.788°	-42.683°	4320 m	Radiolarian Biostratigraphy	Diekmann et al. (2000)
PS2495-3	Gravity Corer	-41.288°	-14.500°	3135 m	Benthic Foraminifera δ^{18}O	Mackensen et al. (2001)
PS2498-1	Gravity Corer	-44.153°	-14.228°	3783 m	Benthic Foraminifera δ^{18}O	Mackensen et al. (2001)
PS2515-3	Piston Corer	-53.556°	-45.310°	3522 m	Radiol. Biostrat., Org. Carb. AMS ^{14}C	Diekmann et al. (2000)
PS2564-3	Piston Corer	-46.141°	35.900°	3035 m	Benthic Foraminifera δ^{18}O	Mackensen (unpublished)
PS2813-1	Piston Corer	-66.730°	-50.000°	3499 m	Lithostratigraphy	Michels (unpublished)
PS2821-1	Piston Corer	-40.943°	9.888°	4575 m	Radiolarian Biostratigraphy (0-80 ka)	Cortese and Abelmann (2002)
ODP Site 1089	Advanced Piston Corer	-40.936°	9.894°	4617 m	Benthic Foraminifera δ^{18}O (80-250 ka)	Hodell et al. (2001)
PS2489-2	Piston Corer	-42.873°	8.973°	3794 m	Benthic Foram. δ^{18}O (0-408 ka)	Becquey and Gersonde (2002)
ODP Site 1090	Advanced Piston Corer	-42.913°	8.899°	3700 m	Benthic Foram. δ^{18}O (408-1800 ka)	Venz and Hodell (2002)

Table 1. Sediment-core locations with references to age models used.

record from the same location, which only covers the last 80 kyr, back to 250 ka. The Site 1090 record covers the last 1800 kyr and is spliced with the late Quaternary PS2489 record at 408 ka. Applied age models and assignments of marine isotope stages (MIS) are based on published foraminiferal $\delta^{18}O$ records, biostratigraphy, or lithostratigraphy (Table 1), using the SPECMAP time scale (Imbrie et al. 1984). For the time-slice reconstruction of clay-mineral ratios in marine sediments, the values for the last glacial maximum (LGM) refer to average values found in samples of the recently defined LGM time interval between 23 and 19 ka (calendar years BP), which corresponds to uncorrected ^{14}C ages between 20 and 16.5 ka (Bard 1999). The GeoB sediment cores used for LGM reconstructions are specified on the Internet (www. pangaea.de/Institutes/GeoB/Cores/LGM.html, compiled by H.S. Niebler and S. Mulitza). Stratigraphic information for PS sediment cores is given by Gersonde et al. (in press). The Recent time slice is based on data from sea-bottom samples.

Mass-accumulation rates of terrigenous sediment (MART [g cm^{-2} kyr^{-1}]) were calculated from sediment dry-bulk density multiplied by the relative proportion of the non-biogenic sediment fraction and linear sedimentation rates that depend on the applied age models. The non-biogenic fraction was determined by substracting the measured percentages of biogenic carbonate, biogenic opal and organic carbon (Kuhn and Diekmann 2002b). MART time-series were resampled to equal time increments of 2 kyr. The abundance of gravel clasts >2 mm, estimated from x-radiographs, was used as a parameter of ice-rafted detritus (IRD) (Grobe 1987). Time-series of IRD flux are expressed as clast abundances per area and time unit [# cm^{-2} kyr^{-1}], by integrating counted IRD abundances at equal time increments of 1 kyr. The proportions of non-carbonate silt and clay (referred to as terrigenous mud) were determined by grain-size separation in settling tubes, after sieving the bulk samples through a 63-μm mesh for separation of the sand fraction and dissolution of carbonate with 10-% acetic acid (Brehme 1992; Kuhn and Diekmann 2002a, b). Silt grain-size analyses were conducted on sediment cores from the southern Cape Basin (Site 1089/PS2821) and the northwestern Weddell Sea (PS2515). The non-carbonate silt fraction was dispersed in sodium-polyphosphate solution and measured with a 'Micromeritics Sedi-Graph 5000E' to determine the grain-size distribution in 1/10-Phi steps and hence the proportion of sortable silt (particle sizes 63-10 μm) in the mud fraction (McCave et al. 1995). The bias of the terrigenous grain-size signal caused by biosiliceous particles in the non-carbonate mud fraction is negligible, because of relatively low opal concentrations that range between 3% and 7% in sediment core PS1575 (Bonn et al. 1998) and are ≤15% in the record of Site 1989/PS2821 (Kuhn and Diekmann 2002a, b). Moreover, no correlation exists between opal concentrations and grain-size variability for the latter record with the slightly elevated opal concentrations (Kuhn and Diekmann 2002b).

Clay-mineral data were obtained following standard procedures in marine geology (Biscaye 1965; Ehrmann et al. 1992; Petschick et al. 1996; Vogt 1997). Relative analytical precision for major clay minerals is 6-9% and 8-14% for minor clay minerals (Ehrmann et al. 1992). Although kaolinite and chlorite mostly represent minor components in the studied sediments (<25%), the calculation of kaolinite/chlorite ratios is quite reliable, because the ratio is inferred from the double peak of 3.54Å (chlorite) and 3.58Å (kaolinite) in the x-ray diffractograms (see details in Biscaye 1964; Petschick et al. 1996). Repeated sample preparations and measurements revealed a relative analytical error of ≤10% for kaolinite/chlorite ratios between 0.5 and 2.0 and around 20% for ratios <0.5 and >2.0 (Ehrmann et al. 1992; Vogt 1997). Ratios <0.25 and >4.0 have lower statistical confidence, which are not considered in this paper.

Temporal variation in grain-size spectra of terrigenous mud and the percentage of the four major clay-mineral groups smectite, illite, chlorite, and kaolinite are presented at equal time increments of 2 kyr. The conversion of original data into data at equal time steps was facilitated by the 'AnalySeries' software through linear integration between original data points (Paillard et al. 1996).

Spatial and Temporal Patterns of Terrigenous Sediment Supply

Terrigenous sediments in the study area are dominated by silt- and clay-sized detritus with small proportions of sand and gravel delivered by icebergs and/or sea ice. High concentrations of ice-rafted detritus (IRD) only appear on the shelves off Antarctica, while they range below 15% beyond the shelves (Diekmann and Kuhn 1999). In this section, we will refer to mass-accumulation of total terrigenous sediment (MART) and discuss IRD fluxes during the last 250 kyr.

Mass-Accumulation Rates of Terrigenous Sediment (MART)

The calculation of MARTs quantifies the deposition of terrigenous sediment, mainly of mud, at a given location and thus gives an impression of the detrital input and the terrigenous sediment fluxes in the ocean. MARTs at the selected sediment-core locations show high spatial and temporal variability in the range between 0.1 and 10 g cm^{-2} kyr^{-1}, with exceptionally high values of up to 150 g cm^{-2} kyr^{-1} in the southeastern Weddell Sea (PS1789) (Fig. 3). Compared to the tropical South Atlantic, where terrigenous input is dominated by almost pure aeolian sediment supply with MARTs that rarely exceed values >1.2 g cm^{-2} kyr^{-1} through time (Ruddiman 1997), the often high MARTs in the temperate to polar South Atlantic document a variety of processes that are responsible for terrigenous sediment dispersal.

On average, elevated MART values (>2.5 g cm^{-2} kyr^{-1}) appear off the continental margins because of the proximity to detrital sources. High values are also evident in the distal regions of the Weddell Sea and the Scotia Sea, where the dispersal of fine-grained glacigenic detritus and reworking of marine sediments by the bottom currents of the Weddell Gyre and the Antarctic Circumpolar Current is promoted. The sea floor in the Scotia Sea exhibits widespread scour features, which are often associated with a lack of Quaternary sediment cover (Pudsey and Howe 1998). Sediment accumulation is focused and restricted to current-sheltered depressions, like at core locations PS2319 and PS2515. High accumulation rates also characterize drift deposits in the southern Cape Basin (PS2821/Site 1089), which are influenced by bottom contour currents (Kuhn and Diekmann 2002a). MARTs in the remote eastern ACC region are variable and mostly low (<1.0 g cm^{-2} kyr-1), with slightly increased values at core location PS2498 (>2.0 g cm^{-2} kyr^{-1}) on the Mid-Atlantic Ridge. Spatial variability in average MARTs points to lateral particle advection within the water masses of the ACC, leading to marked winnowing and focusing effects. The importance of lateral sediment transport is also apparent from the calculation of thorium-normalized flux rates (^{230}Th$_{excess}$ data) of biogenic sediment constituents (Frank et al. 1996).

Glacial-interglacial MART fluctuations show an inconsistent pattern. Along Antarctica, MARTs are increased during glacial stages at core locations on drifts and levee deposits in channel systems that incise the continental margins of the Weddell Sea and the Bellingshausen Sea. Two examples are location PS1565 on the continental rise off the Antarctic Peninsula (Hillenbrand 2000) and location PS1789 on the continental slope in the southeastern Weddell Sea. MARTs at the latter location were approximately 15 times higher during the last glacial stage than during the Holocene (Weber et al. 1994). Higher sediment fluxes during cold stages with low sea levels are caused by the "bulldozing effect" of shelf-grounded ice sheets that triggers turbidity currents to the deep-sea (Grobe and Mackensen 1992). The formation of levee structures documents the interaction of contour currents and turbidity currents (Weber et al. 1994; Rebesco et al. 1996; Diekmann and Kuhn 1997; Hillenbrand 2000; Michels et al. 2002). It is assumed that the suspensions of the turbidity currents are deflected by coriolis forcing during their downslope movement through the channels, and that parts of their overspilling sediment loads are entrained in contour currents.

In the Scotia Sea, high MARTs are also related to glacial stages (PS2319 and PS2515). Higher sediment fluxes in the ACC can be explained by strong glacigenic sediment input from the Antarctic Peninsula and from Patagonia in response to extended ice sheets that reach the open ocean and the shelf edges (Diekmann et al. 2000; Walter et

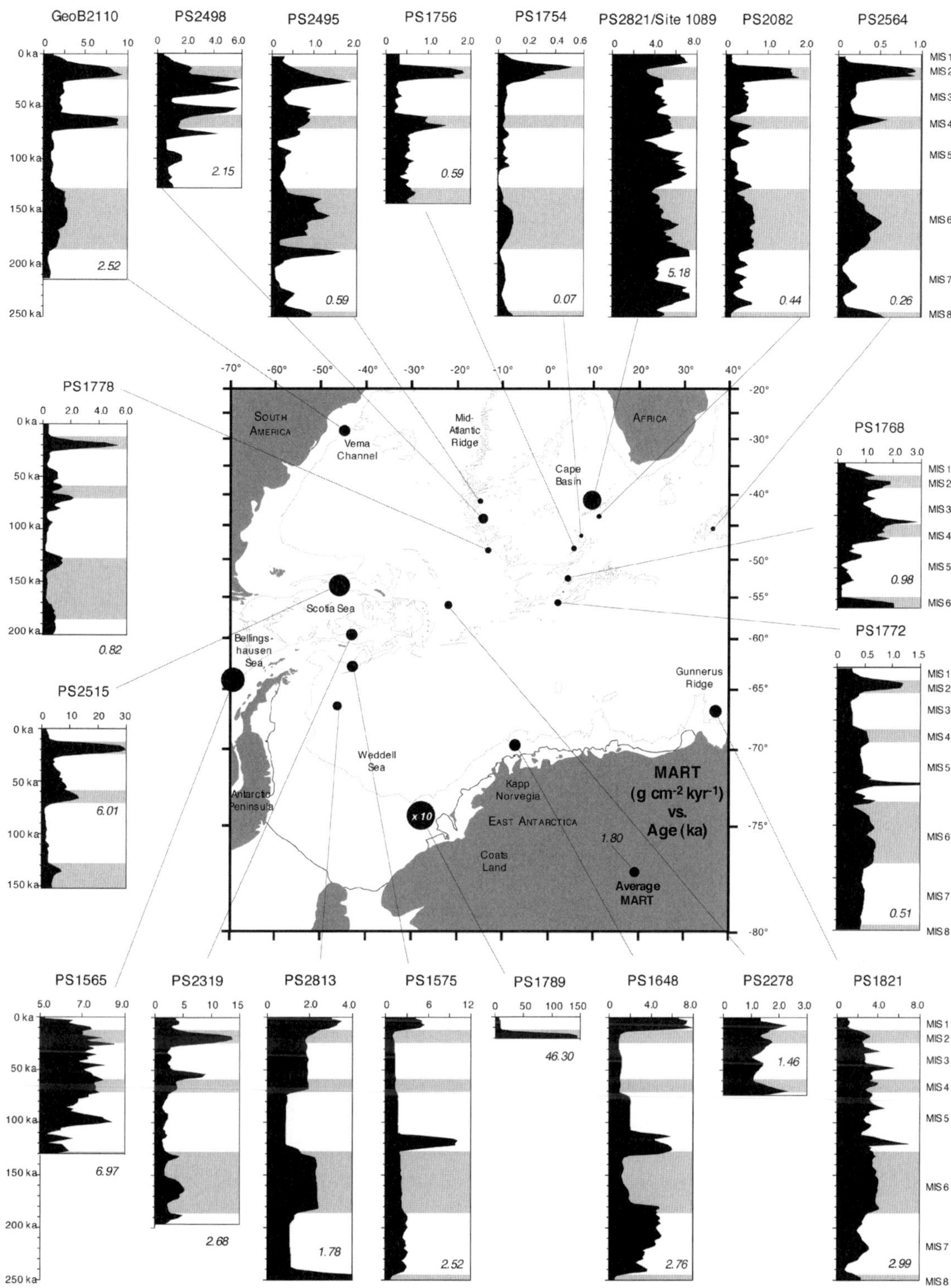

Fig. 3. Temporal and spatial variability of mass-accumulation rates (MART) of terrigenous sediment (MART) recorded in sediment cores of the study area. Diagrams display MARTs in g cm^{-2} kyr^{-1} versus age in ka (last 250 kyr) with indication of average MART value (italic number) and marine isotope stages (glacial stages in grey). Black circles on the map are proportional to average MARTs in the respective sediment cores. Bathymetry is indicated by the 3000-m contour line.

al. 2000). Furthermore, a stronger wind-forcing of the ACC during glacial times enhances transport capacity of the ACC (Pudsey and Howe 1998; Howe and Pudsey 1999). Elevated MARTs during glacial intervals that document increased detrital fluxes in the water masses of the ACC are also evident at many distal core locations farther eastward from the Scotia Sea. This area is characterized by dense nepheloid suspension layers (Biscaye and Eittreim 1977).

It has been suggested that increased detrital fluxes might reflect an increased long-distance supply of air-borne dust from Patagonia due to increased continental aridity and/or enhanced wind strengths of the southern-hemispheric westerlies during glacial times (Martin 1990; Kumar et al. 1995). However, this suggestion is not well supported by provenance analyses taking into account the mineralogical and geochemical composition (major and trace elements, neodymium and strontium isotopes) of marine muds (Bareille et al. 1994; Diekmann et al. 2000; Walter et al. 2000). Clues for stronger atmospheric dust fluxes during glacial times actually arise from increased dust concentrations of Patagonian origin in the ice cores from the East Antarctic ice sheet (Petit et al. 1990; Grousset et al. 1992). The Vostok dust record for example suggests greater dust fluxes by a factor between 15 and 25 for the last glacial maximum (Petit et al. 1990; Mahowald et al. 1999). However, in relation to the low present-day dust fluxes over the Southern Ocean (Duce et al. 1991; Husar et al. 1997), dust fluxes were too low even during glacial times to strongly affect the observed MARTs in the Southern Ocean (Bareille et al. 1994; Diekmann et al. 1999; Maher and Dennis 2001).

In the southern Cape Basin (PS2821), elevated MARTs occur during interglacial stages and are consistent with strong bottom currents that give rise to high sediment fluxes, as discussed below. At the continental margin off the Brazilian coast (GeoB 2110), higher MARTs appear during glacial periods and likely reflect a higher sediment input across the shelf edge, caused by low sea levels (Massé et al. 1996; Diekmann et al. 1999).

Ice-Rafted Detritus (IRD)

Although IRD only forms a small proportion of the terrigenous sediment fraction, its spatial and temporal distribution provides important insights into ice-sheet dynamics and the extent of cold surface water masses that control the distribution and survival of both icebergs and sea ice. The spatial IRD distribution does not show a simple proximal to distal marine gradient as is more or less displayed by the MARTs (Fig. 4). High IRD fluxes (4.23 n cm^{-2} kyr^{-1}) appear in the southeastern Weddell Sea off Coats Land (PS1789), where prominent ice streams drain the East Antarctic ice sheet, while average IRD fluxes off Dronning Maud Land (PS1648, PS1388, PS1821) are moderate to low (0.04-1.16 n cm^{-2} kyr^{-1}). Moderate IRD fluxes (0.48 n cm^{-2} kyr^{-1}) also appear in the Bellingshausen Sea off the Antarctic Peninsula (PS1565). In the pelagic study area, the Polar Front delineates an area with low IRD fluxes ($\leq$ 0.15 n cm^{-2} kyr^{-1}) and warm surface waters to the north from the remaining area with moderate to high IRD fluxes (>0.2 n cm^{-2} kyr^{-1}), which is occupied by cold water masses. Unusually elevated average IRD fluxes (>2.0 n cm^{-2} kyr^{-1}) appear east of the South Sandwich Islands (PS2278) and near Bouvet Island (PS1768).

Temporal fluctuations reveal systematic regional trends. In the Weddell Sea, apart from location PS1789, higher IRD fluxes are related to interglacial stages (Grobe and Mackensen 1992). Marked IRD fluxes in the Bellingshausen Sea off the Antarctic Peninsula (PS1565) and in the southern (PS2319) and the northern Scotia Sea (PS2515) coincide with glacial terminations and moreover show small short-term spikes during MIS 3 and MIS 2 (Hofmann 1999; Diekmann et al. 2000). In the Bellingshausen Sea, short-term fluctuations in IRD supply for the last glacial period were also noted by Pudsey and Cammerlenghi (1998) and Cofaigh et al. (2001). At the two southern locations (PS1565, PS2319), short-term IRD fluctuations are moreover evident for MIS 5. In the southeastern Atlantic maximum IRD fluxes appear during glacial stages. High-resolution IRD records from ODP

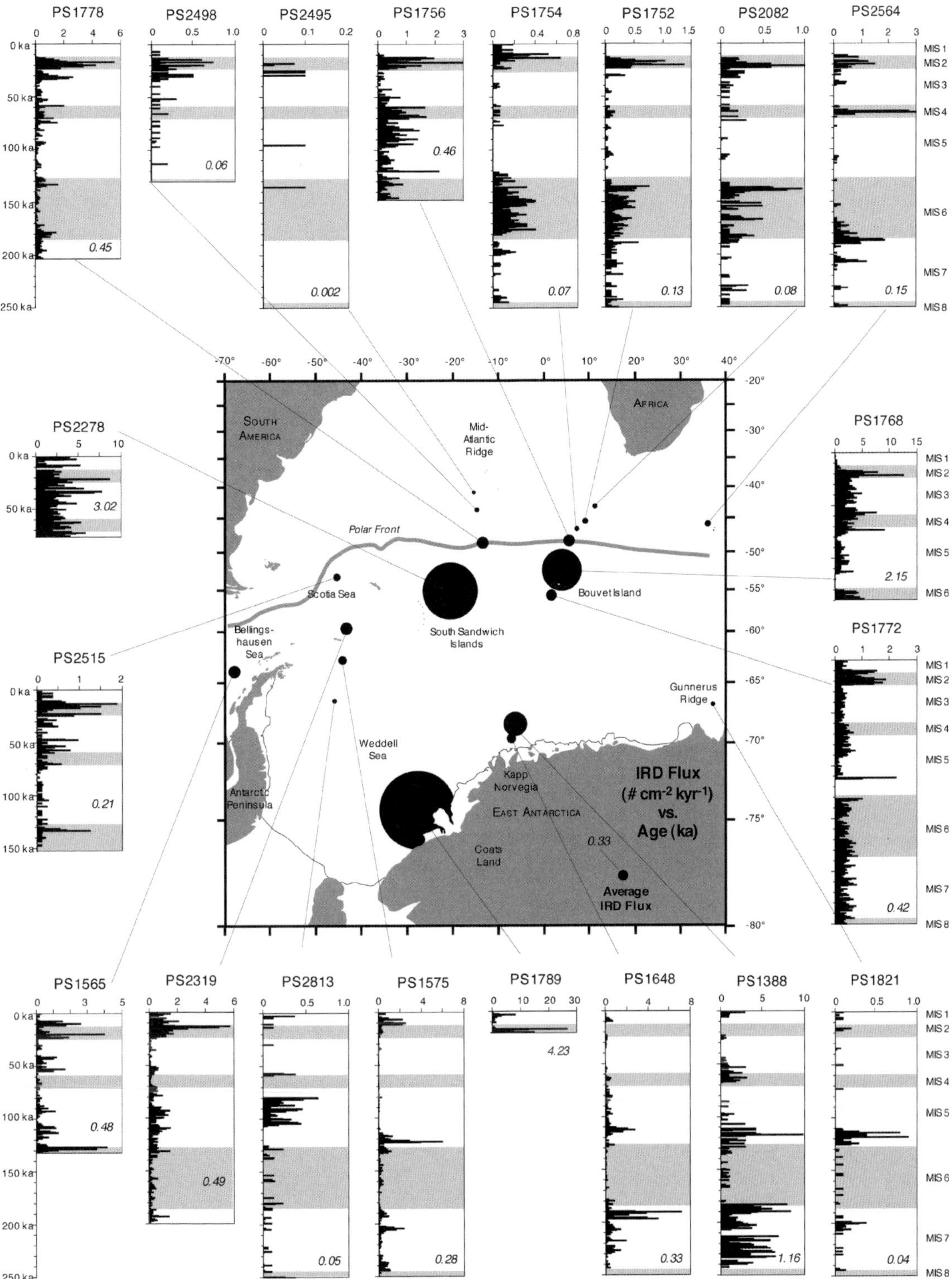

Fig. 4. Temporal and spatial variability of flux rates of ice-rafted detritus (IRD) recorded in sediment cores of the study area. Diagrams display IRD fluxes in # cm^{-2} kyr^{-1} versus age in ka (last 250 kyr) with indication of average IRD value (italic number) and marine isotope stages (glacial stages in grey). Black circles on the map are proportional to average IRD values in the respective sediment cores.

sites in the same area were interpreted to mark short-term IRD events during MIS 3, which likely correspond to interstadials in the northern hemisphere (Kanfoush et al. 2000).

For the interpretation of IRD fluxes, it is important to consider the IRD sources inferred from petrographic and mineralogical studies. In the southern and eastern Weddell Sea, the composition of IRD generally points to nearby source rocks, with abundant clasts derived from East Antarctic crystalline basement rocks, carried by icebergs in the course of the Weddell Gyre (Oskierski 1988; Anderson et al. 1989; Diekmann and Kuhn 1999). In the northwestern Weddell Sea, a minor IRD contribution from the eastern Antarctic Peninsula slightly dilutes the East Antarctic signal (Edwards and Goodell 1969; Diekmann and Kuhn 1999). In the Weddell Sea, the generally higher IRD fluxes during interglacial stages thus document increased calving rates of icebergs from floating ice masses of the East Antarctic ice sheet during high stands of sea level (Grobe and Mackensen 1992).

IRD clasts in the Bellingshausen Sea originate from Mesozoic and Cainozoic intrusive and volcanic rocks as well as meta-sedimentary rocks of the Antarctic Peninsula (Edwards and Goodell 1969; Diekmann and Kuhn 1999). Magmatic rocks also make up a high proportion of IRD in the Scotia Sea (Conolly and Ewing 1965; Diekmann et al. 2000). In the northern Scotia Sea, the supply of IRD from the Pacific sector of the Southern Ocean is supplemented by IRD from the southern Andean magmatic belt during glacial times and terminations, when the Patagonian ice fields coalesce and expand to the open ocean (Diekmann et al. 2000). In contrast to East Antarctica, where intense iceberg calving is related to interglacial stages, maximum iceberg calving rates from Patagonia and the Antarctic Peninsula document the rapid disintegration of marine ice masses during climate amelioration associated with glacial terminations and less pronounced during short-term climate perturbations (Hofmann 1999; Diekmann et al. 2000). Patagonian land records are consistent with these findings and moreover suggest interhemispheric teleconnections with northern-hemispheric climate dynamics (Lowell et al. 1995; Heusser et al. 2000; McCulloch et al. 2000).

The area between the southwest Atlantic and the Indian sector of the Southern Ocean represents an IRD province with high abundances of volcaniclastic glass shards, pumice and volcanic rock fragments mostly derived from the South Sandwich Islands and other islands (Conolly and Ewing 1965; Smith et al. 1983). This fact is reflected by elevated IRD fluxes in the vicinity of such source areas, for example PS2278 northeast of the South Sandwich Islands and PS1768 near Bouvet Island. The association of high IRD fluxes with late Quaternary glacial stages and reduced IRD fluxes north of the Polar Front underlines the major control of sea-surface temperatures on the dispersal of IRD (Smith et al. 1983; Labeyrie et al. 1986; Allen and Warnke 1991; Becquey and Gersonde 2002; Pirrung et al. 2002). The interpretation of IRD fluxes in terms of Antarctic ice-sheet dynamics in this IRD province is hence complicated for several reasons: (1) the temperature effect on the survival of icebergs, (2) IRD sources from offshore Antarctica, (3) the possible sea-ice transport of volcanic ashes (Cooke and Hays 1982; Smith et al. 1983), (4) the impact of episodic volcanic eruptions that are not coupled with climate variability. An attempt to overcome these unpredictable volcanic influences is to calculate the fluxes of ice-rafted quartz, which presumably is solely derived from Antarctic sources (Smith et al. 1983; Labeyrie et al. 1986; Kanfoush et al. 2000). This approach has to be regarded with caution, because quartz as a major rock-forming mineral is supplied from a variety of Antarctic sources and because the temporal variability of IRD supply from East and West Antarctic sources differs markedly. Furthermore, the glacial-interglacial IRD patterns observed in the south-east Atlantic clearly contradict those along the Antarctic continent.

Sortable Silt and Bottom Currents

Apart from the modes of sediment input (glacial, aeolian and fluvial supply), the grain-size distribution of terrigenous muds is determined by bottom-current strengths and its impact on winnowing and focusing effects and selective grain-size transport within nepheloid layers. Granulometric analyses of the terrigenous mud fraction are often used to deduce the influence of current-induced particle transport (Ledbetter 1986; McCave et al. 1995; Höppner and Henrich 1999), although the method is not indisputable (e.g. Anderson and Kurtz 1985). The basic idea is that the terrigenous mud fraction is mostly delivered by lateral sediment supply and susceptible to current sorting by deep-sea currents. In contrast, the grain-size distribution of biogenic particles, in addition to current sorting, is inherited from the modes of biological productivity and mainly originates from the vertical settling of skeletal planktonic remains through the water column, and is also affected and modified by dissolution processes in the water and the sediment column, especially in the South Atlantic (Bickert and Wefer 1996; Diekmann and Kuhn 1997; Dittert et al. 1999). One well established grain-size parameter is the percentage of so-called "sortable silt" (grain size 63-10 µm in diameter) in terrigenous mud. It represents the mud fraction that is transported as single grains, whereas finer grained material is affected and modified by cohesive aggregation of particles (Mc Cave et al. 1995).

In the study area, the reconstruction of palaeo bottom-current speeds on the basis of terrigenous silt grain size provides information about the relative dynamics of Circumpolar Deep Water (CPDW), the deepest water mass of the ACC, and Antarctic Bottom Water (AABW). AABW represents the deepest water mass in the study area, which is confined to the abyssal basins and trenches and directed by both thermohaline circulation and topography (Figs. 1, 2). The reconstruction of AABW dynamics is of palaeoclimatic interest, since its production is linked to Antarctic ice-sheet behaviour and at present depends to a large part on the presence of floating ice shelves. Today, it is formed as a mixture of various polar water masses around the Antarctic margins. A large proportion originates in the Weddell Sea by processes related to sea-ice formation and associated brine release as well as supercooling mechanisms beneath the Filchner-Ronne Ice Shelves (Foldvik and Gammelsrød 1988; Fahrbach and Beckmann 2001). The northward spread of dense AABW contributes to the ventilation of the abyssal world ocean.

Grain-size records of CPDW dynamics from the Scotia Sea basically indicate a stronger current activity during glacial stages, because of a stronger wind-forcing of the ACC by the southern-hemispheric westerlies (Howe and Pudsey 1999; Hofmann 1999). Previous studies on the timing of glacial-interglacial variability in AABW flow, however, have revealed inconsistencies. In the channel systems off the Filchner Trough, which partly drain cold Ice Shelf Water, the hemipelagic muds of levee deposits tend to be coarser grained in glacial intervals (Weber et al. 1994). However, these grain-size records are likely affected by granulometric signals produced by turbidity currents, as revealed by mineralogical provenance analyses of the muds (Diek-mann and Kuhn 1997). In the northwestern Weddell Sea, coarse-grained muds are clearly associated with interglacial intervals (Fütterer et al. 1988; Pudsey 1992; Gilbert et al. 1998). A grain-size record from the southern South Orkney continental slope (PS1575) displays maximum values in substages of MIS 7 and MIS 5 and in the Holocene (Brehme 1992) (Fig. 5). Along the western branch of the AABW outflow path, a maximum in the early Holocene contrasts with minimum values in MIS 2 and the late Holocene in a sediment core from the South Sandwich Trench (Diekmann and Kuhn 1997). Farther north, the Vema Channel between the Argentine Basin and the Brazil Basin represents another focus of AABW outflow, where relatively increased AABW current strengths coincide with interglacial-glacial transitions (Ledbetter 1986; Massé et al. 1994).

The reconstruction of AABW dynamics generally suffers from the poor age constraints of the sediment records under discussion, which all were recovered from water depths below the carbonate lysocline and thus do not permit any age assignments by foraminiferal oxygen-isotope stratigraphy. In the recent past, a well-dated and high-resolution grain-size record was obtained from sediment core PS2821/Site1089, which was recovered on a drift deposit in the southern Cape Basin and which already was addressed for the discussion of MARTs (Kuhn and Diekmann 2002a) (Figs. 1, 3, 5). The carbonate-bearing drift sediments were deposited at high sedimentation rates (15-20 cm kyr⁻¹) and MARTs (3-7 g cm⁻² kyr⁻¹) under the influence of clockwise contour currents that are prevalent in the

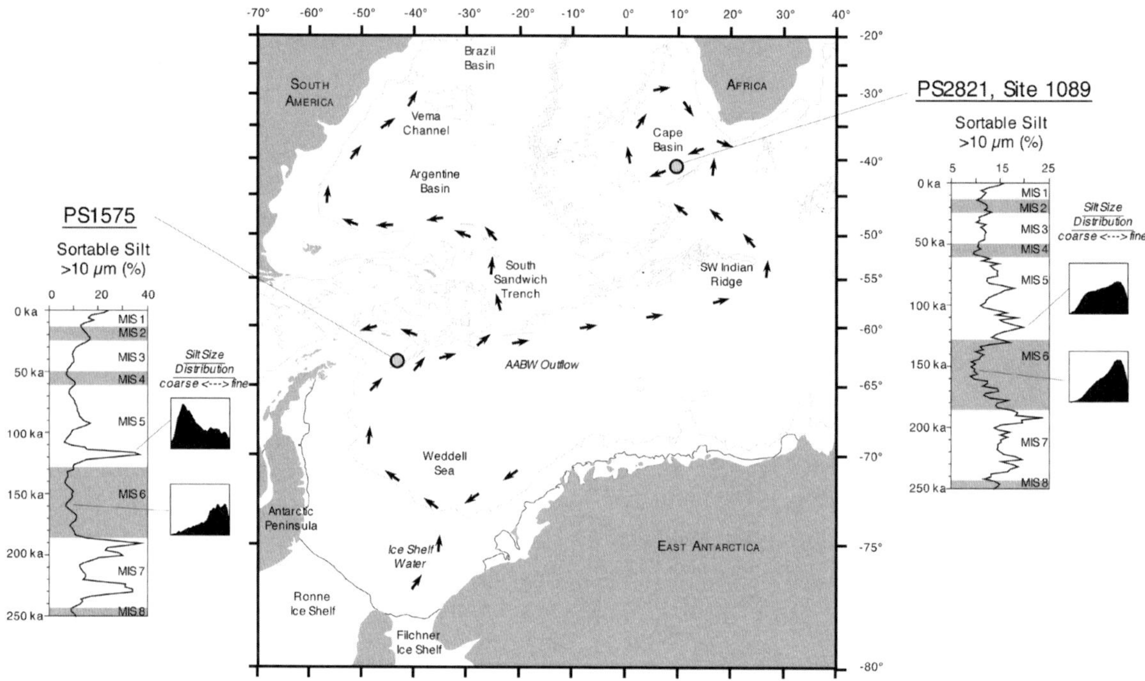

Fig. 5. Temporal variability (last 250 kyr) of percentages of sortable silt in two sediment cores from the northwestern Weddell Sea (PS1575) and the southern Cape Basin (PS2821/Site 1089). Small insets show silt-grain size distribution curves of characteristic glacial and interglacial sediment-core intervals, respectively. Diagrams include indication of marine isotope stages (glacial stages in grey). Black arrows delineate the flow paths of Antarctic Bottom Water. Bathymetry is indicated by the 3000-m contour line.

abyssal Cape Basin. These bottom currents oppose the general eastward ACC jets and originate in the northward outflow of the AABW through gaps of the SW Indian Ridge (Bornhold and Sum-merhayes 1977; Tucholke and Embley 1984) (Fig. 1).

Downcore variations in the percentage of sortable terrigenous silt at location PS2821/Site1089 demonstrate fluctuations with affinities to glacial-interglacial cycles (Fig. 5). Minimum values are related to glacial intervals and point to a relative decrease in abyssal current strengths. Abundant pyrite streaks in the glacial sediment intervals indeed support the interpretation of reduced bottom-water ventilation (Kuhn and Diekmann 2002a). Values of maximum sortable silt correspond to interglacial optimum values, comprising the substages MIS 7.5 and 7.1, MIS 5.5, 5.3 and 5.1 and MIS 1. These maximum values actually exhibit a

good correlation with maximum values recorded in sediment core PS1575 recovered from the northwestern Weddell Sea (Fig. 5), which was dated by lithostratigraphic and chemostratigraphic methods (Bonn et al. 1998). Time-series analyses of the silt grain-size parameter at location PS2821/Site1089 reveal spectral power in the 100-kyr[-1] frequency domain of Milankovitch cycles, but lack significant spectral power in the 40-kyr[-1] and 23-kyr[-1] frequency bands (Kuhn and Diekmann 2002a). This finding is consistent with maximum AABW formation during high stands of sea level that permit the supercooling mechanism of waters beneath floating ice shelves, particularly in the southern Weddell Sea. Since AABW production is strongly coupled with ice-sheet behavior, another conclusion is that the polar ice masses of Antarctica respond less sensitively to short-term climate variability and sea-level fluctuations than the more subpolar ice masses of Patagonia.

Clay Minerals as Tracers of Sediment Provenance and Dispersal

A crucial aspect of terrigenous sediment supply in the study area is the identification of terrigenous sediment sources. Because of the dominance of fine-grained muds, clay mineralogy is an appropriate tool to constrain the origin of terrigenous sediments. The composition of terrestrial clay minerals reflects the development of prevailing weathering regimes in the continental source areas, which depend on the climate zonation that determine the intensity of pedogenesis and chemical and physical weathering (Chamley 1989). In the South Atlantic, late Quaternary clay-mineral assemblages constrain the sources and marine transport paths of clay-sized material in response to climate variability and its impacts on the modes of sediment input and ocean circulation.

Modern Clay-Mineral Provinces

The general distribution pattern of terrigenous clay-mineral assemblages in the whole South Atlantic highlights the sedimentary processes that control their distribution. A detailed clay-mineralogical survey of modern sea-bottom sediments in the South Atlantic was first undertaken by Biscaye (1965) and more recently by Petschick et al. (1996). The latter authors particularly provided an improved knowledge of clay-mineral distributions in the higher latitudes of the South Atlantic. The major clay-mineral provinces are well displayed by clay-mineral assemblages, as defined by the proportions of the three clay-mineral groups illite, kaolinite, and chlorite in a ternary concentration diagram (Fig. 6). The interpretation of smectite minerals is enigmatic, because they can be of both terrigenous and authigenic marine origin (Chamley 1997). The configuration of clay-mineral provinces reveal kaolinite- and chlorite-bearing assemblages to represent contrasting end members of the low and high latitudes, respectively (Fig. 6). Provinces dominated by the kaolinite end member or by mixed kaolinite-illite assemblages appear north of the ACC region, whereas mixed illite-chlorite assemblages are restricted to the south.

In the equatorial regions, kaolinite-dominated assemblages appear off tropical Africa and off the Brasilian Coast, increasingly diluted by illite toward the pelagic South Atlantic. In Africa, the large tropical Congo and Niger rivers as well as trade winds from the Sahel Zone supply detrital particles rich in kaolinite from fossil and modern tropical soils (Aston et al. 1973; Pastouret et al. 1978; Van der Gaast and Jansen 1984; Gingele et al. 1998). Off Brazil, fluvial suspensions, originating from deeply weathered basement rocks (laterites) of the Santos/Sao Paulo Plateaus, provide the bulk of kaolinite in marine sediments (De Melo et al. 1975; Gingele et al. 1999). The western coastal area of southern Africa is characterized by an illite-dominated clay-mineral province that passes to illite-kaolinite assemblages farther offshore. Deserts and semi-arid regions of southern Africa are major contributors of illite, supplied by the southeastern trade winds and to a lesser degree by fluvial input (Bremner and Willis 1993; Gingele 1996). The transition to illite-kaolinite assemblages farther offshore indicates an increasing influence of kaolinite-bearing detritus from northern sources, supplied by deep-water advection (Petschick et al. 1996). Kaolinite is also delivered by the Agulhas Current that brings detrital material from the subtropical regions of eastern Africa to the South Atlantic (Kolla et al. 1976).

Marine sediments underlying the ACC exhibit highest chlorite concentrations of the entire South Atlantic. Chlorite originates from the Andean mobile belts of South America and the Antarctic Peninsula distributed by the ACC, with possibly minor aeolian contributions (Petschick et al. 1996; Diekmann et al. 2000; Hillenbrand and Ehrmann 2001; Hillenbrand et al. 2003). The influence of southern-source water masses on particle advec-tion is well displayed by latitudinal gradients of kaolinite/chlorite ratios in sea-bottom sediments, with decreasing kaolinite/chlorite ratios from north to south. In the southwestern Atlantic, northward particle transport by deep southern-source water masses is also indicated by displaced Anatarctic diatoms (Jones and Johnson 1984) and the pattern of strontium isotope signatures in sea-bottom sediments (Biscaye and Dasch 1971). Another illite

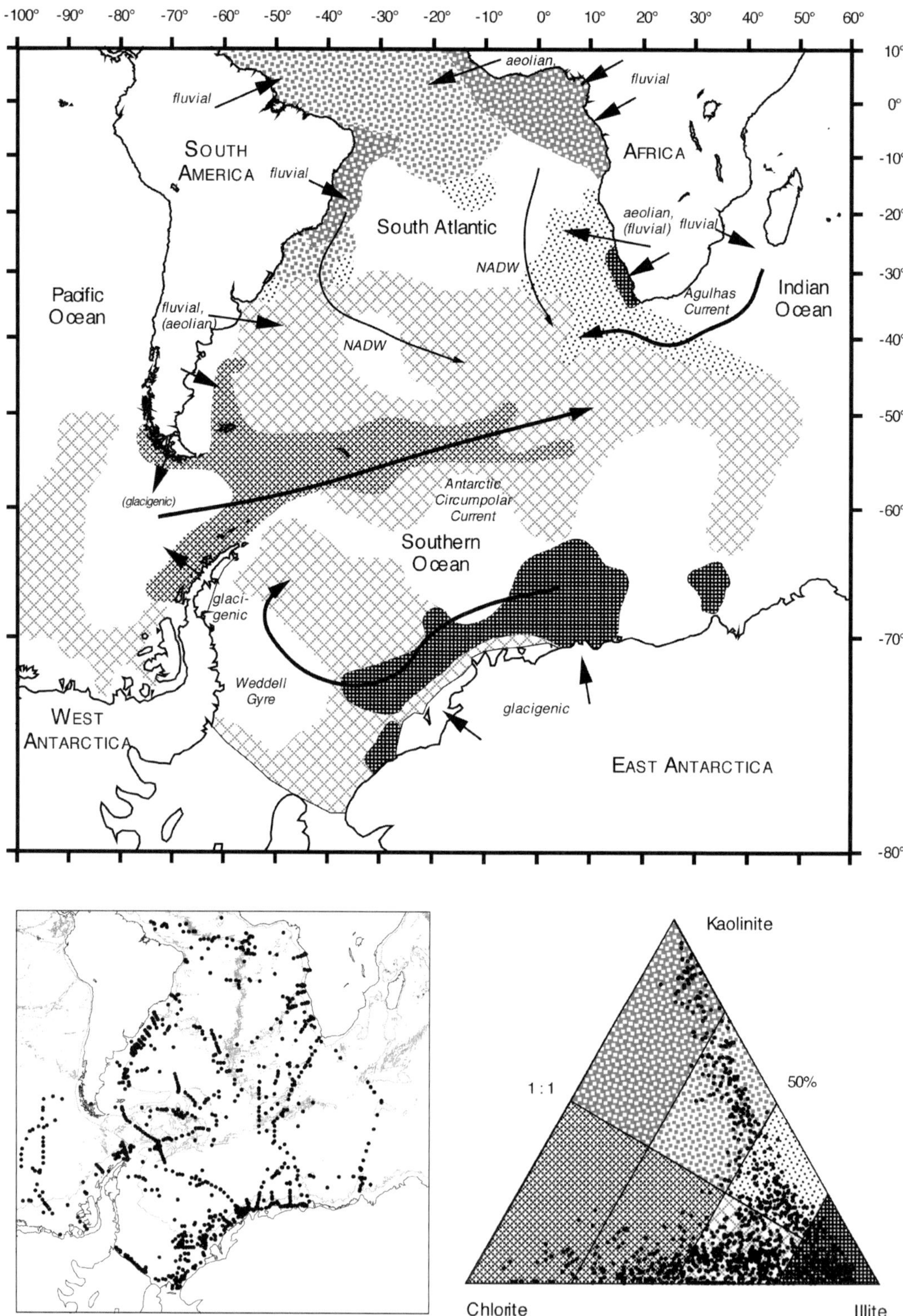

Fig. 6. Modern clay-mineral provinces and modes of terrigenous clay supply and dispersal. Lower left map shows the distribution of sea-bottom samples (black circles) and illustrates the data base (Petschick et al. 1996; Diekmann et al. 2000). The ternary concentration diagram visualizes the definition of clay-mineral asssemblages, according to the proportions of the three terrigenous clay-mineral groups kaolinite, chlorite, and illite.

province fringes East Antarctica, reflecting the widespread presence of crystalline basement rocks (Petschick et al. 1996). In contrast, southern Weddell Sea sediments include mixed clay-mineral assemblages (Ehrmann et al. 1992; Melles et al. 1994).

Temporal Variability of Clay-Mineral Assemblages

Along the Antarctic continental margin, the downcore clay-mineral distributions show distinct glacial-interglacial variations off East Antarctica and the Antarctic Peninsula. A common leit-motif is the alternation of local clay-mineral input during glacial times versus lateral clay-mineral supply from more distant sources during interglacial stages, as indicated by typical key minerals that differ regionally (Fig. 7). The local signal can be attributed to the glacigenic sediment supply by advanced and grounded ice sheets on the shelves that dilute the lateral hemipelagic signal produced by deep and shallow coastal currents (Grobe and Mackensen 1992; Diekmann and Kuhn 1997; Diekmann et al. 2000; Hillenbrand and Ehrmann 2001). Thus, location PS1821 at the Gunnerus Ridge during interglacials is bathed by water masses that carry kaolinite from the Indian sector to the west, while during cold stages it receives more illite from crystalline basement rocks of the hinterland. Off Kapp Norvegia (PS1648), a higher supply of corrensite (mixed-layered smectite-chlorite clay mineral) from local magmatic sources admixes with illite-bearing water masses during glacial stages. In the Bellingshausen Sea (PS1565), chlorite from the central Antarctic Peninsula dilutes the hemipelagic smectite signal during glacial stages. Also in the northern Scotia Sea (PS2515), continental ice dynamics manifested itself as temporal clay-mineral fluctuations (Fig. 7). There, low smectite/chlorite ratios together with elevated MARTs indicate a strong chlorite supply from southern South America in response to extended Patagonian ice fields.

In the northern part of the study area, downcore clay-mineral fluctuations are of lower amplitude than farther south, pointing to relatively constant sediment sources through time (Diekmann et al.

1996). Downcore variations merely concern the proportions of kaolinite and chlorite that originate from distant and opposite sources. Sediment cores from that area (GeoB2110, PS2498, PS2821/Site 1089) show common downcore fluctuations in their kaolinite/chlorite ratios (Fig. 7). These glacial-interglacial contrasts are shown by maps of kaolinite/chlorite ratios in modern sea-bottom sediments and sediments of the last glacial maximum (Fig. 8). Both time slices exhibit pronounced latitudinal gradients of kaolinite/chlorite ratios, which underline the low-latitude character of kaolinite and the high-latitude character of chlorite. Compared to the modern situation, the map of kaolinite/chlorite ratios of the last glacial maximum reveals a general shift of low kaolinite/chlorite ratios towards lower latitudes. The clay minerals kaolinite and chlorite reflect the modes of sediment supply through time and are good tracers of water-mass advection, as revealed by the present distribution of clay-mineral provinces.

Along the Brazilian continental margin (GeoB 2110) (Fig. 7), temporal variations in kaolinite/chlorite ratios mainly monitor changes in glacial-interglacial changes in river runoff and the discharge of kaolinite-bearing sediment plumes to mid-water depths in response to alternating humid and arid climate conditions in the Brazilian hinterland (Gingele et al. 1999). Accordingly, fluvial detrital input is reduced during arid glacial periods, causing a decline in kaolinite supply. Another effect in glacial periods that reduces kaolinite/chlorite ratios is the farther northward and increased vertical extension of chlorite-bearing southern-source deep and bottom water masses, which enhance chlorite advection in the area (Jones 1984; Massé et al. 1996; Diekmann et al. 1999) and is also indicated by higher concentrations of displaced Antarctic diatoms in sediments related to glacial periods (Jones and Johnson 1984). On the time-slice maps, these modified processes are indicated by lower values and less pronounced vertical gradients of kaolinite/chlorite ratios in the Vema Channel region for the last glacial situation (Fig. 8).

Temporal variations of kaolinite/chlorite ratios on the Mid-Atlantic Ridge (PS2498) can be attributed to alternating modes in water-mass advection,

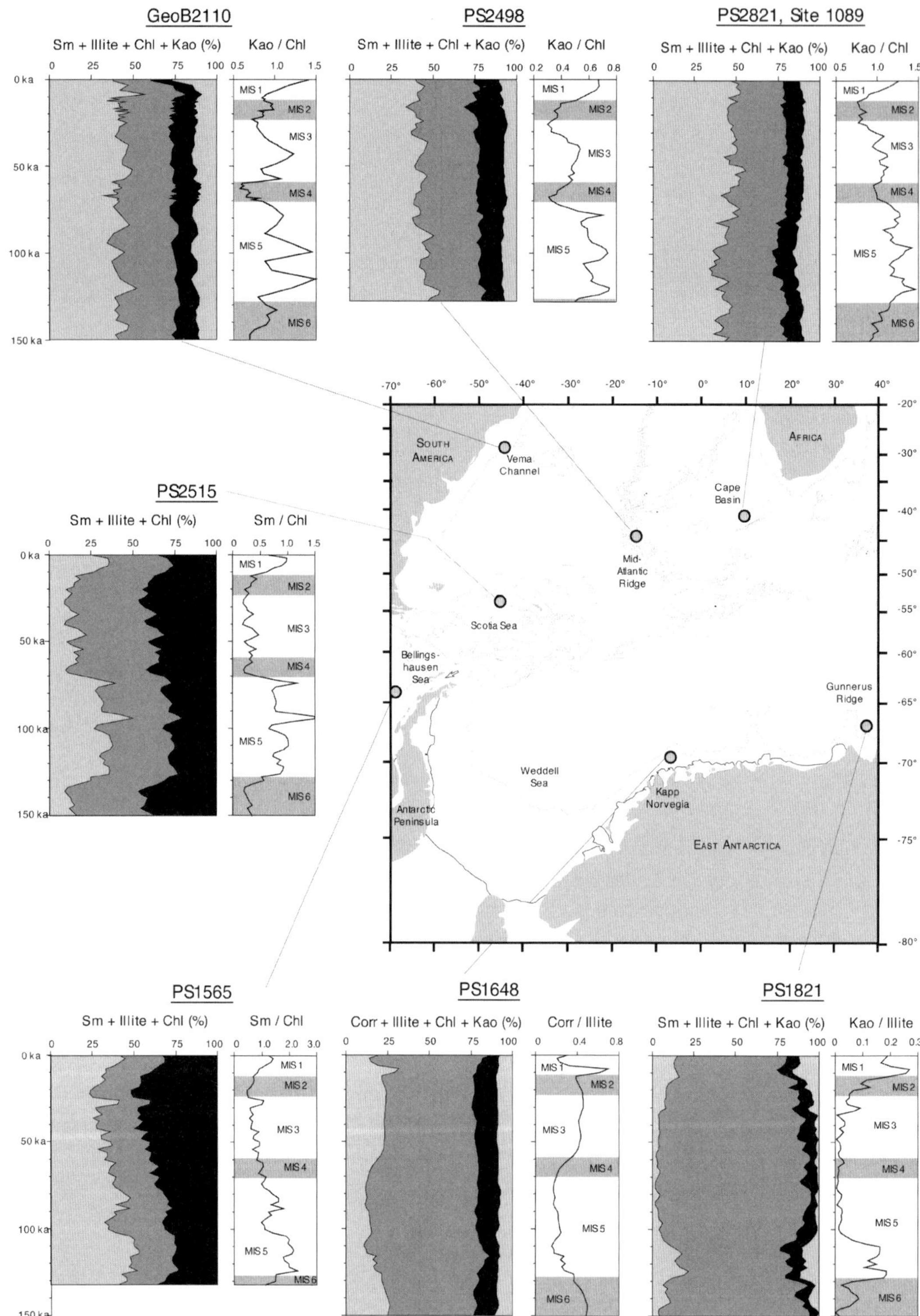

Fig. 7. Temporal (last 150 kyr) and spatial variability of clay-mineral assemblages in sediment cores of the study area, (Sm) smectite, (Kao) kaolinite, (Chl) chlorite, (Corr) corrensite. Diagrams for each sediment core display clay-mineral percentages (left diagram) and clay-mineral ratios (right diagram) with indication of marine isotope stages (glacial stages in grey). Bathymetry on the map is shown by the 3000-m contour line.

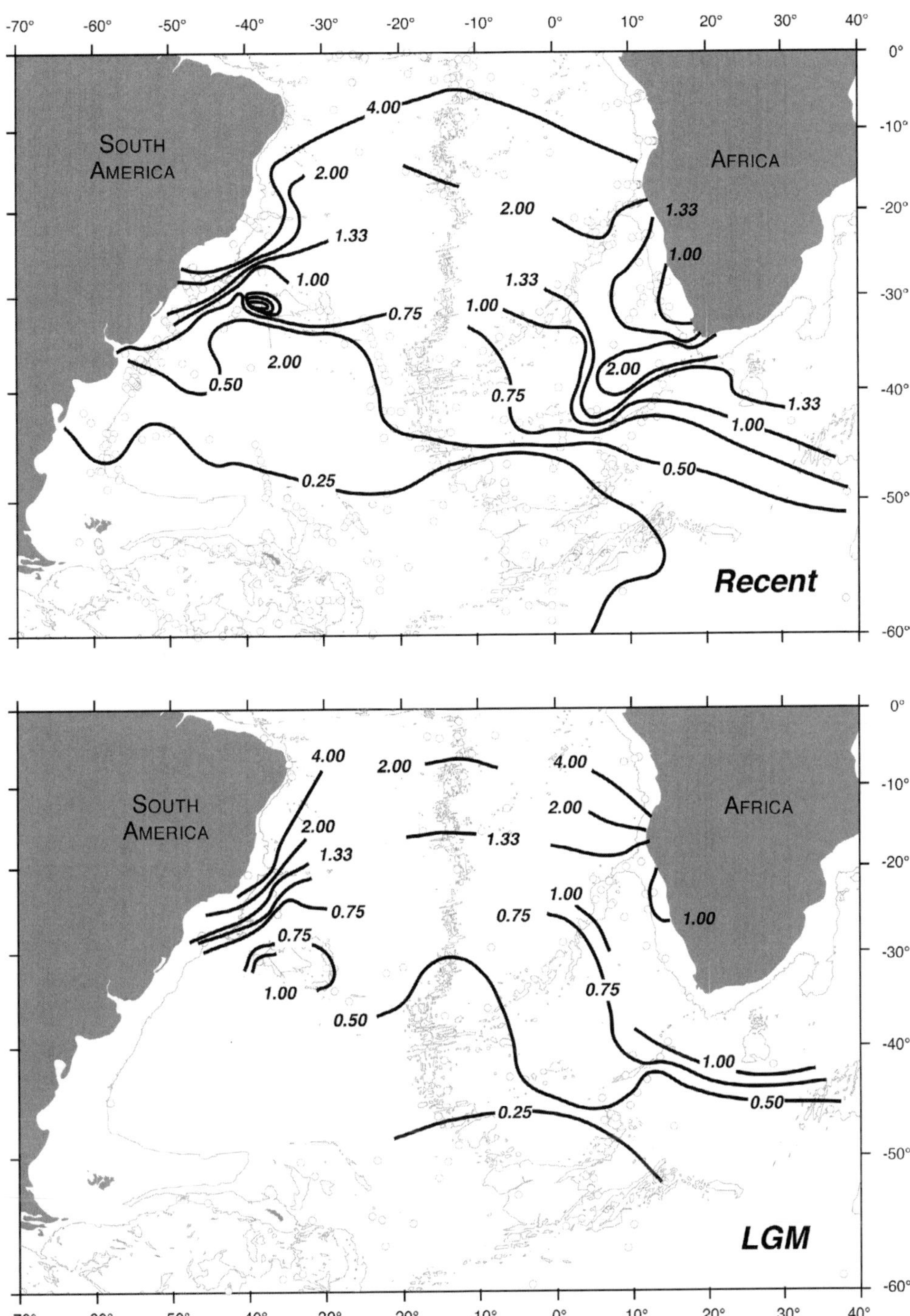

Fig. 8. Time-slice reconstructions of kaolinite/chlorite ratios in modern sea-bottom sediments (Recent, upper map) and marine sediments of the last glacial maximum (LGM, lower map), documenting particle advection in different water masses. Compared to the Recent situation the LGM map shows a displacement of low kaolinite/chlorite ratios to the northeast towards lower latitudes and to shallower water depths on topographic highs, because of a wider extension of chlorite-bearing southern-source water masses and a reduced influx of kaolinite-bearing North Atlantic Deep Water. Moreover, the Recent tongue of high kaolinite/chlorite ratios off southern Africa disappears on the LGM map, which can be explained by a reduced leakage of the kaolinite-bearing Agulhas Current into the South Atlantic. Open circles indicate locations of sediment cores used for reconstruction.

with a wide propagation of chlorite-bearing water masses of the ACC and a reduced influx of kaolinite-bearing NADW into the ACC during glacial periods (Diekmann et al. 1999). This interpretation is supported by the contour-line pattern of kaolinite/chlorite ratios, which exhibit a marked northeastward bulge of low values beyond the Mid-Atlantic Ridge for the last glacial situation (Fig. 8). Beside modified water-mass configurations, the changes in clay-mineral ratios can also be attributed to variable clay-mineral fluxes in the opposite water masses. As pointed out, ACC water masses are fed by glacigenic chlorite during glacial periods, whereas NADW receives abundant kaolinite during interglacials, when kaolinite-bearing NADW passes the Vema Channel region and migrates towards the Mid-Atlantic Ridge. These contrasts may overprint the water-mass-related clay-mineral signal to a certain degree.

The reconstruction of the operational modes of South Atlantic ocean circulation (Figs. 1, 2) in response to cyclic climate changes is of worldwide significance, because the water masses in the South Atlantic present part of the global conveyor circulation, which apart from atmospheric processes is responsible for the interhemispheric teleconnection of climate signals (Broecker and Denton 1989, Berger and Wefer 1996). Particularly in the southeastern South Atlantic, kaolinite/chlorite ratios are a suitable parameter to reconstruct particle advection within southern-source water masses of the ACC, within North Atlantic Deep Water (NADW) that is injected into the ACC, and within water masses of the Agulhas Current that enter the South Atlantic. High kaolinite/chlorite ratios in interglacial intervals document the inflow of kaolinite-bearing suspensions, entrained in the filaments of the Agulhas-Current retroflection and within NADW. Glacial periods are represented by low kaolinite/chlorite ratios, when the region is mainly bathed by ACC water masses. In the time-slice maps of kaolinite/chlorite ratios, the influence of the Agulhas Current on detrital fluxes is displayed by a contour-line tongue of high kaolinite/chlorite ratios for the Recent situation that extends from the Indian Ocean into the South Atlantic (Fig. 8). For the LGM situation, kaolinite/chlorite ratios are lower.

The palaeoclimatic relevance of the temporal variability in regional ocean circulation is underlined by time-series analyses of the clay-mineral parameters (Diekmann et al. 1996; Kuhn and Diekmann 2002a). The results reveal significant spectral power in the 100-kyr^{-1}, 41-kyr^{-1}, and 23-kyr^{-1} frequency bands of orbital eccentricity, obliquity, and precession. Moreover, the inferred palaeoceano-graphic variability is coherent and in phase with changes in (northern-hemispheric) ice volume and NADW production. Apart from Milankovitch cycles, variations in regional ocean circulation are also evident on a millennial time scale, particularly for MIS 3 (Kuhn and Diekmann 2002a). Thus, for interglacial periods, the operation of the modern warm-route conveyor mode is evident, implying a far southward injection of relatively warm and saline NADW into the ACC, compensated to a large extent by the northward outflow of warm surface and intermediate waters, which enter the South Atlantic via the Agulhas Current (Gordon et al. 1992). During glacial stages, the conveyor operates in cold route, which is characterized by prevailing cold southern-source water masses in the study area.

These results support palaeoceanographic reconstructions on the basis of other water-mass proxies. A diminished leakage of the Agulhas Current to the South Atlantic during glacial periods is recorded by changes in nannofossil assemblages in marine sediments off the South African cape region (Flores et al. 1999) and by strontium isotopes in detrital material in marine sediments of the southern Cape Basin (Goldstein et al. 1999). Benthic foraminiferal assemblages (Schmiedl and Mackensen 1997) and benthic foraminiferal δ^{13}C variations in South Atlantic sediment cores (Duplessy et al. 1988; Charles and Fairbanks 1992; Mackensen and Bickert 1999) both suggest a weakened NADW influx to the south during glacial periods. Moreover, benthic foraminiferal δ^{13}C records document a variable NADW influx on a millennial time scale, but on higher frequency as suggested by the clay-mineral proxy (Charles et al. 1996).

From these findings, it can be stated that regional ocean circulation in the South Atlantic is connected with global conveyor circulation and northern-hemispheric climate processes. Other palaeoceanographic changes in the ACC region seem to be dictated by southern-hemispheric processes, as suggested, for example, by variations in sea-surface temperatures that generally lead changes in conveyor circulation (Charles et al. 1996; Brathauer and Abelmann 1999). The coupling of northern-hemispheric processes with the South Atlantic is also evident on a longer time scale, as deduced from the Quaternary oxygen-isotope and clay-mineral record of Site 1090, which was also drilled during Leg 177 of the Ocean Drilling Program in the southeastern South Atlantic (Diekmann and Kuhn 2002) (Fig. 9). Thus the onset of the modern glacial-interglacial pattern in regional circulation at 1200 ka and the development of strong glacial-interglacial contrasts after the mid-Pleistocene around 650 ka match a corresponding temporal pattern in NADW production in the North Atlantic and the adjoining northern seas (Henrich and Baumann 1994; Henrich et al. 2002).

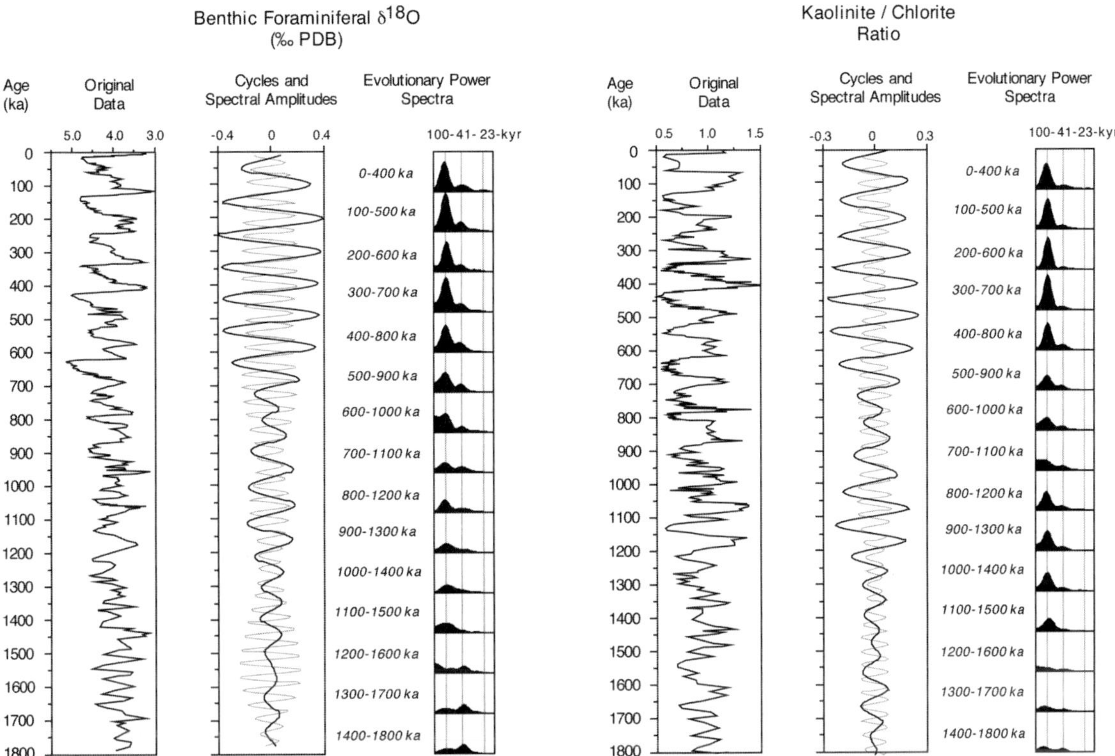

Fig. 9. Temporal fluctuations of benthic foraminiferal $\delta^{18}O$ ratios (Becquey and Gersonde 2002) and kaolinite/chlorite ratios (Diekmann and Kuhn 2002) at ODP Site 1090 in the southeastern South Atlantic. Cyclicity and spectral amplitudes of 100-kyr and 41-kyr periods were deduced from Gaussian filtering. Evolutionary power spectra are based on Blackman-Tuckey time-series analyses on 400-kyr intervals at steps of 100 kyr. Note the development of 100-kyr cycles across the mid-Pleistocene climate transition between 1200 ka and 650 ka and their marked strengthening after 600 ka, displayed by both parameters. These common spectral features underline the close relationship between global ice volume (oxygen isotope parameter) and ocean circulation in the southeastern South Atlantic (kaolinite/chlorite ratios). Modified from Diekmann and Kuhn (2002).

Conclusions

Detrital sources and modes of terrigenous sediment supply in the polar to temperate South Atlantic reflect glacial-interglacial environmental changes in response to late Quaternary climate cycles.

Mass-accumulation rates of terrigenous sediment depend on the detrital influx driven by advancing ice masses, fluvial discharge and aeolian influx and reflect the intensity of bottom-current transport. The spatial and temporal pattern of ice-rafted detritus (IRD) in marine sediments along Antarctica and in the Scotia Sea reflects the calving rates of Antarctic and Patagonian ice masses, while IRD abundances in the southeastern South Atlantic are related to the presence of cold surface waters. High proportions of coarse silt in the mud fraction of terrigenous sediments point to increased bottom-current strengths. Clay minerals indicate the sources of terrigenous mud and trace the dispersal of fine-grained suspensions by water-mass advection within ocean currents.

The combined interpretation of the various terrigenous sediment parameters suggests East Antarctic ice dynamics with dominant 100-kyr cycles and high rates of Antarctic Bottom Water formation and iceberg discharge during interglacial times with floating ice shelves because of high sea level. The more subpolar ice masses of the Antarctic Peninsula and Patagonia respond with maximum iceberg discharges during glacial terminations related to the rapid disintegration of advanced ice masses. In the northern Scotia Sea, increased sediment supply from southern South America points to extended ice masses in Patagonia during glacial times. In the southeastern South Atlantic, changes in regional ocean circulation are linked to global conveyor circulation and are in phase with northern-hemispheric processes of ice build-up and the associated formation of North Atlantic Deep Water, which decreases during glacial times and permits a wider extension of southern-source water masses in the study area.

Acknowledgements

This study was funded by the Deutsche Forschungsgemeinschaft through the Sonderforschungsbereich 261 at Bremen University and at the Alfred Wegener Institute in Bremerhaven and through grant Di-655/2. We appreciate the kind assistance of ship crews and scientists. For sample preparation and technical support during the last years, we are indebted to Rita Froehlking, Silvia Janisch, Norbert Lensch, Walter Luttmer, Helga Rhodes (†), Michael Seebeck, Maren Thomas, Franziska Stenzel, Jutta Vernaleken, and numerous student trainees. Finally we acknowledge the reviews of Carol Pudsey, Steven Hovan and an anonymous referee. This is SFB-261 contribution No. 363 and AWI publication No. 10492. Data are available under www.pangaea.de/Projects/SFB261.

References

Allen CP, Warnke DA (1991) History of ice-rafting at Leg 114 sites, Subantarctic/South Atlantic. Proceedings ODP, Scientific Results 114: 599-607

Anderson JB (1999) Antarctic Marine Geology. Cambridge University Press, Cambridge, 289 p

Anderson JB, Kurtz D (1985) The use of silt grain size parameters as a paleovelocity gauge: A critical review and case study. Geo-Mar Lett 5: 55-59

Anderson JB, Andrews BA, Bartek LR, Truswell EM (1989) Petrology and palynology of Weddell Sea glacial sediments: Implications for subglacial geology. In: Thomson MRA, Crame JA, Thomson JW (eds) Geological Evolution of Antarctica. Cambridge University Press, Cambridge, pp 231-235

Aston SR, Chester R, Johnson LR, Padgham RC (1973) Eolian dust from the lower atmosphere of the eastern Atlantic, and Indian Ocean, China Sea and Sea of Japan. Mar Geol 14: 15-28

Bard E (1999) Ice age temperatures and geochemistry. Science 284: 1133-1134

Bareille G, Grousset FE, Labracherie M, Labeyrie LD, Petit JR (1994) Origin of detrital fluxes in the southeast Indian Ocean during the last climatic cycles. Paleoceanography 9: 799-819

Barrett PJ (1996) Antarctic palaeoenvironment through Cenozoic times - a review. Terra Antartica 3: 103-119

Becquey S, Gersonde R (2002) Past hydrographic and climatic changes in the Subantarctic Zone of the South Atlantic - The Pleistocene record from ODP Site 1090. Palaeogeogr Palaeoclimatol Palaeoecol 182: 221-239

Berger WH, Wefer G (1996) Expeditions into the past: paleoceanographic studies in the South Atlantic. In: Wefer G, Berger WH, Siedler G, Webb DJ (eds) The

South Atlantic: Present and Past Circulation. Springer, Berlin, pp 363-410

Bickert T, Wefer G (1996) Late Quaternary deep water circulation in the South Atlantic: Reconstruction from carbonate dissolution and benthic stable isotopes. In: Wefer G, Berger WH, Siedler G, Webb DJ (eds) The South Atlantic: Present and Past Circulation. Springer, Berlin, pp 599-620

Biscaye PE (1964) Distinction between kaolinite and chlorite in Recent sediments by x-ray diffraction. Am Mineral 49: 1281-1289

Biscaye PE (1965) Mineralogy and sedimentation of recent deep-sea clay in the Atlantic Ocean and adjacent seas and oceans. Geol Soc Am Bull 76: 803-832

Biscaye PE, Dasch EJ (1971) The rubidium, strontium, strontium-isotope system in deep-sea sediments: Argentine Basin. J Geophys Res 76: 5087-5096

Biscaye PE, Eittreim SL (1977) Suspended particulate loads and transports in the nepheloid layer of the abyssal Atlantic Ocean. Mar Geol 23: 155-172

Bonn WJ, Gingele FX, Grobe H, Mackensen A, Fütterer DK (1998) Palaeoproductivity at the Antarctic continental margin: opal and barium records for the last 400 ka. Palaeogeogr Palaeoclimatol Palaeoecol 139: 195-211

Bornhold BD, Summerhayes CP (1977) Scour and deposition at the foot of the Walvis Ridge in the northernmost Cape Basin, South Atlantic. Deep-Sea Res 24: 743-752

Brathauer U, Abelmann A (1999) Late Quaternary variations in sea-surface temperatures and their relationship to orbital forcing recorded in the Southern Ocean (Atlantic sector). Paleoceanography 14: 135-148

Brehme I (1992) Sedimentfazies und Bodenwasserstrom am Kontinentalhang des nordwestlichen Weddellmeeres. Ber Polarforsch, Bremerhaven 110, 127 p

Bremner JM, Willis JP (1993) Mineralogy and geochem-istry of the clay fraction of sediments from the Namibian continental margin and the adjacent hinterland. Mar Geol 115: 85-116

Broecker W, Denton GH (1989) The role of ocean-atmosphere reorganizations in glacial cycles. Geochim Cosmochim Acta 53: 2465-2501

Chamley H (1989) Clay Sedimentology. Springer, Berlin, 623 p

Chamley H (1997) Clay mineral sedimentation in the ocean. In: Paquet H, Clauer N (eds) Soils and Sediments: Mineralogy and Geochemistry. Springer, Berlin, pp 269-302

Charles CD, Fairbanks RG (1992) Evidence from Southern Ocean sediments for the effect of North Atlantic Deep Water flux on climate. Nature 355: 416-419

Charles CD, Lynch-Stieglitz J, Ninnemann US, Fairbanks RG (1996) Climate connections between the hemispheres revealed by deep-sea sediment core / ice core correlations. Earth Planet Sci Lett 142: 19-27

Cofaigh CO, Dowdeswell JA, Pudsey CJ (2001) Late Quaternary iceberg rafting along the Antarctic Peninsula continental rise and in the Weddell and Scotia Seas. Quat Res 56: 308-321

Conolly JR, Ewing M (1965) Ice-rafted detritus as a climatic indicator in Antarctic deep-sea cores. Science 150: 1822-1824

Cooke DW, Hays JD (1982) Estimates of Antarctic Ocean seasonal sea-ice cover during glacial intervals. In: Craddock C (ed) Antarctic Geoscience. University of Wisconsin Press, Madison, pp 1017-1027

Cortese G, Abelmann A (2002) Radiolarian-based paleotemperatures during the last 160 kyrs at ODP Site 1089 (Southern Ocean, Atlantic Sector). Palaeogeogr Palaeoclimatol Palaeoecol 182: 259-286

De Melo U, Summerhayes CP, Ellis JP (1975) Salvador to Vitoria, southeastern Brazil, part IV. In: Milliman JD, Summerhayes CP (eds) Upper Continental Margin Sedimentation off Brazil. Contributions to Sedimentology, Vol 4, Schweizerbart, Stuttgart, pp 78-116

Diekmann B, Kuhn G (1997) Terrigene Partikeltransporte als Abbild spätquartärer Tiefen- und Bodenwasserzirkulation im Südatlantik und angrenzendem Südpolarmeer. Z Deut Geol Ges 148: 405-429

Diekmann B, Kuhn G (1999) Provenance and dispersal of glacial-marine surface sediments in the Weddell Sea and adjoining areas, Antarctica: Ice-rafting versus current transport. Mar Geol 158: 209-231

Diekmann B, Kuhn G (2002) Sedimentary record of the mid-Pleistocene climate transition in the sub-Antarctic South Atlantic (ODP Site 1090). Palaeogeogr Palaeoclimatol Palaeoecol 182: 241-258

Diekmann B, Petschick R, Gingele FX, Fütterer DK, Abelmann A, Brathauer U, Gersonde R, Mackensen A (1996) Clay mineral fluctuations in late Quaternary sediments of the southeastern South Atlantic: Implications for past changes of deep water advection. In: Wefer G, Berger WH, Siedler G, Webb DJ (eds) The South Atlantic: Present and Past Circulation. Springer, Berlin, pp 621-644

Diekmann B, Kuhn G, Mackensen A, Petschick R, Fütterer DK, Gersonde R, Rühlemann C, Niebler H-S (1999) Kaolinite and chlorite as tracers of modern and late Quaternary deep water circulation in the South Atlantic and the adjoining Southern Ocean. In: Fischer G, Wefer G (eds) Use of Proxies in

Paleoceanography: Examples from the South Atlantic. Springer, Berlin, pp 285-313

Diekmann B, Kuhn G, Rachold V, Abelmann A, Brathauer U, Fütterer DK, Gersonde G, Grobe H (2000) Ter-rigenous sediment supply in the Scotia Sea (Southern Ocean): Response to Late Quaternary ice dynamics in Patagonia and on the Antarctic Peninsula. Palaeogeogr Palaeoclimatol Palaeoecol 162: 357-387

Dittert N, Baumann K-H, Bickert T, Henrich R, Huber R, Kinkel H, Meggers H (1999) Carbonate dissolution in the deep-sea: Methods, quantification and pale-oceanographic application. In: Fischer G, Wefer G (eds) Use of Proxies in Paleoceanography: Examples from the South Atlantic. Springer, Berlin, pp 255-284

Duce RA, Liss PS, Merrill JT, Atlas EL, Buat-Menard P, Hicks BB, Miller JM, Prospero JM, Arimoto R, Church TM, Ellis W, Galloway JN, Hansen L, Jickells TD, Knap AH, Reinhardt KH, Schneider B, Soudine A, Tokos JJ, Tsunogai S, Wollast R, Zhou M (1991) The atmospheric input of trace species to the world ocean. Glob Biogeochem Cycl 5: 193-259

Duplessy JC, Shackleton NJ, Fairbanks RG, Labeyrie LD, Oppo D, Kallel N (1988) Deep water source variations during the last climatic cycle and their impact on global deep water circulation. Paleoceanography 3: 343-360

Edwards DS, Goodell HG (1969) The detrital mineralogy of ocean floor surface sediments adjacent to the Antarctic Peninsula. Mar Geol 7: 207-234

Ehrmann WU (1994) Die känozoische Vereisungsgeschichte der Antarktis. Ber Polarforsch, Bremerhaven 137, 152 p

Ehrmann WU, Melles M, Kuhn G, Grobe H (1992) Significance of clay mineral assemblages in the Antarctic Ocean. Mar Geol 107: 249-273

Fahrbach E, Beckmann A (2001) Weddell Sea Circulation. In: Steele JH, Turekian KK, Thorpe SA (eds) Encyclopedia of Ocean Sciences. Academic Press, San Diego, pp 560-570

Flores JA, Gersonde R, Sierro FJ (1999) Pleistocene fluctuations in the Agulhas Current Retroflection based on the calcareous plankton record. Mar Micro-paleontol 37: 1-22

Foldvik A, Gammelsrød T (1988) Notes on Southern Ocean hydrography, sea-ice and bottom water formation. Palaeogeogr Palaeoclimatol Palaeoecol 67: 3-17

Frank M, Gersonde R, Rutgers van der Loeff M, Kuhn G, Mangini A (1996) Late Quaternary sediment dating and quantification of lateral sediment redistribution applying 230-Th-ex: A study from the eastern Atlantic sector of the Southern Ocean. Geol Rundsch 85: 554-566

Fütterer DK, Grobe H, Grünig S (1988) Quaternary sediment patterns in the Weddell Sea: Relations and environmental conditions. Paleoceanography 3: 551-361

Georgi DT (1981) Circulation of bottom waters in the southwestern South Atlantic. Deep-Sea Res 28: 959-979

Gersonde R, Hodell DA, Blum P et al. (1999) Proceedings ODP, Initial Reports, 177 [CD-ROM]. Available from: Ocean Drilling Program, Texas A&M University, College Station, TX 77845-9547, USA

Gersonde R, Abelmann A, Brathauer U, Becquey S, Bianchi C, Cortese G, Grobe H, Kuhn G, Niebler H-S, Segl M, Sieger R, Zielinski U, Fütterer DK (2003) Last glacial sea-surface temperatures and sea-ice extent in the Southern Ocean (Altanic-Indian sector) - A multiproxy approach. Paleoceanography 18: 1061, doi 10.1029/2002PA000809

Gilbert IM, Pudsey CJ, Murray JW (1998) A sediment record of cyclic bottom-current variability from the northwest Weddell Sea. Sediment Geol 115: 185-214

Gingele FX (1996) Holocene climatic optimum in Southwest Africa - evidence from the marine clay mineral record. Palaeogeogr Palaeoclimatol Palaeoecol 122: 77-87

Gingele FX, Müller PM, Schneider RR (1998) Orbital forcing of freshwater input in the Zaire Fan area - clay mineral evidence from the last 200 kyr. Palaeogeogr Palaeoclimatol Palaeoecol 138: 17-26

Gingele FX, Schmieder F, von Dobeneck T, Petschick R, Rühlemann C (1999) Terrigenous flux in the Rio Grande Rise area during the past 1500 ka: Evidence of deepwater advection or rapid response to continental rainfall patterns? Paleoceanography 14: 84-95

Goldstein SL, Hemming SR, Kish S, Rutberg R (1999) Strontium isotopes in South Atlantic detritus: A surface current proxy and tracer of Agulhas leakage. Ninth Annual Goldschmidt Conference. Lunar Planet Inst, Houston, TX pp #7537 [CD-ROM]

Gordon AL, Weiss RF, Smethie WMJ, Warner MJ (1992) Thermocline and intermediate water communication between the South Atlantic and Indian Oceans. J Geophys Res 97: 7223-7240

Grobe H (1987) A simple method for the determination of ice-rafted debris in sediment cores. Polarforschung 57: 123-126

Grobe H, Mackensen A (1992) Late Quaternary climate cycles as recorded in sediments from the Antarctic continental margin. In: Kennett JP, Warnke DA (eds) The Antarctic Paleoenvironment: A Perspective on Global Change, Part One. Antarctic Research Series, Vol. 56, American Geophysical Union, Washington DC, pp 349-376

Grousset FE, Biscaye PE, Revel M, Petit JR, Pye K, Joussaume S, Jouzel J (1992) Antarctic (Dome C) ice core dust at 18 ky BP: Isotopic constraints on origins. Earth Planet Sci Lett 111: 175-182

Henrich R, Baumann KH (1994) Evolution of the Norwegian Current and the Scandinavian Ice Sheets during the past 2.6 my: Evidence from ODP Leg 104 biogenic carbonate and terrigenous records. Palaeogeogr Palaeoclimatol Palaeoecol 108: 75-94

Henrich R, Baumann KH, Huber R, Meggers H (2002) Carbonate preservation records of the past 3 Myr in the Norwegian-Greenland Sea and the northern North Atlantic: Implications for the history of NADW production. Mar Geol 184: 17-39

Heusser CJ, Heusser LE, Lowell TV, Moreira AM, Moreira SM (2000) Deglacial palaeoclimate at Puerto del Hambre, subantarctic Patagonia, Chile. J Quater Sci 15: 101-114

Hillenbrand CD (1994) Spätquartäre Sedimentationsprozesse am Kontinentalrand des nordöstlichen Bellingshausenmeeres (Antarktis). Diploma Thesis, Universität Würzbug, 124 p

Hillenbrand CD (2000) Glazialmarine Sedimentationsentwicklung am westantarktischen Kontinentalrand im Amundsen- und Bellingshausenmeer - Hinweise auf Paläoumweltveränderungen während der quartären Klimazyklen. Ber Polarforsch, Bremerhaven 346, 182 p

Hillenbrand CD, Ehrmann WU (2001) Distribution of clay minerals in drift sediments on the continental rise west of the Antarctic Peninsula, ODP Leg 178, Sites 1095 and 1096. In: Barker PF, Camerlenghi A, Acton GD, Ramsay ATS (eds) Proceedings ODP, Scientific Results 178, pp 1-29

Hillenbrand CD, Grobe H, Diekmann B, Kuhn G, Fütterer DK (2003) Distribution of clay minerals and proxies for productivity in surface sediments of the Bellings-hausen and Amundsen seas (West Antarctica) - Relation to modern environmental conditions. Mar Geol 193: 253-271

Hodell DA, Charles CD, Sierro FJ (2001) Late Pleistocene evolution of the ocean's carbonate system: A detailed record from Ocean Drilling Program (ODP) Site 1089. Earth Planet Sci Lett 192: 109-124

Höppner R, Henrich R (1999) Kornsortierungsprozesse am Argentinischen Kontinentalhang anhand von Siltkorn-Analysen. Zentralblatt für Geologie und Paläontologie Teil 1 1997: 897-905

Hofmann A (1999) Kurzfristige Klimaschwankungen im Scotiameer und Ergebnisse zur Kalbungsgeschichte der Antarktis während der letzten 200 000 Jahre. Ber Polarforsch, Bremerhaven 345, 162 p

Howe JA, Pudsey CJ (1999) Antarctic Circumpolar Deep Water: A Quaternary paleoflow record from the northern Scotia Sea, South Atlantic Ocean. J Sediment Res 69: 847-861

Husar RB, Prospero JM, Stowe LL (1997) Characterization of tropospheric aerosols over the oceans with the NOAA advanced very high resolution radiometer optical thickness operational product. J Geophys Res D102: 16889-16909

Imbrie J, Hays JD, Martinson DG, McIntyre A, Mix AC, Morley JJ, Pisias NG, Prell WL, Shackleton NJ (1984) The orbital theory of Pleistocene climate: Support from a revised chronology of the marine $\delta^{18}O$ record. In: Berger A, Imbrie J, Hays J, Kukla G, Saltzman B (eds) Milankovitch and Climate. NATO ASI SERIES, D Reidel, Dordrecht, pp 269-305

Jones GA (1984) Advective transport of clay minerals in the region of the Rio Grande Rise. Mar Geol 58: 187-212

Jones GA, Johnson DA (1984) Displaced Antarctic diatoms in Vema Channel sediments: Late Pleistocene/Holocene fluctuations in AABW flow. Mar Geol 58: 165-186

Kanfoush SL, Hodell DA, Charles CD, Guilderson TP, Mortyn PG, Ninnemann US (2000) Millennial-scale instability of the Antarctic ice sheet during the last glaciation. Science 288: 1815-1818

Keir RS (1988) On the late Pleistocene ocean geochemistry and circulation. Paleoceanogr 3: 413-445

Kolla V, Henderson L, Biscaye P (1976) Clay mineralogy and sedimentation in the western Indian Ocean. Deep-Sea Res 23: 949-961

Kuhn G, Diekmann B (2002a) Late Quaternary variability of ocean circulation in the southeastern South Atlantic inferred from the terrigenous sediment record of a drift deposit in the southern Cape Basin (ODP Site 1089). Palaeogeogr Palaeoclimatol Palaeoecol 182: 287-303

Kuhn G, Diekmann B (2002b). Data report: Bulk sediment composition, grain size, clay and silt mineralogy of Pleistocene sediments from ODP Leg 177 Sites 1089 and 1090. In: Gersonde R, Hodell DA, Blum P (eds) Proc. ODP, Sci. Results, 177 [Online]

Kumar N, Anderson RF, Mortlock RA, Froelich PN, Kubik P, Dittrich-Hannen B, Suter M (1995) Increased biological productivity and export production in the glacial Southern Ocean. Nature 378: 675-680

Labeyrie LD, Pichon JJ, Labracherie M, Ippolito P, Duprat J, Duplessy JC (1986) Melting history of Antarctica during the past 60,000 years. Nature 322: 701-706

Ledbetter MT (1986) A late Pleistocene time-series of bottom-current speed in the Vema Channel. Palaeogeogr Palaeoclimatol Palaeoecol 53: 97-105

Lisitzin AP (1996) Oceanic Sedimentation. American Geophysical Union, Washington DC, 400 p

Locarnini RA, Whitworth III, T, Nowlin Jr WD (1993) The importance of the Scotia Sea on the outflow of Weddell Sea Deep Water. J Mar Res 51: 135-153

Lowell TV, Heusser BJ, Anderson BG, Moreno PI, Hauser A, Heusser LE, Schlüchter C, Marchant DR, Denton GH (1995) Interhemispheric correlation of Late Pleistocene glacial events. Science 269: 1541-1549

Mackensen A, Bickert T (1999) Stable carbon isotopes in benthic foraminfera. In: Fischer G, Wefer G (eds) Use of Proxies in Paleoceanography: Examples from the South Atlantic. Springer, Berlin, pp 229-254

Mackensen A, Rudolph M, Kuhn G (2001) Late Pleistocene deep-water circulation in the subantarctic eastern Atlantic. Glob Planet Change 30: 197-229

Maher BA, Dennis PF (2001) Evidence against dust-mediated control of glacial-interglacial changes in atmospheric CO_2. Nature 411: 176-180

Mahowald N, Kohfeld K, Hansson M, Balkanski Y, Harrison SP, Prentice IC, Schulz M, Rodhe H (1999) Dust sources and deposition during the last glacial maximum and current climate: A comparison of model results with paleodata from ice cores and marine sediments. J Geophys Res D104: 15895-15916

Martin JH (1990) Glacial-interglacial CO_2 change: The iron hypothesis. Paleoceanography 5: 1-13

Massé L, Faugères JC, Bernat M, Pujos A, Mézerais ML (1994) A 600,000-year record of Antarctic Bottom Water activity inferred from sediment textures and structures in a sediment core from the southern Brazil Basin. Paleoceanography 9: 1017-1026

Massé L, Faugères JC, Pujol C, Pujos A, Labeyrie LD, Bernat M (1996) Sediment flux distribution in the southern Brazil Basin during the late Quaternary: the role of deep-sea currents. Sedimentology 43: 115-132

McCave IN, Manighetti B, Robinson SG (1995) Sortable silt and fine sediment size/composition slicing: Parameters for palaeocurrent speed and palaeoceano-graphy. Paleoceanography 10: 593-610

McCulloch RD, Bentley MJ, Purves RS, Hulton NRJ, Sudgen DE, Clapperton CM (2000) Climatic inferences from glacial and paleoecological evidence at the last glacial termination, southern South America. J Quat Sci 15: 409-417

Melles M, Kuhn G, Fütterer DK, Meischner D (1994) Processes of modern sedimentation in the southern Weddell Sea, Antarctica - evidence from surface sediments. Polarforschung 64: 45-74

Michels KH, Kuhn G, Hillenbrand CD, Diekmann B, Fütterer DK, Grobe H, Uenzelmann-Neben G (2002) The southern Weddell Sea: Combined contourite-turbidite sedimentation at the southeastern margin of the Weddell Gyre. In: Stow DAV, Pudsey CJ, Howe JA, Faugères JC, Viana Ar (eds) Deepwater Contourite Systems: Modern Drifts and Ancient Series, Seismic and Sedimentary Characteristics. Geol Soc, London, Mem 22, London, pp 305-323

Niebler HS (1995) Rekonstruktion von Paläo-Umweltparametern anhand von stabilen Isotopen und Faunenvergesellschaftungen planktischer Foraminiferen im Südatlantik. Ber Polarforsch, Bremerhaven 167, 198 p

Orsi AH, Whitworth III, T, Nowlin Jr, WD (1995) On the meridional extent and fronts of the Antarctic Circum-polar Current. Deep-Sea Res 42: 641-673

Oskierski W (1988) Verteilung und Herkunft glazialmariner Gerölle am Antarktischen Kontinentalrand des östlichern Weddellmeeres. Ber Polarforsch, Brermhaven 47, 166 p

Paillard D, Labeyrie L, Yiou P (1996) Macintosh program performs time-series analysis. EOS Transactions AGU 77: 379

Pastouret L, Chamley H, Delibrias G, Duplessy J-C, Thiede J (1978) Late Quaternary climate changes in western tropical Africa deduced from deep-sea sedimentation off the Niger delta. Oceanol Acta 1: 217-232

Peterson RG, Stramma L (1991) Upper-level circulation in the South Atlantic Ocean. Prog Oceanogr 26: 1-73

Petit JR, Mounier L, Jouzel J, Korotkevich YS, Kotlyakov VI, Lorius C (1990) Palaeoclimatological implications of the Vostok core dust record. Nature 343: 56-58

Petschick R, Kuhn G, Gingele FX (1996) Clay mineral

distribution in surface sediments of the South Atlantic: sources, transport, and relation to oceanography. Mar Geol 130: 203-229

Pirrung M, Hillenbrand CD, Diekmann B, Fütterer DK, Grobe H, Kuhn G (2002) Magnetic susceptibility and ice-rafted debris in surface sediments of the Atlantic sector of the Southern Ocean. Geo-Mar Lett 22: 170-180

Pudsey CJ (1992) Late Quaternary changes in Antarctic Bottom Water velocity inferred from sediment grain size in the northern Weddell Sea. Mar Geol 107: 9-33

Pudsey CJ, Camerlenghi A (1998) Glacial-interglacial deposition on a sediment drift on the Pacific margin of the Antarctic Peninsula. Antarctic Science 10: 286-308

Pudsey CJ, Howe JA (1998) Quaternary history of the Antarctic Circumpolar Current: Evidence from the Scotia Sea. Mar Geol 148: 83-112

Rebesco M, Larter RD, Barker PF (1996) Giant sediment drifts on the continental rise west of the Antarctic Peninsula. Geo-Mar Lett 16: 65-75

Ruddiman WF (1997) Tropical Atlantic terrigenous fluxes since 25,000 yrs BP. Mar Geol 136: 189-207

Schmiedl G, Mackensen A (1997) Late Quaternary paleo-productivity and deep water circulation in the eastern South Atlantic Ocean: Evidence from benthic foraminifera. Palaeogeogr Palaeoclimatol Palaeoecol 130: 43-80

Smith DG, Ledbetter MT, Ciesielski PF (1983) Ice-rafted volcanic ash in the southeast Atlantic sector of the Southern Ocean during the last 100,000 years. Mar Geol 53: 291-312

Tucholke BE, Embley RW (1984) Cenozoic regional erosion of the abyssal sea floor off South Africa. In: Schlee JS (eds) Interregional Unconformities and Hydrocarbon Accumulation. AAPG Memoir 36, American Association of Petroleum Geologists, Tulsa, pp 145-164

Van der Gaast SJ, Jansen JHF (1984) Mineralogy, opal and manganese of middle and late Quaternary sediments of the Zaire (Congo) deep-sea fan: Origin and climatic variation. Netherland J Sea Res 17: 313-341

Venz KA, Hodell DA (2002) New evidence for chnages in Plio-Pleistocene deep water circulation from Southern Ocean ODP Leg 177, Site 1090. Palaeogeogr Palaeoclimatol Palaeoecol 182: 197-220

Vogt C (1997) Zeitliche und räumliche Verteilung von Mineralvergesellschaftungen in spätquartären Sedimenten des Arktischen Ozeans und ihre Nützlichkeit als Klimaindikatoren während der Glazial/Interglazialwechsel. Ber Polarforsch, Bremerhaven 251, 309 p

Walter HJ, Hegner E, Diekmann B, Kuhn G, Rutgers van der Loeff MM (2000) Provenance and transport of terrigenous sediment in the South Atlantic Ocean and their relations to glacial and interglacial cycles: Nd and Sr isotope evidence. Geochim Cosmochim Acta 64: 3813-3827

Weber ME, Bonani G, Fütterer DK (1994) Sedimentation processes within channel-ridge systems, southeastern Weddell Sea, Antarctica. Paleoceanography 9: 1027-1048

Wefer G, Berger WH, Bijma J, Fischer G (1999) Clues to ocean history: A brief overview of proxies. In: Fischer G, Wefer G (eds) Use of Proxies in Paleoceanography: Examples from the South Atlantic. Springer, Berlin, pp 1-68

Fluxes at the Benthic Boundary Layer -
A Global View from the South Atlantic

C. Hensen[1*], K. Pfeifer[2], F. Wenzhöfer[3], A. Volbers[4], S. Schulz[2], J. Holstein[2], O. Romero[2] and K. Seiter[2]

[1]*GEOMAR, Forschungszentrum für Marine Geowissenschaften, Wischhofstr. 1-3, 24148 Kiel, Germany*
[2]*Universität Bremen, Fachbereich Geowissenschaften, Klagenfurter Strasse, 28359 Bremen, Germany*
[3]*Marine Biological Laboratory, University of Copenhagen, Strandpromenaden 5, 3000 Helsingør, Denmark*
[4]*Bundesanstalt für Geowissenschaften und Rohstoffe, Stilleweg 2, 30655 Hannover, Germany*
* corresponding author (e-mail): chensen@geomar.de*

Abstract: Fluxes between the ocean waters and the sediments are key regulation processes for the marine biogeochemical cycles and, thus, their quantification is of crucial importance. At this transition it is ultimately determined how much of a primary particulate signal is preserved or mineralized and hence recycled. Our review summarizes two major approaches how to use spatial information obtained from surface sediments: (1) In the first part we summarize the state-of-the-art regarding the use of biogenic barium as a proxy for primary productivity. We discuss the possibilities and limitations of this approach mainly based on the results of a recent study in the South Atlantic. The general outcome of this study was that the spatial pattern of primary productivity can well be traced back by calculating (sub-)recent accumulation rates of biogenic barium and applying available and newly formulated empirical equations. Most of those equations, however, fail to give the really observed magnitude of today's productivity values. The main reasons for this are mostly the uncertainty of the C_{org}/Ba_{bio} depth relation, which differs between distinct ocean regions, dynamic sedimentary processes at ocean margins combined with badly constrained values of terrigeneous barium input, and the effect of barite dissolution due to subsequent anoxic diagenesis. To improve the quality of prognoses for past productivity multi proxy approaches are recommended to bypass the uncertainty in predictions from a single proxy. (2) The more extensive second part is based on the large amount of studies that aimed at the quantification of benthic fluxes of nutrients and oxygen, which are good measures for the amount of reactive particulate material being mineralized at the seafloor and thus returned into the marine cycle. Those results enabled us to give profound calculations of the benthic oxygen consumption and the release of nitrate, phosphate, and silicate at the seafloor of the South Atlantic and give upscaled estimates for the global area of the sea floor. Additionally, we discuss more detailed studies focusing on control parameters for benthic fluxes like primary production and lateral advection along the ocean margins off Southwest Africa and Argentina. A very conspicuous result was obtained by calculating mass balances for biogenic opal in those regions indicating a dramatic underestimation of accumulation fluxes of opal by "conservative" methods, which is believed to be of global significance. The last section mainly focuses on the effect of benthic mineralization on the dissolution of calcium carbonate even above the chemical lysocline. This process is in discussion since more than two decades. A number of studies have been performed, mainly using *in situ* devices, to determine $CaCO_3$ dissolution. We summarize and discuss the results obtained from the South Atlantic and use a recently developed empirical algorithm to show the worldwide distribution of supralysoclinal $CaCO_3$ dissolution fluxes in marine surface sediments and give an estimate of their total amount. Finally, a table for benthic fluxes of major constituents is provided on ocean wide and global scales.

From WEFER G, MULITZA S, RATMEYER V (eds), 2003, The South Atlantic in the Late Quaternary: Reconstruction of Material Budgets and Current Systems. Springer-Verlag Berlin Heidelberg New York Tokyo, pp 401-430

Introduction

Because of their inherent importance in global biogeochemical cycles, the investigation of the uppermost sediment layers staying in contact and exchanging with the deep-ocean water is still a central theme in marine biogeochemical research. Important unsolved questions in this regard are whether and under which circumstances marine sediments can act as a final sink for anthro-pogenically produced CO_2. At this transition, early diagenetic processes are most intense giving the possibility to study various pathways of organic matter decay and their complex interaction with other redox processes. To what degree early dia-genesis masks the primary sedimentary signal arriving at the sea floor is a function of a number of parameters, such as the supply and the composition of organic material, the availability of oxidants for microbial activity (comprising e.g. the oxygen level in the bottom water), the sedimentation rate, and the effect of sediment mixing (bioturbation) and "non-local" pore water exchange (bioirrigation) by macrofaunal activity. Apart from the understanding of the various microbial and chemical processes underneath, the quantification of resulting material fluxes is essential in connection with the calculation of mass budgets within the oceans. The cycling of carbon (organic material plus $CaCO_3$) and nutrients constitute a key issue in global calculations. This comprises mainly the fluxes between the different compartments of the oceanic system determining the amount of organic carbon fixed by primary producers in the surface waters, the organic carbon fraction exported to the deep ocean and the sea floor, and the fraction of organic carbon arriving at the seafloor, which is finally buried or released back into the bottom water. In connection with carbon fluxes, it is widely accepted that about 90% of the annual primary productivity (PP) is recycled within the surface waters and only about 1% of the remaining 10% of organic matter, which escape as export production to the deep ocean, is finally buried in deep-sea sediments (i.e. de Baar and Suess 1993). The important link between burial of organic matter and the enhancement of primary production by nutrient supply is to a certain amount determined by the biogeochemical mineralization efficiency in the deep waters and in the surface sediments.

Closely related to the recycling of organic carbon on the sea floor is the cycle of $CaCO_3$. The dissolution of calcium carbonate in marine sediments is controlled by two major factors, the undersaturation of calcite or aragonite in the bottom water and the (local) undersaturation in the pore water due to the production of carbon dioxide by oxic respiration. Metabolically driven carbonate dissolution in marine sediments has not been considered to play an important role on a global scale for a number of years, although Emerson and Bender already pointed out its importance for the preservation of calcium carbonate in marine sediments in 1981. The understanding whether dissolution of calcium carbonate is driven by one or the other process is very important to correctly interpret the accumulation of calcium carbonate in sediments over time. For example, calcium carbonate preservation at a given site may either be determined by the calcite compensation depth (CCD) or a change in the rain ratio of organic carbon to calcium carbonate controlling the CO_2 release within the sediments. In this regard, Archer and Maier-Reimer (1994) demonstrated that a shift to higher rain ratios could explain both reduced pCO_2-levels during the last glacial and sedimentary calcium carbonate concentrations in deep-sea sediments. Thus, the calcite dissolution by oxic respiration of organic matter can mask effects of changes in carbonate productivity and deep-water chemistry in the sedimentary carbonate record (Martin and Sayles 1996).

In summary, it is obvious that the processes at the benthic boundary layer are critical in many instances for the investigation of mass transfer and mass fluxes in the ocean. Thus, we will address a number of aspects concerning regional-quantitative and model approaches centered on the results from the South Atlantic over the past 12 years. This comprises the (1) preservation of the primary sedimentation signal (the reconstruction of the recent primary productivity pattern using biogenic barium accumulation rates), (2) the quantification of benthic fluxes on regional and global scales, (3) regional aspects of the oxygen consumption and benthic nutrient fluxes in the South Atlantic, and (4) the quantitative effect of organic matter decomposition on the dissolution of calcium carbonate.

Signals of Primary Productivity: Biogenic Barium in Surface Sediments

To reconstruct present and past primary productivity, sediment proxies are required which reflect the primary productivity in relation to its distribution in the sedimentary record (e.g. organic matter, carbonate or opal). Most important is the knowledge of the burial efficiencies related to these proxies, which can be affected by temporal and spatial changes depending on variations of the biogeochemical environment. However, the main biogenic phases can be applied only under certain conditions to calculate productivity fluctuations. For example, large uncertainties exist when C_{org} is used as a proxy in oligotrophic open-ocean regions. This is due to its very low concentrations in the sediment and varying degrees of preservation. On the other hand, the primary productivity signal can be overprinted along continental margins by the input of terrestrial C_{org}, lateral sediment transport, and mixing through the action of bottom currents. Calcium carbonate, as an alternative proxy for productivity in oligotrophic ocean areas, can only be used above the lysocline (Rühlemann et al. 1999) and where dissolution by metabolically produced CO_2 is expected to be negligible (see Supralysoclinal dissolution of $CaCO_3$). Opal is considered to have a robust preservation potential, but until today it has not been possible to relate biogenic opal production in surface waters to opal accumulation in the sediment. Thus, additional proxies are required which are associated with a main biogenic phase and are well preserved during transport through the water column and within the sediments.

Barium has often been suggested as a geochemical proxy to reconstruct past and present primary productivity (i.e. Dymond et al. 1992; Dymond and Collier 1996; Gingele and Dahmke 1994; Francois et al. 1995), however, its general applicability is still under discussion. Calculations of the new production (P_{new}) from biogenic barium (Ba_{bio}) are based on the assumption that barite formation is associated with the decomposition of or-

ganic matter where it precipitates in microenvironments due to the sulfate released from organic components (Dehairs et al. 1980; Bishop 1988) or dissolution of acantharian derived celestite as a source of barium and sulfate (Bernstein et al. 1992). Analyses of C_{org}/Ba_{bio} ratios obtained from sediment trap material revealed highest ratios in the western Atlantic and lowest in the Pacific (Dymond et al. 1992). Generally, the ratio decreases with increasing water depth caused by increasing Ba-fluxes and the decay of organic material. In a recent study Pfeifer et al. (2001) used a data set of about 190 surface sediment samples in the South Atlantic, in order to investigate the suitability of sedimentary barium accumulation for the reconstruction of primary productivity, to our knowledge the first time on a basin-wide scale. The hypothesis was that the spatial distribution of Ba_{bio} at the sediment surface should reflect the known, present-day primary productivity pattern of the South Atlantic Ocean. Since terrigeneous Ba (Ba_{terr}), associated with the alumosilicate fraction of the sediment, cannot be attributed to the productivity in surface waters, a normative analysis was applied to account for this fraction. Ba_{bio} can then be calculated by subtracting the concentration of terrigeneous Ba from the total Ba concentration (Ba_{tot}):

$$Ba_{bio} = Ba_{tot} - (Al \times Ba/Al_{alumosilicate}) \qquad (1)$$

where Ba_{bio} is the biogenic barium content or barite, Ba_{tot} is the total barium, Al is the total aluminum content in the sediment, and $Ba/Al_{alumosilicate}$ denotes the terrigeneous Ba/Al ratio, which has to be estimated either from tables of average crustal element abundances or from the composition of the known alumosilicate source.

Pfeifer et al. (2001) used a uniform lithogenic Ba/Al ratio of 0.004 for all samples to estimate the content of Ba_{bio} in the sediment (see original study for more details). The recent accumulation rates of biogenic barium (AR-Ba_{bio} in g cm^{-2} kyr^{-1}) are shown in Figure 1a. AR-Ba_{bio} were calculated to consider regionally varying sediment mass accumulation rates (MAR). Most MARs used for the calculations of AR-Ba_{bio} were based on the data published by Cwienk (1986; derived from Jahnke 1996).

However, it should be noted that MAR calculations are critical with respect to continental margin deposits since the available data set does not reflect the spatial heterogeneity in those areas. From Figure 1a it is obvious that three different regions with increased accumulation rates can be identified in the South Atlantic: (1) the upwelling region off Southwest Africa (Cape Basin) with AR-Ba_{bio} of up to 3 mg cm^{-2} kyr^{-1}, (2) the Argentine Basin with rates of up to 3.6 mg cm^{-2} kyr^{-1}, and (3) the equatorial divergence zone with rates between 0.25 and 0.5 mg cm^{-2} kyr^{-1}. Generally, a high primary production in the surface waters correlates with the high AR-Ba_{bio} in the surface sediments confirming the qualitative correctness of the approach.

To calculate new and primary productivity from Ba-accumulation additional assumptions have to be made that are expressed in the empirical formulations listed below. The new production from AR-Ba_{bio} or Ba_{bio}-fluxes, respectively, was estimated according to Dymond et al. (1992) with:

$$P_{new} = (FBa_{bio} \cdot 0.171 \cdot Ba_{diss}^{2.218} \cdot z^{0.476-0.00478\,Ba_{diss}} / 2056)^{1.504}$$

with Ba_{diss} =55 nmol kg^{-1}
(eq. South Atlantic, Chan et al. 1977)   (2)

where P_{new} is the new production, FBa_{bio} is the flux of Ba_{bio} to the sediment surface, Ba_{diss} is the concentration of dissolved Ba in sea water, and z is the water depth.

This formulation is based on the assumption that barite formation is associated with the decomposition of organic matter and hence the flux of C_{org}. To estimate the Ba flux on the basis of Ba-accumulation rates (Eq. 3), the amount of preserved Ba_{bio} (%$Ba_{pres.}$) has to be calculated, which can be related to the MAR (in mg cm^{-2} kyr^{-1}) of the sediment by Equation 4 (Dymond et al. 1992).

$$FBa_{bio} = AR\text{-}Ba_{bio} / 0.209 \log (MAR) - 0.213$$
$$(3)$$

$$\% Ba_{pres} = 20.9 \log (MAR) - 21.3 \qquad (4)$$

Approaches similar to Dymond et al. (1992) have also been published by Nürnberg (1995) and Francois et al. (1995). However, their equations generally lead to an underestimation of the new production of 40-60 % compared to results calculated after Dymond et al. (1992). Finally, primary production (PP in gC m^{-2} yr^{-1}) was calculated according to Eppley and Peterson (1979):

$$PP = 20 \, (P_{new})^{0.5} \qquad (5)$$

The resulting primary productivity map is shown in Figure 1b. Accordingly, the regions displaying an increased primary productivity are the southwest coast of Africa, the Argentine Basin and the equatorial divergence zone. Using Equation (2), the estimates are generally comparable with those published by Berger (1989), but are distinctly lower than the more recent model results obtained by Behrenfeld and Falkowski (1997). The application of the equation by Francois et al. (1995), however, resulted in primary productivity values between 16 and 81 gC m^{-2} yr^{-1} in the Cape Basin, which is much lower than those derived from recent surface water estimates (Table 1).

These results suggest that there is a strong regional component between the water column fluxes of C_{org} (FC_{org}) and biogenic Ba (FBa_{bio}). Thus, the data obtained from the Cape Basin were used to develop a new equation for this region applying a different C_{org}/Ba_{bio} ratio. The ratio of FC_{org} and FBa_{bio} was estimated from the amount of Ba_{bio} and C_{org} in the surface sediments, by calculating the amount removed by dissolution and decay at the sediment/water interface. The proportion of C_{org} mineralization was estimated by relating O_2 fluxes into the sediment to the decay of organic matter (Hensen et al. 2000). The total C_{org} flux (FC_{org}) to the sediment could then be determined by Equation 6

$$FC_{org} = AR \, C_{org} + FC_{ox} \qquad (6)$$

where AR C_{org} (gC cm^{-2} kyr^{-1}) is the amount of carbon buried in the sediment and FC_{ox} (gC cm^{-2} kyr^{-1}) is the portion of oxidized carbon. For FC_{ox}, it was assumed that the amount of oxygen is completely consumed for the decomposition of C_{org} (see discussion below). The function was fitted to the C_{org}/Ba_{bio} ratios (Eq. 7) and applied to calculate

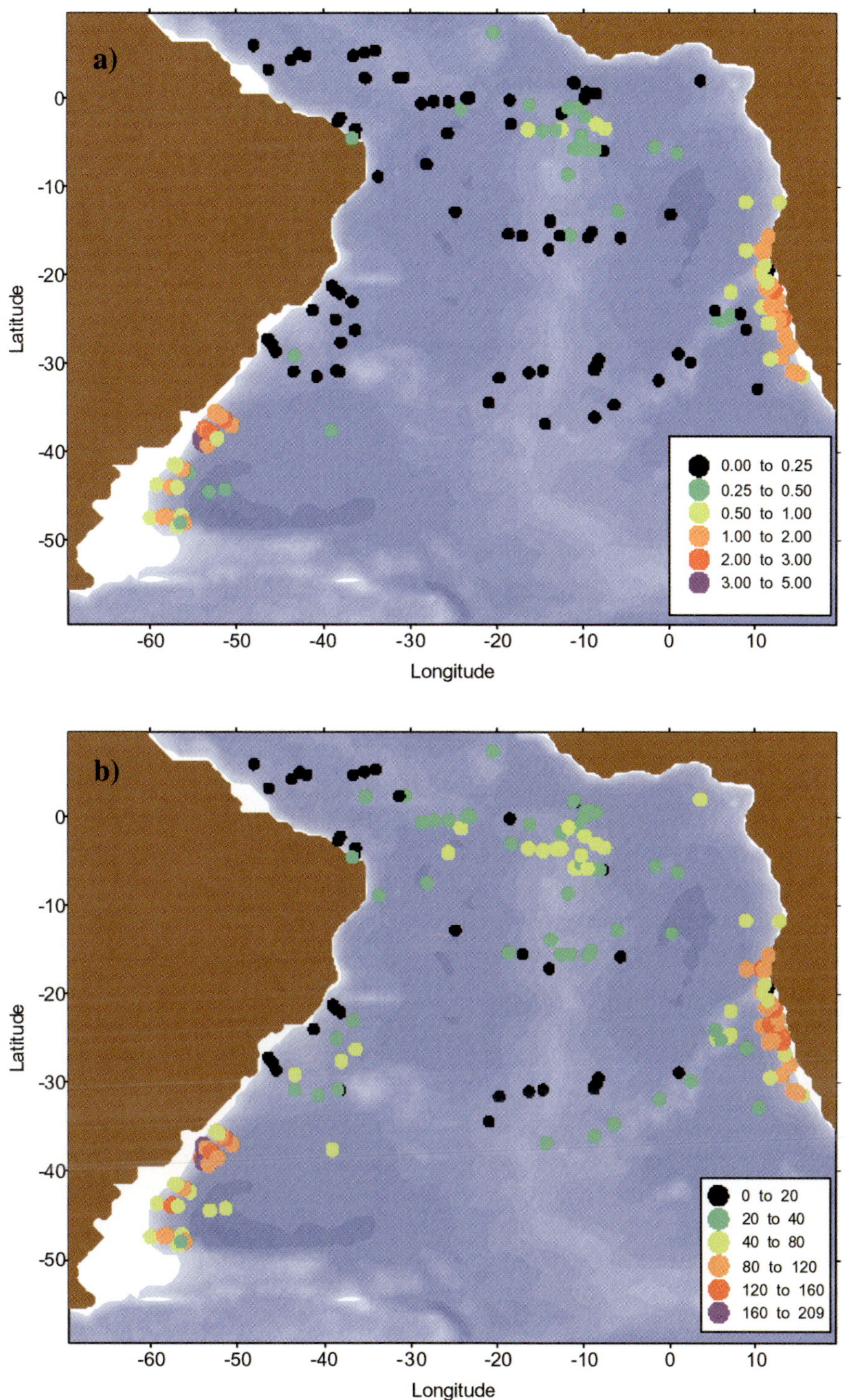

Fig. 1. a) Estimated accumulation rates of biogenic barium in mg cm^{-2} kyr^{-1} and **b)** the resulting primary productivity in gC m^{-2} yr^{-1} using Eq. 2 and 5 for various sites in the South Atlantic (redrawn from Pfeifer et al. 2001).

Region	PP (g C m^{-2} yr^{-1})				
	Behrenfeld and Falkowski (1997)	from P_{new} (Equ. 2)	from P_{new}: Francois et al. (1995) $P_{new}=1.95(FBa_{bio})^{1.41}$	from P_{new}: Dehairs et al. (2000) $P_{new}=22.7 (FBa_{bio})^{1.504}z^{-0.139}$	from P_{new}: Pfeifer et al. (2001) $P_{new}=683.34 (FBa_{bio})^{1.504}z^{-0.482}$
Equatorial Divergence	~ 50 - 250	2 - 69	1 - 37	2 - 72	3 - 105
Cape Basin	~ 50 - 400	29 - 135	16 - 81	28 - 181	36 - 286
Argentine Basin	~ 75 - 400	28 - 209	17 - 110	32 - 257	45 - >400

Table 1. Estimated primary productivity (gC m^{-2} yr^{-1}) from accumulation rates of biogenic barium (AR-Ba$_{bio}$) in different regions of the South Atlantic.

new and primary production. Primary productivity values between 36-286 gC m^{-2} yr^{-1} in the Cape Basin were obtained by this method (Table 1), which come fairly close to the estimate of Behrenfeld and Falkowski (1997) of 50 to 400 gC m^{-2} yr^{-1}.

$$FC_{org}/FBa_{bio} = 155.753\ z^{-0.8745} \qquad (7)$$

A similar transfer function based on barium data derived from sediment traps deployed in the Northeast Atlantic margin was defined by Dehairs et al. (2000) to generally relate the FBa$_{bio}$ to the export production characteristics of margin areas (Eq. 8).

$$FC_{org}/FBa_{bio} = 16.360\ z^{-0.646} \qquad (8)$$

Using this relation to calculate the new production for the locations in the Cape Basin again reveals much lower primary productivity values, these being in the range of those based on P_{new} values of Dymond et al. (1992). This result (documented in Table 1) emphasizes the limited applicability of such empirical equations, especially in ocean margin regions with enhanced biological activity in surface waters. However, applying Equation (7) to other regions in the South Atlantic reveals realistic results, although the correctness of the applied method is questionable. Overall, two major problems of estimating new production and primary productivity remain, when Ba is used as

a proxy. The first and major problem is the uncertainty of the C$_{org}$/Ba$_{bio}$ depth relation, which differs between distinct ocean regions. As pointed out by Dehairs et al. (2000) for margin areas, P$_{new}$ values calculated by the FC$_{org}$/FBa$_{bio}$ ratio in sediment traps are generally lower than those derived from direct estimates of new production. This is possibly due to either the conversion of new production to dissolved organic matter or lateral, advective transport of organic matter into the open ocean instead of sinking vertically to the sea floor. This would mean that the equation relating the flux of organic carbon in the water column to new production (Sarnthein et al. 1988) is not suitable in these regions. The second problem concerns the sedimentary Ba content and the burial efficiency in the sediment. The preservation of the Ba signal is believed to depend on the MAR (Eq. 4) and the total amount of sedimentary Ba (Dymond et al. 1992). Combined sediment trap, surface sediment, and model studies revealed different degrees of barite preservation in deep-sea locations in the Pacific and Atlantic Ocean: Dymond et al. (1992) and Paytan and Kastner (1996) predict about 30% preservation, McManus et al. (1994) 30-80% and Kumar et al. (1996) only 20% preservation of the primary FBa$_{bio}$ at the sediment surface.

Interesting and still being discussed is, however, the influence of early diagenesis on the preservation of Ba$_{bio}$. It is well known that sulfate reduction (with total depletion of pore water sulfate) results in the dissolution of barite and a subsequent redistribution of Ba inventories. However, suboxic diagenesis (i.e. reduction of Mn- and Fe-oxides, without depletion of sulfate) is also discussed to be responsible for the dissolution of barite (McManus et al. 1998). In a recent study, Schenau et al. (2001) proposed that the saturation state of barite in bottom water is probably the main environmental factor regulating the burial efficiency (which is only 0–10% at 500 m increasing with water depth) in Arabian Sea surface sediments.

Taking all the evidence into consideration much work remains to be done in order to identify the processes in surface sediments leading to barite dissolution and the control of the preservation efficiency. A promising alternative might be, however, the use of several proxies (multi-proxy approach) to diminish the uncertainty of predictions based on a single proxy and improve the understanding of dissolution and mineralization processes in surface sediments (i.e. Kumar et al. 1996). One main interrelation for the application of such proxies regarding the complex pore water / solid-phase interactions is a better quantitative understanding of benthic fluxes on regional and global scales, which shall be studied in the following sections.

Quantification of Benthic Fluxes

General Strategies and Approaches

Benthic fluxes are a measure for the amount of an element that is consumed in the sediments or released back to ocean water in dissolved form at the sediment-water interface, and therefore provide us with a tool to estimate the supplementary amount not to be preserved in the sediment. Thus, as stated above, the quantification of the total mass fluxes across the sediment-water interface is of central importance to estimate (their role in the) global carbon and nutrient budgets. Since benthic fluxes of dissolved oxygen or nutrients as well as the burial of organic carbon or biogenic opal show an overall correlation with the export of biogenic particles to the sea floor (i.e. Henrichs 1992; Canfield 1993; Ragueneau et al. 2000), benthic flux estimations should, if possible, be based on empirical relationships of primary productivity data and the depth dependence of organic matter export (i.e.

Betzer et al. 1984; Berger et al. 1987). To follow such an approach, oxic consumption, which is the major driving force of benthic respiration in deep-sea sediments with oxygen as the ultimate electron acceptor (Jørgensen 1983; Canfield 1993), can be related directly to benthic mineralization rates (e.g. Smith and Hinga 1983; Thamdrup and Canfield 2000) and thus converted in terms of total flux of organic material. Because of all subsequent mineralization processes occurring below, this is not strictly correct. Rather, a complete net-reoxidation of all reduced species produced during anoxic diagenesis by oxygen would then be required (Pamatmat 1971; Berner 1982). Any burial of reduced species by formation of sulfides or similar processes, or the flux of reduced species into the bottom water, would lead to an underestimation of the total respiration (e.g. Canfield 1993; Hensen and Zabel 2000). Strictly speaking, such an approach is only valid when virtually no organic carbon is preserved, which is only applicable in the strictly oligotrophic areas of the ocean. In highly productive areas and coastal environments, the carbon input exceeds the oxidation potential of the sediments by far. Here, and in areas where bottom waters are even anoxic, other pathways of organic matter decomposition dominate, clearly limiting this approach to the deep-ocean realm. Although the simple relation between the amount of organic carbon fixation by phototrophic algae in the surface water and benthic recycling and burial rates may as a matter of principle be correct, numerous geochemical and oceanographic studies give evidence that this link is much more complex in detail. Substantial uncertainties might arise from a lateral input of organic matter in the sediment due to downslope mass fluxes along continental margins (i.e. Jahnke et al. 1990; Rowe et al. 1994). Productivity can be enhanced by upwelling of nutrient rich water masses especially along continental shelves and slopes. Wherever a more complex current-system favors lateral transport processes, a significant decoupling between the pattern of productivity and the distribution of benthic remineralization rates can often be observed. In these regions simple transfer functions for export production usually fail to determine the decomposition of organic matter expressed as a function of water depth. Lateral drift during the process of particle settling, as well as re-suspension and re-deposition of sediments result in the development of so called "depocenters" at continental slopes or rises (Walsh 1991) which are generally attributed to a cross-shelf export of fine-grained organic material. Since it is very difficult to assess the influence of this process on the quantification of benthic fluxes, the handling of this problem will be discussed in the subsequent sections (see Opal budget in specific regions of the South Atlantic) and deserves much more attention in future studies (Zabel and Hensen 2002).

A fundamental understanding of biogeochemical processes in sediments enclosing the benthic boundary layer requires a quantitative assessment of process rates. *In situ* measurements have provided an effective way to quantify benthic fluxes across the sediment-water interface for quite a number of dissolved species (i.e. O_2, NO_3, PO_4, TIC, Si, etc.). The success of *in situ* techniques is generally attributed to the fact that sediment disturbance is minimized. While laboratory measurements require recovery of the sediment, benthic lander measurements leave the sediment in place and thereby the natural gradients of metabolites and substrates remain intact. Since this problem was identified, a number of studies have targeted discussion on the usability of *ex situ* data for single compounds. For example, studies by Glud et al. (1994, 1999), Sauter et al. (2001), and Wenzhöfer and Glud (2002) revealed that laboratory measurements generally overestimate the oxygen flux *in situ*, while the oxygen penetration depth (OPD) measured *ex situ* is much lower than *in situ*. Generally, the reasons for differences between in and *ex situ* measurements are most likely attributed to (1) increased microbial activity caused by transient heating during core recovery, (2) depressurization leading to a higher availability of labile organic matter released during lysis of barophilic or psycrophilic microbiota and/or precipitation of carbonate minerals, and (3) the treatment of the sediment to extract the pore water by sectioning, centrifugation or pressure squeezing (e.g. Glud et al. 1994, 1999; Berelson et al. 1990; Martin et al. 1996; Hensen et al. 1998; Sauter et al. 2001). However, a major problem in this context is that

for a large number of important parameters (nutrients, CO_2, Ca, etc.) *in situ* measurements either do not exist or the databases are insufficient, and thus, *ex situ* data have to be used.

In situ solute exchange mechanisms and rates can either be measured with benthic chambers (Smith et al. 1976; Glud et al. 1994), which measure the total benthic solute flux (e.g. total oxygen uptake - TOU), with micro-profiling instruments (Reimers 1987; Gundersen and Jørgensen 1990; Wenzhöfer et al. 2001a, b), which measure the solute flux supported by molecular diffusion (e.g. diffusive oxygen uptake - DOU) or with coring devices for tracer injection (Greeff et al. 1998), which measure process rates within the sediment. For studies focusing on the dynamics of sediment oxygen (and the related mineralization of organic matter), it is necessary to measure the diffusive as well as the total uptake rates. TOU measurements with benthic chambers can be taken as a measure for the total carbon mineralization in deep-sea sediments, however, they consider the enclosed sediment as a black box. In contrast, oxygen sensors resolve a one-dimensional picture of the oxygen distribution at a single spot neglecting benthic fauna activity and sediment topography, which can alter the oxygen penetration and uptake rate. On the other hand, it is possible to define the oxygenated sediment horizon with oxygen profiles and model consumption rates, which can be used to determine distinct zones of different microbial activities. The difference between TOU and DOU is furthermore a measure for the non-diffusive uptake (also called fauna-mediated uptake), including the metabolic respiration of the fauna as well as the sediment uptake through bioirrigation and biodiffusion.

Disregarding these obvious differences between TOU and DOU (or *in situ* and *ex situ* measurements) the crucial problem of quantifying benthic fluxes is their total availability and heterogeneity of distribution on a global scale. Major approaches to circumvent this problem consist in the application of geostatistical interpolation methods, the development of empirical equations on more widely available parameters, like the primary production or solid-phase compounds in the surface sediments, and, at least, the use of coupled diagenetic models. Focusing on the studies carried out in the South Atlantic, the next section will present the most important results obtained during the past five years.

Benthic Oxygen Fluxes

Up to now, there are two global (Jahnke 1996; Christensen 2000) and two other large-scale (Schlüter et al. 2000; Wenzhöfer and Glud 2002) estimates of the benthic oxygen consumption and related carbon mineralization. Due to the limited availability of *in situ* microelectrode measurements and benthic chamber incubations, the study of Jahnke (1996) was based on a set of 64 data points restricted to certain areas of the Pacific and Atlantic Ocean. For reasons of extrapolation considering such a strongly limited data set, he determined an empirical function, which relates the oxygen flux across the sediment/water interface to the more widely available data sets of organic carbon and $CaCO_3$ in the surface sediments and corresponding accumulation rates. In this procedure, oxygen depletion is a function of the organic carbon burial rate. The accumulation of $CaCO_3$ is calculated and subtracted from the total accumulation rate, in order to compensate for the variations of dilution by non-reactive particles. This simply means that a higher input of non-reactive particles increases the rate of accumulation and thus favors organic carbon burial above remineralization. The application of this method came out with a relation between the organic carbon burial rate (corrected for calcium carbonate accumulation) and the benthic oxygen flux, with a maximum deviation from measured values of about a factor of 2:

$$FO_2 = 0.134 \cdot (COB)^{0.467} \qquad (9)$$

where FO_2 is the benthic oxygen flux and COB is the rate of C_{org}-burial corrected for $CaCO_3$.

The result obtained from this regression is a grid (2° resolution) of global oxygen consumption rates between 60°S and 60°N for water depths below 1000 m (Jahnke 1996). As expected, the structures roughly follow the pattern of primary production and thus organic matter export. The resulting total rate of global benthic oxygen consumption calculated by this method is 5.4×10^{14} moles yr^{-1}, corresponding to an oxidized flux rate of particulate organic carbon of about 3.3×10^{13} moles yr^{-1} in the deep-sea. This value is equivalent to 45% of the POC-flux over the 1000m depth horizon (7.2×10^{13} moles C yr^{-1}), estimated after the empirical expression of Berger et al. (1987).

More recently, Christensen (2000) compiled benthic oxygen fluxes derived from *in situ* and *ex situ* microprofile measurements and *in situ* chamber incubations (in total 63 measurements) to achieve a global relation between primary productivity (PP), water depth (z), and benthic oxygen demand (BOD):

$$BOD = 19.6537 \cdot z^{-0.935411} \cdot PP^{1.139487} \qquad (10)$$

Based on the results of a diagenetic model study, Christensen found that a linear relationship between FCorg and BOD only exists below a critical value of 9 gC m^{-2} yr^{-1}. Thus, he only used data from locations below this critical value of carbon input in regression analysis. The result estimates the global amount of mineralized carbon in the order of 739 MT C yr^{-1}, which equals 32% of the carbon flux at 1000m. The annual global oxygen consumption is estimated to be 50 % higher compared to the global compilation of Jahnke (1996). Christensen (2000) explained this difference with the fact that more organic matter reaches the sediments than is predicted by sediment trap studies, and that his relation was based on calculations at a higher spatial resolution.

Concentrating on regional aspects, Schlüter et al. (2000) and Wenzhöfer and Glud (2002) applied similar approaches of relating the flux of particulate organic carbon to benthic oxygen fluxes in the northern section of the North Atlantic and the South Atlantic. Schlüter et al. (2000) calculated rain rates of organic carbon to the seafloor on the basis of multiple regression analyses of *in situ* oxygen fluxes (based on microsensor measurements), primary productivity and water depth which also showed a reasonable agreement with the data obtained from sediment trap deployments and the *ex situ* determination of benthic oxygen fluxes. Accordingly, they could define the organic carbon flux (FC_{org}) to the seafloor (>500 m water depth, z) for the northern part of the North Atlantic between 60°N and 80°N by:

$$FC_{org} = PP^{1.873} \cdot z^{-1.172} \qquad (11)$$

For the investigated area of about $2 \cdot 10^6$ km^2, with a primary production of $232 \cdot 10^6$ tC yr^{-1}, an organic carbon flux arriving at the seafloor was calculated to be 2.72 10^6 t yr^{-1} (1.2% of the total primary production; Table 2). Since COB values were not calculated in this study and there is no spatial overlap to the study of Jahnke (1996), both results are difficult to compare. However, the results of Jahnke (1996) indicate that a similar amount of organic carbon (~1.1% of primary production) is recycled on the seafloor between 50°N and 60°N, showing a good agreement between both approaches (see Schlüter et al. 2000 or Sauter et al. 2001).

Wenzhöfer and Glud (2002) compiled a data set of 45 stations, where the benthic oxygen uptake was measured by *in situ* devices in the South Atlantic. Altogether, the data comprise 43 determinations of DOU and 21 determinations of TOU (based on a total number of 139 micro-profiles and 25 chamber incubations) along with *ex situ* measurements. The stations are spread over an area located between 30°N and 40°S. *In situ* measurements of oxygen uptake rates show that both DOU and TOU decrease with increasing water depth from the continental slope to the deep-sea (Figure 2a,b). Both *in situ* DOU and TOU reflect a distinct separation between the western and eastern South Atlantic. However, the *in situ* TOU was higher than the *in situ* DOU at the shallower sites on the continental slope. The ratio between TOU and DOU was generally ≥ 1, with high ratios of 3 - 4 in shallow and productive areas (Figure 2c). This difference between *in situ* TOU and DOU correlates with the dry weight of macrofauna and is a measure of the fauna-mediated oxygen uptake (Glud et al. 1994). Jahnke (2001) also showed a trend of increasing ratios between the total oxygen flux and the quotient of total and diffusive fluxes. The higher the total oxygen fluxes the greater is the ratio. Thus, macrobenthic activity can have a major influence on the oxygen uptake of the sediment in specific regions of high C input, such as

the continental upwelling (eastern South Atlantic: Glud et al. 1994; Wenzhöfer and Glud 2002; eastern South Pacific: Glud et al. 1999) and upper continental slope and shelf areas (eastern North Pacific: Archer and Devol 1992). In contrast to the decreasing uptake rates, the *in situ* oxygen penetration depth (OPD) increases with increasing water depth (Figure 2d). Although there is a significant correlation of the *in situ* DOU, TOU, and OPD with water depth, this is not the case for the laboratory measurements and the determined data differ strongly from the *in situ* measurements (Wenzhöfer and Glud 2002); a finding which is attributed to the processes of sediment recovery and pore water processing (see section General strategies and approaches).

Given the fact that the *in situ* TOU and DOU exhibited different rates, especially at shallow or active sites, Wenzhöfer and Glud (2002) defined two correlations between surface water primary productivity (PP), water depth (z) and organic carbon mineralized at the sea floor:

$$C\text{-}DOU = PP^{0.7358} \cdot z^{-0.3306} \qquad (12)$$

$$C\text{-}TOU = PP^{1.0466} \cdot z^{-0.4922} \qquad (13)$$

where C-DOU and C-TOU is the equivalent amount of mineralized carbon (applying Redfield ratio).

The benthic carbon flux in the South Atlantic derived from both relationships (Eq. 12 and 13) is shown in Figure 3a and 3b. The patterns of the C-DOU (which can be regarded as a minimum value) and the C-TOU both reflect the primary production in the surface water and revealed highest fluxes at about 7 and 14 g C m^{-2} yr^{-1} along the West African coast. The different approaches revealed almost identical flux rates for the open ocean, but in total, considerably lower values of approximately 2 gC m^{-2} yr^{1} were obtained. These fluxes are generally higher than the previous estimates made by Jahnke (1996). The high fluxes in the productive coastal area of the eastern South Atlantic are, however, comparable to measurements carried out in the upwelling area off central Chile (Glud et al. 1999). Comparing these relationships with those of Christensen (2000) it shows that the fluxes calcu-

Parameter	Area	Flux	Source
Oxygen	Global	54	Jahnke (1996)
	Global	80	Christensen (2000)
	Global	78 - 98	after Wenzhöfer and Glud (2002)
	North Atlantic (60°N - 80°N)	0.33	after Schlüter et al. (2000)[4] and Sauter et al. (2001)[4]
	South Atlantic (30°N - 50°S)	13	Christensen (2000)
	South Atlantic (35°N - 50°S)	14.5 - 18.2	Wenzhöfer and Glud (2002)
Silicate	South Atlantic (10°N - 50°S)	2.1	Hensen et al. (1998)
	Antarctic/Atlantic (40°S - 80°S)	4.28	Schlüter et al. (1998)
	Global	19.5	Hensen et al. (1998)
		23	Tréguer et al. (1995)
Phosphate	South Atlantic (10°N - 50°S)	0.035	Hensen et al. (1998)
	Global	0.31	Hensen et al. (1998)
Nitrate	South Atlantic (10°N - 50°S)	1.55	Hensen et al. (1998)
	Global	14.9	Hensen et al. (1998)
Respiratory CO_2[1]	Global	60 - (110)	Schneider et al. (2000)
	Global	40	after Jahnke (1996)
	South Atlantic (10°N - 50°S)	6 - (11)	Schneider et al. (2000)
	South Atlantic (30°N - 50°S)	10	after Christensen (2000)
	South Atlantic (35°N - 50°S)	11.1 - 13.9	after Wenzhöfer and Glud (2002)
	North Atlantic (60°N - 80°N)	0.23	after Schlüter et al. (2000)
Calcite Dissolution	Global	27 - 54	Archer (1996)
	Global	13	Milliman (1993), Milliman et al. (1999)
	Global (supralysoclinal)	5.7	this study
	Global	22 - 81	this study
Alkalinity / TCO_2	Pacific	55	Berelson et al. (1994)
	Indo-Pacific	91	Berelson et al. (1994)[2]
	Global	100	after Berelson et al. (1994)
		120	after Berelson et al. (1994)[2]
		120	Mackenzie et al. (1993)
DOC	Atlantic [3]	4	Otto (1996)
	Global [3]	18	Otto (1996)

[1] Estimated as oxic respiration of organic matter.

[2] Using data of Broecker and Peng (1987).

[3] Including water depths < 1000m.

[4] Water depths > 500 m.

Table 2. Estimated benthic fluxes from deep-sea sediments (> 1000 m water depth) in 10^{12} mol yr^{-1}.

Hensen et al.

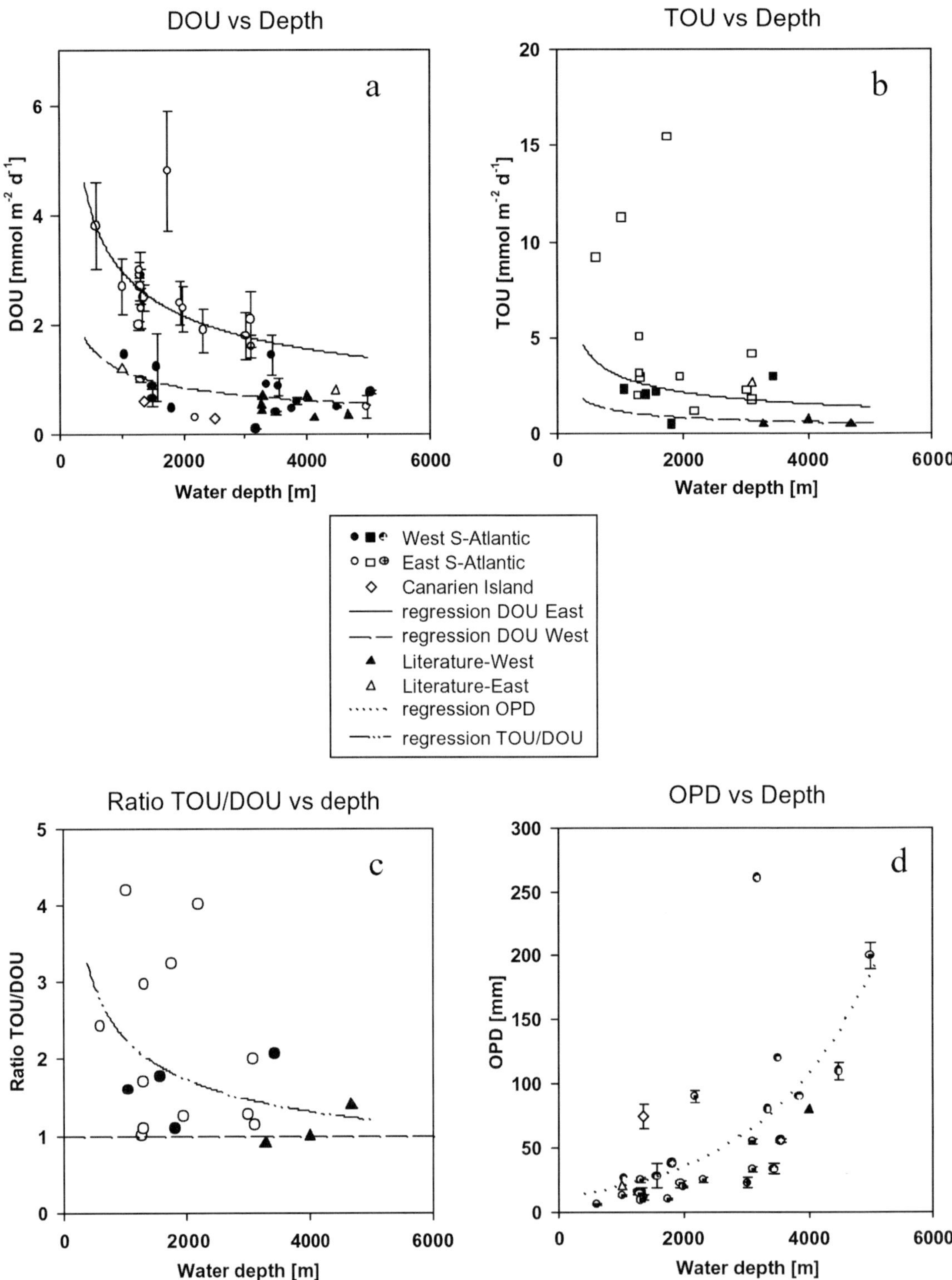

Fig. 2. Various plots of oxygen consumption patterns in relation to water depth in the South Atlantic showing **a)** the diffusive oxygen uptake (DOU), **b)** the total oxygen uptake (TOU), **c)** the ratio of total and diffusive oxygen uptake, and **d)** the oxygen penetration depth (OPD) at the sites investigated (from Wenzhöfer and Glud 2002).

lated by the latter are lower in areas with water depths above 3000 m and when primary productivity is below 200 gC m^{-2} yr^{-1}, but higher in shallow areas of high productivity. As mentioned above, the difference in TOU and DOU is a measure for the flux mediated by the benthic macrofauna. Using this interrelation Wenzhöfer and Glud (2002) determined fluxes related to the benthic fauna in the South Atlantic as given in Figure 3c. It is obvious that the overall contribution of the benthic fauna is highest in the high-productive areas (up to 50% higher TOU than DOU). Differences concerning

open ocean areas are within the accuracy of the flux measurements, which, however, does not exclude that the benthic fauna can enhance the uptake of oxygen and thus the mineralization of carbon rate in these areas.

Another correlation is determined by Wenzhöfer and Glud (2002), which relates DOU and OPD, so that the OPD can be estimated by the C-DOU distribution for the South Atlantic according to Equation 14

$$OPD = 114.6968 \cdot \text{C-DOU}^{-0.7541} \qquad (14)$$

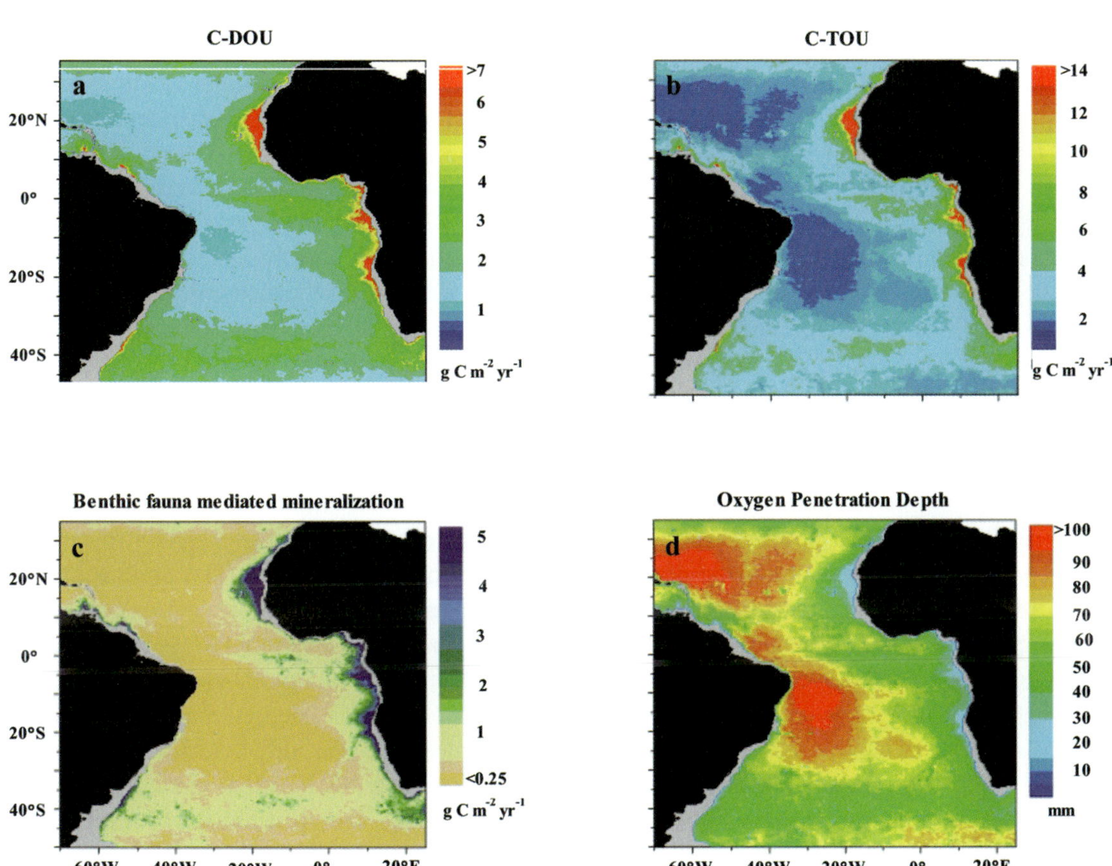

Fig. 3. Oxygen fluxes (expressed as carbon mineralized, gC m^{-2} yr^{-1}) and oxygen penetration depths for the South Atlantic between 30°N and 50°S. Diffusive **a)** and total **b)** fluxes were calculated using Eq. 12 and 13, respectively. **c)** Impact of the benthic (macro-) fauna on the total mineralization determined from the difference between total and diffusive fluxes. **d)** Estimated oxygen penetration depth (OPD) according to Eq. 14 (from Wenzhöfer and Glud 2002).

In combination with the correlation of C-DOU with primary productivity and water depth (Eq. 12), the oxygen penetration can be further described as a function of primary productivity and water depth

$$OPD = 114.6968 \cdot PP^{-0.555} \cdot z^{0.249} \qquad (15)$$

Compiling oxygen fluxes from the northeast Pacific Ocean, Cai and Reimers (1995) found that the benthic O_2 consumption is quantitatively related to the oxygen concentration in bottom water (BW_{O2}). Considering this regulatory function of oxygen availability (Cai and Sayles 1996) for the diffusive uptake of oxygen, Wenzhöfer and Glud (2002) proposed two further equations to describe the relation between OPD, C-DOU, BW_{O2}, PP and z:

$$OPD = BW_{O2}^{0.8619} \cdot C\text{-}DOU^{-0.7341} \qquad (16)$$

$$OPD = BW_{O2}^{0.8619} \cdot PP\text{-}0.54\ z^{0.243} \qquad (17)$$

The distribution of the oxygen penetration depth for the South Atlantic is shown in Figure 3d. Since the relation between DOU and OPD is very sensitive in low activity areas (Cai and Sayles 1996; Wenzhöfer and Glud 2002), it is only possible to calculate the penetration of oxygen to a maximum depth of 100 mm. While oxygen penetrates only into few millimeters of the sediment at shallow sites of the continental margin with high C input, it exceeds several centimeters in oligotrophic areas. *In situ* oxygen optode measurements carried out in the western South Atlantic exhibited penetration depths in the range of 8 to 26 cm (Wenzhöfer et al. 2001a). Integrating the benthic fluxes of both relations (Eq. 12 and 13) for the investigated area revealed a total mineralization rate of benthic carbon amounting to 133.5 and 168.1×10^{12} gC yr⁻¹, respectively, which equals 1.7-2.1 % of the surface primary productivity. The two equations also predict that 50–63 % of the total carbon flux over the 1000 m depth horizon is mineralized in deep-sea sediments. These high seafloor mineralization rates are consistent with values from Schlüter et al. (2000), but are higher than estimates from Jahnke (1996) and Christensen (2000). Assuming that the relations for the South Atlantic are also valid for the entire Atlantic,

Wenzhöfer and Glud (2002) estimated an annual flux of 18.9 and 23.8×10^{12} mol O_2. This is in the same range as estimates by Christensen (2000), but are 1.3-1.6-times higher than values reported by Jahnke (1996). A detailed overview of basin and global wide benthic fluxes is given in Table 2.

Benthic Nutrient Fluxes

It is difficult to assess to what extent estimations of the benthic oxygen flux can be used in predictions of the rates of benthic nutrient recycling, in which phosphate and nitrate are released during the degradation of organic material. The released amounts of nitrate and phosphate strongly depend on the C:N:P ratio of the organic material mineralized. Furthermore, adsorption on solid-phase surfaces (phosphate), denitrification, and ammonium oxidation may play important roles in influencing the amounts of nitrate and phosphate released from the sediments. Similar problems exist in the derivation of benthic silicate fluxes. A number of factors, which vary from place to place, govern the preservation of biogenic opal. These include the de-coupling of the biogeochemical cycles of Si and C, differences in composition and silicification of predominantly diatom assemblages, or the large variability of export and sedimentary opal fluxes between open ocean and the coastal zone (Ragueneau et al. 2000). Owing to uncertainties resulting from the empirical correlations and a general lack of information on nutrient release from the deep ocean sediments, a few other approaches were used to estimate the recycling rates of nutrients at the sediment/water interface on a basin wide scale (Zabel et al. 1998; Hensen et al. 1998; Schlüter et al. 2000). Hensen et al. (1998) compiled pore water data from surface sediments and bottom water data from 180 locations in the South Atlantic and applied the geostatistical methods of variogram analysis and kriging to create distribution maps of diffusive benthic release rates for nitrate, phosphate, and silicate for water depths >1000 m. The distribution pattern of all investigated constituents is generally expected to reflect the coupling to surface water productivity. Figure 4 shows a comparison between the distribution of benthic phosphate fluxes with the primary produc-

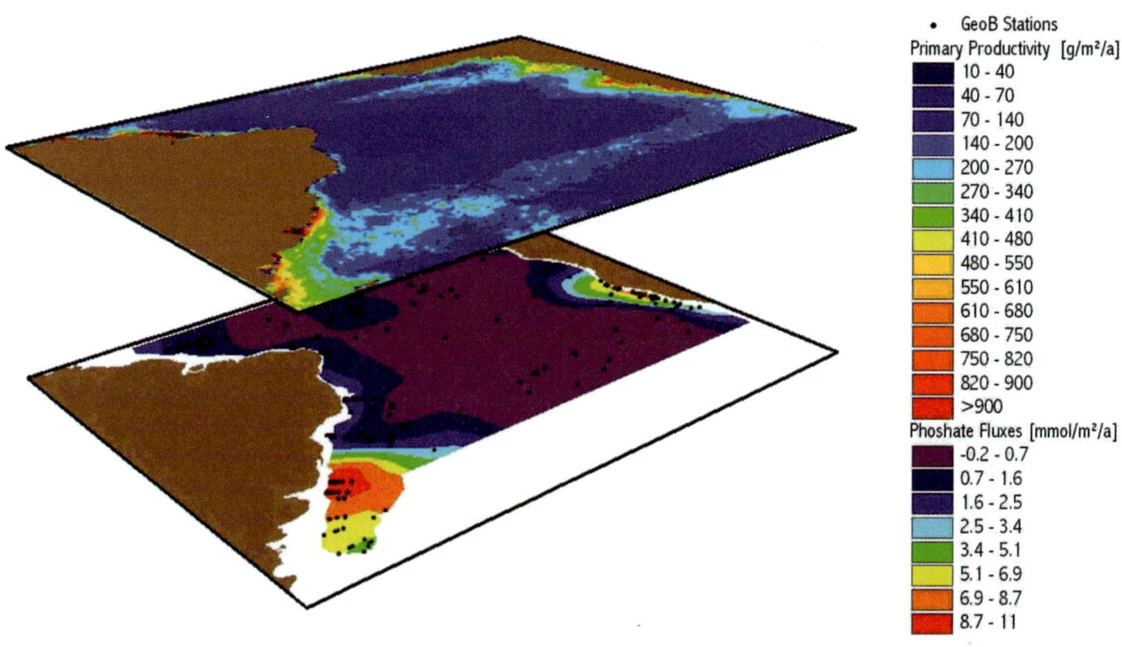

Fig. 4. Comparison of the regional patterns of the primary productivity (Behrenfeld and Falkowski 1997) and benthic nutrient fluxes (here phosphate) from Hensen et al. (1998). Essentially, good correlations exist in the Argentine and the Cape Basin, whereas strong differences are obvious in the eastern equatorial Atlantic.

tivity pattern as derived from Behrenfeld and Falkowski (1997). A general agreement is obvious except for the open-ocean part of the eastern equatorial divergence region, where comparatively low benthic fluxes were calculated. A prominent oxygen minimum zone, thus favors the notion of an intense organic matter recycling within the water column. Additionally, a striking feature along the continental margin off Argentina is that benthic fluxes extend far into the deep Argentine Basin. The calculated fluxes are highly variable between the highly productive and the open ocean regions for nitrate (10-180 mmol m^{-2} yr^{-1}) phosphate, (-2-11 mmol m^{-2} yr^{-1}), and silicate (0 0.65 mol m^{-2} yr^{-1}). The fluxes compiled for the total investigated area are shown in Table 2. Open ocean areas with moderate to low fluxes contribute to about 70-80% of the total fluxes estimated. As stated above, all flux rates used for that study were calculated from *ex situ* pore water analyses. They might, however, be subject to uncertainties due to artifacts result-

ing from core retrieval or pore water extraction procedures (see above). Due to such artifacts diffusive fluxes of phosphate are believed to be a lower end estimate (even negative fluxes occur in some areas), thus, the total area estimation is based on the assumption that fluxes are not lower than 0.25 mmol m^{-2} yr^{-1} at any location within the study area. In the same sense, nitrate fluxes represent an estimate at the upper end of the scale. However, anticipating that the South Atlantic is representative of the global ocean, annual global release rates are given in Table 2. The only other large-scale distribution map of benthic nutrient fluxes so far has been published by Schlüter et al. (1998). In their study, they applied an approach based on GIS to regionalize a set of surface sediment and pore water samples from 130 stations in order to determine the sedimentary silica balance for the Atlantic sector of the Southern Ocean. According to their calculation, this area situated between ~40°-80°S and 60°W-20°E (9.5·10^6 km^2) is responsible for a benthic dissolution flux of Si amounting to $4.28 \cdot 10^{12}$ mol yr^{-1} and a burial rate of $0.84 \cdot 10^{12}$ molSi yr^{-1} (in form of biogenic opal), resulting in a total flux of $5.12 \cdot 10^{12}$ molSi yr^{-1} to the seafloor. A main outcome was that their estimate of the burial of biogenic silica is about 50% lower than previous estimates (e.g. DeMaster 1981), thereby decreasing the discrepancy between opal production and burial as compared to other regions displaying a high production of biogenic silica.

Apart from the general pattern and the quantification of diffusive fluxes of benthic nutrients in the South Atlantic, some interesting differences in the regional patterns of the Cape and Argentine Basin were observed. Representatively, diffusive nutrient fluxes compiled from several transects across the continental slope in front of the Rio de la Plata mouth and from the Cape / S-Angola Basin are plotted in Figure 5a. In the Argentine Basin, the highest release rates of nitrate and phosphate were detected at intermediate and low depths of the slope, whereas silicate fluxes tend to increase with depth and thus with increasing distance from the continent. Although there is much scattering of data presented from the Cape / S-Angola Basin, the decrease of the flux rates with increasing water depth more or less reflects an expected correlation. This observation illustrates the different spatial extension of areas with high release rates into the basin. Hensen et al. (2000) were able to show that the situation in the Cape / S-Angola Basin widely corresponds to a "normal" situation where the mineralization processes in the surface sedi-ments are predominantly fostered by the input of degradable material coming from a "vertical" flux component (export production). As stated above, a downslope distribution of benthic fluxes as it is found in the Argentine Basin is also not uncommon. For example, compared to a situation in the Northeast Pacific (Cai and Reimers 1995), where oxygen limitation is the reason for reduced benthic fluxes, the bottom water in the Argentine Basin is well oxygenated, suggesting that intense transport processes deliver large amounts of sediments and organic matter downslope. Since nitrate and phosphate fluxes are reasonably well correlated with the organic matter content in the surface sediments, obviously, most of the degradable material is de-posited in mid-slope depths, whereas the upper slope surface sediments are partly depleted of organic carbon (Hensen et al. 2000). The release of dissolved silica is usually also well correlated with the organic carbon content, since the deposition of opaline tests takes place in form of detrital aggregates that consist to a large part of organic tissue. In general, the dissolution of biogenic opal depends on the degree of undersaturation, the temperature, and the surface exposed to the interstitial water, hence, the removal of organic coatings around the tests markedly intensifies the dissolution of biogenic opal as it was recently shown by Bidle and Azam (1999). A reasonable correlation could only be found in the sediments of the Cape / S-Angola Basin. A significant difference, however, is observed for silicate where both, the flux pattern related to water depth and the correlation to C_{org} (Figure 5b) differ from the other nutrients. This suggests an additional source of silica being significantly dissolved in the sediments below a water depth of 4000 m. Both the regional distribution of benthic silica reflux (Hensen et al. 1998) and literature data indicated that the reason for this is a long distance disposal of Antarctic diatoms by the Antarctic Bottom Water (AABW, Jones and Johnson 1984; Hensen et al. 2000; Romero and Hensen 2002). To support this hypothesis, surface sediments were analyzed for their biogenic silica content and diatom assemblages (Romero and Hensen 2002). Figure 6 shows a transect of 3 stations across the slope off the Rio de la Plata mouth. The general trend indicates a 3-10 wt.% increase of the opal content. The origin of diatoms is more or less well balanced at the shallowest location (~1800 m), whereas the most dominant species are of neritic origin or come from the deeper stations (~3500-4500 m) of the Antarctic underlining predominantly lateral input pathways, namely the downslope transport from the shelf and by the Antarctic Intermediate and Bottom Water. Below a water depth off 4000 m Antarctic species contribute to more than 50% of the total diatom assemblages. These results suggest that a more detailed view on the budget of biogenic silica in these areas is required and will therefore be presented below.

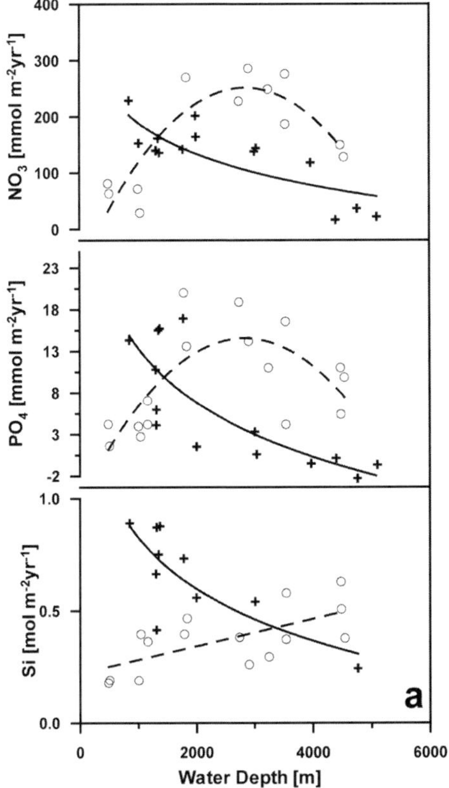

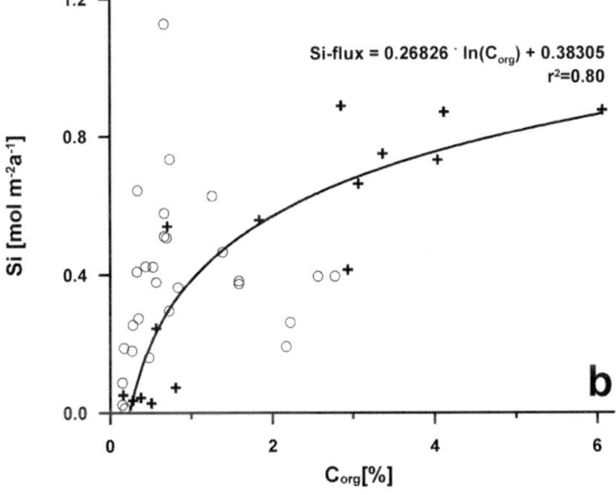

Fig. 5. Plots of benthic nutrient fluxes related to water depth and organic carbon from transects across the Rio de la Plata mouth (Argentine Basin) and the continental slope off Namibia (Cape and S-Angola Basin), illustrating the differences of the regional control parameters. Plot **a)** shows the nutrient flux vs. water depth in the Cape and the Argentine Basin, respectively, and **b)** the relation between silicate fluxes and the organic carbon content in the surface sediments of both regions. Crosses and solid regression lines indicate stations of the Cape Basin, open symbols and broken line those of the Argentine Basin.

Opal Budget in Specific Regions of the South Atlantic

In comparison to the diffusive benthic silica fluxes in the Cape / S-Angola Basin and the Argentine Basin, another interesting result was revealed by the determination of opal accumulation rates in surface sediments (Figure 7a), when additional data from both regions and the N-Angola / Guinea Basin were combined (Schulz 2000; Holstein 2002). Opal accumulation rates (expressed as mg Si cm^{-2} yr^{-1}) were calculated by multiplying the Si-content in the upper first centimeter of the sediment (80-85% porosity) by the sedimentation rate and diffusive fluxes from Hensen et al. (1998). The graphical comparison shows that the flux of biogenic silica is underestimated by more than one order of magnitude compared to the dissolution flux. This imbalance is most prominent at the Cape / S-Angola Basin locations. Supposing the correctness and an approximate invariability of the profile of dissolved silica over time (Sayles et al. 1996; Schlüter et al. 1998), the flux of opal to the sediment surface is much underestimated by this approach. Since sedimentation rates were mostly derived from fairly rough compilations or biostratigraphic age models comprising time spans of ten thousands to hundred thousands of years (i.e. Cwienk 1986; Summerhayes et al. 1995), which are not corrected for compaction or any kind of decomposition/dissolution process, such data, in fact, must lead to an underestimation of the real fluxes to the sediment surface. Additionally, opal dissolution starts right at the sediment-water interface, which gives reason to the assumption that an integrated value for the opal content also results in an underestimation of the flux. Thus, a major outcome of this comparison is that the "conservative" method to calculate a specific sediment accumulation rate – although widespread in use - fails to reflect the true picture. As the dissolution of opal starts at the very sediment surface, it follows that a much higher flux of opal to the sediment is required. To better constrain the "real" particle flux to the sediment surface, a model study with the numerical model CoTReM (Adler et al. 2001) was performed, pre-supposing that the dissolved silica

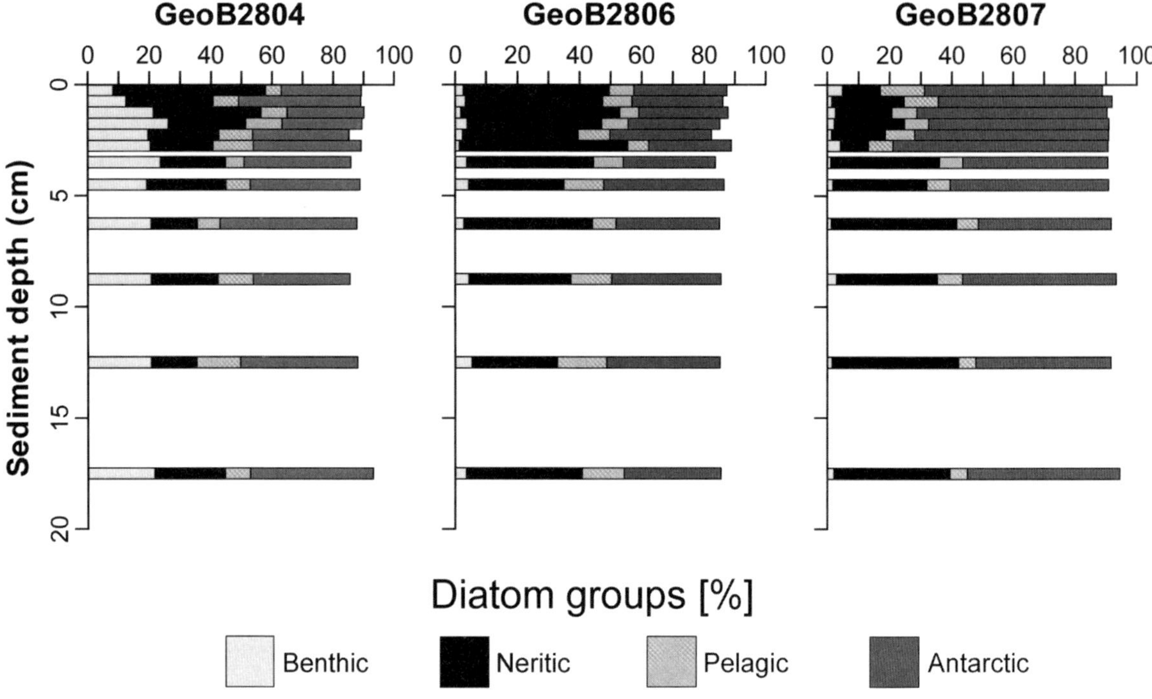

Fig. 6. Distribution of diatom species (sorted by 4 main groups) on a transect across the Rio de la Plata mouth. A simultaneous increase of the total amount of opal and the amount of Antarctic diatom species is clearly visible, supporting the theory that long-distance disposal of Antarctic diatoms has a major impact on opal sedimentation and regeneration in this area.

profile reflects a steady-state situation. For a number of key-stations, the percentage of opal in the total solid-phase input into the surface sediment, the sedimentation rate, and the bioturbation rate were varied until a reasonable approximation to the solid-phase profile of biogenic silica was obtained (Schulz 2000; Holstein 2002). We determined regional sedimentation rates from the model runs and multiplied them by the residual opal content below the zone of opal dissolution in order to estimate the biogenic opal burial at each station. The sum of the burial rates and the dissolution fluxes gives the overall required accumulation rates of biogenic silica for all 23 stations as shown in Figure 7b. Opal preservation is obviously much better in the Argentine Basin (14-30 %) than in the Cape / S-Angola Basin (4-9 %) or the Guinea / N-Angola Basin (2-9 %). Regional exceptions only exist at station GeoB 3714 in the S-Angola Basin, where the biogenic opal content is high (>10 wt%), and about 15 % of the biogenic Si-flux is buried;

and at GeoB 2804, where the opal preservation is significantly lower compared to the surrounding stations. Averaged over all sites the amount of opal required in the particle rain arriving at the seafloor is 21.5 %, which is clearly at the upper end and above compared to results of sediment trap studies conducted in modern high productivity and upwelling areas (e.g. Fischer and Wefer 1996; Berelson et al. 1997; Koning et al. 1997). These most likely reflected the effects of lateral advection. Although the obtained results on Si-accumulation rates are exclusively based on dissolved silica profiles and biogenic opal in the sediments, they do agree fairly well with those from other continental upwelling areas (DeMaster 1981), the Antarctic (DeMaster 1981; Schlüter et al. 1998) and NW Indian Ocean (Koning et al. 1997). Much lower estimates based on a few sediment trap results from the Angola Basin (Jansen and van der Gaast 1988), the Walvis Ridge (Wefer and Fischer 1993), and the southern Cape Basin (Fischer et al.

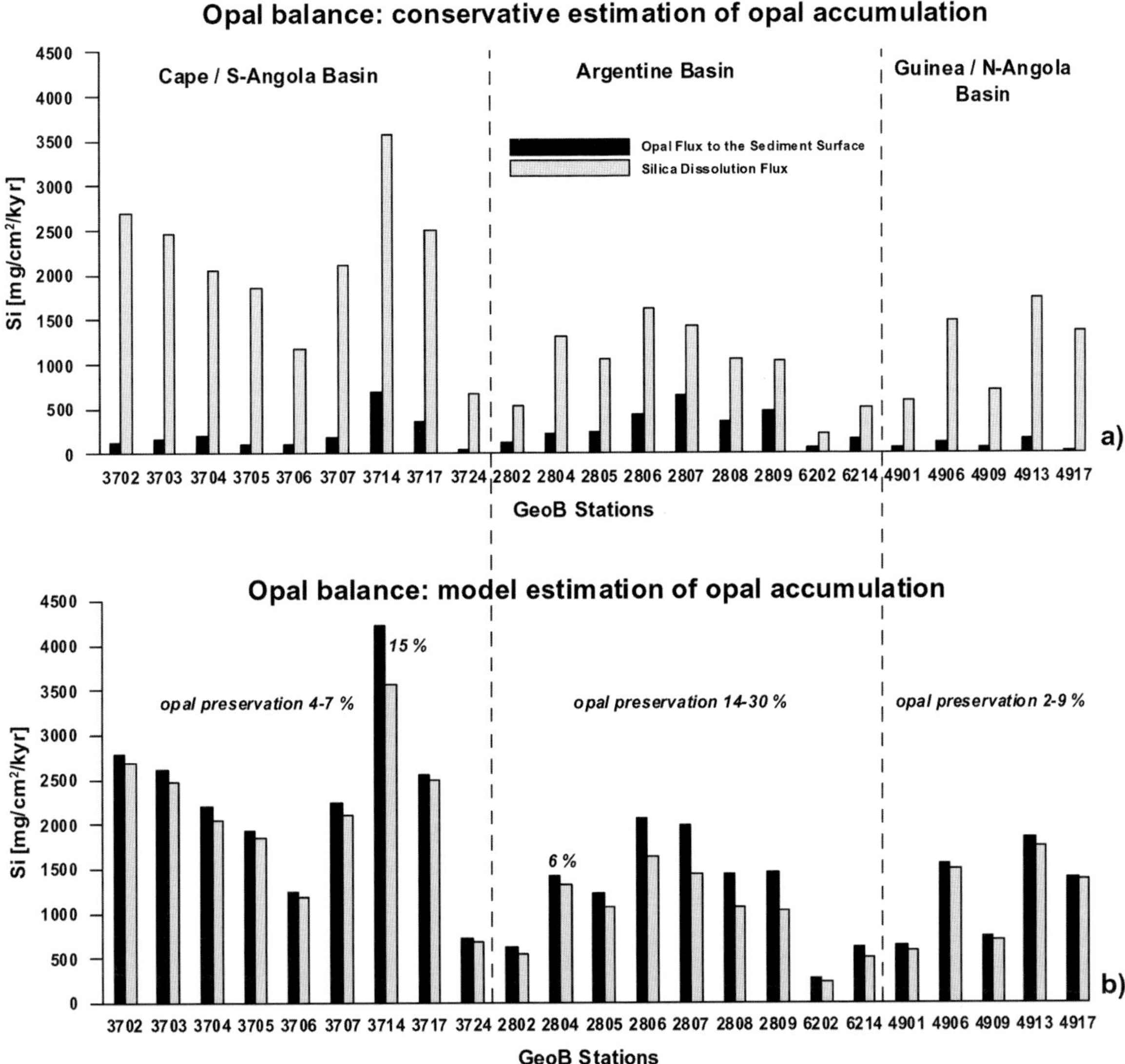

Fig. 7. Opal balances for 14 stations in the Cape and the Argentine Basin. Diffusive benthic Si-fluxes are compared with **a)** a conservative estimation of opal accumulation as derived from the opal content in the upper 1 cm of the sediment and the sedimentation rate, and **b)** a model derived accumulation rate which is necessary for obtaining the measured opal concentration profile, assuming steady-state dissolution conditions. Obviously, conservative methods underestimate the opal accumulation in the Cape Basin by more than one order of magnitude and a factor of 3-4 in the Argentine Basin. On average, biogenic opal is much better preserved in the Argentine than in the Cape Basin (see text for details).

2000) question either how representative the available water column fluxes are for larger oceanic regions and again the correctness of the assumption of a predominantly "vertical" supply of biogenic material as discussed by Hensen et al. (2000).

The dimension of the observed discrepancies based on the conservative approach clarifies, however, that coupled sediment trap and sediment-pore water studies are required to have a better time and space resolution to resolve this problem on a global scale, since most of the areas around the world distinguished by a significant productivity of biogenic opal (except Antarctica) are underrepresented in terms of silica mass balances.

Sedimentary CaCO₃ Dissolution by Organic Matter Decomposition

The decomposition of organic material is also the driving process for many secondary reactions in the sediment / pore water system, such as mineral dissolution and precipitation. Increased attention was paid to the process of metabolically driven $CaCO_3$ dissolution and its quantification in deep-sea sediments as an important supplement to dissolution resulting from bottom water undersaturation. The main reason for this is a still existing uncertainty in determining global benthic calcite dissolution fluxes resulting in a large information gap concerning the C-balance in the global ocean (Milliman 1993; Milliman et al. 1999). Beginning with a recent carbonate production in the surface waters of about $60\text{-}90\times10^{12}$ mol yr^{-1} and a particulate carbonate export of 24×10^{12} mol yr^{-1} at about 1000 m water depth (determined by long-term sediment trap monitoring), Milliman et al. (1999) suggested strong carbonate dissolution in the water column at depths above 1000m. A remaining sedimentary re-dissolution flux of 13×10^{12} mol yr^{-1} is based on the assumption that only 20% of the calcium carbonate settling above the lysocline dissolves (Milliman 1993). This, however, seems to be unlikely regarding the results of Archer (1996b) who predicts a combined re-dissolution flux by benthic respiration above and below the lysocline and by bottom water undersaturation of $27\text{-}54\times10^{12}$ mol yr^{-1} contributing to the alkalinity of the ocean.

One major reason for this shortcoming is again the limited availability of high quality, sensitive data from *in situ* deployments, which, however, became increasingly abundant by the end of the past decade. The use of benthic lander systems to study calcite dissolution in deep-sea sediments started - subsequent to the invention of benthic chamber and oxygen-microsensors - with the development of highly sensitive electrodes and optodes able to

record high resolution pore water profiles of pH, pCO_2, and calcium in the deep-sea (Archer et al. 1989; Hales et al. 1994; Hales and Emerson 1996, 1997a; Cai and Reimers 1993; Wenzhöfer et al. 2001b). Benthic chamber incubations were also used to determine total fluxes of DIC, alkalinity, and calcium from the sediment into the bottom water (Berelson et al. 1990, 1994; Jahnke et al. 1994, 1997). The kinetics of calcium carbonate dissolution has been investigated in laboratory and field experiments regarding reagent grade and natural calcite. The most widely applied kinetic rate law to describe calcite dissolution in sea and in pore waters is given in Equation 18 as suggested by Keir (1980).

$$DR = k \cdot (1 - \Omega)^n \qquad\qquad (18)$$

where DR is the calcite dissolution rate, k the dissolution rate constant, Ù the saturation state of the porewater with respect to calcite, and n is the reaction order.

The most extensively used reaction order for modeling calcite dissolution is 4.5 (as suggested by Keir 1980). A compilation of reaction orders from model and experimental studies is given by Cai et al. (1995). However, the discussion concerning the "correct" reaction order still continues. More recently, Hales and Emerson (1997b) found that dissolution in calcite-rich sediments could be better described by first order dissolution kinetics. In contrast, upon studying calcite dissolution kinetics in calcite-poor sediments of the equatorial Atlantic, Adler et al. (2001) again favored higher reaction orders. In this sense, the observed dissolution rate constants are highly variable, which seems to be mainly dependent on differences in the physical (e.g. surface area) and chemical properties (high/low Mg-calcite) of the calcite mineral phase. The values of k reported so far range over several orders of magnitude from 0.005-0.16 % d^{-1} (Hales and Emerson 1996) up to laboratory values of 10-1000 % d^{-1} (Keir 1980, 1983).

Usually, transport and reaction models were used to explain the interrelation between measured microprofiles, to predict overall calcite dissolution rates by defining the dissolution rate constants, and to distinguish between dissolution driven by organic matter oxidation and by the undersaturation of the bottom water. It is generally reported that dissolution at and above the saturation horizon is solely attributed to the oxidation of organic matter. The importance of oxidation-related dissolution decreases with increasing undersaturation of the bottom water, in other words, the efficiency by which C_{org} oxidation drives $CaCO_3$ dissolution increases with increasing undersaturation of the bottom water (Martin and Sayles 1996). In the same sense, Berelson et al. (1994) suggested that the undersaturation of the bottom water is more important for better soluble calcite phases (high k-values), whereas when calcite phases are more resistant to dissolution (low k-values), the rain rates of C_{org} to the seafloor, and thus mineralization, become the driving force for calcite dissolution. This can be understood as an alteration process, where dissolution starts at the sediment-water interface, right after deposition of the calcite phase, and becomes more resistant to dissolution during burial. In the zone of oxic respiration the removal of organic coatings around calcite grains might also play an important role, by exposing a larger surface area to the pore water, thus triggering dissolution processes.

Supralysoclinal Dissolution of $CaCO_3$

A number of studies in the South Atlantic targeted the process of calcite dissolution in surface sediments, in order to elucidate problems arising from the gap of knowledge in the oceanic mass balance and concerning the preservation potential of $CaCO_3$ in marine sediments. In the following we will give two examples, in which supralysoclinal calcite dissolution is indicated by completely different investigation methods, and finally try to quantify the process on a global scale.

From a geochemical point of view, the hydrographical lysocline is determined either by chemical analyses of water or pore water samples. It is the water depth where the *in situ* CO_3^{2-} concentration falls below the CO_3^{2-} concentration of calcite saturation. This depth horizon is mainly attributed to the interface of water masses, which differ in their physical and chemical properties. In contrast, micropaleontologists and paleoceano-graphers

usually use the sedimentary calcite lyso-cline, which means the depth at which a significant $CaCO_3$ dissolution is observed. An established method for a qualitative classification of the degree of dissolution is the detailed ultrastructural investigation of planktic foraminiferal tests (Henrich 1986; Dittert and Henrich 2000; Volbers and Henrich 2002a,b). These tests are characterized by relatively low species diversities and usually occur in extraordinary high individual numbers. In particular, species that thrive in almost all different hydrographic regimes, as e.g. *Globigerina bulloides* in the South Atlantic Ocean, serve as ideal proxies to determine the degree of calcium carbonate dissolution within a sediment sample. The *Globigerina bulloides* dissolution index (BDX') detects the onset of significant calcite dissolution within South Atlantic sediments and was successfully employed to determine the depth of the sedimentary calcite lysocline (Volbers and Henrich 2002b).

According to the authors, investigations of surface sediments from the eastern South Atlantic reveal that significant calcite dissolution seems to occur at depths, which are 600 to 1600 m shallower than predicted from hydrographic data. The variable interplay of organic matter degradation and sediment accumulation seems to account for the different positions of the sedimentary calcite lysocline. Particularly, in the vicinity of the Congo river, there are strong indications for the oxidation of organic matter based on an examination of the water mass properties at around 11°S: The anomaly of oxygen, phosphate, and nitrate is most pronounced just at the upper continental rise, reflecting fluxes of oxygen and nutrients across the sediment-water interface, probably resulting from the decomposition of plankton (Warren and Speer 1991). The sedimentary calcite lysocline in this region is encountered at around 3400 m, and is at its shallowest position within the eastern South Atlantic Ocean (Figure 8). Strong dissolution seems to occur below 3400 m in the sediment, although the surrounding North Atlantic Bottom Water (NADW) is still supersaturated with respect to calcite.

Based on *in situ* microsensor measurements of O_2, pH, pCO_2, and Ca in the eastern South Atlantic carried out by means of an autonomous profiling

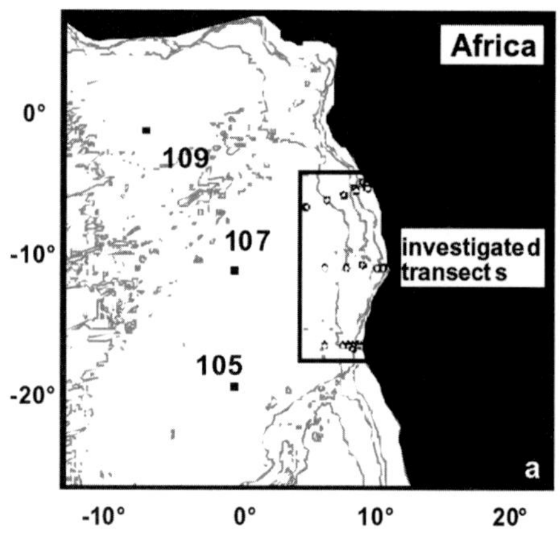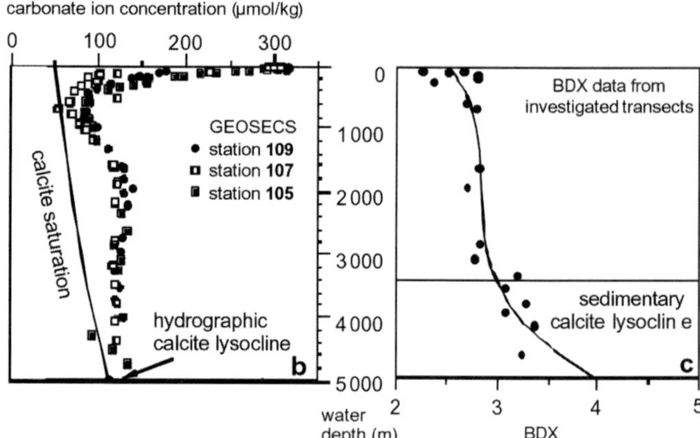

Fig. 8. Use of the *Globigerina bulloides* dissolution index (BDX) for the determination of the sedimentary lysocline. **a)** Location of the investigated GEOSECS stations and GeoB transects in the eastern South Atlantic. **b)** Measured carbonate concentration in seawater and the calculated saturation concentration indicate that the hydrographic lysocline is being reached at the GEOSECS sites at a water depth of about 5000 m. **c)** Applying the BDX the sedimentary lysocline begins below a water depth of 3400 m.

lander system, a combined geochemical and modelling approach was made. Subject of investigation were surface sediments from the upper continental slope off Gabon (GeoB 4906 and GeoB 4909) at a water depth of approximately 1300 m (Wenzhöfer et al. 2001b; Adler et al. 2001; Pfeifer et al. 2002). The bottom water at both stations is slightly oversaturated with respect to calcite (Ω = 1.07 and 1.10, Adler et al. 2001), thus, dissolution must exclusively be mediated by metabolically produced CO_2. For the first time, *in situ* measurements of all parameters describing the carbon dioxide system, combined with O_2 microprofiles, permitted the quantification of the amount of C_{org} mineralization relative to $CaCO_3$ dissolution. Thus, the calcite dissolution flux could be quantified in two different ways: (1) by calculating the Ca^{2+}-gradient from the sediment into the bottom water of the Ca-microprofile (Wenzhöfer et al. 2001b) and (2) by the model derived dissolution rates. The numerical model CoTReM was applied to investigate the depth dependent effects of respiration and redox processes related to $CaCO_3$ dissolution (Adler et al. 2001; Pfeifer et al. 2002; Wenzhöfer et al. 2001b). Simultaneously, the interdependence of the measured *in situ* profiles was evaluated by defin-

ing the dissolution rate constant of calcite in the kinetic rate law. The general model strategy was to come out with an optimal fit to the pH depth profile, which is the most sensitive parameter (for detailed description of the model approach see Adler et al. 2001). Following the model approach, Ca-fluxes resulted in 35.8 µmol cm^{-2} yr^{-1} at site GeoB 4906 and 33 µmol cm^{-2} yr^{-1} at site GeoB 4909 (calcite dissolution fluxes calculated from Ca-microprofiles are distinctively lower with 20.1 and 21.2 µmol cm^{-2} yr^{-1}, respectively, which is attributed to scattering data and inconsistencies in the measured profiles compared to the model results; see Pfeifer et al. 2002 for details). These rates were added to a compilation of literature-derived calcite dissolution fluxes and C_{org} mineralization rates from deep-sea sediments located above the saturation horizon and slightly below (Ω~0.8; Table 3). It is obvious that GeoB 4906 and 4909 exhibit high fluxes for both parameters. The correlation of the mean estimates of the calcite dissolution fluxes and the C_{org} mineralization rates (Pfeifer et al. 2002) produced a reasonable correlation of r^2 = 0.77, thus, the calcite dissolution rate in the sediment can be estimated by

$$DR_{CaCO_3} = 1.1 \cdot MR_{C_{org}} - 9.3 \qquad (19)$$

where DR_{CaCO_3} is the calcite dissolution flux and $MR_{C_{org}}$ is the C_{org} mineralization rate (both in µmol cm^{-2} yr^{-1}).

To our knowledge the only significant deviation from this trend has been reported by Jahnke et al. (1994) who found hardly any evidence for calcite dissolution at or above the saturation horizon at deep-sea sites on the Ontong Java Plateau and off NW-Africa. Using Eq. 19 we calculated an estimate of the global calcite dissolution flux above the hydrographic lysocline by application of a GIS-system (Figure 9). To define the position of the hydro-graphic lysocline, we used the gridded global field of bottom water ΔCO_3^{2-} data (Archer 1996a) and calculated the C_{org}-mineralization rates from the grid of oxygen consumption rates after Jahnke (1996), assuming a Redfield ratio of decomposed organic material. The global area above the lysocline was estimated to be at about 7.4×10^7 km^2 (about 20% of the total ocean area) and sums up to an annual benthic dissolution flux of 5.7×10^{12} mol $CaCO_3$. Approximately 30% of the surface area in the South Atlantic is located above the lysocline, which makes up 20% of the global area above the lysocline with a reflux of 1.0×10^{12} mol $CaCO_3$ yr^{-1}. Taking into account the more recent estimates for the global oxygen demand of the seafloor (Christensen 2000; Wenzhöfer and Glud 2002) respiratory driven calcite dissolution can be expected to be at maximum 50% higher. Compared to Archer's modeling result for the global seafloor calcite dissolution flux in 1996, this adds up to 10-20% of the total, again pointing out the obvious importance of this process in determining the preservation potential of calcitic remains in marine sediments, and hence the application of calcium carbonate as a proxy for paleoproductivity. However, this result allows us to draw some further conclusions: (1) In accordance with the study of Archer (1996b), we can reasonably state that the

Area	CaCO$_3$-dissolution	C$_{org}$-mineralization	Reference
	(µmol cm^{-2} yr^{-1})		
E-equatorial Pacific	8 - 20	16 - 24	Berelson et al. (1994)
W-equatorial Pacific (Ontong-Java Plateau)	3.5 - 6	7.5 - 16.2	Hales and Emerson (1996)
W-equatorial Atlantic (Ceará Rise)	4 - 16	8 - 18	Hales and Emerson (1997a)
W-North Atlantic	10 - 24	15 - 35	Hales et al. (1994)
W-North Atlantic	52 - 61	40 - 47	Jahnke & Jahnke (2000)
W-equatorial Atlantic (Ceará Rise)	4 - 11	18 - 26	Martin and Sayles (1996)
E-equatorial Atlantic	20 - 36	28 - 56	Wenzhöfer et al. (2001b); Pfeifer et al. (2002)

Table 3. Calcite dissolution versus organic carbon mineralization rates of supralysoclinal sediments (modified after Pfeifer et al. 2002).

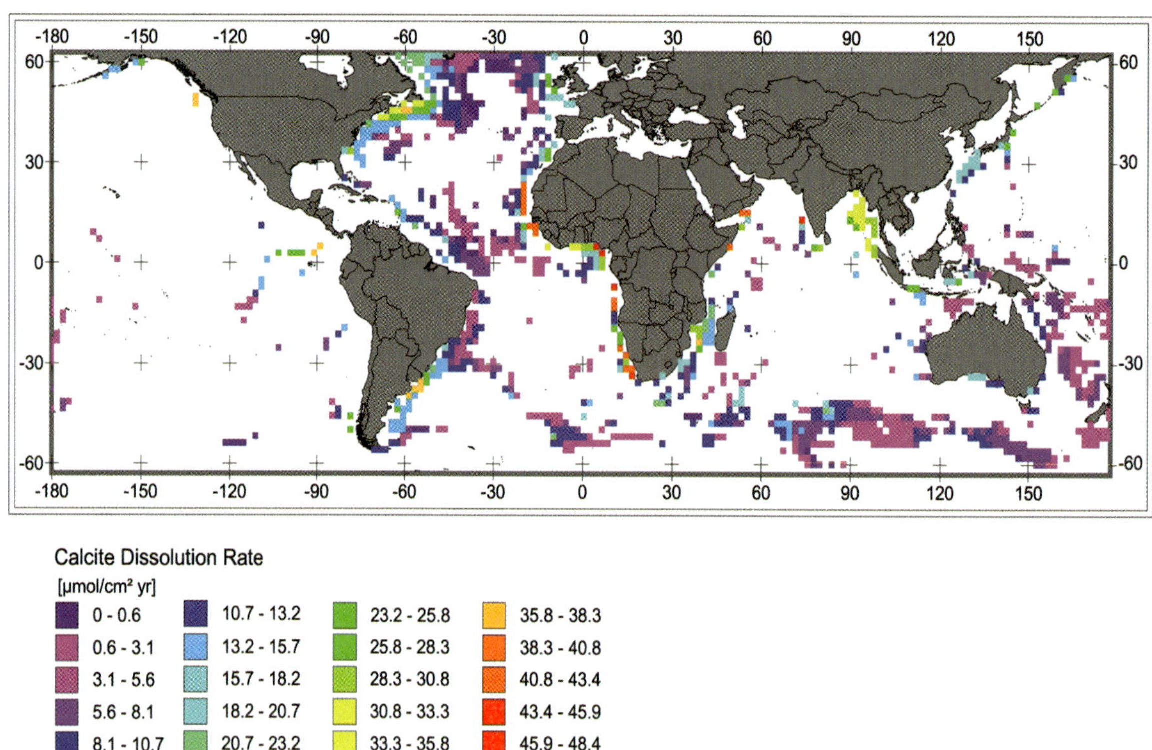

Fig. 9. Map showing the global distribution of supralysoclinal calcite dissolution by applying Equation (19), based on global grids of bottom water ΔCO_3^{2-} (Archer 1996a) and oxygen consumption rates (Jahnke 1996).

estimates of Milliman (1993, 1999) clearly underestimate the total benthic $CaCO_3$-dissolution. (2) Even though Equation (19) is a simple tool to predict the metabolically induced $CaCO_3$-dissolution in surface sediments above the saturation horizon, its application to the global seafloor should at least result in a minimum estimate of the total flux. The reasons for this underestimation are clearly the direct effect of the undersaturated water column in the major part of the deep ocean, but also the fact that stronger undersaturation increases the effect of metabolically induced $CaCO_3$-dissolution (see above, Berelson et al. 1994, Martin and Sayles 1996). The application of Equation (19) to C_{org}-mineralization rates derived from the various estimates of the global oxygen consumption in deep-sea sediments (Table 2) results in global fluxes of $CaCO_3$-dissolution lying between 14.8×10^{12} mol yr^{-1} (after Jahnke 1996), 33.9×10^{12} mol yr^{-1} (after Christensen 2000), and 32.4-47.2×10^{12} mol yr^{-1}

(after Wenzhöfer and Glud 2002). A global ocean area of $3.1 \cdot 10^8$ km^2 (Kennett 1982) was assumed for the calculation of the average C_{org} mineralization rate per cm^{-2} yr^{-1}. Combining these results with those published by Archer (1996a), who predicted a global benthic dissolution (neglecting respiratory dissolution) between 7-34×10^{12} mol yr^{-1}, this makes a global $CaCO_3$-dissolution ranging from 22 to 81×10^{12} mol yr^{-1}. This result is interesting in two respects. On the one hand, it confirms Archer's results based on a number of newly obtained *in situ* measurements from different deep-sea locations, and, on the other hand, it suggests that calcite dissolution fluxes to the seafloor might be further corrected upwards, approximately by ~50% as compared to Archer's maximum estimation of 54×10^{12} mol yr^{-1}, when the higher benthic mineralization rates derived from Christensen (2000) and Wenzhöfer and Glud (2002) are favored.

Outlook

The present manuscript aimed at presenting an overview of the state-of-the-art in the field of the quantifying diffusive benthic fluxes. Due to the specialized scope of this book, examples focus on the South Atlantic. Table 2 is a compilation of, to our knowledge, all recently available data estimated for larger regions or on a global scale. All these results are necessary constituents to approach a more consolidated material budget of the world ocean and to constrain inputs and outputs for coupled biogeochemical models in future studies. Apart from the fact that it clearly shows the progress obtained over the past decade, this compilation also exposes gaps and inconsistencies that still exist. These originate from a poor database and the need of extrapolation with the aid of empirical formulations that are often developed for certain regions and, thus, it is often at least questionable if those were crudely transferred to larger scales. One example in this respect is the comparison of the "data based" maps of nutrient releases from the sediments of the South Atlantic (Hensen et al. 1998) with the "empirically" developed maps of benthic oxygen uptake in the South Atlantic by Wenzhöfer and Glud (2002). Although the general patterns are driven by primary productivity and are consequently consistent, regional deviations from this general pattern caused by lateral transport, or other oceanographic phenomena, the comparatively small database of high-quality oxygen measurements obviously does not produce the necessary detailed resolution. The fact that the large *ex situ* data sets used for regionalization might be affected during sampling and processing shows the dilemma, but should also encourage to enlarge *in situ* data sets and to work more thoroughly on the development of empirical relationships for the transfer from regional to global scales. One important step in this regard might become the development of a global system of biogeochemical provinces, following the concept of Longhurst et al. (1995) on global ocean primary productivity. Presently, several groups of geochemists, biologists, and oceanographers are working on this task by collecting and compiling the available data sets, mainly to figure out how benthic fluxes can be best quantified by a number of con-trol parameters in certain key areas of the world ocean, and to define similarities and discrepancies of those. The outcome of this approach will surely help to constrain the estimates of benthic reflux of dissolved carbon and major nutrients, but also provide an improved tool for the prediction of the preservation potential of key elements in specific regions. It requires, however, additional efforts to obtain a larger coverage of benthic flux measurements in space and time. Together with the establishment of a reliable database of MARs for surface sediments - especially at ocean margins, as the example from the Cape Basin shows - this will be one major task in studying present and past marine biogeochemical cycles. The successful application of proxy elements for the reconstruction of paleoproductivity, like Ba_{bio}, will strongly depend on these developments. However, much work also needs to be done concerning the coupling of the Ba-cycle to those of the major elements. McManus et al. (1999) found a constant ratio of benthic Ba and alkalinity fluxes in sediments of the equatorial Pacific and the continental margin off California, but no functional relationship between Ba-fluxes and organic matter mineralization. This implies that the process that causes the similarity of barium and alkalinity fluxes is not identified, but differs from the likely coupling to oxic respiration and $CaCO_3$-dissolution. In a similar manner, the cycles of Si and Ba seem to be decoupled during early diagenesis (McManus et al. 1994), thus fundamental principles are yet not understood and also have to be overcome by detailed process studies in the future, specifically focusing on the coupling and interactions of elements.

Acknowledgements

We would like to thank the captains and the crews of the RV *Meteor* on many cruises in the South Atlantic. For technical assistance on board and countless pore water analyses we are indebted to Sigrid Hinrichs, Karsten Enneking and Susanne Siemer. For technical assistance with the lander systems we likewise thank Stephan Meyer, Axel Krack, Gabriele Eickert, Anja Eggers, Vera Hübner and Volker Meyer. Rick Jahnke and David Archer are thanked for their helpful and constructive comments of the manuscript. This research was funded by the German Science Foundation (contribution no. 367 of Special Research Project SFB 261 at the University of Bremen). Data are available under www.pangaea.de/Projects/SFB261.

References

Adler M, Hensen C, Wenzhöfer F, Pfeifer K, Schulz HD (2001) Modeling of calcite dissolution by oxic respiration in supralysoclinal deep-sea sediments. Mar Geol 177: 167-189

Archer DE (1996a) An atlas of the distribution of calcium carbonate in sediments of the deep-sea. Glob Biogeochem Cycl 10: 159-174

Archer DE (1996b) A data-driven model of the global calcite lysocline. Glob Biogeochem Cycl 10: 511-526

Archer DE, Devol AH (1992) Benthic oxygen fluxes on the Washington shelf and slope: A comparison of *in situ* microelectrode and chamber flux measurements. Limnol Oceanogr 37: 614-629

Archer DE, Maier-Reimer E (1994) Effect of deep-sea sedimentary calcite preservation on atmospheric CO_2 concentration. Nature 367: 260 - 263

Archer D, Emerson S, Reimers C (1989) Dissolution of calcite in deep-sea sediments: pH and O_2 microelectrode results. Geochim Cosmochim Acta 53: 2831-2845

Behrenfeld MJ, Falkowski PG (1997) Photosynthetic rates derived from satellite-based chlorophyll concentration. Limnol Oceanogr 42: 1-20

Berelson WM, Hammond DE, O'Neill D, Xu XM, Chin C, Zukin J (1990) Benthic fluxes and pore water studies from sediments of central equatorial north Pacific: Nutrient diagenesis. Geochim Cosmochim Acta 54: 3001-3012

Berelson WM, Hammond DE, McManus J, Kilgore TE (1994) Dissolution kinetics of calcium carbonate in equatorial Pacific sediments. Glob Biogeochem Cycl 8: 219-235

Berelson WM, Anderson RF, Dymond J, DeMaster D, Hammond DE, Collier R, Honjo S, Leinen M, McManus J, Pope R, Smith C, Stephens M (1997) Biogenic budget of particle rain, benthic remineralization and sediment accumulation in equatorial Pacific. Deep-Sea Res II 44: 2251-2282

Berger WH (1989) Appendix global maps of ocean productivity. In: Berger WH, Smetacek V, Wefer G (eds) Productivity of the Ocean: Present and Past.

Dahlem Konferenzen. J Wiley & Sons, Chichester, pp 429-455

Berger WH, Fischer K, Lai C, Wu G (1987) Ocean productivity and organic carbon flux. I. Overview and maps of primary production and export production. Univ California, San Diego, SIO Reference 87-30, 67 p

Berner RA (1982) Burial of organic carbon and pyrite sulfur in the modern ocean: Its geochemical and environmental significance. Am J Sci 282: 451-473

Bernstein RE, Byrne RH, Betzer PR, Greco AM (1992) Morphologies and transformations of celestite in seawater: The role of acantharians in strontium and barium geochemistry. Geochim Cosmochim Acta 56: 3273-3279

Betzer PR, Showers WJ, Laws EA, Winn CD, DiTullio GR, Kroopnick PM (1984) Primary productivity and particle fluxes on a transect of the equator at 153°W in the Pacific Ocean. Deep-Sea Res 31: 1-11

Bidle KD, Azam F (1999) Accelerated dissolution of diatom silica by marine bacterial assemblages. Nature 397: 508-512

Bishop JKB (1988) The barite-opal-organic carbon association in oceanic particulate matter. Nature 332:341-343

Broecker WS, Peng T-H (1987) The role of $CaCO_3$ compensation in the glacial to interglacial atmospheric CO_2 change. Glob Biogeochem Cycl 1: 15-29

Canfield DE (1993) Organic matter oxidation in marine sediments. In: Wollast R, Mackenzie FT, Chou L (eds) Interactions of C, N, P, and S in Biogeochemical Cycles and Global Change. NATO ASI Series 14, Springer, Berlin, pp 333-363

Cai WJ, Reimers CE (1993) The development of pH and pCO_2 microelectrodes for studying the carbonate chemistry of pore waters near the sediment-water interface. Limnol Oceanogr 38: 1762-1773

Cai WJ, Reimers CE (1995) Benthic oxygen flux, bottom water oxygen concentration and core top organic carbon content in the deep northeast Pacific Ocean. Deep-Sea Res 42: 1681-1699

Cai WJ, Reimers CE, Shaw T (1995) Microelectrode studies of organic carbon degradation and calcite dissolution at a California Continental rise site. Geochim Cosmochim Acta 59: 497-511

Cai WJ, Sayles FL (1996) Oxygen penetration depths and fluxes in marine sediments. Mar Chem 52: 123-131

Chan LH, Drummond D, Edmond JM, Grant B (1977) On the barium data from the Atlantic GEOSECS expedition. Deep-Sea Res 24: 613-649

Christensen JP (2000) A relationship between deep-sea

benthic oxygen demand and oceanic primary production. Oceanol Acta 23: 65-82

Cwienk DS (1986) Recent and glacial age organic and biogenic silica accumulation in marine sediments. MS thesis, School of Oceanography, Univ Rhode Island, 237 p

DeBaar HJW, Suess E (1993) Ocean carbon cycle and climate change - An introduction to the interdisciplinary Union Symposium. Glob Planet Change 8: VII - XI

Dehairs F, Chesselet R, Jedwab J (1980) Discrete suspended particles of barite and the barium cycle in the open ocean. Earth Planet Sci Lett 49: 528-550

Dehairs F, Fagel N, Anita AN, Peinert R, Elskens M, Goeyens L (2000) Export production in the Bay of Biscay as estimated from barium - barite in settling material: A comparison with new production. Deep-Sea Res I 47: 583-601

DeMaster DJ (1981) The supply and accumulation of silica in the marine environment. Geochim Cosmochim Acta 47: 1715-1732

Dittert N, Henrich R (2000) Carbonate dissolution in the South Atlantic Ocean: Evidence from ultrastructural breakdown in *Globigerina bulloides*. Deep-Sea Res I 47: 603-620

Dymond J, Collier C (1996) Particulate barium fluxes and their relationships to biological productivity. Deep-Sea Res II 43: 1283-1308

Dymond J, Suess E, Lyle M (1992) Barium in Deep-Sea Sediments: A Geochemical Proxy for Paleoproduc-tivity. Paleoceanography 7: 163-181

Emerson S, Bender M (1981) Carbon fluxes at the sediment-water interface of the deep-sea; calcium carbonate preservation. J Mar Res 39: 139-162

Eppley R, Peterson BJ (1979) Particulate organic matter flux and planktonic new production in the deep ocean Nature 282: 677-680

Fischer G, Wefer G (1996) Long-term observations of particle fluxes in the eastern Atlantic: Seasonality, changes of flux with depth and comparison with the sediment record. In: Wefer G, Berger WH, Siedler G, Webb D (eds) The South Atlantic: Present and Past Circulation. Springer, Berlin, pp 325-344

Fischer G, Ratmeyer V, Wefer G (2000) Organic carbon fluxes in the Atlantic and the Southern Ocean: Relationship to primary production compiled from satellite radiometer data. Deep-Sea Res II 47:1961-1997

Francois R, Honjo S, Manganini SJ, Ravizza GE (1995) Biogenic barium fluxes to the deep-sea: Implications for paleoproductivity reconstruction. Glob Biogeo-chem Cycl 9: 289-303

Gingele FX, Dahmke A (1994) Discrete barite particles and barium as tracers of paleoproductivity in South Atlantic sediments. Paleoceanography 9: 151-168

Glud RN, Gundersen JK, Jørgensen BB, Revsbech NP, Schulz HD (1994): Diffusive and total oxygen uptake of deep-sea sediments in the eastern South Atlantic Ocean: *In situ* and laboratory measurements. Deep-Sea Res 41: 1767-1788

Glud RN, Gundersen JK, Holby O (1999) Benthic *in situ* respiration in the upwelling area off central Chile. Mar Ecol Prog Ser 186: 9-18

Greeff O, Glud RN, Gundersen JK, Holby O, Jørgensen BB (1998) A benthic lander for tracer studies inthe sea bed: *In situ* measurements of sulfate reduction. Cont Shelf Res 18: 1581-1594

Gundersen JK, Jørgensen BB (1990) Microstructure of diffusive boundary layers and the oxygen uptake of the sea floor. Nature 345: 604-607

Hales B, Emerson S (1996) Calcite dissolution in sediments of the Ontong-Java Plateau: *In situ* measurements of pore water O_2 and pH. Glob Biogeochem Cycl 10: 527-541

Hales B, Emerson S (1997a) Calcite dissolution in sediments of the Ceara Rise: *In situ* measurements of porewater O_2, pH, and CO_2(aq). Geochim Cosmochim Acta 61: 501-514

Hales B, Emerson S (1997b) Evidence in support of first-order dissolution kinetic of calcite in seawater. Earth Planet Sci Lett 148: 317-327

Hales B, Emerson S, Archer D (1994) Respiration and dissolution in the sediments of the western North Atlantic: estimates from models of *in situ* microelectrode measurements of porewater oxygen and pH. Deep-Sea Res 41: 695-719

Henrich R (1986) A calcite dissolution pulse in the Norwegian-Greenland Sea during the last deglaciation. Geol Rundsch 75: 805-820

Henrichs SM (1992) Early diagenesis of organic matter in marine sediments: progress and perplexity. Mar Chem 39: 119-149

Hensen C, Landenberger H, Zabel M, Schulz HD (1998) Quantification of diffusive benthic fluxes of nitrate, phosphate and silicate in the Southern Atlantic Ocean. Glob Biogeochem Cycl 12: 193-210

Hensen C, Zabel M (2000) Early diagenesis at the benthic boundary layer: Oxygen and nitrate in marine sedi-ments. In: Schulz HD, Zabel M (eds) Marine Geo-chemistry. Springer, Berlin, pp 209-231

Hensen C, Zabel M, Schulz HD (2000) A comparison of benthic nutrient fluxes from deep-sea sediments off Namibia and Argentina. Deep-Sea Res II 47: 2029-2050

Holstein J (2002) Effekt benthischer Mineralisations-prozesse auf die Lösung von biogenem Opal. Unpubl Diploma Thesis, University of Bremen

Jahnke RA (1996) The global ocean flux of particulate organic carbon: Areal distribution and magnitude. Glob Biogeochem Cycl 10: 71-88

Jahnke RA (2001) Constraining organic matter cycling with benthic fluxes. In: Boudreau BP and Jørgensen BB (eds) The Benthic Boundary Layer. Oxford University Press, New York, pp 302-319

Jahnke RA, Jahnke DB (2000) Rates of C, N, P and Si recycling and denitrification at the US Mid-Atlantic continental slope depotcenter. Deep-Sea Res 47: 1405-1428

Jahnke RA, Reimers CE, Craven DB (1990) Intensification of recycling of organic matter at the sea floor near ocean margins. Nature 348: 50-54

Jahnke RA, Craven DB, Gaillard J-F (1994) The influence of organic matter diagenesis on $CaCO_3$ dissolution at the deep-sea floor. Geochim Cosmochim Acta 58: 2799-2809

Jahnke RA, Craven DB, McCorkle DC, Reimers CE (1997) $CaCO_3$ dissolution in California continental margin sediments: The influence of organic matter remin-eralization. Geochim Cosmochim Acta 61: 3587-3604

Jansen FJH, van der Gaast SJ (1988) Accumulation and dissolution of opal in quaternary sediments of the Zaire deep-sea fan. Mar Geol 83: 1-7

Jones GA, Johnson DA (1984) Displaced antarctic diatoms in Vema Channel sediments: Late Pleistocene/Holocene fluctuations in AABW flow. Mar Geol 58: 165-186

Jørgensen BB (1983) Processes at the sediment-water interface. In: Bolin B, Cook RB (eds) The Major Bio-geochemical Cycles and their Interactions. SCOPE, J Wiley & Sons, New York, pp 477-515

Keir RS (1980) The dissolution kinetics of biogenic calcium carbonates in seawater. Geochim Cosmochim Acta 44: 241-252

Keir RS (1983) Variation in the carbonate reactivity of deep-sea sediments: Calcium carbonates in seawater. Deep-Sea Res 30: 279-296

Kennett JP (1982) Marine Geology. Prentice Hall Inc, Englwood Cliffs, USA

Koning E, Brummer GJ, van Raaphorst W, van Bennekom J, Helder W, van Iperen J (1997) Settling, dissolution and burial of biogenic silica in the sediments off Somalia (northwestern Indian Ocean). Deep-Sea Res II 44: 1341-1360

Kumar N, Anderson RF, Biscaye PE (1996) Remineralization of particulate authigenic trace metals in the Middle Atlantic Bight: Implications for proxies of export production. Geochim Cosmochim Acta 60: 3383-3397

Longhurst A, Sathyendranath S, Platt T, Caverhill C (1995) An estimate of global primary production in the ocean from satellite radiometer data. J Plankton Res 17: 1245-1271

Mackenzie FT, Ver LM, Sabine C, Lane M, Lerman A (1993) C, N, P, S global biogeochemical cycles and modeling of global change. In: Wollast R, Mackenzie FT and Chou L (eds) Interactions of C, N, P and S Biogeochemical Cycles and Global Change. NATO ASI Series, I 4, Springer, Berlin, pp 1-61

Martin WR, Sayles FL (1996) $CaCO_3$ dissolution in sediments of the Ceará Rise, western equatorial Atlantic. Geochim Cosmochim Acta 60: 243-263

McManus J, Berelson WM, Klinkhammer GP, Kilgore TE, Hammond DE (1994) Remobilization of barium in continental margin sediments. Geochim Cosmochim Acta 58: 4899-4907

McManus J, Berelson WM, Klinkhammer GP, Johnson KS, Coale KH, Anderson RF, Kumar J, Burdige DJ, Hammond DE, Brumsack HJ, McCorkle DC, Rushdi A (1998) Geochemistry of barium in marine sediments: Implications for its use as a paleoproxy. Geochim Cosmochim Acta 62: 3452-3473

McManus J, Berelson WM, Hammond DE, Klinkhammer GP (1999) Barium cycling in the North Pacific: Implications for the utility of Ba as a paleoproduc-tivity and paleoalkalinity proxy. Paleoceanography 14: 53-61

Milliman JD (1993) Production and Accumulation of Calcium Carbonate in the Ocean: Budget of a Nonsteady State. Glob Biogeochem Cycl 7: 927-957

Milliman JD, Troy PJ, Balch WM, Adams AK, Li Y-H, Mackenzie FT (1999) Biologically mediated dissolution of calcium carbonate above the chemical lysocline? Deep-Sea Res I 46: 1653-1669

Nürnberg CC (1995) Bariumfluss und Sedimentation im südlichen Südatlantik - Hinweise auf Produktivitäts-änderungen im Quartär. Geomar Report 38, Kiel, 105 p

Otto S (1996) Die Bedeutung von gelöstem organischen Kohlenstoff (DOC) für den Kohlenstoffffluß im Ozean. PhD Thesis, Ber Fachber Geowiss, Univ Bremen 87, 150 p

Pamatmat MM (1971) Oxygen consumption in the sea-bed. IV. Shipboard and laboratory experiments. Limnol Oceanogr 16: 536-550

Paytan A, Kastner M (1996) Benthic Ba fluxes in the central Equatorial Pacific, implications for the oceanic Ba cycle. Earth Planet Sci Lett 142: 439-450

Pfeifer K, Kasten S, Hensen C, Schulz HD (2001) Reconstruction of primary productivity from barium contents in surface sediments of the South Atlantic Ocean. Mar Geol 177: 13-24

Pfeifer K, Hensen C, Adler M, Wenzhöfer M, Weber B, Schulz HD (2002) Modeling of subsurface calcite dissolution regarding respiration and reoxidation processes in the equatorial upwelling off Gabon. Geochmim Cosmochim Acta 66: 4247-4259

Ragueneau O, Tréguer P, Leynaert A, Anderson RF, Brzezinski MA, DeMaster DJ, Dugdale RC, Dymond J, Fischer G, François R, Heinze C, Maier-Reimer E, Martin-Jézéquel V, Nelson DM, Quéguiner B (2000) A review of the Si cycle in the modern ocean: recent progress and missing gaps in the application of biogenic opal as a paleoproductivity proxy. Glob Planet Change 26: 317-365

Reimers CE (1987) An *in situ* microprofiling instrument for measuring interfacial pore water gradients: methods and oxygen profiles from the North Pacific Ocean. Deep-Sea Res 34: 2019-2035

Romero O, Hensen C (2002) Oceanographic control of biogenic opal and diatoms in surface sediments of the South Western Atlantic. Mar Geol 186: 263-280.

Rowe GT, Boland GS, Phoel WC, Anderson RF, Biscaye PE (1994) Deep-sea floor respiration as an indication of lateral input of biogenic detritus from continental margins. Deep-Sea Res II 41: 657-668

Rühlemann C, Müller PJ, Schneider RR (1999) Organic carbon and carbonate as paleoproductivity proxies: Examples from high and low productivity areas of the tropical Atlantic. In: Fischer G, Wefer G (eds) Use of Proxies in Paleoceanography: Examples from the South Atlantic. Springer, Berlin, pp 315-344

Sarnthein M, Winn K, Duplessy J-C, Fontugne MR (1988) Global variations of surface ocean productivity in low and mid latitudes: Influence on CO_2 reservoirs of the deep ocean and atmosphere during the last 21,000 years. Paleoceanography 3:361-399

Sauter EJ, Schlüter M, Suess E (2001) Organic carbon flux and remineralization in surface sediments from the northern North Atlantic derived from pore-water oxygen microprofiles. Deep-Sea Res I 48: 529-553

Sayles FL, Deuser WG, Goudreau JE, Dickinson WH, Jickells TD, King P (1996) The benthic cycle of biogenic opal at the Bermuda Atlantic Time Series site. Deep-Sea Res 43: 383-409

Schenau SJ, Prins MA, DeLange GJ, Monnin C (2001) Barium accumulation in the Arabian Sea: Controls on barite preservation in marine sediments. Geochim Cosmochim Acta 65: 1545-1556

Schlüter M, Rutgers van der Loeff MM, Holby O, Kuhn G (1998) Silica cycle in surface sediments of the South Atlantic. Deep-Sea Res I 45: 1085-1109

Schlüter M, Sauter EJ, Schäfer A, Ritzrau W (2000) Spatial budget of organic carbon flux to the seafloor of the northern North Atlantic (60°N – 80°N). Glob Biogeochem Cycl 14: 329-340

Schneider RR, Schulz HD, Hensen C (2000) Marine Carbonates: Their formation and destruction. In: Schulz HD, Zabel M (eds) Marine Geochemistry. Springer, Berlin, pp 283-307

Schulz S (2000) Frühdiagenese von Opal in Oberflächen-sedimenten des Südatlantiks – Versuch einer Bi-lanzierung. unpubl Diploma Thesis, University of Bremen

Smith KL jr, Clifford CH, Eliason AH, Walden B, Rowe GT, Teal JM (1976) A free vehicle for measuring benthic community metabolism. Limnol Oceanogr 21: 164-170

Smith KL jr, Hinga KR (1983) Sediment community respiration in the deep-sea. In: Rowe GT (ed) The Sea. Vol. 8, J Wiley & Sons, Chinchester, pp 331-370

Summerhayes CP, Kroon D, Rosell-Melé A, Jordan RW, Schrader H-J, Hearn R, Villanueva J, Grimalt JO, Eglington G (1995) Variability in the Benguela Current upwelling system over the past 70,000 years. Prog Oceanogr 35: 207-251

Thamdrup B, Canfiled DE (2000) Benthic respiration in aquatic sediments. In: Sala OE, Jackson RB, Mooney HA, Horwarth RW (eds) Methods in Ecosystem Science. Springer, Berlin, pp 86-103

Tréguer P, Nelson DM, van Bennekom AJ, DeMaster DJ, Leynaert A, Quéguiner B (1995) The silica balance in the World Ocean: A reestimate. Science 268:375-379

Volbers ANA, Henrich R (2002a) Late Quaternary variations in calcium carbonate preservation of deep-sea sediments in the northern Cape Basin: results from a multiproxy approach. Mar Geol 180:203-220

Volbers ANA, Henrich R (2002b) Present water mass calcium carbonate corrosiveness in the eastern South Atlantic inferred from ultrastructural breakdown of *Globigerina bulloides* in surface sediments. Mar Geol 186: 471-486

Walsh JJ (1991) Importance of continental margins in the marine biogeochemical cycling of carbon and nitrogen. Nature 350: 53-55

Warren B, Speer KG (1991) Deep circulation in the eastern South Atlantic ocean. Deep-Sea Res 38: 281-322

Wefer G, Fischer G (1993) Seasonal patterns of vertical particle flux in equatorial and coastal upwelling areas of the Eastern Atlantic. Deep-Sea Res 40: 1613-1645

Wenzhöfer F, Glud RN (2002) Benthic carbon mineralization in the Atlantic: A synthesis based on *in situ* data from the last decade. Deep-Sea Res 49: 1255-1279

Wenzhöfer F, Holby O, Kohls O (2001a) Deep penetrating benthic oxygen profiles measured *in situ* with oxygen optodes. Deep-Sea Res 48: 1741-1755

Wenzhöfer F, Adler M, Kohls O, Hensen C, Strotmann B, Boehme S, Schulz HD (2001b). Calcite dissolution driven by benthic mineralization in the deep-sea: *In situ* measurements of Ca^{2+}, pH, pCO_2, O_2. Geochim Cosmochim Acta 65: 2677-2690

Zabel M, Hensen C (2002) The importance of mineralization processes in surface sediments at continental margins. In: Wefer G, Billet D, Hebbeln D, Jørgensen BB, Schlüter M, Van Weering Tj (eds) Ocean Margin Systems. Springer, Berlin, pp 253-267

Zabel M, Dahmke A, Schulz HD (1998) Regional distribution of diffusive phosphate and silicate fluxes through the sediment-water interface: The eastern South Atlantic. Deep-Sea Res 45: 277-300

Processes and Signals of Nonsteady-State Diagenesis in Deep-Sea Sediments and their Pore Waters

S. Kasten[1*], M. Zabel[1], V. Heuer[1] and C. Hensen[1,2]

[1] Universität Bremen, Fachbereich Geowissenschaften, Postfach 33 04 40, 28334 Bremen, Germany
[2] GEOMAR – Forschungszentrum für Marine Geowissenschaften, Wischhofstr. 1-3, 24148 Kiel, Germany
* corresponding author (e-mail): skasten@uni-bremen.de

Abstract: Nonsteady-state conditions – induced by changes in the fluxes of electron donors and acceptors and environmental conditions – are shown to have been and to be still widespread in sediments of the equatorial and South Atlantic Ocean. Typical diagenetic phenomena initiated under such nonsteady-state conditions comprise the fixation and downward progression of redox boundaries and reaction fronts. Intervals most severely altered by diagenetic overprint often occur cyclically within the sedimentary record and are mostly associated with full glacial/interglacial transitions. The extent of post-depositional oxidation of organic carbon as well as the dissolution and re-precipitation of minerals across these glacial terminations was shown to depend on the overall sedimentation rate and the magnitude of change encountered in the various depositional and geochemical factors. A sedimentation rate of about 2 cm/kyr was confirmed to be the critical value below which no significant amounts of non-refractory organic carbon are preserved. The influence of climatically induced variations in environmental conditions is not restricted to the geochemical boundaries in the vicinity of the sediment surface (e.g. oxic/post-oxic and Fe redox boundary) but well extends into much deeper sediment sections – namely into the zone of anaerobic oxidation of methane (AOM). In this way, processes within the zone of AOM can produce a further profound diagenetic alteration of the sediment composition up to hundreds of thousands of years after initial deposition and thus a significantly delayed chemical log-in. The long-term utility of all primary and secondary signals – also those formed and initially preserved across the oxic/post-oxic and Fe redox boundaries – is ultimately controlled by the geochemical processes within and below the sulfate/methane transition (SMT). While dissolution of authigenic and productivity-related barite takes place in sulfate-depleted sediment sections, iron sulfides as well as sulfurized organic matter and associated trace elements have a high potential to survive burial below the SMT. Nonsteady-state diagenesis can be triggered not only by changes in conditions at the sediment/water interface like TOC input, sedimentation rate or O_2 content of bottom water but also by processes in the underlying sediment – namely the formation and/or liberation of methane. Apart from the distinct alteration of the solid-phase composition, variations in the upward flux of methane also have a considerable impact on the shape of sulfate pore water profiles. Modelling the effects of such variations in methane flux on sulfate profiles has illustrated that considering possible nonsteady-state situations in the sediment/pore water system is of utmost importance for the interpretation of pore water data.

Introduction

During burial, marine sediment passes through a series of depth intervals where microbially medi- ated reactions can change the chemical composi- tion of solid particulates and surrounding pore

From WEFER G, MULITZA S, RATMEYER V (eds), 2003, *The South Atlantic in the Late Quaternary: Reconstruction of Material Budgets and Current Systems.* Springer-Verlag Berlin Heidelberg New York Tokyo, pp 431-459

water. This "early diagenesis" is typically examined assuming steady state conditions, where reactive inputs and overlying water properties remain constant over time. This assumption is rarely tested and, in many cases, clearly incorrect. Numerous studies published during the past two decades have demonstrated that nonsteady-state depositional and diagenetic conditions are common even in deposits at great water depths and/or with low organic carbon contents (e.g. Thomson et al. 1984, 1996; Wallace et al. 1988; Tarduno 1994; Tarduno and Wilkison 1996; Kasten et al. 1998). Nonsteady-state diagenesis can be initiated by any changes in the fluxes of electron donors and acceptors and environmental conditions. While the sediment/pore water system adjusts to the new depositional and/or diagenetic conditions, the primary sediment composition can be efficiently modified. For example, organic carbon is oxidized to various degrees post-depositionally, (magnetic) minerals are dissolved, and redistribution processes cause distinct depletions and enrichments of elements and minerals in the solid phase.

The imprint that such nonsteady-state processes can leave on the sedimentary record has been extensively studied in turbiditic/pelagic sediment sequences in the Madeira and Nares Abyssal Plains (e.g. Colley et al. 1984; Wilson et al. 1985, 1986) and along sapropel/hemipelagic transitions in the eastern Mediterranean (e.g. Pruysers et al. 1991, 1993; Higgs et al. 1994; Thomson et al. 1995; Van Santvoort et al. 1996, 1997; Passier et al. 1996). Besides redistribution of elements along turbidite/hemipelagic or sapropel/hemipelagic sediment sequences, post-depositional redistributions of elements due to non-constant depositional conditions have also been reported for transitional sediment intervals from glacial to interglacial times. The most recent example is the Pleistocene/Holocene boundary (oxygen isotope stage boundary 2/1 or Termination I) where the diagenetic processes responsible for the formation of distinct solid phase element enrichments were shown to be still active (Thomson et al. 1984, 1990; Wallace et al. 1988; Kasten et al. 1998). Thomson et al. (1996) studied the penultimate full glacial/interglacial transition – i.e. Termination II - in Atlantic sediments as the fossil analogue of the Pleistocene/Holocene boundary. They

demonstrated that similar nonsteady-state diagenetic processes were also active along this penultimate climatic boundary.

Nonsteady-state diagenetic conditions can also be induced by (giant) submarine landslides which were shown to be widespread along continental margins (e.g. De Lange 1983; Mienert et al. 1998; Zabel and Schulz 2001). The mechanisms that trigger such massive sediment movements (tectonics or climate-related destabilization of underlying gashydrates) as well as their feedback to climate (possible release of methane to the atmosphere) are still under dispute. Furthermore the impact of these nonsteady-state events on the shape of pore water profiles and the solid phase composition of sediments is not yet fully understood. Several studies relate the occurrence of authigenic carbonates (Bohrmann et al. 1998) or ikaites to the destabilization of gas hydrates (Schubert et al. 1997). Similarly, changes in the diffusive flux of methane or in advective fluid/gas migration have a strong potential to initiate non-steady state diagenesis from below.

The aim of this contribution is to describe the diagenetic processes which are initiated during non-constant depositional and geochemical conditions and to give an overview of related studies on non-steady-state diagenesis in the frame of the Special Research Project SFB 261 (Sonderforschungsbereich 261) "The South Atlantic in the Late Quaternary: Reconstruction of Material Budgets and Current Systems". We also emphasize the potential of secondary sedimentary signals to reconstruct changes in paleoceanographic, depositional and geochemical conditions, particularly focussing on changes in organic carbon burial and in methane fluxes from deeper sediment strata over time. These examples will demonstrate the potential of nonsteady-state diagenesis to overprint the primary sediment composition – with strongest modifications occurring across the $Fe(II)/Fe(III)$ redox boundary and at the sulfate/methane transition (SMT), respectively. We will also present a modeling approach which gives evidence that particular shapes of sulfate pore water profiles which have been either unexplained until now or interpreted in different ways can result from changes in methane fluxes from below or from the emplacement of turbidites or submarine landslides – thus from a nonsteady-state geochemical or depositional situation.

Redox Zones, Redox Boundaries and Reaction Fronts

The microbial degradation/oxidation of organic matter in aquatic sediments by successive use of different terminal electron acceptors (TEA) leads to the establishment of a typical redox zonation. The concept of redox zonation is based on the assumption that the different electron acceptor processes for the oxidation of organic material are spatially segregated into separate sediment intervals and that the succession of these zones is determined by the overall energy yield of the different reactions. The TEAs are generally used in the sequence: oxygen, nitrate, manganese oxides, iron oxides, sulfate -

followed by methane formation (Froelich et al. 1979). Redox zonation has also been used to classify the different early diagenetic environments in marine sediments (Froelich et al. 1979; Berner 1981) as indicated in Fig.1.

The classical sequence of redox zones in marine sediments is depicted in Fig. 1 including schematized pore water profiles of the different dissolved TEAs (oxygen, nitrate, sulfate) as well as of the products of the particular TEA processes (Mn^{2+}, Fe^{2+}, HS^-). It should be noted that this idealized model represents a specific situation in which the sulfate pore water profile is primarily shaped by the reaction of sulfate with upward diffusing methane, i.e. by anaerobic oxidation of methane (AOM).

However, studies in lacustrine and marine sediments as well as groundwater systems have dem-

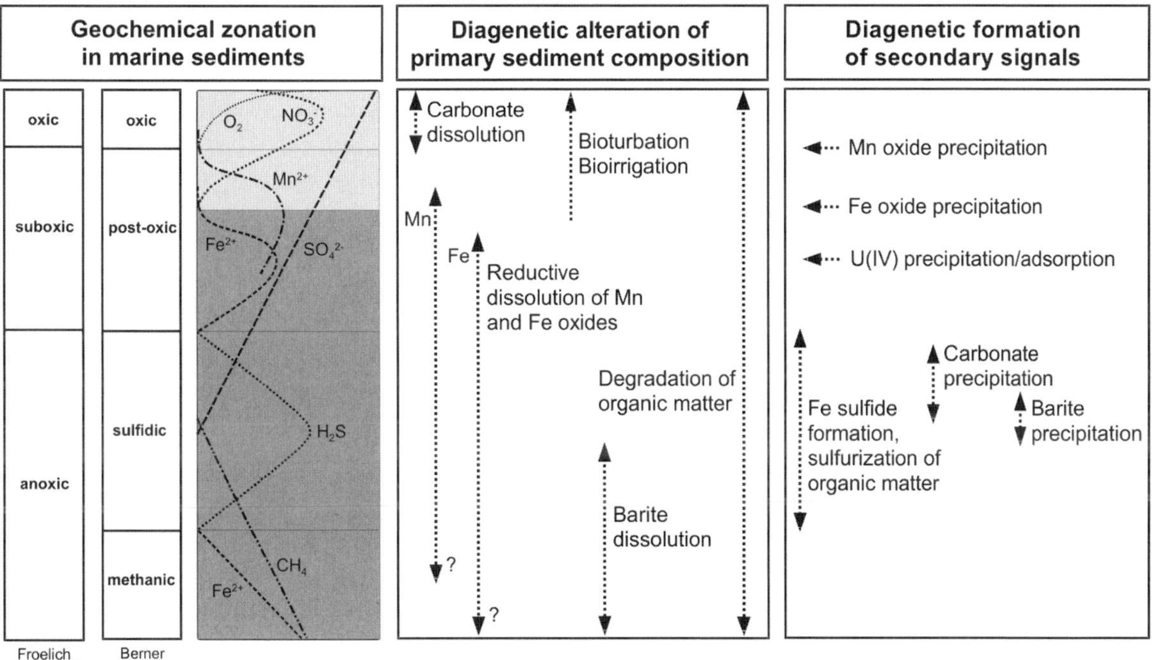

Fig. 1. Schematic representation of redox zonation in marine sediments as well as the most important early diagenetic processes for the alteration of the primary sediment composition and for the formation of secondary signals in the sedimentary record. The classifications of the different early diagenetic environments as given by Froelich et al. (1979) and Berner (1981) are shown on the left side. Pore water profiles of the dissolved terminal electron acceptors (O_2, NO_3^-, SO_4^{2-}) and the products of the different TEA processes (Mn^{2+}, Fe^{2+}, HS^-) are depicted here as well. This illustration represents a particular situation in which the sulfate pore water profile is primarily shaped by the reaction with upward diffusing methane, i.e. by anaerobic oxidation of methane.

onstrated that the different terminal electron acceptor (TEA) processes are not necessarily separated into distinct zones (e.g. Wersin et al. 1991; Canfield et al. 1993; Jakobsen and Postma 1999). Moreover, the different reactions of organic matter degradation can occur simultaneously or the zones can even be reversed (e.g. Wersin et al. 1991). As a general explanation for redox zonation the energy gain of the overall process has been given. In fact it is known that the anaerobic degradation of organic matter proceeds via at least two steps, i.e. a first fermentative step and a second TEA reaction step (e.g. Jørgensen 2000). In the first fermentative step organic material is degraded to small organic molecules like acetate and formate as well as hydrogen. These fermentation products are subsequently consumed by the different TEA processes.

As the TEA processes are much faster than fermentation they could be considered as approaching chemical equilibrium. Based on this line of argumentation Postma and Jakobsen (1996) developed the concept of 'partial equilibrium' in which fermentation determines the overall rate of organic matter oxidation, while equilibrium is approached by the different terminal electron acceptors. In other words: it is not the energy gain of the overall process but of the second step in the process of organic matter mineralization, i.e. the TEA process, that determines the sequence or overlapping of redox zones. In this way they offered a more plausible explanation for the segregation of TEA processes into different redox zones as well as for the often-observed simultaneous occurrence of sulfate reduction and iron reduction or a reversal of different redox zones. Examples of such concurrently occurring degradation processes have been found in sediments of the Amazon Fan where Fe reduction was shown to continue in the methanic zone (Flood et al. 1995; Kasten et al. 1998). These findings are also confirmed by experimental studies of Schinzel et al. (1993) who demonstrated that methanogenesis and iron reduction can take place contemporaneously.

The vertical extension of the different redox zones depends on the input of organic matter to the sediment, the supply of TEAs (e.g. oxygen and sulfate content of the bottom water, Mn and Fe oxihydroxide content of the sediment) as well as the flux of reduced pore water constituents (e.g. methane) from greater sediment depth. The higher the flux of organic detritus settling to the seafloor and/or the higher the flux of reduced components from below the more condensed this zonation generally will be. The exact location of the different redox boundaries – e.g. the Fe and Mn redox boundaries – with respect to the sediment surface therefore is a function of the balance between oxidants and reductants on the one hand and sedimentation rate on the other hand. We emphasize the significance of redox boundaries and reaction fronts in the following discussions because these are the sites where the strongest modifications of the primary sediment composition take place.

The most important early diagenetic processes that lead to an alteration of the primary sediment composition as well as to the formation of secondary signals in the sedimentary record are illustrated in Fig. 1. The most intense redistributions of elements between the solid phase and the pore water including dissolution and precipitation of (magnetic) minerals take place (1) in the surface sediment interval reaching from the oxic/post-oxic (suboxic) boundary (often also referred to as 'redoxcline') down across the Fe(II)/Fe(III) redox boundary and (2) around the sulfate/methane transition where hydrogen sulfide and carbon dioxide (bicarbonate) are produced through the anaerobic oxidation of methane.

Besides bioturbation and bioirrigation which can destroy the initial depth/age relation of sediment components, carbonate dissolution driven by the release of metabolic CO_2 during aerobic degradation of organic matter represents one of the most important diagenetic reactions in surface sediments. The significance of this process is discussed in detail by Wilson and Thomson (1998), Adler et al. (2001), Pfeifer et al. (2002) and Hensen et al. (this volume). Moving deeper into the sediment, the reduction of Mn and Fe oxihydroxides takes place both biotically and abiotically. This process can occur over considerable sediment depth sometimes even continuing within the methanic zone – as has been described above. Besides the redistribution of Fe and Mn in the sediment column and the pronounced alteration of the initial Fe and Mn mineralogy, reductive dissolution of ferric iron components also has a strong impact on the rock magnetic properties of the sediment (e.g. Tarduno 1994; Tarduno and Wilkison 1996; Robinson et al. 2000). Funk et al. (this volume) demonstrate that in sediments of the equatorial Atlantic reductive dissolution below the Fe redox boundary under nonsteady-state conditions has extensively altered the magnetic mineral inventory of these deposits.

With respect to the diagenetic formation of secondary minerals, the sediment interval across the oxic/post-oxic boundary hosts a suite of redox boundaries for different redox-sensitive elements like e.g. Mn, Fe and U (cf. Fig. 1). Correspondingly, the formation of the authigenic solid-phase enrichments of these elements - including precipitation of bacterial magnetite - occurs at different sediment depths.

Within the zone of anaerobic oxidation of methane the primary mineral that is most affected by diagenetic alteration is barite ($BaSO_4$) (e.g. Kasten and Jørgensen 2000). In the sediment interval below the sulfate/methane transition where interstitial sulfate is completely depleted, barite is undersaturated and consequently subject to dissolution. The Ba^{2+} ions liberated into the pore water by this process diffuse upwards and precipitate as secondary barite slightly above the sulfate penetration depth at a so-called authigenic barite front (cf. Fig. 1) (e.g. Brumsack 1986; Von Breymann et al. 1992; Torres et al. 1996; Dickens 2001). This process is of particular significance since the dissolution and re-precipitation of barite limits the use of biogenic Ba as a potential proxy for paleoproductivity (for a review see Gingele et al. 1999). Examples where biogenic or excess barium can be used as a persistent sediment component to trace past fluxes of organic matter to the seafloor are presented below. Besides authigenic barite, carbonates and iron sulfides are two other important secondary minerals precipitated within the sulfate/methane transition zone. Both mineral groups form as a direct consequence of AOM which produces large amounts of hydrogen sulfide and carbon dioxide (bicarbonate) according to equation (1).

$$CH_4 + SO_4^{2-} \rightarrow HCO_3^- + HS^- + H_2O \qquad (1)$$

Boetius et al. (2000) have shown that the anaerobic oxidation of methane by sulfate is mediated by a consortium of methane-consuming archaea and sulfate-reducing bacteria. Hydrogen sulfide liberated into the pore water by this process has at least a two-fold effect on the sedimentary solid phase as it initiates the precipitation of iron sulfides (e.g. Berner 1969, 1984) as well as the sulfurization of organic matter (e.g. Sinninghe Damsté and de Leeuw 1990; Schouten et al. 1994). Moreover, both iron sulfides and organic sulfur compounds affect the retention and accumulation of trace elements in the solid phase of sulfidic sediments (e.g. Huerta-Diaz and Morse 1992; Morse and Arakaki 1993; Heuer et al. 2003a, b).

Nonsteady-State Diagenesis

During steady-state diagenesis redox zones and redox boundaries establish at specific subsurface depths and remain at these constant relative distances with respect to the sediment surface over time. These depths are determined by a balance between fluxes of oxidants and reductants on the one hand and sedimentation rate on the other hand. A permanent steady-state condition - including permanent steady-state diagenesis - requires constant sedimentary and geochemical conditions over a long period of time. Only under such circumstances the relative depth positions of the redox boundaries and reaction fronts with respect to the sediment surface as well as the vertical extension of the redox zones remain constant too. However, as has already been pointed out by Pruysers et al. (1993) the balance between fluxes and sedimentation rate is very delicate, and thus, a steady-state condition, persisting over a long period of time is very unlikely in any sedimentary environment.

A change in at least one of the depositional or geochemical parameters alters the balance of the sediment/pore water system. The time interval during which the system adjusts to the new conditions – in order to obtain a new balance and thus a new steady state – is referred to as a nonsteady-state diagenetic situation. In the following we will show that such periods of nonsteady-state diagenesis have especially strong potential for overprinting the primary sediment composition and for producing distinct secondary signals in the sedimentary record.

Typical depositional and geochemical processes and events which can initiate nonsteady-state dia-genetic conditions comprise (1) changes in organic carbon burial induced by variations in sedimentation rate, organic carbon input to the sediment and/or oxygen content of bottom water, (2) mass transfer events of sediments in the form of turbidites, debris flows or submarine landslides, (3) variations in the upward-directed diffusive flux of methane or other reduced components from greater sediment depth, and (4) fluctuations in the advective flow of pore water and fluids from deeper sources or across the sediment/water interface. A decrease in organic carbon burial or a surplus in the supply of oxidants over time will result in an extension of redox zones. In the opposite case, redox boundaries and reaction fronts will be shifted towards the sediment surface and the redox succession will be condensed.

The effects of increases in the input and burial of organic carbon over time - accompanied by an upward shift of the redox boundaries and reaction fronts – have been described by Finney et al. (1988), De Lange et al. (1994) and Passier et al. (1996). They showed that metastable element enrichments (e.g. Mn hydroxides) can be preserved after a sudden upward shift of the redox boundaries. However, the long-term persistence of these enrich-ments is very restricted. Passier et al. (1996) described downward sulfidization fronts that can lead to the formation of distinct Fe sulfide enrichments below organic-rich layers like sapropels. In this contribution we will, however, focus on the opposite case – i.e. decrease in organic carbon burial and extension of redox zones over time – because it is the scenario which leads to the strongest diagenetic overprint of the primary sediment composition as we will demonstrate by means of the following schematic representations.

The potential of nonsteady-state diagenesis to produce distinct secondary element enrichments in the solid phase is shown in Fig. 2 using Mn, Fe and U as examples. This illustration demonstrates that the redox boundaries of the elements under consideration are arranged in a typical depth succession across the Fe(II)/Fe(III) redox boundary.

Under steady-state diagenetic conditions, the different redox boundaries and thus also the solid-phase enrichments of the elements will remain at constant depths with respect to the sediment surface over time. With ongoing sedimentation the lower parts of the solid-phase enrichments of Fe and Mn move into more reducing conditions and are successively dissolved. This reductive dissolution is followed by preferential upward migration of Mn^{2+} and Fe^{2+} and subsequent re-precipitation as oxihydroxides in the top part of the peak. Uranium has a contrasting geochemical behaviour and is immobilized under post-oxic conditions. The mechanism leading to the authigenic enrichment of U in the solid phase is diffusion of dissolved U species from seawater into the post-oxic part of the sediments followed by reduction and subsequent precipitation and/or adsorption to reactive surfaces. Under steady-state conditions, each sediment layer receives the same amount of authigenic U in addition to the U background content of the sedimenting material. In this way a constant U concentration level evolves in the sediments below the U redox boundary.

In contrast, a decrease in organic carbon burial over time can cause a fixation of the redox boundaries at a particular sediment depth for a prolonged period of time thus producing higher solid-phase concentrations of the metal under consideration than during steady-state. Thus, the occurrence of peaks in the concentration profile of authigenic U and strong enrichments of other elements (e.g. Thomson et al. 1990, 1996) are indicative for nonsteady-state diagenesis to have been active at some time in the past.

Besides the above described fixation of redox boundaries and reaction fronts other typical nonsteady-state processes comprise progressively downward-moving oxidation fronts – a concept developed by Thomson et al. (1984) and Wilson et al. (1985, 1986) – as well as the already mentioned sulfidization fronts below sapropels (Passier et al. 1996). Furthermore, Funk et al. (this volume) demonstrate that redox boundaries can be trapped at or within sediment intervals in which elevated amounts of degradable/non-refractory organic carbon are preserved. In this way, these layers are subject to a much higher degree of diagenetic over-

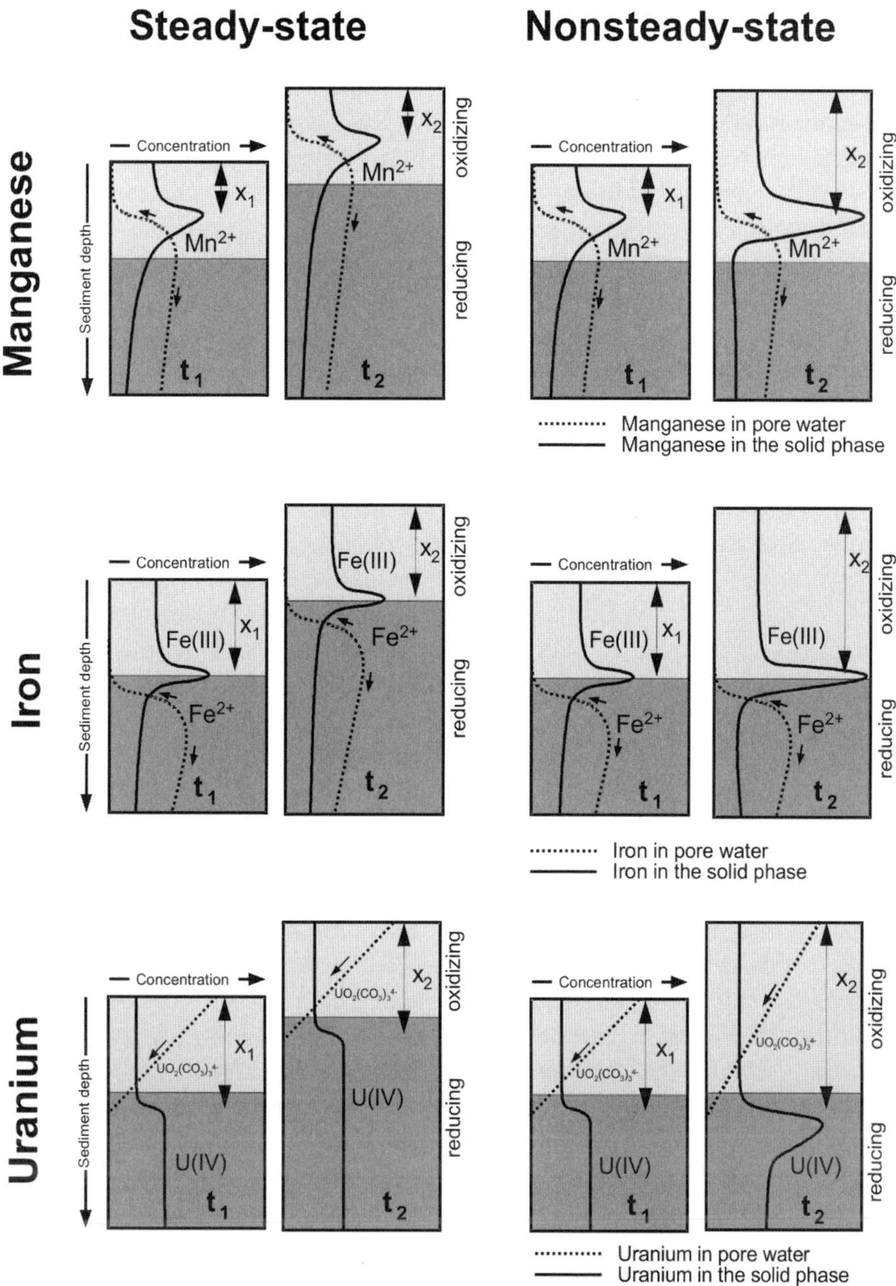

Fig. 2. Model concept for the formation of solid-phase enrichments of manganese, iron and uranium under steady-state (left side) and nonsteady-state (right side) depositional and diagenetic conditions. Dotted lines represent pore water profiles, solid lines show solid-phase concentration profiles. Note that authigenic solid-phase peaks can also form under steady-state conditions if the formation process is faster than the sedimentation rate.

print compared to sediment intervals affected by steady-state diagenesis.

With respect to the long-term burial and preservation of "nonsteady-state" element enrichments more recent works by Mercone et al. (1999) and Crusius and Thomson (2003) have shown that certain less frequently reported elements like selenium, mercury and silver form sharp peaks immediately into anoxic conditions below active oxidation fronts and that these peaks remain immobile in anoxic conditions during long-time burial. Thus, these elements have a considerable diagnostic potential to define relict positions of oxidation fronts. While uranium was important in earlier studies that worked out how oxidation fronts operate (e.g. Thomson et al. 1993, 1995, 1998) it tends to form very diffuse peaks compared with selenium, mercury and silver (Crusius and Thomson 2000).

In the following sections we will present studies of nonsteady-state diagenesis from the equatorial and the South Atlantic Ocean. These works are grouped into those which focus on diagenetic processes across the oxic/post-oxic boundary and the Fe redox boundary and those which illustrate the extent of diagenetic modification within the zone of anaerobic methane oxidation.

Redistribution of Elements Across the Oxic/Post-Oxic Boundary and the Fe(II)/Fe(III) Redox Boundary

High-resolution solid-phase investigations have been performed along glacial Terminations in sediments of the central equatorial Atlantic Ocean (Reitz, unpublished Diplom thesis) and of the upwelling area off Morocco, NW Africa (Jaquet, unpublished Diplom thesis). These sites were selected because we expected strong nonsteady-state diagenetic overprint of the transitional sediment intervals across the glacial/interglacial boundaries. The aim of these studies was to identify the extent of post-depositional modification of the primary sediment composition during a fixation or a progressive downward movement of the redoxcline and the Fe redox boundary induced by deglacial changes in depositional and geochemical conditions. Furthermore we wanted to reveal which of the secondary element enrichments produced are preserved in these sediments on further burial and can thus serve as valuable indicators for changes in organic carbon burial as has been done in detail by Thomson et al. (1996, 1998). For this purpose we compared the last glacial/interglacial transition – i.e. Termination I - where these nonsteady-state processes were shown to be still active with the fossil example – i.e. the penultimate glacial/interglacial transition or Termination II, respectively.

Central Equatorial Atlantic Ocean – Periphery of Equatorial Divergence

Gravity core GeoB 2908-7 (00°06,4'N; 23°19, 6'W) was recovered from a water depth of 3809 m during RV Meteor cruise M 29/3 (Schulz et al. 1995). The core has a total length of 1094 cm and reaches into oxygen isotope stage 10 (OIS 10). Sedimentation rates are on average 4.3 cm/kyr during glacial periods and 2.3 cm/kyr during interglacials (Funk et al. this volume). The two core segments of interest – i.e. the intervals across the OIS boundaries 2/1 and 6/5 - are characterized by color changes from brown to grey/green (cf. Fig. 3). The most pronounced color change is found at a sediment depth of about 46 cm and is located 10 cm below the maximum of Termination I (age: about 13.4 ka). This colour change marks the position of the active Fe(II)/Fe(III) redox boundary at this site. Around the 6/5 OIS boundary, about 15 cm below the maximum of Termination II (age: about 128 ka) a less distinct colour change occurs at about 433 cm which represents the fossil analogue of the modern Fe(II)/Fe(III) redox boundary.

The results of high-resolution geochemical investigations at 1-cm intervals are shown in Fig. 3 and reveal a characteristic succession of the solid-phase peaks of Mn, Fe, V, Cd and U with depth across the active Fe redox boundary (Reitz, unpublished Diplom thesis). Along the fossil Fe redox boundary the same sequence of peaks is observed except for Mn. This depth succession of the en-richments of redox-sensitive elements is a characteristic result of nonsteady-state diagenesis and has also been observed by Thomson et al. (1996) in glacial/interglacial transition sediments of the northeast Atlantic Ocean. We suggest that - similar to the schematic representation shown in Fig. 2 – this depth arrangement of the peaks as well as their magnitude was caused by a sustained localization of the oxic/post-oxic boundary and the Fe redox boundary at particular levels in the sediment. The conditions which are likely to have induced these nonsteady-state diagenetic periods are shifts to lower organic carbon input and sedimentation rates as well as to higher bottom water oxygen contents at the onset of enhanced NADW production during interglacials.

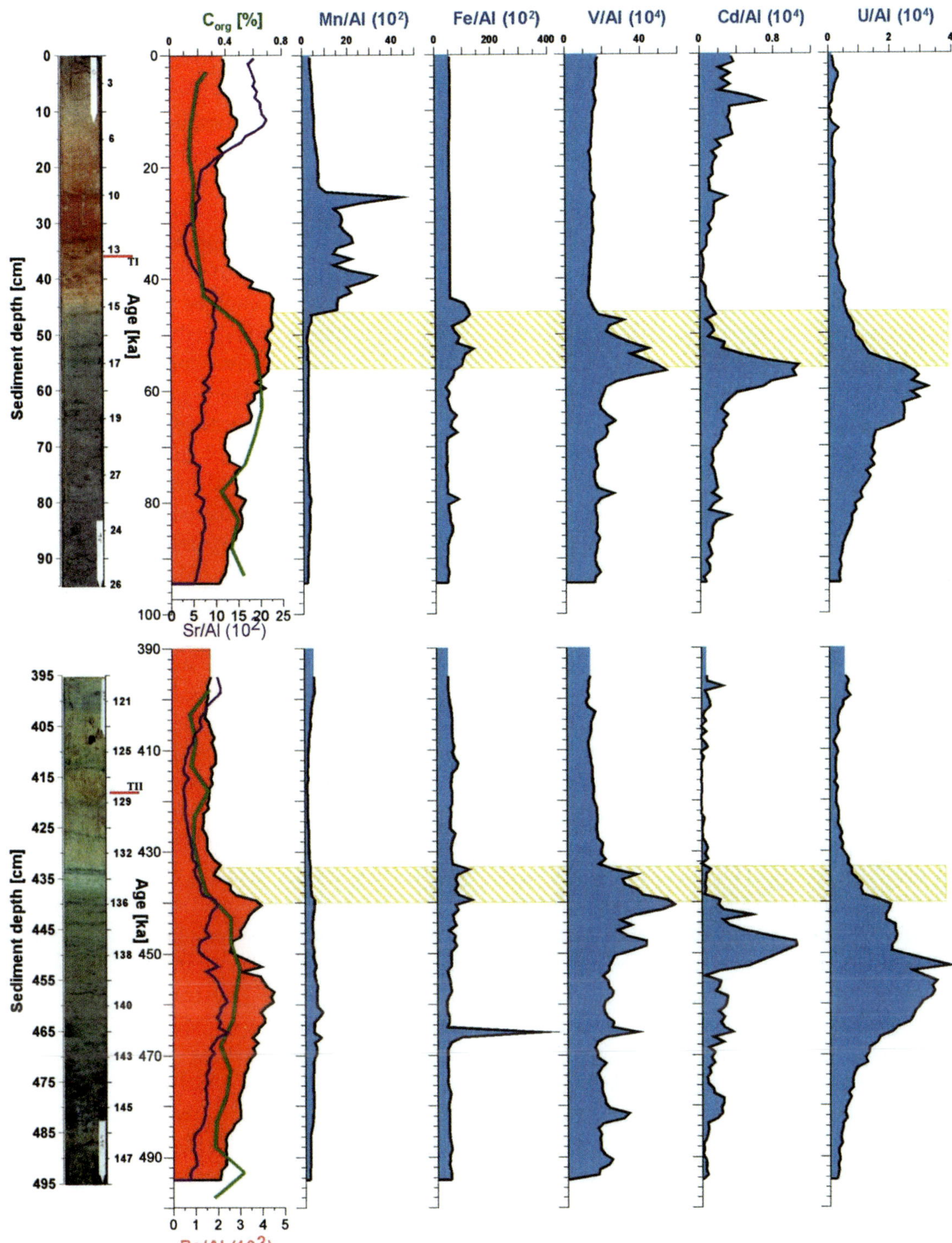

Fig. 3. Al-normalized solid-phase concentrations of Mn, Fe, V, Cd and U versus sediment depth along the active (above) and the fossil (below) Fe redox boundary in gravity core GeoB 2908-7 from the central equatorial Atlantic Ocean. On the left side photographs of the two core segments as well as organic carbon (C_{org}) and Ba/Al and Sr/Al data are shown. Organic carbon data from Funk (unpubl. data). All other data from Reitz (unpubl. Diplom thesis).

The concentration profile of TOC displays higher preserved amounts within glacial intervals and illustrates that the modern Fe redox boundary is trapped at the top of the youngest TOC-enriched sediment interval at about 46 cm depth. This fixation of the active Fe(II)/Fe(III) redox boundary at the upper boundary of the most recent TOC pulse is a ubiquitous phenomenon in the equatorial Atlantic (cf. Funk et al. this volume). Compared to this active Fe redox boundary the peaks of the elements which form authigenic enrichments in post-oxic sediments (V, Cd, U) are still in place along the fossil Fe redox boundary at about 433 cm as also found by Thomson et al. (1996). The Mn spike which was initially located at the fossil oxic/post-oxic boundary above Termination II has been mobilized by reduction during further burial and re-precipitated at the modern redoxcline in the form of a broad enrichment between 24 and 47 cm sediment depth. Although being similarly unstable under post-oxic and anoxic conditions, a relict of the Fe enrichment is still present at the fossil Fe redox boundary due to the slower reduction kinetics of Fe compared to Mn. The distinct Fe spike found at 468 cm depth is not coincident with the Fe(III) peak of the original boundary and is likely to represent an enrichment of authigenic iron sulfides as sulfur (not shown here) displays a pronounced peak at the same depth.

Close examination of the concentration profiles of TOC and solid-phase Ba in the vicinity of the two colour changes reveals a discrepancy between these two components immediately above the TOC-enriched sediment intervals. The relatively higher Ba contents at these depths suggest that there has not only been a fixation but a slight downward progression of the redox boundaries across the glacial/interglacial boundaries. This downward movement is likely to have caused an efficient post-depositional oxidation of the upper part of the TOC-enriched sediment interval leaving Ba as a relict. Based on numerous studies (e.g. Goldberg and Arrhenius 1958; Dehairs et al. 1980; Schmitz 1987; Bishop 1988; Gingele and Dahmke 1994) sedimentary barium – or more precisely biogenic barite - is assumed to be a persistent tracer of present and past productivity as long as pore water sulfate is not completely depleted due to sulfate reduction (e.g. Brumsack 1986; Von Breymann et al. 1992; Torres et al. 1996; Dickens 2001). Although the association between the fluxes of organic carbon and biogenic barite has been confirmed (see above), the exact mechanism of barite formation in sea-water is still a matter of debate (cf. Dehairs et al. 1980; Bishop 1988; Bernstein et al. 1992; Dymond et al. 1992; Francois et al. 1995; Gingele et al. 1999). However, the advantage of barite compared to organic carbon is that it is much better preserved within the sediment and not affected by post-depositional oxidation (e.g. Kasten et al. 2001).

Nonsteady-state diagenesis acting along these glacial terminations has not only led to the formation of distinct secondary element enrichments and the post-depositional oxidation of organic carbon, but has also caused a strong dissolution of magnetic Fe minerals below the Fe redox boundaries. The imprint on the magnetic mineral inventory is described in detail by Funk et al. (this volume) for sediment core GeoB 2908-7 as well as for numerous other sites in the equatorial Atlantic. The high-resolution geochemical and rock magnetic data demonstrate that the extent of diagenetic overprint of the primary sediment composition is proportional to the magnitude of change encountered in depositional and geochemical conditions. For this reason glacial/interglacial transitions, which represent time intervals of the most profound variations in environmental parameters are subject to the highest degree of diagenetic overprint in an undisturbed sedimentary sequence. With respect to the potential use of secondary signals as indicators for changes – or more precisely decreases - in organic carbon burial, those elements which form authi-genic enrichments in post-oxic sediment sections are most promising as they are stable and thus preserved during further burial into more reducing conditions.

Upwelling-Filament Area off Morocco, NW Africa

Site GeoB 4216 is located offshore the upwelling area off Morocco and strongly influenced by Trade

Wind driven filament activity (Freudenthal et al. 2002). The 1117 cm long gravity core GeoB 4216-1 (30°38'N, 12°24'W) was taken from a water depth of 2324 m during RV *Meteor* cruise M 37/1 (Wefer et al. 1997) and reaches into OIS 7. With an average value of 5 cm/kyr (Freudenthal et al. 2002), sedimentation rates at this site are slightly higher than at location GeoB 2908 in the central equatorial Atlantic. However – in accordance with site GeoB 2908 – sedimentation rates and pre-served TOC contents are higher during glacial times than during interglacials.

The records of solid phase Ba and TOC across the 2/1 and the 6/5 oxygen isotope stage boundaries of core GeoB 4216-1 (Jaquet, unpublished Diplom thesis) are depicted in Fig 4. As for site GeoB 2908 presented above, a disagreement between these two parameters is obvious above the glacial/interglacial boundaries whereas a striking positive correlation exists in the sediment interval below. These findings suggest that the TOC signal in these parts of the sediment has suffered from post-depositional degradation due to the action of downward-progressing oxidation fronts during deglacial nonsteady-state depositional conditions. The fact that the discrepancy between TOC and Ba is even more pronounced at site GeoB 4216 - with TOC contents exhibiting a sharp drop - suggests that the contrast in sedimentation rate and/or bottom-water oxygen contents between glacials and interglacials was more intense than at site GeoB 2908. The shape of the Ba profile (Fig. 4) indicates that, rather than increased productivity during glacial periods, maximum productivity seems to have occurred during the most pronounced climatic transitions, i.e. during deglacial periods. This hypothesis is supported by similar Ba peaks found at glacial terminations in sediments of the northwest African margin (Matthewson et al. 1995), the Ontong Java Plateau in the Pacific (Schwarz et al. 1996), the Portuguese margin (Thomson et al. 2000), the Sierra Leone Rise and Ceará Rise in the Atlantic (Kasten et al. 2001) and the upwelling and filament influenced area off Morocco (Moreno et al. 2002; Freudenthal et al. 2002).

In the high sedimentation regime of the Portu-guese margin (Thomson et al. 2000), coincident maxima of Ba, TOC, and diatom abundance are

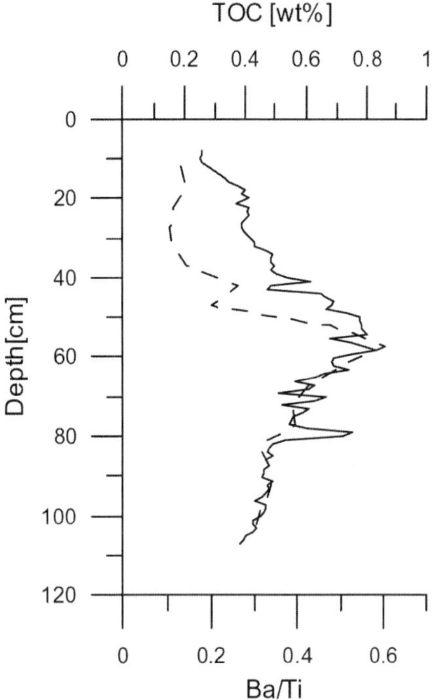

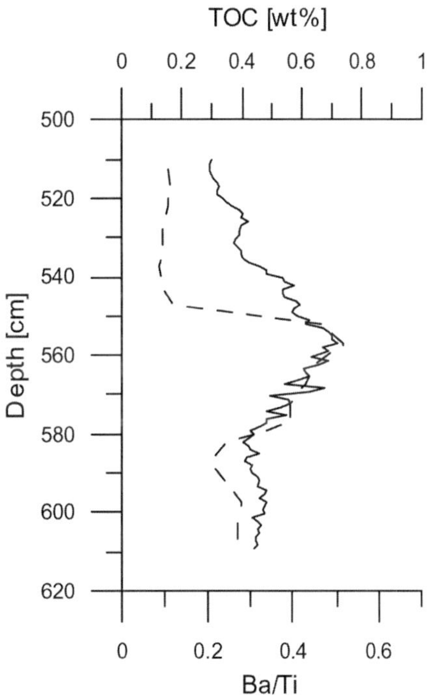

Fig. 4. Organic carbon contents (dashed lines) and solid-phase Ba/Ti ratios (solid lines) versus depth across the 2/1 and the 6/5 OIS boundaries in gravity core GeoB 4216-1 from the upwelling area off Morocco, NW Africa (Jaquet, unpubl. Diploma Thesis).

found at glacial terminations implying that the Ba peaks indeed formed in association with productivity events (see above). In contrast, no correlation between Ba and organic carbon is found in the deglacial sediments of the northwest African margin (Matthewson et al. 1995), the Ontong Java Plateau (Schwarz et al. 1996), or the Sierra Leone Rise and Ceará Rise (Kasten et al. 2001). These locations are all characterized by considerably lower sedimentation rates compared to the Portuguese margin. Sites GeoB 2908 and GeoB 4216 presented here as well as the study sites of Moreno et al. (2002) and Freudenthal et al. (2002) represent intermediate situations where only the upper part of the TOC-enriched intervals has been oxidized post-depositionally whereas a good correlation between Ba and TOC still exists in the lower un-oxidized section (cf. Figs. 3 and 4). The shape and arrangement of the TOC and Ba peaks along the glacial/interglacial boundaries at site GeoB 4216 (Fig. 4) show striking similarities to the depth distribution of these two components along active oxidation fronts at the most recent sapropel, S1, in eastern Mediterranean sediments (Thomson et al. 1995; Van Santvoort et al. 1996; Zonneveld et al. 2001) as well as in older partly oxidized sapropel intervals (e.g. Van Santvoort et al. 1997). This analogy is strong evidence for oxidation fronts to have been active along the 2/1 and the 6/5 stage boundaries in core GeoB 4216 as well (Fig. 4).

As has already been pointed out by Kasten et al. (2001), the degree or depth of "burn-down" – i.e. the post-depositional oxidation of TOC - along glacial/interglacial boundaries is controlled by the overall rate of sedimentation as well as by the magnitude of change in environmental and depositional conditions that occurred during these full climatic transitions. Using a modelling approach, Jung et al. (1997) demonstrated that sedimentation rate is the dominant factor which controls whether a sapropel is oxidized completely or only partly (also cf. Higgs et al. 1994; Van Santvoort et al. 1997). Jung et al. (1997) showed that preservation of elevated concentrations of non-refractory organic carbon is generally improbable when the sedimentation rate is lower than 1-2 cm/kyr. This value corresponds well to the value of about 2 to 2.5 cm/kyr on average for the sites on the Ceará and Sierra Leone Rise where no corrrelation between Ba and TOC at glacial/interglacial transitions has been found. At stations GeoB 2908 and GeoB 4216 which are characterized by slightly higher sedimentation rates only the upper part of the TOC-enriched interval has been affected by post-deposi-tional oxidation.

These studies illustrate the strong variability in diagenetic overprint and thus in degradation/preservation of organic carbon along glacial/interglacial transitions in dependence on (changes in) depositional and bottom water oxygen conditions. Besides the general constraints of TOC as a productivity proxy, they clearly demonstrate that using organic carbon as a productivity indicator within these transitional sediment intervals bears a particularly high risk of misleading interpretation. Various investigations further suggest that other organic sediment components sensible to oxygen might also be affected by burn-down phenomena (e.g. De Lange et al. 1994; Hoefs et al. 1998, 2002; Zonneveld et al. 2001; Sinninghe Damsté et al. 2002). Moreover, as an important secondary effect of aerobic degradation, the extent and species-selectivity of carbonate dissolution driven by the metabolic production of CO_2 during aerobic mineralization of organic matter (e.g. Wilson and Thomson 1998; Pfeifer et al. 2002; Volbers and Henrich 2002; Hensen et al. this volume) has to be considered too within these transitional sediment intervals.

Further detailed investigations concerning the various organic/biogenic sediment components are needed to unravel the extent and the controlling factors of post-depositional overprint along glacial/interglacial transitions. Considering the different degrees in preservation of biogenic components within a sedimentary sequence will help to separate preservation from productivity (to improve environmental reconstructions from sedimentary components) and to solve discrepancies between existing productivity reconstructions. A promising approach to tackle with the difficulties in discriminating between productivity and preservation has recently been presented by Versteegh and Zonneveld (2002) based on the differences in degradation rates of various organic sediment components.

Redistribution of Elements and Formation of Element Enrichments within the Zone of Anaerobic Oxidation of Methane

While numerous studies have shown the effects of reaction fronts acting around the oxic/post-oxic redox boundary only few investigations focus on the sulfate/methane transition zone. Nevertheless, the formation of a number of significant authigenic minerals, namely iron sulfides, barite and carbonates as well as the sulfurization of organic matter at and around this important biogeochemical boundary is well-known (cf. Fig. 1). The following examples from the equatorial and South Atlantic will illustrate the strong modification of the primary sediment composition by the biogeochemical processes acting within the zone of anaerobic methane oxidation. They will demonstrate that overprint of the sedimentary record by diagenetic processes can occur up to hundreds of thousands of years after initial deposition of the particulate material. Furthermore, they highlight the potential of these secondary signals to reconstruct variations in geochemical conditions and in the position of redox boundaries and reaction fronts over time. The nature and extent of diagenetic overprint is strongly influenced by the oceanographic, depositional and geochemical conditions at each particular site. In terrigenous-dominated settings Fe is usually available in high amounts and any hydrogen sulfide produced by anaerobic methane oxidation is immediately converted into Fe sulfides. In depositional environments characterized by high input of biogenic material, hydrogen sulfide is produced in excess of Fe and sulphurization of organic matter becomes the dominating process. As a consequence striking differences in the secondary signals formed as well as in the distribution patterns of trace elements within the sedimentary record evolve as we will show in the following.

Amazon Deep-Sea Fan – Iron-Dominated Setting

The studies by Kasten et al. (1998) and Adler et al. (2000) in sediments of the Amazon Fan can serve as valuable examples for the formation of massive authigenic iron sulfides within the zone of anaerobic methane oxidation under nonsteady-state conditions (Fig. 5). The sedimentary environment on the Amazon deep-sea fan is characterized by strong differences in depositional conditions between glacial and interglacial times. While during glacial sealevel lowstands Amazon river load is directly transported to the deep-sea fan via channels, sedimentation during interglacials is dominated by pelagic carbonaceous material because most of the Amazon-discharged terrigenous material is transported to the Northeast parallel to the coast (Flood et al. 1995). These differences in the sedimentary regimes also result in extreme contrasts in sediment accumulation rates. While glacial sedimentation rates can be as high as several meters per thousand years, interglacial rates average around only 3-4 cm per thousand years (Flood et al. 1995). Therefore, the Amazon Fan represents an extreme nonsteady-state setting with respect to depositional as well as diagenetic conditions. Furthermore, the sedimentary system of the fan is iron-dominated resulting from the high amounts of lateritic whethering products supplied by the Amazon River. As a consequence, although sulfate reduction is occurring at high rates within the zone of anaerobic methane oxidation, hydrogen sulfide is only detectable at micromolar concentrations in the pore water of these sediments.

The 7.1 m long gravity core GeoB 1514 (5°08.4'N, 46°34.6'W) which was recovered from a water depth of 3509 m on the Amazon Fan is characterized by a distinct enrichment of iron sulfides between 6.0 and 6.5 m sediment depth as can be seen from the solid phase concentrations of sulfur and iron (Fig. 5). This iron sulfide peak is located slightly below the current depth of the sulfate/methane transition around 5.3 m. As only minor amounts of free hydrogen sulfide were detectable in the pore water at this site, Kasten et al. (1998) assumed that all HS⁻ produced by anaerobic methane oxidation is immediately fixed in the sediment as iron sulfides. The solid phase sulfur contents within this pronounced peak therefore represent the integrated amount of deep sulfate reduction that occurred within this depth interval. Kasten et al. (1998) further concluded that the condition that caused the magnitude of this Fe sulfide enrichment was the pronounced decrease

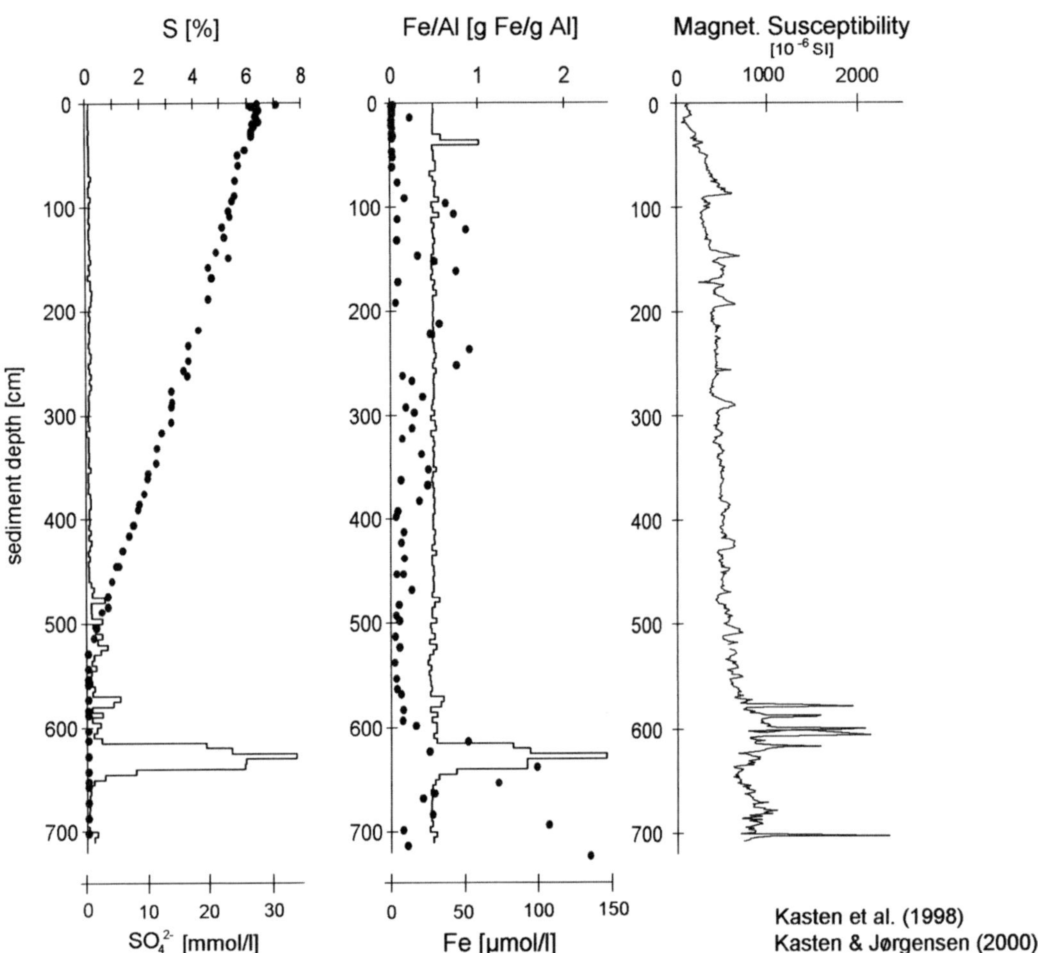

Fig. 5. Solid-phase profiles of total S and Al-normalized concentrations of Fe from total digestion (solid lines), magnetic susceptibiliy (J. Funk unpubl. data) as well as pore-water concentration profiles of SO$_4^{2-}$ and Fe^{2+} (solid circles) for core GeoB 1514-6 from the Amazon deep-sea fan. The SMT is currently located at a depth of 5.3 m. A strong enrichment of Fe sulfides is found between 6 and 6.5 m sediment depth. Please note that the iron sulfide maximum does not directly coincide with the peak in magnetic susceptibility. Modified from Kasten et al. (1998) and Kasten and Jørgensen (2000), pore water data taken from Schulz et al. (1994).

in sedimentation and organic carbon accumulation rates during the transition from the Pleistocene to the Holocene. While methane fluxes from below remained more or less constant, this strong decrease in sedimentation rate caused a fixation of the zone of anaerobic methane oxidation for a prolonged period of time, thus producing the high amount of authigenic Fe sulfides at this depth.

Besides the strong modification of the bulk sediment composition, nonsteady-state diagenesis within the zone of deep sulfate reduction has also generated a distinct secondary magnetic signal within the sedimentary record. A pronounced maximum in magnetic susceptibility is found slightly above the distinct Fe sulfide peak (Fig. 5). As the concentration profile of Fe sulfides does not display a local maximum at the depth of the suscepti- bility peak, Kasten et al. (1998) suggested the pres- ence of an iron sulfide mineral with a high mag- netic potential. X-ray diffraction analyses per- formed on samples from the depth of the suscepti- bility maximum indeed revealed the presence of the magnetic iron sulfide mineral, greigite (Fe_3S_4).

Upwelling Area off Namibia – Sulfide-Dominated System

Site GeoB 3718 is located south of Walvis Bay at the continental slope off Namibia (24°53.6'S, 13°09.8'E, water depth: 1312 m). Due to the lack of large rivers terrigenous input is of minor impor- tance at this site. Solid phase contents of Al and Fe are contributing 3-10 wt% in terms of Al_2O_3 and 1-5 wt% in terms of Fe_2O_3 (Heuer et al. 2003a). In contrast, primary productivity within the surface waters and organic matter export to the sediment

are high since trade winds cause the upwelling of nutrient-rich water masses. As a result, intense remineralization of C_{org} takes place in the sediment and methanogenesis becomes a significant proc- ess even at shallow sediment depths. At station GeoB 3718 organic carbon (C_{org}) makes up 4-11 wt% of the solid phase (Heuer et al. 2003a) and the pore water study of Niewöhner et al. (1998) has shown a distinct sulfate/methane transition at 6 meters below the sediment/water interface (Fig. 6).

The complete consumption of sulfate at this sulfate/methane transition has initiated a redistribu- tion of barite ($BaSO_4$) that can be observed in both the solid phase and the pore water of the 13.7 m long gravity core GeoB 3718-9 (Heuer et al. 2003a). In the solid phase a distinct enrichment of Ba occurs in the vicinity of the sulfate/methane transition at 6.0 mbsf (Fig. 6). Total solid phase contents of barium (Ba_{tot}) invariably exceed 750 ppm in the sediment

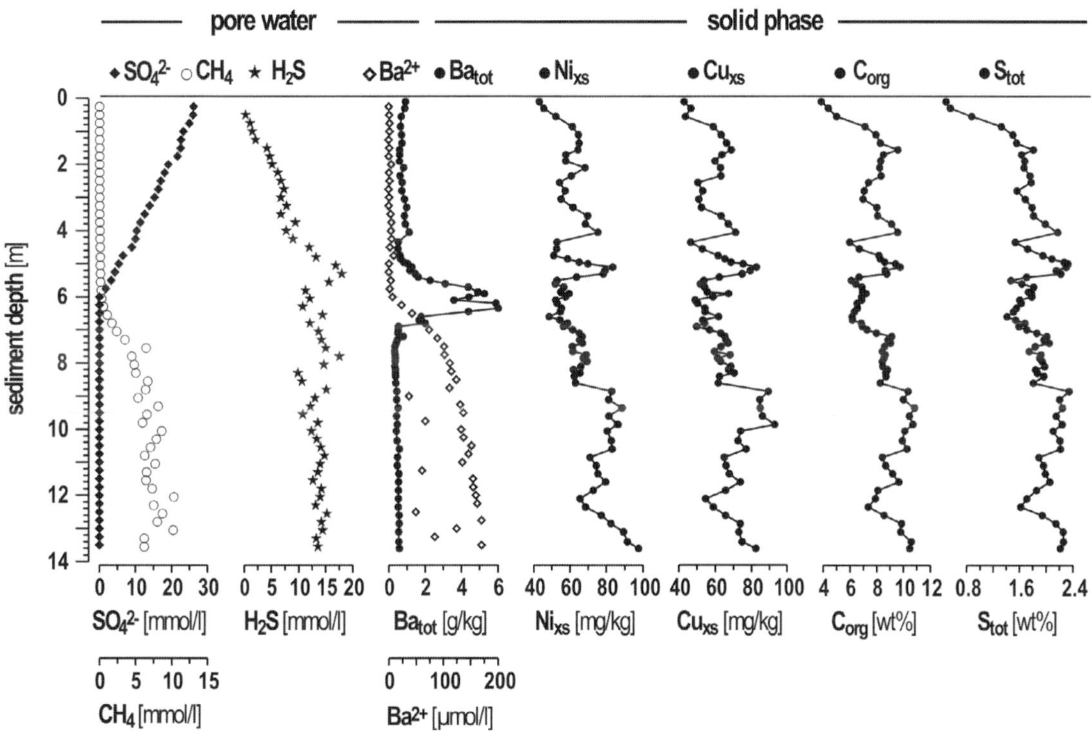

Fig. 6. Pore water (symbols) and solid phase (solid lines) data for gravity core GeoB 3718 from the continental slope off Namibia. The sulfate/methane transition is located at a sediment depth of 6 m. SO_4^{2-}, CH_4 and H_2S data according to Niewöhner et al. (1998). All other data taken from Heuer et al. (2003a).

section between 5.0 and 6.8 mbsf and peak Ba_{tot} concentrations amount up to 6020 ppm at 6.4 mbsf. In contrast, above and below the Ba enrichment Ba_{tot} contents average around 750 ppm and 470 ppm, respectively. Pore water data confirm the precipitation of Ba at the sulfate/methane transition (Fig. 6). While dissolved Ba^{2+} is negligible in the sulfate containing pore water throughout the upper 6.0 m of the sediment, the dissolution of barite has produced up to 170 µmol/l dissolved Ba^{2+} in the sulfate-depleted pore water of the underlying sediment section. The sulfate concentration profile indicates upward diffusion of dissolved Ba^{2+} from sulfate-depleted to sulfate rich pore water where the pore water concentration of Ba^{2+} becomes zero due to the precipitation of barite. The contact of Ba^{2+} and sulfate containing pore water coincides with the distinct enrichment of Ba_{tot} in the solid phase of the sediment.

The linear shape of the sulfate pore water profile indicates that at present consumption of sulfate by anaerobic methane oxidation at this site is at a steady state. Sulfate concentrations decrease (linearly) with depth and the diffusive fluxes of sulfate and methane (30.2 mmol*m^{-2}*a^{-1}) are balanced (Niewöhner et al. 1998). Moreover, the solid-phase Ba enrichment (Fig. 6) suggests that upward Ba^{2+} diffusion has already been exceeding downward barite burial for a considerable period of time. According to Dickens (2001) the precipitation of a diagenetic barite front forms a time-integrated signal of sulfate depletion and can be used to assess changes in the upward methane flux. This approach offers one particular advantage. Barite fronts record a time-integrated signal of sulfate depletion whereas the current pore water profiles of sulfate could have had a complex history.

At site GeoB 3718, the amount of Ba that has been enriched in the barite front totals 0.220 g cm^{-2}, or $1.61*10^{-3}$ moles of Ba for a 1 cm^2 vertical column of sediment (Heuer et al. 2003a). Assuming steady state conditions, the diffusive flux of Ba^{2+} across the sulfate/methane transition can be calculated from the concentration gradient of dissolved Ba^{2+} and Fick's first law of diffusion. With an observed concentration gradient of $75.4*10^{-10}$ mmol*m^{-3}*m^{-1}, the Ba^{2+} flux is $2.14*10^{-8}$ mmol* m^{-2}*s^{-1} or $6.74*10^{-8}$ mol*cm^{-2}*a^{-1}. A Ba^{2+} flux of

this magnitude would allow the build up of the observed Ba enrichment within 23,900 years – with a constant flux assumption. Consequently, the Ba enrichment at site GeoB 3718 suggests that the present methane flux and thus the depth of complete sulfate depletion have been similar for at least 23,900 years (Heuer et al. 2003a).

At the presented sulfate/methane transition, anaerobic oxidation of methane not only consumes all sulfate but also produces high concentrations of pore water sulfide (Fig. 6) (Niewöhner et al. 1998). Thereby methane oxidation provides the means for the sulfurization of organic matter (cf. Sinninghe Damsté and de Leeuw 1990). The reaction of organic matter with dissolved sulfide and polysulfides is of particular importance in sediments with high C_{org} contents and low reactive Fe contents that prevent the complete trapping of sulfide as iron sulfides (e.g. Mossmann et al. 1991; Werne et al. 2000). These conditions are also met at site GeoB 3718 where high amounts of preserved C_{org} occur along with low Fe_{tot} contents. In fact, a close correspondence between the total solid phase contents of sulfur (S_{tot}) and the distribution of C_{org} indicates that organic sulfur represents the dominant sulfur species in the sediment (Fig. 6) (Heuer et al. 2003a).

Furthermore, the sulfurization of organic matter seems to control the distribution of trace elements in the solid phase of gravity core GeoB 3718-9 as well (Heuer et al. 2003a). Based on a normative calculation which divides the total solid phase contents of an element into a lithogenic and an excess fraction, Ni and Cu turned out to have large excess fractions (Ni_{xs}, Cu_{xs}) that match both the solid phase contents of C_{org} and the total sulfur contents (S_{tot}) (Fig. 6). Heuer et al. (2003a) propose that the correspondence between these trace metals and both C_{org} and S_{tot} results from an early diagenetic reaction of Cu and Ni with secondary organo-sulfur compounds. In this context, the incorporation of Ni and Cu into organic sulfur compounds might proceed either via the trace metals' complexation with sulfur containing functional groups of low-molecular-weight organic sulfur compounds or via the incorporation of metal ions into polysulfide bridges of vulcanized organic macromolecules. Additional evidence for a secondary binding of Ni and Cu as well as Zn and Mo to organic sulfur compounds comes from further investigations at the continental slope off Namibia (Heuer et al. 2003b).

Niger Deep-Sea Fan

At the Niger deep-sea fan the current position of the sulfate/methane transition was not reached by the 20.3 m long gravity core GeoB 4901 (02°40.7'N, 06°43.2'E, water depth: 2184 m). Dissolved Mn indicates post-oxic conditions in the upper half of the core (0.1 - 12.5 m) while free hydrogen sulfide is present in pore water below 12.5 m sediment depth (Fig. 7). Nevertheless, a distinct enrichment of Ba occurs in the solid phase of the sediment at 12.7 m sediment depth and this enrichment might be a fossil diagenetic barite front

that formed in the past during a period of increased upward methane flux (Heuer et al. 2003c).

Support for this hypothesis comes from the distribution of trace elements in the solid phase of gravity core GeoB 4901 (Fig. 7) (Heuer et al. 2003c). Within the oxic and post-oxic zone (0-12.5 m), a good correlation is observed between biogenic Ba and the excess fractions of Ni, Cu, and Zn. The latter elements are known for their nutrient-type relation to productivity in the water column. Thus, their correlation with Ba in the solid phase of the sediment supports the reliability of biogenic Ba as a proxy for paleoproductivity. However, the solid phase contents of Ni, Cu, and Zn do neither match the Ba maximum at 12.5 m sediment depth nor the distribution of Ba in the underlying sulfidic sediment section. This discrepancy can be explained by a redistribution of barite at a fossil sulfate/methane transition that

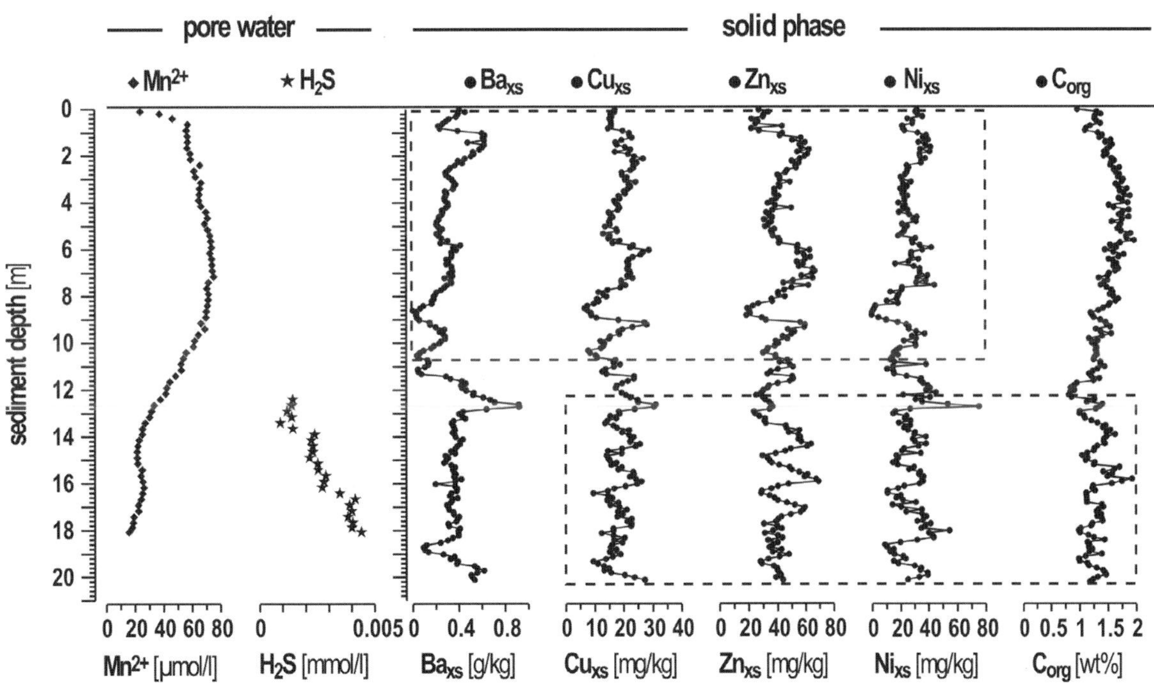

Fig. 7. Pore water (symbols) and solid phase (solid lines) data for gravity core GeoB 4901 from the Niger deep-sea fan. The current depth of the sulfate/methane transition was not penetrated by this core. Heuer et al. (2003c) propose that the SMT was located at a much shallower depth, namely around 12.5 m, in the past. Organic carbon data from Kolonic (unpubl. data). All other data taken from Heuer et al. (2003c).

used to be located at shallower sediment depth than today as proposed by Dickens (2001).

On the other hand, the changing relation between biogenic Ba and the excess fractions of Ni, Cu, and Zn might also result from a redistribution of the latter elements. In the upper 12.5 m of gravity core GeoB 4901, solid phase contents of Ni_{xs}, Cu_{xs}, and Zn_{xs} do not match the distribution of C_{org} since the degradation of organic matter has destroyed the primary signal. However, a close correspondence between Ni, Cu, Zn and C_{org} was identified under sulfidic conditions at greater sediment depth and supports the hypothesis of a secondary binding between trace metals and organic matter in the presence of sulfide (Heuer et al. 2003c).

On the whole, the upward flux of methane has considerable impact on the composition of the solid phase as it is the crucial factor controlling the depth position of the sulfate/methane transition. Within the SMT, diagenesis can cause a dramatic modification of the primary sediment composition several thousands of years after initial deposition of the sediment. In this way, processes within the SMT lead to a sort of delayed chemical log-in of various elements and minerals and ultimately determine which signals are preserved on longer time scales. Up to now, mostly changes in conditions at the sediment/water interface (C_{org} input, bottom water O_2 content, sedimentation rate) have been identified to initiate nonsteady-state diagenetic episodes. The studies presented above, however, clearly demonstrate that not only variations in these "influences from above" but also in the "influences from below" in the form of changing upward methane fluxes can trigger non-constant diagenetic situations. These variations in methane flux are driven by processes in the underlying sediment – namely by the proportion of methanogenesis and/or release from gas hydrates.

The extent and nature of diagenetic overprint within the SMT is dependent on the time the SMT is fixed at a particular sediment layer and the overall geochemical characteristics of the sedimentary setting. In Fe-dominated environments, Fe sulfide rich layers are strong indicators for the prolonged fixation of the zone of anaerobic methane oxidation and thus formation during nonsteady-state diagenetic intervals. Sulfurization of organic mat-

ter in sulfide-dominated settings enhances preservation of organic matter and produces a close association with a number of trace metals. Although these diagenetic processes also take place under steady-state conditions, changes in the association/correlation of trace elements with either Ba or organic carbon over depth may help to reconstruct the migration of reaction fronts – namely of the SMT – during nonsteady-state periods. Both Fe sulfide enrichments as well as sulfurized organic matter and associated trace metals have a high potential of being preserved in the sedimentary record during further burial below the depth of the SMT. Enrichments of authigenic barite formed within the SMT can be used as valuable tracers for higher methane fluxes in the past and a decrease in methane fluxes over time and thus a subsequent downward migration of the SMT (Dickens 2001; Heuer et al. 2003c). However, these secondary barites are only stable as long as they are not subject to sulfate-depleted pore water conditions – i.e. as long as they remain above the depth of the SMT.

Impact of Nonsteady-State Diagenesis on the Shape of Sulfate Pore Water Profiles

Nonsteady-state diagenesis can also significantly influence the shape of pore water profiles. Here we use interstitial sulfate to demonstrate that the consideration of possible nonsteady-state conditions is of utmost importance for the interpretation of pore water profiles.

Biogeochemical transfer reactions have an effect on pore water concentrations of involved reactants. Therefore, changes in pore water concentration gradients – i.e. deviations from the linear shape - are commonly used to identify specific reactions and where they take place (e.g. Schulz 2000). However, it has to be noted that reaction rates deduced from or modeled based on pore water concentration profiles always represent net rates because pore water profiles are the result of all primary and secondary reactions taking place. Gross reaction rates can be obtained using radiotracer techniques - e.g. ^{35}S for the determination of sulfate reduction rates (SRR) (e.g. Fossing and Jørgensen 1989). Comparisons of radiotracer-derived SRR with those calculated from pore water profiles often reveal that modeled rates underestimate the values obtained from direct rate measurements. The largest discrepancies between modeled and measured SRR occur close to the sediment surface where highest rates are determined by the $^{35}SO_4$ technique without any detectable gradient change in the sulfate profile (e.g. Fossing et al. 2000; Kasten and Jørgensen 2000). These discrepancies are attributed to the re-oxidation of reduced sulfur species formed during sulfate reduction (Kasten and Jørgensen 2000).

In contrast, the opposite also can be the case – i.e. that obvious changes in pore water gradients do not necessarily have to indicate a reaction but can be the result of a nonsteady-state condition. Borowski et al. (1999) have described the different shapes of sulfate pore water profiles in gas hydrate bearing sediments of the Carolina Rise and the Blake Ridge. They argued that steep, linear sulfate gradients and shallow depths of the sulfate/methane transition (SMT) are indicators of sites prone to gas hydrate occurrences below. Non-linear sulfate profiles with concave-down curvature were interpreted as background cases for low upward methane flux where a significant amount of sulfate is consumed by dissimilatory sulfate reduction – i.e. for the degradation of organic matter. Figure 8 shows examples of the five different types of sulfate profiles which have been examined in about 88 gravity cores recovered during 1989 and 2000 in the South Atlantic (Schulz et al. 1994; Haese et al. 1997; Kasten et al. 1998; Niewöhner et al. 1998; Kasten and Jørgensen 2000; Schulz 2000; Zabel and Schulz 2001; Hensen et al. 2003; Heuer 2003a, b, c). In general, the sulfate penetration depth is controlled by the upward flux of methane which is produced fermentatively or thermogenically in deeper parts of the sediment or supplied from decomposing gas hydrates. Within the sulfate/methane transition (SMT) methane is consumed by anaerobic oxidation of methane (AOM) with sulfate according to Eq. 1.

As the SO_4^{2-} concentration in bottom waters can be considered constant over time, the position of the SMT is controlled by the amount of dissolved methane and the depth of its release (Borowski et al. 1999). Therefore, both the accumulation and burial rates of reactive organic compounds are

important factors controlling the depth of the SMT. Shallow SO_4^{2-} gradients are generally found in sediments of oligotrophic areas characterized by low organic carbon contents while steep gradients dominate in organic-rich sediments. Apart from variations in the depth position of the SMT, many SO_4^{2-} profiles are characterized by distinct changes in concentration gradients – i.e. they deviate from the linear type of profile (Fig. 8). Various attempts to explain these profiles on the basis of steady-state conditions have failed so far, because either sulfate reduction rates did not occur at sufficiently high rates or the necessary electron acceptor (such as reactive ferric iron) was not available in adequate amounts (Haese et al. 1997; Fossing et al. 2000). However, all of these shapes can be explained by the assumption of nonsteady-state conditions in the pore water system.

To illustrate the effects of changing CH_4 fluxes on SO_4^{2-} pore water profiles we have modeled two simplified scenarios within a 15 m long sediment column by applying Fick's second law of diffusion. Besides diffusive transport (with coefficients D_s of 105 and 164 $cm^2\ yr^{-1}$ for SO_4^{2-} and CH_4, respectively) only the cooperative process of anaerobic methane oxidation at the SMT (Eq. 1) is taken into account. Porosity was assumed to decrease exponentially from 90% at the sediment surface to a constant value of 50% below 4 m. The concentration of SO_4^{2-} within the bottom water was kept constant at 27 mmol l^{-1}. The only variable parameter is thus the CH_4 concentration at the bottom of the model area. The initial situation for scenario A represents the steady-state when the CH_4 concentration at a depth of 15 m is 3 mmol l^{-1} (Fig. 9-A1). The slight gradient change of the SO_4^{2-} profile close to the sediment surface is caused by the given decrease in porosity (cf. Dickens 2001). Figures 9-A2 and 9-A3 show the development of the sulfate pore water profile when the CH_4 concentration at the lower boundary is set to 50 mmol l^{-1}. The SMT moves upward and reaches a new steady-state after about 7000 years. Even if such a long period of constant CH_4 flux seems to be unrealistic, the model run definitely demonstrates the effect of an increasing CH_4 release on the corresponding SO_4^{2-} pore water profile. As a result, transient shapes are of "concave-up"-type as depicted in Figure 9-C.

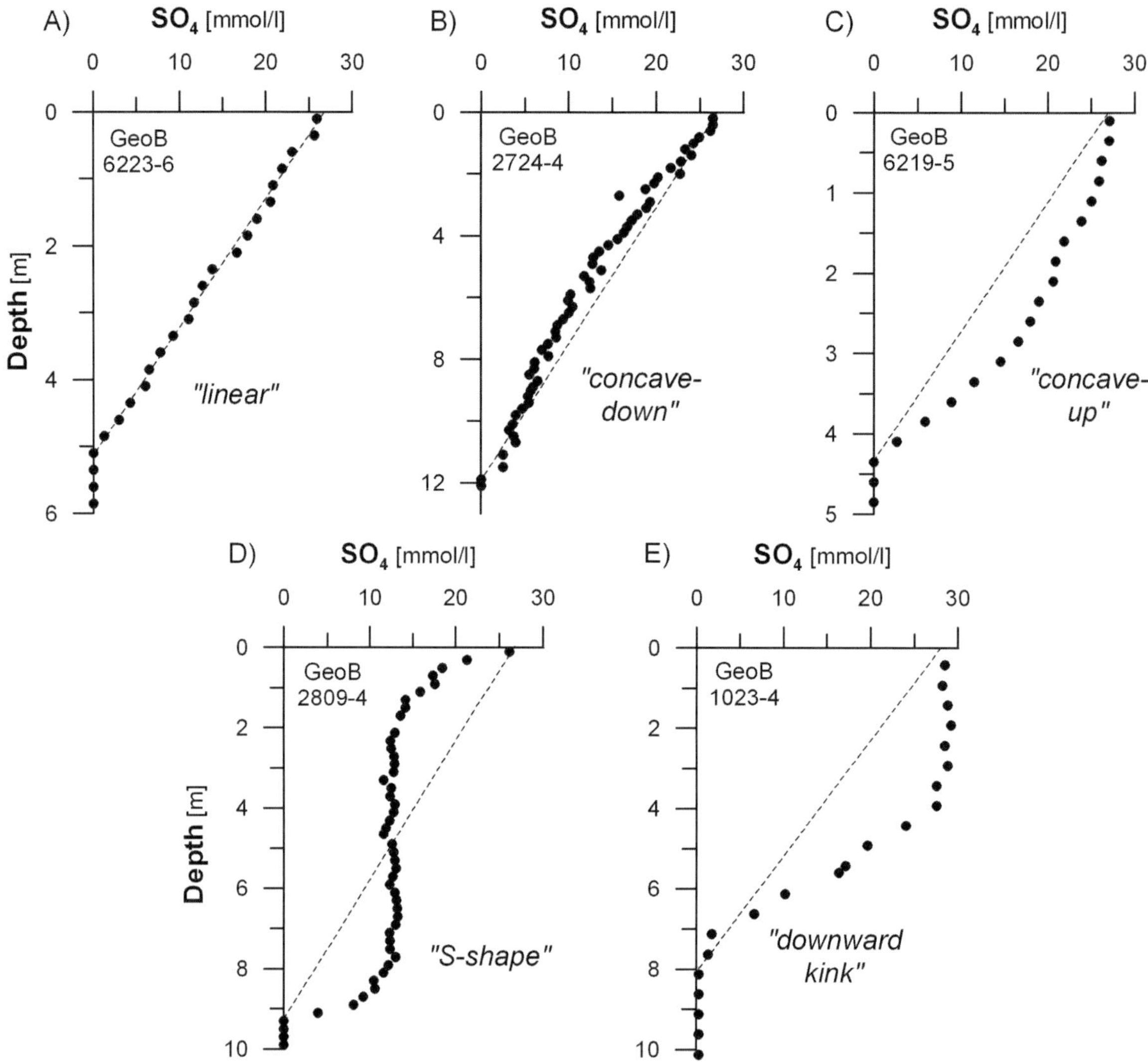

Fig. 8. Examples of the different shapes/types of sulfate pore water profiles as observed in South Atlantic sediments **(a-d)** GeoB 6223-6, GeoB 2724-4, GeoB 6219-5, GeoB 2809-4, all Argentine Basin, **e)** GeoB 1023-4, Angola Basin.

Scenario B (Fig. 9-B1 to B3) describes the opposite case when the CH$_4$ flux decreases over time. Starting with the steady-state situation (A3) this value was reduced to 3 mmol l^{-1} (Fig. 9-B1 to B3). During the subsequent downward movement of the SMT, the SO$_4^{2-}$ profile transiently shows a "concave-down"-shape (cf. Fig. 8-B). Sulfate reduction rates (SRR) derived from the flux rates within the SMT during the model runs (Fig. 10) are in the same range as those determined within the zone of deep sulfate reduction by means of the radio-tracer technique (e.g. Fossing et al. 2000). These values

are about one order of magnitude lower than those measured close to the sediment surface (Ferdelman et al. 1999).

As there is strong evidence that the release of methane in deep-sea sediments is not a continuous process and constant in space and time (e.g. Dickens et al. 1995; Norris and Röhl 1999) we suggest that nonsteady-state conditions in the sulfate/methane system are common and have to be considered when interpreting sulfate profiles. Our modeling approach demonstrates that the often quoted explanation for concave-down sulfate profiles – degrada-

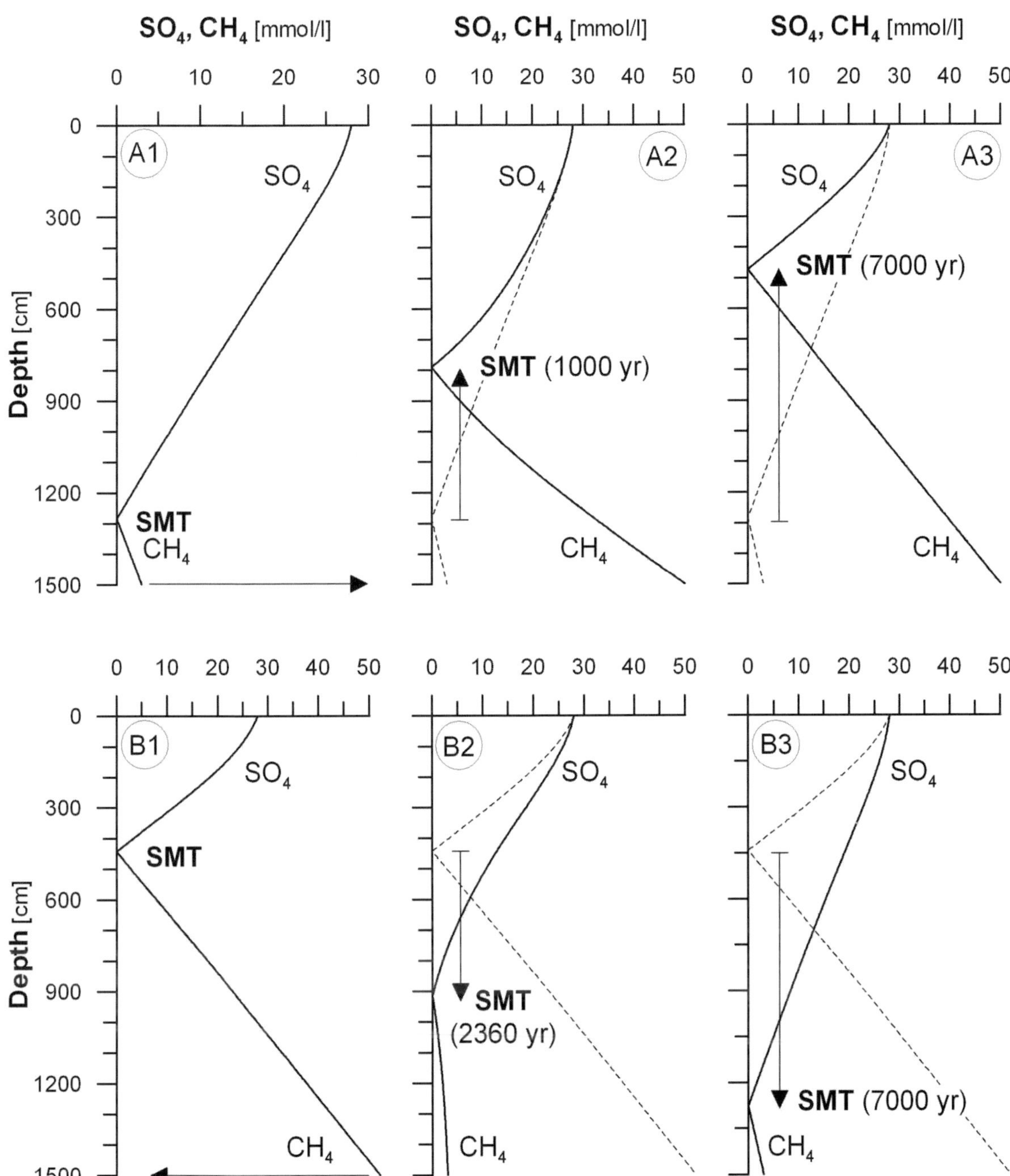

Fig. 9. Modelling of the effects of differing fluxes of methane from below on the shape of sulfate pore water profiles. **a)** Results of a model run showing the development of the sulfate profile when the CH$_4$ concentration at the lower boundary is increased from 3 to 50 mmol l^{-1}. Transiently, SO$_4^{2-}$ profiles are of concave-up type. **b)** Results of a model run where the CH$_4$ concentration was decreased from 50 to 3 mmol l^{-1}. This scenario temporarily produces SO$_4^{2-}$ profiles of concave-down shape.

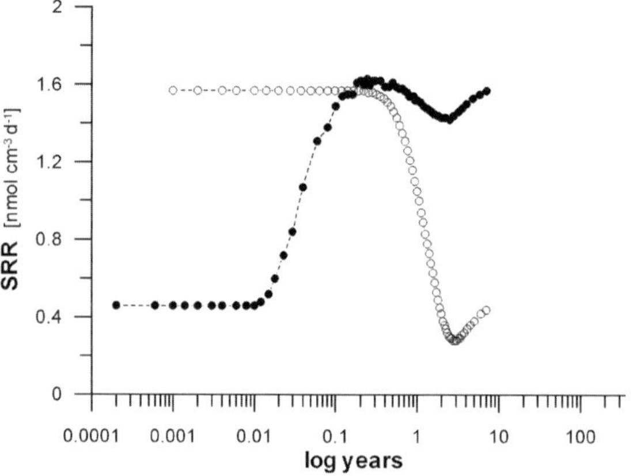

Fig. 10. Development of sulfate reduction rates (SRR) within the sulfate/methane transition (SMT) over time during the model runs A and B shown in Fig. 9. Closed and open symbols represent scenarios A (Figs. 9-A1 to A3) and B (Figs. 9-B1 to B3), respectively. SRR were calculated from the sulfate flux rates at the SMT.

tion of organic matter by dissimilatory sulfate reduction (e.g. Borowski et al. 1999) – does not always hold true, at least never in sediments that contain large amounts of methane. Nonsteady-state conditions seem to be the only conclusive explanation which is consistent with all available data, and underlines the assumption that SO_4^{2-} pore water profiles, showing similar non-linear shapes as described above, do not necessarily have to result from specific (bio)geochemical reactions or non-local pore water transport (cf. Dickens 2001).

Pore water profiles which are characterized by prominent changes (kinks) in the sulfate gradient/ profile (cf. Fig. 8 D) cannot be simulated by this approach. After Zabel and Schulz (2001) such profiles can originate from the displacement of submarine landslides. When water-saturated deposits are rapidly overthrusted by a coherent package of sediment, the pore water systems in both sections start to adapt to each other. Such drastic sedimentation processes do not inevitably set off (bio)geochemical reactions. This depends on several parameters like the difference in sediment compositions and/or of the geochemical environmental condition in both (lithological) units. In the case presented by Zabel and Schulz (2001), the authors give con-

clusive evidence for the sliding of a large 10 m thick coherent block of pelagic sediments down the continental slope, overthrusting on a turbidite sequence. Applying a simple transient model they could demonstrate that effects of submarine landslides on pore water systems are able to explain the observed shapes in SO_4^{2-} profiles (Fig. 11). Accordingly, such pore water profiles represent a transitional stage between two or more steady-state stages which are formerly established in the single sediment units. The period until the measured and modeled profiles correspond to each other may give an indication of the adaptation between sediment systems. For the example shown in Figure 11, Zabel and Schulz (2001) have estimated a period of 300 ± 25 yr.

A slightly different, but nevertheless similar explanation may apply to the rare "S-shape" pore water profiles (Fig. 8-E). Hydroacoustic and physical property data (magnetic susceptibility and density), solid phase analyses of iron speciation, and also AMS ^{14}C dating of an ikaite crystal from 9.4 m depth (26.770 yr) indicate that at site GeoB 2809 at least two slides or sedimentary events must have occurred to establish the present-day situation (Hensen et al. 2003). As a prerequisite for the intermediate and constant pore water SO_4^{2-} concentrations over depth immediately after deposition, complete mixing of the two sediment packages has to be assumed. Hensen et al. (2003) also applied a simple transport/reaction model to reproduce the S-shape of the sulfate profile at this location. Using this modelling approach an optimal fit for the measured data was obtained after 25 years. After about 500 years – and with a constant CH_4 flux from below - an almost linear profile would result.

As also shown for the kink-type, translocated sediment packages may carry their geochemical signature from their origin, thus the accumulation of different packages with originally different pore water signatures should be able to produce any type of sulfate pore water profile, only forced by the position of the SMT and diffusion. However, it is evident that diffusion is very effective in smoothing the concentration differences and that both, kink- and S-type, are obviously very short-lived phenomena.

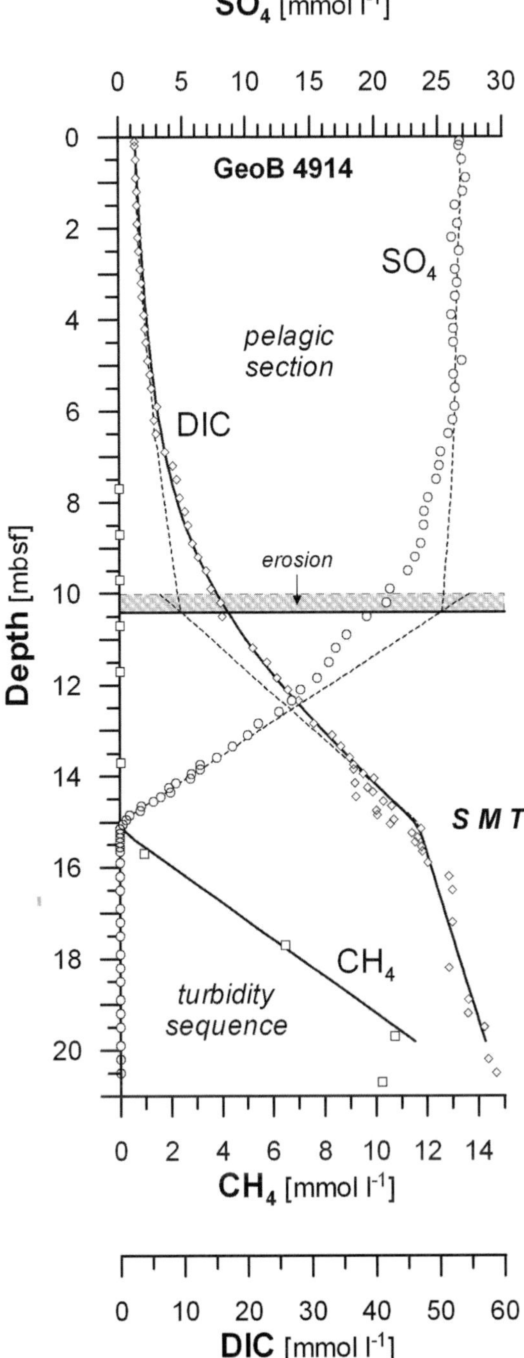

Fig. 11. Submarine landslide in sediments of the Southern Congo Fan (gravity core GeoB 4914). Results of a transient model as compared to measured data. Dashed lines: presumed starting situation representing steady-state conditions in both sediment segments; solid lines: model result after a period of 300 years; shaded depth interval: erosion caused by the overthrusting of the turbidity deposits by the pelagic sequence (after Zabel and Schulz 2001).

Conclusions and Future Perspectives

Periods of nonsteady-state diagenesis are common features in sediments of the equatorial and South Atlantic Ocean – even in slowly accumulating pelagic deposits and/or in sediments with overall low organic carbon contents. Sediment intervals affected by pronounced diagenetic overprint under nonsteady-state conditions often occur cyclically within the sedimentary record and are mostly associated with full glacial/interglacial transitions. At numerous sites these diagenetic processes were shown to be still active. These findings agree very well with studies and principals of nonsteady-state diagenesis elucidated elsewhere. As a consequence we conclude that the sediments over large areas of the entire Atlantic have not yet come to a new equlibrium with the changes that occurred at the glacial/Holocene transition.

The fact that strongest diagenetic alterations of the primary sediment composition are found across glacial/interglacial boundaries is consistent with glacial terminations being the times of the most profound changes in oceanographic, depositional and geochemical conditions. The resulting imbalance in the sediment/pore water system initiates a number of diagenetic phenomena - like fixation or downward progression of redox boundaries and reaction fronts - which have a high potential for modifying the primary sediment composition and producing distinct secondary signals in the solid-phase. The mechanisms that are likely to have triggered these very "effective" diagenetic processes are decreases in sedimentation rate and organic carbon input as well as increase in bottom water O$_2$ content due to enhanced NADW production at the onset of interglacial periods.

The strongest diagenetic modifications within a sedimentary redox succession take place in the interval across the oxic/post-oxic and the Fe(II)/Fe(III) redox boundary as well as within the zone of anaerobic methane oxidation. Across the oxic/post-oxic and Fe redox boundary the most important diagenetic processes are the post-depositional aerobic degradation of organic matter, dissolution of carbonate and ferric iron minerals as well as formation of distinct secondary enrichments of redox-sensitive elements. Of the latter authigenic enrichments, those elements which form their solid-phase peaks under post-oxic conditions – e.g. V, Cd, and U - can serve as valuable indicators for decreases in C_{org} burial over time and thus for periods of nonsteady-state diagenesis. The extent of post-depositional oxidation of organic matter in these transitional sediment intervals was shown to depend on the overall sedimentation rate and the magnitude of change encountered in the various depositional and geochemical factors. A sedimentation rate of about 2 cm/kyr was shown to be the critical value below which no significant amounts of non-refractory organic carbon are preserved. For this reason deglacial productivity pulses and elevated glacial TOC inputs which have been widespread in the equatorial Atlantic have been oxidized to various degrees by oxidation fronts across glacial terminations. In sediments of the Ceará Rise and the Sierra Leone Rise which are characterized by low sedimentation rates around 2 cm/kyr the initially present deglacial pulses of organic carbon have been completely burned-down - leaving only peaks of biogenic barium as relicts.

The influence of climatically induced changes in sedimentation rate and/or organic carbon burial is not restricted to the oxic/post-oxic and the Fe redox boundary, but well extends to deeper sediment sections. This effect has been illustrated for deposits of the Amazon deep-sea fan, where the strong decrease in sedimentation rate and organic carbon accumulation encountered during the Pleistocene/Holocene transition has caused a fixation of the zone of deep sulfate reduction for a considerable period of time. In this way nonsteady-state diagenesis has a high potential for significantly overprinting the sediment composition up to hundreds of thousands of years after initial deposition. The most relevant diagenetic/ geochemical processes within the zone of anaerobic methane oxidation comprise the formation of authigenic barites, carbonates and iron sulfides (most important in iron-dominated systems) as well as the sulfurization of organic matter (most important in organic matter rich, sulfide-dominated systems). Enrichments of authigenic barite formed slightly above the SMT can be used to indicate higher methane fluxes in the past and a decrease in methane fluxes over time and thus a subsequent downward migration of the

SMT. However, these barites are only stable under sulfate-containing pore water conditions.

Recent investigations in sediments of the Niger deep-sea fan and the upwelling area off Namibia have revealed that sulfurization of organic matter in sulfidic parts of the sediment also strongly affects the solid-phase distribution/association of trace metals. While a significant correlation between Ni, Cu, Zn and Ba exists in non-sulfidic sediments, the presence of free sulfide in pore water is likely to cause the incorporation of these trace elements into organic sulfur compounds producing a close correspondence with the amount of organic matter preserved. The association of these trace elements with either Ba or TOC might therefore be a valuable indicator for sulfidic and non-sulfic conditions and to trace the temporal migration of the sulfate/methane transition during nonsteady-state periods.

The long-term usability of all primary and secondary signals – also of those formed and preserved across the oxic/post-oxic and the Fe redox boundary - is ultimately controlled by the geochemical processes within and below the sulfate/methane transition. While dissolution of authigenic barite as well as productivity-related barite takes place during burial into sulfate-depleted sediment sections, iron sulfides as well as sulfurized organic matter and associated trace elements have a high potential to survive burial below the SMT.

Our studies further demonstrate that nonsteady-state situations can be induced not only by changes in conditions at the sediment/water interface like productivity, C_{org} burial or sedimentation rate but also by processes in the underlying sediment – namely the formation and/ or liberation of CH_4. Besides the above described modifications of the solid phase, variations in the upward flux of methane also have a considerable impact on the shape of sulfate pore water profiles. Modelling the effects of such variations in methane flux on sulfate profiles has demonstrated that considering possible nonsteady-state situations in the sediment/pore water system is of utmost importance for the interpretation of pore water data.

To conclude we emphasize that glacial terminations and sediment sections subject to anaerobic methane oxidation are those intervals which are and have been most severely altered by post-depositional diagenetic processes in an otherwise undisturbed sediment sequence. As this distinct overprint often also goes along with a particularly poor preservation of a number of proxies (e.g. TOC, carbonate, Ba) special care has to be taken when interpreting the records of these components within such sediment sections.

Further detailed and high-resolution studies on the fate of a variety of sediment constituents – e.g. calcareous and organic components - during non-steady-state diagenesis are required to improve environmental reconstructions from the various sedimentary components and to solve discrepancies between existing productivity reconstructions. Furthermore, the suite and controlling factors of authigenic mineral formation within the sulfate/methane transition – like e.g. iron sulfides - in natural systems is still a matter of debate. Similarly, the geochemical processes involved in the sulfurization of organic matter and the coupling of trace metals to the biogeochemical cycles of sulfur and carbon are not known in detail yet and await further investigation.

Acknowledgements

We would like to thank the captains and crews of the RV *Meteor* for their support during the numerous cruises. For technical assistance on board and in the home laboratory we are indebted to Sigrid Hinrichs, Karsten Enneking, Susanne Siemer and Silvana Hessler. This research was funded by the Deutsche Forschungsgemeinschaft (contribution no. 369 of Special Research Project SFB 261 at the University of Bremen). Data are available under www.pangaea.de/Projects/SFB261.

References

Adler M, Hensen C, Kasten S, Schulz HD (2000) Computer simulation of deep sulfate reduction in sediments of the Amazon Fan. Int J Earth Sci 88: 641-654

Adler M, Hensen C, Wenzhöfer F, Pfeifer K, Schulz HD (2001) Modeling of calcite dissolution by oxic respiration in supralysoclinal deep-sea sediments. Mar Geol 177: 167-189

Berner RA (1969) Migration of iron and sulfur within anaerobic sediments during early diagenesis. Am J Sci 267: 19-42

Berner RA (1981) A new geochemical classification of sedimentary environments. J Sediment Petrol 51: 259-365

Berner RA (1984) Sedimentary pyrite formation: An update. Geochim Cosmochim Acta 48: 605-615

Bernstein RE, Byrne RH, Betzer PR, Greco AM (1992) Morphologies and transformations of celestite in seawater: The role of acantharians in strontium and barium geochemistry. Geochim Cosmochim Acta 56: 3273-3279

Bishop JKB (1988) The barite-opal-organic carbon association in oceanic particulate matter. Nature 332: 341-343

Boetius A, Ravenschlag K, Schubert CJ, Rickert D, Widdel F, Gieseke A, Amann R, Jørgensen BB, Witte U, Pfannkuche O (2000) A marine microbial consortium apparently mediating anaerobic oxidation of methane. Nature 407: 623-626

Bohrmann G, Greinert J, Suess E, Torres M (1998) Authigenic carbonates from the Cascadia subduction zone and their relation to gas hydrate stability. Geology 26: 647-650

Borowski WS, Paull CK, Ussler III W (1999) Global and local variations of interstitial sulfate gradients in deep-water, continental margin sediments: Sensitivity to underlying methane and gas hydrate. Mar Geol 159: 131-154

Brumsack HJ (1986) The inorganic geochemistry of Cretaceous black shales (DSDP Leg 41) in comparison to modern upwelling sediments from the Gulf of California. In: Summerhayes CP, Shackleton NJ (eds) North Atlantic Palaeoceanography. Vol. 21, Blackwell Scientific Publications, Oxford, pp 447-462

Canfield DE, Thamdrup B, Hansen JW (1993) The anaerobic degradation of organic matter in Danish coastal sediments: Iron reduction, manganese reduction, and sulfate reduction. Geochim Cosmochim Acta 57: 3867-3883

Colley S, Thomson J, Wilson TRS, Higgs NC (1984) Post-depositional migration of elements during diagenesis in brown clay and turbidite sequences in the North East Atlantic. Geochim Cosmochim Acta 48: 1223-1235

Crusius J, Thomson J (2000) Comparative behavior of authigenic Re, U, and Mo during reoxidation and subsequent long-term burial in marine sediments. Geochim Cosmochim Acta 64: 2233-2242

Crusius J, Thomson J (2003) Mobility of authigenic rhenium, silver, and selenium during postdepositional oxidation in marine sediments. Geochim Cosmochim Acta 67: 265-273

Dehairs F, Chesselet R, Jedwab J (1980) Discrete suspended particles of barite and the barium cycle in the open ocean. Earth Planet Sci Lett 49: 528-550

De Lange GJ (1983) Geochemical evidence of a massive slide in the southern Norwegian Sea. Nature 305: 420-422

De Lange GJ, Van Os B, Pruysers PA, Middelburg JJ, Castradori D, Van Santvoort P, Müller PJ, Eggenkamp H, Prahl FG (1994) Possible early diagenetic alteration of palaeo proxies. In: Zahn R, Pedersen TF, Kaminski MA, Labeyrie L (eds) Carbon Cycling in the Glacial Ocean: Constraints on the Ocean's Role in Global Change. NATO ASI Series, pp 225-258

Dickens GR (2001) Sulfate profiles and barium fronts in sediment on the Blake Ridge: Present and past methane fluxes through a large gas hydrate reservoir. Geochim Cosmochim Acta 65: 529-543

Dickens GR, O'Neil JR, Rea DK, Owen RM (1995) Dissociation of oceanic methane hydrate as a cause of the carbon isotope excursion at the end of the Paleocene. Paleoceanography 10: 965-971

Dymond J, Suess E, Lyle M (1992) Barium in deep-sea sediment: A geochemical proxy for paleoproductivity. Paleoceanography 7: 163-181

Ferdelman TG, Fossing H, Neumann K, Schulz HD (1999) Sulfate reduction in surface sediments of the south-east Atlantic continental margin between 15°38'S and 27°57'S (Angola and Namibia). Limnol Oceanogr 44: 650-661

Finney BP, Lyle MW, Heath GR (1988) Sedimentation at Manop site H (eastern Equatorial Pacific) over the past 400,000 years: Climatically induced redox variations and their effects on transition metal cycling. Paleoceanography 3: 169-189

Flood RD, Piper DJW, Klaus A and cruise participants (1995) Proc. ODP, Init Repts 155, College Station, TX (Ocean Drilling Program).

Fossing H, Jørgensen BB (1989) Measurement of bacterial sulfate reduction in sediments: Evaluation of a single-step chromium reduction method. Biogeochemistry 8: 205-222

Fossing H, Ferdelman TG, Berg P (2000) Sulfate reduction and methane oxidation in continental margin sediments influenced by irrigation (South-East Atlantic off Namibia). Geochim Cosmochim Acta 64: 897-910

Freudenthal T, Meggers H, Henderiks J, Kuhlmann H, Moreno A, Wefer G (2002) Upwelling intensity and filament activity off Morocco during the last 250,000 years. Deep-Sea Res II 49: 3655-3674

Francois R, Honjo S, Manganini SJ, Ravizza GE (1995) Biogenic barium fluxes to the deep-sea: Implications for paleoproductivity reconstruction. Glob Bio-geochem Cycl 9: 289-303

Froelich PN, Klinkhammer GP, Bender ML, Luedtke NA, Heath GR, Cullen D, Dauphin P, Hammond D, Hartman B, Maynard V (1979) Early oxidation of organic matter in pelagic sediments of the eastern equatorial Atlantic: Suboxic diagenesis. Geochim Cosmochim Acta 43: 1075-1090

Gingele FX, Dahmke A (1994) Discrete barite particles and barium as tracers of paleoproductivity in South Atlantic sediments. Paleoceanography 9: 151-168

Gingele FX, Zabel M, Kasten S, Bonn WJ, Nürnberg CC (1999) Biogenic barium as a proxy for paleoproduc-tivity: Methods and limitations of application. In: Fischer G, Wefer G (eds) Use of Proxies in Paleoceanography: Examples from the South Atlantic. Springer, Berlin, pp 345-364

Goldberg ED, Arrhenius GOS (1958) Chemistry of Pacific pelagic sediments. Geochim Cosmochim Acta 13: 153-212

Haese RR, Wallmann K, Dahmke A, Kretzmann U, Müller PJ, Schulz HD (1997) Iron species determination to investigate early diagenetic reactivity in marine sediments. Geochim Cosmochim Acta 61: 63-72

Hensen C, Zabel M, Pfeifer K, Schwenk T, Kasten S, Riedinger N, Schulz HD, Boetius A (2003) Control of sulfate pore-water profiles by sedimentary events and the significance of anaerobic oxidation of methane for the burial of sulfur in marine sediments. Geochim Cosmochim Acta 67: 2631-2647

Heuer V, Kasten S, Hensen C, Schulz HD (2003a) Early diagenesis at the sulphate/methane transition: (re)distribution of barium and other trace elements in sediments on the continental slope off Namibia, Southeast Atlantic. Mar Geol, submitted

Heuer V, Kasten S, Schulz HD (2003b) Does sulphuriza-tion create an early diagenetic link between trace elements and organic matter? – Evidence from the upwelling region off Namibia, Southeast Atlantic. Geochim Cosmochim Acta, submitted

Heuer V, Kasten S, Schulz HD (2003c) Trace elements reflecting primary production, degradation and preservation of organic matter in sediments from the Niger deep-sea fan, equatorial Atlantic. Earth Panet Sci Lett, submitted

Higgs NC, Thomson J, Wilson TRS, Croudace IW (1994) Modification and complete removal of eastern Mediterranean sapropels by postdepositional oxidation. Geology 22: 423-426

Hoefs MJL, Versteegh GJM, Rijpstra WIC, De Leeuw JW, Sinninghe Damsté JS (1998) Postdepositional oxic degradation of alkenones: Implications from the measurement of palaeo sea surface temperatures. Palaeoceanography 13: 42-49

Hoefs MJL, Rijpstra WIC, Sinninghe Damsté JS (2002) The influence of oxic degradation on the sedimentary biomarker record I: Evidence from Madeira Abyssal Plain turbidites. Geochim Cosmochim Acta 66: 2719-2735

Huerta-Diaz MA, Morse JW (1992) Pyritization of trace metals in anoxic marine sediments. Geochim Cosmochim Acta 56: 2681-2702

Jakobsen R, Postma D (1999) Redox zoning, rates of sulfate reduction and interactions with Fe-reduction and methanogenesis in a shallow sandy aquifer, Rømø, Denmark. Geochim Cosmochim Acta 63: 137-151

Jørgensen BB (2000) Bacteria and marine biogeochemistry. In: Schulz HD, Zabel M (eds) Marine Geochemistry. Springer, Berlin, pp 173-208

Jung M, Ilmberger J, Mangini A, Emeis K-C (1997) Why some Mediterranean sapropels survived burndown (and others did not). Mar Geol 141: 51-60

Kasten S, Freudenthal T, Gingele FX, Schulz HD (1998) Simultaneous formation of iron-rich layers at different redox boundaries in sediments of the Amazon deep-sea fan. Geochim Cosmochim Acta 62: 2253-2264

Kasten S, Jørgensen BB (2000) Sulfate reduction in marine sediments. In: Schulz HD, Zabel M (eds) Marine Geochemistry. Springer, Berlin, pp 263-282

Kasten S, Haese RR, Zabel M, Rühlemann C, Schulz HD (2001) Barium peaks at glacial terminations in sediments of the equatorial Atlantic Ocean – relicts of deglacial productivity pulses? Chem Geol 175: 635-651

Matthewson AP, Shimmield GB, Kroon D (1995) A 300 kyr high-resolution aridity record of the North African continent. Paleoceanography 10: 677-692

Mercone D, Thomson J, Croudace W, Troelstra SR (1999) A coupled natural immobilisation mechanism for mercury and selenium in deep-sea sediments. Geochim Cosmochim Acta 63: 1481-1488

Mienert J, Posewang J, Baumann M (1998) Gas hydrates along the northeastern Atlantic margin: Possible hydrate-bound margin instabilities and possible release of methane. In: Henriet JP, Mienert J (eds) Gas Hydrates: Relevance to World Margin Stability and Climate Change. Geol Soc London Spec Publ 137, pp 275-291

Moreno A, Nave S, Kuhlmann H, Canals M, Targarona J, Freudenthal T, Abrantes F (2002) Productivity response in the North Canary Basin to climate changes during the last 250,000 yr: A multi-proxy approach. Earth Planet Sci Lett 196: 147-159

Morse JW, Arakaki T (1993) Adsorption and coprecipitation of divalent metals with mackinawite (FeS). Geochim Cosmochim Acta 57: 3635-3640

Mossmann JR, Aplin AC, Curtis CD, Coleman ML (1991) Geochemistry of inorganic and organic sulphur in organic-rich sediments from the Peru Margin. Geochim Cosmochim Acta 55: 3581-3595

Niewöhner C, Hensen C, Kasten S, Zabel M, Schulz HD (1998) Deep sulfate reduction completely mediated by anaerobic methane oxidation in sediments of the upwelling area off Namibia. Geochim Cosmochim Acta 62: 455-464

Norris RD, Röhl U (1999) Carbon cycling and chronology of climate warming during Palaeocene/Eocene transition. Nature 401: 775-778

Passier HF, Middelburg JJ, Van Os BJH, De Lange GJ (1996) Diagenetic pyritisation under eastern Mediterranean sapropels caused by downward sulphide diffusion. Geochim Cosmochim Acta 60: 751-763

Pfeifer K, Hensen C, Adler M, Wenzhöfer F, Weber B, Schulz HD (2002) Modeling of subsurface calcite dissolution - including the respiration and re-oxidation processes of marine sediments in the region of equatorial upwelling off Gabon. Geochim Cosmochim Acta 66: 4247-4259

Postma D, Jakobsen R (1996) Redox zonation: Equilibrium constraints on the Fe(III)/SO$_4$-reduction interface. Geochim Cosmochim Acta 60: 3169-3175

Pruysers PA, de Lange GJ, Middelburg JJ (1991) Geochemistry of eastern Mediterranean sediments: Primary sediment composition and diagenetic alterations. Mar Geol 100: 137-154

Pruysers PA, de Lange GJ, Middelburg JJ, Hydes DJ (1993) The diagenetic formation of metal-rich layers in sapropel-containing sediments in the eastern Mediterranean. Geochim Cosmochim Acta 57: 579-595

Robinson SG, Sahota JTS, Oldfield F (2000) Early diagen-esis in North Atlantic abyssal plain sediments characterized by rock-magnetic and geochemical indices. Mar Geol 163: 77-107

Schinzel U, Dahmke A, Schulz HD (1993) Reaktionen von Eisen(III)-Oxidhydraten während der

Frühdiagenese in marinen Sedimenten: Experimentelle Untersuchun-gen. Z Dt Geol Ges 144: 224-247

Schmitz B (1987) Barium, equatorial high productivity, and the northward wandering of the Indian continent. Paleoceanography 2: 63-77

Schouten S, van Driel GB, Sinninghe Damsté JS, de Leeuw JW (1994) Natural sulphurization of ketones and aldehydes: A key reaction in the formation of organic sulphur compounds. Geochim Cosmochim Acta 58: 511-5116

Schubert CJ, Nürnberg D, Scheele N, Pauer F, Kriews M (1997) ^{13}C isotope depletion in ikaite crystals: Evidence for methane release from the Siberian shelves? Geo-Mar Lett 17: 169-174

Schulz HD, Dahmke A, Schinzel U, Wallmann K, Zabel M (1994) Early diagenetic processes, fluxes and re-action rates in sediments of the South Atlantic. Geochim Cosmochim Acta 58: 2041-2060

Schulz HD, Bleil U, Henrich R, Segl M (1995) Geo Bremen South Atlantic 1994, Cruise No.29, 17 June – 5 September 1994. *Meteor*-Berichte, Universität Hamburg, 95-2, 323 p

Schulz HD (2000) Quantification of Early Diagenesis: Dissolved Constituents in Marine Pore Water. In: Schulz HD, Zabel M (eds) Marine Geochemistry. Springer, Berlin, pp 87-128

Schwarz B, Mangini A, Segl M (1996) Geochemistry of a piston core from the Ontong Java Plateau (western equatorial Pacific): Evidence for sediment redistribution and changes in paleoproductivity. Geol Rundsch 85: 536-545

Sinninghe Damsté JS, De Leeuw JW (1990) Analysis, structure and geochemical significance of organi-cally-bound sulphur in the geosphere: State of the art and future research. In: Durand B, Behar F (eds) Advances in Organic Geochemistry 1989. Org Geochem 16: 1077-1101

Sinninghe Damsté JS, Rijpstra WIC, Reichart G-J (2002) The influence of oxic degradation on the sedimentary biomarker record II: Evidence from Arabian Sea sediments. Geochim Cosmochim Acta 66: 2737-2754

Tarduno JA (1994) Temporal trends of magnetic disso-lution in the pelagic realm: Gauging paleoproduc-tivity? Earth Planet Sci Lett 123: 39-48

Tarduno JA, Wilkison SL (1996) Non-steady state mag-netic mineral reduction, chemical lock-in, and de-layed remanence acquisition in pelagic sediments. Earth Planet Sci Lett 144: 315-326

Thomson J, Wilson TRS, Culkin F, Hydes DJ (1984) Non-steady state diagenetic record in eastern equatorial Atlantic sediments. Earth Planet Sci Lett 71: 23-30

Thomson J, Wallace HE, Colley S, Toole J (1990) Authigenic uranium in Atlantic sediments of the last glacial stage – a diagenetic phenomenon. Earth Planet Sci Lett 98: 222-232

Thomson J, Higgs NC, Croudace IW, Colley S, Hydes DJ (1993) Redox zonation of elements at an oxic/post-oxic boundary in deep-sea sediments. Geochim Cosmochim Acta 57: 579-595

Thomson J, Higgs NC, Wilson TRS, Croudace IW, De Lange GJ, Van Santvoort PJM (1995) Redistribution and geochemical behaviour of redox-sensitive ele-ments around S1, the most recent eastern Medi-terranean sapropel. Geochim Cosmochim Acta 59: 3487-3501

Thomson J, Higgs NC, Colley S (1996) Diagenetic redistributions of redox-sensitive elements in north-east Atlantic glacial/interglacial transition sediments. Earth Planet Sci Lett 139: 365-377

Thomson J, Jarvis I, Green DRH, Green DA, Clayton T (1998) Mobility and immobility of redox-sensitive elements in deep-sea turbidites during shallow burial. Geochim Cosmochim Acta 62: 643-656

Thomson J, Nixon S, Summerhayes C, Rohling EJ, Schönfeld J, Zahn R, Grootes P, Abrantes F, Gaspar L, Vaqueiro S (2000) Enhanced productivity on the Iberian margin during glacial/interglacial transitions revealed by barium and diatoms. J Geol Soc London 157: 667-677

Torres ME, Brumsack HJ, Bohrmann G, Emeis KC (1996) Barite fronts in continental margin sediments: A new look at barium remobilization in the zone of sulfate reduction and formation of heavy barites in diagenetic fronts. Chem Geol 127: 125-139

Van Santvoort PJM, De Lange GJ, Thomson J, Cussen H, Wilson TRS, Krom MD, Ströhle K (1996) Active post-depositional oxidation of the most recent sapropel (S1) in sediments of the eastern Mediter-ranean Sea. Geochim Cosmochim Acta 60: 4007-4024

Van Santvoort PJM, De Lange GJ, Langereis CG, Dekkers MJ, Paterne M (1997) Geochemical and paleomag-netic evidence for the occurrence of "missing" sapropels in eastern Mediterranean sediments. Paleoceanography 12: 773-786

Versteegh GJM, Zonneveld KAF (2002) Use of selec-tive degradation to separate preservation from pro-ductivity. Geology 30: 615-618

Volbers ANA, Henrich R (2002) Present water mass cal-cium carbonate corrosiveness in the eastern South

Atlantic inferred from ultrastructural breakdown of Globigerina bulloides in surface sediments. Mar Geol 186: 471-486

Von Breymann MTK, Emeis KC, Suess E (1992) Water depth and diagenetic constraints on the use of barium as a paleoproductivity indicator. In: Summer-hayes CP (ed) Upwelling Systems: Evolution since the Early Miocene. Geol Soc Spec Publ 64: 273-284

Wallace HE, Thomson J, Wilson TRS, Weaver PPE, Higgs NC, Hydes DJ (1988) Active diagenetic formation of metal-rich layers in NE Atlantic sediments. Geochim Cosmochim Acta 52: 1557-1569

Wefer G and cruise participants (1997) Report and preliminary results of Meteor cruise M 37/1, Lisbon – Las Palmas, 04.12.1996 – 23.12.1996. Ber Fachber Geowiss, Univ Bremen 90, 79 p

Werne JP, Hollander DJ, Behrens A, Schaeffer P, Albrecht P, Sinninghe Damsté JS (2000) Timing of early diagenetic sulfurization of organic matter: A precursor-product relationship in Holocene sediments of the anoxic Cariaco Basin, Venezuela. Geochim Cosmochim Acta 64: 1741-1751

Wersin P, Höhener P, Giovanoli R, Stumm W (1991) Early diagenetic influences on iron transformations in a freshwater lake sediment. Chem Geol 90: 233-252

Wilson TRS, Thomson J (1998) Calcite dissolution accompanying early diagenesis in turbiditic deep ocean sediments. Geochim Cosmochim Acta 62: 2087-2096

Wilson TRS, Thomson J, Colley S, Hydes DJ, Higgs NC, Sørensen J (1985) Early organic diagenesis: The significance of progressive subsurface oxidation fronts in pelagic sediments. Geochim Cosmochim Acta 49: 811-822

Wilson TRS, Thomson J, Hydes DJ, Colley S, Culkin F, Sørensen J (1986) Oxidation fronts in pelagic sedi-ments: Diagenetic formation of metal-rich layers. Science 232: 972-975

Zabel M, Schulz HD (2001) Importance of submarine landslides for non-steady state conditions in pore water systems - lower Zaire (Congo) deep-sea fan. Mar Geol 176: 87-99

Zonneveld KAF, Versteegh GJM, De Lange GJ (2001) Palaeoproductivity and post-depositional aerobic organic matter decay reflected by dinoflagellate cyst assemblages of the Eastern Mediterranean S1 sapropel. Mar Geol 172: 181-195

Late Quaternary Sedimentation and Early Diagenesis in the Equatorial Atlantic Ocean: Patterns, Trends and Processes Deduced from Rock Magnetic and Geochemical Records

J.A. Funk[1*], T. von Dobeneck[1,2], T. Wagner[3] and S. Kasten[1]

[1]*Università Bremen, Fachbereich Geowissenschaften, Postfach 33 04 40,
D-28334 Bremen, Germany*
[2]*Paleomagnetic Laboratory 'Fort Hoofddijk', Faculty of Earth Sciences
Utrecht University, Budapestlaan 17, 3584 CD Utrecht, The Netherlands*
[3]*Woods Hole Oceanographic Institution, Marine Chemistry and Geochemistry Dept.,
Fye Laboratory (MS#4), 360 Woods Hole Rd, Woods Hole, MA 02543-1543, USA
* corresponding author (e-mail): funk@uni-bremen.de*

Abstract: This is an interdisciplinary and synoptic study of Equatorial Atlantic sediment formation in the Late Quaternary aimed at untangling the interlaced signatures of terrigenous and biogenous deposition and early diagenesis. It is based on a stratigraphic network of 16 gravity core records arranged along one meridional and three zonal transects (4°N, 0° and 4°S) crossing the Amazon and Sahara plumes as well as the Equatorial Divergence high productivity region. All newly introduced sediment sequences are collectively dated by their coherent $CaCO_3$ content profiles and two available $\delta^{18}O$ age models. To infer proxy records indicative of individual fluxes and processes, we analyze environmental magnetic parameters describing magnetite concentration, magnetic grain sizes and magnetic mineralogy along with $CaCO_3$, C_{org}, Fe, Mn, Ba and color data. Diagenetically affected layers are identified by a newly introduced Fe/κ index. Reach and climatic variability of the major regional sedimentation systems is delimited from lithological patterns and glacial/interglacial accumulation rate averages. The most prominent regional trends are the N-S decrease in terrigenous accumulation and the Equatorial Divergence high in glacial C_{org} accumulation, which decays much faster south- than northwards. Glacial enrichments in C_{org} and proportional depletions in $CaCO_3$ content appear to reflect sedimentary carbonate diagenesis more than lysoclinal oscillations and dominate temporal lithology changes. Suboxic iron mineral reduction is low at Ceará Rise and Sierra Leone Rise, but more intense on both flanks of the Mid-Atlantic ridge, where it occurs within organic rich layers deposited during oxygen isotope stages 6, 10 and 12, in particular at the terminations. To the equator, these zones reflect a full precessional rhythm with individual diagenesis peaks merging into broader magnetite-depleted zones. Rock magnetic and geochemical data show, that the depths of the Fe^{3+}/Fe^{2+} redox boundary in the Equatorial Atlantic are not indicative of average productivity and were frequently shifted in the past. They are now located just above the topmost preserved productivity pulse. At 4°N, this organically enriched layer coincides with glacial stage 6, at 0° with glacial stage 2. Subsequent oxic and suboxic degradation of organic material entails stratigraphically coincident carbonate and magnetite losses opening new analytical perspectives.

Introduction

The Equatorial Atlantic is an interesting and well established research area for the general discussion of marine sedimentation, land-ocean linkage and early diagenesis through Quaternary climates. Its model character results from pronounced west-east asymmetries with regard to transport pathways

From WEFER G, MULITZA S, RATMEYER V (eds), 2003, The South Atlantic in the Late Quaternary: Reconstruction of Material Budgets and Current Systems. Springer-Verlag Berlin Heidelberg New York Tokyo, pp 461-497

and flux rates of terrigenous matter, which has both lithogenic (e.g. Ratmeyer et al. 1999a, b) and organic (e.g. Wagner 2000; Dupont et al. 2000) components.

In its western part, terrigenous sedimentation is dominated by the massive discharge of the Amazon river (Showers and Angle 1986; Bleil and von Dobeneck this volume). Under modern interglacial conditions, much of the suspended load is deposited on the shelf or carried away by the northwest-ward North Brazil Current (NBC). The sea-level fall in glacial periods causes erosion of the inner continental shelf and channels the discharge directly into the Amazon Canyon (Milliman et al. 1975; Damuth 1977), intensifying mass-flow events on the slopes of the Amazon Fan (Maslin and Mikkelsen 1997). Terrigenous particle fluxes reach the Ceará Rise either by surface transport via a retroflection of the NBC into the North Equatorial Countercurrent (Johns et al. 1990) or by intensified nepheloid transport via the southeastward directed North Atlantic Deep Water (NADW) flow (Kumar and Embley 1977; François and Bacon 1991).

In its eastern part, supply and distribution of terrigenous matter is controlled by eolian dust transport from African source areas. Passat and Harmattan wind systems carry large amounts of dust to the deep-sea in response to glacial-interglacial variations in African aridity (deMenocal et al. 1993; Ruddiman 1997; Sarnthein et al. 1981; Tiedemann et al. 1989). Main glacial source areas for dust reaching the central and eastern Equatorial Atlantic were localized in the semi-desert Sahel Zone surrounding Lake Chad (Ruddiman and Janecek 1989; Bonifay and Giresse 1992). Various paleoclimatic studies have demonstrated that fluctuations in African aridity closely corresponded to changes in high latitude ice volume and North Atlantic sea surface temperatures (Sarnthein et al. 1982; Street-Perrott and Perrott 1990; Tiedemann et al. 1994; see review in Zabel and Wagner this volume). DeMenocal et al. (1993) proposed that African terrestrial climate responded most sensitively to high latitude sea surface temperatures and an intensification of the trade winds at low latitudes during peak glacial conditions while precessional forcing in the tropics was lowest. Both mechanisms are considered to promote stronger African aridity and enhanced dust transport in glacial periods.

Apart from the aspect of land-ocean linkage, the equatorial sector of the Atlantic Ocean plays a key role for the interhemispheric transfer of deep and surface water masses and the complex interchange of African and South American continental climate, which effectively forces the movement of the surface waters of the low latitude current system (e.g Pickard and Emery 1990; Wefer et al. 1996). It is through this wide gap between tropical Africa and South America that young, oxygen-rich NADW flows south into the South Atlantic and further to the Indian and Pacific Ocean, thus conveying oxygen to most major deep ocean basins (Broecker and Denton 1989). Above and below the NADW, Antarctic Intermediate Water (AAIW) and corrosive Antarctic Bottom Water (AABW) flow northwards through the equatorial region into the North Atlantic causing distinct gradients in deep ocean chemistry especially in the western Equatorial Atlantic (Curry 1996; Lutjeharms 1996; Rhein et al. 1996).

The interaction of atmospheric and oceanic circulation patterns controls organic and carbonate carbon production and preservation in the modern Equatorial Atlantic. These boundary conditions recurrently changed in response to orbitally forced Quaternary glacial-interglacial cycles (deMenocal et al. 1993; Duplessy et al. 1996; Raymo et al. 1997; Sarnthein et al. 1994; Verardo and McIntyre 1994) leaving characteristic features in the sedimentological, geochemical and micropaleontological deep-sea record of all low latitude oceans. The coupling of high latitude forced dust supply and productivity-driven organic sedimentation nourishes the assumption that a combination of mechanisms similar to those observed in the modern Arabian Sea and off NW-Africa had control on organic carbon accumulation below the Equatorial Divergence. The persistently high supply of clay-sized lithogenic dust (up to 85% of the bulk lithogenic matter at ODP Site 663 in the central Equatorial Atlantic according to Rath et al. (1999)) probably accelerated the interaction with newly formed labile organic matter.

Climate-related changes in oceanic circulation and productivity led to the cyclic formation of organic-rich layers (ORLs) in marine sediments and can usually be associated with orbital rhythms (Lyle 1988; Rossignol-Strick et al. 1982). ORLs are not only found in high-productivity regions and stagnant basins, but also in presently oligotrophic open oceans. Mesozoic to Quaternary Equatorial Atlantic sediment sections typically include horizons of darker color, which are relatively enriched in organic carbon (Wagner 2002). The Pleistocene to modern layers were formed during glaciations by enhanced organic carbon fluxes from more productive surface waters. The higher productivity of the glacial tropical Atlantic Ocean is attributed to increased oceanic and coastal upwelling as well as to a generally stronger mixing in response to stronger temperature gradients, winds, and currents (Müller et al. 1983). The accumulation of organic carbon was additionally favored by a slow-down of bottom water exchange with better preservation conditions for freshly deposited organic matter (Verardo and McIntyre 1994).

In interglacial periods, the formation of Antarctic bottom waters was reduced (Oppo and Fair-banks 1987) and more young, oxygen-rich NADW moved southwards to the equatorial Atlantic region (Boyle and Keigwin 1985) thereby enhancing organic carbon remineralization in the water column, and, by downward diffusion, also in the underlying suboxic glacial sediments (Kasten et al. 2001). Effects and signatures of such downward propagating oxidation fronts have also been observed at the tops of organic-rich turbidites (Wilson et al. 1985; Wilson et al. 1986) and were shown to significantly alter sapropel units in the eastern Mediterranean (Higgs et al. 1994; van Santvoort et al. 1996; Passier and Dekkers 2002).

Magnetic and geochemical sediment properties react sensitively and irreversibly to temporarily reducing conditions. In the absence of energetically more favorable electron acceptors, detrital magnetite is progressively reduced and subsequently transformed into authigenic iron sulfides (Karlin and Levi 1983; Karlin and Levi 1985; Canfield and Berner 1987; Karlin 1990). Organic-rich turbidite deposits (Robinson et al. 2000), paleooxidation fronts (Sahota et al. 1995), sapropels (Passier et al. 1998; van Santvoort et al. 1997), and pelagic sediment sequences (Tarduno 1994) have been characterized via the processes of reductive magnetite diagenesis. In a methodical companion paper, Funk et al. (this volume) describe a novel non-destructive method to detect the level of diagenetic magnetite depletion and enhancement by combining rock magnetic analytics with X-ray fluorescence spectrometry. Suboxic carbon diagenesis, however, not only alters $Mn^{(IV)}$ and $Fe^{(III)}$ minerals. In an earlier process, aerobic degradation of organic matter liberates CO_2 which in turn dissolves calcium carbonate (Martin and Sayles 1996; Schulte and Bard submitted).

This two-part synthesis addresses the physical mixing and chemical interaction of terrigenous and biogenous sediment fluxes in a temporal and regional context and investigates the distribution of ORLs from a paleoceanographic perspective. For this interdisciplinary interpretation geophysical and geochemical proxy records of Late Quaternary sediment sequences arranged in three zonal core profiles have been combined. The study region links the Ceará Rise in the west, the Sierra Leone Rise in the east, the Equatorial Divergence Zone in the center and the boundary regions to the oligotrophic subtropical gyre in the south.

In the first, sedimentologically oriented part, we take a synoptic view at glacial-interglacial patterns and trends of primary accumulation and secondary alteration of carbon, carbonate and ferric iron minerals. By exploiting the analogy of magnetite and calcite dissolution in the course of early diagenesis, we provide new insights into general and regional formation processes of Equatorial Atlantic sediment records.

In the second, geochemically oriented part, some specific signatures of early organic carbon and iron mineral diagenesis are identified with additional parameters and discussed in detail. We investigate the brown/green subsurface color transition and its varying depth position. A partial oxidation of the organic carbon profiles is detected by magnetic and geochemical proxy records. There is also clear evidence for magnetic mineral dissolution below and neoformation above modern and fossil iron redox boundaries.

Materials

This study combines data of 16 selected gravity cores retrieved in the Equatorial Atlantic on six RV *Meteor* cruises (M6/6, M9/4, M16/1, M16/2, M23/3, M29/3, M38/1) between February 1988 and April 1997, and one hydraulic piston core of ODP Site 663 recovered on Leg 108 (Fig. 1, Tab. 1). The core sites form three zonal transects at 4°N (profile A), 0° (profile B) and 4°S (profile C) and a NW-SE oriented 'quasi-meridional' cross transect (pro-file D) from 4°N to 15°S. From a total of 23 cores newly investigated for this study, nine were found to be of sufficient quality and interest to be presented here in combination with seven previously published cores.

Profile A (GeoB 1523-1, 1505-2, 4311-2, 4312-2, 4313-2, 4315-2, 4317-2, 2910-1) stretches from the Ceará Rise over the western and eastern flanks of the Mid-Atlantic Ridge to the Sierra Leone Rise. This zone is presently oligotrophic ('blue ocean') and receives high fluvial and eolian input from the Amazon River and the Sahara.

Profile B (GeoB 2906-1, 2908-7, 2215-10, 2909-2, ODP 663) follows the western and central parts of the Equatorial Divergence Zone, a mesotrophic ('green ocean') open oceanic up-welling belt. In its western part, this transect is concordant with the boreal winter Intertropical Convergence Zone (ITCZ) and the Southern mar-gin of the Sa-haran winter dust plume (Wagner 2000).

Profile C (GeoB 1117-2, 1112-4, 1041-3) is south of this line, but it equally transects southern and central parts of the Equatorial Divergence Zone.

Profile D (GeoB 4317-2, 2908-7, 1117-2, 1041-3, 1112-4) cuts from the center of the Saharan dust plume across the Equatorial Divergence Zone into the oligotrophic subtropical South Atlantic. It was assembled from representative cores of profiles A-

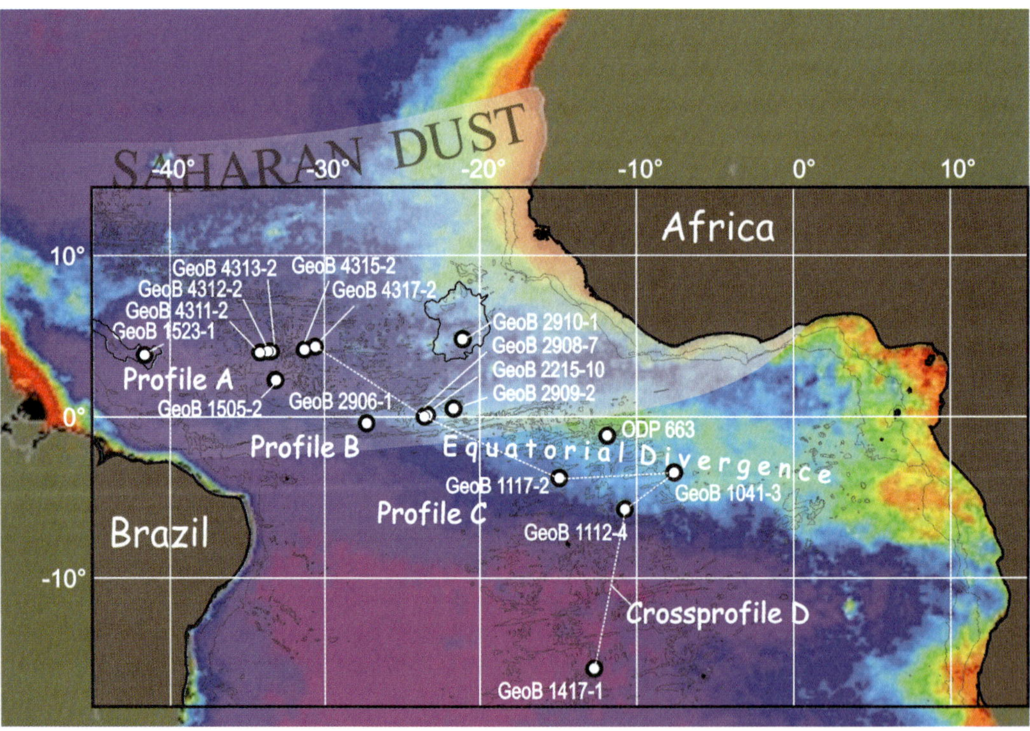

Fig. 1. The sampling sites of cores investigated (white dots) form three zonal transects at 4°N (profile A), 0° (profile B) and 4°S (profile C) and a NW-SE oriented cross transect (profile D, dotted line). The colored graph depicts primary productivity as visualization of the phytoplankton pigment concentration, with increasing values from lilac to red hues (http://seawifs.gsfc.nasa.gov, data collected over the time period November 1978 - June 1986). Saharan dust fall is indicated with a light gray shaded area.

C and extended by an additional RV *Meteor* gravity core from 15°S (GeoB 1417-1).

Profiles A and D form the basis for a statistical analysis of zonal and meridional trends in Equatorial Atlantic sedimentation and diagenesis. All cores of profiles A and B were newly investigated for this study, with the exception of cores GeoB 1523-1

(Rühlemann et al. 1996), 1505-2 (Zabel et al. 1999), 2910-1 (Haese et al. 1998) and ODP 663 (deMenocal et al. 1993). For core GeoB 4317-2 dry bulk densities were provided by P.J. Müller (unpublished, Universität Bremen). For the profiles C and D we refer to literature data published by Bickert (1992) and Thießen (1993) (GeoB 1117-2

Core/ Site	Latitude	Longitude	Core Recovery [m]	Water Depth [m]	Location	Research Cruise, Reference
Profile A						
GeoB 1523-1	3°49,9'N	41°37,3'W	6,65	3292	Ceará Rise	M16/2, [1]
GeoB 1505-2	2°16,0'N	33°00,9'W	8,50	3706	Western flank MAR	M16/2, [1]
GeoB 4311-2	3°59,6'N	34°08,1'W	9,11	4005	Western flank MAR	M38/1, [2]
GeoB 4312-2	4°02,8'N	33°35,6'W	6,01	3438	Western flank MAR	M38/1, [2]
GeoB 4313-2	4°02,8'N	33°26,3'W	8,98	3178	Western flank MAR	M38/1, [2]
GeoB 4315-2	4°10,0'N	31°17,0'W	9,38	3199	Eastern flank MAR	M38/1, [2]
GeoB 4317-2	4°21,3'N	30°36,2'W	12,04	3507	Eastern flank MAR	M38/1, [2]
GeoB 2910-1	4°50,7'N	21°03,2'W	13,91	2703	Sierra Leone Rise	M29/3
Profile B						
GeoB 2906-1	0°24,8'S	27°15,3'W	10,91	3897	Western flank MAR	M29/3
GeoB 2908-7	0°06,4'N	23°19,6'W	10,94	3809	Equatorial MAR	M29/3
GeoB 2215-10	0°00,4'S	23°29,7'W	9,04	3711	Equatorial MAR	M23/3, [3]
GeoB 2909-2	0°29,9'N	21°37,9'W	11,80	4386	Equatorial MAR	M29/3
ODP 663	1°11,9'S	11°52,7'W	153,5	3706	Eastern flank MAR	Leg 108, [4]
Profile C						
GeoB 1117-2	3°48.9'S	14°53.8'W	15,30	3984	Brazil Basin MAR	M9/4, [5]
GeoB 1112-4	5°46,7'S	10°45,0'W	6,91	3125	Guinea Basin MAR	M9/4, [5]
GeoB 1041-3	3°28.5'S	7°36.0'W	11,53	4033	Guinea Basin MAR	M6/6, [6]
Cross Profile D						
GeoB 4317-2	4°21,3'N	30°36,2'W	12,04	3507	Eastern flank MAR	M38/1, [2]
GeoB 2908-7	0°06,4'N	23°19,6'W	10,94	3809	Equatorial MAR	M29/3
GeoB 1117-2	3°48.9'S	14°53.8'W	15,30	3984	Brazil Basin MAR	M9/4, [5]
GeoB 1041-3	3°28.5'S	7°36.0'W	11,53	4033	Guinea Basin MAR	M6/6, [6]
GeoB 1112-4	5°46,7'S	10°45,0'W	6,91	3125	Guinea Basin MAR	M9/4, [5]
GeoB 1417-1	15°32,2'S	12°42,4'W	5,63	2845	MAR	M16/1, [7]

Table 1. List of gravity cores, sampling sites, recovered lengths, depth soundings, locations and cruises. Indices correspond to cruise reports by (1) Schulz et al. (1991), (2) Fischer et al. (1998), (3) Wefer et al. (1994), (4) deMenocal et al. (1993), (5) Wefer et al. (1989), (6) Wefer et al. (1988), (7) (Wefer et al. (1991).

and 1041-3) and Meinecke (1992) (GeoB 1112-4 and 1417-1). Core data and age models are available in the PANGAEA database (www.pangaea.de)

The cores were collected at water depths ranging between 4400 m and 2700 m (Tab. 1). Their records reach 300 to 500 kyr back, covering the full Late Quaternary. A specific chapter is devoted to the chronostratigraphy. The calcareous biogenic sediments consist mainly of foraminiferal and nannofossil oozes (Fischer et al. 1998), have only minor contents of opal (< 5 %) (Gingele and Dahmke 1994; deMenocal et al. 1993) and considerably differing portions of clastic and clayey material (total range 5-70 %). This broad lithological variability is reflected in vivid sediment color changes illustrated for two cores in Fig. 15.

Methods

Rock Magnetic Analyses

Magnetic volume susceptibility κ was measured on split-core archive halves with a *Bartington* MS2 susceptibility meter with MS2F-type spot sensor at 1 cm intervals. Magnetic volume susceptibility consists of ferri- (κ_{fer}), para- (κ_{para}) and diamagnetic (κ_{dia}) contributions. Under oxic marine conditions $\kappa \approx \kappa_{fer}$ is a linear measure of minerals of the (titano) magnetite and -maghemite series. The diamagnetic susceptibility κ_{dia} is a negative, almost constant background signal which was subtracted from κ using an average value of $-15 \cdot 10^{-6}$ for carbonate, quartz and water according to Schmieder et al. (2000). The resulting parameter is termed non-diamagnetic magnetic susceptibility (κ_{nd}) and is proportional to ferri- and paramagnetic mineral contents.

The combined magnetic/geochemical ratio Fe/κ_{nd} has been identified as a quantitative proxy parameter for reductive magnetite diagenesis (Funk et al. this volume) suitable for most open ocean settings. While for pristine sediments this ratio is typically at a constant, source-dependent value, the diagenetic depletion of ferric minerals leads to a marked decrease of susceptibility and increase of Fe/κ_{nd}. Using κ instead of κ_{nd} in this ratio can lead

to singularities in the case of nearly diamagnetic sediments ($\kappa \approx 0$).

By neglecting the influence of paramagnetic and antiferromagnetic components on susceptibility, the magnetite accumulation rate AR_{mag} can be estimated from the magnetic volume susceptibility κ_{nd} by using the equation of Schmieder et al. (2000):

$$ AR_{mag} = \frac{\varkappa_{nd} \cdot SR \cdot \varrho_{mag}}{\varkappa_{mag}} $$

with sedimentation rate SR, a magnetite volume susceptibility κ_{mag} of 3.1 and density ρ_{mag} of 5200 kg/m^3. Accumulation rates are based on linearly interpolated ages and therefore tend to create artificial variability at tie points.

Cores GeoB 4317-2 and GeoB 2908-7 were subsampled at 5 cm intervals for hysteresis measurements using a preparation technique described by von Dobeneck (1996). Hysteresis and backfield measurements to maximum fields of 300 mT were carried out on a *Princeton Measurement Corporation* M2900 alternating gradient force magnetometer. The program 'HYSTEAR' by von Dobeneck (1996) was used for data processing. It provides the ratio of non-ferrimagnetic susceptibility to total susceptibility χ_{nf}/χ_{tot}, a measure of the rela-tive contribution of paramagnetic and diamagnetic sediment constituents to total susceptibility. This hysteresis-based ratio is similar to the above des-cribed diagenesis proxy parameter Fe/κ_{nd} (Funk et al. this volume).

For rock magnetic remanence measurements cubic 6.2 cm^3 single samples were collected from split-core sections at 5 cm intervals. An anhysteretic remanent magnetization M_{ar} was imparted by superimposing a DC field of 0.04 mT over a decaying AC demagnetizing field of 250 mT. M_{ar} particularly quantifies magnetite of single-domain (SD; $0.03<d<0.1$ μm; Dunlop 1973) and pseudo-single-domain (PSD; 0.1 μm$<d<1$-10 μm; Parry 1965) size. An isothermal remanent magnetization M_{ir} was generated in a direct field of 250 mT, which primarily quantifies the total concentration of magnetite. The M_{ar}/M_{ir} ratio is therefore a proxy parameter of magnetite grain-size (Maher 1988). Both M_{ar} and M_{ir} were measured with a *2G*

Enterprises horizontal pass-through SQUID magnetometer.

Saturation isothermal remanent magnetization M_{sir} was acquired in an applied DC field of 2.5 T using a *2G Enterprises* pulse magnetizer. The incremental increase of magnetization from 300 mT to 2.5 T is referred to as M_{hir} and quantifies high coercive magnetic minerals, here essentially hematite. Subsequent to acquisition and measurement of M_{sir}, a back field magnetization M_{bf} at – 300 mT was created to calculate the $S_{-0.3T}$ ratio (Bloemendal et al. 1992) defined as

$$S_{-0.3T} = 0.5 \cdot \left(1 - M_{bf(-0.3T)}/M_{sir}\right) \cong M_{ir(0.3T)}/M_{sir}$$
$$= M_{ir(0.3T)}/\left(M_{ir(0.3T)} + M_{hir}\right)$$

This ratio varies between 0 and 1. A value of 1 represents pure magnetite, lower numbers denote increasingly larger magnetic influence of hematite. The conversion of $S_{-0.3T}$ into relative magnetite and hematite contents involves a highly non-linear and grain-size dependent transfer function.

The rock magnetic data of cores GeoB 1523-1, 1505-2 and 2909-2 were earlier measured using differing field settings and instruments for M_{ar} (0.05 mT direct field, 100 mT and 300 mT alternating fields) and S_{ir} (880 mT, 1000 mT)). Wherever possible, these literature data were converted to recent measurement conditions by using average recalibration factors derived from remanence acquisition curves. This procedure clearly improves the comparability of recent and earlier records, but should not be expected to yield totally consistent data.

X-ray Fluorescence Spectroscopy

Ca, Mn and Fe were determined by X-ray Fluorescence Spectroscopy (XRF) at a depth resolution of 1 cm on surfaces of archive core halves using a non-destructive XRF core logging system especially developed for marine sediments. The general method and its applications are described by Jansen et al. (1998) and Röhl and Abrams (2000). Resulting element data are relative concentrations given in counts per second [cps], but can be calibrated (Funk et al. this volume). Absolute concentrations of aluminum and barium of selected sediment layers were analyzed at 5 cm intervals using a wavelength dispersive *Philips* PW 1400 XRF spectrometer. Prior to analysis, the samples were disintegrated by ultrasonic treatment, dialyzed and washed to remove pore water, ball-milled and pressed.

$CaCO_3$ and TOC Analyses

Analyses of total carbon (TC) and total organic carbon (TOC) of bulk sediments in 5 cm intervals were carried out with a *LECO*-CS 300 infrared analyzer. 100 mg of freeze-dried and homogenized sediment were analyzed with respect to TC. Analytical accuracy was checked using a standard every 10 to 15 samples. For determination of TOC the same amount of sediment was acidified with 6% HCl and measured in the same way. Calcium carbonate content was calculated in weight percentage of the bulk sample by:

$$CaCO_3 \text{ wt.\%} = (TC \text{ wt.\%} - TOC \text{ wt.\%}) \cdot 8.33$$

Color Reflectance Spectroscopy

Color reflectance of split core surfaces was measured (Fischer et al. 1998) with a handheld *Minolta CM – 2002* spectrophotometer at 5 cm intervals and 31 wavelengths within the visible spectrum from 400 to 700 nm. Iron (oxyhydr)oxides like hematite and goethite are highly reflective around 670 nm. $CaCO_3$ as the major matrix component in the sediments is characterized by a high reflectivity in the range of 470 nm. According to Mix et al. (1992) a 670/470 nm ratio of >1.5 characterizes the oxidized zone in sediments.

Chronostratigraphy

The chronostratigraphies of the newly investigated cores of this study are based on the independently established age models of two reference cores, each representing one of the two profiles A and B (Fig. 1). These two cores were selected owing to their nearly continuous and undisturbed stratigraphic records; they also serve here for further detailed investigations. The age model of core GeoB 4317-2 (Profile A, Fig. 2g,h) was established by correlating its variations in $CaCO_3$ content to

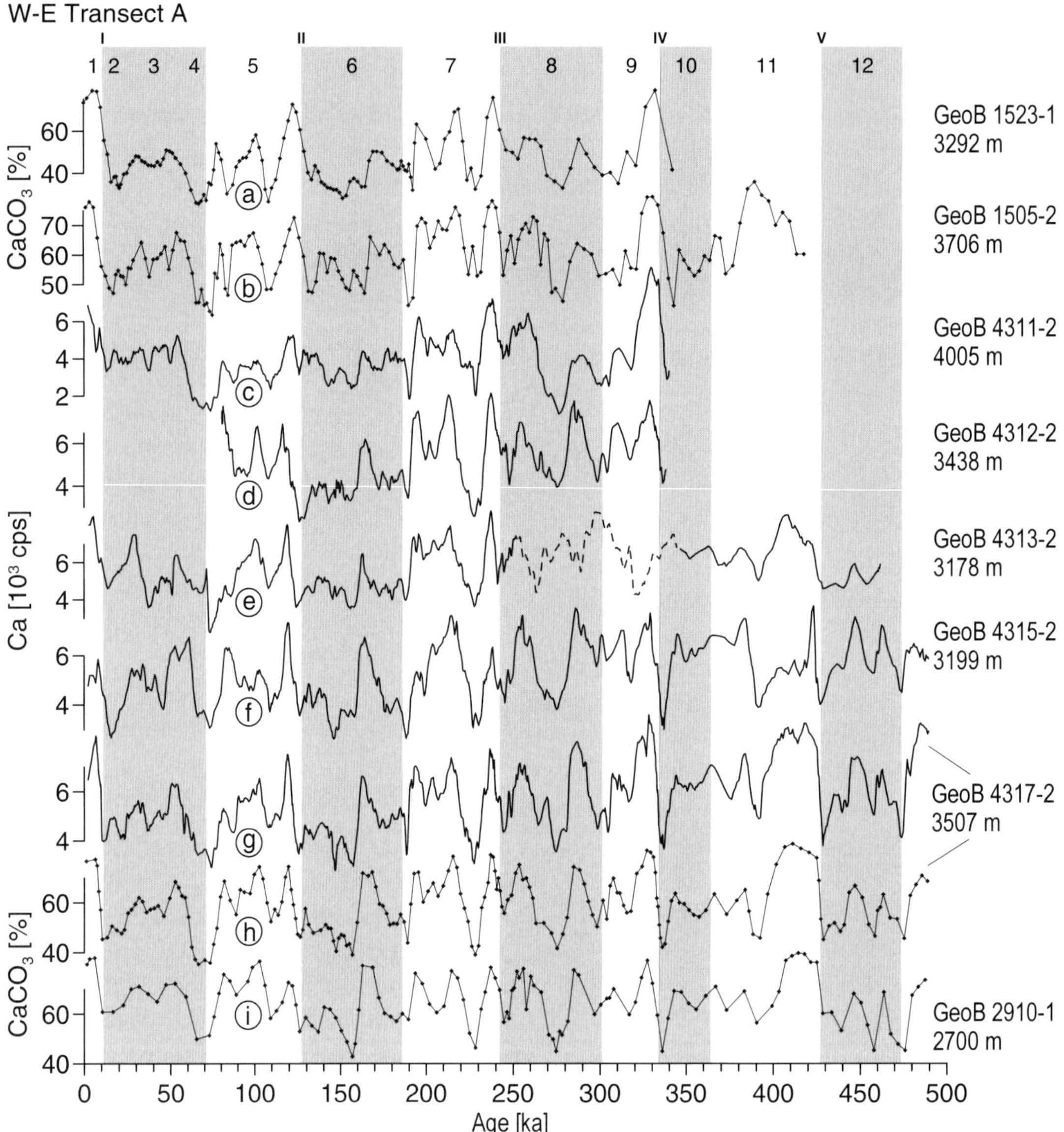

Fig. 2. Variations in CaCO$_3$ (diamonds) and Ca (plain lines) content of cores along profile A (4°N). The prevailing uniform signal pattern in the study area was used to correlate all records (for details and references see text). The δ^{18}O age model of ODP core 663 (Fig.3f,g) was first inherited to core GeoB 4317-2 and subsequently to all others. Ca (**g**) and CaCO$_3$ (**h**) records are clearly equivalent.

that of the δ^{18}O dated (deMenocal et al. 1993) core ODP 663 (Fig. 3f,g). In the same way, core GeoB 2908-7 (Profile B, Fig.3b,c) was correlated to the δ^{18}O dated (Bickert,1992) core GeoB 1117-2 (Profile C, Fig.3h,i). According to these age mo-dels, core GeoB 4317-2 spans the last 500 kyr and core GeoB 2908-7 the last 360 kyr.

The remaining cores of each transect were primarily dated by correlation of calcium (Ca)

records determined with a XRF core scanner in 1 cm intervals. In foraminiferal/nannofossil oozes, Ca profiles are representative of CaCO$_3$ content and allow likewise age modeling by graphic corre-lation, as demonstrated for cores GeoB 4317-2 (Fig. 2g,h) and GeoB 2908-7 (Fig. 3b,c). The ini-tial Ca signal was smoothed by a three-point mov-ing average.

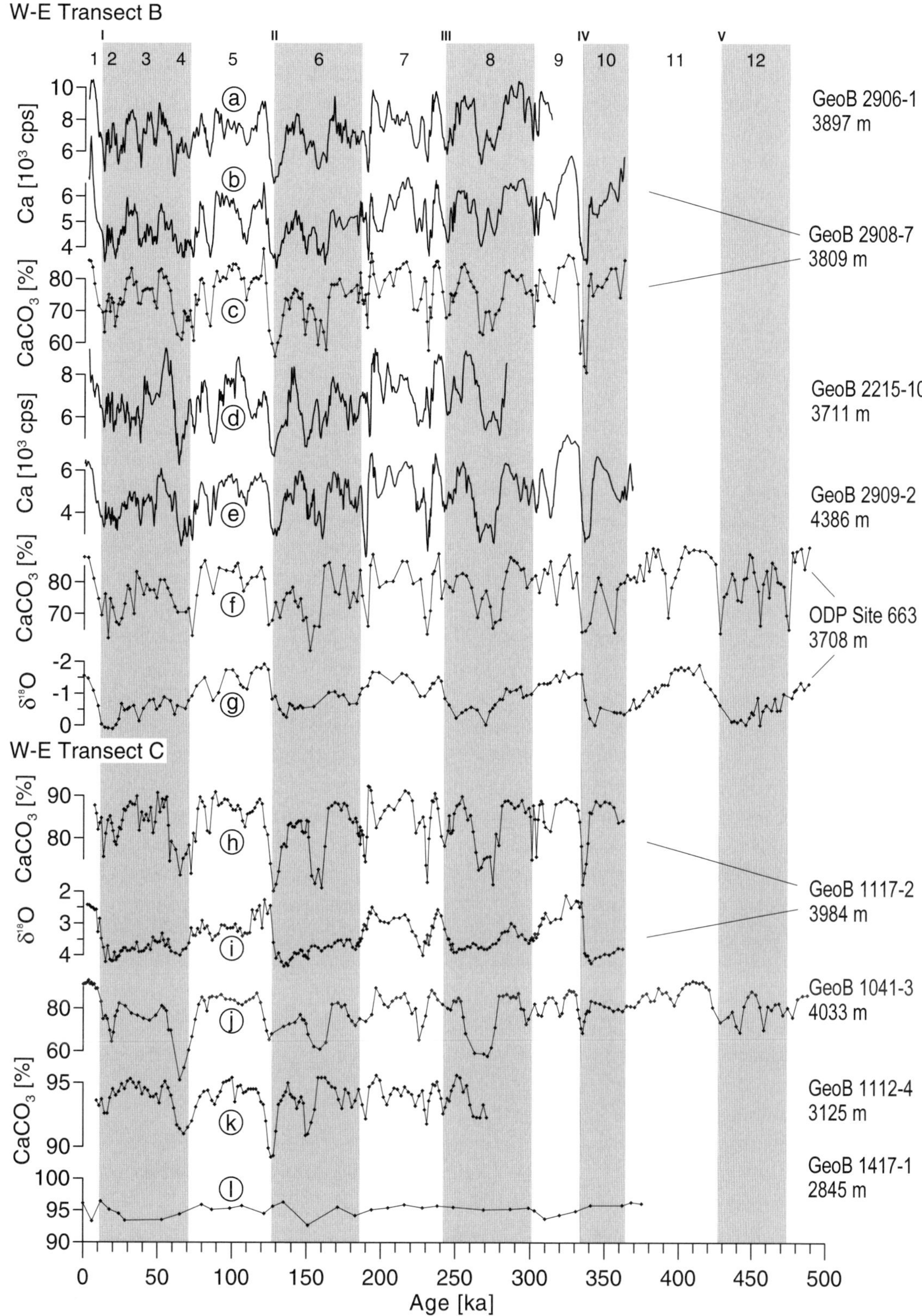

Fig. 3. Same procedure as for profile A has been applied for profile B. Here, core GeoB 2908-7 act as a 'master core' dated by correlation with GeoB 1117-2. Likewise presented are the variations in Ca **b)** and CaCO₃ **c)** content. The remaining cores of profile C and D were dated by former investigators (please see text).

By paralleling the sediment sequences it became obvious that the uppermost section of core GeoB 4312-2 was lost, possibly during coring (Fig. 2d). A strongly disturbed and diagenetically affected layer in core GeoB 4313-2 (Fig. 2e, dotted line) interrupts the otherwise characteristic pattern over 246 cm. Since no age control was available within this layer, the depth-age relation was assumed linear between top and bottom of this section.

The variations in $CaCO_3$ and Ca contents were compared with earlier dated carbonate records from the Equatorial Atlantic (GeoB 1523-1, 1505-2, 2910-1, 1041-3, 1112-4 and 1417-1, see Figs. 2, 3). The close match of the signal patterns down to finest details is striking. The good comparability of individual signal features within a large set of records made a convincing case for a minor readjustment of the earlier age model of core GeoB 2910-1 (Zabel et al. 1999), which resulted in an appreciable smoothing of the sedimentation rate in MIS 9. The now existing carbonate based stratigraphic network of the Late Quaternary Equatorial Atlantic Ocean forms a solid fundament for future work.

While the obvious similarities of the shown $CaCO_3$ signals are primordial for chronostratigraphic purposes, their equally obvious dissimilarities form a promising basis to analyze signal formation mechanisms in a regional and temporal framework. The potential and most likely contributing factors are:

• Variations in primary productivity due to changing nutrient supply to the photic zone. The glacial weakening of the northwesterly North African summer monsoons strengthens the south-westerly trade winds (Kutzbach 1981) causing a rise of the Equatorial Atlantic thermo- and nutricline in the east and a synchronous fall in the west, together with more vigorous mixing and upwelling (Müller et al. 1983).

• Variations in terrigenous dilution of carbonate by enhanced glacial dust flux from the Sahara and Sahel Zone. This material is eroded under conditions of greater aridity and carried by stronger glacial NE trade winds into the eastern Equatorial Atlantic (Sarnthein et al. 1982; deMenocal et al. 1993).

• Variations in terrigenous dilution of carbonate by an enhanced discharge of Amazon detritus into the western Equatorial Atlantic during glacial sea level falls and low stands. At high stands, Amazonian sediments are mainly deposited on the continental shelf (Damuth 1977; François et al. 1990; Bleil and von Dobeneck this volume).

• Variations in carbonate preservation in the water column due to weaker overturning of glacial NADW resulting in a vertical expansion of AABW entailing a rise of the calcite lysocline by several hundred meters (Gardner and Burkle 1975; Boyle and Keigwin 1982; Raymo et al. 1990; Bickert and Wefer 1996). Curry and Lohmann (1986, 1990) demonstrated, that Quaternary carbonates deposited at the Sierra Leone Rise above 3750 m are generally well preserved, while Rühlemann et al. (1996) associate fractionation of foraminifera at 3300 m depth during extreme glacial stages with influence of southern-source deep water.

• Variations in carbonate preservation due to carbonate dissolution by metabolic acids released in the course of aerobic OM decomposition in near-surface sediment layers (Martin and Sayles 1996; Pfeifer et al. 2002; Hensen et al. this volume). This process is particularly active in times of enhanced ocean fertility with enhanced $C_{org}/CaCO_3$ rain ratios. According to Martin and Sayles (1996), the modern sedimentary calcite dissolution rate at the presently oligotrophic and supralysoclinal Ceará Rise amounts to as much as 40-60% of the initially deposited $CaCO_3$, although this estimate should be handled with care (see review on carbonate dissolution in the South Atlantic by Henrich et al. this volume).

Patterns, Trends and Formation Processes of Sedimentary Records

Organic Carbon and Carbonate Records

We approach the multi-factorial problem of sedimentary signal formation in the Equatorial Atlantic by investigating and comparing the patterns and trends of various lithostratigraphic parameters, starting out with the relation of carbonate and organic carbon content.

A compilation of organic carbon and calcium carbonate contents along the meridional cross profile D exhibits very pronounced temporal patterns, regional trends and asymmetries (Fig. 4). Both parameters show pronounced climate cyclicity with large contrasts in organic carbon burial at a glacial/interglacial and substage rhythm.

Glacial C_{org} maxima vary widely between 0.25% and 2%, while interglacial C_{org} minima remain at the same low level of about 0.125% throughout the entire Equatorial Atlantic. This uniformity of recent sedimentary TOC levels contrasts with large modern productivity gradients along profile D indicated by Coastal Zone Color Scanner (CZCS) images (compare to November 1978 – June 1986 chlorophyll pigment concentration in Figure 1). Wagner and Dupont (1999) already observed, that modern, enhanced marine organic carbon fluxes are largely remineralized and therefore not recorded in surface sediments. This situation appears to be a general feature of interglacial conditions. Apparently, only in glacial periods regional productivity pulses have been preserved in the sedimentary archive.

The northernmost core GeoB 4317-2 (4°N, Fig. 4a) is scarcely influenced by the Equatorial Divergence and presently under oligotrophic conditions. On the basis of preserved TOC, there is no evidence for enhanced glacial productivity after termination II. The youngest organic-rich layer (ORL) dates back to MIS 6 reaching the highest observed concentrations of about 0.5% TOC. Similar saw-tooth shaped TOC patterns precede in glacial stages 10 and 12, while MIS 8 compares to the low values encountered in MIS 2 and 4. It will be argued below, that these 'missing ORLs' (in analogy to the term 'missing sapropels' by van Santvoort 1997) may rather be due to a lack of organic carbon preservation than to a lack of deposition. In greater vicinity of the Equatorial Divergence as in core GeoB 2908-7, glacial TOC contents are increasingly higher (Fig. 4b) and fully developed in every glacial stage. A well-defined glacial/interglacial pattern (Lyle 1988) with steeply increasing values is even more evident in cores GeoB 1117-2 and 1041-3 (Bickert 1992) where

individual pre-cessional ORLs merge to broader bands covering entire glacial periods, however, with still clearly separated subpeaks.

In core GeoB 1041-3 these carbon enrichments reach TOC concentrations of *sapropelic layers* according to the definition by Kidd et al. (1978), since they are '*discrete, greater than 1 cm in thickness, set in open marine pelagic sediments and containing between 0.5% and 2.0% or-ganic carbon by weight*'. Our data are consistent with findings of Verardo and McIntyre (1994) who describe a fourfold increase in TOC contents during MIS 2 from the margins to the center of the Equatorial Divergence upwelling. The two southernmost records GeoB 1112-4 and 1417-1 (Fig. 4e,f) show constantly high carbonate contents and accordingly low TOC and terrigenous contents. As shown by the enlarged carbonate record of GeoB 1112-4 in Fig. 3k, a faint climate signature at largely reduced amplitude remains.

It is a general feature of all investigated cores, that C_{org} and $CaCO_3$ records are in a broad sense clearly antisymmetric and anticorrelated. This negative correlation is at maximum at the Equatorial Divergence (r = -0.72), still relatively high in the adjacent northern regions (r = -0.46 to -0.36) and goes down to r = -0.17 in the oligotrophic subtropical gyre (Fig. 5a). This negative correlation is even more pronounced, if sediment sections next to terminations are excluded which obviously have been carbon-depleted by burn-down due to a down-ward progressing postglacial oxidation front (note irregularity in record asymmetry in particular around Termination II).

The inverse pattern of preserved $CaCO_3$ and TOC clearly prohibits the concept of assessing paleoproductivity from carbonate accumulation, which is generally based on the observation, that organic carbon and carbonate fluxes are globally well-correlated in deep-moored sediment traps (Brummer and Van Eijden 1992; Milliman 1993) throughout different oceanic productivity regimes (van Kreveld et al. 1996). Since fluctuations of deep water stratification are not sufficient for explaining the observed effects over the broad range of included core depths, we interpret the observed anticorrelation of $CaCO_3$ and TOC along the lines of Archer and Maier-Reimer (1994) and Martin

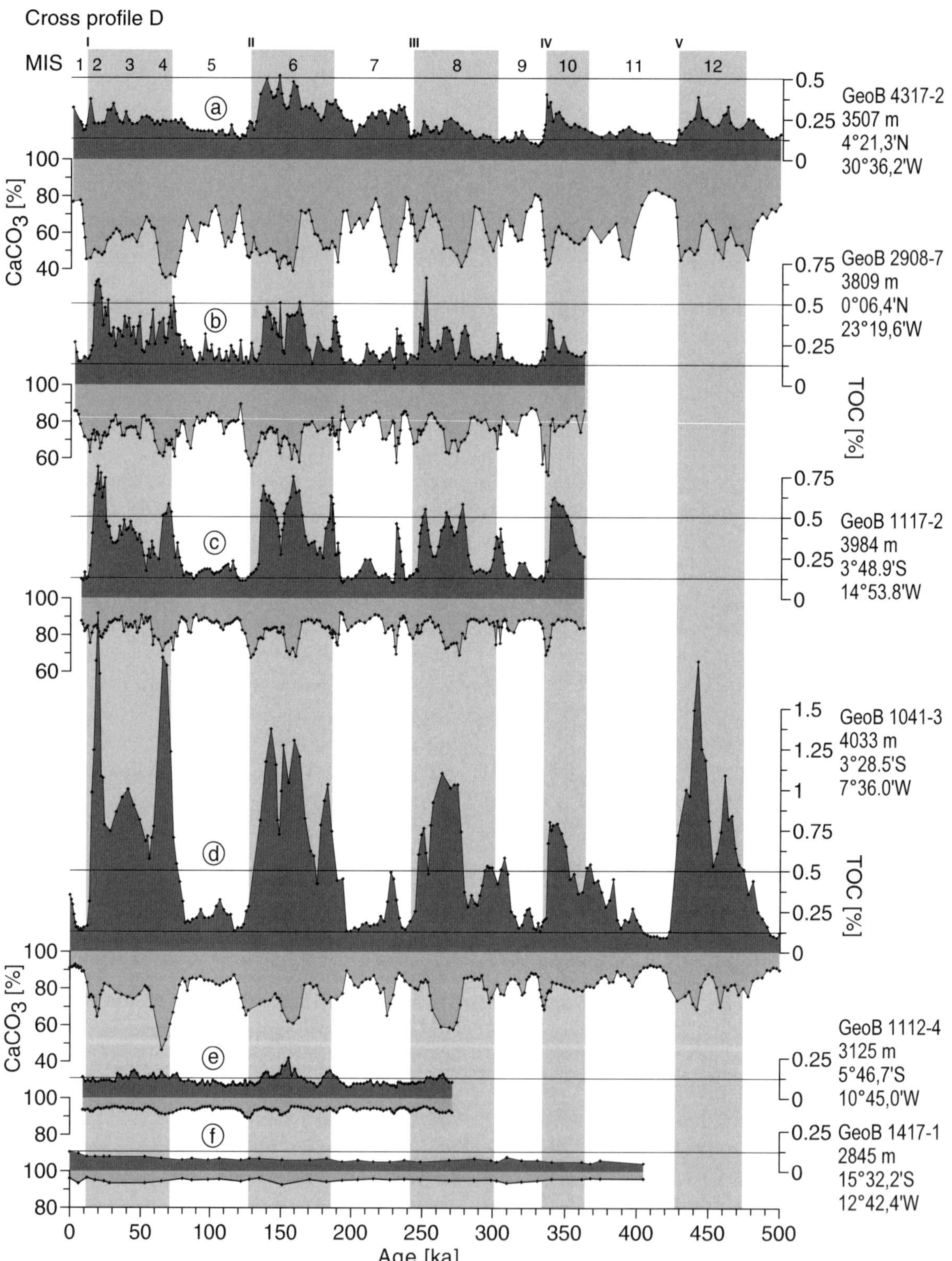

Fig. 4. Anticorrelated records of $CaCO_3$ (dark gray shaded) and C_{org} (light gray shaded) of the meridional cross profile D. Please note offset around terminations probably due to carbon depletion by burn-down in a downward progressing postglacial oxidation front.

and Sayles (1996) as an effect of enhanced glacial $C_{org}/CaCO_3$ rain rates entailing calcite dissolution by metabolic CO_2 release in the course of aerobic organic carbon degradation (Wilson and Thomson 1998; Schneider et al. 2000):

$$\left(CH_2O\right)_{106}\left(NH_3\right)_{16}\left(H_3PO_4\right)+138\ O_2+124\ CaCO_3 \rightarrow$$

$$230\ HCO_3^- +16\ NO_3^- + HPO_4^{2-} +124\ Ca^{2+} +16\ H_2O$$

While from the stoichiometry of the Redfield ratio, the molar loss relationship of organic carbon

and carbonate should not exceed 106:124 (= 1:1.17), ratios above 1:4 were observed in the oxic zones of homogenous turbiditic units (Wilson and Thomson 1998). A plausible explanation for progressing carbonate dissolution in spite of relatively low residual TOC contents is bioturbation. Benthic macrofauna constantly mixes organic material from the sediment-water interface down into the shallow oxic zone over just a few cm depth, where it is metabolized and thereby triggers calcite dissolution. In the present-day Western Equatorial Atlantic, benthic R_{Corg}/R_{CaCO3} rain rate ratios are on the order of 1:1 (0.8 to 1.6) (Martin and Sayles 1996) contrasting drastically with average AR_{Corg}/AR_{CaCO3} ratios documented in our data with numbers between 1:100 and 1:1000 (Fig. 5b). The more than 99% of organic carbon remineralized at around the sediment/water interface are partly available for $CaCO_3$ dissolution by the above depicted bioturbation/-irrigation mechanisms. For the Western Equatorial Atlantic, Martin and Sayles (1996) present model-based estimates of the carbonate dissolution efficiency by aerobic organic carbon degradation. Losses are estimated to reach between 24% and 73% of initial $CaCO_3$ content depending on conditions and assumptions.

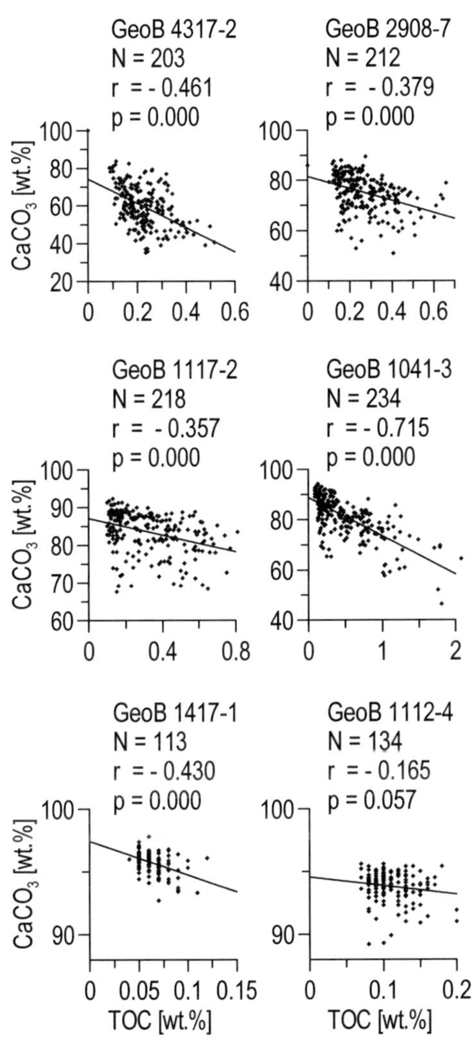

Fig. 5. a) Crossplots of C_{org} and $CaCO_3$ contents of sediment sequences along cross profile D. The various negative Pearson's correlation coefficients at individual core sites probably reflect regional productivity-related calcite dissolution intensities.

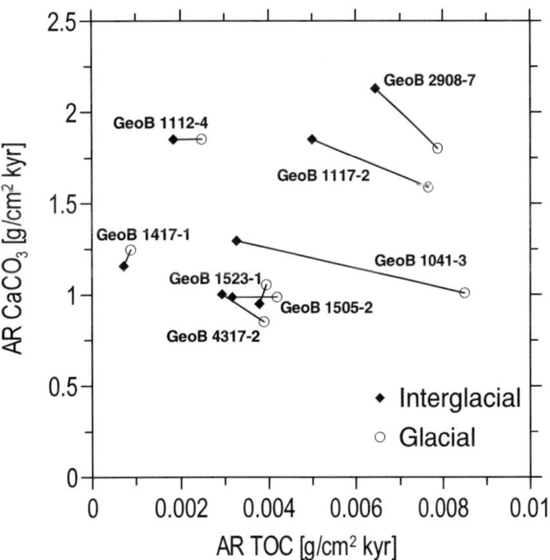

Fig. 5. b) Relation between interglacial (diamonds) and glacial (circles) averages of $CaCO_3$ and TOC accumulation rates.

474   Funk et al.

The tilt of the connection lines between inter-glacial and glacial averages of $CaCO_3$ and TOC accumulation rates (Fig. 5b) is positive (productivity signal) or neutral only in strictly oligotrophic southern regions, while it is negative (dissolution signal) around and north of the Equatorial Divergence. Zonal averages of the AR_{Corg}/AR_{CaCO3} ratio of the central core GeoB 1041-3 show the highest con-trast between glacial (1:125) and interglacial (1:400) conditions. Productivity estimates based on one or the other record would therefore come to very different and certainly contradictory conclusions.

Non-Carbonate, Iron and Magnetic Susceptibility Records

Variations in non-carbonate content (100%-$CaCO_3$%), iron content, and non-diamagnetic susceptibility (κ_{nd}, essentially a measure of ferri- and paramagnetic mineral content) of W-E profiles A

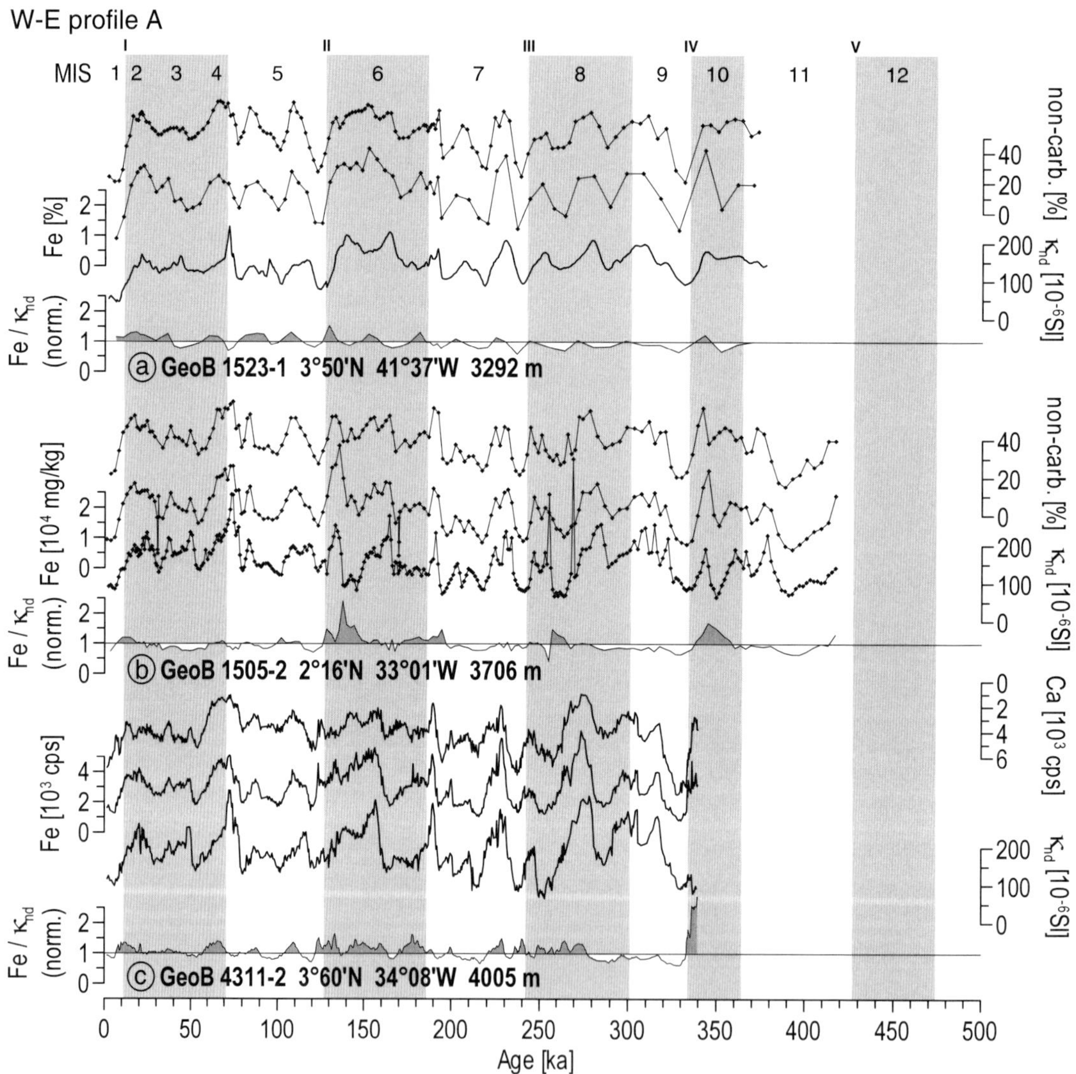

Fig. 6. Variations in non-carbonate content (100%-$CaCO_3$%), iron content (Fe) and non-diamagnetic susceptibility (κ_{nd}) of W-E profile A. Where non-$CaCO_3$ content was not available, an inverse Ca record was used instead. The three parameters show excellent correspondence in unaltered sections, while susceptibility deviates in diagenetically affected layers. This is particularly good to see in cores GeoB 4315-2 and 4317-2 (MIS 6, 10 and 12) as indicated by the magnetite dissolution proxy Fe/κ_{nd}.

and B are presented in Figs. 6a,b and 7 along with a newly introduced proxy parameter Fe/κ_{nd} indicating the extent of diagenetic magnetite dissolution (Funk et al. this volume). Near-baseline values of this very sensitive proxy parameter mark essentially pristine sections of the susceptibility signal, which can be interpreted in terms of mixed biogenic and lithogenic fluxes. Fe/κ_{nd} values exceeding 1 denote diagenetically overprinted rock magnetic signals.

Along transect A at 4°N mean values of κ_{nd} range from 133 to 183·10^{-6} SI, whereas the cores of transect B at 0° latitude exhibit much lower means between 36 and 59·10^{-6} SI. These decreases in magnetic susceptibility and, almost to the same extent, in iron content between the two transects at 4°N and 0° result primarily from two-fold higher carbonate dilution at the equatorial sites and more pervasive diagenesis. In comparison, W-E trends are subordinate and not as easily discernable from this compilation.

In the diagenetically unaltered sections the patterns of non-CaCO$_3$, Fe and κ_{nd} along each of the two latitudinal profiles correspond well. All three parameters basically measure terrigenous content and are therefore mirror images of the comple-

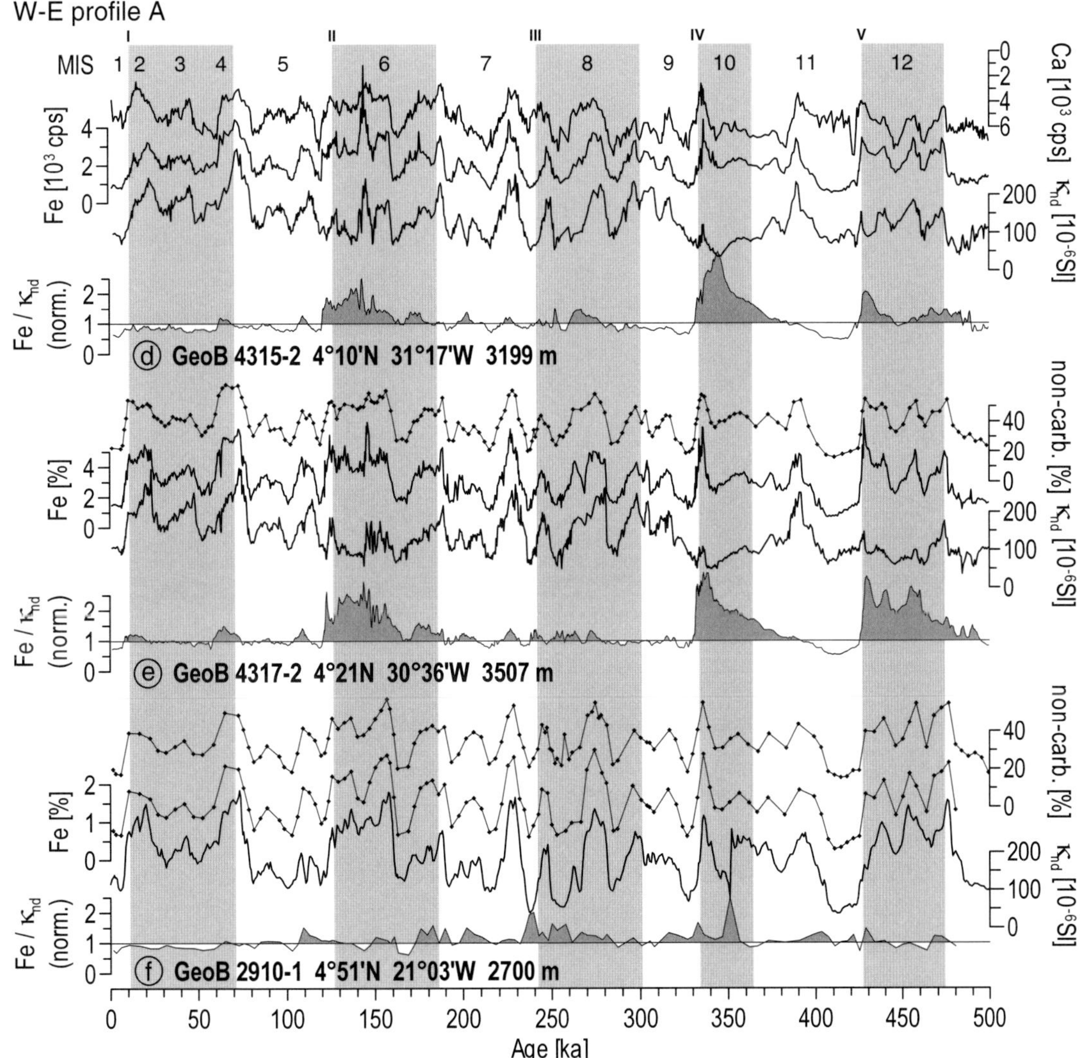

Fig. 6. cont.

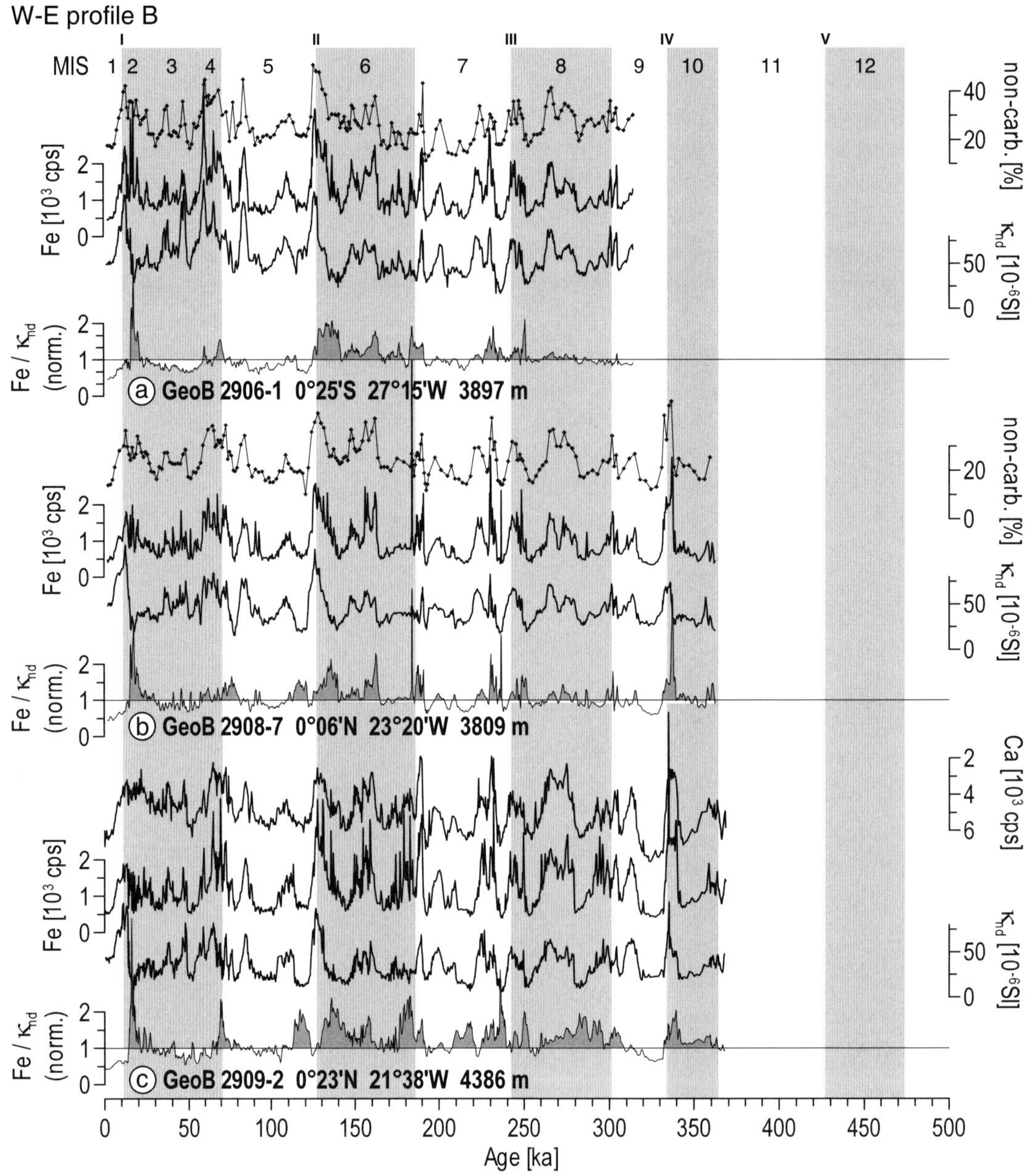

Fig. 7. Variations in non-carbonate content (100%-CaCO$_3$%), iron content (Fe) and non-diamagnetic susceptibility (κ_{nd}) of W-E profile B. Differences of the two profiles A and B are explained in detail in the text.

mentary carbonate contents. Previous remarks on the signal characteristics of the carbonate record therefore apply as well to Fe and κ_{nd} records, albeit in the reverse sense. Sedimentologically and mathematically conjugated by mutual dilution, biogenous and terrigenous components cannot be readily interpreted in terms of flux variations. Calculated accumulation rate records suffer visibly from an insufficient number of absolute age tie

points and are therefore not shown here. However, average glacial and interglacial accumulation rates can be determined with higher precision; they are compiled and discussed below.

At closer inspection of each signal triplet (Figs. 6a,b and 7), it can be seen, that susceptibility recurrently deviates from non-CaCO$_3$ and Fe, in particular at glacial terminations and within glacials (e.g. MIS 6, 10 and 12). In general, these offsets in the susceptibility signal show lower and less modulated values and occur almost simultaneously in every core of each transect. The Fe/κ_{nd} ratio clearly pinpoints this behavior as magnetic mineral dissolution layers. Although dissolution layers in profiles A and B are basically synchronous, they occur more frequently at 0° than at 4° N.

Along the northern profile A situated at 4° N in a recently oligotrophic area (Fig. 6b) a fairly constant baseline value alternates with elevated Fe/κ_{nd} values (gray shaded sections) around major climate transitions, in particular after MIS 6, 10, and 12. These sections coincide well with TOC enrichments shown in Fig. 4. Especially in cores GeoB 4317-2 and 4315-2 the magnetite dissolution peaks form broad sawtooth patterns with steep flanks at the terminations encompassing sediment sections of about 100 cm in thickness. In MIS 2, 4 and 8 dissolution peaks are faint or absent.

Along profile B (Fig. 7) at 0° at the northern margin of the Equatorial Divergence Fe/κ_{nd} peaks are shorter, mostly less intense, but much more frequent. Such a pervasive overprint hampers the use of magnetic signals for chronostratigraphic purposes here. As indicated by pronounced peaks in Fe/κ_{nd}, κ_{nd} goes through extreme minima in late MIS 2 (Termination I) and all preceding climate analogues. Dissolution signatures appear in all glacials and even in some interstadials (5.4, 7.2, 7.4); they respond preferentially to a precessional rhythm (Fig.7). The Fe/κ_{nd} peaks are generally steeper and more symmetric.

With respect to the previously discussed anti-correlation of $CaCO_3$ and C_{org} content, it does not come as a surprise that magnetite dissolution, reduced carbonate content and thereby enhanced terrigenous content go along in organic carbon enriched sections. Subsequent oxic and suboxic degradation of organic matter results in coincident carbonate and magnetite losses. We clearly see this effect in sections deposited during high glacials and terminations, most notably at the terminations II and IV of profile B (Fig. 7).

The high signal conformity of terrigenous contents finds an explanation in a largely similar carbonate dilution regime exerted by uniform, ocean-wide mechanisms of supra- and sub-lysoclinal carbonate dissolution. Vertical lysoclinal displacement was

found to be entirely responsible for the equally high similarity of susceptibility records from the oligotrophic Subtropical Atlantic (Schmieder et al. 2000).

Magnetogranulometric and -Mineralogical Records

So-called 'relational' rock magnetic parameters compare the concentrations of magnetically discernible magnetic mineral and particle size fractions. They are by definition independent of biogenous and terrigenous content and often react very sensitive to changes in source rock geology and transport pathways (Frederichs et al. 1999). Two well established and easily measurable rela-tional parameters are investigated here: the magnetogranulometric M_{ar}/M_{ir} ratio and the magneto-mineralogical $S_{-0.3T}$ ratio, together with their respective basis parameters, M_{ar}, M_{ir} and M_{hir} presented in the following chapter. Details on the expression of redoxomorphic diagenesis in these parameters are described in the complementary paper by Funk et al. (this volume).

M_{ar}/M_{ir} records of the northern profile A (Fig. 8) have the general appearance of ice volume records and correlate highly to the oxygen isotope 'low latitude stack' (Bassinot et al. 1994) in most of their parts (Fig. 8i). These patterns are clear and uniform and lend themselves ideally to cyclostratigraphic purposes. Core mean M_{ar}/M_{ir} values of profile A level around 0.06 (0.055 to 0.08) It is interesting to regard the systematic W-E trend of the total data range. The plateau-like glacial signal sections assume low and therefore 'coarse' levels of 0.04 to 0.05. All interglacial maxima (e.g. MIS 5.5) are well developed and reach their largest, thus 'finest' values of about 0.10 at the westernmost end of profile A. To the east, they take increasingly lower, less fine peak values of 0.07 to 0.08. The individual interglacials are well isolated and extremely spiky; they highly resemble respective $CaCO_3$ records (Fig. 2) and coincide with carbonate-rich layers deposited during periods of sea level high stands, shallow nutricline and hence higher productivity in the west and reduced upwelling with lower productivity in the east. This analogy of M_{ar}/M_{ir} and $CaCO_3$ records cannot be extended to the warm stadials of glacial periods,

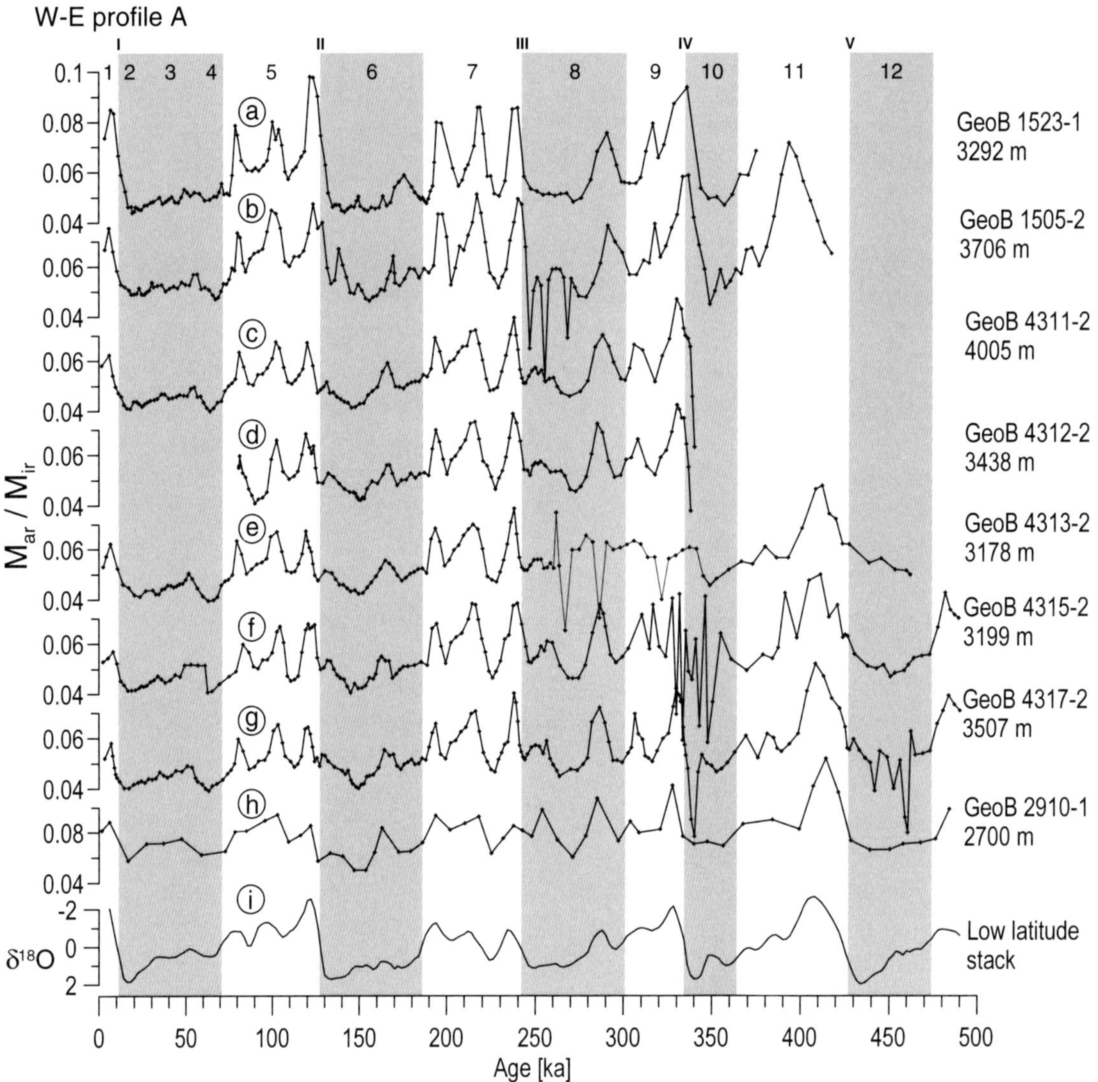

Fig. 8. The magnetic grain size index M_{ar}/M_{ir} exhibit a uniform signal signature along W-E profile A (4°N) with a high correlation to the $\delta^{18}O$ latitude stack (**i**, Bassinot et al. 1994). Sharp signal minima portend to magnetically depleted zones and fall into glacial periods.

where magnetogranulometry shows a smaller variance than carbonate content.

Some glacial sections of the M_{ar}/M_{ir} records exhibit traits of reductive magnetite dissolution similar to those already detected and described for the magnetic susceptibility records. Some extreme-ly peaked M_{ar}/M_{ir} minima within ORLs sharply protrude from the prevalent climate patterns signaling an abrupt coarsening of the magnetic particle assemblage. These features strongly disturb certain sections of cores GeoB 1505-2, 4311-2, 4312-2, 4315-2 and 4317-2, in particular during MIS

8, 10 and 12. Such changes in magnetic grain-size distribution are due to chemical depletion of the finer ferrimagnetic grains, which are more quickly dissolved owing to their large specific surface/volume ratio. This fine particle loss entails a relative enrichment of residual coarser particles and thereby an effective 'coarsening' – even if every grain actually lost volume (Karlin 1990).

In profile B, mean M_{ar}/M_{ir} values vary between 0.068 and 0.079 including altered sections (Fig. 9) and indicate a higher relative proportion of fine magnetic grains in comparison to profile A sedi-

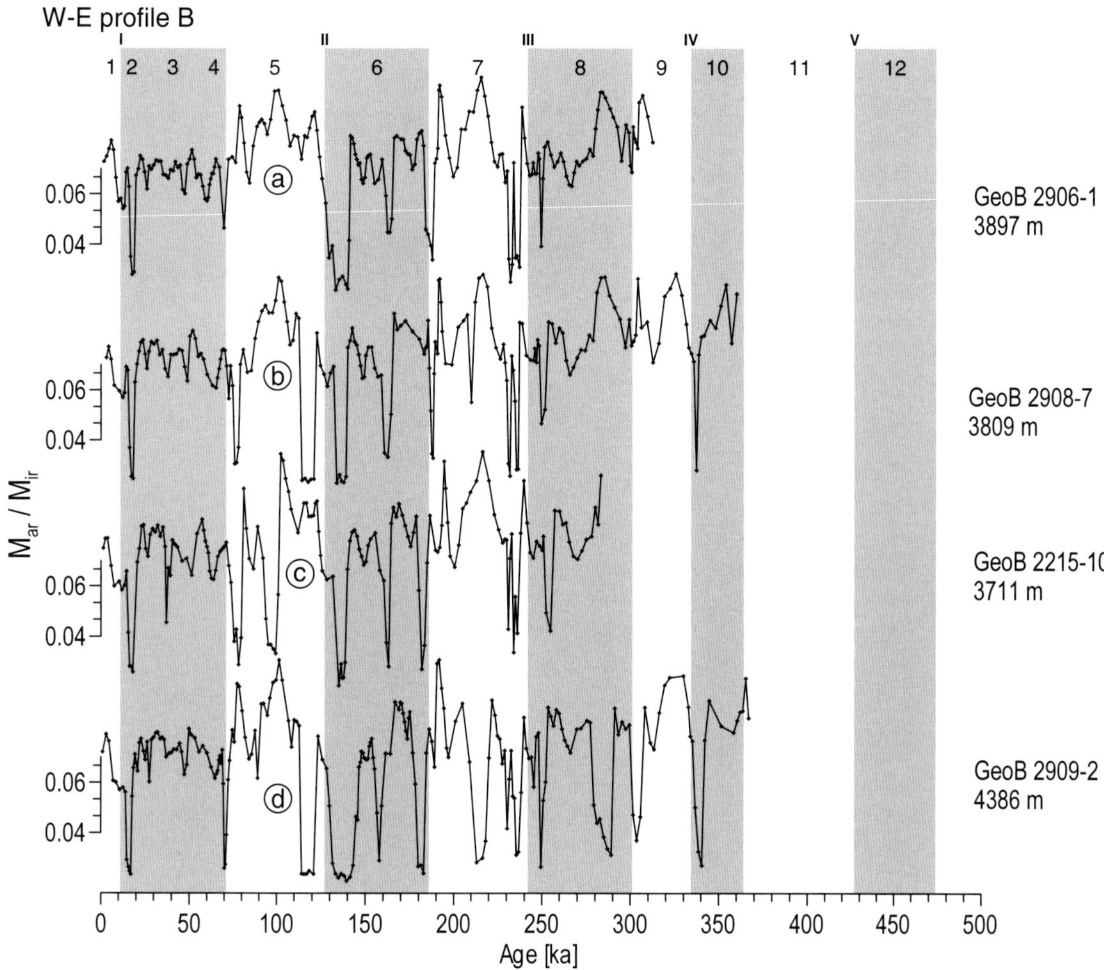

Fig. 9. Along W-E profile B (0°) the M_{ar}/M_{ir} ratio is considerably stronger affected than to the north. The pristine signal is nearly complete masked by extremely deviating 'coarser' values caused by magnetite dissolution.

ments. At a much higher frequency and pervasiveness, reductive dissolution events interrupt the primary signal with extremely deviating and 'coarse' M_{ar}/M_{ir} values around 0.02. In consequence, a primary signature can only be attributed to short, intermittent signal sections. These, nevertheless, seem to correlate to profile A records and the low latitude $\delta^{18}O$ stack. All diagenetically affected intervals show elevated Fe/κ_{nd} values (Figs. 6a,b and 7).

The $S_{-0.3T}$ ratio (Bloemendal et al. 1992) relates hematite to magnetite. In practice, this parameter primarily traces variations between various sources and erosional regimes. The major hematite source of the Equatorial Atlantic region is dust from the arid northwest Africa, followed by an Amazonian

influx distributed eastwards by the Northequatorial Countercurrent (Bleil and von Dobeneck this volume).

As mentioned earlier, the compilation of available $S_{-0.3T}$ records of profiles A and B (Fig. 10) includes data sets produced with different pulse magnetizers and settings. These records follow comparable patterns, but their absolute numbers are not entirely consistent. It is evident at first sight, that all four records of profile A equally represent global climate cycles. Other than the M_{ar}/M_{ir} ratio, they follow glacial more than interglacial dynamics (compare MIS 5 and 6). This precession-dominated, typically 'tropical' signal clearly contrasts with the eccentricity-dominated M_{ar}/M_{ir} pattern. $S_{-0.3T}$ amplitudes and relative hematite content seem to

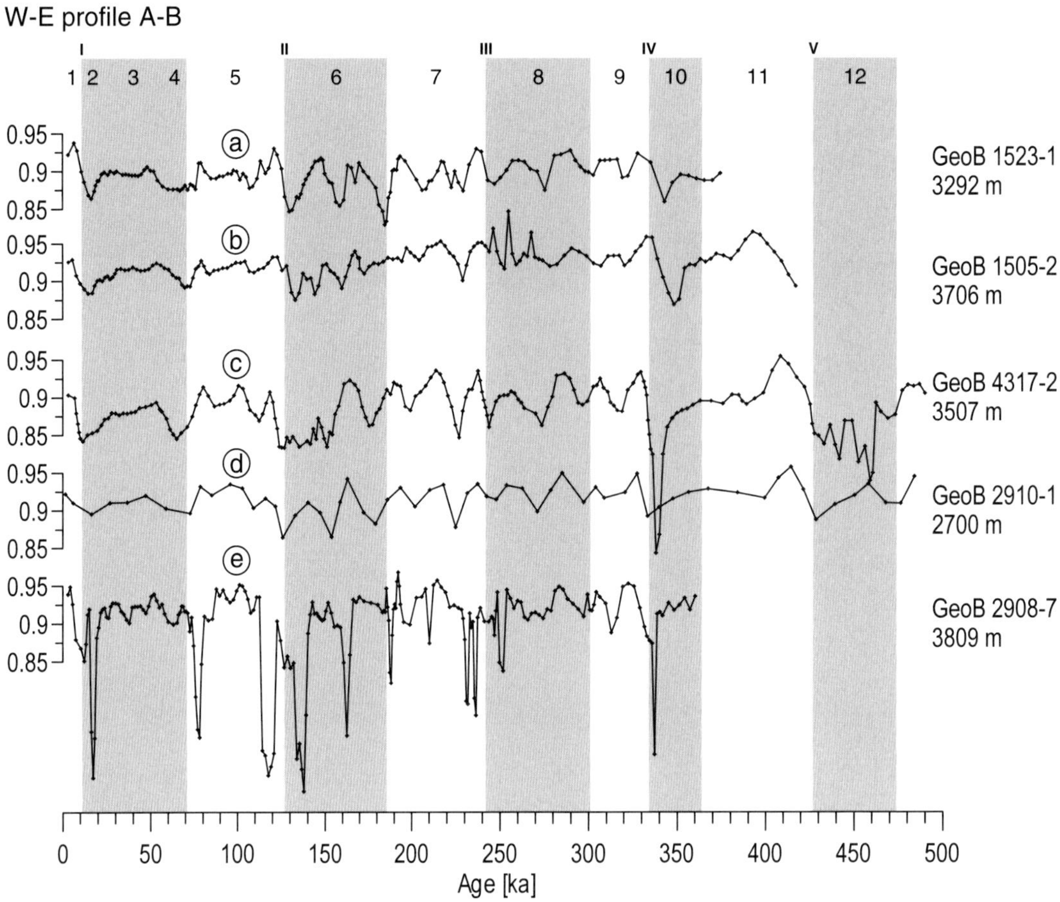

Fig. 10. Compilation of $S_{-0.3T}$ records of profiles A and B. For explanation see text.

increase from GeoB 1523-1 (a) in the west to GeoB 4317-2 (c) at the eastern flank of the mid-Atlantic ridge pointing to a growing influence of Saharan dust. Surprisingly, the record of the easternmost Sierra Leone Rise core GeoB 2910-1 (d) does not continue this trend, but returns to values similar to those of the westernmost Ceará Rise core.

Diagenesis comes again into play at all the previously described dissolution intervals. Its traits are extremely low, seemingly 'hematite-rich' values, which deform the primary signal around termina-ion IV, in MIS 12, and, less significantly, also around termination II. The equatorial core GeoB 2908-7 (e) shows again stronger and more frequent over-printing in all the numerous ORLs at terminations I, II, III and IV as well as in MIS 4 and 6. This change in magnetic composition is primarily related to

magnetite depletion by reductive dissolution, leaving the more reduction-resistant, probably also coarser hematite phases behind as a relict mineral.

Selective Rock Magnetic Records

In order to resolve the climatic controls responsible for the signal characteristics of the two rock magnetic proxies M_{ar}/M_{ir} and $S_{-0.3T}$, it is necessary to take a look at their three defining curves M_{ar}, M_{ir} and M_{hir}, tracking fine-particle (SD) magnetite, total magnetite and hematite content, respectively. The co-compilation of these three parameters (Fig. 11) raises evidence for a regional interpretation.

A detailed inspection of all curves shows, that their patterns are essentially quite similar. However, their dynamics and signal levels vary considerably. These properties are most easily compared on basis

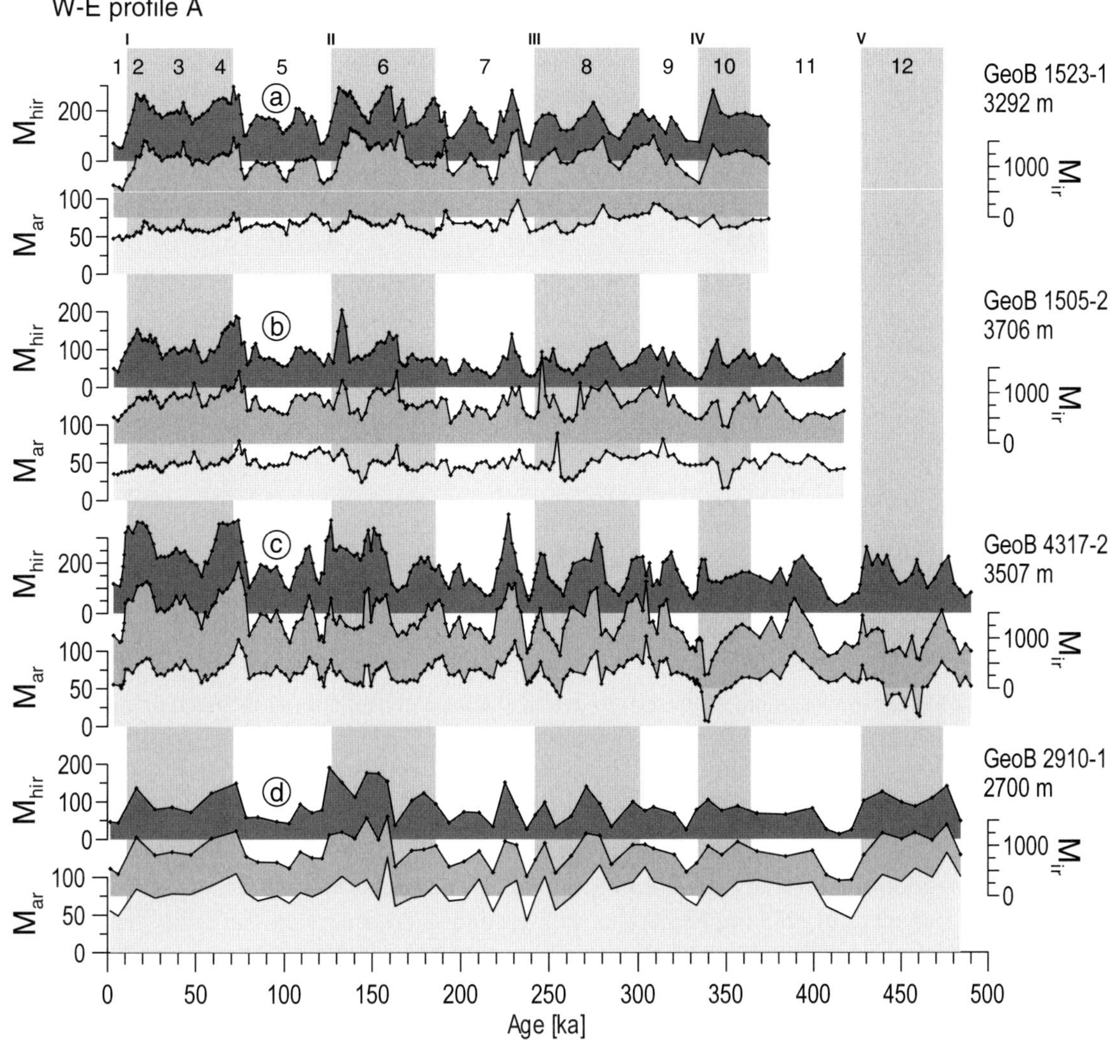

Fig. 11. Compilation of the three rock magnetic parameters M_{ar} (light gray), M_{ir} (gray) and M_{hir} (dark gray) tracking fine-particle (SD) magnetite, total magnetite and hematite content, respectively.

of some statistical quantities compiled in Table 2, comprising arithmetic means, relative standard deviations and interparametric Pearson's correla-tion coefficients. Signal sections severely affected by diagenesis were excluded from this calculation.

Three main trends can be observed:

1. Comparing relative standard deviations of all parameters and sites, we find that hematite vari-ance systematically exceeds coarse magnetite variance and nearly doubles fine magnetite vari-ance. We also note, that all these variances increase from west to east.

2. Correlation coefficients between hematite and coarse magnetite records are always higher than those between fine and coarse magnetite records. Correlations improve from west to east, where both coefficients seem to converge to a sin-gle value.

3. While there is no systematic W-E trend with regard to mean values (partly due to variable carbo-nate dilution), the averages for fine magnetite are regionally more stable than those of hematite and coarse magnetite.

From all three trends it follows, that the fine-grained magnetite phase is to be regarded as a spatially and temporarily more stable phase than hematite and coarse magnetite, which show much higher climatic modulation. According to an en-

Parameter	GeoB 1523-1	GeoB 1505-2	GeoB 4317-2	GeoB 2910-1
$\mu(M_{hir})$ [10^{-3} A/m]	**179**	**83**	**195**	**84**
$\sigma(M_{hir})$ [%]	*32*	*44*	*43*	*50*
$\mu(M_{ir})$ [10^{-3} A/m]	**1183**	**810**	**1426**	**873**
$\sigma(M_{ir})$ [%]	*22*	*28*	*26*	*34*
$\mu(M_{ar})$ [10^{-3} A/m]	**66**	**49**	**73**	**66**
$\sigma(M_{ar})$ [%]	*14*	*22*	*19*	*23*
$\varrho(M_{hir}, M_{ir})$	**0.74**	**0.81**	**0.84**	**0.90**
$\varrho(M_{ar}, M_{ir})$	**0.57**	**0.74**	**0.83**	**0.85**

Table 2. This compilation of the arithmetic means (μ), relative standard deviations (σ) and interparametric Pearson's correlation coefficients (ρ) gives an overview of dynamics and signal levels of the three basic magnetic parameters M_{ar}, M_{ir} and M_{hir} along the W-E profile A. Signal sections severely affected by diagenesis were excluded from this calculation.

vironmental magnetic case study of site GeoB 1505-2 by von Dobeneck (1998) and Frederichs et al. (1999), the latter parameters track terrigenous input from continental origin, while the fine-grained magnetite signal is probably of bacterial origin, or, alternatively, a faint, detrital background signal. In consequence, the magnetogranulometric M_{ar}/M_{ir} ratio deals not with a unimodal magnetic phase changing grain-size over time, but rather with a bimodal system of a coarse detrital and a fine, probably bacterial phase of which only the coarse phase is climate-controlled. The increasing variance of fine magnetite and its correlation to coarse magnetite towards the east implies, that the Saharan dust flux progressively contributes to and thereby modulates this fine magnetite fraction. At the western side, only the coarse magnetite fraction of Amazonian origin shows noticeable climate control, while the fine magnetite fraction barely varies with time. The very high, i.e. fine interglacial values of the M_{ar}/M_{ir} ratio at and near the Ceará Rise imply, that the Amazon particle flux to the pelagial realm is extremely reduced at sea level high-stands. It

finds other sinks in that time, in particular the newly eroded depocenters on the broad north Brazilian shelf (Milliman et al. 1975; Damuth 1977). We conclude, that the climatic information of the M_{ar}/M_{ir} ratio in the Equatorial Atlantic is essentially controlled by the denominator M_{ir}, hence, the terrigenous particle flux, while the more biogenous numerator M_{ar} compensates for carbonate dilution and diagenesis effects. The diminishing amplitudes of the M_{ar}/M_{ir} ratio to the east result from a growing influence of the Saharan source over the fine magnetite fraction. This coupling of coarse and fine magnetite fractions effectively stabilizes their mutual M_{ar}/M_{ir} ratio and reduces its amplitudes.

The mathematical controls of hematite and magnetite on the magnetomineralogical $S_{-0.3T}$ ratio are most easily recognized when expressing it as $S_{-0.3T} = M_{ir}/(M_{ir}+M_{hir})$. We find from the statistical analysis, that the hematite phase consistently shows stronger climate modulation than the magnetite phase. Again the term M_{hir} in the denominator takes a lead in this ratio and overcompensates coherent variations of M_{ir}, while carbonate dilution again cancels out. In other words, we see here a dominance of hematite over magnetite flux variations.

From a sedimentological perspective, this finding suggests, that glacial hematite fluxes were relatively enhanced, either by larger Saharan influence or by different weathering conditions. The lack of a consistent trend in average $S_{-0.3T}$ ratios along profile A is puzzling and possibly an expression of internal variability within the Saharan dust falls in the study area, in particular at their southeastern margins. Records from more northerly locations along the main dust trajectory are needed to study the genuine signature of the southern Saharan magnetic mineral assemblage (Bloemendal et al. 1988).

Synopsis of Glacial/Interglacial Accumulation Rate Averages

For paleoclimate research we need to know individual material fluxes and budgets, which can only be resumed from accumulation rates (AR).In awareness of the known difficulties involved in estimating accumulation rates from records with cyclostratigraphic, i.e. relative age models we decided to follow the approach of Ruddimann (1997) and Wagner (2000), who limited AR errors resulting from dating uncertainties by averaging over larger time intervals. Here we determined ARs over extended glacial (MIS 2+3+4, 6) and interglacial (MIS 1, 5, 7) periods of the 100 kyr cycle. That this approach largely cancels out pre-cessional variations. An additional problem was encountered with the two western cores of profile A, which suffer from noticeable coring induced compression in their lower parts. This problem is often encountered when coring porous, clay-rich and therefore easily deformable sediments. A modest compression has little effect on concentration records if these have been correctly transposed into the time domain. However, the deformation of accumulation records is by far more problematic and not tolerable, when it comes to compare absolute AR values. As a conservative but firm approach, all AR averages of profile A were therefore exclusively derived from the last 200 kyr, where compression (Bleil and von Dobeneck this volume) and early diagenesis seem to have done little damage. Each region is represented by a single sediment core.

For most of the W-E profile (Fig. 12), total sedimentation rates (a) amount to some 2.5 cm/kyr and show little or no consistent glacial/interglacial variability. As sole exception, the Sierra Leone Rise record GeoB 2910-1 has a 40% lower sedimentation rate of about 1.4 cm/kyr. In terms of biogenous (d) and terrigenous (f) content, the latter decreases from about 55% in the west to around 35% in the east. Since AR averages of $CaCO_3$ (e) of roughly 1 g/cm^2 kyr do not show large zonal or temporal gradients, the glacial/interglacial shifts in non-$CaCO_3$ AR averages (g) are largely responsible for lithological variability. We notice a W-E decrease from 1.2 over 0.78 and 0.75 down to 0.35 g/cm^2 kyr. This trend is obviously inverse to the trade wind direction and, unless we take atmospheric focusing of the Saharan dust fall into consideration, also to existing opinions on Saharan prevalence in the central Equatorial Atlantic.

In particular the very low eastern detrital ARs are astonishing and raise some suspect that winnowing may have come into play at this topographic height of the Sierra Leone Rise. On the other hand, we find also the largest, i.e. finest average M_{ar}/M_{ir} ratios at this eastern site (Fig. 9). Magnetic 'fining', however, is contradictive to elevated bottom current velocities. Furthermore, we find also the highest and most climate dependent susceptibilities (h) at this site, in spite of the lower terrigenous contents.

The governing W-E trend from lower to higher carbonate-free susceptibilities is only interrupted by GeoB 4317-2, the only core with relatively lower susceptibility and magnetite AR during glacials (i). This is clearly an effect of the large magnetite losses during glacial MIS 6 as suggested by highest glacial TOC averages (0.30%) among all cores of profile A (b).

Comparing TOC contents (b) and ARs (c), we find, that some subtle shifts towards higher glacial TOC contents observed in the three western cores are at least partly due to carbonate dilution. In reality, TOC accumulation at the Ceará Rise even drops slightly during glacials as to be expected from the shear wind driven seesaw mechanism, which disequilibrates glacial Equatorial Atlantic productivity by lifting the eastern and lowering the western nutricline. Indeed, glacial C_{org} burial rates were

W-E profile A

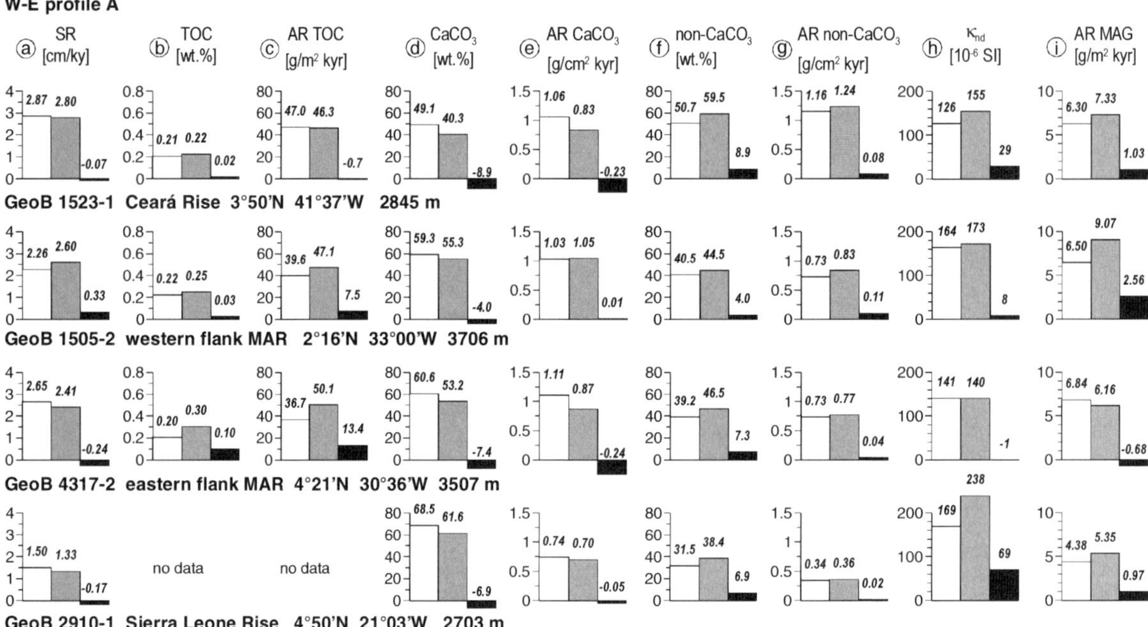

Fig. 12. The 200 ka to present means of sedimentation rates **a)**, TOC content **b)** and accumulation **c)**, CaCO$_3$ content **d)** and accumulation **e)**, non-CaCO$_3$ content **f)** and accumulation **g)**, non-diamagnetic susceptibility **h)** and magnetite accumulation **i)** of W-E profile A. White bars denote averages over interglacial (MIS 1, 5, 7) and gray bars over glacial (MIS 2+3+4, 6) periods, while black bars show their difference.

some 19% higher on the western and 36% higher on the eastern flanks of the MAR, most likely due to larger W-E gradients in ocean productivity.

AR averages of glacial/interglacial C$_{org}$ and CaCO$_3$ shift inversely like the earlier shown signal patterns. GeoB 4317-2, the core with the highest glacial TOC increase and magnetite loss, also features the largest drop to glacial CaCO$_3$ accumulation rates. This drop in carbonate sedimentation is also seen at the shallow Ceará Rise core GeoB 1523-1 (2845 m) whereas the deepest core GeoB 1505-2 (3706 m) shows nearly balanced glacial/interglacial CaCO$_3$ contents and ARs. This again is strong evidence for a lead of sedimentary diagenesis over CCD oscillation in controlling supralysoclinal Equatorial Atlantic carbonate contents.

We now change perspective and proceed to the meridional NW-SE cross profile stretching over an equatorial latitude zone from 4°N to 5°S with an outpost at 15°S (Fig. 13). Here, the interglacial and glacial averages were drawn from total record lengths.

The subtle meridional rise and fall in mean sedimentation rates from some 2.5 cm/kyr (GeoB 4317-2) up to 3.5 cm/kyr (GeoB 2908-7, 1117-2) and back to about 2.5 cm/kyr (GeoB 1041-3, 1112-4) unveils merely a glimpse at the lithological contrasts hidden in this profile (Fig. 13). From the separate compilations of biogenous and terrigenous components and their ARs (b-g), it can be easily inferred, that two large, but mutually opposed meridional trends exist, which counterbalance each other in total sedimentation rates.

The CaCO$_3$ percentages (d) start with merely 57% (GeoB 4317-2) at the northern oligotrophic margin, rise to 76%, 84% and 82% (GeoB 2908-7, 1117-2, 1041-3) in the mesotrophic Equatorial Divergence region and reach nearly pure calcareous values of 94% and 96% at the southern oligotrophic gyre. Consequently, non-CaCO$_3$ content (f) decreases by one order of magnitude from 43% to 4% from north to south. This large gradient results from non-CaCO$_3$ ARs (g), which show vastly decreasing numbers of 0.75, 0.65, 0.33, 0.24, 0.12 and 0.06 g/cm^2 kyr over cross profile D

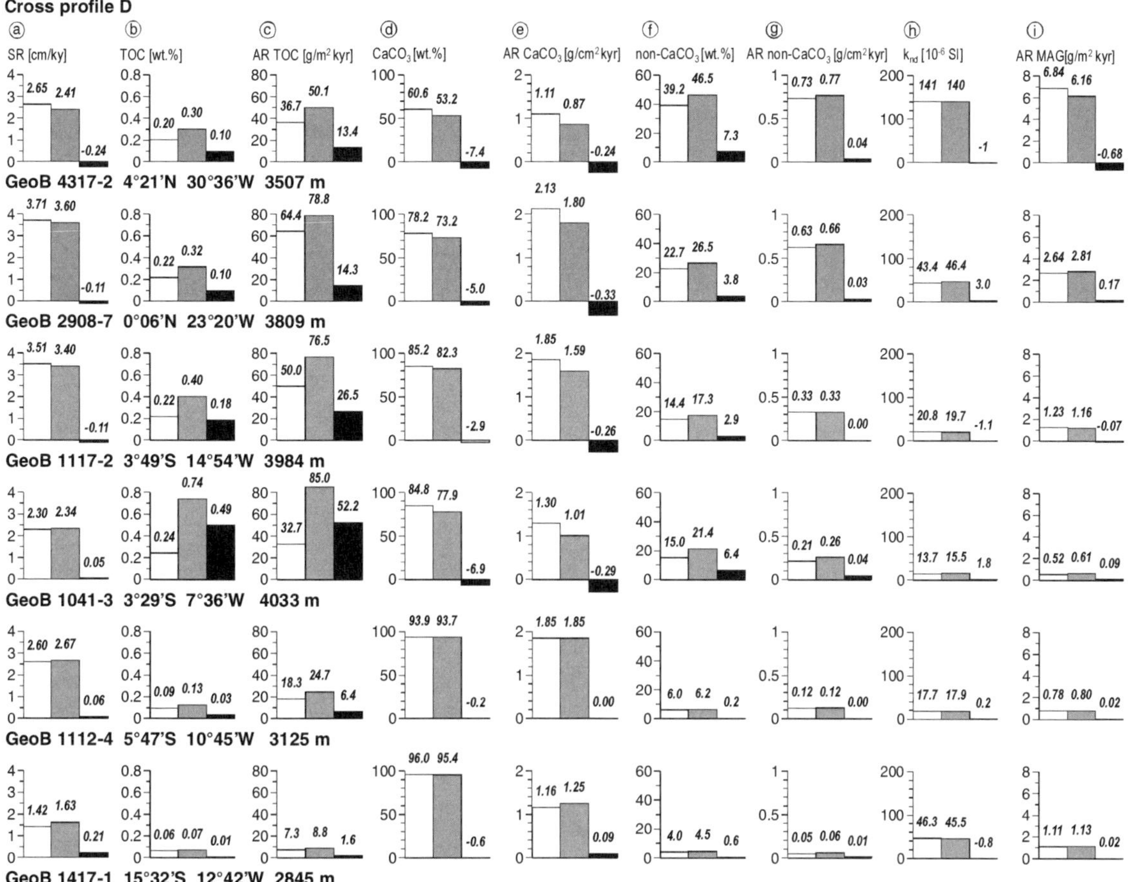

Fig. 13. Total core means of sedimentation rates **a)**, TOC content **b)** and accumulation **c)**, $CaCO_3$ content **d)** and accumulation **e)**, non-$CaCO_3$ content **f)** and accumulation **g)**, non-diamagnetic susceptibility **h)** and magnetite accumulation **i)** of NW-SE cross profile D. White bars denote averages over interglacial, gray bars over glacial periods, while black bars show their difference.

while respective $CaCO_3$ ARs (e) do not vary by more than a factor of two (1-2 g/cm^2 kyr).

Combining profiles A and D we may conclude, that the major regional trends in Equatorial Atlantic lithology are primarily caused by the heterogeneous distribution of continental source regions and transport pathways. Regional productivity variations are clearly subordinate. However, for the temporal trends, i.e. glacial-interglacial variability, we have the opposite regime:

The general trend to lower glacial $CaCO_3$ contents (d) has statistically more to do with lower glacial $CaCO_3$ ARs (e) than with higher glacial non-$CaCO_3$ ARs (g). It should be noted, that the ratio of glacial and interglacial $CaCO_3$ ARs is of nearly equal proportion at 3500, 3800, and 4000 m water

depths. Only at the two southernmost and shallowest (3125 m and 2845 m water depth) stations do we see balanced or even slightly enhanced glacial $CaCO_3$ ARs. As shown earlier for cross profile D (Figs. 4, 5), between 50 and 200% more organic carbon has on average been preserved in glacial than in interglacial sediment sections. The related much higher glacial C_{org} burial rates have left a syndepositional diagenetic overprint in the $CaCO_3$ records as well as a postdepositional signature in the magnetic mineral records. Averaged susceptibilities (h) and magnetite ARs (i) decay at an even steeper rate from N to S than the related non-$CaCO_3$ contents (f) and ARs (g), most likely by an increasingly higher tribute to reductive diagenesis.

In this context, it is interesting to note, that the ratio of C_{org} and $CaCO_3$ ARs is surprisingly high at the northernmost location GeoB 4317-2. Neither productivity nor bottom water chemistry can be hold responsible for this finding, but these superpro-portional C_{org} contents go along with remarkably heavy losses in glacial magnetite accumulation. We may conclude from this situation, that scavenging, i.e. accelerated downward transport by flocculation of organic material with (magnetite-bearing) dust particles, could explain this situation. Intimate contact of particles with a redox partnership within such clusters may lead to particularly high reductive magnetite losses. This would also explain some untypically broad ORLs in the northernmost studied sediments of core GeoB 4317-2.

Specific Signatures of Early Diagenetic Processes

Berner (1987) showed that organic carbon enrichment and degradation is responsible for the early diagenesis of sedimentary iron minerals and resulting changes of rock magnetic signals. The quantity of preserved organic carbon in sediments depends on the organic matter supply from the surface layer of the ocean, water depth, bulk sedimentation rate, and its oxidation by bottom water oxygen and other oxidants (Müller et al. 1988). Subsequent to deposition the organic material is subject to degradation by microorganisms. Froelich et al. (1979) proposed a conceptual model for these microbially mediated processes: Accordingly, the oxidation of organic matter follows a sequential series of terminal electron acceptors in descending order of free energy gain. The oxidants are: dissolved oxygen, nitrate, Mn oxyhydroxides, Fe oxyhydroxides, and sulfate. Oxyhydroxide is a frequently used term for -oxides, -hydroxides, and everything in between these two end -members (van Santvoort et al. 1997). The extent of the reaction sequence depends on the availability and reactivity of both organic matter and reductants and the competitive efficiency of microbial populations (Karlin and Levi 1983). If the amount of organic matter in a sediment is such that suboxic diagenesis involves Fe-reducing bacteria (Froelich et al. 1979), reductive dissolution of detrital iron oxides occurs

(Bloemendal et al. 1992) altering the bulk magnetic properties of the sediment (Robinson et al. 2000).

Subsurface Color Transition

Reductive dissolution of Fe oxides has not only a consequence on magnetic properties, but, due to their important role as pigments, strongly influences the color aspect of a sediment. The prominent subsurface color transition found in all investigated Equatorial Atlantic sediment cores (Fig. 14) represents the modern iron redox boundary (Lyle 1983). Illustrated on the basis of core pictures

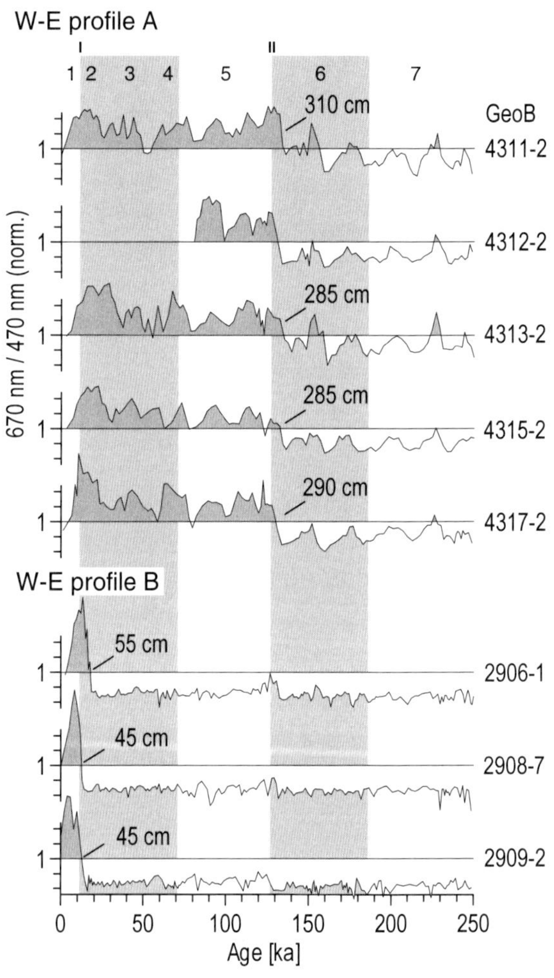

Fig. 14. Reflectance ratio 670nm/470nm along the W-E profiles A and B. The Fe^{2+}/Fe^{3+} redox boundary is indicated by an abrupt color change represented by a drop to values below 1. Note the very different depths of this transition at 4°N and 0°.

(Fig. 15a,i) and color reflectance data the change from hues of reddish browns to greenish grays appears at different depths in cores GeoB 4317-2 (transect A) and 2908-7 (transect B). The oxidized layer is indicated by higher 670 nm/470 nm (red/blue) reflectance ratios (Fig. 15b,j, gray shaded). The color transition in core GeoB 4317-2 (4°N, Fig. 15a) marked by an abrupt drop to values <1 is situated at a depth of 2.9 m, just beneath termination II. Core GeoB 2908-7 (0°) exhibits the redox boundary below termination I at the substantially shallower depth of 0.45 m (Fig. 15b). Remarkably, the color change is almost located at the same depth interval along each longitudinal transect, but varies considerable between profiles A and B. Along profile A the thickness of the oxidized layer (gray shaded) ranges down to sediment depths between 2.85 m and 3.10 m (Fig. 14). In the cores of tran-sect B the depth of the color change is located at only 0.50 m below the sediment surface indicating a remarkable shallowing of the boundary towards the Equatorial Divergence.

Lyle (1983) used the depth position of the transition from Fe(III) to Fe(II) as a measure of the relative supply and degradation rate of organic carbon. He mapped the thickness of the oxidized Fe(III)-bearing surface sediment layer in the eastern tropical Pacific Ocean and correlated thin brown layers to a high accumulation and consumption of organic carbon. (Müller et al. 1988) have also demonstrated a correlation between the thickness of the upper oxidized layer and the productivity of surface waters, organic fluxes and bulk sedimentation rates. In the eastern tropical Pacific they linked hemipelagic sediments dominated by reducing conditions to highly productive waters, and pelagic sediments formed under oxic conditions to less fertile surface waters. On the basis of magnetic hysteresis data Tarduno and Wilkison (1996) defined a progressive downward shift of the modern iron redox boundary due to a decrease in organic carbon input.

Present-day productivity, however, cannot be made responsible for the different depth positions of the iron redox boundary observed in the Equatorial Atlantic. This holds true for all interglacial sections of crossprofile D cores (Fig. 4), where no obvious differences in the sedimentary TOC con-

tents between the north and south are observed. Significant regional variability of TOC is confined to the greatly enhanced glacial productivity, which still today determines the depth position of the redox boundary by forming the topmost ORL (Colley et al. 1984). Upper surfaces of the ORLs therefore coincide both with color and major climate transitions. In the cores under study the iron redox boundary also corresponds to the uppermost magnetite dissolution zone.

The depth of the modern redox boundary is determined by the latest preserved organic carbon enrichments. In profile A at 4°N no productivity pulses are indicated for MIS 2 and 4 (Fig. 15c, GeoB 4317-2). Accordingly, the modern iron redox boundary is located beneath termination II coinciding with the upper limit of the youngest ORL which in turn coincides with the uppermost major magnetite dissolution zone (Fig. 15e). The same holds true for sediments at 0° (Fig. 15B, GeoB 2908-7). Here, the presumable ultimate productivity pulse led to enhanced TOC accumulation during MIS 2 (Fig. 15k). As documented by enhanced Fe/κ_{nd} ratios (Fig. 15m) and earlier shown rock magnetic proxies the topmost magnetically altered section clearly corresponds to the respective ORL. Accordingly, the redox boundary is located beneath termination I (Fig. 15B). These conditions forbid to draw simple links between the color transition depth and modern regional productivity patterns. Never-theless, this position can be used to estimate the historical lateral expansion of the Equatorial upwelling system.

Oxidation of Organic Carbon

At the climate transitions organic matter accumulation decreased and bottom water conditions changed. In a second stage of early diagenesis the oxidative degradation of organic matter proceeded as a descending oxidation front into the sediment (Wilson et al. 1985; Wilson et al. 1986), owing to downward diffusion of O_2 from the overlying bottom water. At the investigated sites, reoxygenation was favoured by O_2 rich North Atlantic Bottom Water at the onset of interglacial periods. In glacial periods pronounced minima in $CaCO_3$ content are attributed to emplacement of corrosive Circumpolar Deep Water (de Menocal et al. 1993; Bickert and Wefer 1996) (Fig. 15f,n). As documented by comparison of TOC profiles with detailed XRF single sample barium measurements the youngest ORLs in cores GeoB 4317-2 and 2908-7 seemingly became progressively thinner during decomposition. The refractory element barium is likewise enriched in the sediment at times of enhanced productivity (Bishop 1988) and considered to be a more reliable proxy for productivity. It is therefore used to re-construct the original TOC profile (Mercone et al. 2000). Methods and limitations of this application were described by Gingele et al. (1999). The total barium content has to be corrected for the detrital fraction by using the following equation:

$$Ba_{excess} = Ba_{total} - \left(Al \cdot Ba/Al_{average}\right)$$

An average Ba/Al ratio of 0.0045 for detrital aluminosilicates as suggested by Kasten et al. (2001) is used here. The Ba_{excess} profiles of cores GeoB 4317-2 and 2908-7 (Fig. 15d,l) do not match the TOC records (Fig. 15c,k). The upper bound-aries of the ORLs were probably steepened by burn-down as described by Thomson et al. (1995) in sapropels and lie now somewhat below their original position. A distinct Ba maximum is found just above the major climate transition II in core GeoB 4317-2 (Fig. 15d) and delineates a relatively higher organic matter input at the onset of inter-glacial conditions. Remarkably, the Ba_{excess} profiles coincide much better with variations of the Fe/κ_{nd} ratio (Fig. 15e). This indicates that the degradation of the today 'missing' TOC was accompanied by dissolution of magnetic minerals. Obviously, the partial magnetite depletion still witnesses the prior TOC accumulation. Also in core GeoB 2908-7

(Fig. 15B) the comparison of the TOC and barium profiles suggests a partial oxidative burn-down of the TOC record (Fig. 15k,l). Again, the former extension of the ORL is still documented by the irreversible reductive loss of detrital magnetite (Fig. 15m).

In the suboxic ORLs below the oxidation front, Mn^{2+} and Fe^{2+} are released by reductive dissolution of Mn and Fe oxides through bacterially mediated degradation of organic matter. These released Mn^{2+} and Fe^{2+} ions diffused upwards and precipi-tated as Mn and Fe (hydr)oxides at the downward moving oxidation front above the youngest ORLs (Fig. 15g,h,o,p).

Scanning electron microscope investigations of core GeoB 2908-7 provide further information about the stage of diagenesis reached in the ORL in MIS 6. Authigenesis of framboidal and octa-hedral pyrite respectively implies intermittent an-oxic conditions of this section (Fig. 16a). In this case, the flux of organic matter to the sediment exceeds the oxygen flow and the degradation pro-ceeds by bacterial sulfate reduction. Diffusive fluxes of Fe, liberated by dissolution of iron oxides such as magnetite, resulted in the formation of pyrite, which was partly oxidized by an oxidation front at later stage (Fig. 16b).

Authigenesis of Magnetic Minerals above the Iron Redox Boundary

Three susceptibility patterns of crossprofile D hold evidence for authigenic magnetic enhancement above the active iron redox boundary (Fig. 17b-d). The initially inverse pattern of magnetic suscepti-bility and carbonate clearly deviates in a positive

---

Right: Fig. 15. Comparison of the uppermost sections of cores GeoB 4317-2 (transect A) and 2908-7 (transect B). The oxidized layer and the depth position of the color change, indicative for the iron redox boundary, identified by eye **(a,i)** and by the ratio of 670 nm to 470 nm **(b,j,** gray shaded). Please note the particular relation to major climate transitions. The color contrast results from the presence of both Fe(II) and organic material. In each case the youngest ORL is located just below the redox boundary **(c,k)**. Comparison of the productivity indices TOC and Ba_{excess} indicates **(d,l)** burn down of the upper reaches of the ORLs. After formation of the ORLs suboxic conditions led to dissolution of magnetite documented by enhanced Fe/κ_{nd} ratios **(e,m,** gray shaded). Pronounced minima in $CaCO_3$ content may also provoked by oxidation of organic matter and metabolically mediated CO_2 **(f,n)**. Downward moving oxidation fronts have oxidized the upper part of the ORLs and have brought into contact with upward diffusing Mn^{2+} and Fe^{2+} ions. This produces Mn and Fe (hydr)oxides at the top of the ORLs **(g,h,o,p,** gray shaded).

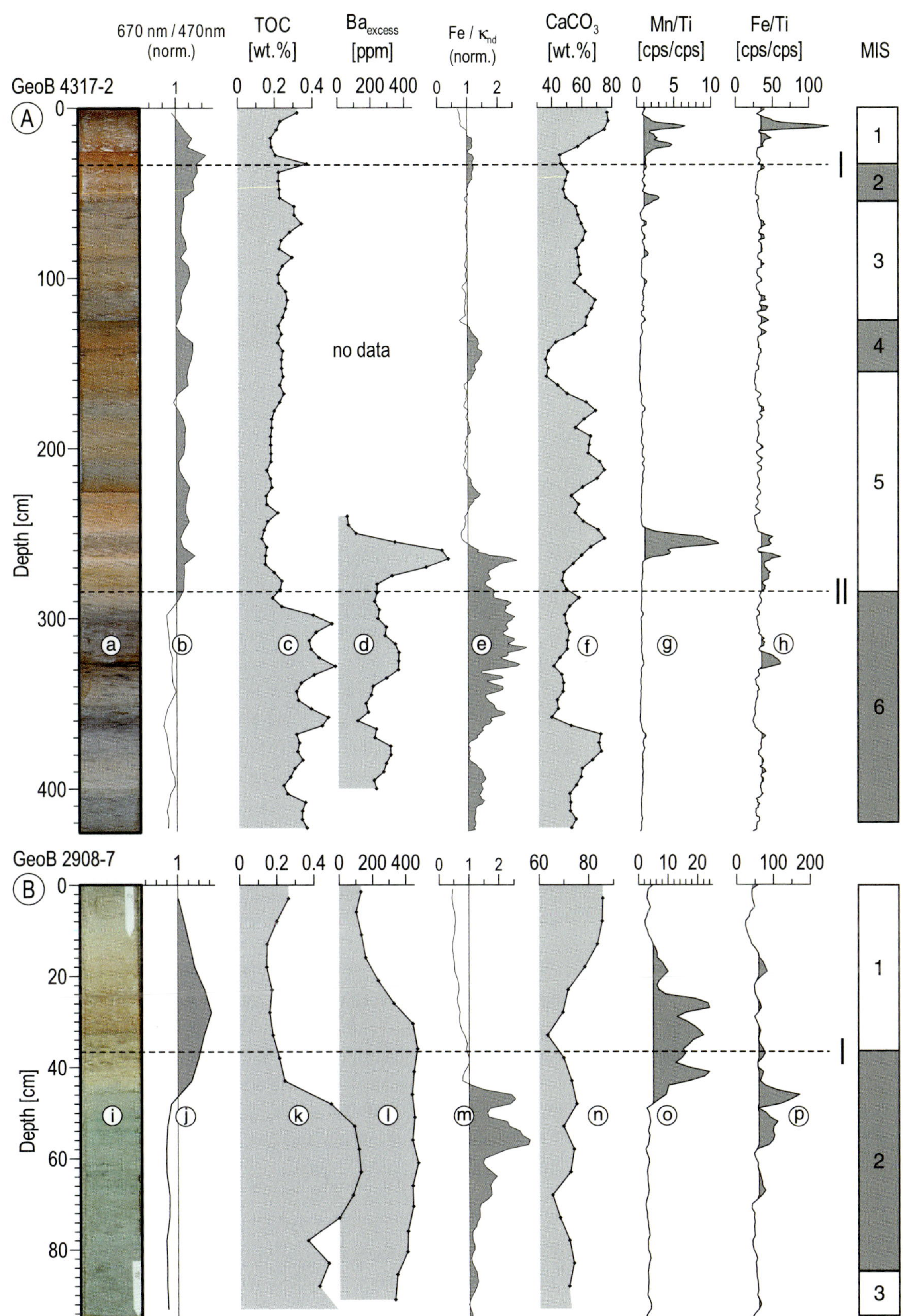
670 nm / 470nm
(norm.)
TOC
[wt.%]
Ba_excess
[ppm]
Fe / κ_nd
(norm.)
CaCO_3
[wt.%]
Mn/Ti
[cps/cps]
Fe/Ti
[cps/cps]
MIS
GeoB 4317-2
A
0
1
0 0.2 0.4
0 200 400
0 1 2
40 60 80
0 5 10
0 50 100
100
200
Depth [cm]
300
400
no data
a
b
c
d
e
f
g
h
1
2
3
4
5
6
I
II
GeoB 2908-7
B
0
1
0 0.2 0.4
0 200 400
0 1 2
60 80
0 10 20
0 100 200
20
40
Depth [cm]
60
80
i
j
k
l
m
n
o
p
1
2
3
I

Fig. 16. Scanning electron microscopy (SEM) micrographs of authigenic pyrite formation in an ORL of core GeoB 2908-7 (MIS 6). **a)** Octahedral and framboidal pyrite indicate intermittent anoxic conditions in the ORL. **b)** Partially dissolved octahedrons mark more oxic conditions at a later time.

direction within a relatively narrow depth range. In core GeoB 4317-2 the expected correlation between the non-$CaCO_3$ fraction and magnetic susceptibility is obvious. Yet, in core GeoB 2908-7, and more sharply defined in cores GeoB 1117-2 and 1041-3, deviations of the two parameters are indicated above the color transition and at termination I respectively. Distinct maxima in the magnetic susceptibility profiles indicate the presence of ferrimagnetic components of probably secondary origin as described by Tarduno et al. (1998). Karlin et al. (1987) reported evidence of metabolic authigenesis of fine-grained magnetite crystals by Fe(II)-oxidizing magnetotactic bacteria close to the color change marking the Fe(III)/Fe(II)-redox transition (Lyle 1983). Magnetotactic bacteria, which may find this optimal zone with greater efficiency by using the inclination of the earth's magnetic field (Stolz 1992) precipitate ultrafine-grained crystals of single domain, ferro-magnetic minerals intracellulary, which become a stable carrier of remanent magnetization. In sediments of the eastern South Atlantic Petermann and Bleil (1993) found highest concentrations of magnetotactic bacteria in a well-defined narrow subsurface layer. The depth of this layer was found to depend on the input and turnover of organic matter, linking a high flux of organic matter with high numbers of living magnetotactic bacteria. This relation is also found here along the NW-SE cross profile C. At the less productive location GeoB 4317-2 no authigenic maximum is indicated by magnetic susceptibility due to a more expanded oxic/anoxic transition zone. Towards the center of Equatorial Divergence increasingly larger magnetite precipitation peaks in the susceptibility signal of cores GeoB 2908-7, 1117-2 and 1041-3 (Thießen 1993) mirror a narrower zonation of terminal electron acceptors (Froelich et al. 1979) due to higher input of organic matter. Biomineralization seems to be restricted to the oxic/anoxic transition zone where iron is available in its soluble form.

Conclusions

As a contribution to the reconstruction of Late Quaternary material budgets and current systems in the Equatorial Atlantic Ocean, we co-interpret basin-wide spatio-temporal stratigraphic information contained in rock magnetic, sedimentological and geochemical records to gain deeper insight into the nature and relevance of the sediment accumulation and diagenetic alteration processes. Four profiles made up of 16 cores dissect the Equatorial Divergence high productivity region, the Saharan dust and the Amazon suspension plume. All investigated C_{org}, $CaCO_3$ and, to some extent, even terrigenous records were shaped or at least overprinted by diagenetic effects. According to our study, early diagenesis in the central Equatorial

Cross profile D

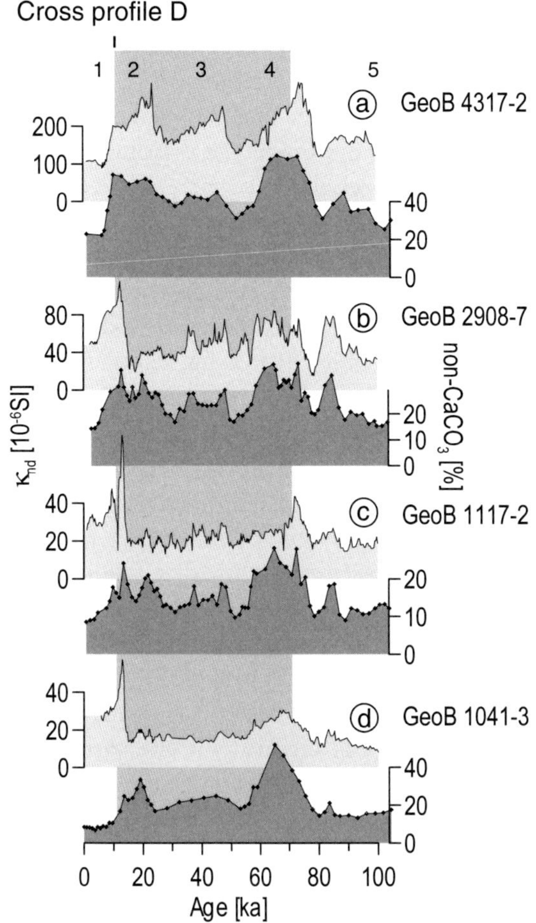

Fig. 17. Non-CaCO$_3$ and magnetic susceptibility records of cross profile D. Sharp maxima in magnetic suscepti-bility above the Fe^{3+}/Fe^{2+} redox boundary are attributed to bacterial magnetite (**b-d,** not present in **a).**

Atlantic is a leading issue for material flux and budget calculations and a serious problem for productivity and primary deposition estimates.

As oxygen isotope records were not available for the more recently collected cores, our ocean-spanning stratigraphic network was based on the variations of Ca and/or CaCO$_3$ content, which were found to be reasonably coherent throughout the entire region. Interestingly, not just the lysoclinal but even the permanently supra-lysoclinal locations show similar carbonate signatures. All CaCO$_3$ records are inversely correlated to the respective C$_{org}$ records with increasing coherency towards the center of the Equatorial Divergence Zone. Both phenomena are likely to be the effect of synsedi-mentary carbonate dissolution which is driven by free CO$_2$ released by aerobic bacterial metabolism in the subsurface sediment. Over time, sufficient quantities of C$_{org}$ are available for this process since typical benthic Equatorial Atlantic C$_{org}$/CaCO$_3$ rain rate ratios of about 1:1 transform into C$_{org}$/CaCO$_3$ accumulation rate ratios of only 1:100 to 1:1000 - this difference offers an enormous potential for subsurface carbonate diagenesis. C$_{org}$/CaCO$_3$ ratios averaged over glacial periods were found to be up to three times higher than those averaged over interglacials. The systematic glacial increase in C$_{org}$% is certainly a consequence of enhanced productivity, quicker burial and lower bottom water oxygenation. The fact that even supralysoclinal CaCO$_3$ accumulation rates systematically decrease in glacial periods is clear evidence for the lead of dissolution over productivity in shaping the CaCO$_3$% records.

Non-carbonate, i.e. terrigenous content is spatially controlled by source distance and shows a strong N-S decline. The temporal variations, however, have apparently more to do with dilution effects of a varying carbonate deposition than with changes in terrigenous accumulation rates. We found non-carbonate (100% - CaCO$_3$%), iron and susceptibility records to be highly collinear down to fine pattern details. Susceptibilities were partly overprinted by reductive diagenesis of magnetite. Magnetically depleted and organically enriched layers coincide perfectly and identify the two solid-state redox reaction partners. The iron to susceptibility ratio Fe/κ proved to be a sensitive proxy for suboxic magnetite diagenesis, indicating pervasive postsedimentary iron reduction within organic rich layers formed during cold periods throughout the Equatorial Divergence region. At the Ceará and Sierra Leone Rises, magnetite dissolution is barely detectable. On both flanks of the Mid-Atlantic ridge we find intense iron reduction during marine oxygen isotope stages 6, 10 and 12, in particular at the terminations. Towards the equator, the diagenesis-affected zones become narrower, but also more frequent and finally reflect a full precessional rhythm until individual diagenesis peaks merge into broader magnetite-depleted zones.

Magnetogranulometric and -mineralogical proxies show highly uniform, SPECMAP type climate signals throughout the study region except for overprinted sections. Interglacials have relatively finer magnetite grain size and relatively lower hematite content. The climate signature is generally better expressed in M_{ar}/M_{ir} as in $S_{-0.3T}$. The amplitude of M_{ar}/M_{ir} variations decreases from west to east as a result of the differing contributions of Amazon and Sahara. We can infer from the three selective rock magnetic parameters M_{ar}, M_{ir}, M_{hir}, that the above discussed magnetic ratios and their climatic variations are determined by mixing of individual sources with specific characteristics. The fine-grained, potentially bacterial or fine detrital magnetite phase has a very stable flux through time, particularly in the western part. The flux of the coarse grained magnetite phase varies strongly with Amazon discharge into the open ocean, with peaks corresponding to glacial sea level fall and low stand. Hematite fluxes show even higher variability. Peaks indicate intensified Saharan dust fall during glacial periods.

A detailed quantitative synopsis of glacial and interglacial accumulation rate averages with regard to total sediment, C_{org}, $CaCO_3$, non-$CaCO_3$, and susceptibility was carried out in order to detect zonal and meridional trends. By far the most prominent features are the 4°N to 4°S decrease in terrigenous accumulation (6:1) and the Equatorial Divergence high in glacial C_{org} accumulation, which decays much faster south- than northwards. The glacial increase in C_{org} and proportional decrease in $CaCO_3$ accumulation clearly reflects sedimentary carbonate diagenesis. This trend does not continue into the supralysoclinal oligotrophic South Atlantic.

A common feature of the Equatorial Atlantic sediment column is a sharp color change from reddish brown to greenish gray hues located at about 3 m depth at 4°N and at about 0.5 m depth at 0°. These transitions mark the modern Fe^{3+}/Fe^{2+} redox boundary and are typically located just above the uppermost ORL. Their depth is obviously not defined by the steady supply and degradation rate of organic carbon. Instead, it appears to be stabilized by the C_{org} reservoir of the topmost preserved productivity pulse. At 4°N, this 'sapropelic' layer coincides with glacial stage 6, at 0° with glacial stage 2. These Fe^{3+}/Fe^{2+} redox depth positions may have previously been shallower than suggested by the present color reflectance and carbon records. An interesting case is core GeoB 4317-2 from a 4°N Mid-Atlantic Ridge location. Not just barium content, but also our magnetite dissolution index indicate, that the productivity pulse in glacial stage 6 lasted longer than indicated by the C_{org} profile. The record was later depressed by 'burn-down', i.e. organic carbon depletion by a downward diffusing oxidation front. Traces of magnetite dissolution during glacials 2 and 4 even suggest, that the Fe^{3+}/Fe^{2+} redox boundary may have been as shallow as 0.35 m during the termination period, before it stepwise retreated to the present depth of 2.85 m with time and increasing bottom water oxygenation. Solid phase records of subsurface Fe^{3+} and Mn^{4+} precipitates agree with this assumption. Authigenic precipitates of relocated ferrous iron can often be detected by magnetic methods. In all sediment cores from the Equatorial Divergence region, a susceptibility spike appears just above the modern redox boundary, always corresponding to an age of about 14 ka. According to rock magnetic characteristics, this magnetite phase is probably constituted of fossil bacterial magnetosomes. Northerly cores with a more expanded redox zonation do not exhibit this magnetic enhancement.

Acknowledgement

We thank officers and crews of RV *Meteor* for their efficient support during cruises M 29/3 and M 38/1. U. Röhl made the XRF scanner measure-ments possible and was very helpful. This study was funded by the Deutsche Forschungsgemein-schaft (Sonderforschungsbereich 261 at Bremen University, contribution No. 386). J.A. Funk was supported by the Deutsche Forschungsgemein-schaft in the framework of Graduiertenkolleg 221. T. von Doberneck is a visiting research fellow at Utrecht University supported by the Netherlands Research Center for Integrated Solid Earth Science (ISES). The two peer reviewers, D. Rey and M. Urbat, have greatly improved this manuscript and are gratefully acknowledged.

References

Archer D, Maier-Reimer E (1994) Effect of deep-sea sedimentary calcite preservation on atmospheric CO_2 concentration. Nature 367: 260-263

Bassinot FC, Labeyrie LD, Vincent E, Quidelleur X, Shakelton NJ, Lancelot Y (1994) The astronomical theory of climate and the age of the Brunhes-Matuyama magnetic reversal. Earth Planet Sci Lett 126: 546-559

Bickert T (1992) Rekonstruktion der spätquatären Bodenwasserzirkulation im östlichen Südatlantik über stabile Isotope bentischer Foraminiferen. Ber Fachber Geowiss, Univ Bremen 27, 205 p

Bickert T, Wefer G (1996) Late Quaternary deep water circulation in the South Atlantic: Reconstruction from carbonate dissolution and benthic stable isotopes. In: Wefer G, WH Berger, Siedler G, Webb DJ (eds) The South Atlantic: Present and Past Circulation. Springer, Berlin, pp 599-620

Bishop JKB (1988) The barite-opal-organic carbon association in oceanic particulate matter. Nature 332: 341-343

Bloemendal J, King JW, Hall FR, Doh S-J (1992) Rock magnetism of late Neogene and Pleistocene deep-sea sediments: Relationship to sediment source, diagenetic processes, and sediment lithology. J Geophys Res 97: 4361-4375

Bloemendal J, Lamb B, King J (1988) Paleoenvironmental implications of rock - magnetic pro-perties of late Quaternary sediment cores from the eastern equatorial Atantic. Paleoceanography 3: 61 -87

Bonifay D, Giresse P (1992) Middle to Late Quaternary sediment flux and post- depositional processes between the continental slope off Gabon and the Mid-Guinean margin. Mar Geol 106: 107-129

Boyle EA, Keigwin LD (1982) Deep circulation of the North Atlantic over the last 200,000 years: Geochemical evidence. Science 218: 784-787

Boyle EA, Keigwin LD (1985) Comparison of Atlantic and Pacific paleochemical records for the last 215,000 years: Changes in deep ocean circulation and chemi-cal inventories. Earth Planet Sci Lett 76: 135-140

Broecker WS, Denton GH (1989) The role of oceanatmosphere reorganisations in glacial cycles. Geochim Cosmochim Acta 53: 2465-2501

Brummer GJA, Van Eijden AJM (1992) 'Blue-ocean' paleoproductivity estimates from pelagic carbonate mass accumulation rates. Mar Micropaleontol 19: 99-117

Canfield DE, Berner RA (1987) Dissolution and pyriti-zation of magnetite in anoxic marine sediments. Geochim Cosmochim Acta 51: 645-659

Colley S, Thomson J, Wilson TRS, Higgs NC (1984) Post-depositional migration of elements during diagenesis in brown clay and turbidite sequences in the North East Atlantic. Geochim Cosmochim Acta 48: 1223-1235

Curry WB, Lohmann GP (1986) Late Quaternary sedimentation at the Sierra Leone Rise (eastern equatorial Atlantic Ocean). Mar Geol 70: 223-250

Curry WB, Lohmann GP (1990) Reconstructing past particle fluxes in the tropical Atlantic Ocean. Paleoceanography 5: 487-506

Curry WB (1996) Late Quaternary deep circulation in the western Equatorial Atlantic. In: Wefer G, Berger WH, Siedler G, Webb D (eds) The South Atlantic: Present and Past Circulation. Springer, Berlin, pp 577-598

Damuth JE (1977) Late Quaternary sedimentation in the western equatorial Atlantic. Geol Soc Am Bull 88: 695-710

deMenocal PB, Ruddiman WF, Pokras EM (1993) Influences of high- and low- latitude processes on African terrestrial climate: Pleistocene eolian records from equatorial Atlantic ocean drilling program Site 663. Paleoceanography 8: 209-242

Dunlop DJ (1973) Superparamagnetic and single-domain threshold sizes in magnetite. J Geophys Res 78: 1780–1793

Duplessy J C, Labeyrie L, Paterne M, Hovine S, Fichefet T, Duprat J, Labracherie M (1996) High latitude deep water sources during the Last Glacial Maximum and the intensity of the global oceanic circulation. In: Wefer G, Berger WH, Siedler G, Webb D (eds) The South Atlantic: Present and Past Circulation. Sprin-ger, Berlin, pp 445-460

Dupont LM, Jahns S, Marret F, Ning S (2000) Vegetation change in equatorial West Africa: Time-slices for the last 150 ka. Palaeogeogr Palaeoclimatol Palaeoecol 155: 95-122

Fischer G and participants (1998) Report and preliminary results of METEOR cruise M38/1, Las Palmas - Recife, 25.1.-1.3.1997. Ber Fachber Geowiss, Univ Bremen 94, 178 p

François R, Bacon MP (1991) Variations in terrigenous input into the deep Equatorial Atlantic during the past 24,000 years. Science 251: 1473-1475

François R, Bacon MP, Suman DO (1990) Thorium 230 profiling in deep-sea sediments: High-resolution records of flux and dissolution of carbonate in the equatorial Atlantic during the last 24000 years. Paleoceanography 5: 761–787

Frederichs T, Bleil U, Däumler K, von Dobeneck T, Schmidt A (1999) The magnetic view on the marine paleoenvironment: Parameters, techniques, and potentials of rock magnetic studies as a key to paleoclimatic and paleoceanographic changes. In: Fischer G, Wefer G (eds) Use of Proxies in Paleoceanography: Examples from the South Atlantic. Springer, Berlin, pp 575-599

Froelich PN, Klinkhammer GP, Bender ML, Luedtke NA, Heath GR, Cullen D, Dauphin P, Hammond D, Hartman B, Maynard V (1979) Early oxidation of organic matter in pelagic sediments of the eastern equatorial Atlantic: Suboxic diagenesis. Geochim Cosmochim Acta 43: 1075-1090

Gardner JV, Burkle LH (1975) Upper Pleistocene Ethmodiscus rex from the eastern equatorial Atlantic. Micropaleontology 21: 236-242

Gingele F, Dahmke A (1994) Discrete barite particles and barium as tracers of paleoproductivity in South Atlantic sediments. Paleoceanography 9: 151-168

Gingele FX, Zabel M, Kasten S, Bonn WJ, Nürnberg CC (1999) Biogenic barium as a proxy for paleoproductivity: Methods and limitations of application. In: Fischer G, Wefer G (eds) Use of proxies in paleoceanography: Examples from the South Atlantic. Springer, Berlin, pp 345-364

Haese RR, Petermann H, Dittert L, Schulz HD (1998) The early diagenesis of iron in pelagic sediments: A multidisciplinary approach. Earth Planet Sci Lett 157: 233-248.

Higgs HC, Thomson J, Wilson TRS, Croudace IW (1994). Modification and complete removal of eastern Medi-terranean sapropels by postdepositional oxidation. Geology 22: 423-426

Jansen JHF, Van der Gaast SJ, Koster AJ (1998) CORTEX, a shipboard XRF-scanner for element analyses in split sediment cores. Mar Geol 151: 143-153

Johns WE, Schott FA, Zantopp RJ, Evans RH (1990) The North Brazil Current retroflection: Seasonal structure and eddy variability, J Geophys Res 95: 22103-22120

Karlin R, Levi S (1983) Diagenesis of magnetic minerals in recent haemipelagic sediments. Nature 303: 327-330

Karlin R, Levi S (1985) Geochemical and sedimentologi-cal control of the magnetic properties of hemipelagic sediments. J Geophys Res 90: 10,373-10,392

Karlin R, Lyle M, Heath GR (1987) Authigenic magnetic formation in suboxic marine sediments. Nature 326: 490-493

Karlin R (1990) Magnetite diagenesis in marine sediments from the Oregon continental margin. J Geophys Res 95: 4405-4419

Kasten S, Haese RR, Zabel M, Rühlemann C, Schulz HD (2001) Barium peaks at glacial terminations in sediments of the equatorial Atlantic Ocean - relicts of deglacial productivity pulses? Chem Geol 175: 635-651

Kidd RB, Cita MB, Ryan WBF (1978) Stratigraphy of eastern Mediterranean sapropel sequences recovered during DSDP Leg 42A and their paleoenvironmental significance. In: Hsü KJ et al. (eds) Initial Reports of the Deep-Sea Drilling Project. US Government Printing Office, Washington, pp 421-443

Kumar N, Embley RW (1977) Evolution and origin of Ceará Rise: An aseismic rise in the western equatorial Atlantic. Geol Soc Am Bull 88: 683-694

Kutzbach JE (1981) Monsoon climate of the early Holocene: Climatic experiment with the Earth's orbital parameters for 9000 years ago. Science 214: 59-61

Lutjeharms JRE (1996) The exchange of water between the South Indian and South Atlantic Oceans. In: Wefer G, Berger WH, Siedler G, Webb D (eds) The South Atlantic: Present and Past Circulation. Springer, Berlin, pp 125-162

Lyle M (1983) The brown-green color transition in marine sediments: A marker of the Fe(III)-Fe(II) redox boundary. Limnol Oceanogr 28: 1026-1033

Lyle M (1988) Climatically forced organic carbon burial in equatorial Atlantic and Pacific Oceans. Nature 335: 529-532

Maher BA (1988) Magnetic properties of some synthetic submicron magnetites. Geophys J 94: 83-96

Martin WR, Sayles FL (1996) CaCO3 dissolution in sediments of the Ceara Rise, western equatorial Atlantic. Geochim Cosmochim Acta 60: 243-263

Maslin M, Mikkelsen N (1997) Amazon fan mass-transport deposits and underlying interglacial deposits: Age estimates and fan dynamics. Proc ODP Sci Results 155: 353-365

Meinecke G (1992) Spätquartäre Oberflächenwassertemperaturen im östlichen äquatorialen Atlantik. Ber Fachber Geowiss, Univ Bremen 29, 181 p

Mercone D, Thomson J, Croudace IW, Siani G, Paterne M, Troelsta S (2000) Duration of S1, the most recent sapropel in the eastern Mediterranean Sea, as indicated by accelerator mass spectrometry radiocarbon and geochemical evidence. Paleoceanography 15: 336 - 347

Milliman JD, Summerhayes CP, Barretto HT (1975) Qua-ternary sedimentation on the Amazon continental margin: A model. Geol Soc Am Bull 86: 610-614

Milliman JD (1993) Production and accumulation of calcium carbonate in the ocean: Budget of a non-steady. Glob Biochem Cycl 7: 927-957

Mix AC, Rugh W, Pisias N G, Veirs S (1992) Color reflec-tance spectroscopy: A tool for rapid charakterization of deep-sea sediments. Proc ODP, Initial Reports 138: 67-77

Müller PJ, Erlenkeuser H, v Grafenstein R (1983) Glacial - interglacial cycles in oceanic productivity inferred from organic carbon contents in eastern north Atlantic sediment cores. In: Thiede J, Suess E (eds) Coastal Upwelling Part B. Plenum Press, New York, USA, pp 365 - 398

Müller PJ, Hartmann M, Suess E (1988) The chemical environment of pelagic sediments. In: Halbach P, Friedrich G, von Stackelberg U (eds) The manganese nodule belt of the Pacific Ocean. Enke, Stuttgart, 254 p

Oppo DW, Fairbanks RG (1987) Variability in the deep and intermediate water circulation of the Atlantic Ocean during the past 25,000 years: Northern Hemis-phere modulation of the Southern Ocean. Earth Planet Sci Lett 86: 1-15

Parry LG (1965) Magnetic properties of dispersed magnetite powders. Philosophical Magazine 11: 303–312

Passier HF, Dekkers MJ, de Lange GJ (1998) Sediment chemistry and magnetic properties in an anomalously reducing core from the eastern Mediterranean Sea. Chem Geol 152: 287-306

Passier HF, Dekkers MJ (2002) Assessment of the formation of iron oxides in the active oxidation front above sapropel S l in the eastern Mediterranean Sea by low- temperature magnetic properties. Geophys J Int, in press

Petermann H, Bleil U (1993) Detection of live magneto-tactic bacteria in South Atlantic deep-sea sediments. Earth Planet Sci Lett 117: 223-228

Pfeifer K, Hensen C, Adler M, Wenzhöfer F, Weber B, Schulz HD (2002) Modeling of subsurface calcite dissolution - including the respiration and re-oxidation processes of marine sediments in the region of the equatorial upwelling off Gabon. Geochim Cosmochim Acta 66: 4247-4259

Pickard GL, Emery WJ (1990) Descriptive Physical Oceanography. Pergamon Press Oxford UK, 320 p

Rath S, Wagner T, Henrich S (1999) Pleistozäne Schwankungen im Staubeintrag und in der Karbo-naterhaltung im äquatorialen Atlantik (ODP Site 663) zwischen 480 und 390 Ka: Hinweise aus Korngrößen-analysen. Zentralblatt Geologie-Paläontologie Teil I H 7-10: 1-19

Ratmeyer V, Balzer W, Bergametti G, Chiapello I, Fischer G, Wyputta U (1999a) Seasonal impact of mineral dust on deep-ocean particle flux in the eastern subtropical Atlantic Ocean. Mar Geol 159: 241-252

Ratmeyer V, Fischer G, Wefer G (1999b) Lithogenic particle fluxes and grain size distributions in the deep ocean off northwest Africa: Implications for seasonal changes of aeolian dust input and downward trans-port. Deep-Sea Res 46: 1289-1337

Raymo ME, Ruddiman WF, Shakleton NJ, Oppo DW (1990) Evolution of Atlantic - Pacific $\delta^{13}C$ gradients over the last 2.5 my. Earth Planet Sci Lett 97: 353-368

Raymo ME, Oppo DW, Curry W (1997) The mid-Pleistocene climate transition: A deep-sea carbon isotopic perspective. Paleoceanography 12: 546-559

Rhein M, Schott F, Fischer J, Send U, Stramma L (1996) The deep water regime in the Equatorial Atlantic. In: Wefer G, Berger WH, Siedler G, Webb D (eds) The South Atlantic: Present and Past Circulation. Springer, Berlin, pp 261-271

Robinson SG, Sahota JTS, Oldfield F (2000) Early diagenesis in North Atlantic abyssal plain sediments characterized by rock magnetic and geochemical indices. Mar Geol 163: 77-107

Röhl U, Abrams LJ (2000) High-resolution, downhole, and nondestructive core measurements from Sites 999 and 1001 in the Caribbean Sea: Application to the Late Paleocene thermal maximum. In: Leckie RM, Sigurdsson H, Acton GD, Draper G (eds) Proc ODP Sci Results. College Station Texas, pp 191-203

Rossignol-Strick M, Nesteroff W, Olive P, Vergnaud-Grazzini C (1982) After the deluge: Mediterranean stagnation and sapropel formation. Nature 295: 105-110

Ruddiman WF (1997) Tropical Atlantic terrigenous fluxes since 25,000 yrs BP. Mar Geol 136: 189-207

Ruddiman WF, Janecek TR (1989) Pliocene-Pleistocene biogenic and terrigenous fluxes at equatorial Atlantic Sites 662, 663, and 664. In: Ruddiman W, Sarnthein M (eds) Proc ODP Sci Results. College Station, Texas pp 211-240

Rühlemann C, Frank M, Hale W, Mangini A, Mulitza S, Müller PJ, Wefer G (1996) Late Quaternary productivity changes in the western equatorial Atlantic: Evidence from 230Th-normalized carbonate and organic carbon accumulation rates. Mar Geol 135: 127-152

Sahota JTS, Robinson SG, Oldfield F (1995) Magnetic measurements used to identify paleoxidation fronts from the Madeira abyssal plain. Geophys Res Lett 22: 1961-1964

Sarnthein M, Tetzlaff G, Koopmann B, Wolter K, Pflaumann, U 1981. Glacial and interglacial wind regimes over the eastern subtropical Atlantic and North-West Africa. Nature 293: 193-196

Sarnthein M, Thiede J, Pflaumann U, Erlenkeuser H, Fütterer D, Koopmann B, Lange H, Seibold E (1982) Atmospheric and oceanic circulation patterns off Northwest Africa during the past 25 million years. In: von Rad U, Hinz K et al. (eds) Geology of the Northwest African Continental Margin. Springer, Berlin, pp 545-604

Sarnthein M, Winn K, Jung SJA, Duplessy J-C, Labeyrie L, Erlenkeuser H, Ganssen G (1994) Changes in east Atlantic deepwater circulation over the last 30,000 years: Eight time slice reconstructions. Paleoceanography 9: 209-267

Schmieder F, von Dobeneck T, Bleil U (2000) The Mid-Pleistocene climate transition as documented in the deep South Atlantic Ocean: Initiation, interim state and terminal event. Earth Planet Sci Lett 179: 539-549

Schneider RR, Schulz HD, Hensen C (2000) Marine carbonates: Their formation and destruction. In: Schulz HD, Zabel M (eds) Marine Geochemistry. Springer, Berlin, pp 283-307

Schulte S, Bard E (2003) Past changes of biologically mediated dissolution of calcite above the chemycal lysocline documented in Indian Ocean sediments. Quater Sci Rev 22: 1757-1770

Schulz, HD and participants (1991) Report and preliminary results of *Meteor* cruise M16/2, Recife - Belem, 28.4.-20.5.1991. Ber Fachber Geowiss, Univ Bremen 19, 149 p

Showers WJ, Angle DG (1986) Stable isotopic characterization of organic carbon accumulation on the Amazon continental shelf. Continent Shelf Res 6: 227-244

Stolz JF (1992) Magnetotactic Bacteria: Biomineralization, Ecology, Sediment Magnetism, Environ-mental Indicator. In: Skinner HGW, Fitzpatrick RW (eds) Biomineralization processes of iron and manganese. Catena-Verl Cremlingen-Destedt, pp 133-145

Street-Perrott FA, Perrott RA (1990) Abrupt climate fluctuations in the tropics: The influence of Atlantic Ocean circulation. Nature 343: 607-612

Tarduno JA (1994) Temporal trends of magnetic dissolu-tion in the pelagic realm: Gauging paleoproductivity? Earth Planet Sci Lett 123: 39-48

Tarduno JA, Tian W, Wilkison S (1998) Biogeochemical remanent magnetization in pelagic sediments of the western equatorial Pacific Ocean. Geophys Res Lett 25: 3987-3990

Tarduno JA, Wilkison SL (1996) Non-steady state magnetic mineral reduction, chemical lock-in, and delayed remanence acquisition in pelagic sediments. Earth Planet Sci Lett 144: 315-326

Thießen W (1993) Magnetische Eigenschaften von Sedimenten des östlichen Südatlantiks und ihre paläozeanographische Relevanz. Ber Fachber Geowiss, Univ Bremen 4, 170 p

Thomson J, Higgs NC, Wilson TRS, Croudace IW, de Lange GJ, van Santvoort, PJM (1995) Redistribution and geochemical behaviour of redox - sensitive elements around S 1, the most recent eastern Mediterranean sapropel. Geochim Cosmochim Acta 59: 3487-3501

Tiedemann R, Sarnthein M, Shackleton NJ (1994) Astronomic timescale for the Pliocene Atlantic $\delta^{18}O$ and dust flux records of Ocean Drilling Program Site 659. Paleoceanography 9: 619-638

Tiedemann R, Sarnthein M, Stein R (1989) Climatic changes in the western Sahara: Aeolo-marine sediment record of the last 8 million years (Sites 657-661) In: Ruddimann W, Sarnthein M et al. (eds) Proc ODP Sci Results. Vol 108, pp 241-277

van Kreveld SA, Knappertsbusch M, Ottens J, Ganssen GM, van Hinte JE (1996) Biogenic carbonate and ice-rafted debris (Heinrich layer) accumulation in deep-sea sediments from a Northeast Atlantic piston core. Mar Geol 131: 21-46

van Santvoort PJM, de Lange GJ, Thomson J, Cussen H, Wilson TRS, Krom MD, Ströhle K (1996) Active post-depositional oxidation of the most recent sapropel (S1) in sediments of the eastern Mediterranean Sea. Geochim Cosmochim Acta 60: 4007-4024

van Santvoort PJM, de Lange GJ, Langereis CG, Dekkers MJ, Paterne M (1997) Geochemical and paleomag-netic evidence for the occurence of 'missing' sapropels in eastern Mediterranean sediments. Paleoceanography 12: 773-786

Verardo DJ, McIntyre A (1994) Production and destruction: Control of biogenous sedimentation in the tropical Atlantic 0-300,000 years BP. Paleoceanography 9: 63-86

von Dobeneck T (1996) A syematic analysis of natural magnetic mineral assemblages based on modelling hysteresis loops with coercivity-related hyberbolic basis functions. Geophys J Int 124: 675-694

von Dobeneck T (1998) The concept of 'partial susceptibilities'. Geologica Carpathica 49: 228-229

von Dobeneck T, Schmieder F (1999) Using rock magnetic proxy records for orbital tuning and extented time series analyses into the Super- and Sub-Milankovitch bands. In: Fischer G, Wefer G (eds)

Use of Proxies in Paleoceanography: Examples from the South Atlantic. Springer; Berlin, pp 601-633

Wagner T (2002) Organic sedimentation in the Equatorial Atlantic: Evolution from Cretaceous to early Quaternary depositional environments. Palaeogeogr Palaeoclimatol Palaeoecol 179: 113-147

Wagner T (2000) Control of organic carbon accumulation in the late Quaternary equatorial Atlantic (Ocean Drilling Program sites 664 and 663): Productivity versus terrigenous supply. Paleoceanography 15: 181-199

Wagner T, Dupont LM (1999) Terrestrial organic matter in marine sediments: Analytical approaches and eolian-marine records in the Central Equatorial Atlantic. In: Fischer G, Wefer G (eds) Use of Proxies in Paleoceanography: Examples from the South Atlantic. Springer, Berlin, pp 547-574

Wefer G, Berger W H, Bickert T, Donner B, Fischer G, Kemle-von Mücke S, Meinecke G, Müller P J, Mulitza S, Niebler H-S, Pätzold J, Schmidt H, Schneider R R, Segl M (1996) Late Quaternary surface circulation in the South Atlantic: The stable isotope record and implications for heat transport and productivity. In: Wefer G, Berger WH, Siedler G, Webb D (eds) The South Atlantic: Present and Past Circulation. Springer, Berlin, pp 461-502

Wefer G and participants (1988) Report and preliminary results of *Meteor* cruise M6/6, Libreville - Las Palmas, 18.2.-23.3.1988. Ber Fachber Geowiss, Univ Bremen 3, 97 p

Wefer G and participants (1989) Report and preliminary results of *Meteor* cruise M9/4, Dakar - Santa Cruz, 19.2.-16.3.1989. Ber Fachber Geowiss, Univ Bremen 7, 103 p

Wefer G and participants (1991) Report and preliminary results of *Meteor* cruise M16/1, Pointe Noir - Recife, 27.3.-25.4.1991. Ber Fachber Geowiss, Univ Bremen 18, 120 p

Wefer G and participants (1994) Report and preliminary results of *Meteor* cruise M23/3, Recife - Las Palmas, 21.3.-12.4.1993. Ber Fachber Geowiss, Univ Bremen 44, 71 p

Wilson TRS, Thomson J, Colley S, Hydes DJ, Higgs NC, Sorensen J (1985) Early organic diagenesis: The significance of progressive subsurface oxidation fronts in pelgic sediments. Geochim Cosmochim Acta 49: 811-822

Wilson TRS, Thomson J, Hydes DJ, Colley S, Culkin F, Sörensen J (1986) Oxidation fronts in pelagic sedi-ments: Diagenetic formation of metal-rich layers. Science 232: 972-975

Wilson TRS, Thomson J (1998) Calcite dissolution accompanying early diagenesis in turbiditic deep ocean sediments. Geochim Cosmochim Acta 62: 2087-2096

Zabel M, Bickert T, Dittert L, Haese RR (1999) Significance of the sedimentary Al:Ti ratio as an indicator for variations in the circulation patterns of the equatorial North Atlantic. Paleoceanography 14: 789-799

The Late Pleistocene South Atlantic and Southern Ocean Surface - A Summary of Time-Slice and Time-Series Studies

R. Gersonde[1*], A. Abelmann[1], G. Cortese[1], S. Becquey[1], C. Bianchi[1], U. Brathauer[1], H.-S. Niebler[1,2], U. Zielinski[1] and J. Pätzold[2]

[1]*Alfred Wegener Institut für Polar- und Meeresforschung, Columbusstrasse, 27515 Bremerhaven, Germany*
[2]*Universität Bremen, Fachbereich Geowissenschaften, Klagenfurter Strasse, 28359 Bremen, Germany*
** corresponding author (e-mail): rgersonde@awi-bremerhaven.de*

Abstract: Central to global climate evolution is the paleoceanographic development of the South Atlantic as it represents the passageway for inter-hemispheric heat exchange within global thermohaline circulation (THC). Processes in the adjacent Southern Ocean regulate the heat import into the South Atlantic via the Agulhas "warm water route"(WWR) and the Drake Passage "cold water route"(CWR), and amplify climate change through various feedback mechanisms and tele-connections. For paleoceanographic reconstruction an inventory of new data sets and methods is now available, allowing for the estimation of Pleistocene sea-surface water temperatures and sea-ice distribution on time-slices and time-series based on the calcareous and siliceous microfossil record. Reconstruction of the Last Glacial Maximum (LGM) reveals distinct cooling in the Southern Ocean (up to 4 - 6 °C) accompanied by an expansion of winter and summer sea ice, cooling in the African upwelling regimes (up to 10°C) and in the Equatorial Atlantic (4 - 5 °C), but the Subtropical Gyre region remains relatively warm and unchanged compared with the present. While the WWR was not strongly altered during the LGM, heat transport via the CWR was most probably much weaker. The reconstruction of time-slices representing a warm climate end-member at the onset of the last climate cycle documents a distinct lead of southern high-latitudes in global climate development that also affects the south-west African upwelling regions. It is at the Marine Isotope Stage (MIS) 6/MIS 5 transition when Southern Ocean surface temperatures reach maximum values and sea ice is at a minimum, marking a period of South Atlantic heat piracy. During the isotopic minimum of MIS 5.5, the tropical South Atlantic was slightly colder than at present, likely the result of an enhanced poleward heat export. Time-series studies from key areas document that climate variability related to orbital forcing is overprinted by THC changes driven by meltwater injections into the North Atlantic and the Southern Ocean, changes in atmospheric circulation and greenhouse gas concentration, as well as sea ice that amplify climate change at global, hemispheric and regional scales. The study of centennial-scale variability during interglacial optima, such as MIS 5.5 and MIS 11, suggests that the presence of large ice sheets, meltwater events, changes in greenhouse gas concentration and sea-ice distribution are not the only prerequisite to trigger millennial-centennial-scale variability, but that another external agent, changes in solar irradiance, must be considered as an important factor in climate development.

Introduction

As a result of Cenozoic plate tectonic opening and closure of oceanic gateways, the South Atlantic has become an unique passageway within the global ocean conveyor belt that links the North Atlantic

From WEFER G, MULITZA S, RATMEYER V (eds), 2003, *The South Atlantic in the Late Quaternary: Reconstruction of Material Budgets and Current Systems.* Springer-Verlag Berlin Heidelberg New York Tokyo, pp 499-529

500 Gersonde et al.

with the North Pacific. The functionality of the global conveyor is crucial for the understanding of regional and global climate development, as it controls the distribution of heat and nutrients in the global ocean. At present, interglacial conditions, the conveyor circulation results in a relatively warm North Atlantic with elevated surface water temperatures expanding to high Northern latitudes. In the southern hemisphere, a broad cold-water belt with surface waters at temperatures below 4°C extends as far north as 50°S, and a large winter sea-ice field is also present. The unique feature of the South Atlantic is that it delivers large amounts of heat to the North Atlantic by a cross-equatorial transport of warm water (Stommel 1980), while in the World Ocean, the net movement of heat and warm water is generally away from the equator (see compilation in Miller and Russell 1989). The asymmetrically driven Atlantic Heat Conveyor (AHC) transports warm water in either upper or intermediate layers from the South into the North Atlantic, where it cools and sinks down to form North Atlantic Deep Water (NADW) which returns south to the Southern Ocean. The AHC has been identified to be vulnerable especially during glacial conditions and to cause high-amplitude abrupt climate transitions, the Dansgaard-Oeschger oscillations, as well as the millennium-scale asynchrony of northern and southern hemisphere climate records, as documented from Greenland and Antarctic ice cores (e.g. Blunier et al. 1998), and marine sediment cores (e.g. Charles et al. 1996; Arz et al. 1998; Vidal et al. 1999). Modelling this vulnerability, for which Broecker (1998) coined the term "bipolar seesaw", revealed a scenario of "heat piracy" in the Atlantic Ocean, steered by variations in the relative amount of deep-water formation in the northern and southern hemisphere (Seidov and Maslin 2001). This scenario includes three modes of operation: (1) the "North Atlantic heat piracy" at full interglacial conditions or Dansgaard-Oeschger interstadials, with an enhanced NADW formation resulting in warming the northern hemisphere and cooling the southern hemisphere, (2) the "South Atlantic heat piracy" at periods of enhanced meltwater input in the North Atlantic (e.g. Heinrich events) that causes a cessation of NADW formation and results in a collapse of the conveyor, and (3) the "no piracy" mode at full glacial conditions characterized by an approximate balance between southern and northern heat budgets and reduced NADW formation (Fig. 1).

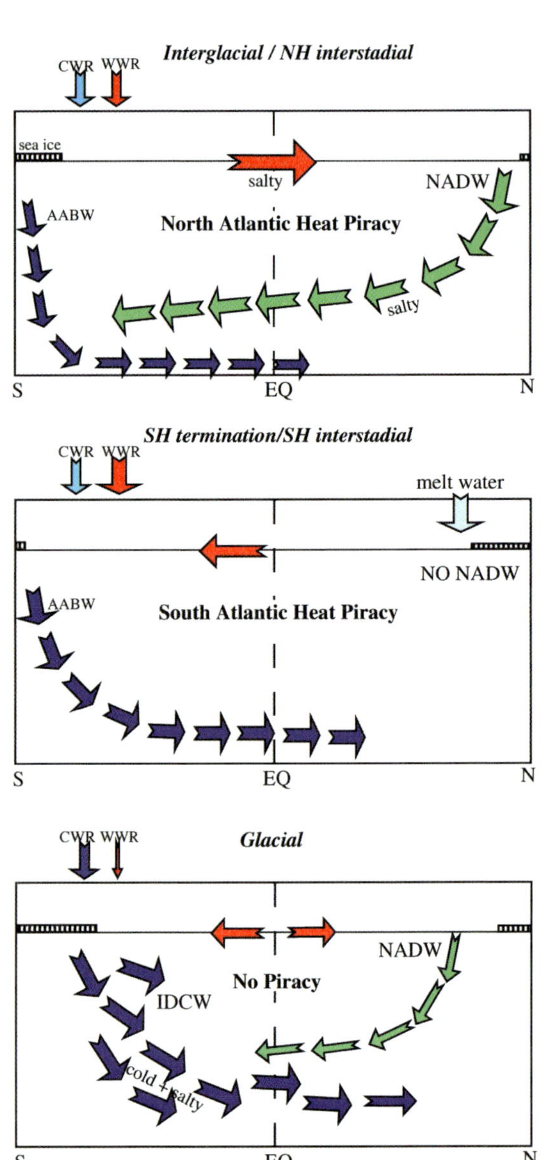

Fig. 1. Schematic representation of three regimes of heat and salt transport in the Atlantic Ocean during different climate states, adapted from scenarios proposed by Seidov and Maslin (2001) and Keeling and Stephens (2001), reconstructions by Duplessy et al. (1988) and Mackensen et al. (2001). NH northern hemisphere, SH southern hemisphere, CWR Cold water route, WWR Warm water route, NADW North Atlantic Deep Water, AABW Antarctic Bottom Water, IDCW Intermediate and deep cold water.

The three-modes-scenario described by Seidov and Maslin (2001) does not require any substantial change in the heat transported by the upper ocean currents (although it does not reject upper-ocean impact), and could be controlled solely by salinity-driven variations in NADW formation and responding Antarctic Bottom Water (AABW) distribution. Other models highlight the importance of alternative mechanisms amplifying climate variability and steering the bipolar seesaw. The extent of sea ice has been proposed to represent a major amplifier, causing the enhanced variability during glacial intervals, because it gears latitudinal thermal gradients and storminess and thus involves the lofting of dust and sea salt into the atmosphere, and in turn the Earth's albedo (Broecker 2001). Keeling and Stephens (2001) hypothesize that Antarctic sea-ice expansion affects not only the amplification of climate variability, but it also steers thermohaline overturning, due to the associated changes in the ocean's salinity structure.

As the AHC is not a closed system, the processes linked with the heat and salt import to the South Atlantic are also in the focus for understanding climate variability. Heat lost by trans-equatorial delivery to the North Atlantic is re-supplied from the Indian and the Pacific Ocean by the global ocean conveyor belt. Two return-flow portals have been identified, the leakage of surface waters around the southern tip of Africa, the so-called "warm water route" (WWR), and the import of waters through the Drake Passage, the "cold water route" (CWR). The WWR was proposed by Gordon (1986) as a major component of the global conveyor belt allowing for the return flow of tropical and subtropical warm and salty surface waters from the Indian and Pacific Ocean to compensate the outflow of NADW. This has been supported by field studies indicating that the major portion of northward heat flux into the Atlantic comes from the Indian Ocean (Döös 1995). The interbasin transfer into the South Atlantic takes place via large Agulhas rings shed at the Agulhas retroflection (Fig. 2), and to a minor proportion in the form of filaments (Lutjeharms 1996). Alternatively, Rintoul (1991) suggested that the return flow of water and heat mainly derived from waters entering via the Drake Passage, equally split between the surface,

intermediate and bottom waters, the CRW. Accepting the presence of the CWR, Gordon et al. (1992) proposed a modification in which CWR waters make an excursion into the Indian Ocean (gaining heat), before passing into the Atlantic via the Agulhas region. More recent modelling studies have been inconclusive, with some supporting the WWR (e.g. Thompson et al. 1997; Blanke et al. 2001), others the CWR (e.g. Drijfhout et al. 1996). Based on inverse modelling, Schlitzer (1996) suggested that the upper layer, i.e. northward flowing waters compensating the NADW outflow from the Atlantic Ocean, mainly consist of intermediate waters entering via the CWR, which are then modified and gradually warmed in the South Atlantic by the process of gaining heat from the atmosphere. Thus, the relative importance of the two routes in the modern thermohaline circulation and the processes involved are still controversial. As the mechanisms governing interbasin exchange are crucial for understanding climate variability on decadal to Milankovitch time scales, determination of the role of the WWR and CWR has become a central question in the "Climate Variability and Predictability" (CLIVAR) implementation plan (see http://www.clivar.org/publications/other_pubs/iplan/iip/pd3.htm).

Past climate variability might be modulated or amplified by changes in the advection of heat, salt and water at the two return-flow portals, related to latitudinal changes of the Southern Ocean cold-water sphere expansion. The WWR is particularly vulnerable, and could be pinched off by a northward movement of the Subtropical Front during glacial periods. While such conditions can be traced with paleoceanographic proxies allowing for the reconstruction of past sea-surface temperatures (SSTs), the lack of appropriate proxies imposes a major dilemma upon paleoceanographers when reconstructing changes that involve the entire water column, or mainly intermediate waters, as required for the CWR. However, hints to variations of the Antarctic Circumpolar Current (ACC) water masses being transferred across the Drake Passage can be derived from SST estimates that provide information as to which extent warmer zonal bands of the ACC are cut off and deflected along the South American coast in the southeast Pacific. This is due

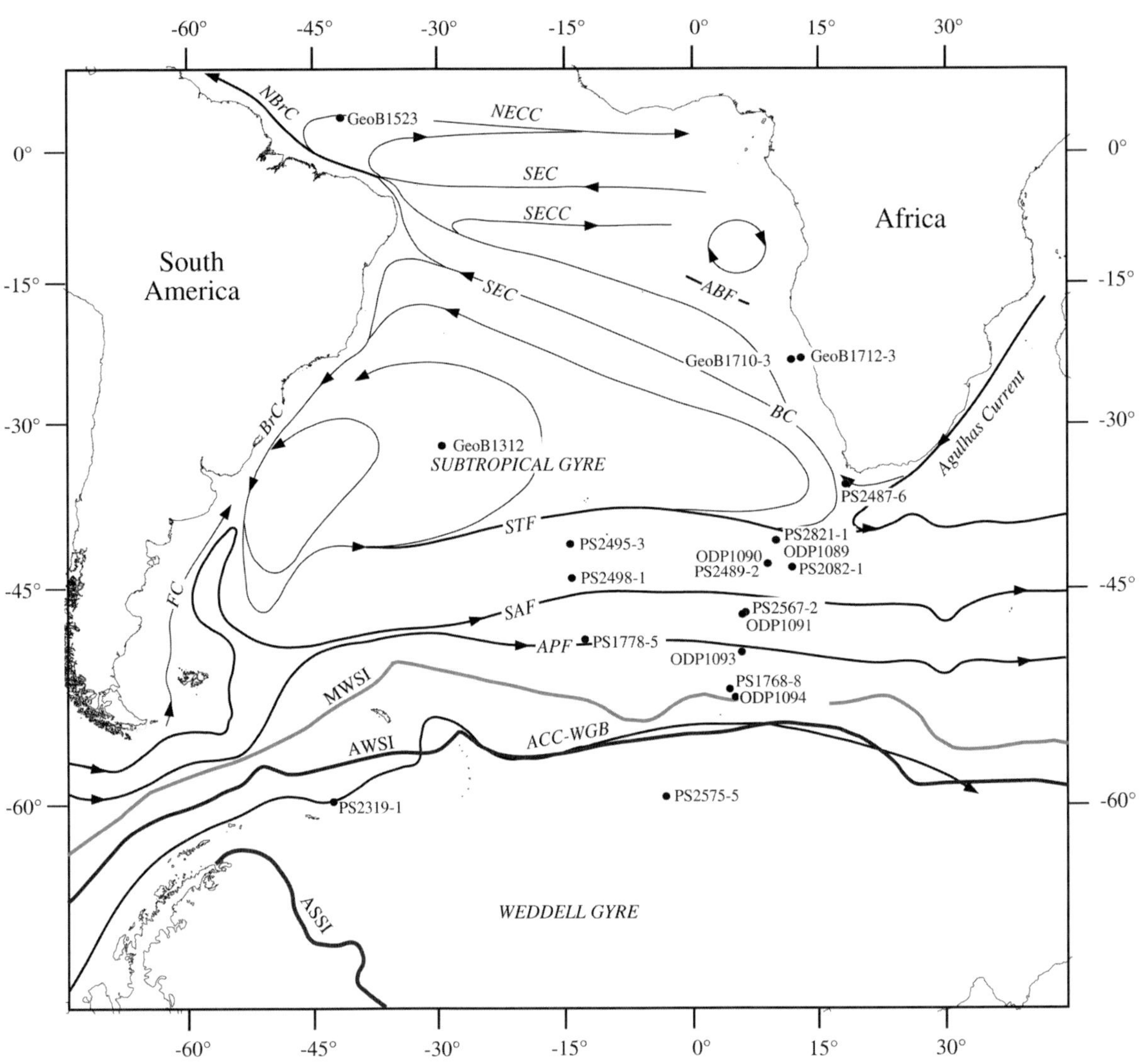

Fig. 2. Schematic representation of the large-scale, upper-level currents and fronts (adapted from Peterson and Stramma 1991; Tomczak and Godfrey 1994), and Antarctic sea-ice distribution (from Naval Oceanography Command Detachment, 1985). Also shown are locations of cores and drill sites discussed in the text. Designation of oceanic fronts and boundaries: ACC-WGB Antarctic Circumpolar Current-Weddell Gyre boundary, APF Antarctic Polar Front, SAF Subantarctic Front, STF Subtropical Front, ABF Angola-Benguela Front. Sea ice boundaries: ASSI average summer sea ice boundary, AWSI average winter sea ice boundary, MWSI maximum winter sea ice boundary. Current systems: FC Falkland-Malvinas Current, BC Benguela Current, BrC Brazil Current, SEC South Equatorial Current, SECC South Equatorial Counter Current, NECC North Equatorial Counter Current, NBrC North Brazil Current.

to the southward incision of the southern tip of South America into the Southern Ocean zonal bands.

Attempts to appropriately reconstruct the changing ocean circulation pattern and its relationship to climate variability on Milankovitch to millennial-centennial time scales should include the reconstruction of salinity and temperature in different layers of the evolving paleoocean. However, the reconstruction of salinity, one of the major drivers of thermohaline circulation (THC), is rather problematic, as summarized by Wolff et al. (1999).

According to these authors, reconstructions of paleosalinity from microfossil assemblages, e.g. foraminifers, yield unrealistic values. The approach to derive paleosalinities based on paired examination of the oxygen isotope ratio of planktic foraminifers and independent SST estimates looks more promising. However, the reliability of the obtained results is affected by the propagation of errors arising from uncertainties in the assumed δ^{18}O-salinity relationship in seawater and uncertainties in the temperature reconstruction ($\pm 1°$C leads to salinity errors between ± 0.5 and ± 1 ‰) (Schmidt, 1999). Nevertheless, the paired δ^{18}O/SST method has now been broadly applied in a global ocean model to simulate Atlantic Ocean water masses and circulation during the Last Glacial Maximum (LGM) (Paul and Schäfer-Neth 2003).

Here, we concentrate on results obtained from the reconstruction of SST, a paleoceanographic parameter whose estimations have undergone significant improvement since their first comprehensive application by the "Climate: Long Range Investigation, Mapping, and Prediction" (CLIMAP) initiative for the reconstruction of winter and summer sea-surface temperatures during the LGM on a global scale. Since the pioneering effort of CLIMAP (1976, 1981), the hydrographic and microfossil reference data sets for the estimation of past SSTs and the water-mass structure have been considerably improved, and new geochemical/stable isotope methods have been developed. Besides large reference bases for the application of planktic foraminifers, new data sets have been established for radiolarians and diatoms, in order to improve temperature reconstruction in the Southern Ocean. New approaches for the reconstruction of sea-ice have also been proposed (see Methods chapter for details). Successful geochemical methods include the alkenone method, using the simplified unsaturation index of C_{37} alkenones ($U^{K'}_{37}$, Prahl and Wakeham 1987), which can be converted to an annual mean mixed-layer SST based on calibration data sets (e.g. Müller et al. 1998). The application of oxygen isotope measurements to planktic foraminifers has also been proposed, including an "isotopic transfer function", which allows for the calculation of average temperature, temperature variability, and the isotopic

composition of seawater, if the oxygen-isotope differences between at least three species are known (Mulitza et al. 1998).

We present surface-water temperature estimates obtained for time slices of two end-members of Pleistocene climate variability, the Last Glacial Maximum and the penultimate climatic optimum during Marine Isotope Stage 5.5 (MIS 5.5) as well as down-core records. The latter were obtained from cores recovered in the South Atlantic and the adjacent Southern Ocean sector during expeditions with RV *Meteor* and RV *Polarstern*, mostly conducted in the scope of projects funded by the Deutsche Forschungsgemeinschaft Sonderforschungsbereich (SFB) 261:"The South Atlantic in the Late Quaternary: Reconstruction of Mass Budget and Current Systems". Additionally, we present results obtained from Pleistocene high-resolution records recovered during Ocean Drilling Program (ODP) Leg 177 on a paleoceanographic transect across the ACC (Gersonde et al. 1999), a leg which significantly profited from pre-site information collected within SFB 261. As many of the South Atlantic down-core records obtained north of the Southern Ocean realm have been summarized and discussed comprehensively in previous articles (e.g. Wefer et al. 1996b; Schneider et al. 1996) presented in Wefer et al. (1996a), we focus on down-core results from the Atlantic Southern Ocean. While this paper summarizes results on Pleistocene surface-water conditions, other contributions will deal with reconstructions of deep and bottom water distribution based on stable isotope (Bickert and Mackensen this issue) and clay mineral distribution (Diekmann et al. this issue).

Methods

Surface-Water Temperature Reconstruction based on the Distribution of Microfossil Assemblages

Reconstructions of the late Quaternary sea surface-temperatures (SST) in the South Atlantic and the adjacent Southern Ocean must cover a broad temperature range from -1.8 to ca. 30°C between the sea-ice covered Antarctic Zone and the tropical region. Thus latitudinal differentiations in the composition of biogenic components, the preservation and species diversity recorded in the various oceanographic regimes must be considered. In the tropical and subtropical realm, but also in the Subantarctic Zone of the Southern Ocean, biogenic sediments above the Carbonate Compensation Depth (CCD) predominantly consist of calcareous nannofossils and foraminifers. In the Subantarctic Zone, biosiliceous components (diatoms, radiolarians) can be prominent in sediments deposited in deep waters and during glacial periods. However diatom assemblages may be affected by dissolution, which reduces their potential for the estimation of paleo-temperature in this area (see below). In the area south of the Subantarctic Front, the biogenic components of the sediments predominantly consist of diatoms and radiolarians. Here, biogenic calcareous components are foraminifers. However, they display assemblages which are of low diversity or even monospecific, and therefore of little value for quantitative estimations of paleo-temperatures. The diatoms are the most important biogenic component between the Antarctic Polar Front and the northern area of the winter sea-ice edge. This is the circumantarctic siliceous belt, the main area of annual silica burial in the World Ocean (Tréguer et al. 1995), where up to 200 x 10^6 diatoms valves per gram dry sediment have been encountered (Zielinski and Gersonde 1997). The diatom assemblages in this area are well preserved and diversified, while radiolarian assemblages tend to display, with decreasing temperatures, lower species diversity. In the seasonal sea-ice zone of the Weddell Basin biosiliceous sediments are strongly affected by dissolution, which biases especially the diatom assemblages towards more strongly silicified species (Zielinski and Gersonde 1997), and precludes the use of diatom-based transfer functions. Consequently, SST estimations for the South Atlantic and the Southern Ocean must be derived from different planktic microfossil groups, the foraminifers, the radiolarians and the diatoms according to their applicability in the different regions.

Statistical methods used for the estimation of microfossil-based SSTs include the classical Imbrie & Kipp Method (IKM, Imbrie and Kipp 1971), the Modern Analog Technique (MAT, Hutson 1980),

and the recently optimised SIMMAX-28 MAT (Pflaumann et al. 2003). The basic assumption of IKM and MAT is that the modern spatial variability of microfossil assemblages in surface sediment samples deposited at known environmental conditions serves as a proxy for past environmental variability documented down-core. By means of factor analysis the IKM resolves microfossil assemblages preserved in surface sediment samples. The resulting varimax factors are calibrated in terms of hydrographic parameters of the surface waters, such as temperature, by using a stepwise multiple regression analysis (which defines both the coefficients of the terms in and the standard error of estimate of the regression equation), which is then applied to down-core assemblages to estimate past hydrographic parameters. The MAT matches down-core with surface sediment assemblages, which possess a similar species composition as measured by a similarity index. The estimates are based on an average of the measurements for the subset of modern samples, weighted by their similarity index, with a calculation of the standard error for each estimate. A user-friendly software package has been developed by Sieger et al. (1999) for the application of both, IKM and MAT.

Planktic foraminifers: Pflaumann et al. (2003) have recently extended a previous data base (Pflaumann et al. 1996) that now includes 947 modern core-top samples covering an area between the Arctic Ocean and the northern Southern Ocean in the Atlantic. This data set allows to estimate temperatures ranging from tropical values to temperatures as cold as −1.2°C for caloric winters, and −0.1°C for caloric summers. For this purpose, the SIMMAX-28 MAT was used (Tab. 1). However, such temperature estimates at the "cold end" of the data matrix rely on a low variability of foraminiferal assemblages, which results in an increased error of the estimate.

Considering that regional data sets may provide more accurate results, Niebler et al. (2003) generated a planktic foraminiferal data set from the >150 µm size fraction obtained from 271 surface sediment samples collected in the Atlantic Ocean between 20°N and 60°S for application in the tropical and subtropical Atlantic (Tab. 1). In order to relatively reduce the massive dominance of few

Application Area	Fossil Group	Method	Equation	Standard dev. of estimate (°C)	Temp. range of reference data set (°C)	Source of hydrogr. ref. (water depth m)	Reference
North and South Atlantic (<40°S-85°N)	Pl. foram. (>150 µm)	IMMAX-28 MA′	F947/38	winter ±0.75 summer ±0.82	-1.2 - 30	WOA (10 m)	Pflaumann et al. (2003)
South Atlantic (>40°S)	Pl. foram. (>150 µm)	IKM	F271/24/4ln	winter ±1.17 summer ±1.21 ann. mean ±1.16	0 - 27	WOA (10 m)	Niebler et al. (2003)
Subtropial- Subantarctic Zone	Pl. foram. (>125 µm)	MAT/IKM	F186/26/5	summer ±1.3	0 - 28	WOA, SOA (10 m)	Niebler & Gersonde (1998) Niebler et al. (in prep.)
	Radiolaria	IKM	R73/24lg/4	summer ±1.16	-1 - 22	WOA, SOA (10 m)	Cortese & Abelmann (2002)
Subantarctic- Antarctic Zone	Radiolaria	IKM	R52/23/4	summer ±1.2	-1 - 21	WOA, SOA (10 m)	Abelmann et al. [1999]
	Diatoms	IKM	D93/29lg/3	summer ±0.66	-1.8 - 21	WOA, SOA (10 m)	Zielinski et al. (1998)
Antarctic Zone (Opal dissolution area)	Diatoms	*Eucampia*-Ind.	EI56/2	summer ±1.1	-1.8 - 5	SOA (10 m)	Gersonde & Censarek (in prep.)

Table 1. Summary of statistical methods and equations, standard deviation of estimates, temperature range of the reference data sets, source of hydrographic reference data set (WOA: World Ocean Atlas, Levitus and Boyer, 1994; SOA: Southern Ocean Atlas, Olbers et al. 1992), and water depth of extracted temperature. Equation designations indicate used fossil or species group (F: foraminifers, R: radiolarians, D: diatoms, E: Eucampia)/number of reference samples/number of taxa or taxa groups/number of factors, lg or ln indicates logarithmic conversion of species abundance data used to compensate the dominance of single taxa.

taxa (e.g. *Neogloboquadrina pachyderma* sin, *Globigerinoides ruber* white) in this data set, and to increase the weight of rarely occurring but paleoecologically significant species, logarithmic conversion of the species counts was applied. Similar to data sets created for application in the subtropical-subantarctic realm (Niebler and Gersonde 1998; Niebler et al. in prep.), deep-dwelling taxa (e.g. *Globorotalia hirsuta, Globorotalia crassaformis, Globorotalia scitula*) have been excluded from statistical treatment, since they do not reflect sea-surface conditions.

The most comprehensive data set available for application in the subantarctic and the adjacent subtropical region is that of Nieber et al. (in prep.), which consists of 186 surface sediment samples, collected in the South Atlantic and Indian sector, between 20°S and 58°S, which includes 26 foraminiferal taxa and taxa groups (Tab. 1). Considering the typically smaller size of polar and subpolar planktic foraminifers, the faunal analysis of cold water assemblages was done with a size fraction >125 µm, rather than >150 µm as is usually done

analysing subtropical and tropical assemblages. This increases the abundances of taxa such as *Turborotalia quinqueloba* and *Globigerina bradyi*, and thus enhances the significance of these taxa in paleoenvironmental reconstruction.

Radiolaria: Radiolarians have been the ultimate proxy for the estimation of Southern Ocean SST in the scope of CLIMAP (1976, 1981, 1984), using IKM on the reference data sets of Lozano and Hays (1976) and Morley (1979). Considering the progress made in radiolarian taxonomy as well as new information on radiolarian autecology obtained from water- column studies (Abelmann and Gowing 1996, 1997), Abelmann et al. (1999) created a data set based on radiolarians, in order to reconstruct sea-surface temperatures of the past Southern Ocean. The data set comprises 23 taxa and taxonomic groups, which are all related to a surface water habitat, obtained from 52 surface sediment samples (Tab. 1). As biogenic particle flux to the seafloor of the Southern Ocean is restricted to the austral summer, including areas not affected by ice coverage (Abelmann and Gersonde 1991;

Abelmann 1992; Gersonde and Zielinski 2000; Fischer et al. 2002), the transfer-function equation used is designed for summer sea-surface temperatures (SSST) estimates only. Cortese and Abelmann (2002) extended this data set, including additional surface sediment assemblages from the Atlantic subtropical zone (up to 30°S) to enable that SSST reconstructions cover the paleo-temperature fluctuations between the Antarctic and subtropical regimes (Tab. 1).

Diatoms: For the reconstruction of SSST based on the diatom record, Zielinski et al. (1998) elaborated a reference data set including 29 taxa and taxa groups from 93 surface sediment samples which were selected from a large surface sediment data set presented by Zielinski and Gersonde (1997) (Tab. 1). In order to compensate for the dominance of the species *Fragilariopsis kerguelensis*, and to relatively increase the weight of paleoceanographically significant species occurring in lower abundances, Zielinski et al. (1998) proposed a continuous ranking system after subjecting the raw data to logarithmic conversion.

SSST estimates derived from either IKM or MAT cannot be accomplished in sequences where the microfossil assemblage composition is altered by dissolution effects. The latter prevents an appropriate comparison with the assemblages contained in the reference data set of surface sediments. To overcome such problem with cores from the Weddell Sea, Gersonde and Censarek (in prep.) developed a method, which allows for the reconstruction of temperature on the basis of the ratio of morphologically different valve types of the coarsely silicified diatom *Eucampia antarctica*, i.e. terminal and intercalary valves. The Eucampia-Index Method (EIM) can only be applied in cold waters below 4°C (Table 1).

Comparing Different Proxies of Sea - Surface Temperatures

Core PS2498-1, recovered in the Subantarctic Zone (Fig. 2, Tab. 2), is hitherto the only known example of a core recovered during RV *Polarstern* expeditions in the Southern Ocean with a late Pleistocene sediment sequence that contains assemblages of foraminifers, diatoms and radiolarians at a state of preservation, which allows to estimate SSSTs on the three microfossil groups using the same set of samples (Fig. 3). Unfortunately, further going comparison with alkenone-derived temperatures was impossible, as the concentration of alkenones in these sediments was too low (P. Müller, personal comm.). In general, the shapes of the three records are rather similar and monitor the broad climatic variations during the past 130 ka. However, a more detailed comparison reveals that the foraminiferal SSSTs are mostly a few degrees colder as compared to the radiolarian and the diatom record. This may be attributed to the dissolution of carbonate biasing the ratio between two dominant taxa of subantarctic assemblages, i.e. the cold *N. pachyderma*$_{sin}$ and the warmer *Globigerina bulloides*, towards colder assemblages. While *N. pachyderma*$_{sin}$, predominantly dwelling in waters colder than 6-8°C (e.g. Niebler and Gersonde 1998), has a robust shell that is more resistant to dissolution, *G. bulloides*, a species that is related to water temperatures between 10-16°C, is more strongly affected by dissolution. The SSST estimates obtained from diatoms and radiolarians display rather similar absolute values and variability. An exception exists in the warm intervals of Marine Isotope Stage (MIS) 5, especially during MIS 5.5 where diatoms indicate colder SSSTs. This pattern can be attributed to biogenic opal dissolution during periods of inferior opal deposition, which, in turn, affects diatoms more strongly than radiolarians (Abelmann and Gersonde 1991). Dissolution causes a bias of the diatom assemblages towards colder assemblages and is related to a relative increase of the coarsely silicified *Fragilariopsis kerguelensis*. During MIS 2, SSSTs based on radiolarians display warmer values as compared to diatom-derived estimates. This is interpreted as being related to the dominant occurrence of *Chaetoceros* spp. in this interval. This diatom group is not considered in the transfer function (see Zielinski et al. 1998) and its high abundance results in a rather exiguous remaining proportion of the considered diatom signal, and thus produces uncertain estimates. A comparable pattern leading to relatively higher radiolarian-based SSST values

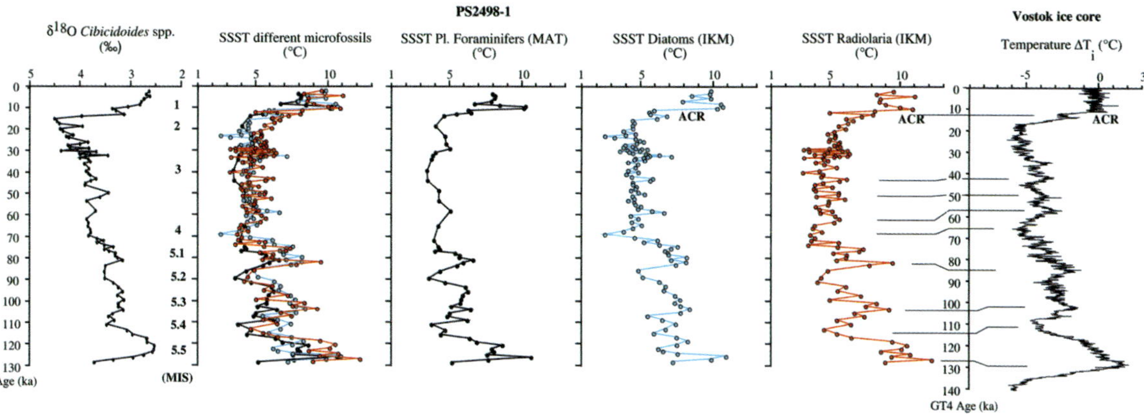

Fig. 3. Comparison of summer sea surface temperatures (SSST) estimates obtained from planktic foraminifers (MAT on F186/26), diatoms (IKM on D93/29lg/3) and radiolaria (IKM on R55/23/4) in Core PS2498-1. Age model based on $\delta^{18}O$ record of the benthic foraminifer Cibicidoides spp. (from Mackensen et al. 2001) and reservoir corrected AMS ^{14}C measurements (from Gersonde et al. 2003). SSST are compared with Vostok ice core atmospheric temperatures at the inversion level (Petit et al. 1999), expressed as deviation from modern temperatures. Vostok age according to the Petit et al. (1999) GT4 time scale. ACR: Antarctic Cold Reversal.

was found in other MIS 2 sections originating from the Subantarctic Zone (Gersonde et al. 2003). In summary, the comparison of the three methods of SSST estimates indicates that the most reliable records from Subantarctic Pleistocene sequences, affected by carbonate dissolution during glacials and opal dissolution during climate optima, can be obtained from estimates based on radiolarians. This is supported by the close correspondence between the PS2498-1 radiolarian record and the independently produced temperature record from the Vostok ice core. Within the limits of stratigraphic resolution and stratigraphic correlation of the age of PS2498-1 derived from marine stable isotopes and AMS ^{14}C and the GT4 age of the Vostok ice core, short-term changes can also be identified (e.g. the Antarctic Cold Reversal, ACR) (Fig. 3). During MIS 3, deposited at higher sedimentations rates and thus allowing enhanced resolution, short-term fluctuations reaching an amplitude of 4°C have been detected, indicating the presence of rapid SSST variability at centennial-millenial time scales, which is comparable to Dansgaard-Oeschger cyles. Only recently, Cortese and Abelmann (2002) have demonstrated the high value of the radiolarian-based method for the reconstruction of past SSST variability in subantarctic cores from another, but more expanded section.

The comparison of radiolarian and diatom-derived SSST estimates obtained from LGM core sections in the PFZ and the AZ, shows close similarity between the results obtained from both fossil groups within the range of the statistical error (Gersonde et al. 2003). This finding indicates coherency between both methods when applied to sediments deposited south of the SAZ.

Sea-Ice Reconstruction

The determination of the past summer and winter sea-ice extent is an important goal for paleoceanographic and paleoclimatic reconstructions, considering the strong impact of this fast changing environmental factor on climate and climate variability. After the qualitative reconstruction of the sea-ice distribution which was mainly based on lithological boundaries (e.g. Cooke and Hays 1982), Crosta et al. (1998a,b) were the first to propose a quantitative reconstruction of sea ice, expressed as "num-ber of months with sea-ice coverage per year", applying a MAT transfer function to a sea-ice distribution and diatom reference data set from the Southern Ocean. Such a quantitative sea-ice estimate would considerably augment our ability to study the impact of sea-ice on past climate, and is urgently requested by paleoclimate modellers.

Core	Latitude	Longitude	SST modWS (°C)	SST LGMWS (°C)	LGM anomaly WS(°C)	Method	Reference
GeoB1008-3	-6,582	10,318	27,25	22,05	-5,2	F	Niebler et al. 2003
GeoB1016-3	-11,77	11,682	26,61	22,61	-4	F	Niebler et al. 2003
GeoB1028-5	-20,103	9,185	20,9	15,1	-5,8	F	Wefer et al. 1996; Niebler et al. 2003
GeoB1031-4	-21,88	7,102	21,69	18,89	-2,8	F	Wefer et al. 1996
GeoB1032-3	-22,915	6,037	22	20	-2	F	Wefer et al. 1996
GeoB1034-3	-21,735	5,422	22,38	20,08	-2,3	F	Niebler et al. 2003
GeoB1041-3	-3,475	-7,6	27,35	20,45	-6,9	F	Niebler et al. 2003
GeoB1101-5	1,658	-10,98	29,24	24,94	-4,3	F	Niebler et al. 2003
GeoB1105-4	-1,665	-12,428	28,33	22,33	-6	F	Wefer et al. 1996; Niebler et al. 2003
GeoB1112-4	-5,778	-10,75	27,91	23,71	-4,2	F	Wefer et al. 1996; Niebler et al. 2003
GeoB1115-4	-3,558	-12,58	28,49	23,99	-4,5	F	Niebler et al. 2003
GeoB1117-2	-3,815	-14,897	28,65	23,85	-4,8	F	Dittert et al. in press / Niebler et al. 2003
GeoB1214-1	-24,69	7,24	21,49	19,29	-2,2	F	Niebler et al. 2003
GeoB1220-1	-24,033	5,307	22,24	18,64	-3,6	F	Wefer et al. 1996
GeoB1306-1	-35,207	-26,763	20,72	18,12	-2,6	F	Niebler et al. 2003
GeoB1309-2	-31,667	-28,667	22,85	21,8	-1,1	F	Gersonde et al. 2003
GeoB1413-4	-15,68	-9,455	25,52	23,02	-2,5	F	Niebler et al. 2003
GeoB1417-1	-15,537	-12,707	25,73	23,43	-2,3	F	Niebler et al. 2003
GeoB1419-2	-15,537	-17,068	26,45	24,45	-2	F	Niebler et al. 2003
GeoB1501-4	-3,678	-32,007	29,09	25,69	-3,4	F	Niebler et al. 2003
GeoB1503-1	2,311	-30,646	29,51	24,61	-4,9	F	Niebler et al. 2003
GeoB1505-1	2,27	-33,01	29,5	25	-4,5	F	Niebler et al. 2003
GeoB1508-4	5,333	-34,025	26,12	24,62	-1,5	F	Niebler et al. 2003
GeoB1515-1	4,238	-43,667	25,98	23,28	-2,7	F	K. U. Gräfe / Niebler et al. 2003
GeoB1523-1	3,832	-41,622	25,87	22,87	-3	F	Hale and Pflaumann 1999
GeoB1701-4	1,95	3,552	29,74	25,74	-4	F	Niebler et al. 2003
GeoB1706-2	-19,562	11,175	19,69	9,69	-10	F	Little et al. 1997
GeoB1711-4	-23,315	12,377	18,02	10,12	-7,9	F	Little et al. 1997
GeoB1722-1	-29,45	11,75	20,7	19,3	-1,4	F	Niebler et al. 2003
GeoB1903-3	-8,675	-11,845	27,59	24,89	-2,7	F	Niebler et al. 2003
GeoB1905-3	-17,143	-13,987	25,58	24,68	-0,9	F	Niebler et al. 2003
GeoB2004-2	-30,87	14,343	20,2	19,4	-0,8	F	Niebler et al. 2003
GeoB2016-1	-31,9	-1,33	21,75	20,05	-1,7	F	Niebler et al. 2003
GeoB2019-1	-36,06	-8,778	19,8	15,7	-4,1	F	Niebler et al. 2003
GeoB2021-5	-36,833	-14,403	18,68	14	-4,7	F	Gersonde et al. 2003
GeoB2109-1	-27,912	-45,882	25,61	23,71	-1,9	F	Niebler et al. 2003
GeoB2125-1	-20,823	-39,863	27,05	25,45	-1,6	F	Niebler et al. 2003
GeoB2202-4	-8,198	-34,265	28,92	26,72	-2,2	F	Niebler et al. 2003
GeoB2204-2	-8,528	-34,022	29,02	26,32	-2,7	F	Hale and Pflaumann 1999
GeoB2215-10	0,007	-23,495	28,9	23,7	-5,2	F	Niebler et al. 2003
GeoB2819-1	-30,848	-38,343	23,69	21,6	-2,1	F	Gersonde et al. 2003
GeoB3104-1	-3,667	-37,717	28,65	26,25	-2,4	F	Niebler et al. 2003
GeoB3117-1	-4,295	-37,092	28,65	26,35	-2,3	F	Niebler et al. 2003
GeoB3175-1	-7,053	-34,463	29,02	26,12	-2,9	F	Niebler et al. 2003
GeoB3176-1	-7,012	-34,442	29,11	26,31	-2,8	F	Niebler et al. 2003
GeoB3603-2	-35,125	17,543	21,12	18,52	-2,6	F	Acheson et al. submitted
GeoB3722-2	-25,25	12,018	18,88	13,48	-5,4	F	Niebler et al. 2003
GeoB3801-6	-29,512	-8,305	23,59	21,89	-1,7	F	Niebler et al. 2003
GeoB3808-6	-30,812	-14,712	24,14	21,44	-2,7	F	Niebler et al. 2003
GeoB3813-3	-32,268	-21,967	23,2	20,4	-2,9	F	Gersonde et al. 2003
GeoB5112-5	-23,825	-16,258	25,75	24,15	-1,6	F	Niebler et al. 2003
GeoB5121-2	-24,183	-12,022	25,67	24,27	-1,4	F	Niebler et al. 2003
GeoB5133-3	-19,087	-10,192	25,54	23,74	-1,8	F	Niebler et al. 2003
EN66-10	6,64	-21,897	25,44	21,84	-3,6	F	Mix 1986
RC09-49	11,183	-58,6	26	24	-2	F	Mix 1986
RC13-184	3,86	-43,31	26,27	23,97	-2,3	F	Mix 1986
RC13-189	1,863	-30	29,27	25,67	-3,6	F	Mix 1986
RC24-01	0,563	-13,65	29,41	24,21	-5,2	F	Mix 1986
RC24-16	-5,038	-10,19	27,76	24,46	-3,3	F	Mix 1986
RC24-27	-5,412	-0,363	28,09	25,09	-3	F	Mix 1986
RC24-7	-1,342	-11,92	28,32	22,92	-5,4	F	Mix 1986
V15-168	0,2	-39,9	28,67	25,57	-3,1	F	Mix 1986
V22-177	-7,75	-14,6	28,13	24,53	-3,6	F	Mix 1986

Table 2.

V22-182	-0,533	-17,27	28,94	24,34	-4,6	F	Mix 1986
V22-38	-9,55	-34,25	28,57	26,47	-2,1	F	Mix 1986
V23-110	17,633	-45,87	24,46	24,06	-0,4	F	Mix 1986
V25-59	1,367	-33,48	28,95	26,45	-2,5	F	Mix 1986
V25-60	3,283	-34,83	25,98	23,58	-2,4	F	Mix 1986
V25-75	8,583	-53,17	26,24	24,74	-1,5	F	Mix 1986
V29-144	-0,2	6,05	29,12	25,12	-4	F	Mix 1986
V30-36	5,35	-27,32	25,76	23,46	-2,3	F	Mix 1986
V30-40	-0,2	-23,15	28,92	23,92	-5	F	Mix 1986
V30-41	0,217	-23,07	28,97	23,97	-5	F	Mix 1986
PS1433-1	-47,5428	15,3648	6,18	3,9	-2,3	D+R/2	Gersonde et al. 2003
PS1444-1	-55,373	9,9777	0,51	0,1	-0,4	D	Gersonde et al. 2003
PS1649-2	-54,9105	3,3077	-0,04	-0,1	-0,1	D	Gersonde et al. 2003
PS1651-1	-53,6308	3,8562	0,44	0,3	-0,2	D+R/2	Gersonde et al. 2003
PS1652-2	-53,664	5,0995	0,24	-0,5	-0,8	D	Gersonde et al. 2003
PS1654-2	-50,1575	5,7225	4,56	1,4	-3,1	D	Gersonde et al. 2003
PS1754-1	-46,771	7,6123	7,31	2,9	-4,4	F	Gersonde et al. 2003
PS1756-5	-48,9	6,7147	5,03	2,7	-2,3	D+R/2	Gersonde et al. 2003
PS1765-3	-51,8315	4,806	2,93	0,8	-2,1	D	Gersonde et al. 2003
PS1768-8	-52,593	4,476	1,47	0,7	-0,8	D+R/2	Gersonde et al. 2003
PS1775-4	-50,9517	-7,505	2,03	0,4	-1,7	D	Gersonde et al. 2003
PS1777-6	-48,2333	-11,04	4,92	2	-3	D	Gersonde et al. 2003
PS1778-5	-49,0133	-12,7	4,81	1,8	-3	D+R/2	Gersonde et al. 2003
PS1779-2	-50,4	-14,08	3,97	1	-3	D+R/2	Gersonde et al. 2003
PS1780-5	-51,704	-15,3	2,86	1,1	-1,7	D	Gersonde et al. 2003
PS1782-5	-55,19	-18,61	1,16	1,4	0,2	D	Gersonde et al. 2003
PS1783-5	-54,9117	-22,71	0,85	0,7	-0,1	D+R/2	Gersonde et al. 2003
PS1786-1	-54,9233	-31,72	1,4	1,1	-0,3	D	Gersonde et al. 2003
PS2082-1	-43,2202	11,7383	11,35	6,6	-4,8	R	Gersonde et al. 2003
PS2089-1	-53,1887	5,3302	1,36	0,4	-1	D+R/2	Gersonde et al. 2003
PS2090-1	-53,1783	5,133	1,36	0,4	-0,9	D	Gersonde et al. 2003
PS2102-2	-53,073	-4,9856	0,55	0	-0,5	D	Gersonde et al. 2003
PS2104-2	-50,7421	-3,2288	3,6	1,9	-1,7	R	Gersonde et al. 2003
PS2250-5	-45,1002	-57,9473	12,1	2,7	-9,4	D+R/2	Gersonde et al. 2003
PS2271-5	-51,5305	-31,3485	3,1	2,7	-0,4	R	Gersonde et al. 2003
PS2276-4	-54,6355	-23,9508	0,89	0,5	-0,4	D	Gersonde et al. 2003
PS2278-3	-55,9687	-22,2248	0,66	0,6	0	D	Gersonde et al. 2003
PS2280-4	-56,8397	-22,3242	0,22	0,8	0,6	D	Gersonde et al. 2003
PS2305-6	-58,7203	-33,0365	1,4	0,2	-1,2	D	Gersonde et al. 2003
PS2307-1	-59,0535	-35,6093	0,58	0	-0,6	D	Gersonde et al. 2003
PS2319-1	-59,7883	-42,6833	0,82	0,5	-0,3	D	Gersonde et al. 2003
PS2489-2	-42,8733	8,9733	10,69	4,7	-6	F	Gersonde et al. 2003
PS2491-3	-44,955	5,97	9,32	5	-4,3	R	Gersonde et al. 2003
PS2492-2	-43,1733	-4,055	11,42	6,5	-4,9	R	Gersonde et al. 2003
PS2493-1	-42,8833	-6,02	11,42	6,6	-4,8	R	Gersonde et al. 2003
PS2495-3	-41,275	-14,49	14,72	5,4	-9,3	F	Gersonde et al. 2003
PS2498-1	-44,1533	-14,2283	11,11	6,2	-4,9	R	Gersonde et al. 2003
PS2499-5	-46,5117	-15,3333	5,54	2,7	-2,8	D	Gersonde et al. 2003
PS2502-2	-50,25	-23,24	4,37	1,2	-3,1	D	Gersonde et al. 2003
PS2515-3	-53,545	-45,2917	4,35	0,8	-3,5	D	Gersonde et al. 2003
PS2561-2	-41,8583	28,5417	16,4	10,9	-5,5	D	Gersonde et al. 2003
PS2563-2	-44,56	34,7867	9,38	4,1	-5,3	D	Gersonde et al. 2003
PS2564-3	-46,1417	35,8983	9,03	4,3	-4,7	D	Gersonde et al. 2003
PS2567-2	-46,935	6,2567	6,9	3,8	-3,1	D+R/2	Gersonde et al. 2003
PS2575-5	-59,4817	-3,2	0,4	-0,8	-1,2	D	Gersonde et al. 2003
PS2603-3	-58,9867	37,6283	1,51	0,9	-0,6	D	Gersonde et al. 2003
PS2606-6	-53,2317	40,8017	3,09	0,5	-2,6	D	Gersonde et al. 2003
PS2608-1	-51,8783	41,6517	3,09	0,1	-3	D	Gersonde et al. 2003
PS2610-3	-50,685	40,1333	3,74	0,6	-3,1	D	Gersonde et al. 2003
PS2821-1	-40,943	9,888	15,3	12,1	-3,2	R	Gersonde et al. 2003

Table 2. cont. Designation, geographic coordinates, modern warm season sea surface temperature (SSTmod WS), estimated warm season sea surface temperature for EPILOG LGM (SST LGMWS), modern/LGM anomaly, applied fossil group for SST estimate (F: foraminifers, R: radiolaria, D: diatoms), and reference of estimate.

However, questions on the reliability of the estimated annual sea-ice duration at a given core location have been raised by Gersonde and Zielinski (2000), based on a combined study of annual diatom fluxes from the Southern Ocean surface to the seafloor using time-series sediment traps and mapping of diatom assemblages preserved in surface sediments. While the approach of Crosta et al. (1998a,b) implies that the annual duration of sea-ice coverage at a given location is reflected by the production and deposition of diatom sea-ice signals, such relationship was disproved by sediment trap experiments. The results of Gersonde and Zielinski (2000) show that the diatom sea-ice signal preserved in the sediment record is produced under austral summer open water conditions from diatom blooms that have been seeded from sea-ice to open water. In addition, Gersonde and Zielinski (2000) claim that the statistical approach may lead to misinterpretations of the duration of sea-ice coverage in areas affected by an enhanced dissolution of opal, e.g. in areas with longer annual sea ice coverage. Until other, more appropriate quantitative estimation techniques have been elaborated and successfully tested, it is recommended to rather rely on qualitative estimates of sea-ice extent than on a quantitative approach, which may produce unrealistic values, as it is only based on a statistical relationship between two parameters sampled on a latitudinal gradient. Gersonde and Zielinski (2000) proposed to use the abundance pattern of the diatom species *Fragilariopsis curta* and *F. cylin-drus*, combined as *F. curta/cylindrus* group, as tracers of the winter sea-ice edge and waters influenced by the spring melt. A relative abundance of the *F. curta/cylindrus* group higher than 3% of the total assemblage is considered as a qualitative threshold between the average presence of winter sea-ice and year-round open waters. The proximity of the summer sea-ice limit was detected by means of higher abundance of *Fragilariopsis obliquecostata*, a taxon limited to very cold waters (<-1°C). This species is relatively thickly silicified and thus not much affected by opal dissolution. It remains a valuable ice cover tracer even in conditions of low sedimentation rates and enhanced opal dissolution.

A South Atlantic LGM Time Slice

About 25 years after the CLIMAP (1976, 1981) pioneer study to reconstruct global sea-surface temperatures and sea-ice extent during the LGM, Niebler et al. (2003) and Gersonde et al. (2003) presented new maps of sea-surface temperature (SST) and the sea-ice extent for this end-member of late Pleistocene climate variability from the South Atlantic and the adjacent Atlantic sector of the Southern Ocean. They mapped SST and sea-ice distribution obtained with planktic foraminifers-, radiolarian- and diatom-based transfer functions from a set of 123 cores between 10°N and 60°S (Tab. 2). The selected LGM time slice covers a period of maximum sea-level low stand (ca. 135 m) (Yokoyama et al., 2000) between 19.5 and 16 ^{14}C ka (equal to 23,000 to 19,000 cal yr. B. P.), as proposed by "Environmental processes of the ice age: land, oceans, glaciers" (EPILOG) (Mix et al. 2001). The effort to reconstruct the South Atlantic LGM is part of the German "Glacial Atlantic Ocean Mapping" (GLAMAP-2000) initiative (Sarnthein et al. 2003), which will be merged into the international EPILOG initiative for global reconstruction of the past ice age. The stratigraphic identification of the EPILOG LGM (ELGM) time slice was accomplished by the combination of AMS ^{14}C measurements, oxygen isotope and (in Southern Ocean cores barren in carbonate) calibrated siliceous microfossil abundance records (e.g. *Cyclado-phora davisiana*, *Eucampia antarctica*). This provides the high level of stratigraphic control needed for the establishment of a global LGM reconstruction. The obtained sea-surface temperatures indicate three major zones of significant cooling during the LGM as compared with modern conditions: the central equatorial Atlantic, where temperatures drop between 4 and 5°C (with a maximum close to 7° C), the Namibia upwelling regime with temperature reductions up to 10°C, and the present Subantarctic Zone where the SSST decreased between 4 and close to 6°C (Fig. 4a). The observed cooling in the central tropical Atlantic is in good agreement with the results presented by Mix et al. (1999) and Pflaumann et al. (2003), but also confirms late glacial numerical simulations (e.g. Ganopolski et al. 1998). It provides further evidence of substantial cooling in the equatorial current systems of the Atlantic Ocean during the LGM. There is considerable debate whether the tropical ocean was only cooled by 1 - 2°C, as concluded by CLIMAP (1976, 1981) (Fig. 4b), or whether the tropical temperatures were colder (e.g. see Crowley 2000 for extensive discussion). A scenario with cooler tropical LGM sea-surface temperatures allows to resolve apparent disagreements between ocean and land data, the latter indicating a significant cooling from different sources (e.g. Stute et al. 1995; Thompson 2000). According to simulations by Seidov and Maslin (2001), the cold tropical LGM scenario shows very little cross-equatorial heat transport into the North Atlantic, and results in an approximate balance between southern and northern heat budgets in the Atlantic Ocean, the "No piracy" mode (Fig. 1). Mix et al. (1999) suggested cooling of the ice-age thermocline to be driven by faster upper ocean ventilation. They speculated that cold mode waters forming in the southern polar region returned to the surface through equatorial and eastern boundary upwelling which is strongly enhanced by strengthened wind fields. Poleward return of the tropical surface water would then complete the advective loop, resulting in additional heat lost in the tropics.

The Mix et al. (1999) scenario is in line with the observation of a strongly cooled Namibia upwelling system (Fig. 4a). Here, SST dropping at maximum to 10°C and at minimum to values around 8°C, have been estimated using a transfer function based on foraminifers (Niebler et al. 2003). The strong cooling may result from an increased upwelling intensity, related to stronger atmospheric gradients. This is considered to be a response to the expansion of the Southern Ocean cold-water sphere, in concert with the presence of very cold temperatures of the upwelled water originating in the polar latitudes. Such import might be signalled by the massive occurrence of the preferentially polar-subpolar foraminifer *N. pachyderma*$_{sin}$ (Niebler et al. 2003). LGM oxygen isotope values of foraminifers dwelling in surface waters support a distinct cooling (about 8-9°C) in the Namibia upwelling (Niebler et al. 2003). In addition, continental climate

reconstructions show a 6-7°C decrease at the southwestern coast of Africa during the LGM (Partridge et al. 1999). However, alkenone-derived temperatures obtained from cores located in the Benguela Coastal Current do not document strong cooling. In contrast, Kirst et al. (1999) and Schneider et al. (1996) report that the Namibia upwelling had been at its maximum before the LGM, between 50 – 35 ka. These authors ascribe relatively warm temperatures to the LGM (Fig. 5, GeoB1712-3) that are explained to result from southward warm-water protrusions of tropical water masses during periods of weakening of the Angola-Benguela Front (similar to recent "Benguela Niño events", resulting from pertur-bations in the trade-wind system over the western tropical Atlantic). The apparent mismatch between temperatures derived from alkenones and fora-minifers could be a consequence of the preferential presence of alkenone-producing algae during warmer and more nutrient-poor conditions that would affect the Namibia upwelling area during periods of tropical-water-mass import.

The cooling in the Southern Ocean, which mainly affected the present Subantarctic Zone, was related to an expansion of the winter sea-ice field by 60-70 % compared to the present (Gersonde et al. 2003). The summer sea-ice edge was also shifted northwards and reached, at least sporadically, as far north as 55°S in the eastern Weddell Sea sector. Assuming that the modern relationship between SSST and the location of Southern Ocean frontal systems can be applied to the ELGM, the Polar and Subantarctic Fronts were shifted northwards by 4-5 degrees in latitude. However, the displacement of the Subtropical Front, blocked to the north by the Subtropical Gyre that did not experience any significant temperature change during the LGM, was minor (2-3 degrees in latitude), likely resulting in a steepening of the thermal gradients in the ELGM Subtropical Front area (Fig. 4a). The observed broad zone of cooling in the present Subantarctic Zone agrees well with simulations obtained from a coupled atmosphere-ocean climate model (Ganopolski et al. 1998). The distinct expansion of the Southern Ocean cold-water sphere and sea-ice are the candidates for the generation of a cold intermediate and deep water mass in the Southern Ocean, as suggested from stable

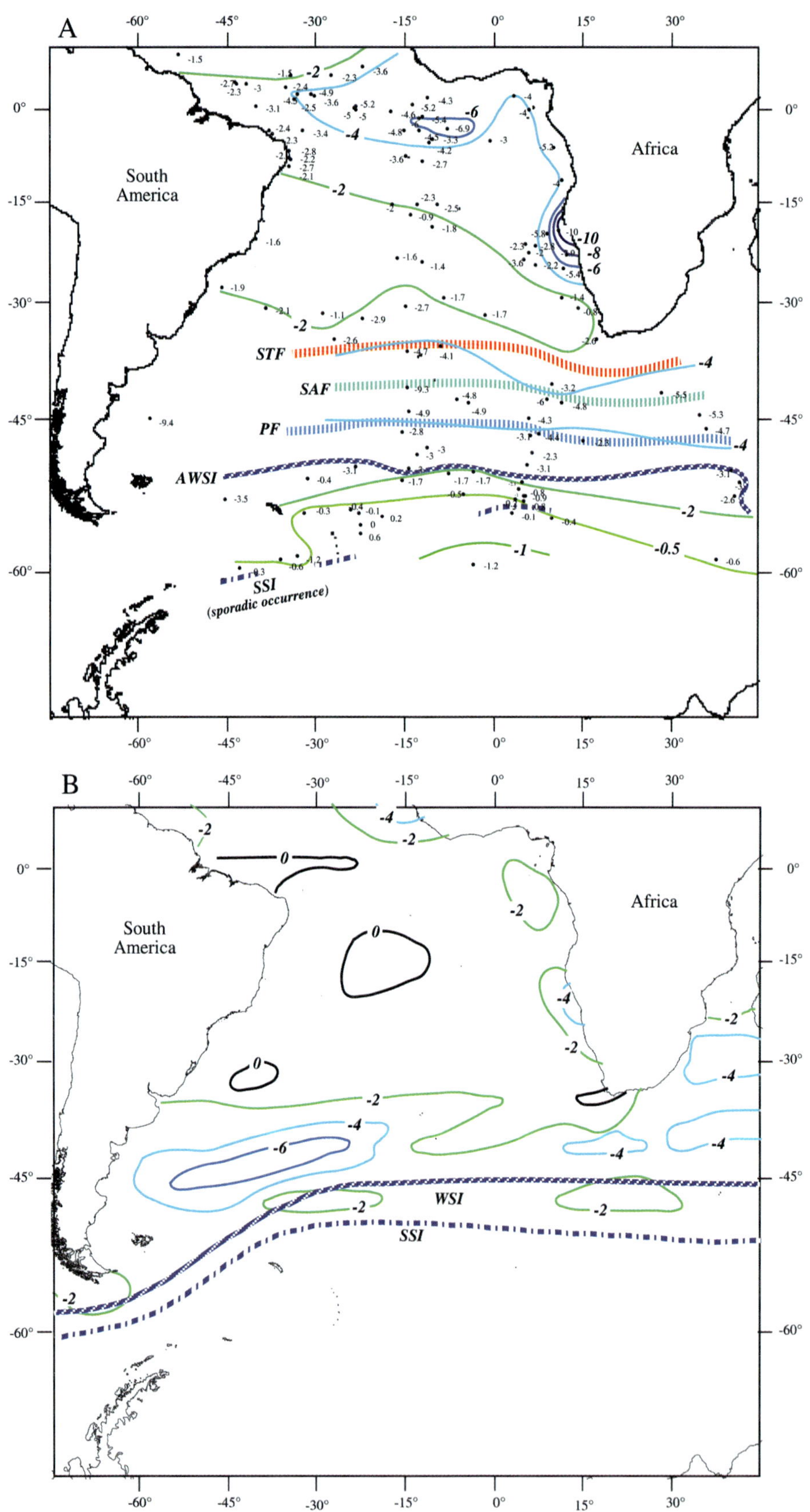
A
South America
Africa
STF
SAF
PF
AWSI
SSI
(sporadic occurrence)
B
South America
Africa
WSI
SSI

Fig. 4. left a) Austral summer sea-surface temperature anomaly (modern/EPILOG LGM) compiled from Niebler et al. (2003) and Gersonde et al. (2003). Also indicated is the location of the estimated position of the ELGM Subtropical Front (STF), Subantarctic Front (SAF), and Polar Front (PF), as well as the boundary of average winter sea-ice (AWSI) extent. Arrows indicate sporadic expansion of summer sea ice (SSI) into the northern Weddell Basin. For core location, derivation and values of estimates, as well as source of estimate, see Tab. 2. Location of modern Southern Ocean fronts and sea ice distribution can be viewed on Fig. 1. **b)** Austral summer sea-surface temperature anomaly (modern/LGM), summer sea ice (SSI), and winter sea-ice (WSI) extent from CLIMAP (1981).

isotope studies for the LGM (Mackensen et al. 2001). This water mass may be the source of the cold waters reaching the surface in the eastern boundary and the equatorial upwelling regions.

Only few cores document the LGM temperature change in the area around the southern tip of Africa, the pathway of the WWR. The data obtained by Niebler et al. (2003) and Gersonde et al. (2003) do not point to a strong decrease in temperature in this area (2 to 3°C average decrease in SST) that would be indicative of a pinch-off of the warm water route due to a northward displacement of the Subtropical Front. Such a finding is in accordance with the CLIMAP (1976, 1981) reconstruction (Fig. 4). However, estimates presented by Pflaumann et al. (2003) indicate a strong seasonality with a cold winter SST from this area, providing a hint to potential seasonal closure of the Cape valve. Clay mineralogical and strontium isotope studies show that Indian-Atlantic bottom water transport was distinctly reduced during the LGM (Diekmann et al. this issue; Goldstein et al. 1999). No direct information is available from the CWR because of the lack of appropriate sediment cores from the Drake Passage. Considering a northward displacement of the Southern Ocean zones, the northern warmer waters passing the Drake Passage at present would be truncated and deflected along the South American coast in the southeast Pacific, due to the southward incision of the southern tip of South America. Such a scenario is corroborated by foraminiferal studies in the eastern tropical Pacific showing that the cooling in the LGM

Equatorial Pacific was related to an increased northward advection of polar waters along the eastern boundary of South America (Feldberg and Mix 2002, 2003). Gersonde et al. (2003) proposed that such a northward deflection of a large proportion of the ACC would have weakened the heat transport from the Pacific to the Atlantic Ocean during the LGM.

Compared with CLIMAP (1981), the reconstruction of Niebler et al. (2003) and Gersonde et al. (2003) results in cooler temperatures for the Equatorial Atlantic, the Namibia Upwelling and a continuous zone of cooling in the present Sub-antarctic Zone area (Fig. 4b). Agreement can be found in the subtropical gyre, where only minor changes have been encountered. The subtropical warm pool may be fostered by an enhanced southward heat transport from the equatorial zone according to Ganopolski et al. (1998). Together with the finding of a close relationship between foramini-feral faunas associated with the eastern boundary currents and the equatorial area (Mix et al. 1999), this suggests enhanced circulation of the subtropical gyre during the LGM. Altogether, the new reconstructions result in a significantly cooler South Atlantic Ocean compared with CLIMAP (1981). However, the broad expansion of the winter and summer sea ice proposed by CLIMAP (1981) could not be confirmed. Gersonde et al. (2003) placed the average winter sea-ice edge at around 50°S, and the maximum winter sea-ice as far north as ca. 48°S, which is in good agreement with the estimate presented by Crosta et al. (1998a,b), but few degrees south of the CLIMAP (1981) winter sea-ice edge (Fig. 4b). More disagreement becomes obvious when the location of the summer sea-ice edge is considered. While CLIMAP (1981) located the summer sea-ice boundary as far as 50°S, thus indicating an extremely expanded permanent sea-ice cover and little sea-ice seasonality, Crosta et al. (1998a,b) speculated that the LGM summer sea-ice distribution was similar to modern conditions, which would result in a stronger seasonality than at present. Gersonde et al. (2003) proposed that the summer sea-ice edge was more expanded than today and may have sporadically reached as far north as ca. 55°S in the eastern Atlantic sector. Relative to numerical model results obtained using

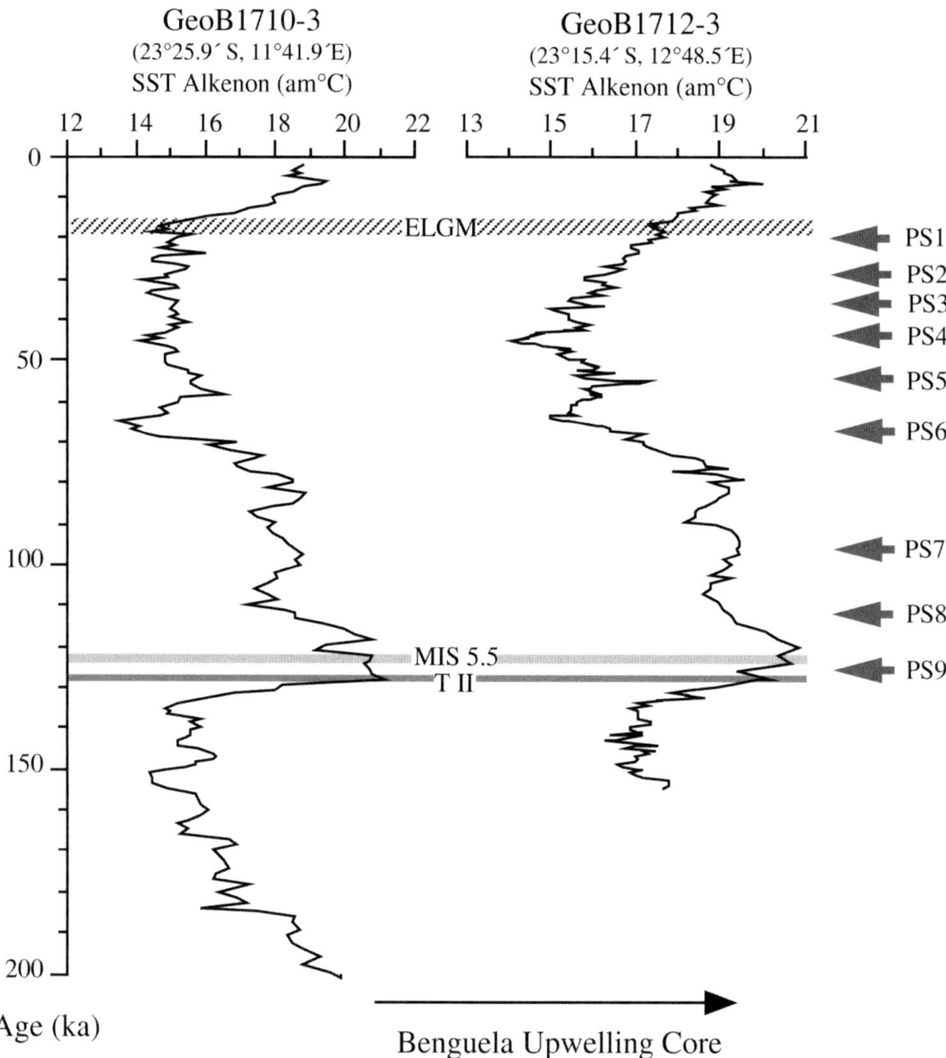

Fig. 5. Alkenone-derived sea surface temperatures (annual mean °C) in cores GeoB 1710-3 and GeoB 1712-3 (data from Kirst et al. 1999) located in the northern Benguela upwelling region (Fig. 2), compared with Benguela upwelling events (PS 1-9), as defined by Little et al. (1997). PS-events may have a duration of several thousand years. Age according to Imbrie et al. (1984) age model.

CLIMAP data, the enhanced cooling but reduced sea ice estimated within the GLAMAP 2000 initiative should modify the simulations of ocean and atmosphere circulation as well as the moisture patterns and the physical impact on CO_2 storage in the ocean.

A South Atlantic MIS 5.5 Time Slice

As a counterpart to the LGM time slice we selected temperature estimates obtained from various sources and a variety of methods, for the climate optimum of the penultimate interglacial (MIS 5). This was done to illustrate surface water conditions during a "warm end" of Pleistocene climate (Fig. 6, Tab. 3). MIS 5.5 was, on global average, slightly warmer than today, with a global sea level 3-6 m higher during the MIS 5 isotopic minimum (MIS 5.5) than at present (Chappel and Shackleton 1986; Stirling et al. 1998). The sea-level highstand has been related to the melt-down of Greenland ice (Körner 1989; Cuffey and Marshall 2000).

Core	Latitude	Longitude	SST Modern	SST MIS 5.5 (°C)	?T MIS 5.5 (°C)	SST Term II	?T Term II	Method	Reference
GeoB1008-	06°34.94'	10°19.12'	25	26	1	26,8	1,8	A	Holmes et al. (1999)
GeoB1016-	11°46.23'	11°40.91'	24,2	26,3	2,1	26,5	2,3	A	Holmes et al. (1999)
GeoB1028-	20°06.2'S	09°11.1'E	19	24,3	5,3	24,6	5,6	A	Müller et al. (1997)
GeoB1105-	01°39.9'S	12°25.7'W	27,1	26,5	26,5	24,3	24,3	F (MAT)	Wefer et al. (1996)
GeoB1112-	05°46.7'S	10°45.0'W	26,7	25,4	-1,3	25,6	-1,1	F	Wefer et al. (1996)
GeoB1710-	23°25.9'S	11°41.9'E	17,5	20,8	3,3	20,7	3,2	A	Kirst et al. (1999)
GeoB1711-	23°18.91'	12°22.6'E	16,6	21,2	4,6	20,7	4,1	A	Kirst et al. (1999)
GeoB1712-	23°15.42'	12°48.5'E	16,6	20,4	3,8	19,4	2,8	A	Kirst et al. (1999)
ODP 1089	40°56'S	9°54'E	12,8	16,8	4	19	6,2	R (IKM)	Cortese & Abelmann (2002)
ODP 1094	53°0.8'S	5°7.8'E	1,4	2,2	0,8	3,8	2,4	D (IKM)	Bianchi & Gersonde (2002)
PS1768-8	52°35.58'	4°28.56'E	1,4	2,4	1	3,7	2,3	D (IKM)	Zielinski et al. (1998)
PS1772-8	55°27.50'	1°09.80'E	0,1	-	-	2,7	2,6	D (IKM)	Bianchi & Gersonde (2002)
PS1778-5	49°00.8'S	12°42'W	4,8	5,4	0,6	1,8	-3	R (IKM)	Brathauer & Abelmann
PS2102-2	53°04.38'	4°59.14'W	0,5	2,8	2,3	2,9	2,4	D (IKM)	Bianchi & Gersonde (2002)
PS2276-4	54°38.13'	23°57.05'	0,9	2,4	1,5	3,1	2,2	D (IKM)	Bianchi & Gersonde (2002)
PS2305-6	58°43.22'	33°02.19'	0,9	0,4	-0,5	1,4	0,5	D (IKM)	Bianchi & Gersonde (2002)
PS2489-2	42°52.40'	8°58.40'E	10,2	9,8	-0,4	11,1	0,9	F (MAT)	Becquey & Gersonde
PS2495-3	41°16.50'	14°29.40'	14,7	15,4	0,7	-	-	F (MAT)	Niebler et al., in prep.
PS2498-1	44°09.20'	14°13.70'	11,1	10,7	-0,4	12,2	1,1	R (IKM)	this study
PS2603-3	58°59.20'	37°37.70'	1,5	3,2	1,7	3,8	2,3	D (IKM)	Bianchi & Gersonde (2002)
RC11-86	35°47'S	18°27'E	21,1	23,8	2,7	-	-	C (IKM)	CLIMAP (1984)
				18,9	-2,2	-	-	F (IKM)	CLIMAP (1984)
RC12-294	37°16'S	10°06'W	17	20,3	3,3	22	5	C (IKM)	CLIMAP (1984)
				19,1	2,1	20,8	3,8	F (IKM)	CLIMAP (1984)
RC13-205	02°17.28'	05°10.98'	28	26,4	-1,6	25,8	-2,2	R (IKM)	CLIMAP (1984)
				27,1	-0,9	25	-3	F (IKM)	CLIMAP (1984)
RC13-228	22°19.8'S	11°11.88'	19,7	22,3	2,6	24,1	4,4	C (IKM)	CLIMAP (1984)
				24,2	4,5	25,1	5,4	R (IKM)	CLIMAP (1984)
RC13-229	25°29.40'	11°18.42'	19,6	23,9	4,2	23,1	3,4	R (IKM)	CLIMAP (1984)
RC24-16	05°02.28'	10°11.40'	26,7	26,5	-0,2	25	-1,7	F (IKM)	McIntyre et al. (1989)
V22-38	09°33.00'	34°15.00'	27,9	25,7	-2,2	25,8	-2,1	C (IKM)	CLIMAP (1984)
				27,3	-0,6	27,9	0	F (IKM)	CLIMAP (1984)
V22-108	43°11'S	3°15'W	11,8	7,9	-3,9	15,5	5,5	R (IKM)	CLIMAP (1984)
V22-174	10°04.20'	12°49.20'	25,8	25,7	-0,1	26	0,2	C (IKM)	CLIMAP (1984)
				26,7	0,9	26,3	0,5	F (IKM)	CLIMAP (1984)
V22-182	00°31.98'	17°16.20'	27,3	26,4	-0,9	26,3	-1	C (IKM)	CLIMAP (1984)
				23,2	-4,1	23,1	-4,2	F (IKM)	CLIMAP (1984)
V25-56	03°33.00'	35°13.80'	27,8	27,5	-0,3	26,8	-1	F (IKM)	McIntyre et al. (1989)
V25-59	01°22.02'	33°28.80'	27,2	27,6	0,4	26,7	-0,5	F (IKM)	McIntyre et al. (1989)
V30-40	00°12.00'	23°09.00'	27,3	25,7	-1,6	25,6	-1,7	F (IKM)	McIntyre et al. (1989)

Table 3. Designation, geographic coordinates, modern sea surface temperature (warm season SST for locations with microfossil based estimates, annual mean temperatures for locations with alkenone-based estimates), SST estimates for MIS 5.5 and Termination II and anomalies, applied method for SST estimate (F: foraminifers, R: radiolaria, D: diatoms, A: alkenones), and reference of estimate. While microfossil derived estimates represent summer sea surface temperatures, alkenone temperatures indicate average annual temperatures.

CLIMAP (1984), which presented the first attempt to compile SST information of the last interglacial ocean, already pointed out that warmest conditions have been reached asynchronously, the southern hemisphere temperature increase leading the global ice-volume draw-down by 3-5 kyr. This pattern has later been corroborated by other sediment and ice core studies (e.g. Charles et al. 1996; Broecker and Henderson 1998; Brathauer and Abelmann 1999; Cortese and Abelmann 2002; Bianchi and Gersonde 2002). Considering the southern lead, we selected the temperatures from two time slices, as proposed by Bianchi and Gersonde (2002). The early time slice is placed in Termination II ("TII-slice"), averaging SSTs from the interval of 127.5-128.5 ka and around 126 ka from cores that have been dated according to Martinson et al. (1987) and Imbrie et al. (1984), respectively. The values reported for the second, the MIS 5.5 time slice represent the average SST calculated on the interval of the isotopic minimum at 122.5-125 ka in cores dated after Martinson et al. (1987), and at around 122 ka in cores dated according to Imbrie et al. (1984). Due to the relatively low number of cores and the uneven distribution of the core locations, the results obtained from 33 core locations do not allow for the generation of an isotherm mapping as was compiled for the ELGM. In all cores from the Southern Ocean, including RC12-294 from the southern subtropical realm, the temperature maximum of the penultimate interglacial, exceeding the modern SSST by up to 5-6°C, have been found to be limited to a period within Termination II (Fig. 6). This clearly documents the lead of the ocean temperature development at southern high latitudes for the MIS 6/MIS 5 transition (Figs. 6, 7). The SSST of the Southern Ocean decreased to values close to modern temperatures during MIS 5.5. A very early response of the SSST also occurred in the Namibia upwelling and the adjacent open ocean areas, but maximum values were only reached during MIS 5.5 in this region. In contrast, most cores from the tropical zone are close to, or slightly below, modern temperatures. This points to a minor but widespread relative cooling during MIS 5.5 in the tropical South Atlantic.

Bianchi and Gersonde (2002) suggest that the 3-4 kyr-temperature lead in the Southern Ocean and the adjacent subtropical zone was initially triggered by precessional changes influencing the summer insolation in high southern latitudes. Rapid re-

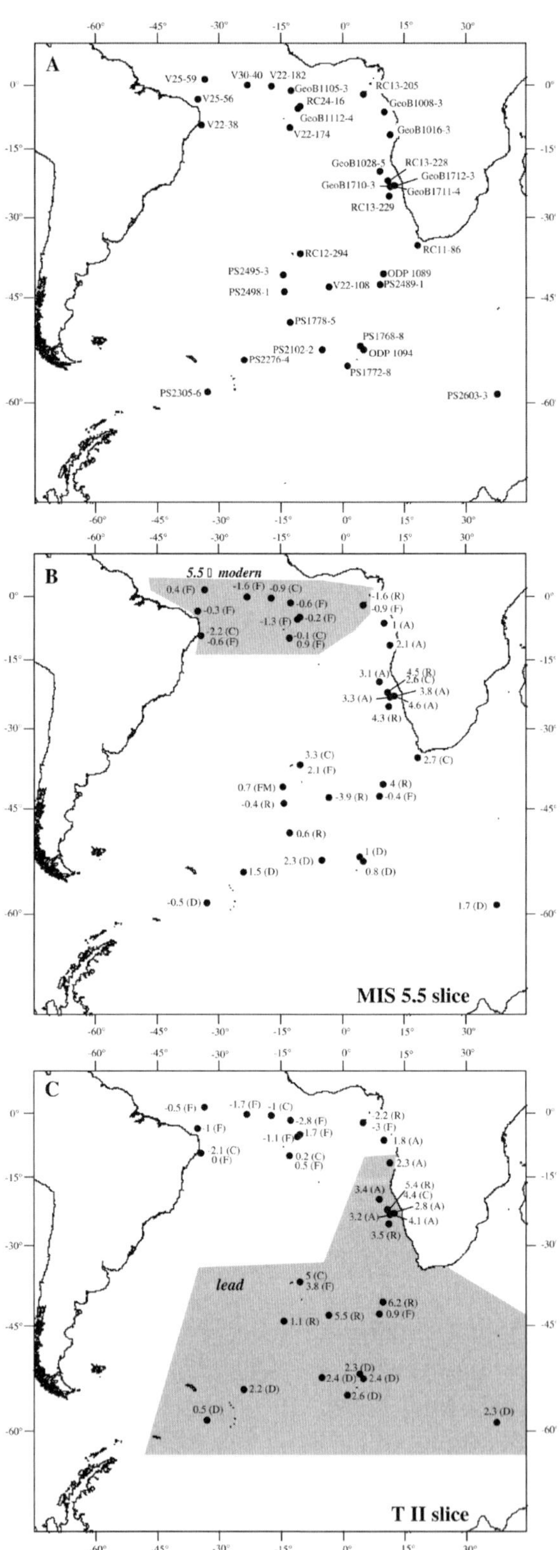

Fig. 6. a) Location and designation of cores selected for South Atlantic MIS 5.5 and T-II time slices. **b)** Summer sea surface temperature anomalies (modern/MIS 5.5 time slice) obtained from foraminifers (F), calcareous coccolith (C), radiolarian (R), and diatom (D) based transfer functions, and annual average temperatures derived from alkenone (A) measurements, compiled from various sources. Shaded area indicates the zone where MIS 5.5 temperatures were similar or slightly colder than present. **c)** Same as in B, but on Termination II time slice. Shaded area indicates the zone where Termination II temperatures exceed MIS 5.5 values. For definition of time slices, see text. For core location, derivation and values of estimates, as well as source of estimate, see Tab. 3.

duction of the Antarctic sea ice to a field less expanded than at present and an increase of the SSST acted as positive feedback mechanisms in reducing the surface albedo and allowing for an enhanced ocean-atmosphere gas exchange and thus release of CO_2 into the atmosphere. Further reinforcement of the SSST during the termination may be related to a collapse or strong reduction in NADW production (see Oppo et al. 1997), leading to southern hemisphere heat piracy (Fig. 1). The resumption of NADW production at the end of Termination II and the subsequent cessation of the southern hemisphere heat piracy could have caused the MIS 5.5 temperature to decrease to conditions only slightly warmer than at present. Assuming that the modern relationship between SSST and the location of Southern Ocean frontal systems can be applied to Termination II, the Southern Ocean oceanic zonal bands were shifted southward by 3 to 5 degrees in latitude (Bianchi and Gersonde 2002), resulting in a wider opening of the WWR and less truncation of warmer components of the ACC bands at the southern tip of South America. Import of warmer waters might have caused an early increase in SST in the Benguela upwelling area, in the presence of a moderate upwelling event caused by intensified South Atlantic trade winds (PS9-event of Little et al. 1997) (Fig. 5). This is followed by another temperature increase during MIS 5.5 and distinct reduction in upwelling-related productivity in this area (Kirst et al. 1999; Little et al. 1997). The lack of temperature increase in the tropical South Atlantic can be interpreted to indicate a heat loss due to a transfer to the north (enhanced North Atlantic heat piracy) but also to the south, as indicated by increased SST in the Subtropical Gyre documented in core GeoB1312 (Hale and Pflau-mann 1999). This might signal an enhanced pole-ward transport via the Brazil Current, which could have been recirculated within the Subtropical Gyre.

The Southern Ocean Pleistocene Temperatures and Climate Variability

The yet most extended SSST time-series of the Southern Ocean has been accomplished by applying a transfer function based on foraminifers (MAT) to a composite core section consisting of core PS2489-2 and sections recovered at the ODP Site 1090 in the Subantarctic Zone (Becquey and Gersonde 2002, 2003). This time-series covering the past 1.83 Ma identifies two main periods of SSST variability and cyclicity, connected by a transitional interval. Between MIS 65 (1.83 Ma) and MIS 22 (0.87 Ma) the record displays a minor glacial/interglacial variability of relatively cold temperatures ranging between 3° and 5°C. This indicates that isotherms characteristic for the present PFZ were shifted to the north by about 7° in latitude, and governed the area of the present central Subantarctic Zone during most of the early Pleistocene. Such a pattern might have resulted in a strong reduction or pinch-off of warm water advection into the South Atlantic via the WWR, and a relative increase in cold-water transfer via the CWR. Power spectra of the SSST record from this period display frequencies that are not in the Milankovitch bands. Forcing of these frequencies (29, 35 and 54 kyr), which have also been reported from low latitudes (e.g. Bassinot et al. 1994) still remains unexplained. The cold period of the Southern Ocean in the early Pleistocene was accompanied by a decrease in SST and an intensification of upwelling in the Benguela region (Marlow et al. 2000), as well as the establishment of strong zonal flow of southern trade winds along the equator (Ruddiman and Janecek 1989). As a response to a major increase in ice volume (Ruddiman et al. 1989), resulting in enhanced glacial/interglacial climate contrasts, the Subantarctic SSST variability started to change between 0.87 and 0.43 Ma (MIS 12), and a 41-kyr and 100 kyr cyclicity was established. While the glacial SSST in the present Subantarctic Zone dropped to values corresponding to those at the present Polar Front, interglacial SSSTs reached values close to present temperatures at the Subantarctic Front, resulting in a steepening of the glacial/interglacial SSST contrast (3°C to 10°C). Following the transitional period, a strong 100 kyr cyclicity at high glacial/interglacial amplitude with SST fluctuating between ca. 4°C and 15°C was established. However, carbonate dissolution can affect SSST estimates by increasing the relative number of cold-water dwellers, such as *N. pachyderma* during glacials, leading to colder estimates and warm-water species resistant to dissolution

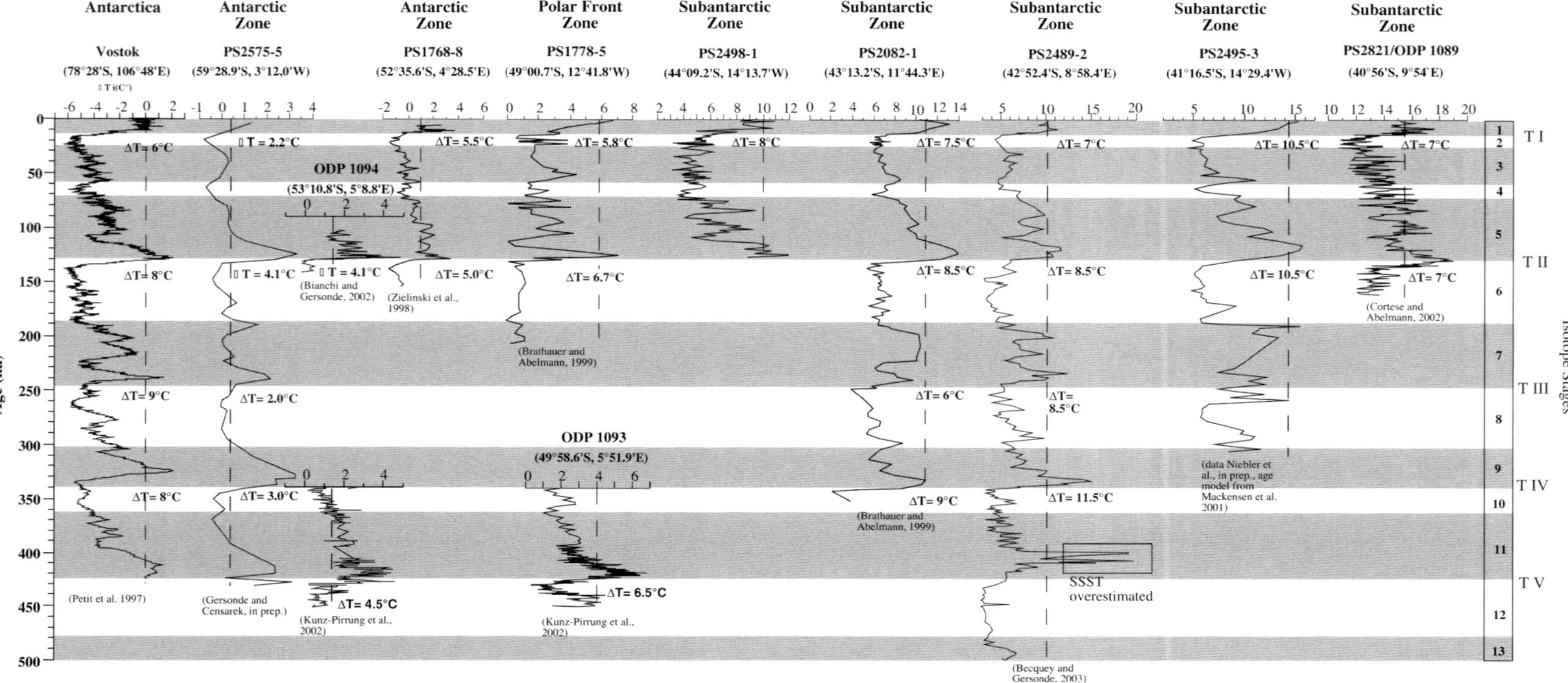

Fig. 7. Comparison of summer sea surface temperatures (°C) time-series obtained for sediment cores located on a S-N transect across the Southern Ocean (for core location, see Figure 2) with the atmospheric temperature at the inversion level (ΔT_i, °C) from the Vostok ice core. The Vostok temperature record is plotted on the GT4 ice core time scale (Petit et al. 1999), the marine records on a SPECMAP-based time scale, causing temperature mismatches in some intervals (e.g. in MIS 9). ΔT_i values indicate maximum temperature change at terminations T I – T V. Stippled lines indicate modern SSST at core location. SSST were calculated based on Eucampia Index Method (PS2575-5), diatom-based IKM (PS 1768-8, ODP 1093, 1094), radiolarian-based-IKM (PS1778-5, PS2567-2, PS2498-1, PS2082-1, PS2821/ODP 1089) and foraminifer-based MAT (PS2489-2, PS2495-3).

(e.g. *Globorotalia inflata*) during warm periods. The latter has been observed during MIS 11, resulting in a broadly overestimated SSST (Becquey and Gersonde 2003) (Fig. 7).

While the Pleistocene SSST record obtained from the PS2489-2/ODP1090 record has a Milankovitch scale resolution, cores recovered at ODP Leg 177 sites 1089, 1093 and 1094 (Fig. 2) allow for the establishment of Pleistocene SSST and sea-ice records at decadal to centennial resolution. Such high-resolution records are the prerequisite for the correlation of atmospheric climate series obtained from continental ice cores and marine records from the North Atlantic, serving as a baseline for better understanding the physical and biological processes governing global climate variability. Yet, SSST records have been only presented from selected intervals of the recovered high-resolution Pleistocene sections (Cortese and Abelmann 2002; Kunz-Pirrung et al. 2002; Bianchi and Gersonde 2002) (Fig. 7). Ongoing studies will soon be completed to present continuous time series at submillennial resolution, covering the past 600 ka from the three high-resolution sites using transfer functions based on radiolarians or diatoms. This period exceeds the time window obtained from the Vostok ice core (Petit et al. 1999), and will be ideal for comparisons with records now being recovered within the European Project for Ice Coring in Antarctica (EPICA). In 2002 the EPICA ice-core recovery expanded the ice-core record beyond MIS 11. It is expected to reach a basal age around 800 ka or older (EPICA Dome C 2001-02 Science and Drilling Teams 2002).

Within the limits of time-resolution and stratigraphic accuracy, the SSST records obtained from cores recovered from a region between the northern Weddell Basin (59°28.9' S) and the Subtropical Front (40°56'S) show good correlation with the atmospheric temperatures recorded in the Vostok ice core (Fig. 7). The SSST amplitudes at the past five terminations range from approximately 4°C in the South to up to 10°C in the Subantarctic cores, compatible with those obtained from Vostok. Highest temperatures exceeding modern values and signalling a southward shift of the Southern Ocean zones, were accompanied by reduction of the sea-ice field as compared with today. They are restricted to the terminations and lead the minimum of benthic $\delta^{18}O$ by several kyr. (e.g. Brathauer and Abelmann 1999; Cortese and Abelmann 2002; Kunz-Pirrung et al. 2002; Bianchi and Gersonde 2002; Becquey and Gersonde 2002, 2003; Mortyn et al. 2003). This pattern should be associated with early changes in atmospheric circulation, impact on ocean/atmosphere CO_2 exchange rates, and the import of warm and salty waters via the WWR and CWR. The lead of Southern Ocean temperature with respect to the global ice volume also occurs in relation to NADW variability, ruling out that NADW, as a transmitter of the insolation effect, triggers climate changes in southern high latitudes on Milankovitch time scales. Brathauer and Abelmann (1999) proposed that the Southern Ocean lead, which has also been reported from tropical cores (Schneider et al. 1995; Pisias and Mix 1997), is related to insolation-modulated changes in the atmosphere and wind fields. These affect the ocean dominated southern hemisphere most efficiently, while the climate change in the northern hemisphere is delayed due to the sluggish response of the large ice sheets. High-resolution records of Termination V (Kunz-Pirrung et al. 2002), Termination II (Bianchi and Gersonde 2002), and Termination I (Cortese and Abelmann 2002) show that the mechanisms triggering the early response include mainly precessional but also obliquity-related orbital forcing. This is amplified by internal processes, such as a rapid reduction of the sea-ice field, reducing the surface albedo, and allowing for an enhanced ocean/atmosphere gas exchange (CO_2 release into the atmosphere), a pattern documented by the close relationship between the steep temperature and CO_2 increase in the ice core record (Cuffey and Vimeux 2001). Further reinforcement of the SSST rise during terminations can be related to changes in the circulation of the global ocean, affected by a collapse or strong reduction in NADW production (e.g. Termination II; Bianchi and Gersonde 2002), and causing South Atlantic heat piracy (Fig. 1).

Changes in THC have been suggested to be associated with the very large millennial-scale changes in SSST observed in a composite section (PS2821-1/ODP 1089) from the northern Subantarctic Zone during MIS 3 and 4. They are almost as great as those observed at Terminations I and II (Cortese and Abelmann 2002) (Fig. 7). This variability can be interpreted to be equivalent to the Dansgaard-Oeschger cycles recognized in the Greenland ice cores, however, operating out of phase relative to the northern hemisphere (Charles et al. 1996; Ninnemann et al. 1999) which is a result of the bipolar seesaw. However, the exact phase relationship between the climate's evolution of the northern hemisphere, the tropical zone, and the southern hemisphere is still under discussion, as it is steered by different processes and amplification mechanisms that vary on different time scales in different regions (Clark et al. 2002). This intricate pattern is highlighted by growing evidence from records documenting climate change during the last deglaciation from different regions on our globe. While the Heinrich 1 (H1) and the Younger Dryas (YD) affect the northern North Atlantic as well as the eastern North Atlantic via the southward advection of cold waters with the Canary Current (Zhao et al. 1995), the temperature development in the western tropical Atlantic covaries with the climate development documented for the Southern Ocean and Antarctic ice cores (Rühlemann et al. 1999) (Fig. 8). This geographical pattern of warm water distribution has been attributed to changes in THC causing South Atlantic heat piracy. However, the finding of a sea-surface temperature change in the tropical Indian Ocean, showing an in-phase development with the northern hemisphere during Termination I, strongly supports the fact that THC variability alone cannot be viewed as the only agent for inter-hemispheric climate tele-connection (Bard et al. 1997). The latter authors speculated that dynamics in the low-latitude atmosphere and related greenhouse forcing caused by changes in atmospheric water vapour content are mighty factors driving climate variability. Recently, Kim et al. (2002) reported an in-phase cooling response during the YD period from the Benguela upwelling region, while the SST displays a continuous warming period during the H1 meltwater event, as was found for the southern high latitudes and the western tropical Atlantic (Fig. 8). Due to a distinctly reduced or halted NADW formation during H1 this response pattern was interpreted to indicate that the change in THC had

caused South Atlantic heat piracy, while, as a result of less pronounced THC modification, the bipolar seesaw effect was distinctly weaker during the YD. Indeed, there is growing evidence that the YD was not related to major meltwater discharge that resulted in a weakened NADW formation (de Vernal et al. 1996; Moore et al. 2000). However, other records point to the contrary (Bond et al. 1997). Kim et al. (2002) suggested that the YD represents a synchronous global phenomenon that had been triggered by atmospheric cooling coupled to alterations in greenhouse gas concentrations. Indeed, the YD is accompanied by a distinct reduction of the atmospheric methane concentration (Chapellaz et al. 1993), while CO_2 strongly increases, after a halt during the period of the Antarctic Cold Reversal (ACR)(Stenni et al. 2001). This highlights the influence exerted by processes on the continents of the northern hemisphere that affect methane variations, whereas the Southern Ocean is proposed to be an important regulator of CO_2 concentrations (Monin et al. 2001). However, although there is further evidence for an atmospherically driven YD cooling from the mid-latitudes of the southern hemisphere from Chilean land records (Moreno et al. 2001), there is yet also no indication for high southern latitudes to be affected by a YD cooling (Rosquist et al. 1999; Singer et al. 1998; Bianchi and Gersonde in prep). Bard et al. (1997) proposed that a decoupling of climate evolution in the high southern latitudes during the YD points to changes in the dynamics of the low-latitude atmosphere, affecting the atmospheric water vapour and its greenhouse effect. This mechanism would be in line with the observed YD reduction in atmospheric methane concentration, which is closely linked with low latitude moisture and temperature. Considering such an interpretation, the mechanisms causing in the western tropical Atlantic a deglacial sequence of events, which is in-phase with the southern high latitudes (Rühlemann et al. 1999) should be reconsidered.

Southern Ocean records document the presence of the Antarctic Cold Reversal (ACR), which is concurrent with the northern hemisphere's warm Bølling/Allerød event (Fig. 8). This points to the bipolar seesaw mechanism and corroborates the climate records obtained from the Byrd, Vostok

Fig. 8. Comparison of climate records of Termination I from the Greenland GRIP and the Antarctic Byrd ice cores with marine surface temperature records from the tropical Atlantic (M35003-4), the eastern South Atlantic (GeoB 1023-5), and the Subantarctic Zone (PS2489-1). The M35003-4 and the GeoB 1023-5 records are alkenone-based estimates representing annual mean surface temperatures. The PS2489-1 record is a radiolarian-based austral summer SST record. Climate period abbreviations: YD Younger Dryas, BA Bølling/Allerød, H1 Heinrich Event 1, ACR Antarctic Cold Reversal.

and EPICA Dome C ice cores. Cold Southern Ocean events are accompanied by an expansion of the Antarctic sea-ice field, which represents an effective amplifier of climate change via its albedo effect (Greenfell 1983), its control over the thermohaline and atmospheric circulation (Comiso and Gordon 1998; Carleton 1989), and CO_2 ventilation (Stephens and Keeling 2000; Morales Marqueda and Rahmsdorf 2002). Rapid variability of the sea-ice field expansions has been documented during the last glacial period (MIS 4-2) from the southern Scotia Sea (core PS2319-1), an ideal area to monitor sea-ice variability, mostly associated to IRD deposition events (Gersonde et al. 2003). The exact timing of these events, and their correlation to the climate of the northern hemisphere, is problematic because of the lack of benthic foraminifers in the sediment records from this area that would allow for the establishment of a $\delta^{13}C$ record for monitoring deep-water circulation. However, combined $\delta^{13}C$ and IRD records from the eastern Atlantic Southern Ocean document that the IRD events are associated with northern hemisphere interstadials and an increased NADW production, suggesting an anti-phase relationship between these regions (Kanfoush et al. 2000). These authors proposed that the linking mechanism may have been sea level rise associated with the melting of ice sheets in the northern hemisphere that ungrounded Antarctic ice shelves, shedding armadas of icebergs to the Southern Ocean during strong inter-stadial events.

Cold reversals as observed during the last deglacial are prominent features of other late and middle Pleistocene terminations documented in

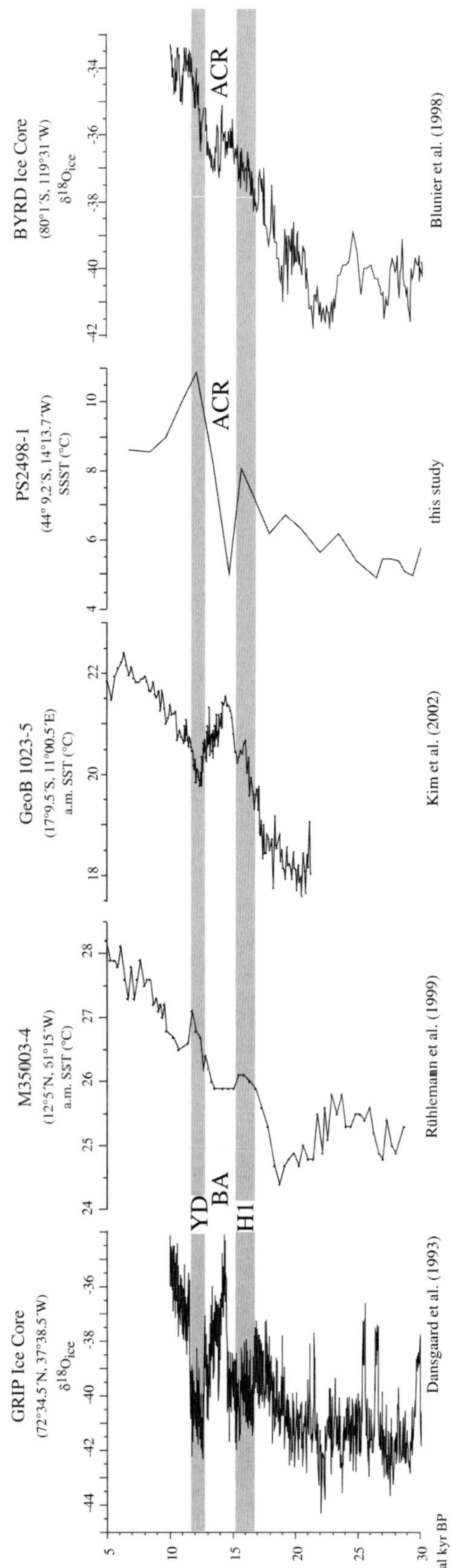

high-resolution Southern Ocean records. The nature of these events, now reported from Termination II (Cortese and Abelmann 2002; Bianchi and Gersonde 2002) and Termination V (Kunz-Pirrung et al. 2002) (Fig. 7) is still under discussion. Possible mechanisms involve northern hemisphere driven changes in THC, but also a discharge of meltwater into the Southern Ocean by a massive production of icebergs. These cooling events entrain a northward expansion of the Southern Ocean cold sphere. Related steepening in thermal gradients might have caused an upwelling event documented in the Benguela upwelling system during Termination II (Little et al. 1997: PS9, Fig. 5).

Our examples of the complexity in interpreting the nature and inter-hemispheric relationship of rapid climate change during glacial and deglacial periods document that, besides global THC changes driven by meltwater injections into the North Atlantic and the Southern Ocean, changes in the atmospheric circulation and greenhouse gas concentration, as well as sea ice, govern and amplify climate change on global, hemispheric and regional scales. The exact interaction of these parameters is however only poorly understood and awaits further studies, including well-dated field observations and their integration in improved Earth system models. The occurrence of millennial-centennial scale SSST variability during interglacial optima, such as MIS 5.5 (Bianchi and Gersonde 2002) and MIS 11 (Kunz-Pirrung et al. 2002), despite being at a lower amplitude than during glacials, suggests that the presence of large ice sheets, meltwater events, changes in the greenhouse gas concentration, and sea ice distribution are not the only prerequisites to trigger millennial-centennial scale variability. As documented for the North Atlantic Holocene (Bond et al. 2001), changes in solar irradiance represent another, external factor inducing such a climate variability.

Constructing a Scenario of South Atlantic Interglacial/Glacial Climate and Circulation Development

Analysing late Pleistocene planktic foraminiferal assemblages from the tropical Atlantic, Mix and Morey (1996) depicted a succession of faunal events, which they used for the reconstruction of the spatio-temporal development of climate-relevant factors in a coupled atmosphere and ocean. They illustrated the reconstructed equatorial upwelling, strength of ocean current system, and the development of the trade-wind field during a climate cycle in the shape of a schematic diagram. Considering the additional new information summarized in the chapters above, we updated the model of Mix and Morey (1996) to construct a more developed scenario for the evolution of the South Atlantic's climate and circulation in the late Pleisto-cene (Fig. 9).

The transition from an interglacial state comparable to modern times (Fig. 9a), governed by a relatively warm northern and a colder southern hemisphere, which resulted from North Atlantic heat piracy, to colder glacial climate conditions is most probably triggered by orbital forcing related annual insolation changes that influence temperatures in the high southern latitudes (Petit et al. 1999). This environmental change is followed by physical and biological processes in the Southern Ocean that entrain changes in atmospheric CO_2, acting as a strong amplifier of climate change (Lorius et al. 1990). Resulting from northward expansion of the Southern Ocean cold-water sphere including summer and winter sea-ice extent, and related increase of greenhouse cooling (Fig. 9b1), thermal gradients on the southern hemisphere become displaced and intensified. This affects the zonal component of the southern trade winds, which, in turn, is the main control mechanism of equatorial and coastal upwelling, and results in an early climatic response in the upwelling areas (Mix and Morey 1996). As a consequence of the new position of the gradients and oceanic frontal systems of the Southern Ocean, the import of warm waters via the WWR was reduced and possibly seasonally pinched off, and the waters imported into the South Atlantic via the CWR across the Drake Passage were relatively colder. This results in a cooler Benguela Current and lowers the temperatures in the southwestern African upwelling (Fig. 9b2). In the equatorial region, enhanced upwelling causes cooling only during the initial part of a glacial cycle. An increased northward advection of the cold Benguela Current water, fed by a cooled CWR

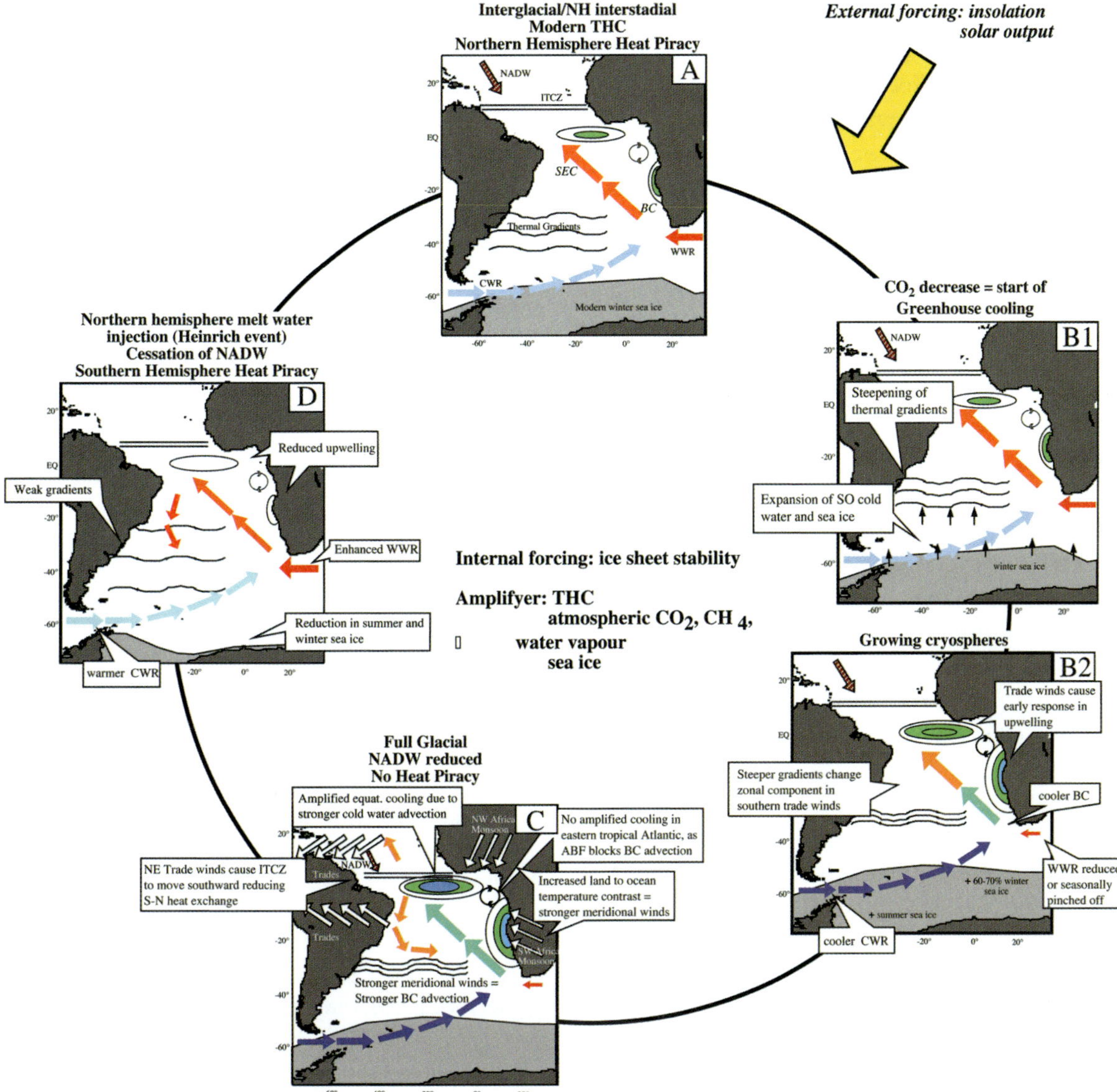

Fig. 9. Schematic representation of the sequence of events involved in the generation of an interglacial/glacial cycle in the South Atlantic. For further explanation, see text. SO Southern Ocean, SEC South Equatorial Current, BC Benguela Current, WWR Warm water route, CWR Cold water route, ABF Angola Benguela Front, NADW North Atlantic Deep Water, ITCZ Intertropical Convergence Zone.

import and by intermediate and deep waters advected from the Southern Ocean, takes over as a cooling agent later in the cycle. The observation that wind-driven upwelling responds faster to orbital forcing than the advection of cold water from the Southern Ocean (also suggested by Mix and Morey 1996) is confirmed in cores from the eastern tropical South Atlantic, where TOC slightly leads SST changes at precessional periodicities, and additionally, by the SST variations in the Angola Basin (cores GeoB1008-3 and 1016-3) which, in turn, precede the ice volume changes (Schneider et al. 1996).

At about this point in the cycle, the intensification of the NE trade winds (generated in the northern hemisphere, and related to the North African monsoon system) pushes the ITCZ to the south (Wefer et al. 1996b), thus working against the exchange of heat between the South and the North Atlantic. Therefore, more heat is trapped in the eastern equatorial South Atlantic. This is an important step towards a new equilibrium status, the No Heat Piracy mode (Figs. 1, 9c). The NADW import is weakened, and the already cooled sea surface, caused by the early response in the upwelling areas, displays a stronger thermal contrast to the adjacent (still warm) continent. This drives intensified meridional winds both close to NW and SW Africa. The result of the meridional winds is a stronger advection of Benguela Current (mainly controlled by meridional winds close to SW Africa) and an enhanced transport of cold polar waters into the upwelling regions, thus amplifying the ongoing cooling. As a result of the more vigorous circulation in the South Atlantic, equatorial cooling is further enhanced by the southward-directed heat export (keeping the Subtropical Gyre relatively warm). This closes the advective loop (Fig. 9c). The eastern tropical Atlantic seems to be excluded from this amplified cooling mechanism (Fig. 4), as the advection of Benguela Current is hindered by the presence of the strong Angola-Benguela Front at 15°S.

Meltwater injections into the North Atlantic terminate the No piracy mode. The related cessation of NADW formation causes a collapse in THC that was still operating, although with reduced strength, during the glacial. As a consequence, heat starts to pile up in the South Atlantic, whereas the northern hemisphere remains in a cold state. Accumulating excess heat melts back the Antarctic sea-ice entraining, via feedback mechanisms, further warming and weakening of the thermal gradients (Fig. 9d). At terminations, this warming may be additionally strengthened by insolation changes that result in a minimum sea-ice distribution, a southward shift of the Southern Ocean's cold-water zones, a wide opening of the WWR, and a warmer CWR. As a consequence, the Benguela upwelling region receives warmer waters during the terminations (Fig. 5). The tropical Atlantic is not warmed up additionally, which is possibly related to enhanced southward export of heat feeding the subtropical region. With the resumption of NADW formation, normal interglacial conditions are re-established.

Acknowledgement

The authors are grateful to Sharon L. Kanfoush and one anonymous reviewer for valuable comments. This study was funded by the Deutsche Forschungsgemeinschaft through the Sonderforschungsbereich 261 at the University of Bremen and the Alfred Wegener Institute at Bremerhaven. This is SFB 261 contribution no. 365. Data are available under www.pangaea.de/Projects/SFB261.

References

Abelmann A (1992) Radiolarian flux in Antarctic waters (Drake Passage, Powell Basin, Bransfield Strait) Polar Biol 12: 357-372

Abelmann A, Gersonde R (1991) Biosiliceous particle flux in the Southern Ocean. Mar Chem 35: 503-536

Abelmann A, Gowing MM (1996) Horizontal and vertical distribution pattern of living radiolarians along a transect from the Southern Ocean to the South Atlantic subtropical region. Deep-Sea Res 43: 361-382

Abelmann A, Gowing MM (1997) Spatial distribution pattern of living polycystine radiolarian taxa: Baseline study for paleoenvironmental reconstructions in the Southern Ocean (Atlantic sector). Mar Micropal 30: 3-28

Abelmann A, Brathauer U, Gersonde R, Sieger R, Zielinski U (1999) A radiolarian-based transfer function for the estimation of summer sea-surface temperatures in the Southern Ocean (Atlantic sector). Paleoceanography 14: 410-421

Arz HW, Pätzold J, Wefer G (1998) Correlated millennial-scale changes in surface hydrography and terrigenous sediment yield inferred from last-glacial marine deposits off northeastern Brazil. Quat Res 50: 157-166

Bard E, Rostek F, Sonzogni C (1997) Interhemispheric synchrony of the last deglaciation inferred from alkenone palaeothermometry. Nature 385: 707-710

Bassinot FC, Beaufort L, Vincent E, Labeyrie LD, Rostek F, Müller PJ, Quidelleur X, Lancelot Y (1994) Coarse fraction fluctuations in pelagic carbonate sediments from the tropical Indian Ocean: A 1,500 kyr record of carbonate dissolution. Paleoceanography 9: 579-600

Becquey S, Gersonde R (2002) Past hydrographic and climatic change in the Subantarctic Zone–The Pleistocene record from ODP Site 1090. Palaeogeogr Palaeoclimatol Palaeoecol 182: 221-239

Becquey S, Gersonde R (2003) A 0.55 Ma paleotempera-ture record from the Subantarctic zone: Implications for Antarctic Circumpolar Current development. Paleoceanography 18: 1014, doi: 10.1029/2000PA 000576

Bianchi C, Gersonde R (2002) The Southern Ocean surface between Marine Isotope Stages 6 and 5d: Shape and timing of climate changes. Palaeogeogr Palaeoclimatol Palaeoecol 187: 151-177

Bianchi C, Gersonde R (in prep) Climate evolution at the last deglaciation: The role of the Southern Ocean

Blanke B, Speich S, Madec G, Döös K (2001) A global diagnostic of interocean mass transfer. J Phys Oceanogr 31: 1623-1632

Blunier T, Chappellaz J, Schwander J, Dallenbach A, Stauffer B, Stocker TF, Raynaud D, Jouzel J, Clausen HB, Hammer CU, Johnsen SJ (1998) Asynchrony of Antarctic and Greenland climate change during the last glacial period. Nature 394: 739-743

Bond G, Showers W, Cheseby M, Lotti R, Almasi P, deMenocal P, Priore P, Cullen H, Hajdas I, Bonani G (1997) A pervasive millennial-scale cycle in North Atlantic Holocene and glacial climates. Science 278: 1257-1266

Bond G, Kromer B, Beer J, Muscheler R, Evans MN, Showers W, Hoffmann S, Lotti-Bond R, Hajdas I, Bonani G (2001) Persistent solar influence on North Atlantic climate during the Holocene. Science 294: 2130-2136

Brathauer U, Abelmann A (1999) Late Quaternary variations in sea surface temperatures and their relationship to orbital forcing recorded in the Southern Ocean (Atlantic sector). Paleoceanogr 14: 135-148

Broecker WS (1998) Paleocean circulation during the last deglaciation: A bipolar seesaw? Paleoceanography 13: 119-121

Broecker WS (2001) The big climate amplifier ocean circulation-sea ice-storminess-dustiness-albedo. In: Seidov D, Haupt BJ, Maslin M (eds) The Oceans and Rapid Climate Change: Past, Present, Future. Geophysical Monograph 126, AGU, Washington, DC, pp 53-56

Broecker WS, Henderson GM (1998) The sequence of events surrounding Termination II and their implications for the cause of glacial-interglacial CO_2 changes. Paleoceanography 13: 352-364

Carleton AM (1989) Antarctic sea ice relationships with indices of atmospheric circulation of the Southern hemisphere. Clim Dyn 3: 207-220

Chappel J, Shackleton NJ (1986) Oxygen isotopes and sea level. Nature 324: 137-140

Chappellaz J, Blunier T, Raynaud D, Barnola JM, Schwander J, Stauffer B (1993) Synchronous changes in atmospheric CH_4 and Greenland climate between 40 and 8 kyr BP. Nature 366: 443-445

Charles CD, Lynch-Stieglitz J, Ninnemann US, Fairbanks RG (1996) Climate connections between the hemispheres revealed by deep-sea sediment core/ice core correlations. Earth Planet Sci Lett 142:19-27

Clark PU, Pisias NG, Stocker TF, Weaver AJ (2002) The role of the thermohaline circulation in abrupt climate change. Nature 415: 863-869

Climate: Long Range Investigation, Mapping, and Prediction (CLIMAP) Project Members (1976) The surface of the ice-age earth. Science 191: 1131-1137

Climate: Long Range Investigation, Mapping, and Prediction (CLIMAP) Project Members (1981) Seasonal reconstructions of the Earth's surface at the Last Glacial Maximum. Geol Soc Am Map and Chart Ser MC-36

Climate: Long Range Investigation, Mapping, and Prediction (CLIMAP) Project Members (1984) The Last Interglacial ocean. Quat Res 21: 123-224

Comiso JC, Gordon AL (1998) Interannual variability in summer sea ice minimum, coastal polynias and bottom water formation in the Weddell Sea. In: Jeffries MO (ed) Antarctic Sea Ice, Physical Processes, Interactions and Variability. Antarctic Res Ser 74, pp 293-315

Cooke DW, Hays JD (1982) Estimates of Antarctic ocean seasonal ice-cover during glacial intervals. In: Craddock C (ed) Antarctic Geoscience. IUGS, Ser B, No. 4, pp 1017-1025

Cortese G, Abelmann A (2002) Radiolarian-based paleo-temperatures during the last 160 kyr at ODP Site 1089 (Southern Ocean, Atlantic Sector). Palaeogeogr Palaeoclimatol Palaeoecol 82: 259-286

Crosta X, Pichon JJ, Burckle LH (1998a) Application of the modern analog technique to marine Antarctic diatoms: reconstruction of maximum sea-ice extent at the Last Glacial Maximum. Paleoceanography 13: 284-297

Crosta X, Pichon JJ, Burckle LH (1998b) Reappraisal of Antarctic seasonal sea-ice at the Last Glacial Maximum. Geophys Res Let 25: 2703-2706

Crowley TJ (2000) CLIMAP SSTs re-revisited. Clim Dyn 16: 241-255

Cuffey KM, Marshall SJ (2000) Substantial contribution to sea-level rise during the last interglacial from the Greenland ice sheet. Nature 404: 591-594

Cuffey KM, Vimeux F (2001) Covariation of carbon dioxide and temperature from the Vostok ice core after deuterium-excess correction. Nature 412: 523-526

Dansgaard W, Johnsen SJ, Clausen HB, Dahl-Jensen D,

Gundestrup NS, Hammer CU, Hvidberg S, Steffensen JP, Sveinbjörnsdottir AE, Jouzel J, Bond G (1993) Evidence for general instability of past climate from a 250-kyr ice-core record. Nature 364: 218-220

de Vernal A, Hillaire-Marcel C, Bilodeau G (1996) Reduced meltwater outflow from the Laurentide ice margin during the Younger Dryas. Nature 381: 774-777

Döös, K (1995) Interocean exchange of water masses. J Geophys Res 100(C7):13,499-13,514

Drijfhout SE, Maier-Reimer E, Mikolajewicz U (1996) Tracing the conveyor belt in the Hamburg large scale geostrophic ocean general circulation model. J Geophys Res 101: 22,563-22,575

Duplessy JC, Shackleton NJ, Fairbanks RG, Labeyrie L, Oppo D, Kallel N (1988) Deep water sources during the last glacial cycle and their impact on the global deepwater circulation. Paleoceanography 3: 343-360

EPICA Dome C 2001-02 science and drilling teams (2002) Extending the ice core record beyond half a million years. Eos 83: 509-517

Feldberg MJ, Mix AC (2002) Sea-surface temperature estimates in the Southeast Pacific based on planktonic foraminiferal species: Modern calibration and Last Glacial Maximum. Mar Micropal 44: 1-29

Feldberg MJ, Mix AC (2003) Planktonic foraminifera, sea surface temperatures, and mechanisms of oceanic change in the Peru and south equatorial currents, 0-150 ka BP. Paleoceanography 18: 1016, doi: 10.1029/2001PA000740

Fischer G, Gersonde R, Wefer G (2002) Organic carbon, biogenic silica and diatom fluxes in the Northern Seasonal Ice Zone in the Polar Front Region in the Southern Ocean (Atlantic Sector): Interannual variation and changes in composition. Deep-Sea Res II 49: 1721-1745

Ganopolski A, Rahmsdorf S, Petoukhov V, Claussen M (1998) Simulation of modern and glacial climates with a coupled global model of intermediate complexity. Nature 391: 351-356

Gersonde R, Hodell DA, Blum P et al. (1999) Proc. ODP, Initial Reports, 177 [CD-ROM]. Available from: Ocean Drilling Program, Texas A&M University, College Station, TX 77845-9547, USA

Gersonde R, Zielinski U (2000) The reconstruction of late Quaternary Antarctic sea-ice distribution-the use of diatoms as a proxy for sea-ice. Palaeogeogr Palaeo-climatol Palaeoecol 162: 263-286

Gersonde R, Abelmann A, Brathauer U, Becquey S, Bianchi C, Cortese G, Grobe H, Kuhn G, Niebler HS, Segl M, Zielinski U, Fütterer DK (2003) Last glacial sea-surface temperatures and sea-ice extent in the Southern Ocean (Atlantic-Indian sector) – A multi-proxy approach. Paleoceanography 18: 1061, doi: 10.1029/2002PA000809

Gersonde R, Censarek B (2003) The Eucampia Index, a method for the reconstruction of cold sea-surface temperatures of the Pleistocene Southern Ocean. Mar Micropaleontol, in prep

Goldstein SL, Hemming SR, Kish S, Rutberg R (1999) Strontium isotopes in South Atlantic detritus: A surface current proxy and tracer of Agulhas leakage. Ninth Annual Goldschmidt Conference, Lunar Planet Inst, Houston, TX, #7537 [CD-ROM]

Gordon AL (1986) Interocean exchange of thermocline water. J Geophys Res 91: 5037-5046

Gordon AL, Weiss RF, Smethie WM Jr, Warner, MJ (1992) Thermocline and intermediate water communication between the South Atlantic and Indian Oceans. J Geophys Res 97: 7223-7240

Greenfell T C (1983) A theoretical model of the optical properties of sea ice in the visible and near infrared. J Geoph Res 88: 9723-9735

Hale W, Pflaumann U (1999) Sea-surface temperature estimations using a Modern Analog Technique with foraminiferal assemblages from western Atlantic Quaternary sediments. In: Fischer G, Wefer G (eds) Use of Proxies in Paleoceanography. Springer, Berlin, pp 69-90

Holmes ME, Eichner C, Struck U, Wefer G (1999) Reconstruction of surface ocean nitrate utilization using stable nitrogen isotopes in sinking particles and sediments. In: Fischer G, Wefer G (eds) Use of Proxies in Paleoceanography - Examples from the South Atlantic. Springer, Berlin, pp 447-468

Hutson WH (1980) The Agulhas Current during the late Pleistocene: Analysis of modern faunal analogs. Science 207: 64-66

Imbrie J, Kipp NG (1971) A new micropaleontological method for quantitative paleoclimatology: Application to a late Pleistocene Caribbean Core. In. Turekian KK (ed) The Late Cenozoic Glacial Age. Yale Univ Press, New Haven, Connecticut, pp 71-181

Imbrie J, Hays JD, Martinson DG, McIntyre A, Mix AC, Morley JJ, Pisias NG, Prell WL, Shackleton NJ (1984) The orbital theory of Pleistocene climate: Support from a revised chronology of the marine $\delta_{18}O$ record. In: Berger A, Imbrie J, Hays J, Kukla G, Saltzman B (eds) Milankovitch and Climate. D Reidel Publishing Company, pp 269-305

Kanfoush SL, Hodell DA, Charles CD, Guilderson TP,

Mortyn PG, Ninnemann US (2000) Millennial-scale instability of the Antarctic ice sheet during the last glaciation. Science 288: 1815-1818

Keeling RF, Stephens BB (2001) Antarctic sea ice and the control of Pleistocene climate instability. Paleoceanography 16: 112-131

Kim JH, Schneider RR, Müller PJ, Wefer G (2002) Interhemispheric comparison of deglacial sea-surface temperature patterns in Atlantic eastern boundary currents. Earth Planet Sci Lett 194: 383-393

Kirst GP, Schneider RR, Müller PJ, von Storch I, Wefer G (1999) Late Quaternary temperature variability in the Benguela Current system derived from alkenones. Quat Res 52: 92-103

Körner RM (1989) Ice core evidence for extensive melting of the Greenland ice sheet through the last glacial-interglacial cycle. Science 244: 964-968.

Kunz-Pirrung M, Gersonde R, Hodell DA (2002) Mid-Brunhes century-scale diatom sea surface temperature and sea ice records from the Atlantic sector of the Southern Ocean (ODP Leg 177, Sites 1093, 1094 and core PS 2089-2). Palaeogeogr Palaeoclimatol Palaeoecol 182: 305-328

Levitus S, Boyer T (1994) World Ocean Atlas, Vol. 4: Temperature, NOAA Atlas NESDIS 4, US dept of Commerce, Washington, DC

Little MG, Schneider RR, Kroon D, Pice B, Summerhayes CP, Segl M (1997) Trade wind forcing of upwelling, seasonality, and Heinrich events as a response to sub-Milankovitch climate variability. Paleoceano-graphy 12: 568-576

Lorius C, Jouzel J, Reynaud D, Hansen J, Le Treut H (1990) Greenhouse warming, climate sensitivity and ice core data. Nature 347: 139-145

Lozano JA, Hays JD (1976) Relationship of radiolarian assemblages to sediment types and physical ocea nography in the Atlantic and western Indian Ocean sectors of the Antarctic Ocean. In: Investigation of Late Quaternary Paleoceanography and Paleoclimatology. Geol Soc Am Mem 145: 303-336

Lutjeharms JRE (1996) The exchange of water between the South Indian and South Atlantic Oceans. In: Wefer G, Berger WH, Siedler G, Webb DJ (eds) The South Atlantic: Present and Past Circulation. Springer, Berlin, pp 125-162

Mackensen A, Rudolph M, Kuhn G (2001) Late Pleistocene deep-water circulation in the subantarctic eastern Atlantic. Glob Planet Change 30: 197-229

Marlow JR, Lange CB, Wefer G, Rosell-Melé A (2000) Upwelling intensification as part of Pliocene-Pleisto-cene climate transition. Science 290: 2288-2291

Martinson DG, Pisias NG, Hays JD, Imbrie J, Moore TC, Shackleton NJ (1987) Age dating and orbital theory of the ice ages: development of a high-resolution 0 to 300,000-year chronostratigraphy. Quat Res 27: 1-29

McIntyre A, Ruddiman WF, Karlin K, Mix AC (1989) Surface water response of the equatorial Atlantic Ocean to orbital forcing. Paleoceanography 4: 19-55

Miller JR, Russell GL (1989) Ocean heat transport during the last glacial maximum. Paleoceanography 4: 141-155

Mix AC (1986) Late Quaternary paleoceanography of the Atlantic Ocean: Foraminiferal faunal and stable-isotope evidence. PhD Thesis, Columbia Univ

Mix AC, Morey AE (1996) Climate feedback and Pleisto-cene variations in the Atlantic South Equatorial Current. In: Wefer G, Berger WH, Siedler G, Webb DJ (eds) The South Atlantic: Present and Past Circulation. Springer, Berlin, pp 503-525

Mix AC, Morey AE, Pisias NG, Hostetler SW (1999) Fora-miniferal faunal estimates of paleotemperature: circumventing the no-analog problem yields cool ice age tropics. Paleoceanography 14: 350-359

Mix AC, Bard E, Schneider R (2001) Environmental processes of the ice age: Land, oceans, glaciers (EPILOG). Quat Sci Rev 20: 627-658

Monin E, Indermühle A, Dällenbach A, Flückiger J, Stauffer B, Stocker TF, Raynaud D, Barnola JM (2001) Atmospheric CO_2 concentrations over the last glacial termination. Science 291: 112-114

Moore JK, Abbott MR, Richman JG, Nelson DM (2000) The Southern Ocean at the Last Glacial Maximum: A strong sink for atmospheric carbon dioxide. Glob Biogeochem Cycl 14: 455-475

Morales Marqueda MA, Rahmsdorf S (2002) Did Antarctic sea-ice expansion cause glacial CO_2 decline? Geophys Res Lett 29: 111-113

Moreno PI, Jacobson GL, Lowell TV, Denton GH (2001) Interhemipheric climate links revealed by a late glacial cooling episode in southern Chile. Nature 409: 804-808

Morley JJ (1979) A transfer function for estimating paleoceanographic conditions based on deep-sea surface sediment distribution of radiolarian assemblages in the South Atlantic. Quat Res 12: 381-395

Mortyn PG, Charles CD, Ninnemann US, Ludwig K, Hodell DA (2003) Deep-sea sedimentary analogs for the Vostok ice core. Geochem Geophys Geosyst 4: 8405, doi: 10.1029/2002GC000475

Mulitza S, Wolff T, Pätzold J, Hale W, Wefer G (1998) Temperature sensitivity of planktic foraminifera and its influence on the oxygen isotope record. Mar Micropal 33: 223-240

Müller PJ, Cepek M, Ruhland G, Schneider RR (1997) Alkenone and coccolithophorid species changes in Late Quaternary sediments from Walvis Ridge: Implications for the alkenone paleotemperature method. Palaeogeogr Palaeoclimatol Palaeoecol 135: 71-96

Müller PJ, Kirst G, Ruhland G, von Storch I, Rosell-Melé A (1998) Calibration of the alkenone paleotemperature index $U^{K'}_{37}$ based on core-tops from the eastern South Atlantic and the global ocean (60°N-60°S). Geochim Cosmochim Acta 62: 1757-1772

Naval Oceanography Command Detachment (1985) Sea-Ice Climatic Atlas. Vol 1, Antarctic, Natl Space Technol Lab, Asheville, N, 131 p

Niebler HS, Gersonde R (1998) Planktic foraminifera reference data set for reconstruction of past paleotem-peratures (southern South Atlantic Ocean). Mar Micropaleontol 34: 213-234

Niebler HS, Mulitza S, Donner B, Arz H, Pätzold J, Wefer G (2003) Sea-surface temperatures in the Equatorial and South Atlantic Ocean during the Last Glacial Maximum (23-19 ka). Paleoceanography 18: 1069, doi: 10.1029/2003PA000902

Niebler HS et al. (in prep) A new foraminiferal-based data set for reconstruction of southern high latitude Pleistocene surface water temperatures.

Ninnemann US, Charles CD, Hodell DA (1999) Origin of global millennial scale climate events: Constraints from the Southern Ocean deep-sea sedimentary record. In: Clark U, Webb RS, Keigwin LD (eds) Mechanisms of Global Climate Change at Millennial Time Scale. AGU, Washington DC, pp 99-112

Olbers D, Gouretski V, Seiß G, Schröter J (1992) Hydrographic Atlas of the Southern Ocean, Alfred Wegener Institute for Polar and Marine Research, Bremerhaven, Germany

Oppo DW, Horowitz M, Lehman SJ (1997) Marine core evidence for reduced deep water production during Termination II followed by a relatively stable substage 5e (Eemian). Paleoceanography 12: 51-63

Partridge TC, Scott L, Hamilton JE (1999) Synthetic reconstruction of southern African environments during the Last Glacial Maximum (21-18 kyr) and the Holocene Altithermal (8-6 kyr). Quat Int 57/58: 207-214

Paul A, Schäfer-Neth C (2003) Modeling the water masses of the Atlantic Ocean at the Last Glacial Maximum. Paleoceanography 13: 1058, doi: 10.1029/2002PA000783

Peterson, RG, Stramma L (1991) Upper-lever circulation in the South Atlantic Ocean. Prog Oceanogr 26: 1-73

Petit JR, Jouzel J, Raynaud D, Barkov NI, Barnola JM, Basile I, Bender M, Chappellaz J, Davis M, Delaygue G, Delmotte M, Kotlyakov M, Legrand M, Lipenkov Y, Lorius C, Pepin L, Ritz C, Saltzman E, Stievenard M (1999) Climate and atmospheric history of the past 420,000 years from the Vostok ice core, Antarctica. Nature 399: 429-436

Pflaumann U, Duprat J, Pujol C, Labeyrie LD (1996) SIMMAX. A modern analog technique to deduce Atlantic sea surface temperatures from planktonic foraminifera in deep-sea sediments. Paleoceanography 11: 15-35

Pflaumann U, Sarnthein M, Chapman M, Funnell B, Huels M, Kieer T, Maslin M, Schulz H, Swallow J, van Kreveld S, Vautravers M, Vogelsang E, Weinelt M (2003) The Glacial North Atlantic: Sea-surface conditions reconstructed by GLAMAP-2000. Paleoceanography 18: 1065, doi: 10.1029/2002PA 000774

Pisias NG, Mix AC (1997) Spatial and temporal oceanographic variability of the eastern equatorial Pacific during the late Pleistocene. Evidence from radiolarian microfossils. Paleoceanography 12: 381-393

Prahl FG, Wakeham SG (1987) Calibration of unsaturated patterns in long-chain ketone compositions for paleotemperature assessment. Nature 330: 367-369

Rintoul SR (1991) South Atlantic interbasin exchange. J Geophys Res 97: 5493-5550

Rosquist GC, Rietti-Shatti M, Shemesh A (1999) Late glacial to middle Holocene climatic record of lacustrine biogenic silica oxygen isotopes from a Southern Ocean island. Geology 27: 967-970

Ruddiman WF, Janecek TR (1989) Pliocene-Pleistocene biogenic and terrigenous fluxes at equatorial Atlantic Sites 662, 663 and 664. In: Ruddiman WF, Sarnthein M, Baldauf J (eds) Proceedings of the Ocean Drilling Program, Scientific Results. Ocean Drilling Program, College Station, TX, pp 211-240

Ruddiman WF, Raymo ME, Martinson DG, Clement BM, Backman J (1989) Pleistocene evolution: Northern Hemisphere ice sheets and North Atlantic Ocean. Paleoceanography 4: 353-412

Rühlemann C, Mulitza S, Müller PJ, Wefer G, Zahn R (1999) Warming of the tropical Atlantic Ocean and slowdown of thermohaline circulation during the last deglciation. Nature 402: 511-514

Sarnthein M, Gersonde R, Niebler HS, Pflaumann U, Spielhagen R, Thiede J, Wefer G, Weinelt M (2003) Overview of Galcial Atlantic Ocean Mapping (GLAMAP 2000). Paleoceanography 18, doi: 10.1029/2002PA000769

Schlitzer R (1996) Mass and heat transports in the South Atlantic derived from historical hydrographic data. In: Wefer G, Berger WH, Siedler G, Webb DJ (eds) The South Atlantic: Present and Past Circulation. Springer, Berlin, pp 305-323

Schmidt, GA (1999) Error analysis of paleosalinity calculations. Paleoceanography 14: 422-429

Schneider RR, Müller PJ, Ruhland G (1995) Late Quaternary surface circulation in the east equatorial South Atlantic. Evidence from alkenone sea surface temperatures. Paleoceanography 10: 197-219

Schneider RR, Müller PJ, Ruhland G, Meinecke G, Schmidt H, Wefer G (1996) Late Quaternary surface temperatures and productivity in the East-Equatorial South Atlantic: response to change in trade/monsoon wind forcing and surface water advection. In: Wefer G, Berger WH, Siedler G, Webb DJ (eds) The South Atlantic: Present and Past Circulation. Springer, Berlin, pp 527-551

Seidov D, Maslin M (2001) Atlantic Ocean heat piracy and the bipolar climate see-saw during Heinrich and Dansgaard-Oeschger events. J Quat Sci 16: 321-328

Sieger R, Gersonde R, Zielinski U (1999) New software package available for quantitative paleoenvironmental reconstructions. EOS 80: 223

Singer C, Shulmeister J, McLea B (1998) Evidence against a significant Younger Dryas cooling event in New Zealand. Science 281: 812-814

Stenni B, Masson-Delmotte V, Johnsen S, Jouzel J, Longinelli A, Monnin E, Röthlisberger R, Selmo E (2001) An oceanic cold reversal during the last deglaciation. Science 293: 2074-2077

Stephens BB, Keeling RF (2000) The influence of Antarctic sea-ice on glacial-interglacial CO_2 variations. Nature 404: 171-174

Stirling CH, Esat TM, Lambeck K, McCulloch MT (1998) Timing and duration of the Last Interglacial: Evidence for a restricted interval of widespread coral reef growth. Earth Planet Sci Lett 160: 745-762

Stommel H (1980) Asymmetry of interoceanic freshwater and heat fluxes. Proc Nat Acad Sci (USA) 77: 2377-2381

Stute M, Forster M, Frischkorn H, Serejo A, Clark JF, Schlosser P, Broecker WS, Bonani G (1995) Cooling of tropical Brazil (5°C) during the last glacial maximum. Science 269: 379-383

Thompson LG (2000) Ice core evidence for climate change in the tropics: Implications for our future. Quat Sci Rev 19: 19-35

Thompson SR, Stevens DP, Döös K (1997) The importance of interocean exchange south of Africa in a numerical model. J Geophys Res 102: 3303-3316

Tomczak M, Godfrey JS (1994) Regional oceanography. An introduction. Pergamon, New York

Tréguer P, Nelson DM, Van Bennekom AJ, deMaster DJ, Leynaert A, Quéguiner B (1995) The silica balance in the world ocean: A reestimate. Science 268: 375-379

Vidal L, Schneider RR, Marchal O, Bickert T, Stocker TF, Wefer G (1999) Link between the North and the South Atlantic during Heinrich events of the last glacial period. Clim Dyn 15: 909-919

Wefer G, Berger WH, Siedler G, Webb DJ (1996a) The South Atlantic: Present and Past Circulation. Springer, Berlin, 644 p

Wefer G, Berger WH, Bickert T, Donner B, Fischer G, Kemle-von Mücke S, Meineke G, Müller PJ, Mulitza S, Niebler HS, Pätzold J, Schmidt H, Schneider RR, Segl M (1996b) Late Quaternary circulation of the South Atlantic. The stable isotope records and implications for heat transport and productivity. In: Wefer G, Berger WH, Siedler G, Webb DJ (eds) The South Atlantic: Present and Past Circulation. Springer, Berlin, pp 461-502

Wolff T, Grieger B, Hale W, Dürkopp A, Mulitza S, Pätzold J, Wefer G (1999) On the reconstruction of paleosalinities. In: Fischer G, Wefer G (eds) Use of Proxies in Paleoceanography. Springer, Berlin, pp 207-228

Yokoyama Y, Lambeck K, de Deckker P, Johnson P, Fifield K (2000) Timing for the maximum of the Last Glacial constrained by lowest sea-level observation. Nature 406: 713-716

Zhao M, Beveridge NAS, Shackleton NJ, Sarnthein M, Eglington G (1995) Molecular stratigraphy of cores off northwest Africa: Sea surface temperature history over the last 80 ka. Paleoceanography 10: 661-675

Zielinski U, Gersonde R (1997) Diatom distribution in Southern Ocean surface sediments (Atlantic sector): implications for paleoenvironmental reconstructions. Palaeogeogr Palaeoclimatol Palaeoecol 129: 213-250

Zielinski U, Gersonde R, Sieger R, Fütterer D (1998) Quaternary surface water temperature estimations: calibration of a diatom transfer function for the Southern Ocean. Paleoceanography 13: 365-383

The Atlantic Ocean at the Last Glacial Maximum: 1. Objective Mapping of the GLAMAP Sea-Surface Conditions

C. Schäfer-Neth* and A. Paul

DFG Forschungszentrum Ozeanränder, Universität Bremen, Postfach 33 04 40, 28334 Bremen, Germany
** corresponding author (e-mail): csn@uni-bremen.de*

Abstract: Recent efforts of the German paleoceanographic community have resulted in a unique data set of reconstructed sea-surface temperature for the Atlantic Ocean during the Last Glacial Maximum, plus estimates for the extents of glacial sea ice. Unlike prior attempts, the contributing research groups based their data on a common definition of the Last Glacial Maximum chronozone and used the same modern reference data for calibrating the different transfer techniques. Furthermore, the number of processed sediment cores was vastly increased. Thus the new data is a significant advance not only with respect to quality, but also to quantity. We integrate these new data and provide monthly data sets of global sea-surface temperature and ice cover, objectively interpolated onto a regular 1°x1° grid, suitable for forcing or validating numerical ocean and atmosphere models. This set is compared to an existing subjective interpolation of the same base data, in part by employing an ocean circulation model. For the latter purpose, we reconstruct sea surface salinity from the new temperature data and the available oxygen isotope measurements.

Aims of this study

Until very recently, ocean-wide - and even more so global - reconstructions of glacial sea-surface conditions suffered from severe limitations:

- Data scarcity: Except for the pioneering CLIMAP (1981) study, all data sets were confined to small areas of the ocean, and the number of sediment cores was relatively small. For example, the reconstruction of Weinelt et al. (1996) was limited to the northeastern part of the Atlantic Ocean and employed only 25 cores with temperature data.
- Chronozone definition: Depending on the type of sediment core data measured by the different groups, the chronozone of the Last Glacial Maximum (LGM) within the core could be identified using a $\delta^{18}O$ criterion, ^{14}C datings, the minimum of reconstructed SST as in CLIMAP, lithological parameters, and other evidence. Thus, at the small scale, even cores in close vicinity could easily yield "LGM" SSTs belonging to considerably different times and introduce unrealistic high variability into

any field interpolated horizontally from the core data. At the large scale, data sets provided by different institutions for different parts of the ocean could not be consistently combined due to offsets between them.

- Methodology: SST estimates may be derived from different proxy data, such as faunal assemblages of foraminifera and dinoflagellates, alkenone concentrations, oxygen isotopes, and more. Results from different proxies are not equivocal. For example, tropical and subtropical SST estimates from alkenones and corals sampled from the western Atlantic Ocean are lower than SSTs derived from faunal assemblages (Guilderson et al. 1994; Rühlemann et al. 1999; Crowley 2000). Reconstructions of glacial land temperatures (Rind and Peteet 1985; Stute et al. 1992; Aeschbach-Hertig et al. 2000; Aeschbach-Hertig et al. 2002) too indicate that the assemblage-based SSTs are in part too high.

From WEFER G, MULITZA S, RATMEYER V (eds), 2003, The South Atlantic in the Late Quaternary: Reconstruction of Material Budgets and Current Systems. Springer-Verlag Berlin Heidelberg New York Tokyo, pp 531-548

• Calibration: Regardless of the method used, it has to be calibrated using the modern relations between SST and the respective proxy. Any change of the modern reference data set will alter the calculated paleo-SSTs. Quite commonly, the atlas by Levitus (1982) and, more recently, the WOA 1994 data set (Levitus and Boyer 1994) have been employed. Increasing the differences further, for some studies the surface values were taken, whereas others were based on the temperatures from 10 meters depth, or a vertical average over the upper 50 meters. In addition, some groups used local data sets with higher spatial resolution that were more appropriate for the region studied than the global compilations (e.g. De Vernal et al. 2000).

All these differences cause inconsistencies between SST reconstructions at individual sediment core sites that make interpolation over a larger region a difficult if not impossible task. However, a reliable and consistent set of sea-surface conditions is still a necessity for driving and validating numerical models of paleoclimate. Being aware of this, a number of the institutions contributing paleo-temperature reconstructions (Table 1) succeeded to ground their reconstructions on a common base, that is, a uniform definition for the LGM chronozone and a standardized set of modern reference data. This cooperation greatly improved both quality and quantity of the SST estimates, and gave an opportunity to produce new, and better, seasonal maps of glacial SST on a regular grid. We realize that the supplemental data we used to fill in the remaining void areas (Table 2) do not fully comply with the new standards. However, we carefully checked these data for inconsistencies with the GLAMAP sets, both in the selection process and by variogram analysis.

Improved LGM SST and SSS Data

The joint effort of the contributing research groups aimed at reducing the discrepancies between individual SST reconstructions and filling the still large undersampled areas of the ocean. Many of the hindrances mentioned in the introduction were overcome within the last few years, at least for the Atlantic Ocean:

• The number of sampled and analyzed sediment cores was vastly increased. Instead of typically some ten or twenty cores per group, now there are nearly 300 SST reconstructions available for the Atlantic ocean (Fig. 1, top), fairly exceeding all previous studies, even CLIMAP (1981) with its almost 180 cores on the global scale. These point-wise data is supplemented by Summer and Winter sea-ice reconstructions for both hemispheres.

• It was agreed on two only slightly different definitions for the LGM: the GLAMAP chronozone (Sarnthein et al. 2003a) between 18000 and 22000 calendar years before present, based on the maximum $\delta^{18}O$ values in the cores, and the EPILOG chronozone (Mix et al. 2001) from 19000 to 23000 years, defined using the minimum glacial sea level. Thus, both definitions overlap in the time interval of 19-22 ka, and SST estimates for both chronozones differ only marginally at a given core location (Pflaumann et al. 2003; Gersonde et al. 2003; Niebler et al. 2003). In fact, both chronozones are subsets of the slightly broader EPILOG level 2

Region, Season, Depth	Organisms[1]	Method[2]	Age[3], ka BP	Authors
SST				
South Atlantic, February, 10 m	F R D	TF MAT	23-19	Gersonde et al. 2003; Niebler and Gersonde 1998; Abelmann et al. 1999; Zielinski et al. 1998;
South to tropical Atlantic, annual mean / seasonality, 10 m	F	TF	23-19	Niebler et al. 2003
Tropical to North Atlantic, Jan-Mar / Jul-Sep, 10 m	F	SIMMAX	22-18	Pflaumann et al. 2003
Ice Cover				
South Atlantic, August	D	A	23-19	Gersonde et al. 2003; Gersonde and Zielinski 2000
North Atlantic, Jan-Mar / Jul-Sep	via SST		22-18	Pflaumann et al. 2003

Table 1. GLAMAP data employed for this study.

Region, Season, Depth		Organisms[1]	Method[2]	Age[3], ka BP	Authors
	SST				
North Atlantic, February / August, 0 m		Dc	MAT	23-19	De Vernal et al. 2000
Atlantic, February / August, 0 m		F	MAT	18[14]C	Prell 1985
Mediterranean, February / August, 0 m		F	TF	18[14]C	Bigg 1994
Global Ocean, February / August, 0 m		F R D C	TF	18[14]C	CLIMAP 1981
	Ice Cover				
North Atlantic, February / August		Dc	MAT	23-19	De Vernal et al. 2000
Global Ocean, February / August		F R D C	TF	18[14]C	CLIMAP 1981
	Planktic foraminiferal δ^{18}O				
South Atlantic	Melles 1991; Duplessy et al. 1996				
North Atlantic	Kellogg et al. 1978; Ruddiman and McIntyre 1981; Jansen and Erlenkeuser 1985; Zahn et al. 1985; Bard et al. 1987; Morris 1988; Jones and Keigwin 1989; Keigwin and Boyle 1989; Jansen and Veum 1990; Vogelsang 1990; Duplessy et al. 1991; Lackschewitz 1991; Veum et al. 1992; Duplessy et al. 1992; Köhler 1991; Jünger 1993; Weinelt 1993; Sarnthein et al. 1995; Weinelt et al. 1996				

Notes on Table 1 and 2:
[1] F: Foraminifera, R: Radiolaria, D: Diatoms, C: Coccolithophores, Dc: Dinocysts.
[2] TF: Transfer functions, MAT: Modern analog technique, SIMMAX: revised MAT with distance weights (Pflaumann et al. 1996), A: Abundance.
[3] Calendar ages, if not marked as [14]C.
[4] Average sea-ice curve.

Table 2. Supplemental data collected for this study.

chronozone between 18000 and 24000 years (Mix et al. 2001).

• For the modern reference data set, the WOA 94 (Levitus and Boyer 1994) temperature at 10 meters depth was chosen.

• These new definitions for chronozone and reference data set were not only applied to newly sampled sediment cores, but older census counts were re-evaluated under the new settings.

To show how significant these improvements are, we compare geostatistical analyses of modern and glacial SST data sets (Fig. 2). The top panel illustrates the benefits of the common chronozone definition and modern reference data. Between 50°N and 80°N, the data available in 1994 (Schulz 1994) exhibit high variances at small distances (dashed line). This is in contrast to the spatial dependencies of modern temperatures (shaded), which show low variance at short distance, as expected. To obtain this variogram, we used only the WOA 98 temperatures sampled at the positions of the 15 '94 SST reconstructions. With the advent of the new GLAMAP data, the situation has been improved substantially (solid line). Even taking only the data for the '94 locations yields a much more reasonable variogram, indicating similar values at close distance, and growing differences as distance increases. In the same manner, the data quality was improved in the South Atlantic Ocean (middle panel). As can be seen from the coincident peaks of the two variograms from the South Atlantic Ocean, sampling density is also an important factor. Using the full GLAMAP data set between 50°N and 80°N (lower panel, shaded) yields a much smoother variogram that is more suitable for fitting and gridding than the variogram (solid line) that is obtained if the same data is reduced to the '94 sites. However, the main advance is the careful refinement of the criteria for including core samples into the LGM reconstruction, and how to calibrate the transfer technique. The underlying methods and proxies to determine SST are still different throughout the research groups (Table 1).

SST

The SST reconstructions employed for the present study are listed in Tables 1 and 2 and are described in detail by Paul and Schäfer-Neth (2003) - termed PSN hereafter. To compile our new Atlantic-wide

Glacial Sea Surface Temperature from Sediment Cores

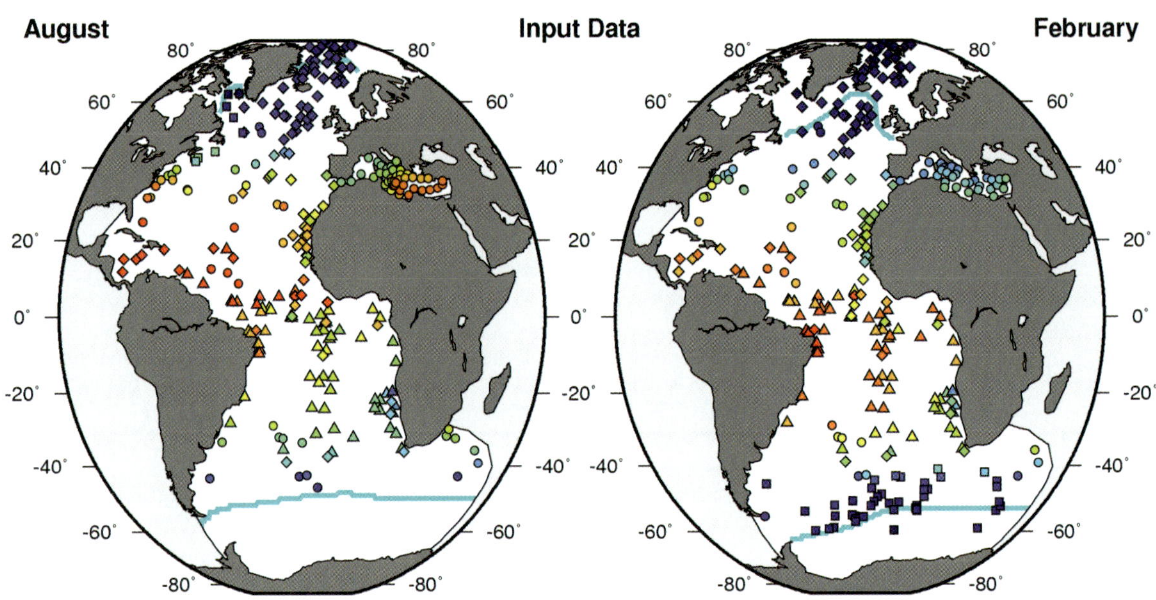

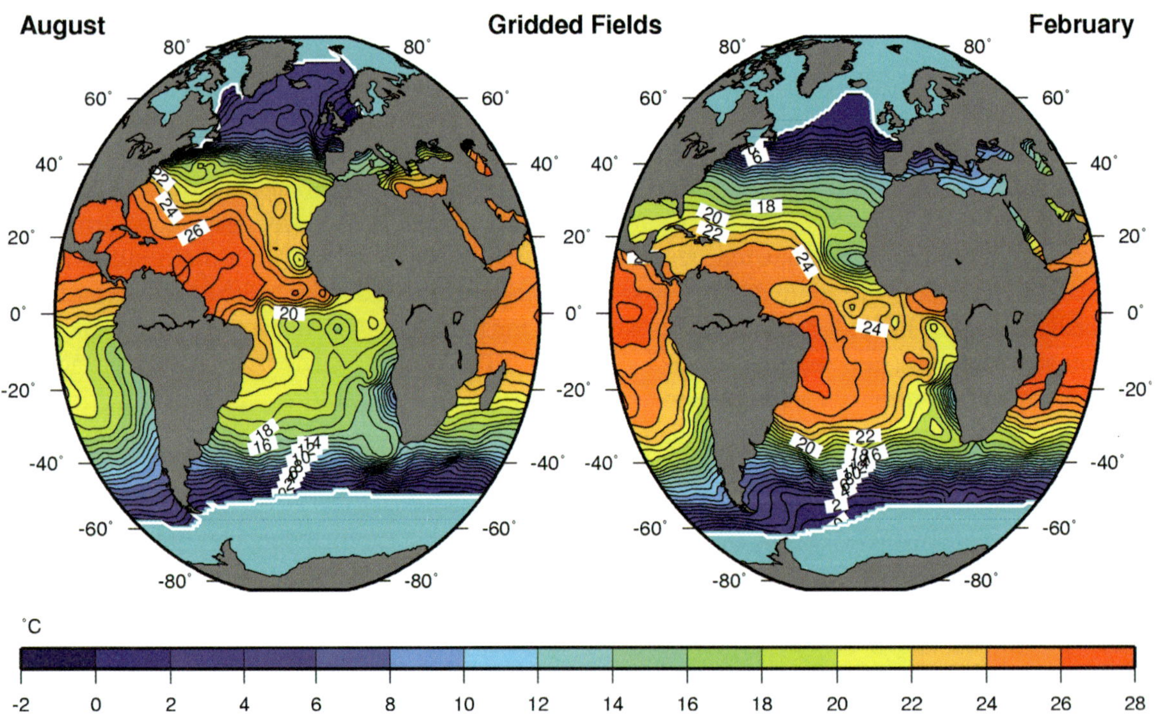

Fig. 1. Glacial sea-surface temperature (°C). Top row: Data base. Triangles: Niebler et al. 2003. Diamonds: Pflaumann et al. 2003. Boxes: De Vernal et al. 2000 (Aug), Gersonde et al. 2003 (Feb). Circles: Prell 1985 (Atlantic), Bigg 1994 (Mediterranean). Blue lines: Ice edges after Pflaumann et al. 2003; De Vernal et al. 2000; Gersonde and Zielinski 2000. Light grey shade: SST data taken from CLIMAP (1981). Bottom row: Resulting 1°×1° gridded fields, white lines denote ice edge.

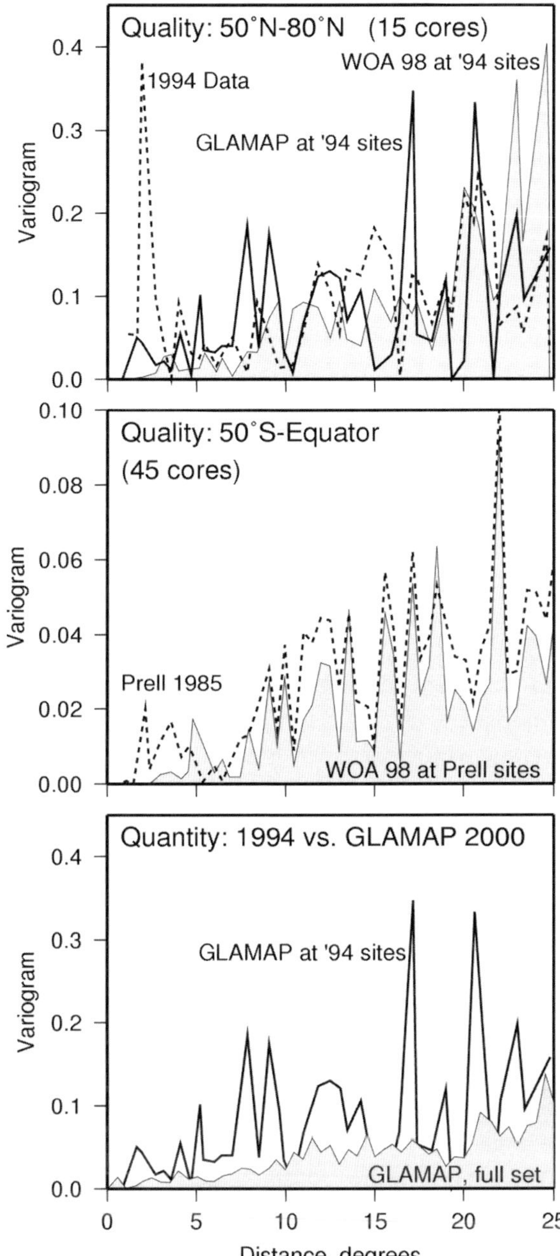

Fig. 2. Empirical variograms employing different present-day and glacial data sets of boreal summer SSTs. Top panel: Effect of the refined chronozone definition and calibration for the North Atlantic Ocean: results for the data available in 1994 (dashed, Schulz 1994) and for the GLAMAP (solid) and WOA 98 (shaded) data at the same core locations. Middle: Like above, but for the South Atlantic Ocean: results from the Prell (1985) data set (dashed) compared to the WOA 98 data (shaded) sampled at the same core sites. Bottom: Effect of the inclusion of new SST estimates: GLAMAP at the 1994 core sites (solid) compared to the full GLAMAP set.

gridded SST fields, we used the following GLAMAP reconstructions:

• The austral summer SSTs from the Atlantic sector of the Southern Ocean by Gersonde et al. (2003), based on faunal assemblages of foraminifera (Niebler and Gersonde 1998), radiolaria (Abelmann et al. 1999), and diatoms (Zielinski et al. 1998).

• The seasonal SSTs computed from the annual means and seasonalities reconstructed by Niebler et al. (2003) for the South Atlantic Ocean.

• The seasonal SSTs for the North Atlantic Ocean estimated by Pflaumann et al. (2003).

Since these new data do not cover the Atlantic Ocean and its marginal seas as a whole, we included the following earlier reconstructions after carefully checking for possible inconsistencies with the GLAMAP data:

• The August summer SSTs for the North Atlantic Ocean derived by De Vernal et al. (2000) from assemblages of dinoflagellate cysts. In the northeastern Atlantic, these SSTs seem unrealistically high, in part even higher than modern temperatures. There are two conceivable causes for this: First, the high SST may indicate a very shallow, light surface layer produced by summer ice melt that is rapidly warmed to these high values. Second, there might be advection of warmer water from lower latitudes. Both scenarios would be accompanied by a strong stratification near the sea surface, leading to differences between temperatures reconstructed from the shallow-dwelling dinoflagellates and the foraminifers dwelling at greater depth. Since the northeastern Atlantic Ocean is well covered by GLAMAP, we did not use the reconstructions of De Vernal et al. (2000) for this area.

• The revised seasonal CLIMAP SSTs by Prell (1985) in the western parts of the Atlantic Ocean and east of South Africa.

• The seasonal SSTs for the Mediterranean Sea as compiled by Bigg (1994) from the reconstructions by Thiede (1978) and Thunell (1979).

• The seasonal CLIMAP (1981) data for the western part of the Mexican Gulf.

However, there are a number of significant differences between the interpolated SST fields of our previous study and those presented here. Most important, the data by Pflaumann et al. (2003) exists in two versions, namely the original core-based pointwise SSTs, and as part of a map of isolines that were constructed based on a subjective interpolation of their own and the Niebler et al. (2003) data. It was this second version we used in our earlier study (PSN).

Despite the advances in data quality that have been demonstrated above, there are still sediment cores that are located closely to each other but nevertheless show considerable differences between the reconstructed temperatures. When interpolating such data by hand, some decision has to be made as to whether take some sort of a mathematically strict average, discard less reliable estimates, include some prior knowledge from other proxy data, or even honor the best-sampled core as the single representative one. Here we investigate the differences between these subjectively interpolated data and an objective interpolation of the core-based SST. In our earlier study, we had to employ additional tie points where the isolines ended in the open ocean to avoid artificial fronts between these ends and the ocean margins. This was not necessary anymore. Furthermore, using now the core-based SSTs only, we did not include the reconstructed ice edges in the gridding process. Instead, SSTs were interpolated 'as is', and the ice covers were superimposed onto the gridded fields.

Like in our previous study, we used variogram analysis and kriging for interpolation of the SSTs. This was done with spherical-coordinate versions (Schäfer-Neth et al. 1998) of the gamv2 and okb2d routines of the GSLIB package (Deutsch and Journel 1992). To account for changing spatial variabilities between the different regions of the Atlantic Ocean, we divided each of the monthly sets into 10 latitude belts 30 degrees wide, each overlapping the next by 15 degrees, and calculated five experimental variograms, one omnidirectional and four in the local meridional, zonal, SW-NE, and SE-NW directions, the latter four with an angular tolerance of 45 degrees. Lag spacing was set to a maximum of 50 lags of 2 degrees. The kriging was carried out for each belt on a regular $1°\times1°$ grid using variogram models fitted to the pair of perpendicular variograms showing maximum and minimum variance. In cases of small overall variance, the omnidirectional variograms were used. Joining the overlapping belt-wise grids by weighted averaging and merging with the global CLIMAP (1981) data sets with an additional 2° moving average yielded the new August and February SST fields (Fig. 1, bottom row). Following the PMIP (1993) guidelines, we constructed a seasonal cycle by first fitting a sinusoidal cycle to the glacial-to-modern anomalies and then adding the modern monthly SSTs (10 m values from WOA 1998).

Ice Cover

In addition to the SST estimates, there are reconstructed August and February ice edges for the Atlantic Ocean (Tables 1 and 2) In the Southern Hemisphere, Gersonde et al. (2003) and Gersonde and Zielinski (2000) derived maximum and average sea-ice extents during austral winter from diatom abundances. For our compilation, we used their average winter curve. Within 2°-3° of latitude, this line corresponds to the line of maximum ice advance by Crosta et al. (1998), except that it indicates a little more ice in the western, but somewhat less ice in the eastern South Atlantic Ocean. The ice extent during glacial austral summer is less well constrained (Gersonde et al. 2003). However, these authors suggest an ice edge north of its position during modern summer in the western, but close to modern winter conditions in the eastern Atlantic Ocean. We therefore chose a line starting at 64°S in the Drake Passage and reaching 62°S south of Africa. The ice edge lines for the South Atlantic Ocean were smoothly joined to the CLIMAP (1981) ice edges in the Pacific and Indian Oceans. Based on a correlation of modern sea-ice extent and SST estimated from core-top foraminifer assemblages, Sarnthein et al. (2003b) placed the summer and winter ice edges along the 3°C and 0.4°C SST isolines in the northeastern Atlantic Ocean. We extended these reconstructions to the west according to the lines derived from dinoflagellate cyst assemblages by De Vernal et al. (2000). From these minimal and maximal ice covers, we constructed five additional monthly ice cover fields, gradually migrating between the two extremes and setting SST to -1.8°C in the ice-covered regions.

SSS

We estimated sea-surface salinity (SSS) following the approach of Schäfer-Neth (1998) and Schäfer-Neth and Paul (2001) by first computing the sea water oxygen isotope ratio $\delta^{18}O_W$ from the newly gridded temperature and the carbonate isotopic composition $\delta^{18}O_C$ of fossil foraminifer shells, and then calculating SSS from $\delta^{18}O_W$. A total of 143 deep-sea sediment cores recovered from the North Atlantic Ocean were available for this purpose (Tables 1 and 2, Fig. 4, top left). The glacial summer SST was sampled from the gridded fields at the locations of the $\delta^{18}O_C$ cores and corrected to the calcification temperature T_C of the foraminifera according to the following empirical relations:

Neogloboquadrina pachyderma sinistral $\quad$ (1)
$T_C = SST - 2.5 \qquad$ if $T > 4.5°C$
$T_C = 0.42\,SST + 0.39 \qquad$ if $T < 4.5°C$
$\qquad\qquad\qquad\qquad$ Weinelt (1993)
Globigerina bulloides
$T_C = SST - 1 \qquad$ Duplessy et al. (1991)

$\delta^{18}O_W$ was then computed according to the Epstein et al. (1953) paleo-temperature equation:

$$\delta^{18}O_W = \delta^{18}O_C - 21.63 + (310.61 + 10\,T_C)^{1/2}, \quad (2)$$

with the carbonate and water oxygen isotope ratios expressed versus the PDB and SMOW standards, respectively. From $\delta^{18}O_W$, we calculated salinity at each sediment core location according to:

$$S = \Delta S_g + A\,[\,\delta^{18}O_W - \Delta\delta^{18}O_g\,] + B, \quad (3)$$

where $\Delta S_g = 1.07$ denotes the global salinity increase due to the lower glacial sea level and $\Delta\delta^{18}O_g = 1.2$ represents the global increase of $\delta^{18}O_W$ due to the storage of ^{16}O in the continental ice sheets. To account for locally different relations between S and $\delta^{18}O_W$, the coefficients A and B (Fig. 3) vary with latitude (Paul et al. 1999); they were derived from the Atlantic GEOSECS data (Östlund et al. 1987) as contained in the GISS $\delta^{18}O$ database (Bigg and Rohling 2000; Schmidt et al. 1999). We used only the GEOSECS data for fitting because the complete $\delta^{18}O_W$ data set includes a large number of very low $\delta^{18}O_W$ and salinity values from the Labrador Sea that are not representative for most of the North Atlantic Ocean and would have considerably biased the result. We tried to use a $\delta^{18}O_W$-salinity relation varying not only with latitude but also with season to account for the local effects of melting and formation of sea-ice, but the database was not sufficient to establish a relation for the boreal winter. Schmidt (1999) demonstrates that an incorrect $\delta^{18}O_W$-SSS relation does not significantly increase the salinity errors when salinity is high - which is the case in the glacial northern North Atlantic. Furthermore, most of the high-latitude GEOSECS data were obtained during the warm season during which the foraminifera build up their shells, that is, both $\delta^{18}O_W$-SSS relation and $\delta^{18}O_W$ data can be in general attributed to the same time of the year. Therefore we adhered to the annual mean relation for our reconstructions.

Because the available $\delta^{18}O_C$ data span only a limited part of the northeastern Atlantic Ocean (Fig. 4, top left), it turned out to be difficult to incorporate the newly gridded data into whatever global data set without artificial gradients. Therefore, we gridded the glacial-to-modern salinity difference referenced to the 10-m-values of the WOA (1998) summer salinity, using the global 1.07 anomaly outside the data-covered region. The gridding was carried out by the same variogram analysis and kriging process that was employed for the temperature data. By adding the gridded anomaly field to the modern monthly salinity data (Schäfer-Neth and Paul 2001), these as well taken from the 10 m WOA (1998) analyses, we finally arrived at the seasonal SSS cycle (Fig. 4, bottom).

SST and SSS Errors

According to Malmgren et al. (2001), SST is reconstructed from the sediment cores with a statistical accuracy in the order of 1°C by the modern analog techniques (MAT, SIMMAX), and with a slightly larger error of about 1.2°C by the transfer function (TF) approach. This is reflected by the SST differences of neighboring sediment cores (Fig. 1, top), even if the estimates were obtained

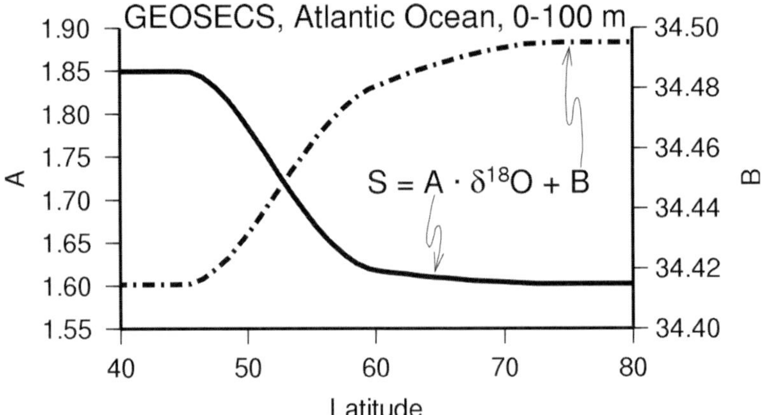

Fig. 3. Latitude-dependent linear fit between salinity and water oxygen isotopic composition obtained from the Atlantic Ocean's GEOSECS measurements.

with the same method. These differences in part exceed the 0.2°C error difference between MAT and TF, and therefore we considered the latter as unimportant for our present study. In addition to the statistical uncertainties, there are a number of possible systematic errors: First, the depth habitats and the growing seasons of the foraminifera might have changed between LGM and present, such that the proxies might not as directly depend on sea surface conditions as MAT and TF assume. Both methods are calibrated to modern data and cannot account for these changes. Second, Malmgren et al. (2001) demonstrate that the inclusion of geographical information obtained from the modern training data might compromise the applicability of SIMMAX to climates very different from the modern. That is, SIMMAX could suffer from this error in regions where, for example, the LGM ocean currents differed from the modern ones. We regard the high northern latitudes of the Atlantic Ocean as most susceptible to this source of uncertainty because of the much larger glacial ice sheets. A comparative application of MAT and TF to the species countings from that region could yield more insight into this problem. However, this has not yet been done. Third, at the cold end of the SIMMAX data, there is a tendency of SST overestimation (Pflaumann et al. 2003) that may amount to more than 1°C. We overcame this problem by including additional information on the

position of the ice edges and setting SST to the freezing point in the ice-covered regions.

Thanks to the recent efforts in refining and standardizing and the acquisition of many new SST estimates, we could not discern any systematic trends or offsets between the contributions of the different research groups. Based on this, the dense spacing of the sediment cores, and the reasonable empirical variograms (Fig. 2), we regard our gridded SST fields as reliable pictures of the robust features at the surface of the Atlantic Ocean at the LGM.

Reconstructions of SSS are even more problematic because several sources of error are involved. To begin with the least important, there are the analytical errors of the parameters SSS is computed from: the SST uncertainty of about 1°C and the $\delta^{18}O_C$ measurement error of typically 0.1. Error propagation through the application of the paleotemperature equation (2) yields a $\delta^{18}O_W$ error of about 0.4, and, with the relation between $\delta^{18}O_W$ and salinity (3), an SSS error around 0.7. This is about 25% of the east-west and north-south contrasts of the reconstructed SSS anomalies in the subtropical and subpolar North Atlantic Ocean (Fig. 4, top left), that is, errors due to false SST and/or incorrect $\delta^{18}O_C$ would not change the salinity patterns very much, especially given the high spatial sampling density. Despite this fairly robust estimate of local SSS gradients we are left with a greater un-

Glacial Sea Surface Salinities

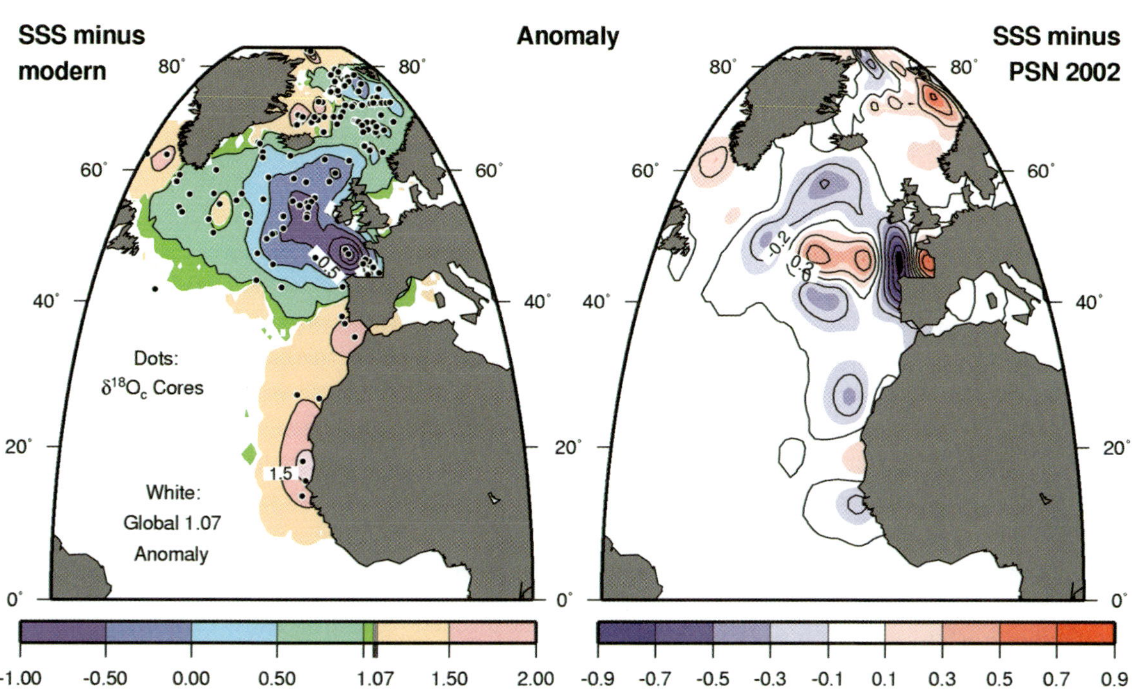

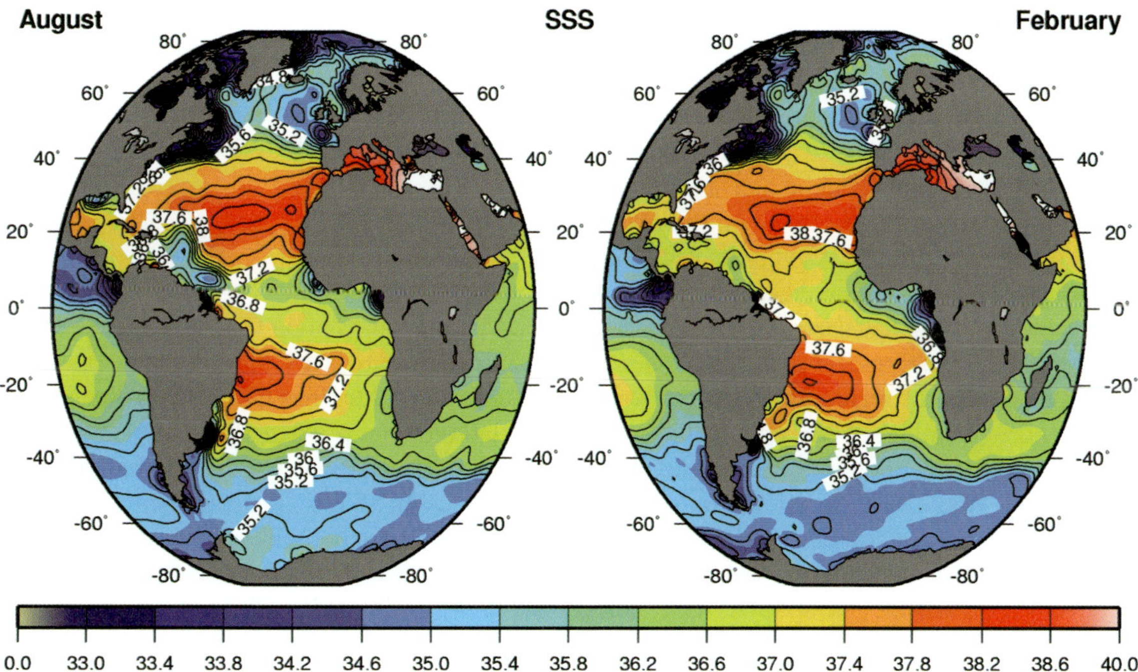

Fig. 4. Glacial sea-surface salinity. Top row: Glacial-to-modern anomaly (left) and difference between salinities based on sediment-core-SST and isoline-SST (right). Bottom row: Gridded seasonal fields.

certainty introduced by the parameters of the empirical relationships between SST, $\delta^{18}O_W$, $\delta^{18}O_C$, and SSS that affects the absolute salinity values. It turns out that the relation between $\delta^{18}O_W$ and salinity is less critical than the paleo-temperature equation (2) that basically links water temperature, water oxygen isotopic composition and the isotope ratio fixed in the carbonate shells of the foraminifera. Thus, Equation (2) is intimately tied to the life cycle of these organisms and varies depending on numerous parameters, such as the prevailing water masses during the season of reproduction and the water depth preferred by the different biota. Even for a single species, there may be different $\delta^{18}O_C$ values depending on when and where the samples were taken. For example, Mulitza et al. (this volume) report a systematic contrast of 1 ‰ between shells of *Neogloboquadrina pachyderma sinistral* that were sampled near the surface ocean (lighter values) and those retrieved from the core-top sediments (heavier due to secondary calcification), whereas there is no such offset for *Globigerina bulloides*. The temperature adjustments given by Equation (1) are just an attempt to compensate for these different offsets, and according to the review of different paleo-temperature calibrations by Bemis et al. (1998), SST adjustments of up to 4-5°C might be necessary, depending on the species, to yield a consistent $\delta^{18}O_W$ data set. Summarizing in a detailed analysis the influence of these effects on the reconstructed salinity, Schmidt (1999) estimates a total salinity error that may easily exceed 1 and reach even values around 1.8 in the tropics. A promising new approach in this respect is to set up an ecological model for the relationship between water $\delta^{18}O_W$ and foraminiferal $\delta^{18}O_C$ on a global scale, as proposed by Schmidt and Mulitza (2002) to quantify the vital effects based on temperature, depth habitat, and calcification processes. Presently, this model is in its initial stage and can not be directly applied to the glacial chronozone for deducing water isotopic composition from the carbonate $\delta^{18}O_C$ data (Schmidt pers. comm. 2002). Ideally, such a model should be combined with a global set of glacial $\delta^{18}O_C$ measurements presently under preparation (Mulitza et al. this volume). For the moment we can only test the effects of recon-

structed SST and regional SSS by numerical model experiments and assess the reconstructions with regard to the resulting water mass and circulation patterns. In our earlier publication (PSN), we therefore considered different additional salinity offsets in the Southern Ocean and their interplay with the reconstructed SSS in the North Atlantic Ocean. Despite all uncertainties of the individual SSS reconstructions, we gain confidence in the overall LGM SSS patterns from the results of recent coupled atmosphere-ocean models. For example, Shin et al. (2003) find local glacial anomalies in the range of our reconstructions. Especially the pattern of anomalies in excess of the global shift in the northwestern Atlantic Ocean and lower than the global shift near the Gulf of Biscay (Fig. 4) is well met in their model, which does not rely on any SST or SSS reconstruction.

Availability

The gridded 1°×1° SST and SSS data are available from the World Data Center for Paleoclimatology, 325 Broadway, Boulder, Colorado; http://www.ngdc.noaa.gov/paleo/paleo.html; email: paleo@noaa.gov

Comparison to CLIMAP

Core-Based and CLIMAP SST

When comparing the newly gridded SST and the CLIMAP reconstruction (Fig. 5, top), the most-prominent feature is that the new data turns out to be generally colder than CLIMAP in the low and mid latitudes but is distinctly warmer in the high latitudes, especially during the summer season.

The lower SST is consistent with the lower temperature based on other proxy data evidence that were discussed by Prell (1985), Guilderson et al. (1994), and Crowley (2000). A glacial Atlantic Ocean cooler than CLIMAP is as well supported by reconstructed land temperatures in eastern North America (Aeschbach-Hertig et al. 2000) and in tropical America (Aeschbach-Hertig et al. 2002; Rind and Peteet 1985; Stute et al. 1995), although indirectly. Only in a narrow band in the tropical

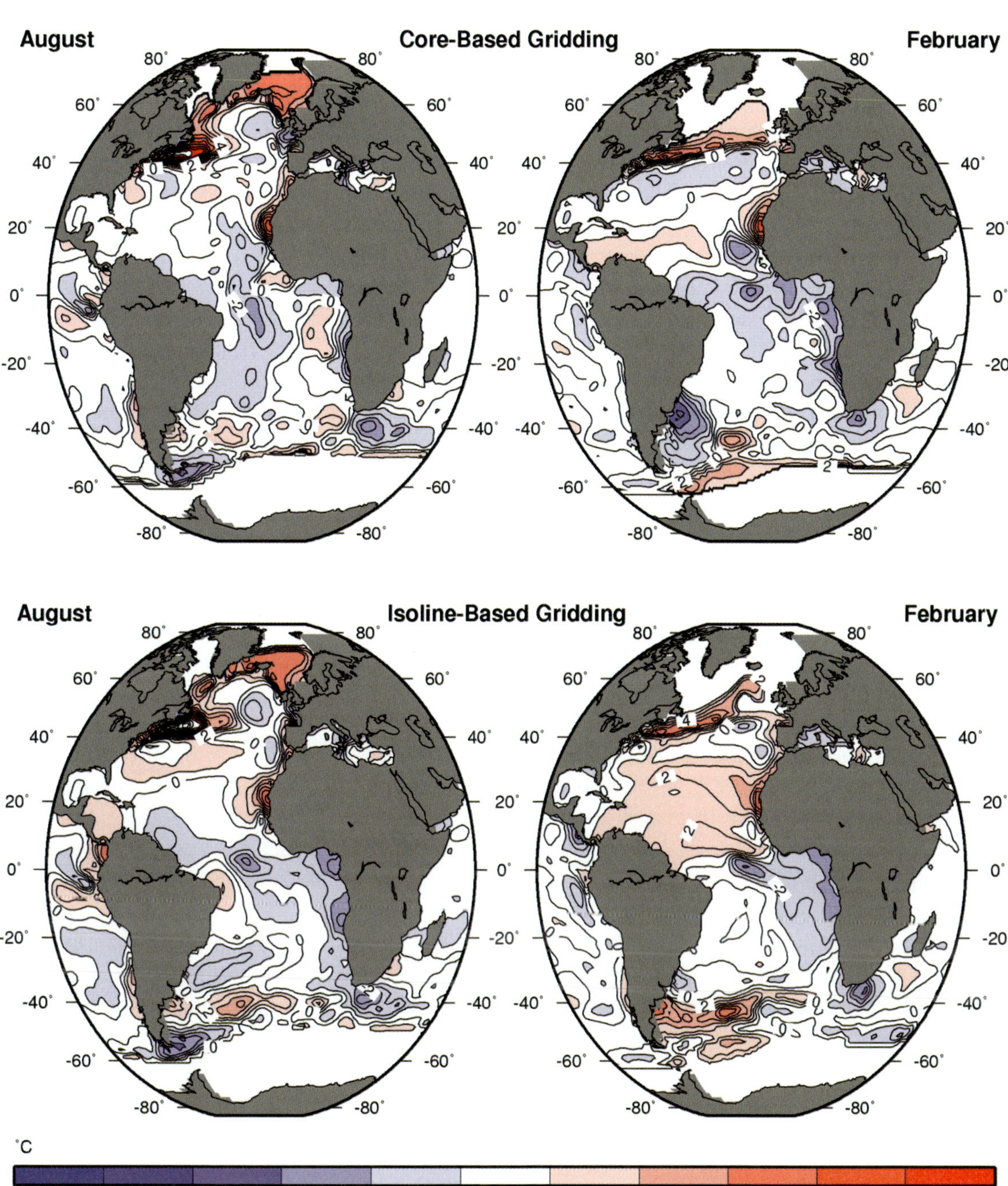

Fig. 5. Differences between the new GLAMAP and the CLIMAP (1981) reconstruction. Top row: Objectively interpolated from the sediment core-based reconstructions. Bottom row: Gridded using the digitized isolines by Pflaumann et al. (2002).

North Atlantic Ocean during February, the new data turned out to be warmer than the CLIMAP reconstruction. This band is supported by numerous sediment cores from different institutions (Fig. 1, top right). Thus the new data set is in good agreement with the modest glacial cooling of 2-2.5°C in the tropics that has been suggested by Crowley (2000) based on his review of the different available SST proxies. According to Hostetler and Mix (1999) and Mix et al. (1999), the tropical cooling was most pronounced in the eastern parts of the oceans where the eastern boundary currents interact with the equatorial circulation. This east-west gradient of glacial cooling is well represented in the core-based temperature fields. The glacial-to-modern anomalies increase from values around 2°C in the western equatorial Atlantic Ocean to values around 5°C in the eastern (not shown here).

The increase of SSTs relative to CLIMAP in the high latitudes is simply caused by the higher seasonality of the new ice cover reconstruction, yielding less ice in summer and as well in winter. The ice-free Nordic Seas are not only supported by the faunal assemblages underlying the present compilation but by temperature reconstructions from alkenone concentrations, too (Rosell-Melé 1997; Rosell-Melé and Comes 1999). There is a small area warmer than CLIMAP in both seasons at about 50°S/40°W which is in an area where no SST estimates are available (Fig. 1, top). We regard this feature as an artifact.

Along the coast of Africa, there are several smaller regions that indicate an SST differing from CLIMAP, in most cases lower, in some cases higher. These are related to the upwelling regions of the eastern Atlantic Ocean and the water masses transported by the Benguela current that are much better resolved and more detailed in the new GLAMAP data set than in CLIMAP. According to recent modeling results of PSN and Paul and Schäfer-Neth (2003), these characteristics indicate changed properties of the upwelled waters, but no distinctly different upwelling rates.

The intense lowering of SSTs in the Argentine Basin in February is linked to a single sediment core (Fig. 1, top) and needs perhaps further investigation.

Isoline-Based and CLIMAP SST

In contrast to the core-based reconstruction, the gridded field derived from the isolines exhibits markedly higher SST than CLIMAP, especially in the tropical and subtropical North Atlantic Ocean and along approximately 40°S during boreal winter, clearly in contrast to other proxy data evidence (Crowley 2000). We identified two main causes for this. First, in latitudes higher than 15° (both hemispheres), the isolines seem to be drawn too far poleward with respect to the core data, which causes the large negative anomaly between the core- and isoline-based grids around 30°N in February (Fig. 6, right). Likewise, the negative anomaly at about 10°S in August (Fig. 6, left) can be attributed to an eastward shift of the isolines. However, it must be noted that we did not find a true mismatch between the core data and the hand-drawn isolines: the shift of the isolines is most prominent in regions where there are large distances between the sediment cores, leaving several possibilities for 'true' interpolating lines. Second, the 26°C and 27°C isolines for February enclose an area of cores indicating generally lower temperatures, but some actually showing precisely these values. In this case, the cores are densely spaced but show widely different temperatures, and drawing the isolines is a matter of judging some cores as more reliable than others. Again, the researcher is left with a good deal of freedom. As a result, the isoline-based SST exceeds the core-based by more than 3°C northeast of northern Brazil (Fig. 6, right). A possible reason for this bias towards higher SST values in the isoline reconstruction might be the fact that the isolines were constructed for the GLAMAP chronozone, which in general is somewhat warmer than EPILOG (Gersonde et al. 2003). Unfortunately, no EPILOG version of the isolines was available for our study.

The strong positive anomaly relative to CLIMAP in the Argentine Basin (Fig. 5, lower right) indicates another problem related to the isoline reconstruction. In order to avoid unrealistic zonal gradients in the gridded SST field, we had (PSN) to extend the isolines by additional tie points into an area with almost no sediment cores (Fig. 1, top). As in the case of the hand-drawn isolines in

Core-Based minus Isoline-Based SST

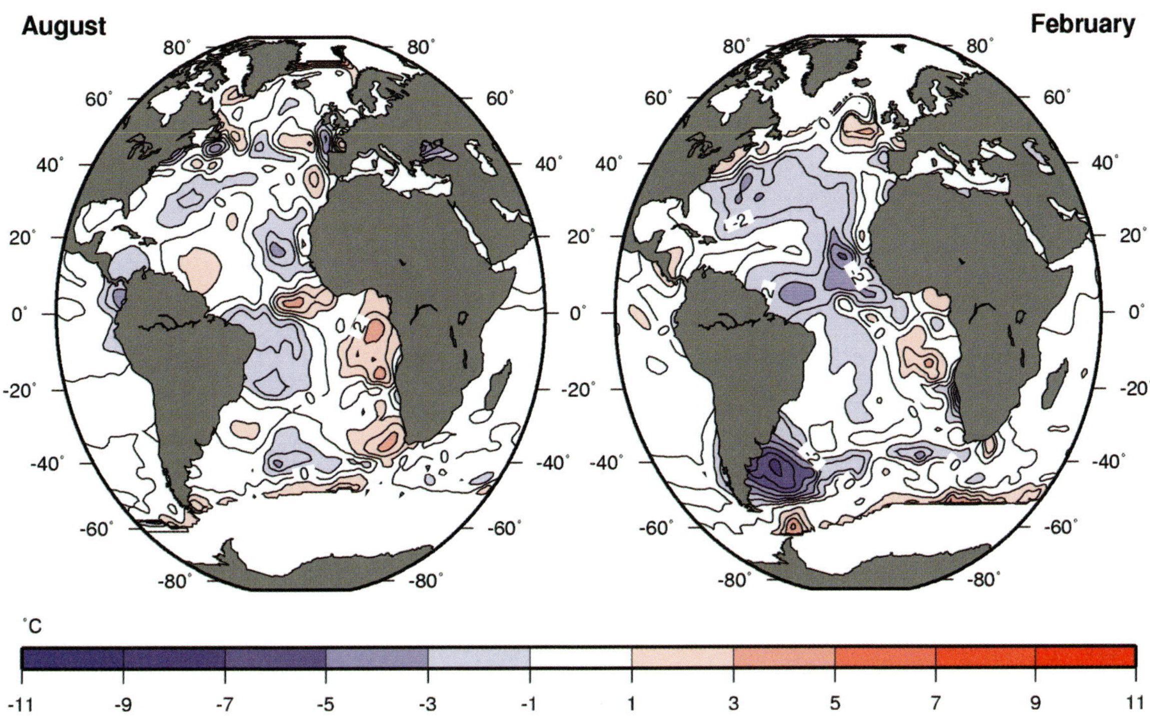

Fig. 6. Differences between the sediment core-based SST reconstruction and the isoline-based.

areas of sparse data, the position of these tie points is arbitrary to some extent, and moving the points could either diminish or enhance the anomaly.

SSS Differences between Isoline-Based and Core-Based SST Reconstructions

Except for a small region west of Ireland, the new SST field yields lower salinities than we reconstructed in our previous study (Fig. 4, top right). However, these changes are smaller than the accuracy of the SSS reconstruction and limited to the northeastern part of the Atlantic Ocean, so that we will focus on the SST differences that extend over almost the entire Atlantic Ocean.

Model Experiments

To further investigate the differences between the two LGM SST reconstructions, we performed a number of experiments with a numerical ocean circulation model that are described in detail in our accompanying publication (Paul and Schäfer-Neth 2003).

For the present study, we compare three model experiments: A control run that was driven by present-day boundary conditions, and two runs forced by the core-based and isoline-based glacial SST fields. The wind stress fields for the model experiments were derived from the control run of the European Centre/Hamburg atmospheric general circulation model (ECHAM3) at T42 resolution and a run that employed the isoline-based gridded SST field as bottom boundary condition (Lorenz and Lohmann, pers. comm.; for a more detailed discussion, see PSN). Both glacial model runs discussed here were driven with this wind field, which in this context allows us to highlight the effect of the different thermohaline boundary conditions.

Figure 7 displays the modeled glacial-to-modern temperature changes along a section through the western Atlantic Ocean. Both glacial runs show a marked cooling of the deep convection area in the northern North Atlantic Ocean as well as of the thermocline waters in low and mid-latitudes. In the run forced by the core-based SST, the low-latitude cooling is much more pronounced than in the run forced by the isoline-based temperature. This corresponds well to the findings of Slowey and Curry (1995) from the western Bahama Banks, who propose a temperature shift of around 4°C for the upper 1000 m, and of Curry et al. (1999), who reconstruct a cooling of 4-6°C at 60°N. Thus, the model results suggest that the core-based SST field is more consistent with the reconstructed meridional structure of the glacial thermocline than the isoline-based SST field.

In their coupled atmosphere-ocean model experiment, Liu et al. (2002) reproduce a similar cool-

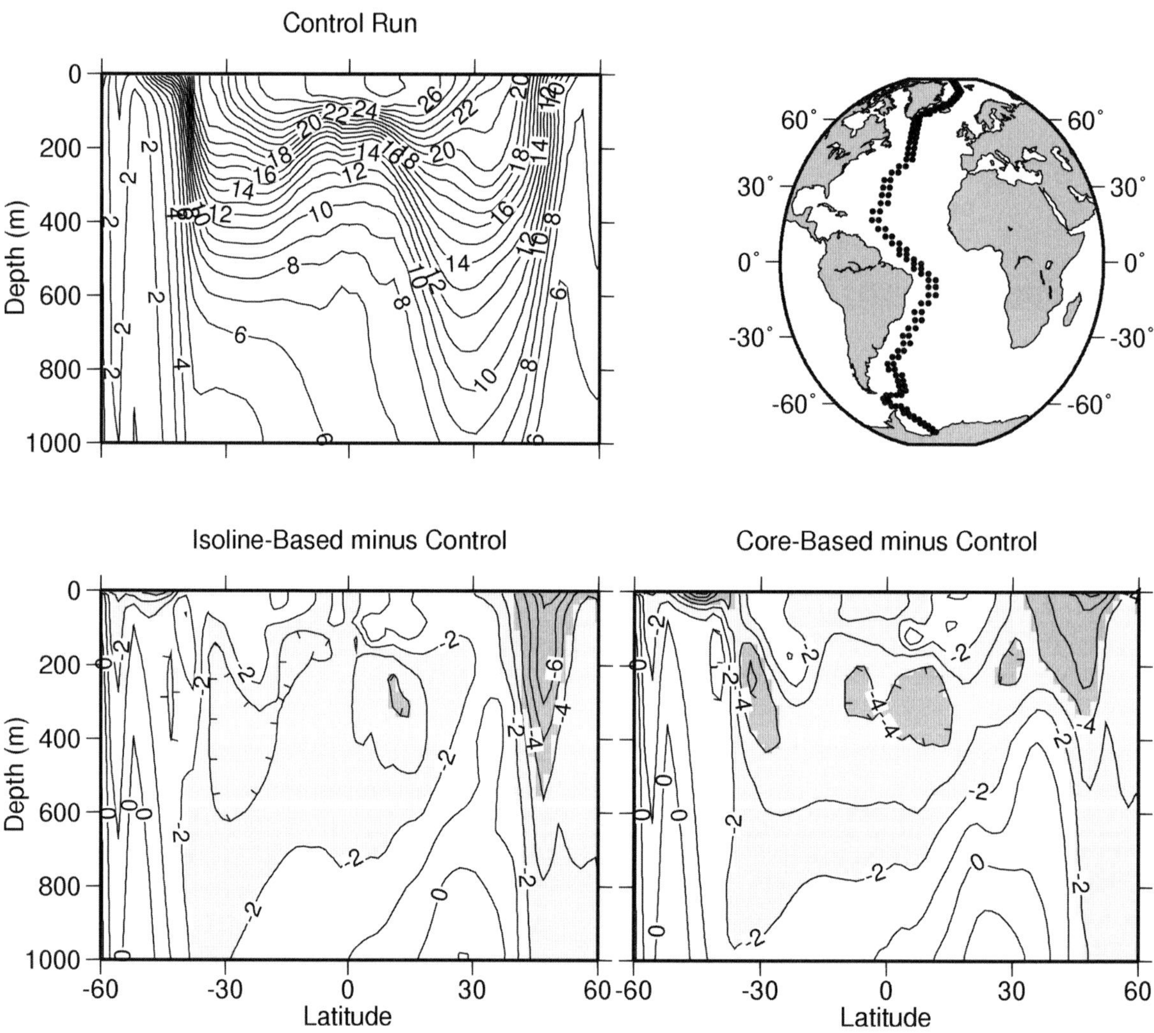

Fig. 7. Modeled temperatures along a section following GEOSECS (top right) through the western Atlantic Ocean. Top: Control run under present-day forcing. Bottom: Differences between the LGM and control runs. Left: LGM run driven by the isoline-based SST reconstruction, right: LGM experiment forced with the core-based SST fields. Light grey: below -2°C, dark grey: below -4°C.

ing signature, not only for the northern hemisphere, but as well for the South Atlantic Ocean.

Summary

• Based on new sediment-core-based SST estimates, we constructed objectively interpolated data sets of February and August SST for the whole Atlantic Ocean.

• These fields were incorporated into the global CLIMAP data set, and a seasonal cycle for every month was constructed according to the PMIP approach.

• Judging from (i) a direct comparison of these data with an earlier set based on a subjective interpolation, (ii) a comparison of both fields with CLIMAP and other paleo-SST evidence, and (iii) the outcome of two experiments with a general ocean circulation model, we regard the objectively interpolated fields as more reliable.

Acknowledgements

We wish to thank S. Mulitza for his help with the application of the different available paleo-temperature equations and the interpretation of the numerous species-dependent relations between temperature and oxygen isotopes. Our paper benefited from the comments of M. Weinelt and an anonymous referee, which we gratefully acknowledge. This research was funded by the Deutsche Forschungs-gemeinschaft (DFG) as part of the Sonderfor-schungsbereich 261, No. 375, and the DFG Research Center "Ocean Margins" of the University of Bremen, No. RCOM 0082.

References

Abelmann A, Brathauer U, Gersonde R, Sieger R, Zielinski U (1999) Radiolarian-based transfer function for the estimation of sea surface temperatures in the Southern Ocean (Atlantic sector). Paleoceanography 14: 410-421

Aeschbach-Hertig W, Peeters F, Beyerle U, Kipfer R (2000) Palaeotemperature reconstruction from noble gases in ground water taking into account equilibration with entrapped air. Nature 405: 1040-1044

Aeschbach-Hertig W, Stute M, Clark JF, Reuter RF, Schlosser P (2002) A paleotemperature record derived from dissolved noble gases in groundwater of the Aquia Aquifer (Maryland, USA). Geochim Cosmo-chim Acta 66: 797-817

Bard E, Arnold, M, Maurice P, Duprat J, Duplessy J-C (1987) Retreat velocity of the North Atlantic polar front during the last deglaciation determined by ^{14}C accelerator mass spectrometry. Nature 328: 791-794

Bemis BE, Spero HJ, Bijma J, Lea DW (1998) Reevalua-tion of the oxygen isotopic composition of planktonic foraminifera: Experimental results and revised paleotemperature equations. Paleoceanography 13: 150-160

Bigg GR (1994) An ocean general circulation model view of the glacial Mediterranean thermohaline circulation. Paleoceanography 9: 705-722

Bigg GR, Rohling EJ (2000) An oxygen isotope data set for marine water. J Geophys Res 105: 8527-8535

CLIMAP Project Members (1981) Seasonal reconstructions of the Earth's surface at the Last Glacial Maximum. Geological Society of America, Map and Chart Series MC-36, 18 p

Crosta X, Pichon JJ, Burckle L (1998) Application of modern analog technique to marine Antarctic diatoms: reconstruction of maximum sea-ice extent at the last glacial maximum. Paleoceanography 13: 284-297

Crowley T (2000) CLIMAP SSTs re-revisited. Clim Dyn 16: 241-255

Curry WB, Marchitto TM, McManus JF, Oppo DW, Laarkamp KL (1999) Millennial-scale changes in ventilation of the thermocline, intermediate, and deep waters of the glacial North Atlantic. In: Clark PU, Webb RS, Keigwin LD (eds) Mechanisms of Global Climate Change at Millennial Time Scales. Geophysical Monograph Series, Vol 112, pp 59-76

De Vernal A, Hillaire-Marcel C, Turon J-L, Matthiesen J (2000) Reconstruction of sea-surface temperature, salinity and sea ice cover in the northern North Atlantic during the Last Glacial Maximum based on dinocyst assemblages. Canadian J Earth Sci 37: 725-750

Deutsch CV, Journel AG (1992) GSLIB, Geostatistical Software Library and User's Guide. Oxford University Press, New York, Oxford

Duplessy J-C, Labeyrie L, Arnold M, Paterne M, Duprat J, van Weering TCE (1992) Changes in surface salinity of the North Atlantic during the last deglaciation. Nature 358: 485-488

Duplessy J-C, Labeyrie L, Juillet-Leclerc A, Maitre F, Duprat J, Sarnthein M (1991) Surface salinity reconstruction of the North Atlantic Ocean during the last glacial maximum. Oceanol Acta 14: 311-324

Duplessy J-C, Labeyrie L, Paterne M, Hovine S, Fichefet T, Duprat J, Labracherie M (1996) High Latitude Deep Water Sources During the Last Glacial Maximum and the Intensity of the Global Oceanic Circulation. In: Wefer G, Berger WH, Siedler G, Webb DJ (eds) The South Atlantic: Present and Past Circulation. Springer, Berlin, pp 445-460

Epstein S, Buchsbaum R, Lowenstam HA, Urey HC (1953) Revised carbonate-water isotopic temperature scale. Geol Soc Am Bull 64: 1315-1325

Gersonde R, Abelmann A, Brathauer U, Cortese G, Fütterer D, Grobe H, Niebler H-S, Segl M, Sieger R, Zielinski U (2003) Last Glacial Maximum sea surface temperature and sea ice extent in the Southern Ocean (Atlantic-Indian sector): A multiproxy approach. Paleoceanography 18: doi: 10.10292002PA000809

Gersonde R, Zielinski U (2000) The reconstruction of late Quaternary Antarctic sea-ice distribution-the use of diatoms as proxies for sea-ice. Paleogeogr Paleoclimatol Paleoecol 162: 263-286

Guilderson TP, Fairbanks RG, Rubenstone JL (1994) Tropical temperature variations since 20,000 years ago: Modulating interhemispheric climate change. Science 263: 663-664

Hostetler SW, Mix AC (1999) Reassessment of ice-age cooling of the tropical ocean and atmosphere. Nature 399: 673-676

Jansen E, Erlenkeuser HH (1985) Ocean circulation in the Norwegian Sea 15000 bp to present. Boreas 14: 189-206

Jansen E, Veum T (1990) Evidence for two-step deglaciation and its impact on North Atlantic deep water circulation. Nature 343: 612-616

Jones GA, Keigwin LD (1989) Evidence from FRAM Strait (78°N) for early deglaciation. Nature 336: 56-59

Jünger B (1993) Tiefenwassererneuerung in der Grönlandsee während der letzten 340000 Jahre. GEOMAR Report No. 35, Kiel Germany 103 p

Keigwin LD, Boyle EA (1989) Late Quaternary chemistry of high-latitude surface waters. Paleogeogr Paleoclimatol Paleoecol 3: 85-106

Kellogg TB, Duplessy J-C, Shackleton NN (1978) Planc-tonic foraminiferal and oxygen isotopic stratigraphy and paleoclimatology of Norwegian deep-sea cores. Boreas 7: 61-73

Köhler, SEI (1991) Spärtquartäre paläo-ozeanographische Entwicklung des Nordpolarmeers anhand von Sauerstoff- und Kohlenstoffisotopenverhältnissen der planktischen Foraminifere *Neogloboquadrina pachyderma* (sin). GEOMAR Report No. 13, Kiel, Germany, 104 p

Lackschewitz KS (1991) Sedimentationsprozesse am aktiven mittelatlantischen Kolbinsey Rücken (nördlich von Island). GEOMAR Report No. 9, Kiel Germany 121 p

Levitus S (1982) Climatological atlas of the World Ocean. NOAA Prof. Paper No. 13, 173 p

Levitus S, Boyer TP (1994) World Ocean Atlas. Volume 4: Temperature. NOAA Atlas NESDIS No. 4, 117 p

Liu Z, Shin S, Otto-Bliesner B, Kutzbach JE, Brady EC, Lee D (2002) Tropical cooling at the last glacial maximum and extratropical ocean ventilation. Geophys Res Lett 29: doi: 10.1029/2001GL013938

Malmgren BA, Kucera M, Waelbroeck C, Nyberg J (2001) Comparison of statistical and artificial neural network techniques for estimating past sea-surface temperatures from planktonic foraminifer census data. Paleoceanography 16: 520-530

Melles M (1991) Late Quaternary paleoglaciology and paleoceanography at the continental margin of the southern Weddell Sea, Antarctica. Berichte zur Polarforschung. Bremerhaven Vol 81, Alfred-Wegener-Institut für Polar- und Meeresforschung, Bremer-haven, Germany

Mix AC, Bard E, Schneider R (2001) Environmental processes of the ice age: Land, oceans, glaciers (EPILOG) Quat Sci Rev 20: 627-658

Mix AC, Morey AE, Pisias NG, Hostetler SW (1999) Fora-miniferal faunal estimates of paleotemperature: Circumventing the no-analog problem yields cool ice age tropics. Paleoceanography 14: 350-359

Morris TH (1988) Stable isotope stratigraphy of the Arctic Ocean: Fram Strait to Central Arctic. Paleogeogr Paleoclimatol Paleoecol 64: 201-219

Niebler H-S, Gersonde R (1998) A planktic foraminiferal transfer function for the southern South Atlantic Ocean. Mar Micropaleontol 34: 213-234

Niebler H-S, Mulitza S, Donner B, Arz H, Pätzold J, Wefer G (2003) Sea-surface temperatures in the equatorial and South Atlantic Ocean during the Last Glacial Maximum (23-19 ka). Paleoceanography 18: doi: 10.1029/2002PA000902

Östlund H, Craig H, Broecker WS, Spencer D (1987) Geosecs Atlantic, Pacific, Indian Ocean expeditions. shorebased data and graphics. GEOSECS Atlas Series Vol. 7, US Government Printing Office, Washington, DC

Paul A, Mulitza S, Pätzold J, Wolff T (1999) Simulation of oxygen isotopes in a global ocean model. In: Fischer G, Wefer G (eds) Use of Proxies in Paleoceanography: Examples from the South Atlantic. Springer, Berlin, pp 655-686

Paul A, Schäfer-Neth C (2003) Modeling the water masses of the Atlantic Ocean at the Last Glacial Maximum. Paleoceanography 18: doi: 10.1029/2002PA 000783

Pflaumann U, Duprat J, Pujol C, Labeyrie LD (1996) SIMMAX: A modern analog technique to deduce Atlantic sea surface temperatures from planktonic foraminifera in deep-sea sediments. Paleoceanography 11: 15-35

Pflaumann U, Sarnthein M, Chapman M, Funnel B, Huels M, Kiefer T, Maslin M, Schulz H, Swallow J, van Kreveld S, Vautravers M, Vogelsang E, Weinelt M (2003) The Glacial North Atlantic: Sea-surface conditions reconstructed by GLAMAP-2000. Paleoceanography 18: doi: 10.1029/2002PA000774

PMIP (1993) Paleoclimate modelling intercomparison project, http://www-pcmdi.llnl.gov/pmip/newsletters/newsletter02.html. Technical report

Prell WL (1985) The stability of low latitude sea surface temperature. An evaluation of the CLIMAP reconstruction with emphasis on the positive SST anomalies. Technical report, Department of Energy, Washington, DC

Rind D, Peteet D (1985) Terrestrial conditions at the last glacial maximum and climap sea-surface temperature estimates: Are they consistent? Quat Res 24: 1-22

Rosell-Melé A (1997) Appraisal of climap temperature reconstruction in the NE Atlantic using alkenone proxies. EOS 78: F28

Rosell-Melé A, Comes P (1999) Evidence for a warm last glacial maximum in the nordic seas, or an example of shortcomings in UK37' and UK37 to estimate low sea surface temperature? Paleoceanography 13: 694-703

Ruddiman WF, McIntyre AA (1981) The North Atlantic during the last deglaciation. Paleogeogr Paleoclimatol Paleoecol 35: 145-214

Rühlemann C, Mulitza S, Müller PJ, Wefer G, Zahn R (1999) Warming of the tropical Atlantic Ocean and slowdown of thermohaline circulation during the last deglaciation. Nature 402: 511-514

Sarnthein M, Gersonde R, Niebler H-S, Pflaumann U, Spielhagen R, Thiede J, Wefer G, Weinelt M (2003a) Overview of Glacial atlantic ocean mapping (GLAMAP-2000) Paleoceanography 18: doi: 10.1029/2002PA00769.

Sarnthein M, Jansen E, Weinelt M, Arnold M, Duplessy J-C, Erlenkeuser H, Flatøy A, Johannessen G, Johannessen T, Jung S, Koç N, Labeyrie L, Maslin M, Pflaumann U, Schulz H (1995) Variations in Atlantic surface ocean paleoceanography, 50°-80° N: A time-slice record of the last 30,000 years. Paleoceanography 10: 1063-1094

Sarnthein M, Pflaumann U, Weinelt M (2003b) Past extent of sea ice in the northern North Atlantic inferred from foraminiferal paleotemperature estimates. Paleoceanography 18: doi: 10.1029/2002PA000771

Schäfer-Neth C (1998) Changes in the seawater-oxygen isotope relation between last glacial and present: Sediment core data and OGCM modelling. Paleoclimatology 2: 101-131

Schäfer-Neth C, Hervada-Sala C, Pawlowsky-Glahn V, Stattegger K (1998) Geostatistical interpretation of paleoceanographic data over large ocean basins - reality and fiction. In: Buccianti A, Nardi G, Potenza R (eds) Proceedings of the Fourth Annual Conference of the International Association for Mathematical Geology. De Frede, Napoli, Italy, pp 111-116

Schäfer-Neth C, Paul A (2001) Circulation of the glacial Atlantic: A synthesis of global and regional modeling. In: Schäfer P, Ritzrau W, Schlüter M, Thiede J (eds) The northern North Atlantic: A changing environment. Springer, Berlin, pp 441-462

Schmidt GA (1999) Error analysis of paleosalinity calculations. Paleoceanography 14: 422-429

Schmidt GA, Bigg GR, Rohling EJ (1999) Global seawater oxygen-18 database. http://www.giss.nasa.gov/data/o18data Technical report, Goddard Institute for Space Studies

Schmidt GA, Mulitza S (2002) Global calibration of ecological models for planktic foraminifera from coretop carbonate oxygen-18. Mar Micropalaentol 44: 125-140

Schulz H (1994) Meeresoberflächentemperaturen im frühen Holozän 10,000 Jahre vor heute. Dissertation, Universität Kiel, Germany

Shin, S, Liu Z, Otto-Bliesner O, Brady EC, Kutzbach JE, Harrison SP (2003) A simulation of the Last Glacial Maximum Climate using the NCAR-CCSM. Clim Dyn 20: 127-151

Slowey NC, Curry WB (1995) Glacial-interglacial differences in circulation and carbon cycling within the upper western North Atlantic. Paleoceanography 10: 715-732

Stute M, Forster M, Frischkorn H, Serejo F, Clark A, Schlosser P, Broecker WS, Bonani G (1995) Cooling of tropical Brazil (5°C) during the Last Glacial Maximum. Science 269: 379-383

Stute M, Schlosser P, Clark JF, Broecker WS (1992) Paleotemperatures in the southwestern United States derived from noble gas measurements in groundwater. Science 256: 1000-1003

Thiede J (1978) A glacial Mediterranean. Nature 276: 680-683

Thunell RC (1979) Eastern Mediterranean sea during the Last Glacial Maximum; an 18,000-years BP reconstruction. Quat Res 11: 353-372

Veum T, Arnold M, Beyer I, Duplessy J-C (1992) Water mass exchange between the North Atlantic and the Norwegian Sea during the last 28000 years. Nature 356: 783-785

Vogelsang E (1990) Paläo-Ozeanographie des Europäischen Nordmeers an Hand stabiler Kohlenstoff- und Sauerstoffisotopen. Berichte aus dem Sonderforschungsbereich 313 der Universität Kiel No. 23, University Kiel, Germany, 136 p

Weinelt M (1993) Veränderungen der Oberflächenzirkulation im Europäischen Nordmeer während der letzten 60,000 Jahre-Hinweise aus stabilen Isotopen. Berichte aus dem Sonderforschungsbereich 313 der Universität Kiel No. 41, University Kiel, Germany, 105 p

Weinelt M, Sarnthein M, Pflaumann U, Schulz H, Jung S, Erlenkeuser H (1996) Ice-free Nordic Seas during the Last Glacial Maximum? Potential sites of deep-water formation. Paleoclimatology 1: 283-309

WOA (1998) World ocean atlas 1998, http://www.nodc.noaa.gov/oc5/woa98.html. Technical report, National Oceanographic Data Center, Silver Spring, Maryland

Zahn R, Markussen B, Thiede J (1985) Stable isotope data and depositional environments in the Late Quaternary. Nature 314: 433-435

Zielinski U, Gersonde R, Sieger R, Fütterer D (1998) Quaternary surface water temperature estimations: Calibration of a diatom transfer function for the Southern Ocean. Paleoceanography 13: 365-383

The Atlantic Ocean at the Last Glacial Maximum:
2. Reconstructing the Current Systems with a Global Ocean Model

A. Paul* and C. Schäfer-Neth

*DFG Forschungszentrum Ozeanränder, Universität Bremen, Postfach 33 04 40,
28334 Bremen, Germany
* corresponding author (e-mail): apau@palmod.uni-bremen.de*

Abstract: We use a global ocean general circulation model (OGCM) with low vertical diffusion and isopycnal mixing to simulate the circulation in the Atlantic Ocean at present-day and the Last Glacial Maximum (LGM). The OGCM includes $\delta^{18}O$ as a passive tracer. Regarding the LGM sea-surface boundary conditions, the temperature is based on the GLAMAP reconstruction, the salinity is estimated from the available $\delta^{18}O$ data, and the wind-stress is derived from the output of an atmospheric general circulation model. Our focus is on changes in the upper-ocean hydrology, the large-scale horizontal circulation and the $\delta^{18}O$ distribution. In a series of LGM experiments with a step-wise increase of the sea-surface salinity anomaly in the Weddell Sea, the ventilated thermocline was colder than today by 2–3°C in the North Atlantic Ocean and, in the experiment with the largest anomaly (1.0 beyond the global anomaly), by 4–5°C in the South Atlantic Ocean; furthermore it was generally shallower. As the meridional density gradient grew, the Antarctic Circumpolar Current strengthened and its northern boundary approached Cape of Good Hope. At the same time the southward penetration of the Agulhas Current was reduced, and less thermocline-to-intermediate water slipped from the Indian Ocean along the southern rim of the African continent into the South Atlantic Ocean; the 'Agulhas leakage' was diminished by up to 60% with respect to its modern value, such that the cold water route became the dominant path for North Atlantic Deep Water (NADW) renewal. It can be speculated that the simulated intensification of the Benguela Current and the enhancement of NADW upwelling in the Southern Ocean might reduce the import of silicate into the Benguela System, which could possibly resolve the 'Walvis Opal Paradox'. Although $\delta^{18}O_w$ was restored to the same surface values and could only reflect changes in advection and diffusion, the resulting $\delta^{18}O_c$ distribution came close to reconstructions based on fossil shells of benthic foraminifera.

Introduction

Current Systems

In the present-day eastern South Atlantic Ocean, the surface waters flow northward with the Benguela Current (Figure 1). They become warmer and, because of the high evaporation in the subtropics, saltier. A fraction crosses the equator and feeds the Gulf Stream and the North Atlantic Drift to eventually reach the subpolar North Atlantic Ocean. Here these waters release their heat to the atmosphere and cool. Because of their high salinity, they become dense enough to sink to great depth in the Labrador and Nordic Seas and thus form deep water.

There are actually two distinct branches of the Benguela Current: the Benguela Oceanic Current (BOC) and the Benguela Coastal Current (BCC). The BOC is part of the South Atlantic subtropical gyre and includes thermocline as well as intermediate waters. At the surface, it is replenished from two sources: There is a supply of Indian Ocean water which originates in the Agulhas Current and

From WEFER G, MULITZA S, RATMEYER V (eds), 2003, The South Atlantic in the Late Quaternary: Reconstruction of Material Budgets and Current Systems. Springer-Verlag Berlin Heidelberg New York Tokyo, pp 549-583

Surface Circulation of the Atlantic Ocean

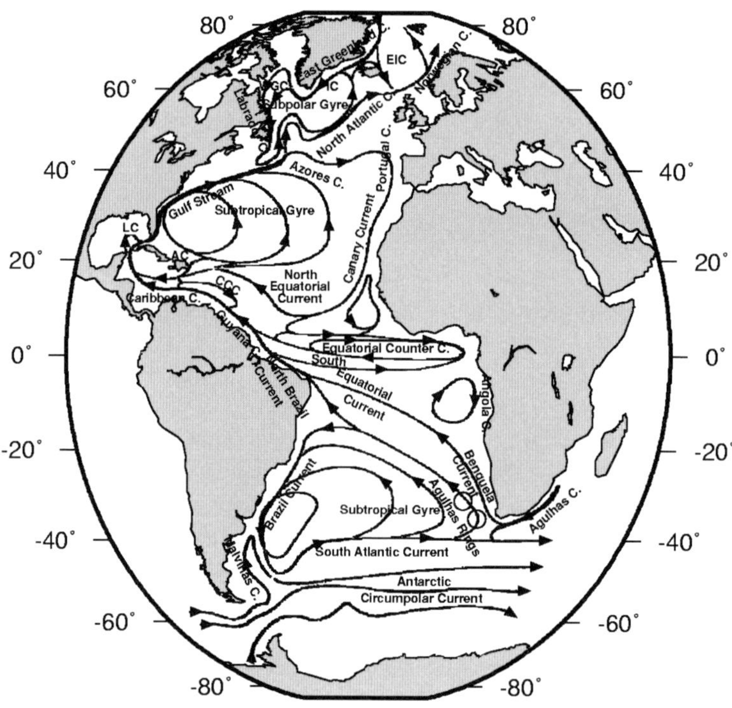

Fig. 1. Surface circulation of the Atlantic Ocean, based on Figure 14.2 of Tomczak und Godfrey (1994), which in turn is adapted from Duncan et al. (1982), Krauss (1986) and Peterson and Stramma (1991). Abbreviations are used for the East Iceland (EIC), Irminger (IC), West Greenland (WGC), Loop (LC) and Antilles (AC) Currents and the Caribbean Countercurrent (CCC). Two Agulhas rings are shown in the Benguela System off the coast of southwest Africa to indicate the role of eddy shedding in the Agulhas leakage.

slips into the South Atlantic Ocean around Cape of Good Hope. Because Indian Ocean water is relatively warm, this source is called the 'warm water route' (Gordon 1986). The remaining water is subantarctic surface water added at the Subtropical Front (STF, Shannon et al. 1989) or intermediate water that enters the South Atlantic Ocean through Drake Passage, upwells into the thermocline, warms up and flows eastward with the South Atlantic Current. This second source largely corresponds to the 'cold water route' first described by Rintoul (1991). Changes in the heat and salt transports from the South to the North Atlantic Ocean may be caused by changes in the relative contributions from these two routes and linked to the climate variations of the Quaternary (e.g. Berger and Wefer 1996; Boyle 2000). The upstream Agulhas Current may have undergone only limited glacial-

interglacial change regarding its temperature, lateral position and latitude of retroflection (Winter and Martin 1990). However, there is evidence, e.g. from molluscs (Pether 1994), for warmer conditions in the Benguela System during the last deglaciation, as a result of enhanced advection close to the coast of water derived from the Agulhas Current and a concomitant suppression of upwelling.

The BCC flows on the south-west African shelf and transports upwelled cold water northwards. At about 15°S it meets the poleward Angola Current. The water that upwells along the coast in the distinct cells of the Benguela System is thermocline water from at most 400 m depth. It comprises Eastern South Atlantic Central Water (ESACW, which originates at the STF by mixing and sinking of subtropical and subantarctic surface water), South Indian central water (advected into the eastern

South Atlantic Ocean in modified form by Agulhas rings and filaments) and tropical South Atlantic Central Water (SACW). The tropical Atlantic Ocean is a major source of thermocline water in the Benguela System. The northern Benguela System in particular is influenced by the Angola Current and the South Equatorial Current-Undercurrent system. Thermocline water and Antarctic Intermediate Water from the tropical Atlantic Ocean are advected polewards into the Benguela System certainly as far as 27°S (i.e. the Lüderitz cell). The oceanography of the southern Benguela System is more strongly affected by the Agulhas Current.

The relatively fresh, oxygen-rich ESACW is poor in nutrients and transported northward with the BOC. However, the saltier, oxygen-depleted SACW is greatly enriched in nutrients and transported poleward with a strong sub-surface flow (Shannon and Nelson 1996). Hence much of the nutrients used in the biological production off the coast of Namibia is brought in by poleward transport, from the high-concentration area north of Walvis Ridge. If this poleward transport were relatively decreased during glacial periods, by a weakening of the deep undercurrent as compared to the BOC, the quality of the upwelled water would be severely affected (Berger and Wefer 2002).

It has indeed been inferred from proxy data that the flow of the Benguela Current was slighty more intense at the Last Glacial Maximum (LGM), which led to a northward shift of the Angola-Benguela Frontal Zone by a few degrees (Jansen et al. 1996). Such a shift could be associated with a reduced import of nutrients (such as silicate) into the northern Benguela System and contribute to the paradoxical decrease of the production of diatoms and their flux to the sea floor during glacial periods ('Walvis Opal Paradox', Berger and Wefer 2002).

Modeling the Paleo-Ocean

This study was initiated by the sub-project C2 "Modeling the Paleo-Ocean" of the Special Research Project (Sonderforschungsbereich, SFB) 261, "The South Atlantic during the Late Quaternary: Reconstruction of mass budget and current systems", funded by the Deutsche Forschungsgemeinschaft (DFG). The goal of this effort was to reconstruct the LGM state of the South Atlantic Ocean. The focii were on the exchange between the South Atlantic and the other oceans:

• Changes in the exchange of water, heat and salt between the South Atlantic Ocean and the North Atlantic, Pacific and Indian Oceans, across the equator and through the circumpolar and Agulhas current systems

• Influence of the Antarctic Intermediate Water (AAIW) and Antarctic Bottom Water (AABW) on the North Atlantic Ocean

• Changes in the distribution of conservative tracers (temperature, salinity and $\delta^{18}O_w$)

• Changes in biologic production and the distribution of biogeochemical tracers (e.g. $\delta^{13}C$)

The LGM and the modern states represent two extremes of the climate variability during the Late Quaternary. They are particularly useful test cases for climate models because they are best documented by (proxy-) data, which enables us to force and validate climate models.

In the SFB 261, we employed two coarse-resolution ocean general circulation models: the large-scale geostrophic (LSG) ocean model, originally developed by the Max Planck Institute of Meteorology in Hamburg, Germany, and modified in the SFB 261, and the Modular Ocean Model (MOM), provided by the Geophysical Fluid Dynamics Laboratory (GFDL) in Princeton, USA. The modifications to the original LSG code included the implementation of a new tracer advection scheme (see Schäfer-Neth and Paul 2001 and Matthies et al. this volume for further details). Overall, the MOM turned out to be more flexible in testing different vertical resolutions and mixing parameterizations.

We used the LSG model to carry out a number of sensitivity experiments, which resulted in a robust circulation pattern for the glacial ocean. This provided the bases for a synthesis of a global and a regional study of the Atlantic Ocean at the LGM (Schäfer-Neth and Paul 2001) as well as for two simulations of the ocean carbon cycle (Matthies et al. this volume; Schulz and Paul 2003).

With the help of the MOM, we simulated the present-day distribution of the oxygen-isotopic content of sea water, in good agreement with the GEOSECS data (Paul et al. 1999).

Here we present the first step towards a similar simulation with the MOM for the last glacial maximum. Furthermore, we discuss glacial-interglacial changes in the South Atlantic Ocean circulation at a regional scale, e.g. in the Brazil and Malvinas currents. In constructing the surface boundary conditions we used the proxy-data acquired by the SFBs 261 and 313 ("Veränderungen der Umwelt: Der nördliche Nordatlantik", Christian-Albrechts-Universität zu Kiel) over a ten-year period. To compute meaningful $\delta^{18}O_w$ values and compare them with $\delta^{18}O$ values of benthic foraminifera, we had to reduce the temperature error in the thermocline. To this end, we used relatively high vertical resolution, low vertical diffusion and state-of-the-art parameterizations of isopycnal mixing and eddy-induced tracer transport.

Our accompanying contribution (Schäfer-Neth and Paul this volume) deals with the details of the underlying sea-surface temperature (SST) reconstruction and assesses different gridding methods. In contrast to our previous study (Paul and Schäfer-Neth 2003), we now use as model forcing the SST fields obtained with the most reliable gridding technique (i.e. variogram analysis and kriging) applied to the proxy-data at the core locations, to simulate the glacial-to-interglacial changes in the hydrology and circulation of the Atlantic Ocean.

In our model, a fraction of the Agulhas Current leaked into the South Atlantic Ocean. Since the horizontal resolution was too coarse to resolve the eddies that are generated in the retroflection region and shed into the South Atlantic Ocean, only a broad westward advective transport was simulated (cf. England and Garçon 1994); in addition some westward diffusive transport of heat was invoked in terms of the isopycnal and eddy transport parameterizations. We find a reduced Indian-South Atlantic Ocean exchange at the LGM and a slightly intensified BOC that flowed in a more northerly direction along the coast off Namibia and Angola. We furthermore report on changes in thermocline ventilation, large-scale circulation and the $\delta^{18}O_w$ and $\delta^{18}O_c$ distribution implied by altered advection and diffusion.

Methods

Data

For a detailed description of the database used for forcing the model experiments, of the gridding process involved, and for a discussion of the errors within the data, the reader is referred to our accompanying publication (Schäfer-Neth and Paul this volume). In the following we give a short overview.

Sea Surface Temperature. We employed the SST estimates for the GLAMAP (Sarnthein et al. 2003) and EPILOG (Mix et al. 2001) chronozones that overlap in a common time span between 19000 and 22000 calendar years before present (BP). This data set was not only subject to a common stratigraphic control, but was also based on the same modern reference data set for the calibration of the transfer techniques, which greatly improved the data consistency despite different transfer or modern analog formulae. It includes more than 220 individual SST samples for boreal winter and more than 170 samples for boreal summer, which cover almost the entire Atlantic Ocean. We filled the remaining gaps by SST estimates from other publications, carefully selected to avoid discrepancies with the GLAMAP data. In addition to the SST reconstructions, new estimates for the extent of the glacial sea-ice cover were provided by the GLAMAP group. This information was included into the SST data base by setting SST to the freezing point over the respective regions. From the raw data, we produced regularly gridded ($1° \times 1°$) summer and winter fields for the entire Atlantic Ocean by variogram analysis and kriging. We then smoothly incorporated these fields into the CLIMAP (1981) data set to produce improved new global SST maps. Following the PMIP (1993) approach, we constructed twelve monthly data sets by fitting a sinusoidal cycle to the winter and summer glacial-to-modern anomalies and then adding the modern monthly SSTs (10 m values from the World Ocean Atlas WOA 1998). Finally, these monthly data sets were interpolated to the model grid.

Sea Surface Salinity. For the northern North Atlantic Ocean, we estimated the sea-surface salinity (SSS) at the LGM from the reconstructed SST and the oxygen isotopic composition of fossil foraminiferal shells $\delta^{18}O_c$ using a method that has been developed over the last decade (Duplessy et al. 1991; Weinelt et al. 1996; Schäfer-Neth 1998; Schäfer-Neth and Paul 2001). In total, we employed 143 $\delta^{18}O_c$ measurements from this region. The oxygen isotope ratio of sea water $\delta^{18}O_w$ was computed according to the paleotemperature equation of Epstein et al. (1953) and the species-dependent SST corrections given in Weinelt et al. (1996). SSS was calculated from $\delta^{18}O_w$ using a linear correlation between these properties. The parameters of this correlation were derived from the GISS $\delta^{18}O$ Database (Schmidt et al. 1999; Bigg and Rohling 2000) and vary with latitude. For the remaining ocean, we used a global glacial-to-modern salinity shift of 1.07 (which, with respect to the whole depth of the ocean and not just its surface, corresponds to a change in relative mean sea level of about 100 m). Following the evidence from the South Atlantic Ocean (Melles 1991; Duplessy et al. 1996), we applied an additional SSS anomaly of 0.55 and 1.0 in the Weddell Sea that went beyond the global shift (Experiments GB and GC in Table 1). The seasonal cycle of SSS was constructed by adding the difference between the gridded glacial reconstruction and the boreal Summer salinity taken from the 10 m (WOA 1998) analyses to the monthly modern salinity fields.

Oxygen Isotopic Composition of Surface Waters. In addition to T and S, our model carried the oxygen isotopic composition of seawater as a third tracer. To force the $\delta^{18}O_w$ at the sea surface, we used the values for the upper 100 m from the GISS $\delta^{18}O$ database (Bigg and Rohling 2000; Schmidt et al. 1999). The forcing was applied as a restoring boundary condition for all model grid cells for which $\delta^{18}O_w$ measurements were available from the database, as indicated by the colored circles in

Figure 16. All other surface points of the model were allowed to evolve freely. The restoring values were computed as the annual mean of all $\delta^{18}O_w$ data from the upper 100 m of the ocean that fell in the respective grid cell. We employed the modern $\delta^{18}O_w$ data for all runs. Thus, although we cannot present a true simulation of glacial $\delta^{18}O_w$, changes in this passive tracer help to discriminate between hydrographic and circulation changes between the control and glacial model experiments.

Wind Stress. The wind stress anomaly for the glacial model experiments was derived from the control run of the European Centre/Hamburg atmospheric general circulation model (ECHAM3) at T42 resolution and a run that employed the GLAMAP SST fields as bottom boundary condition. (Lorenz and Lohmann, pers. comm.; these SST fields were gridded using a different interpolation technique, for a more detailed discussion, see Paul and Schäfer-Neth 2003). Differences to the SST maps used here are discussed in Schäfer-Neth and Paul (this volume). All glacial model runs presented here were driven by these wind fields to isolate the effect of the different thermohaline boundary conditions.

Model

We applied Version 2 of the Modular Ocean Model (MOM 2, Pacanowski 1996) to the global ocean. The resolution, geometry and bottom topography were similar to the coarse-resolution model of Large et al. (1997). The longitudinal resolution was constant at 3.6°, whereas the meridional resolution was 1.8° near the equator, decreased to a minimum of 3.4° away from the equator, then increased in midlatitudes as the cosine of latitude and was finally kept constant at 1.8° poleward of 60°. There were 27 vertical levels with monotonically increasing thickness from 12 m near the surface to 450 m near the bottom. The minimum model depth was 49 m, corresponding to 3 vertical levels, and the maximum model depth was 5900 m.

The modern bottom topography of the model was obtained from the ETOPO5 topography data (NCAR Support Section 1986). To derive the glacial bottom topography, we computed the glacial anomaly from the Peltier (1994) reconstruction and

Experiment	Time period	Additional Weddell Sea SSS anomaly
M	modern	no
GA	LGM	no
GB	LGM	intermediate (0.55)
GC	LGM	large (1.0)

Table 1. List of model experiments.

added it to the modern bottom topography. Either topography was first interpolated to the model grid and then smoothed and adjusted according to the procedure outlined by Large et al. (1997).

We took the vertical diffusion coefficient to be depth-dependent (Bryan and Lewis 1979):

$$A_{tv} = A_0 + \frac{C_r}{\pi} \arctan\left[\lambda\left(z - z_0\right)\right]$$

where we set $A_0 = 0.7 \times 10^{-4}\,m^2\,s^{-1}$, $C_r = 1.25 \times 10^{-4}\,m^2\,s^{-1}$, $\lambda = 4.5 \times 10^{-4}\,m^{-1}$ and $z_0 = 2500$ m. Thus A_{tv} ranges from $0.1 \times 10^{-4}\,m^2\,s^{-1}$ near the surface to $1.3 \times 10^{-4}\,m^2\,s^{-1}$ near the bottom. Above the turnover depth z_0, the vertical diffusion coefficient was smaller than in the previous studies of Paul et al. (1999) and Schäfer-Neth and Paul (2001). According to the few available observational estimates, a low value of $0.1 \times 10^{-4}\,m^2\,s^{-1}$ indeed seems to be appropriate below the permanent pycnocline and away from topography (Ledwell et al. 1993).

As an improvement over the two previous studies, we employed isopycnal mixing and the mesoscale eddy tracer transport parameterization of Gent and McWilliams (1990). The isopycnal diffusion and thickness diffusion coefficients were chosen to be equal, $A_I = A_{ITD} = 0.5 \times 10^3\,m^2\,s^{-1}$, and the horizontal diffusion coefficient A_{hv} was set to zero. The vertical and horizontal viscosity coefficients were $A_{mv} = 16.7 \times 10^{-4}\,m^2\,s^{-1}$ and $A_{mh} = 2.5 \times 10^5\,m^2\,s^{-1}$.

To relax the diffusive stability limit imposed on the model time step by the converging meridians at high latitudes, the horizontal mixing coefficients were tapered near the North Pole (Large et al. 1997). The isopycnal diffusion and thickness diffusion coefficients were reduced to 36% at 88.2°N, and the horizontal viscosity coefficient was reduced to 53% and 20% at 85.5°N and 87.3°N, respectively.

We used cross-land mixing to parameterize the exchange through the Bering Strait and the Strait of Gibraltar. Here the mixing coefficients were set such that the simulated inflow and outflow were equal to 0.8 Sv in the case of the Bering Strait and 1.75 Sv in the case of the Strait of Gibraltar. For the LGM experiments, the exchange through the Bering Strait was set to zero.

For the momentum and barotropic integrations, the time step was 1800 s. The tracer time step was 2.5 d at all levels. All experiments were integrated for at least 2000 tracer years (Experiment M for 5000 tracer years). Fourier filtering was applied to the flow variables south of 71.1°S and north of 71.1°N, and to the tracer variables south of 70.2°S and north of 70.2°N.

The surface momentum flux was provided by the zonal and meridional wind stress components, and the net surface heat and freshwater fluxes were computed by restoring the potential temperature and salinity at the first model level to prescribed monthly SST and SSS with a relaxation time scale of 50 days relative to the upper 50 m. Seawater oxygen-18 ($\delta^{18}O_w$) at the first model level was restored to the sparse modern annual-mean observations with the same relaxation time scale.

Experimental Setup

We carried out four model experiments (Table 1), one for the modern and three for the LGM time period:

• Experiment M was subject to modern sea surface boundary conditions. The sea-surface temperature and salinity fields were taken from the WOA 1998 data for 10 m depth. The wind stress fields were derived from the NCEP reanalysis data covering the four years 1985 through 1988 (Kalnay et al. 1996) as described by Large et al. (1997).

• Experiment GA employed the LGM SST reconstruction, SSS field and wind stress anomaly, all as described above.

• Experiment GB was run under the same conditions as Experiment GA, but with the 0.55 salinity anomaly in the Weddell Sea.

• Experiment GC employed the 1.0 salinity anomaly in the Weddell Sea.

The main difference between the LGM experiments discussed here and those presented in Paul and Schäfer-Neth (2003) is the use of the core-based as opposed to the isoline-based temperature reconstruction (see our accompanying publication Schäfer-Neth and Paul this volume). With respect to $\delta^{18}O_w$, Experiments GA-GC are not true simulations of the LGM, but important sensitivity experiments that isolate the effect of only the circulation changes on the distribution at depth, without changes of the surface hydrology. Our results for $\delta^{18}O_w$ and $\delta^{18}O_c$ are shown here for the first time.

Diagnostics

In diagnosing the model output, we used three different criteria of mixed-layer depth. The most common definition employs a constant density increment (here 0.125 kg m^{-3}; Levitus 1982) and sets the minimum value to the depth of first model layer. Plots based on this definition can be compared to other work on the subduction process (e.g. Huang and Qiu 1994; Williams et al. 1995).

To further investigate the ventilation of the thermocline, we computed the thickness of the surface isothermal layer from the model output by extracting the depth where the temperature differed from the temperature at the surface by more than 0.5°C; this depth is representative of the depth of the seasonal thermocline (Sprintall and Tomczak 1992; Tomczak and Godfrey 1994). In this connection, the thickness of the surface isopycnal layer is defined by the depth where the density is larger than the density at the surface by an amount which corresponds to the temperature change of 0.5°C used in the construction of the surface isothermal layer. To find these two thicknesses, we interpolated the local temperature and density linearly between the model depths. We took the annual maximum of the surface isothermal layer thickness as the lower boundary of the seasonal thermocline.

The depth difference between the surface isothermal and isopycnal layer can be different from zero. Positive differences show the presence of a barrier layer between the halocline and the thermo-

cline that does not allow a heat flux through the bottom of the mixed layer. Negative differences in the subtropics are produced by the subduction process (Tomczak and Godfrey 1994). They are therefore a reliable indicator for the ventilation of the thermocline.

From the distribution of the mixed-layer density and the thickness of the barrier layer, we derived the density criteria for the base of the ventilated thermocline: We overlaid the density and thickness contours and picked the highest density values that fell in regions of negative barrier layer thickness near the poleward boundaries of the subtropical gyres. To define water mass boundaries between intermediate and deep water, and deep and bottom water, we identified the isopycnal surfaces that approximate the lower boundary of the salinity minimum in the South Atlantic Ocean and run through the center of the bottom water cell in the meridional overturning streamfunction for the Atlantic Ocean, respectively.

Calcite $\delta^{18}O$ in equilibrium with ambient seawater was computed from the paleotemperature equation by Mulitza et al. (this volume).

Results

Hydrology

Our four model experiments yielded the global mean temperatures and salinities given in Table 2. As compared to Experiment M, the LGM sea-surface boundary conditions prompted a vertically averaged global mean cooling that ranged from 1.04°C in Experiment GA to 1.87°C in Experiment GC. At the same time, the ocean became saltier by 1.35 to 1.46. At the sea surface, the cooling was 1.36°C

Experiment	T		SST		S		SSS	
	Atlantic	global	Atlantic	global	Atlantic	global	Atlantic	global
M	4.16	3.73	16.77	18.14	34.86	34.65	35.32	34.70
GA	3.71	2.69	14.07	16.78	34.30	36.01	36.48	35.87
GB	3.11	2.32	14.05	16.78	34.27	36.00	36.50	35.88
GC	2.48	1.86	14.04	16.78	34.35	36.11	36.52	35.88

Table 2. Global mean temperatures and salinities, vertically integrated (T, S) and at the sea surface (SST, SSS).

for the global ocean and 2.70-2.73°C for the Atlantic Ocean.

The changes in the surface hydrology were strongly influenced by the changes in the precribed wind stress distribution (Figure 2 a and b). The main characteristics of the glacial anomaly were the more prominent trades in the North Atlantic Ocean and the much stronger westerlies in the Southern Ocean. Consequently, the Ekman downwelling (computed from the divergence of the Ekman transport) increased in the subtropics of both hemispheres (Figure 2 c and d). There was only little change in the Ekman upwelling off the coasts of northwest and southwest Africa because there was hardly any change in the South Atlantic trades. The zero line clearly shows a southward shift of the westerlies in the Southern Ocean.

The winter mixed-layer depth in Experiment M (the mixed-layer depth during March in the northern hemisphere and September in the southern hemisphere, Figure 3) indicates that central water was formed near 40° latitude in the subtropics of both hemispheres, and AAIW between 55° and 60°S in the South Pacific Ocean, consistent with the distribution of Ekman downwelling. In addition, there was a general poleward deepening.

In all glacial experiments, the low surface salinities reconstructed from low surface $\delta^{18}O_w$ values suppressed convection in the northern subpolar region, except for a small cell in the Irminger Sea where still some deep water was formed. A region of convection deeper than 200 m appeared at about 40°W in the northern subtropics. In the southern hemisphere, the band of mixed-layer depths larger than 100 m near 60°S extended from the South Pacific Ocean into the South Atlantic Ocean. In Experiment GC, the deep convection in the Weddell Sea clearly reflected the 1.0 SSS anomaly.

In Experiment M, the winter mixed-layer density (the potential density during March in the northern hemisphere and September in the southern hemisphere, referenced to the surface and averaged over the mixed-layer depth, Figure 4) was basically symmetric about the equator. It had low values in the tropics, became denser with latitude due to the increasing surface buoyancy loss and reached values higher than 27 kg m^{-3} in the sub-

polar oceans of either hemisphere. The contrast between Experiment GC and Experiment M ranged from typically 2 kg m^{-3} units in the low latitudes to 1 kg m^{-3} unit in the high latitudes and was enhanced in the Weddell Sea due the additional salinity anomaly. Experiments GA and GB (not shown) differed from Experiment GC only in the magnitude of the enhanced density contrast in the Weddell Sea.

The additional salinity anomaly in the Weddell Sea was thought to reflect a larger sea-ice formation and export at the LGM. As demonstrated by Toggweiler and Samuels (1995b), the net rate of sea-ice formation and melting can be inferred from the restoring boundary condition on salinity. For the inner Weddell Sea in Experiment M, the implied net freezing rate was 0.061 m a^{-1}, and the corresponding divergence of the net surface freshwater flux was 4.87 mSv (Table 3). These values changed sign in Experiments GA and GB. In Experiment GC, the fluxes were of the same sign as in Experiment M, but considerably larger.

Negative values of barrier layer thickness (Figure 5) occur where the depth of the isopycnal mixed layer exceeds the depth of the isothermal mixed layer and faithfully indicate the subduction of thermocline water (Sprintall and Tomczak 1992; Tomczak and Godfrey 1994).

In the subtropics, we identified the northeastern corner of the North Atlantic Ocean and the southwestern corner of the South Atlantic Ocean as those regions where the highest densities of the mixed layer coincided with the most negative values of barrier layer thickness. These regions determined the lower boundary of the ventilated thermocline. Mixed-layer densities in these regions were about $\sigma_0 = 26.9$ kg m^{-3} in Experiment M and $\sigma_0 = 28.1$ kg m^{-3} in Experiment GC (cf. Figure 3) and were used to trace the depth of the ventilated thermocline equatorward from its origin at the subtropical fronts. The alternating pattern of negative (or near-zero) and positive values of the barrier-layer thickness in the tropical Atlantic Ocean in Experiment M corresponds well to the present-day observations (Sprintall and Tomczak 1992).

The depth of the ventilated thermocline that followed from our density criteria is shown in Figure 6. In Experiment M, it was shallow near the

Sea Surface Wind Stress and Ekman Pumping

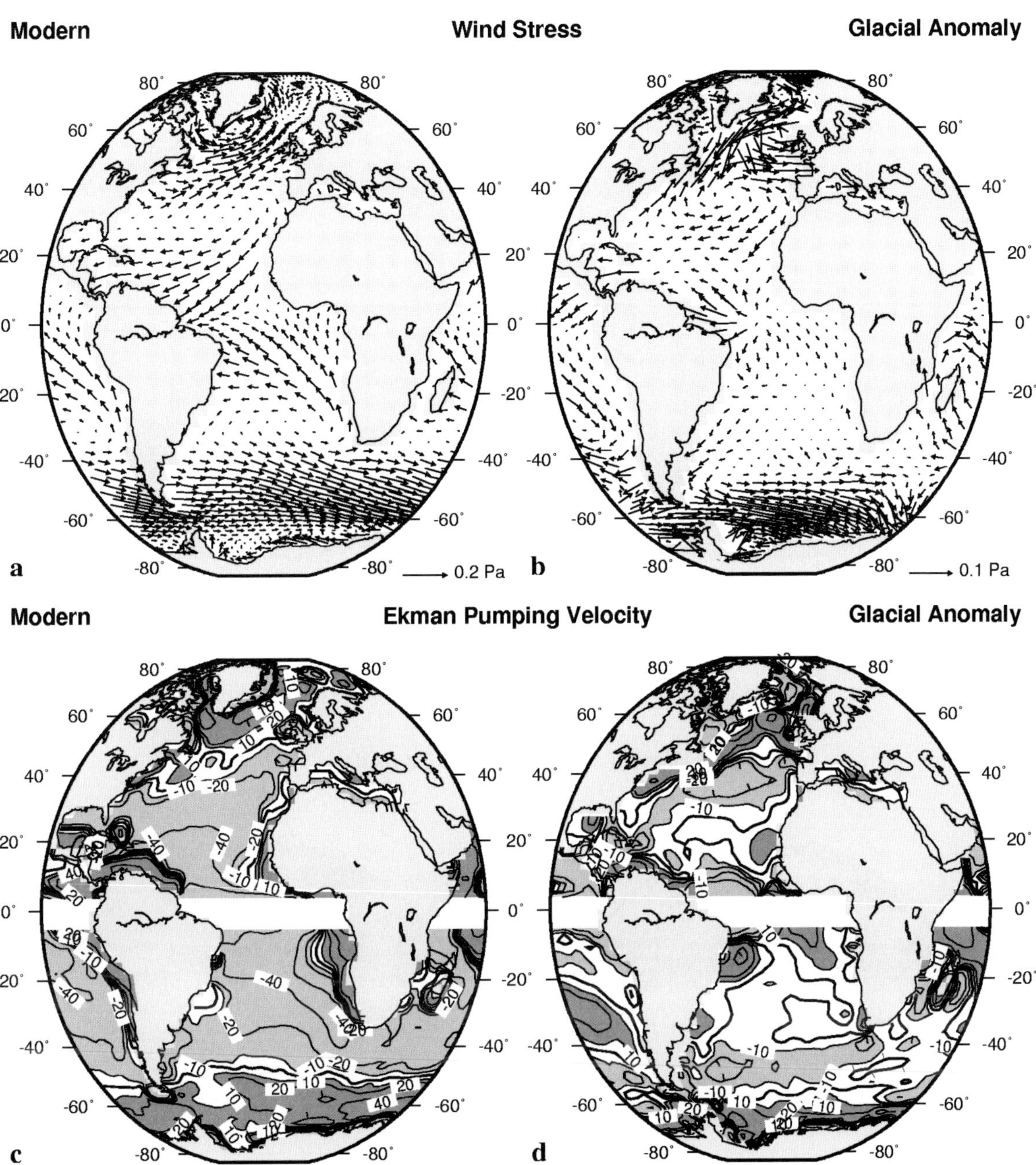

Fig. 2. Annual-mean wind stress and Ekman pumping. **a)** Modern wind stress distribution (Pa), derived from NCEP reanalysis data (Kalnay et al. 1996) and used in Experiment M. **b)** LGM wind stress anomaly (Pa), derived from the ECHAM3 atmospheric general circulation model and used in Experiments GA to GC. Note that the scale is doubled as compared to a. **c)** Modern Ekman pumping velocity, computed from the divergence of the Ekman transport (m a^{-1}, Experiment M). No values are plotted in the region between 5°S and 5°N where the Coriolis parameter tends to zero at the equator and it is impossible to compute a pumping rate. The thick contour denotes the zero line. Dark shading indicates positive, light shading negative values. **d)** LGM Ekman pumping velocity anomaly (m a^{-1}, Experiments GA to GC). Annotation as in c.

Winter Mixed-Layer Depth

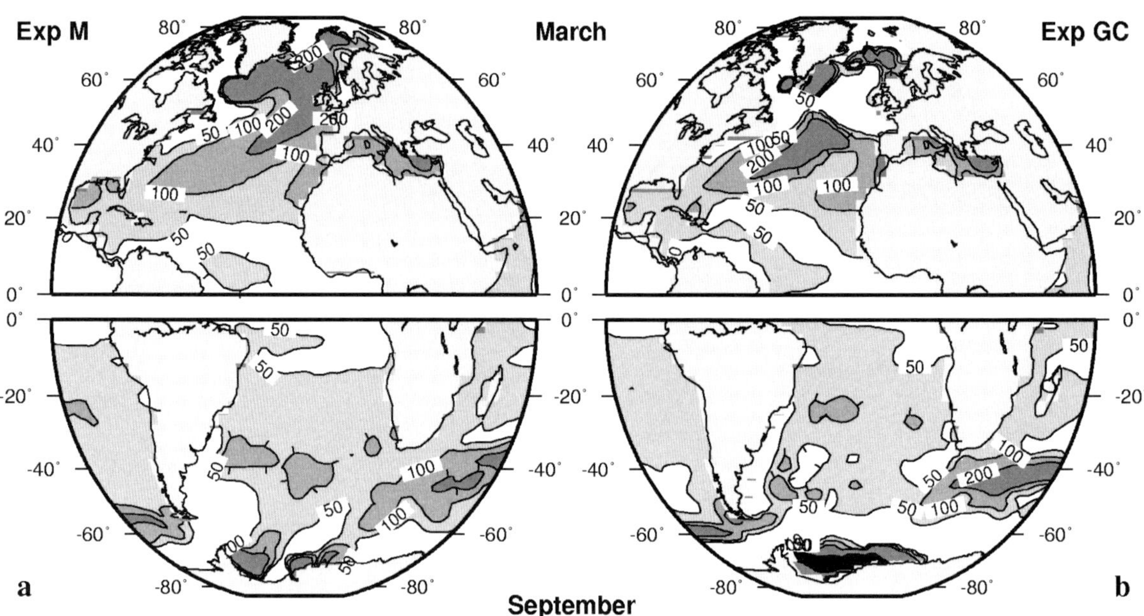

Fig. 3. Winter mixed-layer depth (m) for the Atlantic Ocean (March in northern hemisphere, September in southern hemisphere). **a)** Modern (Experiment M). **b)** LGM (Experiment GC). The mixed-layer depth is based on a constant density contrast ($\Delta\sigma_0 = 0.0125$ kg m^{-3}). The contour interval is 50 m. Values larger than 300 m are colored black.

Winter Mixed-Layer Density

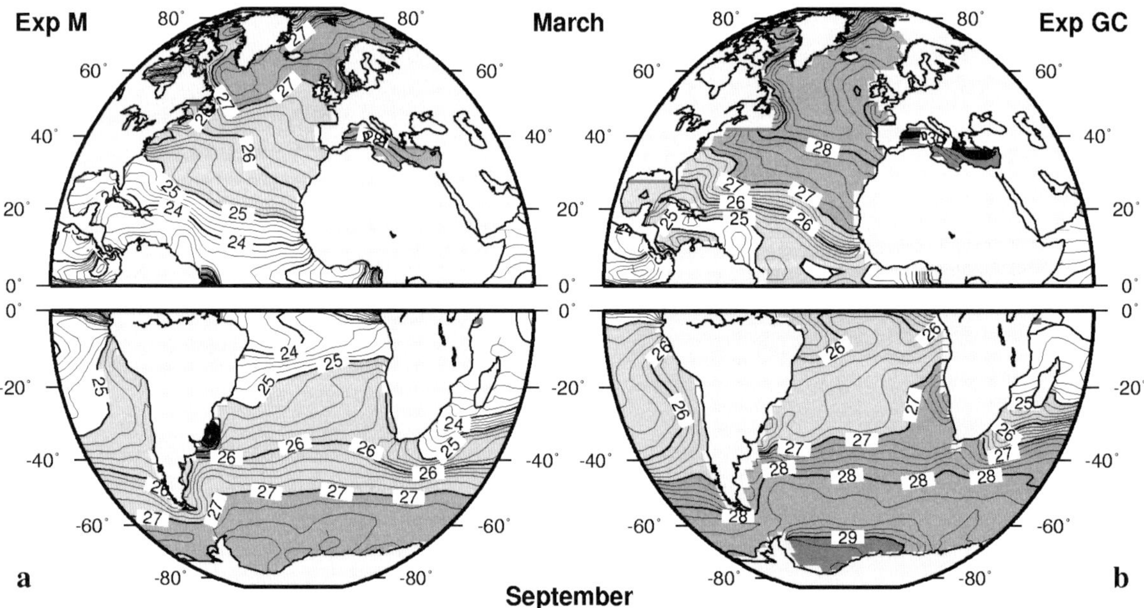

Fig. 4. Winter mixed-layer density (kg m^{-3}) for the Atlantic Ocean (March in the northern hemisphere, September in the southern hemisphere). **a)** Modern (Experiment M). **b)** LGM (Experiment GC). The contour interval is 0.2 kg m^{-3}.

Winter Barrier-Layer Thickness

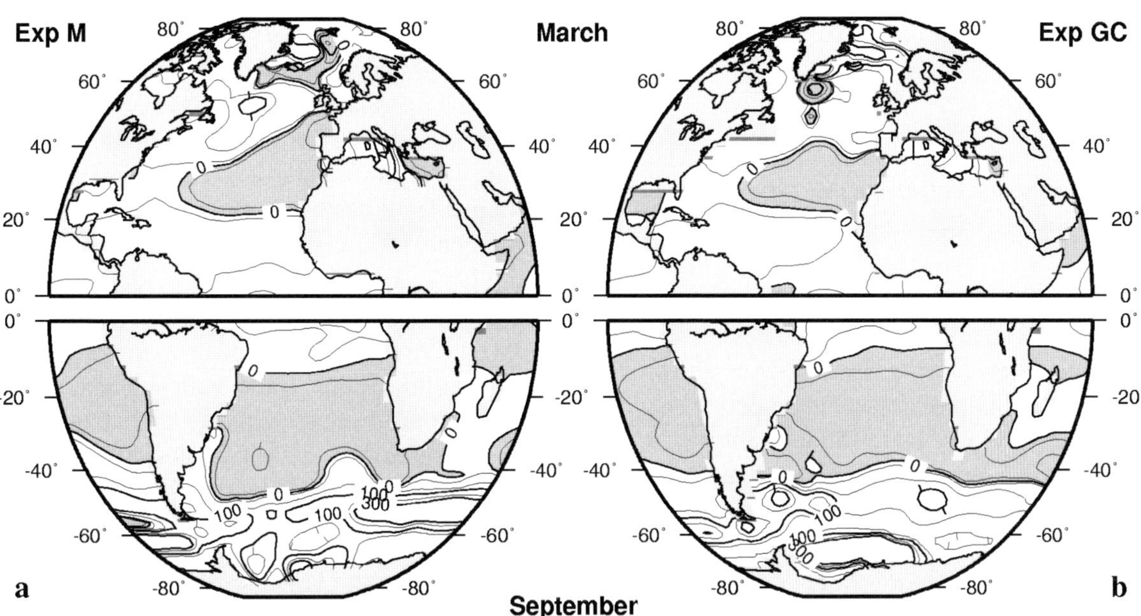

Fig. 5. Barrier layer thickness (depth difference between the isothermal and isopycnal mixed layer, m) for the Atlantic Ocean during winter (March in northern hemisphere, September in southern hemisphere). **a)** Modern (Experiment M). **b)** LGM (Experiment GC). Light shading indicates negative differences. Isolines are drawn at 0, ±5, ±10, ±50, ±100, ±200 and ±300 m.

Ventilated Thermocline Depth

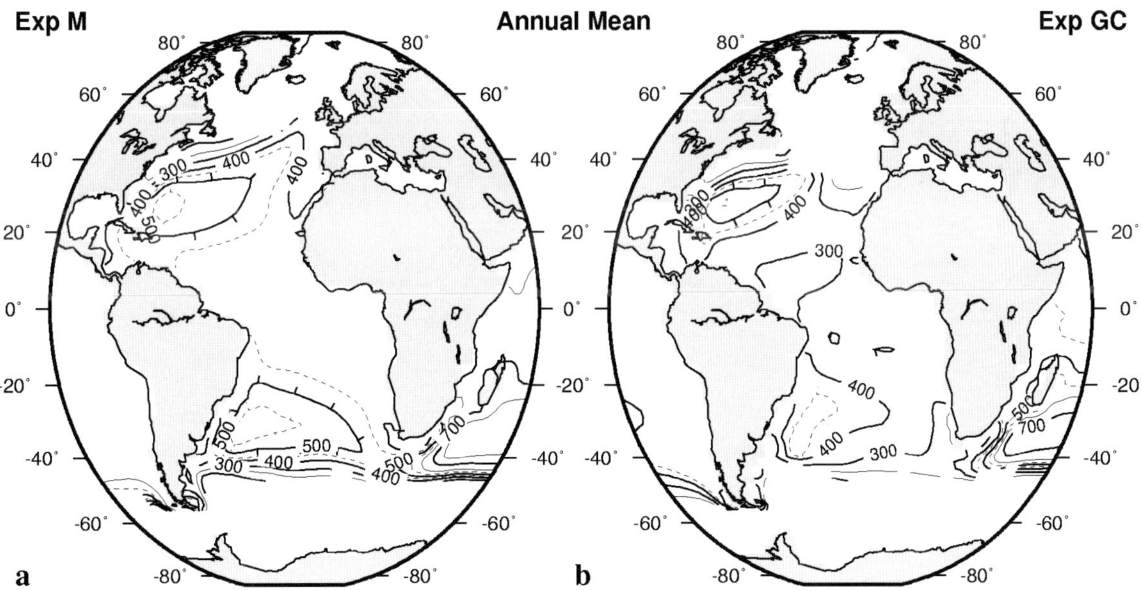

Fig. 6. Annual-mean maximum depth of the ventilated thermocline (m) for the Atlantic Ocean. **a)** Modern (Experiment M, $\sigma_0 = 26.9$ kg m^{-3} isopycnal surface). **b)** LGM (Experiment GC, $\sigma_0 = 28.1$ kg m^{-3} isopycnal surface). The contour interval is 100 m, additional contours are drawn at 450 and 550 m depth (dashed lines).

Characteristic	M	GA	GB	GC
Water flux (m a^{-1})	-0.061	0.227	0.008	-0.592
Water divergence (mSv)	-4.87	13.97	0.51	-36.30
SSS	34.35	35.49	35.94	36.30

Table 3. Implied water flux, its divergence and the sea-surface salinity (SSS) in the inner Weddell Sea (south of 63.9°S). The implied water flux approximates the net sea-ice formation or freezing rate in the ocean model.

equator (400-450 m) and near the outcrop regions, but reached up to 650 m in the centers of the sub-tropical gyres. During the LGM (as shown for Experiment GC), the tropical thermocline between 20°S and 20°N was up to 150 m shallower. At the same time the outcrop regions shifted equatorward in both hemispheres. As a result, the ventilated thermocline was shallower and reached only 500 m depth in the North Atlantic Ocean and 450 m depth in the South Atlantic Ocean. In addition, the contour lines that follow the path of South Atlantic Current (cf. Figure 1) indicate the enhanced advection of colder and denser waters from the western South Atlantic Ocean into the Benguela

System that replaced the inflow of warmer waters from the Indian Ocean.

While the base of the ventilated thermocline shallowed in the equatorial region (5°S–5°N), in the west more so than in the east (Figure 6), the mixed-layer and the permanent thermocline deepened in the west, at least during northern hemisphere summer (Figure 7). Thus we find a larger east-west slope of the mixed-layer depth as well as of the permanent thermocline, in agreement with the study by Wolff et al. (1999). However, it is also evident that this deepening went along with a general cooling by about 2°C. While the deepening was due to a strengthening of the trade winds, the cooling resulted from lower glacial SST and changes in ventilation at extra-tropical latitudes.

Circulation

Figure 8 shows the large-scale, depth-integrated horizontal circulation. The major meridional and horizontal volume transports are given in Table 4. In the southern hemisphere, the volume transport of the Antarctic Circumpolar Current (ACC) was 121 Sv in Experiment M and increased from 123 Sv in

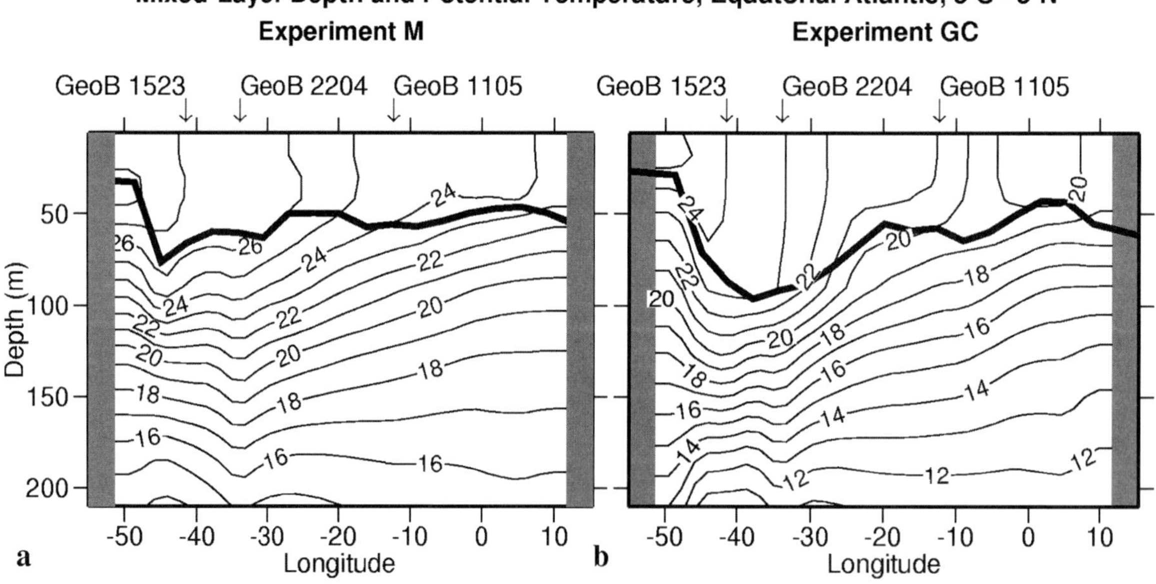

Fig. 7. Potential temperature (°C) distribution in the equatorial Atlantic Ocean (5°S-5°N) during September. The thick black line indicates the isothermal mixed-layer thickness. Arrows indicate the locations of the sediment cores studied by Wolff et al. (2000).

Horizontal Transport Streamfunction

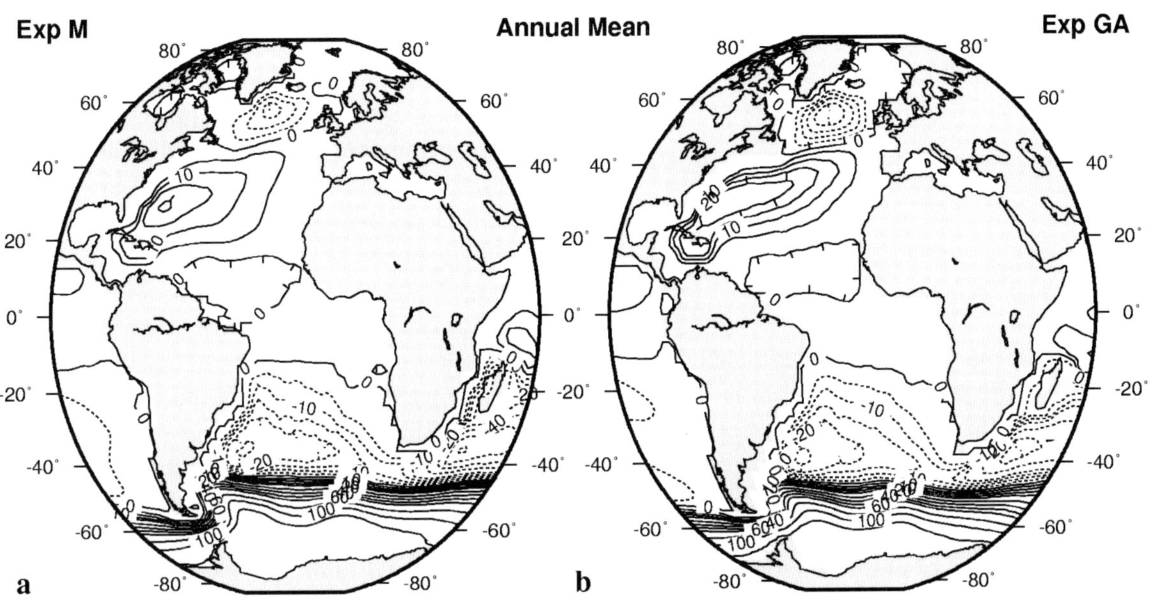

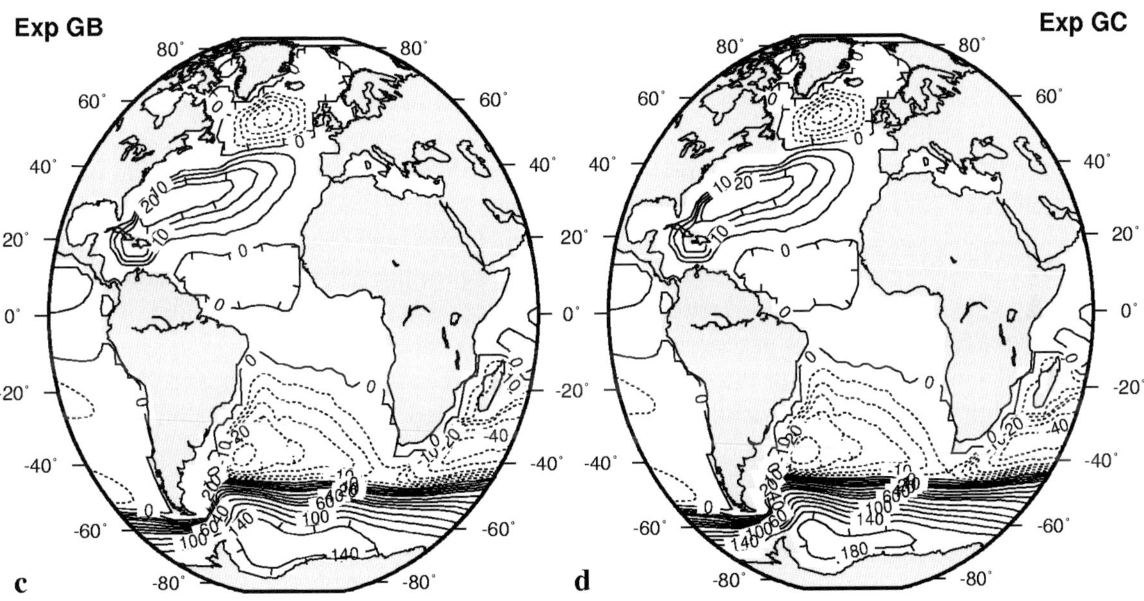

Fig. 8. Annual-mean vertically integrated volume transport (Sv) for the Atlantic Ocean as given by the barotropic streamfunction. The contour interval is 5 Sv up to 20 Sv, 10 Sv between 20 Sv and 80 Sv, and 20 Sv for more than 80 Sv. Dotted lines represent negative contour levels and indicate anti-clockwise circulation. **a)** Modern (Experiment M). **b)** LGM (Experiment GC).

Water mass, current	M	GA	GB	GC
NADW (production)	10	10	10	9
NADW (outflow at 30°S)	8	10	9	9
AABW (inflow at 30°S)	4	2	3	3
Northern subpolar gyre	16.9	24.1	23.6	23.4
Gulf Stream	21.4	27.4	28.0	28.1
Brazil Current	32.6	32.3	34.0	36.4
Agulhas Current	62.3	46.5	48.0	49.0
Agulhas leakage	11.7	12.1	8.1	5.1
ACC	120.7	123.1	138.0	175.1
Weddell Sea gyre	19.1	12.3	14.4	13.5

Table 4. Major meridional and horizontal transports (Sv) simulated in the ocean model.

Experiment GA to 175 Sv in Experiment GC. At the same time, the Agulhas leakage was 11.7 Sv in Experiment M and decreased from 12.1 Sv in Experiment GA to 5.1 Sv in Experiment GC. In all LGM experiments, the Agulhas Current was weaker by about 25% than at present, while the volume transport of the Brazil Current (and correspondingly the Benguela Current) was roughly 10% larger. The confluence of the Brazil and Malvinas Currents shifted southward by one gridpoint. The Weddell Sea Gyre turned out to be weaker under LGM conditions. In the northern hemisphere, the volume transport of the Gulf Stream increased by roughly 30%, and the North Atlantic Drift turned from a northwestward to a westward direction. The northern subpolar gyre intensified by nearly 40% and expanded southward.

The decreasing leakage of Indian Ocean waters from the Agulhas Current into the Atlantic Ocean was clearly reflected by changes in the horizontal transport and temperature over the depth range of the ventilated thermocline in the vicinity of Cape

of Good Hope, as can be seen from Figure 9. This figure details the large-scale horizontal circulation and temperature in five layers: the seasonal thermocline, ventilated thermocline, intermediate water, deep water and bottom water (see Table 5 for the definition of the water mass boundaries). The flow in the seasonal thermocline (whose lower boundary is defined as the annual maximum of mixed-layer depth, Huang and Qiu 1994) again shows the more zonal path of the Gulf Stream and North Atlantic Drift in Experiment GC. The water recirculating with the northern subtropical gyre was eventually entrained into the thermocline, in which the southwest transport gradually increased. The thermocline water cooled by 2–3°C in the North Atlantic Ocean and 4–5°C in the South Atlantic Ocean. While the cooling in the North Atlantic Ocean was due to lower temperatures at the sea surface and enhanced ventilation, in the South Atlantic Ocean, less warm water was imported from the Indian Ocean and more and colder water was subducted at the Subtropical Front. A large cooling was also evident in the intermediate and deep water. In the North Atlantic Ocean, both were ventilated from further south in Experiment GC as compared to Experiment M. The transport of Antarctic Intermediate Water in the southern subtropical gyre and across the equator was slightly weaker. The northward flow of Antarctic Bottom Water was stronger, mainly in the western trough of the Atlantic Ocean, and reached further north; its potential temperature fell below 0°C everywhere.

There was actually no upwelling of intermediate water into the ventilated thermocline in the tropics (between 20°S and 20°N), except for the Caribbean; there was very limited upwelling of deep

	Modern (Experiment M)	LGM (Experiment GC)
Seasonal thermocline	Annual maximum of isothermal mixed-layer depth	Annual maximum of isothermal mixed-layer depth
Ventilated thermocline	$\sigma_0 < 26.9$	$\sigma_0 < 28.1$
Intermediate Water	$26.9 < \sigma_0 < 27.5$	$28.1 < \sigma_0 < 28.8$
Deep Water	$\sigma_0 > 27.5$ and $\sigma_4 < 45.9$	$\sigma_0 > 28.8$ and $\sigma_4 < 47.45$
Bottom Water	$\sigma_4 > 45.9$	$\sigma_4 > 47.45$

Table 5. Definition of water mass boundaries. Potential densities (σ_0 = referenced to the surface, σ_4 = referenced to 4000 m depth) are given in kg m^{-3}.

Horizontal Volume Transport and Temperature

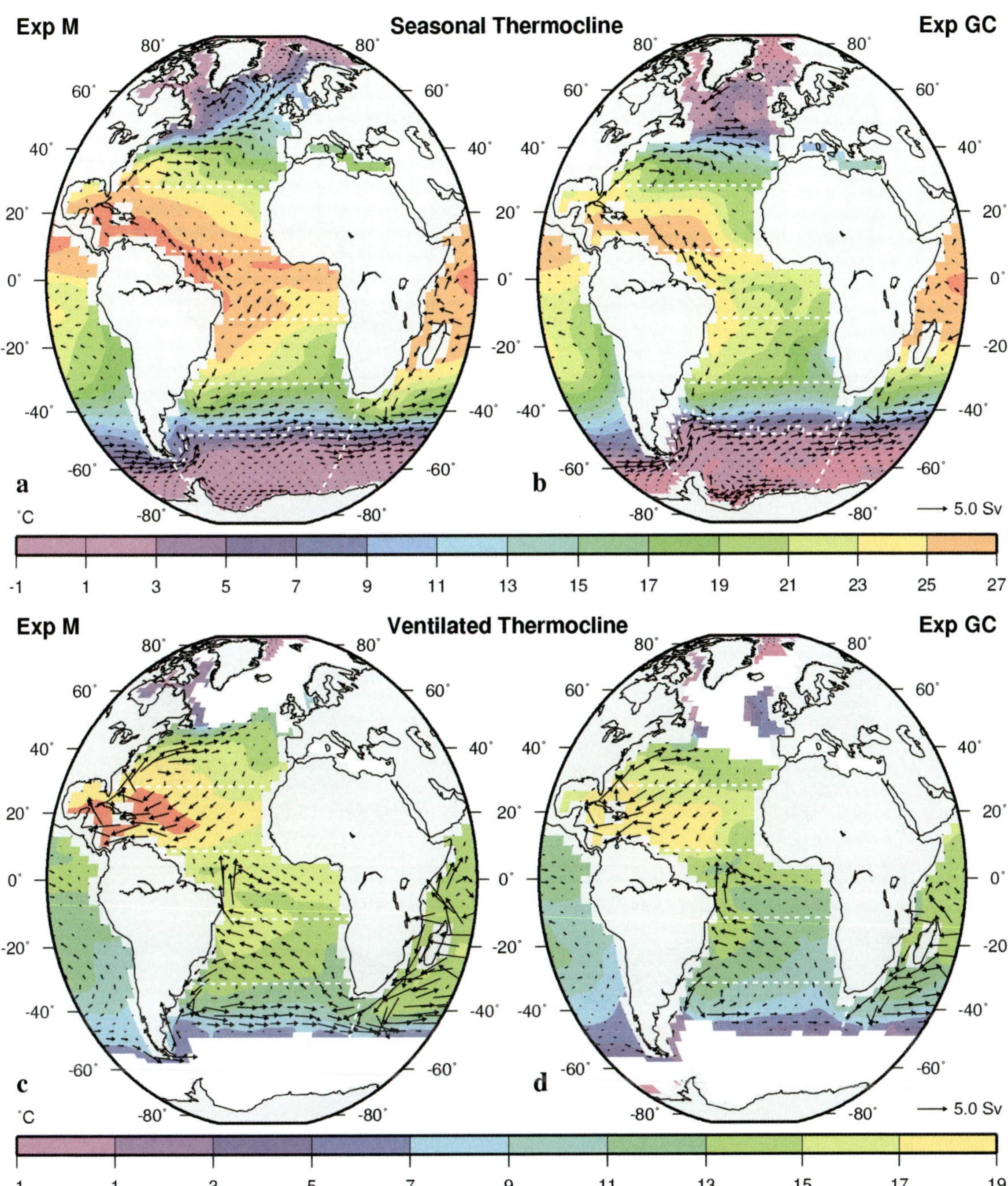

Fig. 9. Annual-mean horizontal volume transport (Sv) and temperature (°C), vertically averaged over the (**a,b**) seasonal thermocline, (**c,d**) ventilated thermocline, (**e,f**) intermediate water, (**g,h**) deep water and (**i,j**) bottom water layers. Left column: Modern (Experiment M). Right column: LGM (Experiment GC). For clarity, we show only one vector in two and arbitrarily truncate arrow lengths to that of the vector that is longer than 92% of all vectors. The water mass boundaries that define the respective layers are given in Table 5. Dashed white lines denote the zonal and meridional sections used in Figures 13-15.

Paul and Schäfer-Neth

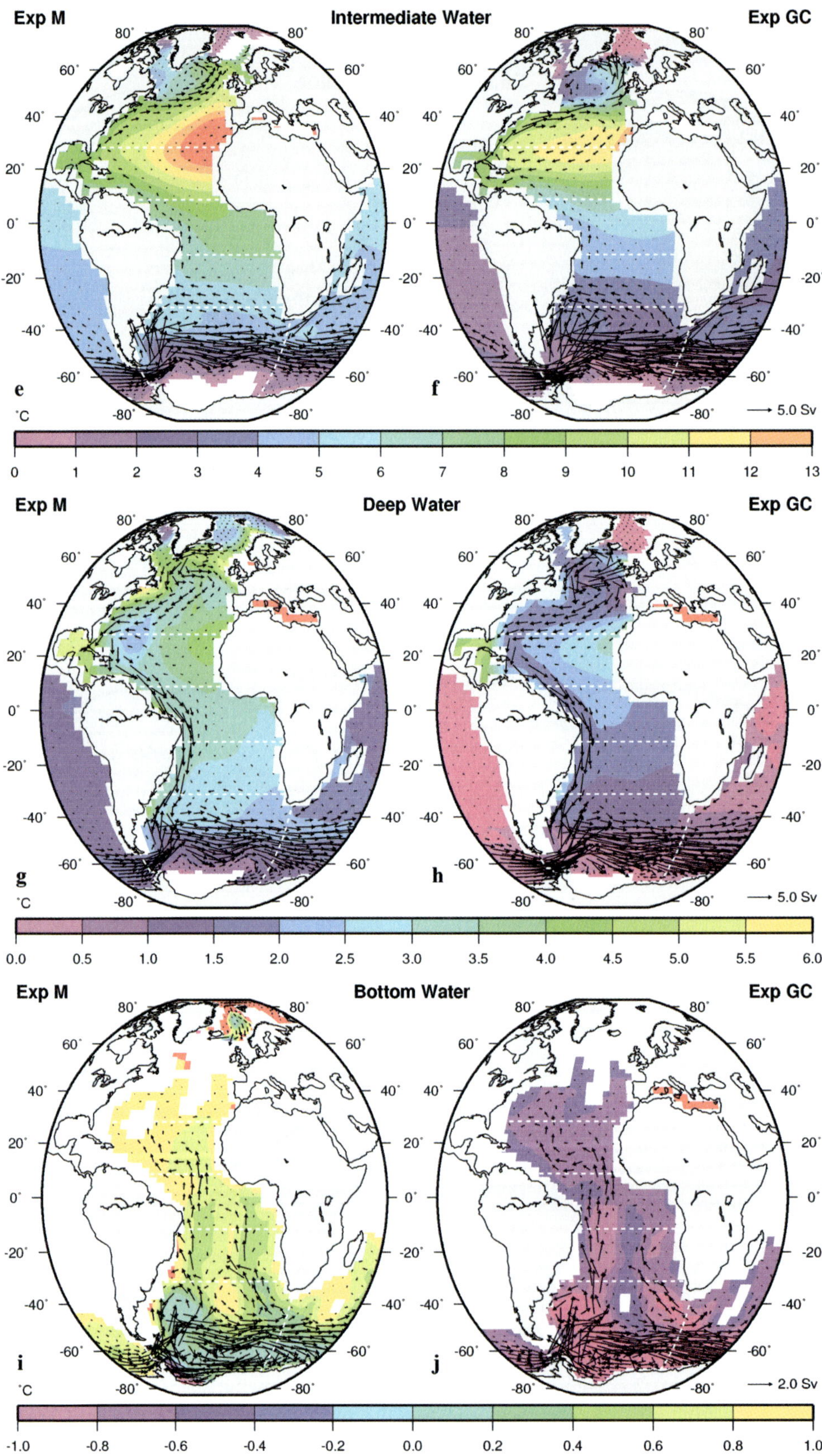

Fig. 9. cont.

Vertical Volume Transport

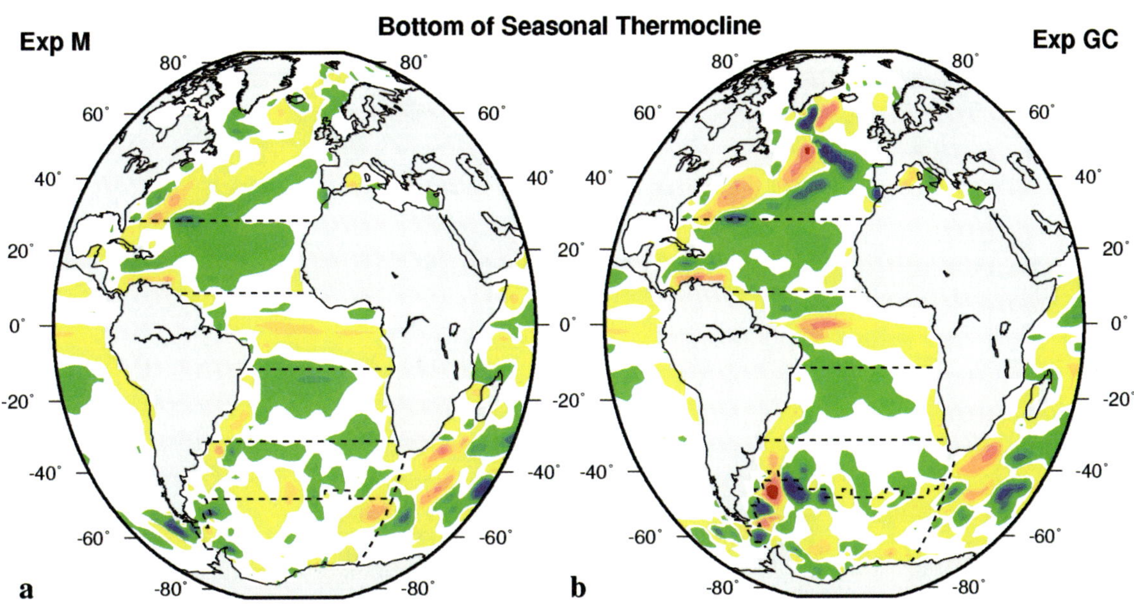

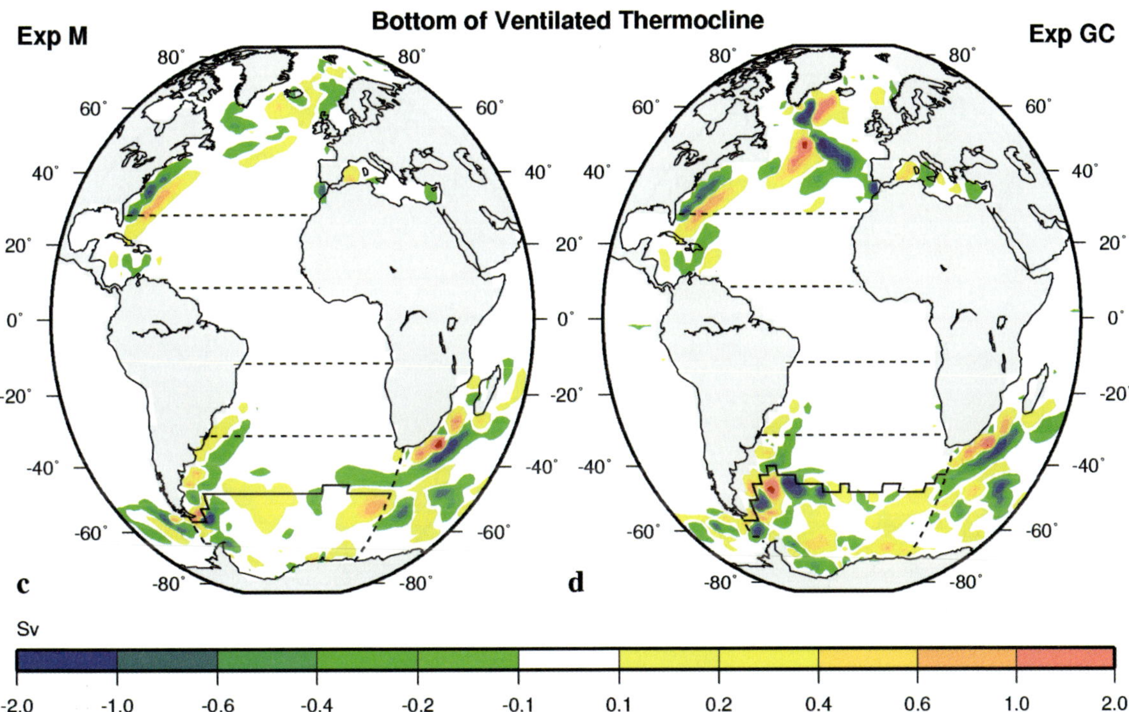

Fig. 10. Vertical volume transports (Sv) at the base of the (**a,b**) seasonal thermocline, (**c,d**) ventilated thermocline, (**e,f**) intermediate water and (**g,h**) deep water layers. Left column: Modern (Experiment M). Right column: LGM (Experiment GC). Positive values (yellow to red) indicate upwelling, negative values (green to blue) indicate downwelling. Whenever a layer outcrops at high latitudes, the vertical volume transport at the base of next deeper layer is shown (cf. Figure 15). The solid lines in c-h indicate the outcrop of the respective layer in the Southern Ocean. The dashed lines denote the zonal and meridional sections used in Figures 13-15 (the southernmost dashed line in a and b reflects the outcrop of the ventilated thermocline layer below).

Exp M
Bottom of Intermediate Water
Exp GC
80°
60°
40°
20°
0°
-20°
-40°
-60°
-80°
e
f

Exp M
Bottom of Deep Water
Exp GC
80°
60°
40°
20°
0°
-20°
-40°
-60°
-80°
g
h
Sv
-2.0   -1.0   -0.6   -0.4   -0.2   -0.1   0.1   0.2   0.4   0.6   1.0   2.0

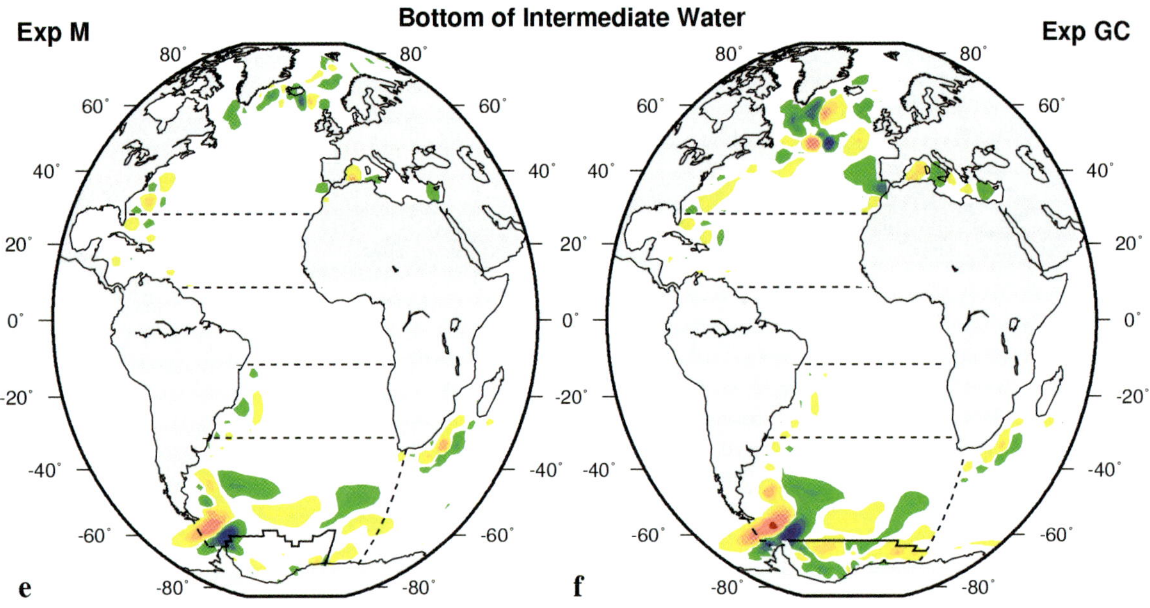

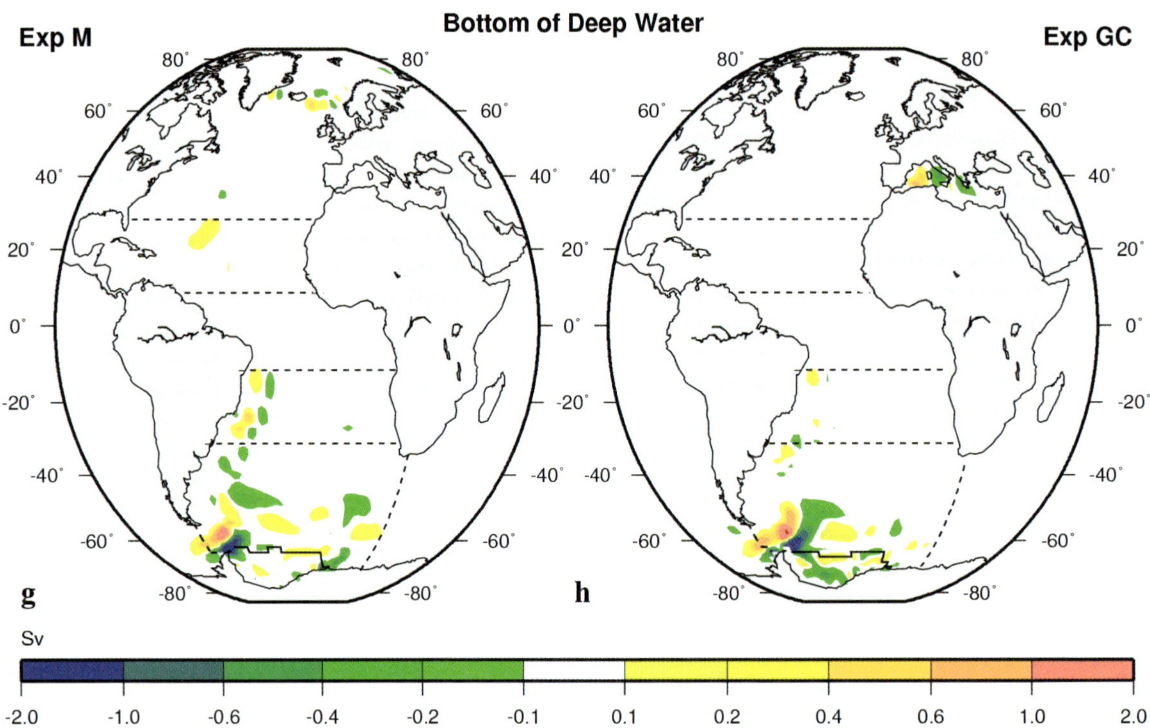

Fig. 10. cont.

water at the western boundary, and none at the equator (Figure 10). In Experiment GC as compared to Experiment M, equatorial upwelling (between 5°S and 5°N) into the mixed layer was stronger by a factor of 3 to 5 (Figure 10a). The equatorial downwelling was also more intense. Coastal upwelling at the LGM slightly increased in the southern Benguela System, but decreased in the northern Benguela System, consistent with the change in the Ekman pumping velocity (Figure 2). In the deeper layers, there was upwelling as well as downwelling south of 40°S.

Figure 11 shows the meridional overturning circulation in the Atlantic Ocean. In addition to three shallow, wind-driven cells, there are two cells at depth indicating the formation and outflow of North Atlantic Deep Water (NADW) and the inflow of AABW. In Experiment M, 10 Sv of new NADW were formed in the northern North Atlantic Ocean, almost 4 of which originated from the Arctic Ocean. Only 1 Sv upwelled north of 30°N and 1 Sv upwelled at the equator, leaving 8 Sv for export across 30°S. Experiment GC shows similar total rates of NADW formation and upwelling in the North Atlantic Ocean as Experiment M, but the formation was shifted to the south and split into two convection areas, one south of 50°N, the other south of 70°N. Thus, even under glacial conditions, about half of the NADW was formed at or slightly north of the Greenland-Iceland-Scotland ridge. In both Experiment M and Experiment GC, the inflow of AABW at 30°S amounted to 4 Sv.

The meridional heat transport in the Atlantic Ocean was northward at all latitudes (Figures 12 and 13). The maximum occurred at about 28°N and amounted to 0.57 PW in Experiment M and 0.55 PW in Experiment GC. In the LGM case with the reduced Agulhas leakage, the northward heat transport in the South Atlantic Ocean was reduced. There was only a very small change of about 0.02 PW in the cross-equatorial heat transport. However, northward of 40°N, the decrease in Experiment GC as compared to Experiment M reached 0.1 PW.

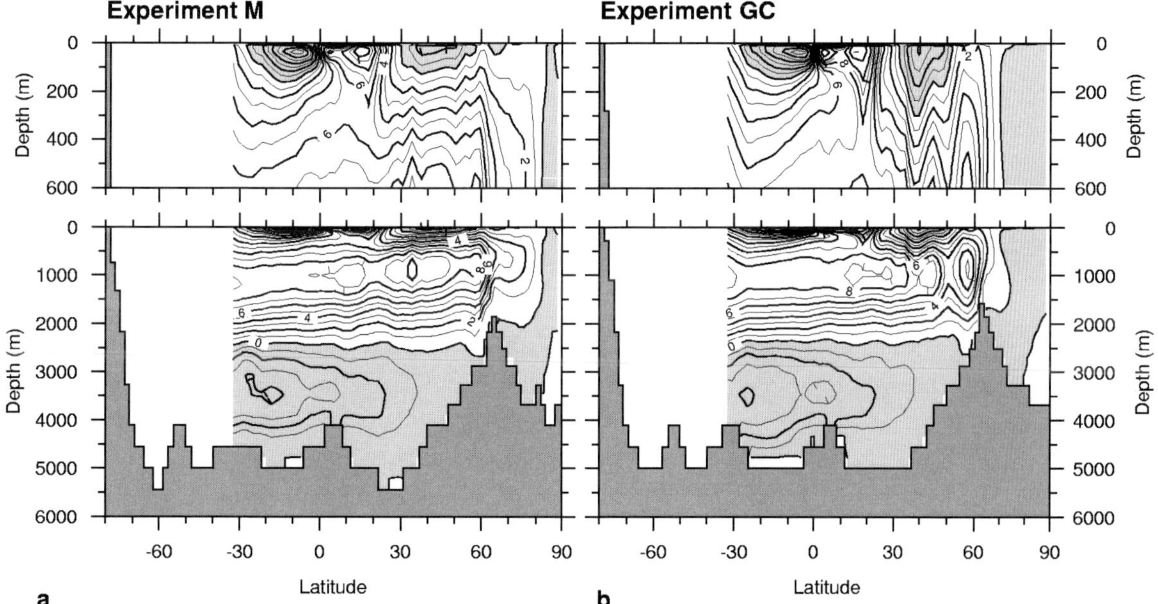

Fig. 11. Annual-mean meridional overturning streamfunction (Sv) for the Atlantic Ocean. **a)** Experiment M. **b)** Experiment GC. The contour interval is 1 Sv. Light shading indicates negative contour levels and anti-clockwise circulation. No streamfunction can be defined at latitudes south of Cape of Good Hope where there is zonal exchange with the Pacific and Indian Oceans.

Northward Heat Transport

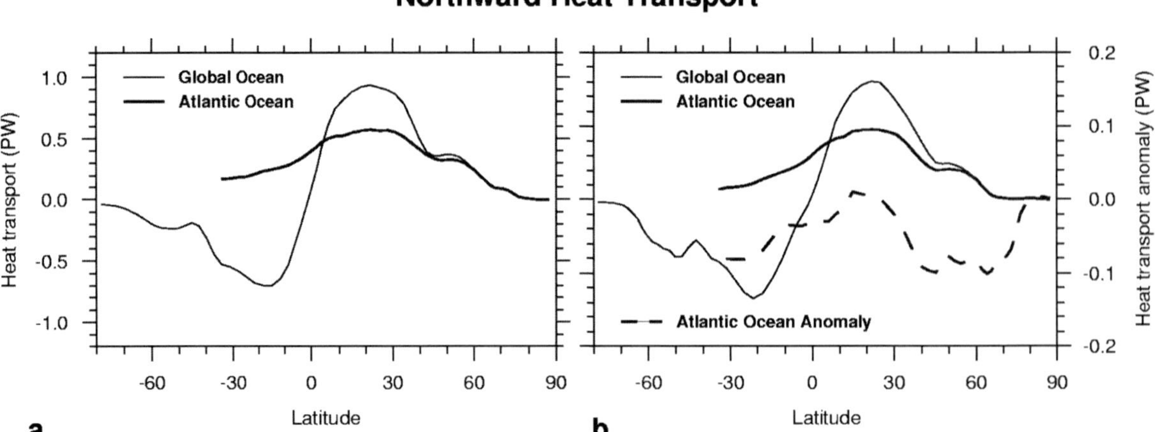

Fig. 12. Annual-mean zonally averaged northward heat transport (PW). **a)** Modern (Experiment M). **b)** LGM (Experiment GC). Thin solid line: global ocean, thick solid line: Atlantic Ocean, thick dashed line: Atlantic Ocean LGM anomaly.

Total Depth-Integrated Volume and Heat Transports

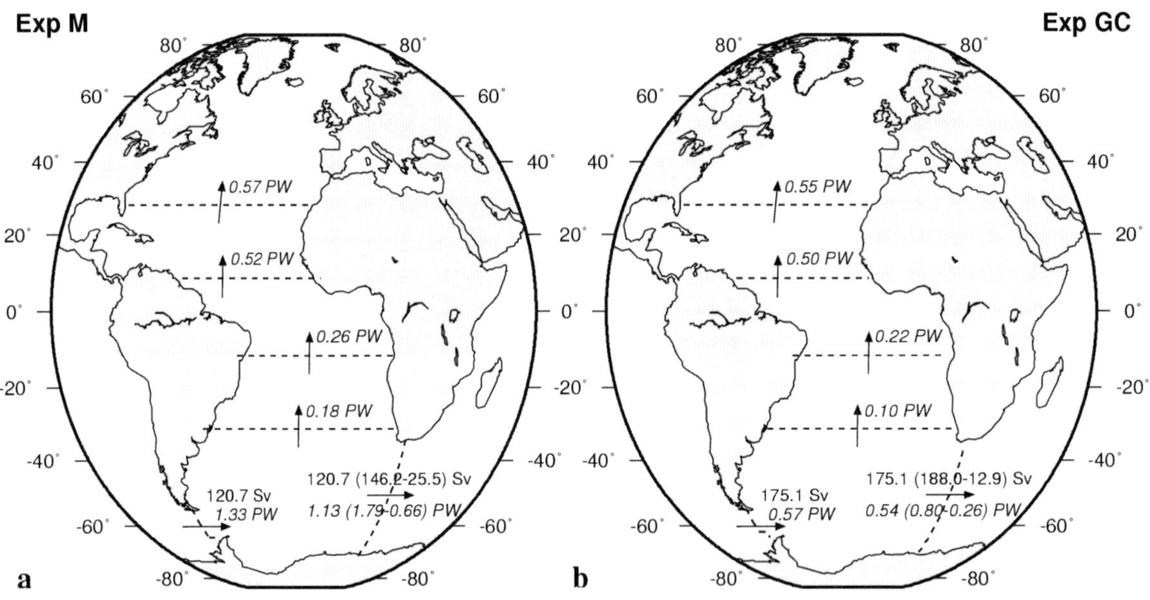

Fig. 13. Schematic representation of the simulated horizontal volume and heat transports, vertically integrated over the whole depth of the water column (cf. Sloyan and Rintoul 2001, Figure 2). **a)** Modern (Experiment M). **b)** LGM (Experiment GC). The dashed lines mark selected zonal sections at ~30°S, ~11°S, ~8°N (close to classic hydrographic sections) and ~24°N (near the southern tip of Florida), and meridional sections at ~68°W (at Cape Horn) and ~20°E (near Cape of Good Hope). Volume transports (Sv) are zero across the zonal sections because there is no advective flow through Bering Strait. Heat transports (PW) are relative to 0°C.

In addition to the heat transports across selected zonal sections at 31.17°S, 11.48°S, 8.55°N and 28.04°N, Figure 13 indicates the volume and heat transports across the meridional sections at 68.4°W (at Cape Horn) and 21.6°E (near Cape of Good Hope) in the southern hemisphere (these model latitudes and longitudes are close to classic hydrographic sections at 30°S, 11°S, 8°N, 24°N, 68°W and 20°E, e.g. Hall and Bryden 1982, Klein et al. 1995, Sloyan and Rintoul 2001). The air-sea heat flux in the South Atlantic Ocean south of 31.17°S was nearly balanced and amounted to -0.02 PW in Experiment M and +0.07 PW in Experiment GC. Thus the heat gained by the South Atlantic Ocean north of 31.17°S was mainly determined by the difference between the heat exported by the ACC and the heat imported by the Agulhas leakage. The heat exchanged between the South Atlantic and Indian Oceans via the Agulhas leakage decreased from about 0.66 PW in Experiment M to about 0.26 PW in Experiment GC.

In Figure 14 we summarize the exact depth-integrated horizontal volume transports within the five layers shown in Figure 9 (cf. England and Garçon 1994 and Sloyan and Rintoul 2001 for similar diagrams of the present-day circulation in the South Atlantic Ocean), and Figure 15 details the vertical mass exchange between these different layers (cf. Sloyan and Rintoul 2001, Figure 13). The most prominent changes between the modern and the glacial experiment occured in the Drake Passage and south of Cape of Good Hope. In Experiment M, the inflow of Indian Ocean water via the Agulhas leakage accounted for 28% of the transport of the Benguela Current in the seasonal thermocline (1.5 out of 5.4 Sv). In the ventilated thermocline, this fraction increased to 55 % (6 out of 11 Sv), and in the Intermediate Water eventually reached as much as 87 % (6 out of 6.9 Sv). In total, 60 % of the Benguela Current in Experiment M were fed by the Agulhas leakage. In Experiment GC, the respective numbers were 20 % (seasonal thermocline, 0.7 out of 3.5 Sv), 11 % (ventilated thermocline, 1 out of 9.5 Sv), and 38 % (Intermediate Water, 4.5 out of 11 Sv), which summed up to 25 % over the three levels, leaving the mixture of waters from Drake Passage, from the Brazil Current, and from the upwelling around 60° S (cf.

Figure 15) as the dominant source for the water masses in the southern Benguela system. In this experiment, the Benguela Current was slightly stronger than in Experiment M and stayed closer to the coast.

Oxygen-18

Sea-water oxygen-18 ($\delta^{18}O_w$) was a passive tracer in our ocean model, restored to the same sparse and scattered modern annual-mean observations in all experiments. For the sea surface, this treatment was similar to the horizontal interpolation scheme by Takahashi et al. (1997), but it allowed for a vertical exchange with the deep ocean. Accordingly, the resulting sea-surface distribution in Experiment M shows a general agreement between model and data (Figure 16). In Experiment GC, there were large local changes, particularly in upwelling areas (cf. Figure 2c and d) where the restoring boundary condition was less effective and the characteristics of the upwelled water dominated. In the south-eastern South Atlantic Ocean, the upwelled water was depleted in oxygen-18 by 0.4‰ and originated from the western South Atlantic Ocean rather than the Indian Ocean (cf. Figure 9). In the North Atlantic Ocean, the changes in the subtropical gyre reflected the more zonally-oriented Gulf Stream and the stronger recirculation.

Despite identical $\delta^{18}O$ restoring data, our model experiments developed distinctly different $\delta^{18}O_w$ patterns with depth, indicating changes in the formation, transport and mixing of the various water masses. In the control run (Figure 17a) there were extremely light values in the high latitudes of the Atlantic Ocean, below -0.2‰ at the surface, that influenced the oxygen isotope ratios of the underlying bottom waters. Consequently, the deep ocean was filled by light AABW < 0‰ and only slightly heavier water north of the Greenland-Scotland ridge. These light values were contrasted by higher ratios > 0.2‰ in the NADW and in the ventilated thermocline, where more than 1.1‰ were reached at about 30°N. AAIW was characterized by values around 0.1–0.2‰.

All glacial experiments were marked by considerably heavier values around 0.5‰ within the NADW tongue, which yielded a much sharper

Exact Depth-Integrated Volume Transports

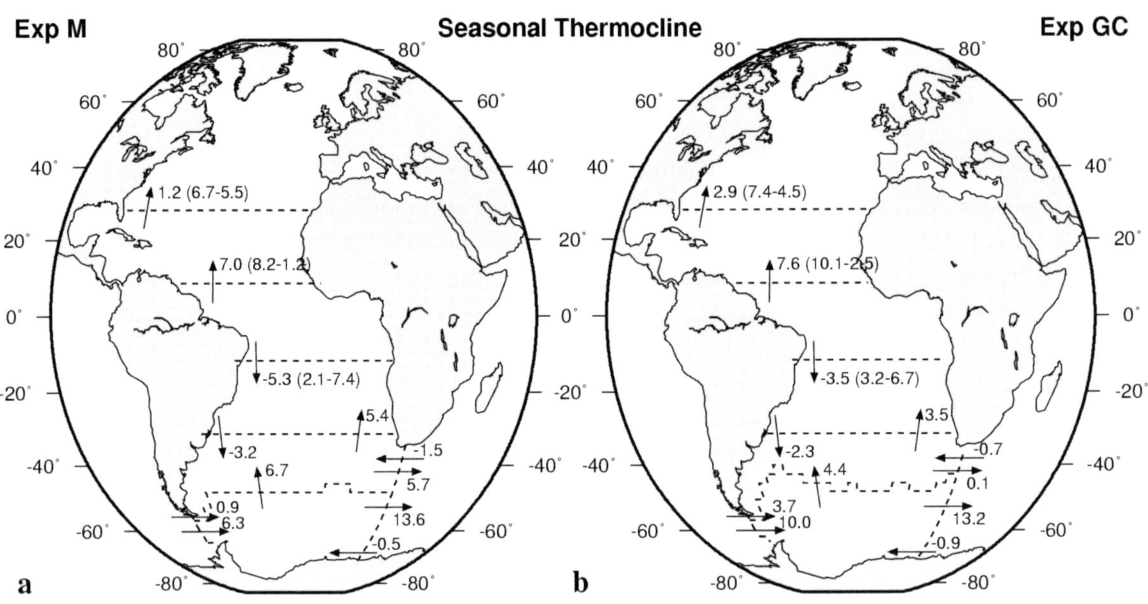

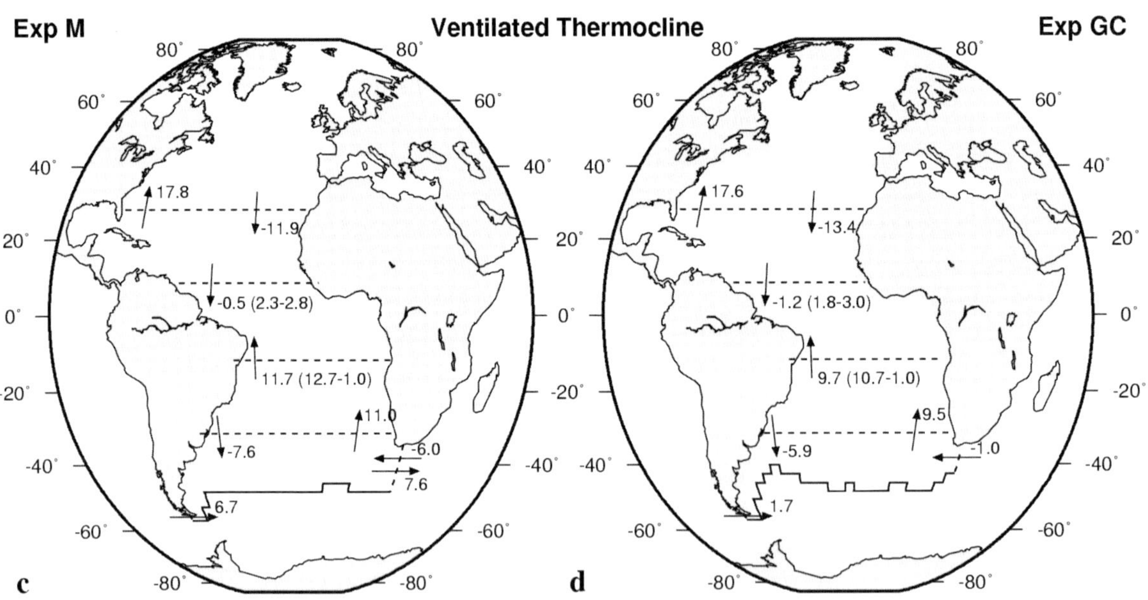

Fig. 14. Schematic representation of the simulated horizontal volume transports (Sv), vertically integrated over the (**a,b**) seasonal thermocline, (**c,d**) ventilated thermocline, (**e,f**) intermediate water, (**g,h**) deep water and (**i,j**) bottom water layers (cf. England and Garçon 1994, Figure 11, and Sloyan and Rintoul 2001, Figure 5). Left column: Modern (Experiment M). Right column: LGM (Experiment GC). The water mass boundaries that define the respective layers are given in Table 5. The solid lines in c-h indicate the outcrop of the respective layer in the Southern Ocean. The dashed lines mark the same zonal and meridional sections as in Figure 13.

Exp M
Intermediate Water
Exp GC
4.2
-1.8
2.5 (3.0-0.5)
1.9 (3.1-1.2)
6.9
-4.5
-6.0
49.9
59.4
-0.3
e
7.7
-5.8
2.2 (2.8-0.6)
1.7 (3.8-2.1)
11.0
-8.5
-4.2
78.8
81.1
f

Exp M
Deep Water
Exp GC
-10.2 (2.1-12.3)
-12.2 (1.6-13.8)
-11.7 (2.3-14.0)
-12.1 (2.4-14.5)
-1.5
50.0
40.0
-3.3
g
-10.7 (1.0-11.7)
-11.7 (1.3-13.0)
-11.3 (2.4-13.7)
-11.0 (2.0-13.0)
-1.0
75.2
69.8
-2.0
h

Exp M
Bottom Water
Exp GC
0.7 (1.0-0.3)
3.2 (3.3-0.1)
3.4 (4.2-0.8)
4.1 (5.5-1.4)
-1.1
19.4
7.4
-5.3
i
1.7 (2.5-0.8)
3.2 (3.4-0.2)
3.4 (4.2-0.8)
3.7 (5.4-1.7)
-2.5
20.6
8.8
-0.5
j

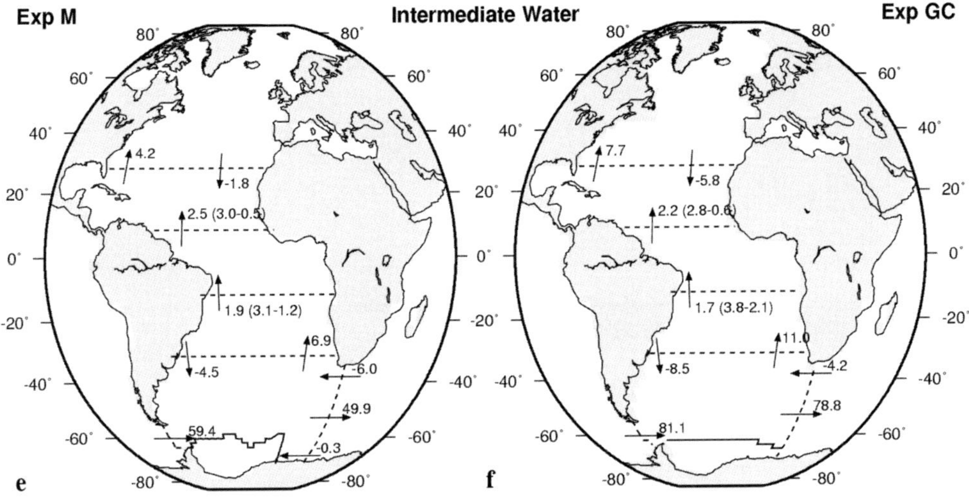

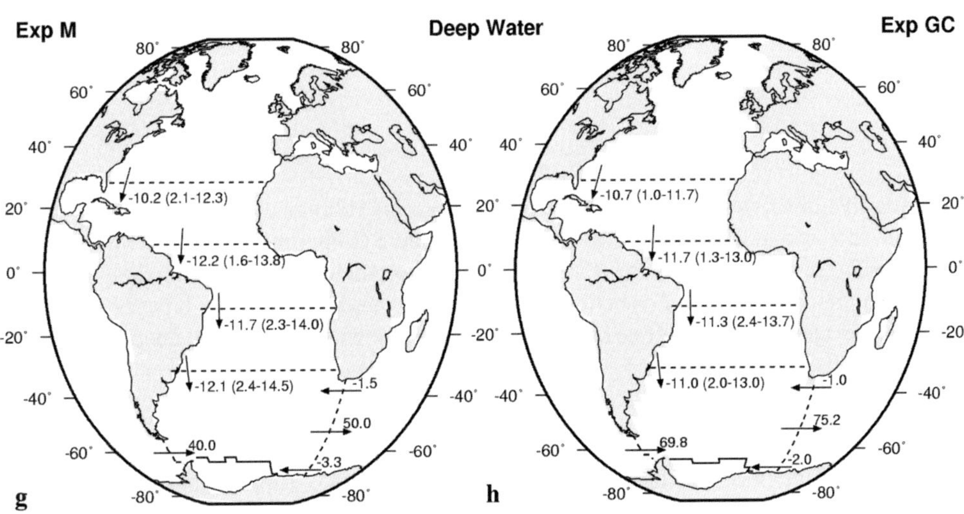

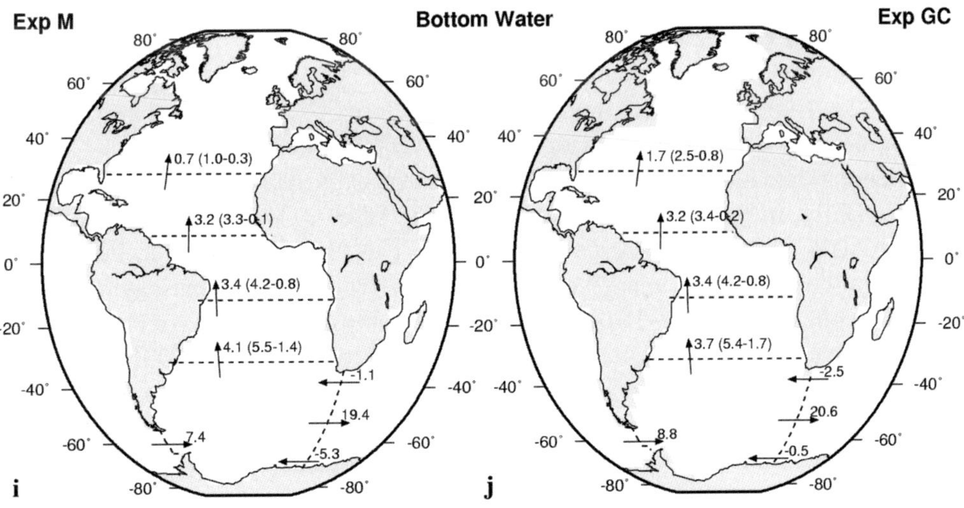

Fig. 14. cont.

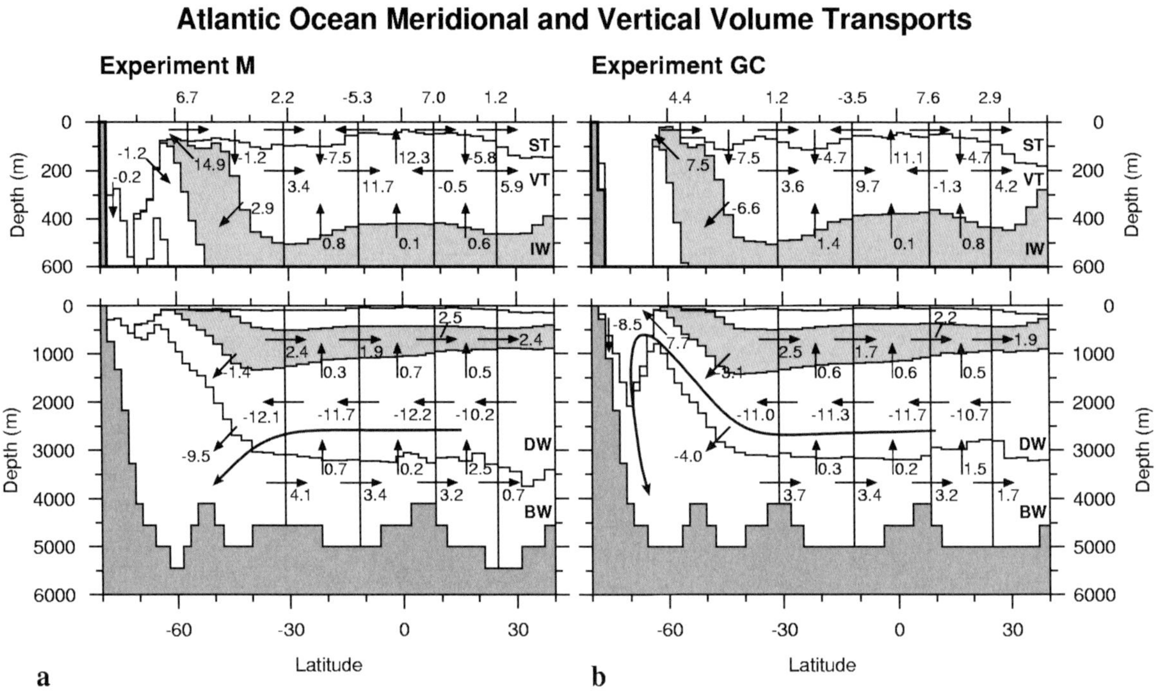

Atlantic Ocean Meridional and Vertical Volume Transports

Fig. 15. Schematic five-layer view of the simulated annual-mean meridional overturning circulation (Sv) for the Atlantic Ocean (cf. Sloyan and Rintoul 2001, Figure 13). **a)** Modern (Experiment M). **b)** LGM (Experiment GC). The five layers are the seasonal thermocline (ST), ventilated thermocline (VT), intermediate water (IW), deep water (DW) and bottom water (BW). For better orientation, the intermediate water layer is shaded in both the upper (0-600 m depth) and lower panels (surface to bottom). The long arrows indicate the path taken by the deep water flow.

contrast to the almost unaltered AAIW isotope ratios. In the deep polar North Atlantic Ocean, there were higher values > 0.5‰ as well, separated by a narrow band of 0.1–0.2‰ from the subpolar Atlantic Ocean.

AABW experienced drastic changes in the course of the changing haline boundary condition in the Weddell Sea. Experiments GA and GB yielded $\delta^{18}O_w$ values that were increased with respect to Experiment M and reached up to to 0.3‰ (GA) and 0.1‰ (GB). In Experiment GC, the $\delta^{18}O_w$ values of AABW $\delta^{18}O_w$ turned negative and were comparable to those of Experiment M.

The distribution of calcite oxygen-18 ($\delta^{18}O_c$) in equilibrium with ambient seawater (Figure 18) was computed from the modeled temperature and $\delta^{18}O_w$ fields and the paleotemperature equation by Mulitza et al. (2003), ignoring any 'vital' or explicit 'ice-volume' effects. It is largely a function of temperature with depth, with the lightest values in the upper ocean. As compared to the $\delta^{18}O_w$ distribution, the AAIW tongue almost vanished in Experiment M and was much less prominent in the glacial experiments.

Discussion

Hydrology

The vertically averaged temperature and salinities for the Atlantic Ocean (4.16°C, 34.86) and the global ocean (3.73°C, 34.65) in Experiment M are much closer to the present-day observations than in the control experiments of Paul et al. (1999) and Schäfer-Neth and Paul (2001). This reflects the more realistic representation of the thermocline, intermediate and deep waters in our present configuration of the MOM, which is mainly due to the low vertical diffusion and the isopycnal mixing parameterization that overcomes the warm and

Sea Surface δ¹⁸O of Sea-Water

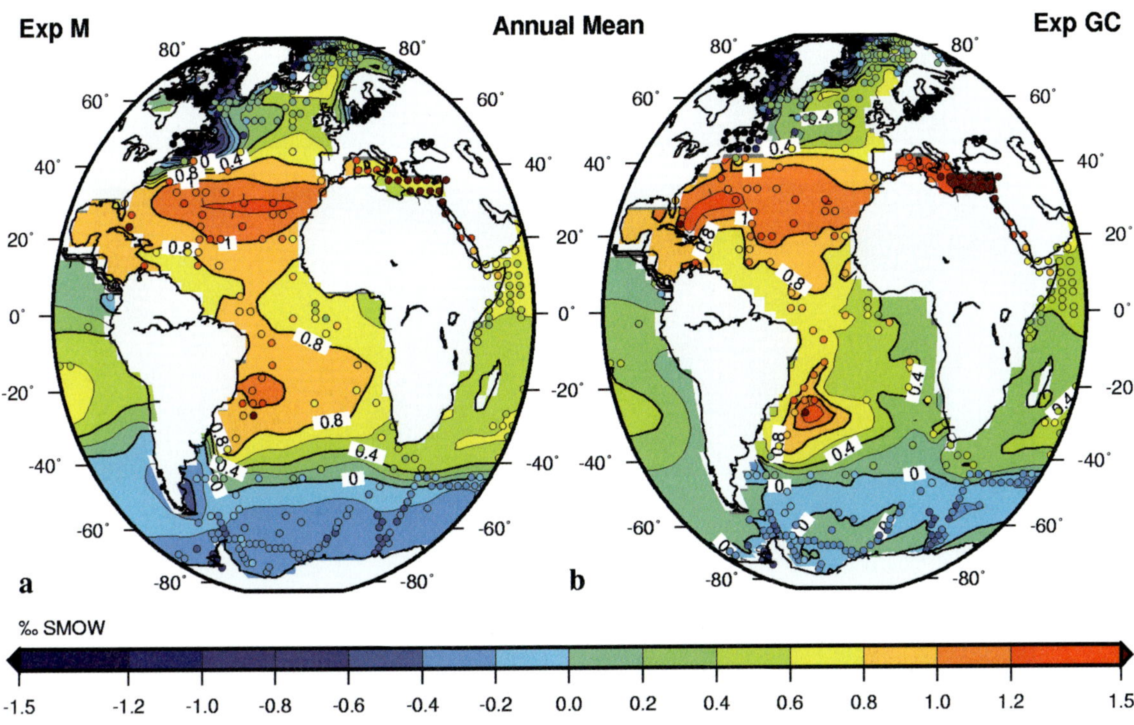

Fig. 16. Sea-surface distribution of annual-mean sea-water $\delta^{18}O$ (‰ SMOW) for the Atlantic Ocean. **a)** Modern (Experiment M). **b)** LGM (Experiment GC). Circles denote observed present-day annual-mean sea-water $\delta^{18}O$ from the GISS database (Schmidt et al. 1999). The contour interval is 0.2‰.

fresh bias of the older simulations with the MOM and LSG models.

A change in the global mean salinity by 1.35 to 1.46 corresponds to a change in relative mean sea level of about 130 m, which is close to recent reconstructions (Clark and Mix 2002). In spite of the reorganization of the deep ocean stratification, the SST changed only very little from Experiment GA to GC. The SSS changed only in the Weddell Sea. This is a peculiarity of the restoring boundary condition that tries to maintain the prescribed sea-surface conditions even against changes of the large-scale circulation.

Following the recipe outlined in Section 2.3, we obtained $\sigma_0 = 26.9$ kg m^{-3} as the lower boundary of the ventilated thermocline in Experiment M (Table 5). This compares well with density criteria derived from modern observations which range be-

tween 26.8 and 27.1 kg m^{-3}. For example, Williams et al. (1995) and Schmid et al. (2000) use $\sigma_0 = 27.0$ kg m^{-3}. Our slightly lower value may reflect that in Experiment M the salinity (and hence the density) of the upper Atlantic Ocean is somewhat too low (not shown).

In our model, the outcrop locations of the thermocline isopycnal surfaces in the North Atlantic Ocean shifted from about 40°–55°N in Experiment M to about 35°–45°N in Experiments GA–GC (Figure 5). As a net result, the thermocline waters cooled by 2–3°C in the North Atlantic Ocean and 4–5°C in the South Atlantic Ocean. The general cooling and shoaling of the ventilated thermocline is in agreement with the reconstruction by Slowey and Curry (1995), which is based on large $\delta^{18}O$ increases in benthic foraminifera on the margins of the Little and Great Bahama Banks.

Atlantic Ocean δ¹⁸O of Sea-Water

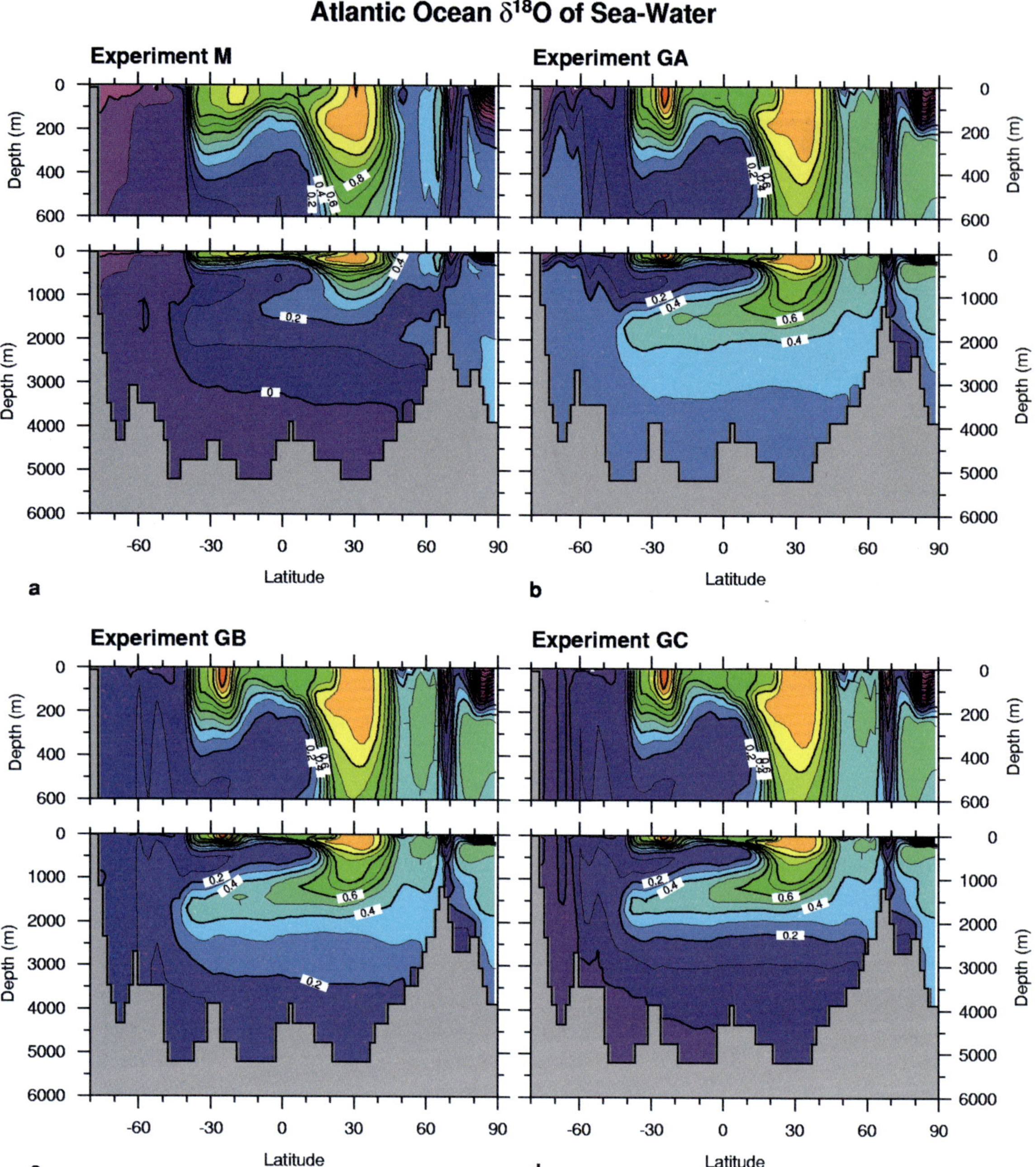

Fig. 17. Annual-mean distribution of sea-water $\delta^{18}O$ (‰ SMOW) along the Western Atlantic Ocean transect. **a)** Experiment M. **b)** Experiment GA. **c)** Experiment GB. **d)** Experiment GC. The contour interval is 0.1‰.

However, we do not find a shallower ventilated thermocline precisely at the location of the Bahama banks. There are two possible reasons for this: First the Caribbean is influenced by complex topography, which is only crudely resolved in our model. Second, the western boundary current region may be governed by different dynamics than the interior of the subtropical gyre.

The net annual ice export of the inner Weddell Sea south of 63°S has been estimated as 50±19 mSv (Harms et al. 2001) from a twenty-year deployment of upward-looking sonar systems and

Atlantic Ocean δ¹⁸O of Foraminiferal Carbonate

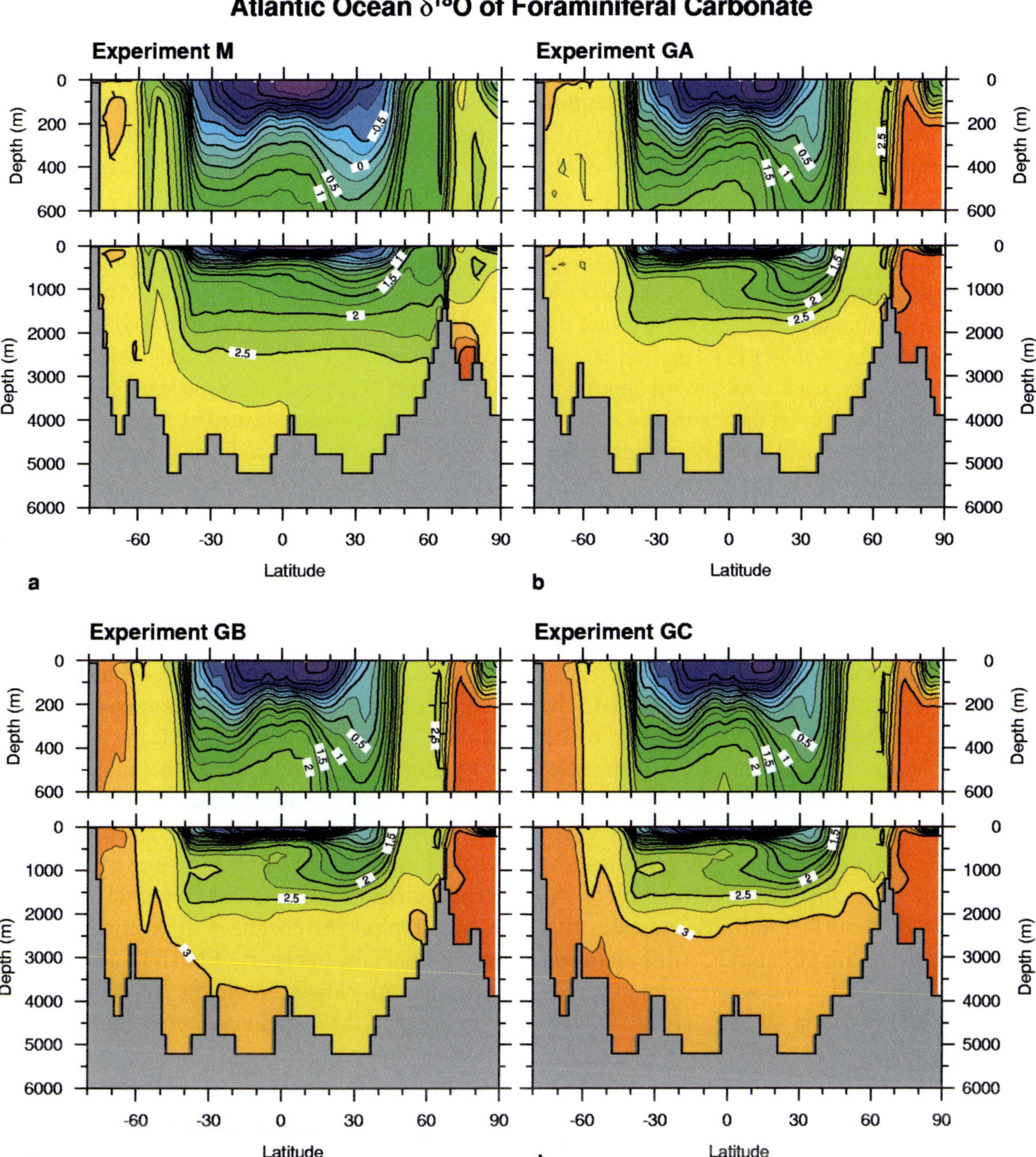

Fig. 18. Annual-mean distribution of calcite $\delta^{18}O$ (‰ PDB) in equilibrium with ambient seawater along the Western Atlantic Ocean transect. **a)** Experiment M. **b)** Experiment GA. **c)** Experiment GB. **d)** Experiment GC. The contour interval is 0.25‰.

satellite microwave measurements. Timmermann et al. (2001) use these data to validate their coupled sea ice-ocean model, which yields an ice export of 42±25 mSv. According to this model, the freshwater loss due to sea-ice formation roughly balances ice-shelf basal melting and net precipitation and amounts to 5±13 mSv in the annual mean. In our case, the net freshwater loss through the surface of the inner Weddell Sea may be inferred from the implied salinity fluxes due to the

SSS restoring boundary condition. For Experiment M, we find nearly 4.9 mSv (Table 3), which is in good agreement with the results of Timmermann et al. (2001). However, it should be kept in mind that the available observations are the residuum of several large terms of different signs, namely snowfall/precipitation, glacial melt from the ice shelves and ice advection, which easily exceed the net freezing rate by an order of magnitude. This is even more the case for the seasonal variations and clearly evident from the large error margin of the model estimate (5±13 mSv). As Timmermann et al. (2001) conclude, sea-ice formation appears to be a necessary condition for the renewal of AABW, and any changes in the Weddell Sea freshwater budget have a significant effect on the global circulation, a result that is also born out by our series of glacial experiments (GA–GC). As can be seen from Table 3, there was a net freshening of the Weddell Sea in Experiments GA and GB. Due to the up-welling of salty NADW, the restoring to lower SSS implied a net melting of sea ice instead of sea ice formation. Only in Experiment GC, the net freshwater flux turned negative again, which indicated the net export of sea ice (freshwater) from the inner Weddell Sea and corresponded to a net formation of sea-ice, at a rate vastly increased with respect to our control run. This nonlinear dependence of the implied sea ice formation on the Weddell Sea sea-surface salinity indicates that between Experiments GB and GC a threshold was passed, beyond which the water column turned unstable and convection was triggered.

Circulation

The transport of the ACC in Experiment M was 121 Sv, which was close to the observed value (e.g. 130±20 Sv - Witworth and Peterson 1985) and much smaller than the value of 225 Sv in the study by Paul et al. (1999). The improvement was due to the effects of the isopycnal mixing and mesoscale eddy transport parameterizations on the isopycnal form stress in the ACC. The transport rates of the western boundary currents tended to be lower than observed, mainly because of the relatively coarse zonal resolution of our ocean model.

This was particularly true for the Gulf Stream transport of 20 Sv that was only half of the observed strength, and for the northern subpolar gyre. With this exception, the horizontal mass transports of our control run agreed reasonably well with the observed current strengths (e.g. England and Garçon 1994).

The cooling of the high northern latitudes and the associated changes in the wind stress distribution (Figure 2b) caused the North Atlantic Drift to take a more easterly direction (Figure 9a), which is a common finding in many modeling studies of the LGM ocean. Changes in the wind stress distribution in the Brazil Basin as indicated by Figure 2b and the southward shift of the zero line of Ekman pumping as implied by Figure 2c also led to the larger Brazil Current transport and southward shift of the Brazil and Malvinas Currents by one grid-point, in spite of the strengthening of the ACC.

In contrast to previous modeling studies of the glacial ocean, we used low vertical diffusion in the upper 2000 m and a relatively high vertical resolution, which yielded a realistic representation of modern AAIW. In Experiment M, the associated $\delta^{18}O_w$ minimum (closely related to a salinity minimum, not shown) could be traced as far as 10°N; it was 750 m deep at 40°S and gradually rose until it reached a depth of 500 m. Furthermore, the low vertical diffusion also yielded very little upwelling into the tropical thermocline in all experiments (Figure 10), thus the transport of NADW across 30°S was only 1-2 Sv lower than across 30°N (Figure 11). In this respect, the meridional circulation resembled the 'reconfigured conveyor belt' of Togg-weiler and Samuels (1995a) with most of the deep water upwelling in the Southern Ocean.

We note that the results in Figure 9 are sensitive to the choice of the water mass boundaries. However, the barrier layer thickness provides a mean to determine the base of the ventilated thermocline in an objective way, for present time as well as during the LGM, independently of the actual sea-water density that may be subject to global and local changes.

An increase in Weddell Sea sea-surface salinity leads to a positive density anomaly, which is transported into the deep ocean. Accordingly, the meridional density gradient grew from Experiment

GA to Experiment GC, and the ACC strengthened (cf. Borowski et al. 2002). In Experiment GC, the additional 1.0 SSS anomaly imposed on the Weddell Sea accelerated the ACC to almost 150% of its modern transport, as compared to Experiment M (Table 4). At the same time, the ACC broadened and its northern boundary approached Cape of Good Hope. The southward penetration of the Agulhas Current was reduced, and consequently, in Experiment GC the Agulhas leakage was diminished to only 40% of that found in Experiment M or GA, in agreement with the theoretical and modeling studies reviewed by de Ruijter et al. (1999). The model resolution was too coarse to generate baroclinic eddies such as Agulhas rings, thus only a broad westward advective transport was simulated, plus some westward diffusive transport of heat due to the isopycnal and eddy transport parameterizations (cf. England and Garçon 1994).

The maximum of the meridional heat transport in the Atlantic Ocean (Figures 12 and 13) amounted to only half of the estimates by e.g. Hall and Bryden (1982) and Klein et al. (1995). This could also be due to the coarse horizontal resolution of our model, which yielded a weak Gulf Stream with only half the observed transport. The injection of warm subtropical Indian Ocean Water into the southeastern corner of the Atlantic Ocean contributed significantly to the northward heat transport across 30°S (cf. Gordon 1986). Our value of 0.66 PW for the zonal heat transport by the Agulhas Current at 20°E in Experiment M compares well with the high-resolution model results of Thompson et al. (1997, -0.51 PW) and Biastoch and Krauss (1999, 0.87 PW). In Experiment GC, the same value was reduced to 0.26 PW or 40%, in agreement with the reduction in the volume transport associated with the Agulhas leakage. Similarly, our values for the integrated heat fluxes across the meridional sections at 68°W and 20°E (Figure 13) come close to those derived from present-day hydrographic data (1.41 and 1.19 PW according to Sloyan and Rintoul 2001), but are too small across the zonal section at 30°S (compared to 0.28 PW after Sloyan and Rintoul 2001).

Since 60% of the water flowing northwestward with the Benguela Current stemmed from the Agulhas Current, the warm water route proposed by Gordon (1986) dominated the NADW return flow in Experiment M. In contrast, in Experiment GC only 25% of the Benguela Current waters were of Indian Ocean origin, which indicates that the cold water route suggested by Rintoul (1991) became the main path for NADW renewal. We note that in our prognostic model the warm water path could be reconciled with a realistic transport of the ACC, which was not possible in the inverse model of Rintoul (1991). Consistent with the cold water path, a significant amount (14.9 Sv in Experiment M and 7.5 Sv in Experiment GC) of intermediate water that entered through Drake Passage was modified to thermocline water in the southeast Atlantic Ocean (Figure 15 - cf. England and Garçon 1994).

In Experiments M and GC, the Atlantic Ocean as a whole converted about the same amount of thermocline and intermediate water to deep and bottom water. Experiment M yielded an inflow from the Pacific Ocean of 73 Sv in the upper three levels (seasonal thermocline, ventilated thermocline and intermediate water), but only 47 Sv in the lower two levels (deep water and bottom water). For the outflow to the Indian Ocean, the respective transports were 62 Sv and 58 Sv. The net downwelling in the Atlantic Ocean of about 11 Sv was roughly equal to the rate of NADW formation (Figure 11). Except for higher horizontal transports with the ACC (96 Sv and 79 Sv inflow, 85 Sv and 90 Sv outflow, 11 Sv downwelling), this scenario applies as well to Experiment GC.

However, in the Southern Ocean, the path taken by the deep water flow was quite different in the two experiments (bold arrows in Figure 15): In Experiment M, most of the NADW directly joined the AABW at about 50°S, but in Experiment GC, it first upwelled into the seasonal thermocline and then downwelled in the Weddell Sea to contribute to the formation of AABW, to eventually fill the deep Atlantic and Indian Ocean basins. This is clearly visible from the areas of upward and downward transport in the Weddell Sea (Figure 10b, d and f): The deep convection caused by the large additional salinity anomaly was balanced by upward motion just northward of the high-salinity region.

Finally, we can speculate that the slighty (~10%) more intense Benguela Current in Experiment GC (Figures 8a, d and 9) could lead to a northward shift of the Angola-Benguela Frontal Zone by a few degrees and a reduced import of nutrients into the northern Benguela System (Berger and Wefer 2002). With regard to silicate, in the South Atlantic subtropical gyre today there is a maximum that can be identified with Upper Circumpolar Water, which lies below AAIW because of its much higher density. This water is enriched in silicate over AAIW because AAIW originates from silicate-poor surface waters, and over NADW because of the admixture of high-silicate Pacific and Antarctic waters (cf. Talley 1996 and references therein). North of 20°S, the density of the silicate maximum is only slightly higher than that of the salinity minimum, because it is truncated from below by NADW. Again we can speculate that today, in a situation that resembles our Experiment M, this water is partly mixed into SACW, possibly in upwelling cells such as those visible in Figure 10c, e and g just east of Cape Horn and within Drake Passage, and is advected with the Benguela, South Equatorial and Angola Currents to become a source for the silicate that is consumed in the Namibian upwelling region. In a situation akin to Experiment GC, the silicate maximum might be truncated from above by the upwelling of NADW in the Southern Ocean, and its silicate might be lost to the circulation with the South Atlantic subtropical gyre. Hence a circulation scheme as depicted in Figure 15b could possibly contribute to the resolution of the Walvis Opal Paradox: the fact that during glacial times of increased upwelling and organic matter supply to the sea floor, the flux of diatoms and other siliceous plankton remains was decreased (Berger and Wefer 2002). It would further resemble the pattern proposed by Michel et al 1995 to explain the glacial $\delta^{13}C$ distribution, in which Subantarctic Mode water filled the deep ocean basins at the LGM.

Oxygen-18

Our model carried $\delta^{18}O_w$ as an additional passive tracer. This tracer was restored to the sparse and scattered modern annual-mean observations by Schmidt et al. (1999). Thus there were no surface fluxes over the vast areas of the ocean that were void of data. As far as the surface ocean is concerned, the model acted like an interpolation or extrapolation method using advective and diffusive fluxes (Takahashi et al. 1997). The $\delta^{18}O_w$ tracer became even more artificial in the glacial experiments, but still served the purpose to highlight the effects of changes in advection and diffusion.

In spite of these limitations, the $\delta^{18}O_w$ distribution in Experiment M compared well to the GEOSECS data (Birchfield 1987). The changes in the $\delta^{18}O_w$ distribution exhibited by the glacial experiments were clearly related to changes in the meridional overturning circulation. For example, NADW was isotopically enriched because at the LGM the bulk of it was formed further to the south (cf. Figures 11 and 16). Similarly, the drastic changes of the $\delta^{18}O_w$ of AABW reflected the changes in the location and intensity of convection in the Weddell Sea. Experiments GA and GB were characterized by only weak convection in the southern Weddell Sea, such that the bottom water $\delta^{18}O_w$ values were dominated by mixing with heavier water that originated from the North Atlantic Ocean. In Experiment GC, the high Weddell Sea SSS anomaly caused intense convection leading to AABW $\delta^{18}O_w$ values below zero, comparable to those of Experiment M. Besides the influence of NADW, these different bottom water $\delta^{18}O_w$ ratios were linked to a latitudinal shift of the AABW formation region between the glacial experiments. Neither the location nor the rate of AAIW formation changed much, and therefore the $\delta^{18}O_w$ of AAIW was about 0.1-0.2‰ in all experiments.

Changes in the $\delta^{18}O_c$ distribution were caused to a large extent by temperature changes. Below 2000 m, Experiments GA to GC reflected the increasingly colder deep and bottom waters. The difference in AABW $\delta^{18}O_c$ between Experiment GC and M amounted to 0.5‰, consistent with a cooling of about 1.0°C (not shown). The differences between Experiments GC and M were roughly consistent with reconstructed glacial-Holo-cene anomalies in benthic foraminiferal $\delta^{18}O$ and a global ice-volume effect of about 1.2‰ (Matsu-moto and Lynch-Stieglitz 1999). In agreement with Matsumoto and Lynch-Stieglitz (2001), they did not show an appreciable shift of the polar front in the Southern Ocean.

Comparison to CLIMAP

Based on the GLAMAP reconstruction, our model experiments resulted in a global mean SST anomaly of about 1.4 °C. According to the CLIMAP reconstruction, the present-day global mean SST is about 1.6°C higher than at the LGM. The smaller difference implied by GLAMAP is due to less sea-ice cover and distinctly higher SST in the high latitudes of the North Atlantic Ocean. In the low and mid latitudes, GLAMAP is colder than CLIMAP (see Figure 5 and the related discussion in Schäfer-Neth and Paul this volume). In conjunction with the available $\delta^{18}O_c$ measurements, the higher SST yielded a higher SSS and sea-surface density over most of the subpolar and polar North Atlantic Ocean, where it may cause deep convection and the formation of a relatively dense NADW. This was indeed the case in all our glacial model experiments.

Other Model Simulations

Combing global and regional modeling, we describe in a previous study (Schäfer-Neth and Paul 2001) experiments based on the temperature reconstruction by Weinelt et al. (1996), with similarly seasonally ice-free Nordic Seas and a sea-surface density field comparable to our present study. With the help of a global carbon-cycle box model, Schulz and Paul (2003) study the effect of these GLAMAP-like sea-surface conditions on the high-latitude $\delta^{13}C$ distribution (Experiment "LGMW"in Schulz and Paul 2003; see also Matthies et al. this volume for a coupling to the Hamburg Ocean Carbon Cycle Model). Compared to $\delta^{13}C$ reconstructions for the eastern North Atlantic Ocean, they find a better correspondance than in a case with CLIMAP ice cover (Experiment "LGMC"), because the water on the southern flank of the Greenland-Iceland-Scotland ridge is still ventilated from the north, although at a smaller rate. In our experiments, roughly half of the NADW was formed at convection sites south of 48°N where the overturning reached only 1500 m depth (Figure 11) and the

convection was limited to 1000–1200 m depth (Figure 3); the rest still sank at or even north of the ridge.

There is a large number of other model simulations of the LGM ocean based on a wide variety of OGCMs, which for example employ air temperature and freshwater flux from an atmospheric model (Lautenschlager and Herterich 1990; Bigg et al. 1998), reconstructions of SST and SSS (Seidov et al. 1996; Winguth et al. 1999) or an energy balance model for the temperature boundary condition (Fieg and Gerdes 2001). Two coupled model studies were carried out with intermediate complexity models (Ganopolski et al. 1998; Weaver et al. 1998). Hewitt et al. (2001) were the first to present a multicentury simulation of the LGM with a three-dimensional atmosphere-ocean general circulation model. Further coupled model experiments were performed by Kitoh et al. (2001) and Shin et al. (2003). Some of these model studies obtain a meridional overturning circulation weaker, others stronger than today. This is consistent with our finding that, given the high sea-surface density in the northern North Atlantic Ocean as implied by the GLAMAP SST, the meridional overturning circulation could have been similar to today. In contrast to the present study, none of these other model simulations have so far focussed on the upper ocean hydrology and circulation.

In Experiments GA–GC, the subpolar sea-surface density gradient (Figure 4) occured over a smaller and more southern range of latitudes than in Experiment M, which reflected the southward shift as well as the more zonal path of the glacial Gulf Stream and North Atlantic Drift. It is this pattern and the still active although reduced exchange of water masses between the northeastern Atlantic Ocean and the Nordic Seas that compares particularly well with the LGM results of the coupled model by Hewitt et al. (2001). They also find that there is still some weak convection in the Norwegian Sea, and in the midlatitudes even more heat than at present is transported by the strong subpolar gyre circulation.

In a different coupled model, the NCAR CCSM by Shin et al. (2003), the tropics show a glacial cooling of 3°C over land and 2°C over the ocean. Thus the SST is about 1°C lower than in the

CLIMAP reconstruction, which also applies to GLAMAP (see again Figure 5 in Schäfer-Neth and Paul this volume). With respect to the SSS in the Weddell Sea, they find a large anomaly of about 1.5, excluding the global salinity increase of 1. As we anticipated in our series of model experiments GA-GC, this salinity anomaly under sea ice is mainly due to the release of brine during winter sea-ice formation and the subsequent sea-ice export to lower latitudes. Its magnitude is even larger than in our most extreme case (Experiment GC with a Weddell Sea SSS anomaly of 1.0).

Conclusions

With its high vertical resolution, low vertical diffusivity and isopycnal mixing, our ocean model realistically simulated the circulation of the thermocline and intermediate waters as well as the characteristics of the deep ocean. In particular, the vertical diffusion of only 0.1 cm^2 s^{-1} in the upper 2000 m allowed for very low rates of upwelling from the deep ocean into the tropical thermocline and thus a more intact conveyor belt. Together with the isopycnal mixing parameterization, we achieved a better representation of the subsurface flow field than in our previous modeling studies (Paul et al. 1999; Schäfer-Neth and Paul 2001).

In our ocean model, the ventilated thermocline was cooler during the LGM than today by 2–3°C in the North and 4–5°C in the South Atlantic Ocean. In the tropics, its depth was reduced by up to 150 m. In the North Atlantic Ocean, the outcrop locations of the thermocline isopycnal surfaces migrated southward by 5°–10°, and the ventilation increased. In the South Atlantic Ocean, the ventilated thermo-cline shoaled to the southwest of Cape of Good Hope. Correspondingly, the mixed-layer and thermocline water masses were dominated by cold water that originated from near Drake Passage as opposed to warm water from the Indian Ocean. In the equatorial region (5°S–5°N), the east-west slope of the depth of the mixed layer and the permanent thermocline increased, which was accompanied by a general cooling of about 2°C.

Given the high density implied by the GLAMAP SST reconstruction for the surface of the northern North Atlantic Ocean, the modeled

formation and export of NADW across the equator was nearly unchanged. The flow of AABW depended strongly on the SSS in the Weddell Sea. In the glacial experiment with the largest Weddell Sea salinity anomaly, it was of similar strength as in the modern experiment, but reached further north. The size of the Weddell Sea salinity anomaly also influenced the strength of the ACC. A stronger ACC flow led to a more intense Agulhas retroflection and a leakage of warm Indian Ocean water westward into the South Atlantic Ocean that was reduced by up to 60% with respect to its modern value. Hence the cold water route became the primary path for NADW renewal.

Based on the comparison of our modern and LGM experiments, we speculate that a slighty more intense Benguela Current might lead to a northward shift of the Angola-Benguela Frontal Zone and, in conjunction with enhanced upwelling of NADW in the Southern Ocean, to a reduced import of silicate into the Benguela System, which could partially resolve the Walvis Opal Paradox (Berger and Wefer 2002).

In the LGM experiment with the largest Weddell Sea SSS anomaly, the AABW $\delta^{18}O_w$ values were comparable to those of the modern experiment. Although $\delta^{18}O_w$ was restored to the same surface values and only reflected changes in advection and diffusion, the resulting $\delta^{18}O_c$ distribution came close to reconstructions based on fossil shells of benthic foraminifera.

Acknowledgments

We thank Guy Delaygue and an anonymous referee for their constructive comments that helped to improve our manuscript considerably. This research was funded by the Deutsche Forschungsgemeinschaft (DFG) as part of the Sonderforschungsbereich 261, No. 374, and DFG Research Center Ocean Margins of the University of Bremen, No. RCOM 0083.

References

Berger WH, Wefer G (1996) Expeditions into the past: Paleoceanographic studies in the South Atlantic. In: Wefer G, Berger WH, Siedler G, Webb DJ (eds) The South Atlantic: Present and Past Circulation.

Springer, Berlin, pp 363-410

Berger WH, Wefer G (2002) On the reconstruction of upwelling: Namibian upwelling in context. Mar Geol 180: 3-28

Biastoch A, Krauss W (1999) The role of mesoscale eddies in the source regions of the Agulhas Current. J Phys Oceanogr 29: 2303-2317

Bigg GR, Wadley MR, Stevens DP, Johnson JA (1998) Simulations of two last glacial maximum ocean states. Paleoceanography 13: 340-351

Bigg GR, Rohling EJ (2000) An oxygen isotope data set for marine water. J Geophys Res 105: 8527-8535

Birchfield GE (1987) Changes in deep-ocean water $\delta^{18}O$ and temperature from the last glacial maximum to the present. Paleoceanography 2: 431-442

Borowski D, Gerdes R, Olbers D (2002) Thermohaline and wind forcing of a circumpolar channel with blocked geostrophic contours. J Phys Oceanogr 32: 2520-2540

Boyle EA (2000) Is the ocean thermohaline circulation linked to abrupt stadial/interstadial transitions? Quat Sci Rev 19: 255-272

Bryan K, Lewis LJ (1979) A water mass model of the world ocean circulation. J Geophys Res 84: 2503-2517

Clarke PU, Mix AC (2002) Ice sheets and sea level of the Last Glacial Maximum. Quat Sci Rev 21: 1-7

CLIMAP Project Members (1981) Seasonal reconstructions of the Earth's surface at the Last Glacial Maximum. Geological Society of America, Map and Chart Series MC-36, 18 p

De Ruijter WPM, Biastoch A, Drijfhout SS, Lutjeharms JRE, Matano RP, Pichevin T, van Leeuwen PJ, Weijer W (1999) Indian-Atlantic interocean exchange: Dynamics, estimation and impact. J Geophys Res C 104: 20,885-20,910

Duncan CP, Schladow SG, Williams WG (1982) Surface currents near the Greater and Lesser Antilles. Int Hydrogr Rev 59: 67-78

Duplessy JC, Labeyrie L, Juillet-Leclerc A, Maitre F, Duprat J, Sarnthein M (1991) Surface salinity reconstruction of the North Atlantic Ocean during the last glacial maximum. Oceanol Acta 14: 311-324

Duplessy J-C, Labeyrie L, Paterne M, Hovine S, Fichefet T, Duprat J, Labracherie M (1996) High latitude deep water sources during the Last Glacial Maximum and the intensity of the global oceanic circulation. In: Wefer G, Berger WH, Siedler G, Webb DJ (eds) The South Atlantic: Present and Past Circulation. Springer, Berlin, pp 445-460

England MH, Garçon V (1994) South Atlantic circulation in a world ocean model. Ann Geophys 12: 812-825

Epstein S, Buchsbaum R, Lowenstam HA, Urey HC (1953) Revised carbonate-water isotopic temperature scale. Geol Soc Am Bull 64: 1315-1325

Fieg K, Gerdes R (2001) Sensitivity of the thermohaline circulation to modern and glacial surface boundary conditions. J Geophys Res 106: 6853-6867

Ganopolski A, Rahmstorf S, Petoukhov V, Claussen M (1998) Simulation of modern and glacial climates with a coupled model of intermediate complexity. Nature 391: 351-356

Gent PR, McWilliams JC (1990) Isopycnal mixing in ocean circulation models. J Phys Oceanogr 20: 150-155

Gordon A (1986) Interocean exchange of thermohaline water. J Geophys Res 91: 5037-5046

Hall MM, Bryden HL (1982) Direct estimates and mechanisms of ocean heat transport. Deep-Sea Res 29: 339-359

Harms S, Fahrbach E, Strass VH (2001) Sea ice transports in the Weddell Sea. J Geophys Res C 106: 9057-9073

Hewitt CD, Broccoli AJ, Mitchell JFB, Stouffer RJ (2001) A coupled model study of the last glacial maximum: Was part of the North Atlantic relatively warm? Geophys Res Lett 28: 1571-1574

Huang RX, Qiu B (1994) Three-dimensional structure of the wind-driven circulation in the subtropical North Pacific. J Phys Oceanogr 24: 1608-1622

Kalnay E et al. (1996) The NCEP/NCAR reanalysis project. Bull Am Met Soc 77: 437-471

Kitoh A, Murakami S, Koide H (2001) A simulation of the Last Glacial Maximum with a coupled ocean-atmosphere GCM. Geohys Res Lett 28:2221-2224

Klein B, Molinari R, Siedler G, Müller TJ (1995) A transatlantic section at 14.5°N: Meridional volume and heat fluxes. J Mar Res 53: 929-957

Krauss W (1986) The North Atlantic Current. J Geophys Res 89: 3407-3415

Lautenschlager M, Herterich K (1990) Atmospheric response to ice age conditions: Climatology near the Earth's surface. J Geophys Res D 95: 22,547-22,557

Large WG, Danabasoglu G, Doney SC, McWilliams JC (1997) Sensitivity to surface forcing and boundary layer mixing in a global ocean model: Annual-mean climatology. J Phys Oceanogr 27: 2418-2447

Ledwell JR, Watson AJ, Law CS (1993) Evidence for slow mixing across the pycnocline from an open-ocean tracer release experiment. Nature 364: 701-703

Levitus S (1982) Climatological atlas of the World Ocean. NOAA Prof. Paper No. 13, 173 p

Matsumoto K, Lynch-Stieglitz J (1999) Similar glacial and Holocene deep water circulation inferred from southeast pacific benthic foraminiferal carbon isotope composition. Paleoceanography 14: 149-163

Matsumoto K, Lynch-Stieglitz J (2001) Similar glacial and Holocene Southern Ocean hydrography. Paleoceanography 16: 445-454

Melles M (1991) Late Quaternary paleoglaciology and paleoceanography at the continental margin of the southern Weddell Sea, Antarctica. Ber Polarforsch, Bremerhaven, Germany, 81, 190 p

Michel E, Labeyrie LD, Duplessy JC, Gorfti N, Labracherie M, Turon JL (1995) Could deep Subantarctic convection feed the world deep basins during the last glacial maximum? Paleoceanography 10: 927-942

Mix AC, Bard E, Schneider R (2001) Environmental processes of the ice age: Land, oceans, glaciers (EPILOG) Quat Sci Rev 20: 627-658

NCAR Data Support Section (1986). NGDC ETOPO5 global ocean depth & land elevation, 5-min. Data Set 759.1, National Center for Atmospheric Research, Boulder, Colorado

Pacanowski RCE (1996) MOM 2. Documentation, User's Guide and Reference Manual. Technical Report 3.2, GFDL Ocean Group, GFDL, Princeton, New Jersey

Peterson RG, Stramma L (1991) Upper-level circulation in the South Atlantic Ocean. Prog Oceanogr 26: 1-73

Paul A, Mulitza S, Pätzold J, Wolff T (1999) Simulation of oxygen isotopes in a global ocean model. In: Fischer G, Wefer G (eds) Use of Proxies in Paleoceanography: Examples from the South Atlantic. Springer, Berlin, pp 655-686

Paul A, Schäfer-Neth C (2003) Modeling the water masses of the Atlantic Ocean at the Last Glacial Maximum. Paleoceanography 18: doi: 10.1029/ 2002PA 000783

Peltier WR (1994). Ice age paleotopography. Science 265: 195-201

Pether J (1994) Molluscan evidence for enhanced deglacial advection of Agulhas water in the Benguela Current off southwestern Africa. Palaeogeogr Palaeoclimatol Palaeoecol 111: 99-117

PMIP (1993) Paleoclimate modelling intercomparison project, http://www-pcmdi.llnl.gov/pmip/newsletters/newsletter02.html.

Rintoul SR (1991) South Atlantic interbasin exchange. J Geophys Res 96: 2675-2692

Sarnthein M, Gersonde R, Niebler S, Pflaumann U, Spielhagen R, Thiede J, Wefer G, Weinelt M (2003) Preface: Glacial atlantic ocean mapping (GLAMAP-2000). Paleoceanography 18: doi: 10.1029/2002PA 00769

Schäfer-Neth C (1998) Changes in the seawater-oxygen isotope relation between last glacial and present: Sediment core data and OGCM modelling. Paleoclimates 2: 101-131

Schäfer-Neth C, Paul A (2001) Circulation of the glacial Atlantic: A synthesis of global and regional modeling. In: Schäfer P, Ritzrau W, Schlüter M, Thiede J (eds) The northern North Atlantic: A changing environment. Springer, Berlin, pp 441-462

Schmid C, Siedler G, Zenk W (2000) Dynamics of intermediate water circulation in the subtropical South Atlantic. J Phys Oceanogr 30: 3191-3211

Schmidt GA, Bigg GR, Rohling EJ (1999) Global seawater oxygen-18 database. http:// www.giss.nasa.gov/data/o18data, Goddard Institute for Space Studies, New York

Schulz M, Paul A (2003) Sensitivity of the ocean-atmosphere carbon cycle to ice-covered and ice-free conditions in the Nordic Seas during the Last Glacial Maximum. Palaeogeogr Palaeoclimatol Palaeoecol, in press

Seidov D, Sarnthein M, Stattegger K, Prien R, Weinelt M (1996) North Atlantic ocean circulation during the Last Glacial Maximum and a subsequent meltwater event: A numerical model. J Geophys Res C 101: 16,305-16,332

Shannon LV, Nelson G (1996) The Benguela: Large scale features and processes and system variability. In: Wefer G, Berger WH, Siedler G, Webb DJ (eds) The South Atlantic: Present and Past Circulation. Springer, Berlin, pp 163-210

Shin S-I, Liu Z, Otto-Bliesner B, Brady EC, Kutzbach JE, Harrison SP (2003) A simulation of the Last Glacial Maximum climate using the NCAR-CCSM. Clim Dyn 20: 127-151

Slowey NC, Curry WB (1995) Glacial-interglacial differences in circulation and carbon cycling within the upper western North Atlantic. Paleoceanography 10: 715-732

Sloyan BM, Rintoul SR (2001) The Southern Ocean limb of the global deep overturning circulation. J Phys Oceanogr 31: 143-173

Sprintall J, Tomczak M (1992) Evidence of the barrier layer in the surface layer of the Tropics. J Geophys Res C 97: 7305-7316

Takahashi T, Feely RA, Weiss RF, Wanninkhof RH, Chipman DW, Sutherland SC, Takahashi TT (1997) Global air-sea flux of CO_2: An estimate based on measurements of sea-air pCO_2 difference. Proceedings of the National Academy of Sciences 94: 8292-8299

Thompson SR, Stevens DP, Döös K (1997) The importance of interocean exchange south of Africa in a numerical model. J Geophys Res 102: 3303-3315

Timmermann R, Beckmann A, Hellmer HH (2001). The role of sea ice in the fresh water budget of the Weddell Sea. Ann Glaciol 33: 419-424

Toggweiler JR, Samuels B (1995a) Effect of Drake Passage on the global thermohaline circulation. Deep-Sea Res 42: 477-500

Toggweiler JR, Samuels B (1995b). Effect of sea ice on the salinity of Antarctic bottom waters. J Phys Oceanogr 25: 1980-1997

Tomczak M, Godfrey JS (1994) Regional Oceanography: An Introduction. Pergamon, Oxford, England

Tsuchiya M (1989) Circulation of the Antarctic Intermediate Water in the North Atlantic Ocean. J Mar Res 47: 747-755

Tsuchiya M, Talley LD, McCartney MS (1994) A western Atlantic section from South Georgia Island (54°S) northward across the equator. J Mar Res 52: 55-81

Weaver AJ, Eby M, Fanning AF, Wiebe EC (1998) The climate of the last glacial maximum in a coupled atmosphere-ocean model. Nature 394: 847-853

Weinelt M, Sarnthein M, Pflaumann U, Schulz H, Jung S, Erlenkeuser H (1996) Ice-free Nordic Seas during the Last Glacial Maximum? Potential sites of deepwater formation. Paleoclimates 1: 283-309

Whitworth III. T, Peterson RG (1985) Volume transport of the Antarctic Circumpolar Current from bottom pressure measurements. J Phys Oceanogr 15: 810-816

Williams RG, Spall MA, Marshall JC (1995) Does Stommel's mixed layer "demon" work? J Phys Oceanogr 25: 3089-3102

Winguth AME, Archer D, Duplessy JC, Maier-Reimer E, Mikolajewicz, U (1999) Sensitivity of paleonutrient tracer distributions and deep-sea circulation to glacial boundary conditions. Paleoceanography 14: 304-323

Winter A, Martin K (1990) Late Quarternary history of the Agulhas Current. Paleoceanography 5: 479-486

WOA (1998) World ocean atlas 1998, http://www.nodc.noaa.gov/oc5/woa98.html. National Oceanographic Data Center, Silver Spring, Maryland

Wolff T, Mulitza S, Rühlemann C, Wefer G (1999) Response of the tropical Atlantic thermocline to late Quaternary trade wind changes. Paleoceanography 14: 374-383

Inverse Modelling of the Glacial Atlantic Circulation under Geostrophic Side Conditions

R. Schlotte[1*] and B. Grieger[2]

[1]*Universität Bremen, Fachbereich Geowissenschaften, Klagenfurter Straße,
28359 Bremen, Germany*
[2]*Max-Planck-Institut für Aeronomie, Max-Planck-Str. 2,
37191 Katlenburg-Lindau, Germany*
* *corresponding author (e-mail): r.schlotte@science-computing.de*

Abstract: The inverse ocean model 2RAIOM is presented. It is designed to determine the mean circulation of the glacial ocean from observations of temperature and salinity. Derived from a model code suited for application to the present day ocean, new terms have been added to the objective function to enhance the model's performance if temperature data are sparse. The effects of the new objective function are studied with present day temperature and salinity data. The model is then applied to the Atlantic Ocean during the Last Glacial Maximum. Although the performance of the new model version is improved considerably, the amount of available data is only sufficient to reconstruct the ocean circulation during the Last Glacial Maximum qualitatively.

Introduction

Paleoceanographic considerations mostly come in one of two flavors: The first kind are investigations of a small number of sediment cores. Using counts of foraminiferal tests or measurements of physical or chemical properties they provide time series of proxy data. Oceanographic parameters like sea surface temperature or salinity may then be computed. Due to the small number of cores involved, this kind of investigation usually concentrates on local or regional aspects of paleoceanography. Examples for this experimental approach are presented by Vink et al. (2001) and Baumann et al. (1996). The second kind of investigations use numerical ocean models, seeking a theoretical understanding of the processes at work in the climate system. Most commonly used are general circulation models (GCMs) which simulate the evolution of the physical system in time, starting from a complete set of start parameters. Because regional models include poorly known lateral boundary conditions, global models are preferred. GCMs consume large amounts of computer resources, so their application generally is limited to the simulation of rather short time slices. Examples of the mod-

elling approach are given by Paul et al. (1999) and Schäfer-Neth and Paul (2000).

The combination of these two kinds of investigations is not straightforward, because the input parameters necessary for ocean models like surface heat flux or wind stress are often hard or impossible to observe by sedimentological methods. Some of the GCM output data like sea surface temperature (SST) or salinity (SSS) are available experimentally, so the question arises whether the ocean models can be inverted: In an *inverse* model, the roles of input and output parameters are - at least partially - interchanged with respect to the corresponding *forward* model. Several ocean GCMs have been inverted in order to assimilate data while simulating the present day ocean, for an example see Giering (1996). The resulting inverse models, however, consume at least an order of magnitude more resources than the forward models they are based on (days to weeks of CPU time). For the application to the past ocean, this is not desirable, because the number of data points that need to be assimilated is very low compared to that of the present day ocean. It is more reasonable to construct

From WEFER G, MULITZA S, RATMEYER V (eds), 2003, *The South Atlantic in the Late Quaternary: Reconstruction of Material Budgets and Current Systems.* Springer-Verlag Berlin Heidelberg New York Tokyo, pp 585-599

simpler inverse models that are better adapted to the paucity of data by neglecting the evolution in time, and reconstruct a steady-state ocean only. LeGrand and Wunsch (1995) show an example of this kind of specialized inverse model.

The model used in this work is derived from an inverse model developed for the application to the present day ocean (Schlitzer 1993). While the original model is able to reconstruct the present day circulation from temperature and salinity data, it is not well suited for reconstructing the ocean of past times (Grieger and Schlitzer 1996). It makes use of the fact that temperature T and salinity S are known for almost the complete volume of the present day ocean: All computations for the principle of geo-strophy (connecting pressure gradients to velocities) are based on the input data; this method will herein be called *static geostrophy*. For the past ocean this is not possible, because sedimentological methods do not provide estimates for temperature and salinity of the whole water column, but only at depths populated by microorganisms, i.e. usually close to the sea surface and occasionally at the bottom. In its internal forward part, however, the inverse model computes temperature and salinity for the whole model domain. Therefore, it was possible to extend the model by introducing the so-called *dynamic geostrophy*, which bases its calculations on the internally computed temperatures and salinities. The use of the T and S input data reduces to serve as comparison values during the minimization of the cost function (see below). Therefore the model can cope with sparse and unevenly distributed data.

This paper describes the inverse model 2RAIOM, presenting in more detail the features that are new in this version. Finally, three numerical experiments are presented, demonstrating the application of the model to the present day and the LGM ocean.

Model Description: 2RAIOM

Grid Geometry

The ocean is approximated by an irregular grid of simple boxes. The resolution may be chosen freely, the geometry used in this work for the Atlantic Ocean has a resolution of 2.5-10° horizontally and 60-500 m vertically and is identical to that of Schlitzer (1993), see Fig. 1. The grid consists of n_{box} = 4375 boxes in n_{col} = 294 columns and n_{lay} = 20 layers, which results in n_{surf} = 11836 permeable surfaces. The present-day topography was also used in the experiments concerning the LGM; the global sea-surface depression of about 130 m was neglected.

Flow velocities are defined at the centers of the box surfaces perpendicular to them; the horizontal velocities $|\vec{v}|$ are the first set of independent parameters (n_v = 7461). Vertical velocities are calculated by balancing the transports for any box. The vertical velocity at the top of a box column is identical to the surface freshwater flux given by the difference of evaporation and precipitation. As the (small) surface freshwater fluxes are computed as residuals of (large) horizontal transports it is necessary to provide observational data for the surface freshwater fluxes at the complete sea surface.

Tracer

2RAIOM calculates just two tracers, potential temperature θ and salinity S. The flux of salt through the sea surface is zero everywhere, the sea surface heat fluxes Q are the second set of independent parameters (n_Q = 294, this is equal to the number of box columns). The equation of continuity is stated for every box by summing up the terms for advection and diffusion over all box surfaces. For the boxes in the top layer, the surface heat and freshwater fluxes are taken into account. For every tracer the equations of continuity form a set of linear equations that can be solved using standard procedures.

Advection is implemented using a control parameter $e = 0.7$, this is a mixture of the centered-in-space advection scheme ($e = 0.5$) and the upstream scheme ($e = 1$). Explicit diffusion is represented by the horizontal and the vertical mixing coefficients k_h and k_v, respectively, that are constant throughout the whole model domain. k_h and k_v form the third set of independent parameters.

Schlitzer (1993) reduces the computational cost of the problem by introducing clusters of horizontally neighbouring boxes. Temperature and salinity are constant within any cluster, so the number of

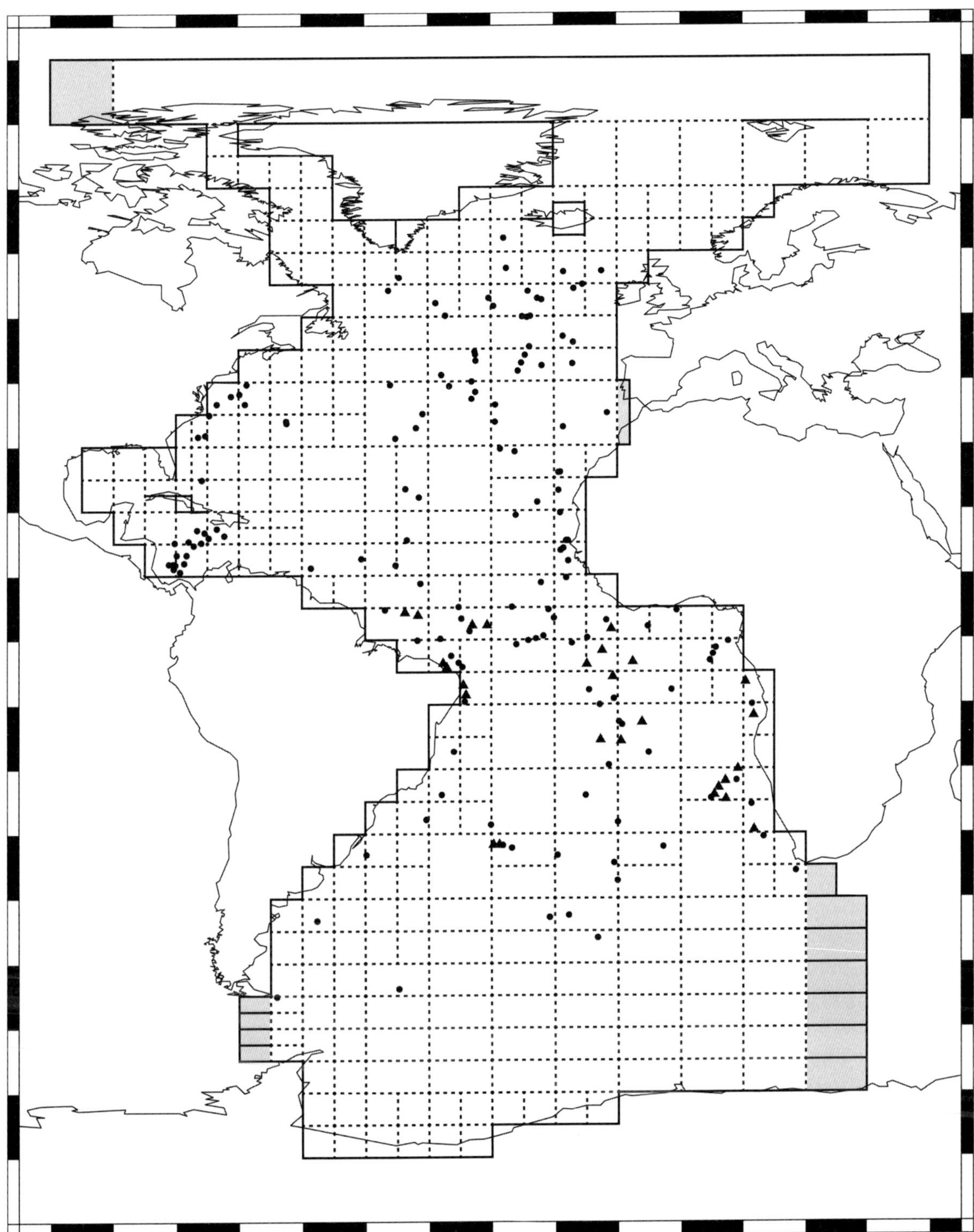

Fig. 1. The model grid used in this work. Solid lines represent box surfaces impermeable to water, dashed ones represent permeable surfaces. Box columns at the open boundaries are shown in gray, boundary conditions for temperature and salinity are necessary there. Dots indicate the locations of the CLIMAP set of temperature data, triangles indicate the locations of additional data provided by Niebler et al. (2003).

unknowns in the equation systems drops from n_{box} = 4375 to $n_{cluster}$ = 1777. The concept of clusters leads to two classes of box surfaces: Those which are located in the interior of a cluster and those which separate two clusters. With this kind of distinction the implementation of the dynamic geostrophy would have been much harder, however, the increase in available computational resources made it possible to drop the concept of clusters.

Objective Function and Optimization

The n_p = 7757 independent parameters $\vec{p}$ (n_v = 7461 horizontal velocities, n_Q = 294 surface heat fluxes, and two mixing coefficients) only implicitly define the ocean state. In the scope of the model the state is completely defined after solving the sets of linear equations for θ and S. Now the quality of the state must be determined by evaluating an objective function $K(\vec{p})$. The many terms of $K(\vec{p})$ may be separated into three classes:

1. measurement of the ocean state's deviation from observed data,
2. measurement of the ocean state's deviation from an inner consistency,
3. additional constraints for stabilization.

Ad 1: Observed data including error estimates for the following quantities may be fed into the model:

- temperature and/or salinity for any grid box,
- surface freshwater flux and/or surface heat flux for any box in the top layer,
- two mixing coefficients,
- vertical shears of transports (for the static geostrophy, see next section).

Additional a priori constraints on mass and heat transports may be imposed, e.g. *transport at Drake Passage is about* 120 Sv. Schlitzer (1993) introduced about 20 a priori constraints, none of them are used in the experiments presented here.

Ad 2: In the original version of 2RAIOM just the linear vorticity balance had to be satisfied approximately. In the current version the approximate satisfaction of the geostrophic equation is demanded, see the next section for details.

Ad 3: The reconstruction problem is mathematically ill-posed (Louis 1989), that means,

- it is not necessary that an ocean state exists that reproduces the observed data (because they may be inconsistent due to measuring errors),
- if such a state exists, it is not necessarily unique (in general, one state must be chosen as the optimum from a manifold of states reproducing the data),
- if a unique state exists, it is possible that it depends very sensitively on the observed data (in this case noise amplification renders the reconstruction useless, even when data errors are small).

For these reasons, a so-called regularization must be performed. In 2RAIOM this is done in two steps: Some of the constraints are weakened, for example, the observed temperature and salinity data do not need to be reproduced exactly, but only within their respective error estimates. However, the conservation laws for mass and energy must still be satisfied exactly. The second step is done by introducing new constraints to the solution: Some quantities like vertical component of velocity, surface freshwater flux and surface heat flux are required to be smooth. This is accomplished by adding terms to the objective function that penalize large second derivatives of the quantities mentioned above.

The problem of reconstructing the ocean state is now reduced to minimizing the objective function $K(\vec{p})$. For very simple (linear) problems this can be done analytically; 2RAIOM, however, requires a numerical-iterative method: Starting from a first guess set of parameters $\vec{p}_0$, the objective function is evaluated and the parameters are changed slightly again and again, until an optimum set of parameters $\vec{p}_{opt}$ is found (see Fig. 2). As a minimization algorithm a quasi-Newton algorithm is used that evaluates not only the objective function $K(\vec{p})$, but also its derivatives with respect to the parameters (Gilbert and Lemaréchal 1989). As a disadvantage this method requires the objective function to be encoded as well as a second function for computing the derivatives. However, the advantage is a large gain in performance: The computational cost of evaluating the derivatives with a dedicated function is in the order of evaluating $K(\vec{p})$ itself (about 5 times slower than $K(\vec{p})$); computing the derivative by a secant method would

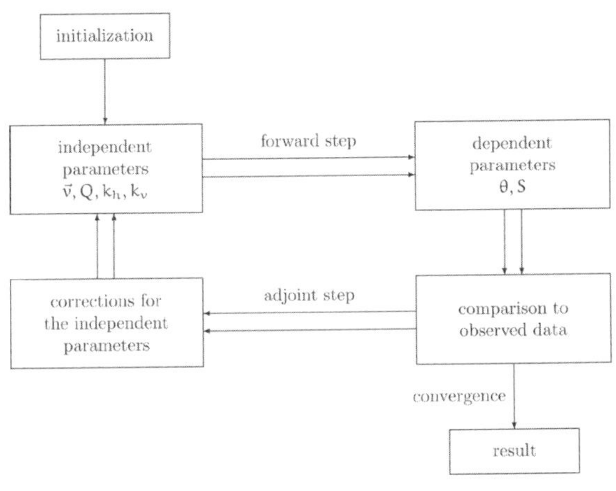

Fig. 2. 2RAIOM's mode of operation: After an initialization the forward step and the adjoint step are repeated again and again, until the comparison to observed data shows acceptable agreement.

require $n_p = 7757$ evaluations of the objective function. Hence, encoding and using the first derivative gains a factor of about 1000 in runtime speed.

Contrary to the method of Schlitzer (1993), all experiments in this work started with an ocean at rest, i.e. at the beginning of the iterations all velocities and the surface heat fluxes are equal to zero, the start values for the mixing coefficients are the data values: $k_h = 5 \cdot 10^2$ m^2/s, $k_v = 0.5 \cdot 10^{-4}$ m^2/s. The first reason for this is a practical one: Determination of a special state to start from as performed by Schlitzer (1993) needs δ and S data for the whole model domain; these data are not available for the ocean of past times. Generally, it is possible to start the LGM runs from an ocean state precomputed with recent data; however, this poses a more theoretical problem: How does the model result depend on the initial state? A perfect optimizer should find the global minimum of the objective function even in the presence of additional, local minima; the algorithm used here cannot accomplish that. It is far more probable that it finds a local minimum not too far away from the initial state; one could say that the model has some kind of memory for the state from which it started the minimization. To avoid problems with this behavior, it was decided to always start from a neutral ocean, the one at rest.

All numerical experiments were conducted with a fixed number of 25,000 iterations, this number being a compromise between time consumption and accuracy. Inspection of the values of the objective function shows that it always decreases by several orders of magnitude very rapidly after the beginning of the optimization, but decreases very slowly just before aborting the iterative procedure, so this procedure seems justified.

Geostrophy

The original version of the model by Schlitzer performs geostrophic calculations exactly once (static geostrophy): Before starting the optimization of the independent parameters, the observed values of θ and S (which have to be available for the whole model domain) are used to compute dynamic heights. Differences in dynamic height between neighboring water columns are then used to calculate pressure gradients. These gradients can be easily transformed to geostrophic velocities. However, because the reference velocity (or the level of no motion) is unknown, the geostrophic velocities may not be compared directly to the flow velocities. Schlitzer solves this problem by computing geostrophic shears, i.e. differences of geostrophic velocities in vertically adjacent box surfaces, and comparing them to the analogously computed shears of flow velocities. The geostrophic shears are treated in similar way as the temperature and salinity data as they are all constant throughout the procedure of optimization.

This flow of computation is possible only if temperature and salinity data are available for (almost) the whole model domain; in particular, the data must be available for the whole water column. This is almost the case for the present day ocean, however, there are only very few observations for the ocean of past times, and these are mostly available for the sea surface only, because they rely on microorganisms living near the sea surface.

The model itself, however, computes in its forward step temperature and salinity within the whole model domain. These values can be used to perform geostrophic calculations anew during every single iteration. 2RAIOM has been extended in this respect by implementing a new term in the objective function:

Salinity, potential temperature and pressure are used to compute the density of sea water according to the equation of state as used by Blumberg and Mellor (1987) in the Princeton Ocean Model (POM). The pressure is calculated by a simple integration over depth in every box column, starting at the sea surface where the pressure is assumed to be zero. Now, for every pair of adjacent box columns the following calculations are performed:

(a) Compute the pressure difference for every pair of horizontally adjacent boxes, i.e. for each depth level. (b) Compute the geostrophic velocity for every depth level, resulting in a velocity profile for a pair of box columns. (c) Compute the misfit between this velocity profile and the profile resulting from the flow velocities in the two columns, allowing for a velocity offset. The velocity offset is chosen to minimize the misfit. This offset bears similarity to the reference level velocity in usual geostrophic calculations; it is constant for a pair of adjacent box columns, however, it may vary from one pair to the next pair, and from one iteration of the optimizer to the next. (d) The misfit is multiplied by the cross section of the two box columns.

The sum of all misfits is multiplied by a weight factor and added to the objective function.

This construction ensures that any deviations from the geostrophic equilibrium are penalized, however, the geostrophic equation does not need to be satisfied exactly. Because the calculations are done repeatedly using the temperature and salinity currently determined by the model itself, the method is called *dynamic geostrophy*, in contrast to the *static geostrophy* of the original version of the model. It is also possible to think of *internal* and *external* geostrophy, respectively, because the former uses the internally computed values for θ and S, the latter, however, requires the input of observed data for the tracers.

Correlation with the Wind
Preliminary experiments showed that 2RAIOM produces completely unrealistic results if all the following conditions are met simultaneously:
- Optimization starts from an ocean at rest.
- No a priori constraints are imposed.
- Dynamic geostrophy is used instead of static geostrophy.
- The number of observed θ- and S-data is small.

In this case the subtropical gyres are lacking, the Equatorial Current flows to the east. The reason for this behaviour is an interaction between the minimization algorithm and the structure of the objective function near the starting state, which leads to a local minimum far from the global one.

To overcome this problem, another new term of the objective function K_W was constructed. The basic idea is that the oceanic surface flows are mostly wind-driven, hence the surface (water) velocity is roughly parallel to the wind direction. (That is not really a contradiction to Nansen's observation and Ekman's explanation of the surface flow being deflected *cum sole* from the wind direction; border effects are neglected in Ekman theory.) The computation works as follows: First the (zonal or meridional) transport is determined for all vertical box surfaces in the top layer, resulting in a set of 545 water transports $\vec{T}_W$; the values of this set change over the course of the optimization. Now the respective component of the 2 m-wind at the center of the box surface is taken and multiplied with the area of the surface, resulting in a second set of 545 fictive air transports $\vec{T}_A$; this set remains constant during the optimization. Then the correlation coefficient $r_{W,A}$ of the two data sets is computed. To get a positive contribution to the objective function that decreases with increasing correlation the sign is changed and 1 is added. Finally a weight factor is applied:

$$K_W = w_W \cdot (1 - r_{W,A}).$$

By this construction the wind velocity does not influence the result directly, only quotients of velocities do. Even if the wind velocities have been higher in the Last Glacial Maximum because of higher temperature gradients, this fact does not influence the model's result; only changes in the wind *patterns* may do so.

Numerical Experiments and Results

The three experiments presented in this work were carried out with the objective function terms for dynamic geostrophy enabled, and the terms for static geostrophy and the explicit a priori transports of Schlitzer (1993) disabled. Values for temperature and salinity at the open boundaries are taken from the recent data set of Schlitzer (1993), which was compiled from several sources. Surface heat fluxes originate from the COADS data set (Woodruff et al. 1987; Oberhuber 1988). Surface freshwater fluxes are taken from Hellerman (1973). For horizontal and vertical mixing coefficients first guess values of $k_h = 5 \cdot 10^2$ m²/s, $k_v = 0.5 \cdot 10^{-4}$ m²/s are used (Olbers et al. 1985; Olbers and Wenzel 1989). By choosing rather large weights for the terms penalizing deviations from these values, it was guaranteed that the mixing coefficients stayed almost constant throughout the optimization.

The purpose of the first experiment, henceforth called FULL, is to prove the concept of dynamic geostrophy. The data set from Schlitzer (1993) was used, containing temperature and salinity data for (almost) the whole model domain; obviously this is only possible for the recent ocean. The term *correlation with the wind* was not used.

The resulting surface circulation is shown in Fig. 3 (left panel). The Antarctic Circumpolar Current (ACC), the Equatorial Current and the subtropical gyre of the South Atlantic are clearly visible. The Gulf Stream and the northern subtropical gyre are less pronounced. The transport at the Drake Passage amounts to 41.1 Sv.

The meridional stream function (see Fig. 4, top panel) shows an overturning in the North Atlantic of about 10.5 Sv and an export into the Southern ocean of about 4 Sv. An inflow of Antarctic Bottom water is visible below 3500 m. Deep water production in the North Atlantic reaches from 55° N up to the North Pole.

The modeled temperatures deviate from the data by 0.90 °C (rms), this is more than the mean error of the temperature data of 0.13 °C.

Experiment FULL shows the right features qualitatively. Quantitatively, however, the circulation is slower than in reality: The North Atlantic overturning is actually in the order of 20 Sv,

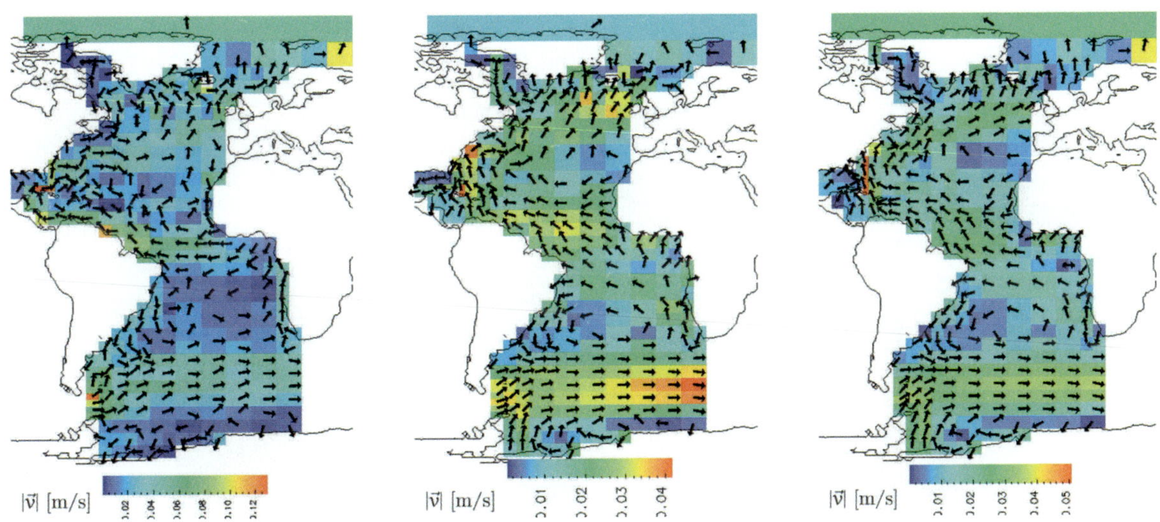

Fig. 3. 2RAIOM's reconstruction of the surface circulation of the Atlantic Ocean. The arrows (all of equal length) show the direction of the motion; the color represents the velocity (note the different scales). **Left:** Experiment FULL for the present day ocean. **Center:** Experiment REC with a reduced amount of data. **Right:** Experiment LGM using data of the Last Glacial Maximum.

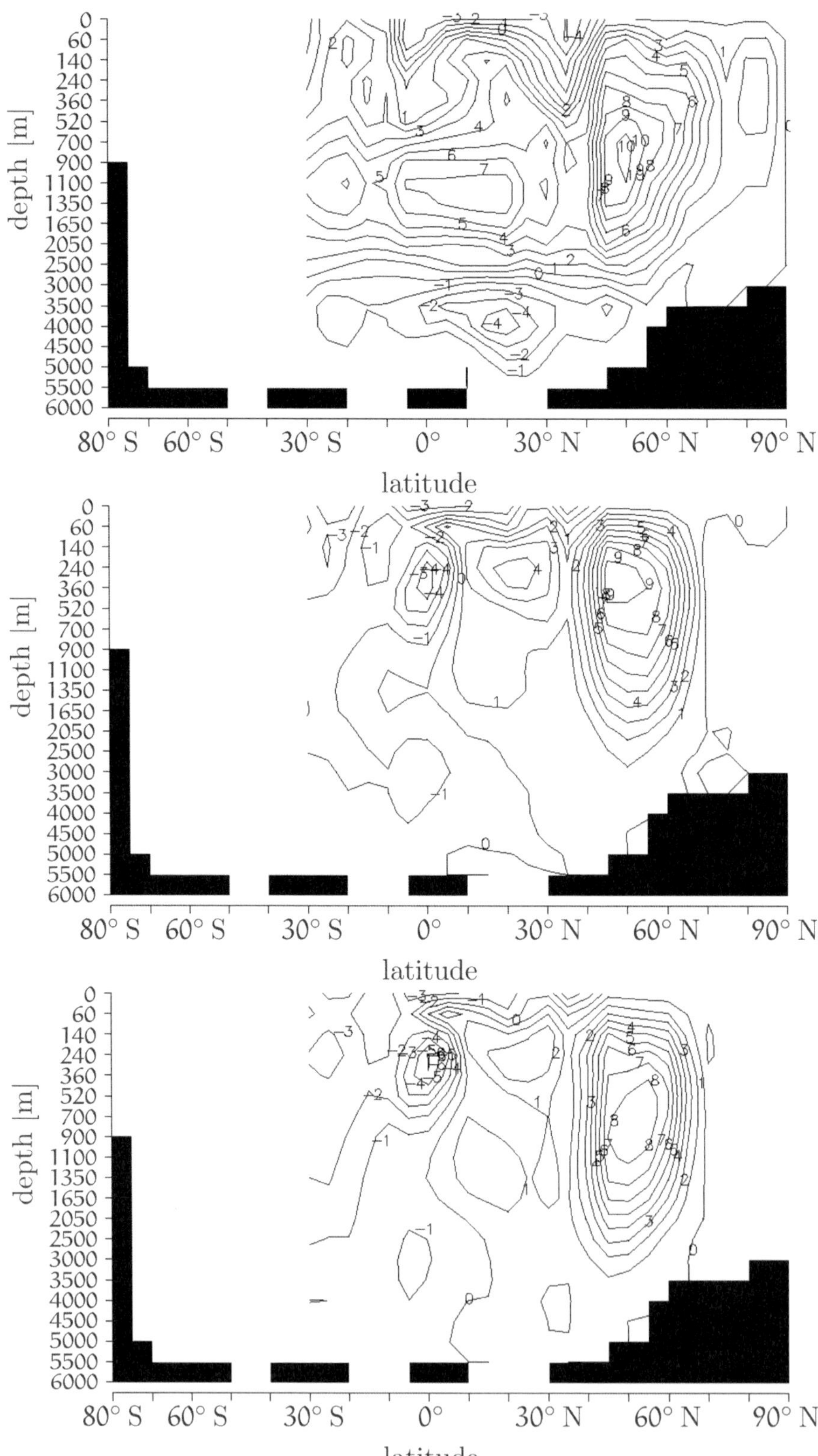

Fig. 4. Meridional stream function of the Atlantic Ocean. Top panel: experiment FULL. Center panel: experiment REC. Bottom panel: experiment LGM.

whereas the transport in the Drake Passage amounts to about 120 Sv. This result is not unexpected: The regularizing terms of the objective function are constructed in order to favor smooth circulations without pronounced features. Together with the start of the optimization with an ocean at rest, the model is bound to yield the smoothest and slowest circulation that is consistent with the data. On the other hand, the transport at Drake Passage of about 40 Sv is quite high compared to numerical experiments using static geostrophy instead of dynamic geostrophy (not shown here): Using static geostrophy, the transport at Drake Passage always stays rather low (a few Sverdrups), unless an a priori constraint is imposed on it. Model run FULL shows that these terms might be unnecessary.

Experiment REC uses the same recent data set for temperature and salinity as experiment FULL, however, the amount of data is reduced to the amount that is available for the LGM from the CLIMAP (1981) data set. No salinity data are input (except at the open boundaries), temperature data are given for only 83 boxes instead of 4345. These 83 boxes are distributed unevenly over the model domain. They are all located in the top layer and in tropical to mid latitudes due to methodical and technical limitations (see Fig. 1). The error estimates of the temperature data are replaced by the estimates used in the third (LGM) run, see below. On account of the paucity of data and the change in the error estimates, the weight factor for the term *deviation from temperature data* of the objective function had to be increased by 100. The term *correlation with the wind* is used in experiment REC, the wind field stems from a model run of the atmospheric general circulation model ECHAM (Lorenz et al. 1996). In all other respects experiment REC is identical to FULL. The resulting top layer circulation is shown in Fig. 3 (center panel). The dominant feature is the ACC; the subtropical gyres are slightly more pronounced than in experiment FULL. The Equatorial Current has an evident component to the north.

The meridional stream function (see Fig. 4, centre panel) shows a slight decrease of the North Atlantic overturning to 9.5 Sv. The North Atlantic Deep Water (NADW) production is now limited to latitudes lower than 70° N. Neither an export to

the Southern Ocean, nor an inflow of AABW are visible. Altogether, there are less features visible than in experiment FULL.

The modeled temperatures now deviate from the data by 0.34 °C (rms), indicating a slightly too strong adjustment to the data: The mean error estimate of the data amounts to 0.74 °C, the error of the Modern Analogue Technique (MAT) itself is at about 0.95 °C (Wefer et al. 1999). These numbers show that it is not desirable to increase the weight factor penalizing deviations from temperature data any further: The model would then adapt the solution to the current set of measurement errors instead of finding an optimum overall solution.

The reduction of the circulation in experiment REC is a consequence of the reduced amount of data: As there are less structures in the input data to reproduce, the model can satisfy all constraints with a solution that shows less structures itself. In particular, the deep ocean's circulation is not explicitly constrained by any data located there. The surface circulation, however, is now bound to the patterns of the wind field, the subtropical gyres therefore are better reconstructed than in experiment FULL.

The experiment LGM is an application of 2RAIOM to the Last Glacial Maximum. The setup is identical to experiment REC, except for the recent temperature and wind field data being replaced by data applying to the Last Glacial Maximum. The main data source is CLIMAP (1981) as reanalyzed by Prell (1985) using the Modern Analogue Technique. 153 of CLIMAP's 261 cores lie within the model domain. This data set is slightly enlarged by the 26 temperature estimates provided by Niebler et al. (2003) and Grieger and Niebler (2003). The total of 179 core locations fall into 86 different grid boxes. The arithmetic mean of the temperature values was calculated for boxes containing more than one core location. The obvious method to obtain the essential error estimates is to calculate the standard deviation of the temperature values in each single box. However, the standard deviation is a proper error estimate only if the values are distributed normally (Gaussian), and if the number of data values is large enough. These conditions are not very well met here. Therefore, the error estimate for any box providing temperature data is set to the reciprocal value of the number of core locations belonging to the box. By this method, any box containing a single core location gets an error estimate of 1 °C, which corresponds to the error estimate of the Modern Analogue Technique that is about ±0.95 °C.

The resulting surface circulation of experiment LGM is very similar to that of experiment REC (see Fig. 3, right panel), however, the North Atlantic Current is directed more to the east than in experiment REC, the Gulf Stream is located nearer to the coast of North America and reaches a higher speed. An Angola Current is indicated in experiment LGM but not visible in experiment REC.

The meridional stream function of experiment LGM (see Fig. 4, bottom panel) shows only small differences as compared to experiment REC. NADW production is reduced to 8.8 Sv.

The reconstructed sea-surface temperatures (SSTs) of all model runs are shown in Figure 5; the SST is assumed to be identical with the temperature in the top layer with a thickness of 60 m.

Experiment FULL (left panel) shows that the model reproduces the data well, except for some boxes with temperatures below -1.8 °C: The effect of sea water freezing to ice is not included in 2RAIOM. Therefore, the model reacts to the large

heat transport into the atmosphere by cooling the water below its freezing-point.

Experiment REC (center panel) yields an SST that is on large scales rather similar to that of FULL. More details are visible in the difference plot (Fig. 6, left panel): Over large areas of the open ocean, the SST difference lies in the range -4 to +6 °C. On the one hand, these numbers are quite large compared to the error estimates for SST reconstruction methods like MAT; on the other hand, the amount of temperature data fed into the model is by far smaller than at experiment FULL, so larger differences do not come unexpectedly.

In the Gulf of Mexico, Baffin Bay and Norwegian Sea much larger SST differences of -30 to +20 °C are evident. No temperature data are prescribed in these regions (see Fig. 1). Additionally, these are margin basins not well connected to the large scale circulation, so the advective heat transport in the basins is small. The model still reproduces the surface heat flux data rather well (see below); the only way to accomplish this task is by diffusive transport, i.e. large temperature differences.

The reconstructed SST difference of experiments LGM and REC (see Fig. 6, right panel) mainly represents the differences in the input tem-

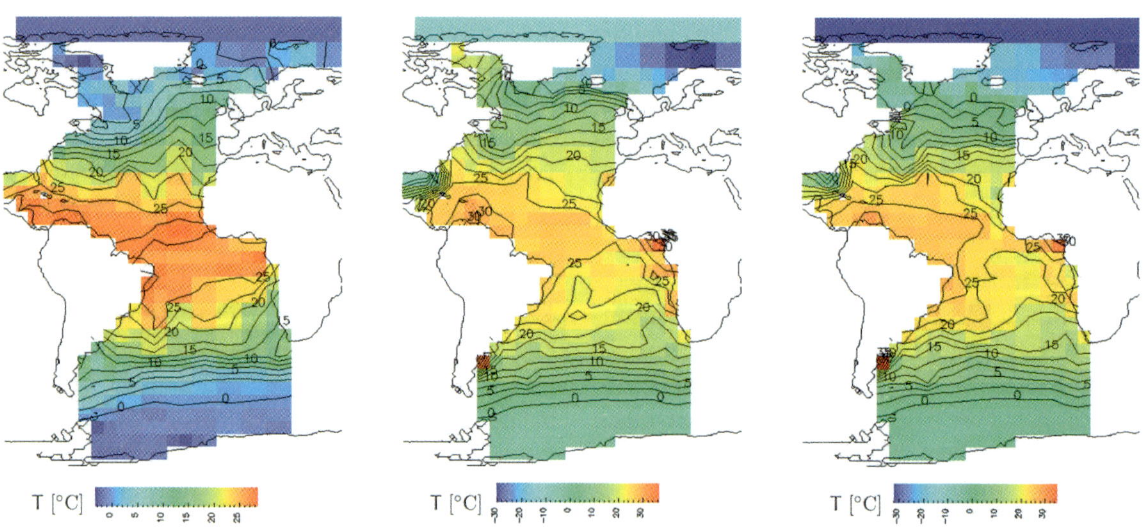

Fig. 5. 2RAIOM's reconstruction of the sea-surface temperature of the Atlantic Ocean. The distance of isotherms is 2.5 °C. **Left:** Experiment FULL applies to the present day ocean. **Center:** Experiment REC with reduced amount of data. **Right:** Experiment LGM with data of the Last Glacial Maximum. Note the different scales.

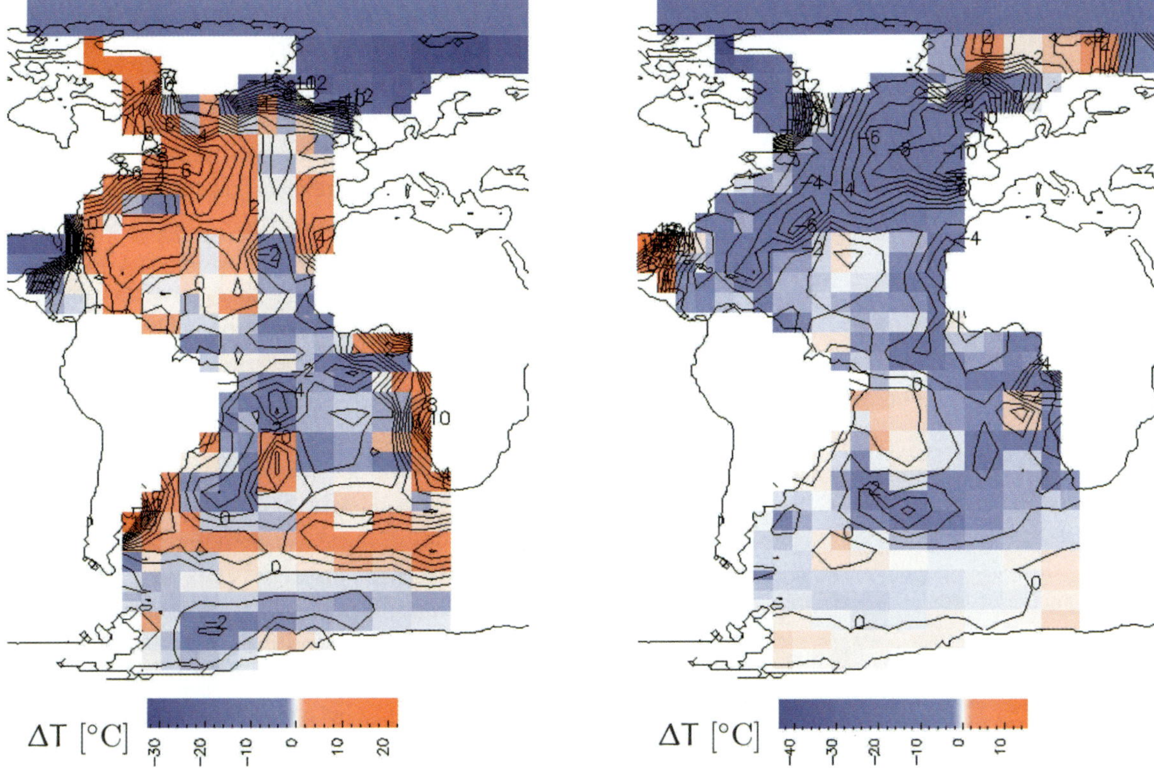

Fig. 6. Differences of reconstructed sea surface temperatures. **Left:** Experiment REC minus FULL. **Right:** Experiment LGM minus REC.

perature data: In the LGM experiment the North Atlantic is colder by up to 10 °C. In the South Atlantic, the SST differences are much smaller, they mostly stay in the range of -2 to +2 °C. The SST difference prescribed by the input data is interpolated smoothly into regions where no input data are available, e.g. in large parts of the Western North Atlantic.

The extrapolation, however, fails in experiment LGM similarly to experiment REC, in that the Norwegian Sea and the Gulf of Mexico are modeled by far too cold and the Baffin Bay too warm. In contrast to this finding, the SST difference in the Southern Ocean is almost negligible, where no data are available, either: This is a consequence of the open boundaries at Drake Passage and toward the Indian Ocean, where temperature data are given for the whole water column.

The zonally integrated meridional heat transport in the Atlantic Ocean can be computed by integrating the surface heat flux, starting at the North Pole. Neglecting the possible (small) offset due to heat transport through the Bering Strait, results for all three experiments are shown in figure 7 together with the input data derived from Oberhuber (1988).

North of 40° N, the agreement of the modeled heat transport with the data is very good for all experiments; south of about 30° N, the modeled transport is too low by an offset which is almost constant for each experiment. Obviously, the disagreement is localized to a small band between 30° N and 40° N. Inspection of the heat flux data shows that the Gulf Stream's sharp maximum in the surface heat flux is not adequately represented in the modeled data. This is a consequence of a regularization term in the cost function, penalizing deviations from a smooth surface heat flux. It seems advisable to reduce this term's weight factor in future experiments.

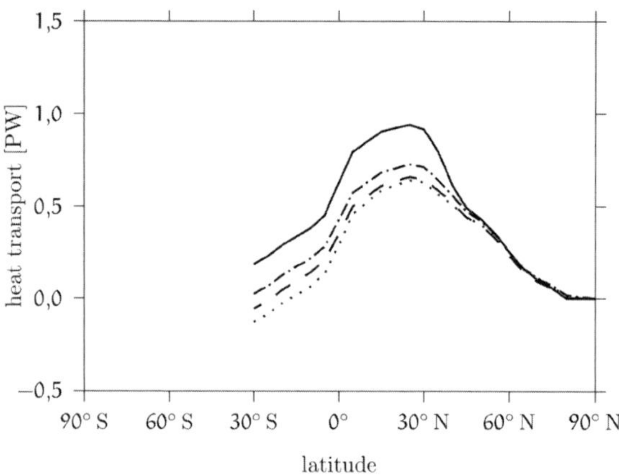

Fig. 7. Meridional heat transport in the Atlantic Ocean. Full line: input data from COADS. Dotted: Experiment FULL. Dashed: Experiment REC. Dot-dashed: Experiment LGM.

Discussion

Reconstructing the oceanic circulation from observational data poses a challenge: Even for the present day ocean, observations of temperature and salinity are not numerous enough to enable the direct calculation of the circulation from the density field. Only a coarse approximation of the density field is known; the problem is mathematically ill-posed, and results are subject to large uncertainties. For the ocean of the geological past, the problem is still harder because of the paucity of data and the even larger error of each single observation.

The numerical experiments performed with 2RAIOM show that it is nevertheless possible to reconstruct an approximation of the ocean with its circulation and distributions of temperature and salinity. It is not necessary to incorporate the deviation from an ocean state near the one expected into the objective function as suggested by Grieger and Schlitzer (1996): The *dynamic* implementation of the geostrophic principle introduced in 2RAIOM proves as a strong constraint which effectively prevents completely unrealistic results, even if the optimization is started from an ocean at rest. In spite of the fact that the physically relevant process of convection is not included in the model, the greatest densities appear at the bottom of the ocean. At the same time, however, dynamic geostrophy works as a "brake" for the circulation, because a homogenic ocean at rest perfectly satisfies the geostrophic condition. The regularizing terms in the cost function have the same effect. In this respect, the model works as constructed: The reconstructed ocean state just shows the features that are necessary to reproduce the observational data; the absence of any feature in the numerical result does not imply its absence in the real ocean, it just indicates that the available data do not contain enough information.

As an aid, the model needs information about the wind stress at the ocean surface, especially if temperature and salinity data for less than the whole model domain are provided. On the one hand, this necessity just represents the fact that the ocean circulation is driven by a combination of thermohaline effects and the surface wind stress. On the other hand, this finding proves that the small number of LGM observations alone does not suffice to constrain even the large scale features of oceanic circulation: In an experiment not shown here observational temperature data for the whole top layer were input to 2RAIOM, this is equivalent to about three times the number of LGM data. Even in this optimistic case did the reconstructed circulation not show any export of NADW into the southern ocean. We conclude that the currently available data do not suffice to answer questions of present day interest like *How much NADW was produced during the LGM?* with a model like 2RAIOM. These findings and conclusions are in agreement with the results of LeGrand and Wunsch (1995): The inverse model used there is structurally similar to 2RAIOM, although it uses a coarser resolution and covers the Atlantic only from 10 to 50° N and below a depth of 1000 m. Additionally the model is designed as a diagnostic model because it necessarily starts the optimization process from a realistic oceanic circulation. LeGrand and Wunsch (1995) showed that both modern circulation and a modern circulation weakened by a factor of 2 are consistent with LGM tracer data. They conclude that the available paleo-data are sufficient to constrain the distribution of water masses, but not sufficient to estimate mass transports.

As stated above, all experiments presented in this work were started with an ocean at rest, making an additional term in the cost function necessary. The weight factor of this term was kept constant during the optimization. It may also be possible to reduce the weight factor, e.g. set to zero after one half of the iterations. The result would then be less influenced by the wind field, for which no good proxies are available.

A different approach is to use a realistic ocean circulation as a starting point, for example the modern circulation (Grieger and Schlitzer 1996), and not to use the term *correlation with the wind*. Using the modern circulation also matches better with the use of modern temperatures and salinities at the open boundaries. However the model results have to be interpreted differently: They show the smallest changes in the circulation that are necessary to render the circulation consistent with the temperature and salinity data.

Another approach is to construct different guesses for the circulation, and use them as starting points. The extent to which the circulation changes in the course of optimization and the resulting values of the cost function then show how well the guesses match the data. The drawback of this approach is an additional burden for the scientist having to make the guesses and evaluate the results. Ideally, this work is better left to the combination of cost function and optimizer.

All these approaches are reasonable as well, but beyond the scope of this work.

2RAIOM seems more suitable as a tool for interpolating punctually available sea-surface temperature data. Whether the inclusion of physical laws justifies the enormous expense on computer resources, in comparison to simple interpolation methods remains to be seen. However, the *extrapolation* into regions where no data are available is prone to large errors. There are several possible solutions to this problem: The obvious one is to increase the amount of available data. The model itself may be enhanced by introducing the notion of freezing: this could be accomplished fairly easily by adding a term to the cost function drastically penalizing any occurrence of boxes with a temperature below the freezing point of sea water. In the context of an inverse model and a long term mean state, the processes of freezing and melting with their influence on salt and heat budgets do not need to be modeled in detail.

Paleoceanographic proxies often are difficult to interpret, because they do not depend on a single oceanic property but on several ones. For example, the oxygen isotope ratio of foraminiferal calcite $\delta^{18}O_c$ is linked by a thermodynamic effect to the isotope ratio of the ambient water $\delta^{18}O_w$ and the temperature T; approximately the equation

$$T = 16{,}5 - 4{,}3 \cdot (\delta^{18}O_c - \delta^{18}O_w) + 0{,}14 \cdot (\delta^{18}O_c - \delta^{18}O_w)^2$$

holds, where T is the temperature in $^\circ$C, $\delta^{18}O_c$ and $\delta^{18}O_w$ are the isotopic composition of calcite and water, respectively, relative to the PDB standard (Urey 1947; Wolff et al. 1999). Now, the value of $\delta^{18}O_w$ is linked to the salinity S, because atmospheric water vapour and precipitation are depleted in ^{18}O and simultaneously free of salt. Consequently $\delta^{18}O_c$ appears as a function of temperature *and* salinity:

$$\delta^{18}O_c = \delta^{18}O_c(T, S).$$

Therefore, any single observation $\delta^{18}\hat{O}_c$ of the isotopic composition does not constrain the possible values of temperature or salinity, it cannot be used without further assumptions for forward ocean models which usually need both parameters. In the context of an inverse model like 2RAIOM, however, a considerable advantage is granted by the freedom in choosing the cost function: It is possible to construct a term penalizing any deviation from the observational value $\delta^{18}\hat{O}_c$ by computing the dependent parameter $\delta^{18}O_c(T, S)$ from the internally known parameters T and S. With this approach, it is possible to utilize by far more proxy data into inverse models than into forward models.

Acknowledgements

We thank Reiner Schlitzer for a copy of the original model code and many helpful hints. We are grateful to our colleagues at the working group *Paleoceanographic Modelling*, esp. André Paul and Michael Matthies. The paper benefited from a review by Pascal LeGrand. This research was funded by the Deutsche Forschungsgemeinschaft (Sonderforschungsbereich 261 at Bremen University, contribution no. 377).

References

Baumann KH, Cepek M, Kinkel H (1996) Coccolithophores as indicators of ocean water masses, surface-water temperature, and paleoproductivity - examples from the South Atlantic. In: Wefer G, Berger WH, Siedler G, Webb DJ (eds) The South Atlantic: Present and Past Circulation. Springer, Berlin, pp 117-144

Blumberg AF, Mellor GL (1987) A description of a three-dimensional coastal ocean circulation model. In: Heaps N (ed) Three-Dimensional Coastal Ocean Models. American Geophysical Union, pp 1-16

CLIMAP Project Members (1981) Seasonal reconstructions of the Earth's surface at the Last Glacial Maximum. Geological Society of America, Map and Chart Series, MC-36: 1-18

Giering R (1996) Erstellung eines adjungierten Modells zur Assimilierung von Daten in ein Modell der globalen ozeanischen Zirkulation. Ph D Thesis, Universität Hamburg

Gilbert JC, Lemaréchal C (1989) Some numerical experiments with variable-storage quasi-Newton algorithms. Mathematical Programming 45: 407-435

Grieger B, Niebler HS (2003) Glacial South-Atlantic surface temperatures interpolated with a semi-inverse ocean model. Paleoceanography 18: 1056, doi: 10.1029/2002PA000773

Grieger B, Schlitzer R (1996) Inverse modelling of the glacial Atlantic circulation system: Investigation of data requirements. In: Wefer G, Berger WH, Siedler G, Webb DJ (eds) The South Atlantic: Present and Past Circulation. Springer, Berlin, pp 411-422

Hellerman S (1973) The net meridional flux of water by the oceans from evaporation and precipitation estimates. Unpublished manuscript

LeGrand P, Wunsch C (1995) Constraints from paleotracer data on the North Atlantic circulation during the last glacial maximum. Paleoceanography 10: 1011-1045

Lorenz S, Grieger B, Helbig P, Herterich K (1996) Investigating the sensitivity of the atmospheric general circulation model ECHAM 3 to paleoclimatic boundary conditions. Geol Rundsch 85: 513-524

Louis AK (1989) Inverse und schlecht gestellte Probleme. Teubner, Stuttgart

Niebler H-S, Arz HW, Donner B, Mulitza S, Pätzold J,

Wefer G (2003) Sea surface temperatures in the equatorial and South Atlantic Ocean during the Last Glacial Maximum (23–19 ka). Paleoceanography 18: 1069, doi: 10.1029/2003PA000902

Oberhuber JM (1998) An atlas based on the COADS data set: The budgets of heat, buoyancy and turbulent kinetic energy at the surface of the global ocean. Report 15, Max-Planck-Institut für Meteorologie, Hamburg

Olbers DJ, Wenzel M, Willebrand J (1985) The inference of North Atlantic circulation patterns from clima-tological hydrographic data. Rev Geophys 23: 313-356

Olbers D, Wenzel M (1989) Determining diffusivities from hydrographic data by inverse methods with applications to the Circumpolar Current. In: Anderson DLT, Willebrand J (eds) Oceanic Circulation Models: Combining Data and Dynamics, volume 284 of NATO ASI Series C: Mathematical and Physical Sciences. Kluwer Academic Publishers, Dordrecht, pp 95-139

Paul A, Mulitza S, Pätzold J, Wolff T (1999) Simulation of oxygen isotopes in a global ocean model. In: Fischer G, Wefer G (eds) Use of Proxies in Paleoceanography: Examples from the South Atlantic. Springer, Berlin, pp 655-686

Prell WL (1985) The stability of low-latitude sea surface temperatures: An evaluation of the CLIMAP reconstruction with emphasis on the positive SST anomalies. Spec Pub TRO25, US Dep of Energy, Washington DC

Schäfer-Neth C, Paul A (2000) Circulation of the glacial Atlantic: A synthesis of global and regional modeling. In: Schäfer P, Ritzrau W, Schlüter M, Thiede J (eds) The Northern North Atlantic: A Changing Environment. Springer, Berlin, pp 441-462

Schlitzer R (1993) An Adjoint Model for the Determination of the Mean Oceanic Circulation, Air-Sea Fluxes and Mixing Coefficients. Habilitation thesis, University of Bremen

Urey HC (1947) The thermodynamic properties of isotopic substances. J Chem Society, pp 562-581

Vink A, Rühlemann C, Zonneveld KAF, Mulitza S, Hüls M, Willems H (2001) Shifts in the position of the North Equatorial Current and rapid productivity changes in the western Tropical Atlantic during the last glacial. Paleoceanography 16: 479-490

Wefer G, Berger WH, Bijma J, Fischer G (1999) Clues to ocean history: A brief overview of proxies. In: Fischer G, Wefer G (eds) Use of Proxies in Paleoceanography: Examples from the South Atlantic. Springer, Berlin, pp 1-68

Wolff T, Grieger B, Hale W, Dürkoop A, Mulitza S, Pätzold J, Wefer G (1999) On the reconstruction of paleosalinities. In: Fischer G, Wefer G (eds) Use of Proxies in Paleoceanography: Examples from the South Atlantic. Springer, Berlin, pp 207-228

Woodruff SD, Slutz RJ, Jenne RL, Steurer PM (1987) A comprehensive ocean-atmosphere data set. Bull Am Met Soc 68: 1239-1250

Palaeoceanographic Changes in the Northern Benguela Upwelling System over the last 245.000 Years as Derived from Planktic Foraminifera Assemblages

A.N.A. Volbers[1,3]*, H.-S. Niebler[3], J. Giraudeau[2], H. Schmidt[3] and R. Henrich[3]

[1]*Bundesanstalt für Geowissenschaften und Rohstoffe (BGR), Stilleweg 2, 30655 Hannover, Germany*
[2]*DGO - UMR 5805 EPOC, Université Bordeaux I, Avenue des Facultés, 33405 Talence, France*
[3]*Universität Bremen, Fachbereich Geowissenschaften, Klagenfurter Strasse, 28359 Bremen, Germany*
* *corresponding author (e-mail): a.volbers@bgr.de*

Abstract: Planktic foraminiferal records from six sediment cores recovered from the Walvis Ridge and the northern Cape Basin indicate changes in the spatial and temporal variability in the degree of upwelling during the past 245 kyrs. During periods of intensified upwelling, northern Benguela upwelling cells were displaced westward and increased in size, covering areas at least three times larger than present day. Distinct upwelling events were recognized during oxygen isotopic stage (OIS) 2 and 3 and oxygen isotopic event (OIE) 4.2, 5.2, 5.4, 5.53, 6.2, 6.4/6.5, and 7.4. During OIS 3, OIE 5.4 and 7.4 the maximum upwelling was recorded around the Namibia/Walvis Bay cells and during OIE 3.1, 5.4, and 6.2 at around Walvis Bay/Lüderitz. During OIE 5.1 and 5.51, upwelling was at its minimum. A good correlation between upwelling events in the northern Benguela region and increases in equatorial seasonality implies that both regions respond to the same mechanism, i.e. probably changes in the trade wind intensity.

Introduction

Equatorial currents, in particular where they interact with the eastern boundary current systems, may have significantly contributed to the bulk of the last glacial maximum (LGM) cooling within the tropics (Mix et al. 1999). During the LGM coastal and equatorial upwelling seem to have intensified compared to today (Sarnthein et al. 1988). In glacial periods, trade winds are assumed to strengthen and to induce intensification of the wind-driven upwelling systems coupled with an increase of global productivity (Sarnthein et al. 1988). This process draws down CO_2 from the atmosphere to the ocean and then contributes to global cooling.

Sediments which underlie the major upwelling centers preserve important information on past variations in the strength and the areal extent of upwelling. Such information is of great relevance to studies on global carbon budgets and their role in climate change (Peterson et al. 1995). Biotic parameters indicate an extension of upwelling and divergence in the equatorial and eastern boundary current regions during the LGM (CLIMAP 1981; Mix 1989), a feature also observed in the sediments from beneath the Benguela Upwelling System (BUS), which this study focuses on. It is one of the four main upwelling regions of the world ocean and plays a major role in global climate, primarily through heat exchange and primary production (e.g. Christensen and Giraudeau 2002). It is driven by atmospheric conditions controlling trade wind intensity and

From WEFER G, MULITZA S, RATMEYER V (eds), 2003, *The South Atlantic in the Late Quaternary: Reconstruction of Material Budgets and Current Systems.* Springer-Verlag Berlin Heidelberg New York Tokyo, pp 601-622

zonality (e.g. Chang et al. 1999). Winds from the southeast blow parallel to the coast and displace warm surface water northward which is being deflected to the left by the Coriolis force. It is replaced by cold and nutrient-rich subsurface water, creating a high productivity regime with characteristic cold-water planktic foraminifera assemblages. As even regional modulations of the upwelling process are clearly reflected by the distribution and frequency of individual taxa (Giraudeau 1993) we use these marine protozoa to reconstruct the spatial and temporal variability in the degree of upwelling during the past 245 kyrs.

Hydrographic Setting

The Benguela Current (BC) forms the eastern boundary current of the South Atlantic Subtropical Gyre and plays a crucial role in the thermohaline circulation. It lies at a major choke point in the global conveyor belt, transporting warm and salty surface-water from the southern to the northern hemisphere. The BC is mainly fed by the South Atlantic Current, which carries subtropical surface waters (Shannon 1985; Stramma and Peterson 1989). It extends from 17°S to 35°S, interacting with the warm Angola Current in the north and the Agulhas Current in the south (Fig. 1). Delivery of warm eddy influx from the Indian Ocean was reported by Nelson and Hutchings (1983) and Lutjeharms (1996).

At about 30°S, the BC splits into the Benguela Oceanic Current (BOC) which flows to the west, and the Benguela Coastal Current (BCC) which moves northward (Stramma and Peterson 1989). The BOC turns to the northwest and feeds the South Equatorial Current (SEC; Stramma and Peterson 1989). The SEC is forced by the trade winds: In boreal summer (June-September) southern trades are strong and SEC speed, equatorial divergence are at maximum. This coincides with lowest temperatures in the Benguela region due to the enhanced coastal upwelling of cold and nutrient-rich water.

A well developed thermal front around the shelf edge separates the BCC from the warmer waters of the subtropical Atlantic (Shannon 1985). The BCC follows the coast until it encounters the An-gola Current which carries warm water southwards (Angola-Benguela Front, ABF; Hart and Currie 1960). During a "Benguela Nino" the tropical eastern Atlantic becomes anomalously warm as a consequence of relaxation in the trade winds. Benguela Ninos occur on average every ten years, when tropical water moves eastwards and southwards along the Namibian coast, displacing the ABF to around 25°S and reducing upwelling at Namibia (Shannon et al. 1986; Boyer et al. 2000; Shannon and O'Toole 1999).

The prevailing southerly and southeasterly winds induce coastal upwelling of cold and nutrient-rich South Atlantic Central Water (SACW; Shannon 1985) in which a small proportion of cold and nutrient-rich Antarctic Intermediate Water (AAIW) might be entrained (Giraudeau 1993). AAIW originates by sinking at the Polar Front and occurs between 450-900 m in the study area (Stramma and Peterson 1989). Additionally, upwelling at the shelf edge was reported by Hart and Currie (1960).

The BCC is subdivided into a northern (17°S to 25°S) and a southern part (25°S to 35°S) which differ in upwelling intensity (Shannon 1985; Lutjeharms and Meeuwis 1987). Upwelled waters originate from a shallower depth in the north compared to the south (Shannon 1985). In the northern Benguela region a decrease in wind stress reduces the intensity of the upwelling compared with the coastal areas of 25°S to 31°S that experience the most intense upwelling and lowest SSTs (Shannon 1985; Lutjeharms and Meeuwis 1987). The Lüderitz upwelling cell (around 25°S, Fig. 2) is regarded as the center of upwelling and displays an average seaward extension of 270 km away from the coast (Lutjeharms and Meeuwis 1987). Compared to the Lüderitz cell, upwelling at Walvis Bay (at around 22°S, average seaward extension 240 km) was at least two times lower during a three year period of study. Changes in the strength and direction of longshore winds can, if prolonged over several months or more, result in periods of anomalously low temperatures along the coast due to increased upwelling or warmer conditions associated with reduced upwelling or downwelling (Shannon and O'Toole 1999). Abnormally cool shelf waters were reported from the northern Benguela region in the early 1980s associated with a prolonged period of stronger than

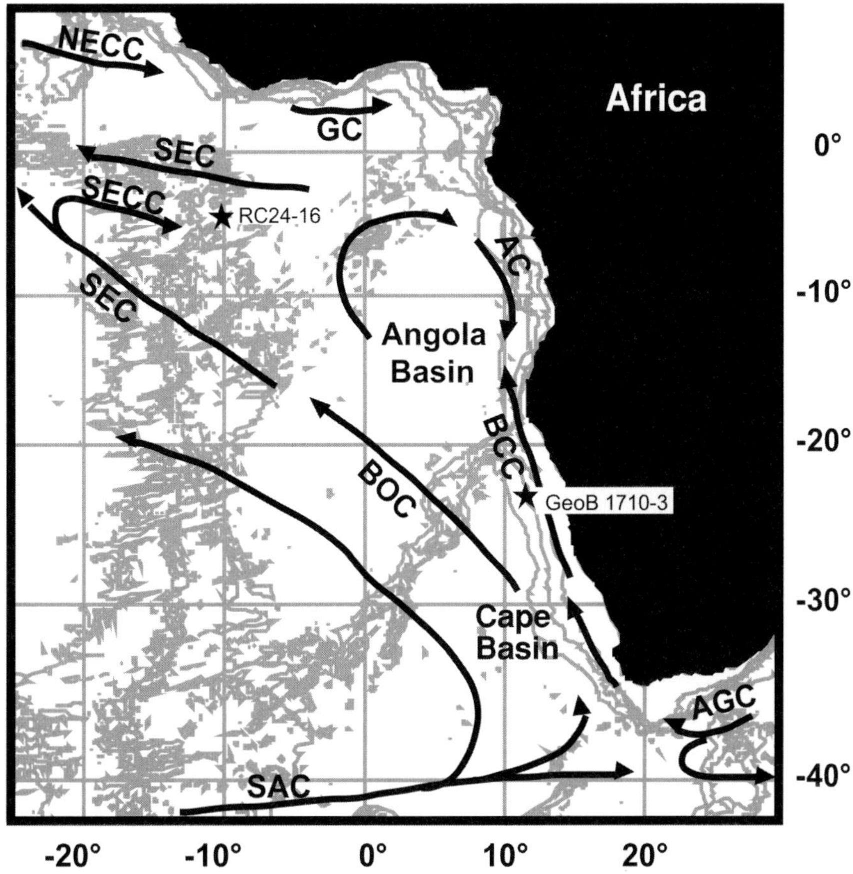

Fig. 1. Map of the South Atlantic surface circulation pattern and position of sediment cores GeoB 1710-3 and RC24-16. (NECC) North Equatorial Counter Current, (SEC) South Equatorial Current, (SECC) South Equatorial Counter Current, (GC) Guinea Current, (AC) Angola Current, (BCC) Benguela Coastal Current, (BOC) Benguela Oceanic Current, (AGC) Agulhas Current, (SAC) South Atlantic Current.

normal longshore winds. Enhanced upwelling in the southern Benguela region during 1993-1994 was a consequence of stronger south-easterly winds during summer and coincided with a wide-spread negative SST anomaly in the South Atlantic (Shannon and Nelson 1996; Shannon and O'Toole 1999).

A broad mixing area, which is an important component of the total upwelling regime today, extends more than 600 km seaward (Lutjeharms and Stockton 1987). Filaments of cold and nutrient enriched waters may even increase the area of high productivity. Hence, the total offshore extent of upwelling may cover a distance up to 1000 km from the coastal sites (Lutjeharms et al. 1991). Sur-

face-water masses of the mixing zone are relatively nutrient-enriched and colder in comparison to the surrounding low productivity regimes of the open ocean (Lutjeharms and Stockton 1987; Shillington et al. 1990).

Distribution of Modern Planktic Foranminiferal Assemblages

Already 100 years ago it was recognized that species distribution of planktic foraminifera is closely linked to surrounding water temperature (Murray and Renard 1891; Murray 1897). Following this

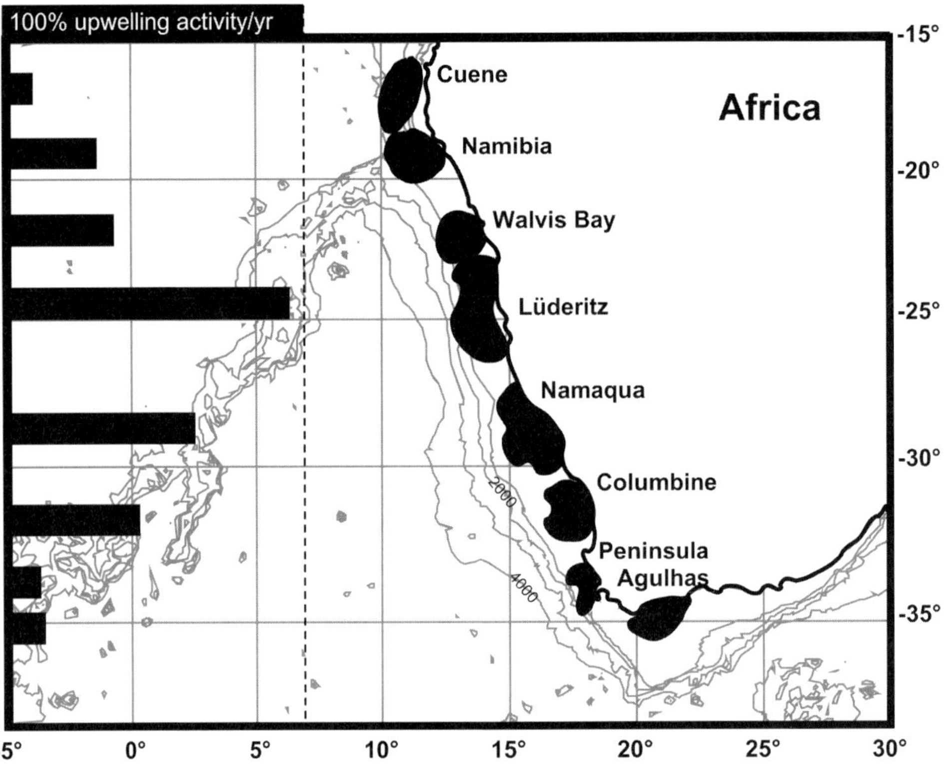

Fig. 2. Geographic location and annual frequency of occurrence of eight identified upwelling cells within the Benguela System (redrawn from Lutjeharms and Meeuwis 1987).

observation, the oceans were subdivided into five planktic foraminiferal provinces: tropical, subtropical, transition, subpolar, and polar (Bè and Tolder-lund 1971). When it became obvious that recent and fossil planktic foraminiferal assemblages varied significantly over time, Imbrie and Kipp (1971) calibrated recent planktic foraminferal assemblages on modern sea-surface temperatures (SSTs) to determine paleo SSTs. Within the CLIMAP project (1976, 1981, 1984) paleo-SSTs of the world ocean were calculated for distinct time periods. These micropaleontologic studies were mostly restricted to the tropics and subtropics.

Niebler and Gersonde (1998) developed a factor model that covered the tropics and the high latitudes of the South Atlantic. However, the coastal upwelling regions were not covered, since the planktic foraminiferal assemblages in the Benguela regions could not be described by this model. As

presented by Volbers et al. (1999), the reconstructed paleo-SSTs for the Benguela region would be <5°C during OIS 3, if their model was used. This is because the upwelling fauna consists of high percentages of *N. pachderma* sin. (Giraudeau 1993; Ufkes and Zachariasse 1993) which was previously regarded as a typical polar species restricted to surface waters below 7°C (e.g. Bé and Tolderlund 1971). North of the Arctic Front, the abundance of this species is greater than 90% in all surface sediment assemblages (Pflaumann et al. 1996). The massive occurrence of *N. pachyderma* sin. in Benguela coastal waters with SSTs above 15°C is surprising (Ufkes and Zachariasse 1993). Usually, the subpolar species *G. bulloides* was found to occur in high abundances in upwelling regions (e.g. NW Africa, Arabian Sea; Thiede, 1975; Prell and Curry, 1981) and although relative percentages >40% were reported locally, it is not the dominant species in surface sediments beneath the Benguela System (on average 18%; Giraudeau 1993). Instead, *N. pachyderma* sin. dominates the foraminiferal assemblages beneath the most intense and frequent upwelling cells with up to 78% (Giraudeau 1993; Lutjeharms and Meeuwis 1987). *G. bulloides* and *N. pachyderma* dex. characterize the mixed environment between the coastal upwelling and offshore oligotrophic regimes (Giraudeau 1993), whereas *Globorotalia inflata* reflects the warmer outer limits of the BUS.

The occurrence of *N. pachyderma* sin. in upwelling regions is not limited to the BUS. Minor percentages of this species were also reported from the Oman and Somalia upwelling areas, where its growth and reproduction seems to be tied to the upwelling period (Ivanova et al. 1999). As mentioned by Ivanova et al. (1999), there is no general consensus on factors controlling the distribution of *N. pachyderma* sin. At least for the Benguela region, Giraudeau and Rogers (1994) report a missing correlation of the dominant planktic foraminifera *N. pachyderma* sin. and *N. pachyderma* dex. with respect to phytoplankton biomass, suggesting that despite the overall high availability of food compared to the open ocean regime, their distribution is primarily controlled by temperature. Its massive abundances of up to >80% in the late Quaternary sediments, called PS events (Little et al. 1997a) were attributed to intensified upwelling, increased productivity, and lower SSTs in the Benguela region (Oberhänsli 1991; Schmidt 1992; Little et al. 1997b; Ufkes et al. 2000).

Factor Analysis

Since global factor models and temperature transfer functions concentrate mainly on oligotrophic or mesotrophic ecological conditions, they commonly exclude upwelling regions because enhanced biological productivity could bias temperature-related transfer function models (Molfino et al. 1982; Chen and Prell 1998; Watkins and Mix 1998). For this reason, upwelling regions are covered by local factor models and local transfer functions (Giraudeau and Rogers 1994; this study). Relevant for this approach in the Benguela region is that the distribution of planktic foraminifera in this area is mainly

temperature-controlled as stated by Giraudeau and Rogers (1994). In this paper we extended the surface data set on which the model of Giraudeau and Rogers (1994) is based.

Our Benguela Model (for details please refer to the material and methods section) was then applied on planktic foraminiferal counts of five cores from the Walvis Ridge in order to determine the degree of upwelling at Namibia and Walvis Bay upwelling cell. To examine upwelling intensity of Walvis Bay and the Lüderitz upwelling cell we introduce planktic foraminiferal counts of GeoB 1710-3 from northern Cape Basin.

Materials and Methods

The Material Used

With respect to the oceanographic configuration mentioned above, the presented core sites from Walvis Ridge are located as follows: DSDP Site 532 is located beneath the Namibia upwelling cell (Fig. 2 and Fig. 3), which is at least three times weaker than the Lüderitz cell today (Lutjeharms and Meeuwis 1987). All other stations (GeoB 1028-5, GeoB 1031-2, GeoB 1032-3, GeoB 1220-1) are located on the Walvis Ridge, an area not directly affected by today's coastal upwelling. Gravity core GeoB 1710-3 was retrieved 280 km off the Namibian coast, beneath the broad-mixing filamentous domain induced by the Walvis Bay and Lüderitz upwelling cells (Fig. 2).

Planktic foraminiferal assemblages of DSDP 532 were counted by Oberhänsli (1991), with at least 300 tests in the >125 µm fraction. *N. pachyderma-N. dutertrei* intergrates (PDI) were counted as *N. pachyderma* dex. Census counts from GeoB 1028-5, GeoB 1031-4, GeoB 1032-3, and GeoB 1220-1 were provided by Schmidt (1992), who examined at least 400 tests >150 µm. PDI were added to *N. pachyderma* dex. Planktic foraminiferal counts from RC13-229 were taken from CLIMAP (1981). For further information please refer to Table 1.

In this study we present planktic foraminiferal census counts for GeoB 1710-3 from northern Cape Basin. For micropaleontological investigations, all samples were freeze-dried, weighed, and

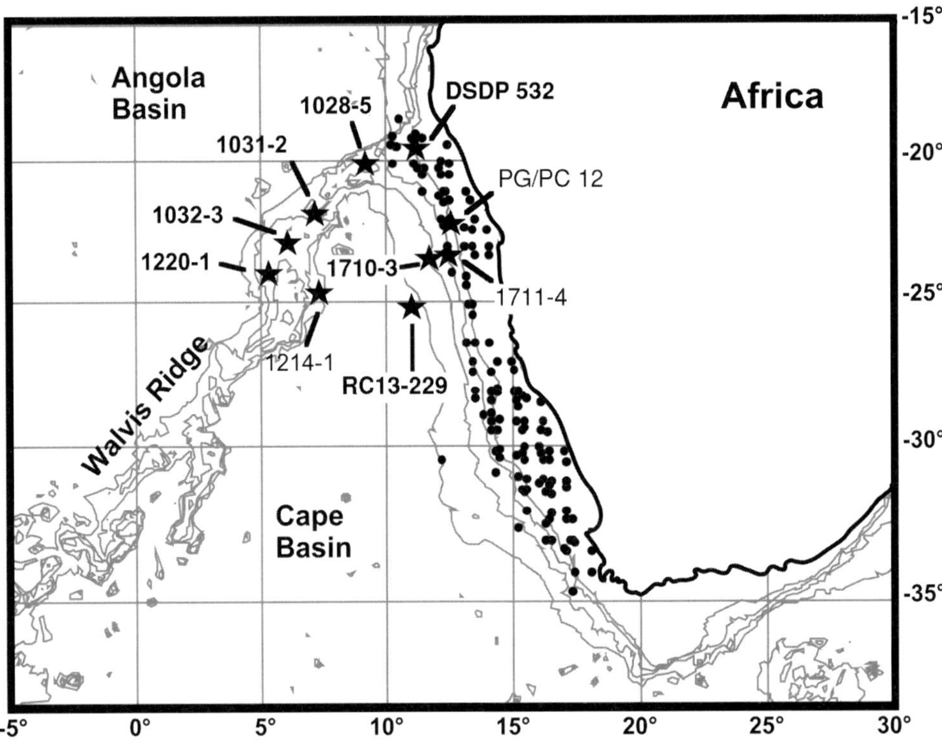

Fig. 3. Stars outline geographic location of investigated (bold) gravity cores. Dots represent surface sediment census counts of planktic foraminifera used in this study (Giraudeau 1993; Niebler 1995).

	Core Location	Water Depth (m)	Age Model, taken from	Census Counts
Walvis Ridge				
DSDP 532	19.44°S, 10.31°E	1331	δ^{18}O *G. bulloides*, distribution of *G. lacunosa*; Oberhänsli 1991	Oberhänsli 1991
GeoB 1028-5	20.11°S, 9.19°E	2209	δ^{18}O *G. ruber* (white) and *G. bulloides*; Müller et al. 1997	Schmidt 1992
GeoB 1031-4	21.88°S, 7.10°E	3105	δ^{18}O *G. ruber* (white) and *G. inflata*; Schmidt 1992	Schmidt 1992
GeoB 1032-3	22.92°S, 6.04°E	2505	δ^{18}O *C. wuellerstorfi*; Bickert 1992	Schmidt 1992
GeoB 1220-1	24.03°S, 5.31°E	2265	δ^{18}O *G. ruber* (white) and *G. inflata*; Schmidt 1992	Schmidt 1992
Cape Basin				
GeoB 1710-3	23.43°S, 11.7°E	2987	δ^{18}O *C. wuellerstorfi*; Bickert and Wefer 1999	this study
RC13-229	25.30°S, 11.18°E	4194	CLIMAP 1982; Oppo and Rosenthal 1994	CLIMAP 1982

Table 1. Sediment cores used in this study.

washed through a 63 µm sieve. The samples were then dry-sieved into the fractions 63-125 µm, 125-250 µm, 250-500 µm, and >500 µm. Census counts were conducted on the >125 µm fraction in order to include small specimen of *N. pachyderma* and *Turborotalita quinqueloba* (Peeters et al. 1999). All samples were split into subsamples using a microsplitter. A minimum of at least 300 non-fragmented planktic foraminiferal specimen were identified using an OLYMPUS SZ 40 microscope. All planktic foraminiferal species were identified using the taxonomic concepts of Hemleben et al. (1989). For the purpose of this paper, right-coiling *N. pachyderma* and *N. pachyderma-N. dutertrei* intergrates (PDI) were grouped as *N. pachyderma* dex. whereas species with a wider opening and at least five chambers were counted as *N. dutertrei.*

Benguela Factor Model

The Benguela Factor Model was based on 135 surface samples from the southwest African continental margin from 17°S to 35°S and 10°E to 18°E (Fig. 3). The surface samples were evenly distributed on the sea-floor and contained twenty four taxa of planktic foraminifera. As usual, 300 individuals from the fraction >125 µm were counted by Giraudeau (1993) and Niebler (1995), these surface samples comprise the ideal reference data set. Species that made up less than 2% in any sample were excluded. We also excluded *Globorotalia hirsuta, Globorotalia scitula,* and *Globorotalia crassaformis* because they generally prefer deep habitats (Lohmann 1992; Ravelo and Fairbanks 1992).

The transfer function was calculated with the program package CABFAC of Klovan and Imbrie (1971) and Imbrie and Kipp (1971). The factor analysis combines a large number of species yielding a smaller number of assemblages (factors). This step provides two matrices: the varimax factor loading matrix, which explains the importance of the individual factor in each sample, while the varimax factor score matrix explains the species importance in each factor. According to Backhaus et al. (1989), loadings >0.4 and scores >0.2 are significant.

The Benguela factor model consists of three planktic foraminiferal assemblages, called "factors", which explain 95,3% of the total variance. The great majority of the samples displays communalities >0.9. Factor 1 with 38.6% of the total variance is defined by *G. inflata* (Table 2). This factor characterizes the outer shelf and the slope of the Benguela region and the Walvis Ridge (open-oceanic assemblage) (Fig. 4). Factor 1 is absent between 22°S and 25°S. The second factor is dominated by *N. pachyderma* sin. with minor contribution of *N. pachyderma* dex. and 34.37% of the total variance. This factor (upwelling assemblage) characterizes the innermost shelf between 21°S and 34°S and underlies the most active upwelling cells (Walvis Bay, Lüderitz, Namaqua, Columbine). Factor 3 consists of *N. pachyderma* dex. and *G. bulloides,* with 22.35% of the total variance. This

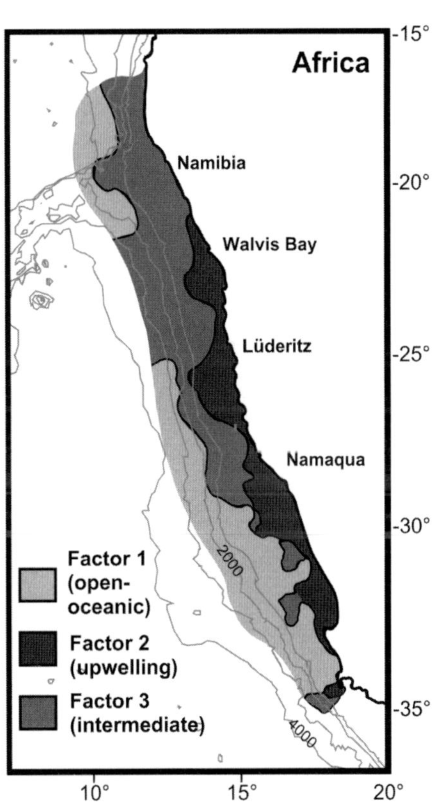

Fig. 4. Composite map of the dominant biogeographic areas of planktic foraminiferal factors produced by Q-mode factor analysis.

Taxa	Factor 1 (open-oceanic)	Factor 2 (upwelling)	Factor 3 (intermediate)
G. siphonifera	0.019	-0.008	-0.008
G. anfracta	0	0	-0.001
G. bulloides	0.124	-0.057	**-0.571**
N. dutertrei	0.065	0.015	-0.05
G. falconensis	0.014	0.017	-0.005
G. glutinata	0.014	-0.008	-0.013
G. inflata	**0.985**	0.033	0.066
P. obliquiloculata	0.002	0.001	0.001
***N. pachyderma* sin.**	-0.021	**0.964**	0.138
***N. pachyderma* dex.**	-0.009	**0.218**	**-0.779**
T. quinqueloba	-0.034	0.122	-0.063
G. ruber	0.043	-0.015	-0.01
G. sacculifer	0.026	-0.003	-0.004
G. truncatulinoides	0.067	-0.01	0
O. universa	-0.018	-0.056	-0.193
Variance (%)	38.6	34.37	22.35
Cummulative variance (%)	38.6	72.97	95.32

Table 2. Varimax factor score matrix derived from Q-mode factor analysis of the planktic foraminiferal census data (135 sediment samples, 13 taxa).

assemblage characterizes the middle shelf and upper slope environment between 17°S to 29°S. South of 29°S it is restricted to small patches associated with Factor 1. This factor represents the mixing of warmer oligotrophic waters with colder waters from coastal upwelling (intermediate assemblage).

Results

As sediment cores taken from beneath the Benguela Upwelling region display characteristic variations in species distribution through time, we show the relative percentages of the most important planktic foraminifera species in an overview (Fig. 5 a, b).

GeoB 1710-3 fauna consisted of eight species which comprise 95% of all planktic foraminiferal species of which *N. pachyderma* dex., *N. pachyderma* sin., and *G. bulloides* are the most abundant. *N. pachyderma* sin. showed rapid oscillations throughout the last 245 kyrs marked by thirteen PS events (Fig. 6). *G. bulloides* recorded the third-highest abundance in the core, and was slightly enhanced during interglacial periods (OIE 5.1 and 5.5). *T. quinqueloba* was most abundant during OIS 6 and 7, whereas *G. inflata* showed two pronounced maximum abundances during warm OIE 5.1 and 5.5. *Globigerinella calida* remained relatively independent of glacial-interglacial conditions, with maximal abundances occurring during the Holocene and during OIS 4, whereas increased values of *Neogloboquadrina dutertrei* rising up to 20% were observed during interglacials. Apart from an abundance peak at the beginning of the Holocene, *Globigerinita glutinata* was evenly distributed in the course of the last 245 kyrs.

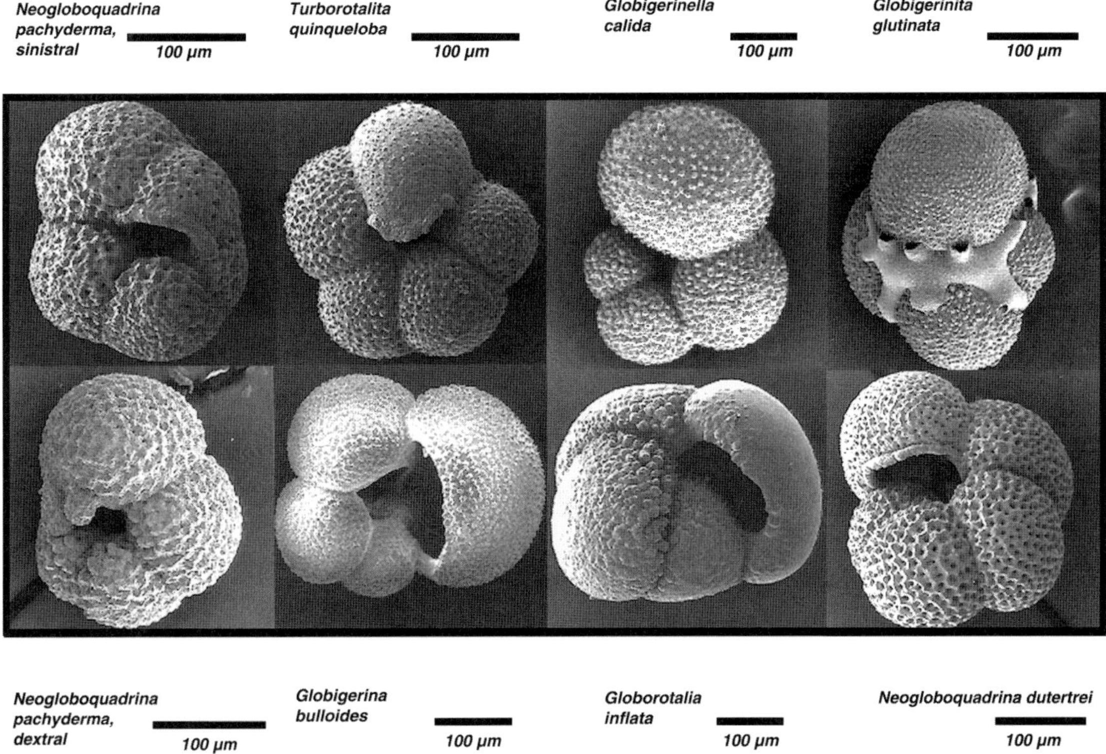

Fig. 5. a) REM photos of the most abundant planktic foraminifera.

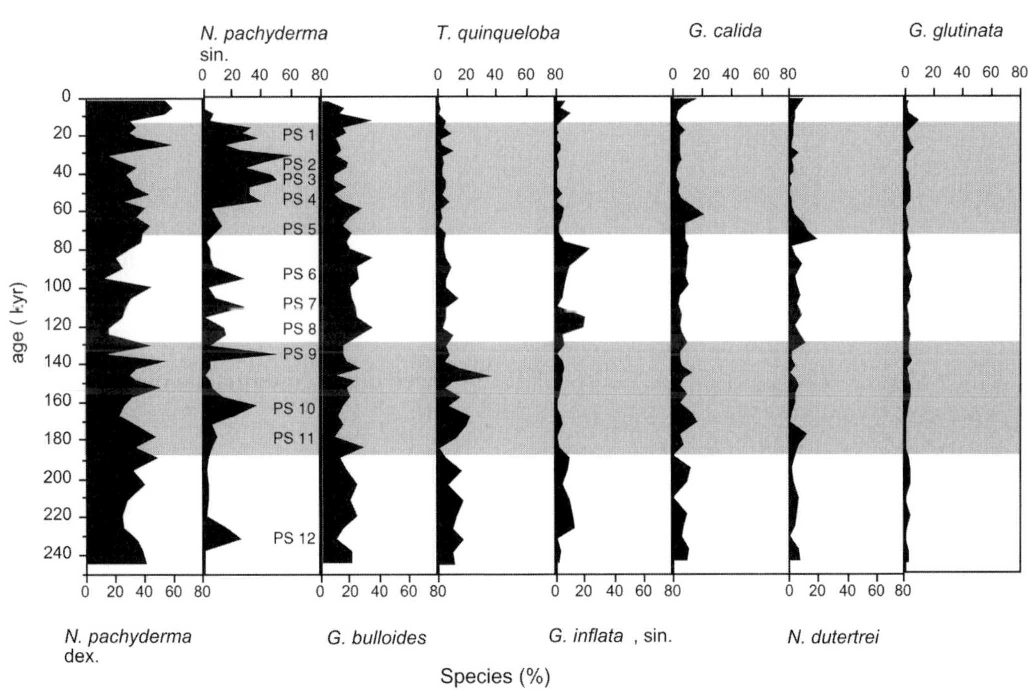

Fig. 5. b) relative abundances of the main planktic foraminifera species of GeoB 1710-3 in %. Peak abundances of *Neogloboquadrina pachyderma* sin. ("PS events") as introduced by Little et al. (1997a) are labelled PS1 – PS12. Grey bars represent glacial periods.

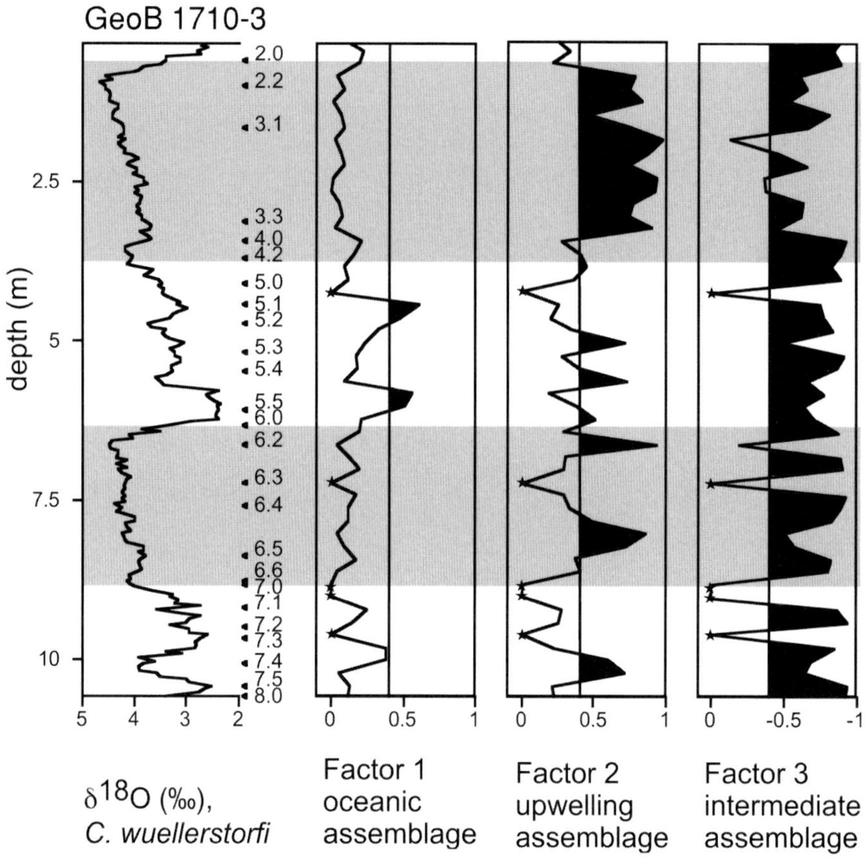

$\delta^{18}O$ (‰),
C. wuellerstorfi

Fig. 6. Graphic representation of GeoB 1710-3 parafactor loading, stars represent no-analog situations. Age model after Bickert and Wefer (1999). Grey bars represent glacial periods.

Sediment Cores from the Cape Basin

Downcore samples which have no core-top equivalents show up as non-analogous-situations. For purposes pursued in this study, non-analogous situations with <20% of one species were accepted, as they still may be interpreted with caution. As carbonate dissolution can strongly alter planktic foraminiferal assemblages, carbonate preservation in all GeoB 1710-3 samples was examined by applying several independent dissolution proxies. Sample 883 cm was strongly altered by carbonate dissolution and therefore excluded from the data set (Volbers and Henrich 2002). RC 13-229 data are limited to interval 2.5-3.19 m, the individual results are not graphically shown.

The intermediate assemblage (Factor 3) explains most associations of GeoB 1710-3 faunal in the last 245 kyrs (Fig. 6). The upwelling assemblage (Factor 2) is the second important factor dominating in during OIS 2 and 3. It is significant during OIE 4.2, 5.2, 5.4, 5.53, 6.2, 6.4/6.5, and 7.4. The upwelling assemblage is also significant at location RC13-229 in OIE 5.4 (2.76 m level, Oppo and Rosenthal 1994) and OIE 5.53 (3.08 m level).

In OIE 5.3 and 5.51 the oceanic assemblage (Factor 1) was significant whereas at RC13-229, the oceanic assemblage dominated over the intermediate assemblage in OIE 5.51 (2,9 m level, CLIMAP 1984) as well.

Sediment Cores from the Walvis Ridge

The five sediment cores from Walvis Ridge comprised a W-E transect, with GeoB 1220-1 furthest to the west and DSDP 532 to the east, only 200 km off the coast of Angola (Fig. 3). DSDP 532 contained sediments ranging up to OIS 13 (Fig. 7). Time resolution compared to all other cores was rather low, therefore high fluctuations of characteristic planktic foraminiferal assemblages may not necessarily be reflected. The upwelling assemblage was most important during OIS 2-4, parts of OIS 5 and throughout OIS 10 to 13. The intermediate assemblage (Factor 3) was strongest during OIS 5 and 6, whereas the oceanic assemblage (Factor 1) was only significant during short time periods of OIS 5 and mid-6. Due to the overwhelming occurrence of *G. falconensis* OIS 7 to 9 are marked by non-analogous situations.

In contrast to DSDP 532, GeoB 1028-5 fauna was dominated by the intermediate assemblage throughout the last 280 kyrs (Fig. 8). The oceanic assemblages were also significant during interglacial periods. The upwelling assemblage was restricted to OIS 2-3, and OIE 5.2/5.3, 5.4, 5.53, 6.2, and 6.5.

GeoB 1031-2 displayed the increasing importance of the oceanic assemblage (Factor 1) throughout the last 250 kyrs. However, the intermediate assemblage remained dominant with exception of OIE 7.4/7.5 (Fig. 9) whereas the upwelling assemblage remained restricted to OIE 3.3 and 5.4, and 7.4.

Further to the west, the upwelling assemblage was absent during the last 280 kyrs (Fig. 10) and foraminiferal assemblages of GeoB 1032-3 were dominated by the intermediate assemblage. The oceanic assemblage was significant during OIE 2.2, 5.1/5.2, 5.4, 7.2.

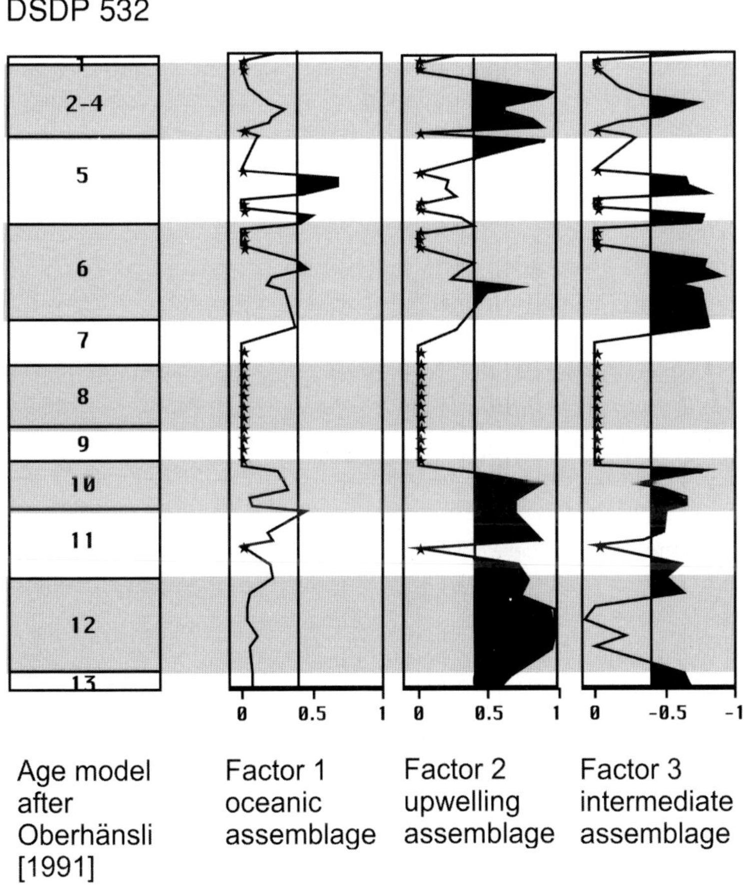

Fig. 7. Graphic representations of DSDP 532 parafactor loading matrix, stars represent no-analog situations. Age model from Oberhänsli (1991). Grey bars represent glacials.

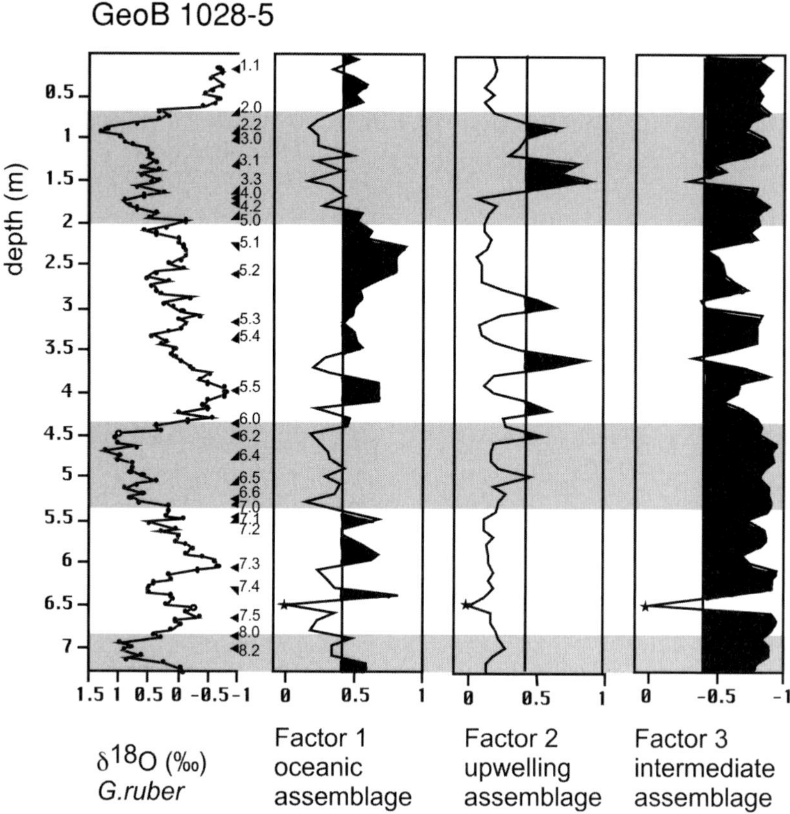

Fig. 8. Graphic representations of GeoB 1028-5 parafactor loading matrix, stars represent no-analog situations. Age model after Müller et al. (1997). Grey bars represent glacial periods.

The increasing number of no-analogoue situations (compare Fig. 10 and 11) clearly shows the limitations of the Benguela Model. GeoB 1032-3 fauna displayed increasing abundances of *Globigerinoides ruber, Globorotalia truncatulino-ides, Globigerinoides sacculifer,* and *Pullenia-tina obliquiloculata.* Furthest to the west, GeoB 1220-1 fauna consisted of even higher percentages of these subtropical-tropical species. Only during glacial periods some (minor) influence of the coastal upwelling was indicated by the significance of the intermediate assemblage. As displayed in Fig. 11, the Benguela Model was only party useful to explain conditions so far away from today's coastal upwelling.

Reconstruction of Northern Benguela Upwelling Intensity

Over the last 245 kyrs, the surface waters over the core locations were predominantly derived from variable mixing of the warm oligotrophic offshore waters with the cool, upwelled waters from the coast. This is reflected by the overwhelming dominance of the intermediate assemblage (Factor 3, compare Fig. 6-Fig. 11). Over long periods in the past 245 kyrs, the areal extent of northern Benguela upwelling cells was mainly restricted to the proximity of the African coast due to the low upwelling intensity seen today. During OIE 5.3, and 5.51 the upwelling assemblage was not even significant in the innermost cores.

In contrast, increased upwelling intensity displaced the boundary between the eutrophic and mesotrophic water masses to the west as the upwelling cells widened. During periods of increased upwelling the upwelling assemblage (Factor 2) was also significant in the cores retrieved from a more westward section of Walvis Ridge. Our data indicate that these distinct intervals of considerably larger extension were independent from glacial-interglacial cycles. We focus on the

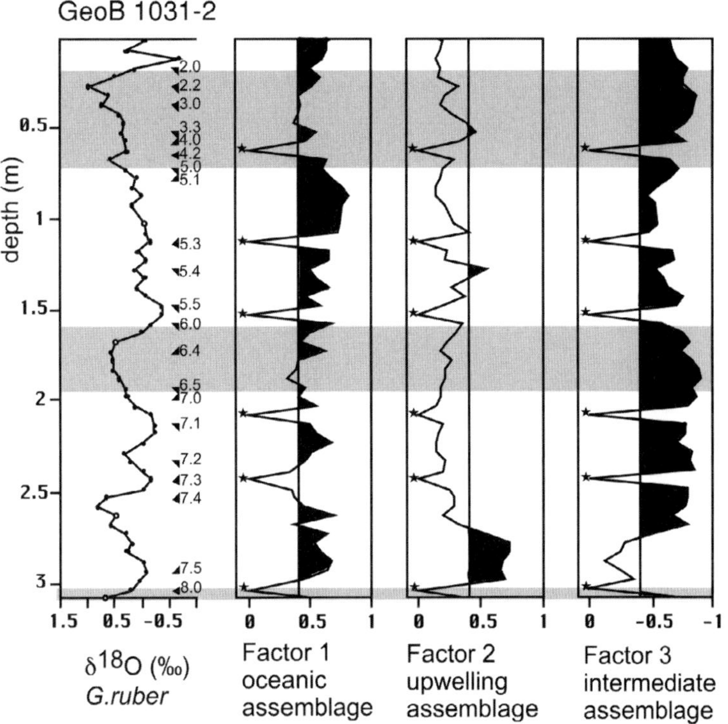

Fig. 9. Graphic representation of GeoB 1031-2 parafactor loading matrix, stars represent no-analog situations. Age model from Schmidt (1992). Grey bars represent glacial periods.

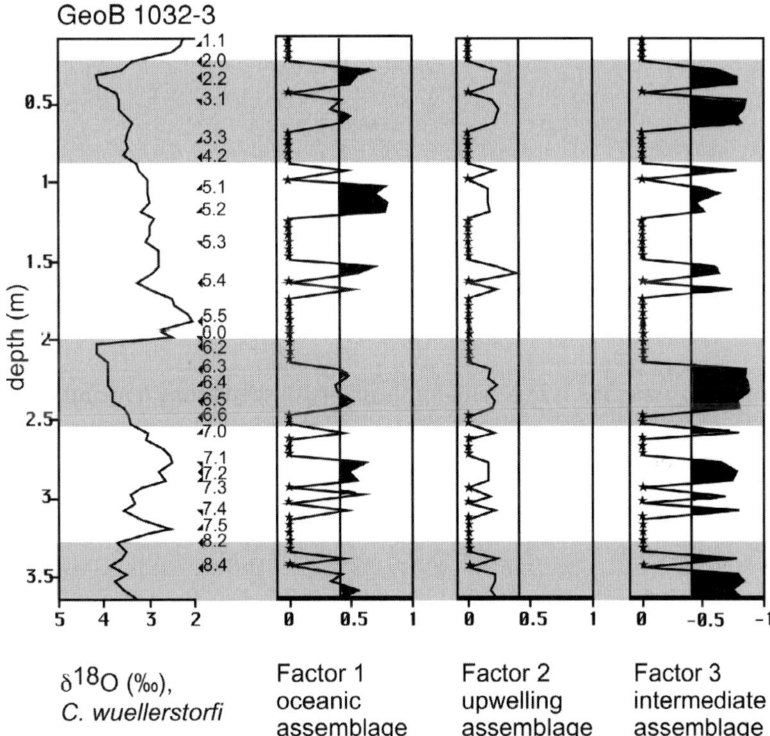

Fig. 10. Graphic representation of GeoB 1032-3 parafactor loading matrix, stars represent no-analog situations. Age model after Bickert (1992) and Bickert and Wefer (1996). Grey bars represent glacial periods.

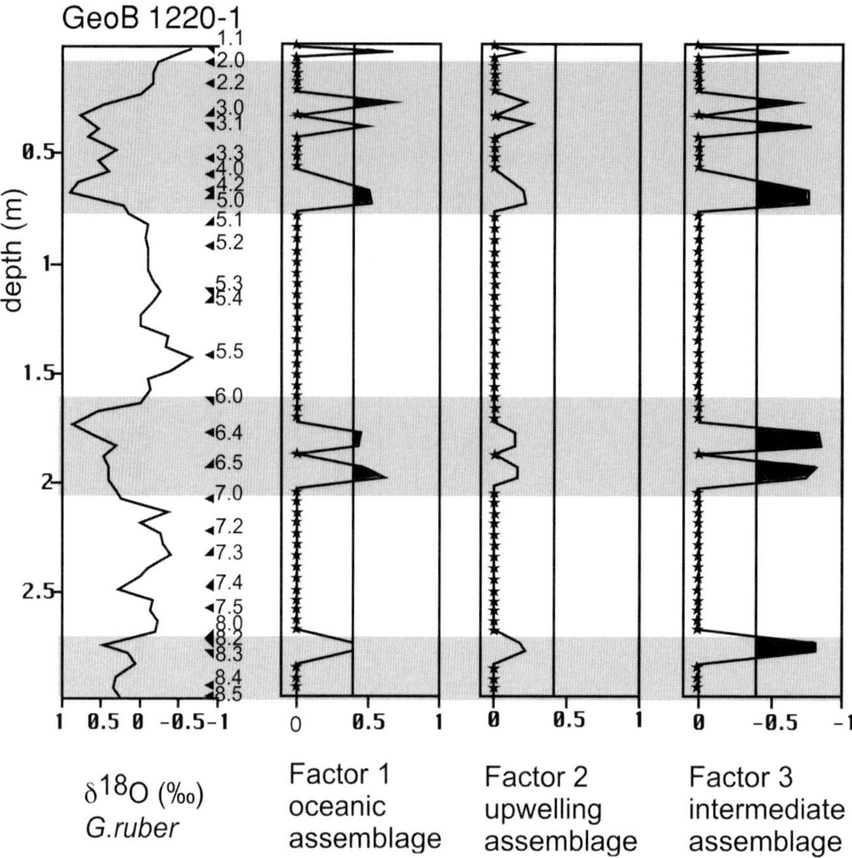

Fig. 11. Graphic representation of GeoB 1220-1 parafactor loading matrix, stars represent no-analog situations. Age model from Schmidt (1992). Grey bars represent glacial periods.

most characteristic time slices to determine the maximum and minimum extent of upwelling during the past 245 kyrs.

Walvis Bay/Lüderitz Upwelling Cell

Walvis Bay and Lüderitz upwelling cells periodically extended GeoB 1710-3 location. Increased upwelling was reflected by the upwelling assemblage during OIS 2-3, and OIE 4.2, 5.2, 5.53, 5.4, 6.2, 6.4/6.5, and 7.4. The maximum westward extension of the Walvis Bay/Lüderitz upwelling cell was found during OIE 3.1 and 6.2 (Fig. 6). A westward extension of more than 280 km off the coast was indicated by the exclusive significance of the upwelling assemblage during these time periods. GeoB 1031-4 indicated maximum westward exten-

sion of the Walvis Bay/Lüderitz upwelling cell also during OIE 7.4, when the northernmost upwelling cell was rather small (compare Fig. 12j). This was supported by the presence of the high productivity benthic foraminifera fauna of GeoB 1214-1 (Schmiedl and Mackensen 1997).

In OIS 2, and OIE 5.2, 5.4, 6.4/6.5, and 7.4 the areal extent of Walvis Bay/Lüderitz cell must have been somewhat smaller compared to its extension in OIS 3 and OIE 6.2, as was indicated by the co-existence of the intermediate assemblage (Fig. 12). Its areal extent must have been considerably smaller in OIE 4.2 and 5.53. The dominance of the intermediate assemblage at RC13-229 location indicated that this core was at the outer edge of the Lüderitz upwelling cell in OIE 5.4 (compare Fig. 12), whereas it was beneath the upwelling cell during OIE 5.53. Our findings are in good agreement with the work of Schmiedl and Mackensen (1997) on benthic foraminifera which suggested increased primary productivity during OIS 2 to 4, and during OIE 5.2 and 5.4. In addition, a high-productive benthic foraminiferal fauna was found in OIS 2-3, and OIE 6.1-6.5, and 7.4.

In contrast, the areal extent of the Walvis Bay upwelling cell was smallest in OIE 5.1 and 5.51, when the open-ocean assemblage yielded highest values. Waters of the open ocean invaded the BUS during these periods of rather low upwelling intensity.

Namibia/Walvis Bay Upwelling Cell

The areal extent of the Namibia upwelling cell can be reconstructed in much greater detail. During periods of increased upwelling intensity, the upwelling cells off Namibia and Walvis Bay cells might have been connected to one another (Fig. 12). DSDP 532, closest to the coast of Angola, showed intense upwelling during OIS 2-4, parts of OIS 5, and from OIS 10 to 13 (Fig. 7). High abundances of *N. pachyderma* sin. during OIS 12 and 13 might be explained as an accumulation of this resistant species resulting from severe carbonate dissolution (Diester-Haas 1985). *N. pachyderma* sin. is regarded as a dissolution-resistant planktic foraminiferal species which accumulates during progressive dissolution. However, there is also evidence for high productivity during OIS 12 in the benthic foraminiferal fauna recovered in GeoB 1214-1 (Fig. 3; Schmicdl and Mackcnscn 1997). Although data are limited to two core positions, the situation during OIS 12 might have been similar to the one reconstructed for OIE 7.4 (Fig. 12j).

For purposes pursued in this study, we concentrate on the last 245 kyrs. From OIS 2 to 4 and during parts of OIS 5 and 6, Namibia/Walvis Bay upwelling cell proceeded DSDP 532 location. GeoB 1028-5 data indicated, that GeoB 1028-5 was beneath the upwelling cell during OIS 3 and OIE 5.4, whereas GeoB 1031-4 was located at the edge of Namibia/Walvis Bay. During these time periods, the maximum areal extent of the Namibia/Walvis Bay upwelling cell was displayed (Fig. 12e). During OIS 2, 4, and OIE 5.2, 5.5, 6.2, and 6.5 upwelling was less intense.

The significance of the upwelling assemblage in the west is restricted to the location of GeoB 1031-4. The sediment cores further to the west were exclusively dominated by the intermediate assemblage and the open-oceanic assemblage which showed the outer limits of coastal upwelling.

The restricted significance of the oceanic assemblage to parts of OIS 5 and mid-OIS 6 indicates that waters of the open ocean invaded the Benguela region exclusively during these time periods. The areal extent of the Namibian upwelling cell must have been relatively small. According to data from DSDP 532 and GeoB 1028-5, the open-oceanic assemblage was significant in the Holocene, most of OIS 5 and 7 (Fig. 8), indicating periods of reduced upwelling intensity. The open-ocean assemblage was the dominant factor at the position of GeoB 1028-5 particularly during OIE 5.1, reflecting the most intense open-ocean character during this period of time.

Paleoceanographic Implications

According to Shannon and O'Toole (1999), modern inter-annual variability in the northern Benguela region is associated with changes in the zonal winds in the equatorial Atlantic and the variability in the Pacific El Nino/La Nina, documenting its importance in global climate forcing, although the linkages and mechanisms are not understood. The BUS is therefore considered to be an important site for the early detection of global climate change (Shannon and O'Toole 1999).

Regarding the past 245 kyrs (Fig. 12), and even further back in time (DSDP 532, Fig. 7), our data reveal periods of severe upwelling intensification in the northern Benguela region. As monitored by the irregular occurrence of 12 PS events during this time period, fluctuations in upwelling intensity were independent from the classic glacial-interglacial cycle, indicating an increased intensity and zonality of the South Atlantic trade winds controlling the BUS (compare Little et al. 1997a,b). This is consistent with the findings of Volbers (2001) who demonstrated that paleo-SSTs determined from planktic foraminiferal assemblages GeoB 1710-3

Volbers et al.

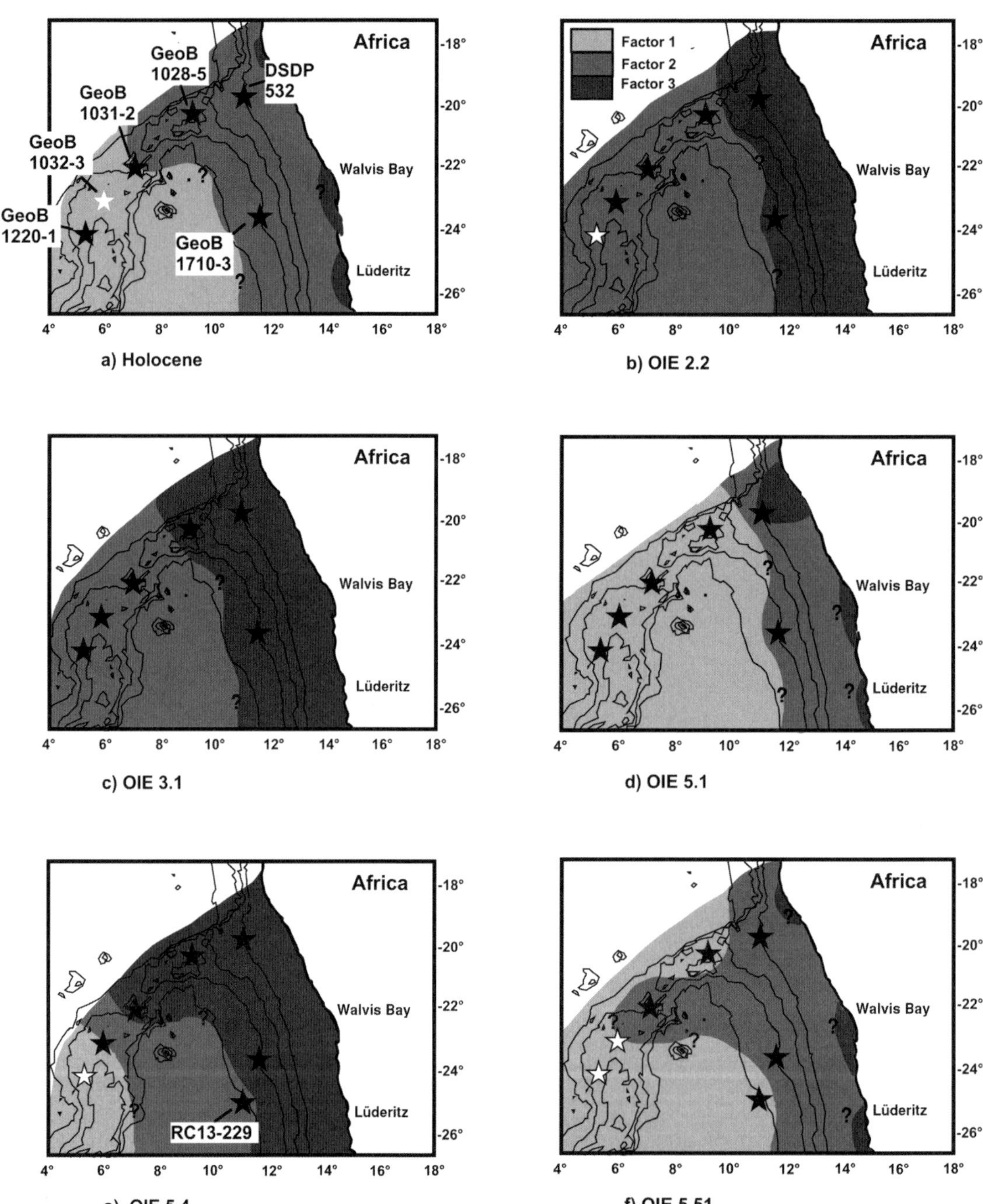

Fig. 12. Graphic interpretation of parafactor loading matrix of all investigated cores during distinct time periods **a-j)** to reconstruct areal extent of northern Benguela upwelling cells (no analog situations are not illustrated). **Figure 12j** also shows location of GeoB 1214-1 because investigation of benthic foraminifera indicate a high productivity fauna during OIE 7.4 (Schmiedl et al. 1997).

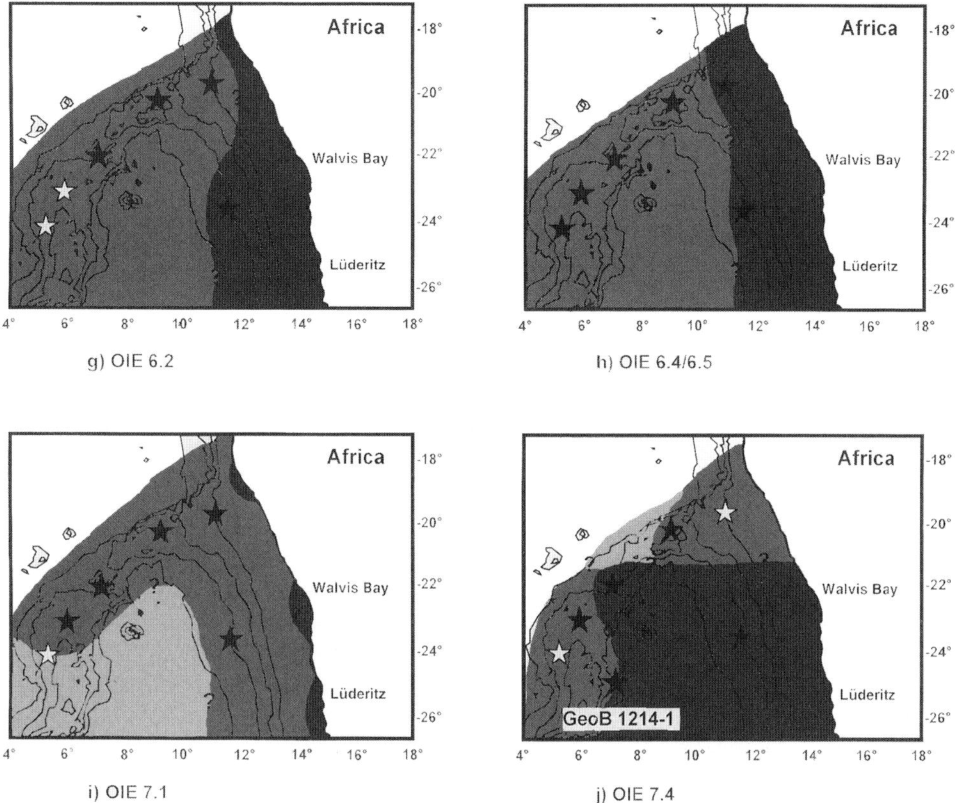

Fig. 12. cont.

do not simply follow a glacial-interglacial pattern as was implied by paleo-SSTs derived from alkenones (e.g. Kirst et al. 1998), but relate to changes in upwelling intensity. Generally, the low paleo SSTs of GeoB 1710-3 coincided with PS events on a sub-Milankovitch scale (Volbers 2001).

Similar fluctuations in upwelling were reported from the eastern equatorial Atlantic (McIntyre et al. 1989). In this region, trade-wind zonality is strongly related to seasonality. The modern strong seasonal variation in the trade winds produces a fluctuating equatorial system with low SSTs in the boreal summer (strong southern trades) and high SSTs, minimum divergence, and SEC speed during the boreal winter. McIntyre et al. (1989) determined paleo SSTs in order to reconstruct upwelling history in the eastern equatorial region for the past 250 kyrs. They provided a temperature equation for the cold (Tc) and the warm season (Tw) derived from planktic foraminifera. Results

from RC24-16 (Fig. 1) pointed to an unstable eastern equatorial region, with high fluctuations in the Tc curve.

Equatorial seasonality was calculated from the differences in the Tc and Tw curves of RC24-16 (Fig. 13). Low seasonality (small SST differences) was associated with periods of low upwelling intensity, whereas high seasonality coincided with increased upwelling. GeoB 1710-3 PS events and RC24-16 equatorial seasonality show high fluctuations in the degree of upwelling in both regions on a similar time scale, substantiating a linkage of the Benguela region and the equatorial Atlantic (Little et al. 1997b, this study).

There are several possibilities to explain coupling of the Benguela upwelling region and the equatorial Atlantic in the past. According to Pokras (1987) and Schneider (1991), the BC might have expanded further into the equatorial region, cooling the eastern equatorial waters during glacials.

However, reconstructions of northward shifts of the Angola-Benguela Front (ABF) showed that its northernmost position was either 12°S during the late Quaternary (Shi et al. 1997) or north of 9°S during OIS 3-4 (Jansen et al. 1996). Although the ABF was north (Jansen et al. 1996) or at its present position (Schneider et al. 1995) for most of the past 220 kyrs, the BC did not penetrate into the Gulf of Guinea (Jansen et al. 1996). Therefore, waters from the BC do not seem to have been entrained into the equatorial region along the African coast.

Shannon et al. (1989) reported episodic input of subantarctic water into the Benguela region and suggested that these cold intrusions might be important for the transfer of heat and water between the hemispheres. Therefore, another way to cool eastern equatorial waters is by transporting colder waters from higher latitudes towards the equator, lowering SSTs in this region. If this had occurred,

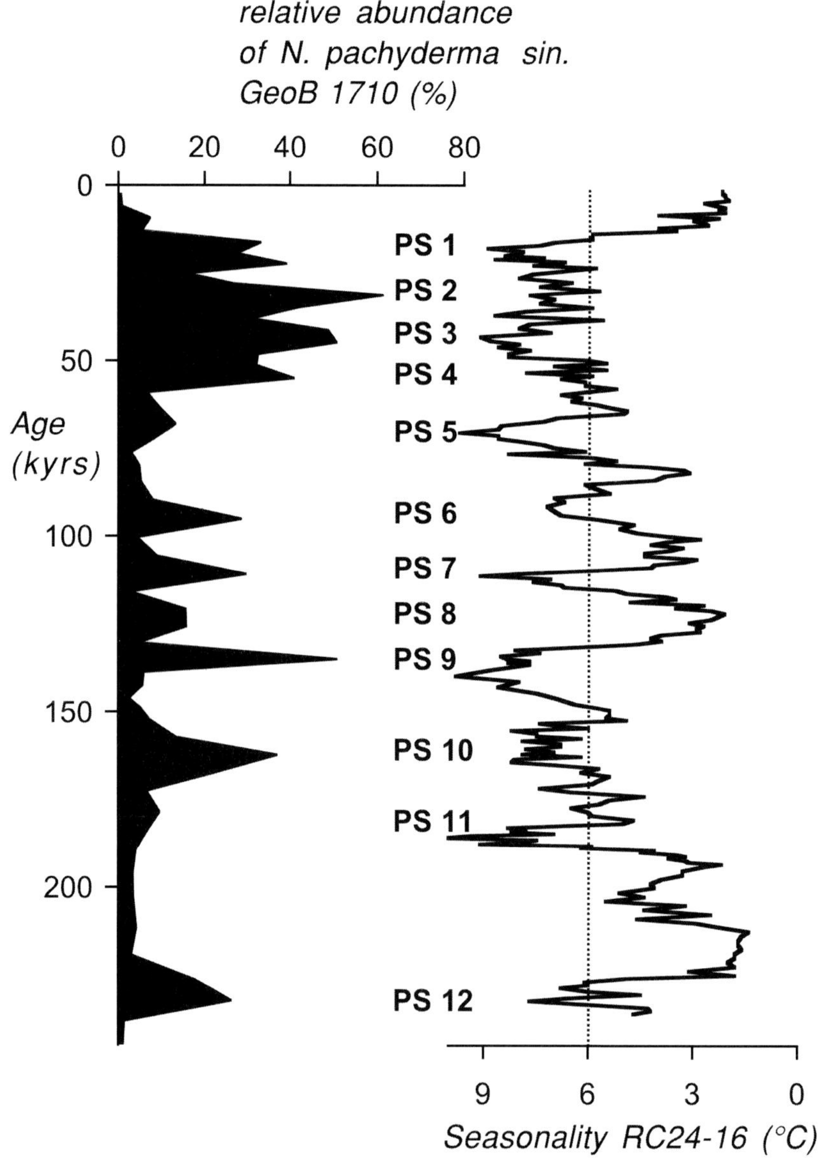

Fig. 13. Comparison of northern Benguela upwelling events indicated by *N. pachyderma* sin. Maximal abundances (PS events) from GeoB 1710-3 (Cape Basin) with equatorial seasonality (°C) (seasonality after McIntyre et al. 1989).

higher abundances of cold water planktic foraminifera, like *N. pachyderma* sin., would have been discovered in the south compared to the north. The low relative abundances of *N. pachyderma* sin. in the southern Cape Basin and high relative abundances of this species in the northern part found by Little et al. (1997b) provide strong evidence against cold-water advection of sub-Antarctic waters from the subtropical convergence. It is more likely that cool waters from the Benguela coastal upwelling were transported towards the equator. Mix et al. (1999) proposed that the BS contributes to the equatorial cooling during LGM (PS 1). Although the BC is currently not strong enough to cause extreme cooling in the central equatorial Atlantic (Mix and Morey 1996), the cold waters from the Benguela System may have turned westward between 10° and 20°S (around 17°S (Diester-Haass et al. 1992) and finally entered the equatorial zone during glacial time (Schneider et al. 1995). However, as demonstrated in this study, both northern Benguela upwelling and equatorial seasonality did not follow the classic glacial-interglacial pattern but varied on a sub-Milankovitch scale. The scenario described above was obviously not limited to glacial periods as has been presumed previously. Instead, the BS contributes to the equatorial cooling not only during LGM (PS 1) but also during eleven other time intervals during the past 245 kyrs.

So far, mechanisms capable of increasing the intensity and zonality of SE trade winds were directly coupled to the glacial-interglacial cycle. The polar front was assumed to move northward coupled with an increase in the meridional pressure gradients during glacial periods, finally resulting in intensified SE trade winds. However, according to our observation, increases in the intensity and zonality of SE trade winds varied on a sub-Milankovitch scale, and pointed out to a more variable atmospheric and oceanic circulation pattern than had been previously assumed for the late Quaternary.

Conclusions

Fluctuations in characteristic planktic foraminifera assemblages derived from 6 cores on the Walvis Ridge and from the Cape Basin indicated significant changes in the degree of upwelling in the BUS over the past 245 kyrs.

• GeoB 1710-3 reflected twelve distinct upwelling events (PS 1 to PS 12) during OIS 2 and 3, and OIE 4.2, 5.2, 5.4, 5.53, 6.2, 6.4/6.5, and 7.4.

• During periods of intensified upwelling, northern Benguela upwelling cells increased in size covering areas at least three times larger than the present day.

• Upwelling in the northern Benguela region was at maximum during OIE 5.4 and 7.4. During OIE 5.1 and 5.51 upwelling was minimal and open oceanic waters invaded the Benguela region.

• Periods of increased upwelling intensity were not linked to the glacial-interglacial cycle but occurred erratically on a sub-Milankovitch scale. Attributing to this observation, increases in the intensity and zonality of the trade winds were not limited to glacial periods as previously described.

• Simultaneous increases in eastern equatorial seasonality demonstrate the close connection of both regions, confirming the outstanding role of the BUS in oceanic circulation and global climate.

Acknowledgments

We thank H. Oberhänsli who kindly provided planktic foraminifera counts and the stratigraphy of DSDP 532 to us. We would like to acknowledge R. R. Schneider and S. Gerhardt for discussion, and R. Sieger for providing his software package PaleoToolBox/WinTransfer and his help with the database PANGAEA (www.pangaea.de). Helpful suggestions and reviews by Frank Peeters and Michal Kucera are gratefully acknowledged. C. Devey kindly corrected the English. This research was funded by the Deutsche Forschungsgemeinschaft (Sonderforschungsbereich 261 at Bremen University, Contribution No. 352). Data are available under www.pangaea.de/Projects/SFB261.

References

Backhaus K, Erichson B, Plinke W, Schuchard-Ficher C, Weiber R (1989) Multivariate Analysemethoden. Springer, Berlin, vol 5, 418 p

Bé AWH, Tolderlund DS (1971) Distribution and ecology of living planktonic foraminifera in surface waters of the Atlantic and Indian Oceans. In: Funnell DM, Riedel WR (eds) The Micropaleontology of Oceans. Cambridge University Press, Cambridge, pp 105-149

Bickert T (1992) Rekonstruktion der spätquartären Bodenwasserzirkulation im östlichen Südatlantik über stabile Isotope benthischer Foraminiferen. Ber Fachber Geowiss, Univ Bremen, vol 27, 205 p

Bickert T, Wefer G (1999) South Atlantic and benthic foraminifer $\delta^{13}C$-deviations: Implications for reconstructing the Late Quaternary deep-water circulation. Deep-Sea Res 46: 437-452

Boyer D, Cole J, Bartholome C (2000) Southwestern Africa: Northern Benguela Current region. Mar Poll Bull 41: 123-140

Chang P-Y, Chang C-C, Wang L-W, Chen M-T, Wang C-H, Yu E-F (1999) Planktonic foraminiferal sea surface temperature variations in the southeast Atlantic Ocean: A high-resolution record MD962085 of the past 400,000 years from the IMAGES II-NAUSICAA Cruise. TAO 10: 185-200

Chen M-T, Prell WL (1998) Faunal distribution patterns of planktonic foraminifers in surface sediments of the low-latitude Pacific. Palaeogeogr Palaeoclimatol Palaeoecol 137: 55-77

Christensen BA, Giraudeau J (2002) Neogene and Quaternary evolution of the Benguela upwellig system. Mar Geol 180: 1-2

CLIMAP (1976) The surface of the ice-age Earth. Science 191: 1131-1137

CLIMAP Project Members (1981) Seasonal reconstructions of the Earth's surface at the last glacial maximum. Geol Soc Am Map and Chart Ser, MC-36, Boulder, Colorado, 18 p

CLIMAP Project Members (1984) The last interglacial ocean. Quat Res 21: 123-224

Dieser-Haass L (1985) Late Quaternary sedimentation on the eastern Walvis Ridge, SE Atlantic (HPC 532 and 4 piston cores). Mar Geol 65: 145-189

Dieser-Haass L, Meyers PA, Rothe P (1992) The Benguela Current and associated upwelling on the southwest African margin: A synthesis of the Neogene-Quaternary sedimentary record at DSDP sites 362 and 532. In: Summerhayes CP, Prell WL, Emeis KC (eds) Upwelling Systems: Evolution since the early Miocene. Geol Soc Spec Publ, vol. 64, Geol Soc London, pp 331-342

Giraudeau J (1993) Planktonic foraminiferal assemblages in surface sediments from the southwest African continental margin. Mar Geol 110: 47-62

Giraudeau J, Rogers J (1994) Phytoplankton biomass and sea-surface temperature estimates from sea-bed distribution of nannofossils and planktonic foramini-fera in the Benguela upwelling system. Micro-paleontology 40: 275-285

Hart TJ, Currie RI (1960) The Benguela Current. Discovery Rep 31: 123-298

Hemleben C, Spindler M, Anderson OR (1989) Modern Planktionic Foraminifera. Springer, New York, 363 p

Imbrie J, Kipp NG (1971) A new micropaleontological method for quantitative paleoclimatology: Application to a late Pleistocene Caribbean core. In: Turekian KK (ed) The Late Cenozoic Glacial Ages. Yale Univ Press, New Haven, pp 71-181

Ivanova EM, Conan SM-H, Peeters FJC, Troelstra SR (1999) Living *Neogloboquadrina pachyderma* sin and its distribution in the sediments from Oman and Somalia upwelling areas. Mar Micropal 36: 91-107

Jansen JHF, Ufkes E, Schneider RR (1996) Late Quaternary movements of the Angola-Benguela-Front, SE Atlantic, and implications for advection in the equatorial ocean. In: Wefer G, Berger WH, Siedler G (eds) The South Atlantic: Present and Past Circulation. Springer, Berlin, pp 553-575

Kirst GJ, Schneider RR, Müller PJ, von Storch I, Wefer G (1999) Late Quaternary temperature variability in the Benguela Current system derived from alkenones. Quat Res 52: 92-103

Klovan JE, Imbrie J (1971) An algorithm and FORTRAN IV Program for large scale Q-mode factor analysis and calculation of factor scores. Math Geol 3: 61-77

Little MG, Schneider RR, Kroon D, Price B, Summerhayes CP, Segl M (1997a) Trade Wind forcing of upwelling, seasonality, and Heinrich events as a response to sub-Milankovitch climate variability. Paleoceano-graphy 12: 568-576

Little MG, Schneider RR, Kroon D, Price B, Bickert T, Wefer G (1997b) Rapid palaeoceanographic changes in the Benguela Upwelling System for the last 160,000 years as indicated by abundances of planktonic foraminifera. Palaeogeogr Palaeoclimat Palaeo-ecol 130: 135-161

Lohmann GP (1992) Increasing seasonal upwelling in the subtropical South Atlantic over the past 700,000 yr: Evidence from deep-living planktonic foraminifera. Mar Micropal 19: 1-12

Lutjeharms JRE (1996) The exchange of water between the South Indian and South Atlantic Oceans. In: Wefer G, Berger WH, Siedler G, Webb DJ (eds) The South Atlantic: Present and Past Circulation. Springer, Berlin, pp 125-162

Lutjeharms JRE, Meeuwis JM (1987) The extent and variability of South-East Atlantic upwelling. S Afr J Mar Sci 5: 51-62

Lutjeharms JRE Stockton PL (1987) Kinematics of the upwelling front off Southern Africa. S Afr J Mar Sci 5: 35-49

Lutjeharms JRE, Shillington FA, Duncombe R (1991) Observations of extreme upwelling filaments in the Southeast Atlantic Ocean. Science 253: 774-776

McIntyre A, Ruddiman WF, Karlin K, Mix AC (1989) Surface water response of the equatorial Atlantic Ocean to orbital forcing. Paleoceanography 4: 19-55

Mix AC (1989) Pleistocene Paleoproductivity: Evidence from organic carbon and foraminiferal species. In: Berger WH, Smetacek VS, Wefer G (eds) Productivity of the Ocean: Present and Past. J Wiley & Sons, Chichester, pp 313-336

Mix AC, Morey AE (1996) Climate feedback and Pleistocene variations in the Atlantic South Equatorial Current. In: Wefer G, Berger WH, Siedler G, Webb DJ (eds) The South Atlantic: Present and Past Circulation. Springer, Berlin, pp 503-525

Mix AC, MoreyAE, Pisias NG, Hostetler SW (1999) Foraminiferal faunal estimates of paleotemperature: Circumventing the no-analog problem yields cool ice age tropics. Paleoceanography 14: 350-359

Molfino B, Kipp NG, Morley JJ (1982) Comparison of foraminiferal, coccolithophorid, and radiolarian paleotemperature equations: Assemblage coherency and estimate concordancy. Quat Res 17: 279-313

Müller PJ, Cepek M, Ruhland G, Schneider RR (1997) Alkenone and coccolithophorid species changes in late Quaternary sediments from the Walvis Ridge: Implications for the alkenone paleotemperature method. Palaeogeogr Palaeoclimat Palaeoecol 135: 71-96

Murray J (1897) On the distribution of the pelagic foraminifera at the surface and on the floor of the ocean. Nat Sci 11: 17-27

Murray J, Renard AF (1891) Deep-sea deposits based on the specimens collected during the voyage of HMS Challenger in the years 1872-1876. Reports Voyage Challenger, London, 525 p

Nelson G, Hutchings L (1983) The Benguela upwelling area. Prog Oceanogr 12: 333-356

Niebler HS (1995) Rekonstruktion von Paläo-Umweltparametern anhand von stabilen Isotopen und Faunen-Vergesellschaftungen planktischer Foraminiferen im Südatlantik. Ber Polarforsch, Alfred-Wegener-Institut, Bremerhaven 167, 198 p

Niebler HS, Gersonde R (1998) A planktic foraminiferal transfer function for the southern South Atlantic Ocean. Mar Micropaleontol 34: 213-234

Oberhänsli H (1991) Upwelling signals at the Northeastern Walvis Ridge during the past 500,000 years. Paleoceanography 6: 53-71

Oppo DW, Rosenthal Y (1994) Cd/Ca changes in a deep Cape Basin core over the past 730,000 years: Response of circumpolar deepwater variability to northern hemisphere ice sheet melting? Paleoceanography 9: 661-675

Peeters FJC, Ivanova EM, Conan SM-H, Brummer GJA, Ganssen GM, Troelstra SR, van Hinte JE (1999) A size analysis of planktic foraminifera from the Arabian Sea. Mar Micropaleontol 36 : 31-61

Peterson LC, Abbott MR, Anderson DM, Caulet J-P, Conté MH, Emeis K-C, Kemp AES, CP Summerhayes, Group Report (1995) How do Upwelling Systems vary through time? In: Summerhayes CP, Prell WL, Emeis KC (eds) Upwelling in the Ocean. Modern Processes and Ancient Records. J Wiley & Sons, Chichester, pp 285-311

Pflaumann U, Dupra J, Pujol C, Labeyrie L (1996) SIMMAX: A modern analog technique to deduce Atlantic sea surface temperatures from planktonic foraminifera in deep-sea sediments. Paleoceanography 11: 15-35

Pokras EM (1987) Diatom record of late Quaternary climatic change in the eastern equatorial atlantic and tropical Africa. Paleoceanography 2: 273-286

Ravelo AC, Fairbanks RG (1992) Oxygen isotopic composition of multiple species of planktonic foraminifera: Recorders of the modern photic zone temperature gradient. Paleoceanography 7: 815-831

Sarnthein M, Winn K, Duplessy J-C, Fontugne M-R (1988) Global variations of surface ocean productivity in low and mid latitudes: Influence on CO2 reservoirs of the deep ocean and atmosphere during the last 21,000 years. Paleoceanography 3: 361-399

Schmidt H (1992) Der Benguela-Strom im Bereich des Walfisch-Rückens im Spätquartär. Ber Fachber Geowiss, Univ Bremen 28, 172 p

Schmiedl G, Mackensen A, Müller PJ (1997) Recent benthic foraminifera from from the eastern South Atlantic Ocean: Dependence on food supply and water masses. Mar Micropalenotol 32: 249-287

Schneider RR (1991) Spätquartäre Produktivitätsänderungen im östlichen Angola-Becken: Reaktion auf Variationen im Passat-Monsun-Windsystem und in der Advektion des Benguela-Küstenstrom. Ber Fachber Geowiss, Univ Bremen 21, 198 p

Schneider RR, Müller PJ, Ruhland G (1995) Late Quaternary surface circulation in the east equatorial South Atlantic: Evidence from alkenone sea surface temperatures. Paleoceanography 10: 197-219

Shannon LV (1985) The Benguela ecosystem part I. Evolution of the Benguela. Physical features and processes. Oceanogr Mar Biol Ann Rev 23: 105-182

Shannon LV, Boyd AJ, Brundrit GB, Taunton-Clark J (1986) On the existence of an El Nino-type phenomenon in the Benguela system. J Mar Res 44: 495-520

Shannon LV, Lutjeharms JRE, Agenbag JJ (1989) Episodic input of Subantarctic water into the Benguela region. S Afr J Sci 85: 317-322

Shannon LV, Nelson G (1996) The Benguela: Large scale features and processes and system variability. In: Wefer G, Berger WH, Siedler G, Webb DJ (eds) The South Atlantic: Present and Past Circulation. Springer, Berlin, pp 163-210

Shannon LV, O'Toole MJ (1999) Integrated overview of the oceanography and environmental variability of the Benguela Current region. Synthesis and assessment of information on the Benguela Current large marine ecosytem (BCLME), Thematic report NO. 2, http://www.ioinst.org/bclme/factfig/oceanography.htm

Shi N, Dupont LM, Beug H-J, Schneider R (1997) Vegetation and climate history of SW Africa: A marine palynological record of the last 300,000 years. Veg Hist Archaeobot 6: 117-131

Shillington FA, Peterson WT, Hutchings L, Probyn TA, Waldron HN, Agenrag JJ (1990) A cool upwelling filament off Namibia, southwest Africa: Preliminary measurements of physical and biological features. Deep-Sea Res 37: 1753-1772

Stramma L, Peterson RG (1989) Geostrophic transport in the Benguela Current Region. J Phys Oceanogr 19: 1440-1448

Thiede J (1975) Distribution of foraminifera in surface waters of a coastal upwelling area. Nature 253: 712-714

Ufkes E, Zachariasse W-J (1993) Origion of coiling differences in living neogloboquadrinids in the Walvis Bay region, off Namibia, southwest Africa. Micropaleontology 39: 283-287

Ufkes E, Jansen JHF, Schneider RR (2000) Anomalous occurrences of *Neogloboquadrina pachyderma* (left) in a 420-ky upwelling record from Walvis Ridge (SE Atlantic). Mar Micropal 40: 23-42

Volbers A (2001) Planktic foraninifera as paleoceanographic indicators: Production, preservation, and reconstruction of upwelling intensity. Impllications from late Quaternary South Atlantic sediments. Ber Fachber Geowisss, Univ Bremen 184, 122 p

Volbers A, Henrich R, Niebler H-S (1999) Examiniation of late Quaternary planktic foraminiferal assemblages and investigations of carbonate preservation in the sediment core GeoB 1710-3 from the eastern South Atlantic. Workshop on Paleoceanography, Bern, Switzerland

Volbers ANA, Henrich R (2002) Late Quaternary variations in calcium carbonate preservation of deep-sea sediments in the northern Cape Basin: Results from a multiproxy approach. Mar Geol 180: 203-220

Watkins JM, Mix AC (1998) Testing the effects of tropical temperature, productivity, and mixed-layer depth on foraminiferal transfer functions. Paleoceanography 13: 96-105

Carbon Isotopes of Live Benthic Foraminifera from the South Atlantic: Sensitivity to Bottom Water Carbonate Saturation State and Organic Matter Rain Rates

A. Mackensen[*] and L. Licari

Alfred Wegener Institute for Polar and Marine Research, Columbusstrasse, 27568 Bremerhaven, Germany
** corresponding author (e-mail): amackensen@awi-bremerhaven.de*

Abstract: Live (Rose Bengal stained) and dead benthic foraminifera of surface and subsurface sediments from 25 stations in the eastern South Atlantic Ocean and the Atlantic sector of the Southern Ocean were analyzed to decipher a potential influence of seasonally and spatially varying high primary productivity on the stable carbon isotopic composition of foraminiferal tests. Therefore, stations were chosen so that productivity strongly varied, whereas conservative water mass properties changed only little. To define the stable carbon isotopic composition of dissolved inorganic carbon ($\delta^{13}C_{DIC}$) in ambient water masses, we compiled new and previously published $\delta^{13}C_{DIC}$ data in a section running from Antarctica through Agulhas, Cape and Angola Basins, via the Guinea Abyssal Plain to the Equator. We found that intraspecific $\delta^{13}C$ variability of all species at a single site is constantly low throughout their distribution within the sediments, i.e. species specific and site dependent mean values calculated from all subbottom depths on average only varied by ±0.09 ‰. This is important because it makes the stable carbon isotopic signal of species independent of the particular microhabitat of each single specimen measured and thus more constant and reliable than has been previously assumed. So-called vital and/or microhabitat effects were further quantified: (1) $\delta^{13}C$ values of endobenthic *Globobulimina affinis*, *Fursenkoina mexicana*, and *Bulimina mexicana* consistently are by between -1.5 and -1.0 ‰ VPDB more depleted than $\delta^{13}C$ values of preferentially epibenthic *Fontbotia wuellerstorfi*, *Cibicidoides pachyderma*, and *Lobatula lobatula*. (2) In contrast to the Antarctic Polar Front region, at all stations except one on the African continental slope *Fontbotia wuellerstorfi* records bottom water $\delta^{13}C_{DIC}$ values without significant offset, whereas *L. lobatula* and *C. pachyderma* values deviate from bottom water values by about -0.4‰ and -0.6‰, respectively. This adds to the growing amount of data on contrasting cibicid $\delta^{13}C$ values which on the one hand support the original 1:1-calibration of *F. wuellerstorfi* and bottom water $\delta^{13}C_{DIC}$, and on the other hand document severe depletions of taxonomically close relatives such as *L. lobatula* and *C. pachyderma*. At one station close to Bouvet Island at the western rim of Agulhas Basin, we interpret the offset of -1.5 ‰ between bottom water $\delta^{13}C_{DIC}$ and $\delta^{13}C$ values of infaunal living *Bulimina aculeata* in contrast to about -0.6 ± 0.1 ‰ measured at eight stations close-by, as a direct reflection of locally increased organic matter fluxes and sedimentation rates. Alternatively, we speculate that methane locally released from gas vents and related to hydrothermal venting at the mid-ocean ridge might have caused this strong depletion of ^{13}C in the benthic foraminiferal carbon isotopic composition. Along the African continental margin, offsets between deep infaunal *Globobulimina affinis* and epibenthic *Fontbotia wuellerstorfi* as well as between shallow infaunal *Uvigerina peregrina* and *F. wuellerstorfi*, $\delta^{13}C$ values tend to increase with generally increasing organic matter decomposition rates. Although clearly more data are needed, these offsets between species might be used for quantification of biogeochemical paleogradients within the sediment and thus paleocarbon flux estimates. Furthermore, our data suggest that in high-productivity areas where sedimentary carbonate contents are lower than 15 weight %, epibenthic and endobenthic foraminiferal $\delta^{13}C$ values are strongly influenced by ^{13}C enrichment probably due to carbonate-ion undersaturation, whereas above this sedimentary carbonate threshold endobenthic $\delta^{13}C$ values reflect depleted pore water $\delta^{13}C_{DIC}$ values.

From WEFER G, MULITZA S, RATMEYER V (eds), 2003, *The South Atlantic in the Late Quaternary: Reconstruction of Material Budgets and Current Systems.* Springer-Verlag Berlin Heidelberg New York Tokyo, pp 623-644

Introduction

Benthic foraminifera are the only abundant and ubiquitous benthic marine protists of the deep-sea that, due to their mostly calcareous tests, have a great potential to become fossilized. They record paleoenvironmental conditions directly in the trace elemental and isotopic composition of their calcareous test, and indirectly by particular faunal compositions specifically adapted to environmental parameters such as food supply, oxygen content, hydrodynamic and physicochemical conditions of bottom water masses. For paleoceanographic purposes it is most interesting to know what specific environmental conditions are recorded by benthic foraminifera and how sensitive their response is to changes in these parameters.

Since the work of Corliss (1985) and the detailed analyses of Gooday (1986) it is agreed that specific deep-sea benthic foraminifera do live in stratified depths within the sediment down to a depth of 10-15 cm below the seafloor. After a decade of intensive research and collecting high-quality samples using multiple corers, there is no doubt that food availability and oxygen content of the interstitial waters most substantially control the benthic foraminiferal microhabitat (e.g. Mackensen and Douglas 1989; Gooday and Turley 1990; Bernhard 1992; Jorissen et al. 1992; Rathburn and Corliss 1994; Alve and Bernhard 1995; Wollenburg and Mackensen 1998). Generally, it is assumed that the oxygen penetration depth controls the maximum habitat depth as long as food is available. Consequently, a model was proposed that determines the microhabitat depth in eutrophic environments by a critical oxygen level within the sediment; and by a critical level of food supply in case of oligotrophic environments (Corliss and Emerson 1990; Jorissen et al. 1995). It was further suggested that the ver-tical foraminiferal distribution of species within the sediments reflects the presence of various populations of anaerobic and sulfate- and nitrate-reducing bacteria, which the foraminifera are thought to feed on directly. Otherwise they might selectively depend on the state of bacterial organic matter degradation (Caralp 1989; Jorissen et al. 1998). Since the pore-water oxygen content and food availability, including bacterial communities, are coupled to

seasonal fluctuations in the ocean's surface productivity, this implies that microhabitat depth preferences do not only vary between different species, but also within a single species, depending on season and food supply (cf. Linke and Lutze 1993).

In most oceans, a linear correlation is observed between the stable carbon isotopic composition of the dissolved inorganic carbon ($\delta^{13}C_{DIC}$) values and nutrient contents of deep and bottom water masses because the distribution of both are controlled by the interaction of biological uptake at the sea surface and decomposition in deeper water masses with the general circulation of the ocean (Kroopnick 1980, 1985). This is important because the carbon isotopic composition of foraminiferal carbonate exhibits consistent relationships with the isotopic composition of dissolved inorganic carbon in ambient waters at the time of precipitation (Woodruff et al. 1980; Graham et al. 1981). Thus the $\delta^{13}C$ signal of the water masses is recorded in epibenthic foraminifera and as such is extensively used as a nutrient proxy to reconstruct deep ocean paleocirculation (e.g. Duplessy et al. 1984; Curry et al. 1988; Oppo et al. 1990; Raymo et al. 1990; Boyle 1992; Sarnthein et al. 1994; Mackensen et al. 2001).

Some species assumed to live within the sediment were detected to reflect by their carbon isotopic test composition the amount of organic carbon fluxes to the seafloor (Woodruff and Savin 1985; Zahn et al. 1986; Loubere 1987; McCorkle et al. 1994). Other species known for their epi-benthic lifestyle, such as *Fontbotia wuellerstorfi*, were shown to record the isotope signal of dissolved inorganic carbon in bottom water (Duplessy et al. 1984; Grossman 1987). Even the carbon isotopic composition of epibenthic *F. wuellerstorfi*, however, if collected from areas prone to seasonal plankton blooms and subsequent rapid sedimentation and the development of a phytodetritus layer at the sea floor, significantly deviates from the bottom water isotope signal, thus making the interpretation of fossil tests more difficult and ambiguous (Mackensen et al. 1993b; Mackensen and Bickert 1999).

Generally, only few species of the marine calcareous macrobenthos and meiobenthos precipitate calcite in equilibrium with the isotopic composition of dissolved inorganic carbon in ocean bottom water (Wefer and Berger 1991). Basically two mechanisms are responsible for carbon isotope disequilibria seen in benthic foraminifera: so called vital (physiological) and microhabitat effects. Vital effects can be further subdivided into two categories: metabolic and kinetic isotope effects (Mc Connaughey 1989a, 1989b). Metabolic effects reduce the $\delta^{13}C$ values due to the incorporation of respired CO_2 into the foraminiferal test (Spero and Lea 1996; McConnaughey et al. 1997; Wilson-Finelli et al. 1998). In addition, kinetic fractionation can occur during stages of rapid shell calcification and would result in even more depleted $\delta^{13}C$ values. The term 'microhabitat effect' summarizes both equilibrium as well as kinetic fractionation during calcification within the sediment porewater or other specific microhabitats, for instance, within a phytodetrital layer directly on the sediment surface (see above). Generally, porewater $\delta^{13}C_{DIC}$ values rapidly decrease with increasing depth in the sediment within the top centimeter. Calcitic tests of infaunal species are therefore expected to generally display low $\delta^{13}C$ values (Grossman 1984b; McCorkle et al. 1985). Studies of Rathburn et al. (1996) and Mackensen et al. (2000), however, revealed that most taxa caught over a depth range within the sediments do not show consistent gradients in $\delta^{13}C$ as would be expected if carbon isotopic composition were influenced only by the average porewater environments in which they had been found. The saturation state of the ambient water with respect to carbonate may also influence the isotopic signal recorded in the benthic foraminiferal shell. Culturing experiments revealed that the stable isotopic composition of planktic foraminiferal tests responds to changes in seawater carbonate ion concentration (Spero et al. 1997). Recently, this 'Carbonate Ion Effect' was applied to interpret severe deviations of $\delta^{13}C_{DIC}$ values of planktic foraminifera *Neogloboquadrina pachyderma* from surface water $\delta^{13}C_{DIC}$ values in the Okhotsk Sea (Bauch et al. 2002). If the responses of benthic and planktic foraminifera are similar, a decrease in carbonate ion concentration by 10 µmol/kg would be equivalent to an increase in calcitic test $\delta^{13}C$ of 0.1 ‰ (Lea et al. 1999).

Relatively few studies from deep-sea environments are available that relate live (Rose Bengal stained) benthic foraminiferal test isotopic composition to ambient bottom and pore water dissolved inorganic carbon isotopes (Grossman 1984a, 1984b, 1987; Mackensen and Douglas 1989; McCorkle et al. 1990; Mackensen et al. 1993b; Rathburn et al. 1996; McCorkle et al. 1997; Mackensen and Bickert 1999; Mackensen et al. 2000). Therefore, in this study we further address the relationship between $\delta^{13}C$ values of live (Rose Bengal stained) benthic foraminifera and their specific microhabitat preferences and environments to increase the reliability and substantiate the use of benthic fora-miniferal test $\delta^{13}C$ as a proxy for paleoceano-graphic reconstructions. Benthic $\delta^{13}C$ data, due to their dependency on ocean surface productivity as well as air-sea gas exchange, provide crucial insight into the links between upper ocean climate and deep water circulation.

Material

Live (Rose Bengal stained) and dead benthic foraminifera of surface and subsurface sediments from 25 stations in the eastern South Atlantic Ocean and the Atlantic sector of the Southern Ocean were analyzed in order to decipher a potential influence of seasonally and spatially varying high primary productivity on the stable carbon isotopic composition of foraminiferal tests (Fig. 1; Table 1). We have chosen two sample sets out of more than 250 samples from the South Atlantic Ocean, most of which have been investigated earlier for their benthic foraminiferal content and ecological preferences (Mackensen et al. 1990; Mackensen et al. 1993a; Mackensen et al. 1995; Harloff and Mackensen 1997; Schmiedl et al. 1997; Mackensen et al. 2000; Schumacher 2001; Licari et al. 2003).

One set of nine surface sediment samples is from a transect across a mid-ocean ridge close to Bouvet Island, roughly between 48° and 55°S (Mackensen et al. 1993a; Mackensen et al. 2000). Ecologically this area is characterized by a locally and seasonally highly varying ocean surface productivity associated with the Antarctic Polar Frontal Zone between the Subantarctic and the Polar Fronts (Peterson and Stramma 1991; Orsi et al. 1995).

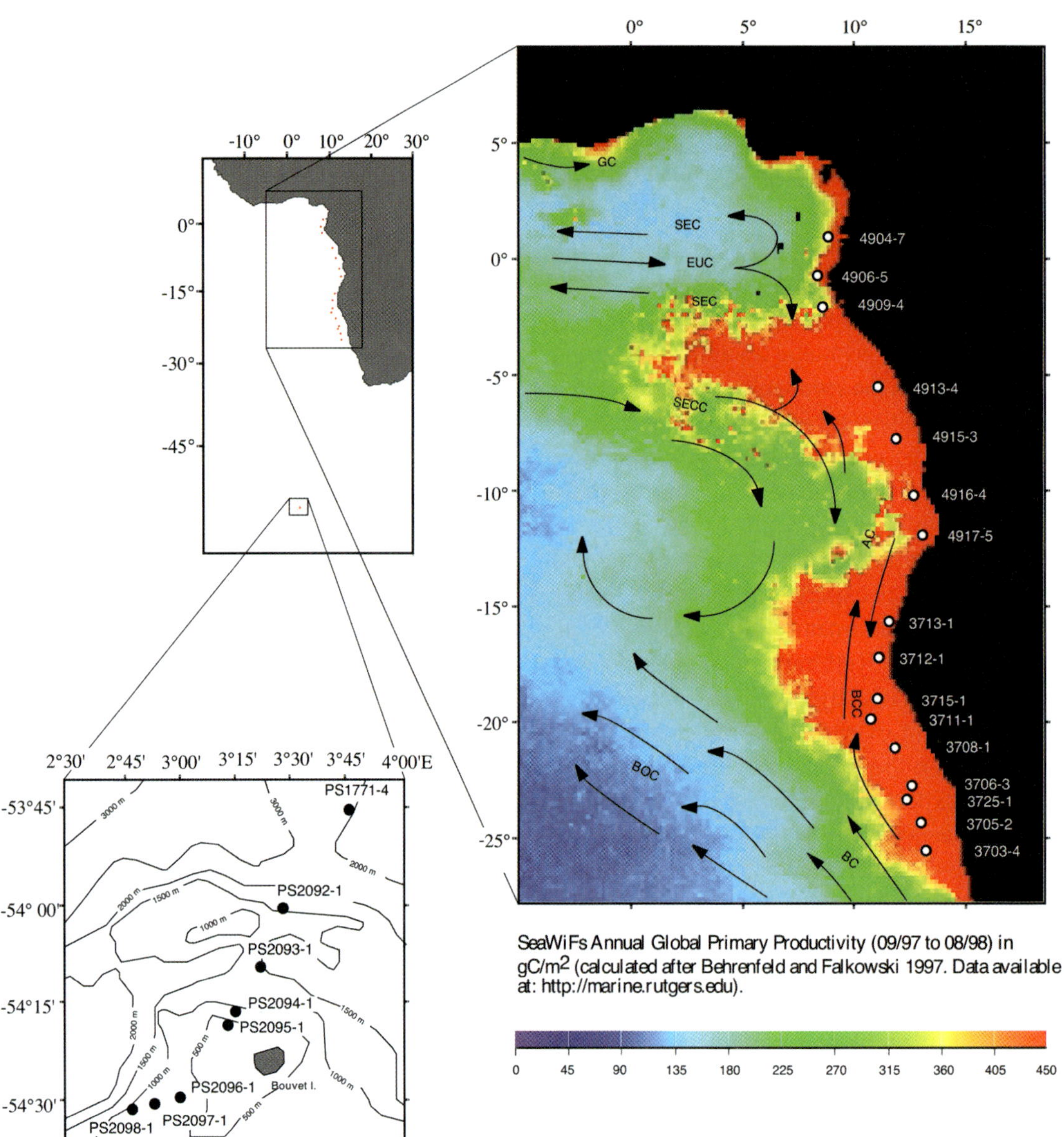

Fig. 1. Positions of sites investigated in this study close to Bouvet Island in the Southern Ocean and in the South Atlantic Ocean off the southwest African continental margin. Bathymetry is after GEBCO and surface hydrography according to Voituriez and Herbland (1982) and Peterson and Stramma (1991). AC = Angola Current, BC = Benguela Current, BCC = Benguela Coastel Current, BOC = Benguela Oceanic Current, EUC = Equatorial Under Current, GC = Guinea Current, SEC = South Equatorial Current, SECC = South Equatorial Counter Current.

Sample	Latitude [°]	Longitude [°]	Water depth [m]	Stocks [#/50cm²]	TOC [%]	Carbonate [%]	$\delta^{13}C_{DIC}$ [‰VPDB]
GeoB4904-7	0.96	8.88	1341	315	1.2	*12.9*	0.53
GeoB4906-5	-0.69	8.38	1277	166	1.5	*16.5*	0.70
GeoB4909-4	-2.07	8.63	1313	279	1.3	*5.2*	0.67
GeoB4913-4	-5.50	11.07	1300	48	1.7	*4.0*	0.65
GeoB4915-3	-7.75	11.87	1305	156	1.3	*20.5*	*0.63*
GeoB4916-4	-10.17	12.69	1300	357	2.5	*15.8*	*0.61*
GeoB4917-5	-11.90	13.07	1299	930	2.2	*15.5*	0.63
GeoB3713-1	-15.63	11.58	1330	827	1.6	5.1	*0.35*
GeoB3712-1	-17.19	11.13	1242	1645	2.4	5.7	*0.25*
GeoB3715-1	-18.95	11.06	1204	332	4.5	25.6	*0.30*
GeoB3711-1	-19.84	10.77	1214	726	3.8	66.7	*0.30*
GeoB3708-1	-21.09	11.83	1283	994	4.9	52.6	*0.35*
GeoB3706-3	-22.72	12.60	1313	580	3.7	55.2	*0.40*
GeoB3725-1	-23.32	12.37	1980	283	2.1	76.4	*0.40*
GeoB3705-2	-24.30	13.00	1305	391	4.5	66.0	*0.45*
GeoB3703-4	-25.52	13.23	1376	746	8.1	55.3	*0.45*
PS1777-7	-48.23	11.03	2556	595	0.4	63.0	0.64
PS1771-4	-53.77	3.78	1811	205	0.6	1.6	0.28
PS2092-1	-54.02	3.48	1897	275	0.3	2.7	0.32
PS2093-1	-54.17	3.38	1441	345	0.2	3.2	0.45
PS2094-1	-54.28	3.27	937	880	0.4	3.9	0.33
PS2095-1	-54.32	3.23	486	800	0.1	3.8	0.33
PS2096-1	-54.5	3.02	502	1695	0.2	3.2	0.27
PS2097-1	-54.52	2.9	1020	335	0.3	3.1	0.21
PS2098-1	-54.53	2.8	1500	180	0.3	2.2	0.38

Table 1. Station list with sample numbers, geographical position, water depth, standing stocks of benthic foraminifera, as well as total organic carbon and carbonate contents of the sediment, and stable carbon isotopic composition of dissolved inorganic carbon of bottom water. Numbers in italics indicate extrapolation from measurements at nearby stations. Sample PS1777-7 is omitted in Figure 1 for graphical reasons. Sediment data are from Pangaea@awi-bremerhaven.de, originally published by Mollenhauer et al. (2002) and Wagner et al. (submitted).

Another set of 16 samples is from the South African continental slope between the equator and about 30°S, from a water depth of approximately 1300 m, chosen to assure that conservative water mass characteristics change only little and most sample positions are bathed by Antarctic Intermediate Water or Upper Circumpolar Deep Water, except one dominated by North Atlantic Deep Water (Mackensen et al. 1995; Schmiedl et al. 1997; Licari et al. 2003). The sample positions in the eastern South Atlantic are ecologically dominated by several systems of permanent and seasonal upwelling cells which generate productivities that vary significantly in time and space (Lutjeharms and Meeuwis 1987).

At the equator, strong seasonal variations in the trade winds produce a highly fluctuating ocean surface productivity system. At 6°S, in addition to an enhanced nutrient supply from Congo River outflow, the rapidity of this freshwater outflow forced by a narrow estuary, induces upwelling at the river mouth by entrainment of cool subsurface oceanic waters which are rich in phosphate and nitrate (Van Bennekom et al. 1978). Between 10° and 15°S off Angola, a complex system of fronts, gyres, and thermal domes induces highly seasonal oceanic upwelling (Stramma and Schott 1999). Between 15°-20°S, southerly and southeasterly trade winds induce permanent intense coastal upwelling cells (Lutjeharms and Meeuwis 1987).

This general scheme, schematically illustrated in Figure 1, is in good agreement with global maps of primary productivity estimates (calculated after Behrenfeld and Falkowski (1997), based on recent satellite chlorophyll data (SeaWIFs chlorophyll data from September 1997 to August 1998; http://marine.rutgers.edu/opp)). According to these maps, annual primary productivity values increase from 180 gC m^{-2} y^{-1} in the northern part of the Guinea Basin to 450 gC m^{-2} y^{-1} in the Angola Basin.

Remineralization of organic material below the high productivity areas and off the river mouths leads to the formation of an oxygen minimum zone (OMZ), which reaches its largest extension off the Cunene River (around 17°S), where it extends between water depths of 50 and 1200 m with an oxygen minimum value of 1ml l^{-1}. The intensity of the OMZ decreases progressively northward (Chapman and Shannon 1985, 1987; Schulz and Cruise participants 1992).

Methods

Surface sediment samples were taken with the aid of a multiple corer down to a subbottom depth of 15 cm, preserved in a mixture of alcohol and Rose Bengal (1 g Rose Bengal dissolved in 1L 96%-ethanol) and kept cool at 4 °C until further treatment (Walton 1955; Lutze 1964; Mackensen and Douglas 1989). From each station, at least four of twelve multiple corer subcores (6 cm diameter), or two of ten (10 cm diameter) were analyzed, selected according to quality, i.e. clarity of the supernatant water, and position, i.e. maximum distance between subcores. This way we achieved to obtain replicates from the surface area covered by a multiple corer, i.e. of about one m^2, by just one deployment of the sampling gear. This allows for a reasonable treatment of patchiness in benthic foraminiferal abundance patterns on a spatial scale from tens of centimeters to about one meter. On shore, samples were wet sieved over 63 µm and the dried fraction >125 µm was investigated for its benthic foraminiferal content. Live (stained) and dead assemblages were counted separately. The stable isotopic composition of live and dead benthic foraminifera was determined with a Finnigan MAT251 isotope ratio gas mass spectrometer directly coupled to an automatic carbonate preparation device (Kiel I) and calibrated via NIST 19 to the VPDB (Vienna Pee Dee Belemnite) scale. The number of specimens analyzed in a single measurement varied between two and five for cibicids and uvigerinids, between five and eight for *Bulimina aculeata*, and nine and 25 for all other species. All values are reported as δ-notation versus VPDB. The overall precision of the measurements based on repeated analyses of a laboratory standard (Solnhofen limestone) over a one year period was better than 0.06 ‰ and 0.09 ‰ for carbon and oxygen, respectively.

The water column was sampled with the aid of Niskin bottles mounted on a water-sampling rosette. The bottom water from directly above the sediment/water interface was sampled from one of the multiple corer tubes. Although sub-sampled immediately after recovery on board, the bottom water in the sub-cores of the multiple corer was nevertheless influenced by shipboard handling that may have led to the release of porewater into the bottom water (Mackensen et al. 1993b). Dissolved inorganic carbon was either manually extracted in a preparation line following Mook et al. (1974), or automatically extracted with a Finnigan Gas Bench using gas chromatography to purify CO_2. The stable isotopic composition of resulting purified CO_2 was determined with a Finnigan MAT252 isotope ratio gas mass spectrometer and calibrated to the VPDB scale. The precision of carbon measurements, including CO_2 extraction, is better than 0.1 ‰. All samples including gas extraction were processed in duplicate.

Results

Water Column $\delta^{13}C_{DIC}$

To define the range of water depths in which conservative water mass properties and $\delta^{13}C_{DIC}$ values vary only slightly, we compiled new $\delta^{13}C_{DIC}$ data measured in this study and data published earlier in a section from Antarctica through the Agulhas Basin, crossing the middle oceanic ridge at Bouvet Island, the Walvis Ridge between Cape and Angola Basins and eventually, via the Guinea Abyssal Plain, the Equator (Oestlund et al. 1987; Mackensen et al. 1993b; Mackensen et al. 1996; Bickert and Wefer 1999; Mackensen 2001). All of the stations investigated in this study are bathed by waters with $\delta^{13}C_{DIC}$ values ranging from 0.2 to 0.7 ‰ VPDB, i.e. AAIW and UCDW (Fig. 2, Table 1).

Microhabitats and Foraminiferal $\delta^{13}C$

In accordance with the distribution of benthic foraminiferal faunal provinces characterized by specific assemblage compositions (Mackensen et al. 1993a; Mackensen et al. 1995; Schmiedl et al. 1997; Schumacher 2001; Licari et al. 2003), we present and discuss in this paper faunal and isotopic data in the order of the following geographic positions of sites (Fig. 1, Table 1): (1) samples from sites PS1771 to PS2098 from near Bouvet Island between 48 and 55°S (hereinafter referred to as Agulhas Basin samples), (2) samples south of

Walvis Ridge including GeoB3711-1 (hereinafter referred to as Cape Basin samples), (3) samples from north of Walvis Ridge from site GeoB3715 to site GeoB4916 (hereinafter referred to as Angola Basin samples), and (4) north of 10°S (hereinafter referred to as Guinea Abyssal Plain samples).

The number of live (stained) benthic foraminifera in the >125 µm size fraction generally decreases from Bouvet Island in the south at about 50°S, via the Cape Basin and Angola Basin to the Guinea Abyssal Plain in the north near the equator (Fig. 3). Interestingly, in Cape Basin and Angola Basin samples, highest standing stock values are found between one and two cm below the sediment surface, in contrast to the expected exponential decrease from the sediment surface down to 11 cm subbottom depth as found off Bouvet Island and at the Guinea Abyssal Plain continental slope.

Agulhas Basin: At the mid-ocean ridge, which Bouvet Island is part of, the benthic foraminiferal fauna is characterized by the presence of *Bulimina aculeata* and *Angulogerina angulosa* (Macken-sen et al. 1994). At all stations except one the number of live specimens of these two species exponentially decreases from the uppermost surface sediment down to about five centimeters subbottom depth (Fig. 4). At station PS2094, however, live specimens are found at a water depth of 937 m down to a subbottom depth of 11 cm, and surface values decrease from 59 individuals per 10 cm^3 in the top centimeter to four in the second sediment slice, then increasing to a second subsurface maximum at a subbottom depth of 3-5 cm (Fig. 4).

The stable carbon isotope offset of preferentially infaunal *B. aculeata* from bottom water $\delta^{13}C_{DIC}$ at all stations investigated rather constantly varies around -0.6 ± 0.1 ‰ throughout the sediment, except at station PS2094 where the $\delta^{13}C$ values are consistently lower, varying around -1.5 ± 0.2 ‰ (Mackensen et al. 2000). The $\delta^{13}C$ values of live and dead specimens of *A. angulosa* vary around the same mean within the standard deviation of the *B. aculeata* values (Fig. 5). *Angulogerina angulosa* does not thrive in water depths beyond 800 m in this area and thus is not found at station PS2094. The stable carbon isotopic com-

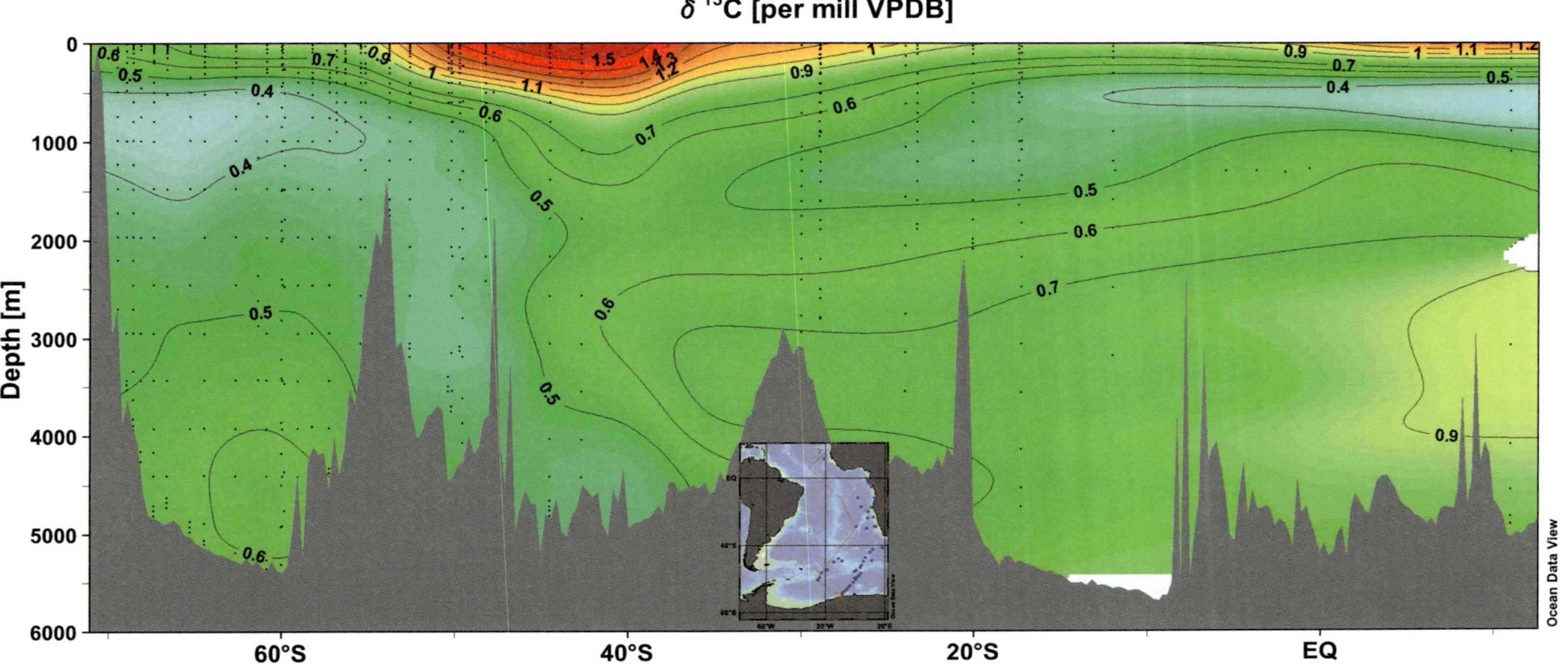

Fig. 2. Distribution of $\delta^{13}C_{DIC}$ (‰ VPDB) on a section from 70°S at the Antarctic continental margin along the African continental slope to 10°N. Data are compiled from Mackensen et al. (1993b, 1996, 2001), Bickert and Wefer (1999), and Oestlund et al. (1987) using Ocean Data View visualization software (Schlitzer 2002).

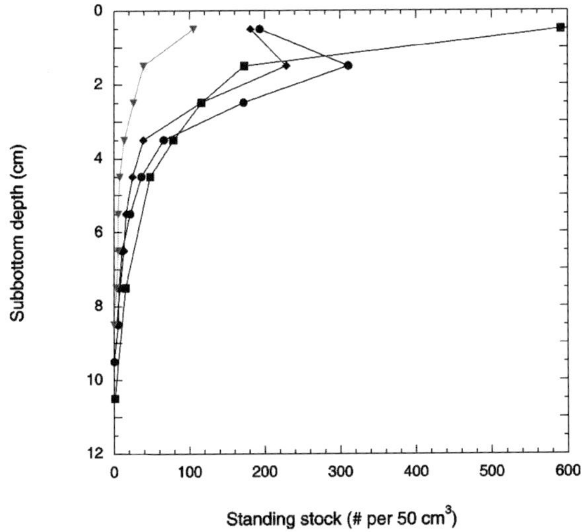

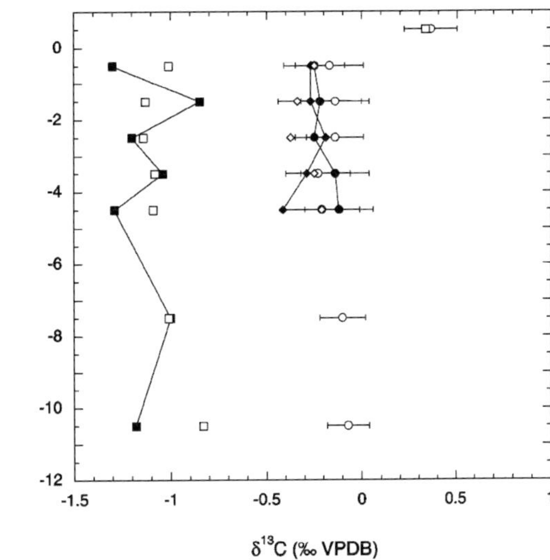

Fig. 3. Mean total benthic foraminiferal standing stocks of nine stations from Agulhas Basin (squares), and of six and five stations from the African continental slopes of Cape (diamonds) and Angola Basins (dots), respectively, as well as five stations from Guinea Abyssal Plain slope (triangles).

Fig. 5. δ^{13}C values of live and dead *Bulimina aculeata* (filled and open circles, respectively) from eight stations and one single station PS2094 off Bouvet Island (squares) are plotted versus sediment depth, as well as δ^{13}C values of live and dead *Angulogerina angulosa* from two of the eight stations for comparison (filled and open diamonds, respectively). Error bars indicate the standard deviation of the mean of eight and two stations, respectively. In addition, the δ^{13}C$_{DIC}$ values of the bottom water are indicated, as determined in multiple corer water simultaneously sampled (for graphic reasons plotted 0.5 cm above sediment/water interface).

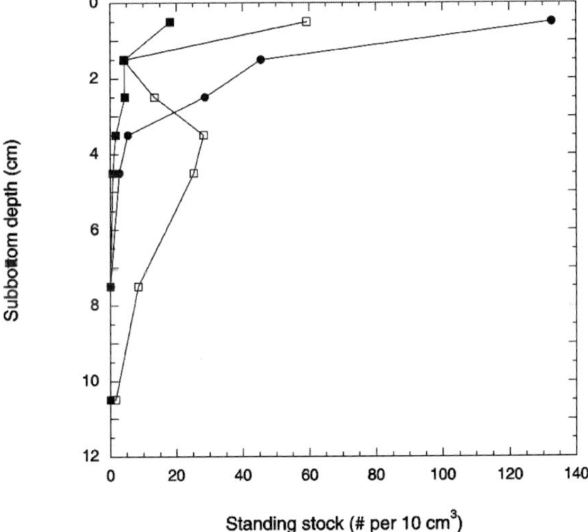

position of dissolved inorganic carbon was determined in multiple corer water and checked with δ^{13}C$_{DIC}$ measurements from the water column (Mackensen et al. 1993b; Mackensen et al. 1996). According to this, δ^{13}C$_{DIC}$ values from multiple corer water may underestimate the actual bottom water δ^{13}C$_{DIC}$ by 0.2 ‰ at the most (see Methods section).

Cape and Angola Basins: The live benthic foraminiferal fauna at the African upper continental slope between 30 and 10°S is characterized by assemblages either dominated by *Fontbotia wuellerstorfi* or *Uvigerina auberiana*, mainly depending on substrate composition and organic matter fluxes (Schmiedl et al. 1997). Further north *Uvigerina peregrina* becomes a more dominant component of the live fauna, in addition to species such as *Globobulimina turgida* and *G. affinis*, as well as *Fursenkoina earlandi* and *F. mexicana*

Fig. 4. Mean standing stocks of live *Bulimina aculeata* (filled squares) from eight stations and one single station PS2094 (open squares) off Bouvet Island are plotted versus sediment depth. Mean standing stock of live *Angulogerina angulosa* (dots) from two of the eight stations is given for comparison.

632   Mackensen and Licari

that were found associated discontinuously along the entire continental slope at this water depth (Mackensen et al. 1995; Schmiedl et al. 1997; Licari et al. 2003). Although total benthic foraminiferal standing stocks at the slopes of both Cape Basin and Angola Basin clearly show a shallow subsurface maximum between one and two centimeters within the sediment (Fig. 3), the preferentially epibenthic living *F. wuellerstorfi* mostly is restricted to the first centimeter and is on average equally distributed within the top two sediment centimeters at all stations in Cape Basin. On the contrary, deeply infaunal species such as *G. affinis* and *F. mexicana* exhibit a subsurface distributional maximum between five and seven, and three and six centimeters, respectively (Fig. 6).

All mean $\delta^{13}C$ values of the various epifaunal and infaunal species under investigation do not significantly vary or obviously decrease with increasing depth within the sediment (Fig. 7, Table 2). The data vary only slightly around a mean value for each single species at each single site, no matter what habitat the species prefers and at what subbottom depth the particular specimens were

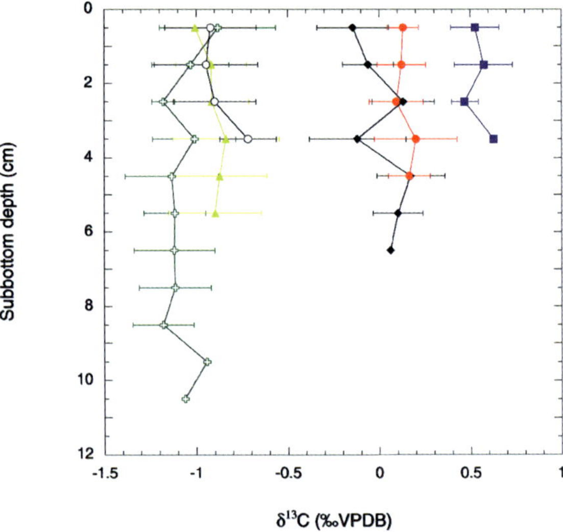

Fig. 7. Mean $\delta^{13}C$ values of live *Fontbotia wuellerstorfi* (squares), *Cibicidoides pachyderma* (diamonds), and *Lobatula lobatula* (dots) from five stations in Cape Basin, as well as $\delta^{13}C$ values of live *Globobulimina affinis* (crosses), *Fursenkoina mexicana* (triangles), and *Bulimina mexicana* (open circles) from seven, four, and ten stations, respectively, from Cape and Angola Basins plotted versus sediment depth. Error bars give standard deviations of means of stations used for the particular species. Bottom water $\delta^{13}C_{DIC}$ values (not plotted) at stations lie exactly within the range of *F. wuellerstorfi* $\delta^{13}C$, i.e. between .45 and .65 ‰.

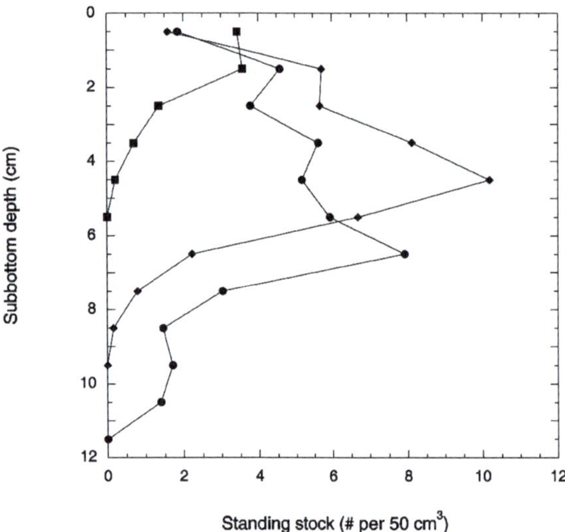

Fig. 6. Mean standing stock of *Fontbotia wuellerstorfi* (squares) from four stations in the Cape Basin, of *Globobulimina affinis* from six stations in Cape and Angola Basins (dots), and of *Fursenkoina mexicana* (diamonds) from two stations in Angola Basin plotted versus sediment depth.

caught. The preferentially epibenthic living species *F. wuellerstorfi*, *Lobatula lobatula* and *Cibicidoides pachyderma*, however, can be clearly differentiated by $\delta^{13}C$ values more than 1 ‰ higher than the values applying to the deep and shallow infaunal *G. affinis*, *F. mexicana*, and *B. mexicana*.

Guinea Abyssal Plain: Standing stock values at Guinea Abyssal Plain are the lowest of all stations investigated in this study (Fig. 3). Consequently, mean standing stock values of single taxa are low as well, and only few specimens are available for stable isotope determinations. This is in particular the case for preferentially infaunal species. One of the most common taxa in these samples is *Uvigerina peregrina* (Licari et al. 2003). Although generally considered as a shallow infaunal species, it clearly has its distribution maximum within the top centimeter of the sediment. In the

Sample	L. lobatula	stdev	C. pachyderma	stdev	F. wuellerstorfi	stdev	B. mexicana	stdev	G. pacifica	stdev	F. mexicana	stdev	U. auberiana	stdev	U. peregrina	stdev
GeoB4904-7															-0.48	-0.95	0.12
GeoB4906-5							-1.05								-0.49	-0.94	0.03
GeoB4909-4															-0.19	-0.66	0.08
GeoB4913-4																-0.76	
GeoB4915-3			0.44		0.63				-0.93							-0.55	
GeoB4916-4							-1.25	0.03									
GeoB4917-5							-1.22	0.05	-1.02	0.13			-1.03	0.09	-0.38	0.16
GeoB3713-1					1.01		-0.82	0.07	-0.86	0.08					0.03	
GeoB3712-1							-0.44	0.11			-0.49	0.03	-0.63	0.08		
GeoB3715-1							-0.85	0.08	-0.98	0.24	-0.84	0.10	-0.93	0.05		
GeoB3711-1	0.13	0.14	-0.01	0.09	0.39	0.02	-1.08	0.03			-1.06	0.03	-1.21	0.03	-0.71	
GeoB3708-1	0.00	0.08	-0.03	0.03	0.48	0.05	-0.92	0.14	-1.27	0.04	-1.04	0.02	-1.15	0.01		
GeoB3706-3	0.26	0.05	0.12	0.06	0.64	0.11	-0.74	0.10	-1.15	0.20						
GeoB3725-1	0.14		0.10	0.40	0.60	0.07			-1.30	0.13					-0.86	
GeoB3705-2	0.15		-0.28	0.13	0.45		-0.98	0.07								
GeoB3703-4							-1.04	0.06	-1.00	0.14			-1.09	0.14		
GeoB4904-7																-0.62	0.11
GeoB4906-5																-1.03	0.17
GeoB4909-4																-0.64	0.10
GeoB4915-3					0.75				-1.33							-0.28	
GeoB4916-4							-1.14	0.06									
GeoB4917-5							-1.09	0.04	-1.03	0.16					-0.44	0.01
GeoB3713-1							-0.84	0.06	-0.95	0.32					-0.20	0.10
GeoB3712-1							-0.75	0.07			-0.55	0.07	-0.77	0.08		
GeoB3715-1							-0.96	0.02	-1.22	0.20	-0.85	0.04	-0.98	0.14		
GeoB3711-1	0.04	0.11	0.01	0.16	0.37	0.17	-0.98	0.04			-0.93	0.06	-1.09	0.05		
GeoB3708-1	-0.01	0.04	-0.07	0.10	0.55	0.05	-1.10	0.06	-1.28	0.15	-1.07	0.05	-1.31	0.06		
GeoB3706-3	0.15	0.14	0.09	0.09	0.58	0.17	-1.03	0.05	-1.34	0.26						
GeoB3725-1	0.25		0.17	0.23	0.34	0.22			-1.14	0.19						
GeoB3705-2			-0.18	0.31	0.53	0.15	-1.06	0.03								
GeoB3703-4									-0.89	0.13						

Table 2. Mean $\delta^{13}C$ (‰ VPDB) values and standard deviations of live (upper part) and dead (lower part) species from the sediment surface and all subbottom depths at every single station.

second and third centimeter within the sediment, however, about half of the total number of live specimens is still found (Fig. 8). The $\delta^{13}C$ values of stained and empty *U. peregrina* tests vary between stations by ±0.25 ‰, regardless of the bottom water $\delta^{13}C_{DIC}$ values, but they do not significantly change with increasing subbottom depths (Fig. 8, Table 2).

Discussion

Intraspecific $\delta^{13}C$ Variability at Single Sites

In all four series of samples from stations in Agulhas Basin, Cape Basin and Angola Basins as well as the Guinea Abyssal Plain, $\delta^{13}C$ values of all species

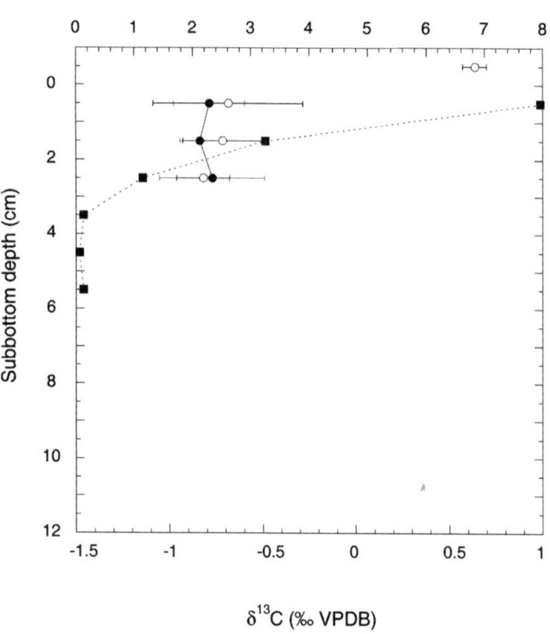

Fig. 8. Mean $\delta^{13}C$ values of live and dead *Uvigerina peregrina* (dots and open circles, respectively), as well as standing stocks (squares) from five stations at the Guinea Abyssal Plain continental slope are plotted versus sediment depth. Error bars give standard deviations of means of stations used. $\delta^{13}C_{DIC}$ values of the bottom water are indicated, as determined from simultaneously sampled Multiple corer water (for graphic reasons plotted 0.5 cm above sediment/water interface).

vary on average ±0.09 ‰ (range of standard deviations: 0.01 – 0.40 ‰) around a species specific and site dependent mean value calculated from all subbottom depths (Figs. 5, 7, 8; Table 2). This simply reflects that at each station live specimens of each particular species have about the same stable carbon isotopic composition, regardless of the actual sediment depth they were caught in. This in turn means that no matter which habitat depth the different species prefer, all specimens of the same species record the same $\delta^{13}C$ signal. This was shown earlier for some of the species discussed here (Mackensen and Douglas 1989; McCorkle et al. 1990; McCorkle and Keigwin 1994; Rathburn et al. 1996; Mackensen et al. 2000), but to our knowledge it has not been clear yet that obviously most, if not all benthic foraminiferal species behave like this. This means further that if specimens of particular species vertically migrate within the sediment, as is known from culture experiments (Severin 1987; Moodley 1992; Ernst et al. 2002) they obviously do not record different $\delta^{13}C$ signals in accordance with the stratified and downwards rapidly decreasing $\delta^{13}C$ values of pore water DIC. Based on a limited data set comprising values merely pertaining to *B. aculeata* , it was suggested that by moving within the sediment, individuals of infaunal species either record an average isotope signal of the pore water or the signal of one specific calcification depth level (Mackensen et al. 2000). Now more data from different species being at hand, it is safe for paleoceanographers to interpret fossil benthic foraminiferal $\delta^{13}C$ values as representing one signal at one particular site at one particular time. In other words, if there is strong variation within fossil specimens of the same species at one site in a sample of one particular age, mechanisms other than migration of specimens and subsequent recording of different $\delta^{13}C$ signals during lifetime have to be considered as an explanation. The most common possibilities should include reworking of older sediments, or lateral and down-slope transport from elsewhere.

Interspecific $\delta^{13}C$ Differences at Single Sites

Generally, at all sites along the African continental slope, $\delta^{13}C$ values of live and dead *Fontbotia wuellerstorfi, Cibicidoides pachyderma*, and *Lobatula lobatula* differ consistently and significantly from $\delta^{13}C$ values of live and dead *Globobulimina affinis, Fursenkoina mexicana*, and *Bulimina mexicana* by 1.5 to 1.0 ‰ (Fig. 7, Table 2). This finding is in good agreement with the hypothesis that lower $\delta^{13}C$ values of *G. affinis, B. mexicana* and *F. mexicana* — species generally considered to dwell in deep or shallow infaunal microhabitats (Kitazato 1994; Jorissen et al. 1998) — simply reflect a calcification depth within the sediment and its depleted $\delta^{13}C_{DIC}$ porewater values (Woodruff et al. 1980; Grossman 1984b; Mackensen and Douglas 1989; McCorkle et al. 1990; Rathburn et al. 1996). Moreover, depleted $\delta^{13}C$ values of *C. pachyderma* and *L. lobatula* additionally reflect that cibicids and in particular *C. pachyderma* are not strictly epifaunal in contrast to *F. wuellerstorfi* (Wollenburg and Mackensen 1998). Alternatively, different taxa may feed on different diets or may even live in symbiosis with bacteria, instead of eating them (Bernhard and Bowser 1992; Jorissen et al. 1995; Kitazato and Ohga 1995; Bernhard et al. 2000). This eventually, via the metabolism of the foraminifera, affects the stable carbon isotopic composition of their calcareous tests.

It also appears that *F. wuellerstorfi* records bottom water $\delta^{13}C_{DIC}$ values almost perfectly, without any significant offset, whereas *L. lobatula* and *C. pachyderma* values deviate from bottom water values by about -0.4‰ and -0.6‰, respectively (Fig. 7). This adds to the growing data on contrasting cibicid $\delta^{13}C$ values which on the one hand supports the original calibration between *F. wuellerstorfi* and bottom water $\delta^{13}C_{DIC}$ (Duplessy et al. 1984; McCorkle and Keigwin 1994), and on the other hand documents severe depletions of *L. lobatula* which was shown just recently to reliably record $\delta^{13}C_{DIC}$ values in the Nordic seas (Hald and Aspeli 1997; Mackensen et al. 2000). Moreover, the confirmation of *F. wuellerstorfi* as a reliable recorder of bottom water $\delta^{13}C_{DIC}$ values at the Cape Basin continental slope supports the view that explains negative deviations in *F. wuellerstorfi* tests of up to -0.6 ‰ close to the Antarctic Polar Front (Mackensen et al. 1993b; Mackensen et al. 2001) and in other high sedimentation rate environments

(McCorckle pers. Comm.) as being mainly due to a highly seasonal ocean surface production and the subsequent deposition of ephemeral phyto-detritus layers. A closer look on a satellite-based primary productivity map of the ocean at the positions of the seven stations investigated here for *F. wuellerstorfi* $\delta^{13}C$ values, shows that most of them are situated not below the belt of highly seasonal coastal upwelling, but are rather influenced by sustained ocean surface productivity and constant rain rates (Fig. 1) (Behrenfeld and Falkowski 1997; Schmiedl et al. 1997). However, because of the paramount importance of *F. wuellerstorfi* and other epibenthic cibicids for paleoceanographic reconstruction more research is demanded to assess the $\delta^{13}C$ fidelity of these taxa.

Intraspecific $\delta^{13}C$ Differences between Sites

Generally, site specific $\delta^{13}C$ values of epibenthic foraminiferal species can be influenced by (1) the $\delta^{13}C_{DIC}$ values of the bottom water mass (Duplessy et al. 1984; Curry et al. 1988), (2) the saturation state of bottom water with respect to calcite, which may influence the isotopic signal either by *post mortem* dissolution or by altering the fractionation during calcification (McCorkle et al. 1995; Boyle and Rosenthal 1996; Spero et al. 1997; Mackensen et al. 2000), and (3) highly seasonal ocean surface productivity with high-sedimentation rates and the subsequent development of an ephemeral phytodetritus layer at the sea floor. This phytodetrital layer may influence the isotopic signal either by inducing the animals to calcify test chambers when plenty of food is available but fluffy layer $\delta^{13}C_{DIC}$ values are depleted in ^{13}C due to increased organic decomposition rates, or by altering fractionation coefficients during the more rapid calcification processes (Mackensen et al. 1993b; Mackensen and Bickert 1999; Mackensen et al. 2000).

In addition to the environmental conditions that affect $\delta^{13}C$ values of epibenthic foraminifera, endobenthic species further can be influenced by (1) the $\delta^{13}C_{DIC}$ gradients in the pore water, which in turn depend on the decomposition rate of sedimentary organic matter, which again is largely driven by the rain rate of particulate organic matter to the sea floor (Belanger et al. 1981; Woodruff and Savin 1985; Zahn et al. 1986; McCorkle et al. 1990), and (2) by the release of methane into the pore water, which either significantly lowers the pore water $\delta^{13}C_{DIC}$ values via the pathway of bacterial oxidation to CO_2, or which favor benthic foraminifera to feed directly on methane consuming bacteria and thus alter the isotopic fractionation in the course of their metabolism (Wefer et al. 1994; Kennett et al. 2000; Bernhard et al. 2001).

In the Agulhas Basin data, we interpret the offset of 1.5 ± 0.2 ‰ between bottom water $\delta^{13}C_{DIC}$ and $\delta^{13}C$ values of *B. aculeata* at one single station in contrast to about 0.6 ± 0.1 ‰ at all other stations, as a direct reflection on locally increased organic matter fluxes and sedimentation rates (Fig. 9). This particular station PS2094 exhibiting lower $\delta^{13}C$ values in live and dead *Bulimina aculeata* calcite at about 1 ‰ compared to the rest of the stations, displays similar total organic carbon contents of the sediments but shows a specific high-productivity/low-oxygen conditions indicating deep infaunal *Fursenkoina earlandi* component and very high standing stocks (Mackensen et al. 1993a;

Mackensen et al. 2000). Alternatively, from the one station with low values in contrast to eight stations close-by with significantly higher values, all of them overlain by the same general productivity regime, one may speculate that this strong depletion of ^{13}C in the composition of isotopic carbon in benthic foraminifera may have been caused by methane which could have been locally released from gas vents, such as reflecting hydrothermal activities at the mid-ocean ridge (cf. Rathburn et al. 2000). It is worth mentioning that the $\delta^{18}O$ values of *B. aculeata* from station PS2094 do not differ from the $\delta^{18}O$ values of *B. aculeata* at the rest of the stations.

We plotted mean $\delta^{13}C$ values of live *Fontbotia wuellerstorfi*, *Lobatula lobatula*, *Cibicidoides pachyderma*, *Bulimina mexicana*, *Globobuli-mina affinis*, *Fursenkoina mexicana*, *Uvigerina auberiana*, and *U. peregrina* as well as $CaCO_3$ contents of the sediments from all stations along the African continental margin versus latitude (Fig. 10). Mean $\delta^{13}C$ values of live and dead preferentially epibenthic *F. wuellerstorfi* reduced by station specific $\delta^{13}C_{DIC}$ values, vary between 0 and 0.2‰, i.e. well within the ±0.2 ‰ range of variation around the expected 1:1 relationship between foraminiferal calcite and bottom water $\delta^{13}C$ values (Figs. 10, 13), except at station GeoB3713 at 15°S where $\delta^{13}C$ values exceed $\delta^{13}C_{DIC}$ values of bottom water by 0.7‰ (Fig. 10).

Comparing $\delta^{13}C$ values of all species from all stations in a more generalized view, it becomes obvious that all species exhibit significantly higher values between 19 and 15°S, centered at stations Geob 3712 and 3713 (Figs. 10, 13). This particular area off Angola coincides with low $CaCO_3$ and organic carbon contents of the sediments. Further north, both sedimentary carbonate and organic carbon contents increase again between 16 and 7°S, to finally decrease to lowest values in the Guinea Abyssal Plain samples. Below the equator, carbonate values of the continental slope in a water depth of 1300 m then increase to about 17 weight% (Fig. 11). The low sedimentary carbonate contents are at least partially related to the oxidation of particulate organic matter within the water column, which is confirmed by a well developed oxygen minimum zone, as well as by the organic matter decomposition in the sediments

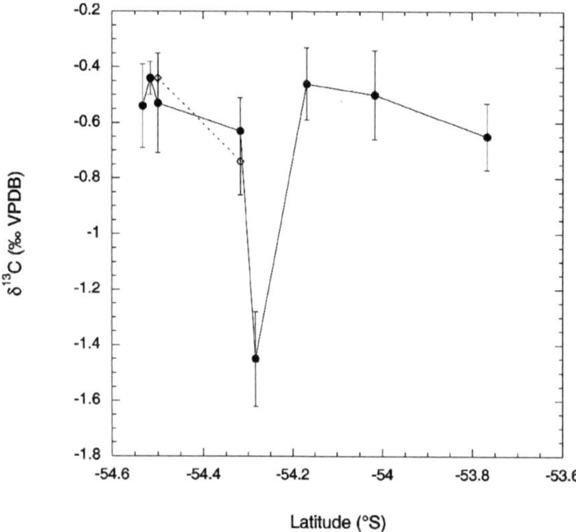

Fig. 9. Mean $\delta^{13}C$ values of live *Bulimina aculeata* (dots) and *Trifarina angulosa* (diamonds) from all stations in Agulhas Basin plotted versus latitude. Error bars indicate standard deviations of means of all downcore values at each single station. Bottom water $\delta^{13}C_{DIC}$ values at each of the stations are subtracted.

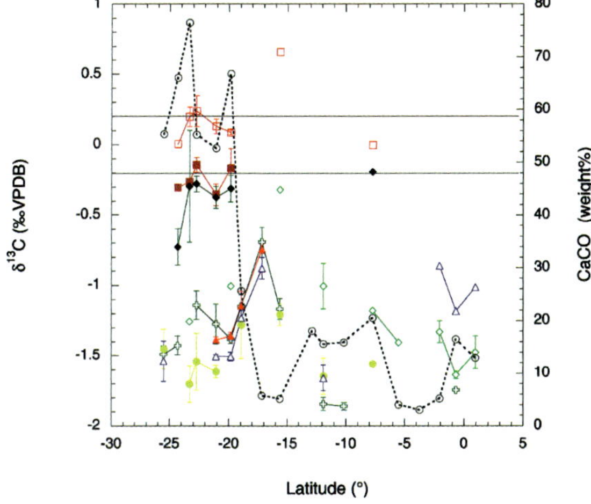

Fig. 10. Mean δ¹³C values of live *Fontbotia wuellerstorfi* (open squares), *Lobatula lobatula* (filled squares), *Cibicidoides pachyderma* (filled diamonds), *Bulimina mexicana* (crosses), *Globobulimina affinis* (dots), *Fursenkoina mexicana* (filled triangles), *Uvigerina auberiana* (open triangles), and *U. peregrina* (open diamonds), as well as CaCO₃ contents (open circles) of the sediments from all stations along the African continental margin plotted versus latitude. Error bars indicate standard deviations of means of all downcore values at each single station. Bottom water δ¹³C$_{DIC}$ values are subtracted. Horizontal lines at ±0.2 ‰ give range of 1:1 relationship between δ¹³C values of bottom water DIC and foraminiferal calcite as commonly tolerated for paleonutrient reconstructions.

environment. We, therefore, interpret the enriched calcite δ¹³C values as mainly reflecting a severe undersaturation of bottom water and pore waters.

Although there is some scattering of data due to the lack of sufficient specimen numbers applied in the stable isotope determinations of all species in each sample, *U. auberiana,* and partly *U. peregrina* , corroborate our interpretation as both exhibit higher values which coincide with the low carbonate contents of the sediment (Figs. 10, 13). Furthermore, the mean δ¹⁸O value of live *F. wuellerstorfi* at station GeoB3713 is enriched by 0.47 ‰, relative to the lowest value of Station GeoB3705 from the same water depth and displaying about the same conservative water mass properties, and by 0.22‰ as compared to the mean of all stations. This supports the interpretation of an isotope fractionation effect caused by low carbonate ion concentrations as was first discussed by Lea et al. (1999) with regard to benthic foraminifera and which up to now has been observed under natural conditions only for planktonic foraminifera (Bauch et al. 2002).

A scatter plot of the sedimentary carbonate content versus the mean δ¹³C values of *F.*

of the continental slope beneath. Both processes cause the water to become more acidic and lower the carbonate saturation state through the consequent decrease in [CO₃²⁻], the carbonate ion concentration, which results in highly carbonate aggressive bottom and pore waters. A low carbonate ion concentration results in enhanced isotopic fractionation during calcification of foraminiferal tests in this environment (Spero et al. 1997; Bijma et al. 1999), or an initial dissolution of calcitic tests as early as during the lifetime of the organisms. Although both of these processes induce a shift toward higher δ¹³C values, the latter process, however, seems rather unlikely, since most benthic foraminifera protect their calcitic tests with an outer organic lining against dissolution in a carbonate aggressive micro-

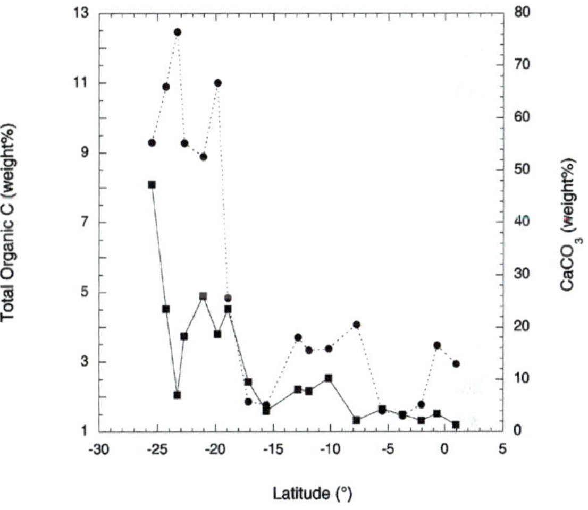

Fig. 11. Sediment CaCO₃ (dots) and total organic carbon (squares) contents from along the African continental slope of surface sediment samples investigated in this study plotted versus latitude.

wuellerstorfi, F. mexicana and *U. auberiana* further illustrates that, below a threshold of about 15% sedimentary carbonate content, $\delta^{13}C$ values are enriched by up to 0.7‰ relative to values of samples containing 50-80% carbonate (Fig. 12). Above we discussed a significant depletion in ^{13}C of *B. aculeata* tests at one single station off Bouvet Island with respect to all other stations in this area as probably being caused by an enhanced local deposition of organic matter and subsequent low pore water $\delta^{13}C_{DIC}$ values. On the contrary, on the African continental slope between 19 and 15 °S, we have shown that if the decomposition of organic matter within the intermediate water column and the interstitial waters is not compensated for by a sufficient supply of carbonate from the oceanic surface layer, the carbonate ion concentration decreases below a threshold that finally results in a

^{13}C enrichment of epibenthic *F. wuellerstorfi* and different preferentially infaunal species, although pore water $\delta^{13}C_{DIC}$ values must still be low. This is no contradiction, but just the reflection of processes with opposing effects on the stable isotopic composition of benthic foraminiferal tests in areas with generally high ocean surface productivity and organic matter rain rates. This may be best illustrated by the shallow infaunal *Bulimina mexicana*: Mean $\delta^{13}C$ values from stations between 26 and 20 °S with >50 % sedimentary $CaCO_3$ content, vary between -1.2 and -1.5 ‰, between 19 and 15 °S with <5% $CaCO_3$ values increase to -0.7 and $-1,2$ ‰, and between 12 °S and the equator with >15 % carbonate, values decrease again below -1.7 ‰ (Fig. 10). As suggested above, in areas with a generally high productivity below the sedimentary carbonate threshold of 15 weight%, epibenthic and endobenthic $\delta^{13}C$ values are predominantly influenced by ^{13}C enrichment due to carbonate ion undersaturation, whereas above this threshold $\delta^{13}C$ values of endobenthic species reflect the influence of the strongly depleted pore water $\delta^{13}C_{DIC}$ values.

Although the dead assemblage represents a longer period of time than the live assemblage, good agreement between live and dead benthic stable isotope values has often been observed (McCorkle et al. 1990; Mackensen et al. 1993b). To broaden the data base and help to assess the persistence of the features observed in the living assemblage, we plotted mean $\delta^{13}C$ values of dead *Fontbotia wuellerstorfi, Lobatula lobatula, Cibicidoides pachyderma, Bulimina mexicana, Globobuli-mina affinis, Fursenkoina mexicana, Uvigerina auberiana*, and *U. peregrina* as well as $CaCO_3$ contents of the sediments from all stations along the African continental margin versus latitude (Fig. 13). Although the variability within some species at single stations tends to be increased, the isotopic composition of empty tests generally corresponds perfectly to the pattern observed within the live data set. This corroborates our speculation that postmortem calcite dissolution does not play a major role in shaping the variability of benthic $\delta^{13}C$ values between the stations.

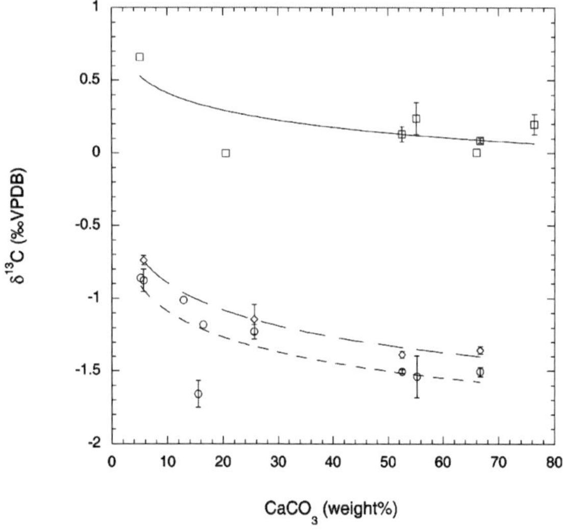

Fig. 12. Mean $\delta^{13}C$ values of live *Fontbotia wuellerstorfi* (squares), *Fursenkoina mexicana* (diamonds), and *Uvigerina auberiana* (circles) are plotted versus $CaCO_3$ contents of the sediments from all stations along the African continental margin. Error bars indicate standard deviations of means of down-core values at each single station. Bottom water $\delta^{13}C_{DIC}$ values are subtracted. Curve fit is logarithmic with y=0.81-0.39*log(x) and R=0.74 for *F. wuellerstorfi*, y=-0.28-0.62*log(x) and R=0.99 for *F. mexicana*, and y=-0.49-0.59*log(x) and R=0.81 for *U. auberiana*, reflecting virtually no response in $\delta^{13}C$ values on decreasing carbonate values below the threshold of 15 % $CaCO_3$.

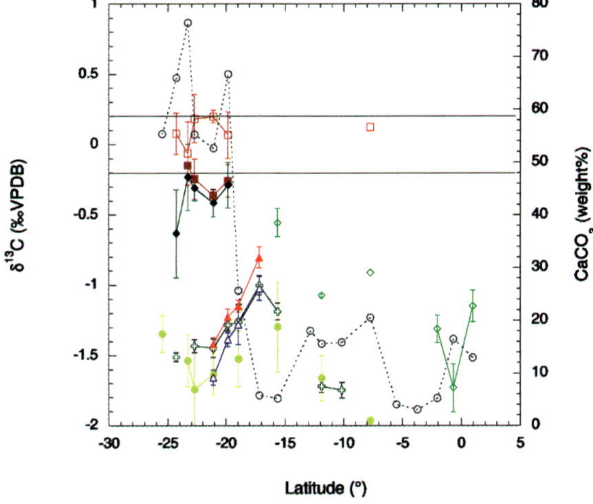

Fig. 13. Mean $\delta^{13}C$ values of dead *Fontbotia wuellerstorfi* (open squares), *Lobatula lobatula* (filled squares), *Cibicidoides pachyderma* (filled diamonds), *Bulimina mexicana* (crosses), *Globobulimina affinis* (dots), *Fursenkoina mexicana* (filled triangles), *Uvigerina auberiana* (open triangles), and *U. peregrina* (open diamonds), as well as $CaCO_3$ contents (open circles) of the sediments from all stations along the African continental margin plotted versus latitude. Error bars indicate standard deviations of means of downcore values at each single station. Bottom water $\delta^{13}C_{DIC}$ values are subtracted. Horizontal lines at ±0.2 ‰ show the range of 1:1 relationship between $\delta^{13}C$ values of bottom water DIC and foraminiferal calcite, which is commonly tolerated for paleonutrient reconstructions.

$\delta^{13}C$ Differences between Epi- and Endo-Benthic Species at Different Sites

Since Zahn et al. (1986) it is known that infaunal benthic $\delta^{13}C$ values depend on the organic carbon decomposition rate within the sediment and thus on the organic matter rain rate from the upper ocean to the sediment surface. On the other hand, epibenthic species such as *F. wuellerstorfi* in general reliably record the bottom water $\delta^{13}C_{DIC}$ ratios in their tests (Duplessy et al. 1984). Consequently, it has long been suggested that decreasing rain rates should be mirrored by decreasing differences between epi- and infaunal benthic $\delta^{13}C$ values, and *vice versa* (Woodruff and Savin 1985; Zahn et al. 1986; McCorkle et al. 1994). However, it was

argued that because of the expected down-core ^{13}C depletion of live specimens of infaunal taxa corresponding with the gradient of organic matter decomposition, the difference between epifaunal and infaunal $\delta^{13}C$ values randomly depends on how many specimens per sample were analyzed in the mass spectrometer and how sedimentation rates varied. This new study now corroborates earlier findings of Rathburn et al. (1996) and Mackensen et al. (2000) in suggesting that indeed most, if not all infaunal species have relatively consistent $\delta^{13}C$ values throughout their distribution within the sediments. Thus, we investigated whether the offsets between epifaunal cibicids and deeply infaunal globobuliminids or shallow infaunal uvigerinids vary with latitude (Fig. 14). The variation with latitude of the offsets between *G. affinis* as well as *U. peregrina* and *F. wuellerstorfi* $\delta^{13}C$ values around means of −1.77 ±0.13 and −1.18 ±0.20 ‰, respectively, is not significant, given the widely accepted overall margin of reliability of ±0.2 ‰ typical of this kind of investigation. However, there might be a visible trend indicating an increase of absolute dif-

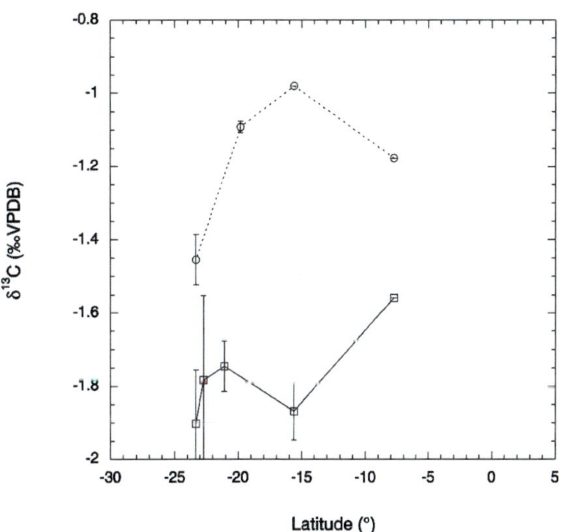

Fig. 14. Offset between deep infaunal *G. affinis* and epifaunal *F. wuellerstorfi* (squares) and between shallow infaunal *U. peregrina* and *F. wuellerstorfi* (circles) mean $\delta^{13}C$ values plotted versus latitude. Error bars give propagated standard deviations of differences between means of downcore values at each single station. Bottom water $\delta^{13}C_{DIC}$ values are subtracted.

ferences from the north to the south and, in particular, at the southernmost station which would comply with the hypothesis of increasing differences with increasing organic matter rain rates and subsequent decomposition rates, if corroborated by further investigations (Figs. 11, 14).

Conclusions

Summarizing the above discussion in terms of paleoceanographic applicability, we conclude:
• All species investigated display relatively consistent $\delta^{13}C$ values throughout their distribution within the sediments.
• The $\delta^{13}C$ values of *Bulimina aculeata* from a location close to Bouvet Island reflect ocean surface water productivity in the sense that high organic carbon fluxes result in low $\delta^{13}C$ values of tests of live specimens. Alternatively, methane locally released from gas vents and related to hydrothermal venting at the mid-ocean ridge may have caused the strong depletion of ^{13}C in the benthic foraminiferal carbon isotopic composition.
• If the decomposition of organic matter within the Oxygen Minimum Zone of the intermediate water column and within the interstitial waters is not compensated for by a sufficient supply of carbonate from the oceanic surface production, the carbonate ion concentration drops below a threshold that finally results in a ^{13}C enrichment of epibenthic *F. wuellerstorfi* and even of preferentially infaunal species, although pore-water $\delta^{13}C_{DIC}$ values are low.
• The offsets between deep endobenthic *Globobulimina affinis* and *Fontbotia wuellerstorfi* as well as between shallow endobenthic *Uvigerina peregrina* and *F. wuellerstorfi* $\delta^{13}C$ values tend to increase from north to south, in parallel with generally increasing rates of organic matter decomposition. Although clearly more data are needed, these offsets between species might be used for the quantification of biogeochemical paleogradients within the sediment and thus for paleocarbon flux estimates as was previously suggested.

Generally this study confirms that further detailed investigations of quantifying the dependence of $\delta^{13}C$ values of infaunal species from organic carbon fluxes are urgently needed to derive reliable paleoproductivity proxies.

Acknowledgements

We are grateful to the crews and shipboard scientific parties of RV *Meteor* for the recovery of virtually undisturbed surface sediment samples, as well as to M. Warnkroß and H. Röben for sample preparation and picking of stained specimens. We thank G. Traue, M. Matura and G. Meyer for running and supervising the mass spectrometers. The final version of this manuscript profited from thorough reviews and helpful suggestions of F. Jorissen and S. Heß. This is Special Research Project 261 publication no. 370. Data are available under www.pangaea.de/Projects/SFB261.

References

Alve E, Bernhard JM (1995) Vertical migratory response of benthic foraminifera to controlled oxygen concentrations in an experimental mesocosm. Mar Ecol Prog Ser 116: 137-151

Bauch D, Erlenkeuser H, Winckler G, Pavlova G, Thiede J (2002) Carbon isotopes and habitat of polar planktic foraminifera in the Okhotsk Sea: The 'Carbonate Ion Effect' under natural conditions. Mar Micropaleontol 45: 83-99

Behrenfeld MJ, Falkowski PG (1997) Photosynthetic rates derived from satellite-based chlorophyll concentration. Limnol Oceanogr 42: 1-20

Belanger PE, Curry WB, Matthews RK (1981) Coretop evaluation of benthic foraminiferal isotopic ratios for paleo-oceanographic interpretations. Palaeogeogr Palaeoclimatol Palaeoecol 33: 205-220

Bernhard JM (1992) Benthic foraminiferal distribution and biomass related to porewater oxygen content: Central California continental slope and rise. Deep-Sea Res 39: 585-605

Bernhard JM, Bowser SS (1992) Bacterial biofilms as a trophic resource for certain benthic foraminifera. Mar Ecol Prog Ser 83: 263-272

Bernhard JM, Buck KR, Farmer MA, Bowser SS (2000) The Santa Barbara Basin is a symbiosis oasis. Nature 403: 77-80

Bernhard JM, Buck KR, Barry JP (2001) Monterey Bay cold-seep biota: Assemblages, abundance, and ultrastructure of living foraminifera. Deep-Sea Res I 48: 2233-2249

Bickert T, Wefer G (1999) South Atlantic and benthic

foraminifer $\delta^{13}C$ deviations: Implications for reconstructing the Late Quaternary deep-water circulation. Deep-Sea Res II 46: 437-452

Bijma J, Spero H, Lea DW (1999) Reassessing foraminiferal stable isotope geochemistry: Impact of the oceanic carbonate system (experimental results). In: Fischer G, Wefer G (eds) Use of Proxies in Paleoceanography: Examples from the South Atlantic. Springer, Berlin, pp 489-512

Boyle EA (1992) Cadmium and $\delta^{13}C$ paleochemical ocean distributions during the stage 2 glacial maximum. Annu Rev Earth Planet Sci 20: 245-287

Boyle EA, Rosenthal Y (1996) Chemical hydrography of the South Atlantic during the last glacial maximum: Cd vs. $\delta^{13}C$. In: Wefer G, Berger WH, Siedler G, Webb DJ (eds) The South Atlantic, Present and Past Circulation. Springer, Berlin, pp 423-443

Caralp MH (1989) Size and morphology of the benthic foraminifer *Melonis barleeanum*: Relationships with marine organic matter. J Foraminiferal Res 19: 235-245

Chapman P, Shannon LV (1985) The Benguela ecosystem Part II. In: Barnes M (ed) Oceanography and Marine Biology: An Annual Review. Aberdeen University Press, vol 23, pp 183-251

Chapman P, Shannon LV (1987) Seasonality of the oxygen minimum layer at the extremities of the Benguela system. In: Payne AIL, Gulland JA, Brink KH (eds) The Benguela and Comparable Ecosystems. S Afri J Mar Sci 5: 85-94

Corliss BH (1985) Microhabitats of benthic foraminifera within deep-sea sediments. Nature 314: 435-438

Corliss BH, Emerson S (1990) Distribution of Rose Bengal stained deep-sea benthic foraminifera from the Nova Scotian continental margin and Gulf of Maine. Deep-Sea Res 37: 381-400

Curry WB, Duplessy JC, Labeyrie LD, Shackleton NJ (1988) Changes in the distribution of $\delta^{13}C$ of deep water ΣCO_2 between the last glaciation and the Holocene. Paleoceanography 3: 317-341

Duplessy JC, Shackleton NJ, Matthews RK, Prell W, Ruddiman WF, Caralp MH, Hendy CH (1984) ^{13}C record of benthic foraminifera in the last interglacial ocean: implications for the carbon cycle and the global deep water circulation. Quat Res 21: 225-243

Ernst SR, Duijnstee IAP, van der Zwaan GJ (2002) The dynamics of the benthic foraminiferal microhabitat: Recovery after experimental disturbance. Mar Micro-paleontol 46: 343-361

Gooday AJ (1986) Meiofaunal foraminiferans from the bathyal Porcupine seabight (northeast Atlantic): Size structure, standing stock, species diversity and vertical distribution in the sediment. Deep-Sea Res 33: 1345-1373

Gooday AJ, Turley CM (1990) Responses by benthic organisms to inputs of organic material to the ocean floor: A review. Phil Trans R Soc Lond 331: 119-138

Graham DW, Corliss BH, Bender ML, Keigwin LD Jr (1981) Carbon and oxygen isotopic disequilibria of Recent deep-sea benthic foraminifera. Mar Micropaleontol 6: 483-497

Grossman EL (1984a) Stable isotope fractionation in live benthic foraminifera from the southern California borderland. Palaeogeogr Palaeoclimatol Palaeoecol 47: 301-327

Grossman EL (1984b) Carbon isotopic fractionation in live benthic foraminifera - comparison with inorganic precipitate studies. Geochim Cosmochim Acta 48: 1505-1512

Grossman EL (1987) Stable isotopes in modern benthic foraminifera: A study of vital effect. J Foram Res 17: 48-61

Hald M, Aspeli R (1997) Rapid climatic shifts of the northern Norwegian Sea during the last deglaciation and the Holocene. Boreas 26: 15-28

Harloff J, Mackensen A (1997) Recent benthic foraminiferal associations and ecology of the Scotia Sea and Argentine Basin. Mar Micropaleontol 31: 1-29

Jorissen FJ, Barmawidjaja DM, Puskaric S, Van Der Zwaan GJ (1992) Vertical distribution of benthic foraminifera in the northern Adriatic Sea: The relation with the organic flux. Mar Micropaleontol 19: 131-146

Jorissen FJ, de Stigter HC, Widmark JGV (1995) A conceptual model explaining benthic foraminiferal micro-habitats. Mar Micropaleontol 26: 3-15

Jorissen FJ, Wittling I, Peypouquet JP, Rabouille C, Relexans JC (1998) Live benthic foraminiferal faunas off Cape Blanc, NW-Africa: Community structure and microhabitats. Deep-Sea Res I 45: 2157-2188

Kennett JP, Cannariato KG, Hendy IL, Behl RJ (2000) Carbon isotopic evidence for methane hydrate instability during Quaternary interstadials. Science 288: 128-133

Kitazato H (1994) Foraminiferal microhabitats in four marine environments around Japan. Mar Micropaleontol 24: 29-41

Kitazato H, Ohga T (1995) Seasonal changes in deep-sea benthic foraminiferal populations: Results of long-term observations at Sagami Bay, Japan. In: Sakai H, Nozaki Y (eds) Biogeochemical

Processes and Ocean Flux in the Western Pacific. Terra Scientific Publishing Company (Terrapub), Tokyo, pp 331-342

Kroopnick P (1980) The distribution of ^{13}C in the Atlantic Ocean. Earth Planet Sci Lett 49: 469-484

Kroopnick PM (1985) The distribution of ^{13}C of total CO_2 in the world oceans. Deep-Sea Res 32: 57-84

Lea DW, Bijma J, Spero HJ, Archer D (1999) Implications of a carbonate ion effect on shell carbon and oxygen isotopes for glacial ocean conditions. In: Fischer G, Wefer G (eds) Use of Proxies in Paleoceanography. Springer, Berlin, pp 513-522

Licari L, Schumacher S, Wenzhoefer F, Zabel M, Mackensen A (2003) Communities and microhabitats of living benthic Foraminifera from the tropical east Atlantic: Impact of different productivity regimes. J Foram Res, in press

Linke P, Lutze GF (1993) Microhabitat preferences of benthic foraminifera - a static concept or a dynamic adaptation to optimize food acquisition? Mar Micro-paleontol 20: 215-134

Loubere P (1987) Late Pliocene variations in the carbon isotope values of North Atlantic benthic foraminifera: Biotic control of the isotopic record? Mar Geol 76: 45-56

Lutjeharms JRE, Meeuwis JM (1987) The extend and variability of South-East Atlantic upwelling. In: Payne AIL, Gulland JA, Brink KH (eds) The Benguela and Comparable Ecosystems. S African J Mar Sci 5: 51-62

Lutze GF (1964) Zum Färben rezenter Foraminiferen. Meyniana 14: 43-47

Mackensen A (2001) Oxygen and carbon stable isotope tracers of Weddell Sea water masses: New data and some paleoceanographic implications. Deep-Sea Res I 48: 1401-1422

Mackensen A, Douglas RG (1989) Down-core distribution of live and dead deep-water benthic foraminifera in box cores from the Weddell Sea and the California continental borderland. Deep-Sea Res 36: 879-900

Mackensen A, Bickert T (1999) Stable carbon isotopes in benthic foraminifera: Proxies for deep and bottom water circulation and new production. In: Fischer G, Wefer G (eds) Use of Proxies in Paleoceanography: Examples from the South Atlantic. Springer, Berlin, pp 229-254

Mackensen A, Grobe H, Kuhn G, Fütterer DK (1990) Benthic foraminiferal assemblages from the eastern Weddell Sea between 68 and 73°S: distribution, ecology and fossilization potential. Mar Micropaleontol 16: 241-283

Mackensen A, Fütterer DK, Grobe H, Schmiedl G (1993a) Benthic foraminiferal assemblages from the eastern South Atlantic Polar Front region between 35° and 57°S: Distribution, ecology and fossilization potential. Mar Micropaleontol 22: 33-69

Mackensen A, Hubberten H-W, Bickert T, Fischer G, Fütterer DK (1993b) The δ^{13}C in benthic foraminiferal tests of *Fontbotia wuellerstorfi* (Schwager) relative to the δ^{13}C of dissolved inorganic carbon in southern ocean deep water: Implications for glacial ocean circulation models. Paleoceanography 8: 587-610

Mackensen A, Grobe H, Hubberten H-W, Kuhn G (1994) Benthic foraminiferal assemblages and the δ^{13}C-signal in the Atlantic sector of the southern ocean: Glacial-to-interglacial contrasts. In: Zahn R, Pedersen T, Kaminski M, Labeyrie L (eds) Carbon Cycling in the Glacial Ocean: Constraints on the Ocean's Role in Global Change. Springer, Heidelberg, NATO ASI Series I: Global Environmental Change, vol 17, pp 105-144

Mackensen A, Schmiedl G, Harloff J, Giese M (1995) Deep-sea Foraminifera in the South Atlantic Ocean: ecology and assemblage generation. Micropaleontology 41: 342-358

Mackensen A, Hubberten H-W, Scheele N, Schlitzer R (1996) Decoupling of δ^{13}CΣCO_2 and phosphate in Recent Weddell Sea Deep and Bottom Water: Implications for glacial southern ocean paleoceanography. Paleoceanography 11: 203-215

Mackensen A, Schumacher S, Radke J, Schmidt DN (2000) Microhabitat preferences and stable carbon isotopes of endobenthic foraminifera: Clue to quantitative reconstruction of oceanic new production? Mar Micropaleontol 40: 233-258

Mackensen A, Rudolph M, Kuhn G (2001) Late Pleistocene deep-water circulation in the subantarctic eastern Atlantic. Glob Planet Change 30: 195-226

McConnaughey T (1989a) ^{13}C and ^{18}O isotopic disequilibrium in biological carbonates: II. *In vitro* simulation of kinetic isotope effects. Geochim Cosmochim Acta 53: 163-171

McConnaughey T (1989b) ^{13}C and ^{18}O isotopic disequilibrium in biological carbonates: I. Patterns. Geochim Cosmochim Acta 53: 151-162

McConnaughey T, Burdett J, Whelan JF, Paull CK (1997) Carbon isotopes in biological carbonates: Respiration and photosynthesis. Geochim Cosmochim Acta 61: 611-622

McCorkle DC, Keigwin LD (1994) Depth profiles of δ^{13}C in bottom water and core-top *C. wuellerstorfi* on the Ontong-Java Plateau and Emperor Seamounts. Paleoceanography 9: 197-208

McCorkle DC, Emerson SR, Quay PD (1985) Stable carbon isotopes in marine porewaters. Earth Planet Sci Lett 74: 13-26

McCorkle DC, Keigwin LD, Corliss BH, Emerson SR (1990) The influence of microhabitats on the carbon isotopic composition of deep-sea benthic foraminifera. Paleoceanography 5: 161-185

McCorkle DC, Veeh HH, Heggie DT (1994) Glacial-Holo-cene paleoproductivity off western Australia: a comparison of proxy records. In: Zahn R, Pedersen T, Kaminski M, Labeyrie L (eds) Carbon Cycling in the Glacial Ocean: Constraints on the Ocean's Role in Global Change. Springer, Heidelberg, NATO ASI Series, vol I 17, pp 443-480

McCorkle DC, Martin PA, Lea DW, Klinkhammer GP (1995) Evidence of a dissolution effect on benthic foraminiferal shell chemistry: $\delta^{13}C$, Cd/Ca, Ba/Ca, and Sr/Ca results from the Ontong Java Plateau. Paleoceanography 10: 699-714

McCorkle DC, Corliss BH, Farnham CA (1997) Vertical distributions and stable isotopic compositions of live (stained) benthic foraminifera from the North Carolina and California continental margins. Deep-Sea Res I 44: 983-1024

Mollenhauer G, Schneider RR, Müller PJ, Spieß V, Wefer G (2002) Glacial/interglacial variability in the Benguela upwelling system: Spatial distribution and budgets of organic carbon accumulation. Glob Biogeochem Cycl doi: 10.1029/2001GB001488

Moodley L (1992) Laboratory experiments on the infaunal activity in benthic foraminifera. In: Moodley L (ed) Experimental Ecology of Benthic Foraminifera in Soft Sediments and its (Paleo) Environmental Significance. Ph D Thesis, Vrije Universiteit, Amster-dam, pp 89-106

Mook WG, Bommerson JC, Staverman WH (1974) Carbon isotope fractionation between dissolved bicarbonate and gaseous carbon dioxide. Earth Planet Sci Lett 22: 169-176

Oestlund HG, Craig C, Broecker WS, Spencer D (eds) (1987) GEOSECS Atlantic, Pacific, and Indian Ocean Expeditions, Shorebased Data and Graphics. Government Printing Office, Washington DC, 200 p

Oppo DW, Fairbanks RG, Gordon AL, Shackleton NJ (1990) Late Pleistocene Southern Ocean $\delta^{13}C$ variability. Paleoceanography 5: 43-54

Orsi AH, Whitworth III T, Nowlin Jr WD (1995) On the meridional extent and fronts of the Antarctic Circum-polar Current. Deep-Sea Res I 42: 641-673

Peterson RG, Stramma L (1991) Upper-level circulation in the South Atlantic Ocean. Prog Oceanogr 26: 1-73

Rathburn AE, Corliss BH (1994) The ecology of living (stained) deep-sea benthic foraminifera from the Sulu Sea. Paleoceanography 9: 87-150

Rathburn AE, Corliss BH, Tappa KD, Lohmann KC (1996) Comparisons of the ecology and stable isotopic compositions of living (stained) benthic foraminifera from the Sulu and South China Seas. Deep-Sea Res I 43: 1617-1646

Rathburn AE, Levin LA, Held Z, Lohmann KC (2000) Benthic foraminifera associated with cold methane seeps on the northern California margin: ecology and stable isotopic composition. Mar Micropaleontol 38: 247-266

Raymo ME, Ruddiman WF, Shackleton NJ, Oppo DW (1990) Evolution of Atlantic-Pacific $\delta^{13}C$ gradients over the last 2.5 my. Earth Planet Sci Lett 97: 353-368

Sarnthein M, Winn K, Jung SJA, Duplessy J-C, Labeyrie L, Erlenkeuser H, Ganssen G (1994) Changes in east Atlantic deepwater circulation over the last 30,000 years: Eight time slice reconstructions. Paleoceano-graphy 9: 209-267

Schlitzer R (2002) Ocean Data View. http://www.awi-bremerhaven.de/GEO/ODV

Schmiedl G, Mackensen A, Müller PJ (1997) Recent benthic foraminifera from the eastern South Atlantic Ocean: Dependence on food supply and water masses. Mar Micropaleontol 32: 249-288

Schulz HD, Cruise participants (1992) Bericht und erste Ergebnisse über die Meteor-Fahrt M20/2. Ber Fachber Geowiss, Univ Bremen 25: 1-173

Schumacher S (2001) Mikrohabitatansprüche benthischer Foraminiferen in Sedimenten des Süd-atlantiks. Ber Polarforsch, Alfred-Wegener-Institut, Bremerhaven 403: 1-151

Severin KP (1987) Laboratory observations of the rate of subsurface movement of a small miliolid foraminifer. J Foram Res 17: 110-116

Spero HJ, Lea DW (1996) Experimental determination of stable isotope variability in Globigerina bulloides: implications for paleoceanographic reconstructions. Mar Micropaleontol 28: 231-246

Spero HJ, Bijma J, Lea DW, Bemis BE (1997) Effect of seawater carbonate concentration on foraminiferal carbon and oxygen isotopes. Nature 390: 497-500

Stramma L, Schott F (1999) The mean flow field of the tropical Atlantic Ocean. Deep-Sea Res II 46: 279-303

Van Bennekom AJ, Berger GW, Helder W, de Vries RTP (1978) Nutrient distribution in the Zaire estuary and river plume. Netherlands J Sea Res 12: 296-323

Voituriez B, Herbland A (1982) Comparaison des systèmes productifs de l'Atlantique Tropical Est: dômes thermiques, upwellings côtiers et upwelling équatorial. Rapports et Procès-verbaux des Réunions (Conseil international pour l'Exploration de la Mer) 180: 114-130

Wagner T, Holtvoeth J, Schubert CJ (2003) Sources and distribution of terrigenous organic carbon in surface sediments of the central to eastern Equatorial Atlantik: A multiparameter approach including lignin. Org Geochem, submitted

Walton WR (1955) Ecology of living benthonic foraminifera, Todos Santos Bay, Baja California. J Paleontol 29: 952-1018

Wefer G, Berger WH (1991) Isotope paleontology: growth and composition of extant calcareous species. Mar Geol 100: 207-248

Wefer G, Heinze P-M, Berger WH (1994) Clues to ancient methane release. Nature 369: 282

Wilson-Finelli A, Chandler GT, Spero HJ (1998) Stable isotope behavior in paleoceanographically important benthic foraminifera: Results from microcosm culture experiments. J Foraminiferal Res 28: 312-320

Wollenburg J, Mackensen A (1998) On the vertical distribution of living (Rose Bengal stained) benthic foraminifers in the Arctic Ocean. J Foraminiferal Res 28: 268-285

Woodruff F, Savin S (1985) $\delta^{13}C$ values of Miocene Pacific benthic foraminifera: correlations with sea level and biological productivity. Geology 13: 119-122

Woodruff F, Savin S, Douglas RG (1980) Biological frac-tionation of oxygen and carbon isotopes by recent benthic foraminifera. Mar Micropaleontol 5: 3-11

Zahn R, Winn K, Sarnthein M (1986) Benthic foraminiferal $\delta^{13}C$ and accumulation rates of organic carbon: *Uvigerina peregrina* group and *Cibicidoides wuellerstorfi*. Paleoceanography 1: 27-42

Carbonate Preservation in Deep and Intermediate Water Masses in the South Atlantic: Evaluation and Geological Record (a Review)

R. Henrich[*], K.-H. Baumann, S. Gerhardt, M. Gröger and A. Volbers

*Universität Bremen, Fachbereich Geowissenschaften, Postfach 330440,
D- 28334 Bremen, Germany*
** corresponding author (e-mail): henrich@uni-bremen.de*

Abstract: Evaluation of conventional dissolution proxies in South Atlantic surface sediments revealed broad applicability only in far offshore, rather oligotrophic regimes in the western basins. In contrast, they fail or produce misleading and incorrect results in the more productive eastern South Atlantic basins, due to the combined effects of variable dilution by non-carbonate material and fluctuating ecological conditions. Much more promising are the results from new dissolution proxies on the planktic foraminifer *Globigerina bulloides* (BDX') and the pteropod *Limacina inflata* (LDX) which were calibrated with carbonate saturation as indicated by GEOSECS data. In the western South Atlantic, the sedimentary calcite lysocline is encountered by the BDX' at the transition between AABW and LNADW. However, it rises up into the LNADW close to the equator due to additional supralysoclinal dissolution. In the eastern South Atlantic basins, supralysoclinal dissolution results in an elevation of the sedimentary calcite lysocline of several hundred metres to a maximum of 1600 m as compared to the position of the hydrographic lysocline, with aragonite preservation in the eastern South Atlantic being even poorer. At most sites investigated, the surface sediments are void of pteropods and thus LDX failure is indicated. However, in the western South Atlantic the LDX displays a double lysocline for aragonite, the upper lysocline at a water depth of 750 m and the lower at 2500 m. Aragonite and calcite preservation profiles indicate much weaker stratification of the water during the LGM. With 3200 m, the position of the calcite lysocline is encountered at the same level in the southern parts of the eastern and western basins dropping to 4000 m near the equator. Along the western continental margin no indication for aragonite-corrosive glacial AAIW was found, providing clear evidence for a strengthened GNAIW flow along the Brazil margin. The long-term history of carbonate dissolution in the equatorial Atlantic was reconstructed by a multiproxy approach combining benthic foraminifer stable isotopes and new proxies from silt analysis. For the first time, this allows a reconstruction of the chemical (nutrient content, carbonate corrosiveness) and physical (bottom current strength) properties of deep and intermediate water masses. The terrigenous silt records of ODP Site 927 at the Ceara Rise show rapid shifts from low to very high bottom-currents speeds for nearly all the isotopic transitions in the Brunhes epoch, indicating subsequent phases of shutdown and rapid reinstatement of LNADW circulation. A drastic reduction of glacial bottom-current strength at Site 927 is inferred after 2.75 Ma, synchronous with the first occurrence of larger continental ice shields and with a drastic decrease in deep convection in the Norwegian-Greenland Sea. After the mid-Pleistocene climate transition, progressively weaker bottom currents and poorer carbonate preservation during glacials indicate a progressive reduction of LNADW from the Late Pliocene to Pleistocene. On the contrary, an opposite trend with progressive improvement of preservation during glacials from Late Pliocene to the Pleistocene is observed in the Caribbean at Site 999. This indicates a contemporaneous progressive increase in the contribution of UNADW to the Atlantic in glacial periods. Altogether, a progressive weakening of the circulation in the LNADW loop and a contemporaneous strengthening of the UNADW loop are evident since the mid Pleistocene transition.

From WEFER G, MULITZA S, RATMEYER V (eds), 2003, *The South Atlantic in the Late Quaternary: Reconstruction of Material Budgets and Current Systems.* Springer-Verlag Berlin Heidelberg New York Tokyo, pp 645-670

Introduction

Constraints on the Marine Carbonate System in the South Atlantic

In the focus of the ongoing climate discussion, much effort has been devoted to gain a better understanding of the carbon cycle on a global basis. However, most of the research has been concerned with the organic carbon cycle and less attention has been paid to the inorganic carbon cycle, although their corresponding net fluxes are comparable in size. Since there is over 60 times more CO_2 in the world oceans than in the atmosphere (Broecker and Peng 1982), the oceans play an important role in the global CO_2 budget, in particular during Quaternary climatic oscillations. To evaluate the role of the ocean in the global exchange of CO_2 requires a sound knowledge of the complex processes involved in the marine carbonate system, as does developing budgets for marine carbonate production and dissolution under various oceanic regimes. There is general consensus that in the present-day cycle more than half of the global carbonate budget accumulates on the shelves even though their surface area represents only 7.5 % of the global ocean (Milliman 1993; Wollast 1994). This pattern is clearly related to interglacial sea-level highstands. However, the database used for budgeting modern open-ocean carbonate production and dissolution rates is sparse, and estimates differ significantly between various models (Milliman and Droxler 1996). Geochemists have calculated a mean global pelagic carbonate production of 21-24 g/ m² per year by modelling alkalinity anomalies and mean residence times of various water masses (Morse and MacKenzie 1990). In contrast, the global average fluxes of pelagic carbonate ranging from only 8 g/m² per year (Milliman 1993) to 10-12 g/m² per year (Milliman and Droxler 1996) were derived from long-term sediment trap measurements. In order to balance the large differences between pelagic productivity and global carbonate flux a considerable loss of carbonate has been assumed to occur by dissolution (Milliman and Droxler 1996). In particular, significant dissolution above the hydrographic lysocline has been proposed. This process is induced by the degradation of organic mat-

ter close to or at the sediment/water interface, e.g. "respiratory dissolution" or "supralysoclinal dissolution" (Archer et al. 1989; Jahnke et al. 1994). Despite the significant progress that has been achieved in modelling these very early diagenetic effects of dissolution (Emerson and Bender 1981; Emerson and Archer 1990; Archer and Meier-Reimer 1994), there is still an ongoing discussion regarding magnitude and intensity of supralysoclinal dissolution at various levels of the oceans. Recently, Milliman et al. (1999) suggested that as much as 60 –80% of the originally produced carbonate might be dissolved in the upper water masses by biological mediation. However, they also stressed that the database still is insufficient to seriously evaluate global carbonate budgeting models which take this effect into account. There is an evident need to extend the database, in particular, along the highly productive continental margins with respect to production and dissolution rates at various depth levels of the ?

The South Atlantic is an ideal natural laboratory to study the various processes of pelagic carbonate production and preservation (Fig. 1a-c). Carbonate production rates measured in sediment traps vary strongly (Fischer and Wefer 1996). They range from 8.2 g/m² per year at the northern boundary of the central oligotrophic gyre in the southern Guinea basin to 13.1 g/m² per year in the equatorial upwelling zone at the northern Guinea basin, and to 27.5 g/m² per year in the coastal upwelling area offshore Namibia. In addition, organic carbon fluxes (Fischer and Wefer 1996) and estimates of primary productivity (Berger 1989) are highly variable. Oligotrophic conditions are registered in the western and central South Atlantic (e.g. 1.7 g C_{org}/ m² per year in the southern Guinea basin; Fischer and Wefer 1996). In contrast, high productivity regimes dominate along the eastern South Atlantic continental margin (e.g. 4.6 g C_{org}/ m² per year in the coastal upwelling area offshore Namibia, Fischer and Wefer 1996). Hence, the susceptibility to supralysoclinal dissolution might also differ accordingly in various regions of the South Atlantic. In addition, sediments at the sea floor of the South Atlantic are bathed in completely different deep and intermediate water regimes (Fig. 1b,c). This includes the carbonate-saturated Lower North Atlantic Deep Water (LNADW), the carbonate under-

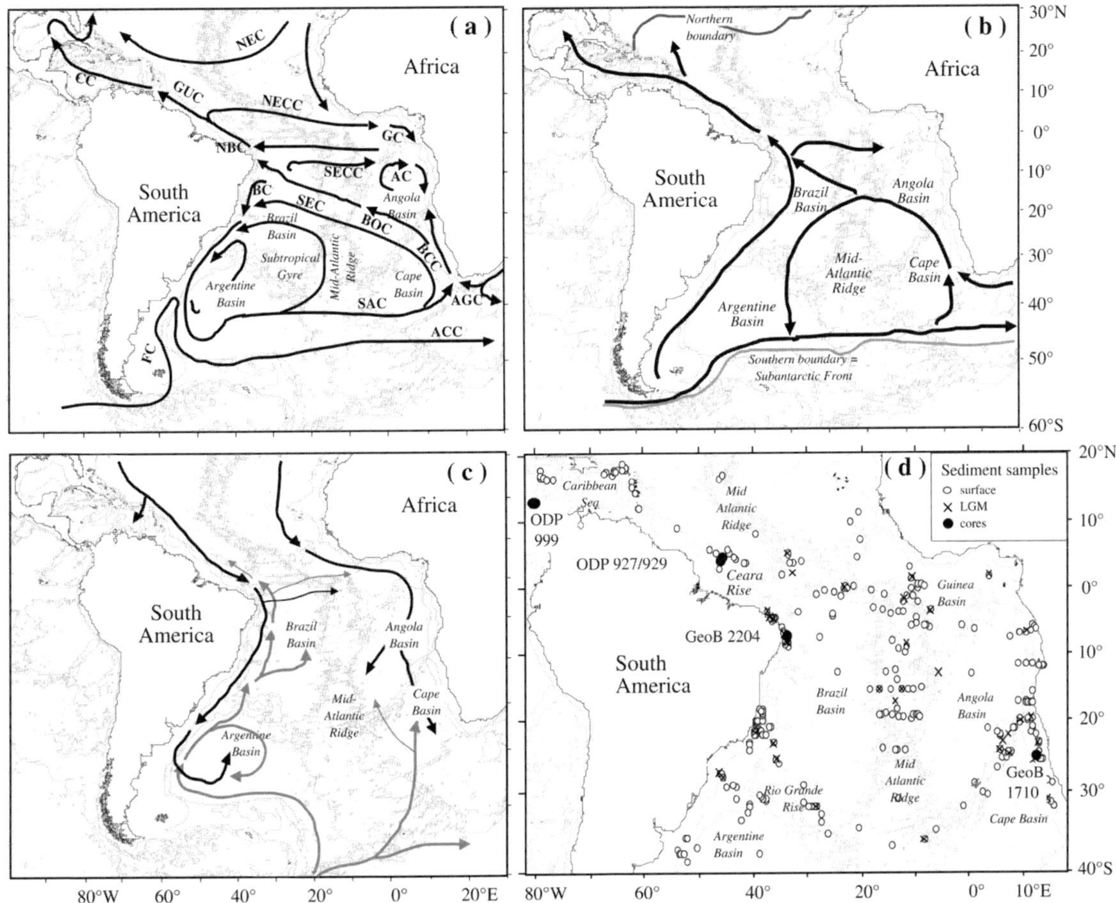

Fig. 1. a) Recent South Atlantic and Caribbean surface water circulation: Angola Current (AC), Antarctic Circumpolar Current (ACC), Agulhas Current (AGC), Brazil Current (BC), Benguela Coastal Current (BCC), Benguela Oceanic Current (BOC), Caribbean Current (CC), Falkland Current (FC), Guinea Current (GC),Guyana Current (GUC), North Brazil Current (NBC), North Equatorial Current (NEC), North Equatorial Counter Current(NECC), South Atlantic Current (SAC), South Equatorial Current (SEC), South Equatorial Counter Current (SECC) (modified after Peterson and Stramma, 1991). **b)** Recent South Atlantic and Caribbean Antarctic Intermediate Water (AAIW) distribution paths. Grey lines mark salinities at the AAIW salinity minimum (modified after Talley 1996). Black arrows mark AAIW flow directions (inferred from the salinity distribution after Talley 1996 and modified after Gordon 1986 and Gordon et al. 1992). **c)** Recent South Atlantic and Caribbean deep-water distribution paths. Grey arrows mark the main and subordinate flow directions of Antarctic Bottom Water; black arrows indicate the main and subordinate flow directions of North Atlantic Deep Water (modified after Wüst 1935 and Reid 1994, 1996).

saturated Antarctic Bottom Water (AABW), the carbonate-saturated Upper North Atlantic Deep Water (UNADW), and the slightly carbonate under-saturated Antarctic Intermediate Water (AAIW).

In this article we will summarise the results from our studies on carbonate preservation in South Atlantic sediments of the Quaternary. We will start with an evaluation of carbonate dissolution proxies by testing their applicability and limitations in different oceanic regimes. We will then investigate the variability of carbonate preservation in surface sediments from the South Atlantic, using new SEM-derived dissolution proxies for aragonite and calcite, which are the only proxies that are not influenced by other processes, like the dilution by non-carbonate material or variable ecological conditions. The results from the modern regimes will be compared to those attained for the time slice of the Last Glacial Maximum (LGM) in order to highlight the most prominent changes in deep and intermediate water carbonate chemistry and to discuss the potential implications for the LGM paleoclimate. Finally, we demonstrate the applicability of proxies from silt analyses in order to reconstruct the long-term record of carbonate preservation and bottom current strength of intermediate and deep-water masses in the equatorial Atlantic during the past 3 myrs.

Review of Methods

Evaluation of Carbonate Dissolution Proxies

In general, the state of preservation of carbonate particles in marine environments depends on a variety of factors. They include:
• Dissolution kinetics in sea water of the respective carbonate mineralogy,
• Overall size, wall thickness, crystal size, surface texture and porosity of the particles,
• pCO_2 and CO_3^{2-} concentration in the water mass, which varies according to its formation mode, age, and the actual depth level at which the water mass is situated,
• Sedimentation rate, and thus the duration of exposure or burial of particles,
• Carbon rain ratio (Berger and Keir 1984), which is defined as the ratio between organic carbon and carbonate, and
• Diagenetic regime in pore waters.

The effects and relative significance of these factors have been tested in various dissolution experiments carried out in the laboratory (Bé et al. 1975; Sliter et al. 1975; Adelseck and Berger 1975; Adelseck 1978) as well as under natural conditions (Peterson 1966; Berger 1967; Milliman 1975; Henrich and Wefer 1986). It is always a combination of different factors that determines the preservation of individual carbonate particles, and the preservation of assemblages of particles of variable composition. Therefore, many of the commonly used criteria to judge the state of preserva-

tion of calcium carbonate remain qualitative. Furthermore, most of the proxies are influenced by other, additional processes (e.g. dilution by non-carbonate material and/or variable ecological conditions), and thus can produce misleading or incorrect results.

In order to test and compare the applicability of the dissolution proxies most commonly used, we have carried out a dissolution experiment in the South Atlantic (Dittert et al. 1999). Three depth transects into the Brazil and the Cape Basin, including areas above and below the calcite lysocline, were studied. The state of carbonate preservation in surface sediments was determined using the different dissolution proxies, and the results were then compared with the state of carbonate saturation indicated by hydrographic data. The set of dissolution proxies tested included:
• *Bulk sediment parameters:* Carbonate content, weight percentages of the sand fraction (Berger et al. 1982; Wu and Berger 1991), the carbon rain ratio (Berger and Keir 1984).
• *Microfossil parameters:* Planktic foraminiferal concentration, the fragmentation index (Fi) of planktic foraminifers (Berger 1973), radiolarian to planktic foraminiferal ratio (r/pf) (Diester-Haas and Rothe 1987), benthic to planktic foraminiferal ratio (bf/pf) (Parker and Berger 1971; Diester-Haas and Rothe 1987), percentage of dissolution- sensitive planktic foraminifer taxa (FDX, Berger 1973) and foraminiferal dissolution index defined according to variable dissolution resistance of species (Berger 1979).
• *SEM dissolution indices:* As dissolution proceeds, progressive steps of ultra-structural breakdown of planktic foraminiferal skeletons occur (Be et al. 1975), an attribute that has been used to determine the intensity of dissolution semiquantita-tively. Henrich (1989) developed the first concept on SEM dissolution indices for the planktic fora-minifer species *N. pachyderma*, which is the planktic foraminifer species most resistant to dissolution. As a polar to subpolar species it displays widespread distribution in the vicinity of the polar oceans in both hemispheres. In addition, it is also observed in cold-water upwelling regimes in the South Atlantic (Little et al. 1997). This paleoceanographic tool has been successfully used to char-

acterise carbonate preservation in various modern oceanic regimes in the North Atlantic and the Norwegian-Greenland Sea (Huber et al. 2000) and the Quaternary (Henrich and Baumann 1994; Baumann and Meggers 1996, Henrich et al. 2002). SEM preservation studies have also been carried out on the spinose foraminifera *Globigerina bulloides* which reveals a high- ranking dissolution susceptibility, a consequence of its more porous test structure (Thunell and Honjo 1981). *G. bulloides* is prevalent throughout the South Atlantic, with particularly high present-day abundances in the productive surface waters of the eastern South Atlantic (Kemle- von Mücke and Oberhänsli 1999). Based on pioneer SEM studies by van Krefeld-Alfane (1996), Dittert developed the *Globigerina bulloides* Dissolution Index (BDX - Dittert et al. 1999; Dittert and Henrich 2000). The BDX method was then improved and slightly modified by Volbers and yielded the BDX' (Volbers and Henrich 2002a). Fig. 2 presents a graphical illustration of the BDX', with six stages of preservation (decreasing preservation from 0 to 5 - for details on the analytical procedure and calculation of the BDX' see Volbers and Henrich 2002a). The high precision of the BDX' to reconstruct foraminiferal carbonate preservation in the modern South Atlantic and its power as a paleoceanographic tool in the fossil record will be demonstrated in the next chapter.

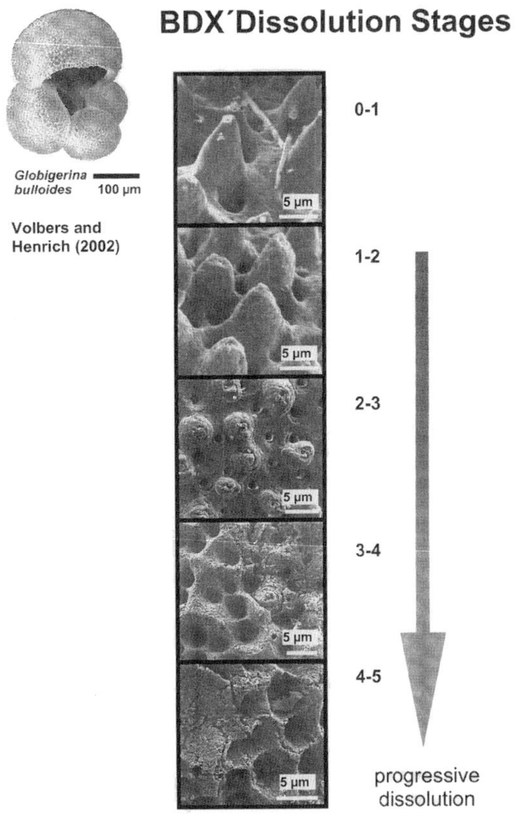

The basic rationale behind all these bulk sediment and microfossil parameters is the progressive breakdown of carbonate particles in the course of dissolution by the following mechanisms: (1) The amount of carbonate is diminished (e.g. bulk carbonate contents, sand percentages and the grain size of the particles decreases, whereas the number of fragments increases). (2) Planktic foraminiferal species assemblages are modified by preferentially removing surface dwellers with thin porous walls and enriching deep-dwelling thick-walled more robust shells. (3) The more soluble planktic foraminifers are preferentially removed, passively enriching benthic foraminifers and radiolarians.

However, in many cases the above-mentioned conventional proxies fail, or produce incorrect and misleading results. This has been documented by Dittert et al. (1999) in the South Atlantic dissolution experiment, and was even more clearly elaborated in a down-core study of the highly productive eastern Atlantic marginal coastal upwelling regimes off Namibia (Fig. 3; Volbers and Henrich 2002b). The processes responsible for this failure are manifold: (1) The bulk sediment parameters might be affected by dilution with non-carbonate material (e.g. input of dust material, river supply of terrigenous fine-grained sediments), downslope resuspension and lateral advection, or by winnowing of sediment by bottom currents. (2) The microfossil parameters might respond to changes in ecology in the course of time time.

Fig. 3 displays different dissolution proxy records in sediment core GeoB 1710 retrieved from

Fig. 2. Calcite dissolution proxy BDX' (*Globigerina bulloides* dissolution index): six steps of progressive breakdown in the ultrastructure of *Globigerina bulloides* tests. As dissolution proceeds pores widen, the inter-pore area is etched, ridges and spines become denuded until tests finally break down (modified from Volbers and Henrich 2002a).

the marginal upwelling zone off Namibia. Note the discrepancies between the conventional dissolution proxies that are commonly used in many studies and the BDX'. The r/pf index, (radiolarian to planktic foraminifer ratio), and the bf/pf index, (benthic to planktic foraminifer ratio) display various peaks which are not recorded in the BDX'. This is due the fact that both the r/pf and the bf/pf index are sensitive to changes of ecological conditions induced by variations in upwelling intensity, whereas the BDX' is not affected. These results are in accordance with the studies of Schmiedl and Mackensen (1997) who showed that changes in benthic foraminiferal assemblages and the accu-

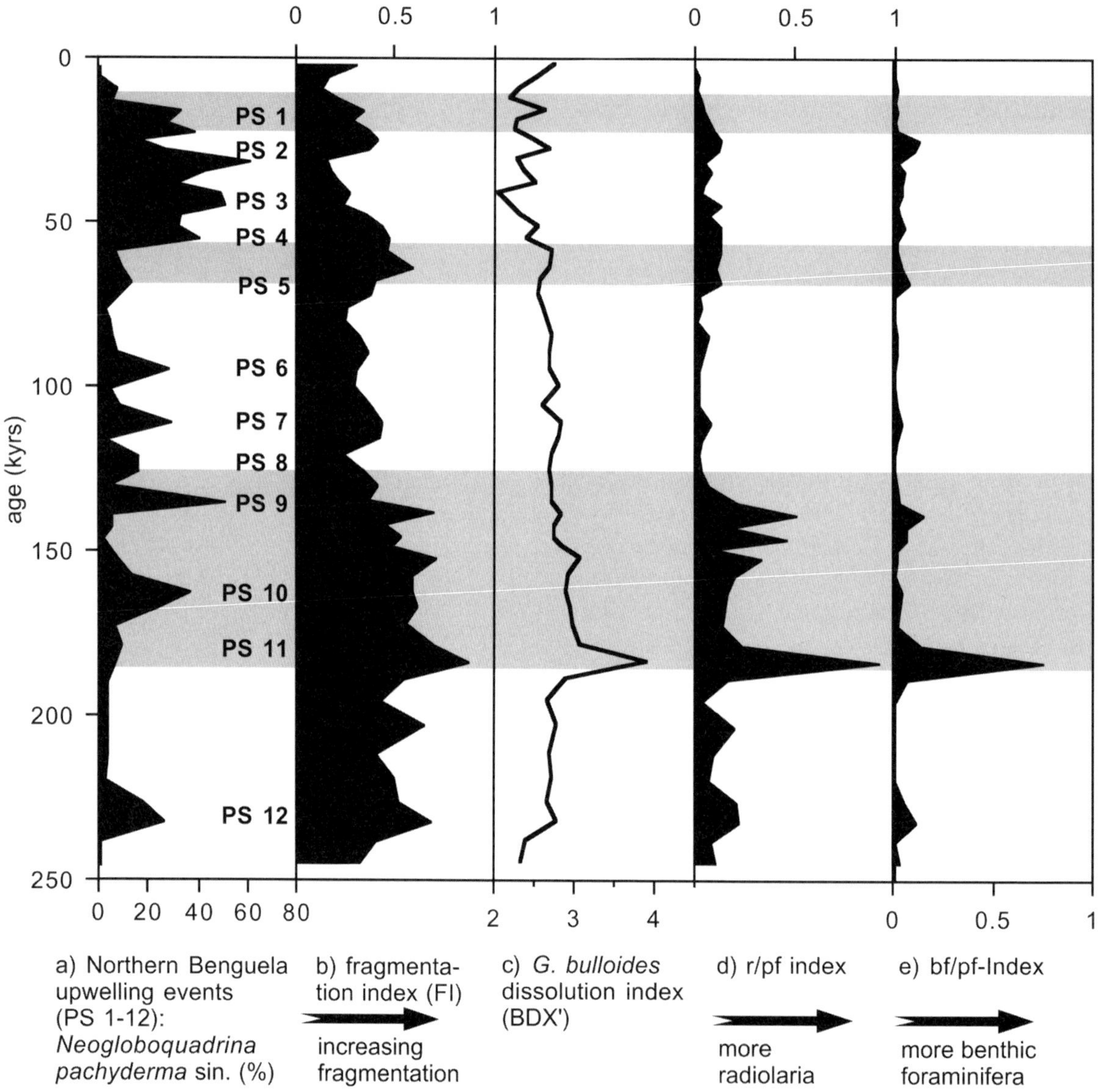

Fig. 3. Compilation of different dissolution proxy records in sediment core GeoB 1710 retrieved from the marginal upwelling zone off Namibia. Please note the discrepancies between the conventional dissolution proxies that are commonly used in many studies and the BDX'. The r/pf index, the radiolarian to planktic foraminifer ratio, and the bf/pf index, the benthic to planktic foraminifer ratio, display various peaks which are not recorded in the BDX'. This is due the fact that both, the r/pf and the bf/bf index, are very sensitive to changing ecological conditions induced by variations in upwelling intensity; whereas the BDX' is not affected (modified from Volbers and Henrich 2002b).

mulation rates were clearly related to upwelling events. In addition, Diester-Haas (1977) concluded from her studies in the Benguela upwelling region that the r/pf index might better serve as a fertility indicator than a reliable dissolution proxy. Other dissolution proxies like the FDX, and the percentages of species sensitive to dissolution cannot be applied to this region, because most of these species are adapted to tropical and subtropical oligotrophic environments, and are thus only occasionally found in cool nutrient-rich surface water (Kemle- von Mücke and Oberhänsli 1999).

As a first conclusion from our evaluation of conventional calcite dissolution proxies we recommend to apply them only with caution. It should always be assured that the proxies under consideration were not influenced by the other factors discussed above. In addition, when studying fossil records, the assumption that other factors did not significantly vary through time should be tested and verified. As a final remark, we stress the point that the SEM foraminiferal dissolution indices are the only proxies that determine the state of calcite preservation on at least semi-quantitative scales, and are not affected by other factors apart from dissolution.

New calcite dissolution proxies are derived from silt grain-size analysis, in particular, of the calcareous silt fraction. Recently, the potential to deduce the state of carbonate preservation from silt parameters was independently discovered by Gröger et al. (2003) and Stuut et al. (2002). Silt grain-size analysis by Göger et al. (2003) was determined with a Micromeritics SediGraph 5100, after separating the sand fraction by wet sieving and removing the clay fraction according to Stoke's law with Atterberg settling tubes. Measurements were carried out within the given size spectrum ranging from 100 to 0.1 μm. The grain size distribution of the 10–63 μm fraction and the total silt (2-63 μm) was calculated separately from the raw data, with each spectrum re-scaled to 100%. The calcareous silt fraction consisted of (1) coccoliths, which were exclusively present in the fine silt with a maximum around 3 μm, and in the clay fraction, and (2) fragments of foraminifers and juvenile foraminifers preferentially abundant in the 10-63 μm silt fraction. Gröger et al. (2003) showed by comparison with planktic

foraminiferal dissolution proxies that the overall calcareous silt content and the coarse calcareous silt mode decreases with progressive dissolution and that values for the mean and modal grain size in the coarse silt fraction diminishes consistently. In parallel to this, the amount of calcareous clay increased indicating a continuous transfer of particles from coarser to fine grain sizes. These patterns clearly indicate that foraminifers in the silt fraction are less resistant to carbonate dissolution than coccoliths. Hence statistical parameters of the calcareous silt fraction provide an additional reliable tool to determine carbonate dissolution. In order to avoid effects introduced by dilution with non–carbonate material, all parameters were determined in the calcareous silt fraction re-scaled to 100%. It should be mentioned that shifts in the statistical parameters of the silt fraction might also reflect changes in ecological conditions. However, the effects introduced by a variable composition of the planktic foraminifer associations are rather small in the silt fraction as compared to the sand fraction, because the differences in the average particle size and thickness of the tests are less prominent. In addition, changes in the production rates of coccoliths and planktic foraminifers will certainly be portrayed in the ratio of clay and fine silt versus coarse silt. However, the statistical parameters may be not considerably influenced by the coarse silt mode, and hence these ecological effects might be recognised. In conclusion, the new dissolution proxies derived from silt grain- size analysis are much more sensitive than conventional dissolution parameters. However, since an overprint by ecological effects cannot be excluded, in particular, if the sedimentary sections display drastically changing environments (e.g. oligotrophic versus highly productive regimes), their reliability should be tested and evaluated against SEM dissolution proxies. Case studies demonstrating the usefulness of these new dissolution proxies are presented below.

Stuut et al. (2002) used a laser-particle sizer in their measurements. They defined a new dissolution index as the log-ratio of two coarse modes, mode A (25-90 μm) and mode B (> 90 μm). Similar to Gröger et al. (2003), they related the decrease in the coarsest mode to progressive carbonate dissolution. An obvious disadvantage of the laser-particle sizer measurements is the low and unreliable resolution in the fine silt and clay size spectra.

The calcite dissolution proxies discussed above are sensitive indicators of deep-water carbonate chemistry and supralysoclinal dissolution (Volbers and Henrich 2002a). However, due to the decreased dissolution kinetics of calcite at lower pressures, their sensitivity is much lower at intermediate water levels. Since aragonite is roughly 1.5 times more soluble than calcite (Morse et al. 1980), dissolution proxies derived from aragonitic skeletons might fill this gap. However, establishing reliable aragonite dissolution proxies is much more difficult than one might initially assume. All conventional approaches like simple fragmentation indices or ratio with other groups, fail to produce reliable results (compare Fig. 4, Gerhardt et al. 2000). The main reason for this failure is the high fragility of certain pteropod species, whereas others are much more robust. These highly fragile species produce a high amount of fragments by simple mechanical breakdown even in perfectly preserved sediment samples that have not undergone any dissolution. During incipient dissolution, these delicate fragments are preferentially dissolved, thus decreasing the fragmentation index in a sample. Hence, conventionally this decrease in fragmentation would be misinterpreted as improved preservation, which is certainly not the case. In addition, all other potential proxies, like ratios between planktic foraminifers and pteropods, or the ratio of calcite and aragonite contents, are strongly influenced by dilution phenomena as well as variable production rates of the different groups in surface waters. A first successful attempt to escape from this dilemma was made by Haddad and Droxler (1996) by establishing a composite dissolution index (CDI) of % Mg calcite, pteropod abundance, % whole pteropods, and % clear pteropods. This approach was improved and simplified by Gerhardt, who established dissolution indices for various species of the more robust pteropod genera *Limacina* by distinguishing characteristic features of progressive structural breakdown of tests under the light microscope (Fig. 4, Gerhardt et al. 2000). The *Limacina inflata* dissolution index (LDX) turned out to be the most

practicable and useful in this group (Gerhardt and Henrich 2001). Fig. 5 presents a graphical illustration of the six steps of progressive dissolution of *Limacina inflata*. As dissolution proceeds tests become milky and lustreless/opaque, display additional damage, and finally break down.

The calibration of the LDX to the modern environments in the South Atlantic and its application as a paleoceanographic tool will be discussed below.

Results and Discussion

Present Intermediate and Deep-Water Masses in the South Atlantic: Circulation Mode, Carbonate Saturation and Carbonate Preservation Patterns in Surface Sediments

Carbonate preservation in surface sediments in the South Atlantic primarily reflects changes in the carbonate chemistry of intermediate and deep waters. This primary signature may be overprinted by different secondary effects, among which supralysoclinal dissolution is by far the most prominent process. Since supralysoclinal dissolution is most efficient under highly productive surface waters, we observe a strong west to east contrast in carbonate preservation of surface sediments. Accordingly we will discuss the preservation patterns for the western and eastern South Atlantic basins separately. We will start with the predominantly oligotrophic basins of the western South Atlantic. Fig. 6 displays a cross-section of the western Atlantic basins with the state of carbonate preservation in surface sediment samples (collected between 55°W to 5 °W (Fig. 1d)) for calcite (e.g. BDX' values), and aragonite (e.g. LDX values). All data are plotted into a single longitudinal transect and are displayed relative to latitude and water depth. In addition, the measured state of saturation in the water column expressed in values of carbonate- ion concentration as derived from the GEOSECS data set (Bainbridge 1981) is shown. Values below 100 µmol $[CO_3^{2-}]$ /l indicate slight undersaturation of aragonite, whereas undersaturation with respect to calcite is reached when values are below 85 µmol $[CO_3^{2-}]$ /l. The

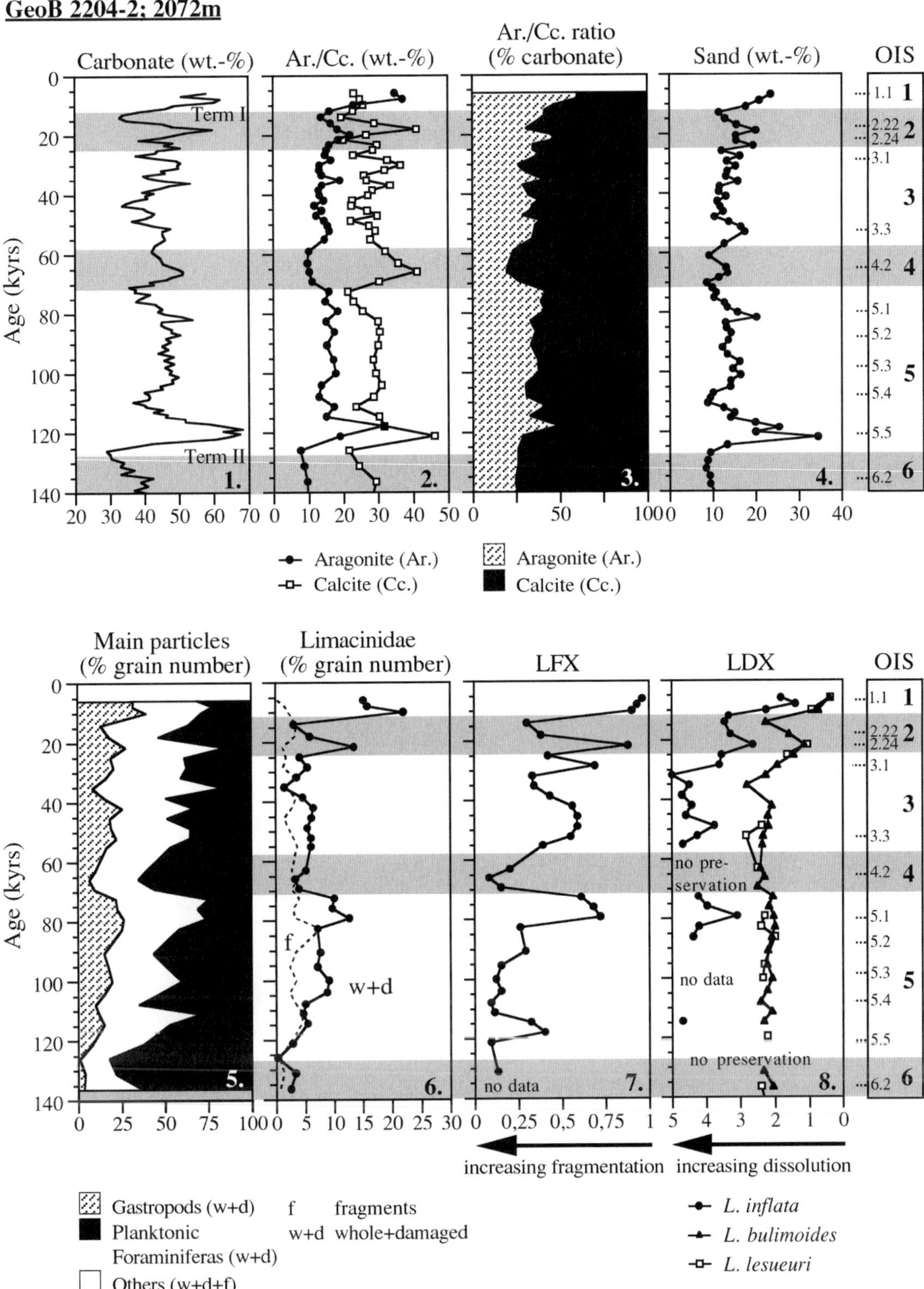

Fig. 4. Aragonite preservation indices for the last 140 ka as recorded in the gravity core GeoB 2204-2 (Brazil continental margin- Pernambuco Plateau, 2072 m water depth) as a function of time. Carbonate data partly and sand data completely from Rühlemann (1996); OIS = Oxygen Isotope Stages. Shaded areas mark glacial time intervals.

a) *Limacina inflata*: cross-section (SEM)

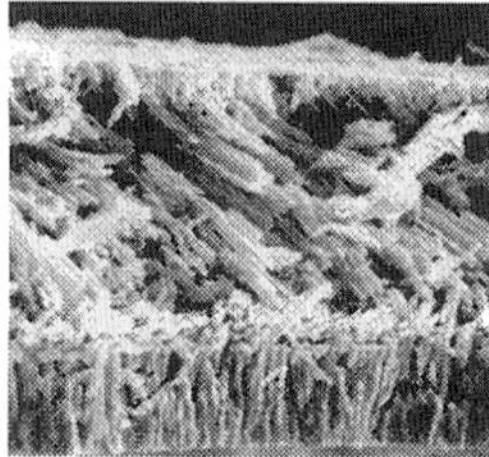

b) *Limacina inflata* Dissolution Index

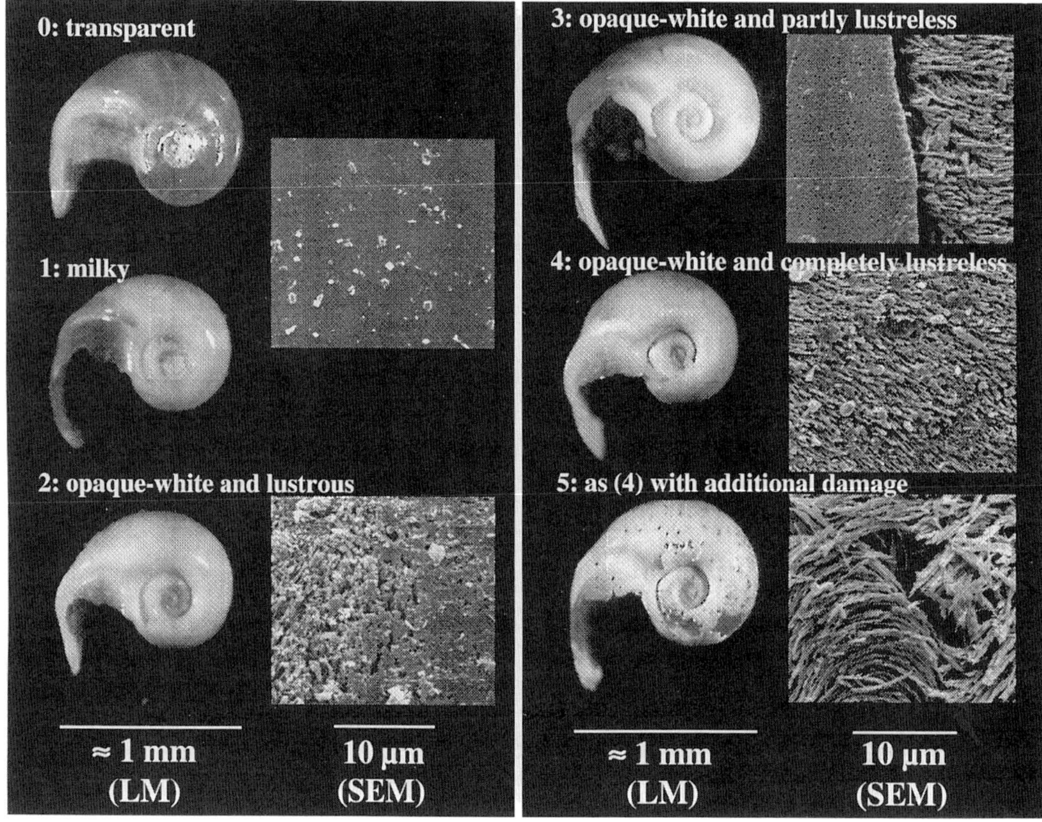

Fig. 5. Aragonite dissolution proxy LDX (*Limacina inflata* dissolution index): six steps of progressive dissolution of *Limacina inflata*. As dissolution proceeds tests become milky and lustreless, display additional damage and finally break down (modified from Gerhardt and Henrich 2001).

deep South Atlantic contains two strongly contrasting water masses (Fig. 1c, 6). The young, oxygenrich and carbonate-saturated NADW with a high $[CO_3^{2-}]$ /l is formed in the northern N-Atlantic (Dickson and Brown 1994) and invades the South Atlantic (from the north). In contrast, along the Antarctic continental margin much older and carbonate-corrosive waters, such as the Lower Circumpolar Deep Water (LCDW) and the Weddell Sea Deep Water (WSDW), originate from brines formed during sea- ice formation admixed with old recycled NADW (Reid 1996; Siedler et

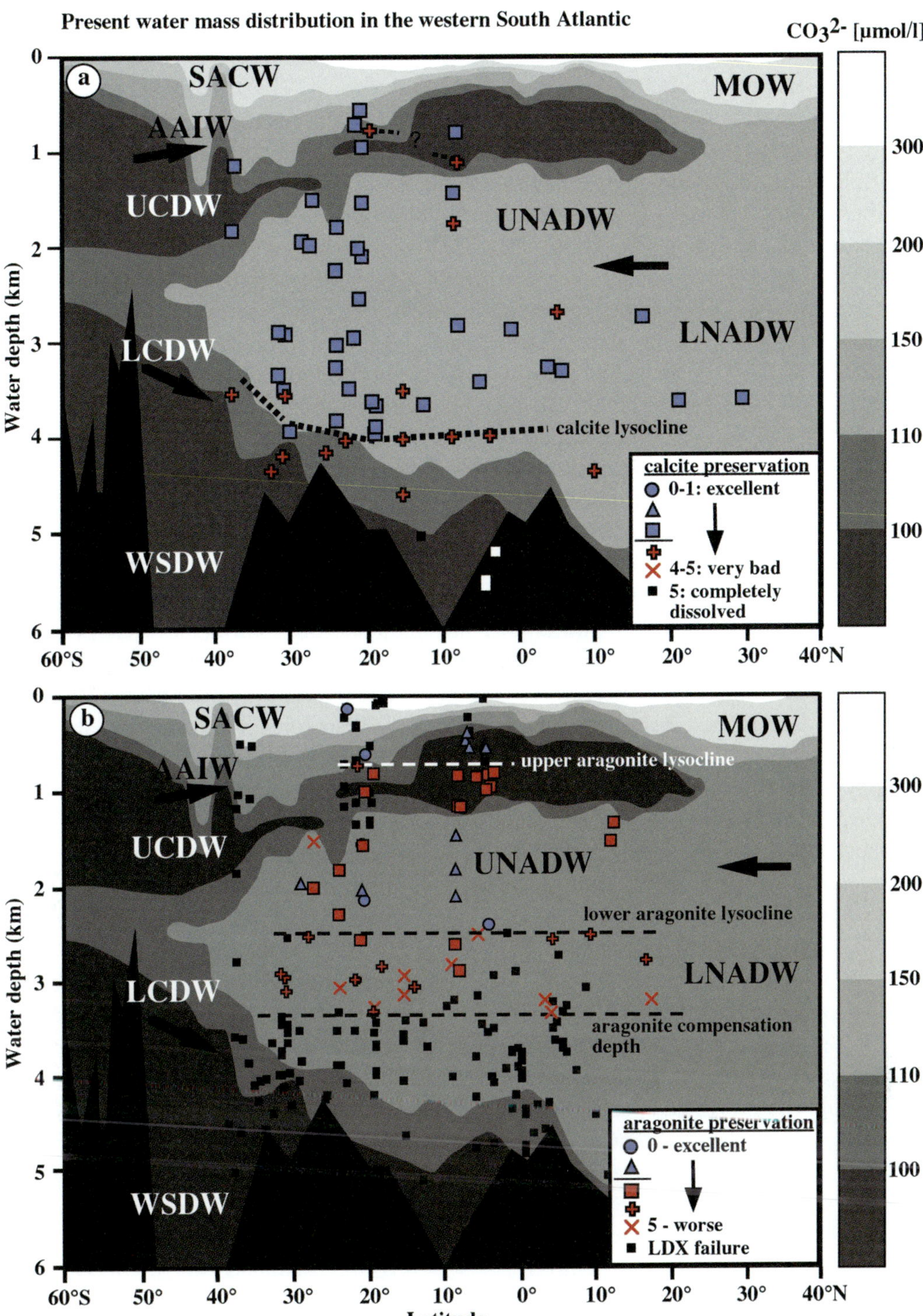

Fig. 6. Intermediate- and deep-water mass distribution, carbonate saturation state and reconstruction of the sedimentary lyscline for aragonite **a)** and calcite **b)** in the western South Atlantic today (LDX and BDX' data compiled from Gerhardt and Henrich (2001), and Volbers and Henrich (2002a). The main surface, intermediate and deep-water masses are indicated by their commonly used abbreviation.

al. 1996). The WSDW flows northward through the Argentine basin and enters the Brazil Basin through the Vema and the Hunter channels. Above the WSDW the LCDW spreads northward, with contours following the main paths of the underlying the WSDW. Both water masses are often linked together and then called Antarctic Bottom Water (AABW). At intermediate depths (Fig. 1b, 6), the northward flowing Upper Circumpolar Deep Water (UCDW) separates NADW from the Antarctic Intermediate Water (AAIW), which can be clearly distinguished from the UCDW by its low-salinity and high-oxygen values (Siedler et al. 1996). In the depth interval from 600 m to 1300 m, which is occupied by the AAIW and the UCDW, a minimum in the $[CO_3^{2-}]/l$ (e.g. values < 100 µmol) is recorded in the GEOSECS profiles. Finally, the South Atlantic Central Water (SACW) covers the AAIW. The Mediterranean Outflow Water (MOW) spreads west and southwestward from the Straits of Gibraltar at 800 m to 1500 m (Kaese and Zenk 1987).

The BDX' results are plotted in the upper graph of Fig. 6 (Volbers and Henrich 2002b). They reflect the hydrography very well, accurately displaying the state of carbonate saturation in the modern South Atlantic. In the AAIW, incipient calcite dissolution is indicated north of 25° S. Overall good preservation in the NADW is recorded by BDX' values of 2. The calcite lysocline is encountered at a water depth of 3,500 m at 35° S. It gradually increases in depth e.g. down to 3,900 m at 25° S, and coincides with the LNADW/ LCDW transition. Near the equator, the lysocline is uplifted into the LNADW. This might be explained by an additional supralysoclinal dissolution due to increased organic productivity in the equatorial upwelling zone (Rühlemann et al. 1999). The lower graph of Fig. 6 displays the results of the aragonite proxy LDX (modified from Gerhardt and Henrich 2001). Excellent preservation corresponds to the high $[CO_3^{2-}]/l$ within the SACW. Within the low carbonate ion concentration core of the AAIW and CDW at first very good preservation is observed until a transition to moderate preservation indicates an upper aragonite lysocline at a water depth of approximately 750 m. Below this lysocline we again observed good preservation of pteropod tests within

the UNADW until the lower aragonite lysocline is encountered at a water depth of 2,500 m. Broecker and Takahashi (1978) determined an average depth of 2,750 m for the aragonite lysocline in the western Atlantic from hydrographic data, which is about 250 m deeper than our data implies. Weak supralysoclinal dissolution affecting the surface sediments might explain the difference. The aragonite compensation depth is recorded at a water depth of 3,400 m as indicated by LDX failure in all samples below this depth level. Surprisingly, LDX failure is also found at shallow water depths along the western continental margin in supersaturated waters (Gerhardt and Henrich 2001). LDX failure is observed at 5°-7° S (23-565 m water depth), close to the Victoria Trinidade Ridge between 18°-23°S (23-1,330 m water depth), and near the mouth of the Rio de la Plata between 35°-38°S (490-1,836 m water depth). We relate these failures of the LDX to supralysoclinal dissolution, which was induced by enhanced productivity due to supply of organic matter and nutrients by the Amazon River and the Rio de la Plata (compare Frenz et al. this volume). This is consistent with the findings of Berger (1978) that pteropod tests are completely absent in the surface sediments of this region.

The eastern South Atlantic basins show a much more differentiated water mass structure than the western South Atlantic. This is predominately due to pronounced topographic barriers that exert a strong influence on the spreading of deep-water masses. The Walvis Ridge hinders the AABW to proceed northward into the South Atlantic. Only small quantities can enter the Angola Basin via the Walvis Passage (Shannon and Chapman 1991). The submarine barrier of the mid-Atlantic Ridge prevents an overflow of AABW and allows only penetration into the Guinea Basin via the Romanche Fracture Zone (Warren and Speer 1991). As a consequence, the Angola Basin and the Guinea Basin are mostly filled with NADW. This results in a very deep position of the hydrographic calcite lysocline below 4700 m to 5000 m, whereas in the Cape Basin the hydrographic calcite lysocline is encountered at a water depth of 4,400 m which is close to the NADW/AABW interface. This is in accordance with former studies that allocated the sedimentary calcite lysocline in the

Guinea and Angola Basin to a water depth of approximately 4,700m , on the basis of variations of carbonate content and the conventional planktic foraminifer fragmentation index (FI, Thunell 1982). Fig. 7 presents calcite preservation profiles identified by the BDX' proxy along various depth-tran-sects into the different eastern South Atlantic basins (modified from Volbers and Henrich 2002b). In all profiles a significant decrease in *G. bulloides* preservation marks the sedimentary calcite lyso-cline, which is encountered by BDX' values above 3. However, the character of this shift changes from rather abrupt in the offshore transects (e.g. Walvis Ridge–Cape Basin, Walvis Ridge–Angola Basin) to more gradual at the mid- Atlantic Ridge and in the continental margin transects (e.g. Namibian coast, Angola coast). In addition, the sedimentary lysocline identified by the BDX' values in all profiles is much shallower than the hydro-graphic calcite lysocline. Within the northern Cape Basin, the sedimentary calcite lysocline is 400 m shallower than the hydrographic lysocline and coincides with the NADW/AABW interface at a water depth of approximately 4,000 m. The sedimentary lysocline rises even more towards the

continental margin, being encountered at a water depth of 3,600 m in the costal Cape Basin and thus 400 m shallower than at the open-ocean North Cape Basin. The reason for this rise of the sedimentary calcite lysocline of 400 m to 800 m as compared to the hydrographic lysocline is due to degradation of organic matter which causes additional supra-lysoclinal dissolution. The degree of this process depends on the amount of organic material in the sediments and can thus be estimated by the organic carbon contents that are still preserved. The second important factor is the accumulation rate which determines the time span during which supra-lysclinal dissolution is effective in the uppermost centimetres of the sediment. In general, increased accumulation rates eventually stop the process (Volbers and Henrich 2002b). The gap between the sedimentary and the hydrographic lysocline is even more pronounced in the Guinea and Angola Basin (e.g. 1,000-1,600 m shallower, compare Fig. 7). In addition, the BDX' values display a more gradual increase with depth (e.g. a more hyperbolically shaped profile) in these coastal and equatorial high productivity areas, indicating that supralysoclinal dissolution is progressively stronger affecting fora-

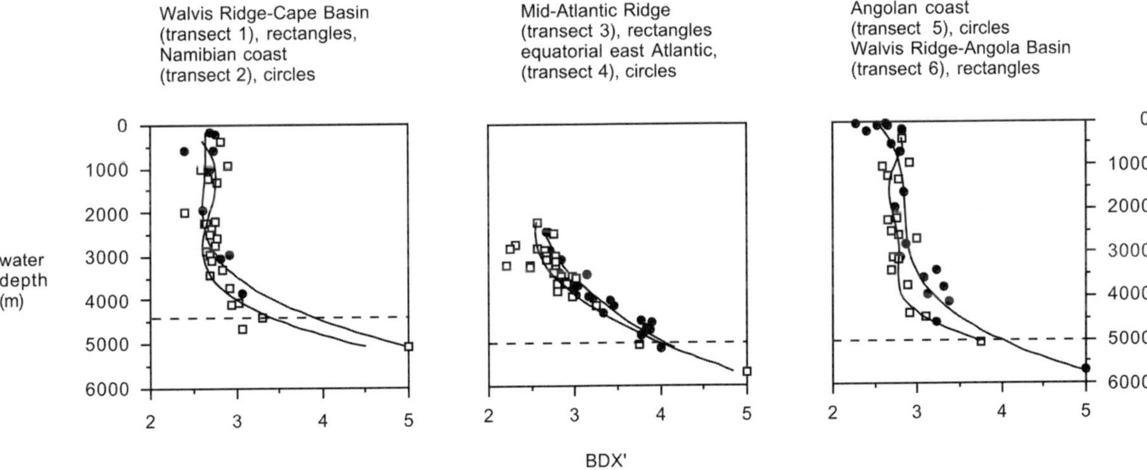

Fig. 7. Determination of calcite preservation along different transects through the eastern South Atlantic basins as indicated by BDX' preservation values in surface sediments. The depth level at which the hydrographic lysocline is recorded indicated by the dashed line. Note the much higher position of the sedimentary lysocline in all profiles, e.g. where the BDX' value increases above 3 (modified from Volbers and Henrich 2002b).

miniferal preservation over a much broader depth interval than in the open ocean. A special case is the mid -Atlantic Ridge transect where the sedimentary lysocline is encountered at a water depth of 4,000 m, 700 m above the hydrographic lysocline. Here, the BDX' profile is also hyperbolically shaped despite a continuously low organic carbon content in the sediments. Due to the very low sedimentation rates we assume that the particles were exposed and affected by weak supralysoclinal dissolution over a much longer period.

Due to the increased influence of supralysoclinal dissolution, wide areas of the eastern South Atlantic are characterised by strong aragonite dissolution. Berger (1978) previously mentioned the nearly complete absence of pteropod tests in surface sediments of the eastern South Atlantic (compare Baumann et al. this volume). Hence, only at a few sites at the mid-Atlantic Ridge could the LDX be applied and then revealed poor to very poor preservation. All other locations are void of pteropods and thus indicate failure of the LDX (Gerhardt and Henrich 2001).

From our results we come to the following conclusions: (1) Although the Atlantic has been regarded as the largest present-day carbonate sink of the deep-sea (Milliman 1993), the carbonate preservation potential in the eastern South Atlantic is considerably lower than would be expected from previous studies. (2) Because of a strong secondary overprint by supralysoclinal dissolution, the carbonate preservation pattern in the eastern South Atlantic cannot be used to reconstruct the deep and intermediate water mass structure. This stands in clear contrast to the western South Atlantic, where carbonate preservation profiles are much more clearly related to water mass characteristics.

Intermediate and Deep Water Masses in the South Atlantic during the Last Glacial Maximum (LGM- 23-19cal-ka-BP): Reconstruction of Carbonate Preservation Patterns

As a general consensus from various LGM benthic $\delta^{13}C$ studies, a significant suppression of LNADW contribution to the South Atlantic is indicated (Oppo

and Fairbanks 1987; Sarnthein et al. 1994). In contrast, the production of less dense UNADW, often referred to as Glacial North Atlantic Intermediate Water (GNAIW), was enhanced (Oppo et al. 1995). As a consequence, southern-source deep water (e.g. glacial AABW or Southern Component Water (SCW) – Bickert 1992) could have expanded to shallower depths compensating for the LNADW in the western and eastern Atlantic basins (Duplessy et al. 1988; Curry 1996; Bickert and Wefer 1996; Bickert et al. 1997a,b). Based on benthic $\delta^{13}C$ evidence, Bickert and Wefer (1996) divided the SCW into lower (LSCW) and upper (USCW) levels. Using conventional dissolution proxies (e.g. carbonate and sand contents), they reconstructed the LGM calcite lysocline at a water depth of 3,800 m near the equator at identical levels in the western and eastern South Atlantic basins, from where they slightly rise towards the Southern Ocean. Recent by, rapid shifts in the contribution of glacial northern-source deep and intermediate waters were recognised (Venz et al. 1999), whereas much less is known about the glacial fluctuation of southern-source intermediate waters. Oppo and Horrowitz (2000) inferred from LGM benthic $\delta^{13}C$ and Cd/Ca records that the GNAIW extended at least as far south as to 28°S in the western South Atlantic, replacing the AAIW and the UCDW. According to the results of Haddad and Droxler (1996), the AAIW and the UCDW were not present in the Caribbean Sea during the LGM. However, the details of the LGM intermediate water- mass structure farther south are still unknown. We have tested the different LGM conceptual models by comparing them with the results from our carbonate preservation studies in the LGM time slice. Details on the chronostratigraphy of the sample set used in this study are given by Arz et al. (1999) and Niebler et al. (2003).

Fig. 8 shows the carbonate preservation in LGM sediment samples of the western South Atlantic, the LDX for aragonite preservation (Gerhardt and Henrich submitted) and the BDX' for calcite preservation (Volbers and Henrich submitted) have been applied. LDX and BDX' values are plotted in the same way as in the modern transect. Also shown in the graph is the LGM water- mass distribution as derived from the benthic foraminifer δ^{13}C records of Bickert (1992). Extreme shifts in the LGM water- mass structure are recognised in comparison to the modern setting. The main feature is an increase of southern deep water and an almost complete shutdown of the LNADW. This pattern is excellently confirmed by the calcite preservation as documented by the BDX' (Fig. 8a). Preservation in the LGM samples is good to moderate throughout the water column till the calcite lysocline is encountered. We observed the LGM calcite lysocline to be at 3,300 m at 40°S and 3,600 m at 15°S, which indicates a rise of more than 500 m on average compared to modern settings. Interestingly, the calcite lysocline deepens below a water depth of 4,000 m in the LGM between 15°S and the equator, whereas it is uplifted in this region today. We attribute this deepening to an increase in coccolith production because of the higher intensity of LGM upwelling (Kinkel et al. 2003). The resultant higher supply of carbonate to the sea floor accounts for the better preservation. Another interesting difference is the rather gradual increase in dissolution below the lysocline in the LGM, which contrasts with the much steeper gradient today.

Fig. 8b displays the results of aragonite preservation obtained by application of the LDX to LGM samples from the western South Atlantic. Very good to incipiently moderate preservation is found in intermediate waters above 2,000 m water depth, indicating the presence of a water mass that is equal to modern UNADW in preservation character. Most interesting, there is no indication for an upper lysocline, giving evidence that UCDW/AAIW was almost completely substituted by the GNAIW in the intermediate water layer. This pattern agrees well with the δ^{13}C and Cd/Ca reconstruction of Oppo and Horrowitz (2000) indicating the extension of GNAIW as far south as 28° S latitude. At water depths below 2,000 m, LDX failure is indicated, except for one location with poor preservation at water depth of 2,974 m. Hence, the aragonite lysocline was more than 500 m shallower during the LGM than it is today, indicating an expansion of the SCW. The aragonite lysocline was situated above the GNAIW/SCW transition, which is considered to mark the ACD at that time.

Fig. 8c displays the BDX' results obtained for the LGM time slice in the eastern South Atlantic. BDX' values vary only slightly within the upper 3,000 m of the water column and do not decrease below the calcite lysocline as rapidly as observed in the modern ocean. The position of the calcite lysocline was entcountered in the southern part at a water depth of approximately 3,200 m and drops towards the equator to 4,000 m depth. In the northern Guinea Basin, the calcite lysocline deepens to a water depth of 4,000 m.

Intermediate and Deep Water Carbonate Production and Preservation: Implications of the Atmospheric pCO_2 in the LGM

The aragonite and calcite preservation profiles indicate that the South Atlantic was less stratified in the intermediate and upper deep-water level during the LGM than under modern conditions. Because the position of the calcite lysocline was at the same depth level in the eastern and western South Atlantic, we infer that the modern west-east asymmetry in the South Atlantic did not exist during the LGM. This is interpreted to result from an expansion of southern deep waters in the water column, which were able to penetrate into the Angola and Guinea Basin. The effect on carbonate preservation caused by the shift in the primary water- mass structure seems to have dominated during the LGM, whereas secondary overprinting by supralysoclinal dissolution appears to have been surprisingly less significant than under modern conditions. Francois et al. (1990) reconstructed carbonate fluxes to the sea floor using high- resolution 230 Th$_{ex}$ profiling of the sedimentary sections deposited at various depths of the Ceara and Sierra Leone Rises. In these areas of the equatorial Atlantic, which are not directly influenced by upwelling, carbonate fluxes of cores estimated to be above the lysocline were significantly lower during the LGM than today. Francois et al. (1990) interpreted these lower fluxes to record lower carbonate production during the LGM. This corresponds well with our finding of generally good preservation at intermediate water depths during the LGM. Hence, the commonly expressed assumption that

LGM water mass distribution in the western South Atlantic

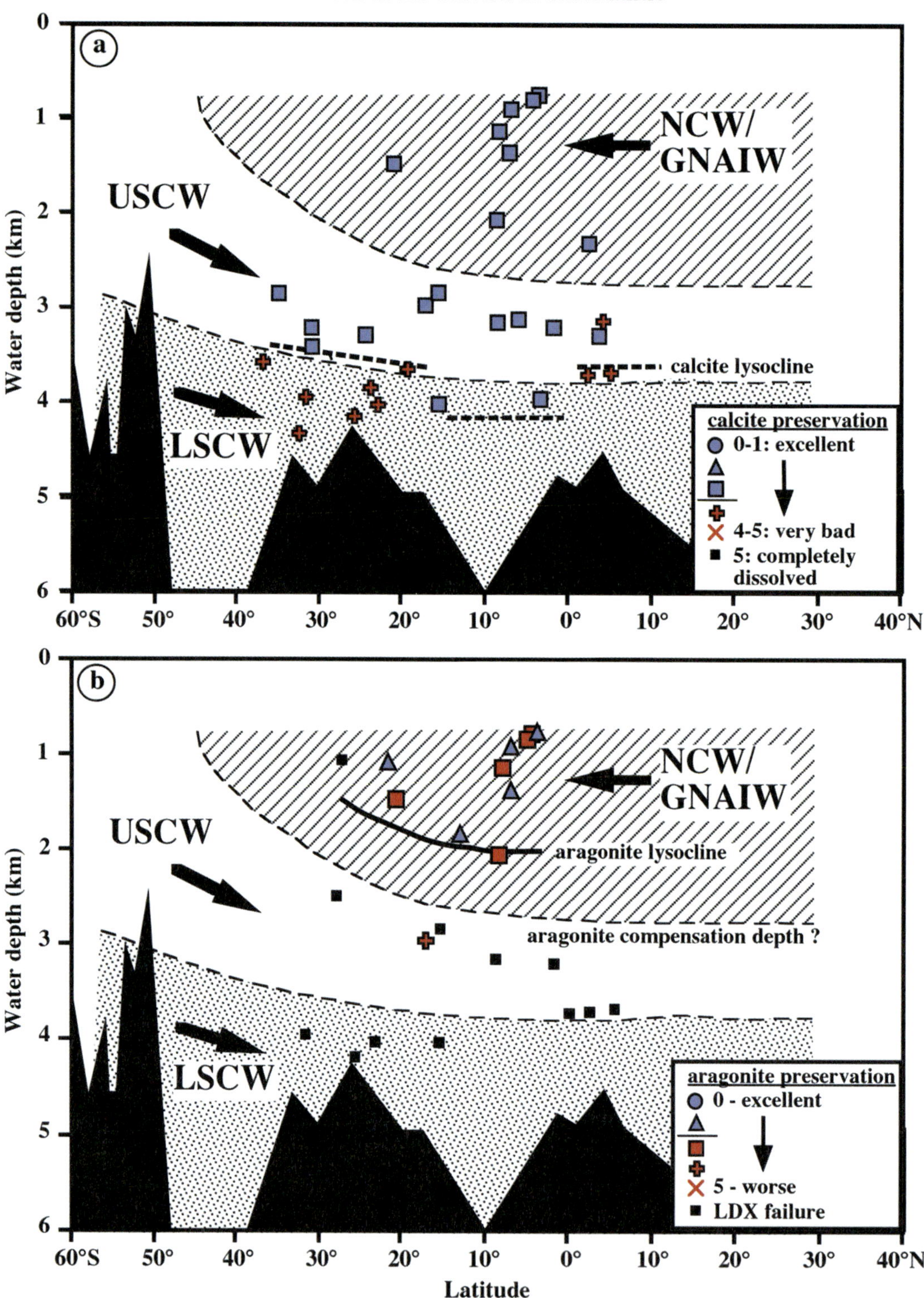

Fig. 8. Distribution of intermediate and deep water masses during the LGM reconstruction from benthic foraminifer δ^{13}C values by Bickert (1992) and aragonite (LDX) and calcite (BDX') preservation pattern (LDX data compiled from Gerhardt and Henrich 2003; BDX' data compiled from Volbers and Henrich 2003). **a)** LGM – calcite preservation in the western South Atlantic, **b)** LGM – argonite preservation in the western South Atlantic.

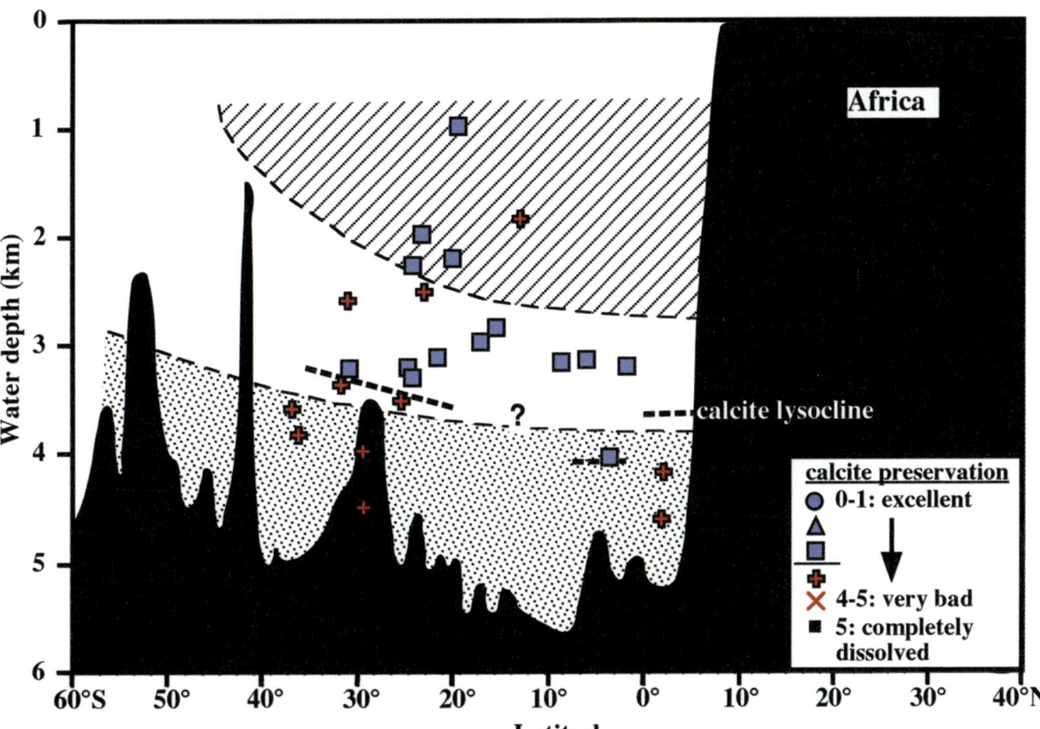

Fig. 8. c) LGM – calcite preservation in the eastern South Atlantic.

the low carbonate contents in glacial sediments at intermediate water depths may have resulted from the dissolution of carbonate can be ruled out. However, the low carbonate values in LGM sediments from water depths deeper than 4000 m are clearly related to an increased carbonate dissolution as indicated by significantly elevated BDX' values. In conclusion, the overall good preservation at intermediate water depths in the LGM testifies the presence of the GNAIW possessing a low nutrient content. Intermediate waters are the main source for nutrients to be supplied to the surface waters. Hence, in the LGM, the low nutrient content of the GNAIW may have depressed new production (Sarnthein et al. 1988), in particular, in the oligotrophic central gyres of the Atlantic. This might explain the low carbonate production rates (Francois et al. 1990), which in turn might have contributed to the low atmospheric pCO$_2$ measured in polar ice cores. In contrast, the overall increase in carbonate corrosiveness in deep waters of the

South Atlantic (e.g. below a water depth of 3500 - 3800 m) indicate significantly lower [CO$_3^{2-}$] values in deep waters as compared to today. Following the argumentation by Archer et al. (2000) and Sigman and Boyle (2000), a shoaling of the calcite lysocline in the deep ocean indicates a decrease in oceanic alkalinity and thus would have the effect of increasing atmospheric CO$_2$ levels. Hence, the observed carbonate dissolution in South Atlantic deep waters provides a counteracting mechanism to lowering the atmospheric CO$_2$ level.

The Geological Record of Intermediate and Deep Water Masses in the Western Equatorial Atlantic during the past 3.4 Ma: Aspects from Carbonate Preservation Patterns

In the last chapter we will compare the records of two ODP Sites to study long-term trends in the carbonate preservation of deep and intermediate waters in the equatorial Atlantic. Site 927 on the north-eastern slope of the Ceara Rise at a water depth of 3,300 m is well positioned in the core of the LNADW (Fig. 1). Site 999 is located in the Colombian Basin in the Caribbean Sea at a water depth of 2800 m (Fig. 1). Since the connection between the Caribbean Sea and the Atlantic is restricted by a sill front to depths shallower than 1600-1800 m, the area is an ideal monitor for variations of intermediate water circulation in the equatorial Atlantic. Today, a mixture of AAIW and UNADW enters the Caribbean Sea.

To evaluate the long-term evolution of physical and chemical parameters of deep and intermediate waters in the equatorial Atlantic, we will use a combination of proxies including benthic foraminifer stable-isotope records, different parameters from silt analysis and planktic foraminifer fragmentation indices. Gröger et al. (2003a) have demonstrated that the late Quaternary records of mean grain size of the terrigenous "sortable silt" can be used to reconstruct a history of highly variable bottom current speeds at the Ceará Rise (for details of methodological approach see Gröger et al. 2003a). Fig. 9 displays the results obtained for the period from 0.85 Ma to 0.25 Ma. A good correlation of decreased benthic foraminifer $\delta^{13}C$ values, minimum values of the mean grain size of terrigenous sortable silt and low percentages of whole planktic foraminifers is observed during glacial stages. This clearly indicates poor ventilation, a high carbonate corrosiveness, and a sluggish exchange of bottom waters, thus evidencing a considerably weakened circulation of the LNADW at the Ceara Rise. In contrast to this, high benthic-foraminifer $\delta^{13}C$ values, maximum values of the mean grain size of terrigenous sortable silt and high percentages of whole planktic foraminifers during interglacial periods indicate a strengthening of the circulation and much better ventilation in the LNADW-loop. One might speculate whether the LNADW might have been considerably "older" compared to interglacials periods due to the weakening circulation during glacials. This is corroborated by a strong negative relationship between the mean sortable silt and $\delta^{13}C$ records. Thus, gradients in $\delta^{13}C$ and $[CO_3^{2-}]$ between the LNADW and southern-source

deep waters might have been less extreme during glacials as compared to interglacials.

Apart from these general glacial/interglacial shifts, the record displays additional interesting phenomena. The most striking changes in paleo-current speeds, inferred from records of the mean grain size of terrigenous sortable silt, occur during glacial-to-interglacial transitions. This shift from the lowest to very high bottom-currents speeds occurs during the transitions 18/17, 14/13 and 10/9 (Fig. 9). This indicates a rapid reinstatement of the lower NADW circulation relatively soon after stages of "shutdown". As a consequence, this pattern evidences a widespread re-balancing of western Atlantic deep-water circulation during the Late Quaternary glacial to interglacial transitions.

In addition, a long-term trend becomes apparent: An abrupt shift to generally poorer preservation of planktic foraminifers, together with the highest current speeds inferred from the mean sortable silt record, is recognised at the onset of glacial stage 12. This marks the beginning of the mid-Brunhes dissolution cycle (Diester Haas and Rothe 1987; Farrel and Prell 1989; Bassinot et al. 1994). In the western equatorial Atlantic. This phenomenon might be interpreted as a consequence of climatic changes in the source region of the NADW. After the onset of glacial stage 12 (~480 ka), a much higher variability is recorded in the foraminiferal preservation during both glacial and interglacial periods. This indicates that changes in carbonate preservation at the Ceará Rise from then on not only occurred in response to glacial-interglacial climatic changes, but also to sub-orbital time scales.

We will now compare the long-term evolution of deep and intermediate water circulation in the western equatorial Atlantic during the past 3.4 Ma, by comparing the results from Site 927 at the Ceara Rise and Site 999 in the Caribbean Sea (Gröger et al. 2003b). At Site 927, the terrigenous record of the mean grain size of sortable silt (terrigenous mean(SS)) displays a stepwise reduction of glacial bottom-current strength from the Late Pliocene to the Pleistocene. This is interpreted to reflect a stepwise weakening of the circulation in the LNADW cell, with main shifts occurring at 2.75 Ma and 0.95 Ma. The most stable

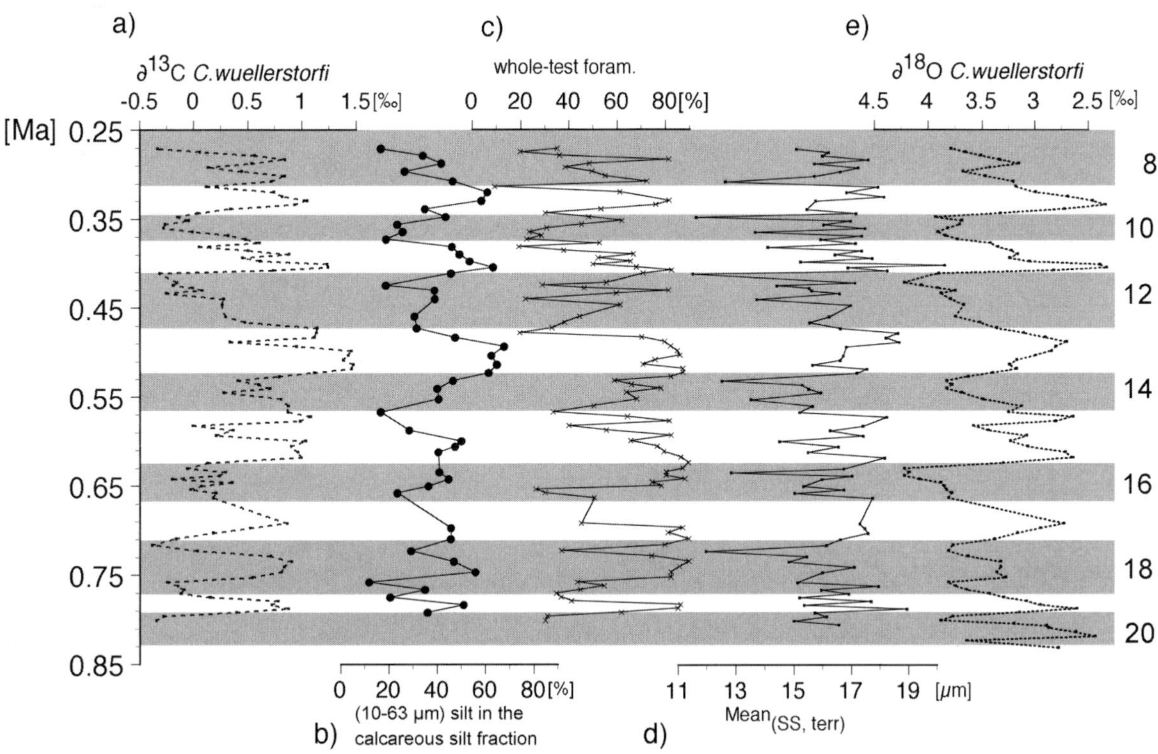

Fig. 9. Glacial/ interglacial variability (800 ka to 300 ka) of geochemical and physical properties in the lower circulation loop of North Atlantic Deep Water in the equatorial Atlantic (ODP Site 927, Ceara Rise). Benthic foraminifer $\delta^{18}O$ and $\delta^{13}C$ values are displayed against different carbonate preservation proxies, e.g. percentage of calcareous sortable silt, percentage of whole foraminifer tests, and a proxy for bottom -current strength, e.g. mean diameter of terrigenous sortable silt (modified from Gröger et al. 2003).

circulation in the LNADW cell, as inferred from terrigenous mean(SS) and planktic foraminifer preservation, is found during the initial phase of Northern Hemisphere Glaciation between ~3.2 and ~2.75 Ma. This is consistent with the "super-conveyor model" of Raymo et al. (1996). However, our data do not support this model in the time span from 3.5 to 3.2 Ma. We propose that variations in the LNADW flux prior to 3.2 Ma were mainly controlled by variations in the surface-water salinity in the Atlantic Ocean, whereas they are controlled by ice-sheet forcing after 2.75 Ma. A drastic reduction of glacial bottom-current strength at Site 927 is inferred after 2.75 Ma (Fig. 10). This shift is synchronous with the first occurrence of larger continental ice shields and a drastic decrease in

deep convection in the Norwegian-Greenland Sea (Henrich et al. 2002). An additional impact on the Atlantic deep-water circulation is registered during the mid-Pleistocene climate transition around 1.1 to 1.0 Ma. It is indicated by drastic decreases in bottom-current strength and preservation of planktic foraminifers (Fig. 10). In agreement with the history of bottom-current strength inferred from the terrigenous mean(SS) proxy data, a stepwise worsening in the preservation of planktic fora-minifers during glacials at the Ceará Rise indicates a diminished influence of the LNADW at the Ceará Rise from the Late Pliocene to Pleistocene. This in-dicates a general decrease in the contribution of the LNADW to the Atlantic. In contrast, an opposing trend is recognised in the Caribbean at Site

999, showing poorer preservation of planktic foraminifers during the Late Pliocene interval moving toward better preservation during the Pleistocene (Fig. 10). This indicates an increasing overturn of northern-source deep waters through the Caribbean sill front from the Atlantic Ocean into the Caribbean basins, indicating an increase in the contribution of the UNADW to the Atlantic.

In summary, the combination of a conventional planktic foraminifer fragmentation index with proxies derived from silt grain size analysis has proved to be a powerful tool to decipher long and short-term changes in carbonate corrosiveness and current strength of deep and intermediate water masses. However, as a final remark, we underline the necessity to ensure that the proxies under consideration were not influenced by other factors as discussed in the first chapter of this article. For the investigated site this test was positive (see Gröger et al. 2003a, Gröger and Henrich 2003 for more details).

Conclusions

Our evaluation of conventional calcite dissolution proxies in South Atlantic surface sediments revealed that their overall broad applicability is limited to oligotrophic far-offshore environments. Many of these proxies only allow for a qualitative characterisation of the stage of preservation. In continental margin settings, in particular under high-productivity regimes, most of these conventional proxies either fail or produce incorrect or misleading results. This is due to effects of variable dilution with non-carbonate material and the variable ecological conditions. The only exception is the conventional fragmentation index on planktic foraminifers. However, the sensitivity of this proxy is much lower than those of newly developed SEM dissolution indices for planktic foraminifers. We recommend to use conventional dissolution proxies only with caution. In particular, when studying fossil records, the assumption that other factors did not significantly vary through time should be tested and verified.

New sedimentological carbonate dissolution proxies are derived from silt analysis. During progressive dissolution the overall calcareous silt con-tent and the coarse calcareous silt mode decrease, whereas the values for the mean and modal grain size in the coarse calcareous silt fraction diminish consistently. Parallel to this, the amount of calcareous clay increases, indicating a continuous shift of particles from coarse to fine grain-size classes. These patterns clearly indicate that foraminifers in the silt fraction are less resistant to carbonate dissolution as compared to coccoliths.

New micropaleontological dissolution proxies allow a semi-quantitative determination of calcite and aragonite preservation by distinguishing progressive steps of (ultra)-structural breakdown of the skeletons of planktic foraminifers (*Globigerina bulloides* Dissolution Index – BDX') and pteropods (*Limacina inflata* dissolution index - LDX).

The applicability of the BDX' and LDX was tested in South Atlantic surface sediments and calibrated against the state of carbonate saturation indicated by GEOSECS data. In the western South Atlantic basins, the LDX and BDX' predominantly reflect the state of carbonate corrosiveness of the different surface-, intermediate- and deep-water masses. The sedimentary calcite lysocline is encountered at the transition between the AABW and the LNADW. However, it rises up into the LNADW close to the equator due to additional supralysoclinal dissolution caused by an enhanced degradation of organic matter in the equatorial upwelling regime. In the eastern South Atlantic basins the secondary overprint by supralysoclinal dissolution is even more prominent, particularly under the influence of the continental margin high-productivity regimes. This results in an elevation of the sedimentary calcite lysocline of several hundred metres to a maximum of 1,600 m as compared to the position of the hydrographic lysocline. Aragonite preservation in the eastern South Atlantic is even poorer. At most sites investigated the surface sediments are void of pteropods and thus LDX failure is observed. However, in the western South Atlantic the LDX provides an excellent tool to trace the only slightly undersaturated AAIW/UCDW core. This results in a double lysocline for aragonite, the upper lysocline being encountered at a water depth of 750 m, whereas the lower lysocline is traced by the LDX at a water depth of 2500 m.

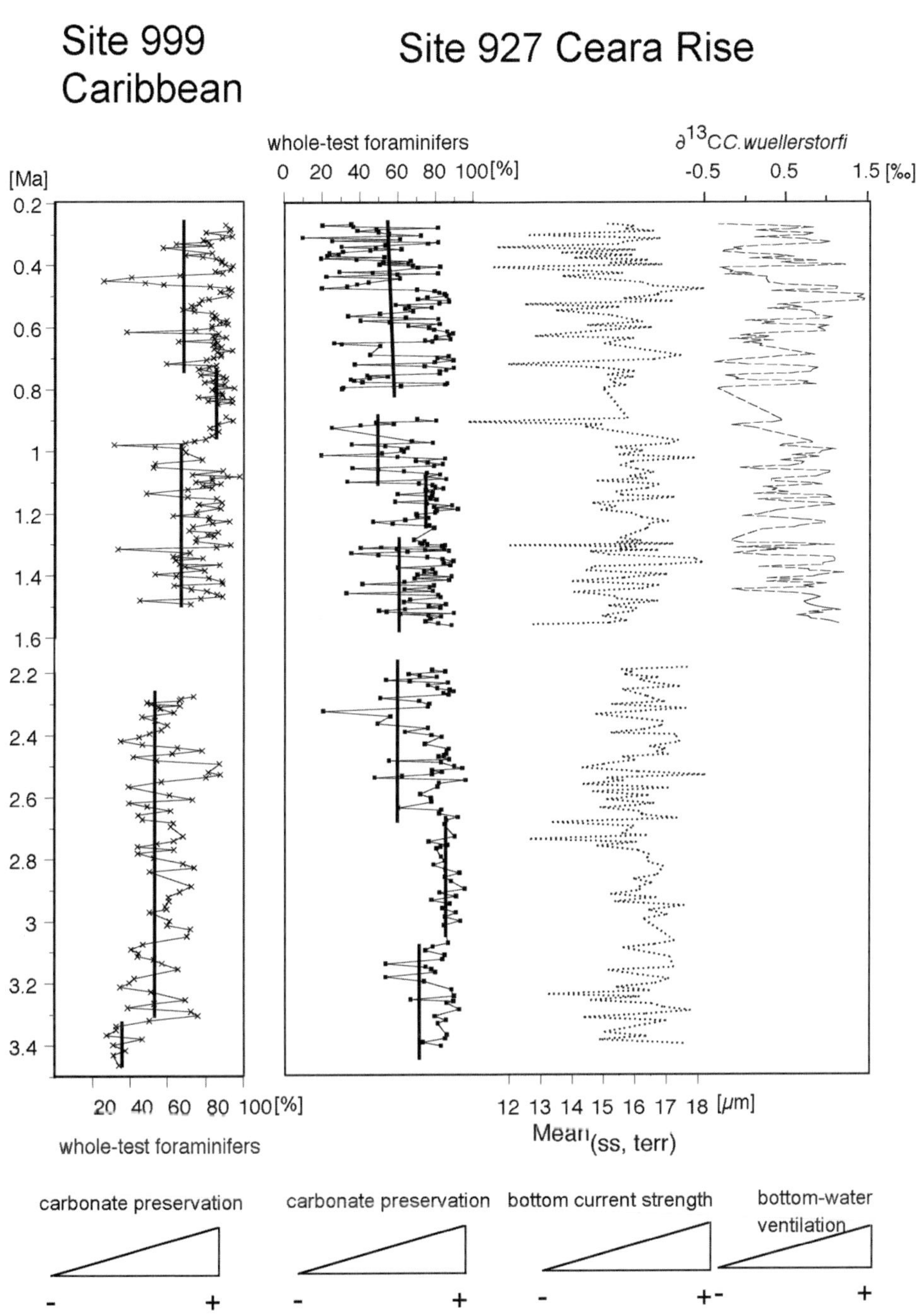

Fig. 10. Comparison of the conventional planktic foraminifer fragmentation index, the benthic foraminifer $\delta^{13}C$ record and the mean grain size of terrigenous sortable silt (terrigenous mean(SS)) in the equatorial Atlantic at the Ceara Rise (ODP Site 927) and the Caribbean Sea (ODP Site 999) (modified from Gröger and Henrich 2003).

Aragonite and calcite preservation profiles indicate that the South Atlantic was less stratified during the LGM. The modern west-east asymmetry in the South Atlantic was not developed. The position of the sedimentary calcite lysocline is encountered at exactly the same level in the eastern and western basins at around 3,200 m in the southern part, dropping to 4,000 m near the equator. Carbonate preservation patterns seem to be dominated by the primary water- mass corrosiveness during the LGM, whereas secondary overprinting by supra-lysoclinal dissolution appears to have been surprisingly less significant than under modern conditions. Along the western continental margin no indication for aragonite-corrosive glacial AAIW was found providing clear evidence for a strengthened GNAIW flow along the Brazil margin.

Results from a multiproxy study combining records of benthic foraminifer stable isotopes, planktic foraminifer fragmentation indices, and mean grain size in the terrigenous sortable silt fraction provide interesting new insights into the geological evolution of deep and intermediate water masses in the western equatorial Atlantic. By this combination, chemical (nutrient content, carbonate corrosiveness) and physical (bottom- current strength) properties can be estimated for the first time. The most striking changes in paleo-current speed inferred from variations in mean grain size of terrigenous sortable silt are recorded at nearly all the isotopic transitions in the Brunhes epoch at ODP Site 927 at the Ceara Rise. Here, rapid shifts from low to high bottom-current speeds are recognised indicating subsequent phases of shutdown and rapid reinstatement of LNADW circulation. In the longterm the comparison of the records of ODP Site 927 at Ceara Rise and Site 999 in the Caribbean Sea display stepwise shifts of LNADW and UNADW circulation. The most stable circulation in the LNADW cell, inferred from terrigenous mean(SS) and planktic foraminifer preservation, is found during the incipient phase of Northern Hemisphere Glaciation between ~3.2 and ~2.75 Ma. A drastic reduction of glacial bottom-current strength at Site 927 is inferred after 2.75 Ma, synchronous with the first occurrence of larger continental ice shields, and with a drastic decrease of deep convection in the Norwegian-Greenland Sea. During the mid-Pleistocene climate transition around 1.1 to 1.0 Ma, another significant decrease of bottom-current strength and preservation of planktic foraminifers is found in the Ceara Rise record. Since then a stepwise deterioration (in the preservation) of planktic foraminifers during glacials indicate a progressive reduction of LNADW from the Late Pliocene to Pleistocene intervals. In contrast, an opposite trend is recognised in the Caribbean at Site 999, showing poorer preservation of planktic foraminifers during the Late Pliocene improving during the Pleistocene. This indicates a contemporaneous progressive increase in the contribution of UNADW to the Atlantic in glacial periods. Altogether, since the mid- Pleistocene transition a progressive weakening of the circulation in the LNADW loop and a contemporaneous strengthening of the UNADW loop are observed.

Acknowledgements

We are grateful to G. Wefer and his working group who provided surface sediment samples to us. B. Donner kindly contributed *G. bulloides* from sediment traps. We thank Renate Henning, Helga Heilmann and numerous student workers for laboratory assistance and technical support. Review comments by J. Milliman and H. Westphal improved the manuscript and are gratefully acknowledged. This research was funded by the Deutsche Forschungsgemeinschaft (DFG-Grant He 1671/ 10; Grant for Sonderforschungbereich 261 at the University of Bremen, Contribution No. 366). Data are available under www.pangaea.de/Projects/SFB261.

References

Adelseck CG, Berger WH (1975) On the dissolution of planktonic and associated microfossils during settling and on the seafloor. Cushman Foundation Spec Pub 13: 70-81

Adelseck CG (1978) Dissolution of deep-sea carbonate: Preliminary calibration of preservational and morphologic aspects. Deep-Sea Res 25: 1167-1185

Archer D, Emerson S, Reimer M (1989) Dissolution of calcite in deep-sea sediments: pH and O_2 microelectrode results. Geochim Cosmochim Acta 53: 2831-2845

Archer DE, Maier-Reimer E (1994) Effect of deep-sea sedimentary calcite preservation on atmospheric CO_2 concentration. Nature 367: 260-264

Archer D, Winguth A, Lea D, Mahowald N (2000) What caused the glacial/interglacial atmospheric pCO_2 cycles? Rev Geophys 38: 159-189

Arz HW, Pätzold J, Wefer G (1999) The deglacial history of the western tropical Atlantic as inferred from high resolution stable isotope records off northeastern Brazil. Earth Planet Sci Lett 167: 105-117

Bainbridge AE (1981) GEOSECS Atlantic Expedition, Hydrographic Data 1972-1973. National Science Foundation

Bassinot FC, Beaufort L, Vincent LE, Labeyrie LD, Rostek F, Müller PJ, Quidelleur X, Lancelot Y (1994) Coarse fraction fluctuations in pelagic carbonate sediments from the tropical Indian Ocean: A 1500-kyr record of carbonate dissolution. Paleoceano-graphy 9: 579-600

Baumann KH, Meggers H (1996) Paleoceanographical changes in the Labrador Sea during the last 3.1 My: Evidence from calcareous plankton records. In: Moguilevsky A, Whatley R (eds) Microfossils and Oceanic Environments. The University of Wales, Aberystwyth Press, pp 131-153

Berger WH (1967) Foraminiferal ooze: Solution at depths. Science 156:383-385

Berger WH (1973) Deep-sea carbonates: Pleistocene dissolution cycles. J Foram Res 3: 187-195

Berger WH (1978) Deep-sea carbonate: Pteropod distribution and the aragonite compensation depth. Deep-Sea Res 25: 447-452

Berger WH (1979) Preservation of foraminifera. In: Lipps JH, Berger WH, Buzas M, Douglas A, Ross RG, Cushman G (eds) Preservation of Foraminifera. Foundation Spec Pub 13: 105-155

Berger WH, Bonneau MC, Parker FL (1982) Foraminifera on the deep-sea floor: Lysocline and dissolution rate. Oceanol Acta 5: 249-258

Berger WH (1989) Global maps of Ocean Productivity. In: Berger WH, Smetacek VS and Wefer G (eds) Productivity of the Ocean: Present and Past. Springer, Berlin, pp 429-455

Berger WH, Keir R (1984) Glacial-Holocene changes in atmospheric CO_2 and the deep-sea record. In: Broecker WS, Hansen J, Takahasi T (eds) Climate Processes and Climate Sensitivity. Am Geophys Union, pp 337-351

Bé AWH, Morse JW, Harrison SM (1975) Progressive dissolution and ultrastructural breakdown of planktonic foraminifera. In: Sliter WV, Bé AWH, Berger WH (eds) Progressive Dissolution and Ultrastructural Breakdown of Planktonic Foraminifera. Cushman Foundation Spec Pub 13: 27-55

Bickert T (1992) Rekonstruktion der spätquartären Bodenwasserzirkulation im östlichen Südatlantik über stabile Isotope benthischer Foraminiferen. Ber Fachber Geowiss, Univ Bremen 27, 205 p

Bickert T, Wefer G (1996) Late Quaternary deep-water circulation in the South Atlantic: Reconstruction from carbonate dissolution and benthic stable isotopes. In: Wefer G, Berger WH, Siedler G (eds) The South Atlantic: Present and Past Circulation. Springer, Berlin, pp 599-620

Bickert T, Cordes R, Wefer G (1997a) Late Pliocene to mid-Pleistocene (2.6-1.0 M y) carbonate dissolution in the western equatorial Atlantic: Results of Leg 154, Ceara rise. Proc ODP Sci Results 154: 229-237

Bickert T, Curry WB, Wefer G (1997b) Late Pliocene to Holocene (2.6-1.0 Ma) western equatorial Atlantic deep-water circulation: inferences from benthic stable isotopes. Proc ODP Sci Results 154: 239-254

Broecker WS, Takahashi T (1978) The relationship between lysocline depth and *in situ* carbonate ion concentration. Deep-Sea Res 25: 65-95

Broecker WS, Peng TH (1982) Tracers in the sea. Eldigio Press, 689 p

Curry WB (1996) Late Quaternary deep circulation in the western equatorial Atlantic. In: Wefer G, Berger WH, Siedler G, Webb DJ (eds) The South Atlantic: Present and Past Circulation. Springer, Berlin, pp 577-598

Dickson RR, Brown J (1994) The production of North Atlantic deep water: Sources, rates, and pathways. J Geophys Res 99: 12319-12341

Diester-Haass L (1977) Radiolarian/planktonic foraminiferal ratios in a coastal upwelling region. J Foram Res 7: 26-33

Diester-Haass L, Rothe P (1987) Plio-Pleistocene sedimentation on the Walvis Ridge, Southeast Atlantic (DSDP Leg 75, Site 532) - Influence of surface currents, carbonate dissolution and climate. Mar Geol 77: 53-85

Dittert N, Baumann KH, Bickert T, Henrich R, Huber R, Kinkel H, Meggers H, Müller PJ (1999) Carbonate dissolution in the deep-sea: Methods, quantification and paleoceanographic application. In: Fischer G, Wefer G (eds) Use of Proxies in Paleoceanography. Springer, Berlin, pp 255-284

Dittert N, Henrich R (2000) SEM ultrastructure of a spinose planktonic foraminifer, *Globigerina bulloides* (d'Orbigny), and its paleoceanographic power. Deep-Sea Res I 47: 603-620

Duplessy JC, Shackleton NJ, Fairbanks RG, Labeyrie L, Oppo D, Kallel N (1988) Deepwater source variations during the past climatic cycle and their impact on the global deepwater circulation. Paleoceanography 3: 343-360

Emerson S, Bender M (1981) Carbon fluxes at the sediment-water interface of the deep-sea: Calcium carbonate preservation. J Mar Res 39:139-162

Emerson SR, Archer D (1990) Calcium carbonate preservation in the ocean. Phil Trans R Soc Lond 331: 29-40

Farrell JW, Prell WL (1989) Climatic change and $CaCO_3$ preservation: An 800,000 year bathymetric reconstruction from the Central Equatorial Pacific Ocean. Paleoceanography 4: 447-466

Fischer G, Wefer G (1996) Long-term observations of particle fluxes in the eastern Atlantic: Seasonality, changes of flux with depth and comparison with the sediment record. In: Wefer G, Berger WH, Siedler G, Webb DJ (eds) The South Atlantic: Present and Past Circulation. Springer, Berlin, pp 325-344

Francois RM, Bacon M, Suman DO (1990) Thorium 230 profiling in deep-sea sediments: High-resolution records of flux and dissolution of carbonate in the equatorial Atlatic during the last 24,000 years. Paleoceanography 5: 761-787

Gerhardt S, Groth H, Rühlemann C, Henrich R (2000) Aragonite preservation in late Quaternary sediment cores on the Brazilian Continental Slope: Implications for intermediate water circulation. Int J Earth Sci 88: 607-618

Gerhardt S, Henrich R (2001) Shell preservation of *Limacina inflata* (pteropoda) in surface sediments from the central and South Atlantic ocean: A new proxy to determine the aragonite saturation state of water masses. Deep-Sea Res I 48: 2051-2071

Gerhardt S, Henrich R (2003) Intermediate water circulation during the last glacial maximum in the South Atlantic Ocean inferred from changes in the shell preservation of *Limacina inflata* (Pteropoda). Deep-Sea Res I, submitted

Gröger M, Henrich R (2003) Deep water circulation during the Pleistocene (0.8-0.25 Ma): Interference from near bottom-current flow variability and deep water chemistry in the western equatorial. Mar Geol, submitted

Gröger M, Henrich R, Bickert T (2003a) Glacial-interglacial variability in the lower circulation loop of North Atlantic deep water: Interference from silt grain size analysis and carbonate preservation studies at Ceara Rise, western equatorial Atlantic. Mar Geol 201: 321-332

Gröger M, Bickert T, Henrich R (2003b) Variability of silt grain size and planktic foraminifer preservation in the Plio/Pleistocene sediments from the western equatorial Atlantic (ODP Site 927) and Caribbean (ODP Site 999): Implications for the history of Atlantic deep and intermediate water circulation. Mar Geol 201: 307-320

Gordon AL (1986) Interocean exchange of thermocline water. J Geophys Res 91:5037-5046

Gordon AL, Ray FW, Smethie Jr WM, Warner MJ (1992) Thermocline and intermediate water communication between the South Atlantic and Indian Oceans. J Geophys Res 97C: 7223-7240

Haddad GA, Droxler WA (1996) Metastable $CaCO_3$ dissolution at intermediate water depths of the Caribbean and western North Atlantic: Implications for intermediate water circulation during the past 200,000 years. Paleoceanography 11: 701-716

Henrich R (1989) Glacial/interglacial cycles in the Norwegian Sea: Sedimentology, paleoceanography, and evolution of late Pliocene to Quaternary northern hemisphere climate. Proc ODP Sci Results 104: 189-232

Henrich R, Wefer G (1986) Dissolution of biogenic carbonates: Effects of skeletal structure. Mar Geol 71: 341-362

Henrich R, Baumann KH (1994) Evolution of the Norwegian current and the Scandinavian Ice Sheets during the past 2.6 my: Evidence from ODP Leg 104 biogenic carbonate and terrigenous records. Palaeogeogr Palaeoclimatol Palaeoecol 108: 75-94

Henrich R, Baumann KH, Huber R, Meggers H (2002) Carbonate preservation records of the past 3 Myr in the Norwegian-Greenland Sea and the northern North Atlantic: Implications for the history of NADW production. Mar Geol 184: 17-41

Huber R, Meggers H, Baumann KH, Henrich R (2000) Recent and Pleistocene carbonate dissolution in sediments of the Norwegian-Greenland Sea. Mar Geol 165: 123-136

Jahnke RA, Craven DB, Gaillard JF (1994) The influence of organic matter diagenesis on $CaCO_3$ dissolution at the deep-sea floor. Geochim Cosmochim Acta 58: 2799-2809

Kaese RH, Zenk W (1987) Reconstructed Mediterranean salt lens trajectories. J Phys Oceanogr 17: 158-163

Kemle-von Mücke S, Oberhänsli H (1999) The distribution of living planktic foraminifera in relation to Southeast Atlantic oceanography. In: Fischer G, Wefer G (eds) Use of proxies in paleoceanography. Examples from the South Atlantic. Springer, Berlin, pp 91-115

Kinkel H, Dittert N, Henrich R (2003) Calcareous plankton record of the Equatorial Atlantic: A 300 kyrs record of climate feedback, productivity and dissolution. Palaeogeogr Palaeoclimatol Palaeoecol, submitted

Little MG, Schneider RR, Kroon D, Price B, Bickert T, Wefer G (1997) Rapid palaeoceanographic changes in the Benguela Upwelling System for the last 160,000 years as indicated by abundance of planktonic foraminifera. Palaeogeogr Palaeoclimatol Palaeoecol 130: 135-161

Milliman JD (1975) Dissolution of aragonite, Mg-calcite, and calcite in the North Atlantic Ocean. Geology 3: 461-462

Milliman JD (1993) Production and accumulation of calcium carbonate in the ocean: Budget of a non steady state. Glob Biogeochem Cycl 7: 927-957

Milliman JD, Droxler WA (1996) Neritic and pelagic carbonate sedimentation in the marine environment: ignorance is not bliss. Geol Rundsch 85: 496-504

Milliman JD, Troy PJ, Balch WM, Adams AK, Li YH, Mackenzie FT (1999) Biologically mediated dissolution of calcium carbonate above the chemical lysocline? Deep-Sea Res I 46: 1653-1669

Morse JW, Mackenzie FT (1990) Geochemistry of sedimentary carbonates. Developments in Sedimentology 48:1-706

Morse JW, Mucci A, Millero FJ (1980) The solubility of calcite and aragonite in seawater at various salinities, temperatures and 1 atmosphere total pressure. Geochim Cosmochim Acta 44: 85-94

Niebler H-S, Mulitza S, Donner B, Arz H, Pätzold J, Wefer G (2003) Sea-surface temperatures in the equatorial and South Atlantic Ocean during the Last Glacial Maximum (23-19 ka). Paleoceanography 18: doi: 10.1029/2002PA000902

Oppo DW, Fairbanks RG (1987) Variability in the deep and intermediate water circulation of the Atlantic Ocean during the past 25,000 years: Northern Hemisphere modulation of the Southern Ocean. Earth Planet Sci Lett 86: 1-15

Oppo DW, Horowitz M (2000) Glacial deep water geometry: South Atlantic benthic foraminiferal Cd/Ca and δ13C evidence. Paleoceanography 15: 147-160

Oppo DW, Raymo ME, Lohmann GP, Mix AC, Wright JD, Prell WL (1995) A $\delta^{13}C$ record of Upper North Atlantic Deep Water during the past 2.6 million years. Paleoceanography 10: 373-394

Parker FL, Berger WH (1971) Faunal and solution pattern of planktonic foraminifera in surface sediments of the south Pacific. Deep-Sea Res 18: 73-107

Peterson MNA (1966) Calcite: Rates of Dissolution in a Vertical Profile in the Central Pacific. Science 154: 1542-1544

Peterson RG, Stramma L (1991) Upper-level circulation in the South Atlantic Ocean. Prog Oceanogr 26: 1-73

Raymo ME, Grant B, Horowitz M, Rau GH (1996) Mid-Pliocene warmth: Stronger greenhouse and stronger conveyor. Mar Micropaleontol 27: 313-326

Reid JL (1994) On the total geostrophic circulation of the North Atlantic Ocean: Flow patterns, tracers, and ransports. Prog Oceanogr 33: 1-92

Reid JR (1996) On the circulation of the South Atlantic Ocean. In: Wefer G, Berger WH, Siedler G, Webb DJ (eds) The South Atlantic: Present and Past Circulation. Springer, Berlin, pp 13-44

Rühlemann C (1996) Akkumulation von Carbonat und organischem Kohlenstoff im tropischen Atlantik: Spätquartäre Produktivitäts-Variationen und ihre Steuerungsmechanismen. Ber Fachber Geowiss, Univ Bremen 84, 139 p

Rühlemann C, Müller PJ, Schneider R (1999) Organic carbon and carbonate as paleoproductivity proxies: Examples from high and low productivity areas of the tropical Atlantic. In: Fischer G, Wefer G (eds) Use of proxies in paleoceanography: Examples from the South Atlantic. Springer, Berlin, pp 315-344

Sarnthein M, Winn K, Duplessy JC, Fontugne MR (1988) Global variations of surface ocean productivity in low and mid latitudes: Influence on CO_2 reservoirs of the deep ocean and the atmosphere during the last 21,000 years. Paleoceanography 3: 361-399

Sarnthein M, Winn, K, Jung SJA, Duplessy JC, Labeyrie L, Erlenkeuser H, Ganssen G (1994) Changes in east Atlantic deepwater circulation over the last 30,000 years: Eight time slice reconstructions. Paleoceano-graphy 9: 209-267

Schmiedl G, Mackensen A (1997) Late Quaternary paleo-productivity and deep water circulation in the eastern South Atlantic Ocean: Evidence from benthic foraminifera. Palaeogeogr Palaeoclimatol Palaeoecol 130: 43-80

Shannon LV, Chapman P (1991) Evidence of Antarctic Bottom Water in the Angola Basin at 32°S. Deep-Sea Res 38: 1299-1304

Siedler G, Müller TJ, Onken R, Arhan M, Mercier H, King BA, Saunders PM (1996) The Zonal WOCE Sections in the South Atlantic. In: Wefer G, Berger WH, Siedler G, Webb DJ (eds) The South Atlantic: Present and Past Circulation. Springer, Berlin, pp 83-104

Sigman DM, Boyle EA (2000) Glacial/interglacial variations in atmospheric carbon dioxide. Nature 407: 859-869

Sliter WV, Bé AWH, Berger WH (1975) Distribution of deep-sea carbonates. Cushman Foundation of Foraminiferal Research Special Publication 13: 1-159

Stuut JB, Prins MA, Jansen HF (2002) Fast reconnaissance of carbonate dissolution based on the size distribution of calcareous ooze on Walvis Ridge. Mar Geol 190: 581-589

Talley LD (1996) Antarctic Intermediate Water in the South Atlantic. In: Wefer G, Berger WH, Siedler G, Webb DJ (eds) The South Atlantic: Present and Past Circulation. Springer, Berlin, pp 219-238

Thunell RC, Honjo S (1981) Calcite dissolution changes in the South Atlantic: Response to initiation of northern hemisphere glaciation. Paleoceanography 4: 565-583

Thunell RC (1982) Carbonate dissolution and abyssal hydrography in the Atlantic. Mar Geol 47:165-180

van Kreveld-Alfane S (1996) Late Quaternary sediment records of mid-latitude northeast Atlantic calcium carbonate production and dissolution. PhD Thesis University of Amsterdam, 212 p

Venz KA, Hodell DA, Stanton C, Warnke DA (1999) A 1.0 Myr record of Glacial North Atlantic Intermediate Water variability from ODP site 982 in the northeast Atlantic. Paleoceanography 14: 42-52

Volbers A, Henrich R (2002a) Late Quaternary Variations in calcium carbonate preservation of deep-sea sedi-ments in the northern Cape Basin: Results from a multiproxy approach. Mar Geol 180: 203-220

Volbers A, Henrich R (2002b) Present water mass calcium carbonate corrosiveness in the eastern South Atlantic inferred from ultrastructural breakdown of *Globigerina bulloides* in surface sediments. Mar Geol 186: 471-486

Volbers A, Henrich R (2003) Calcium carbonate corrosiveness in the South Atlantic during the Last Glacial Maximum as inferred from changes in the preservation of of *Globigerina bulloides*: A proxy to determine deep-water circulation pattern? Mar Geol, submitted

Warren B, Speer KG (1991) Deep circulation in the eastern South Atlantic Ocean. Deep-Sea Res 39: 281-322

Wollast R (1994) The relative importance of biomineralization and dissolution of $CaCO_3$ in the global carbon cycle. In: Doumenge F (ed) Past and Present Biomineralization Processes: Considerations about the Carbonat Cycle. Bulletin de l'Institut Oceanographique, Monaco, pp 16-36

Wu G, Berger WH (1991) Pleistocene $\delta^{18}O$ record from Ontong Java Plateau: Effects of winnowing and dissolution. Mar Geol 96: 193-209

Wüst G (1935) Schichtung und Zirkulation des Atlantischen Ozeans. Deutsche Atlantische Expedition *Meteor*, 1925-1927, Wissenschaftliche Ergebnisse 2

Last Glacial to Holocene Changes in South Atlantic Deep Water Circulation

T. Bickert[1*] and A. Mackensen[2]

[1]*Università* *Bremen, Fachbereich Geowissenschaften, Klagenfurter Strasse,*
28359 Bremen, Germany
[2]*Alfred-Wegener-Institut für Polar- und Meeresforschung, Columbusstrasse,*
27515 Bremerhaven, Germany
* corresponding author (e-mail): bickert@uni-bremen.de*

Abstract: A set of 55 benthic foraminiferal stable carbon and oxygen isotope time series, including 28 new records, is presented from the South Atlantic Ocean between 6°N and 47°S. We compiled these records with published data of the eastern North Atlantic to reconstruct the Atlantic deepwater circulation for the Last Glacial Maximum (19-23 ka) and the Late Holocene (0-4 ka) times. To better understand the spatial distribution of deep and bottom water masses, we assigned these records to three North-South sections representing the western South Atlantic, the central Atlantic east of the Mid-Atlantic Ridge, and the eastern marginal Atlantic. Corrections of up to +0.4‰ are suggested for several benthic $\delta^{13}C$ values of cores located in high-productivity areas, to adjust for phytodetritus-induced depletion of especially glacial values. As a result of this new compilation, no shift of NADW to intermediate depth during the last glacial maximum is evident in the eastern and western marginal Atlantic. Instead, the core of an ^{13}C-enriched water mass spreading southward to at least 30°S between 1200 and 1900 m points to a source of this water mass close to the Isthmus of Gibraltar, indicated by $\delta^{13}C$-values of up to 1.8‰. Therefore, we interpret this layer as an extended tongue of the Mediterranean Outflow Water. Below, a layer of glacial NADW is shown to flow southward at about the same depth interval or even deeper than it does today, although slightly depleted in ^{13}C and less extended in water column. The admixing of NADW into the circumantarctic deepwater belt occurred a few degrees farther north than today, marked by a steep gradient in glacial $\delta^{13}C$ between 30° and 40° S. From these gradients we derive a local formation of Southern Ocean deep water in the zone of extended winter sea-ice coverage south of the polar front. The spreading of this newly formed water mass, however, is restricted to the Atlantic basins south of Walvis Ridge and Rio-Grande Rise, where only a small amount of nutrient-enriched deep water passes across these barriers into the northern basins. Converted into nutrient concentrations, the new carbon isotope data set gives only a slight increase in the nutrient inventory of the deep Atlantic, in good agreement with previously published Cd/Ca data.

Introduction

Since the first measurements of gas inclusions in ice cores by Berner et al. (1978) it is well known that the concentration of carbon dioxide in the atmosphere was lower during ice ages than it is today. However, as yet there is no broadly accepted explanation for this difference. Most theories for low glacial atmospheric CO_2 concentrations focus either on an increase in the strength of the biological pump as a result of changes in the supply or utilization of nutrients or light, or on an increase in the ocean's alkalinity due to coral reef dissolution or carbonate sediment interactions (for recent summaries see Archer et al. 2000; Sigman and Boyle 2000). However, models that invoke biological pump

From WEFER G, MULITZA S, RATMEYER V (eds), 2003, *The South Atlantic in the Late Quaternary: Reconstruction of Material Budgets and Current Systems.* Springer-Verlag Berlin Heidelberg New York Tokyo, pp 671-695

changes are generally inconsistent with the magnitude or direction of glacial/interglacial changes in $^{13}C/^{12}C$ ratios and nutrients (Charles and Fairbanks 1990; Boyle 1992), and with the lack of widespread glacial deepwater anoxia. On the other hand, models invoking alkalinity changes predict large increases in the depth of the glacial lysocline which are not observed (e.g. Farell et al. 1989; Bickert and Wefer 1996; Henrich et al. this volume).

In search of a better explanation than the given hypotheses, Stephens and Keeling (2000) proposed a mechanism that is not based on productivity or alkalinity increases, but on changes in the rate of air-sea gas exchange as a result of increased sea-ice cover at high southern latitudes. By significantly limiting the sea-to-air CO_2 flux in the primary region for deepwater ventilation, expanded Antarctic sea ice during glacial times may trap relatively more carbon in the deep ocean, thereby reducing atmospheric CO_2 concentrations. However, the sea ice-driven regulation of CO_2 outgassing is effective only if the wintertime sea-ice area fraction southward of the Antarctic Polar Front rises to 99- 100%. At present, strong circumpolar winds pull the ice apart, and intense oceanic heat fluxes from deep vertical mixing hinder ice growth to a point that sea ice covers no more than 80-90% of the total winter ice-pack area. Morales Maqueda and Rahmstorf (2002) presented simulations with a coupled sea ice/upper ocean model indicating that the CO_2 sequestration under glacial ice coverage could account for at most 15-50% of the total glacial CO_2 decline.

Sigman and Boyle (2000) presented a hypothesis that focuses on both the biology and physics of the open ocean surrounding Antarctica. In this hypothesis, a cooler climate caused a northward shift in the maximum westerlies that drive upwelling and northward surface flow in the modern Antarctic. This shift caused a decrease in the upwelling of deep water into the Antarctic surface, replacing it with upwelling of intermediate-depth water into the subantarctic surface. Subsequently, a stable, fresh, frequently ice-covered surface layer developed, further reducing the deep ocean ventilation in the open Antarctic. The increased stratification that resulted from these changes lowered the rate of nutrient supply to the Antarctic surface and reduced

CO_2 outgassing. On the other hand, the hypothetical subantartic was more productive due to a combination of higher export production and a reduction in these nutrients from the Antarctic surface. This hypothesis is supported by nitrogen isotope data which suggest that during the last ice age nitrate utilization was twice or more its current value (Francois et al. 1997). A box model calculation predicts that 25-40% higher nitrate utilization (that is, 50-65% during the last ice age compared to 25% during the present interglacial) could lower atmospheric CO_2 by the full glacial-interglacial amplitude (Sigman et al. 1999). Because paleoceano-graphic proxy data suggest that Antarctic export production was lower during the last ice age, Sigman and Boyle (2000) infered that more complete nitrate utilization in the Antarctic was due to a lower rate of nitrate supply from the subsurface, implying that the fundamental driver of the CO_2 change was an ice age decrease in the ventilation of deep waters at the surface of the Antarctic.

However, a study by Yu et al. (1996), based on measurements of the ^{213}Pa / ^{230}Th ratios in glacial and modern sediments indicates that the the export of ^{213}Pa from the Atlantic into the Southern Ocean continued at roughly the modern rate during the LGM. They infer from these results that the export of deep water formed in the North Atlantic into the Southern Ocean continued at a comparable rate during LGM. If this was true, it would call the hypothesized collapse of NADW formation during glacial cold periods (Keeling and Stephens 2001; Sigman and Boyle 2000) into question.

To adress the role of deep water ventilation as the driver for CO_2 changes, we present a new data set of benthic stable isotope records from 55 sediment cores in the South Atlantic that allows a more detailed reconstruction of the deep water circulation in this region sensitive to climate changes. Combined with published data from the North Atlantic, we assigned these records to three North-South sections representing the western South Atlantic, the central Atlantic east of the Mid-Atlantic Ridge, and the eastern marginal Atlantic, to better understand the spatial distribution of deep and bottom water masses. We compared our results to studies using other proxies for deep ocean circulation like Cd/Ca and carbonate dissolution indices.

Modern Hydrography

The deep South Atlantic is a crucial place to observe changes in the depth distribution of deep water properties because it reveals the mixing zone between NADW and southern-source intermediate and deep water masses (Reid 1989). Today the circulation in the deep western South Atlantic, the main flow path of deep water, is dominated by interactions between North Atlantic Deep Water (NADW) flowing toward the south and Circumpolar Deep Water (CDW) flowing to the north. The density characteristics of these water masses are such that at about 45°S the NADW divides the southern water mass into an upper and lower branch. Consequently the relatively warm and saline NADW occupies the depth interval between 1500 and 4000 m, while below 4000 m lower CDW (LCDW) is encountered. The mixing zone between NADW and LCDW is marked by gradients in temperature, salinity, nutrient concentrations, and in the corrosiveness of the water with respect to calcium carbonate. For the eastern basins of the South Atlantic, i.e. the Guinea and the Angola Basins, the inflow of LCDW is restricted by the Mid-Atlantic Ridge in the west and the Walvis Ridge in the south to only small quantities passing the sills through the Romanche Fracture Zone (Van Bennekom and Berger 1984; Warren and Speer 1991) and the Walvis Passage (Shannon and Chapman 1991). Therefore the deepest parts of these basins are filled almost exclusively by NADW. The Cape Basin, although located eastwards of the Mid-Atlantic Ridge, is dominated by LCDW below 4000 m due to a bottom water passage, which allows LCDW to enter the basin from the south.

The modern asymmetry in bottom water distribution is traced by the pattern of $\delta^{13}C_{DIC}$ (Kroopnick 1985; Mackensen et al. 1996; Bickert and Wefer 1999; Mackensen 2001). Below 4000 m water depth in the western Atlantic basins and in the Cape Basin the $\delta^{13}C$-values show the signature of the circumantarctic deep water of about 0.4‰, whereas in the domain of NADW between 1500 and 4000 m in the west and below 1500 m down to the bottom in the Guinea and Angola Basins, these values are higher than 0.9‰, slightly decreasing along the flow path of NADW to the south.

Methods

Sediment Samples

For reconstructing the late Quaternary deep circulation in the South Atlantic, we focus on 55 gravity corers and the associated giant box corers or multicorers retrieved during several *Meteor, Polarstern,* and *Victor Hensen* cruises in the South Atlantic (Fig.1 and Table 1). These pelagic sediment cores are distributed from about 6°N to about 47°S in water depths ranging from 767 to 4675 m. Sediment sequences represent different depositional realms of the South Atlantic, which comprise carbonate oozes of eutrophic areas along the eastern equatorial upwelling, the upwelling off Namibia and close to the oceanic fronts between 38° and 50° S, carbonate muds from areas characterized by higher terrigenous input like the Ceara Rise and along the continental margin off South Africa, and low-accumulating nannofossil oozes retrieved from the oligo-trophic open ocean areas.

For the 28 newly presented cores, as in previously published South Atlantic records, the epibenthic taxon *Cibicidoides wuellerstorfi* was chosen to determine the stable carbon and oxygen isotopic composition using standard procedures. Two to five specimens of this species were picked from the >250 µm size fraction. The samples were analyz-ed using a Finnigan MAT 251 micromass-spectro-meter coupled with a Finnigan automated carbonate device at the University of Bremen. The carbonate was reacted with orthophosphoric acid at 75°C. The reproducebility of the measurements, as refered to an internal carbonate standard (Solnhofen limestone), is ± 0.07 ‰ and ± 0.05 ‰ (1σ over a one year period) for oxygen and carbon isotopes, respectively. The conversion to the VPDB-scale was performed using the international standard NIST 19.

Stratigraphy, Time Intervals and Transect Construction

The age models for all cores are based on graphic correlation of the $\delta^{18}O$ records to the SPECMAP standard record. The graphic correlations were aided by applying the AnalySeries Software Ver-

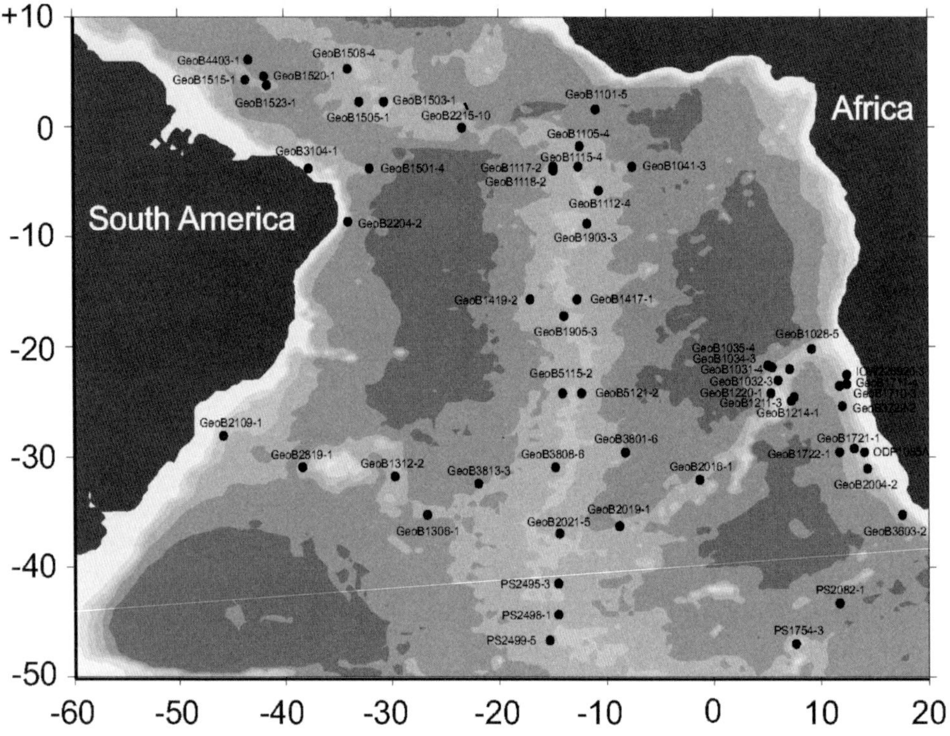

Fig. 1. Bathymetry of the South Atlantic with the core sites presented in this study.

sion 1.2 (Paillard et al. 1996). To ensure to have an undisturbed record of the uppermost part of each sediment column, a composite record was established by using the sediment sequences from both box corer and gravity corer at many sites. The cores were sampled at 5 cm spacing (box cores at 3 cm) throughout their entire length, two of the low-resolution cores of the Mid-Atlantic Ridge at about 24°S in 1 cm intervals (GeoB 5115-2, GeoB 5121-2). Some age models were checked by AMS ^{14}C dates over the last 30.000 years (see Tab.1) and were found to be in good agreement with reservoir-corrected and to calendar-year converted ^{14}C ages.

The sedimentation rates range between 2.5 and 11.3 cm/ky in the eutrophic areas along the eastern equatorial upwelling, the upwelling off Namibia and close to the subantarctic ocean fronts between 38° and 50° S, between 2.6 and 9.7 cm/ky in areas characterized by higher terrigenous input, and between 0.7 and 2.5 cm/ky in the cores retrieved from the oligotrophic open ocean areas. Although

many sequences extend far back in time, only the last 30.000 years of each record are presented here. The complete data sets here discussed are available from the PANGAEA world data center (http://www.pangaea.de/).

To better understand the spatial distribution of deep and bottom water masses, we added published data from the eastern North Atlantic compiled by Sarnthein et al. (1994) and from the western marginal South Atlantic (Oppo and Horrowitz 2000) to the new data set of the South Atlantic, and assigned all these records to three North-South sections representing the western South Atlantic, the central Atlantic east of the Mid-Atlantic Ridge, and the eastern marginal Atlantic. Due to the still insufficient number of data points per section, the contouring of the geochemical data (Figs 3 and 4) was done manually instead of using mapping routines.

To compare glacial-to-interglacial variation in deep water distribution, we selected two time intervals each spanning 4000 years: The Late Holocene (0-4 ka, calendar year scale, 0-3.6 ky ^{14}C scale)

Core	Latitude	Longitude	Water depth (m)	Section	Cruise	Sed. rates (cm / ky)	14C-ages	Reference benth. isotopes
GeoB1028-5	-20.10	9.19	2209	E	M 6-6	3.10	yes	this study
GeoB1031-4	-21.88	7.10	3105	E	M 6-6	1.26		this study
GeoB1032-3	-22.92	6.04	2505	E	M 6-6	1.51		Bickert and Wefer 1996
GeoB1034-3	-21.74	5.42	3772	E	M 6-6	1.17		Bickert and Wefer 1996
GeoB1035-4	-21.59	5.03	4456	E	M 6-6	1.34		Bickert and Wefer 1996
GeoB1041-3	-3.48	-7.60	4033	M	M 6-6	2.13		Bickert and Wefer 1996
GeoB1101-5	1.66	-10.98	4588	M	M 9-4	1.41		Bickert and Wefer 1996
GeoB1105-4	-1.67	-12.43	3225	M	M 9-4	4.39	yes	Bickert and Wefer 1996
GeoB1112-4	-5.78	-10.75	3125	M	M 9-4	2.63		Bickert and Wefer 1996
GeoB1115-4	-3.56	-12.56	2921	W	M 9-4	5.21		Bickert and Wefer 1996
GeoB1117-2	-3.82	-14.90	3984	W	M 9-4	4.11		Bickert and Wefer 1996
GeoB1118-2	-3.56	-14.90	4675	W	M 9-4	2.67		Bickert and Wefer 1996
GeoB1211-3	-24.47	7.53	4084	E	M 12-1	1.25		Bickert and Wefer 1996
GeoB1214-1	-24.69	7.24	3210	E	M 12-1	1.22		Bickert and Wefer 1996
GeoB1220-1	-24.03	5.31	2265	E	M 12-1	1.29		this study
GeoB1306-1	-35.21	-26.76	4057	W	M 15-2	0.82		this study
GeoB1312-2	-31.66	-29.65	3436	W	M 15-2	0.78		this study
GeoB1417-1	-15.54	-12.71	2845	M	M 16-1	1.42		this study
GeoB1419-2	-15.54	-17.07	4024	W	M 16-1	1.21		this study
GeoB1501-4	-3.68	-32.01	4257	W	M 16-2	2.65		this study
GeoB1503-1	2.31	-30.65	2306	W	M 16-2	1.32		this study
GeoB1505-1	2.27	-33.01	3705	W	M 16-2	2.49	yes	Zabel et al. 1999
GeoB1508-4	5.33	-34.03	3682	W	M 16-2	1.63		this study
GeoB1515-1	4.24	-43.67	3129	W	M 16-2	3.25	yes	Vidal et al. 1999
GeoB1520-1	4.59	-41.93	3911	W	M 16-2	2.60		this study
GeoB1523-1	3.83	-41.62	3291	W	M 16-2	2.57		this study
GeoB1710-3	-23.43	11.70	2987	E	M 20-2	5.05	yes	Bickert and Wefer 1999
GeoB1711-4	-23.32	12.38	1967	E	M 20-2	7.77	yes	Bickert and Wefer 1999
GeoB1721-1	-29.17	13.08	3044	E	M 20-2	2.89	yes	Mollenhauer et al. 2002
GeoB1722-1	-29.45	11.75	3973	E	M 20-2	1.57		this study
GeoB1903-3	-8.68	-11.85	3161	M	SO 84	2.72		this study
GeoB1905-3	-17.14	-13.99	2974	M	SO 84	2.07		this study
GeoB2004-2	-30.87	14.34	2569	E	M 23-1	5.02		this study
GeoB2016-1	-31.90	-1.33	3385	M	M 23-1	1.08		this study
GeoB2019-1	-36.06	-8.78	3825	M	M 23-1	2.31		this study
GeoB2109-1	-27.91	-45.88	2504	W	M 23-2	3.82	yes	Vidal et al. 1999
GeoB2204-2	-8.53	-34.02	2072	W	M 23-3	2.49		this study
GeoB2215-10	0.01	-23.50	3711	W	M 23-3	4.04		this study
GeoB2819-1	-30.85	-38.34	3435	W	M 29-2	0.84		this study
GeoB3104-1	-3.67	-37.72	767	W	JOPSII-6	9.07	yes	Arz 1998
GcoB3603-2	-35.13	17.54	2840	E	M 34-1	3.09		this study
GeoB3722-2	-25.25	12.02	3506	E	M 34-2	3.55	yes	Mollenhauer et al. 2002
GeoB3801-6	-29.51	-8.31	4546	M	M 34-3	1.20		this study
GeoB3808-6	-30.81	-14.71	3213	W	M 34-3	2.52		this study
GeoB3813-3	-32.27	-21.97	4331	W	M 34-3	0.67		this study
GeoB4403-1	6.13	-43.44	4503	W	M 38-2	4.68		this study
GeoB5115-2	-24.14	-14.04	3291	W	M 41-3	0.91		this study
GeoB5121-2	-24.17	-12.36	3486	M	M 41-3	0.84		this study
IOW226920-3	-22.45	12.36	1683	E	M 48-2	11.31	yes	Mollenhauer et al. 2002
ODP1085A	-29.37	13.99	1713	E	ODP 175	4.42		Vidal (unpubl.)
PS1754-3	-46.77	7.60	2476	M	ANT VIII-3	1.51		Mackensen et al. 1994
PS2082-1	-43.22	11.75	4661	M	ANT IX-4	4.71		Mackensen et al. 1994
PS2495-3	-41.28	-14.49	3134	M	ANT XI-3	2.94	yes	Mackensen et al. 2001
PS2498-1	-44.15	-14.49	3783	W	ANT XI-3	8.93	yes	Mackensen et al. 2001
PS2499-5	-46.51	-15.33	3175	W	ANT XI-3	3.45	yes	Mackensen et al. 2001

Table 1. Core positions and references for South Atlantic benthic stable isotope records presented in the text.

as representative for modern conditions, and the Last Glacial Maximum (19-23 ka, calendar year scale, 16.1-19.5 ky ^{14}C scale) which was defined following the recommendations of the EPILOG group (Mix et al. 2001). Glacial carbon isotope data from Sarnthein et al. (1994) have been chosen from the time interval between 21.5 and 23.5 ka without any change. Differences to late glacial values have been shown to be less than 0.1‰.

Results

Late Pleistocene to Holocene $\delta^{18}O$-Variability

In Figure 2a we present the last 30.000 years of the 28 yet unpublished benthic $\delta^{18}O$-records of the South Atlantic. Most cores show the typical glacial- to interglacial shift in the order of 1.4 to 2.1‰. This shift is to a large part related to ice volume changes, with an additional effect of deep water temperature and salinity changes. To prove the quality of the presented records, they are compared with the $\delta^{18}O$ curve of core GeoB 1117, which has been shown to serve as a good reference for deep South Atlantic oxygen isotope records (Bickert and Wefer 1996). Although many of the cores have a low time resolution, and even in some cores the upper Holocene sediment sequence is missing, the $\delta^{18}O$ records reflect the full glacial-to-interglacial amplitude range. For some cores, where the benthic oxygen isotope record seemed to be to low in resolution (GeoB 1220-1, GeoB 2019-1, GeoB 3801-6, GeoB 3813-3) or not to extend far back enough to be sure to register the LGM (GeoB 5115-2, GeoB 5121-2), their stratigraphy has been checked with a planktic oxygen isotope record (Niebler et al. 2003). Therefore, we assume the selected cores to reveal the correct pattern for last Glacial to Holocene $\delta^{13}C$ changes.

Late Pleistocene to Holocene $\delta^{13}C$-Variability

In Figure 2b the new epibenthic $\delta^{13}C$-records of the 28 additional South Atlantic cores are presented. Most records reveal lower values during the LGM compared to the Late Holocene. However, there are large differences in amplitudes ranging from 0.35‰ in many cores of the oligotrophic western South Atlantic to 1.1‰ in the deep cores of the abyssal western equatorial Atlantic. From the higher-resolution records it is evident that the minimum in $\delta^{13}C$ occurs clearly after the end of the defined LGM interval. This phenomenon has been described from many other deep Atlantic sites (e.g. Sarnthein et al. 1994) and is interpreted to reflect the low overturning of the deep ocean circulation at the beginning of Termination I. The lowest glacial $\delta^{13}C$ values are found at site GeoB 1306, which is located in the deep Argentine Basin just south of the Rio-Grande Rise. Interestingly, the record of GeoB 3813 in a corresponding position north of the ridge exhibits glacial $\delta^{13}C$ values about 0.5‰ higher, which reflects the barrier effect of the bottom topography in this region.

Discussion

Epibenthic $\delta^{13}C$-Record of Late Holocene (0-4 ka) Atlantic Deep Water Mass Distribution

In Fig. 3a, epibenthic $\delta^{13}C$ mean values for the last 4000 years are presented from the North Atlantic data by Sarnthein et al. (1994) and the newly compiled data set from the South Atlantic (for references see Table 2). Most features of modern Atlantic deep water circulation are displayed by the $\delta^{13}C$ pattern along the three sections. The most prominent feature is the layer of high $\delta^{13}C$ values representing the oxygen-rich, but nutrient-depleted NADW, which spreads in water depths between 1500 and 3500 m towards the south. In the western South Atlantic, this water mass is characterized by values of >1‰ as far as 40°S, where it is encountered into the CDW. In the western South Atlantic, NADW is underlain by the ^{13}C-depleted lower branch of the CDW (LCDW), which spreads from the circum-antarctic realm towards the north. Along its path through the western Atlantic basins, this low-oxygenated, but nutrient-rich water mass is partly reflected at each zonal-oriented ridge. This is evident at the Rio-Grande-Rise, where only the upper part of LCDW enters the Brasil Basin through the Vema and Hunter Chan-

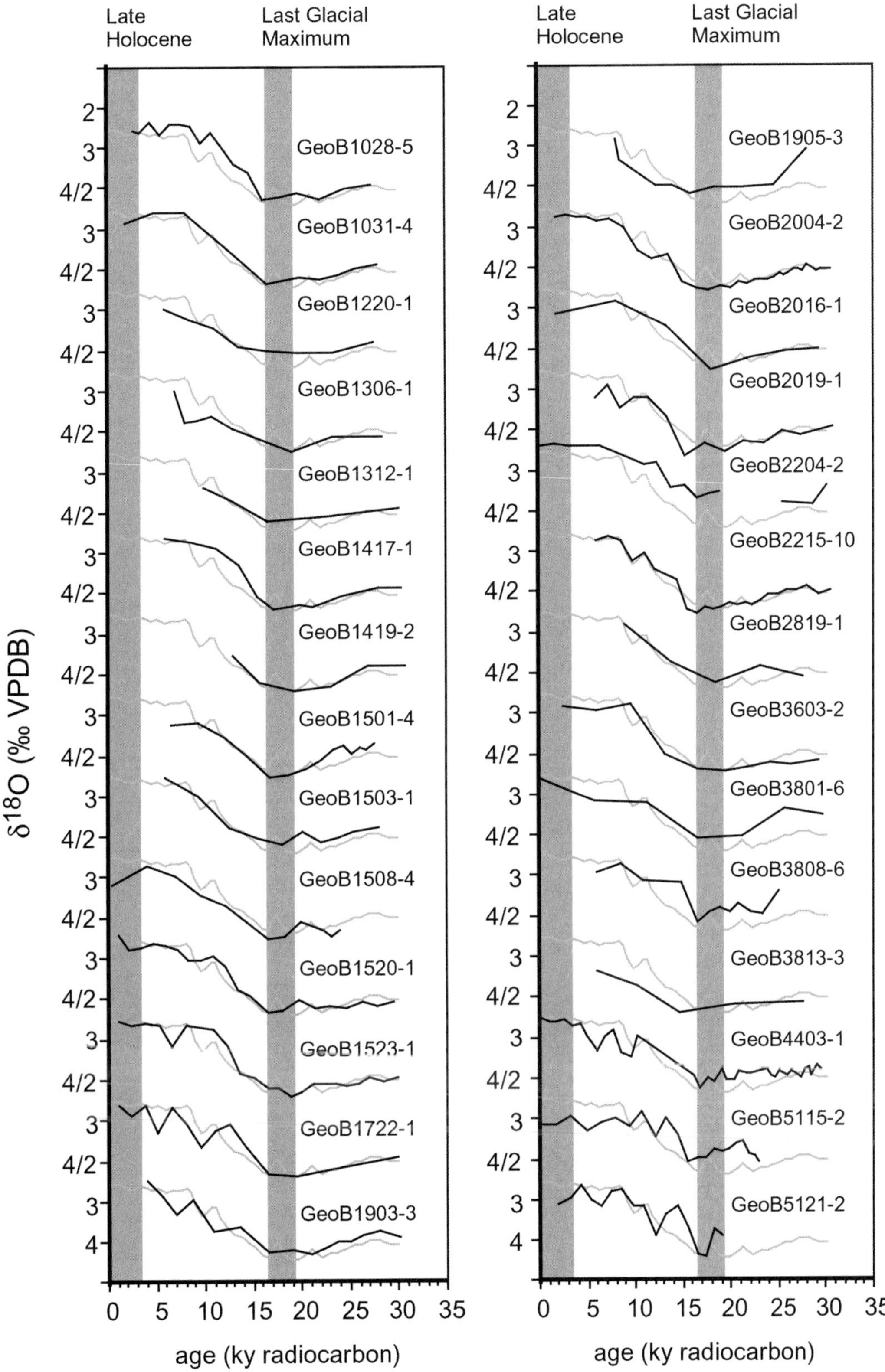

Fig. 2. a) Plot of oxygen isotope records, measured on the benthic foraminifer taxon *C. wuellerstorfi* , for the last 30,000 years in order of core number. The epibenthic $\delta^{18}O$ record of deep equatorial Atlantic site GeoB1117 is added in gray to each record for guidance.

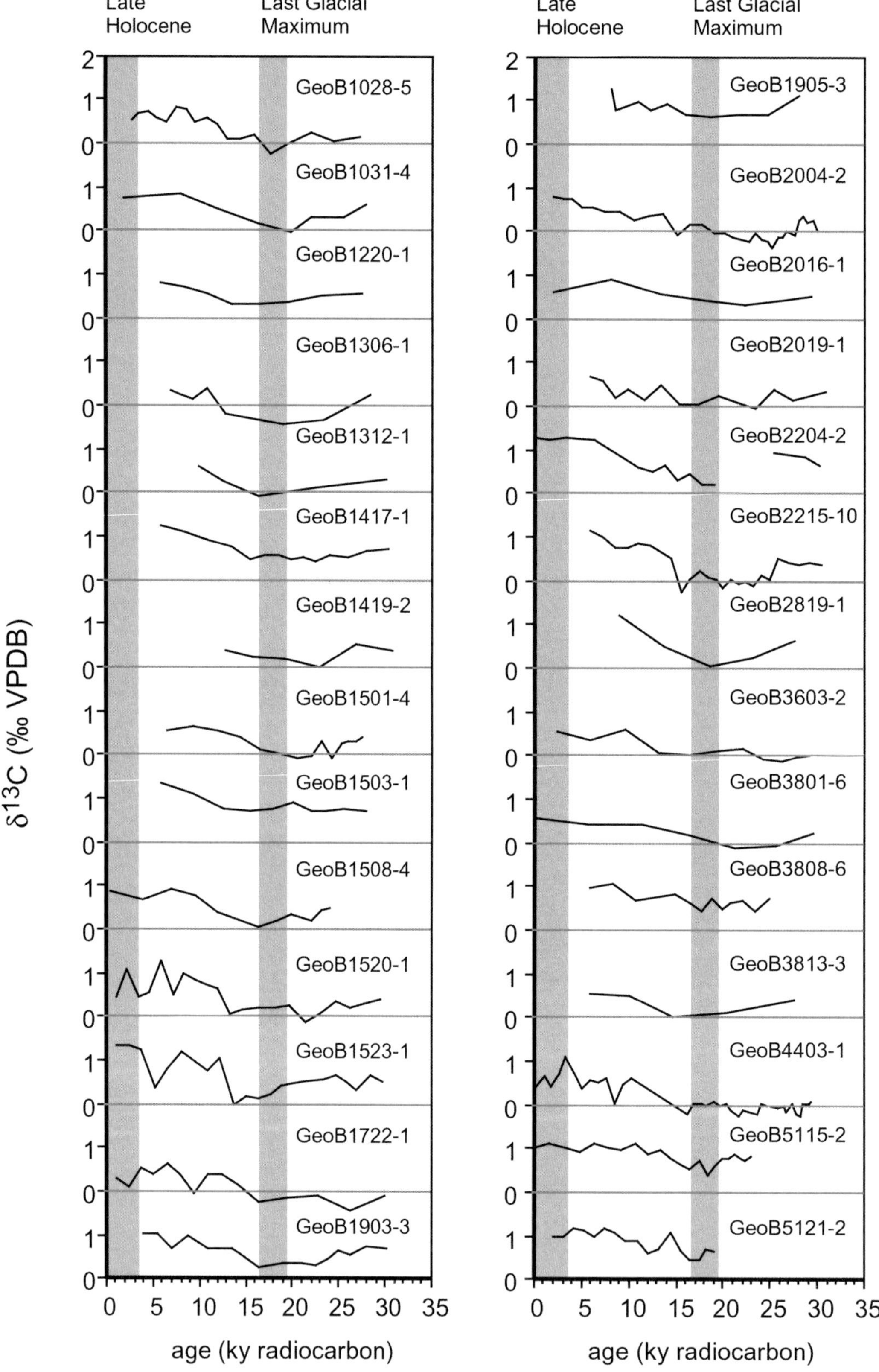

Fig. 2. b) Plot of stable carbon isotope records, measured on the benthic foraminifer taxon *C. wuellerstorfi* , for the last 30,000 years in order of core number.

nels (Warren and Speer 1991), which leads to a decline of the 0.5‰- and 0.25‰-isolines by 500 m on the northern side of the ridge. The same is true for the overflow of the Ceara Rise, where another declining of the 0.75‰-isoline indicates the hold back of most of the LCDW south of the barrier. In the eastern Agulhas and Cape Basins, $\delta^{13}C$ values <0.5‰ also indicate an inflow of LCDW, filling these basins from the bottom up to about 3500 m water depth. The Walvis Ridge prevents the spreading of LCDW further north due to a sill depth of about 3100 m (Connary and Ewing 1974). Therefore, in the eastern basins north of the Walvis Ridge the NADW fills the deeper parts of the basins. Only small amounts of ^{13}C-depleted LCDW enters the eastern Sierra Leone, Guinea and Angola Basins through the Romanche Fracture Zones across the Mid-Atlantic Ridge at 1°N, and the Canary Basins through the Vema Fracture Zone at 11°N. As a consequence, the $\delta^{13}C$ values of the Guinea Basin below 3500 m are as high as 0.78‰ to 0.90‰, and even in the Canary and Angola Basins these values are as high as 0.55‰ to 0.75‰. The slightly lower values in the Angola Basin indicate a small, but considerable amount of LCDW entering this basin through the narrow Walvis Passage with a bottom depth of 4200 m (Connary and Ewing 1984). Above the NADW, the decrease in $\delta^{13}C$ values indicate the upper branch of the CDW (UCDW), which extends as a nutrient rich, but oxygen depleted water mass towards the north. Along its path it mixes with the overlying Antarctic Intermediate Water (AAIW), which is characterized by relatively low salinity values (Reid et al. 1989). A special feature is the local maximum of $\delta^{13}C$ values in the eastern marginal Atlantic transect at 36°N and 800 m water depth. $\delta^{13}C$ values of up to 1.27‰ here indicate the outflow of nutrient-depleted Mediterranean deep water (MOW), extending from the isthmus of Gibraltar towards the south to at least 25°N (Zahn et al. 1987).

However, at several core locations, low $\delta^{13}C$ values occur, which are substantially lower than the values of surrounding cores. These cores are all located in high-productive areas, like the cores PS2498, PS2499 (late Holocene mean $\delta^{13}C$ values of 0.31‰ and 0.22‰) in the area of the western subantartic frontal system, PS2082 (0.15‰) below the eastern subantarctic front, and IOW226920 (0.22‰) in the central Namibia upwelling zone. These epibenthic $\delta^{13}C$ values have been shown to deviate from the local DIC values up to 0.4‰ (Bickert and Wefer 1999; Mackensen et al. 2001). Since these sites are exposed to seasonally high fluxes of organic carbon to the seafloor, the $\delta^{13}C$ values are interpreted to reflect the degradation of a seasonal rapidly deposited phytodetritus layer.

However, no such deviations are evident in cores beneath the eastern equatorial upwelling zone, although the annual primary production has been shown to be more than twice of that in the subantarctic frontal area (Mollenhauer et al. 2002). As has been shown by Bickert and Wefer (1996), deviations related to the phytodetritus effect occur in these sediments only during glacial times. The same is true for a few sites along the NW African and Portuguese continental margins, described by Sarnthein et al. (1994).

Phytodetritus-Effect on Epibenthic $\delta^{13}C$-Values

The phytodetritus-effect is a crucial point in the LGM deep ocean circulation reconstruction by using epibenthic $\delta^{13}C$. Therefore, a separate paragraph is added to this paper, although this issue has been adressed in many studies before (for a summary see Mackensen and Bickert 1999).

The benthic foraminifer species *Cibicidoides wuellerstorfi* and related taxa are assumed to secrete calcite very close to the carbon isotope values of the ambient bottom-water $\Sigma CO2$. However, it was noticed that some single $\delta^{13}C$ values of *Cibicidoides* sp. were much lower relative to the expected or measured values of the ambient seawater (Sarnthein et al. 1988). From paired water/sediment measurements on a South Atlantic transect, Mackensen et al. (1993) deduced that such deviations in foraminiferal $\delta^{13}C$ are associated with high-productivity areas, where even epibenthic specimens calcifiy their tests in an environment strongly depleted in ^{13}C due to the respiration of organic matter. The depletion might not only result from low $\delta^{13}C$ of the interstitial water in the fluffy layer, but from a stronger metabolic isotope fractionation due to rapid growth of the foraminifer individuals (McConnaughey et al. 1997). However, although the phytodetritus effect only occurs in epibenthic foraminifer tests collected in high-productive areas, there is no clear relationship between the accumulation of organic material and the deviation from $\delta^{13}C_{DIC}$ of the ambient bottom water. For example, as described above, the recent epibenthic $\delta^{13}C$ values at the core locations PS 2498 and 2499, which are today located below the western subantarctic front, do show deviations up to -0.24‰ (Mackensen et al. 2001), whereas the recent epibenthic $\delta^{13}C$ values at the core location GeoB 1105, located below the center of eastern equatorial upwelling with even higher accumulation rates of TOC, do not show any deviation today. However, especially for the equatorial Atlantic site it has been shown by Bickert and Wefer (1996) that there is a considerable deviation of epibenthic $\delta^{13}C$ in the order of 0.4‰ during glacial times, if the record of GeoB 1105 is compared to the record of the nearby station GeoB1112, which is in about the same water depth, but clearly outside of the equatorial upwelling area. The same is true for cores at the edge of the Namibia up-welling zone, like cores GeoB 1710 and 1711, but also RC 13-228 and 13-229 (Bickert and Wefer 1999). It has been suggested by Mackensen et al. (2001) that the annual mean organic carbon flux is not the important parameter causing $\delta^{13}C$ deviations. Instead, seasonally high production is leading to a phytodetritus layer by rapid deposition of particulate organic matter. Hence, a $\delta^{13}C$ gradient within this layer above the sediment-water interface is developing. Since there is yet no parameter to quantify the seasonal flux and degradation of organic matter for past times, we tried to identify core locations which are supposed to be influenced by the phytodetritus effect by simply comparing glacial $\delta^{13}C$ values of neighboured cores from in- and outside of high productive areas. And indeed, there is a good agreement between the glacial $\delta^{13}C$ values of neighboured cores high and low productive areas, if an average constant of 0.4‰ is added to the $\delta^{13}C$ values from those cores supposed to be influenced by the phytodetritus effect. We believe that this phytodetritus effect correction will help to solve contradictory results in the glacial Southern

Ocean using cadmium as another nutrient proxy (Boyle 1992; Rosenthal et al. 1997). Whereas benthic $\delta^{13}C$ values are suggested to reveal the glacial Southern Ocean deep water to have higher nutrient concentrations, glacial maximum Cd data suggested nutrient concentrations comparable to today. We will adress this issue in the discussion below.

Reconstruction of Last Glacial Maximum (19-23 ka) Deep Atlantic Circulation

In Fig. 3b, epibenthic $\delta^{13}C$ mean values of the Last Glacial Maximum between 19 and 23 ka (calender years) are presented from the North Atlantic data by Sarnthein et al. (1994) and the newly compiled data set from the South Atlantic. To better compare the glacial $\delta^{13}C$ values to those of the Late Holocene, a correction for the global $\delta^{13}C$ shift has been made by adding 0.32‰ (Duplessy et al. 1988). Furthermore, at those locations which are supposed to be influenced by high seasonal fluxes of organic material, a constant value of 0.4‰ has been added to correct for the phytodetritus effect, as discussed above. These locations belong to the eastern equatorial upwelling zone, to several coastal upwellling cells along the south western and north western African continental margin, and to the subantarctic ocean frontal system (see Tab. 2). The phytodetritus-corrected values are indicated by italic letters in Fig. 3b.

The general features of modern Atlantic deep water circulation are evident in the glacial Atlantic Ocean as well, but with significant changes as indicated by the $\delta^{13}C$ pattern. The Atlantic Ocean north of 40°N was dominated by downwelling of a glacial NADW, indicated by a well defined layer of high $\delta^{13}C$ values along the Mid-Atlantic Ridge section. Further along the flow path of NADW, the benthic $\delta^{13}C$ values are decreasing from 1.14‰ to 1.38‰ at 50°N and 2500 m water depth to values between 0.74‰ and 0.96‰ at 20°S. Although these values are slightly lower by about 0.25‰ compared to the modern values, the overall decrease along the flow path of the NADW is comparable to today. This might indicate only a small glacial decrease in downwelling of a surface-derived deep water at least in the eastern North

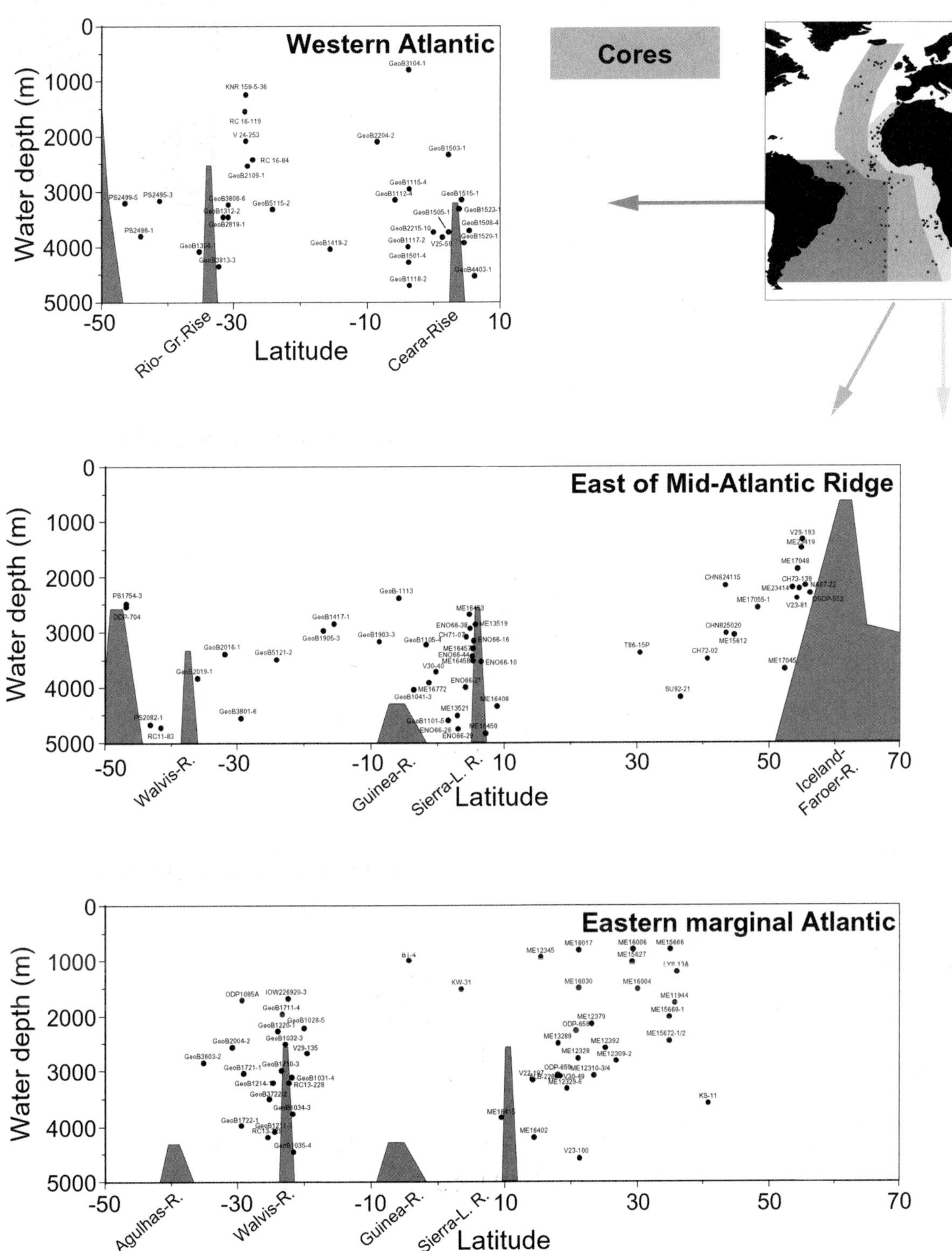

Fig. 3. Sediment cores along the three vertical sections representing the western South Atlantic, the central Atlantic east of the Mid-Atlantic Ridge, and the eastern marginal Atlantic.

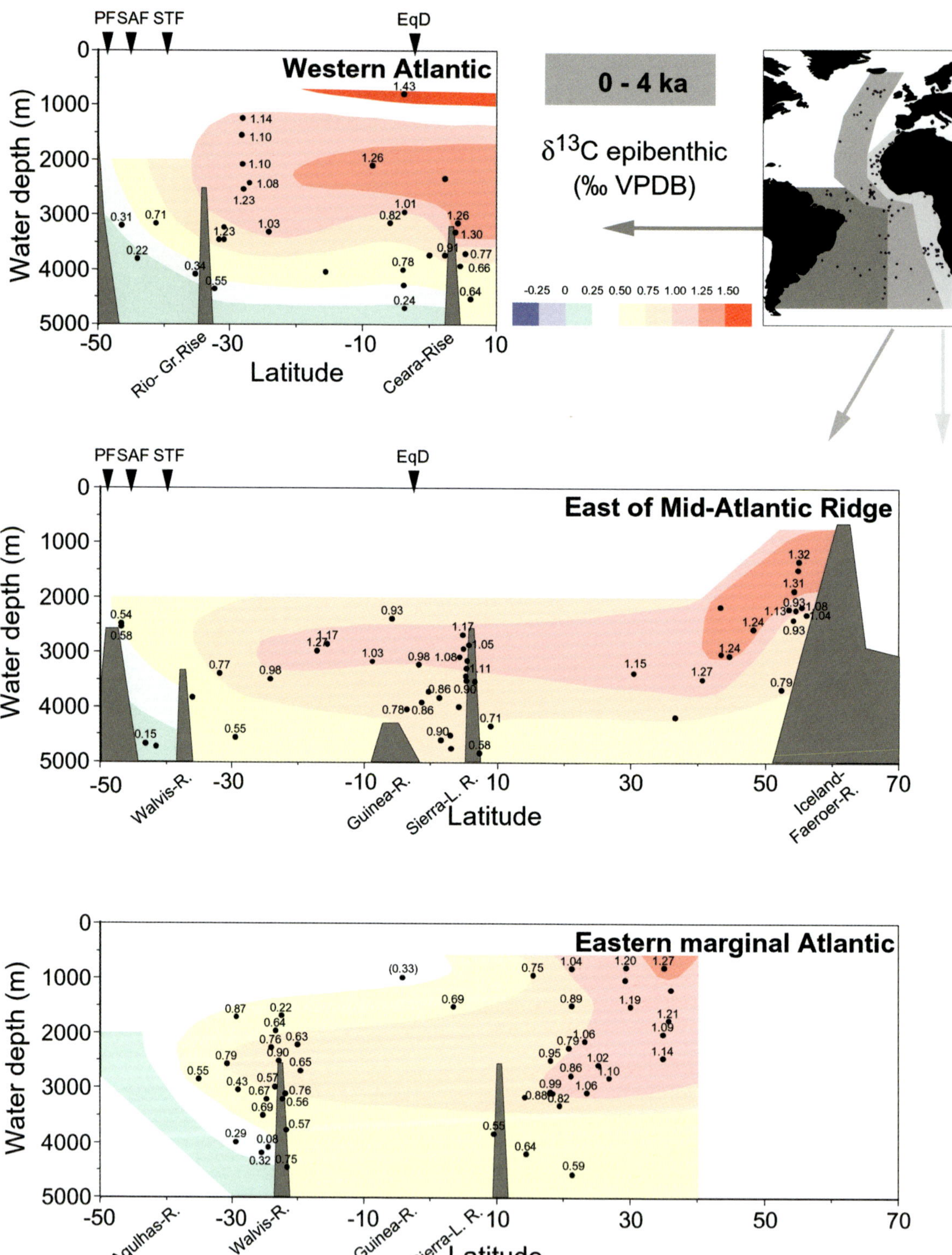

682   Bickert and Mackensen

Fig. 3. a) Late Holocene (0-4 ka) epibenthic δ¹³C distribution along three vertical sections representing the western South Atlantic, the central Atlantic east of the Mid-Atlantic Ridge, and the eastern marginal Atlantic. The data are from the North Atlantic data set of Sarnthein et al. (1994) and from the compiled data from the South Atlantic (this study). PF polar front, SAF subantarctic front, STF subtropical front, EqD equatorial divergence.

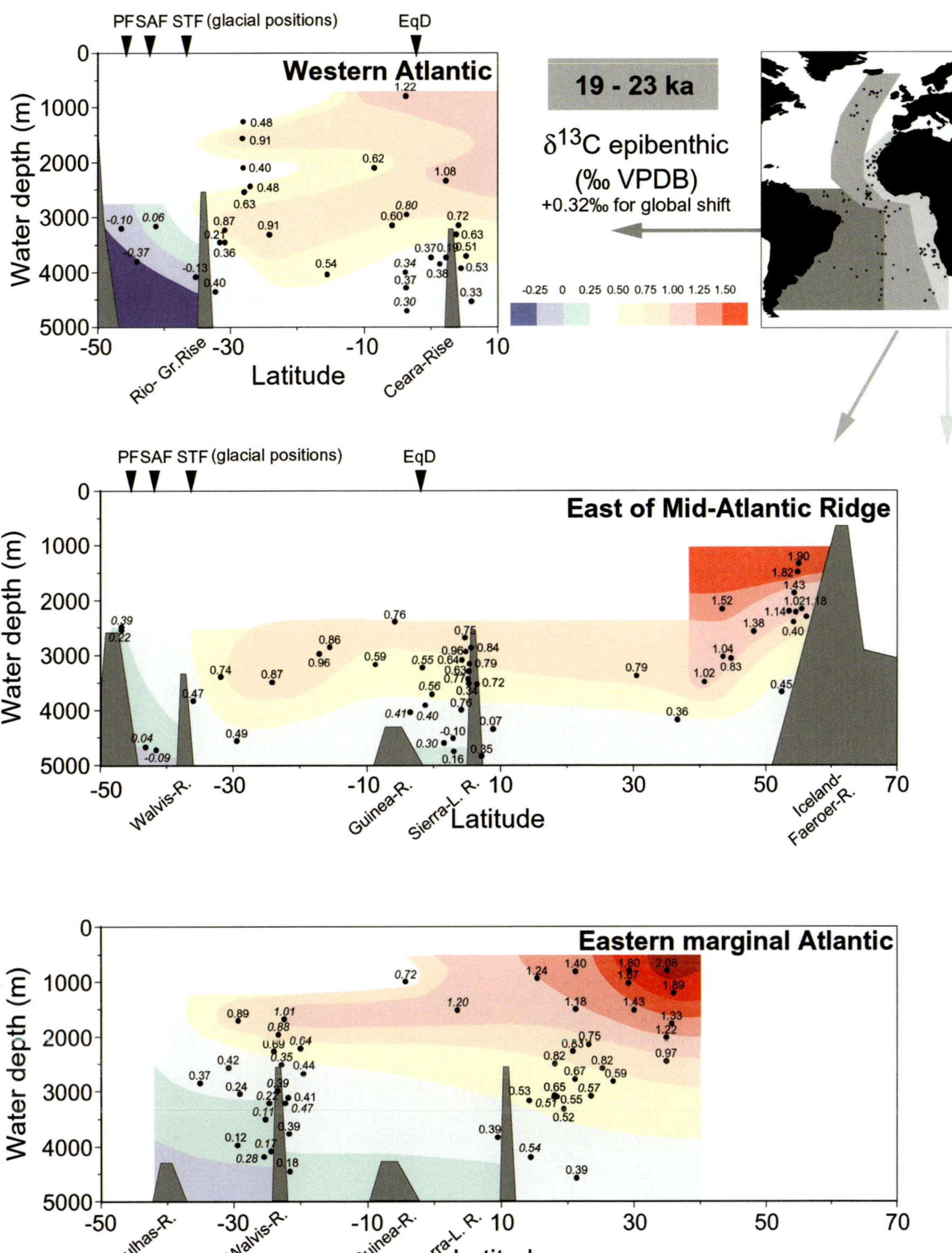

Fig. 3. b) Last Glacial Maximum (19-23 ka) epibenthic $\delta^{13}C$ distribution along three vertical sections representing the western South Atlantic, the central Atlantic east of the Mid-Atlantic Ridge, and the eastern marginal Atlantic. The data are from the North Atlantic data set of Sarnthein et al. (1994) and from the compiled data from the South Atlantic (this study). All values are corrected for the global ocean shift by adding 0.32‰ (Duplessy et al. 1988). An additional correction for the phytodetritus-effect at some locations is indicated by values in italic lettern.

Core	Latitude	Longitude	Water depth (m)	Section	Depositional realm	δ^{13}C (0 - 4 ka) (‰ VPDB)	δ^{13}C (19-23 ka) (‰ VPDB)	phytodetr-corr	global-corr	Reference
LYII-13A	35,97	-7,82	1201	E	continental margin		1,57	1,57	1,89	Sarnthein et al. 1994
ME11944	35,65	-8,06	1765	E	continental margin	1,21	1,01	1,01	1,33	Sarnthein et al. 1994
ME15666	34,96	-7,12	803	E	continental margin	1,27	1,76	1,76	2,08	Sarnthein et al. 1994
ME15669-1	34,89	-7,82	2022	E	continental margin	1,09	0,90	0,90	1,22	Sarnthein et al. 1994
ME15672-1/2	34,86	-8,12	2455	E	continental margin	1,14	0,65	0,65	0,97	Sarnthein et al. 1994
ME16004	29,98	-10,65	1512	E	continental margin	1,19	1,11	1,11	1,43	Sarnthein et al. 1994
ME16006	29,27	-11,50	796	E	continental margin	1,20	1,48	1,48	1,80	Sarnthein et al. 1994
ME15627	29,17	-12,09	1024	E	continental margin		1,35	1,35	1,67	Sarnthein et al. 1994
ME12309-2	26,84	-15,11	2820	E	continental margin	1,10	0,27	0,27	0,59	Sarnthein et al. 1994
ME12392	25,17	-16,85	2575	E	continental margin	1,02	0,50	0,50	0,82	Sarnthein et al. 1994
ME12310-3/4	23,50	-18,72	3080	E	coastal upwelling	1,06	-0,07	0,33	0,65	Sarnthein et al. 1994
ME12379	23,14	-17,75	2136	E	continental margin	1,06	0,43	0,43	0,75	Sarnthein et al. 1994
ME16017	21,25	-17,80	812	E	continental margin	0,99	1,08	1,08	1,40	Sarnthein et al. 1994
ME16030	21,24	-18,06	1500	E	continental margin	0,89	0,86	0,86	1,18	Sarnthein et al. 1994
V23-100	21,30	-21,68	4579	E	continental margin	0,59	0,07	0,07	0,39	Sarnthein et al. 1994
ME12328	21,15	-18,57	2778	E	continental margin	0,86	0,35	0,35	0,67	Sarnthein et al. 1994
ODP-658	20,75	-18,58	2263	E	continental margin	0,79	0,51	0,51	0,83	Sarnthein et al. 1994
ME12329-6	19,37	-19,93	3320	E	continental margin	0,82	0,20	0,20	0,52	Sarnthein et al. 1994
V30-49	18,43	-21,08	3093	E	continental margin		0,23	0,23	0,55	Sarnthein et al. 1994
ODP-659	18,08	-21,03	3069	E	continental margin	0,99	0,33	0,33	0,65	Sarnthein et al. 1994
ME13289	18,07	-18,01	2490	E	continental margin	0,95	0,50	0,50	0,82	Sarnthein et al. 1994
ALB-226	17,95	-21,05	3100	E	continental margin	0,88	0,19	0,19	0,51	Sarnthein et al. 1994
ME12345	15,48	-17,36	945	E	continental margin	0,75	0,92	0,92	1,24	Sarnthein et al. 1994
ME16402	14,42	-20,57	4203	E	coastal upwelling	0,64	-0,18	0,22	0,54	Sarnthein et al. 1994
V22-197	14,17	-18,58	3167	E	continental margin		0,21	0,21	0,53	Sarnthein et al. 1994
ME16415	9,57	-19,11	3841	E	continental margin	0,55	0,07	0,07	0,39	Sarnthein et al. 1994
KW-31	3,52	5,57	1515	E	coastal upwelling	0,69	0,48	0,88	1,20	Sarnthein et al. 1994
BT-4	-4,33	10,43	1000	E	coastal upwelling	0,33	0,00	0,40	0,72	Sarnthein et al. 1994
V29-135	-19,60	8,88	2675	E	open ocean	0,65	0,12	0,12	0,44	Sarnthein et al. 1994
GeoB1028-5	-20,10	9,19	2209	E	coastal upwelling	0,63	-0,10	0,30	0,62	this study
GeoB1035-4	-21,59	5,03	4456	E	open ocean	0,75	-0,14	-0,14	0,18	Bickert and Wefer 1996
GeoB1034-3	-21,74	5,42	3772	E	open ocean	0,57	0,07	0,07	0,39	Bickert and Wefer 1996
GeoB1031-4	-21,88	7,10	3105	E	open ocean	0,76	0,09	0,09	0,41	Sarnthein et al. 1994
RC13-228	-22,33	11,20	3204	E	coastal upwelling	0,56	-0,25	0,15	0,47	this study
IOW226920-3	-22,45	12,36	1683	E	coastal upwelling	0,22	0,39	0,79	1,11	Mollenhauer et al. 2002
GeoB1032-3	-22,92	6,04	2505	E	open ocean	0,90	0,03	0,03	0,35	Bickert and Wefer 1996
GeoB1711-4	-23,32	12,38	1967	E	coastal upwelling	0,64	0,16	0,56	0,88	Bickert and Wefer 1999
GeoB1710-3	-23,43	11,70	2987	E	coastal upwelling	0,57	-0,33	0,07	0,39	Bickert and Wefer 1999
GeoB1220-1	-24,03	5,31	2265	E	open ocean		0,37	0,37	0,69	this study
GeoB1211-3	-24,47	7,53	4084	E	coastal upwelling	0,08	-0,35	0,05	0,37	Bickert and Wefer 1996
GeoB1214-1	-24,69	7,24	3210	E	coastal upwelling	0,67	-0,30	0,10	0,42	Bickert and Wefer 1996
GeoB3722-2	-25,25	12,02	3506	E	coastal upwelling	0,69	-0,21	-0,21	0,11	Mollenhauer et al. 2002
RC13-229	-25,50	11,30	4194	E	coastal upwelling	0,32	-0,44	-0,04	0,28	Sarnthein et al. 1994
GeoB1721-1	-29,17	13,08	3044	E	continental margin	0,43	-0,08	-0,08	0,24	this study
ODP1085A	-29,37	13,99	1713	E	continental margin	0,87	0,57	0,57	0,89	Vidal (unpubl.)
GeoB1722-1	-29,45	11,75	3973	E	continental margin	0,29	-0,20	-0,20	0,12	Mollenhauer et al. 2002
GeoB2004-2	-30,87	14,34	2569	E	continental margin	0,79	0,10	0,10	0,42	this study
GeoB3603-2	-35,13	17,54	2840	E	continental margin	0,55	0,05	0,05	0,37	this study
NA87-22	55,50	-14,70	2161	M	open ocean	1,08	0,86	0,86	1,18	Sarnthein et al. 1994
V29-193	55,00	-19,00	1326	M	open ocean	1,32	1,58	1,58	1,90	Sarnthein et al. 1994
ME23419	54,96	-19,76	1491	M	open ocean		1,50	1,50	1,82	Sarnthein et al. 1994
CH73-139	54,63	-16,35	2209	M	open ocean	0,93	0,70	0,70	1,02	Sarnthein et al. 1994
ME17048	54,31	-18,18	1859	M	open ocean	1,31	1,11	1,11	1,43	Sarnthein et al. 1994
V23-81	54,25	-16,83	2393	M	open ocean	0,93	0,08	0,08	0,40	Sarnthein et al. 1994
ME23414	53,54	-20,29	2196	M	open ocean	1,13	0,82	0,82	1,14	Sarnthein et al. 1994
ME17045	52,43	-16,66	3663	M	open ocean	0,79	0,13	0,13	0,45	Sarnthein et al. 1994
ME17055-1	48,22	-27,06	2558	M	open ocean	1,24	1,06	1,06	1,38	Sarnthein et al. 1994
ME15612	44,69	-26,54	3050	M	open ocean	1,24	0,51	0,51	0,83	Sarnthein et al. 1994
CHN825020	43,50	-29,87	3020	M	open ocean		0,72	0,72	1,04	Sarnthein et al. 1994
CHN824115	43,37	-28,23	2151	M	open ocean		1,20	1,20	1,52	Sarnthein et al. 1994
CH72-02	40,60	-21,70	3485	M	open ocean	1,27	0,70	0,70	1,02	Sarnthein et al. 1994
SU92-21	36,57	-23,74	4170	M	open ocean		0,04	0,04	0,36	Sarnthein et al. 1994
T86-15P	30,43	-37,07	3375	M	open ocean	1,15	0,47	0,47	0,79	Sarnthein et al. 1994

Table 2. Benthic stable carbon and oxygen isotope mean values for the Holocene and the Last Glacial Maximum, as discussed in the text.

Core	Latitude	Longitude	Water depth (m)	Section	Depositional realm	δ^{13}C (0 - 4 ka) (‰ VPDB)	δ^{13}C (19-23 ka) (‰ VPDB)	phytodetr-corr	global-corr	Reference
ME16408	9,01	-21,50	4336	M	open ocean	0,71	-0,25	-0,25	0,07	Sarnthein et al. 1994
ME16459	7,28	-26,19	4835	M	open ocean	0,58	0,03	0,03	0,35	Sarnthein et al. 1994
ENO66-10	6,65	-21,90	3527	M	open ocean		0,40	0,40	0,72	Sarnthein et al. 1994
ME13519	5,66	-19,85	2862	M	open ocean	1,05	0,52	0,52	0,84	Sarnthein et al. 1994
ENO66-16	5,46	-21,14	3152	M	open ocean		0,47	0,47	0,79	Sarnthein et al. 1994
ME16457	5,39	-21,72	3291	M	open ocean	1,11	0,31	0,31	0,63	Sarnthein et al. 1994
ME16458	5,34	-22,06	3518	M	open ocean	0,90	0,02	0,02	0,34	Sarnthein et al. 1994
ENO66-44	5,26	-21,71	3428	M	open ocean		0,45	0,45	0,77	Sarnthein et al. 1994
ENO66-38	4,92	-20,50	2931	M	open ocean		0,64	0,64	0,96	Sarnthein et al. 1994
ME16453	4,73	-20,95	2675	M	open ocean	1,17	0,43	0,43	0,75	Sarnthein et al. 1994
CH71-07	4,38	-20,87	3083	M	open ocean	1,08	0,32	0,32	0,64	Sarnthein et al. 1994
ENO66-21	4,23	-21,63	3995	M	open ocean		0,12	0,12	0,44	Sarnthein et al. 1994
ENO66-26	3,09	-20,02	4745	M	open ocean		-0,16	-0,16	0,16	Sarnthein et al. 1994
ME13521	3,02	-22,03	4504	M	open ocean		-0,42	-0,42	-0,10	Sarnthein et al. 1994
ENO66-32	2,47	-19,73	5003	M	open ocean		-0,20	-0,20	0,12	Sarnthein et al. 1994
ENO66-29	2,46	-19,76	5104	M	open ocean		-0,16	-0,16	0,16	Sarnthein et al. 1994
GeoB1101-5	1,66	-10,98	4588	M	equatorial upwelling	0,90	-0,42	-0,02	0,30	Bickert and Wefer 1996
V30-40	-0,20	-23,15	3706	M	equatorial upwelling		-0,16	0,24	0,56	Sarnthein et al. 1994
ME16772	-1,21	-11,96	3912	M	equatorial upwelling	0,86	-0,32	0,08	0,40	Sarnthein et al. 1994
GeoB1105-4	-1,67	-12,43	3225	M	equatorial upwelling	0,98	-0,17	0,23	0,55	Bickert and Wefer 1996
GeoB1041-3	-3,48	-7,60	4033	M	equatorial upwelling	0,78	-0,31	0,09	0,41	Bickert and Wefer 1996
GeoB-1113	-5,75	-11,04	2374	M	open ocean	0,93	0,44	0,44	0,76	Sarnthein et al. 1994
GeoB1903-3	-8,68	-11,85	3161	M	open ocean		0,27	0,27	0,59	this study
GeoB1417-1	-15,54	-12,71	2845	M	open ocean		0,54	0,54	0,86	this study
GeoB1905-3	-17,14	-13,99	2974	M	open ocean		0,64	0,64	0,96	this study
GeoB5121-2	-24,17	-12,36	3486	M	open ocean	0,98	0,55	0,55	0,87	this study
GeoB3801-6	-29,51	-8,31	4546	M	open ocean	0,55	0,17	0,17	0,49	this study
GeoB2016-1	-31,90	-1,33	3385	M	open ocean	0,77	0,42	0,42	0,74	this study
GeoB2019-1	-36,06	-8,78	3825	M	open ocean		0,15	0,15	0,47	this study
RC11-83	-41,60	9,72	4718	M	subtropical front		-0,81	-0,41	-0,09	Sarnthein et al. 1994
PS2082-1	-43,22	11,75	4661	M	subantarctic front	0,15	-0,68	-0,28	0,04	Mackensen et al. 1994
PS1754-3	-46,77	7,60	2476	M	subantarctic front	0,54	-0,33	0,07	0,39	Mackensen et al. 1994
ODP-704	-46,88	7,42	2532	M	subantarctic front	0,58	-0,50	-0,10	0,22	Sarnthein et al. 1994
GeoB4403-1	6,13	-43,44	4503	W	open ocean	0,64	0,01	0,01	0,33	this study
GeoB1508-4	5,33	-34,03	3682	W	open ocean	0,77	0,19	0,19	0,51	this study
GeoB1520-1	4,59	-41,93	3911	W	open ocean	0,66	0,21	0,21	0,53	this study
GeoB1515-1	4,24	-43,67	3129	W	open ocean	1,26	0,40	0,40	0,72	Vidal et al. 1999
GeoB1523-1	3,83	-41,62	3291	W	open ocean	1,30	0,31	0,31	0,63	this study
GeoB1503-1	2,31	-30,65	2306	W	open ocean		0,76	0,76	1,08	this study
GeoB1505-1	2,27	-33,01	3705	W	open ocean	0,91	-0,13	-0,13	0,19	Zabel et al. 1999
V25-59	1,37	-33,48	3824	W	open ocean	0,86	0,06	0,06	0,38	Sarnthein et al. 1994
GeoB2215-10	0,01	-23,50	3711	W	open ocean		0,05	0,05	0,37	this studay
GeoB1118-2	-3,56	-14,90	4675	W	equatorial upwelling	0,24	-0,42	-0,02	0,30	Bickert and Wefer 1996
GeoB1115-4	-3,56	-12,56	2921	W	equatorial upwelling	1,01	0,08	0,48	0,80	Bickert and Wefer 1996
GeoB3104-1	-3,67	-37,72	767	W	continental margin	1,43	0,90	0,90	1,22	Arz et al. 1998
GeoB1501-4	-3,68	-32,01	4257	W	continental margin		0,05	0,05	0,37	this study
GeoB1117-2	-3,82	-14,90	3984	W	equatorial upwelling	0,78	-0,38	0,02	0,34	Bickert and Wefer 1996
GeoB1112-4	-5,78	-10,75	3125	W	open ocean	0,82	0,28	0,28	0,60	Bickert and Wefer 1996
GeoB2204-2	-8,53	-34,02	2072	W	continental margin	1,26	0,30	0,30	0,62	this study
GeoB1419-2	-15,54	-17,07	4024	W	open ocean		0,22	0,22	0,54	this study
GeoB5115-2	-24,14	-14,04	3291	W	open ocean	1,03	0,59	0,59	0,91	this sudy
V24-253	-26,95	-44,68	2069	W	continental margin	1,10	0,08	0,08	0,40	Oppo+Horowitz 2000
RC16-84	-26,70	-43,33	2438	W	continental margin	1,08	0,16	0,16	0,48	Oppo+Horowitz 2000
KNR159-5-36	-27,51	-46,47	1268	W	continental margin	1,14	0,16	0,16	0,48	Oppo+Horowitz 2000
RC16-119	-27,70	-46,52	1567	W	continental margin	1,10	0,59	0,59	0,91	Oppo+Horowitz 2000
GeoB2109-1	-27,91	-45,88	2504	W	continental margin	1,23	0,31	0,31	0,63	Vidal et al. 1999
GeoB3808-6	-30,81	-14,71	3213	W	open ocean	1,01	0,55	0,55	0,87	this sudy
GeoB2819-1	-30,85	-38,34	3435	W	open ocean		0,04	0,04	0,36	this sudy
GeoB1312-2	-31,66	-29,65	3436	W	open ocean		-0,09	-0,09	0,23	this sudy
GeoB3813-3	-32,27	-21,97	4331	W	open ocean	0,55	0,08	0,08	0,40	this sudy
GeoB1306-1	-35,21	-26,76	4057	W	open ocean		-0,45	-0,45	-0,13	this sudy
PS2495-3	-41,28	-14,49	3134	W	subtropical front	0,71	-0,66	-0,26	0,06	Mackensen et al. 2001
PS2498-1	-44,15	-14,49	3783	W	subantarctic front	0,22	-1,09	-0,69	-0,37	Mackensen et al. 2001
PS2499-5	-46,51	-15,33	3175	W	subantarctic front	0,31	-0,82	-0,42	-0,10	Mackensen et al. 2001

Table 2. cont.

Atlantic east of the Mid-Atlantic Ridge (Sarnthein et al. 1994). However, in the source region south of the Iceland-Faeroer Ridge above 1500 m water depth, the $\delta^{13}C$ values are as high as 1.90‰ and 1.82‰ at stations V29-193 and M23419, respectively, and therefore about 0.55‰ higher than those of the Late Holocene time interval. Below, there is a sharp decrease in $\delta^{13}C$ to the "normal" values of this region. We assume that these high $\delta^{13}C$ values are related to the $\delta^{13}C$ values as high as 2.2‰ in the easternmost section which point to a source at 36°N and might be, therefore, interpreted as the extending tongue of the glacial MOW. Zahn et al. (1987) showed the glacial MOW to spread down from the isthmus of Gibraltar south to 20°N, approximately as far as today. In the eastern marginal Atlantic section of our study, $\delta^{13}C$ values of >1.0‰ are found in water depth between 1000 and 2000 m as far south as the Walvis Ridge, suggesting an extended spreading of the glacial MOW towards at least 30°S. And even in the western South Atlantic, a layer of higher $\delta^{13}C$ values spreads in about the same intermediate water depth towards the south. Therefore we suggest that due to the decreased downwelling of NADW the influence of MOW became more evident in the glacial North as well as South Atlantic. This layer can be clearly distinguished from the glacial NADW below which started near Rockall Plateau in the northern Atlantic.

The slightly ^{13}C-depleted, i.e. less oxygenated glacial NADW can be traced southwards between 2000 and 3500 m to about 30°S, where it mixes into the glacial CDW. However, south of 10°N this glacial NADW appears to be less extended in the water column between 2500 and 3500 m, slightly shifting downwards on its way south of the equator. A similar pattern is seen in the Western Atlantic section, if the 0.5‰-isoline is assumed to outline the contour of the glacial NADW. The reason for this thinning in glacial NADW close to the equator is related to the bulge of ^{13}C-depleted bottom water in the Mid and West Atlantic sections, which spreads in water depths below 3500 m towards the south as well as to the north, as indicated by the polward declining isolines in both sections. Such a feature can only be explained by an increased advection of low-oxygenated and nutrient-enriched

CDW during the LGM. And indeed, south of the Rio-Grande Rise and Walvis Ridge, the low $\delta^{13}C$ values representing CDW are higher in the water column compared to today and shifted a few degrees towards the north. Obviously, the isolines of this glacial CDW decline from the west to the east, which might indicate a source region west of the Mid-Atlantic Ridge and an eastward downwelling of this water mass. Furthermore, in the section of the Eastern Marginal Atlantic, this southern water mass flew well across the Walvis Ridge and spread along the eastern margins of the Angola, Guinea and Canary Basins. Here, it replaced the modern NADW filling of these basins. From the vertically much higher positioned 0.5‰-isoline in the eastern equatorial compared to the central and western equatorial Atlantic we conclude that this southern water mass flew from the east through the fracture zones towards the western Brasil Basin, the opposite direction compared to the bottom water flow today. Nevertheless, the much lower $\delta^{13}C$ values below a water depth of 3200 m south of Rio-Grande Rise and Walvis Ridge compared to the corresponding values at the nothern slopes of these ridges indicate the barrier effect of these ocean floor morphologies. The most ^{13}C-depleted lower part of the glacial CDW is restricted to the southern Argentine and Cape Basins.

What is the difference of the presented glacial deep Atlantic circulation reconstruction compared to earlier published ones? Previous reconstructions that used glacial $\delta^{13}C$ distributions (Boyle and Keigwin 1987; Curry et al. 1988; Duplessy et al. 1988; Sarnthein et al. 1994; Michel et al. 1995; Oppo and Horowitz 2000) suggested a downward shift of the nutrient maximum in the Atlantic water column, most likely due to a northward expansion of nutrient-rich, southern source bottom water. A shoaling of glacial NADW and a greater stratification of the water column compared to the modern ocean was suggested. This interpretation implies that a weakened turn-over was active during the LGM. However, all these scenarios relied on only a few sites south of the equator, and most of these southern sites were located in areas characterized by high productivity and phytodetritus influences on epibenthic $\delta^{13}C$ values. The extension of the glacial southern water mass has therefore been overestimated. A shift of glacial NADW to intermediate depth, which does not extend further south than 20°S (Sarnthein et al. 1994), can no longer be retained. Indeed, a kind of "glacial Atlantic Intermediate Water" (Duplessy et al. 1988) can be seen in the western South Atlantic as well as in the Eastern Marginal Atlantic section, where the layer of highest $\delta^{13}C$ values shifted from 2500 m today up to 1500 m during the LGM. But this layer has been shown to represent the stronger importance of the glacial MOW.

The slightly decreased vertical extension of glacial NADW, as well as the by 0.25‰ reduced $\delta^{13}C$ values and the diminished spreading towards the south, indicate a reduction of the NADW formation in glacial times, as suggested by many physical ocean model experiments. Numerical experiments by Schäfer-Neth and Paul (2000) reveal a 50% reduction of glacial NADW to the Southern Ocean, if the overall flow strength is adapted to LGM sea-surface temperature and salinity data sets from the North Atlantic Ocean. Consequently, the compensating cross-equatorial heat flow was decreased by 0.1 PW compared to 0.35 PW today. However, the modelled glacial NADW does not extend as deep and as far south as the data reveal. This can be obtained by the $\delta^{13}C$-tracer experiments of Matthies et al. (this volume) using a carbon cycle model coupled to the general circulation model of Schäfer-Neth and Paul (2000). The observed discrepancy between model output and paleodata may be explained in part by the general underestimation of NADW as already evident in the control run for the modern ocean (Matthies et al. this volume). It could also mean that the glacial NADW reduction of 50% is overestimated to produce a deep layer of glacial NADW.

A futher noticeable feature of the glacial $\delta^{13}C$ distribution is the much steeper $\delta^{13}C$ gradient in the mixing zone of glacial NADW and CDW south of Walvis Ridge and Rio-Grande Rise. While today in this region the $\delta^{13}C_{DIC}$ values range uniformly between 0.4‰ and 0.5‰ (Mackensen et al. 1996; Mackensen 2000), the glacial values range between 0.15‰ and 0.55‰, in spite of correcting for the phytodetritus effect. Assuming that the applied corrections are correct, the observed gradients can only be explained by a deep water forming in the

circumantarctic belt. Michel et al. (1995) explains the steep $\delta^{13}C$ gradients by a deep water formation along the Subantarctic convergence. As obtained from box model experiments, stronger wind stirring and enhanced vertical convection would be needed to cause a downwelling of the subantarctic mode water deep enough to fill even the deep ocean basins. However, an increased surface ocean mixing due to higher wind stress during glacials would also enhance the air-sea gas exchange in a way that $\delta^{13}C$ values of the surface ocean would increase, opposite to the ^{13}C depletion observed in this region. Mackensen et al. (2001), therefore, proposed a deep water forming in the broad seasonally sea-ice covered zone considerably south of the southern polar front. In this area, away from the influence of the westwind system in the north, and also in a far distance from the katabatic winds at the continental ice edge in the south, the thermodynamic imprint due to air-sea gas exchange should have been weak. Salt rejection and brine release during sea-ice formation resulted in a deep convection of nutrient-enriched Southern Ocean deep water mass without changing the low $\delta^{13}C$ signal. This scenario is corroborated by the evaluation of $\delta^{18}O$ values showing highest values in the assumed zone of downwelling south of the polar front, which most likely is explained by a considerable salinity increase due to extended sea-ice formation (Mackensen et al. 2001).

Wherever this glacial Southern Ocean Deep Water (SODW) has been produced, it is evident from all three South Atlantic $\delta^{13}C$ sections that most of this downwelled water mass is prevented from flowing into the northern basins by the barriers of the Rio-Grande Rise in the west and Walvis Ridge in the east. Only the upper part of the glacial SODW, characterized by values between 0.18‰ and 0.44‰ can flow across Walvis Ridge, filling the deepest part of the eastern Angola and Guinea Basins, spreading from there in the western and the northern basins, as mentioned above. Only a small amount of upper SODW, characterized by values between 0.35‰ and 0.49‰ can flow directly across Rio-Grande Rise into the southern Brasil Basin.

How does the presented reconstuctions fit to those obtained from Cd/Ca measurements (Boyle and Keigwin 1987; Boyle 1992; Rosenthal et al. 1997)? To better compare our results with those derived from Cd/Ca ratios, we converted both proxies into phosphate concentrations. For converting $\delta^{13}C$ we used the equation by Broecker and Maier-Reimer (1992), who showed that if there were no air-sea gas exchange, the relationship between $\delta^{13}C$ and PO_4 in the ocean would be

$$\delta^{13}C - \delta^{13}C_{M.O.} =$$
$$\Delta photo/\Sigma CO_{2M.O.} * C/P_{org} * (PO_4 - PO_{4M.O.})$$

When reasonable values are substituted ($\delta^{13}C_{M.O.} = 0.3‰$, $\Delta photo = -19‰$, $\Sigma CO_{2\,M.O.} = 2200\ \mu mol\ kg^{-1}$, $C/P_{org} = 128$, $PO_{4\,M.O.} = 2.2\ \mu mol\ kg^{-1}$), the predicted relationship closely matches the relationship for waters in the deep Indian and Pacific Oceans ($\delta^{13}C = 2.7 - 1.1 \times PO_4$). This is to be expected, as the effect of air-sea exchange should be constant for these deep water masses due to the homogeneity of source waters for the deep Indian and Pacific Oceans. Prior to calculating phophate concentrations, we applied the correction for the global $\delta^{13}C$ shift by adding 0.32‰ (Duplessy et al. 1988), and at some locations a constant value of 0.4‰ to correct for a phytodetritus effect on locations influenced by high seasonal fluxes of organic material. For converting Cd_w values, we used the global relationship between Cd_w and phospate as presented by Boyle (1988).

As a result, it is evident from Fig. 4 that there is an excellent agreement between the nutrient concentrations obtained from the two proxies. Most of the previously described discrepancy between $\delta^{13}C$- and Cd-based reconstructions of glacial deep water circulation does not remain. However, a few data points are not congruent, especially the phosphate value of 1.71 $\mu mol\ kg^{-1}$ in core RC12-267 at about 38.7°S in the deepest western South Atlantic section. Since two more cores from the Indian Ocean sector at about the same latitude reveal similarly low Cd_w and hence phophate concentrations (Rosenthal et al. 1997), we do not believe that the former value represents an outlier. Instead, we follow the argumentation by Mackensen et al. (2001) that in

addition to the phytodetritus effect the discrepancy between $\delta^{13}C$ and Cd_w derived nutrient concentrations in this particular area of the Southern Ocean might be related to a reduced thermodynamic imprint, which decreases the preformed $\delta^{13}C$ value of a downwelling SODW by the sea-ice coverage of the source region. A reduction or even shut-down of the southwestern Wedell Sea Bottom Water, which today is marked by high preformed $\delta^{13}C_{DIC}$ values (Mackensen et al. 1996), could in addition contribute to the observed $\delta^{13}C$-Cd_w discrepancy. Furthermore, in the southwestern Weddell Sea a minimum of bottom current activity during the LGM is indicated by grain size spectra of the terrigenous sediment fraction (Pudsey 1992). On the other hand, no significant reduction in bottom current velocities is deduced from sortable silt analyses in the southern central and southeastern South Atlantic (Diekmann and Kuhn 1997). Both observations suggest a reduced bottom water formation in the southwestern Weddell Sea and a deep water formation south of the polar front instead (Michel et al. 1995, Mackensen et al. 2001).

Role of the Deep South Atlantic for Late Pleistocene Atmospheric CO_2 Changes

What can the presented reconstruction of glacial to interglacial changes of the South Atlantic deep water circulation add to the discussion on late Pleistocene atmospheric CO_2 changes? As one mechanism to explain the glacial decreases in atmospheric CO_2, an increase in the deep ocean nutrient reservoir is proposed (e.g. Broecker 1982). However, it was recognized that large changes in the nutrient reservoirs would be required to produce the entire observed amplitude of CO_2 change. Since the glacial phosphate concentrations from the new data set of the deep Atlantic has been shown to be in good accordance with the nutrient distribution obtained from published Cd_w data, we just can confirm the given results by Boyle (1992), who estimated the change in the deep Atlantic Ocean's nutrient inventory to be as small as 14.5%. If we re-estimate the inventory by taking the new $\delta^{13}C$ values from the South Atlantic, leaving all other estimations unchanged, then we obtain only a small increase in the nutrient inventory of the deep

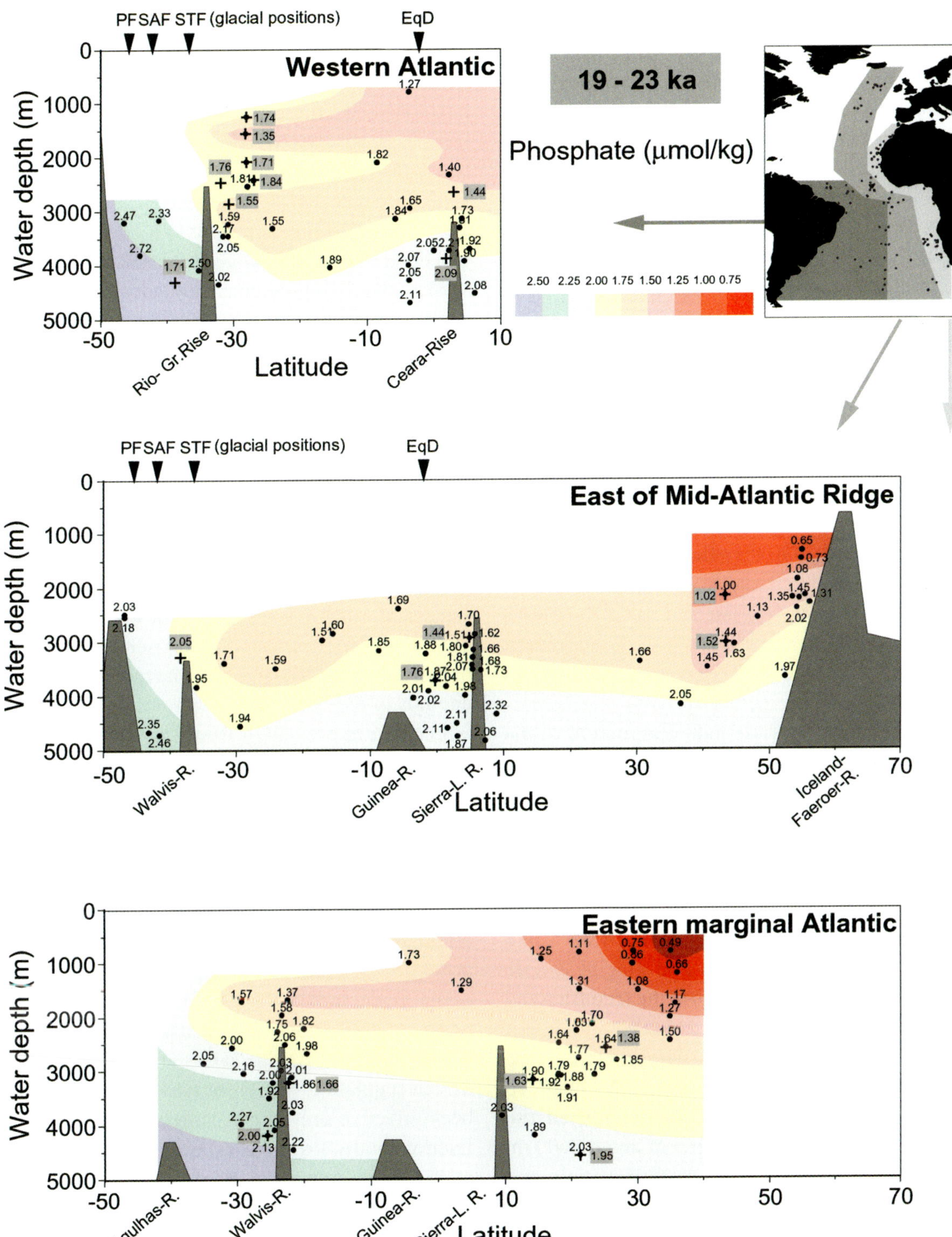

Fig. 4. Last Glacial Maximum (19-23 ka) phosphate distribution calculated from epibenthic $\delta^{13}C$ (this study) and from Cd_w (Boyle 1992, shaded values) along three vertical sections representing the western South Atlantic, the central Atlantic east of the Mid-Atlantic Ridge, and the eastern marginal Atlantic. Prior to calculating phosphate concentrations $\delta^{13}C$ values were corrected for the global ocean shift and occasionally for the phytodetritus-effect as indicated in Fig. 3b.

Atlantic Ocean by 5.9% compared to the modern situation, close to the previous results from Cd_w (Boyle 1992). According to our calculations, the nutrient inventory during LGM decreased in the depth range between 1000 and 3000 m by -2.3%, and increased below 3000 m by 18.4% relative to the Holocene inventories. This is also in well accordance with the modelling results of Matthies et al. (this volume), who obtained by using a carbon cycle model coupled to the general circulation model of Schäfer-Neth and Paul (2000) that under the given glacial conditions of a 50% reduction of NADW and an unchanged advection of southern component water, there is only a small increase in the deep Atlantic nutrient inventory. Balanced on a global scale, the nutrient increase in the deep ocean below 3000 m derived from $\delta^{13}C$ is only 8.1%, and even negative (-18%) according to results from Cd_w. On the other hand, an increase of more than 50% is neccessary to lower atmospheric CO_2 by 80 ppm, and still 30% to 40% nutrient increase is necessary, if a coincident decrease in $CaCO_3$ export is assumed (Sigman and Boyle 2000).

Also changes in carbonate distribution of the deep Atlantic are not suitable to explain the decrease in the CO_3^{2-} pump as a second mechanism. Detailed studies of dissolution proxies (see Henrich et al. this volume, for a summary) reveal a rise of the calcite lysocline in the South Atlantic to a depth of 3700 to 3800 m in the eastern as well as in the western basins, slightly rising south of 20°S to a depth of 3500 m at 30°S. The glacial calcite lysocline was, therefore, close to the reconstructed boundary between glacial NADW and upper SODW (Henrich et al. this volume). Only in the region of equatorial upwelling, the enhanced production of biogenic carbonate pushed the lysocline down to a water depth of about 4000 m. Although especially in the eastern Angola and Guinea Basins the replacement of NADW by the southern component water caused a remarkable vertical shift of the lysocline, it is the opposite direction to result in an increased pumping of CO_2 into the deep-sea. Archer et al. (2000) argue that only a substantial deepening of the lysocline could cause a larger decrease in atmospheric CO_2. However, the positions of the lysoclines were at a comparable level in the glacial oceans (e.g. Bickert and Wefer 1996). As a result, the carbonate chemistry of the deep Atlantic and Pacific, today distinguished from each other by the formation of freshly ventilated NADW, are assumed to have been more homogenous during glacial times.

A third mechanism proposed to explain the glacial pCO_2 reduction is related to a decrease in the rate of air-sea gas exchange as a result of increased sea-ice cover at high southern latitudes (Stepens and Keeling 2000). By significantly limiting the sea-to-air CO_2 flux in the primary region for deepwater ventilation, expanded Antarctic sea ice during glacial times may trap relatively more carbon in the deep ocean, thereby reducing atmospheric CO_2 concentrations. As discussed in the previous chapter, there might be an evidence for a deep water forming of a cool, high-saline and nutrient-enriched water mass like that proposed in the LGM scenario by Stephens and Keeling (2000). However, if their suggested water mass is congruent with the glacial SODW evident from our construction, the distribution of this water mass might not have been extended enough to explain the CO_2 pumping, because a larger part is prevented from spreading into the equatorial and north Atlantic by the barriers of Walvis Ridge and Rio Grande Rise. On the other hand, a melting of the assumed winter sea-ice during summer times may have caused a salinity stratification necessary to prevent CO_2 outgassing during the warm season, when the sea-ice extent is minimal. Morales Maqueda and Rahmstorf (2002) presented simulations with a coupled sea ice/upper ocean model indicating that the CO_2 sequestration under glacial ice coverage could account for at most 15-50% of the total glacial CO_2 decline. A sea ice-driven regulation of CO_2 outgassing would have been effective only if the wintertime sea-ice area fraction southward of the Antarctic Polar Front rises to 99 - 100%. At present, strong circumpolar winds pull the ice apart, and intense oceanic heat fluxes hinder ice growth to a point that sea ice covers no more than 80-90% of the total winter ice-pack area. However, although these winds are assumed to blow even stronger during glacial times, this might not be the case in the region south of the polar front, where the glacial SODW forming is assumed, as stated above.

Conclusions

A set of 55 benthic foraminiferal stable carbon and oxygen isotope time series, including 28 new records, is presented from the South Atlantic Ocean between 6°N and 47°S to reconstruct the Atlantic deepwater circulation for the Last Glacial Maximum (19-23 ka) and the Late Holocene (0-4 ka) times. Compiled with published data of the eastern North Atlantic, we assigned these records to three North-South sections representing the western South Atlantic, the central Atlantic east of the Mid-Atlantic Ridge, and the eastern marginal Atlantic, to better understand the spatial distribution of deep and bottom water masses. Corrections of +0.4‰ are suggested for several benthic $\delta^{13}C$ values of cores located in high-productivity areas, to adjust for phytodetritus-induced depletion of especially glacial values.

As a result of this new compilation, the following pattern for the glacial deep Atlantic circulation have been obtained:

• No shift of NADW to intermediate depth during the last glacial maximum is evident in the glacial Atlantic. Instead, the core of an ^{13}C-enriched water mass spreading southward to at least 30°S between 1200 and 1900 m points to a source of this water mass close to the Isthmus of Gibraltar, indicated by $\delta^{13}C$-values of up to 1.8‰. This layer is interpreted as an extended tongue of the Mediterranean Outflow Water.

• In the central and western South Atlantic, glacial NADW is shown to flow southward at about the same depth interval or even deeper than it does today, although slightly depleted in ^{13}C and less extended in the water column. The admixing of NADW into the circumantarctic deepwater belt occurred a few degrees further north than today.

• Steep gradients in glacial $\delta^{13}C$ between 30° and 40° S suggest a local formation of Southern Ocean deep water in the zone of extended winter sea-ice coverage south of the polar front. The spreading of this newly formed water mass, however, is restricted to the Atlantic basins south of Walvis Ridge and Rio-Grande Rise, where only a small amount of nutrient-enriched deep water passes across these barriers into the northern basins.

• Converted into nutrient concentrations, the new carbon isotope data set gives only a slight increase in the nutrient inventory of the deep Atlantic, in good agreement with previously published Cd/Ca data.

Acknowledgements

We thank the crews and scientific parties of several *Meteor* and *Polarstern* cruises for successful ventures at sea. We are indebted to B. Hellwig, who investigated the benthic isotope measurements of several cores within her diploma thesis, and M. Segl and her team, who carefully supervise the operation of the mass spectrometers of the Fachbereich Geowissenschaften, University of Bremen. We would like to thank E. Michel and an anonymous reviewer for valuable comments to this manuscript. We acknowledge financial support from the Deutsche Forschungsgemeinschaft (Sonderfor-schungsbereich 261). This is SFB 261 contribution no. 362. Data are available under www.pangaea.de/Projects/SFB261.

References

Archer D, Winguth A, Lea D, Mahowald N (2000) What caused the glacial/interglacial atmospheric pCO$_2$ cycles? Rev Geophys 38: 159-189

Berner W, Stauffer B, Oeschger H (1978) Past atmospheric composition and climate, gas parameters measured on ice cores. Nature 276: 53-55

Bickert T, Berger WH, Burke S, Schmidt H, Wefer G (1993) Late Quaternary stable isotope record of benthic foraminifera at sites 805 and 806, Ontong Java Plateau. Proc ODP, Sci Results 130: 411-420

Bickert T, Wefer G (1996) Late Quaternary deep water circulation in the South Atlantic: Reconstruction from carbonate dissolution and benthic stable isotopes. In: Wefer G, Berger WH, Siedler G, Webb D (eds) The South Atlantic: Present and Past Circulation. Springer, Berlin, pp 599-620

Bickert T, Wefer G (1999) South Atlantic and benthic foraminifer $\delta^{13}C$-deviations: Implications for reconstructing the Late Quaternary deep-water circulation. Deep-Sea Res I 46: 437-452

Boyle EA (1988) Cadmium: Chemical tracer of deepwater paleoceanography. Paleoceanography 3: 471-489

Boyle EA (1992) Cadmium and $\delta^{13}C$ palechemical ocean distributions during the stage 2 glacial maximum. Annu Rev Earth Planet Sci 20: 245-287

Boyle EA, Keigwin L (1987) North Atlantic thermohaline circulation during the past 20,000 years linked to high-latitude surface temperature. Nature 330: 35-40

Broecker WS (1982) Ocean chemistry during glacial time. Geochim Cosmochim Acta 46: 1689-1705

Broecker WS, Maier-Reimer E (1992) The influence of air and sea exchange on the carbon isotope distribution in the sea. Glob Biogeochem Cycl 6: 315-320

Charles CD, Fairbanks RG (1990) Glacial to interglacial changes in the isotopic gradients of the Southern Ocean surface water. In: Bleil U,Thiede J (eds) Geological History of the Polar Oceans: Artic versus Antarctic. Kluwer, Dordrecht, pp 519-538

Charles CD, Wright JD, Fairbanks RG (1993) Thermodynamic influences on the marine carbon isotope record. Paleoceanography 8: 691-697

Connary SD, Ewing M (1974) Penetration of Antarctic Bottom Water from the Cape Basin into the Angola Basin. J Geophys Res 79: 463-469

Curry WB, Duplessy JC, Labeyrie LD, Shackleton NJ (1988) Changes in the distribution of $\delta^{13}C$ of deep water TCO_2 between the last glaciation and the Holo-cene. Paleocanography 3: 317-341

Curry WB, Lohmann GP (1983) Reduced advection into Atlantic Ocean deep eastern basins during last glaciation maximum. Nature 306: 577-580

Curry WB, Lohmann GP (1990) Reconstructing past particle fluxes in the tropical Atlantic Ocean. Paleoceanography 5: 487-506

Diekmann B, Kuhn G (1997) Terrigene Partikeltransporte als Abbild spätquartärer Tiefen- und Bodenwasser-zirkulation im Südatlantik und angrenzenden Süd-polarmeer. Z DGG 148: 405-429

Duplessy JC, Shackleton NJ, Fairbanks RG, Labeyrie L, Oppo D, Kallel N (1988) Deepwater source variations during the last climatic cycle and their impact on the global deepwater circulation. Paleoceanography 3: 343-360

Farrell JW, Prell WL (1989) Climatic change and $CaCO_3$ preservation: An 800,000 year bathymetric reconstruction from the central equatorial Pacific Ocean. Paleoceanography 4: 447-466

Flower BP, Oppo DW, McManus JF, Venz KA, Hodell DA, Cullen JL (2000) North Atlantic intermediate to deep water circulation and chemical stratification during the past 1 Myr. Paleoceanography 15: 388-403

Imbrie J, Hays JD, Martinson DG, McIntyre A, Mix AC, Morley JJ, Pisias NG, Prell WL,Shackleton NJ (1984) The orbital theory of Pleistocene climate: Support from a revised chronology of the marine $\delta^{18}O$ record. In: Berger A, Imbrie J, Hays J, Kukla G,Saltzman B (eds) Milankovitch and Climate. Part ID Reidel, Dordrecht, pp 269-305

Keeling R, Stephens BB (2001) Antarctic sea ice and the control of Pleistocene climate instability. Paleoceano-graphy 16: 112-131 and 330-334 (corrections).

Kroopnick P (1985) The distribution of ^{13}C of T CO_2 in the world oceans. Deep-Sea Res 32: 57-84

Labeyrie LD, Duplessy JC, Blanc PL (1987) Variations in mode of formation and temperature of oceanic deep waters over the past 125,000 years. Nature 327: 477-482

Mackensen A (2001) Oxygen and carbon stable isotope tracers of Wedell Sea water masses: New data and some paleoceanographic implications. Deep-Sea Res 48: 1401-1422

Mackensen A, Hubberten HW, Bickert T, Fischer G, Fütterer DK (1993) $\delta^{13}C$ in benthic foraminiferal tests of *Fontbotia wuellerstorfi* (Schwager) relative to $\delta^{13}C$ of dissolved inorganic carbon in Southern Ocean deep water: implications for glacial ocean circulation models. Paleoceanography 8: 587-610

Mackensen A, Grobe H, Hubberten HW, Kuhn G (1994) Benthic foraminiferal assemblages and the ^{13}C-signal in the Atlantic sector of the Southern Ocean: glacial-to-interglacial contrasts. In: Zahn R, Pedersen TF, Kaminski MA, Labeyrie L (eds) Carbon Cycling in the Glacial ocean: Constraints on the Ocean's Role in Global Change. Springer, Berlin, pp 105-142

Mackensen A, Hubberten HW, Scheele N,Schlitzer R (1996) Decoupling of $\delta^{13}C_{?CO2}$ and phosphate in recent Weddell Sea Deep and Bottom Water: Implications for glacial Southern Ocean paleoceanography. Paleoceanography 11: 203-215

Mackensen A, Rudolph M, Kuhn G (2001) Late Pleistocene deep-water circulation in the subantarctic eastern Atlantic. Glob Planet Change 30: 197-229

McConnaughey TA, Burdett J, Whelan JF, Paull CK (1997) Carbon isotopes in biological carbonates: Respiration and photosynthesis. Geochim Cosmochim Acta 61: 611-622

Michel E, Labeyrie LD, Duplessy JC, Gorfti N, Labracherie M, Turon J-L (1995) Could deep subantarctic convection feed the world deep basins during the last glacial maximum? Paleoceanography 10: 927–942

Mix AC (1989) Pleistocene paleoproductivity: Evidence from organic carbon and foraminiferal species. In: Berger WH, Smetacek VS, Wefer G (eds) Productivity of the Ocean: Present and Past. J Wiley & Sons, Chichester, pp 313-340

Mix AC, Bard E, Schneider RR (2001) Environmental processes of the ice age: Land, oceans, glaciers (EPILOG). Quat Sci Rev 20: 627-657

Mollenhauer G, Eglinton TI, Ohkouchi N, Schneider RR, Müller PJ, Grootes PM, and Rullkötter J (2003) Asynchronous Alkenone and Foraminifera Records from the Benguela Upwelling System. Geochim Cosmochim Acta, in press

Mollenhauer G, Schneider RR, Müller PJ, Spieß V, and Wefer G (2002) Glacial/interglacial variability in the Benguela Upwelling System: Spatial distribution and budgets of organic carbon accumulation. Glob Bio-geochem Cycl 16-4: 81(1-15)

Morales Maqueda MA, Rahmstorf S (2002) Did Antarctic sea-ice expansion cause glacial CO_2 decline? Geophys Res Lett 29: 111-113

Niebler H-S, Mulitza S, Donner B, Arz H, Pätzold J, Wefer G (2003) Sea surface temperatures in the equatorial and South Atlantic ocean during the Last Glacial Maximum (23-19 ka) Paleoceanography, in press

Ninnemann US, Charles CD (2002) Changes in the mode of Southern Ocean circulation over the last glacial cycle revealed by foraminiferal stable isotopic variability. Earth Planet Sci Lett 201: 383-396

Oppo DW, Fairbanks RG (1990) Atlantic Ocean thermo-haline circulation of the last 150.000 years: Relation-ship to climate and atmospheric CO_2. Paleoceano-graphy 5: 277-288

Oppo DW, Rosenthal Y (1994) Cd/Ca changes in a deep Cape Basin core over the past 730,000 years: Response of circumpolar deep water variability to northern hemisphere ice sheet melting? Paleoceanography 9: 661-676

Östlund HG, Craig C, Broecker WS, Spencer D (1987) GEOSECS Atlantic, Pacific, and Indian Ocean Expedition, Shorebased data and graphics (GEOSECS Atlas Ser. vol. 7. US Government Printing Office, Washington, 200 p

Paillard D, Labeyrie L and Yiou P (1996) Macintosh program performs time-series analysis. Eos Trans AGU 77: 379

Pudsey CJ (1992) Late Quaternary changes in Antarctic Bottom Water velocity inferred from sediment grain size in the northern Wedell Sea. Mar Geol 107: 9-33

Reid JL (1989) On the total geostrophic circulation of the South Atlantic Ocean: Flow patterns, tracers, and transports. Prog Oceanogr 23: 149-244

Rosenthal Y, Boyle EA, Labeyrie L (1997) Last glacial maximum pleochemistry and deepwater circulation in the Southern Ocean: Evidence from foraminiferal cadmium. Paleoceanography 12: 787-796

Sarnthein M, Winn K, Jung SJA, Duplessy JC, Labeyrie L, Erlenkeuser H, Ganssen G (1994) Changes in east Atlantic deepwater circulation over the last 30,000 years: Eight time slice reconstructions. Paleoceanography 9: 209-268

Schäfer-Neth C and Paul A (2000) Ciculation of the Glacial Atlantic: A synthesis of global and regional modelling. In: Schäfer P, Ritzrau W, Schlüter M, Thiede J (eds) The Northern North Atlantic: A Changing Environment. Springer, Berlin, pp 441-462

Shackleton NJ (1977) Tropical rainforest history and the equatorial Pacific carbonate dissolution cycles. In: Anderson NR, Malahoff A (eds) Fate in Fossil Fuel CO_2 in the Oceans. Plenum, New York, pp 401-427

Shannon LV, Chapman P (1991) Evidence of Antarctic Bottom Water in the Angola Basin at 32°S. Deep-Sea Res 38: 1299-1304

Sigman DM, Altabet MA, Francois R, McCorkle DC, Gaillard J-F (1999) The isotopic composition of diatom-bound nitrogen in Southern Ocean sediments. Paleoceanography 14: 118-134

Sigman DM, Boyle EA (2000) Glacial/interglacial variations in atmospheric carbon dioxide. Nature 407: 859-869

Stephens BB, Keeling RF (2000) The influence of Antarctic sea ice on glacial-interglacial CO_2 variations. Nature 404: 171-174.

Van Bennekom AJ, Berger GW (1984) Hydrography and silica budget of the Angola Basin. Neth J Sea Res 17: 149-200

Warren BA, Speer KG (1991) Deep circulation in the eastern South Atlantic Ocean. Deep-Sea Res 38, Suppl.1: S281- S322

Winguth AME, Archer D, Duplessy JC, Maier-Reimer E, Mikolajewicz U (1999) Sensitivity of paleonutrient tracer distributions and deep-sea circulation to glacial boundary conditions. Paleoceanography 14: 304-323

Yu E, Francois R, Bacon MP (1996) Similar rates of modern and last-glacial ocean thermohaline circulation inferred from radiochemical data. Nature 379: 689-694

Zahn R, Sarnthein M (1987) Benthic isotope evidence for changes of the mediterranean outflow during the late Quartenary. Paleoceanography 2: 543-559

Last Glacial δ^{13}C Distribution and Deep-Sea Circulation in the Atlantic Ocean: A Model - Data Comparison

M. Matthies*, T. Bickert and A. Paul

*Universität Bremen, Fachbereich Geowissenschaften, Klagenfurter Straße,
28334 Bremen, Germany*
** corresponding author (e-mail): mmat@palmod.uni-bremen.de*

Abstract: We used a carbon cycle model (HAMOCC2) coupled to a general ocean circulation model (LSG) to explore the δ^{13}C distribution in the glacial Atlantic Ocean. We compared the simulated δ^{13}C pattern with a new data set of benthic carbon isotopes of the Western and Eastern Atlantic from the Last Glacial Maximum (18,000 to 20,000 ^{14}C years or 21,000 - 23,500 calendar years before present). The model output fits the δ^{13}C distribution derived from sediment samples, when the glacial export of NADW to the Southern Ocean was reduced by 50 % and the inflow of glacial AABW was held constant. In most cases, the modeled δ^{13}C pattern matched the paleodata within a range of ±0.2 ‰. Furthermore, the asymmetry between the glacial NADW distribution in the South Atlantic basins was reproduced by the coupled ocean circulation and carbon cycle models. No additional increase of the nutrient inventory in the deep ocean was necessary to reproduce the paleodata. Hence we conclude that a significant increase in biological pumping during glacials may not be necessary to explain the reconstructed δ^{13}C distribution in this region. The results are discussed with respect to other scenarios for the decrease of global atmospheric pCO_2.

Introduction

The partial pressure of atmospheric carbon dioxide (pCO_2) in the in the Last Glacial Maximum (LGM), about 21,000 years ago is estimated to about 200 ppm, about 80 ppm less compared to the preindustrial value. The cause of this variation in carbon dioxide (CO_2) has not yet been identified (Broecker and Henderson 1998; Archer et al. 2000a; Sigman and Boyle 2000). In the recent literature, there are many hypotheses trying to explain this strong reduction in atmospheric pCO_2. These hypotheses involve changes in the nutrient inventory, in the oceanic pH, in the ocean circulation, or a combination of these factors.

Biological Pump Scenarios

One group of mechanisms to lower the glacial pCO_2 is to increase the rate of biological produc-tivity in the surface ocean, whereby carbon is trans-ferred from the surface ocean to the deep-sea in the form of sinking particles. Either an increase in the ocean inventory of the nutrients phosphate (PO_4^{3-}) and nitrate (NO_3^-) or a change in the ratio of nutrients to carbon in phytoplankton could have stimulated the biological pump in this way. The observation that iron availability limits phytoplankton growth in some parts of the ocean such as the Southern Ocean provides a mechanism by which the biological pump in high latitudes could have intensified in a dustier, more iron-rich glacial climate. Broecker and Henderson (1998) suggested that an enhancement of glacial iron supply to the surface ocean stimulates the rate of nitrogen fixation, causing an increase in the NO_3^- : PO_4^{3-} ratio of the deep-sea and an increase in the effective nutrient reservoir of the ocean.

From WEFER G, MULITZA S, RATMEYER V (eds), 2003, *The South Atlantic in the Late Quaternary: Reconstruction of Material Budgets and Current Systems.* Springer-Verlag Berlin Heidelberg New York Tokyo, pp 695-722

Ocean pH Scenarios

A second class of mechanisms to lower the glacial pCO_2 is to change the pH or alkalinity of the whole ocean, converting seawater CO_2 into bicarbonate (HCO_3^-) and carbonate (CO_3^{2-}) (e.g. Archer et al. 2000a). The pH in the ocean is controlled by the mechanism of $CaCO_3$ compensation. Any imbalance between the influx of dissolved $CaCO_3$ originating from chemical weathering on land and the removal of $CaCO_3$ by burial in the deep-sea sediments will act to change the ocean pH until the flux balance is restored. It is more difficult to preserve $CaCO_3$ in an acidic ocean. If the glacial rate of weathering were higher and the $CaCO_3$ deposition currently occurring in shallow waters were shifted to the deep-sea, or the rate of $CaCO_3$ production decreased, the $CaCO_3$ burial efficiency would increase and the ocean would become more basic. If organic carbon production increased, its degradation in sediments would also promote calcite dissolution, further increasing the pH of the ocean. $CaCO_3$ compensation may also affect the pCO_2 response to the biological pump scenarios described above.

Circulation Changes

Toggweiler (1999) reduced the deep water ventilation in a box model that treats the boundary between mid-depth and deep water masses as a chemical barrier, separating the water with a low CO_2 content above from the water with high contents below. He generated reduced ventilation by decreasing the vertical exchange between deep and Antarctic surface water. This mechanism invokes only a physical process or a set of related physical processes, that stratify the interior while reducing the ventilation of the ocean's deepest water. These processes bottle up remineralised CO_2 at depths below 3,000 m and reduce the CO_3^{2-} in the deep water in contact with most of the ocean's $CaCO_3$ sediments. Through their control of the burial of $CaCO_3$, these ventilation and stratification processes are able to drive atmospheric CO_2 up and down on timescales of several thousands up to ten thousand years. The mechanism works without any changes in biological activity and requires only minimal changes in the distribution of nutrients.

Another mechanism to reduce the atmospheric CO_2 concentration without changing productivity or alkalinity has been proposed by Stephens and Keeling (2000). They suggest that a ventilation change was driven by limitations of air-sea gas-exchange due to an increase of sea-ice in the Southern Ocean. The mechanism that is proposed to explain the observed synchronisation between Antarctic temperature and atmospheric CO_2 calls for dramatic changes in the structure of the deep ocean during cold periods, including a collapse of NADW formation, an expansion of the volume of the deep ocean filled with AABW, and the cooling of this deep layer virtually down to the freezing point (Keeling and Stephens 2001).

However, a study by Yu et al. (1996), based on measurements of the $^{231}Pa/^{230}Ta$ ratios in glacial and modern sediments, indicates that the export of ^{231}Pa from the Atlantic into the Southern Ocean continued during the LGM at roughly the modern rate. They drew the conclusion that the export of deep water formed in the North Atlantic also continued to move into the Southern Ocean at a comparable rate during the LGM. If this were true, it would falsify the hypothesized collapse of NADW formation during glacial cold periods, as first suggested by Duplessy et al. (1988) and also stated as a requirement by Keeling and Stephens (2001).

In the study presented here, we used a model-data comparison of $\delta^{13}C$ changes in the deep Atlantic, in order to test the role of deep water overturning as a main potential driving mechanism for glacial pCO_2 reduction. A new data set of benthic stable isotopes from sediment cores in the South Atlantic (Bickert and Mackensen this volume) allows a more detailed reconstruction of the glacial nutrient distribution in this region. We will discuss the results in view of the various hypotheses described above.

Model Descriptions

Ocean General Circulation Model

The physical ocean model used in this study is known as Hamburg Large-Scale Geostrophic Ocean Model (LSG) (Maier-Reimer et al. 1993). The glacial and modern fields of the velocity, temperature, salinity and convective adjustments were derived from the global model runs of Schäfer-Neth and Paul (2001) (labeled GM for the modern and G1 for a slightly modified version of the glacial simulation) and represented the input fields for the carbon cycle model. The modifications with respect to Maier-Reimer et al. (1993) include:

• The air temperature advection has been removed. This procedure was originally used to bring the simulated North Atlantic Deep Water (NADW) production to an appropriate magnitude of the present-day climate (see Schäfer-Neth and Paul 2001).

• The advection routine for temperature and salinity is formulated in accordance with the QUICK scheme of Leonard (1979), in order to compensate for the strong reduction of NADW formation due to the neglect of air temperature advection. This scheme results in a larger fraction of newly formed NADW that flows across the equator into the South Atlantic. The price to be paid is a smaller time-step, which is about four times less than the monthly time-step used by Maier-Reimer et al. (1993) and necessitates a linear interpolation of the monthly mean sea-surface temperature and salinity.

• The sea-ice model of the original version of the LSG has not been included. Instead, the grid cells covered with sea-ice are set to the freezing point (-1.8°C). Thus the model sea-ice cover has rather been driven by monthly mean sea-surface temperatures.

• A convection scheme, which makes the water column completely stable after every time-step, has been added.

• Depth-dependent horizontal diffusivity values between 8×10^6 cm^2 s^{-1} in the top layer and $4 \times$ 10^6 cm^2 s^{-1} in the bottom layer are employed. Values between 0.3 cm^2 s^{-1} and 1.1 cm^2 s^{-1} respectively have been assumed for the vertical diffusivity.

The forcing data for the simulations under present-day and LGM boundary conditions are specified in Table 1. The modified LSG ocean model has the same resolution, land-sea mask and bathymetry as the original model, except for the glacial experiment in which the Bering Strait is closed.

The zonally integrated meridional overturning in the modern Atlantic Ocean is shown in Figure 1, left. NADW is produced at a rate of 14 Sv, 8 Sv of which is exported across the equator to the Southern Ocean at depths below 1,500 m. There is a deep inflow of Antarctic Bottom Water (AABW) which amounts to 4 Sv. The total quantity of deep water (NADW and AABW) which flows into the Southern Ocean is 12 Sv, approximately 30 % less than the value of 18 Sv given by Schmitz (1995). The water mass boundary between NADW and AABW at a depth of 2,500 m is shallower than estimated from observations by approximately 1,500 m, which corresponds to one model layer thickness. The reason for this is probably that the newly formed NADW is too warm (1 to 2°C) above 3,000 m in the northern North Atlantic Ocean and thus not dense enough to sink to an appropriate depth. However, this is a general problem of coarse resolution models. For a detailed description see Schäfer-Neth and Paul (2001).

In the glacial Atlantic (Fig 1 b), the NADW export to the Southern Ocean is 4 Sv, a reduction of 50 % compared to the modern; the glacial NADW production amounts to 6 Sv. The inflow of glacial AABW remains unchanged at 4 Sv and the bound-

Parameter	Present-day	LGM
Sea-surface temperature	Shea et al. (1990)[1]	Weinelt et al. (1996)
Sea-surface salinity	Levitus et al. (1994)[2]	Schäfer-Neth and Paul (2001)
Wind	Lorenz et al. (1996)	Lorenz et al. (1996)

[1] Surface temperature was set to -1.8 °C over the ice covered regions from Shea et al. (1990).

[2] Winter surface salinities in the Ross and Weddell Seas are adjusted according to Johns et al. (1997).

Table 1. Ocean model forcing data for present-day and LGM boundary conditions.

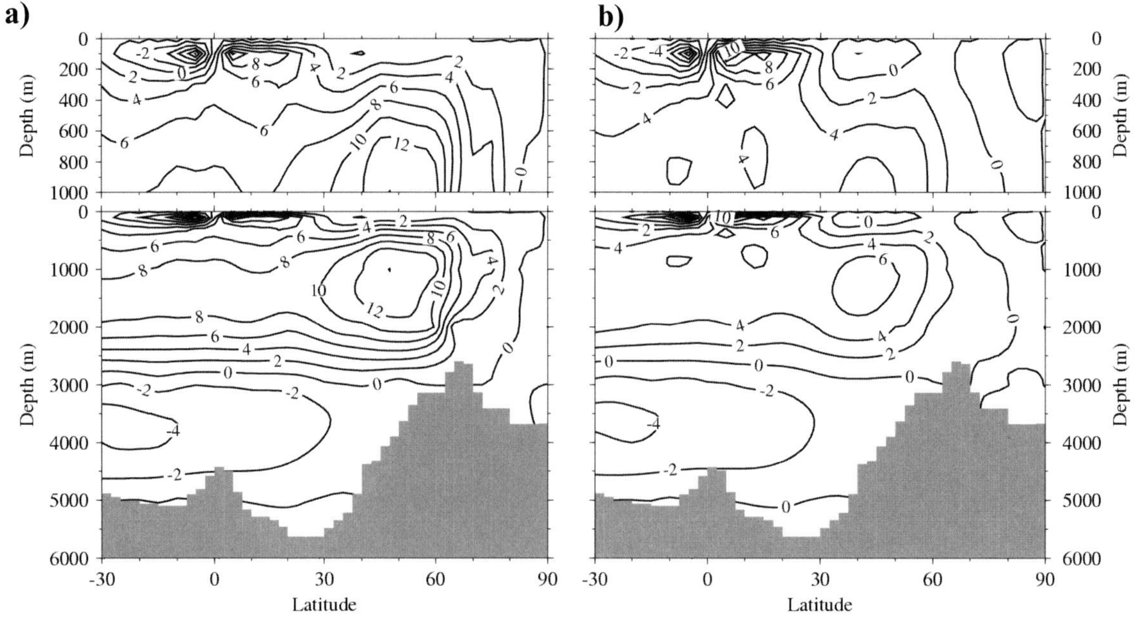

Fig. 1. Annual mean of zonally averaged Atlantic meridional circulation. Contour interval is 2 Sv (1 Sv = $10^6\,\mathrm{m^{-3}\,s^{-1}}$). Negative contour lines indicate anti-clockwise circulation, whereas positive values indicate clockwise circulation. **a)** Simulation corresponds to the present-day boundary conditions. **b)** Simulation corresponds to the LGM boundary conditions.

ary between the Deep Current System and the Bottom Current System is only slightly shallower than in the modern ocean.

Carbon Cycle Model

The model used here is the Hamburg Model of the Oceanic Carbon Cycle (HAMOCC2) (Maier-Reimer and Hasselmann 1987; Heinze et al. 1991). Only a brief outline and modifications to Heinze et al. (1991) are given in the following. The model runs on a global 72 x 72 E-grid (Arakawa and Lamb 1977), with a horizontal resolution of approximately 3.5° x 3.5° and uses 11 vertical layers with layer interfaces at 0, 50, 112.5, 200, 350, 575, 850, 1,500, 2,500, 3,500, 4,500, and 6,000 m, and has a time-step of 1 year. A quasi steady state is reached after 20,000 model years. The model tracer variables include atmospheric pCO$_2$, ΣCO$_2$ (total dissolved inorganic carbon), total alkalinity (including the contribution of borate), PO$_4^{3-}$ (as biolimiting nutrient),

oxygen, POC (particulate organic carbon, plankton soft tissue), and particulate CaCO$_3$ (plankton calcareous shells). For all tracers that include carbon, the three carbon isotopes (the stable ^{12}C and ^{13}C and the radiocarbon ^{14}C) were considered. The organic matter (C$_{org}$ and C$_{CaCO3}$) produced in the uppermost model layer, depending on the nutrient and light availability, is immediately redistributed within the water column because the time-step of 1 year is long relative to the time scale of sinking particulate organic matter (about 100 m day^{-1}, e.g. Suess 1980). The prescribed profiles for the organic carbon flux through the water column are different for POC and CaCO$_3$ and decrease exponentially with depth (with an e-folding depth of 800 and 4,000 m, respectively).

After each model run, the total inventory of ^{13}C was calibrated to a preanthropogenic atmospheric δ^{13}C value of -6.5 ‰, according to Friedli et al. (1986). The model configuration of Heinze et al. (1991) was modified in several respects:

• A modified profile of the vertical distribution of POC was employed. Originally, about one-third of the new production was assumed to fall immediately to the bottom layer to account for organic material coated by hard shells. This formulation was changed in such a manner that no fraction of the new production falls immediately to the bottom layer, leading to some improvements in the tracer distributions. First, this formulation yielded a greater new production (from 6.5 to 6.7 GtC year^{-1}) and lower $CaCO_3$ production (from 1.3 to 1.1 GtC year^1), accompanied with a corresponding decline in the atmospheric pCO_2 from 304 ppm to the preindustrial value of 281 ppm. Second, it reduced the models overestimation of the POC water inventory from 223 to 177 GtC year^{-1} and that of the sediment pool (from 90 to 24 GtC year^{-1}). Third, this profile led to a better tracer distribution especially in the bottom layer of the model, where the original profile yielded an overestimation of the concentrations in the northern parts of the Atlantic and Pacific basin, as compared to the GEOSECS data. By use of this profile, the simulated δ^{13}C values in the western Atlantic basin were reduced from 0.6 to 0.4 ‰ in the source areas of the AAIW, and the water mass characteristics were more pronounced for AAIW, NADW and AABW. Consequently, a better quantitative comparison with the GEOSECS data was accomplished.

• An explicit vertical diffusion is included with the same slight depth dependency as in the ocean model (see above), to make the carbon cycle model more consistent with the ocean model. In general the vertical diffusion tends to reduce the vertical gradients, especially in the deep water column. By using the explicit vertical diffusion, the global new production increased from 6.1 to 6.7 GtC year^{-1}, and $CaCO_3$ production increased from 0.9 to 1.1 GtC year^{-1} in the control run. The atmospheric pCO_2 rose from 278 to 281 ppm. This leads to advantages for some tracers, but others show a worse distribution than without explicit vertical diffusion. The advantages were most obvious in the zone of the phosphate maximum and oxygen minimum in the north and equatorial Pacific, where a consistence with GEOSECS could be reached. However, beneath this zone, the tracer values were higher in δ^{13}C and phosphate and lower in oxygen,

as compared to GEOSECS. Implications of the δ^{13}C concentration in the Atlantic consisted in a small rise of 0.1 ‰ in the northern part (north of 30°N) of the bottom water, and a same enhancement in the region between 30°S and 30°N above a water depth of 800 m.

• A constant Redfield stoichiometry for POC is used, with C:N:P:O equivalent to 117:16:1:-170 according to Anderson and Sarmiento (1994).

• A rain ratio (C_{CaCO3} to C_{org}) is prescribed which is lower than the commonly used value of 0.20 - 0.25. Here, we used a maximum rain ratio value of 0.15, following Bacastow and Maier-Reimer (1990), cf. Yamanaka and Tajika (1996). Due to this change in the rain ratio, all tracers which included $CaCO_3$ were influenced, especially the $CaCO_3$ production which is reduced from 1.7 to 1.1 GtC year^{-1}, if a value of 0.24 is applied. As a result, surface alkalinity and the carbonate ion concentration increased and the atmospheric pCO_2 decreased from 313 to 281 ppm. The effects on the δ^{13}C distribution in the Atlantic were negligible, the extension of the δ^{13}C minimum to the south was enlarged. The effect on the carbonate ion concentration was a better match of the calcite lysocline depth in the Southern Ocean with GEOSECS, and resulted in a deeper lysocline of about 400 m in this region.

The effects of these parameterisations on the model inventories and mean values are summarized in Table 2 and 3, respectively.

Controls on δ^{13}C of Marine Carbon

In the marine carbon cycle, two fractionation processes are important. First, during photosynthesis, organisms preferentially take up the lighter carbon isotope ^{12}C, increasing δ^{13}C in the surface ocean dissolved as inorganic carbon (DIC). When isotopically light organic matter is remineralised, δ^{13}C of marine ΣCO_2 decreases. This fractionation of about -20 ‰ of marine photosynthesis leaves the surface ocean depleted in nutrient with a high content of δ^{13}C, whereas the nutrient-rich deep waters display low δ^{13}C values. Through this mechanism, known as biological pump, carbon and nutrients are transported to the deep-sea and thereby reduce the CO_2 in the surface and atmosphere.

Inventory	Units	POC profile	No explicit vertical diffusion	Rain ratio of 0.24	Control run
Atmospheric pCO_2	Ppm	304	278	313	281
Total primary production	GtC/Year	10.2	7.8	8.6	8.6
New production	GtC/Year	6.5	6.1	6.7	6.7
$CaCO_3$ production	GtC/Year	1.3	0.9	1.7	1.1
POC ocean water pool	GtC	223	174	177	177
$CaCO_3$ ocean water pool	GtC	53.2	52.7	51.1	53.2
ΣCO_2 ocean water pool	GtC	39494	39656	39486	39708
Corg sediment pool	GtC	90.0	22.8	24.3	24.3
$CaCO_3$ sediment pool	GtC	4029	4035	4133	3972

Table 2. Implications of sensitivity experiments under various model parametrisations on the tracer inventories for the modern ocean after 20,000 integrations.

Tracer	Units	POC profile	No explicit vertical diffusion	Rain ratio of 0.24	Control run
ΣCO_2	$\mu mol\,kg^{-1}$	2250	2259	2259	2262
Alkalinity	$\mu equiv\,kg^{-1}$	2367	2366	2355	2373
Phosphate	$\mu mol\,kg^{-1}$	2.03	2.09	2.09	2.09
Oxygen	$\mu mol\,kg^{-1}$	191	199	183	183
$\delta^{13}C$	‰	0.53	0.41	0.40	0.38
CO_3^{2-} ion	$\mu mol\,kg^{-1}$	90.9	86.7	84.8	88.6
Salinity	Psu	34.57	34.57	34.57	34.57
Temperature	°C	4.16	4.16	4.16	4.16

Table 3. Global mean values of the modern ocean for the sensitivity experiments above.

Second, seawater $\delta^{13}C$ can also be altered via exchange with the atmospheric CO_2, without any associated changes in nutrient concentrations. Isotope equilibrium of surface ocean waters with atmospheric CO_2 will cause higher $\delta^{13}C$ in cold surface waters, and lower $\delta^{13}C$ in warm surface waters (Broecker and Maier-Reimer 1992). The difference between the equilibration of CO_2 and the isotopes is that the isotopes require more time to reach equilibrium (about 10 years compared to about 1 year, depending on the depth of the mixed layer and the gas-exchange rate). Because surface waters move about and are replaced on faster timescales than this, there is no region of the ocean where surface water carbon is in complete isotopic equilibrium with the atmosphere (Broecker and Maier-Reimer 1992). To extract the effect of the gas-exchange signature from biological induced alterations, the artificial tracer ($\delta^{13}C_{as}$) could be constructed according to Broecker and Maier-Reimer (1992). If there were no air-sea exchange, the relationship between $\delta^{13}C$ and PO_4^{3-} in the ocean would be

$$\delta^{13}C - \delta^{13}C_{mean} = \frac{\Delta_{photo}}{\Sigma CO_{2\,mean}} \cdot \frac{C}{P} \cdot (PO_4 - PO_{4\,mean}).$$

$$(1)$$

The suffix mean denotes average ocean concentrations and Δ_{photo} and C/P describe the mean $\delta^{13}C$ in particulate organic matter and the Redfield ratio between organic carbon and phosphorus, respectively. When reasonable values are substituted ($\delta^{13}C_{mean} = 0.3$ ‰, $\Delta_{photo} = -20$ ‰, $\Sigma CO_{2\,mean} = 2200$ $\mu mol\,kg^{-1}$, C/P = 128 and $PO_{4\,mean}^{3-} = 2.2$ $\mu mol\,kg^{-1}$), the predicted relationship between $\delta^{13}C$ and

PO_4^{3-} closely matches the relationship for waters of the deep Indian and Pacific Oceans (^{13}C = 2.7 - 1.1 x PO_4^{3-}). This is to be expected, as the effect of air-sea exchange should be constant for theses deep waters on account of the homogeneity of source waters of the deep Indian and Pacific Oceans. The degree to which air-sea exchange processes have affected the surface ocean δ¹³C can determined by subtracting the δ¹³C value predicted from biological cycling exclusively from the actual δ¹³C,

$$\delta^{13}Cas = \delta^{13}C - (2.7 - 1.1 \cdot PO_4). \qquad (2)$$

By definition, water with $\delta^{13}C_{as}$ of 0.0 ‰ has the same air-sea exchange signature as the mean ocean deep water. A positive value of $\delta^{13}C_{as}$ means that the water reveals more of an influence of air-sea exchange at cold temperatures and less at warmer temperatures, whereas a negative $\delta^{13}C_{as}$ value implies less δ¹³C enrichment due to air-sea exchange than for average deep ocean water. However, this formulation will only represent the true effects of air-sea exchange if the assumptions about constant C/P ratios and constant δ¹³C of organic material and reasonable biologically induced changes in oceanic carbon content are sufficiently accurate. For example, Antarctic surface water, where δ¹³C of organic matter is substantially lower than in the rest of the ocean, $\delta^{13}C_{as}$ will lead to an overestimation of the effects of the air-sea exchange. For the glacial ocean, we used the same formulation according to Winguth et al. (1999), but assume the same Redfield slope as for the modern ocean,

$$\delta^{13}Cas = \delta^{13}C - (2.5 - 1.1 \cdot PO_4). \qquad (3)$$

The effect of the ocean circulation on the δ¹³C distribution, appart from the advection of tracers, depends on the residence time of a water mass. The longer a water mass resides in the deep-sea, the more organic material degrades within it, consuming oxygen and depleting ^{13}C. Circulation changes were accompanied with changes in the temperature and salinity which influences the solubility and dissociation constants of CO_2.

A further effect on the marine δ¹³C came from the biosphere where the organic δ¹³C is lower (about -25 ‰) than the marine δ¹³C (about -20 ‰). Changes in the vegetation, soil composition and transport of carbon via continental fluxes to the ocean could influenced the concentration of δ¹³C. Estimations of this effect on the LGM δ¹³C ranged from 0.3 to 0.4 ‰ (Curry et al. 1988; Duplessy et al. 1988).

Carbon Cycle Modelling Results

Modern Atlantic (Control Run)

The reference run (control run) of the carbon cycle model yielded a global primary production rate of 8.62 GtC year^{-1} and a new production rate (defined as the amount of newly formed organic matter which is transferred from the surface layer into deeper layers each year and is thereby removed from the seasonal ocean-atmospheric exchange cycle) of 6.7 GtC year^{-1} (see Table 2). The magnitudes of these values were within the range of results of other global biogeochemical models (e.g. Yamanaka and Tajika 1997; Marchal et al. 1998; Murnane et al. 1999). Estimates from observed data of primary and new production were difficult to compare directly with model outputs, because they included different methods for determining the various kinds of production. Estimations of the global new production ranged between 5.0 and 22.0 GtC year^{-1} (Murnane et al. 1999). The preindustrial atmospheric pCO_2 became stationary at 281 ppm and was in good agreement with estimates from ice cores . The global mean δ¹³C value amounted to 0.38 ‰ (see Table 3), and was slightly lower than the value of 0.50 ‰ reported by Broecker and Maier-Reimer (1992).

The simulated present-day δ¹³C distributions along the Geochemical Ocean Section Study (GEOSECS) sections in the Western and Eastern Atlantic are shown in Figure 2. The GEOSECS data set has been corrected according to Kroopnick (1985). For the Eastern Atlantic transect, we extended the GEOSECS data set to the south using the data set of Mackensen et al. (1996). The GEOSECS transect in the Western Atlantic remained unchanged. The main features of the dif-

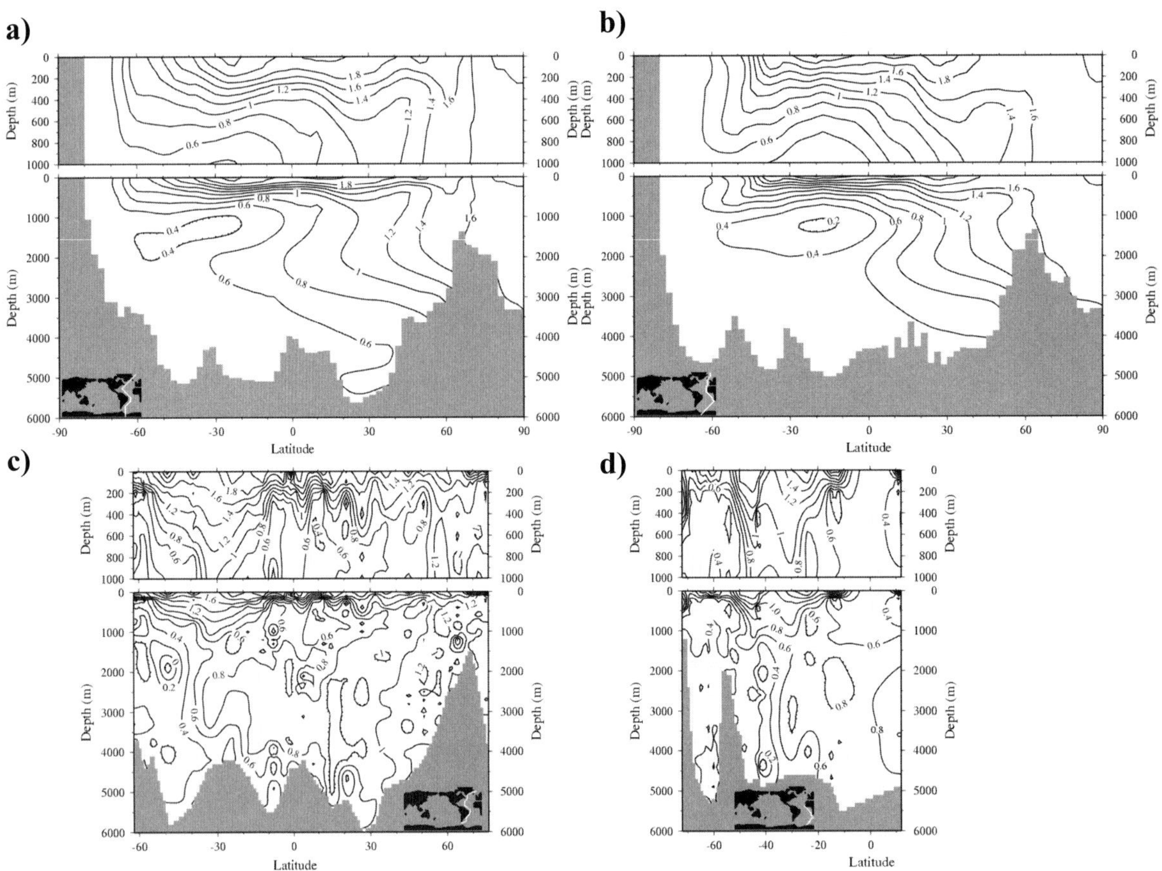

Fig. 2. Meridional distribution of $\delta^{13}C$ in the modern Atlantic Ocean. Contour interval is 0.2 ‰, see small panel for location of the section. **a)** and **b)** Simulated distribution in the Western and Eastern Atlantic, respectively. **c)** Observed $\delta^{13}C$ distribution in the Western Atlantic (GEOSECS), corrected according to Kroopnick (1985). **d)** GEOSECS distribution of $\delta^{13}C$ in the Eastern Atlantic. The GEOSECS transect is extended to the south according to Mackensen et al. (1996).

ferent water masses are adequately reproduced by the model. NADW and AABW can be clearly distinguished, with high and low $\delta^{13}C$ values, respectively. A shortcoming in the simulated $\delta^{13}C$ distribution was the shallower level of the NADW water mass. The core of this water mass was about at 2,500 m depth in the Western Atlantic, whereas in GEOSECS it was about 500 m deeper, but as mentioned above, this represents one model layer at this depth level. This discrepancy was primarily a result of the lower than observed meridional overturning. The simulated minimum of $\delta^{13}C$ in the Antarctic Intermediate Water (AAIW) compares well with data, but the values in the AABW seemed to be up to 0.2 ‰ higher, especially in the Western Atlantic.

In Table 4 we calculated the root mean square error between the control run and GEOSECS data for the entire Atlantic. Table 4 reveals a greater discrepancy of the simulated $\delta^{13}C$ distribution in the modern northern Atlantic compared to the south. This was a further implication for the underestimation of the simulated NADW production.

The contribution of the gas-exchange signature ($\delta^{13}C_{as}$) in the eastern Atlantic is shown in Figure 3. At the ocean surface, the contribution of the $\delta^{13}C_{as}$ signature is mainly due to the temperature dependence of the gas-exchange formulation (see Heinze et al. 1991). Cold regions like the Southern Ocean show higher $\delta^{13}C_{as}$ values than the warm waters of the equator and subtropical gyres, with mostly negative values relative to the deep-sea.

	Control run	Experiment 1	Experiment 2	Experiment 3
75 °N - 60 °N	0.53	-	-	-
60 °N - 45 °N	0.43	0.28	0.33	0.27
45 °N - 30 °N	0.36	0.19	0.51	0.19
30 °N - 15 °N	0.52	0.25	0.35	0.25
15 °N - 0 °	0.42	0.20	0.61	0.20
0 ° - 15 °S	0.40	0.12	0.63	0.12
15 °S - 30 °S	0.29	0.28	0.82	0.28
30 °S - 45 °S	0.26	0.26	0.67	0.26
45 °S - 60 °S	0.33	-	-	-
60 °S - 71 °S	0.27	-	-	-
Entire Atlantic	0.38	0.23	0.57	0.23

Table 4. Root mean square error between simulated δ¹³C values for the control run, glacial experiment 1, 2, 3 and observed data in the entire Atlantic Ocean. Modern data: From GEOSECS, corrected according Kroopnick (1985) and Mackensen et al. (1996). LGM data: From Sarnthein et al. (1994) north of the equator and south of the equator, see Bickert and Mackensen this volume.

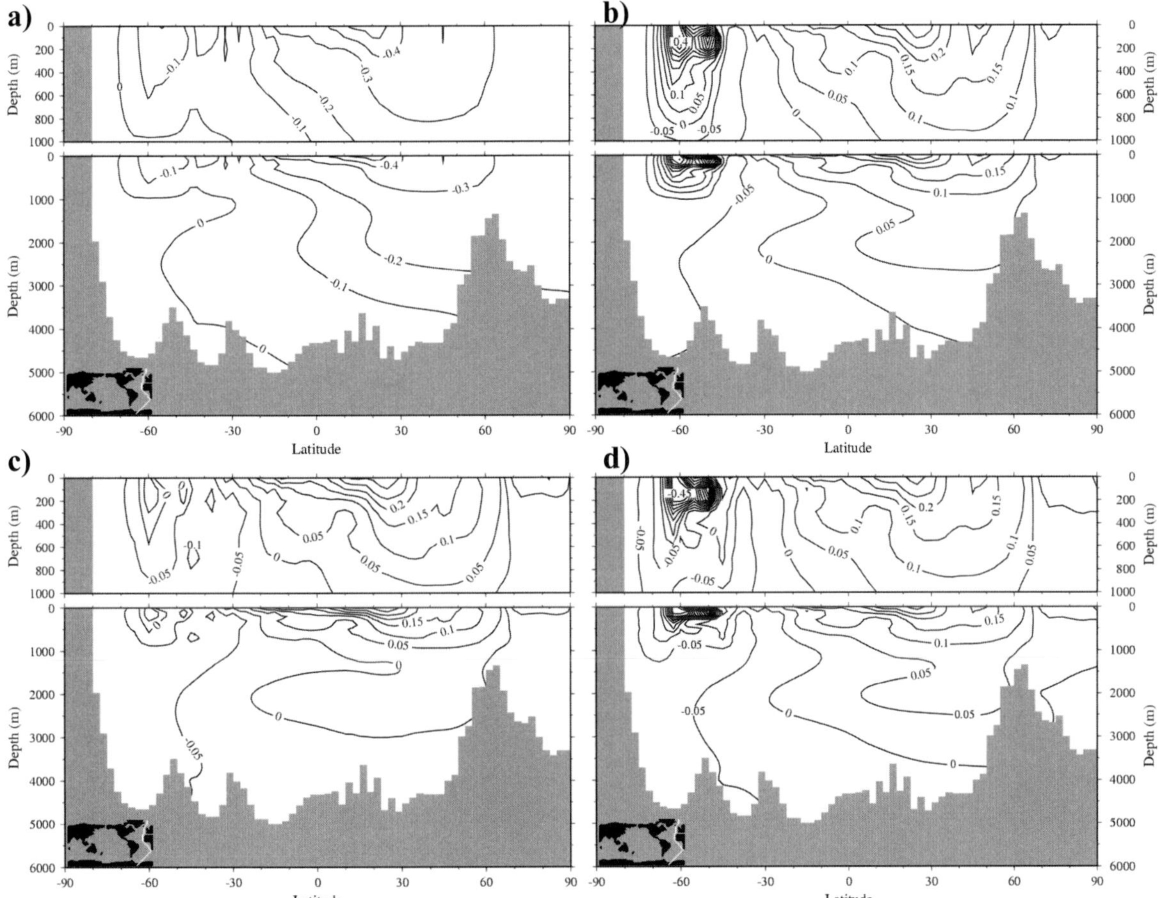

Fig. 3. Meridional section of the gas exchange signature (δ¹³Cas) in the Eastern Atlantic, see small panel for location of the section. **a)** Latitude-depth section of δ¹³Cas for the control run. Contour interval is 0.1 ‰. **b)** Anomaly between glacial experiment 1 and control run. **c)** Anomaly between glacial experiment 2 and control run. **d)** Anomaly between glacial experiment 3 and control run. Contour interval for the anomalies are 0.05 ‰.

Thus the Southern Ocean surface water is enriched in $\delta^{13}C$ from the gas-exchange with the atmosphere at cold temperatures, whereas the warm water regions are reduced in $\delta^{13}C$. However, in the surface water of the Southern Ocean, upwelling of deep water ($\delta^{13}Cas = 0.0$ ‰) influences the gas-exchange signature and the observed low $\delta^{13}C$ in organic matter would tend to overestimate the contribution of the gas-exchange signature in this area. The high $\delta^{13}Cas$ values of about 0.5 ‰ in the surface water of the Weddell Sea - according to Lynch-Stieglitz et al. (1995) - were not reproduced by the model. One reason could be the use of a constant gas-exchange rate, but Lynch-Stieglitz et al. (1995) have shown that its influence was low. It is most likely that the simulated surface $\delta^{13}C$ values were underestimated, whereas the PO_4^{3-} concentration remained in the range of the observations. The deep water column in the north reflected the transport and mixing of NADW carried the gas-exchange signature to the south. The NADW should have a $\delta^{13}C_{as}$ of about -0.4 ‰ according to Broecker and Maier-Reimer (1992), whereas our simulation reached a value of about -0.2 ‰.

Glacial Atlantic (Experiment 1)

To investigate the response of the $\delta^{13}C$ distribution in the Atlantic Ocean to climate change, we performed three model experiments for the glacial Atlantic. Experiment 1 used a modified temperature, salinity, circulation field and more extended ice distribution as driven by glacial ocean circulation. Experiment 2 introduced, in addition, an initial nutrient inventory which was increased by 30 %. To reduce the pCO_2 in the atmosphere without significant changes in the $\delta^{13}C$ values we setup an model run with a chemical glacial ocean according to Sanyal et al. (1995). As a result, experiment 1 showed a reduced new production rate of 5.9 GtC year^{-1} compared to 6.7 in the control run, and the change in the global mean of $\delta^{13}C$ amounted to 0.15 ‰. The average atmospheric pCO_2 was lowered by 1 ppm and attains a value of 280 ppm. The differences in $\delta^{13}C$ in the Atlantic are shown in Figure 4 for the Western and in Figure 5 for the Eastern Atlantic. In the deep Atlantic, there was a re-

duction of 0.2 ‰ as compared to the modern state, whereas the upper ocean north of the equator showed an increase in $\delta^{13}C$. A possible cause of the simulated decrease of $\delta^{13}C$ in the deep ocean is aging of the deep water masses due to the reduced NADW formation and export to the Southern Ocean. The influence of the gas-exchange signature on the $\delta^{13}C$, due to the lower glacial temperatures, showed only minor changes in the deep water column compared to the control run (Fig. 3). Only the surface ocean is significantly affected by the air-sea exchange, especially in the Southern Ocean near the sea-ice edge. The effect of changes in the pattern of the remineralisation of organic matter on the decrease of the $\delta^{13}C$ should be minor, since the global new production is lower in this simulation.

In summary, the simulated distribution of $\delta^{13}C$ in the Atlantic Ocean showed a lowering in the $\delta^{13}C$ of 0.2 ‰ in the deep water column, which was only due to a changed thermohaline circulation field. However, the effect of this circulation on the pCO_2 in the atmosphere was too small to explain the concentrations measured in ice cores. This supports previous findings of Heinze et al. (1999) and Winguth et al. (1999). Therefore, we setup a model run with an intensified biological pump.

Glacial Atlantic (Experiment 2)

In this experiment, we increased the initial nutrient inventory of the glacial ocean by 30 %. Together with the glacial circulation field, this yields a new production rate of 7.5 GtC year^{-1}. As a result, the atmospheric pCO_2 was reduced to 221 ppm. This approaches the observed pCO_2 value of approximated 200 ppm for the LGM (e.g. Barnola et al. 1987; Petit et al. 1999). The global mean $\delta^{13}C$ value is reduced to -0.47 ‰. The enhancement of the new production is accompanied by an additional reduction of the $\delta^{13}C$ in the whole water column of the Atlantic Ocean due to the remineralisation of organic matter (Fig. 6 and 7). Under these conditions, large parts of the deep water column reveal negative $\delta^{13}C$ values. The gas-exchange signature shows no significant changes compared to experiment 1, except in the upper ocean of the Southern Ocean, but this has no influences on the

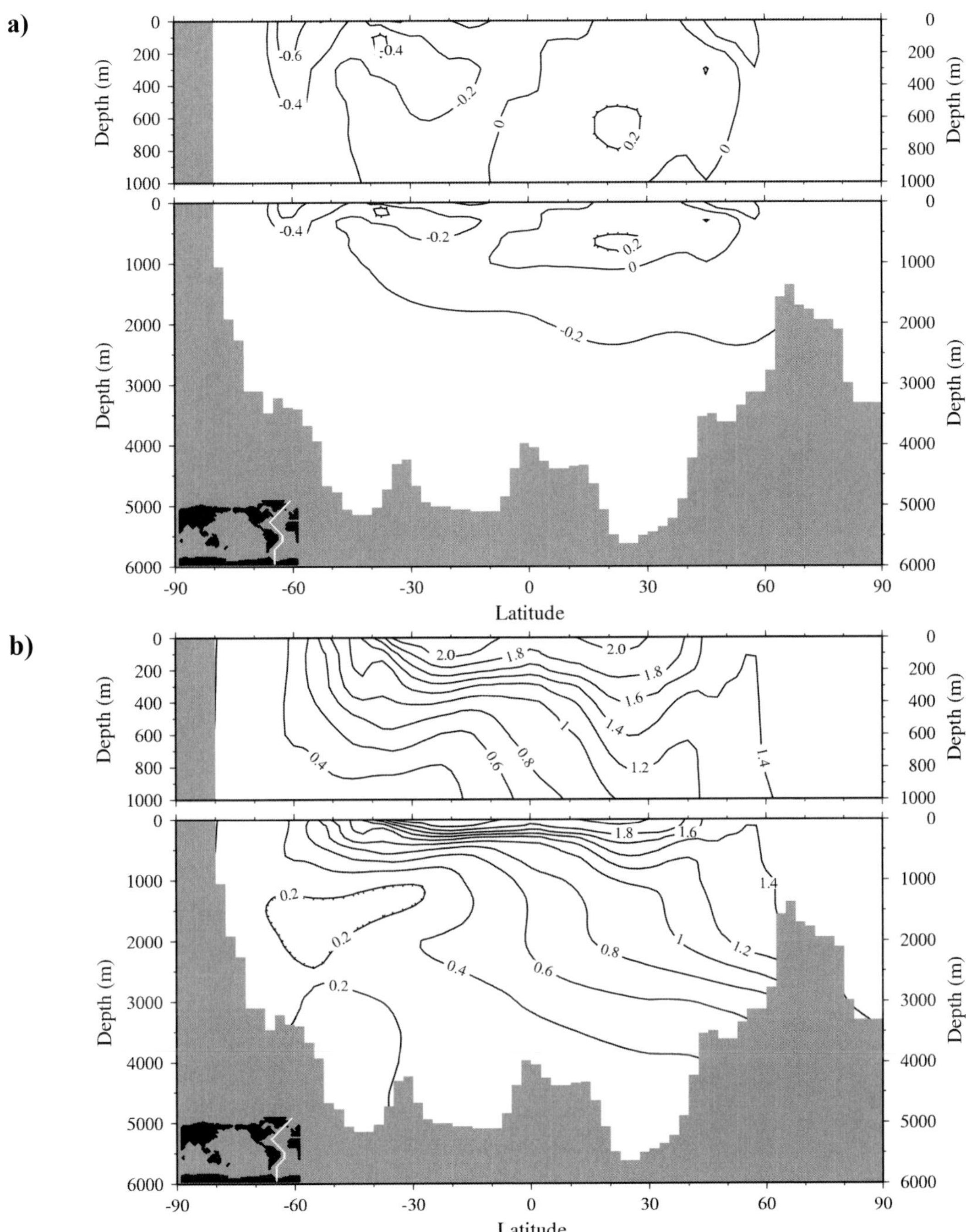

Fig. 4. Meridional distribution of $\delta^{13}C$ in the Western Atlantic for glacial experiment 1. Contour interval is 0.2 ‰.
a) Anomaly between experiment 1 and control run. **b)** Simulated $\delta^{13}C$ distribution.

Matthies et al.

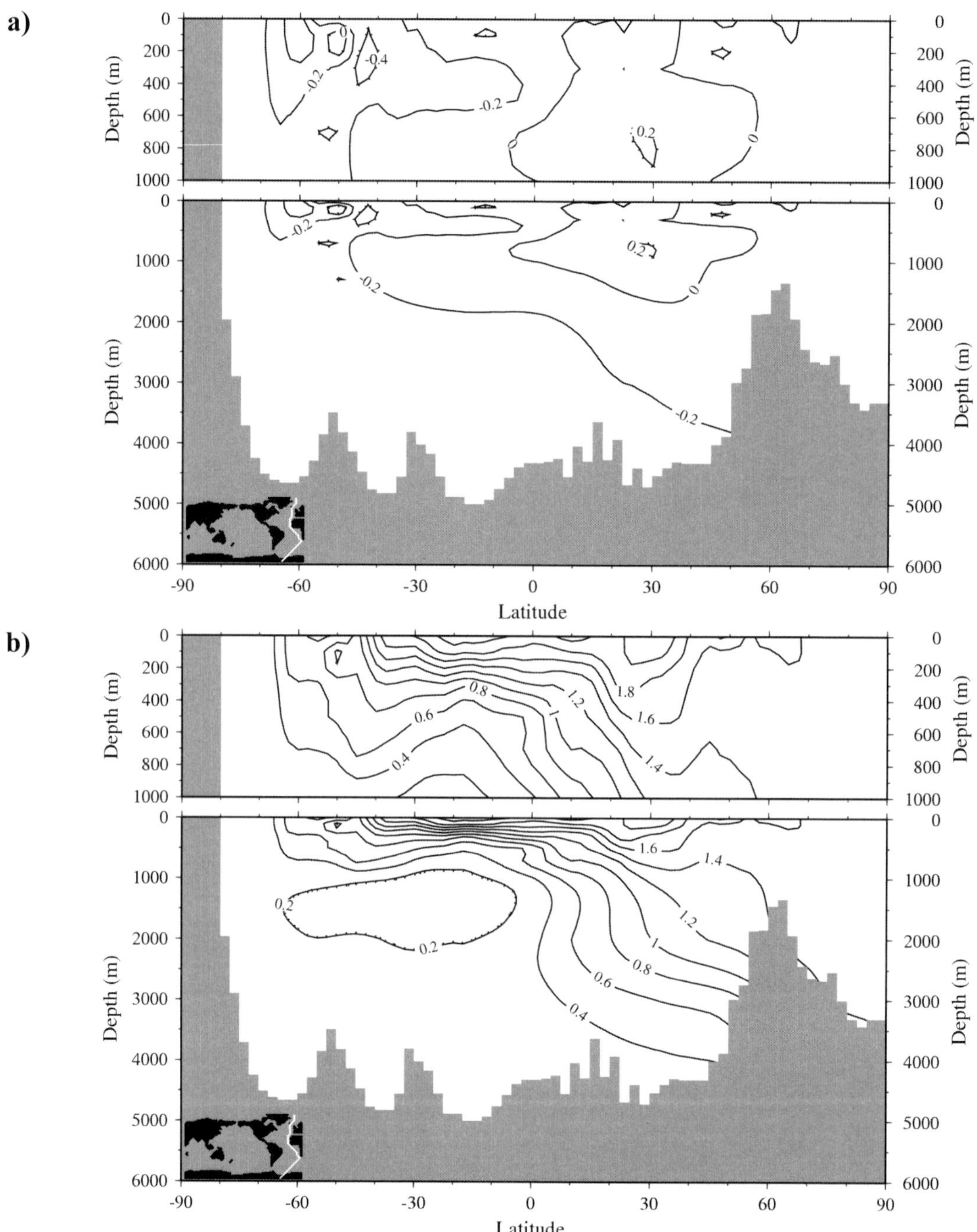

Fig. 5. Meridional distribution of $\delta^{13}C$ in the Eastern Atlantic for glacial experiment 1. Contour interval is 0.2 ‰.
a) Anomaly between experiment 1 and control run. **b)** Simulated $\delta^{13}C$ distribution.

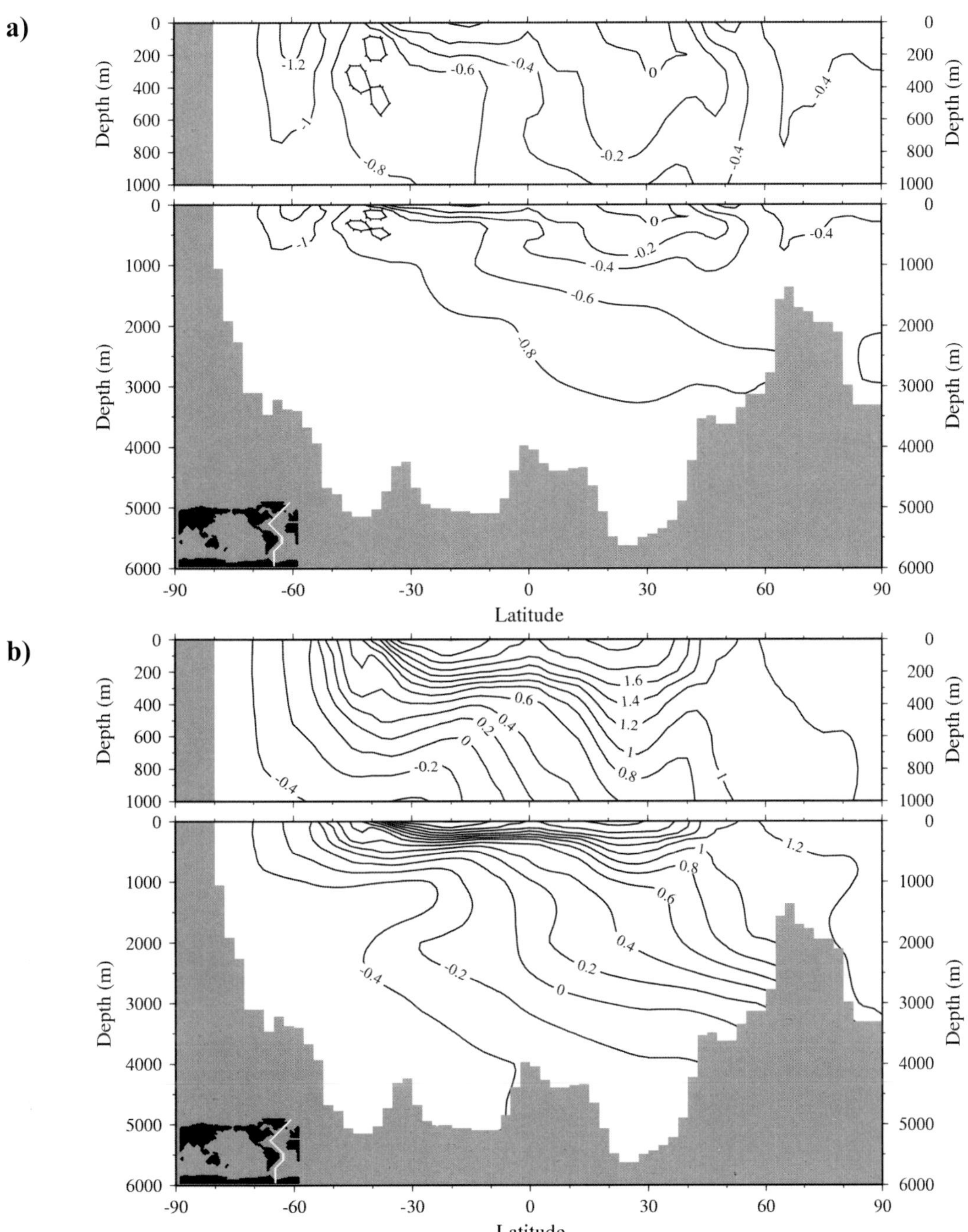

Fig. 6. Meridional distribution of δ¹³C in the Western Atlantic for glacial experiment 2. Contour interval is 0.2 ‰.
a) Anomaly between experiment 2 and control run. **b)** Simulated δ¹³C distribution.

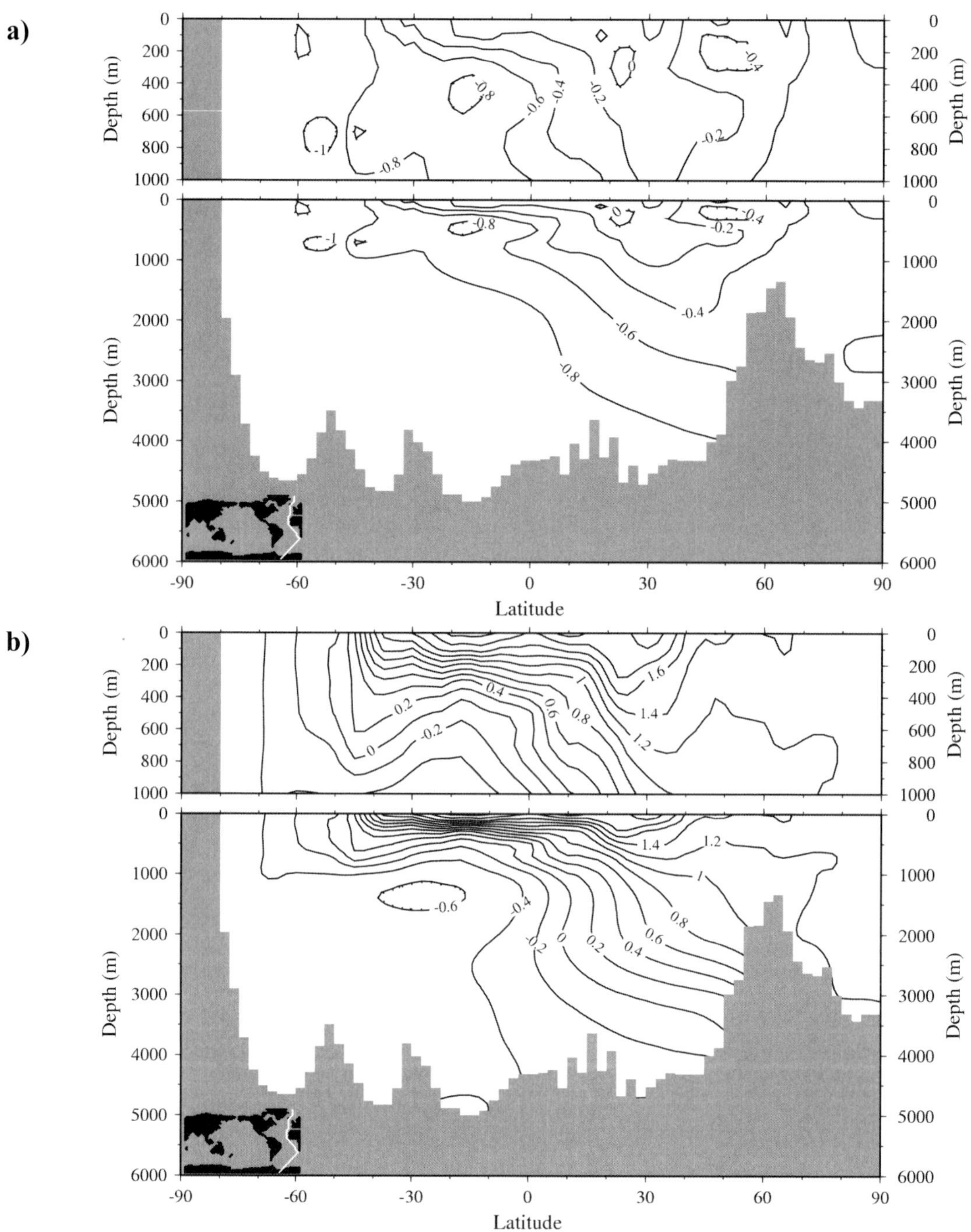

Fig. 7. Meridional distribution of $\delta^{13}C$ in the Eastern Atlantic for glacial experiment 2. Contour interval is 0.2 ‰.
a) Anomaly between experiment 2 and control run. **b)** Simulated $\delta^{13}C$ distribution.

deep water column in this region (compare Fig. 3b with 3c).

Glacial Atlantic (Experiment 3)

Experiment 3 was based on the observations of Sanyal et al. (1995) that the pH values of the glacial deep waters in the Atlantic and Pacific are 0.3 ±0.1 higher as compared to the Holocene period. If the observed higher glacial pH was the result of excess calcite dissolution, then 0.157 x 10^{-3} moles of $CaCO_3$ must have been added to each kilogram of sea water. This corresponds to 0.161 x 10^{-3} mol l^{-1}, if a mean density of 1.025 kg l^{-1} is assumed. We translated these conditions to an increase of the initial model concentrations of 161 µmol l^{-1} ΣCO_2 and 2 x 161 µequiv l^{-1} alkalinity. To mirror an increase of C_{org} in the sediment we halved the maximal rain ratio to 0.075.

It must be considered that this experiment does not take into account the complex interactions between the sediment and the overlying water. In this experiment we introduced the effects of these interactions according to the assumptions of Sanyal et al. (1995), because the model has only a very simple sediment module with a constant exchange rate between the bottom water and sediment layer. The new production is the same as in experiment 1, only the $CaCO_3$ production is lower because of the reduced rain ratio (from 0.15 to 0.075). The implication on the $\delta^{13}C$ distribution is minimal (Fig. 8 and 9) and the pCO_2 decrease to 250 ppm due to the weakening of the carbonate pump. The proposed rise in the pH value of 0.3 in the glacial deep Atlantic according to Sanyal et al. (1995) is not predicted by the model. A maximum elevation of 0.2 in the pH value is found in the northern Indic Ocean at a water depth of 1,500 m, however, in the Atlantic we only reached a maximal increase of 0.1 at about 1,000 m. The reduction of the atmospheric pCO_2 is purchased at the expense of a deepening of the lysocline in the glacial Atlantic, which contradicts observations that show an upward movement of the lysocline. The simulated gas-exchange signature shows roughly the same anomaly patterns as in the other experiments, with an exception in the Southern Ocean, where $\delta^{13}C_{as}$

Comparison of Model Results with Paleoclimate Data

In Figures 10 through 13 we compare the distributions of $\delta^{13}C$ in glacial simulations and in the paleoclimate records of the Western and Eastern Atlantic. A new data set of benthic stable isotopes from sediment cores in the South Atlantic (Bickert and Mackensen this volume) combined with the data of Sarnthein et al. (1994) for the eastern North Atlantic, allows a more detailed reconstruction of the glacial nutrient distribution in this region most sensitive to circulation changes. For the South Atlantic, only values of sites outside of upwelling areas have been chosen for the reconstruction of $\delta^{13}C$ distributions in the past. Several studies from different locations in the South Atlantic have shown that an additional depletion in the $\delta^{13}C$ of epibenthic foraminifer calcite occurs in areas characterized by high surface water productivity and hence high organic matter supply to the seafloor (e.g. Mackensen et al. 1993; Bickert and Wefer 1999). This depletion is most likely explained to be caused by the decay of organic matter, reducing $^{13}C/^{12}C$ ratio in the pore water, which influences to some degree the carbon isotopic composition of the *C. wuellerstorfi* shells of high-productive areas. Therefore, we removed the values of such locations prior to the model-paleodata comparison (see Bickert and Mackensen this volume, for a detailed discussion).

As a result, there is a reasonable fit of the model output of experiment 1 to the $\delta^{13}C$ distribution derived from sediment samples, when the glacial export of NADW to the Southern Ocean is reduced by 50 % and the inflow of glacial AABW is held constant (Fig. 10 and 11, Table 4). The obtained $\delta^{13}C$ pattern from the model experiment matches the paleodata within a range of ±0.2 ‰ for most of the data. Furthermore, the asymmetry between the glacial NADW distribution in the South Atlantic basins is reproduced by the coupled ocean circulation and carbon cycle model. However, some data points in the mid-depth Western Atlantic between 15° and 30° south would suggest the glacial

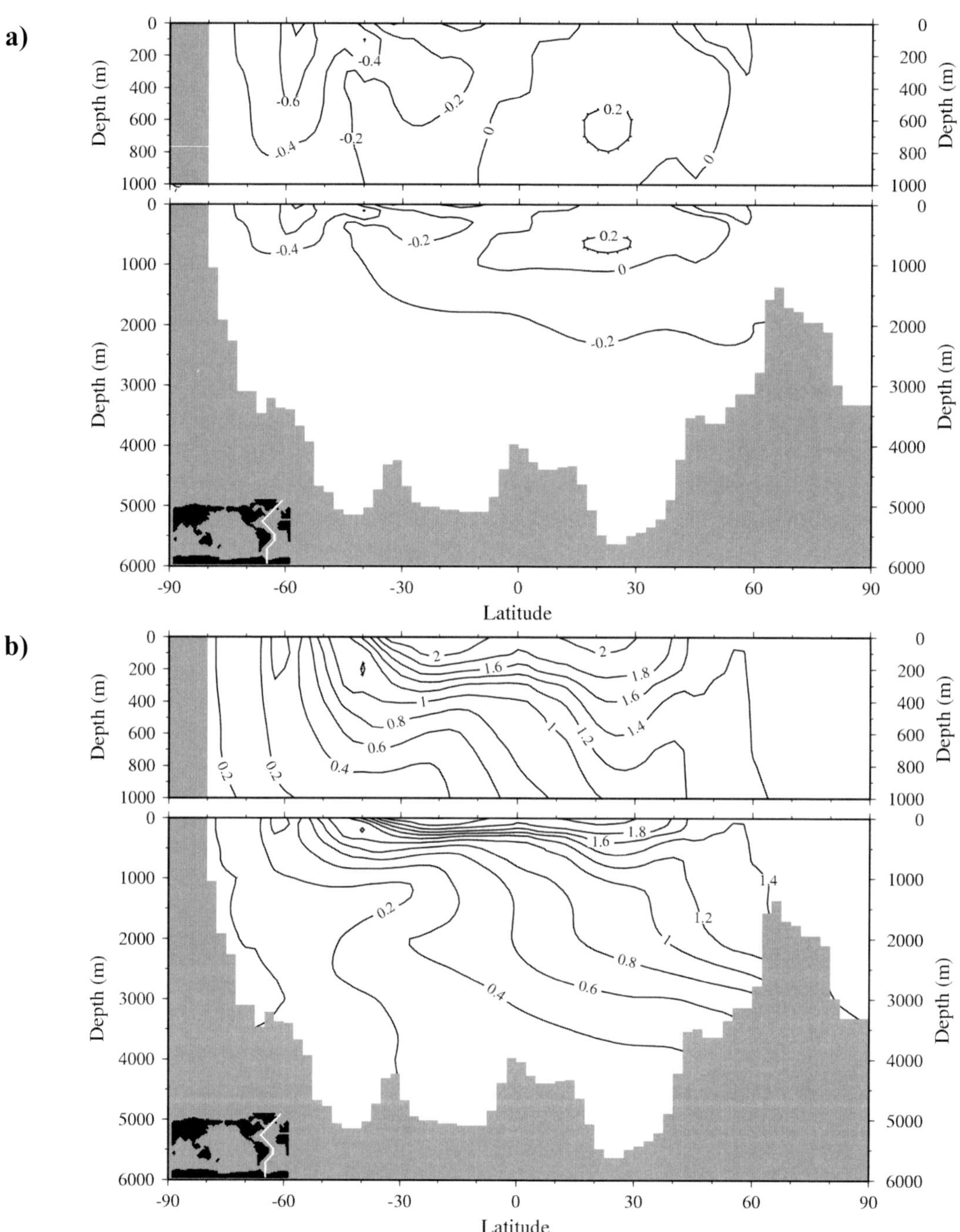

Fig. 8. Meridional distribution of $\delta^{13}C$ in the Western Atlantic for glacial experiment 3. Contour interval is 0.2 ‰. **a)** Anomaly between experiment 3 and control run. **b)** Simulated $\delta^{13}C$ distribution.

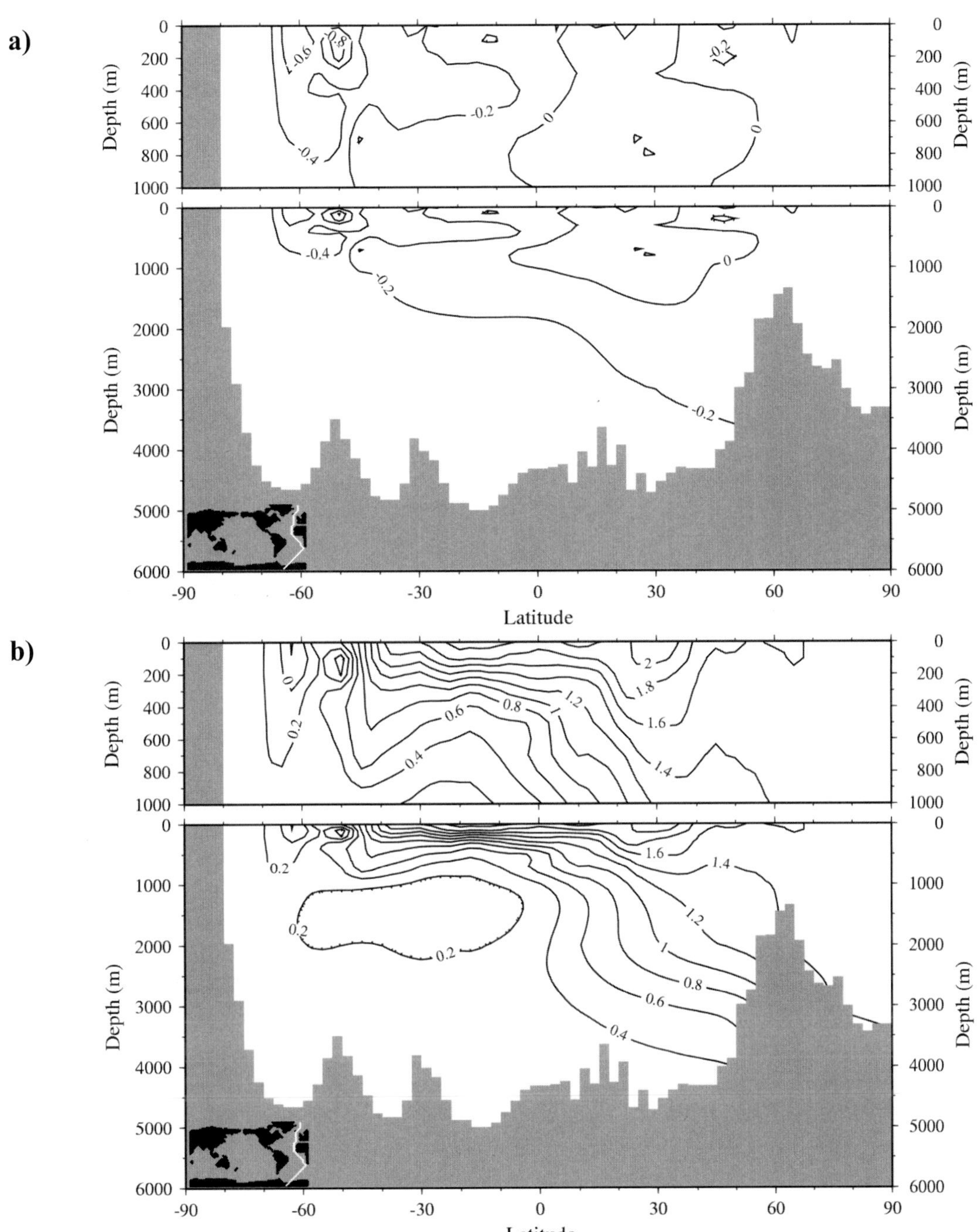

Fig. 9. Meridional distribution of $\delta^{13}C$ in the Eastern Atlantic for glacial experiment 3. Contour interval is 0.2 ‰. **a)** Anomaly between experiment 3 and control run. **b)** Simulated $\delta^{13}C$ distribution.

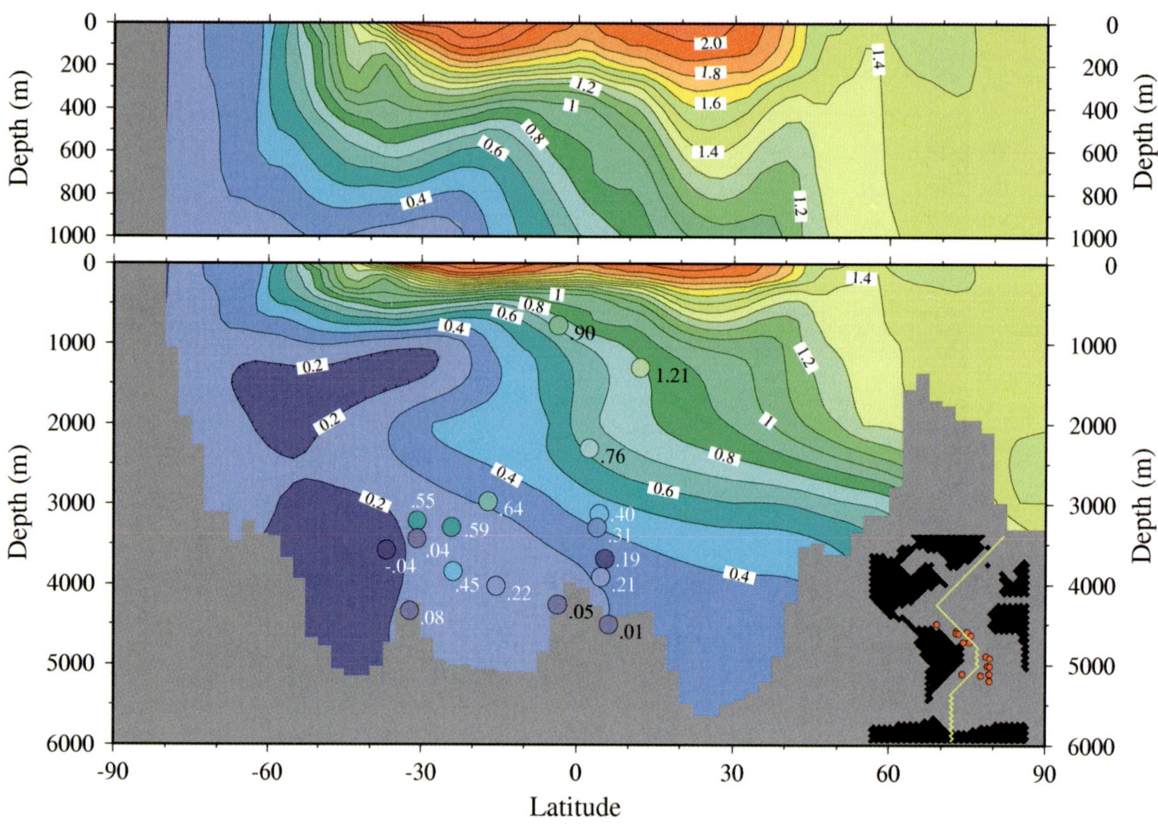

Fig. 10. Comparison between the glacial $\delta^{13}C$ distribution of experiment 1 and observations (data set of SFB 261) in the Western Atlantic. Contour interval is 0.1 ‰, see small panel for locations and data points. The colours of the data points (cycles) corresponds to the colour values in the simulation. If possible, the observed $\delta^{13}C$ values are written next to the corresponding data point.

NADW to extend deeper and further south than the model does. This might be related to the underestimation of the production of NADW as already discussed for the control run of the modern ocean.

The much larger differences between the model output of experiment 2 and the paleodata (Fig. 12 and 13) indicate that there is no additional increase of the nutrient inventory of the deep ocean necessary to reproduce the paleodata. From this observation we derive that no significant increase in biological pumping occurred during glacials. As a consequence, the drop-down of atmospheric CO_2 could not be explained by the mechanism of an intensified biologic pumping related to an enhanced export production. This is in accordance with recent observations that the export production in the glacial Southern Ocean shows no significant changes

(François et al. 1997; Nürnberg et al. 1997). However, our results would even not support the recently presented hypothesis by Stephens and Keeling (2000) that the decrease in atmospheric CO_2 was driven by a reduced outgassing of the upwelling deep water due to an extended sea-ice in the glacial Southern Ocean. Using a box model with deep water upwelling confined to south of 55°S, they suggested that low glacial atmospheric CO_2 levels might result from reduced deep water ventilation associated with either year-round Antarctic sea-ice coverage, or wintertime coverage combined with ice-induced stratification during the summer. As a result, their box model reproduces 67 ppm of the observed glacial-interglacial CO_2 difference and is generally consistent with the additional observational constraints.

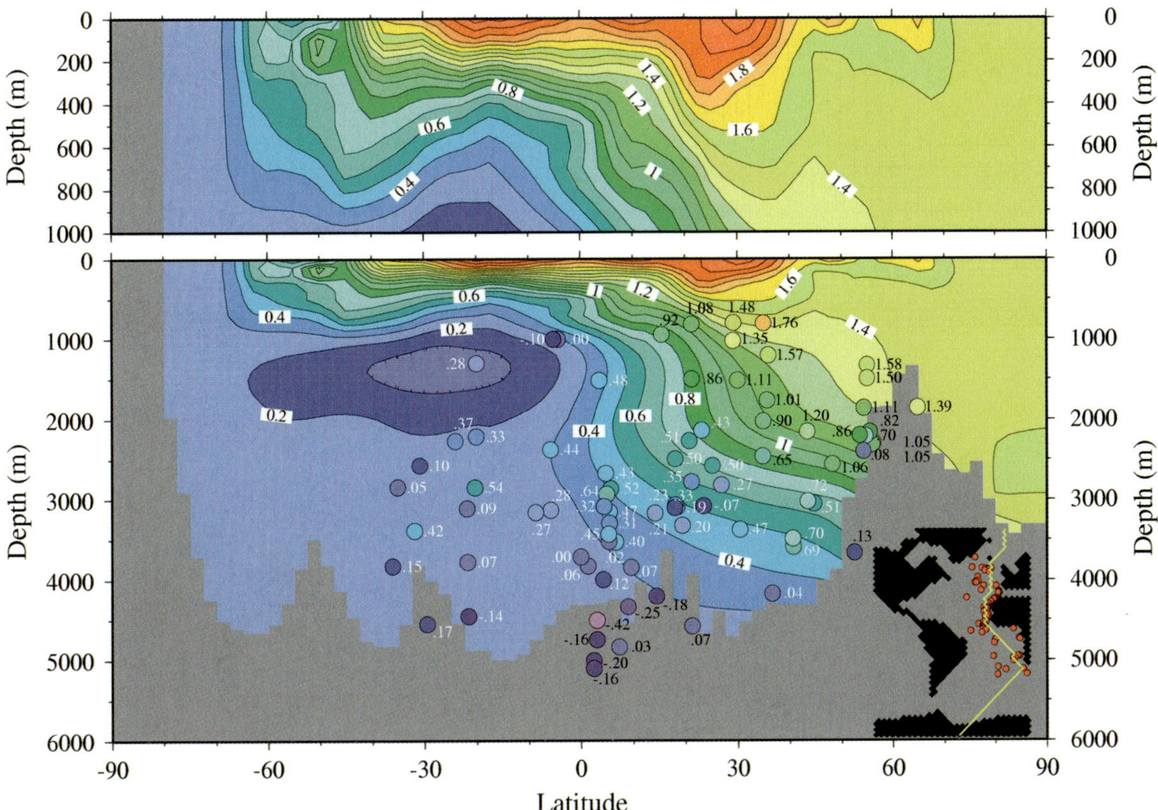

Fig. 11. Comparison between the glacial δ¹³C distribution of experiment 1 and observations (data set of SFB 261 and Sarnthein et al. 1994) in the Eastern Atlantic. Contour interval is 0.1 ‰, see small panel for locations and data points. The colours of the data points (cycles) corresponds to the colour values in the simulation. If possible, the observed δ¹³C values are written next to the corresponding data point.

Figure 14 shows the pCO₂ difference between ocean and atmosphere for our control run and for the glacial experiment 1 as well as the anomaly between both simulations. Negative values indicate that the ocean is a sink for atmospheric CO₂ and positive values indicate that the ocean is a CO₂ source. In both simulations the Southern Ocean represents a CO₂ sink.

In Table 5 we used the same area-weighted zones for the average ΔpCO₂ according to Takahashi et al. (1997). For the control run the Southern Ocean represented a strong CO₂ sink and the equatorial region a strong source. The simulated source and sink ΔpCO₂ pattern is comparable to Takahashi et al. (1997). Quantitative differences between simulations and observations are in the high latitudes of both hemispheres and primarily

due to the fact that the observations are normalized to the year 1990 and the data coverage in the Southern Ocean is scarce. On the other side, the control run represented the preindustrial ocean. The simulated strong CO₂ sink in the Southern Ocean is supported by the observations of Schlitzer (2002) who predicted a significant stronger export production in this region as satellite observations indicated. This should result in a stronger CO₂ input to the surface ocean. In the Atlantic sector of the Southern Ocean this sink seems to be stronger in the glacial run compared to the control run. The minimum value in this region amounts to -80 ppm, whereas in the control run it amounts to -60 ppm. However, the area of the pCO₂ sink is reduced compared to the modern conditions due to the extended sea-ice coverage. This is also taken from

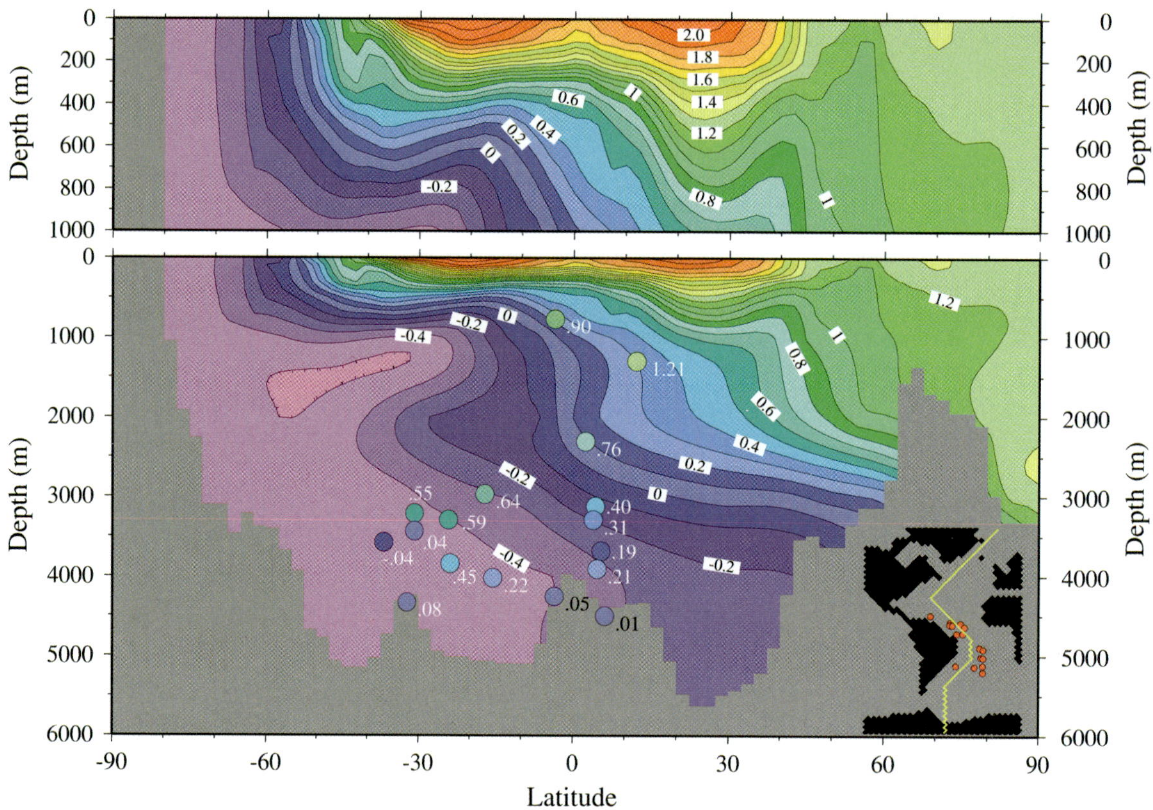

Fig. 12. Comparison between the glacial $\delta^{13}C$ distribution of experiment 2 and observations (data set of SFB 261) in the Western Atlantic. Contour interval is 0.1 ‰, see small panel for locations and data points. The colours of the data points (cycles) corresponds to the colour values in the simulation. If possible, the observed $\delta^{13}C$ values are written next to the corresponding data point.

Table 5 where the average ΔpCO_2 values show a reduction in the CO_2 sink between 50°S and 90°S due to the sea-ice expansion compared to the control run. An intensification is predicted in the zone between 14°S and 50°S. Together with the biological new production which is only slightly stronger in the vicinity of the sea-ice edge in this region, the contribution to lowering the global mean atmospheric pCO_2 appears to be small. Based on our model simulations the extended sea-ice cover in the LGM could not have been responsible for the strong reduction of the global atmospheric CO_2.

The reason for the obvious discrepancy between the results of Keeling and Stephens (2001) and our model output might be that box models and general circulation models (GCM) show a strong different behaviour to changes in the high latitudes

(Archer et al. 2000b). As an example the GCM simulation of Archer et al. (2000a) under conditions of a more efficient nutrient utilization in the glacial Southern Ocean as a result of a higher iron supply shows no sufficient atmospheric CO_2 reduction, whereas box models (see references in Archer et al. 2000b) reaches the glacial CO_2 concentration. This different behaviour of the models to changes in high latitudes was discussed in Archer et al. (2000b) and they concluded that box models overestimate high latitude sensitivity of the real ocean. This uncertainty should be examined in more detail.

The dramatic changes in the circulation and associated exchange processes in the Southern Ocean as proposed by Toggweiler (1999) are always accompanied with changes in the tempera-

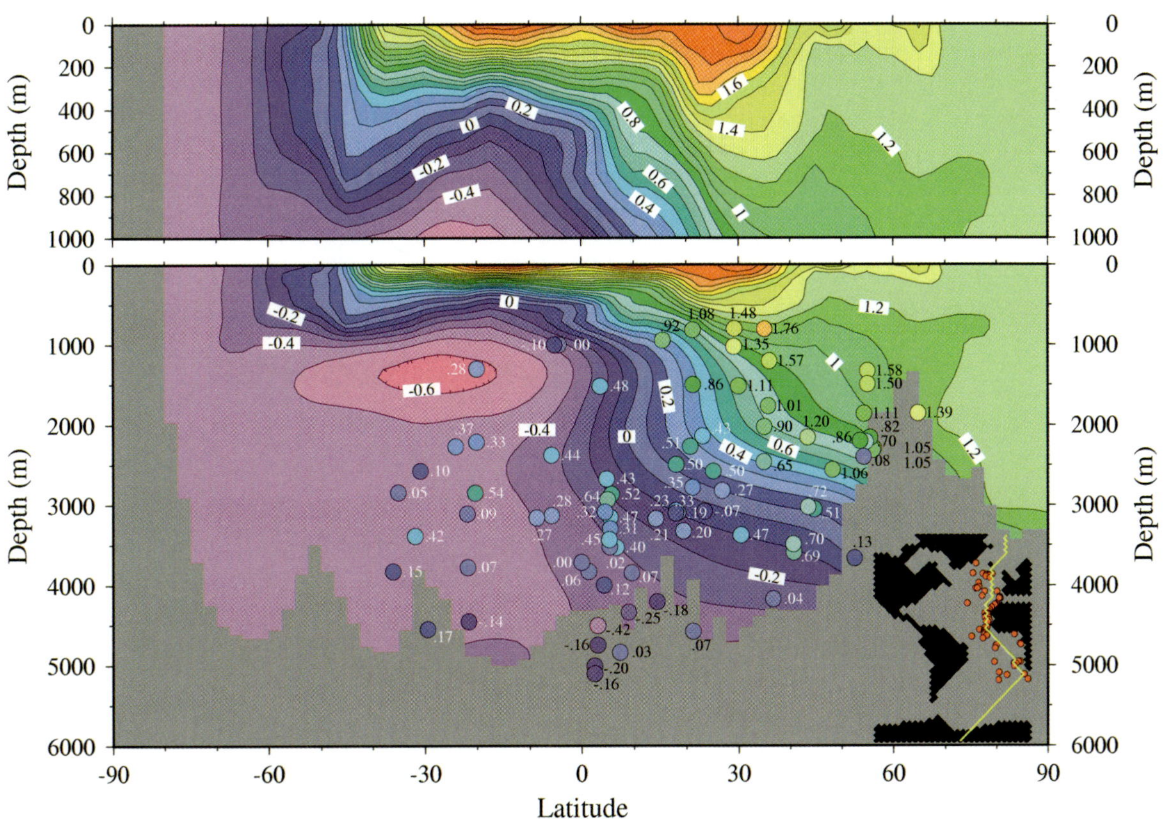

Fig. 13. Comparison between the glacial $\delta^{13}C$ distribution of experiment 2 and observations (data set of SFB 261 and Sarnthein et al. 1994) in the Eastern Atlantic. Contour interval is 0.1 ‰, see small panel for locations and data points. The colours of the data points (cycles) corresponds to the colour values in the simulation. If possible, the observed $\delta^{13}C$ values are written next to the corresponding data point.

Latitude zone	Control run	Experiment 1	Experiment 2	Experiment 3
90 °N - 50 °N	-23.86	-10.55	-8.62	-9.76
50 °N - 14 °N	-3.50	-6.67	-5.61	-6.15
14 °N - 14 °S	+27.64	+23.88	+20.61	+22.26
14 °S - 50 °S	-4.18	-10.62	-9.30	-9.91
50 °S - 90 °S	-32.60	-10.22	-7.55	-8.87
Global mean	-0.24	+0.06	+0.24	+0.16

Table 5. Global annual mean values of sea-air pCO_2-difference (ΔpCO_2, in ppm) for the control run and three glacial experiments for five latitude zones (area-weighted averages). The five latitude belts were selected according to Takahashi et al. (1997).

ture and salinity. Such rearrangements in the water column required, in particular, a change in the salinity that is unfortunately poorly documented for a few locations in the glacial Southern Ocean.

Ultimately, the influence of increased sea-ice in the Southern Ocean and/or a northward shift of glacial wind field on the reduction of the atmospheric CO_2 concentration, as was suggested by Elderfield and Rickaby (2000) and Sigman and Boyle (2000), is related to the different behaviour of the two model types. Winguth et al. (1999) have shown that an increasing glacial wind in the high

Matthies et al.

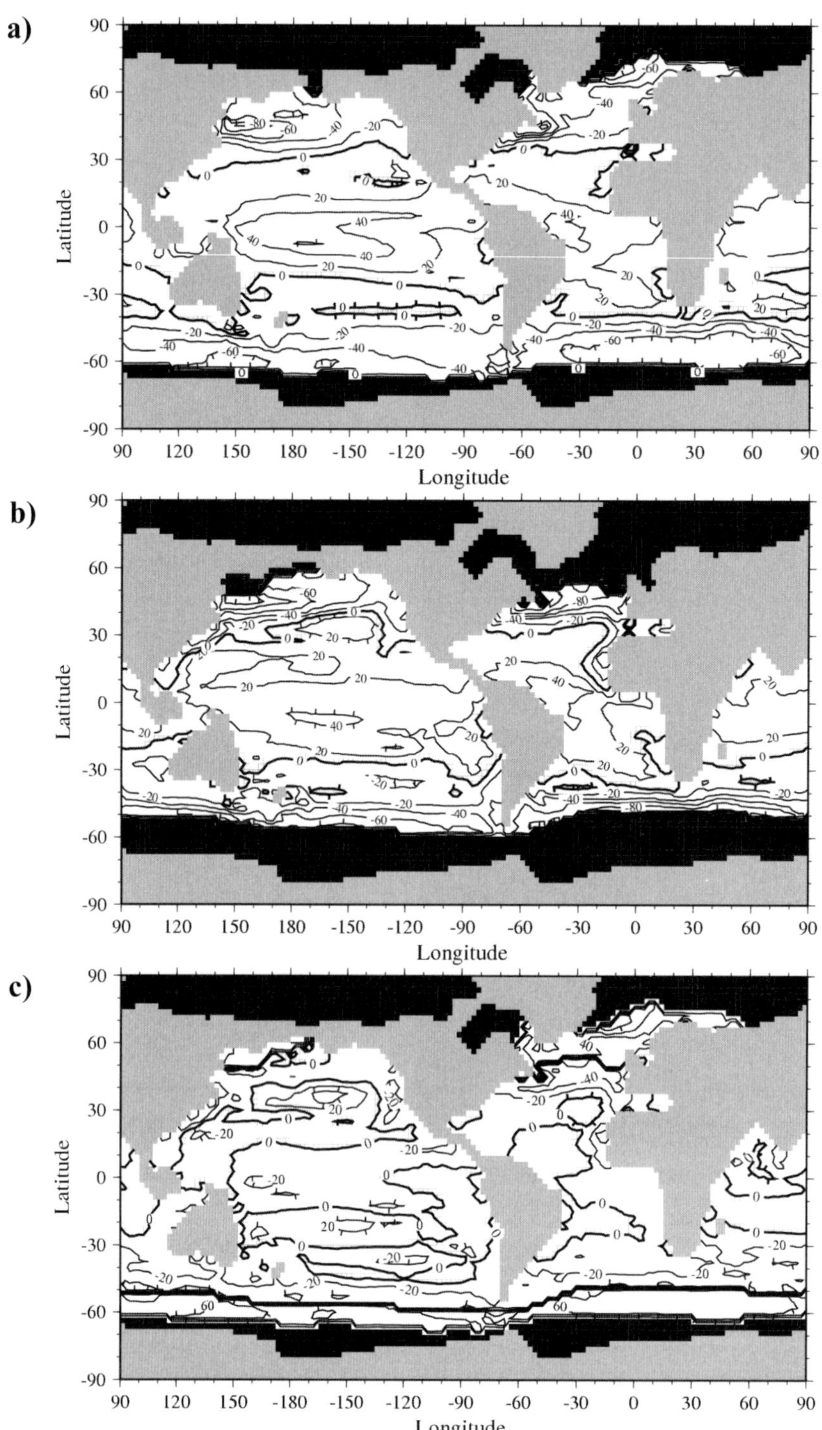

Fig. 14. Annual mean sea-air difference in the CO_2 partial pressure (ΔpCO_2). Contour interval is 20 ppm. Positive values indicate that the ocean is a source for atmospheric CO_2, negative values indicate CO_2 sinks. **a)** Horizontal distribution of ΔpCO_2 for the control run. **b)** Distribution of ΔpCO_2 for the glacial experiment 1. **c)** Difference between experiment 1 and control run. In the two upper panels, black filled areas denote the simulated annual mean sea-ice distribution.

southern latitudes has no significant effect on the $\delta^{13}C$ pattern and the atmospheric pCO_2.

In our opinion, the complex mechanism of carbonate compensation represents a common way of reducing the atmospheric pCO_2, without producing a strong reduction of the $\delta^{13}C$. However, Archer et al. (2000a) have shown in a model simulation which included a complex sediment module that this process works in the right direction, but violates other paleoevidences from sediments, e.g. changes in the depth of the lysocline and the $CaCO_3$ distribution on the seafloor.

In experiment 3, we mimiced the mechanism of the carbonate compensation according to Sanyal et al. (1995) and could reduce the atmospheric pCO_2 to 250 ppm with a $\delta^{13}C$ pattern like in experiment 1 (see Figs. 8, 9 and Table 4). However, the advantage of the pCO_2 reduction in the atmosphere and the reasonable reproduction of the $\delta^{13}C$ distribution is obtained at the expense of carbonate distributions which are not in agreement with observations. For example, the glacial lysocline in the Western Atlantic is deeper (500 to 1,000 m) as compared to the modern situation, whereas observation indicated a rise of the glacial lysocline (Henrich et al. this volume). The estimated pH reduction according to Sanyal et al. (1995) in the glacial Atlantic was not reproduced by the model.

Figure 15 shows another comparison between modelled results and observations in terms of standard deviation and correlation coefficient.

After interpolating modelled data to the observed data levels, the standard deviation and the correlation coefficient were calculated for both data sets (model experiments and observations). These data pairs were transferred as polar coordinates into a scatter diagram according to Taylor (Gates et al. 1998; Taylor 2001). The procedure of the conversion of the value pair in the Taylor diagram was done in three steps:
1. Normalization of the standard deviation of the model simulation and observation was defined as
$$\sigma_{norm} = \sigma^2_{model} \times \sigma^{-2}_{observation}.$$
2. The polar angle, α, was calculated as follows: $\alpha = \arccos r$. The parameter r denotes the correlation coefficient.
3. The determination of the values for the x- and y-axis (the polar coordinates) was calculated by using the equations $x = \sigma_{norm} x \cos \alpha$ and $y = \sigma_{norm} x \sin \alpha$.

The Taylor diagram showed the degree of correspondence between simulated and observed $\delta^{13}C$ fields in the Atlantic Ocean. By definition, all observations showed a normalized standard deviation of 1 and an angle of zero degrees (i.e. the correlation coefficient is equal to 1). This point was accepted as reference point which determines the locality of the various observed data sets used in this diagram and always have the same position. The distance of the simulated tracer point from the origin of the coordinate system is equal to the standard deviation (normalized by the observed standard deviation), whereas the distance from the reference point (observations) is equivalent to the rms (root mean square) pattern difference between the observed and modeled data fields (again normalized by the observed standard deviation). The cosine of the polar angle corresponds to the correlation between the simulated and observed data sets. Thus, a model that is relatively accurate would lie near the dashed arc (indicating it had the correct variance) and close to the observed reference point indicating a small rms error and a high correlation coefficient. The outer circular arc in Figure 15 describes the correlation coefficient. In Figure 15 a), we show the comparison between the model simulations (control run, glacial experiment 1 and 2. Please note that experiment 3 shows the same position as experiment 1 and remains unmentioned in the legend of Figure 15) and the observed data sets (modern: GEOSECS, glacial: Sarnthein et al. 1994). The glacial modeled $\delta^{13}C$ fields are more closely resembled the observation fields (in terms von standard deviation, correlation coefficient and rms error) than the modern modeled $\delta^{13}C$ field. The reason may probably be attributed to the previous finding of low NADW production. The glacial simulation of experiment 2 substantially presents a better agreement between the observed $\delta^{13}C$ pattern of Sarnthein et al. (1994) than the simulation of experiment 1 (and experiment 3).

As shown in Figure 15 b), we combined the observed $\delta^{13}C$ data set of GEOSECS with the data of Mackensen et al. (1996). The observed glacial data set include data of Sarnthein et al. (1994), relating to the north of the equator, otherwise, $\delta^{13}C$

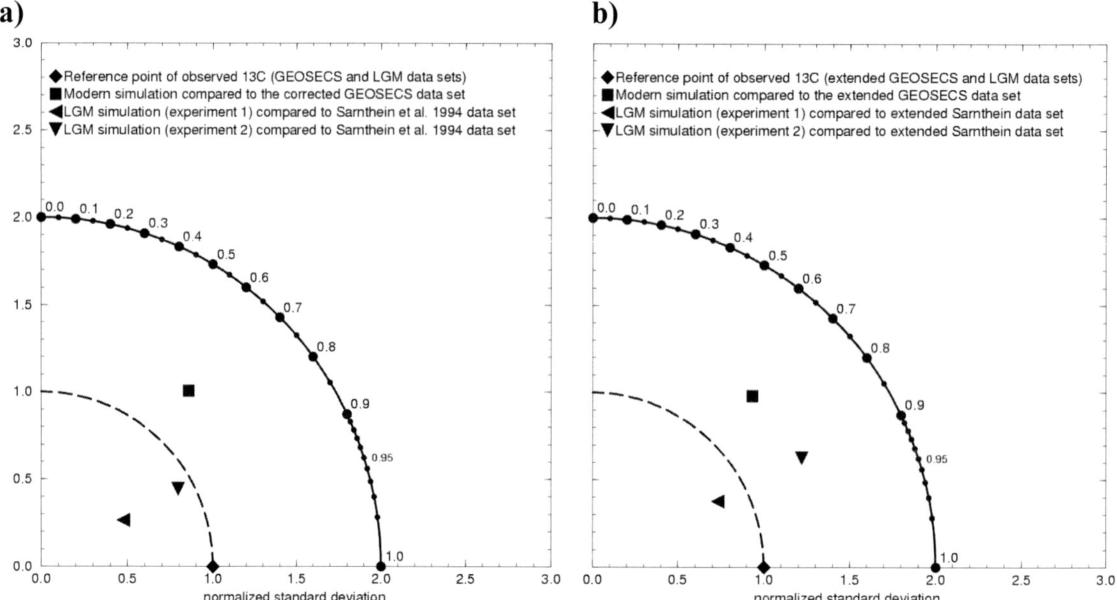

Fig. 15. Comparison between simulated $\delta^{13}C$ pattern with observed $\delta^{13}C$ in the entire Atlantic Ocean. The graphical representation is according to Taylor (Gates et al.1998, Taylor 2001), see text for details. **a)** Simulated $\delta^{13}C$ pattern for the control and the both glacial runs (experiment 1 and 2) in comparison with the GEOSECS observations and the data of Sarnthein et al. (1994). Note that experiment 3 in this diagram shows the same position as experiment 1 and remains unmentioned in the legend. **b)** The same as left, but for extended data sets. For the modern case the GEOSECS data are replenished to the south according to Mackensen et al. (1996). For the comparison with the glacial simulations, we used the data of Sarnthein et al. (1994) north of the equator, otherwise the SFB 261 data set is used.

data measured by the research program of SFB 261 were used. For the modern situation, this led to a slightly improved conformity with these data sets at least as far as the correlation coefficient was concerned. The situation of the glacial simulation showed an opposite behaviour in the $\delta^{13}C$ pattern between experiment 1 (and 3) and 2. The simulation in experiment 1 (and 3) now reveals a better conformity with the observation than could be archieved in experiment 2.

Summary and Conclusions

The new paleodata set of the LGM shows - especially in the South Atlantic - that the reduction of $\delta^{13}C$ in the deep water column is less pronounced than previously assumed. In particular, the Western Atlantic shows no negative $\delta^{13}C$ values. This corresponds favourably with the glacial simulation of experiment 1, in which only the thermohaline

circulation conditions have changed. As compared to the modern ocean, the data in the glacial Eastern Atlantic show generally very low $\delta^{13}C$ concentrations in the deep water column between 20°N and the equator. In the Eastern Atlantic, the simulated $\delta^{13}C$ concentrations correspond well to the observations. In general, deviations are in the range of ±0.2 ‰, with an exception around 20°N where the model produced 0.4 ‰ higher concentrations as compared to the observations. However, in spite of the good correlation between glacial simulation and observation the reduction of the atmospheric CO_2 partial pressure is too low. Thus, the model could reproduce the observed $\delta^{13}C$ pattern, but it was not able to simulate the glacial/interglacial pCO_2 change of about 80 ppm. The distribution of the new production in the Atlantic, not shown, indicated a slight increase in the productivity in the ice free polar regions, whereas the tropical and subtropical ocean showed no significant changes compared to the modern distribution. This simulated pattern is supported by observations that new or export production showed no significant changes in the glacial Southern Ocean (François et al. 1997; Nürnberg et al. 1997) and oligothrophic open ocean regions (Sarnthein et al. 1988; Mix 1989). However, evidences of higher paleoproductivity were observed in the glacial Atlantic, particularly in the eastern boundary upwelling regions (Sarnthein et al. 1988; Schneider et al. 1996) and in the equatorial zones (Mix 1989; Lyle et al. 1992; Schneider et al. 1996). A decrease of paleoproductivity was found in the Artic Ocean (Sarnthein et al. 1988; Schubert and Stein 1996; Knies and Stein 1998).

In model experiment 2 (implications of a stronger biological pump), simulated $\delta^{13}C$ concentrations revealed that the reduction of carbon isotopes was highly overestimated as compared to the observations in the whole Atlantic. However, with respect to the atmospheric CO_2 concentration, the decrease amounted to 60 ppm as compared to the control run and was in relative good accordance with observations from ice cores. The export production in this simulation showed a strong increase in the whole ice free ocean. This was supported by observations made at the glacial subantarctic Atlantic Ocean (Kumar et al. 1995), equatorial Atlantic (Sarnthein et al. 1988; Mix 1989; Lyle et al. 1992) and upwelling regions of the eastern Atlantic (Sarnthein et al. 1988; Schneider et al. 1996).

Experiment 3 displayed a relative strong reduction of the glacial atmospheric pCO_2, in combination with a good reproduction of the $\delta^{13}C$ distribution in the Atlantic. However, the result does not agree with observations of changed lysoclines. An improved model performance could be archieved by accounting for the following considerations:
• As mention above, the physical ocean model showed a too shallow southward extension of the NADW. A higher vertical resolution of the model could result in a better resolution of the various main water masses.
• In the carbon cycle model, the $\delta^{13}C$ of particulate organic matter was calculated with a constant carbon isotope fractionation factor of –20 ‰. However, observations from the Southern Ocean showed values lower than –30 ‰ in $\delta^{13}C$ of organic

matter in the surface water. These low organic carbon isotope values were not reproduced by the model. Rau et al. (1991) attribute the low $\delta^{13}C$ of organic material in this region to the increased pool of aqueous CO_2 in these cold waters. A CO_2-dependent carbon isotope fractionation could possibly correct this discrepancy between simulation and oberservation in the Southern Ocean.
• A more complex production formulation including more different biological species could also improve model performance. However, the functional relationships between the various species are not well understood yet.
• A more complex sediment module should be included which would enhance the performance in the carbonate chemistry.
• A module of the land biosphere should be included which would supply more information on the effect of the marine carbon isotopes.

However, these recommendations are only helpful, if the functional relationships are sufficiently known, ideally, on a global or basin wide scale.

We conclude on the base of our model simulations that no additional increase of the deep ocean's nutrient inventory is necessary to reproduce the $\delta^{13}C$ distribution in the glacial Atlantic, which indicate no significant increase in biological pumping during the LGM. This would exclude all hypotheses that involve changes in the nutrient inventory and/or Redfield ratios to explain the atmospheric glacial-interglacial CO_2 reduction. Our results would rather support other scenarios for the decrease of glacial atmospheric pCO_2, such as the mechanism of carbonate compensation.

Acknowledgments

The models LSG and HAMOCC have been developed and put at our disposal by the Max-Plank-Institut für Meteorologie. Special thanks go to Andreas Manschke for his technical support. The careful comments of A. Winguth and R. Zahn were considerably helpful in improving the manuscript. We kindly acknowledge financial support from Deutsche Forschungsgemeinschaft (Sonderforschungsbereich 261). Data are available under www.pangaea.de/Projects/SFB261.

References

Anderson LA, Sarmiento JL (1994) Redfield ratios of remineralization determined by nutrient data analysis. Glob Biogeochem Cycl 8: 65-80

Arakawa A, Lamb VR (1977) Computational design of basic dynamical process of the UCLA general circulation model. Meth Comput Phys 16: 173-283

Archer D, Winguth A, Lea D, Mahowald N (2000a) What Caused the Glacial/Interglacial Atmospheric pCO_2 Cycles? Rev Geophys 38: 159-189

Archer DE, Eshel G, Winguth A, Broecker WS, Pierrehumbert R, Tobis M, Jacob R (2000b) Atmospheric pCO_2 sensitivity to the biological pump in the ocean. Glob Biogeochem Cycl 14: 1219-1230

Bacastow R, Maier-Reimer E (1990) Ocean-circulation model of the carbon cycle. Clim Dyn 4:95-125

Barnola JM, Raynaud D, Korotkevich YS, Lorius C (1987) Vostok ice core provides 160,000-year record of atmospheric CO_2. Nature 329: 408-414

Bickert T, Wefer G (1999) South Atlantic and benthic foraminifer $\delta^{13}C$-deviations: Implications for reconstructing the Late Quaternary deep-water circulation. Deep-Sea Res 46: 437-452

Broecker WS, Maier-Reimer E (1992) The influence of air and sea exchange on the carbon isotope distribution in the sea. Glob Biogeochem Cycl 6: 315-320

Broecker WS, Henderson GM (1998) The sequence of events surrounding Termination II and their implications for the cause of glacial-interglacial CO_2 changes. Paleoceanography 13: 352-364

Curry WB, Duplessy JC, Labeyrie LD, ShackletonPNJ (1988) Changes in the distribution of $\delta^{13}C$ of Deep Water CO_2 between the Last Glaciation and the Holocene. Paleoceanography 3: 317-341

Duplessy JC, Shackleton NJ, Fairbanks RG, Labeyrie L, Oppo D, Kallel N (1988) Deepwater source variations during the last climatic cycle and their impact on the global deepwater circulation. Paleoceanography 3: 343-360

Elderfield H, Rickaby REM (2000) Oceanic Cd/P ratio and nutrient utilization in the glacial Southern Ocean. Nature 405: 305-310

François R, Altabet MA, Yu EF, Sigman DM, Bacon MP, Frank M, Bohrmann G, Bareille G, Labeyrie LD (1997) Contribution of Southern Ocean surface-water stratification to low atmospheric CO_2 concentrations during the last glacial period. Nature 389: 929-935

Friedli H, Lötscher H, Oeschger H, Siegenthaler U, Stauffer B (1986) Ice core record of the $^{13}C/^{12}C$ ratio of atmospheric CO_2 in the past two centuries. Nature 324: 237-238

Gates WL, Boyle JS, Covey CC, Dease CG, Doutriaux CM, Drach RS, Fiorino M, Gleckler PJ, Hnilo JJ, Marlais SM, Phillips TJ, Potter GL, Santer BD, Sperber KR, Taylor KE, Williams DN (1998) An Overview of the Results of the Atmospheric Model Intercomparison Project (AMIP). Report 45, The Program for Climate Model Diagnosis and Inter-comparison, Livermore, http://www-pcmdi.llnl.gov/pcmdi/pubs/ab45.html

Heinze C, Maier-Reimer E, Winn K (1991) Glacial pCO_2 reduction by the world ocean: Experiments with the Hamburg Carbon Cycle Model. Paleoceanography 6: 395-430

Heinze C, Maier-Reimer E, Winguth AME, Archer D (1999) A global oceanic sediment model for long-term climate studies. Glob Biogeochem Cycl 13:221-250

Johns TC, Carnell RE, Crossley JF, Gregory JM, Mitchell JFB, Senior CA, Tett SFB, Wood RA (1997) The second Hadley Centre coupled ocean-atmosphere GCM: Model description, spinup and validation. Clim Dyn 13: 103-134

Keeling RF, Stephens BB (2001) Antarctic sea ice and the control of Pleistocene climate instability. Paleoceanography 16: 112-131 and (corrections) 330-334

Knies J, Stein R (1998) New aspects of organic carbon deposition and its paleoceanographic implications along the northern Barents Sea margin during the last 30,000 years. Paleoceanography 13: 384-394

Kroopnick PM (1985) The distribution of ^{13}C of CO_2 in the world oceans. Deep-Sea Res 32: 57-84

Kumar N, Anderson RF, Mortlock RA, Froelich PN, Kubik P, Dittrich- Nannen B, Suter M (1995) Increased biological productivity and export production in the glacial Southern Ocean. Nature 378: 675-680

Leonard BP (1979) A stable and accurate convective modeling procedure based on quadratic upstream interpolation. Comp Meth Appl Mech Eng 19: 59-98

Levitus S, Russell B, Boyer TP (1994) World Ocean Atlas 1994. Volume 3: Salinity. Technical report, National Oceanic and Atmospheric Administration, Washington DC

Lorenz S, Grieger B, Helbig P, Herterich K (1996) Investigating the sensitivity of the atmospheric general circulation model ECHAM 3 to paleoclimatic boundary conditions. Geol Rundsch 85: 513-524

Lyle MW, Prahl FG, Sparrow MA (1992) Upwelling and productivity changes inferred from a temperature record in the central equatorial Pacific. Nature 355: 812-815

Lynch-Stieglitz J, Stocker T, Broecker WS, Fairbanks RG (1995) The influence of air-sea exchange on the isotopic composition of oceanic carbon: Observations and modeling. Glob Biogeochem Cyc 9: 653-665

Mackensen A, Hubberten HW, Bickert T, Fischer G, Fütterer DK (1993) The $\delta^{13}C$ in benthic foraminiferal tests of *Fontbotia Wuellerstorfi* (Schwager) relativ to the $\delta^{13}C$ of dissolved inorganic carbon in Southern Ocean Deep Water: Implications for glacial ocean circulation models. Paleoceanography 8: 587-610

Mackensen A, Hubberten HW, Scheele N, Schlitzer R (1996) Decoupling of $\delta^{13}CTCO_2$ and phosphate in Recent Weddel Sea deep and bottom water: Implications for glacial Southern Ocean paleoceanography. Paleoceanography 11: 203-215

Maier-Reimer E, Hasselmann K (1987) Transport and storage of CO_2 in the ocean - an inorganic ocean-circulation carbon cycle model. Clim Dyn 2: 63-90

Maier-Reimer E, Mikolajewicz U, Hasselmann K (1993) Mean circulation of the Hamburg LSG OGCM and Its Sensitivity to the Thermohaline Surface Forcing. J Phys Oceanogr 23: 731-757

Marchal O, Stocker TF, Joos F (1998) Impact of oceanic reorganizations on the ocean carbon cycle and atmospheric carbon dioxide content. Paleoceanography 13: 225-244

Mix AC (1989) Influence of productivity variations on long-term atmospheric CO_2. Nature 337: 541-544

Murnane RJ, Sarmiento JL, Le Quéré C (1999) Spatial distribution of air-sea CO_2 fluxes and the interhemispheric transport of carbon by the oceans. Glob Biogeochem Cycl 13: 287-305

Nürnberg CC, Bohrmann G, Schlüter M, Frank M (1997) Barium accumulation in the Atlantic sector of the Southern Ocean: Results from 190,000-year records. Paleoceanography 12: 594-603

Petit JR, Jouzel J, Raynaud D, Barkov NI, Barnola JM, Basile I, Bender M, Chappellaz J, Davis M, Delaygue G, Delmotte M, Kotlyakov VM, Legrand M, Lipenkov VY, Lorius C, Pépin L, Ritz C, Saltzman E, Stievenard M (1999) Climate and atmospheric history of the past 420,000 years from the Vostok ice core, Antarctica. Nature 399: 429-436

Rau GH, Takahashi T, Des Marais DJ, Sullivan CW (1991) Particulate organic matter $\delta^{13}C$ variations across the Drake Passage. J Geophys Res 96: 15131-15135

Rau GH, Riebesell U, Wolf-Gladrow D (1997) CO_2aq-dependent photosynthetic ^{13}C fractionation in the ocean: A model versus measurements. Glob Biogeochem Cycl 11: 267-278

Sanyal A, Hemming NG, Hanson GN, Broecker WS (1995) Evidence for a higher ph in the glacial ocean from boron isotopes in foraminifera. Nature 373: 234-236

Sarnthein M, Winn K, Duplessy JC, Fontugne MR (1988) Global variations of surface ocean productivity in low and mid latitudes: Influence on CO_2 reservoirs of the deep ocean and atmosphere during the last 21,000 years. Paleoceanography 3: 361-399

Sarnthein M, Winn K, Jung SJA, Duplessy JC, Labeyrie L, Erlenkeuser H, Ganssen G (1994) Changes in east Atlantic deepwater circulation over the last 30,000 years: Eight time slice reconstructions. Paleoceanography 9: 209-267

Schäfer-Neth C, Paul A (2001) Circulation of the Glacial Atlantic: A Synthesis of Global and Regional Modeling. In: Schäfer P, Ritzrau W, Schlüter M, Thiede J (eds) The Northern North Atlantic: A Changing Enviroment. Springer, Berlin, pp 441-462

Schlitzer R (2002) Carbon export fluxes in the Southerrn Ocean: Results from inverse modelling and compaison with satellite-based eastimates. Deep-Sea Res 49: 1623-1644

Schmitz WJ (1995) On the Interbasin-Scale Thermohaline Circulation. Rev Geophys 33: 151-173

Schneider RR, Müller PJ, Ruhland G, Meinecke G, Schmidt H, Wefer G (1996) Late quaternary surface temperatures and productivity in the east-equatorial south atlantic: Response to changes in trademonsoon wind forcing and surface water advection. In: Wefer G, Berger W, Siedler G, Webb DJ (eds) The South Atlantic: Present and Past Circulation. Springer, Berlin, pp 527-551

Schubert CJ, Stein R (1996) Deposition of organic carbon in Arctic Ocean sediments: Terrigenous supply versus marine productivity. Org Geochem 24: 421-436

Shea DJ, Trenberth KE, Reynolds RW (1990) A global monthly sea surface temperature climatology. NCAR Technical Note NCAR/TN 345, NCAR, Boulder, Colorado

Sigman DM, Boyle EA (2000) Glacial/interglacial variations in atmospheric carbon dioxide. Nature 407: 859-869

Stephens BB, Keeling RF (2000) The influence of Antarctic sea ice on glacial-interglacial CO_2 variations. Nature 404: 171-174

Suess E (1980) Particulate organic carbon flux in the oceans-surface productivity and oxygen utilization. Nature 288: 260-263

Takahashi T, Feely RA, Weiss R, Wanninkhof R, Chipman DW, Sutherland SC, Takahashi TT (1997) Global air-sea flux of CO_2: An estimate based on measurements of air-sea pCO_2 difference. Proc Natl Acad Sci 94: 8292-8299

Taylor KE (2001) Summarizing multiple aspects of model performance in a single diagram. J Geophys Res 106: 7183-7192

Toggweiler JR (1999) Variation of atmospheric CO_2 by ventilation of the ocean's deepest water. Paleoceanography 14: 571-588

Weinelt M, Sarnthein M, Pflaumann U, Schulz H, Jung S, Erlenkeuser H (1996) Ice-free Nordic Seas during the Last Glacial Maximum? Potential sites of deepwater formation. Paleoclimatology 1: 283-309

Winguth AME, Archer D, Duplessy JC, Maier-Reimer E, Mikolajewicz U (1999) Sensitivity of paleonutrient tracer distributions and deep-sea circulation to glacial boundary conditions. Paleoceanography 14: 304-323

Yamanaka Y, Tajika E (1996) The role of the vertical fluxes of particulate organic matter and calcite in the oceanic carbon cycle: Studies using an ocean biogeochemical general circulation model. Glob Biogeochem Cycl 10: 361-382

Yamanaka Y, Tajika E (1997) Role of dissolved organic matter in the marine biogeochemical cycle: Studies using an ocean biogeochemical general circulation model. Glob Biogeochem Cycl 11: 599-612

Yu EF, François R, Bacon MP (1996) Similar rates of modern and last-glacial ocean thermohaline circulation inferred from radiochemical data. Nature 379: 689-694

Printing and Binding: Stürtz AG, Würzburg